LINCOLN CHRISTIAN COLLEGE AND SEMINARY P9-CML-813

CANNOT BE CHECKED OUT

For Reference

Not to be taken

from this library

ENCYCLOPEDIA OF EVOLUTION

Editorial Board

EDITOR IN CHIEF

Mark Pagel
University of Reading

EDITORS

Steven A. Frank
University of California at Irvine

Charles Godfray
Imperial College at Silwood Park

Brian K. Hall
Dalhousie University

Kristen Hawkes
University of Utah

David M. Hillis
University of Texas at Austin

Astrid Kodric-Brown
University of New Mexico

Richard E. Lenski
Hannah Professor, Michigan State University

Andrew Pomiankowski
University College London

ADVISORS

Sir Walter Bodmer

L. Luca Cavalli-Sforza

Richard Dawkins

Jared M. Diamond

Sarah B. Hrdy

David L. Hull

Richard Leakey

John Maynard Smith

Edward O. Wilson

Encyclopedia of

Evolution

Mark Pagel

EDITOR IN CHIEF

VOLUME 1

OXFORD

UNIVERSITY PRESS

2002

OXFORD

UNIVERSITY PRESS

Oxford   New York
Auckland   Bangkok   Buenos Aires   Cape Town   Chennai
Dar es Salaam   Delhi   Hong Kong   Istanbul   Karachi   Kolkata
Kuala Lumpur   Madrid   Melbourne   Mexico City   Mumbai   Nairobi
São Paulo   Shanghai   Singapore   Taipei   Tokyo   Toronto

and an associated company in Berlin

Copyright © 2002 by Oxford University Press, Inc.

Published by Oxford University Press, Inc.
198 Madison Avenue, New York, New York 10016
www.oup.com

Oxford is a registered trademark of Oxford University Press

All rights reserved. No part of this publication may be reproduced,
stored in a retrieval system, or transmitted, in any form or by any means,
electronic, mechanical, photocopying, recording, or otherwise,
without the prior written permission of Oxford University Press.

Library of Congress Cataloging-in-Publication Data

(The Oxford) encyclopedia of evolution / Mark Pagel, editor in chief.
p.   cm.
Includes bibliographical references (p.).
ISBN 0–19–512200–3 (set)
ISBN 0–19–514864–9 (vol. 1)
ISBN 0–19–514865–7 (vol. 2)
1. Evolution (Biology)—Encyclopedias.   I. Pagel, Mark D.
QH360.2.O83   2002
576.8'03—dc21   2001021588

1   3   5   7   9   8   6   4   2

Printed in the United States of America
on acid-free paper

Contents

ENCYCLOPEDIA OF EVOLUTION

OVERVIEW ESSAYS: STATES OF EVOLUTION

ALPHABETICAL ENTRIES

109695

Editorial and Production Staff

Commissioning Editor
Sean Pidgeon

Development and Managing Editor
Mark Mones

Project Editor
Mark Gallagher

Copyeditors
Laura Daly   Jane McGary   Melissa Dobson

Proofreaders
Sue Gilad   Carol Holmes

Index
Impressions Book and Journal Services, Inc.

Manufacturing Controller
Genieve Shaw

Book Designer
Joan Greenfield

Publisher
Karen Day

List of Articles

The list of articles that follows reflects the sequence of overview essays and alphabetical entries; for particular terms not noted here, refer to the index, as well as the blind entries (e.g., "**AIDS**. See Acquired Immune Deficiency Syndrome.") which appear selectively as headwords between alphabetical entries.

Topical Outline of Articles

The entries in the Encyclopedia of Evolution are conceived according to the general conceptual categories listed in this topical outline. Some entries are listed more than once because the conceptual categories are not mutually exclusive. Entries in the encyclopedia proper are organized alphabetically, with the exceptions of the nine Overview Essays (which appear at the beginning of Volume 1) and the subarticles that comprise broader (composite) encyclopedia entries. These subarticles are here listed in italics.

OVERVIEW ESSAYS

History of Evolutionary Thought
The Major Transitions in Evolution
Macroevolution
Culture in Chimpanzees
Human Genetic and Linguistic Diversity
Motherhood
Darwinian Medicine
Genomics and the Dawn of Proteomics
The New Replicators

FUNDAMENTAL CONCEPTS, DEFINITIONS, AND THEORY

Classic Evidence for Evolution
    Evolution
    Artificial Selection
    Convergent and Parallel Evolution
    Geological Periods
    Geology
    Homology
    Origin of Life
        The First Fossils
    Paleontology
    Stratigraphy
    Symbiosis
Natural Selection
    Natural Selection
        An Overview
    Character Displacement
    Darwin's Finches
    Galapagos Islands
    Group Selection
    Kin Selection
    Levels of Selection
    Meme
    Mimicry
    Peppered Moth
    Pesticide Resistance

Selfish Gene
Species Selection

Sexual Selection and Signalling
    Sexual Selection
        An Overview
    Signalling Theory

Alternatives to Evolution by Natural Selection
    Creationism
    Directed Protein Evolution
    Lamarckism

Adaptation
    Adaptation
    Coevolution
    Comparative Method
    Constraints on Adaptation
    Exaptation
    Niche Construction
    Optimality Theory
        An Overview
    Red Queen Hypothesis

Fitness
    Fitness

Life History Theory
    Life History Theory
        An Overview
    Demography
    Population Dynamics

Speciation and Species Concepts
    Speciation
    Species Concepts
    Haldane's Rule
    Hybrid Zones

Macroevolution
    Overview Essay *on* Macroevolution
    Cope's Rule
    Extinction
    Mass Extinctions
    Microevolution
    Punctuated Equilibrium

NONBIOLOGICAL APPLICATIONS OF BIOLOGICAL CONCEPTS

Preface

Biological evolution is the cause of the breathtaking diversity of species that make up the organic world. Humans owe to biological evolution their big brains, acute intelligence, and virtuosity with language, lions owe their teeth, leopards their spots, weaver birds their complex nests, plants their stunning flowers, fungi their potent toxins, and bacteria and viruses their vexing abilities to wriggle out of the nets humans cast to catch and control them. We now have greater knowledge than ever before of the workings of evolution and natural selection, and we are beginning to understand the genetic basis of many of our own traits and behaviors. Theodosius Dobzhansky's famous aphorism that "nothing in biology makes sense except in the light of evolution" is overused to the point of exhaustion but has never been more apt than today.

The *Encyclopedia of Evolution* collects together in one place what we have learned about this process that makes sense of biology. This turns out to be a considerable task. Evolutionary biologists are probing the origins of life, finding genes influencing social behaviors and causing disease, uncovering genetic ancestry, documenting biodiversity, sequencing whole genomes, cloning, investigating how animals develop, studying antibiotic resistance, and are involved with genetic modification and genetic screening. To report on all this the *Encyclopedia* presents articles on fundamental concepts, definitions, and theories of evolution, evolutionary and population genetics, molecular evolution, bioinformatics and genomics, systematics and taxonomy, behavioral ecology and social evolution, cell and developmental biology, human cultural and biological evolution, paleontology, mathematical modelling, and the evolutionary basis of disease. Many unexpected topics find their way in among these broad areas: we have, for example, articles on evolutionary economics, prions, the evolution of art, self-deception, and grandmothers and human longevity.

To my knowledge, no other single source has this breadth. The knowledge we have gained from studying evolution is sometimes shocking and salutary—such as the discovery that humans and chimpanzees are on average about 99 percent similar genetically; at times humbling—such as the realization from the human genome project that we humans have scarcely more different genes than a well-known plant and only about double that of a worm; and at other times plainly astonishing—little known relatives of insects, the tardigrades, can live at temperatures between $-272C$ and $+340C$, at pressures ranging from a vacuum to 160,000 pounds per square inch (normal is 14.7), and can persist in the environment as cryptospores for up to 100 years. If life on Earth came from Outer Space (as some have suggested), it was probably a tardigrade.

Our aim is to present these and other topics in a way that is suitable for those new to evolutionary biology and yet technical and up-to-date enough for the *Encyclopedia of Evolution* to be a reference source even for researchers. Drawing on Oxford University Press's great experience in producing reference works, the *Encyclopedia of Evolution* adopts a system of articles designed for audiences of varying backgrounds. *Overview* articles are informative of a broad topic, written at a level suitable for advanced high school or undergraduate students, and yet sophisticated enough even for people with further training. *Technical* and *topical* articles are for readers with greater background, or for those seeking it. *Biographies* and *definitional* articles live up to their designations. *Composite* articles group together two or more articles on closely related topics. The *Encyclopedia of Evolution* provides introductions to technical areas of genetics, evolutionary theory, and evolutionary modelling, as well as accounts of the taxonomy of major groups of plants and animals. These articles are informative in their own right and provide the background to other articles in the *Encyclopedia*. The *topical outline* provided at the front of the *Encyclopedia* is a helpful way to see at a glance the main areas of our coverage and how these areas are linked to articles from other areas of the *Encyclopedia*. This is especially important for a subject such as evolution in which the real connections among topics are extensive.

The content and features of the *Encyclopedia* should make it a useful reference book for courses on introductory biology, evolutionary genetics, developmental biology, animal behavior, phylogenetics and systematics, human biological and cultural evolution, evolutionary psychology, and disease. Those with more training in evolution and biology can use the *Encyclopedia* to learn about topics outside their own expertise.

As Editor in Chief I have read every one of the 365 articles in the *Encyclopedia*. The list of authors reads like a "Who's Who" of evolutionary biology. In addition, a unique feature of this work is the set of overview essays at the beginning of the first volume entitled *States of Evolution*. Here nine of the best-known authors currently writing and thinking about evolutionary biology have contributed their thoughts in extended essays. The articles were chosen to show the breadth and freshness of topics to which evolutionary thinking has been applied, and to which it will be applied in the future: these are the "states of evolution" of this section. The articles include thoughts on such subjects as "the new replicators," "motherhood," "culture in animals," and "genomics and the dawn of proteomics." We think the section will not fail to teach, entertain, and provoke. My notes to that section introduce the articles.

The idea of producing this *Encyclopedia* can be traced back several years to Sean Pidgeon who was then a commissioning editor at Oxford University Press's Scholarly and Professional Reference Department in New York. We began the project by assembling a team of Area Editors and Advisors (listed on the front pages). The Area Editors are authorities in their fields, chosen to oversee the articles in seven broad and overlapping areas of biology: fundamental concepts and ideas (Charles Godfray), evolutionary genetics (Andrew Pomiankowski), developmental biology (Brian K. Hall), systematics and taxonomy (David M. Hillis), behavioral ecology (Astrid Kodric-Brown), human evolution and palaeontology (Kristen Hawkes), disease evolution (Richard E. Lenski), and mathematical modelling and biographies (Steven A. Frank). The Area Editors and I worked together to define the content of the *Encyclopedia*, literally mapping out each article. Our Advisory Board eventually approved the intended content. We are fortunate in having an eminent advisory board, and their comments and contributions have been most helpful.

Then began the hard work of identifying and approaching authors, commissioning articles, and seeing through all of the tasks required to get the articles in. This is an immense, difficult, and often exasperating task. It requires skill and diplomacy, tact, perseverance, obsessive organization, and a deep understanding of the subject. And it does not end with the articles in hand. Someone must oversee, literally take responsibility, for seeing the project not just as a collection of articles, but as a coherent, meaningful, and marketable whole. I cannot emphasize how fortunate I have been to have Mark Mones, Managing Editor of OUP's Scholarly and Professional Reference Department, at the helm steering the project through all of these waters. Ever the diplomat, ever optimistic, and with "encyclopedic" knowledge, I can only despair at the thought of where we would have been without him. About halfway through the project Mark Gallagher assumed responsibility as Project Editor, involving the seemingly endless tasks of chasing recalcitrant authors, turning their articles into finished entries, and fielding demands from the Editor in Chief. A droll sense of humor saw him through. [Having (including myself) three "Mark"s on the project created its own attractions. Once, when dealing with an author also called Mark, we received the following message from an Area Editor: "So Mark says to Mark he e-mailed it to Mark (though did he also copy it to Mark)? Yes, all perfectly clear!"] John Sollami, the Editorial, Design, and Production Director of OUP's Reference Division, oversaw the particulars of our production schedule and helped to polish the work's content as we went into print. Rebecca Seger, Director of Marketing for the Reference Division, formulated a creative plan to publicize and market the finished work. Most significantly, Karen Day, Vice President and Publisher of OUP's Reference Division, applied the discipline and the enthusiasm to keep us to time and budget, but equally, the resources and the insights to dig ourselves out of one or two holes. Outside of OUP, Ruth Mace gave useful advice on the content and structure of several articles. Andrew Meade generously gave his time to help on several technical issues. Masters Thomas and William, although mostly unaware of the project, enforced a welcome line of demarcation between work and play.

We present the *Encyclopedia of Evolution*, 365 articles from 330 different authors, running to 730,000 words.

—MARK PAGEL, EDITOR IN CHIEF
Reading, United Kingdom
December 2001

OVERVIEW ESSAYS: STATES OF EVOLUTION

INTRODUCTION

Mark Pagel

Biology is the physics of the twenty-first century, capturing the public's imagination by undertaking large publicly and privately funded projects, and by tackling some of the most revolutionary and at times disturbing issues of our age. But it is also a subject whose governing theory of evolution genuinely provides a synthetic view of a vast range of phenomena. We invited nine eminent scientists and philosophers working in the field of evolutionary biology to contribute essays to this extraordinary section of the *Encyclopedia of Evolution*. Their essays present a selection of contemporary topics to which evolutionary thinking is bringing what are often challenging and remarkable new insights. They also highlight the provocative ideas and issues (such as whole genome sequencing) that continue to emerge from applying the theories of evolution and natural selection and the technologies of genetics. My purpose here is to introduce the articles of this section, and describe their historical or philosophical context; all nine of these overviews can be read alone or together as an introduction to the rest of the *Encyclopedia*.

David L. Hull's article, *History of Evolutionary Thought*, explains the history of thinking about evolution, up to the present. Hull is a philosopher who has long been interested in philosophical questions surrounding evolution and natural selection. His acclaimed *Science as a Process* (1988) is one of the few philosophical treatments of science that is admired by both scientists and philosophers. In his essay, Hull describes the philosophical insights that allowed scientific thinking about evolution to progress from its essentialist roots in Aristotle up through the Lamarckians and finally on to our current understanding of heredity, evolution, and adaptation by natural selection. The latter themes define what we think we are in the modern age: collections of selfish genes that dupe us into making offspring?—a mere genetic pastiche of our parents? Hull puts the issues and concepts into their historical context and provides a useful introduction to the *Encyclopedia* as a whole.

John Maynard Smith's *The Major Transitions in Evolution* summarizes the main arguments from his scientific monograph of the same name. Maynard Smith has been a leading theorist of evolutionary biology since the 1960s, known for his work on game theory, the evolution of sex, sexual selection, and signalling, and is the author of several books. He is one of the few (several others are in this section) people who have defined the way many of us think about evolution. One of the most difficult issues of evolutionary biology is to understand how life first evolved and then how and why it became more complex. Why do bodies exist? The challenge arises from the fact that at every level of organized life—from naked genes or replicators in the so-called primordial soup, to collections of genes literally stuck together in chromosomes, to simple cells, to primitive organisms, groups, and social systems—the replicating entities at the lower level have to give up some of their autonomy to participate at the higher level. Why, for example, do well over 99.999 percent of all of the cells in your body forgo their own reproduction simply to provide a vehicle to carry around the small number of beknighted germ cells that will form the next generation? How does the body placate or subdue this potentially selfish and uncooperative herd? Maynard Smith provides a unifying answer.

One is easily mesmerized by the power of evolution by natural selection to produce the varied forms that inhabit the Earth. But is the collection of species we now see the outcome solely of this process? Or, must we invoke other mechanisms to explain why, for example, we no longer have saber-toothed tigers but we do have tigers (just), or why about 65 million years ago we lost the dinosaurs, but not the mammals. Stephen Jay Gould of Harvard's Museum of Comparative Zoology is an influential and sought-after speaker and writer, known for his popular books, scientific monographs, and essays on natural history. His views, reported in his essay *Macroevolution*, challenge the received wisdom of some quarters and have shaped the way most people think about the causes of diversity on Earth. Gould's emphasis on "contingency" questions our self-proclaimed status as the pinnacle of evolution. Instead, Gould is fond of pointing out that, but for a large number of highly improbable events, we might never have walked the Earth. The same holds for other species. The upshot is that to understand the current diversity and forms on Earth we must grasp the nature, causes, and occurrences of chance events and be humble about our own existence.

Jane Goodall is known around the world for her work with chimpanzees in Tanzania, beginning in the 1960s and continuing to the present. Her book *Chimpanzees of Gombe* recounts her researches on these species' social and other behaviors. She now travels extensively, raising money for chimpanzee research and conservation. Here Dr. Goodall writes on *Culture in Chimpanzees*. For years social scientists and animal behaviorists have insisted that true culture is the province only of humans. This conceit is now under attack as study after study reports how animals learn local customs and behaviors (local bird song dialects are a good example) by imitating or being taught by older members of their population. Not surprisingly, some of the more remarkable examples of real cultural traditions, fads, and practices are found in chimpanzees and Goodall has been documenting them for decades. Her careful, possibly unmatched, field observations reveal with greater certainty than for any other species the reality of true cultural inheritance in chimpanzees and shortens just a little bit more the distance between us and our intelligent sister species.

L. Luca Cavalli-Sforza is one of the founders of modern human population genetics. Beginning in the 1960s Cavalli-Sforza used patterns of human blood group polymorphisms (genetic markers that vary considerably among individuals) to draw human genealogical histories, tracing human origins and the relationships among human groups around the world. Here Cavalli-Sforza in his essay *Human Genetic and Linguistic Diversity* reports on the latest results of genetic studies of modern humans. The picture that is emerging is one of a species that is young (perhaps as little as 60,000 years) and that quickly upon appearing in Africa walked out of that continent and around the world in perhaps 15,000 years. We know this because they littered their trail with genetic and linguistic clues.

Sarah Blaffer Hrdy's book *The Woman That Never Evolved* almost single-handedly established in the 1980s the view that females had played a far more active role in evolution than previously believed. This view is now becoming mainstream, although small cabals of resistance still exist. Professor Hrdy's essay *Motherhood* is taken from her magisterial book, *Mother Nature*, and documents the theoretical arguments and empirical facts showing how females and their matrilines fundamentally shape evolution, and especially hominid evolution.

Stephen Stearns' theoretical work on life history evolution in the 1970s produced one of a small number of influential papers that give this difficult topic a consistent mathematical grounding. More recently, Stearns has contributed to the field of Darwinian medicine. His essay, entitled *Darwinian Medicine*, describes the many insights into disease evolution that emerge from applying theories (most notably life history theory) of evolution to disease-causing organisms. The simple insights that organisms evolve to exploit their hosts, and that they can normally evolve far quicker than their hosts, explain the features of many infectious diseases. Intriguingly, ideas from evolution are also being applied to attempt to understand human mental illness and Stearns recounts some of these results.

M. J. Bishop is the Director of Bioinformatics at the Human Genome Mapping Project Centre in Cambridge, England. Bishop's work on the statistical analysis of gene sequences is widely used by researchers. Bishop's essay, *Genomics and the Dawn of Proteomics*, describes these emerging fields. Genomics is the large-scale study of whole genomes. Its offshoot, bioinformatics, underpins the biotechnology and pharmaceutical industries' efforts to mine genomes to find genes with medical or other value. Proteomics is the still inchoate field that attempts to understand how proteins interact with each other in our cells. It may sound purely technical, but this field may produce more startling technologies and Nobel prizes in the next twenty years than any other in the life sciences.

Daniel Dennett is one of the leading theorists of the philosophy of mind, on which he has written extensively and entertainingly since the 1970s. Throughout the 1990s Dennett increasingly thought about issues of Darwinism. Dennett's article for this section, entitled *The New Replicators*, sets out a way of thinking about replicators other than our genes. The best known of these are memes, but there are others, including self-replicating computer viruses and, possibly one day, even robots. Are we on the brink of having to hand over power to these self-replicating devices? Are they a form of life?

Nearly ten years ago the doyen of evolutionary biology, Ernst Mayr, wrote a book about Darwinism with the leading title of *One Long Argument*. The one long argument is, of course, Darwin's, but it is also the radical argument of evolution by natural selection. This message is that a blind and aimless process can explain some of the most magnificent creations of Nature—the origin of life itself, the apparent design of organs of perfection, and possibly even our predilections and desires. The juggernaut has not stopped with biology but rumbles on, taking in new ideas and concepts—anthropology, economics, medicine, psychology, and linguistics are embracing elements of Darwinism. Philosophers, and even novelists, are thinking about it. We hope that these nine essays on the *States of Evolution* convey a sense of the profound range of intellectual topics onto which evolutionary thinking has been brought to bear. And, much more follows in the remaining 365 contributions to the *Encyclopedia of Evolution*.

—MARK PAGEL, EDITOR IN CHIEF

HISTORY OF EVOLUTIONARY THOUGHT

David L. Hull

A belief in the evolution of species has had a long and tortuous history. In the development of Western thought from the ancient Greeks to at least the nineteenth century, biological species were construed as being as eternal and immutable as geometric figures and the physical elements. One species could not be transformed into another any more than lead could be transformed into gold or triangularity become rectilinearity. Here and there an early thinker did suggest that species might "evolve," but usually what was meant by this term was that an organism belonging to one species could change into an organism belonging to another. For example, Aristotle (384–322 B.C.) speculated that a water snake might become a viper when the marshes dried up. But an occasional change of species is a far cry from species themselves changing. Certainly, centuries of alchemists tried to transmute samples of cheaper metals into gold, but this process is not the same thing as lead itself or gold itself changing. Finally, a wire triangle might be reshaped into a square, but in the process, triangularity and rectilinearity remain unchanged.

From Archetypes to Ancestors

The view that species of plants and animals are eternal and immutable can be traced back to Plato and his Platonic ideas. However, it reached its peak in the eighteenth and nineteenth centuries in France and Germany. This continental worldview came in many guises and was called by an equal variety of names—idealism, ideal morphology, transcendentalism, rational morphology, typology, and so on. According to such authors as Georges Cuvier (1769–1832), the embryologist Karl Ernst von Baer (1792–1876), the naturalist Louis Agassiz (1807–1873), and the poet Goethe (1749–1832), science consisted in extracting general forms from the variety of patterns that are discernible in nature. For each natural group, an archetype exists, and the goal of science is to discover these archetypes. Just as there is the ideal fish, ideal vertebrate, and ideal plant, there is the ideal mineral. Each of these archetypes can be discovered by stripping away all accidental features to lay bare the essence of the thing. Although some kinds of organisms resemble the ideal of their kind more closely than others, archetypes do not exist in the form of single organisms. Even though the ideal fish looks very much like a perch, no actual perch could be identical with the ideal fish. Archetypes are highly generalized patterns, not actual organisms. The problem that evolutionists saw was how to convert archetypes into ancestors. The task turned out to be much more complicated than anyone at the time supposed.

Richard Owen (1804–1892) was one of the few British scientists of the era who could read and understand the sort of science being done on the Continent. Owen was highly respected in England for his descriptive work but not for his more metaphysical musings, for the simple reason that they were metaphysical. Naturalism, the view that all natural phenomena must be explained naturally, did not seem like a metaphysical view to most Victorian scientists; ideal forms did. It was not surprising that Owen was highly critical of Darwin's theory of evolution, but his contemporaries were a bit taken aback when he claimed priority for the idea of the evolution of species. Why, prior to 1859, had not Owen openly championed the notion of the transmutation of species?

The answer is that he had—but not in language that his British contemporaries were likely to understand. For example, the mechanism that Owen postulated for the transmutation of species involved an antagonism between the Platonic idea (or specific organizing principle) and a general polarizing force.

Idealism was not so much addressed and refuted by naturalist Victorian scientists but ignored with a puzzled shake of the head. Because the idealists lost in the war over the proper way to construe the world, their views remain all but incomprehensible to this day. Only a few present-day scientists advocate anything like an idealist worldview (see, for example, G. Nelson and N. Platnick, *Systematics and Biogeography: Cladistics and Vicariance*, New York, 1981; and G. Webster and B. Goodwin, *Form and Transformation*, Cambridge, 1996).

Jean Baptiste de Lamarck and Robert Chambers

Even though an occasional thinker through the years might suggest that species evolve, none went into any great detail about the mechanisms supposedly responsible for this evolution. The first serious attempts to produce a theory of evolution were published by Jean Baptiste de Lamarck (1744–1829) and Robert Chambers (1802–1871). In his *Philosophie zoologique* (1809) Lamarck postulated two (later three) branching trees of life, beginning with the simplest organisms and culminating with the most complex. For example, one of these trees begins with *les vers* (worms) and ends with mammals. Lamarck also postulated three different mechanisms that brought about this transformation. At the base of each tree, very simple organisms are produced by spontaneous generation. Then, under the influence of a force that inclined all beings toward increased complexity, organisms through successive generations worked their way up their respective trees. Instead of forming a great chain of being, for Lamarck, species formed a great escalator of being, as organisms lower down in a tree are ineluctably forced upward.

The third mechanism operative in Lamarckian evolution has come to be known as "Lamarckian inheritance." The term suggests that Lamarck invented the concept, but a belief that changes in the body of an organism can be transferred somehow to the hereditary material of an organism and thus passed on to its progeny was commonplace at the time. For Lamarck, the inheritance of acquired characteristics was primarily responsible for the branchings in his trees, as organisms confront differences in their environments. Birds living by the seashore would evolve one set of characteristics, whereas relatives of this same species living inland would develop a different set of characteristics.

Most anglophone scientists got whatever knowledge they had of Lamarck's work through a translation of the eulogy written by one of his chief enemies, Cuvier, and criticisms of Lamarck's transmutation theory published by the geologist Charles Lyell (1797–1875) in his *Principles of Geology* (1830–1833). Although Lyell rejected Lamarck's transformation theory, he treated it a good deal more fairly than did Cuvier. In fact, Lyell's exposition was instrumental in convincing several of his contemporaries to accept the evolution of species, including Robert Chambers, a Scottish editor and publisher. In his *Vestiges of the Natural History of Creation* (1844), Chambers rejected the usual parody of Lamarckian inheritance and instead postulated an embryological mechanism according to which a change in the developmental program of an organism might result in its descendants taking a novel turn. He also marshaled all the evidence available at the time in support of evolution.

Because Chambers feared a hostile reaction to his book, he published it anonymously, giving rise to a popular parlor game of trying to guess who the author of this infamous book actually was. The identity of the author became generally known only after his death. Chambers's book sold extremely well and was read by large segments of the general public. However, it was derided by professional scientists. Amateurs had a long history of contributing to science, but by the middle of the nineteenth century, considerable effort was being expended to professionalize science, especially biology, and *Vestiges of the Natural History of Creation* was not the product of a professional scientist. Not only was it highly speculative, but it was filled with errors, some of them of the most rudimentary sort.

Charles Darwin, Charles Lyell, and A. R. Wallace

Although nearly all scientists at the time were highly critical of Chambers's book, it did influence two important personages—Charles Darwin (1809–1882) and A. R. Wallace (1823–1913). By 1844 Darwin had been working on his theory of evolution for almost a dozen years. He did not need to be convinced that species evolve, but the negative reception of Chambers's theory made Darwin hesitant to make his views known until he had developed his theory as fully as possible. When Wallace read Chambers, he was persuaded to adopt a belief in the evolution of species. Most importantly he accepted the evolution of species before he went on his voyage of discovery, while Darwin did not come to adopt this idea until he had returned from his years on the *Beagle*. Wallace had the good fortune of being able to investigate nature with this basic belief in the evolution of species already in place, while Darwin had to reinterpret his observations in retrospect.

Anyone who reads the first edition of *On the Origin of Species by Means of Natural Selection* (1859) gets only an appreciation of what Darwin believed at that time. Darwin changed his mind as he worked on this theory, both before 1859 and after. As is the case with all scientists, Darwin used the work of his predecessors and contemporaries, accepting this part, rejecting that. However, the authors who influenced Darwin most fundamentally were not his precursors, not the occasional author who claimed that species evolve, but the predecessors who rejected the evolution of species. In particular, Darwin's argument in the *Origin of Species* was designed to refute William Paley (1743–1805) and his argument from design. In Paley's view, someone who found a watch lying in a field would assume that the object had been made. Something as complicated and precisely organized as the watch could not arise from purely naturalistic forces. A designer was required. The same conclusion followed for organisms. In this case the only possible designer was God.

Lyell had an equally influential impact on Darwin. According to Lyell, changes in the face of the globe take place gradually over long periods of time, much longer than anyone had supposed before him. In addition, these changes are produced by totally naturalistic processes. One of the best ways to document these changes is by the fossil record. However, Lyell, like Cuvier before him, did not think that the extinct species he discovered in the rocks had given rise to later descendant species. Instead, Lyell was saddled with the belief that some as-of-yet unknown but thoroughly naturalistic mechanism is responsible for the origin of species.

A belief common at the time among geologists was that fossil species exhibit progressive development. Whether God introduced species sequentially through time, or some natural mechanism was responsible, later species are more highly developed than earlier species. Although this view was widespread at the time, it was not universal, and Lyell was one of the few dissenters. Contrary to superficial appearances, Lyell thought that the succession of fossils found in the geological strata did not exhibit any sort of progressive development. Instead they reoccur in cycles. Darwin adopted much of Lyell's thought, in particular a strong predilection for gradual, natural processes and the great expanses of time that Lyell postulated, but he did not share Lyell's views on species.

For Darwin, the evolution of species through time was a highly contingent affair. The conditions of existence did not specify the outlines of predetermined hierarchies the way that they did for Lamarck. Lamarck's trees of life were eternal and immutable. If all species were killed off, the same three trees of life would reemerge, because new, spontaneously generated organisms would confront the same habitats that existed in the past. Nor was evolutionary change a matter of programmed development, as it was for Chambers. It was not predetermined at all. In these respects Darwin's view of evolution differed significantly from his predecessors. However, he did share one belief common at the time. Although he wavered on the issue, he seemed inclined toward the view that in some sense evolution is "progressive." On average, later organisms are more highly developed than earlier organisms.

The elements of Darwin's theory—his mechanism of evolution—are well-known. Darwin thought that the variations operative in evolution are slight and do not occur in any definite direction. His contemporaries tended to read him as claiming that variations occur "by chance" or "at random," but Darwin did not intend such a radical position. He thought that variations are caused—totally caused. However, no mechanism existed for increasing the likelihood that organisms might get the variations that they need. Variations occur at random, not random in any global sense but random with respect to needs. Darwin did acknowledge Lamarckian inheritance but thought of it as only a minor, supplemental mechanism in the origin of species.

Darwin's most novel contribution was natural selection, a process that Herbert Spencer (1820–1903) was later to term "survival of the fittest." Organisms produce more offspring than can possibly survive. For a while, one species might go on producing itself exponentially but only at the expense of other species and only for a while. Eventually the carrying capacity of that environment is reached. In general, certain organisms are so structured that they are better able to exploit their environments than their competitors. As a result, their numbers increase, and the adaptations responsible for this success become more highly developed.

Darwin was aware that others before him had suggested that species evolve, but his principle of natural selection was uniquely his. It was "his" theory. However, contrary to his later critics, he did not think that natural selection is the only mechanism influencing the course of evolution. He also accepted several auxiliary mechanisms, such as sexual selection, use and disuse, and even artificial selection. What his theory did not include was any supernatural directing force, either in the generation of variations or in their selection.

John Herschel and William Whewell

One of the major issues with respect to Darwin's theory of evolution concerned the nature of science. Darwin was not influenced by continental philosophers and scientists about the nature of science as they saw it. Instead, Darwin got his ideas about the nature of science from his compatriots, notably John Herschel (1792–1871) and William Whewell (1794–1866). They were the ones who taught Darwin how proper science was to be conducted. Thus, Darwin was dismayed when both of these highly influential men came out against his theory. Both Herschel and Whewell emphasized the "inductive" character of science, citing Francis Bacon (1561–1626) and Sir Isaac Newton (1642–1727) as patron saints of the inductive method, but what all of these men meant by "induction" differed significantly, and much of it hardly warranted being called "inductive."

Herschel termed Darwin's theory "the law of higgledy-piggledy," and compared it to the Laputan method of composing books—in modern terms, the notion that thousands of monkeys typing at random could produce the likes of Shakespeare. Herschel viewed nature as operating by laws, but these laws must be of the sort that the divine Creator might lay down. God would never produce the regularities observable in nature by chance variation and natural selection. As much as Whewell differed from Herschel on a host of counts, he agreed that nature must be governed by divine law—with the single exception of biological species. Species are produced seriatim through time by divine miracles. It is remarkable that Herschel, Whewell, and later John Stuart Mill (1806–1873), the leading experts on the nature of science at the time, all rejected Darwin's theory of evolution. More surprising still is the fact that the three of them together published only a few pages on one of the most significant advances in the history of science. The literature on Darwin and the evolution of species was huge. The three men most able to evaluate it remained all but silent.

Evolutionary theory served several functions in the second half of the nineteenth century. One of these was to force philosophers and scientists to reexamine the nature of science. Explicit references to God had been gradually removed from physics and chemistry, but they were still acceptable in the biological sciences and remained acceptable for quite a while longer. God was the primary cause. He laid down the basic laws that govern nature, the secondary laws. The problem with Darwin's theory was that God as people of the day conceived of him would not have instituted the sort of secondary laws that Darwin postulated. His contemporaries were willing to accept nature governed by laws but not "blind" laws.

The common claim that Darwin succeeded in getting his theory accepted by his contemporaries is misleading. Certainly a high percentage of scientists and the educated public came to accept the belief that species evolve, keeping in mind that a variety of things might be meant by the term *evolve*. But other elements of his theory were not nearly so successful, in particular those parts that are most original with Darwin, namely the nature of variation and the effects of natural selection. Many of his contemporaries called themselves "Darwinians," but almost none of them accepted all or even most of Darwin's most fundamental ideas. The differences between Darwin's theory and what his successors termed "Darwinism" are so great that at least one Darwin scholar proposed to replace the common designation "the Darwinian Revolution" with "the Non-Darwinian Revolution" (Bowler, 1988).

The Eclipse of Darwinism

Numerous scientists both in Darwin's day and after termed themselves "Darwinians." By this they meant that they identified with Darwin or took his views as a starting place for their own, while retaining the right to disagree with the great man. When

Darwin received Wallace's paper in 1858, he was amazed at how similar it was to his own theory. However, the two men disagreed with each other on several counts, not the least of which was the analogy between artificial selection and natural selection. T. H. Huxley (1825–1895) is considered with some justification Darwin's most effective advocate, but he did not think that evolution was as gradual as Darwin thought that it was. The versions of evolutionary theory that became widely accepted in Darwin's day tended to be developmental. Species evolve according to some inner or preordained program. The combination of ideas that became most prevalent were much closer to the views of Herbert Spencer than Charles Darwin. As in the case of Owen, Spencer claimed priority with respect to evolutionary theory. After all, had he not argued that homogeneity is transformed into heterogeneity? Strangely enough, the man who had the greatest claim to priority—Wallace—never claimed it.

The idea that species might not be eternal and immutable but evolve through time was in itself sufficiently upsetting to Victorians. The extension of this idea to human beings was even more disturbing. Darwin was well aware that his theory had to apply to all species, including to the human species, but he was uneasy about making these views public. He did not publish *The Descent of Man and Selection in Relation to Sex* until 1871, a dozen years after the *Origin of Species*, while from the start Spencer emphasized the messages for human societies of the competitive nature of the evolutionary process—a set of beliefs that came to be known as social Darwinism. In nature, only the fittest organisms survive to reproduce, but in human societies people who would die in the state of nature were kept alive and allowed to reproduce. If the less fit are kept alive and allowed to breed, eventually the human species would deteriorate. Hence, society should not do much if anything to ameliorate the unhappy lot of such inferior people as the Irish.

The development of Darwin's theory did not end in 1859. Darwin and his fellow Darwinians modified this theory, sometimes modifying it so extensively that it ceased to look all that Darwinian. Throughout the second half of the nineteenth century, Darwin and Wallace continued their correspondence over the evolutionary process. Darwin thought that reproductive isolation was merely a side effect of other processes, while Wallace thought that it was an adaptation. Natural selection reinforces reproductive barriers between diverging populations. Moritz Wagner (1813–1887) joined in the dispute, insisting that geographic isolation is necessary for speciation to take place. However, an equal or greater amount of energy was spent in reconstructing phylogenies, especially in Germany. Ernst Haeckel (1843–1919) was one of Darwin's most colorful converts, but his phylogenetic reconstructions were based on developmentalist views.

By the turn of the century, Darwinism was widely held to be an extremely flawed theory. Morphologists on the Continent remained wedded to some form of ideal morphology. Paleontologists tended to be Lamarckian. Evolution was held to occur in abrupt steps (saltation), not gradually. But most importantly, the newly rediscovered principles of Mendelian genetics seemed to be at odds with Darwin's theory. Several mathematically adept biologists were trying to make evolutionary theory more precise, preferably mathematical. One of these, the biometricians, interpreted Darwin as holding that variation is continuous. Mendelians, to the contrary, emphasized characters that are fairly discrete (e.g., wrinkled versus smooth seed coat). Of course, everyone was well aware that some characters vary continuously (e.g., flower petals from red through shades of pink to white), but at the time nearly everyone believed that Mendelian genetics and Darwinian evolution were incompatible.

Anyone reading the literature of this period in the history of evolutionary thought is likely to become frustrated very quickly. The issues were inherently complicated, and the terminology used to describe them was both vague and confused. However, one observation can be made with respect to this period. August Weismann (1834–1914) is credited with refuting the inheritance of acquired characteristics (Lamarckian inheritance) once and for all by means of a series of experiments as well as by distinguishing between the germ plasm and somatoplasm. However, these experiments were not quite as decisive as one might think, because numerous biologists continued to hold some form of Lamarckism well into the next century, not the least of these being Ernst Mayr (b. 1904). Except for a belief in evolution, the dominant view of the evolutionary process was diametrically opposed to just about everything that Darwin himself had thought. Darwinism was clearly in eclipse.

The Modern Synthesis

The rebirth of a truly Darwinian version of evolutionary theory was begun by a triumvirate of three mathematically minded biologists—R. A. Fisher (1890–1962), J. B. S. Haldane (1892–1964), and Sewall Wright (1889–1988). Fisher was a brilliant, arrogant, irascible visionary who founded both modern statistics and mathematical population genetics. He was also an enthusiastic supporter of eugenics (not then a discredited science). Haldane was in the best tradition of a British upper-class eccentric—a mathematical biologist, a popularizer, and an ardent Communist. Wright, to the contrary, was a fairly conventional American academic whose life includes few fascinating stories. His contributions to population genetics were no less fundamental on that account.

These three biologists showed the theoretical limits of the evolutionary process—what could and could not happen given certain assumptions. For example, evolutionary biologists at the time were convinced that very small selection pressures could not have much of an impact on evolution. Fisher showed that these intuitions were mistaken. Haldane was the first to investigate what influence kinship might have on behavior, explaining how kin can act altruistically toward one another. One simplifying assumption that was made at the time was that the populations in nature were large enough that they could be treated as if they were infinite. Wright took just the opposite tack, investigating what effect small populations might have on evolution.

Because the writings of these three men were so technical, they had relatively little impact on other biologists. A second triumvirate served to explain and expand these mathematical formulations—Theodosius Dobzhansky (1900–1975), George Gaylord Simpson (1902–1984), and Ernst Mayr (b. 1904). Dobzhansky took on the task of finding ways to test the theoretical musings of Fisher, Haldane, and Wright. These mathematical formulations certainly were impressive, but were they true? Dobzhansky served to inspire a whole generation of population geneticists to combine mathematical formulations with empirical investigations. He also brought with him from Russia knowledge resulting from the activity of several active groups of Russian population geneticists.

As a paleontologist, Simpson took on the task of refuting several of the beliefs so firmly entrenched in his discipline, mainly Lamarckism, saltation as accepted by ideal morphologists, and all forms of orthogenesis (directed evolution). Mayr took on the influential ideas of Richard Goldschmidt (1878–1958). Like so many before him, Goldschmidt viewed evolutionary change as akin to embryological development. New species arose as the result of systematic mutations that reorganized the genome. Such "hopeful monsters" were the founders of new species. Mayr assumed the task of refuting Goldschmidt and in the process showed what implications the Modern Synthesis had for systematics. He is still most widely known for his biological definition of species in terms of isolating mechanisms.

After the Modern Synthesis and Beyond

Most people view scientific theories as introduced, articulated, and then hardened into stone. The history of evolutionary theory does not fit this view at all. Prior to 1859 Darwin worked on his theory. When he received Wallace's paper, he was forced to produce an abstract of his big book on species, making it as coherent and complete as his current state of understanding allowed. Thereafter, Darwin, his contemporaries, and his intellectual descendants continued to work on evolutionary theory until it suffered its eclipse at the turn of the century. It then reemerged in a form more nearly like Darwin's own version than any of its predecessors. The purpose of the Modern Synthesis was to emphasize the fundamental agreement of all those working on evolutionary theory.

However, this synthesis did not go unchallenged. According to Darwin and later syntheticists, nearly all mutations are slight. Any major change in an organism is likely to be lethal. A few are neutral. Only a very few turn out to enhance survival and reproduction. Natural selection is not the only mechanism involved in evolution, but it is very keen sighted. The first serious challenge to the synthetic theory of evolution stemmed from the writings of Motoo Kimura in the 1960s. According to Kimura, the main cause of evolutionary change at the molecular level is the random fixation of selectively neutral or nearly neutral mutations. For Kimura, most mutations are neutral, while only a few are positive or negative in their effects. As biologists came to better understand genetic material, they discovered that much of an organism's DNA plays no role in the physiology of present-day organisms—it is noncoding. In addition, alleles (alternative copies of the same gene) of genes may be neutral with respect to the proteins produced. In general, findings at the molecular level, in particular at the level of DNA rather than proteins, support many of the arguments of the neutralist school. Although Kimura's theory was treated initially as a competitor to the synthetic theory of evolution, eventually the synthetic theory was modified to include many aspects of the neutralist school.

Niles Eldredge and Stephan Jay Gould introduced another challenge to the synthetic theory in 1972—the punctuated equilibrium model of evolutionary development. As in the case of Kimura, Eldredge and Gould introduced their ideas in a rather extreme form—as requiring a whole new theory of evolution—but as the controversy progressed, their punctuationist ideas were gradually transformed and assimilated into the synthetic theory. Advocates of the synthetic theory had long acknowledged that speciation can in very special circumstances occur rather abruptly. However, in general, speciation occurs very gradually over numerous generations. Eldredge and Gould noticed something about the fossil record. New species appear quite abruptly and continue largely unchanged until they go extinct just as abruptly. More traditional paleontologists had also noticed this phenomena but thought that it was due to the incompleteness of the fossil record. More careful study would produce the requisite intermediary forms.

Eldredge and Gould took the opposite tack. They claimed that the fossil record accurately reflects the speciation process. New species arise abruptly and remain largely unchanged until they go extinct just as abruptly. Mayr's founder principle might explain the abruptness of the appearance of new species. According to Mayr, a common method of speciation is the isolation of small populations on the periphery of their species. In most cases, these small populations go extinct or merge back into the larger population, but on rare occasions one of these peripheral isolates becomes a new species. The net effect is the abrupt appearance of new species. The existence of species over long periods of time largely unchanged was much harder to explain. Perhaps genomes are so finely balanced that slight changes are rejected.

When Eldredge and Gould introduced their views on speciation, they were greeted with considerable resistance, leading Gould to complain that the synthetic theory had hardened. New ideas were being rejected out of hand with little comprehension. As hardened as the synthetic theory might have become, it was sufficiently plastic to incorporate the punctuated equilibrium model of species as one possible mode of speciation. The punctuated equilibrium model was incorporated into the synthetic theory just as smoothly as the neutral theory had been before it.

An issue that has always taxed evolutionary biologists concerns the level or levels at which selection can occur. The most conspicuous levels of organization are genes, organisms, and species, but numerous intermediary levels also exist. Early on, some authors thought that species are not simply the effects of selection occurring at lower levels of organization but can themselves function as units of selection. G. C. Williams (1966) raised serious doubts about selection occurring at levels more inclusive than single organisms or kinship groups. His work fitted nicely with that of William D. Hamilton (1964) and John Maynard Smith (1964). In the process of integrating the theoretical views of these men, the synthetic theory became even more fundamentally modified. Organisms can increase their fitness not only by having offspring of their own but also by caring for the offspring of close relatives. What matters is not just fitness but inclusive fitness. This line of argument has been pursued by some evolutionary biologists to the extreme position that genes and only genes can function as units of selection.

Throughout the history of evolutionary theory, development has been largely left out. Time and again biologists viewed the evolutionary process as basically developmental in character, but they had no theory of development up to the task. Time and again embryologists predicted great advances in development. Time and again they were disappointed. Until quite recently the vast literature on embryology contained massive amounts of data on numerous species but little in the way of overarching theory. In the 1980s Susan Oyama and her colleagues began promising a fundamentally new version of evolutionary theory that at long last incorporates development. The advocates of "evo-devo," as it is called, predict that the synthetic theory will have to be so reworked in the face of increased knowledge of embryological development that it will cease to exist, and have to be replaced by a new theory. More likely is that development will be incorporated into the synthetic theory while that theory continues to exist and be called Darwinian.

If the history of Western thought has anything to teach us, it is the protean nature of all theories, including evolutionary theory. Scientific theories can be and are rejected, but this process is far from easy. Any criticism of the synthetic theory that turned out to have some substance was subsumed in a modified version of this theory. Instead of being a weakness, this ability to change is one of the chief strengths of the synthetic theory of evolution. As in the case of species, scientific theories evolve.

[*See also* Neo-Darwinism.]

BIBLIOGRAPHY

Bowler, P. J. *The Eclipse of Darwinism.* Baltimore, 1983. Darwin's theory of evolution was held in ill repute by the turn of the twentieth century, only to reemerge in the first few decades of the century.

Bowler, P. J. *The Non-Darwinian Revolution: Reinterpreting a Historical Myth.* Baltimore, 1988. Soon after 1859, numerous scientists came to accept some version of evolutionary theory. However, these versions had little in common with Darwin's theory. It was these non-Darwinian theories that resulted in what is commonly termed the Darwinian Revolution.

Burkhardt, R. W. *The Spirit of System: Lamarck and Evolutionary Biology.* Cambridge, Mass., 1977. An excellent explanation of Lamarck's theory.

Depew, D. J., and B. H. Weber. *Darwinism Evolving: Systems Dynamics and the Genealogy of Natural Selection.* Cambridge, Mass., 1995. An exhaustive examination of changes in evolutionary thought from Darwin to the present.

Desmond, A. *Archetypes and Ancestors: Palaeontology in Victorian London, 1850–1857.* Chicago, 1982. Desmond shows exactly how difficult it was to transmute archetypes into ancestors.

Gayon, J. *Darwin's Struggle for Survival: Heredity and the Hypothesis of Natural Selection.* Cambridge, 1992. Gayon guides the reader through the complicated relationships between theories of heredity and evolutionary theory.

Gould, S. J. *Ontogeny and Phylogeny.* Cambridge, Mass., 1977. A history of the changing relationships between views of ontogeny and phylogeny.

Hamilton, W. D. "The Genetical Evolution of Social Behavior." *Journal of Theoretical Biology* 7 (1964): 1–52. The classic paper that worked out the notion of Kin selection in great detail.

Maynard Smith, J. "Group Selection and Kin Selection: A Rejoinder." *Nature* 201 (1964): 1145–1147. Maynard Smith distinguishes between kin selection and the selection of kin groups and shows the implications for group selection.

Mayr, E. *The Growth of Biological Thought.* Cambridge, Mass., 1982. An encyclopedic history of biology, concentrating on evolutionary theory.

Oyama, S. *The Ontogeny of Information: Developmental Systems and Evolution.* Cambridge, Mass., 1985. The source of subsequent works on the role of development in evolution.

Richards, R. J. *Darwin and the Emergence of Evolutionary Theories of Mind and Behavior.* Chicago, 1987. Darwin's theory was a biological theory, but it also had implications for theories of mind and behavior.

Ruse, M. *The Darwinian Revolution: Science Red in Tooth and Claw.* 2d ed. Chicago, 1999. Ruse shows the effects that Darwin's predecessors and contemporaries had on his formulation of evolutionary theory.

Secord, J. A. *Victorian Sensation.* Chicago, 2000. A full explanation of the development and reception of Robert Chambers's *Vestiges of the Natural History of Creation.*

Sterelny, K., and P. E. Griffiths. *Sex and Death: An Introduction to Philosophy of Biology.* Chicago, 1999. Two philosophers examine evolutionary theory.

Williams, G. C. *Adaptation and Natural Selection.* Princeton, N.J., 1966. The book that roused evolutionary biologists from their group-selectionist slumbers.

THE MAJOR TRANSITIONS IN EVOLUTION

John Maynard Smith

The first living things must have been very simple, probably replicating molecules, not linked together on chromosomes or contained within cells. Today, some organisms are immensely complex: even if we find it hard to measure complexity, animals and plants are surely more complex than their protist ancestors, and protists more complex than bacteria. Yet there is no evolutionary law asserting that evolutionary lineages necessarily become more complex with time, and in fact many do not do so. So how and why has complexity increased? One approach to this problem is based on the notion of information. The complexity of existing organisms, in their biochemistry and morphology, depends on information in their genomes. Ignoring noncoding DNA, the total information in the genome of higher organisms amounts to some 10^8 bases, or 2×10^8 bits, equivalent to the information that could be contained in some ten volumes the size of this encyclopedia. Where did all this information come from?

The short answer, of course, is that it arose by natural selection, but there is more to be said. The change from isolated replicating molecules to multicellular organisms living in communities whose members can communicate with one another has required a number of "major transitions" in the way in which information is stored and transmitted: a list of these transitions is given below. None of these was inevitable: although the earliest and simplest forms of life are now extinct, there are still many organisms that have never made the transition from prokaryote to eukaryote, or from eukaryote to multicellular organism. Also, although each of the transitions listed below was a necessary precondition for further increase in complexity, this fact does not explain the transitions: evolution does not happen because of its future consequences. To explain the transitions, we must point to the immediate selective advantages they conferred on the replicating entities that then existed.

A living entity is best defined as one with the properties of multiplication, variation, and heredity, necessary for evolution. Multiplication means that the entities can make copies of themselves; variation means that the copies are not identical; and heredity means that information is transmitted "vertically" from parent to offspring. The justification for the definition of a living entity is as follows. Once entities with these three properties exist, they will evolve the adaptations that are the characteristic feature of living things: it is adaptedness that is the unique feature of life. The first entities with these properties were, in all likelihood, simple polymers, probably nucleic acids, able to replicate by complementary base pairing. Without specific replicases (enzymes that make replication more accurate), replication would have been very inaccurate, and the maximum size of the entities that could replicate correspondingly small. Increase in complexity, then, depended on the following transitions:

1. From replicating molecules to populations of molecules in "compartments," that is, vesicles or primitive cells
2. From unlinked replicators to chromosomes—strings of replicators linked end to end
3. From RNA as both gene and enzyme to a division of labor between DNA (carrying

information) and proteins (enzymes, structural proteins, etc.), requiring the origin of the genetic code

4. From prokaryotes to eukaryotes; origin of organelles, nucleus, and intracellular membranes, mitosis
5. From asexual clones to sexual populations; origin of meiosis
6. From protists to animals, plants, and fungi; origins of cell differentiation and cell heredity, and elaboration of gene regulation
7. From solitary individuals to colonies; origin of nonreproductive castes
8. From primate societies to human societies; origin of language as a nongenetic means of storing and transmitting information

All these transitions involve changes in the storage and transmission of information. Two other features are common to a number of them. First, entities that before the transition were capable of independent replication could only replicate as part of a larger whole after it. For example, a gene that is part of a chromosome can replicate only when the chromosome replicates; eukaryotic organelles (mitochondria, chloroplasts) were once independently replicating prokaryotes; ants can reproduce only as part of a colony, and in practice humans can reproduce only as part of a group. The second common feature is the division of labor, between different enzymes in a cell, between cells in a multicellular organism, and between individuals in a colony or society.

Some of the transitions were, almost certainly, unique: for example, the origin of the genetic code, of meiotic sex, and of language with grammatical rules. Others have occurred more than once. Animals, plants, and fungi, all with highly differentiated cells, evolved independently from single-celled ancestors. Eusociality, involving nonreproductive castes, has evolved repeatedly. The implication is that the unique transitions were in some sense difficult, whereas the origins of multicellular bodies, and of eusociality, required only the presence of suitable ecological conditions.

There are two features of the genetic system that are required for the evolution of complexity. The first is "unlimited heredity," that is, a genetic system capable of carrying an indefinitely large number of messages, or, equivalently, of existing and replicating in an indefinitely large number of different forms. A DNA molecule of n nucleotides can exist in 4^n different forms: if n is large, this is effectively an infinite number. There are several other biological objects that can exist and be transmitted between generations in different forms—prions and cortical patterns of cilia are examples— but the number of different transmissible structures is very limited. Such systems with limited heredity are interesting, but only systems with unlimited heredity—in effect, the genetic system and human language—can be the basis for continuing change. A second requirement is accurate replication. The error rate in DNA replication is of the order of 1 in 10^9, an accuracy that requires both "proofreading" and "mismatch repair." If the error rate was substantially higher than this, errors in the DNA would inevitably accumulate generation by generation.

The Emergence of Compound Entities

The evolution of individuals composed of entities that, ancestrally, were capable of independent replication raises the following problem: why does selection between the lower-level entities not disrupt cooperation at the higher level? That this is a real problem is illustrated by the following examples from existing organisms.

1. Genes in plant mitochondria sometimes mutate to cause male sterility in hermaphrodites; such mutations are favored because mitochondria are transmitted in ovules but not in pollen. By suppressing pollen production, a mitochondrial gene can increase the number of ovules produced by the plant, and hence improve its own chances of transmission.

2. Chromosomal genes may obtain "unfair" representation in future generations by meiotic drive or transposition. For example, a gene that can cause the chromosome of which it is part to finish up in the egg pronucleus, rather than in a polar body, is more likely than its allele to be transmitted.

3. Somatic cells are sometimes malignant; rather than ensuring the survival of the body of which they are part, they multiply at the expense of the body.

4. Worker bees sometimes lay eggs (although, being unmated, they can only produce sons).

5. Genes in a mammalian fetus are selected to acquire more nutrients than genes in the mother are selected to provide.

6. Antisocial and criminal behavior is not unknown in humans.

These examples of "selfishness" survive even after the higher-level individuals have existed for many millions of generations. The problems must have been more severe when such individuals first evolved. Two conditions were needed. First, the advantages of cooperation through the division of labor must have been substantial. Second, if the entities composing the compound individuals—the cells of the body, or the individuals in a community—were genetically similar, cooperation would be more likely to evolve, essentially because, if entities A and B are related, a gene in A causing A to help B may thereby help the survival of a gene identical to itself present in B.

Considering first the advantages of cooperation, it is not hard to think of examples. Several different enzymes in a pathway can synthesize a product that could not be synthesized by any single enzyme. DNA and protein molecules together can ensure biochemical versatility and accuracy of replication better than RNA molecules alone. A body with bone, muscle, and cartilage cells can move more efficiently than a body with only one kind of cell. A group of humans can kill a prey that would be safe from a single hunter. Theoretical models support the conclusion. For example, a model of the evolution of chromosomes shows that selection will favor an AB "chromosome" over independently replicating A and B genes, even if the independent genes replicate more rapidly than the chromosome within a cell, provided that a cell containing both A and B genes grows substantially faster than one lacking one or the other gene. At the other end of the evolutionary scale, the evolution of sociality requires that the average success of individuals in a group be higher than their success if living alone.

The second condition favoring the origin of higher-level individuals is the relatedness of the cooperating entities. This in turn will depend on the way in which genes are transmitted. Cooperation between the lower-level entities is most likely if they are genetically identical. Multicellular organisms typically develop from a single egg cell, so that all the cells are genetically identical except for somatic mutation. The possibilities of intragenomic conflict are therefore restricted, arising usually from some peculiarity of the sexual process. In contrast, the members of animal societies are genetically different, but, as W. D. Hamilton pointed out, they are usually related, so that a gene causing "altruistic" behavior is favoring the propogation of genes identical to itself. [*See* Hamilton, William D.]

In the case of organelles such as the mitochondrion, the evolution of higher-level entities—the eukaryotes—did involve cooperation between genetically unrelated cells. Whether a symbiont will evolve as a mutualist or a parasite depends on how it is transmitted. A symbiont that is vertically transmitted—for example, in the egg—will be selected to ensure the survival of its host, because its own future depends on that of its host. Such a symbiont may evolve into an organelle (*Buchnera*, a bacterium that enables aphids to synthesize amino acids absent in plant sap, is egg-transmitted and is in effect an organelle). In contrast, a horizontally transmitted symbiont is more likely to evolve as a parasite, particularly if a single host is often multiply infected.

Mutualism between genetically different organisms is not confined to that between a host organism and an egg-transmitted symbiont. For example, leaf-cutting ants depend on the presence of specialized fungi able to convert the leaves they have cut into food the ants can digest, and termites depend on a varied population of gut microorganisms able to digest wood. Such symbioses often involve vertical transmission of the symbiont: for example, larval termites acquire their gut fauna by licking the anus of genetically related adults. But there are many examples of mutualistic associations in which transmission is not vertical: for example, between nitrogen-fixing bacteria and plants, or between the animals of deep-sea vents and their symbiotic sulfur-metabolizing bacteria. In both these cases, the association is formed anew in each generation between effectively random partners. This is important, because it shows that, if the benefit to both members of a symbiosis is sufficiently great, it can be maintained in the absence of vertical transmission.

Changes in Hereditary Mechanisms

Three major changes have occurred in the means whereby information is transmitted: the origin of the genetic code, the origin of cell heredity, and the origin of language.

Today, heredity depends on the accurate replication of DNA, and function depends primarily on proteins: the two are connected by the genetic code. It is widely accepted that this pattern was preceded by an "RNA world," in which RNA molecules were both replicators and enzymes (so-called ribozymes). Some ribozymes survive in present-day organisms, representing a trace of the RNA world. How could RNA-based organisms evolve into organisms with DNA, protein, and a translating machinery connecting the two via the genetic code? In seeking an answer, two points are central. First, the assignment of a particular codon to a particular amino acid depends on a specific transfer RNA molecule. This is an RNA molecule that has two essential features: first, an "anticodon"—that is, a triplet of bases complementary to the codon in the messenger RNA—and second, a recognition site to which the appropriate amino acid is attached by an assignment enzyme. It is the presence of these two features in the same molecule—anticodon and recognition site—that ensures that a particular codon on the DNA is translated into a particular amino acid.

The second point is that, during the transition, the enzyme function had to be maintained. This can most easily be explained if each ribozyme was gradually transformed into a protein enzyme with the same function, through the gradual replacement of nucleotides by amino acids.

Szathmáry has proposed a scenario based on these two points. He suggests that the first step was the binding of a specific amino acid to a ribozyme to act as a "cofactor," increasing enzymatic efficiency, just as various substances bind to existing proteins as cofactors. A stepwise evolution of the code then became possible, while retaining enzymic function: first a number of different amino acid cofactors attached to different ribozymes, then several linked amino acids per ribozyme, and finally replacement of ribozymes by protein enzymes.

Multicellular organisms are constructed of a number of different cell types, as different from one another as unrelated protists, but all containing identical genes. This required the evolution of a dual inheritance system. Both in the body and in tissue culture, cells "breed true"—fibroblasts give rise to fibroblasts, epithelial cells to epithelial cells, and so on, despite the fact that all contain the same genes. This cell heredity depends on "labels" attached to the DNA (e.g., methylation) and copied when the DNA is replicated. These labels determine which genes are active, and so determine the phenotype of the cells. The labeling is altered during development in a programmed way. Thus, there is a dual system of inheritance: offspring resemble their parents, and, within a single individual, there are many cell lineages, with cells resembling their cellular ancestors. The development of a complex adult, with differentiated cells, depends on this dual inheritance system, but the dual system itself depends on the transmission of the appropriate information through the sexual process. In view of the complexity of the process, it seems surprising that cellular differentiation has evolved independently on three occasions. The explanation seems to be that the basic biochemical mechanisms were already present in prokaryotes: indeed, there are multicellular bacteria (e.g., myxobacteria) with a degree of cellular differentiation.

In humans, a new mechanism able to transmit information between generations has evolved: language. Human languages differ from communication systems in other animals by possessing grammatical rules. This makes it possible for a speaker with a finite vocabulary to express an indefinitely large number of meanings: language is a system of unlimited heredity. It is accepted that our ability to learn and use language is unique, in the sense that it is distinct from general learning ability, and is absent in other animals. It is a unique, genetically specified competence, making human society and history possible.

Thus, human language is a novel means whereby information can be transmitted. It differs from genetic information in that its transmission is not only from parent to offspring. However, it resembles DNA in being a system of unlimited heredity. It is intriguing that both systems are "digital" rather than "analog" (DNA is a sequence of discrete nucleotides, and language of discrete phonemes), and both are "symbolic" (codon assignments are historically arbitrary, as are the meanings of words). Historical change, unlike evolutionary change, depends on changes in linguistically transmitted information. It is interesting that one can recognize analogues of the major transitions in human history, that is, major changes in the way linguistic information was transmitted, permitting subsequent increase in social complexity. The invention of writing, without which laws cannot be codified or taxes collected, made large-scale civilizations possible. Printing was a necessary precondition for the industrial revolution. We are today living through a third major transition—the electronic storage and transmission of information—whose consequences are impossible to predict.

BIBLIOGRAPHY

Maynard Smith, J., and E. Szathmáry. *The Origins of Life*. Oxford, 1999.

Szathmáry, E., and J. Maynard Smith. "The Major Evolutionary Transitions." *Nature* 374 (1995): 227–232.

MACROEVOLUTION

Stephen Jay Gould

Definition and Standard Darwinian View

In a famous letter, written in 1844, Charles Darwin remarked, for once without false modesty but with genuine awe and affection for his intellectual guru, that "I always feel as if my books came half out of [Charles] Lyell's brain." After showing little concern for the natural world beyond a childhood passion for beetle collecting, Darwin fledged in science primarily as a geologist. (His first three books of the 1840s—preceded only by his popular travelogue, *The Voyage of the Beagle*, published in 1839—all centered on geological subjects: his successful theory for the genesis of coral atolls, and his books on volcanic islands and on the geological history of South America.) From this professional center, Darwin embraced the procedures, and particularly the defining uniformitarian worldview, of England's preeminent geologist, Charles Lyell, whose three-volume textbook, *Principles of Geology*, published between 1830 and 1833, had virtually defined the emerging field. (Lyell's status surely did not derive from his limited skills and experience in field work, but primarily from his unparalleled gifts as a great prose stylist and rhetorician; he was, after all, a lawyer by professional training.)

As a touching anecdote of Lyell's importance to Darwin, the young naturalist carried volume 1 of Lyell's *Principles* (all that had been published) when the *Beagle* sailed in 1831. His greatest recorded joy in receiving parcels from home came not from cherished letters of his family, but when volume 2 of the *Principles*, published in 1832, arrived by mailboat as the *Beagle* docked in Montevideo. (Incidentally, volume 2 opens with a brilliantly fair and trenchantly critical account of J. B. Lamarck's evolutionary theory—until then, the fullest account of the new transformational biology ever published in English.)

Darwin fully grasped the beauty and utility of Lyellian uniformitarianism—and his own decisions about a host of distinctively Darwinian concepts, including gradualism, extrapolationism, and denial of catastrophic mass extinction, follow with almost deductive necessary from the logic of uniformitarian practice and belief. The recognition and designation of a problem behind such a concept as macroevolution—indeed, the very definition of the topic as a distinct subject—flow directly from the extrapolationist central premise of Darwinian theory, and would probably not have arisen at all, or certainly not in the same way, under many competing views of life (Cuverian catastrophism, or the saltationism of Robert Chambers's *Vestiges of the Natural History of Creation*, the popular book on evolution that became such a cause célèbre with its anonymous publication in 1844).

Before we explicate this decisive issue of whether or not, as Darwin advocated in a central premise of his work, the theory of natural selection can explain the causes of all large-scale patterns in the history of life, we must clarify a purely definitional issue that has no empirical content per se, but that often has sown confusion and fruitless debate because antagonists invoke the same term in such different and mutually incomprehensible ways. Shall we use *macroevolution* as a purely descriptive term for all evolutionary patterns at the interspecific level and higher—that is, not for anagenetic changes within populations, but for the splitting of populations into two distinct and noninterbreeding entities (i.e., two separate species), and for all patterns arising within clades and biotas as the result of multiplication and extinction of species? In this descriptive sense, we make no claim about mechanisms by using the word *macroevolution*. These supraspecific patterns may arise entirely as extrapolated consequences of natural selection working, in the conventional Darwinian manner, on organisms within populations (microevolution); or they may arise by mechanisms that only become apparent or relevant at the level of speciation or above. The pattern of change will still be described as macroevolution in either case.

Or shall we invoke the term *macroevolution* in a causal sense—that is, only when we wish to claim that patterns of evolutionary change at or above the species level have been caused by mechanisms that do not operate to produce intraspecific changes within populations (microevolution); or that work in such distinct ways among species (as opposed to within species) that a different causal status must be recognized? I strongly advocate, and shall employ in this entry, the first, or purely descriptive, definition. If we stick to this descriptive usage, then we can specify the empirical pattern of macroevolution before we ask the key question: do these patterns require causes different from those regulating Darwinian change within populations? But if we adopt the causal definition, then we have made a contentious, and as yet unproven, claim simply by so invoking the word itself. I would rather use *macroevolution* as a neutral and descriptive term, so that we can achieve clarity about the nature and features of an empirical pattern before we start wrangling about the causes.

Interestingly, the twentieth century's most famous exponent of distinct causes for macroevolution, Richard Goldschmidt in his 1940 book, *The Material Basis of Evolution*, also defended the purely descriptive definition of macroevolution—and for the same reasons outlined above. That is, Goldschmidt wanted the opportunity to reach agreement about a descriptive pattern before he ruffled everyone's feathers by proposing his controversial and iconoclastic ideas about the causes of these patterns.

The causal issue became distinctive and contentious in the history of evolutionary thought because our standard explanatory paradigm, Darwinian natural selection at the organismal level, so explicitly rejects—for the Lyellian reasons outlined in the first paragraph—any separate realm of macroevolutionary causality, and so forcefully claims that intraspecific processes observable at small scales and short times in modern populations can, by the steady accumulation of tiny increments through the immensity of geological time, produce the full panoply of observed evolutionary patterns on life's full tree.

The power of this central extrapolationist credo links Darwinian evolutionism in biology to Lyellian uniformitarianism in geology for two compelling reasons of scientific preference: (1) methodologically, by locating all causality at a spatial and temporal level that can be grasped in frames accessible to direct observation and experiment in modern populations; and (2) theoretically, by situating all causality (or at least all the major mechanisms) at one level of a potential hierarchy in nature (organisms "struggling" for differential reproductive success), thus reducing an admittedly complex resulting phenomenology of levels of organization and tiers of time to a single locus of causal production.

To cite two examples of this "standard" claim for no distinct macroevolutionary causality—the first from the prime architect of modern neo-Darwinism, the second from a leading paleontologist, agreeing with this claim in the next generation of research—Mayr (1963, p. 586) opened his chapter on "species and transspecific evolution" by stating: "The proponents of the synthetic theory [of modern Darwinism] maintain that all evolution is due to the accumulation of small genetic changes, guided by natural selection, and that transspecific evolution is nothing but an extrapolation and magnification of the events that take place within populations and species." Hoffman (1989, p. 39) concurs: "The neodarwinian paradigm therefore asserts that the history of life at all levels—including and even beyond the level of speciation and species extinction events, embracing all macroevolutionary phenomena—is fully accounted for by the processes that operate within populations and species."

Claims for Distinctive Macroevolutionary Causality

The Noncontroversial Issue of Underprediction and Historical Contingency.
In asking whether observed patterns of macroevolution can be fully explained by principles and mechanisms governing intraspecific change within populations (microevolution), we encounter two distinctly different issues, the first uncontroversial and the second setting a basis for some of the most intense discussion and disagreement within modern evolutionary biology. The first issue, recording a general consensus among scientists, addresses the relationship between any general theory in the historical sciences and the actual results of the unique chronology of complex events, in this case the history of lineages through geological time. In other words, do we expect our general theories to predict the origin and extinction of dinosaurs (as opposed to some other conceivable group of dominant terrestrial creatures in the Mesozoic era), or the rise of hominids in Africa at a definite time?

Even the most committed defender of the full sufficiency of microevolutionary mechanisms (as in the quotes from Mayr and Hoffman above) would not expect their general theory to specify the precise course of unique events in life's history on the earth—for the contingencies of a plethora of unpredictable particulars, from effectively random geographic and climatic changes to the accidental deaths of genuinely fittest individuals, place any realized series of occurrences beyond the power of general theory to specify (just as we do not ask any overall theory of embryology or physiology to explain why the spermatozoon that made me, rather than my nonexistent sister, first penetrated the ovum responsible for my other genetic half). In this sense, any evolutionary theory must underpredict the actual suite of historical events that define so much of our interest and concern for the subject matter of paleontology and macroevolution. Thus, explaining the phenomenology of macroevolution—that is, the actual pathways of phylogeny—requires that we know and specify a large suite of particular historical influences necessarily lying outside the predictive powers of any general theory.

Incidentally, Darwin's own version of evolutionary theory—the principle of natural selection, still the foundation of modern views—fully embraces this necessary underprediction of realized macroevolutionary patterns, and even emphasizes this theme of historical contingency far more forcefully than many competing versions of evolutionary explanation once popular in Darwin's own time. For example, many alternatives posited overarching laws of progress or directionality that could impart a definite and predictable order to life's history. But Darwinian natural selection, based entirely on a principle of local adaptation, specifies no grand and predictable development for the geological totality. In Darwinian theory, "survival of the fittest" only designates better adaptation to particular local conditions that inevitably change in an unpredictable fashion through time, not any concept of global or general improvement. An anatomically degenerate parasite, reduced by natural selection to little more than a bag of reproductive tissue in the body of a host, may be just as well adapted to its local environment as the most admirably evolved organic machine for running, swimming, or flying.

The Controversial Issue of Independent Macroevolutionary Theory. In asking whether the descriptive pattern of macroevolution requires general theory at its own level, beyond extrapolation from the principles of Darwinian microevolution, we finally encounter the most contentious issue of this subject. In defining what would count as separate macroevolutionary theory, we must recall that Darwinian microevolution attributes change to natural selection leading to local adaptation and powered by differential reproductive success of fitter organisms within populations. To arise in this Darwinian manner, the substantial change of macroevolution must therefore be gradual, accumulative, and adaptive.

Consider, for clarity of illustration, two macroevolutionary proposals, starkly contrary to Darwinism, and both formerly popular but now effectively disproven on empirical grounds. First, various forms of saltationism, or immediate origin of new species by mutations of substantial effect (as in the "macromutations" popularized by Hugo De Vries early in the twentieth century), would invalidate natural selection because new forms would then arise by the single "lucky" step of an occasional internal impetus, and not by gradual and adaptive change mediated by natural selection. Second, several late-nineteenth-century paleontologists argued that the phylogeny of lineages follows "life cycles" akin to the ontogeny of organisms, with evolutionary stages of youth, maturity, and senility. This false theory, though still depicting evolution as gradual, contravenes Darwinism by denying control to natural selection and rejecting the adaptive basis of change. Lineages in phyletic "senility" would decline for macroevolutionary reasons of their large-scale inherent "fate," and would therefore fall beyond any possibility of "rescue" by natural selection.

Once we discard these naive (and often basically silly) proposals of past generations, and consider modern advances in genetics, paleontology, development, and philosophical rigor of theoretical formulation, we recognize several serious and important claims for genuine macroevolutionary theory—but in a style far more congenial and compatible with classical Darwinism, and not so starkly destructive or oppositional, as the older notions outlined and rejected above. Consider just two domains of legitimate macroevolutionary theory with strong claims to empirical importance and high promise of interesting integration with classical Darwinian microevolutionary mechanisms.

1. Trends explained by higher levels of selection, with species as macroevolutionary individuals. In conventional microevolutionary accounts, the classical trends of macroevolution—growing mean complexity of ammonite sutures or increasing hominid brain volumes, for example—must be explained as extrapolated consequences of natural selection based on advantages of these traits to competing organisms within populations. Because selection works not only at the organismal level (as Darwin argued) but also on a full hierarchy of genuine biological individuals from gene, to cell lineage, to organism, to deme, to species (see Gould, 2002; Gould and Lloyd, 1999; Sober & Wilson, 1998, for an explication of this claim), such macroevolutionary trends might also be caused by selection among species (treated as stable individuals) and not by extension of the transformation of populations by selection among organisms. [*See* Punctuated Equilibrium.]

Suppose, for example, that a trend to smaller average body size among the species of a clade occurs not because smaller bodies confer adaptive advantages on organisms, but because species composed of small organisms tend to manifest properties at the species level that enhance their rate of producing new species. In that case, decrease in average body size would spread as a trend through the clade by "hitchhiking" on the correlated species-level trait of high speciation rates, and not because small bodies confer Darwinian advantages on organisms. In fact, smaller body size might well be neutral or even slightly detrimental to organisms in competition with larger-bodied individuals of related species, but such competition might occur so rarely that the species-level advantage of higher speciation rates could overcome the organismal detriment of smaller bodies. In any case, such a trend, if driven by selection at the species level, unaided by (or even counteracted by) conventional organismal selection, would represent an irreducibly macroevolutionary process not explainable by microevolutionary mechanics.

2. Catastrophic mass extinction and nonselective reorganization of biotas. Darwin clearly recognized the threat to his theory posed by the apparently sudden destruction of faunas, as indicated by major changes between the fossils of successive strata at several prominent boundaries of the geological record. Darwin therefore argued that the apparent suddenness of mass extinction arose as an artifact of an imperfect fossil record, and that the extinctions, although perhaps uncharacteristically intense and rapid in these episodes (as a consequence of unusual environmental change), still occurred over a considerable interval of time, and by the ordinary mechanisms of competition and natural selection. We now have strong evidence, however, that at least one of the five major mass extinctions—the Cretaceous-Tertiary event (sixty-five million years ago) that marked the demise of dinosaurs and a majority of marine invertebrate species at the same time—was triggered by the truly sudden and globally catastrophic impact of an extraterrestrial object. Extinctions of this rapidity and magnitude not only play a major and obvious role in structuring the history of life (and must therefore be encompassed within any fully satisfactory theory of evolution), but also fall outside the purview of Darwinian microevolution and into the domain of macroevolutionary theory.

At least two important explanations for differential removal of major taxonomic groups in mass extinctions require the formulation of macroevolutionary theory. First, some groups of small membership (but not otherwise destined for extinction by ordinary Darwinian competition and selection) might disappear for random reasons based upon the luck of the draw. For example, only two lineages of trilobites remained when the greatest of all mass dyings, the Permo-Triassic event, removed more than 90 percent of all marine invertebrate species 250 million years ago. The few surviving trilobite species may all have fallen, for reasons of "bad luck" rather than "bad genes" (to use David Raup's somewhat facetious terminology), thus leading to random removal of the entire taxon.

Second, and far more importantly, groups may perish for definite and nonrandom reasons related only to their unfortunate possession of traits ill-suited to the immediate catastrophe, and not for any selective disadvantage in ordinary Darwinian competition. In fact, these fatal traits may have arisen for good Darwinian reasons in normal times—and the group's catastrophic extinction may therefore record an unlucky macroevolutionary consequence of their former microevolutionary success. As a conjectural example, dinosaurs prevailed over mammals for more than 100 million years, presumably for Darwinian reasons. Perhaps the substantially larger size of all dinosaur species spelled their doom in the Cretaceous catastrophe, but provided a major component of their previous and long-sustained Darwinian success. In this sense, any explanation for the current existence and (at least temporary) success of humans today—for mammals only achieved their opportunity for domination after the fortuity of dinosaurian extinction—requires an important input from macroevolutionary theories of mass extinction.

Both these examples—the explanation of trends by higher-level selection and the catastrophic basis of at least some mass extinctions—pose no threat or challenge to the importance or validity, but only to the exclusivity, of the conventional microevolutionary theory of Darwinian natural selection. But these and other examples do indicate that both the historical contingencies of any complex chronology and the different rules and predictions of genuinely macroevolutionary principles must also play a major part in any fully adequate theory of evolutionary mechanisms and the pageant of life's stunning variety and history.

[*See also* Species Selection.]

BIBLIOGRAPHY

Darwin, C. *The Origin of Species*. London, 1859.

Gould, S. J. *The Structure of Evolutionary Theory*. Cambridge, Mass., 2002.

Gould, S. J., and E. A. Lloyd. "Individuality and Adaptation Across Levels of Selection: How Shall We Name and Generalize the Unit of Darwinism?" *Proceedings of the National Academy of Sciences USA* 96 (1999): 11904–11909.

Hoffman, A. *Arguments on Evolution: A Paleontologist's Perspective*. New York, 1989.

Mayr, E. *Animal Species and Evolution*. Cambridge, Mass., 1963.

Sober, E., and D. S. Wilson. *Unto Others*. Cambridge, Mass., 1998.

CULTURE IN CHIMPANZEES

Jane Goodall and Elizabeth Vinson Lonsdorf

On 8 November 1960, three and a half months after beginning the long-term study of the wild chimpanzees of Gombe National Park, Tanzania (the Gombe Stream Game Reserve in Tanganyika at the time), I made my first observation of tool-using behavior. I saw a dark shape squatting on a termite mound, and, peering through the leaves, I saw that it was the male chimpanzee I had named David Greybeard, the first to lose his fear of me. He was, quite clearly, using a piece of grass to fish termites from their underground passages. A few days later, I saw him "fishing" at a different mound, accompanied by a second male. Occasionally, one would pick a leafy twig and strip off the leaves—modifying a natural object to achieve a specific goal. These chimpanzees were not only using but also making tools.

This was a breakthrough observation, as, at that time, it was thought that humans, and only humans, used and made tools. "Man the toolmaker" was a common definition of our species. My mentor, the late Louis Leakey, remarked: "Ah! Now we must redefine man, redefine tool, or accept chimpanzees as humans." Many scientists, however, were reluctant to accept my observations—I had no university training at the time. Moreover, most new data that challenge human uniqueness—such as the cognitive ability to reason, make abstractions and generalizations, transfer information from one sensory modality to another, and recognize self—typically provoke a storm of criticism.

As the years went by, we observed more examples of chimpanzee tool use at Gombe. And, as other scientists began studying communities of chimpanzees in other parts of their range in Africa, it became apparent that not only were there major differences in the type of objects selected as tools and the contexts in which they were used at the different sites, but that there were other behavioral differences as well.

The concept of culture in a nonhuman animal is still controversial. Yet it was first discussed by Dr. Kawai in 1953. He described how a new behavior, first observed in a one-and-a-half-year-old Japanese macaque named Imo, gradually spread through the group. These monkeys were fed regularly to entice them onto the beach for better observation. One day, Imo was seen washing her sweet potato in the sea, thus getting rid of the gritty sand and possibly improving the taste. Three years later, Imo introduced another novel feeding behavior—rather than picking up individual wheat grains that were scattered on the ground and rubbing the sand off on the other wrist, she threw a handful into the sea. The heavier sand sank to the bottom, and Imo skimmed off the floating (and clean) wheat with her hand. Both these behaviors were acquired, gradually, by the rest of her troop.

In their original publications, the Japanese scientists referred to these new behaviors, passed from one individual to another within the group, as subhuman, "protohuman," or "precultural" behavior. These observations paved the way for a new interest in cultural behavior in nonhuman animals. By 1973, it was possible to identify thirteen kinds of tool use and eight social behaviors that appeared to differ between the Gombe chimps and those being studied elsewhere (Goodall, 1973). But the notion of chimpanzee culture did not gain widespread credibility until 1999. This was when the prestigious journal *Nature* published the results of a collaboration between the long-term chimpanzee research sites, initiated by Andrew Whiten (Whiten et al., 1999).

Definitions of Culture

There are many definitions of culture used by scientists in different disciplines. Rendell and Whitehead (2001), for example, list fifteen. Some cultural anthropologists believe that there can be no culture without language. This means that, unless we accept American sign language (ASL) or one of a variety of computer languages that have been designed for the various ape language acquisition studies, no nonhuman animal can show cultural behavior. However, culture is also defined as behavior that is transmitted repeatedly through social or observational learning (Nishida et al., 1987). If we accept this definition, then cultural behavior is theoretically possible in any species in which individuals are able to learn by observing performances in others.

Cultural Behavior in Chimpanzees

Results from this collaborative effort, involving researchers from the seven longest running (eight to thirty-eight years) field studies—Gombe; Mahale (M group and K group); Kibale Forest; Bossou, Guinea; Tai Forest, Ivory Coast; and Budongo Forest—provided most of the data for the 1999 paper. After a good deal of discussion, filling in questionnaires and writing letters back and forth between the different researchers, a list of thirty-nine behaviors was compiled, all of which were thought to be transmitted culturally. Seven of the behaviors originally proposed were rejected because they were common to all the sites, sixteen because they had been seen only a few times at any site, and three because of unique environmental conditions at the sites where they occurred. (This does not mean, of course, that those behaviors are not transmitted culturally—most of them almost certainly are.) The field-site researchers then rated each of the thirty-nine behaviors as (1) customary (occurs in all or most able-bodied members of at least one age-sex class); (2) habitual (occurs repeatedly in several individuals); (3) present (is definitely known to occur, but is neither customary nor habitual); two categories of absent: (4) not recorded—with no obvious environmental reason as to why it is absent and (5) not recorded—with obvious environmental explanation (e.g., cannot fish for termites if no termite mounds are present); and (6) unknown, when researchers had not seen the behavior but could not be certain that it did not occur. It was for this reason that other study sites, where chimpanzees have been observed for only a relatively short time, were not included in the main analysis. It is not until chimpanzees have become well habituated, and the study has continued for at least seven or eight years, that it is possible to be reasonably sure that a behavior does not occur.

The use of stones or wood as a hammer to crack open hard-shelled fruit or nuts has been observed only in West African populations. It is customary at Tai and Bossou. A very early account of using rocks as hammer and anvil to crack open palm nuts was described in Liberia (Beatty, 1947). It does not occur in any of the East African sites. At Gombe, tennis ball–sized fruits are frequently cracked open by banging them against rocks or tree trunks. Thus, the hammering pattern exists there, but it has not been used to crack open nuts despite an unlimited supply of rocks and nuts, and despite the fact that the chimpanzees habitually eat the flesh of the palm nut, the kernel of which is cracked open at Bossou. Good evidence for cultural rather than genetic transmission of nut hammering was provided by Boeshe et al. (1994): within the range of *Pan troglodytes verus*, nut hammering is customary on one side of a large river, and quite absent on the other.

Chimpanzees in different parts of Africa sometimes use different methods to achieve the same goal. Thus, the chimpanzees of Gombe and Tai use different techniques when capturing driver (or army) ants from their underground nests. At Gombe, the chimpanzees open up a nest with frantic digging movements of one hand, which causes the ants to swarm out. A long stick is picked and partly peeled of bark so that it is smooth. This is then pushed down into the nest and left for a moment until the ants have swarmed up en masse to within about 9 inches of the chimpanzee's hand. The tool is then withdrawn, swept rapidly through the free hand, and the mass of ants transferred very quickly to the mouth. At Tai, the chimpanzees gently move the top layer of soil from the nest, disturbing the ants as little as possible, put the end of a short stick a small way into the nest, then take out the stick and pick the ants off with their lips. The only other place where the Gombe technique (which is much more efficient) has been occasionally observed is Bossou, although during a study in Sierra Leone, tools that had been abandoned at driver ant nests were long, peeled, and very similar to those used by the Gombe chimpanzees (Alp, 1993). In these examples, the differences do not seem to be determined by environmental factors. The Mahale chimpanzees, situated farther down the lakeshore from Gombe, for example, do not capitalize on their abundant supply of driver ants at all.

Leaf clipping with the teeth—noisily tearing pieces off one or more leaves—is used in different contexts in different places. At Budongo and Mahale, it is used by males to attract the attention of females in courtship displays. (Gombe males achieve the same purpose with distinctive jerky shaking of leafy twigs.) At Tai, leaf-clipping precedes buttress drumming and also signals frustration. At Bossou, it occurs in frustration and play.

It seems most logical to assume that the differences in behavior are cultural differences, passed from one generation to the next through observational learning and practice. Moreover, the behaviors discussed did not include cultural variation in feeding behavior, because existing data have not yet been systematically analyzed. Nevertheless, some quite striking differences have been listed (Goodall, 1986). The chimpanzees of Gombe make extensive use of the oil nut palm, feeding on the fruit, pith, dried flower stems, and dried or rotten wood fibers. However, they do not eat the kernel inside the palm nut, as do the chimpanzees of Bossou (and Liberia, as reported by Beaty, 1951), using stone tools. The Tai chimpanzees occasionally eat palm pith, but they do not crack open palm nuts or eat their flesh. At Mahale, the chimpanzees eat no parts of the palm. (In Budongo and Kibale, there are no oil palms.) Gombe and Mahale national parks are both situated on the eastern shore of Lake Tanganyika, and there is considerable overlap in much of the flora and fauna. Yet many foods eaten by the Gombe chimpanzees are not eaten in Mahale, even when the same species of plant grew there, and vice versa. At Mahale, the chimpanzees sometimes chew or lick the dry wood of two species of tree, visiting some so regularly that huge caves are formed. When feeding, the chimpanzees creep into these holes so that only their rumps protrude. Nothing like this occurs at Gombe, although both species of tree are present and the chimpanzees feed on the fruits. Of special interest is the fact that stones and rocks from the lakeshore and streambeds are often picked up and licked by the Mahale chimpanzees. This has never been seen in the chimpanzees of Gombe, although it is very common in the baboon troops. Many other differences were revealed in the preliminary comparison, and again most of these are likely to be transmitted culturally. Infants taste the food eaten by their mothers and watch carefully when others are eating. Often, when another starts to feed, a juvenile will approach, watch for a while, then start to feed on the same food. Moreover, when infants try a food that is not part of the repertoire of the community, an adult may remove the offending item.

How Did These Cultures Originate?

Next, we should consider how cultural variations have evolved from a basic, species-specific set of behaviors. Chimpanzees are capable of insightful problem solving. A good example that involved the use of a stick for a new purpose occurred at Gombe when an adult male, back in the early 1980s, was afraid to take a banana from my hand. In his frustration, hair bristling, he shook a clump of tall, dry grass. One piece touched my hand. Instantly, he let go of the grass, broke off a very thin twig, glanced at it, dropped it, broke off a stout stick, and with it knocked the banana from my hand. He used the same technique immediately when I held out another banana. Chimpanzees are intensely curious and watch any unusual action with close attention. Thus, a novel performance of this sort may be passed on to others in the group.

On two occasions at Gombe, juveniles "invented" new patterns. One of them, Fifi, suddenly started wrist shaking, a pattern that had appeared spontaneously in a captive chimp (Gardner and Gardner, 1969). Fifi used it when threatening an older female. A younger individual, Gilka, was with Fifi at the time. The following week, not only Fifi but also Gilka used the gesture. Gilka continued to wrist shake frequently in a variety of contexts. Fifi also continued to use the new gesture, but only seldom. Over the course of the next twelve months, the wrist shake was used by both less and less often, and finally vanished from their repertoire. In the other instance, a male infant was seen to "inspect" a female's genital area by poking at it with a little twig, then sniffing the end, instead of using his finger in the usual way. Within two weeks, another infant was doing the same. For a few months, both used the new pattern frequently, but eventually it was abandoned. If these innovations had been of adaptive value, they might have persisted and gradually spread through the community.

Infants and juveniles are the most likely candidates for introducing new patterns into their communities. Their behavior is more flexible, and they are more likely to try new things than their more conservative elders. When practicing newly acquired skills, such as the use of tools for termite fishing, infants sometimes do so in new contexts. One four-year-old "fished" with a blade of grass in a rain-filled water bowl. As he repeatedly sucked water from the end of the grass, it became crumpled. In the end, he had a minute sponge—this is the kind of behavior that could have led to the "invention" of the leaf-drinking sponge in the first place (the leaf sponge is used at all the sites surveyed). At Gombe, the only instance of the use of a rock hammer was when an infant repeatedly smashed an insect on the ground. He also hit at an insect with a wooden stick. Thus, the patterns necessary for the development of nut-cracking skills are present in the Gombe chimpanzees, who, as mentioned, crack open hard-shelled fruits by hitting them against rocks or tree trunks. It is not impossible that a Gombe chimp might, in the future, "invent" the hammer and anvil technique of the West African chimps.

Young chimpanzees everywhere investigate holes in the ground or tree trunks, using sticks or twigs. At Gombe, infants show such investigatory behavior far more often than adults. One infant poked a stick into a hole in a tree branch from which a stream of fierce black carpenter ants emerged. The infant rushed away, but his mother, who had been watching, immediately approached and ate them. Carpenter ants are not normally eaten by the Gombe chimpanzees, although the Mahale chimpanzees habitually fish for them with twig probes. It is not difficult to see how the actions of the Gombe mother–infant pair could lead to the development of a new cultural pattern in the community.

In fact, some years after we began to habituate the chimpanzees of a second community (Mitumba) at Gombe, to the north of our main study community (Kaeakela), I observed an adolescent female fishing for ants. This seemed very exciting, but the "habituators" said that they had seen this behavior in many Mitumba individuals. They had not realized that ant fishing had not been seen in the Kasakela community in more than thirty years. Just prior to this, a very shy, hard to observe Mitumban adolescent female had transferred into the Kasakela community, presumably taking her culture with her. About two years after this transfer, a Kasakela infant was observed ant fishing—we assume she had learned from the transfer female. Many other individuals of the Kasakela community have now acquired this new technique, which seems likely to become incorporated into the cultural repertoire (Wallauer and Goodall, in press).

How Are These Cultures Passed On?

Infant chimps, like infant humans, undoubtedly learn a good deal as a result of trial and error, stimulus enhancement (when the attention of an observer is attracted to a relevant object, such as a stick or rock), and direct observation and imitation. There have been numerous investigations into the cognitive ability of chimpanzees in a wide variety of captive situations. Home-raised chimps have successfully learned actions as complex as sewing, opening bottles and pouring out drinks, and digging with spades— all of the normal behaviors that are carried on around a house. Roger Fouts and colleagues (1982) described how infant Loulis, housed with three adult signing chimpanzees, acquired thirty-nine signs by the time he was four and a half years old, although he was not actually taught any signs by humans. Loulis continues to learn new signs as he gets older.

Some scientists believe that only when an individual is capable of true imitation can we describe the learned behavior as culturally acquired. Here again, there is evidence that chimpanzees are capable of imitation—even if we take Thorpe's (1958) extraordinarily narrow definition—that an individual is capable of true imitation only if he or she can acquire, by watching another, a motor pattern not in the repertoire of the species. The home-raised Viki satisfied this criterion when she pursed her lips (not part of her innate repertoire) to apply lipstick. More recently, experiments were designed to compare the relative immitative abilities of chimpanzee and human children (Nagel et al.). The experimenters found that human children precisely imitated the way the experimenter handled a rake (with tines up, tines down, or by using the handle), even when this was less efficient. The chimpanzees had no difficulty in using the rake to pull in the food, but they did not often copy the exact demonstration. They complied with Wolfgang Koehler's (1925) definition of imitation—they had the ability to "understand and intelligently grasp what the action of the other means"—described as goal emulation. As a result, their performances were often more efficient than those of the human children: they got the reward more quickly. Whiten (1998) designed artificial "fruits"—complex boxes that could be opened in different ways. He found that after repeated demonstrations, chimpanzees were able to copy different methods of manipulating and opening the "fruit" to get at the rewards inside.

Tuition

There is absolutely no question that chimpanzees can acquire sometimes extremely complex behavior when taught by humans. In the wild, however, there are only two examples, both observed in Tai, of active tuition. Once a mother seemed to demonstrate, for her daughter who was having difficulty cracking a nut, a better grip on a stone tool, and once a mother repositioned a nut on an anvil (Boesche, 1991). There are examples in captive chimpanzees, such as when the adult female Washoe was observed teaching ASL to her adoptive son, Loulis. She herself had been taught by a combination of molding—when her teacher placed her hands in the correct position—and demonstration (Gardner and Gardner, 1969). Once, when a piece of candy was held up, Washoe spontaneously took Loulis's hand and molded it into the correct position until Loulis learned this sign. Another time, in a similar context, she made the sign for *gum* with her hand on his body. The two food signs were appropriated into Loulis's repertoire (Fouts et al., 1982).

Culture in Other Nonhuman Animals

It is not the purpose of this short essay to review reports on culture in other animals. However, now that mainstream science seems increasingly prepared to consider evidence for cultural performances in creatures other than ourselves, more researchers are investigating this fascinating sphere of behavior. The problem is that few animal behavior studies, especially those involving different populations of the same species, have lasted long enough to enable researchers to properly evaluate the data. However, there is good evidence for a variety of cultural behaviors in whales and dolphins. Migrant populations of orcas along the Pacific coast of North America differ culturally from residents, not only in song dialects but also in social structure and hunting behavior (Rendell and Whitehead, 2001). Two kinds of tool use in a community of Sumatran orangutans were observed that have never been observed in long-term studies in Borneo (van Schaik, and Knott, 2001, 1996) Japanese monkeys, in addition to potato and wheat washing, have acquired another new behavior: they will rub, pile up, pick up, scatter, roll, carry, cuddle, drop, toss, and push around pebbles that they have collected from the ground (Huffman, 1996). This was first seen in 1979 and had spread to the majority of the group by the late 1980s. Stone handling has now been seen in five other geographically distinct locations. Studies of capuchin monkeys suggest that there may be many differences in their behavior that are culturally transmitted (Perry and Fragazy, in press). Finally, there is good evidence that there are differences in dialect across populations of songbirds (Jenkins and Baker, 1984). Many other species of animals undoubtedly have rich repertoires of cultural behavior, such as the other apes, many monkeys, elephants, wolves, and hyenas—indeed, all social mammals and birds with complex brains and complex societies.

Conclusion

There is a growing body of evidence to suggest that cultural transmission of diverse behaviors contributes to real (culturally) heritable differences among chimpanzee populations. It is important to point out that the cultural elements described here undoubtedly represent only a fraction of the total variation that can be expected if their behavior is further analyzed. We expect that there will be many differences in vocal repertoires, the precise use of the postures and gestures of nonverbal communication, feeding behavior, the use of medicinal plants, and nest building, among other behaviors. Analysis of the many hours of video that have been collected at all of the sites of chimpanzee field studies will be critical to future exploration of the more subtle cultural behaviors of the different communities.

The other thing that must be emphasized is that results have been presented from very few field studies. In the future, information will be available from Lopez, in Gabon, and it is to be hoped that other longitudinal data will become available from more sites as the years pass. However, the tragedy is that we can never know the full extent of cultural variability in the chimpanzee, our closest living relative. Since I began my study in 1960, hundreds of thousands of chimpanzees have disappeared as a result of habitat loss and hunting. In the 1960s and early 1970s, the live animal trade was responsible for killing thousands of mothers to capture their infants for sale to the international entertainment and medical research industries or as pets for the local market. Today, the bushmeat trade—the commercial sale of wild animals—threatens to cause chimpanzees (and many other animals) to become extinct throughout much of their range in the Congo basin and parts of West Africa.

As more and more communities—indeed, whole populations—are wiped out, the individuals and their cultures, their capacity for innovation, are wiped out, too. Those who hope to increase their understanding of cultural evolution in chimpanzees would do well to assist the international conservation community in its efforts to protect more of their forest habitats before it is utterly too late.

BIBLIOGRAPHY

Alp, R. "Meat Eating and Dipping by Wild Chimpanzees in Sierra Leone." *Primates* 3 (1993): 463–468. Description of early observations in a study that had to be terminated because of civil war.

Beatty, H. "A Note on the Behaviour of the Chimpanzee." *Journal of Mammalogy* 32 (1951): 118. One of the very first observations of a chimpanzee using tools in the wild.

Boesch, C. "Teaching Among Wild Chimpanzees." *Animal Behavior* 41 (1991): 530–532. An account of active teaching by chimpanzees in the wild.

Boesch, C., et al. "Is Nut Cracking in Wild Chimpanzees a Cultural Behavior?" *Journal of Human Evolution* 26 (1994): 325–338. Description of a study showing that cultural variation can exist within a subspecies.

Fouts, R. S., D. H. Fouts, and A. D. Hirsch. "Cultural Transmission of a Human Language in a Chimpanzee Mother Infant Relationship." In *Psychobiological Perspectives*, edited by H. E. Fitzgerald, J. A. Mullins, and P. Gage, pp. 159–193. Child Nurturance Series, vol. 3, New York, 1982. A description of how a young chimpanzee learned signs from other signing chimpanzees.

Gardener, R. A., and B. T. Gardner. "Teaching Sign Language to a Chimpanzee." *Science* 165 (1969): 664–672. Describes how, for the first time, a chimpanzee was taught the signs of American Sign Language (ASL).

Goodall, J. "Cultural Elements in a Chimpanzee Community." In *Precultural Primate Behavior*, edited by E. W. Menzel Basel, 1973. Fourth IPC Symposia Proceedings. The earliest compilation of cultural behaviors in the Gombe chimpanzees.

Goodall, J. *The Chimpanzees of Gombe: Patterns of Behavior*. Cambridge, 1986. The definitive work on wild chimpanzee behavior. Currently out of print, but available in most libraries.

Huffman, M. A. "Acquisition of Innovative Cultural Behaviors in Nonhuman Primates: A Case Study of Stone Handling, a Socially Transmitted Behavior in Japanese Macaques." In *Social Learning in Animals. The Roots of Culture*, edited by C. M. Heys and B. G. Galef, Jr. San Diego, 1996. A comprehensive description and analysis of the innovation and spread of stone-handling behavior in Japanese macaques.

Jenkins, P. F., and A. J. Baker. "Mechanisms of Song Differentiation in Introduced Populations of Chaffinches *Fringilla Coelebs* in New Zealand." *Nature* 223 (1984): 100–101.

Kawai, M. "Newly Acquired Pre-Cultural Behavior of a Natural Troop of Japanese Monkeys on Koshima Island." *Primates* 6 (1965): 1–30. One of the earliest works on cultural behavior in primates. This describes the sweet-potato washing and wheat-washing behaviors of the Japanese macaques.

Koehler, W. *The Mentlity of Apes*. London, 1925. A marvellous account of the ways in which Koehler tested the cognitive skills of his chimpanzee group.

McGrew, W. C. *Material Culture: Implications for Human Evolution*. Cambridge, 1992. One of the best overviews on the subject of chimpanzee tool use and culture.

Nagel, K., J. Ikguin, and M. Tomasello. "Proceses of Social Learning in the Tool-Use of Chimpanzees (*Pan troglodytes*) and Human Children (*Homo sapiens*)." *Journal of Comparative Psychology* 107 (1993): 2 pp. 174–186. A description of experiments on the imitative capabilities of humans and chimpanzees.

Nishida, T. "Local Traditions and Cultural Transmission." In B. B. Smuts, D. L. Cheney, R. M. Seyfarth, R. W. Wrangham, and T. T. Struhsaker, pp. 462–474. Chicago, 1987.

Perry, S., and D. M. Fragaszy, eds. In press. *The Biology of Traditions: Models and Evidence*. Cambridge University Press. Describes cultural behavior in Capuchin monkeys.

Rendell, L., and H. Whitehead. "Culture in Whales and Dolphins." *Behavioral and Brain Sciences* 24 (2000): 309–382. A comprehensive overview of culturally transmitted behaviors in cetaceans.

Thorpe, W. H. *Learning and Instinct in Animals*. London, 1956. For a long time the definite book on this subject.

van Schaik, C. R., and C. D. Knott. "Geographic Variation in Tool Use on *Neesia* Fruits in Orangutans." *American Journal of Physical Anthropology* 114 (2001): 331–342. Evidence for culture in wild orangutans.

Whiten, A., J. Goodall, W. C. McGrew, T. Nishida, V. Reynolds, Y. Sugiyama, C. E. G. Tutin, R. W. Wrangham, and C. Boesch. "Cultures in Chimpanzees." *Nature* 399 (1999): 682–685.

Wrangham, R. W., W. C. MGrew, F. B. M. de Waal, and P. Heltne, eds. *Chimpanzee Cultures*. Cambridge, 1994. An excellent collection of chapters on various aspects of chimpanzee culture.

ONLINE RESOURCES

"Chimpanzee Cultures Online." http://cguo,st-and.ac.uk/cultures. Maintained at the University of St. Andrews, Scotland. This website provides a searchable, graphical database that describes the cultural variations identified in a 1999 *Nature* paper and shows their distribution across the long-term African study sites. Additional photographic and video documentation of these behaviors will be included as material of appropriate quality becomes available.

HUMAN GENETIC AND LINGUISTIC DIVERSITY

L. Luca Cavalli-Sforza

It is agreed that chimpanzees are the nearest cousins of humans, and that the time of their evolutionary separation from humans goes back to about five million years ago. This date was first suggested on the basis of molecular evolution, using protein data, and confirmed by observations on a variety of DNA studies. The next branching date on the fossil line to humans is that of *Australopithecus afarensis* (Lucy), from the Afar region in eastern Ethiopia, 3.2 million years ago. Lucy shows considerable progress toward erect posture, and somewhat increased brain size, as measured by the internal skull volume. It is considered by many the probable progenitor of the human line and also of a line of now extinct australopithecines, found in the region extending from East Africa to South Africa along the Rift Valley, although potential earlier competitors to Lucy's privileged position have recently been described. The first fossil that most paleoanthropologists agree to consider a member of the genus *Homo*, species *habilis* is around 2.5 million years old. The name *Homo erectus* is usually given to humans with a slightly greater brain that shortly after two million years ago begins to appear out of Africa, in various parts of Eurasia. Other names are acquiring supporters, but for our level of descriptive depth, these two will suffice (Figure 1). As far as we can tell today, *H. erectus* never expanded outside the Old World, never having been found in America and Oceania. The brain expanded more or less continuously in size until it reached values comparable with those of modern humans (1,300–1,400 ml) about 300,000 years ago. This is close to four times the values of the great apes, the living primates nearest to humans, and is the mark of our species, *Homo sapiens*. Where exactly *H. sapiens* developed is not clear, but Africa is the continent where transitional forms to "anatomically modern humans" and the first truly modern humans (called *H. sapiens sapiens*) appear around 100,000 years ago. At similar dates they are also found in the Middle East, but here they seem to be replaced between 80,000 and 60,000 years ago by human types previously found exclusively in Europe, and somewhat later in northwestern Asia, labeled Neanderthal from the place in northern Germany where the first was found. It was hypothesized that between 100,000 and 50,000 years ago modern humans who had entered the Middle East from Africa withdrew or somehow disappeared from the Middle East, and were replaced by Neanderthals, believed to have moved south.

Modern humans expanded from Africa to Asia between 50,000 and 60,000 years ago. From there they spread to the other three continents. They are found in Europe in rapidly increasing numbers, starting around 42,000 to 43,000 years ago: at the same time that the previous inhabitants, Neanderthals, tend to disappear or withdraw to refuge areas, the last Neanderthal being dated at about 30,000 years ago. There is a clear distinction at the beginning between the tool set used by all Neanderthals, which, like that of the very first anatomically modern humans in Africa and the Middle East, is of the musterian type, while the first modern humans who spread from Africa to Europe and Asia use a more advanced tool set, aurignacian, starting around 50,000 years ago. They have been called behaviorally modern humans. Their arrival in China may have been as early as 67,000 years ago, based on a dating (of unknown standard

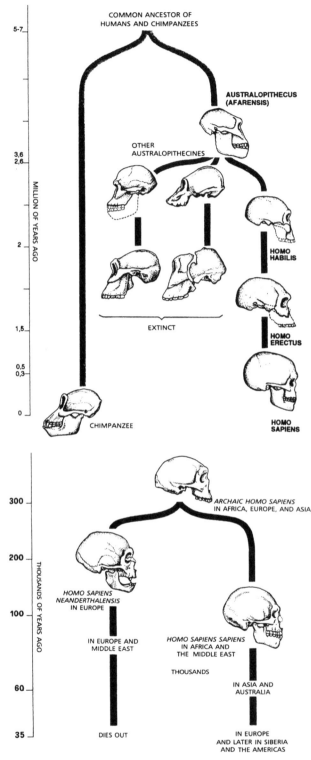

FIGURE 1. Australopithecine Family Tree Showing the Change in Skull Shape and Size in Primates.
Luigi Luca Cavalli-Sforza and Francesco Cavalli-Sforza. *The Great Human Diasporas: The History of Diversity and Evolution.* Addison Wesley, 1995.

error) of the earliest modern human skull in East Asia, and there is now a claim of an early human fossil in Oceania, physically modern, of uncertain origin.

Much evidence, both archaeological and genetic, has today accumulated in favor of the hypothesis that modern humans developed in Africa and expanded to the other continents between 60,000 and 40,000 years ago, but somewhat later from northeastern Asia to America. This hypothesis is called Out of Africa 2 to distinguish it from the expansion out of Africa of *Homo erectus* almost two million years before. The competing "multiregional" hypothesis suggests that *H. erectus* evolved into modern humans, in such a way that currently observed differences in modern humans developed in each continent over the last two million years. Support for the latter hypothesis is claimed on the basis of evidence of morphological continuity in Africa, Europe, and Asia, for the whole period, but recent accurate anthropometric analysis has disputed it. Its strongest bastion was the relative morphological similarity between Neanderthals and modern humans. But this is in total disagreement with molecular evidence on differences between living modern humans that shows their recent origin from a single African population. Moreover, recent evidence from mitochondrial DNA of three Neanderthals has indicated that the separation of European and African ancestry is definitely older than the appearance of modern humans in Africa. The conclusion from these investigations is that European ancestors developed into Neanderthals, now extinct, and the African ancestors into modern, living humans. Out of Africa 2 seems to have reached nearly general consensus as the explanation of the genesis of modern humans. The real proof that would give some credibility to the multiregional hypothesis would be the finding of a modern living human with a definitely archaic DNA. So far, none has been found.

It is clear that after modern humans spread to the whole inhabitable earth, there was more or less continuous growth, until a local saturation of population was reached in many areas. It became clearly necessary to enrich the diet with food production, and domestication of local plants and animals began. It is remarkable that this happened almost at the same time, beginning about 10,000 years ago, in some very distant temperate areas like the Middle East, East Asia, and the Mexican plateau. Very different local plants and animals were domesticated in each area. Certain cereals used as staple diet by local hunter-gatherers before the transition to agriculture were later cultivated (wheat and barley in the Middle East, rice in Southeast Asia, millet in Northeast Asia, and corn in Mexico, along with many other plants). The relative coincidence in dates of agricultural development in different parts of the world may also have been determined by global climatic changes, which altered the local flora and fauna. Agriculture allowed bringing crops closer to settlements, increasing sedentariness and population density. Especially in Europe, North Africa, and parts of China, farming began as a mixed economy, with the development of both plant and animal domesticates. The greater availability of food made possible by food production supplemented the former custom of food collection and stimulated population growth. This may have caused fairly rapid local saturation to new, higher densities and motivated spreading of the farmers to nearby areas favorable for cultivation, and grazing by animal domesticates. Growth and close-range migration thus brought about slow expansion over wider extensions, to the limit of ecological conditions where domesticates could prosper (demic diffusion). Shortly after the development of agriculture, there begins urbanization and development of complex society.

Food production by domestication of plants and animals has permitted as much population growth during the last 10,000 years as in the preceding 100,000 years. If the population was of the order of a few thousands at the beginning of the modern human expansion, it was between 1 million and 15 million at the end of the Paleolithic, about 10,000 years ago, and has since grown by a factor of about 1,000.

Methods of Analyzing the Genetic History of Modern Humans

In the last eighty years, much work has been dedicated to the genetic variation of the human species and to the reconstruction of human evolution on the basis of genetic data on modern humans. Early studies of polymorphisms (common genetic variants) focused on blood groups like ABO, and later MN, RH, HLA, and many others, using immunological methods, but later it became possible to study directly the variation of many other proteins, for example, enzymes, thanks to the introduction of electrophoretic methods in 1948. Hundreds of thousands of data of relative frequencies of alleles of hundreds of proteins were thus accumulated in thousands of populations (Cavalli-Sforza et al., 1994; Mourant et al., 1976; Roychoudhury and Nei, 1988). Alleles of proteins easily scored by these methods are usually few per gene, and it was only when researchers became able to study genetic variation directly at the level of DNA, in the early 1980s, that very substantial progress became possible. With the introduction of the polymerase chain reaction, which allows researcher to amplify specific segments of DNA, and of automated sequencing, the analysis of DNA individual variation became much easier, but in practice this approach became general only in the 1990s. We now know that, in human DNA, one finds variation for one DNA nucleotide about every thousand nucleotide pairs, as order of magnitude. Because there are a number of 3.1 billion nucleotides, the most common type of polymorphism, variants for a single nucleotide (also called single nucleotide polymorphisms, or SNPs) are expected to number about 3 million. Polymorphisms are defined by relatively frequent variants, not less than, say, 1 percent. There is an unknown multitude of rarer variant. SNPs are especially useful in many ways, and are the most common type of genetic variant, but there are other types of variants. Certain types of highly repetitive variants have special merits. For example, in DNA segments formed by the repetition of short sequences of nucleotides, for example, cytosine adenine; or CA (as in CACACACA), mutations increasing or decreasing the number of repeated units (usually by one, sometimes more units) occur with relatively high frequency. These sequences are called microsatellites, and as their mutation rate is high and can be estimated, they are useful markers for dating genetic events. Other more complex, highly repeated DNA segments are also useful for evolutionary studies and for forensic identification of subjects.

Until recently, evolutionary analysis was carried by comparing a set of populations for which allele frequencies of many genes were known, by calculating genetic distances between pairs of populations, based on differences of frequencies of specific alleles. Many methods for calculating distances were suggested; they are all highly correlated, although for specific purposes and specific types of genes, some distances are better than others. Distances between all pairs of n populations were collected in $n \times n$ matrices and analyzed in a variety of ways. Since the very beginning (Cavalli-Sforza and Edwards, 1963, 1967), evolutionary analysis used methods resulting in evolutionary trees or simplified multidimensional representations of the populations. There are a great variety of methods of both classes.

Trees describe the evolution of n populations according to models of $n - 1$ successive, usually binary splits of an original ancestral population, resulting from fission followed by genetic isolation of the two fission products. In the simplest evolutionary model, population fissions happen randomly in time, and differentiation among populations after fission is the result of random genetic drift. The rate of evolutionary change will be constant in the simplest model only if population size is constant in the various branches. The validity of reconstructed trees is affected by deviations from the simple theory as a result of different rates of drift, differences in natural selection in the various environments in which populations evolve after splits, and genetic exchanges between populations after the split. In general, the best results are obtained when populations are wide apart, of not too dissimilar sizes, and distances are calculated from the average of many genes. Tests by bootstrap or other resembling procedures help testing significance of the various branches. The best guarantee of validity is the recovery of the same tree with entirely different sets of genes. The choice of populations that are wide apart or otherwise known to be substantially isolated (having had little or no intermigration) is useful because it is known that genetic migration, whether caused by a change of residence of at least one of the spouses because of marriage or other reasons, falls rapidly with geographic distance. Therefore, one can expect a close correlation between the genetic and the geographic distance of two populations. At great distances, the irregularities of geography and ecology frequently generate barriers (major rivers, sea tracts, mountains, deserts) that severely limit cross-migration. In practice, trees between geographically close populations tend to represent their recent emigrational exchanges as affected by local geography and customs, whereas those between geographically distant populations are more likely to represent the history of fissions.

An alternative method of analysis of matrices populations by gene frequencies is the use of statistical methods of multivariate analysis, which simplify the matrices by reducing a large number of genes to very few variables, with minimum loss of information. The simplest of these is the analysis of principal components (PCs; Hotelling, 1933). It amounts simply to a change of coordinates by rotation and translation of the axes, directed by the correlations existing between the variables (the gene frequencies of the populations). Evolution by successive fissions, as in a genetic tree, introduces a specific pattern of correlations between the gene frequencies. This makes it possible to reduce the number of variables greatly with minimum loss of information, making graphical analyses especially informative. When the major factor of genetic correlations among populations is migration, and this is mostly determined by geographic distance, the map of the first two principal components of a number of populations tends to reproduce their geographic map. If populations differentiate by a sequence of fissions without major migratory exchange between branches thus formed, the order of principal components corresponds exactly to the successive fissions of the tree, and the two methods tend to give identical results. Another precious property of principal components analysis is that when there have been major population expansions at different times and places, which is frequently the case because major technological innovations have stimulated major demographic and geographic expansions of populations starting from the place of origin of the innovation, each principal component separates streams due to different population expansions. The method thus operates an analysis of the effects of evolution by successive fissions and by successive major expansions, of which there have been many in human history.

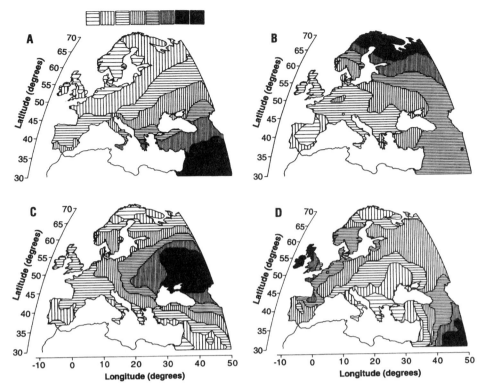

FIGURE 2. Principal Component (PC) Analysis of European Protein Data.
The first PC explains 26–28 percent of the variation and shows a radiation from the Middle East, which corresponds very closely to the archaeological picture of the spread of agriculture 10,000 years ago. It has been interpreted as due to a demic diffusion of farmers. The second PC was misunderstood at first but is likely to pool two expansions that followed a similar route in opposite directions: a major and older one from the southwest to the northeast, the postglacial expansion of western mesolithics, and from the northeast to the southwest, and a minor and later one of Uralic language speakers. The third PC indicates a radiation from southern Russia and is strongly reminiscent of the expansion of Indo-European language speakers from the Kurgan area between the Don and Volga rivers, as postulated by archaeologist M. Gimbutas. The fourth PC is highly suggestive of the Greek migrations in the first millennium BCE. The inferred area of origin of the expansion is the darkest region in A and C, and the lightest in B and D.   Reprinted with permission from L. L. Cavalli-Sforza, P. Menozzi, and A. Piazza. "Demic Expansions and Human Evolution." *Science* 259 (1993): 639–646. © 1993 American Association for the Advancement of Science.

The analysis of protein data with PCs indicated that there were many important expansions, some of which could be given an archaeological interpretation. The first application was made to European data, plotting individual PCs on a geographic map in the guise of an altitude measurement. The first PC of ninety-five genes in Europe (Figure 2A) showed a pattern that was almost identical to that of the radiocarbon dates of the spread of farming in Europe, giving support to the model of demic diffusion of farmers. Lower PCs of Europe (Figures 2B–D) indicated other expansions, which it was possible to confirm and extend by analysis of DNA data.

DNA Analysis

The sequencing of large DNA segments of many individuals permits new, radically different approaches. Data can still be studied by the same methods used for gene frequencies of protein alleles, but one can with greater ease analyze single individuals and reconstruct their pedigree. The products are still shaped as trees, but it is more

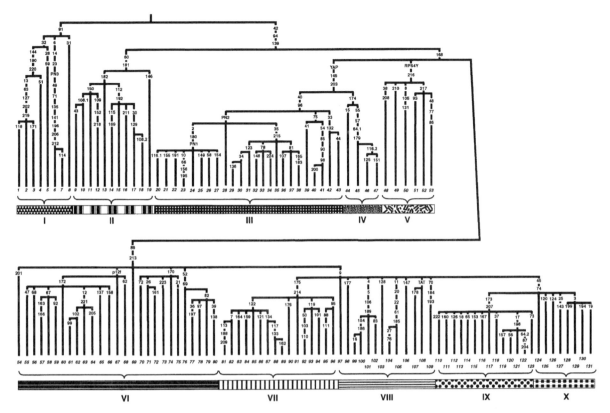

FIGURE 3. Genealogy of NRY (Nonrecombinant Portion of the Y Chromosome) Based on Mutations Detected in a Worldwide Sample of Over 1,000 Indigenous Individuals.
Haplotypes detected were clustered in ten haplogroups.   Reprinted with the permission of Cambridge University Press.

accurate to call them "genealogies." This is especially true of two types of markers, those of mitochondrial DNA (mtDNA) and the nonrecombinant portion of the Y chromosome (NRY). The first is transmitted only by mothers to all children, and the second only by fathers to sons. Thus, these generate real genealogies of individuals, much simpler than those of autosomal genes, which are transmitted by both parents and therefore show potentially more complicated networks.

MtDNA was the first to be used. Thanks to the pioneering work of Allan Wilson, it gave rise to the notion of "African Eve." It is sufficiently widely known that it is worth citing it here, although it is an unfortunate misnomer, because there does not seem to ever have been a time in which there lived only one woman. In 1987, Cann and colleagues identified the most common ancestor (MRCA) of all human mitochondria and dated it to about 200,000 years ago (TMRCA = timing of the MRCA). This work was criticized later because the data on which it was based were weak, but further research proved that it is in the right ballpark. Another great success of mtDNA was that it helped prove that Neanderthals are not direct human ancestors. It has also made it possible to see that autosomal genes, which have been adequately analyzed so far, are expected to be four times higher than that of the NRY or mtDNA. The most common alleles of such genes, those that are usually considered especially when few individuals are analyzed, refer to dates that are much earlier than those relevant to Out of Africa 2. Data from single or few populations are also irrelevant to world history.

World Distribution of Y Chromosome Haplogroups

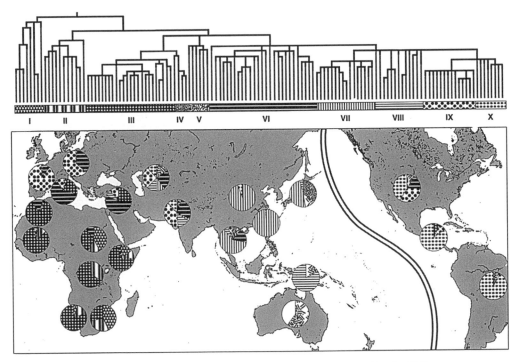

FIGURE 4. Geographic Distribution of the Ten Haplogroups Reconstructed in the Y Chromosome Geneology.
World distribution of Y chromosome Haplogroups.   Reprinted with the permission of Cambridge University Press.

The analysis of the Y chromosome genealogy can be summarized by collecting all the different genetic types (haplotypes) observed (158 in mid-2001), on the basis of more than 200 mutations into ten haplogroups (Figure 3; Underhill et al., 2000). When the ten haplogroups are plotted on the world map with their relative frequencies observed in about thirty populations (Figure 4; Underhill et al., 2001), one sees that the first and oldest haplogroup (considering the origin of the tree, as indicated by the distribution of the mutations in the closest primate species) contains only Ethiopians and Khoisan (Bushmen), almost exclusively from the Great Rift Valley. The second haplogroup extends a little bit further this area, to include Pygmies of central Africa and a few other Africans. The third haplogroup is generated only after some time, expands to all of Africa, and begins to show up in the Middle East. The fourth is found only in Japan; intermediate stages of this migration have left only few descendants en route, in the small sample of individuals examined so far, except for a few in Southeast Asia, indicating that this migration may have happened along the southern coast of Asia. Other migrations, perhaps independently, left East Africa by the southern coast of Asia, reaching Southeast Asia, and from there turning south to Oceania and north to China and Japan—later continuing, perhaps to America, again along the coastal route. It is clear that the expansion to Asia occurred also via northeastern Africa, the Middle East, and central Asia, where the major expansion, probably connected with the Aurignacian culture, may have originated and generated migrational streams again in all directions, also westward to Europe, northward to Siberia, and from there to America.

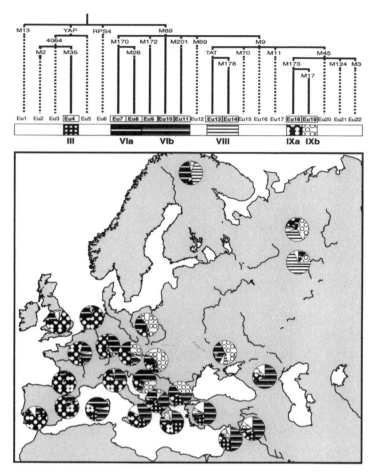

FIGURE 5. Origin of Contemporary Europeans on the Basis of Their Y Chromosome Haplotype. White sectors of circles correspond to rare haplotypes that could not be indicated in the drawing. Reprinted with permission from Semino, O. et al. "The Genetic Legacy of Paleolithic Homo sapiens sapiens in Extant Europeans: A Y Chromosome Perspective." *Science* 290 (2000): 1155–1159. © 2000 American Association for the Advancement of Science.

European data are at the moment more abundant than elsewhere, and Figure 5 shows a study of the peopling of Europe after 42,000 years ago, using the most informative NRY markers. Many African and Asian ancestors settled in Europe, as indicated in the simplified tree above the map. An analysis of the data showed that the first NRY migration in Paleolithic times occurred via the Middle East but was very limited, whereas the major Paleolithic Asian expansion was made of descendants of a later mutation, most probably arisen in central Asia. Figure 5 distinguishes the western and eastern part of these Paleolithic people, who differentiated one from the other most probably after the last glaciation, lasting from 25,000 to 13,000 years ago, forced Europeans to withdraw into two major southern refugia, a western one in Southwestern France/Iberia and an eastern one in the southern Ukraine and the Balkans. After the withdrawal of ice, the Paleolithics (now called by archaeologists Mesolithics because of their distinctive Late Paleolithic culture) expanded toward northwestern Europe, while the Paleolithics expanded to northeastern Europe. Other later expansions are visible in the map: one from the Middle East, which was especially strong when Neolithic farmers spread from the area of origin to the whole Mediterranean, and, to a somewhat lesser extent, toward central and eastern Europe. The latest expansion, probably of reindeer hunters and herders, took place in the last 3,000 to 5,000 years in northeastern Europe across the Urals.

The analysis of mtDNA also indicates three major migrations from East Africa, the first to all of Africa, and two other later ones to Asia, one via the southern coast of Asia to SouthEast Asia, Oceania, and East Asia, and the other to central Asia and from there to Siberia, Europe, and America. Thus, the pictures of expansion at the global level are qualitatively in good agreement for males and females, with discrepancies at the local level, where one finds quantitative and even qualitative differences between dominant male and female migrations. Current genetic methodology will undoubtedly clarify considerably the details of the history of settlement of the earth by modern humans.

Genetic History of Out of Africa 2

The model of migration of modern humans in the last 100,000 years is mostly based on the NRY genealogy, but in principle it is in good agreement with mtDNA data and with earlier conclusions obtained by tree and PC analysis of populations using protein data, as well as with the less abundant data on DNA of other autosomal and X chromosome genes. Yet protein data have inevitably less power, and the study of DNA of one autosomal gene, analyzed separately from the rest, can provide more limited information. Some genes show a complex history, strongly affected by specific natural selection. An autosomal chromosome in the nucleus exists, moreover, in four more copies as compared with DNA transmitted by uniparental inheritance, as NRY and mtDNA. Hence, the TMRCA is calculated from those few differences in the pattern of migration (Seielstad et al., 1998) and variation of reproductive success of males and females (Shen et al., 2000). The amount of mtDNA unfortunately is very small (about 16,600 nucleotides), and the mutation rate is very high, which in part compensates for the small size by providing more mutants, but also generates noise that makes it more difficult to reconstruct its genealogy in complete assurance.

NRY has a lower mutation rate and could give rise to satisfactory genealogies only when its SNPs could be detected in sufficient numbers by appropriate methods provided with higher resolution (Underhill et al., 1998). The genealogies produced with NRY for the world (Underhill, 2000, 2001), Europe (Semino et al., 2000), and Southeast Asia (Li Jin et al., 2000) have greatly contributed to specifying the current model of modern human expansion from Africa. In principle, there are strong similarities between the pattern shown by the study of the male and the female lineage, but there are also some interesting differences. For example, the male MRCA is younger than the female. They both are an upper bound of the first expansion of modern humans out of Africa, and the difference between the MRCA and the actual beginning of the expansion has never been directly estimated. There is, however, an alternative method for estimating the time of beginning of an expansion (Harpending and Rogers; Slatkin), and unpublished observations with mtDNA indicate in fact at least three expansions of women: one within Africa, one from Africa to Asia and to the rest of the world, and one from Asia to Europe, at dates around 100,000, 50,000, and 15,000 years ago, respectively. These dates are very approximate. With NRY, one can see more clearly the second date, again with a younger date than for mtDNA, and the first and third expansions give lighter signals. These male–female differences are again to be expected because of higher mortality of males, higher short-range migration of females, and higher variance of reproductive success of males. Moreover, the initial population size of the population responsible for the major expansion of Out of Africa 2 was rather small (about 1,000 individuals).

The Role of Innovations in the Expansions of Modern Humans

It is likely that many essential innovations were instrumental in making these expansions possible. There is some indirect archaeological evidence that modern humans were linguistically mature before the beginning of Out of Africa 2. G. Isaac suggested that language was responsible for the considerable cultural differentiation observed in Africa in the period 100,000 to 50,000 years ago. The development of art to incredible levels of sophistication is typical of behaviorally modern humans in the last 40,000 years. Although the richest findings are found in Europe, undoubtedly because more research has gone into it, art with similar characteristics is associated with early modern humans almost everywhere.

Demographic inferences from the genealogies of mtDNA and NRY indicate that a small group of early modern humans from East Africa is responsible for the earliest expansions and is ancestral to all other modern humans. Most probably this ancestral group had an already fully developed language at the beginning of Out of Africa 2, and this may be the original language from which all existing languages developed. This is in agreement with the fact that all living humans have languages of similar degrees of grammatical, phonological, and lexical complexity. Although language development may have been gradual or in steps, beginning perhaps with *H. habilis*, it seems likely that the most dramatic developments of language were very late, just prior to Out of Africa 2, and made possible by some new genetic acquisitions characteristic of modern humans. One clear-cut predisposition, which is typically human and essential for the capacity to acquire modern languages, is the presence of a critical period in the first five or six years of age that almost compels children to learn to speak. If a language is not learned in this period, it cannot be acquired afterwards, or only unsatisfactorily. This early development must contribute in a major way to the neural basis needed for the ability to speak. An important part of it, presumably also of genetic origin, must be the capacity to learn grammar and syntax that seems unique of modern humans. Communication requires at least two individuals provided with the specific capacity; if this is transmitted genetically, it is most likely to be found in more than one member of a family where the mutation originated. Thus, a mutation favoring communication will be easily shared by members of the same family, and a family group may be the first social nucleus in which it will easily give its fruits. From this first family it could expand fairly rapidly in successive generations to a closely knit, small social group. If skills connected with language, and, in general, communication, are the result of specific genetic novelties, it seems reasonable to assume that they could spread fast to all the members of the social group in which they originated, and give the group considerable power of demographic and geographic expansion. The usefulness of a highly developed language as an aid in the spread to unknown territory is easy to understand, thinking of how it may have eased communication between members of hunting bands exploring new territory, probably by sending scouting parties that reported to the group. It has been noted that the sharing of information between individuals of a group is different from the sharing of other resources, as it involves lesser costs than other forms of sharing, and is thus more likely to spread rapidly (Lachmann et al., 2000), but if it depends on a skill that is determined genetically, its dynamics will be the slower one of a genetic trait, and not the fast one of a culturally transmitted trait. Thus, it must be a kin group, not a loose social group, that can profit from it. This supports the hypothesis, originated from an evolutionary analysis of NRY data, that it is a small, genetically closely knit population that is at the root of the expansion of modern humans.

Linguists are divided on the issue of whether there was a single original language, but the most common belief is that the problem is insoluble. There may be no compelling linguistic reason in favor of the existence of one or many original languages, but perhaps the best support may come from genetics, if it will be confirmed that the size of the original population responsible for Out of Africa 2 was very small. In other words, there may have been many different independent origins of language, but if the population responsible for Out of Africa 2 was quite small, all surviving languages must come from a single one, or at most a few related ones, spoken by this highly localized and privileged community. The only evidence of a strong difference in original languages, the presence of special sounds known as clicks in the first two haplogroups, in addition to all other sounds common to all languages, and the absence of clicks in all other haplogroups, apart from cases of borrowing by a few neighbors who have had much exchange with the original click speakers, speaks for at most two partially independent origins of both linguistic and genetic peculiarities of modern humans and does not make a double origin necessary.

Another innovation, which may have contributed to Out of Africa 2, is navigation. It is practically certain that New Guinea and Australia, from Southeast Asia, were settled first by modern humans, most likely 40,000 years ago, but possibly 10,000 to 20,000 years earlier, and that it was necessary to cross several tracts of sea requiring some, however primitive, rafts or canoes. Archaeological traces of these are unlikely to be found, and the oldest boats now known are much later (from the Early Neolithic). Although the only strong evidence in favor of navigation is the settlement of Oceania from Southeast Asia, it is accepted by several paleoanthropologists that navigation may have helped the expansion from East Africa to South Asia, and even from East Asia to America. Thus, not only new tools but also greatly improved communication and means of transportation by water must have greatly aided the expansion of behaviorally modern humans, beginning about 50,000 to 60,000 years ago.

If innovations such as new hunting techniques and the domestication of local plants and animals are likely to be later responses to local overcrowding, early expansions to neighboring, insufficiently exploited areas and also by colonization of more remote places may also be natural responses to the same stimulus. Additionally, curiosity and the hope of better environments, or the desire of a group to distance itself from a hostile group by going to a remote territory, are all potentially powerful stimuli for migration and expansion. Migrating groups cannot have been too small because of the danger of inbreeding, but even a very small group of one or two dozen people may have been able to give rise to a growing, successful population. There are historical examples of this, as in the case of the mutineers of the HMS *Bounty*, who in 1789 settled in the Pitcairn Islands. A potential aid to success was the multiethnic composition of the population, which could have decreased the dangers of excessive inbreeding.

How long did it take to settle the world? The rate of advance must have been very different in different areas, and there were many difficult passages because of mountains, high passes, deserts, and other barriers. Where territory is relatively homogeneous, the rate of advance may have been fairly constant. There have been some calculations of rates of advance in relatively homogeneous environments. One archaeological example of an estimation of the rate of advance in virgin territory for hunter-gatherers is from the refugium in southwestern France, where occupants of northwestern Europe had retired during the last glaciation. As ice was withdrawing in this region, around 13,000 years ago, the advance northward was calculated to be about 0.6 kilometers per year as the crow flies (Housely et al., 1997), but it may have been constrained by the rate of ice withdrawal. Rates of advance of populations settling a virgin or almost virgin territory were calculated from archaeological data for agricultural expansions, being about 1 kilometer per year in Europe, and somewhat higher for the Bantu expansion in Africa. It was higher (3–6 km/yr on average) in the Pacific during the Malayo-Polynesian expansion, which started about 5,000 years ago and continued until about 1,000 years ago. In this expansion, the spread took place almost entirely by water and happened in stages from island to island, located often at great distance from each other.

An expansion is caused by continuous population growth accompanied by emigration to the nearest underpopulated area, when local saturation has been reached. The rate of radial advance of an expansion can be predicted by R. A. Fisher's model, and is proportional to the geometric mean of the growth and the migration rates. The model assumes that these rates are constant over time and space. The population growth curve is logistic, beginning exponentially and reaching an upper limit fixed by the carrying capacity of the land, as determined by the environment and the mode of generation of food, but the population growth rate to be used for estimating radial advance is exponential at the growing fringe, which can be very high. The maximum rate of human population growth is about 3–4 percent per year, doubling in 23.5 to 17.6 years. Such rates are observed today in many parts of the world other than Europe and North America. They were also observed in the spread of European farmers in South Africa and in French Canada in the last three centuries, as well as in populations of small islands. These rates of growth assume birth and mortality rates values that seem unrealistically high compared with those observed in living hunter-gatherers, but the latter birth rates may represent an adaptation to the present limitation of territory. Also, death rates may be smaller in truly virgin territory with low population density, especially at the growing fringe of an expansion, where microbial and parasitic diseases are likely to be a lesser evil. The migration rate to be used for predicting the radial rate of expansion is the individual dispersal rate, that is, the mean square distance between birthplaces of parent and offspring. If we accept estimates of dispersal rates calculated from modern mating distances of hunter-gatherers, the rate of advance of hunter-gatherers could be 1 kilometer a year or greater, assuming growth rates at the growing fringe of the expansion were not far from the maximum. These rates of advance as the crow flies would bring hunter-gatherers from East Africa to Southeast Asia and New Guinea about 10,000 to 15,000 years ago, which seems compatible with the observed times of first settlement of the various continents. The expansion rates observed for farmers are in reasonable agreement with those predicted. Agriculture and animal breeding were the major stimuli to population increase in the last 10,000 years and to the origin of towns, cities, and complex societies.

Linguistic Diversity and Its Evolution

It seems unavoidable to conclude that the origin of language has been a major propellant of the enormous cultural changes witnessed in modern humans, by favoring communication. But languages evolve fast, in comparison with genes. The needs of communication force a language to remain sufficiently homogeneous so that all speakers clearly understand each other; however, especially in older times, speakers were relatively small social groups of individuals living at relatively short distances from each other. There is no communication without mutual understanding, and speech training allows communication to remain almost unchanged for at least a few generations—at least for individuals who live at the same time. There is no such constraint of constancy over space, except within the range of normal social intercourse, and the rate of change is such that considerable local differentiation develops. Today there are about 5,000 languages, most of them spoken in very small groups, and the needs of communication among people speaking different languages are served by lingua francas. This must have been a common custom since the very beginning of linguistic differentiation. Polyglottism is common especially where fractionation of languages is high. Language change is faster at the level of lexicon and of sounds, and slower for grammar and syntax, constraints for which seem to be greater. Linguists have created a taxonomy of existing and known extinct languages. Until a short while ago, it was believed that it was impossible to reconstruct the history of language differentiation prior to 5,000 to 6,000 years ago, but the application of more rigorous methods of analysis and the recognition that some words, often those learned most early, are more highly conserved than others are helping to recognize more remote relationships. Most linguists have shown little interest in this type of research, but taxonomy is notoriously the source of disputes in almost every field. Nevertheless, consensus is slowly building. The most recent classification by Joseph Greenberg groups all languages in twelve superfamilies, implying that human languages had a largely common origin. There is an undeniable similarity between what is known about the history of genetic differentiation of modern humans and the major linguistic families, although it can hardly be perfect, because we know many historical examples of radical replacement of a language by another in a relatively short time. This is, however, a result of political events that have become much more common with the development of armies and the frequency of large-scale military interventions and mass migrations, especially after the beginning of transoceanic travel.

It is easier to see a correlation between languages and genes by looking at linguistic superfamilies, families, and their major branches (Figure 6), also because the differences among families or superfamilies must have been established in parallel with earlier, major genetic events, which took place between 50,000 and 5,000 years ago. These events were major migrations, resettlements, and expansions that inevitably affected all aspects of life and were sustained by genetic subsets of populations, each of which had closer cultural, and therefore linguistic, internal connections. One can also consider the correlation of genes and languages in another way, as the result of a common mode of transmission. Genes are clearly transmitted from parents to children. In the absence of schools, however, which are a late development, language

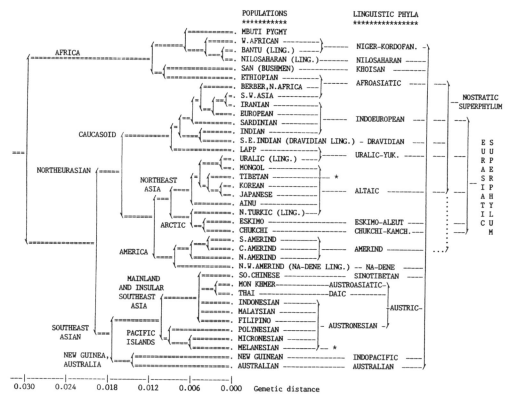

FIGURE 6. Comparison of the Tree of Fourty-Two Human Populations, Based on 110 Protein Genes, and Tree of Languages.
Late language replacement is responsible for the few disagreements (e.g., Sino-Tibetan languages spread from northern China to southern China at the time of the unification of China under the Han Dynasty).   L. L Cavalli-Sforza, P. Menozzi, A. Piazza, and J. Mountain. "Reconstruction of Human Evolution: Bringing Together Genetic, Archaeological, and Linguistic Data." *Proceedings of the National Academy of Sciences of the United States of America* 85 (1988): 6002–6006.

teaching took place in the family or in a very small social group, and was largely done by mothers. This is referred to as vertical cultural transmission, and it has been shown to have a dynamics similar to that of genes (Cavalli-Sforza and Feldman). This means that the common cause responsible for the correlation of genes and languages must be the similarity of the mechanism of genetic and of (vertical) cultural transmission across generations. The same phenomenon (a similarity with the genetic evolutionary pattern of modern humans) is also observed for parasites partly transmitted vertically, including infectious diseases such as hepatitis B and papilloma viruses.

Races

Marriage customs generate enough admixtures with neighbors, even if mating distances are, on average, small. Thus, every human group, unless it is very small and highly isolated, has a high degree of heterozygosity; that is, heterozygotes are very numerous and in most sampled populations are rarely above the frequency expected under random mating. Heterozygous advantage, or the greater Darwinian fitness of a heterozygote compared with the fitness of the two homozygotes, is also not rare, although its existence has been clearly shown only for a few genes. The average genetic difference between two random individuals from the whole world, or the probability that two individuals of a population are genetically different, is higher only by about one-sixth than that of two individuals from the same, average population. This is also expressed by saying that the average difference between two individuals of a single population is 85 percent that between two random individuals of the whole species (Barbujani et al.; Lewontin et al.). It is likely that such a small genetic variation among populations is also in part a consequence of the history of modern humans, which we believe originated from a relatively small population that had a regular genetic variation. In the relatively small period elapsed since the origin, between 2,000 and 4,000 generations ago, new variation must have accumulated among populations because of fresh mutations, and it cannot be large. This is enough for understanding that clearly distinguishable races could not have formed. In fact, it was noted already by Charles Darwin that it is impossible to classify humankind in sharply distinguishable races, and all taxonomists who have attempted racial classifications disagree widely. This is substantially different from the greater intraspecific variation observed in many other species, including primates closest to humans.

Nevertheless, one observes some average differences between continents, or large geographic areas, mostly in skin, hair, and eye color, and in some other clearly visible morphological traits. They allow us to distinguish easily the geographic origin of individuals, and have helped in generating the superficial impression that there are races, and that they are "pure," in the sense that they are homogeneous for a few of these traits. This phenomenon seems to be mostly confined to traits that have adaptive value with respect to the enormous climatic differences that exist on inhabited lands. Traits sensitive to climatic differences must frequently affect surface characteristics of the body, which are almost bound to be especially conspicuous. The impression of strong differences among races, and of racial homogeneity for traits showing strong differences among races, is therefore largely a perceptual error. Genetic differences for nonconspicuous traits among populations only very rarely reach the strong levels observed for these conspicuous traits.

Cultural differences develop easily among isolated populations or social layers and cause important variation in behavior of individuals and social groups. They are usually of a nongenetic nature, but they are often erroneously perceived as having genetic origin. In reality, it is usually very difficult to prove if differences in behavior are genetic or cultural (the result of customs, environment, etc.). Nevertheless, it is very likely that most behavioral differences noted in different societies are of cultural origin. Racism, the belief that such differences are genetically determined, is therefore a very common mistake. It is easily reinforced by our attachment to habits, customs, and beliefs imprinted on us by upbringing and by everyday life, as well as by intolerance and xenophobia, easily elicited by the superficial observation of differences in habits and customs among different cultures. Racism creates serious difficulties in the harmonious life of social groups and frequently reaches levels that make it inevitable to consider it as a serious social disease.

One corollary of the pattern of evolution of modern humans that can be inferred from the evolution of the NRY is about the skin color of earliest modern humans. The two most ancient human haplogroups are not black, but brown. They include East Africans, especially Ethiopians, Khoisan, and African Pygmies. From East Africa one of the paths followed in the settlement of Asia must have been by the Nile Valley or, across the Red Sea, to Saudi Arabia, where archaeological data are scanty. From central Asia, North and East Asia and eventually Europe and America were reached. These are either brown or, in the case of Europe, white. White skin is usually attributed to cereal consumption, a custom largely developed in the last 10,000 years. The darkest complexions are found in a branch of haplogroup III, which has a relatively late origin, and are mostly spread in sub-Saharan Africa. They are also observed by independent mutations in the expansion of other haplogroups along the southern coast of Asia and to Oceania, wherever people stayed closer to the equator. The darkest complexions are thus most probably a derived characteristic, and the original color was most probably brown.

Genetics and the Future of Humanity

The rate of communication keeps increasing rapidly, and the trend to globalization can only bring about an increase in migration, acculturation, and genetic admixture of groups that were previously more or less completely isolated. The global genetic variation is not likely to change dramatically—it is highly conserved in Mendelian systems—but its distribution among groups and within groups will change, with a decrease of the former and an increase of the latter. Ratios of existing differences between the numbers of genetically different groups such as Europeans and non-Europeans are currently changing rapidly, and the low birthrate of the former compared with the latter will cause a progressive browning of humankind, and will be more easily observed in countries with European majorities. Other factors may, however, act in other directions. The much higher spread of certain highly lethal infectious diseases such as AIDS is at the moment ravaging Africa and Asia more than the other continents.

Medicine has become a very efficient, much respected, and desired factor of change in mortality. It also allows the survival of certain types of disabilities, which are bound to increase their incidence. Curable genetic diseases and handicaps are likely to increase in numbers, both relative and absolute. It is difficult to predict the efficiency of future medical technologies, but it is likely that the cost of medicine will increase considerably, not only because of the increase in complexity and cost of medical treatment but also because of the increase in the number of successfully treatable, transmissible diseases. Moreover, many of the greatest advances of medicine, such as the phenomenal decrease of mortality and morbidity due to infectious diseases that occurred in the last 150 years, are increasingly threatened by the evolution of parasites toward increasing antibiotic resistance. Humans are the most highly cultural animals, and cultural evolution has taken the lead over their genetic evolution, but it is also affecting the evolution of all other living organisms.

MOTHERHOOD

Sarah Blaffer Hrdy

Evolutionary biology has come a long way as far as mothers are concerned. Nineteenth-century evolutionists like Herbert Spencer correctly observed that mammalian females ovulate, gestate, bear young, and lactate. They then concluded (incorrectly) that the diversion of so much energy into reproduction inevitably led to "an earlier arrest of evolution in women than in men" (Spencer, 1873). Females were thought of as unvarying breeding machines, designed to bear as many young as they could and to selflessly commit to nurturing each one. The absence of variation between females supposedly limited the opportunity for natural selection to operate on the female sex, which is why Charles Darwin and other early evolutionists assumed that women's intellectual and emotional faculties were inferior to those of men.

An unfortunate byproduct of a century-long delay in correcting wrong assumptions about the female sex was that by the last quarter of the twentieth century, as evolutionary biologists began to recognize the full extent of variation among females and to take selection pressures on females into account, many social scientists, and feminists in particular, had long since closed their minds to evolutionary arguments that they viewed as hopelessly biased. Much of the antagonism characterizing relations between the natural sciences and those in the humanities at the end of the twentieth century—sometimes referred to as the "science wars"—might have been avoided if both sides had been better informed. Evolutionary theory was not the problem here so much as long-standing biases about females that colored starting assumptions. By the last quarter of the twentieth century, however, sociobiologists had begun to incorporate the full range of selection pressures on females—that is, to take female perspectives into account—and to expand evolutionary theory to include both sexes.

Evolutionary paradigms about females began to shift almost imperceptibly from the 1970s onward with the realization that even though Darwinians talk about the origin of species, Darwinian natural selection rarely acts at the level of the group or species. Mothers did not evolve to sacrifice themselves for the benefit of the species, but to translate such reproductive effort as they could muster into progeny who would themselves survive to reproduce. This new view presupposed variable responses from individual mothers, depending on their physical condition and local circumstances.

Prior to this shift, most empirical research on maternal behavior had been by comparative psychologists working with captive animals. Mothers were studied outside of the social and ecological settings in which they evolved, while research protocols relying on "check sheets" preordained what would be noted. Observers counted how often a mother rat, dog, or monkey approached, licked, or suckled her infants. Because food was provided *ad libitum* and no other animals were in the cage besides the infants, little attention was paid to the need for mothers to "make a living," or to interact with other animals that might either help or hinder infant rearing. There was not much else for mothers to do except nurture young, and if they did do something unexpected (e.g., eat the young), their behavior was dismissed as "abnormal."

"Maternal behavior" recorded in this narrowly prescribed way provided convenient operational categories that permitted quantification of nurturing behaviors. But the methods used reinforced prior stereotypes that equated maternal behavior with nurturing. Initial preconceptions shaped the interpretations of primatologists and other field workers when they finally set out (beginning around 1960) to observe mothers in the wild. These early researchers took it for granted that all females become mothers and then instinctively devote themselves to rearing each offspring.

Confronted by an array of challenges and constrained by local history and ecology, different females were likely to respond very differently. There was no one way to be a mother. With sociobiology's recognition of selection at the level of individuals came the belated realization of how much variation there was among mothers, more nearly strategic planners and jugglers than breeding machines. Here was a sex wide open to natural selection.

The Ecology of Motherhood

The reproductive ecologist David Lack was among the first evolutionists to study breeding females as individuals. He recognized that both individual mothers and their circumstances would differ, and that many mothers might well be selected to breed below capacity, or to tolerate brood reduction (e.g., tolerate older, stronger chicks, and eliminate weaker ones), so as to pull at least a few offspring through, or to increase the mother's chances of surviving to breed again, perhaps under better circumstances. This logic paved the way for thinking about maternal behavior in terms of fitness trade-offs made by each female over the course of her life. New models for understanding maternal behavior assumed that mothers traded off reproduction in the present against the possibility of doing better in the future. How one female fares in terms of another in terms of lifetime reproductive success depends on how well a mother handles a series of trade-offs that she encounters over her lifetime. Consider golden hamsters, adapted to irregular rainfall and erratic food supply in arid habitats in the Middle East. In addition to building nests and licking and suckling pups—maternal behaviors in line with early stereotypes—a mother hampster may adjust her litter size by eating a few. Among mice, mothers may cull the smallest pups to enhance litter quality. Occasionally, mother mice, bears, or lions abandon whole litters, if times are tough or if the number born falls below an acceptable threshold.

For each mother, life is a series of such "decisions" about how best to allocate resources over her lifetime, although, except among humans, these decisions are not conscious. When is the best time for the body to stop growing and mature, to divert energy otherwise available for body maintenance and continued growth, to reproduce? When should she ovulate a fertile egg? When should she spontaneously abort, as some wild horses and monkeys do when confronted with dangerous social conditions? Juggling somatic versus reproductive effort represents the foremost trade-off in the life of every mother. The second trade-off is between quality and quantity—whether to produce many offspring, investing little in each one, or whether to produce only a few, investing a great deal (Figure 1).

Many resolutions to these recurring trade-offs are determined over the life of a specific individual. Others are the outcome of solutions forged cumulatively by many individuals, over evolutionary time, leading to species differences. For example, a few prosimians such as the Madagascar sifakas still produce litters that mothers stash in nests. But most primates bear single young, widely spaced, heavily invested in by a mother who carries her semicontinuously nursing infant with her.

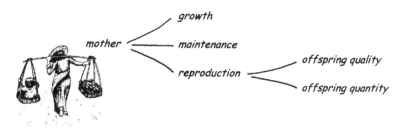

FIGURE 1. Mother's Main "Life History Tradeoffs."
Courtesy of author.

In animals where maternal fecundity or offspring survival is significantly enhanced by large maternal body size, there may be selection on mothers to delay reproduction and keep growing larger. Should a female put all her effort into growing big, then breed in one fecund burst, like the salmon who forage in the ocean, swim upstream, spawn, and die (semelparity)? Or should she bear few offspring, with births spaced at intervals over a long life (iteroparity), the way most primates do?

A survey across species reveals how unusual it is to find the innately self-sacrificing mothers envisioned by Spencer and his predecessors, men who were more nearly moralists than naturalists, projecting onto nature what was essentially wishful thinking. Self-sacrificing mothers exist, but they only evolve as species-typical universals under stringent circumstances. Most often, self-sacrificing mothers are found in highly inbred groups, when mothers are near the end of their reproductive careers, or in semelparous species that breed just once. The prize for what might be called "extreme maternal care" goes to various matriphagous (mother-eating) spiders. After laying her eggs, an Australian social spider (*Diaea ergandros*) continues to store nutrients in a new batch of eggs—odd, oversized eggs, far too large to pass through the mother's oviducts, and lacking genetic instructions. For these eggs are not for laying—they are for eating.

As the spiderlings mature and begin to mill about, the mother becomes strangely subdued. She turns mushy, not in a sentimental, but in a liquefying way. As her tissue melts, it is literally sucked up by the ravenous young, who devour both the mother and the protein-rich eggs dissolving within her.

Antisocial as this cannibalism seems, matriphagy is the key ingredient in the evolution of these spiders' uncharacteristically gregarious lifestyle. By having the bad manners to eat their mother, sated spiderlings are rendered less likely to eat one another, and thus socialize instead. Furthermore, even selfless mothers are not all equal. The more efficient a mother is at capturing prey, the bigger she grows, the more of her there is for her babies to consume, the less inclined her progeny will be to eat one another, the greater their capacity to reap the benefits of a social existence.

The Art of Iteroparity

Few mammals breed in one semelparous burst the way social spiders do. Most are iteroparous, breeding sequentially over a long life. Mother mammals produce litters of variable size, or—as in most primates—single young, spaced at intervals of variable duration over her lifetime. An iteroparous mother who overshoots the optimal clutch size for her circumstances, or who breeds too fast, may lose her entire brood to starvation, or end up so weak (through maternal depletion) that she does not breed the next season when conditions improve. Worse, she may succumb herself, which, if she is a mammal, means her offspring die with her.

Mothers then must engage in a perilous balancing act, allocating time and energy between making a living, resting, producing young, and rearing them. In social mammals, successful rearing of young often depends on the mother's social status, and her ability to elicit tolerance or actual assistance from other group members, activities which take time, energy, and connections. Jeanne Altmann's classic monograph *Baboon Mothers and Infants* (1980) chronicled tight maternal time budgets among mothers living on the arid Amboseli plains of Kenya. Baboon mothers spend 70 percent of daytime just staying fed. Costs of reproduction are high. For example, the death rate among adult females without infants was 0.08 per year; for mothers with infants, the rate was 0.15—nearly twice as high. These primate mothers are under heavy selection pressure to adjust reproductive effort to their circumstances.

Environmentally Sensitive FeedBack Loops

All reproduction is costly, but for most mammals, lactation is the costliest phase. Typically, it is more advantageous for infants than for mothers to prolong this "free meal," leading to mother–offspring conflict over when to wean. [*See* Parent–Offspring Conflict.] Theoretically, then, it would not be prudent for mothers to rely on infant demand alone to set the stopping point for lactational amenorrhea, because it will be in the infant's interest to demand more milk (and also to delay conception of sibling competitors) than it might be optimal for the mother to deliver. Evidence from primates conforms to this theoretical expectation. Intensity of suckling by infants turns out to be an important factor delaying the next conception, but not the only one. Resumption of cycling and subsequent fertile ovulations also depend on the mother's nutritional condition, workload, and energy budget.

Ongoing research among primates is helping to unravel the complex and dynamic system that governs ovarian functioning. This environmentally sensitive feedback loop, involving the hypothalamus, pituitary, and ovaries, is best, albeit still not completely, understood for humans (Ellison, 2001). Right after birth, a rapid rise in circulating levels of the hormone prolactin signals onset of milk production. Thereafter, sucking on the nipples by the baby signals the hypothalamus, lowering secretion of dopamine and triggering increased prolactin secretion. Continued sucking on the nipples causes prolactin levels to spike higher, increasing fifteen-fold above prenursing levels, before returning to base level in about three hours—unless the baby sucks again. In women whose infants nurse frequently, prolactin levels remain continually elevated, suppressing ovulation. On average, eighty minutes of suckling per day spread over a minimum of six bouts should suppress menstrual cycling for eighteen months. But this is not the whole story. It takes far more sucking to suppress ovulation in a sedentary mother (who may menstruate again several months after birth) than in one who is walking long distances carrying heavy loads. In women nutritionally depleted by recent food shortages or from breast feeding the last baby, ovarian cycling may resume but without the production of fertile eggs.

Maternal Effects

Although most models in evolutionary biology focus on genes contributed by parents to their offspring, what a mother is and does, the bodily resources she can impart, the social status she has achieved, and the territory she defends will be critical for infant prospects, with important evolutionary consequences. These maternal effects constitute all the nongenetic influences of the mother's phenotype or local environment on the phenotypes of her offspring.

Biologically, the development of each individual begins as a maternal effect, for the maternal germ cell has already begun dividing prior to contact with sperm. Subsequently, one of these cells continues as the oocyte, whereas others become "nurse cells," manufacturing nutrients and hormones that will be transmitted through the cytoplasm and shape the continuing course of development (West Eberhard, in press).

For mammals, the mother is the most important feature of the environment during the most perilous phases of any animal's existence in utero and right after birth. The mother provides food, security, and, in the case of creatures like primates that carry their young, mobility. Her fortunes, the social and ecological niche she has constructed for herself, as well as her ability to cope with her world—its scarcities, predators, and pathogens, as well as the conspecifics in it—influence the survival chances, and sometimes future reproductive prospects, of her infants. For example, the mother's bodily resources may determine offspring condition and size at birth, which, in humans at least, have long-term implications for brain development and later health. After birth, immatures confront pathogens with immunological defenses imparted in mother's milk.

In most mammals, direct maternal investment ends at weaning. Among some, most notably humans, direct investment in offspring continues as long as the mother lives. Typically, one or both parents continue to provision children through a long period of juvenile and adolescent partial dependency. Continued parental involvement in education, negotiating marriages, and setting up homesteads extends beyond their own life spans, especially if resources are transmitted across generations, as in the transfer of customary rights or other inheritances. Some of the most stunning maternal effects are produced by information about the world communicated by the mother early in life. Such information can be transmitted chemically (e.g., molecules in mother's milk can affect subsequent food choices), through observation, or, in the human case, through linguistically transmitted memes.

For many animals, maternally acquired social statuses or territories not only contribute to the survival of the offspring but also may pass to one or more offspring after death, generation after generation, with profound implications for the relative reproductive success of different matrilines, as has been demonstrated for humans and some other primates, such as cercopithecine monkeys and chimpanzees. "Old Flo," for example, the chimp mother made famous in *National Geographic* films about the primatologist Jane Goodall, had a number of sons and daughters, one of whom, Fifi, inherited her mother's secure, food-rich territory in the center of the Gombe Stream community. Flo's daughter Fifi produced seven surviving offspring, a record number for a wild chimp. Fifi's firstborn daughter in turn remained near her mother. Advantaged by her matriline's safe and productive larder and by a network of well-placed kin (brothers and uncles who were dominant males in the community), this wild chimp matured at the unusually early age of eight.

It is still common in Western society to equate "maternal instincts" with a desire to bear offspring, and many people still equate ambition with being a "bad mother." They are overlooking our primate heritage. There need never have been selection on ape females to "want" to bear children. Rather, there was selection on females to strive for local status and for access to the necessary resources that would allow them to keep such infants as they did bear alive. For any early hominid plump and mature enough to ovulate, pregnancy followed as a consequence of intercourse, not necessarily from a conscious desire for children per se. Hence, we should not be surprised that among modern women—who for the first time in our evolutionary history can consciously choose whether or not to become pregnant—many opt to delay conception, or even to forgo childbearing altogether, in order to strive for status or greater security.

The Optimal Number of Mates

In some cases, maternal effects have more to do with males than with material resources. As Darwin pointed out, female choice of mates is of extraordinary evolutionary importance because offspring may inherit their father's advantageous traits. [*See* Cryptic Female Choice.] Less often considered, however, are the nongenetic consequences of a mother's mating choices. Should she mate once or many times? With one male or many?

If the female is monandrous and her mate is as well, the father's reproductive success is equivalent to the mother's. Confronted with an infant who needs his help, such a father—certain of paternity and lacking other parenting options—should be predisposed to provide it. This is the case among titi monkeys (*Callicebus molloch*). In these monogamous South American monkeys, the father carries the infant 93 percent of the time (based on observations during the first two weeks of life), returning the baby to its mother only to nurse. Interestingly, the mother is more stressed by being separated from her mate than by being separated from her baby. Because it is the father who does the carrying, motherhood in titi monkeys has more to do with giving birth, lactating, and driving away rival females, who might divert her mate's attentions from their offspring, than with infant care.

When male care is neither so costly nor so exclusive as in the titi case, even males less than certain of paternity may protect or provision offspring, or, at least, refrain from harming offspring possibly their own. Hence, in species in which infanticide by strange males is a problem, as is the case with many primates, females solicit and mate with multiple males, possible, as a tactic to confuse paternity. [*See* Infanticide.]

To date, the best evidence for how mothers increase survival of offspring by manipulating information about paternity, comes from birds, especially dunnocks (*Prunella modularis*), studied by ornithologist Nick Davies. Dunnock females typically mate polyandrously, males (if they can) polygynously. Alpha and beta males calibrate mouthfuls of food carried back to nestlings in proportion to how often they managed to copulate with the mother when she was last fertile. Consistent with the hypothesis that it behooves mothers to thus manipulate information about paternity in this species, DNA fingerprinting reveals that provisioning males are often but not always genetic progenitors of the infants that they feed. Among *Prunella collaris*, a close relatives of dunnocks with a similar breeding system, fertile females exhibit primatelike red swellings in the cloacal region, presumably for the same reason as in baboons and chimps—to attract males and to increase a mother's chances of drawing multiple males into the web of possible paternity. Among dunnocks, langur monkeys, and humans, where the behavior of males has a critical impact on infant survival, effects of maternal mating decisions extend beyond genes to include the ways that a mother's recent sexual history affects the behavior of males likely to be in the vicinity of her infants.

Mothers and Allomothers

In addition to fathers and possible fathers, mothers living in social groups may elicit assistance from a range of other group members. Individuals other than the parents who help rear offspring are known as alloparents, from the Greek prefix *allo* ("other than") (Wilson, 1975). However, without DNA tests, it is often impossible to identify fathers, so it is more accurate to talk about assistance by individuals other than the mother, or allomothers. One of the surprises of the postsociobiology era has been the realization of both how common and how evolutionarily important allomaternal assistance can be.

For example, in a number of primates, including vervet monkeys and most species belonging to the Colobine subfamily, mothers are less possessive of their infants than are other Old World monkeys. Mothers in these infant-sharing species permit other females in the group to hold and carry their infants, thus freeing the mother to forage more efficiently. A comparative analysis across primates by John Mitani and David Watts revealed that mothers in infant-sharing species give birth after shorter intervals. The anthropologist Barry Hewlett has done a similar analysis for humans living in hunter-gatherer societies, among whom there is great variation in how much mothers rely on other group members for help with child care and provisioning. For humans, as among other cooperatively breeding animals, the availability of such assistance alters the terms of maternal fitness trade-offs, because mothers with help reproduce faster without excessive maternal depletion or offspring mortality (see Hrdy, 1999, for review).

Allomaternal assistance, ranging from moderate to extensive, is widespread in nature, now documented for a diverse array of insects, birds, and mammals, always with similar outcomes: divisions of labor permit animals to subsist and breed in habitats previously not available to them, and cooperative breeding allows mothers to breed faster, rear larger young, or more young, or rear young that (as in humans) mature slowly and remain dependent on adults for nutritional subsidies for a very long time. The cooperatively breeding Florida scrub jay provides a classic example of how cooperative breeding allows animals to move into habitats that would not be available without a division of labor. These jays can breed in relict patches of stunted forest where other jays can not successfully breed because young, pre-reproductive helpers serve as lookouts while parents forage. This extra assistance and division of labor helps parents defend their young against relentless predation from snakes and raptors in this highly exposed habitat (Woolfenden and Fitzpatrick, 1990). A shift toward cooperative breeding among Pleistocene humans may have been critical for withstanding periodic food shortage and enabled populations of *Homo ergaster* to expand. Hawkes and colleagues (1998) stressed prolongation of the human life span long past menopause, which would have made available a particularly well-qualified class of alloparents, postreproductive group members who were both experienced and had no better reproductive option than to help provision the offspring of kin. The maternal ancestors of modern humans must have been gambling on having such help. Otherwise, how could there have been selection on the earliest human mothers to produce slow maturing young so beyond any woman's means to rear by herself?

FIGURE 2. Among Cooperatively Breeding Tamarins, Allomaternal Assistance Permits a Rapid Pace of Reproduction.
On left, a golden lion tamarin mother passes the twins to one of her former mates to carry, below right, a pre-reproductive helper catches a beetle for the nearly weaned young to eat.
Drawing by Sarah Landry.

In cooperative breeders generally, mothers have been selected to behave so as to increase the availability of allomaternal assistance by remaining near kin, by mating with several males, by eliminating offspring sired by rival females, or by delaying reproduction until such assistance is available. Among cooperatively breeding Callitrichid primates (tamarins and marmosets), mothers give birth to single, twin, or triplet offspring, as often as twice a year. The combined weight of twins may weigh up to 20 percent of the mother's body weight—a staggering reproductive load. Fortunately, allomothers carry the infant much of the time, when the mother is not actually suckling them. The allomother in charge of carrying the infants is usually the male, or one of several males, with whom the mother mated. Other males and immatures—often siblings of the infant—catch insects to supplement their diet around the time of weaning. For three species, *Saquinus mystax*, *Callithrix jacchus*, and *Leontopithecus rosalia*, the golden lion tamarins (Figure 2), a correlation is observed between infant survival and the number of adult males in the group helping to care for them. Learning how powerfully allomaternal assistance affects a mother primate's reproductive success changes the questions sociobiologists ask about maternal behavior. Instead of assuming that mothers always care for babies, the question now arises: why don't mothers delegate care more often? The answer is that reliable, willing, competent alloparents are in short supply, and, for most primates, using allomothers is not a safe option.

Allomaternal and Maternal Commitment (Ultimate Causation)

In 1963, William D. Hamilton sought to explain at a theoretical level why nonreproductive workers in social insects devote themselves to caring for the offspring of the queen. He devised a simple equation to explain the altruism of these allomothers, known as Hamilton's rule:

$$c < rb$$

Altruistic caretaking should evolve whenever the cost to the provider (c), is less than the fitness benefits (b) obtained by helping another individual related by r, a term designating the proportion of genes these two individuals share by common descent. In many of the cooperative breeders, including wild dogs, wolves, hyenas, dingos, dwarf mongooses, and marmosets, as well as in some of the wasps and other social insects, the breeding female—always socially dominant over other females in her group—may kill such infants as a subordinate bears (Digby, 2000). Thus, in species where offspring are too costly for a mother to rear by herself, the dominant female essentially alters potential payoffs to subordinate females from becoming mothers. Under these circumstances, subordinate females make the best of adverse circumstances by delaying reproduction, suppressing their own ovulation. In cooperatively breeding canids such as wolves, subordinate females do ovulate, but instead of becoming pregnant, they undergo pseudopregnancy, swelling up as if the female were pregnant and beginning to produce milk, which is then available for the dominant female's litter.

Callitrichids exhibit full-fledged cooperative breeding. Help from allomothers means that mothers can sustain a heavy reproductive load. However, mothers are thus dependent on help, and their dependency explains why—compared to other primates—maternal commitment in tamarins is unusually sensitive to social circumstances, compared to the social systems of most other primates, in which mothers care for offspring by themselves. If the tamarin mother's mate disappears, or if there are no pre-reproductive tamarins in the group to help, the cost of caretaking exceeds the potential payoff or benefit to the mother from bearing young, so that in a high proportion of such cases, the mother abandons her infants in the first seventy-two hours after birth. Human mothers, whose own highly dependent and often (for an ape) closely spaced offspring require assistance to rear, are similarly sensitive to social support.

Even though all mothers are related to their offspring by roughly the same amount, variation in the cost of caretaking and in the potential payoff means that Hamilton's rule can be insightfully employed to explain the ultimate causation of maternal commitment among mothers as well as allomothers. The proximate causation of nurturing behaviors can also be quite similar for both mothers and allomothers, although for various reasons the threshold for initially responding to infant cues is almost invariably lower in the mother than in fathers or other allomothers.

Eliciting Maternal Commitment (Proximate Causation)

When a virgin rat encounters pups, she may ignore, avoid, or eat them. Only after multiple trials and much exposure is a female conditioned or "primed" to tolerate pups. At that point, she may begin to lick them, crouch over them, even retrieve them in her mouth and consolidate them in a nest. By contrast, a pregnant rat responds within minutes. Furthermore, if a virgin female is injected with blood from a rat who has just given birth—as was done in a now famous 1968 experiment by Joseph Terkel and Jay Rosenblatt—there is a marked reduction in how much time it takes to prime a female to respond in a nurturing way.

A great deal has since been learned about the endocrinological and neural underpinnings of maternal responses. During the last third of pregnancy, a cascade of endocrinological events lowers a mother's threshold for responding to pups in a nurturing way. Prominent in this maternal cocktail are the steroid hormones estrogen and progesterone, manufactured by the placenta and essential to maintaining pregnancy. Because the placenta is delivered along with the babies, progesterone and, a bit later, estrogen, levels decline around parturition. Hence, by themselves, these hormones do not account for nurturing behavior.

Enter two hormones essential for milk production, prolactin and oxytocin. Prolactin, the mother's work order signaling production of more milk, is a very ancient and versatile molecule also linked to nurturing and protective behavior by males, both in mammals and in birds. Oxytocin however, is a quintessentially mammalian hormone. Oxytocin (from the Greek for "swift birth") evolved in mammals as a muscle contractor that, among other things, promotes the uterine contractions leading to birth as well as contractions in mammary glands causing milk ejection. Present when the mother first greets her emerging offspring right after birth, oxytocin continues to be released whenever a mother nurses. Because oxytocin promotes calm and positive social responses, the effect on the mother can be compared to candlelight and soft music for her first date with tiny strangers that suddenly materialize near her nest. Because high levels of oxytocin are also found in mother's milk, there is also the possibility that this soothing hormone helps make the mother's growing attachment to her infant mutual.

As important as hormones can be in preparing mothers to respond to their infants, they do not act in a deterministic fashion. Hormones both affect and are affected by a mother's behavior and her experience. For example, the act of caring for pups leads to reorganization of neural pathways in a mother's brain, making her more likely to respond more quickly to pups the next time. This is one reason why experienced females tend to be more responsive to pups than first-time mothers. This is especially true in primates, where learning is critical for competent caretaking.

Although hormones like prolactin are frequently considered "maternal" hormones, in fact elevated levels of this hormone can be found in both sexes and have been documented for both avian and primate allomothers engaged in carrying, protecting, or provisioning infants. Recent research on men living with pregnant women revealed that prolactin levels rise near term, as do cortisol levels. The most significant effect was the 30 percent drop in testosterone in men right after birth. The more responsive to infants the men were, the more likely it was that their testosterone levels would continue to drop (Wynne-Edwards and Reburn, 2000). No doubt about it, hormonal changes during pregnancy are more pronounced in mothers than in nearby fathers. The point, however, is that primates of both sexes are to some extent, primable, and proximity to and experience with infants matter. This helps explain why adoptive parents can become so deeply committed at a biological level to the infants that they care for, and why a fully engaged father in close contact with his infant can sometimes be even more committed than a detached mother.

However, given that the mother is invariably present at birth, and already hormonally primed to respond to sensory cues from the infant (olfactory, auditory, and visual), the mother is the likeliest candidate to become most attached to her offspring, and the infant to her. Within days of birth, the human mother learns to recognize her own infant's smell, and by this time too—if she is breast feeding—lactation is under way with all the attendant endocrinological consequences. From this point on, infants have lactating mothers on an endocrinological leash as the bond between a mother and an infant develops during the weeks and months after birth.

BIBLIOGRAPHY

Altmann, J. *Baboon Mothers and Infants*. Chicago, 1980. Essential background for researchers.

Clutton-Brock, T. H. *The Evolution of Parental Care*. Princeton, 1991. Essential background for researchers.

Digby, L. "Infanticide by Female Mammals: Implications for the Evolution of Social Systems." In *Infanticide by Males and Its Implications*, edited by Carel van Schaik and Charles Janson, pp. 423–446. Cambridge, 2000.

Ellison, P. T. *On Fertile Ground: A Natural History of Human Reproduction*. Cambridge, Mass., 2001.

Hawkes, K., J. F. O'Connell, N. G. Blurton Jones, H. Alvarez, and E. L. Charnov. "Grandmothering, Menopause, and the Evolution of Human Life Histories." *Proceedings of the National Academy of Sciences USA* 95 (1998): 1336–1339.

Hrdy, S. B. *Mother Nature: A History of Natural Selection, Mothers and Infants*. New York, 1999.

West Eberhard, M. J. *Developmental Plasticity and Evolution*. Oxford, in press.

Wilson, E. O. *Sociobiology: The New Synthesis*. Cambridge, Mass., 1975.

Wynne-Edwards, K. E., and C. J. Reburn. "Behavioral Endocrinology of Mammalian Fatherhood." *Trends in Ecology and Evolution* 15.11 (2000): 464–468.

DARWINIAN MEDICINE

Stephen C. Stearns

Evolutionary thought has long had problems of two sorts with cognate disciplines. Some resisted the application of evolutionary ideas simply because they were defending intellectual territory. Others resisted the reduction of complex, beautiful phenomena to material processes for philosophical, romantic, or spiritual reasons. Recently, however, appreciation that evolutionary thought has both abstract beauty and explanatory power has increased in economics, anthropology, psychology, political science, and medicine. In these disciplines, evolutionary ideas are contributing important insights not accessible from other perspectives. The focus of this article is on the contributions of evolutionary thinking to medical science.

Evolutionary thinking contributes two kinds of ideas to medical science. The first focuses on the rapid, dynamic consequences of natural selection, on adaptations in humans and their pathogens. The second focuses on reconstructing the evolutionary histories of humans and their pathogens, and on the consequences of particular evolutionary histories for health and disease. Because the concepts of natural selection and of a history of descent with modification are both drawn from the work of Charles Darwin, this approach is called evolutionary or Darwinian medicine. Both kinds of ideas contribute to the insights summarized below.

The Evolution of Virulence

Virulence describes how adversely a pathogen affects its host. A virulent pathogen kills its host and kills it quickly. Two major evolutionary mechanisms are believed to shape the level of virulence. In the first, the mode of transmission plays a key role. Diseases that are primarily horizontally transmitted, from host to unrelated host, should evolve higher levels of virulence. Diseases that are primarily vertically transmitted, from host to offspring, should evolve lower levels of virulence, for their transmission then depends on the survival of the host until it can reproduce. If virulence alters the transmission probability of the pathogen, then it will be selected to change intensity in the direction that increases transmission probability. Whether that change is an increase or a decrease depends on how symptoms affect transmission probability.

In the second mechanism, competition of parasites or pathogens within the host plays the key role. Virulence is selected to increase if several pathogen genotypes infect the host at once. The pathogen that wins the competition will be the one that exploits the host most thoroughly and most quickly before the other competing pathogens can rob it of resources.

Thus, the level of virulence that is expected to evolve depends on the conditions that influence the relative importance of competition within hosts and competition for new hosts, that is, the type and frequency of multiple infections and the transmission costs imposed by early host death induced by pathogens. Natural vertical transmission, in which the pathogen must allow the host to survive to ensure its own transmission, can lead to low virulence. In contrast, serial passage experiments, in which the pathogen is transmitted by the experimenter and its success in transmission is not affected by its virulence, produce high virulence. These cases are extremes along a continuum.

Most diseases evolve under a mixture of intermediate conditions to an intermediate level of virulence.

The evolution of virulence also has an historical dimension. The reconstruction of the evolutionary history of pathogens reveals that virulent organisms have accumulated factors associated with virulence in a stepwise fashion. The stepwise pattern suggests that each step represented an adaptive advance for the pathogen. For example, some *Escherichia coli*, a common human intestinal bacteria, have become virulent pathogens responsible for epidemics of food poisoning. Several lineages of *E. coli* acquired the same virulence factors in parallel. The parallel evolution suggests that selection has favored an ordered acquisition of genes that progressively built up the molecular mechanisms that increase virulence.

Close study of the molecular evolution of pathogens can also help to identify the strains that will become problems. Changes in DNA sequences are of two types: those that produce changes in the amino acids in the protein encoded in the sequence (non-synonymous substitutions), and those that do not change the amino acid in the protein (synonymous substitutions). A high rate of nonsynonymous substitutions in a gene suggests that selection on that gene is strong. Just such a high rate of nonsynonymous substitutions has been found in genes associated with virulence in influenza viruses. The clone with the highest proportion of nonsynonymous substitutions was the one that became epidemic the next year, presumably because it had evolved the greatest selective advantage against host defenses. With such information, one could predict which flu strain in this year's flu season will cause next year's epidemic and adjust vaccination policy accordingly.

The Evolution of Antibiotic Resistance

Pathogenic microorganisms rapidly evolve resistance to chemotherapy. Before 1930, we knew that malaria evolves resistance to quinine. Soon after the introduction of sulfa drugs in the 1930s and true antibiotics in the 1940s, many resistant strains evolved. Pathogens that resist several antibiotics in combination have become increasingly common, and some infectious diseases, such as tuberculosis, and some agents of acute infection, such as *Staphylococcus aureus*, that had almost ceased to be problematic have reemerged in resistant form. At the same time the rate of discovery of new antibiotics has decreased. Some strains are close to evolving resistance to all known antibiotics. For infections caused by such pathogens, there will be no cure unless new antibiotics are discovered more rapidly than pathogens evolve resistance to them.

The evolution of antibiotic resistance raises an issue with strategic and moral dimensions. Should one treat the individual patient, or should one treat the entire human population? If one limits antibiotic therapy to slow the evolution of resistance, the individual patient may suffer, but future patients may benefit from simpler therapy and have a quicker recovery. The costs and benefits of antibiotic therapy for a whole population cannot currently be estimated because processes that are poorly measured mediate the population consequences of individual therapy. It is clear, however, that antibiotics should never be used unnecessarily, and when they are used, they should be applied so thoroughly, in large enough doses and for a long enough time, that no mutants with intermediate resistance can survive.

Mechanisms for the evolution of resistance include mutations in chromosomal genes. Do bacterial clones that encounter antibiotics increase their mutation rate, which would allow them more rapidly to produce a mutation that conferred resistance? It has been suggested that bacteria have two different sets of genes: housekeeping genes used for basic metabolism and structure that mutate at low frequency, and highly mutable contingency genes important for adaptation to changing environments. In addition, some bacterial cells with a mutator phenotype have a mutation rate that is from ten to ten thousand times higher than that of normal cells because they have a defective DNA repair system. Contingency loci and mutator phenotypes would allow a bacterial lineage rapidly to accumulate many alleles, some of which could evade host defenses or antibiotics. The large sample of mutations would allow both infectivity and resistance to evolve faster, giving contingency genes and mutator phenotypes a selective advantage for pathogens living in humans under treatment for infectious disease.

Although there is some evidence that the mutation rate of bacteria challenged with antibiotics is elevated in human hosts, an alternative view comes from experiments on mice. During the experimental colonization of mouse guts, a high mutation rate does give an initial benefit because it allows faster adaptation, but this benefit soon disappears. Mutator cells accumulate mutations that are neutral in the mouse gut but that reduce competitiveness during transmission to new hosts. They have a short-term advantage but a long-term disadvantage; they could exist in a population at low or intermediate frequency but would not take over completely.

The Concept of a Selection Arena

A selection arena is a selection process that occurs inside an entity, such as a reproductive female, that is a unit of selection in its own right at a higher level. It has characteristics suggesting that it is an adaptation of the higher level. Examples include the selection of zygotes in the mammalian reproductive tract through selective abortion, where the zygotes are the lower level; the females containing them are the higher level. Selection on variation in reproductive performance of the organisms—the higher level—has shaped the internal selection process on the variation among zygotes at the lower level. Next, we discuss ideas involving selection arenas.

Oocytic Atresia and Menopause

Total loss of oocytes, cells that develop into eggs, occurs by the end of the reproductive life span in humans (menopause) and in at least two other primates, rhesus macaques and bonobos (pygmy chimpanzees). The process, however, starts much earlier. By the third month of pregnancy, ovaries have developed in the female human embryo. They contain about seven million oocytes. By birth, that number has been reduced to less than a million and by menarche, the onset of menses, to less than five hundred. This process of oocyte destruction is called oocytic atresia; it occurs in most mammals.

Is atresia selective, and does it have an evolutionary explanation? One idea starts by noting that mitochondria reproduce asexually and pass regularly through bottle-necks of small population size. Mitochondria are maternally inherited organelles found in the female's oocytes. Such organisms cannot avoid accumulating deleterious mutations. Eventually, all mitochondria would be damaged. The problem can be avoided, however, if the mitochondria with the deleterious mutations can be segregated and discarded. This would happen if a small number of mitochondria were introduced into each of many oocytes. The oocytes with defective mitochondria would advertise that fact in their biochemical profile, giving the maternal tissue a signal that could be used to decide from which oocytes nourishment should be withdrawn.

This hypothesis makes at least three predictions: First, if ovaries are sampled from early embryos to birth, the percentage of defective mitochondria should decline toward birth. This has not been confirmed. Even if such a correlation were demonstrated, it would not establish that the only reason for the destruction of the oocytes is the presence of damaged mitochondria, for signals emitted by other types of damage could be correlated with the presence of damaged mitochondria. Second, the number of mitochondria allocated to a primordial oocyte should be small, ideally just one. The number of mitochondria with which an oocyte starts life is not yet known; some evidence suggests that it is small. The mechanism would not work if the number were large, for then the biochemical signals given off by the oocytes with defective mitochondria would be masked by the signals given from the healthy mitochondria, and oocytes could not be eliminated selectively. Estimating the number of mitochondria with which an oocyte starts life is thus important. Third, the signal that initiates selective oocyte destruction should reliably indicate the presence of damaged mitochondria. The nature of such a signal is not yet known. Establishing its nature is a research priority. When identified, that signal might point to damage in the oocyte that is not in the mitochondria.

Is there a link between atresia and menopause? Menopause has several adaptive evolutionary explanations, but there is also a nonadaptive explanation: menopause may simply be a byproduct of atresia. The variation in age of onset of menopause would then result from random variations in the number of oocytes destroyed prior to menarche. Slight variations in proportion destroyed translate into differences of tens of oocytes available at menarche and into differences of years in onset of menopause. (The rate of oocyte atresia does correlate with the age of "menopause" in mice: high rates of atresia lead to early menopause.) Under this hypothesis, menopause occurs when the female runs out of oocytes to ovulate. The number of oocytes with which the female starts menses is determined by atresia; atresia exists to screen out damaged oocytes. This explanation neatly connects processes at both the beginning and the end of life. It needs tests against alternatives.

Mate Choice and Disease Resistance

The role of mate choice in human reproductive biology developed for years along two independent tracks before merging recently. One track was evolutionary; the other was medical. The evolutionary track began in work on birds and fish with the suggestion that partners select mates on the basis of honest signals of genetically based resistance to parasites and pathogens. If mates vary in genetic quality, and that variation can be detected in the phenotype, then it pays to select a high-quality mate. That selection could occur through mate choice, or the mother's reproductive tract or the ovum itself could select sperm before, during, and after zygote formation. Here again we encounter selection arenas. As we will see next, one operates in humans.

While the evolutionary ideas were being developed, a long-term medical study of the Hutterite communities in South Dakota revealed some remarkable facts. The Hutterites, who moved to North America from Switzerland in the nineteenth century, are a small community that has become relatively inbred. They are notable for their communal lifestyle and their limited number of different versions of the genes (HLA alleles) involved in the immune response. Those genes are known for all couples. Some Hutterite women suffer from recurrent spontaneous abortions, and women whose husbands have similar HLA loci are more likely to suffer spontaneous abortions than women married to men with different HLA alleles. Picking a mate with similar HLA loci is not a good idea.

Do partners avoid potential mates with similar immune genes? Significantly fewer matches of HLA alleles between spouses are observed than would be expected at random: Hutterites avoid mating with partners with the same HLA allele. Among couples whose alleles match, where the wrong choice was made, the matched allele is inherited more often from the father than the mother: men make more mistakes than women. Thus, humans, like mice, may detect variation in immune-response genes and use that information to choose mates. Mice use olfactory signals; the nature of the signals used by humans is not yet well established. Preliminary evidence suggests that they may also be olfactory.

In large, outbred human populations with many HLA alleles, where the probability of finding a partner with the same allele is very low, factors other than immune-response genes, such as social status, income, and ethnicity, are probably more important. But in preagricultural times, in the environment in which these traits may have been selected, humans existed in small groups of hunter-gatherers closely related by endogamy. If that had been the environment of evolutionary adaptedness (see below), conditions would have resembled those now encountered by Hutterites.

Rapid Development of Organs and Increased Virulence of Genetic Diseases

Reproductively active human females in hunter-gatherer societies have children at intervals of about 3.5 years. When a fetus or young infant dies, the mother produces another child in much less than 3.5 years. This replacement of dead infants is called reproductive compensation. Reproductive compensation creates selection at the level of individuals on the selection arenas operating within individuals. A selection arena has a greater advantage when the defective gametes, zygotes, or offspring are discarded early in development before the parents have invested much in them. This may explain the rapid early development of the human embryo, in which most organ systems function within ten weeks of conception. Organ systems may differentiate and develop rapidly so that their performance can be tested as soon as possible. If the embryo lacks, for example, an essential liver enzyme, then the sooner it is discarded, the sooner a potentially healthy sibling can be born.

Reproductive compensation also creates selection on the age of expression of genetic diseases. If the early death of a genetically damaged infant accelerates the birth of a healthy sibling, then the infant mortality rate will itself evolve. Genetic diseases will be selected to become more virulent by being expressed and by killing at earlier ages, to allow the mother sooner to conceive a healthy sibling.

These ideas have not yet been properly tested against alternatives.

Molecular Detective Work: Insights from Phylogenetics

Molecular systematics gives us new insights into the structure of pathogen populations. Such results shed light on many issues, including the potential for rapid horizontal transfer of antibiotic resistance and the identification of the individuals responsible for the spread of very rapidly evolving diseases. The method involves inferences from phylogenetic trees, the reliability of whose construction can become a matter of life or death in a criminal case. A phylogenetic tree describes the pattern of relationships among a group of organisms, similar to a genealogy, which describes a family. In one case involving possible transmission of the human immunodeficiency virus (HIV), a reliable phylogenetic tree identified the person—a dentist—responsible for infecting others with HIV: the dentist's strain of HIV was more closely related to the HIV in his patients than expected of a random strain of HIV. That study also identified an unexpected mechanism of transmission: dental therapy (Hillis and Huelsenbeek, 1994). Here evolutionary methods are so well tested that they have moved from basic research into everyday technology and are being used in criminal laboratories worldwide, where reliability is at a premium.

One can more reliably infer the ancestral relationships of a set of clones than a set of quasi-sexual organisms that exchange genes frequently. Are bacterial populations strictly asexual clones, or do they exchange genes? The answer to this question is also critical for understanding the spread of resistance genes and the potential of pathogen populations to respond quickly to host evolution and to vaccines. Molecular phylogenies have now shown that the degree of clonality varies among species: some are strictly clonal, whereas others exchange genes frequently by any of several mechanisms, including conjugation and transformation. Even the most clonal bacteria are genetic chimeras containing chromosomal genes and portions of genes from different ancestries. Thus, bacteria vary in their potential for horizontal genetic transfer; some behave practically like outcrossing sexual eukaryotes, and all manage some exchange of information.

When pathogens reproduce asexually and form sets of clones related by clear lines of descent, the patterns of relatedness can be used to establish the genetic origin of epidemics. For example, HIV entered the human population from at least two sources. One, in Central Africa, was probably a chimpanzee (HIV-1); the source of another, less virulent form in West Africa (HIV-2) was probably a sooty mangabey, a ground-dwelling monkey.

Intragenomic Conflict in Pregnancy

If not mated for life, the father and mother of an embryonic infant are in conflict with each other over the rate at which the infant should grow and over the amount of resources that it should extract from its mother's body. Because the father could have children by other females, he should want his offspring to extract more resources from this female than she is prepared to give, for any damage done to her will not affect offspring he has by other females. She, on the other hand, should want to hold some resources in reserve for her future offspring and not damage her chances of further reproduction too much.

Evidence for this conflict comes from an unexpected source: genetic imprinting. Genetic imprinting occurs in the germ line of the parents, and it happens only to a very few genes. The imprinted genes are chemically altered so that they will not be expressed in the offspring early in development. In mice and in humans, the genes that are imprinted are the genes that control embryonic growth. In mice, when the father's imprinted genes are activated in the embryo, the embryo grows more slowly and is born at a smaller size. When the mother's imprinted genes are activated, the embryo grows more rapidly and is born at a larger size. The genes that are imprinted in the paternal and maternal germ lines are not the same; they have opposite effects on growth. The genes imprinted in the paternal germ line would, if expressed, decrease growth rate. Those imprinted in the maternal germ line would, if expressed, increase growth rate. Thus, each parent turns off the genes whose expression would run counter to its interests. The state of the system is determined by an evolutionary history of parental conflict over the allocation of resources, a history recorded in the molecular genetic control of embryonic growth.

Evo-Devo: Hope for Nerve and Limb Regeneration

Evo-devo, the label now given to evolutionary developmental genetics, studies the evolution of the major developmental control genes first identified and sequenced in fruit flies, worms, and mice. Comparisons of the DNA sequences among these organisms revealed that genes with similar DNA sequences also shared function to an astounding degree. The genes that initiate brain, eye, and heart formation in fruit flies have DNA sequences similar to those of the genes that do the same in mammals. Their protein products are so similar that when a transgenic mouse gene is expressed in a developing fruit fly, it induces the formation of eyes wherever it is expressed, even on legs. Recently, rapid progress has been made in the study of limb development and nerve growth. We are still a long way from being able to use gene therapy to cause a severed forelimb to regenerate a functional hand or a severed spinal cord to reconnect well enough to restore function. However, never before have we had such good reason to think that such treatments may be possible. If they are to be realized, we will have to trace where in phylogenetic history the ability to regenerate limbs and nerves was lost, and for what reason. To do so, we will need more efficient approaches to the comparative study of developmental control genes in an explicit phylogenetic context. Then we will have to develop the new model systems so identified, models that span the critical losses of function.

The Evolutionary Biology of Mental Disease and Drug Abuse

Why is our neurobiology organized in such a way that we can become addicted to certain chemicals? One idea is that the chemical structures of addictive drugs are an unfortunate coincidence. They hijack pathways that evolved to increase fitness, for example, by providing a reinforcing reward through the production of the subjective impression of pleasure, but that are intrinsically vulnerable to novel drugs with chemical structures that mimic the signals that in the past promised a fitness gain. Thus, susceptibility to addictive drugs is a nonadaptive byproduct of structures and processes evolved for other reasons.

There are also evolutionary hypotheses for the existence of emotional moods. Depression might be adaptive if it could cause avoidance of risky or dangerous situations in which the goal could not be obtained. When it is better to lay low and do nothing, depression could inhibit types of activity that lower fitness. This is a plausible explanation for a moderate level of depression but not of serious depression leading to suicide. If a selection process had repeatedly encountered the problem of deep depression in people young enough to have some remaining reproductive potential, one would expect countermeasures to have evolved to prevent a mood from worsening too far in a dangerous direction.

Changes from the Environment of Evolutionary Adaptedness

Many hypotheses in evolutionary medicine posit a past environment that differed markedly from the present one, an environment in which selection shaped human physiology or behavior in a strikingly different way. They conclude that the problem under analysis results from modern deviations from an ancestral lifestyle characterized as preindustrial, as preagricultural, as Stone Age hunter-and-gatherer, or as something earlier. There is, however, rarely enough information on past environments and past lifestyles to make a strong assertion about the environment of evolutionary adaptedness. Our current evolutionary state has been integrated over a long succession of past environments that have left their traces on us like a moving average weighted toward the recent. Nevertheless, such hypotheses are interesting and worth further exploration, even if, at the moment, few of them have withstood risky experimental tests and exposure to plausible alternatives. Among these hypotheses are the following: Morning sickness is an evolved hypersensitivity to embryo-damaging toxins. Childhood asthma is an autoimmune response to the unnatural lack of exposure to parasitic worms. Sudden infant death syndrome is caused by the unnatural separation of the sleeping infant from the mother. Infant crying and colic are caused by the unnatural separation of the waking infant from the mother. Cardiovascular diseases are reactions to unnaturally low levels of exercise and dietary fiber and high levels of dietary fat. Dental caries are reactions to the novel high-carbohydrate diet that began with agriculture.

The following hypothesis, which belongs to this group, may have immediate benefit.

Menstruation Frequency and Reproductive Cancers

In preindustrial societies, women spent a much greater proportion of their lives pregnant or lactating than they now do. As a result, they went through many fewer menstrual cycles per lifetime than do women who use contraception. Because they cycled less often, their breasts, ovaries, and uteri went through fewer episodes of differentiation and dedifferentiation and were therefore exposed to fewer chances for mistakes in genetic control over the cell cycle. Under this hypothesis, the probability of breast, ovarian, and uterine cancer rises as the number of menstrual cycles per lifetime increases. Women in preindustrial societies with traditional reproductive patterns should be at lower risk, and women in postindustrial societies using types of contraception that do not block the menstrual cycle should be at higher risk. In response to this idea, contraceptive pills that only allow menstruation a few times a year are in development. If the results demonstrate a decreased frequency of reproductive cancers with lower frequency of menstruation, then this will become a case in which a concrete benefit has resulted from the idea that our bodies are adapted to a preindustrial or preagricultural environment.

What Is an Adaptation?

Some evolutionary biologists indulge in adaptive just-so stories without considering nonadaptive alternatives, and many such stories have been told in evolutionary medicine. They are particularly common in speculations about the environment of evolutionary adaptedness and the nonadaptive consequences of postindustrial society. To make discussion of adaptation acceptable in polite company and to lend rigor to an area that has often been relaxed, we will consider four criteria with which to evaluate the claims of adaptation that characterize much of evolutionary medicine. Arranged in rough order of reliability, they are all variations on a single theme: to be accepted as an adaptation, a trait state or a change in trait state must be shown to increase the reproductive success of the organisms that carry it.

The Selection Criterion. Natural selection on a trait is the correlation between variation in the trait and variation in reproductive success. A response to selection occurs when some of the variation in the trait is heritable. If you observe the process and document heritable change in the trait that results from the correlation of trait state with reproductive success, then the change in the trait is an adaptation. This best and strongest criterion—selection and the response to selection both observed—is fulfilled in studies of antibiotic resistance and in serial passage experiments on virulence.

The Perturbation Criterion. This criterion accepts as an adaptation the state of a trait whose optimum has been predicted by a model and tested by experiments. The experiments use mutations, phenocopies, transgenesis, hormones, surgical manipulation, or some other method to perturb the phenotype from the optimal state and, with appropriate controls, to demonstrate that the fitness of the perturbed phenotypes is lower than the fitness of the optimal state. This criterion has been met in studies of clutch size in birds and body size in lizards. For both ethical and practical reasons, it is not often applied to human subjects.

The Functional Criterion. One can also define an adaptation as a change in a phenotype that occurs in response to a specific environmental signal with a clear functional relationship to that signal, resulting in improved survival or reproduction. Under other circumstances, where it imposes a cost, it is not expressed. Examples include induced responses to pathogens, parasites, and predators. Such claims of adaptation are more convincing when they survive tests against at least two alternatives: (1) the host response is an adaptation of the pathogen, not of the host; (2) the host response is pathological, a reflection of damage done to the host by the pathogen.

The Design Criterion. Several experts have suggested that we can recognize an adaptation by its complexity and by its resemblance to something that an engineer might design. To meet this criterion, at least four questions should be answered in the affirmative: (1) Has the performance of the trait in the fulfillment of the function been compared in experiments with alternative states of the trait? (2) Is the state claimed to be an adaptation repeatedly associated with the kind of natural selection needed to produce that adaptation? (3) Is the condition of the trait a byproduct of selection on other traits? and (4) Has the trait been analyzed as a component part of the organism, or might the analysis be confused by an inappropriate abstraction of a piece of the organism from the larger whole in which it is naturally embedded?

The State of the Field

It is hoped that this brief discussion convinces renders that there is no problem in principle with claims of adaptation. Problems arise only when claims are not supported by convincing evidence, when plausible alternatives have not been examined and rejected. In some parts of the field of Darwinian medicine, that is unfortunately still the case, but it is a problem that is diminishing, not increasing, in importance.

Evolutionary biology has brought fresh ideas into medical science. More of them have implications for medical research than for clinical practice, but some important ideas do have clinical implications. It has contributed useful ideas that would not be suggested by other perspectives, and that flow of ideas is increasing.

[*See also* Adaptation; Coronary Artery Disease; *articles on* Disease; Host Behavior, Manipulation of; Nutrition and Disease; Senescence; Virulence.]

BIBLIOGRAPHY

Ebert, D. "Experimental Evolution of Parasites." *Science* 282 (1998): 1432–1435. An overview of serial passage experiments, including those used to produce live attenuated vaccines, in which evolutionary technology is applied to great advantage.

Ewald, P. W. *Evolution of Infectious Diseases.* Oxford, 1994. A stimulating exploration of the idea that virulence evolves primarily in response to mode of transmission.

Flaxman, S. M., and P. W. Sherman. "Morning Sickness: A Mechanism for Protecting Mother and Embryo." *Quarterly Review of Biology* 75 (2000): 113–148.

Haig, D. "Genetic Conflicts in Human Pregnancy." *Quarterly Review of Biology* 68 (1993): 495–532. One of the most striking applications of evolutionary thought to a common human medical problem—the symptoms associated with pregnancy.

Hillis, D. M., and J. P. Huelsenbeck. "Support for Dental HIV Transmission." *Nature* 369 (1994): 24–25. Pioneering application of phylogenetic methods to a problem in forensic medicine.

Krakauer, D. C., and A. Mira. "Mitochondria and Germ-Cell Death." *Nature* 400 (1999): 125–126. A good entry to the idea that oocyte atresia evolved to purge the germ line of defective mitochondria.

Levin, B. R., Lipsitch, M., and S. Bonhoeffer. "Population Biology, Evolution and Infectious Disease: Convergence and Synthesis." *Science* 283 (1999): 806–809. An overview of recent work on the evolution of resistance and of virulence.

Nesse, R. M. "Is Depression an Adaptation?" *Archives of General Psychiatry* 57 (2000): 14–20.

Nesse, R. M., and K. Berridge. "Psychoactive Drug Use in Evolutionary Perspective." *Science* 277 (1997): 63–65.

Nesse, R. M., and G. C. Williams. *Why We Get Sick: The New Science of Darwinian Medicine.* New York, 1995. One book that set the field in motion.

Ober, C., L. R. Weitkamp, N. Cox, H. Dytch, D. Kostyu, and S. Elias. "HLA and Mate Choice in Humans." *American Journal of Human Genetics* 61 (1997): 497–505. Presents the evidence that Hutterites have chosen their spouses in part on the basis of their immune genes.

Short, R. V. "The Evolution of Human Reproduction." *Proceedings of the Royal Society of London B* 195 (1976): 3–24. The first statement that modern women may suffer higher cancer risk because they menstruate many more times in their life than was normal in primitive societies.

Stearns, S. C., ed. *Evolution in Health and Disease.* Oxford, 1998. A critical summary of the state of the field that gives space to both adaptationist and historical perspectives.

Stearns, S. C., and D. Ebert. "Evolution in Health and Disease: Work in Progress." *Quarterly Review of Biology* 76 (2001). A detailed article with many references, of which this article is a précis.

Strassmann, B. I. "The Evolution of Endometrial Cycles and Menstruation." *Quarterly Review of Biology* 71 (1996): 181–220. An analysis of menstruation from the evolutionary perspective notable for its clarity and critical acumen.

Trevathan W. R., E. O. Smith, and J. J. McKenna, eds. *Evolutionary Medicine.* Oxford, 1999. A summary of adaptationist explanations of many human medical conditions, with considerable space given to anthropologists.

Williams, G. C., and R. M. Nesse. "The Dawn of Darwinian Medicine." *Quarterly Review of Biology* 66 (1991): 1–22. An early paper that gave the field strong impetus.

GENOMICS AND THE DAWN OF PROTEOMICS

M. J. Bishop

We now have the ability to sequence the DNA of the genome of any living organism. For the small genomes of archaea and bacteria, this can be accomplished in a short time. For the larger genomes of eukaryotes, it still requires a number of years to sequence the DNA to a satisfactory finished state. The genomes of yeast, nematode, and fruit fly are finished, those of human and puffer fish are in draft, and sequencing of the mouse genome is ongoing.

Why is the genome sequence so important? The sequence must encode all the RNA and protein products that are deployed at all times and in all places during the life of the organism, even if the product exists in a limited quantity, for a limited time, and in a limited location. It would be pleasing to examine the sequence and to predict what all these products are as a first step in understanding how each functions and interacts with others. In archaea and bacteria, this hope can be largely realized, although the function of up to half the genes in an individual species may not presently be understood. In eukaryotes, it is very difficult or impossible to predict all the genes from the DNA sequence. Further, one gene may encode many protein products by the process of alternative splicing of the messenger RNA (mRNA). Proteins may undergo post-translational modification to produce a variety of active forms. Neither splice variants nor modifications can be predicted from the DNA sequence.

Because of the impossibility of predicting all the genes in a eukaryote, large-scale projects to sequence mRNA have been undertaken. The mRNA molecules are formed when genes are transcribed. The population of RNA molecules has been called the transcriptome. These sequences help to give confidence in gene predictions and yield information about splice variants (genes that can produce alternative products) and single nucleotide polymorphisms (SNPs). Once the mRNAs have been identified, studies of gene expression become possible. Representative RNA products or synthetic oligonucleotides can be used to study mRNA populations and to follow changes in expression over time or in different conditions.

Gene expression as revealed by the transcriptome is not the end of the story. Studies have indicated that gene expression may be poorly correlated with protein expression. Proteomics is the name given to the study of all the proteins of an organism (the proteome) by analogy with genomics, which studies all the genes. Proteins are formed by translation of the mRNA sequences. To characterize the proteome, the proteins must be separated and identified. The techniques to do this are becoming easier and faster so that protein expression studies are not beyond our reach. The separation stage may be two-dimensional gel electrophoresis. The identification stage may be by cleavage of the protein to peptides and fingerprinting by mass spectroscopy.

The mRNAs and proteins that we have identified can readily be related back to the DNA sequence of the genome from which they are derived. We can eventually be confident that we have identified all the genes of even the most complex genomes. We will also understand how the genome is organized. The relative positions of genes in the genome may be very important. The case of the bacterial operons is well known in which particular genes are expressed together and their order of appearance in the genome is crucial. Comparative studies of related bacteria demonstrate that genes in operons must preserve their order, whereas genes not in operons do not need to. This helps to confirm that we have correctly identified the operons.

In eukaryotes, we know of gene clusters that do have functional significance, such as globins and *Hox* (homeobox) genes. However, for the majority of genes, there is at present no evidence that order is significant. Vertebrate evolution provides a natural test bed for the truth of this statement. Chromosome evolution leads to numerous break points that can be identified by comparing genome sequences or even maps of genes in a variety of species. In large multigene families, there are tricky issues in identifying those genes that are orthologs and those that are paralogs. Orthologous genes are those homologs (genes of a common origin) in two different species that are thought to have descended from an ancestral gene in a common ancestor of the two species. Paralogous genes are those homologs in a species that are thought to be derived by gene duplication from an ancestral gene within a single species. Both orthologs and paralogs may be subsequently lost in the various lineages leading to extant species. Some multigene families therefore show considerable complexity. Gene order data can help to increase our confidence in these assignments because orthologs or paralogs tend to retain their context within their unrelated neighboring genes.

Another important aspect of genome organization is the role of repetitive families in genome evolution. The complete genome sequence enables us to place all the genes and repeats in their context. In mammals, there are longer repetitive families (LINEs) of about 1,000 to 6,000 nucleotides and shorter repetitive families (SINEs) of about 350 nucleotides. The copy number of some families can reach hundreds of thousands. There are also simple repeats of a few nucleotides that can reach several hundred nucleotides in length, for example, AC (adenine and cytosine) repeated fifty or more times. These can also influence genome evolution and be responsible for human genetic diseases. Comparison of related species helps us to understand the significance of all these complex repetitive families. It has been recognized for many years that these sequences may have profound implications for genome evolution, but the details remain to be unraveled. In fact, recent estimates suggest that at least 50 percent of the human genome is made up of repeat elements.

Comparative genomics, as described, is a richly informative endeavor that will lead to a far deeper understanding of phylogeny and evolution, and it relies heavily on the relatively new field of bioinformatics. Living organisms manage biological information using DNA, RNA, and protein; bioinformatics manages biological information using computers. Statistical models of gene-sequence evolution such as hidden Markov models (HMMs) can encode complete probabilistic descriptions of nucleic acid and amino acid sequences and their relationships. Within a genome, such models can help to identify the different kinds of DNA such as introns, exons, promoters, and intergenic DNA. Coding for protein and the bias in codon usage helps to identify genes within a genome. When combined with cross-species comparisons, this becomes the strongest predictive method we have. Computerized databases of protein families such as PFAM also rely on HMMs to construct their classifications. Members of known families are readily detected in newly sequenced DNA. These models provide the most powerful methods of gene and protein identification. The additional dimension of protein structure is necessary for complete protein classification and to assist in the understanding of protein function.

Genomics and proteomics are producing dramatic advances in biotechnology and medicine. Bioethics and regulation struggle to keep up, as the issues are often very complex. The furor over genetically modified crops is one example of the biotechnology industry getting ahead of the regulators. The issues surrounding the patenting of inventions relating to genes are not resolved. Fears about racism are stultifying studies of ethnicity. Is genetic screening of large sections of the population desirable, and how do we handle genetic testing and counseling? Is it better not to know, as knowledge can bring sorrow as well as joy? What are the dangers of gene therapy and transfer and stem cell therapy? Are animal to human transplants fraught with hidden dangers? Informed consent for the use of human tissue in genetics research is in conflict with the desire to use the material for any purpose, yet unforeseen, in the future. These are all serious issues, and there is a need to become more proactive in anticipating and dealing with them.

The basic aim of understanding the complete development and functioning of an organism at the molecular level is probably achievable by the methodologies that are being developed today. These endeavors are relatively uncontroversial, and skepticism that it would ever be possible to understand the processes of life at the molecular level is evaporating. Anatomy and physiology will be understood in greater depth, and the molecular basis will be apparent. One of the difficulties in describing function is to have a common language that can be used for related processes. The Gene Ontology Consortium is attempting to develop such a language for the processes common to yeast, fruit fly, and mouse.

Controversy begins when we compare human individuals, rather than pretending there is a model normal person whose biology we are describing. It is variation that is the stuff of evolution and it is also variation that leads to bioethical dilemmas. Biotechnology and medicine cannot ignore the advances in molecular biology, but they must exploit these advances carefully and with humanity.

Postgenomics can successfully catalog the building blocks of the central dogma of molecular biology, that is, that DNA make RNA makes protein, by exhaustive enumeration of the genome, transcriptome, and proteome. Knowing the building blocks does not tell us how life operates as a system. We need to move from the molecules to an understanding of their interaction networks in a cell. Higher organisms are formed of a complex network of interacting cells with molecular signaling between them. The organism exists in an environment with which it interacts and has adaptations to deal with harsh or unusual environmental conditions. Finally, organisms exist in ecosystems that are composed of not only the physical environment but also all that other organisms of the same and different species that are present. Postgenome informatics must integrate and analyze all these complex interactions (Kanehisa, 2000).

There are many fascinating questions to be tackled as a result of this understanding. One example is the differences between man and the nearest primate relative, the chimpanzee. As far as we know, the gene sequences and the proteins are about 99 percent identical. Subtle changes in the timing and amount of gene expression and in molecular interactions may underly human–chimpanzee differences rather than the use of different genes—that is, there do not appear as yet to be genes that are uniquely human.

Has anything really changed in the postgenome era? We have been studying genetic variation for more than a century and characterizing genes and proteins for forty years.

The big difference is that knowing the entire genome sequence makes it possible to identify all the genes, transcripts, and proteins. Having the sequences of all transcripts makes it possible to fabricate arrays that allow studies of the expression of all genes simultaneously in high throughput. Knowing all the proteins enables protein and peptide arrays to be used as major tools in functional studies. Arrays of immobilized antibodies may be used to assay binding partners in complex mixtures. Biochemical functions such as enzyme activity or DNA binding can be studied in parallel. Methods of studying protein–protein interactions such as yeast-two hybrid and phage displayed combinatorial peptides can also work in high throughput.

The methods for gene and protein expression commonly work on two-dimensional arrays, known as DNA chips. DNA chip technology allows scientists to investigate which genes are expressed in an organism, at what times they are expressed during development, and in what tissues. The concept of integrating many functions onto a single chip revolutionized the electronics industry. Similarly, biochemical analysis is being revolutionized by miniaturization not only in these two-dimensional arrays but also by microfluidics. A microfluidics instrument is a glass or plastic substrate on the surface of which is a network of extremely fine channels. Analysis or separation is performed by guiding samples and reagents through these channels. Electronic circuits may also be included for detection. We can expect a very wide range of powerful applications to be developed in the future.

These technology developments are awesome, but we must not lose sight of the objectives that are to move from an understanding of the molecules, their interactions, and functions to an integrated picture of the functioning of the cells, the organisms, and ultimately to extract as much information as possible about their evolution.

The biological literature remains one of the largest biological data sets. Medline alone catalogs twelve million entries. Many of the inferences about molecular function are described in the literature and have yet to be fully integrated into computerized databases. Several groups are attempting the automated retrieval of this information. Statistical approaches are based on the frequency of word occurrences within each individual entry in a large body of text. Significant associations may be used to characterize the biological functions being discussed. Computational linguistics methods extract syntactic information and use internal dependencies in the text. Frame-based approaches use previously defined templates in a combination of the statistical and linguistic methods.

The collections of nucleotide and amino acid sequence data probably exceed the literature in sheer volume. More than sixty fully sequenced genomes are now available, and many more genomes are being determined. The EMBL Nucleotide Sequence Database contains (September 2001) 14.5 billion nucleotides including expressed sequence tags (ESTs), sequence tagged sites (STSs), and genome survey sequences (GSSs). At present, it is common to compare a newly determined DNA sequence against all known sequences using a program such as BLAST or FASTA. Computer architectures and the algorithmic difficulties of dealing with mismatches and indels (insertions/deletions) conspire to make this a fact of life. It is dubious how much longer we can go on searching the entirety of known sequences as their volume expands exponentially. The more information we have, the easier it should be to look up related sequences to our query without having to search the universe of sequences.

Sequence variation is a major source of further data with important application in medicine. About 1.4 million human single nucleotide polymorphisms (SNPs) have been mapped, which represents an SNP about every 1,900 bases of DNA. This knowledge helps to define haplotype variation across the genome. It will be very useful to identify allelic variants that are associated with disease. Another major application is in the area of pharmacogenetics that assesses the suitability of particular drugs for individual patients. At present, this has been mainly applied to variants of drug metabolizing enzymes such as the cytochrome P450s. SNPs may also help in defining the genetic basis of varying susceptibility to infectious diseases.

The most rapidly growing type of data today, and one that has the potential to overtake the sequence data in volume, is gene expression data. There are a variety of forms of gene expression data including EST libraries, Serial Analysis of Gene Expression (SAGE) libraries, and DNA microarray data. There are more than eight million EST sequences in dbEST, with their tissue of origin, as well as two million SAGE measurements in the SAGEmap database. The DNA microarrays are of two main types, gridded PCR products from EST clones or synthetic oligonucleotides. Probably the oligonucleotides are the most powerful because they can be chosen to distinguish similar members of multigene families and splice variants of a single gene. The PCR products may hybridize readily to messenger RNAs (mRNAs) that are in fact distinct. The ability to generate data is staggering if one considers that there are over 300 distinct human cell types and that experiments can study time courses over an infinite variety of conditions. It is also important that adequate replicates are performed to ensure statistical validity of the measured expression differences and the interpretation of the results in terms of covariant groups of transcripts.

A major application of expression profiling is to cancer. The results from tumors are highly informative for classification and diagnosis. Correct classification will be invaluable for epidemiological studies. In clinical trials, the effect of treatments will be placed on a much surer foundation.

A further type of gene expression data is that produced by in situ hybridization, which is particularly valuable for studies of embryonic development in organisms such as zebra fish, *Xenopus*, and mouse. The first requirement is to establish a standard anatomy onto which the hybridization patterns of the probes can be mapped. In the case of the mouse, the embryonic anatomy of the twenty-eight Theiler stages (standard sets of stages describing mouse development) needs to be carefully reconstructed from serial sections. In situ patterns of gene expression can then be mapped onto the anatomy. Similar work is being undertaken with human fetal brain. Early analysis of these data indicates that genes with similar expression patterns often have related functional roles.

Protein expression studies are equally as important as gene expression studies but are technically more difficult at present. Protein interactions can also be studied and form the first step in understanding protein interaction networks. For yeast proteins, about 4,000 different interactions have been observed. Similar information needs to be accumulated for human proteins. The Database of Interacting Proteins (DIP) lists 5,958 proteins and 10,782 interactions. Just because two proteins interact in an in vitro experiment does not mean that they necessarily are involved in a pathway in vivo.

A number of attempts are being made to define pathways databases. TRANSFAC is concerned with the regulation of gene expression. Metabolic pathways databases include KEGG and EcoCyc/MetaCyc. Signaling pathways databases such as SPAD are at the early stage of development.

To understand cellular processes, it is also necessary to study the small molecule metabolites that are present; this may be called metabolic profiling. Nuclear magnetic resonance (NMR) spectroscopy provides an aggregate characterization of cell phenotype and enables comparisons to be made with varying conditions. Gas chromatography/mass spectroscopy (GC/MS) is more laborious but enables individual components to be characterized and quantified.

High throughput methodologies yield more data than we can easily handle. They may be hypothesis driven in that they set out to test a specific model of behavior. More commonly they are used to amass data in the hope that some meaning can be teased from these by data mining. Eventually we have to convince ourselves that a protein has been correctly characterized, that its structure has been determined, and that its biochemical role is clearly understood. Further, we understand how variation in this protein can contribute to a disease process. When we set ourselves the task of acquiring precise knowledge about every protein in an organism, we can appreciate why we are at the dawn of proteomics. Genomics and proteomics are technologies more than sciences, and when their time course is run, we return to calling our knowledge genetics and biochemistry.

We have the ability to molecularly dissect, understand the function, and alter the properties of any living organism that we please. The biotechnology revolution has infinite possibilities limited only by our imagination of what we would wish to do and the ethical and legal constraints on what is allowable. We must use this newfound power wisely. In medicine the possibilities for diagnosis and cure are apparent, but for cures the progress may be tantalizingly slow. We could run genetic screens on all patients of a medical practice and advise them on how to live their lives. The social, ethical, and economic implications are great. We must ask ourselves if this is really what we want.

BIBLIOGRAPHY

Kanehisa, M. *Post-Genome Informatics*. Oxford and New York, 2000.

THE NEW REPLICATORS

Daniel Dennett

It has long been clear that, in principle, the process of natural selection is *substrate-neutral*. That is, evolution will occur whenever and wherever three conditions are met: replication, variation (mutation), and differential fitness (competition).

In Darwin's own terms, if there is "descent [i.e., replication] with modification [variation]" and "a severe struggle for life" [competition], better-equipped descendants will prosper at the expense of their competitors. We know that a single material substrate, DNA (with its surrounding systems of gene expression and development), secures the first two conditions for life on earth; the third condition is secured by the finitude of the planet as well as more directly by uncounted environmental challenges. We also know, however, that DNA established its monopoly position over early variations that have left their traces and ongoing exemplars, such as the RNA viruses and prions. Have any other completely different evolutionary substrates arisen on this planet? The best candidates are the brainchildren, planned or unplanned, of one species: *Homo sapiens*.

Darwin himself proposed words as an example: "The survival or preservation of certain favoured words in the struggle for existence is natural selection" (*Descent of Man*, 1871, p. 61). Billions of words are uttered (or inscribed) every day, and almost all of them are replicas—in a sense to be discussed below—of earlier words perceived by their utterers. Replication is not perfect, and there are many opportunities for variation or mutation in pronunciation, inflection, or meaning (or spelling, in the case of written words). Moreover, words are roughly segregated into lineages of replication chains; for instance, we can trace a word's descendants from Latin to French to Cajun. Words compete for air time and print space in many media; words become obsolete and drop out of the word pool, while other words spring up and flourish. We discover *conTROVersy* going to fixation—surviving while all competing pronunciations go extinct—in some regions and *CONtroversy* going to fixation in others, while the original meaning of "begs the question" is supplanted in some quarters by a variant. The detectable historical changes in languages have been studied from one Darwinian perspective or another since Darwin's own day, and a great deal is known about patterns of replication, variation, and competition in the processes that have yielded the diverse languages of today. Some of the investigative methods of modern evolutionary biology—in bio-informatics, for instance—are themselves descended from pre-Darwinian researches conducted by paleographers and other early scholars of historical linguistics. As Darwin noted, "The formation of different languages and of distinct species, and the proofs that both have been developed through a gradual process, are curiously the same" (1871, p. 59).

Words, and the languages they populate, are not the only culturally transmitted variants that have been proposed. Other human acts and practices that spread by imitation have been identified as potential replicators, as have some of the habits of nonhuman animals. The physical substrates of these media are various indeed, including sounds and all manner of visible, tangible patterns in the behavior of the vector organisms. Moreover, behaviors often produce artifacts (paths, shelters, tools, weapons, and signs or symbols) that may serve as better exemplars of replication than the behaviors that produce them, being relatively stable over time and hence in some ways easier to copy, as well as being independently movable and storable. One human artifact, the computer, with its prolific copying ability, has recently provided a distinctly new substrate in which both deliberate and inadvertent experiments in artificial evolution are now burgeoning, taking advantage of the emergence of gigantic networks of linked computers that permit the swift dispersal of propagules made of nothing but bits of information. These *computer viruses* are simply sequences of binary digits that can have an effect on their own replication. Like macromolecular viruses, they travel light, being nothing more than information packets—including a phenotypic overcoat that tends to gain them access to replication machinery wherever they encounter it. Finally, researchers in the new field of artificial life aspire to generate both virtual (simulated, abstract) and real (robotic) self-replicating agents that can take advantage of evolutionary algorithms to explore the adaptive landscapes in which they are situated, generating improved designs by processes that meet the three defining conditions while differing from carbon-based life forms in striking ways. At first glance these phenomena may appear to be only models of evolving entities, thriving in modeled environments, but the boundary between an abstract demonstration and an application in the real world is more easily crossed by these evolutionary phenomena than by others, precisely because of the substrate-neutrality of the underlying evolutionary algorithms. Artificial self-replicators can escape from their original environments on researchers' computers and take on a "life" of their own in the rich new medium of the Internet.

All these categories of new replicators are dependent, like viruses, on replicative machinery that is built and maintained directly or indirectly by the parent process of biological evolution. Were all DNA life forms to go extinct, all their habits and meta-habits, their artifacts and meta-artifacts, would soon die with them, lacking the where-withal (both the machinery and the energy to run the machinery) to reproduce on their own. For the time being, our computer networks and robot fabrication and repair facilities require massive supervision and maintenance by us, but it has been suggested by the roboticist Hans Moravec (1988) that silicon-based electronic (or photonic) artifacts could become entirely self-sustaining and self-replicating, weaning themselves from their dependence on their carbon-based creators. This improbable and distant eventuality is not a requirement for evolution, however, or for life itself. After all, our own self-replication and self-maintenance are entirely dependent on the billions of bacteria without which our metabolisms would fail; and if our artifactual descendants similarly have to enslave armies of our biological descendants to keep their systems up and running, this would not detract from their claim to be a new branch on the tree of life. We have reached a point where we can no longer survive on this planet without our artifacts, just as they can no longer survive without us. This codependency is in some important regards like the codependency of RNA and DNA, and the co-dependency of multicellular hosts and their symbionts—an emerging fact of life.

As with many taxonomies in evolutionary theory, there are controversies and puzzles about how to draw the branchings and how to name them. Some of these puzzles are substantive, and some are merely disagreements about which terms to use. The zoologist Richard Dawkins coined the term *meme* in his 1976 book, *The Selfish Gene*, and the term has caught on. He opened his discussion of these "new replicators" with

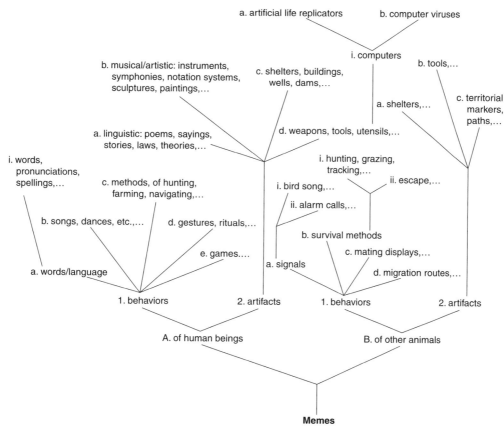

FIGURE 1. The New Replicators.
A Simple Taxonomy of the New Replicators.   Courtesy of Daniel Dennett.

a discussion of bird song, but others who have adopted the term have wanted to restrict memes to human culture. Should such evolving animal traditions as alarm calls, nest-building methods, and chimpanzee tools also be called memes? Researchers concentrating on cultural transmission in animals, such as John Tyler Bonner (1980) and Eytan Avital and Eva Jablonka (2000), have resisted the term, and others writing on human cultural evolution, such as Luca Cavalli-Sforza and Marcus Feldman (1981), and Robert Boyd and Peter Richerson (1985) have also chosen to use alternative terms. But since the word *meme* has secured a foothold in the English language, appearing in the most recent edition of the *Oxford English Dictionary* with the definition "an element of culture that may be considered to be passed on by non-genetic means," we may conveniently settle on it as the general term for any culturally based replicator—if such there are. Those who are squeamish about using a term whose identity conditions are still so embattled should remind themselves that similar controversies continue to swirl around how to define its counterpart, *gene*, a term that few would recommend abandoning.

Memes include not just animal traditions, then, but also computer-based replicators, for two reasons: not only do computers and their maintenance and operation depend on human culture, but the boundaries between computer viruses and more traditional human memes have already been blurred. Simple computer viruses in effect carry the instruction *copy me*, which is directed to the computer in machine language and is entirely invisible to the computer's user. Like the toxins unwittingly ingested by people who catch and eat freshwater fish, such a computer virus, though an element of the users' environment, is arguably not part of their *cultural* environment. However, at least as widespread and virulent as such "proper" computer viruses are bogus computer virus warnings, directed to the computer user in natural language. These warnings, which depend directly on a comprehending (but duped) human vector to get themselves replicated on the Internet, are definitely within the intended understanding of memes; intermediate cases are the computer viruses that depend on enticing human users to open attachments (thereby triggering the invisible copying instruction) by promising some amusing or titillating contents. These too depend on human comprehension; one written in German will not spread readily to the computers of monoglot English speakers. (This pattern may change if users avail themselves regularly of on-line translation services.) In the arms race between virus and anti-virus, ever more elaborate exploitations of human interests are to be expected, so it seems best to include all these replicators under the rubric of memes. Note, however, that some of them make only indirect use of human vectors and hence are only indirectly elements of human culture. We are beginning to see this porous boundary crossed in the other direction as well: it used to be true that the differential replication of such classic memes as songs, poems and recipes depended on their winning the competition for residence in human brains, but now that a multitude of search engines on the Web have interposed themselves between authors and their (human) audiences, competing with one another for a reputation as high-quality sources of cultural items, significant fitness differences between memes can accumulate independently of any human appreciation or cognizance. The day may soon come when a cleverly turned phrase in a book gets indexed by many search engines, and thereupon enters the language as a new cliche without anybody human having read the original book.

Problems of Classification and Individuation

Some problems of classification are substantive, depending in part on historical facts that are not well established, and others are tactical problems for the theorist: What divisions of the phenomena will prove most perspicuous? Are all computer viruses properly descended from the earliest forays into artificial life, or should at least some of them be shown as arising independently of that intellectual movement? Not all computer hackers are A-Life hackers, but there is also the unanswered tactical question of how to characterize what is copied. If one hacker gets the *general idea* of a computer virus from somebody else and then goes on to make an entirely new kind of computer virus, is that new virus properly a descendant, with modifications, of the virus that inspired its creation? What if the hacker adapts elements of the original virus's design in the new type? How much sheer mindless copying must there be, or alternatively, how much comprehending inspiration *may* there be, in an instance of replication? (More on this question below.) Is there cross-species meme-copying in the animal world? Polar bears build a den that includes a raised snow shelf that permits cold air to drain out the depressed opening of the den. Is this wise trend in arctic technology entirely innate (now), or do bear cubs have to copy their mother's example? The same snow shelf is found in an Inuit snow shelter. Did the Inuit copy this tradition from the polar bear, or was it an independent invention? Does it ever happen that one species begins attending to the alarm calls of another and then develops an alarm call tradition of its own? Does the *alarm call* meme spread from species to species, or should we consider the intraspecific alarm calls and their variants as entirely independent lineages (Reader and Laland, 1999)?

Exacerbating these problems are other problems of meme individuation. Should the (English) word "windsurfing" be seen as distinct from the (language-neutral) windsurfing *meme*? Are these two memes or one? Do styles, such as *punk* and *grunge*, count as memes before they have names? Why not? Joining forces with a name-meme is no doubt an excellent fitness advantage for almost any meme. (An exception could be a meme that depends on spreading insidiously; the coining of a name-meme, such as *male chauvinism*, may actually hinder the spread of male chauvinism by sensitizing something like an immune reaction in potential vectors.) It is probably true that as soon as any human meme becomes salient enough in the environment to be discerned, it will thereupon be named by one of its discerners, tightly linking the two memes thereafter: the name and the named, which typically have a shared fate, but not always. (The musical characteristics identifiable as *the blues* include many robust instances that are not *called* "the blues" by those who play and listen to them.) Undiscerned memes can also flourish. For instance, changes in the pronunciation or meaning of a word can move to fixation in a large community before any sharp-eared linguist or other cultural observer takes note. There are more than a few people—comedians as well as anthropologists and other social scientists—who earn their living by detecting and commenting on evolving trends in cultural patterns that have heretofore been at best dimly appreciated.

Until these and other problems of initial theoretical orientation are resolved, skepticism about memes will continue to be widespread and heartfelt. Many commentators are deeply opposed to any proposals to recast questions in the social sciences and humanities in terms of cultural evolution, and this opposition is often expressed in terms of a challenge to prove that "memes exist":

> Genes exist [these critics grant] but what are memes? What are they made of? Genes are made of DNA. Are memes made of neuron-patterns in the brains of enculturated people? What is the material substrate for memes?

There are some proponents of memes who have argued in favor of an attempt to *identify* memes with specific brain-structures—a project still entirely uncharted. On current understandings of how the brain might store cultural information, it is unlikely that any independently identifiable common brain structures, in different brains, could ever be isolated as the material substrate for a particular meme. Although some genes for making eyes do turn out to be identifiable whether they occur in the genome of a fly, a fish, or an elephant, there is no good reason to anticipate that the memes for wearing bifocals might be similarly isolatable in neuronal patterns in brains. It is vanishingly unlikely, that is, that the brain of Benjamin Franklin, who invented bifocals, and the brains of those of us who wear them should "spell" the *idea* of bifocals in a common brain-code. Besides, this imagined path to scientific respectability is based on a mistaken analogy. In his 1966 book *Adaptation and Natural Selection*, the evolutionary theorist George Williams offered an influential definition of a *gene* as "any hereditary information for which there is a favorable or unfavorable selection bias equal to several or many times its rate of endogenous change." As he went on to stress in his 1992 book, *Natural Selection: Domains, Levels, and Challenges*, "A gene is not a DNA molecule; it is the transcribable information coded by the molecule."

Genes—genetic recipes—are all written in the physical medium of DNA, using a single canonical language, the nucleotide alphabet of adenine, cytosine, guanine, and thymine, triplets of which code for amino acids. Let every strand of smallpox DNA in the world be destroyed; if the smallpox genome is preserved (translated from *nucleotides* into the *letters* A, C, G, and T and stored on hard disks on computers, for instance), smallpox is not truly extinct; it could have descendants someday because its genes *still exist* on those hard disks, as what Williams calls "packages of information."

Memes—cultural recipes—similarly depend on one physical medium or another for their continued existence (they aren't magic), but they can leap around from medium to medium, being translated from language to language, from language to diagram, from diagram to rehearsed practice, and so forth. A recipe for chocolate cake, whether written in English in ink on paper, or spoken in Italian on videotape, or stored in a diagrammatic data structure on a computer's hard disk, can be preserved, transmitted, translated, and copied. Since the proof of the pudding is in the eating, the likelihood of a recipe getting *any* of its physical copies replicated depends (mainly) on how successful the cake is. How successful at doing what—at getting a host to make another cake? Usually, but even more important is getting the host to make another copy of the recipe and passing it on. That's all that matters, in the end. The cake may not enhance the fitness of those who eat it; it may even poison them, but if it first somehow provokes them to pass on the recipe, the meme will flourish.

This is perhaps the most important innovation in outlook permitted by recasting traditional investigations in terms of memes: they have their own fitness as replicators, independent of any contribution they may or may not make to the genetic fitness of their hosts, the human vectors. Dawkins put it this way: "What we have not previously considered is that the cultural trait may have evolved in the way that it has, simply because it is *advantageous to itself*" (1976, p. 200). The anthropologist F. T. Cloak, (1975), put it this way: "The survival value of a cultural instruction is the same as its function; it is its value for the survival/replication of itself or its replica."

Those who question whether memes exist because they cannot see what material thing a meme could be should ask themselves if they are equally dubious about whether words exist. What is the word *cat* made of? Words are recognizable, re-identifiable products of human activity; they come in many media, and they can leap from substrate to substrate in the process of being replicated. Their standing as real things is not at all impugned by their abstractness. In the proposed taxonomy, words are but one species of memes, and the other species of memes are the same kind of things that words are—you just can't pronounce or spell them. Some of them you can dance, and some of them you can sing, or play, and others you can follow by making something out of the various building materials the world provides. The word *cat* isn't made out of some of the ink on this page, and a recipe for chocolate cake isn't made of flour and chocolate.

There is no single proprietary code, parallel to the four-element code of DNA, that can be used to anchor meme-identity the way gene-identity can be anchored, for most practical purposes. This is an important difference, but one of degree. If the current trend of language extinctions continues at its present pace, every person on the earth may someday speak the same language, and it will then be difficult to resist the temptation (which should still be resisted!) to *identify* memes with their (now practically unique) verbal labels. But so long as there are multiple languages, to say nothing of the multiple media in which nonlinguistic cultural items can be replicated, we are better off keeping strictly to the abstract, code-neutral understanding of a meme as a "package of information," bearing in mind that for high-fidelity replication to occur, there must always be some "code" or other. Codes play a crucial role in all systems of high-fidelity replication, since they provide finite, practical sets of norms against which relatively mindless editing or proofreading can be done. But even in the clearest cases of codes, there are often multiple levels of norms. Suppose Tommy writes the letters "SePERaTE" on the blackboard, and Billy "copies" it by writing "seperate." Is this really copying? The normalization to all lower-case letters shows that Billy is not slavishly copying Tommy's chalk-marks, but rather being triggered to execute a series of canonical, normalized acts: *make an "s," make an "e,"* and so on. It is thanks to these letter-norms that Billy can "copy" Tommy's word at all. But he does copy Tommy's spelling error, unlike Molly, who "copies" Tommy by writing "separate," responding to a higher norm at the level of word spelling. Sally then goes a step higher, "copying" the phrase "separate butt equal"—all words in good standing in the dictionary—as "separate but equal," responding to a recognized norm at the phrase level. Can we go higher? Yes. Anybody who, when "copying" the line in the recipe "Separate three eggs and beat the yolks until they form stiff white peaks," would replace "yolks" with "whites" knows enough about cooking to recognize the error and correct it. Above spelling and syntactic norms are a host of semantic norms as well.

Norms can both hinder and help replication. The anthropologist Dan Sperber (2000) has distinguished copying from what he calls "triggered reproduction" and has noted that, in cultural transmission, "the information provided by the stimulus is complemented with information already in the system." This complementing tends to absorb mutations instead of passing them on. Evolution depends on the existence of mutations that can survive the proofreading processes of replication intact, but it does not specify the level at which this survival must occur. A brilliant cooking innovation might indeed get corrected away by an all-too-knowing chef in the course of passing on the recipe, but other "errors" might get through and replicate indefinitely. Meanwhile, the correction of other varieties of noise at other levels, responding to spelling norms or others, must be ongoing in order to keep the copying process faithful enough so that multiple exemplars of each innovation can be tested against the environment. As Williams puts it, "A given package of information (codex) must proliferate faster than it changes, so as to produce a genealogy recognizable by some diagnostic effects" (1992, p. 13). Recognizable, that is, to the unfocused, independently varying environment, so it can yield probabilistic verdicts of natural selection that have some likelihood of identifying adaptations of projectible fitness. It is possible, for instance, that encodings of the same meme in different media will differentially compete and differentially mutate, so that they should be considered different memes for some purposes. When a Welsh folk song transmitted orally for hundreds of years is transcribed and transliterated by an ethnomusicologist and then adapted by a composer in a choral work that becomes a popular concert item for amateur choruses whose members know no Welsh, the change in encoding makes a difference to the susceptibility to mutation in subsequent replication streams, which bifurcate into code-dependent trajectories.

Just how big or small can a meme be? A single musical tone is not a meme, but a memorable melody is. Is a symphony a single meme, or is it a system of memes? A parallel question can be asked about genes, of course. No single nucleotide or codon is a gene. How many notes or letters or codons does it take? The answer in both cases tolerates blurred boundaries: a meme, or a gene, must be large enough to carry information worth copying. There is no fixed measure of this, but the bountiful system of case law on copyright and patent infringements indicates that verdicts on particular cases form a relatively trustworthy equilibrium that is stable enough for most purposes.

Other objections to memes seem to exhibit an inverse relationship between popularity and soundness: the more enthusiastically they are championed, the more ill-informed they are. They have been patiently rebutted again and again by proponents, but those who are appalled by the prospect of an evolutionary account of anything in human culture don't seem to notice. A common mistake by critics is to imagine that memes must be more like genes than they need to be for the three conditions to be met. It has been observed, for instance, that when an individual first acquires some encountered cultural item, this is typically not a case of imitating a single instance of it. (If I take up the practice of wearing my baseball cap backwards, or add a new word to my working vocabulary, am I copying the first instance of it I ever noticed, or the most recent instance, or am I somehow averaging over all of them?) This embarrassment of riches in the search for the parent of the new offspring does complicate the model of cultural replication, but it does not in itself disqualify the process as one of replication. For instance, the ultra-high-fidelity copying of computer files depends in many instances on error-correcting code-reading systems that in effect let "majority rule" determine which of several candidate exemplars should count as canonical. In such cases, no single vehicle of the information can be identified as the source—but this is undeniably an instance of replication. Darwin's trio of requirements is both substrate-neutral and implementation-neutral to a degree that is not always appreciated.

Is Cultural Evolution Darwinian?

Marking these unresolved problems of nomenclature and individuation, we can turn to the more fundamental and important question: Do any of these candidates for Darwinian replicator actually fulfill the three requirements in ways that permit evolutionary theory to explain phenomena not already explicable by the methods and theories of the traditional social sciences? Or does this Darwinian perspective provide only a relatively trivial unification? It would still be important to conclude that cultural evolution obeys Darwinian principles in the modest sense that nothing that happens in it contradicts evolutionary theory, even if cultural phenomena are best accounted for in other terms. In *The Origin of Species*, Darwin himself identified three processes of selection: "methodical" selection by the foresighted, deliberate acts of farmers and others intent on artificial selection; "unconscious" selection, in which human beings have engaged in activities that have unwittingly contributed to the differential survival and reproduction of species, mostly on their way to domestication; and "natural" selection, in which human intentions have played no role at all. To this list we can add a fourth phenomenon, genetic engineering, in which the intention and foresight of human designers plays a still more prominent role. All four of these phenomena are Darwinian in the modest sense. Genetic engineers do not produce counterexamples to the theory of evolution by natural selection, any more than plant breeders over the eons have done; they produce novel fruits of the fruits of the fruits of evolution by natural selection. The idea of memes promises similarly to unify under a single perspective such diverse cultural phenomena as deliberate, foresighted scientific and cultural inventions (memetic engineering), such authorless productions as folklore, and even such unwittingly redesigned phenomena as languages and social customs themselves. As we enter the age of deliberate, purportedly foresighted tinkering with our own genomes and the genomes of other species, we face the prospect of strong interactions between genetic and memetic evolution, including many that may take off without having been foreseen at all. It behooves us to investigate these possibilities with the same vigor and attention to detail that we devote to the investigation of the evolution of carbon-based pathogens and to the swift disappearance of natural barriers that have structured the biosphere until very recently.

We should also remind ourselves that, just as population genetics is no substitute for ecology—which investigates the complex interactions between phenotypes and environments that ultimate yield the fitness differences presupposed by genetics—no one should anticipate that a new science of memetics would overturn or replace all the existing models and explanations of cultural phenomena developed by the social sciences. It might, however, recast them in significant ways and provoke new inquiries in much the way genetics has inspired a flood of investigations in ecology. The books listed below explore these prospects in some detail, but still at a very programmatic and speculative level. At this time, there are still only a few works that might be listed as pioneering empirical investigations in specialized branches of memetics: Hull (1988), Lynch, (1996b), Pocklington and Best (1997), and Gray and Jordan (2000).

BIBLIOGRAPHY

Aunger, R., ed., *Darwinizing Culture: The Status of Memetics as a Science.* Oxford, 2000.

Aunger, R. *The Electric Meme: A New Theory of How We Think and Communicate.*

Avital, E. and E. Jablonka. *Animal Traditions: Behavioural Inheritance in Evolution.* Cambridge, 2000.

Blackmore, S. *The Meme Machine.* Oxford, 1999.

Bonner, J. T. *The Evolution of Culture in Animals.* Princeton, 1980.

Boyd, R., and P. Richerson. *Culture and the Evolutionary Process.* Chicago, 1985.

Brodie, R. *Virus of the Mind: The New Science of the Meme.* Seattle, 1996.

Cavalli-Sforza, L., and M. Feldman. *Cultural Transmission and Evolution: A Quantitative Approach.* Princeton, 1981.

Cloak, F. T. "Is a Cultural Ethology Possible?" *Human Ecology* 3 (1975): 161–82.

Dawkins, R. *The Selfish Gene.* Oxford, 1976.

Dennett, D. *Darwin's Dangerous Idea.* New York, 1995.

Dennett, D. "The Evolution of Culture." *Monist* 84 (2001).

Dennett, D. "From Typo to Thinko: When Evolution Graduated to Semantic Norms."

Durham, W. *Coevolution: Genes, Culture and Human Diversity.* Stanford, Calif., 1991.

Gray, R. D., and F. M. Jordan. "Language Trees Support the Express-Train Sequence of Austronesian Expansion." *Nature* 405 (2000): 1052–1055.

Hofstadter, D. R. "Epilogue: Analogy as the Core of Cognition." In *The Analogical Mind: Perspectives from Cognitive Sciences*, edited by D. Gentner, K. J. Holyoak, and B. N. Kokinov. Cambridge, Mass., 2001.

Hull, D. *Science as a Process*. Chicago, 1988.

Laland, K., and G. Brown. *Sense and Nonsense: Evolutionary Perspectives on Human Behaviour*. Oxford, 2002.

Lynch, A. *Thought Contagion: How Belief Spreads through Society*. New York, 1996a.

Lynch, A. "The Population Memetics of Birdsong." In *Ecology and Evolution of Acoustic Communication in Birds*, edited by D. E. Kroodsma and E. H. Miller, pp. 181–197. Ithaca, 1996b.

Moravec, H. *Mind Children: The Future of Robot and Human Intelligence*. Cambridge, Mass., 1988.

Pocklington, R. "Memes and Cultural Viruses." *Encyclopedia of the Social and Behavioral Sciences*.

Pocklington, R., and M. L. Best. "Cultural Evolution and Units of Selection in Replicating Text." *Journal of Theoretical Biology* 188 (1997): 79–87.

Reader, S. M., and K. N. Laland. "Do Animals Have Memes?" *Journal of Memetics* (1999).

Ruhlen, Merritt. *The Origin of Languages*. New York, 1994.

Sperber, D. "An Objection to the Memetic Approach to Culture." In *Darwinizing Culture*, edited by Robert Aunger. 2000.

Williams, G. *Adaptation and Natural Selection*. Princeton, 1966.

Williams, G. *Natural Selection: Domains, Levels, and Challenges*. Oxford, 1992.

ALPHABETICAL ENTRIES

A

ACANTHOCEPHALANS. *See* Animals.

ACQUIRED IMMUNE DEFICIENCY SYNDROME

[*This entry comprises two articles. The introductory article provides a discussion of the origins and phylogeny of HIV; the companion article focuses on different modes of HIV transmission between individuals. For related discussions, see* Immune System *and* Viruses.]

Origins and Phylogeny of HIV

Acquired immune deficiency syndrome (AIDS) was first described in 1981 when clusters of young American men were found to be suffering from the cancer Karposi's sarcoma and pneumonia caused by infection with the fungus *Pneumocystis carinii*. It was soon realized that these usually rare conditions were in fact caused by an immune dysfunction, so that patients were unable to prevent a myriad of microbial infections from taking hold. Although immune deficiency syndromes had been recognized before, AIDS was unique in that its spread was rapid and its associated mortality was effectively 100 percent. Two years later, the causative agent, a retrovirus designated the human immunodeficiency virus (HIV), was isolated from a French AIDS patient. We now know that there are in fact two different forms of HIV circulating in human populations: the vast majority of infections worldwide arise from HIV-1, and a smaller number of people of West African origin are infected with a related virus, HIV-2. Although AIDS was initially recognized in clusters of gay men, it was quickly evident that HIV could be spread through a variety of body fluids and that globally the most common mode of transmission is through heterosexual intercourse.

Since its discovery, HIV has spread with frightening rapidity. By late 2000, more than thirty-six million people worldwide were living with HIV infection, with over fourteen thousand new infections each day, many of which were children, and a cumulative death toll of more than twenty-two million. Worst affected has been sub-Saharan Africa, which accounts for 70 percent of all HIV infections and where the average prevalence in adults reaches 25 percent in some countries.

Basic Biology of HIV. HIV is a retrovirus consisting of two single-stranded RNA genomes of approximately 11 kilobases in length contained within a single virion. Like all retroviruses, replication takes place through a DNA intermediate stage, known as the provirus, that integrates into a host's own DNA, making eradication an extremely difficult problem.

HIV infection may be divided into three stages. The first, primary infection, describes the period in which the virus enters a new host and begins to replicate. Because it takes several weeks for the immune system to recognize that HIV is present, the virus is able to replicate unabated and so reaches very high population sizes ($\sim 10^8$–10^{10}). After an active immune response to HIV is established (seroconversion), the viral population size declines dramatically and the host enters the second, asymptomatic stage, which may last from a few months to many years. Although it is tempting to think of the virus as latent at this time, viral replication is in fact extensive, with some 10^{10} virions produced each day. The final stage of HIV infection is when the host immune system becomes so compromised after years of viral replication that AIDS itself develops.

In many ways, the most important aspect of HIV is that it infects cells of the immune system. The major preference is for macrophages and T lymphocytes, to which the virus attaches using the CD4+ receptor and a variety of chemokine coreceptors, most notably CCR5 and CXCR4. Those HIV-1 strains that prefer to replicate in macrophages (M-tropic strains) use the CCR5 receptor and dominate during the early years of HIV infection. Other HIV strains preferentially replicate in T cells, in which case the CXCR4 receptor is used. These T-tropic strains usually appear later in infection and are associated with a higher rate of CD4+ cell loss, so that AIDS develops faster.

Genetic Diversity of HIV. Like most viruses with RNA genomes, HIV mutates extremely rapidly. This is largely because of the high error rate of the enzyme reverse transcriptase, which the virus uses to make the DNA copy of the RNA genome. Approximately one error is made during each genome replication, which, coupled with the rapid replication of the virus, means that its rate of evolution is about one million times greater than that of human nuclear DNA. Such rapid evolution clearly gives HIV enormous adaptive potential, and resistance to individual antiviral drugs can evolve within months.

The rapid pace of HIV evolution also means that the virus is very variable worldwide, which hinders the development of vaccines that will be effective on a global scale. Three groups of HIV-1 strains can be recognized

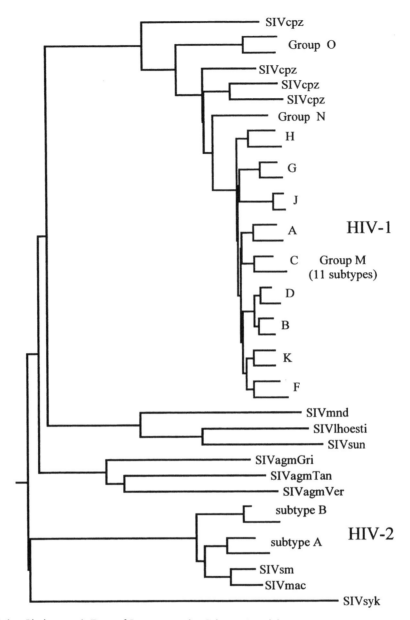

FIGURE 1. Neighbor-Joining Phylogenetic Tree of Representative Primate Lentiviruses.
The tree was constructed using an amino acid alignment of thirty-eight polymerase (*pol*) gene sequences, 884 residues in length. HIV lineages are shown, although only two subtypes of HIV-2 are included here. The tree is midpoint rooted, and all branch lengths are to scale. Abbreviations for the simian immunodeficiency viruses (SIVs) are as follows: SIVcpz = chimpanzee, SIVmnd = mandrill, SIVlhoesti = L'Hoest monkey, SIVsun = solatus monkey, SIVagmGri = grivet monkey, SIVagmTan = tantalus monkey, SIVagmVer = vervet monkey, SIVsm = sooty mangabey, SIVmac = macaque, SIVsyk = Sykes' monkey.
E. C. Holmes, "On the Origin and Evolution of the Human Immunodeficiency Virus (HIV)" *Biological Reviews* 76, 2001, figure 2 (modified). Cambridge University Press.

on phylogenetic trees (see Figure 1). The main group (group M) is globally distributed and accounts for the vast majority of AIDS cases; groups O (outlier) and N (new) are much less common and restricted to West Africa. Group M can also be split into eleven smaller phylogenetic clusters known as subtypes (named A to K), which have differing geographical distributions. The greatest range of subtypes is found in sub-Saharan Africa; other localities usually have a more restricted range of subtypes, such as the predominance of subtype B (the first described) in North America and Europe. Phylogenetic subtypes have not yet been observed in groups O and N, whereas seven subtypes (A to G) have been described in HIV-2, all exclusive to West Africa.

Not only does HIV show extensive variation world-wide, but genetic diversification is so rapid that different viral strains are recovered from patients who have infected one another, and also within single patients. Although this poses a challenge for antiviral therapy, such variation does allow us to retrace with great accuracy the spread of the virus using phylogenetic analysis. This, in turn, enables some key epidemiological questions to be investigated, such as the major routes and mechanisms of viral spread within populations and the source of particular outbreaks, including those involving health care workers (see Vignette; Figure 2).

Although very informative, presenting HIV diversity in the form of a simple phylogenetic tree with discrete clusters of sequences can give a misleading picture of how this virus evolves. In particular, phylogenetic trees ignore the process of recombination among viral RNA strands, which is rife in HIV. Recombination appears to take place during reverse transcription and has been documented between viruses at all phylogenetic levels—from strains that have diversified within single individuals to viruses isolated from different primate species. Intersubtype recombination in HIV-1 has been particularly well characterized and can easily be detected when different gene regions produce different (incongruent) phylogenetic trees. For example, recombination is clearly frequent among the subtypes circulating in Africa, sometimes resulting in complex mosaic sequences comprised of multiple subtypes that are known as circulating recombinant forms. The high incidence of intersubtype recombination means that mixed infections must also occur with a high frequency and that HIV infection with one strain probably confers little immune protection against infection with other strains, an observation that has important implications for vaccine development.

The Origins of HIV. HIV is a member of a group of retroviruses known as the lentiviruses, which cause a range of well-known veterinary diseases in ungulates (grazing animals). More recently, lentiviruses have been identified in felid species and nonhuman primates, the latter of which are particularly important as they are the closest relatives of both HIV-1 and HIV-2. To date, more than twenty primate lentiviruses (PLVs) have been described, and more may be isolated in the future. Because all these viruses infect simian primates, they are also known as the simian immunodeficiency viruses (SIVs). With the exception of the viruses found in humans and macaques, all primate lentiviruses cause asymptomatic infections in their host species. The SIV found in macaques is an interesting exception; the hosts are Asian monkeys, whereas all other SIV-infected nonhuman primates are African in origin, and the virus (SIV mac) induces an AIDS-like illness. However, SIVmac appears to be an example of cross-species transfer in a captive en-

THE FORENSIC ANALYSIS OF HIV SEQUENCE DATA

The extraordinarily rapid evolution of HIV means that gene sequence data from the virus itself can be used to retrace its spread through populations. This property has been utilized in a number of high-profile forensic cases, where the phylogenetic analysis of viral sequence data provided important evidence about the precise pathways of HIV transmission.

Perhaps the most famous of these cases involved a dentist from Florida who, in 1990, was accused of infecting a number of his former patients with the virus. To investigate whether this was true, researchers sampled viral strains from the dentist, eight of his former patients who were HIV infected, and a number of local controls—HIV-infected individuals living within a ninety-mile radius of the dentist's surgery and who were assumed to be carrying viral strains representative of this locality. Phylogenetic analysis of these sequences revealed that those from five of the patients were more closely related to the dentist's viruses than they were to those from the local controls, strongly suggesting that they had indeed been infected by the dentist (see Figure 2). In contrast, three other patients had viral sequences that were more closely related to those from the local controls than they were to the dentist, indicating that they had become infected by some other route. How the dentist infected his former patients is unknown.

—EDWARD C. HOLMES

vironment, because macaques are not infected in the wild. Furthermore, SIVmac is closely related to SIVsm from the sooty mangabey (*Cercocebus torquatus atys*), an African monkey that was kept with macaques in primate research centers.

Although both HIV-1 and HIV-2 are clearly related to the SIVs found in nonhuman primates, it is important to note that they are actually rather distantly related to each other (see Figure 1). Like the macaque viruses, the closest relatives of HIV-2 are the viruses found in sooty mangabeys. SIVsm seems a likely ancestor for HIV-2 because, first of all, the two viruses group closely on molecular phylogenies, as expected if the former had produced the latter, and the geographical distribution of HIV-2 infection overlaps with the range of sooty mangabeys in West Africa. More strikingly, the phylogenies display a mixing of lineages from SIVsm and HIV-2 so that some subtypes of HIV-2 are more closely related to strains of SIVsm than they are to each other. This strongly suggests that there have been multiple transfers of virus from monkeys to humans. Second, the preva-

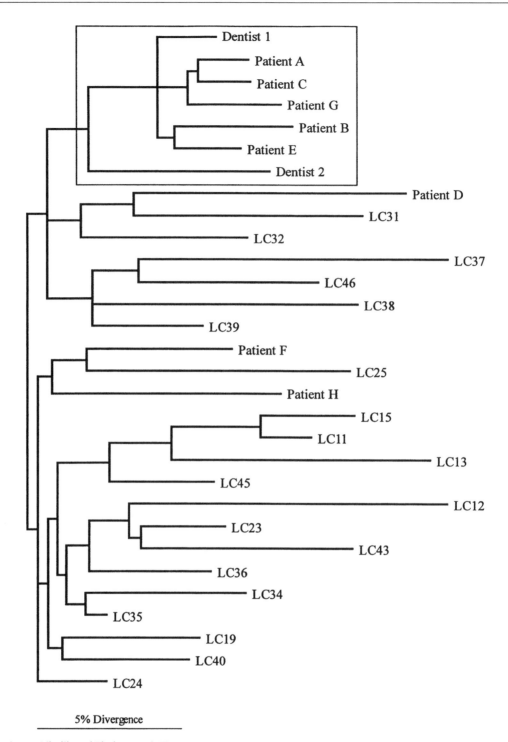

5% Divergence

FIGURE 2. Maximum Likelihood Phylogenetic Tree.
Shown are the relationships among HIV envelope (*env*) gene sequences from the Florida dentist, eight of his former patients (denoted A to H), and twenty randomly sampled local controls (labeled LC). The viral sequences from patients A, B, C, E, and G group closely with the dentist (boxed on the tree), suggesting that they were infected by him, whereas those from patients D, F, and H are more closely related to the local control strains. The viral population within the dentist was so diverse that two of his strains were used in the analysis. The tree is midpoint rooted, and all branch lengths are drawn to scale.   Edward C. Holmes.

lence of SIV in sooty mangabeys is high, but there is no evidence of disease. This might suggest that the virus has been associated with sooty mangabeys for a long time. Finally, sooty mangabeys are occasionally kept as pets or eaten by humans, so that there are plausible routes of transmission from monkey to human populations.

Reconstructing the origin of HIV-1 is more problematic. The closest relatives of HIV-1 are the viruses found in chimpanzees (SIVcpz) (see Figure 1). Because viruses from human and chimpanzee are mixed on phylogenetic trees, with the latter occupying more divergent positions, the obvious implication is that HIV-1 groups M, N, and O might have originated in different geographical populations of the common chimpanzee (*Pan troglodytes*). Once again, such a scenario would mean that there have been multiple transfers of viral strains into humans. Another similarity with the HIV-2–SIVsm story is that there is an overlap of geographic ranges; common chimpanzees are found in various localities in West-Central Africa, and this region contains the M, N, and O groups of HIV-1. Finally, chimpanzees are also hunted by humans, again providing a possible mechanism for viral transmission. The puzzling aspect is that SIVcpz seems to be rarely found in wild-caught chimpanzees. This could mean that both humans and chimpanzees acquired their viruses from a third primate species, as yet unidentified. Alternatively, SIVcpz may occur at high frequencies in some populations of chimpanzee that have been poorly surveyed to date.

The timing of the cross-species transfer events that gave rise to HIV has proven even more controversial. Most debate has focused on the first appearance of the M group of HIV-1 because this accounts for the vast majority of AIDS cases. Although AIDS was not described until the early 1980s, the long asymptomatic stage means that HIV may have existed more than ten years prior to this. Most informative is strain ZR59, obtained from a sample dating to 1959 from what is now the Democratic Republic of the Congo. The existence of ZR59 means that the M group must date from at least 1959.

A variety of methods have been used to infer times of HIV origin from these sequences, particularly using the interval between the sampling times of the strains. For example, the observation that ZR59 falls some distance from the root of the M group tree strongly suggests that this group originated a number of years before the late-1950s. So far, the best estimates are that the M group first appeared in the 1920s or 1930s, which means that HIV must have spread at low levels for much of its history. Such an early date is also important evidence against the theory that the M group emerged in humans following an oral polio vaccination campaign that took place in West-Central Africa from 1957 to 1960. This theory alleges that kidneys from chimpanzees were used to culture the vaccine strains and in so doing contaminated the vaccine stocks. If the HIV-1 group M originated before the polio vaccination campaign, then it must be that multiple viral isolates were transferred from chimpanzees to humans during vaccination, which seems unlikely, and all the remaining vaccine stocks analyzed to date have tested negative for the virus. However, it is equally clear that the divergence times of the various HIV lineages are difficult to infer, particularly with such a small sample of chimpanzee viruses and in the face of frequent recombination, so that the timescale of HIV evolution is not yet fully resolved.

The Evolution of the Primate Lentiviruses: Cospeciation or Cross-Species Transmission? One of the most important issues in the evolution of the primate lentiviruses is whether the predominant mode of viral transmission is cospeciation, so that the viral populations diverge as their host species do over timescales of millions of years, or whether viruses frequently jump species boundaries (cross-species transmission) as has clearly happened for HIV-1 and HIV-2. Evidence for cospeciation is that most primate lentiviruses are depicted as species-specific clusters on phylogenetic trees. As a case in point, each of the four species of African green monkey (genus *Chlorocebus*) has its own phylogenetically distinct immunodeficiency virus, collectively referred to as SIVagm. Furthermore, the viral phylogeny often matches the phylogeny of the host species as expected under long-term host–virus cospeciation. For example, L'Hoest (*Ceropithecus lhoesti lhoesti*) and solatus monkeys (*Ceropithecus solatus*) are closely related, as are the SIVs they harbor, but live in different locations in Africa, thereby preventing any species contact that might enable cross-species viral transmission.

Confusingly, other pieces of evidence suggest that the primate lentiviruses have evolved more recently. First, the match between the host and virus phylogenies is not always apparent. For example, the viruses from chacma baboons (*Papio ursinus*) and yellow baboons (*Papio cynocephalus*) both cluster next to those from vervets, a species of African green monkey living in the same geographical area and that presumably donated its virus to these species in the recent past. Moreover, species specificity could simply be an artifact of the small sample of viruses from each host species: a wider sample, presenting the full geographical range of infected animals, might eventually reveal more interspecies mixing.

Perhaps a bigger problem is that the cospeciation theory predicts that viral divergence times should match those of their host species as their evolutionary histories are intertwined. This is clearly not the case. For example, if SIV agm evolves at the same rate as HIV-1, then the split between the African green monkey hosts would have occurred only several hundred years ago and a long way from the many thousands (perhaps millions) of years during which the host species have been separate. Although it is possible that the viral sequences have been analyzed with unrealistic models of sequence evolution (for example, by ignoring the effects of RNA secondary structure and recombination), it is difficult to imagine that the error could extend to several orders of magnitude. There is also little evidence that rates of mutation and replication are lower in SIVs compared to HIVs, both of which could potentially reconcile species and virus divergence times. The evolutionary timescale of the primate lentiviruses in general, like HIV-1 in particular, therefore remains a puzzle.

[*See also* Disease, *article on* Infectious Disease; Emerging and Re-Emerging Diseases; *articles on* Immune System; Transmission Dynamics; Vaccination; Virulence; Viruses.]

BIBLIOGRAPHY

Barr-Sinoussi, F., J. C. Chermann, F. Rey, M. T. Nugeytre, S. Chamaret, J. Gruest, C. Dauguet, C. Axler-Blue, F. Vezinet-Brun, C. Rouzioux, W. Rozenbaum, and L. Montagnier. "Isolation of a T-Lymphotropic Retrovirus from a Patient at Risk for Acquired Immune Deficiency Syndrome (AIDS)." *Science* 220 (1983): 868–871. First isolation and description of HIV.

Beer, B. E., E. Bailes, R. Goeken, G. Dapolito, C. Coulibaly, S. G. Norley, B. Kurth, J.-P. Gautier, A. Gautier-Hion, D. Vallet, P. M. Sharp, and V. M. Hirsch. "Simian Immunodeficiency Virus (SIV) from Sun-Tailed Monkeys (*Ceropithecus solatus*): Evidence for Host-Dependent Evolution of SIV within the *C. lhoesti* Superspecies." *Journal of Virology* 73 (1999): 7734–7744. Presents important evidence for long-term cospeciation between primate lentiviruses and their host species.

Gao, F., D. L. Robertson, C. D. Carruthers, Y. Y. Li, E. Bailes, L. G. Kostrikis, M. O. Salminen, F. Bibollet-Ruche, M. Peeters, D. D. Ho, G. M. Shaw, P. M. Sharp, and B. H. Hahn. "An Isolate of Human Immunodeficiency Virus Type 1 Originally Classified as Subtype I Represents a Complex Mosaic Comprising Three Different Group M Subtypes (A, G and I)." *Journal of Virology* 72 (1998): 10234–10241. Study that shows the frequency and importance of intersubtype recombination in HIV-1.

Hahn, B. H., G. M. Shaw, K. M. de Cock, and P. M. Sharp. "AIDS as a Zoonosis: Scientific and Public Health Implications." *Science* 287 (2000): 607–614. The most comprehensive review of the origins of HIV.

Holmes, E. C. "On the Origin and Evolution of the Human Immunodeficiency Virus (HIV)." *Biological Reviews* (2001). Comprehensive review of various topics relating to the evolution of HIV.

Jin, M. J., J. Rogers, J. E. Phillips-Conroy, J. S. Allan, R. C. Desrosiers, G. M. Shaw, P. M. Sharp, and B. H. Hahn. "Infection of a Yellow Baboon with Simian Immunodeficiency Virus from African Green Monkeys: Evidence for Cross-Species Transmission in the Wild." *Journal of Virology* 68 (1994): 8454–8460. Interesting example of cross-species transmission among primate lentiviruses.

Korber, B. T., M. Muldoon, J. Theiler, F. Gao, R. Gupta, A. Lapedes, B. H. Hahn, S. Wolinsky, and T. Bhattacharya. "Timing the Ancestor of the HIV-1 Pandemic Strains." *Science* 288 (2000): 1789–1796. Most up-to-date analysis of the likely time of origin of HIV-1.

Ou, C.-Y., C. A. Ciesielski, G. Myers, C. I. Bandea, C.-C. Luo, B. T.

M. Korber, J. I. Mullins, G. Schochetman, R. L. Berkelman, A. N. Economou, J. J. Witte, L. J. Furman, G. A. Satten, K. A. MacInnes, J. W. Curran, H. W. Jaffe, Laboratory Investigation Group, and Epidemiologic Investigation Group. "Molecular Epidemiology of HIV Transmission in a Dental Practice." *Science* 256 (1992): 1165–1171. Description of the famous Florida dentist case of HIV transmission.

Perelson, A. S., A. U. Neumann, M. Markowitz, J. M. Leonard, and D. D. Ho. "HIV-1 Dynamics in vivo: Virion Clearance Rate, Infected Cell Life-span, and Viral Generation Time." *Science* 271 (1996): 1582–1586. Important analysis of the extraordinary replicatory power of HIV within patients.

Zhu, T. F., B. T. Korber, A. J. Nahmias, E. Hooper, P. M. Sharp, and D. D. Ho. "An African HIV-1 Sequence from 1959 and Implications for the Origin of the Epidemic." *Nature* 391 (1998): 594–597. Describes the evolutionary relationships of the key ZR59 strain of HIV-1, which comes from a sample collected in 1959.

— EDWARD C. HOLMES

Dynamics of HIV Transmission and Infection

During the two decades of the human immunodeficiency virus (HIV) pandemic, extensive data have become available concerning the epidemiology of the infection, as well as the within-host dynamics of the disease process. According to recent estimates (2001), some thirty million people are living with HIV worldwide, with at least 16,000 new infections occurring each day. The majority of HIV cases and new infections occur in developing countries, with about two-thirds of infected people living in sub-Saharan Africa. In industrialized nations, the use of combination antiretroviral therapy has resulted in a strong decline in incidences and deaths related to acquired immune deficiency syndrome (AIDS). In contrast, HIV continues to spread in regions where antiretroviral therapy is not available, such as large parts of southern Africa, Asia, and Latin America. The HIV epidemic is having striking effects across society in Africa. High mortality rates among young people significantly decrease the average life expectancy, which in turn weakens the structure of society. AIDS orphans, who have lost both parents to AIDS, now make up a substantial proportion of the population in Africa. Associated with the HIV epidemic are epidemics of opportunistic infections, most notably tuberculosis.

Method and Routes of Transmission. HIV is transmitted between individuals via the exchange of body fluids. Especially in industrialized countries, male homosexuals, intravenous drug users, and recipients of blood transfusions (such as hemophiliacs) were those most often affected by the initial spread of the disease. More recently, heterosexual contact has become a more prevalent route of transmission in industrialized countries. In Africa, heterosexual contact is the major route of HIV transmission, and rates of mother to child (vertical) transmission are generally higher in Africa than in many industrialized nations, probably because of prolonged breast feeding in this region. Major efforts have been directed at health education to change the behavior of people and to interrupt the transmission cycle of HIV. Promotion of condom use and safer sex has played a major role in this context. In addition, the increased control of other sexually transmitted diseases (STDs) is required, because the presence of STDs increases the chances of HIV transmission due to the existence of sores in the genital areas. Further efforts include strengthened HIV surveillance, control of blood safety, widespread testing and early diagnosis, and increased care for HIV-infected patients. Ultimately, improved drug therapy regimes and the construction of an efficient vaccine would play a major role in reducing the incidence of HIV infection. Such measures require a detailed knowledge about the dynamic interactions between HIV replication and the immune system, which are the focus of the remainder of this article.

Mechanics of HIV. HIV infects a variety of cell types, mostly cells of the immune system. The enzyme reverse transcriptase copies the viral ribonucleic acid (RNA) into deoxyribonucleic acid (DNA), which can be integrated into the genome of the host cell. Once the HIV genome has been integrated, the cell is infected for the duration of its life. The integrated HIV genome provides the template to build new virus particles, which eventually leave the host cell by a budding process (Figure 1). It is thought that two types of immune cells are the most important targets for HIV: CD4 T helper cells and antigen presenting cells, such as macrophages and dendritic cells. Both cell types are involved in the induction of virus-specific effector mechanisms, such as CD8 T killer cell and antibody responses. Early studies demonstrated that the CD4 receptor is used by HIV to enter and infect a cell. Later it was revealed that specific coreceptors also play an important role in determining the tropism and pathogenicity of HIV. Although the list of known coreceptors used by the virus is expanding, special importance has been given to CCR5 and CXCR4 coreceptors. CCR5 is expressed on antigen presenting cells and on primary CD4 T cells. CXCR4 is expressed on CD4 T cells only. CCR5 tropic HIV variants are thought to be slowly replicating and weakly cytopathic. In contrast, CXCR4 tropic variants tend to be faster replicating and more cytopathic, which might involve the formation of syncytia. It has been found that early in the infectious process, during the acute phase of the infection, CCR5 tropic variants dominate the virus population. Later in the disease process, HIV expands its coreceptor usage. The virus may become dual tropic, and in about 50 percent of the patients exclusively CXCR4 tropic variants emerge, a factor that is associated with the development of AIDS. It has been observed that individuals lacking a functional CCR5 coreceptor, caused by a deletion in the encoding gene, are characterized by a significantly reduced susceptibility to HIV infection.

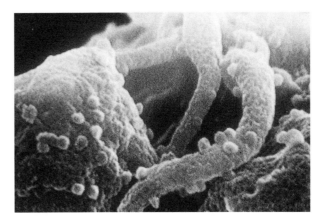

FIGURE 1. HIV Budding from a White Blood Cell ($\times$ 275,000). Visuals Unlimited.

About 1 percent of the Caucasian population carries this mutation. One reason for this resistance could be that antigen presenting cells, expressing the CCR5 coreceptor, are required to carry the virus from the site of infection into the bloodstream. In addition, CD4 T cells, carrying the CXCR4 coreceptor, need to be activated in order to become infected, and a sufficient level of activation is attained only once the virus population has achieved relatively high levels.

During the acute phase of the infection, HIV has been observed to grow to very high numbers. In this early phase of the infection, the virus population is relatively homogeneous. Subsequently, the virus load is reduced and settles at a postacute level, which marks the beginning of the asymptomatic phase. Analysis of drug treatment data, in conjunction with mathematical models, have revealed that the virus is not latent in the asymptomatic phase, but that the virus population is characterized by a relatively high turnover rate. The half-life of infected T cells has been estimated to be about two days, whereas the half-life of free virus particles is on the order of hours. Long-term studies of HIV-infected patients and macaques infected with simian immunodeficiency virus (SIV) indicate that the level of the postacute viral population determines the speed of disease progression. The higher the viral load, the faster the disease progression. In the course of the asymptomatic phase, the virus evolves to become faster replicating and more diverse, and increases its range of coreceptor usage. This results in a gradual increase in virus load and a decrease in the overall CD4 T-cell count. AIDS is defined when the CD4 T-cell count falls below 200 cells/ml, which is accompanied by very high virus loads. The host becomes infected with opportunistic pathogens, and this process culminates in death.

Immune Responses to HIV. Immune responses to the virus have been extensively analyzed in HIV-infected

NAIROBI SEX-WORKERS WITH HEIGHTENED IMMUNITY TO HIV

Despite continued high-level exposure to HIV-1, some sex-workers in Nairobi, Kenya, have remained uninfected for more than ten years. Understanding the nature of their resistance could be crucial for the development of effective vaccines against HIV-1. Various immune responses, including T helper cell responses, CTL responses, and mucosal and plasma antibody responses, have been associated with long-term protection. Particular emphasis has been placed on CTL responses against HIV, because broad and cross-reactive CTL have been detected in highly exposed, persistently sero-negative women. It is thought that repeated transient exposures to HIV-1 in sero-negative individuals could have boosted their specific CTL, contributing to immunity and protection from infection. The precise nature of the boosted responses, however, remains an open question. Our understanding of protection against HIV is further complicated by the observation that a fraction of individuals who were previously classified as resistant to the virus eventually sero-converted, indicating infection. These women might have been infected with "escape" strains (HIV mutants not recognized by the CTL), although there is no direct evidence for this hypothesis. Counterintuitively, late sero-conversion correlated with reduced sexual encounters. Hence, continued exposure to low levels of HIV might be required to repeatedly boost the immune responses. Temporary lack of exposure might thus lead to waning of immunity. In support of this hypothesis, persistently sero-negative women with frequent sexual encounters showed higher levels of HIV-specific CTL than those who had stopped sex work. With respect to vaccine development, the evidence so far suggests that the immune responses to HIV must be continuously boosted in order to be protective in the long term. Such a result might be achieved either by repeated vaccination events or by vaccinating with long-persisting antigen.

—Dominik Wodarz

patients and SIV-infected macaques. CD8 T cells or cyto-toxic T lymphocytes (CTLs) are thought to be most important for limiting HIV load. The role of neutralizing antibodies is less clear. The initial decline in virus load following the acute phase of the infection correlates with the rise of specific CTLs. These CTLs remain during the asymptomatic phase and collapse only toward the end stage of the disease. Depletion of CD8 T cells in SIV-infected macaques has been shown to result in a strong increase in virus load. However, despite the presence of CTLs, the immune system fails to control the infection in the long term. One reason could be the high mutation rate of HIV, resulting in continuing evolutionary escape from immune responses. Escape variants have repeatedly been demonstrated in HIV-infected patients, and mathematical models suggest that the mutation rate of HIV is optimized for the evolution of antigenic escape.

Recent studies indicate an even more profound defect in HIV-specific immunity. In typically progressing patients, significant levels of HIV-specific CD4 T helper cell responses are absent, even during the acute phase of the infection. Because activated CD4 T cells are targets for the virus, it is possible that HIV-specific helper cells become infected and killed. Alternatively, the high virus load attained during acute infection could induce anergy (inactivation) in the HIV-specific T helper cell population. Experiments have revealed that CD4 T-cell help is required for the establishment of a sustained memory CTL response and for long-term control of viral infections. Hence, the CTL responses arising in typical HIV-infected patients in the absence of helper cell responses are likely to be compromised. For example, their antiviral activity might be reduced, or they may fail to be sustained at low levels of antigenic stimulation. In contrast to typically progressing patients, significant levels of HIV-specific CD4 T helper cell responses are observed in long-term nonprogressors. Such patients also maintain high numbers of effective CTLs despite very low virus loads.

Recent research suggests that the dynamics (i.e., rate and growth per unit of time) during the initial acute phase of the infection are an important determinant of subsequent disease progression. If the virus replicates extensively during the acute phase of the infection, CD4 T helper cell responses fail to become established, resulting in the development of compromised CTL responses. This allows for only partial control of the virus. Continuous virus replication, together with the high mutation rate of HIV, results in progressive disease and the development of AIDS. If CD4 T helper cell responses are preserved during early infection, however, an efficient and sustained CTL response develops that keeps virus load at very low levels. This results in reduced opportunities for the virus to mutate and evolve, and virus control is maintained in the long term.

Drug Therapy Regimes. Knowledge of these dynamics helps to improve drug therapy regimes for HIV-infected patients. Currently, a combination of different drugs is administered that can efficiently suppress virus replication to a minimum. At the moment (2001), mainly two types of drugs are used: reverse transcriptase inhibitors, preventing reverse transcription and thus infection of new cells; and protease inhibitors, preventing the generation of infectious virus. Research is under way to develop further drugs targeting different stages of the viral life cycle. In many patients, the drugs currently available have resulted in sustained suppression of virus load below the level of detection. However, therapy is associated with many problems. Clinical data have revealed the existence of infected cells that are very long lived. This, together with the existence of latently infected cells, suggests that eradication of HIV from the host by drug therapy is not feasible at this time. Upon cessation of drug therapy, virus load reemerges to pretreatment levels. Drug therapy therefore has to be maintained for life.

This continuity is difficult to achieve because of severe side effects of the drugs and the ability of the virus to evolve resistance quickly, especially if the patient does not adhere to the complex regimes required. Research is continuing in an effort to find ways to strengthen HIV-specific immunity, which might lead to long-term immunological control of the infection absent continuous treatment. Experiments with SIV-infected macaques and data from HIV-infected patients indicate that this might be achieved by treating early during the acute phase of the infection. Early treatment reduces the amount of initial virus growth. This preserves virus-specific CD4 T helper cell responses, resulting in the development of an efficient and sustained CTL response, which can control the infection in the long term. Data further suggest that so-called structured therapy interruptions, in which therapy is stopped for a brief period of time and then restarted, could augment this effect. These therapy regimes are in principle similar to vaccinating the patient with his or her own virus by providing a limited antigenic stimulus while preventing impairment of CD4 T-cell help. In general, the current trend is to search for treatment regimes that combine the use of antiretroviral drugs with immunotherapy to induce long-term immunological control of the infection in the absence of lifelong drug administration. At this stage, long-term control of the infection seems more practical and feasible than eradication of the virus from the host.

[See also Antibiotic Resistance, article on Strategies for Managing Resistance Coevolution; Disease, article on Infectious Disease, Emerging and Re-Emerging Diseases; Immune System, article on Microbial Countermeasures to Evade the Immune System; Mathematical Models; Red Queen Hypothesis; Transmission Dynamics; Vaccination; Virulence.]

BIBLIOGRAPHY

Chun, T. W., L. Stuyver, S. B. Mizell, L. A. Ehler, J. A. Mican, M. Baseler, A. L. Lloyd, M. A. Nowak, and A. S. Fauci. "Presence of an Inducible HIV-1 Latent Reservoir during Highly Active Antiretroviral Therapy." Proceedings of the National Academy of Sciences USA 94 (1997): 13193–13197. Key paper on the importance of latently infected cells.

Cock, K. M., and H. A. Weiss. "The Global Epidemiology of HIV/AIDS." Trop Med Int Health 5 (2000): A3–9. Good review on HIV epidemiology.

Dalgleish, A. G., P. C. Beverley, P. R. Clapham, D. H. Crawford, M. F. Greaves, and R. A. Weiss. "The CD4 (T4) Antigen Is an Essential Component of the Receptor for the AIDS Retrovirus." Nature 312 (1984): 763–767. Important paper on CD4 as a receptor for HIV.

Dean, M., M. Carrington, C. Winkler, G. A. Huttley, M. W. Smith, R. Allikmets, J. J. Goedert, S. P. Buchbinder, E. Vittinghoff, E. Gomperts, S. Donfield, D. Vlahov, R. Kaslow, A. Saah, C. Rinaldo, R. Detels, and S. J. O'Brien. "Genetic Restriction of HIV-1 Infection and Progression to AIDS by a Deletion Allele of the CKR5 Structural Gene." Hemophilia Growth and Development Study, Multicenter AIDS Cohort Study, Multicenter Hemophilia Cohort Study, San Francisco City Cohort, ALIVE Study [see comments] [published erratum appears in Science 274 (1996): 1069]. Science 273 (1996): 1856–1862. Key paper discussing resistance to HIV based on coreceptor usage.

Ho, D. D., A. U. Neumann, A. S. Perelson, W. Chen, J. M. Leonard, and M. Markowitz. "Rapid Turnover of Plasma Virions and CD4 Lymphocytes in HIV-1 Infection." Nature 373 (1995): 123–126. Key paper demonstrating rapid turnover of HIV.

Lifson, J. D., M. A. Nowak, S. Goldstein, J. L. Rossio, A. Kinter, G. Vasquez, T. A. Wiltrout, C. Brown, D. Schneider, L. Wahl, A. L. Lloyd, J. Williams, W. R. Elkins, A. S. Fauci, and V. M. Hirsch. "The Extent of Early Viral Replication Is a Critical Determinant of the Natural History of Simian Immunodeficiency Virus Infection." Journal of Virology 71 (1997): 9508–9514. Shows how the early viral replication rate can influence disease progression.

Lifson, J. D., J. L. Rossio, R. Arnaout, L. Li, T. L. Parks, D. M. Schneider, R. F. Kiser, V. J. Coalter, G. Walsh, R. Imming, B. Fischer, B. M. Flynn, M. A. Nowak, and D. Wodarz. "Containment of SIV Infection: Cellular Immune Responses and Protection from Rechallenge Following Transient Post-inoculation Antiretroviral Treatment." Journal of Virology 74 (2000): 2584–2593. Shows how timing and duration of therapy during acute infection can crucially determine outcome of SIV infection.

McMichael, A. J., and R. E. Phillips. "Escape of Human Immunodeficiency Virus from Immune Control." Annual Review of Immunology 15 (1997): 271–296. Excellent review of antigenic escape.

Moore, J. P., A. Trkola, and T. Dragic. "Co-receptors for HIV-1 Entry" [see comments]. Current Opinion in Immunology (London) 9 (1997): 551–562. Discusses coreceptor usage by HIV.

Nowak, M. A., R. M. Anderson, A. R. McLean, T. F. W. Wolfs, J. Goudsmit, and R. M. May. "Antigenic Diversity Thresholds and the Development of AIDS." Science 254 (1991): 963–969. Shows how antigenic variation can affect HIV disease progression.

Perelson, A. S., A. U. Neumann, M. Markowitz, J. M. Leonard, and D. D. Ho. "HIV-1 Dynamics in Vivo—Virion Clearance Rate, Infected Cell Life Span, and Viral Generation Time." Science 271 (1996): 1582–1586. Important paper quantifying HIV dynamics.

Rosenberg, E. S., M. Altfeld, S. H. Poon, M. N. Phillips, B. M. Wilkes, R. L. Eldridge, G. K. Robbins, R. T. D'Aquila, P. J. Goulder, and B. D. Walker. "Immune Control of HIV-1 after Early Treatment of Acute Infection" [In Process Citation]. Nature 407 (2000): 523–526. Shows how early therapy in HIV-infected patients can result in improved viral control.

Schmitz, J. E., M. J. Kuroda, S. Santra, V. G. Sasseville, M. A. Simon, M. A. Lifton, P. Racz, K. Tenner-Racz, M. Dalesandro, B. J. Scallon, J. Ghrayeb, M. A. Forman, D. C. Montefiori, E. P. Rieber, N. L. Letvin, and K. A. Reimann. "Control of Viremia in Simian Immunodeficiency Virus Infection by CD8(+) Lymphocytes" [In Process Citation]. Science 283 (1999): 857–860. Shows the potential of CD8 T cells to control SIV.

Thomsen, A. R., A. Nansen, J. P. Christensen, S. O. Andreasen, and

O. Marker. "CD40 Ligand Is Pivotal to Efficient Control of Virus Replication in Mice Infected with Lymphocytic Choriomeningitis Virus." *Journal of Immunology* 161 (1998): 4583–4590. Discusses the effect of helper deficiency on viral dynamics.

Wei, X. P., S. K. Ghosh, M. E. Taylor, V. A. Johnson, E. A. Emini, P. Deutsch, J. D. Lifson, S. Bonhoeffer, M. A. Nowak, B. H. Hahn, M. S. Saag, and G. M. Shaw. "Viral Dynamics in Human-Immunodeficiency-Virus Type-1 Infection." *Nature* 373 (1995): 117–122. Important paper showing rapid turnover of HIV.

Wodarz, D., R. M. May, and M. A. Nowak. "The Role of Antigen-independent Persistence of Memory CTL." *International Immunology* 12 (2000): 467–477. Uses mathematical models to analyze correlates of CTL-mediated virus control.

Wodarz, D., and M. A. Nowak. "Specific Therapy Regimes Could Lead to Long-term Control of HIV." *Proceedings of the National Academy of Sciences USA* 96 (1999): 14464–14469. Gives a theoretical framework for understanding the effect of early therapy and structured therapy interruptions.

— DOMINIK WODARZ

ADAPTATION

Adaptation refers to the many properties of organisms that enable those organisms to stay alive and reproduce. Adaptation is one of the most important abstract concepts in biology, and its informal meaning is widely agreed. However, its exact meaning can be controversial. Three main criteria have been suggested to explain more exactly what adaptation means. The first criterion is design. On this criterion, an adaptation is any attribute of an organism that appears to be designed to enable it to survive and reproduce. The eye is a clear example. An eye has an internal structure that is designed to enable it to form visual images. Likewise, hearts are designed to pump blood, wings are designed for flight, and so on. The design criterion for an adaptation is an engineering criterion: adaptations can be recognized as the sort of structure that an engineer would have designed, given the purpose of the structure. In fact, adaptations evolve by the unconscious process of natural selection; there is not really a conscious designer who creates adaptations. One additional feature of the design criterion is that adaptations refer to attributes of organisms that benefit the organism, but only in so far as the benefit is more than can be predicted simply from the laws of physics and chemistry. G. C. Williams, in his book *Adaptation and Natural Selection* (1966), discussed the example of a flying fish. A flying fish can launch itself out of the water and fly some distance through the air. It then returns to the water, under the influence of the law of gravity. The return of the fish to the water benefits the fish; it helps the fish to stay alive. (Indeed, the fish would soon asphyxiate if it did not return to water.) But the return of the fish to the water is not an adaptation. It is a simple consequence of the law of gravity; natural selection has not worked on flying fish to make them obey that law. But for a true adaptation, such as an eye, it is not the case that any fish will automatically develop an eye. It has required the special workings of natural selection for eyesight to evolve.

A second criterion of adaptation is provided by selection in action. For example, birds lay a certain number of eggs in a clutch. If the birds lay too many eggs, they will produce more chicks than they can feed, and the whole brood may starve. If they lay too few eggs, they will produce fewer offspring than they were able to. We can imagine the birds of a species laying a range of clutch sizes: 2, 3, 4 . . . 10. (Indeed, we do not have to imagine it: experiments have been done in which birds are given artificial clutches of a range of sizes.) The total number of offspring finally produced (i.e., the number of chicks fledged) will be low for small clutches, increase for larger clutches, then decrease again for excessively large clutches. The optimum number is some intermediate. Natural selection will favor the clutch size that maximizes the reproductive output (or fitness) of the parents. In this way, we can define an adaptation as that attribute among a specified set of alternatives that is favored by natural selection. The criterion is closely related to the design criterion. An attribute that is favored by natural selection will also be one that is well designed for life. The advantage of the selection-in-action criterion is that it is objective and operational. If we perform the operation of measuring what natural selection favors, we can unambiguously recognize (on this criterion) the adaptations. The design criterion can be more ambiguous. A supporter of the design criterion, however, would point out that we do not need to measure what natural selection favors in order to say that an eye is an adaptation. The selection-in-action criterion does not easily apply to some attributes, such as eyes, that biologists generally agree to be adaptations.

A third criterion is historical and defines an adaptation as an attribute that benefits an organism and that evolved in the past in order to fulfill the same function as it now has. This criterion will recognize adaptations in much the same way as the other two criteria, but it differs from them in two cases. One is when an organ changes its function during evolution with little or no change in its structure. For example, feathers are now used by birds for flight. But some fossil evidence suggests that feathers first evolved in flightless dinosaurs. If so, feathers first evolved for some other function, such as display or thermoregulation, then later acquired the function of flight. On the historical criterion, it is wrong to call a feather an adaptation for flight, because it did not first evolve for flight: something can only be called an adaptation if it still performs its original function. A second case is as follows. Eyes evolved from ancestrally eyeless creatures but have been secondarily lost: some cave-dwelling fish, for example, are eyeless, but they have evolved from ancestors that had eyes. On the historical criterion, we call eyelessness an adaptation in the secondarily eyeless cave fish. In these creatures, natural selection has acted to favor eyelessness. Yet we do not call eyelessness an adaptation in the originally eyeless creatures. Natural selection has not specifically transformed them to produce the condition of eyelessness. These primitively eyeless fish are "adapted" in the sense that eyes are not needed. But their lack of eyes is not an adaptation on the historical criterion. Some biologists add the historical criterion to their definition of adaptation. For them, an adaptation must not only produce a benefit for the organism, but must also have specifically evolved to produce that benefit. Other biologists concentrate on the function of organs now, independent of the organ's history. In some cases, such as feathers that may have historically changed their function, there can be disagreement about what is properly called an adaptation. All three criteria are attempts to solve a common core problem, and in most cases biologists agree on whether an attribute of an organism is an adaptation, whichever criterion of adaptation the biologists favor.

One difference between the biological use of the word *adaptation* and its everyday unscientific use is worth noting. Biologists use *adaptation* to refer to a state of an organism, independent of time. Eyes, clutch size, or feathers are observable attributes of an organism at any one time. In nonbiological usage, *adaptation* more often refers to a change in which a person adjusts (or a thing is adjusted) to suit the circumstances. Adaptations, in the biological sense, do evolve via a change: organisms with eyes evolved from organisms without eyes. But *adaptation* in biology is generally used to refer to the outcome of that evolutionary change rather than the change itself. (Some biologists might refer to the evolutionary change in which an eye evolves as *adaptation*, and the usage is not wrong. But most often when the word is used in evolutionary biology, it refers to a timeless state.)

The Cause of Adaptation. Adaptations evolve only by means of natural selection. When natural selection acts on a population, it is an almost automatic result that the members of the population become better adapted. (Natural selection also acts to maintain adaptations, once they have evolved. If an attribute ceases to be advantageous, it is soon randomized by mutation. Every adaptation has to be maintained against mutational decay.) Natural selection is not the only force of evolution, however: evolutionary change is also driven by random drift. Random drift does not cause the evolution of adaptation. We can distinguish between adaptive evolution—which is mainly evolution by natural selection—and nonadaptive evolution—which is mainly evolution by genetic drift. Thus, although natural selection is the cause of all adaptation, it is not the cause of all evolution. Genetic drift is unable to cause adaptation, because it is a random process, whereas adaptation is a highly nonrandom state of nature. Something like an eye could not be assembled during evolution, by random genetic change. However, once an adaptation has evolved, it is often (probably always) the case that more than one sequence of nucleotide letters in the DNA can code for it. Evolution can then undergo random genetic drift between the different, equally well-adapted DNA sequences.

Environmental "Instruction" and Adaptation. The main evidence for adaptation comes from the fit of organisms to their environments. Biologists have from time to time suggested that this organism–environment fit could arise not by the evolution of adaptation but by various kinds of instruction from the environment. For example, bird species have beaks that are the right size and shape to eat the food that each bird species in fact eats. Some species have big beaks and eat big seeds; others have small beaks and eat small seeds. If the beak of each species is an adaptation, then natural selection has adjusted the beak size within each species over time. However, it might be that information from the environment somehow directly causes the birds to acquire the right kind of beak. The fit of organism and environment would be by environmental instruction rather than natural selection. Most biologists do not consider this an important process, but it is worth understanding what it means.

For example, beak size might evolve up or down in different species by chance. Then some species would have big beaks and others small beaks. Suppose now that the birds then behaviorally select an environment appropriate to their beaks. The birds with small beaks go to where the small seeds are. Alternatively, the birds with large beaks where there are small seeds could die out. Either way, we end up with a fit between bird beak and environmental seed supply. That fit might be thought not to be caused by natural selection.

For a simple characteristic such as beak size, the "instructionist" explanation is just about possible (though in fact it can be refuted by direct observations of natural selection in action). However, the explanation is impossible for complex adaptations, such as the eye. Beak size could evolve up and down at random, and the birds might always find an environment where their beaks "work": some environments do have small seeds and others big seeds. But in the evolution of the eye there will not be a range of environments where all the random possibilities could work. If lens shape evolves at random, the focus of the visual image will alter. Yet the real world lacks a range of environments where some species could prosper with round lenses and others with flat lenses. Eyes are much more likely to have evolved as adaptations by natural selection than to have evolved

ADAPTATION AND ARCHITECTURAL CONSTRAINT

In architecture, an arch may be positioned within a rectangular surrounding structure, or two arches may be positioned next to each other. In both cases, a triangular space is necessarily created: either between the curve at the top of the arch and the top corners of the rectangle, or between the curves of the two arches. These spaces are called spandrels, in technical language (and to be more exact, the spandrels between two arches are called pendentives). The triangular space is an inevitable geometric consequence of the decision to use an arch in a certain way. The arch is the basic supporting structure of the building, and the spandrel is a minor byproduct.

In a famous paper titled "The Spandrels of San Marco" (1979), S. J. Gould and R. C. Lewontin argued that many attributes of organisms are analogous to spandrels. The human chin, for instance, may have evolved as a byproduct of evolutionary changes in the growth gradients of the face. Attributes of this kind, they argued, are not adaptations, but reflect architectural constraints (or developmental constraints, in the biological case). Natural selection, if Gould and Lewontin are right, did not specifically favor the human chin, and it would be a mistake to look for a function for the chin. Critics would note, however, that no research has been done on how, or whether, natural selection acts on the chin. Research is required before an attribute can be said to be nonadaptive. In other cases (allometry, for example), in which it had been suggested that one attribute is an automatic developmental byproduct of some second attribute, research has shown that natural selection guides the developmental relation between the two traits. It remains unknown how many attributes of organisms are nonadaptive consequences of developmental architecture.

—MARK RIDLEY

at random, with the organisms then selecting an environment where their eyes work. There is no infinity of worlds, such that any shape of eye works somewhere.

Environmental instruction does make some contribution to the fit of organisms and environment that we see. Any example of learning would illustrate the process. However, the way organisms select their environments and the way they learn are themselves adaptations that have evolved by natural selection. We can ask whether any particular piece of behavior was acquired by learning or natural selection, but environmental instruction and adaptation are not deep theoretical alternatives.

The Extent of Adaptation. Many of the attributes of organisms are adaptations. This is not only true of organs such as eyes, hearts, and wings, but also of behavior and of molecular attributes such as enzymes. The structure of every enzyme is a highly nonrandom sequence of amino acids, enabling it to catalyze a specific metabolic reaction. It is also true that many of the attributes of organisms are not adaptations. One reason may be that the attribute is an inevitable consequence of a law of chemistry or physics: for example, obeying the law of gravity is not an adaptation. Or it may be because the attribute evolved by genetic drift. For example, because of the redundancy of genetic code, many nucleotide changes are "synonymous"; they make no difference to the organism. Synonymous nucleotide changes mainly evolve by genetic drift, and the specific DNA sequence (among a set of synonyms) used by an organism is therefore nonadaptive.

Although biologists agree that some attributes of organisms are adaptive and some are nonadaptive, they do not agree on the relative importance of these two categories. Part of the disagreement is about the relative importance of drift and selection in molecular evolution. There is also disagreement over which anatomical characteristics of whole organisms may inevitably follow from nonbiological laws. Those who doubt the importance of adaptation point to "self-organizing" properties of complex molecular systems, in which apparent order can arise from the properties of chemistry. Others have suggested that apparent adaptations are not really adaptations at all, but the inevitable consequences of geometry or deeper architectural principles. (The Vignette discusses one famous suggestion of this kind.) There is sufficient uncertainty in the definition of adaptation that biologists can make mistakes about what is and is not an adaptation. It is a research problem to discover the true extent of adaptation in life, as distinct from random, physically inevitable, and self-organizing features.

The Genetics of Adaptation. In the early twentieth century, there were two main schools of thought about the genetic changes that underlie adaptive evolution. One school, associated with the German-American biologist Richard Goldschmidt, suggested that new adaptations arise via extraordinarily large genetic changes, called macromutations. The other school, associated with the British biologist and statistician R. A. Fisher, suggested that adaptations evolve via many small genetic changes. The difference between the two schools concerns not only the magnitude of genetic changes but also the source of genetic variation. A population at any one time shows a range of forms with respect to any particular attribute such as size: there is typically a normal distribution (or Bell curve) of forms, and that distribution partly reflects the many different genes that influence the attribute. Fisher, following Charles Darwin, suggested that evolution uses the genetic variation that is normally found within a population. Goldschmidt, following many of Darwin's critics, suggested that evolution proceeds via extraordinary genetic variation, outside the range of the normal distribution.

Fisher's argument against the Goldschmidtian school has been highly influential. Fisher pointed out that the observed rates of macromutation are too low to explain evolution, and that the observed amount of normally distributed variation within populations is adequate to explain evolution. He also gave a theoretical argument to suggest that adaptive evolution will proceed in small steps. He explained it in an analogy with a microscope. When focusing a microscope, most of the moves on the focusing knob are small, to improve the focus or slightly adjust where the focus is. A large move would be unlikely to improve the focus. It would either send the lens crashing down through the slide or bring it up far away from the focal layer. The analogy assumes that living creatures are already fairly well adapted (that is, the lens is already somewhere near the focus layer). This is reasonable, because living creatures are, after all, alive, and this will require some degree of adaptation to the environment. Adaptive evolution will then mainly consist of fine-tuning adjustments when the environment changes. Fisher's account of adaptive evolution has been widely accepted, and Goldschmidt had few supporters after the 1940s. (Goldschmidt did enjoy a small revival with the theory of punctuated equilibrium in the 1970s and 1980s, but the orthodox theory of punctuated equilibrium did not posit macromutations.)

A few exceptions exist to the Fisherian account. If it is impossible to evolve from one form to another without maladaptive intermediate stages, then the transformation may occur by a relatively large genetic step. It is controversial whether any examples exist, but the most likely and best studied possible example is mimicry in butterflies. There is more convincing research on another kind of exception. Fisher's argument assumes that the population is already well adapted, and the environment changes by a small amount. However, at least in some human examples, this assumption can be violated. Pesticides have been developed and used against insect pests. When the pesticides are first used, they represent a huge environmental deterioration for the insect. The insect is suddenly poorly adapted. The insects then evolve resistance to the pesticide, and the first step in this evolution is often a relatively large genetic change. That large change compromises the insect, reducing its quality in other respects. Natural selection still favors the change, because the selective force imposed by the pesticide is so large. The next steps then consist of small adjustments to the insect's physiology, to restore it to its earlier functional level, but with the pesticide-resistance machinery in place. In this example, the genetic changes during adaptation consist of a large initial step, followed by smaller later steps.

A full theory of the magnitude of genetic change during adaptation needs to consider four factors: (1) how well adapted the adaptation is to the current environment, (2) the relative frequency of large and small mutations, (3) the chance that a large or a small mutation is advantageous, and (4) the amount of selective advantage in the event that a mutation is advantageous. Fisher assumed that organisms are well adapted, but this may not always be true, as the pesticide example illustrates. Fisher's theory considered factors 2 and 3. Motoo Kimura later pointed out that factor (4) might act in the opposite direction. Although a large mutation is less likely to be advantageous than a small mutation (factor 3—explained by the microscope analogy), in the rare case that a large mutation is advantageous, it can have a larger selective advantage, and natural selection is more likely to establish it in the population. Recent theoretical work is thus adding to Fisher's theory.

Modern genetic work aims to take Fisher's argument from the theoretical to the empirically tested phase. Until recently the theory could not be tested because the genetic factors underlying adaptation were unknown except in rare examples. New molecular methods allow the genetic factors underlying evolution to be identified. Experiments can be performed in which a certain kind of evolution takes place, for instance in a microbial system. The genetic changes underlying that evolution can then be detected. This allows an empirical description of the magnitudes of the genetic effects. An experiment by C. L. Burch and L. Chao (1999) with viruses broadly supported Fisher's theory, but many other kinds of evolution should be testable in further experiments of this kind.

Adaptation and Progress/Improvement. When a new adaptation evolves in a species, it may cause an increase in the survival and reproductive success of individual members of the species. It may also make the species less likely to go extinct. For example, when zebra evolve the ability to run faster, then individual zebra may survive better (being less eaten by carnivores) and the zebra species may persist longer.

However, it is not evolutionarily inevitable that after a new adaptation has evolved, either individuals or species will be any better off than before. The reason is that nature is competitive, and any temporary advantage that a new adaptation produces may quickly (over evolutionary time) be lost by other evolutionary changes within the species or in other species.

On average, an individual in a sexual species must leave two successful offspring. (If the number is on average less than two, the species will soon go extinct. If the number is on average more than two, the species will soon become infinitely abundant, which is absurd.) Suppose an average member of a species produces 100 fertilized eggs. That means on average 98 of them fail to survive and reproduce. Now an advantageous mutation arises within the species. The mutant bearers are superior in some way and produce 200 eggs on average. We can assume the eggs are as good as the other eggs produced by the species. On average, then, 4 of the 200 will survive and reproduce. Natural selection favors the mutation. But as the mutation spreads through the species, the chance of survival per egg must decrease. The fecundity of the species cannot permanently be increased, given the limited supply of environmental resources. Once every member of the species is producing 200 fertilized eggs, 198 on average will die, rather than the previous 98. The improvement has been canceled out by higher mortality. Natural selection still favors the mutation, and the ability to reproduce 200 eggs (rather than 100) is still a good example of adaptation. Yet the example shows that adaptation does not have to mean improvement.

Likewise, the evolution of competing species may reduce any adaptive improvements. When zebra evolve to run fast, lions may go hungry and zebra survive better. But natural selection on lions now favors improved hunting skills, restoring the former chance that any one zebra is eaten by lions. The evolution of adaptive faster running in zebra has not led to an improved chance of zebra survival. This view of evolution, in which adaptive improvements in one species are soon balanced by adaptive improvements in competing species, is sometimes referred to as the "Red Queen theory." Van Valen (1973) put forward the Red Queen theory, to explain a paleontological observation that the chance of extinction of a species is approximately constant over time.

The Adaptationist Research Program. Much biological research aims to understand how the attributes of organisms are adaptive. This kind of research, however, has been criticized. One criticism is that it tends to assume that the attribute under research is an adaptation, whereas in fact it may not be. For instance, we might be studying coloration in snails. We first hypothesize that the color pattern is an adaptation for camouflage, to reduce predation. We do an experiment, perhaps measuring predation rates on painted snails. Suppose we find coloration has no effect on predation rate. We might then proceed to hypothesize that it is an adaptation for thermoregulation. But what if we falsify that hypothesis? According to critics, the research program proceeds from one hypothesis about adaptation to another, without testing whether the attribute is an adaptation. Not all evolution is adaptive, and the attribute may not be adaptive at all.

Biologists who do research on adaptation generally dismiss the criticisms as either confusions about the goals of their research or as using general philosophical points about all science as if they were particular criticisms of research on adaptation. Research on how an attribute is adaptive is different from research on whether an attribute is adaptive, and confusion results if the former kind of research is mistaken for the latter. In science, there is a general question—not peculiar to research on adaptation—about what you should do after a hypothesis is falsified. You can either slightly tune the hypothesis and retest it, or you can move on to test a completely different kind of hypothesis. It makes sense to test repeated hypotheses about adaptation if you are assuming that it is an adaptation: all scientific research makes assumptions.

Convergence As Evidence of Adaptation. In evolution, a similar attribute often evolves repeatedly when different species occupy the same environment. For instance, fish, the extinct reptiles called ichthyosaurs, and mammals such as dolphins and whales have all evolved to swim in the sea. Birds, pterodactyls, and bats have all evolved to fly in the air. In both examples, the various forms have convergently evolved a similar design: a hydrodynamic shape and propulsion by an undulating tail, or an aerodynamic shape and propulsion by flapping wings. The physical properties of the environment result in the evolution of a particular adaptive design. Convergence is found in behavior, in body organs, and in the shapes of enzymes: leaf-eating monkeys, for example, have convergently evolved a similar stomach enzyme to one found in cows and other ruminants. Convergence provides particularly clear evidence for the importance of adaptation in life.

BIBLIOGRAPHY

Barton, N., and L. Partridge. "Limits to Natural Selection." *Bioessays* 22 (2000): 1075–1084. Survey of the factors that limit the perfection of adaptation.

Burch, C. L., and L. Chao. "Evolution by Small Steps and Rugged Landscapes in the RNA Virus Φ6." *Genetics* 151 (1999): 921–927. Lab test of Fisher's theory, with microbial system.

Dawkins, R. *The Blind Watchmaker*. London, 1986. Popular book about the importance of adaptation and its explanation by natural selection.

Dawkins, R. *Climbing Mount Improbable*. London, 1996. Popular book about "uniformitarian" evolution of adaptations.

Fisher, R. *The Genetical Theory of Natural Selection*. Oxford, 1998. Contains his theory that adaptations evolve in many small stages.

Gould, S. J., and R. C. Lewontin. "The Spandrels of San Marco and the Panglossian Paradigm: A Critique of the Adaptationist Program." *Proceedings of the Royal Society of London B* 205 (1979): 581–598.

Harvey, P. H., and M. D. Pagel. *The Comparative Method in Evolutionary Biology*. Oxford, 1991. Research-level book on one method of studying adaptation; discusses allometry.

Lewontin, R. C. *The Triple Helix*. Cambridge, Mass., 2000. Criticizes the way adaptation is studied and suggests some directions for future research.

Nesse, R. M., and G. C. Williams. *Why We Get Sick*. New York, 1995. Popular book, using the theory of adaptation to look at human medicine.

Reeve, H. K., and P. W. Sherman. "Adaptation and the Goals of Evolutionary Research." *Quarterly Review of Biology* 68 (1993): 1–32.

Ridley, M., ed. *Evolution*. Oxford, 1997. Extracts from the literature, with a section on adaptation that includes many of the entries in this bibliography.

Turner, J. R. G. "The Hypothesis That Explains Mimetic Resemblance Explains Evolution: The Gradualist-Saltationist Schism." In *Dimensions of Darwinism*, edited by M. Grene, pp. 129–169. Cambridge, 1983.

Williams, G. C. *Adaptation and Natural Selection*. Princeton, 1966. Argues that adaptations exist for the benefit of organisms (or their genes) and criticizes other approaches.

— MARK RIDLEY

ADAPTED MIND IDEAS. *See* Human Sociobiology and Behavior, *article on* Evolutionary Psychology.

ADOPTION. *See* Alloparental Care.

AGRICULTURE

[*This entry comprises two articles. The first article provides an account of the times and places of earliest centers of domestication in the Near East, Middle America, Eastern North America, East Asia, New Guinea, and Africa; the second article discusses the turning point in human life from hunter-gatherers to food producers. For a related discussion, see* Human Foraging Strategies, *article on* Subsistence Strategies and Subsistence Transitions.]

Origins of Agriculture

Just after the end of the Pleistocene epoch, about 12,000 years ago, the archaeological record reveals one of the most important changes in human history. After thousands of years foraging for wild foods, people began intentionally to encourage the growth of a suite of preferred plants. Gradually, they began to clear land of competing plants, plant seeds, and weed, and, importantly, began to select which seeds would produce the next generations of plants. V. Gordon Childe (1953) labeled this change the "Neolithic Revolution." Over the succeeding 8,000 years, an instant in human evolutionary history, the process would be repeated independently in Southeast Asia, Mesoamerica, Africa, South America, and North America. The adoption of agriculture has been one of the dominant subjects in anthropology over the last fifty years.

The Near East. The earliest, and probably the most extensively studied, sequence from foraging to farming is found in the Near East. Domesticated emmer and einkorn wheat recovered from Aswad, Syria, Ali Kosh, Iran, and Çayönü, Turkey, have been dated to more than 10,000 years ago. Early domesticated barley, though not clearly distinguishable from wild barley, is reported from Gilgal and Netiv Hagdud in the Levant. Early pulses, including lentils and peas, are found at Çayönü, Jericho, and Aswad. Each of these early domestication sites lies within the geographic range of the related wild plant.

Sedentary communities precede agriculture in the Levant. Ofer Bar-Yosef (1998) describes social and economic conditions that led the Natufian culture (beginning approximately 13,000 years ago) to play a major role in the emergence of Neolithic farming communities. Natufian is a prime example of a Mesolithic complex forager society. The Natufian communities exploited a broad spectrum of wild resources. Base camps were concentrated in gallery woodlands where oak and pistachio predominated, with high concentrations of wild cereal grasses in the undergrowth. Wear pattern studies of "polished" stone sickles, along with the presence of pounding and grinding tools, indicate the harvesting of wild cereal grasses. Seed remains, though rarely preserved at Natufian sites, include pulses, cereals, almonds, acorns, and other fruits. Faunal remains include gazelle, deer, wild bovids, wild boar, and migratory waterfowl. There is also evidence of freshwater fishing. Natufian village sites, such as Ain Mallaha and Wadi Hammeh 27, include pit houses, possible ritual structures, and an elaborate funerary tradition, but little evidence of social differentiation. Storage facilities are rare.

The earliest, well-documented farming communities of the region, such as Çayönü and Jericho, were much larger than their Natufian predecessors. Neolithic settlement focused on the Mediterranean woodland zone, which was rich in wild plant and animal resources. Pit houses continued to predominate but were more substantial, with stone foundations and superstructures of unbaked mud brick. Subsistence patterns show much continuity from the Mesolithic era. Early Neolithic villagers continued to hunt and to gather wild plant resources. Pounding tools and polished stone sickles continued to be common. Early Neolithic village sites contain many more food storage facilities than Natufian villages. Also, unlike Natufian sites, they have yielded high frequencies of carbonized plant remains, including barley, wheat, and legumes. According to Bar-Yosef (1998), the overall subsistence pattern is a broad-spectrum strategy much like that of the Natufians, but with the addition of plant cultivation.

Bar-Yosef argues that climatic fluctuations created local population pressure among semisedentary Natufian peoples, triggering adaptive shifts that led to cultivation. Bar-Yosef's central factors are sedentism and population growth under optimal conditions followed by resource stress caused by resource degradation and population movements. Climatic improvements around 13,000 years ago increased already rich food resources. Populations grew, and sedentary hamlets were established. Colder and dryer conditions during the Younger Dryas (c.11,000 to 10,300 years ago) decreased both the production and the geographic distribution of wild cereals. The first experiments in systematic cultivation probably occurred at this time.

Africa. By about 7,000 years ago, agriculturalists were planting barley, emmer wheat, lentils, and chickpeas, and raising sheep, goats, and cattle along the Nile in Egypt (Figure 1). It is unclear how many of these domesticates were introduced from Southwest Asia and how many were indigenous. South of the Sahara the independent adoption of agriculture is more clear, though details are lacking. Jack Harlan, in *The Living Fields* (1995), identifies pearl millet, sorghum, millet, and chickpea as early domesticates south of the Sahara, but goes on to argue that the archaeological record tells us little about the timing and initial centers of domestication. He finds a mosaic of crops and farming practices that seem to him characteristic of noncenters. Most domestication seems to focus on the sub-Saharan belt from the Atlantic to the Indian Ocean, with most domesticates, including sorghum, pearl millet, and *Oryza barthii*, the progenitor of African cultivated rice, native to the savannas of the region.

Harlan distinguishes three distinct prehistoric agricultural complexes for the African Neolithic. The Savanna complex focuses on pearl millet in dry regions and sorghum in the wetter broadleaf savanna. The forest margin complex includes a heavy reliance on yams, which were native to the forest edge, and oil palms as well as Guinea millet. The Ethiopian complex includes finger millet, teff, and enset. Enset (*Ensete ventricosum*) is a relative of the banana; its stem base is nearly pure starch.

East Asia. The earliest evidence of domesticated rice dates to 11,500 years ago along the middle Yangtze River in China (Normille, 1997). This predates the earliest rice cultivation in Southeast Asia (Non Nok Tha, Thailand, c.5,500 years ago) by several thousand years. The period from 11,500 to 8,500 years ago when small village communities were established is not well documented archaeologically, but agricultural villages are well established by the Peiligang culture, active between 8,500 and 7,000 years ago. Chang (1986) reports that Peiligang villagers farmed millet and Chinese cabbage, and kept domesticated pigs, dogs, and chickens.

Higham (1995) sees climatic fluctuations of the Younger Dryas creating similar conditions in the Yangtze

FIGURE 1. By 7,000 Years Ago, Early Egyptian Agriculturalists Began Planting Barley, Wheat, Lentils, and Chickpeas in the Nile Valley.
Visuals Unlimited, Inc.

valley as those described by Bar-Yosef (1998) in the Levant. The earliest known villages probably were established during a cold period, which was followed by a long period of greater warmth. During this warm period, wild rice would have been able to colonize lakes and marshes of the lower and middle Yangtze. Sedentary communities appear to have developed at this time. Early agriculture communities, such as Pengtoushan, which appears to have been cultivating rice by 7,800 years ago, were established in the succeeding cold period. This period would have seen declines in wild grasses, similar to those in the Levant, which might have encouraged the cultivation of millet and rice.

Mesoamerica. Squashes were the earliest Mesoamerican domesticates. Remains of domesticated squash, recovered from the Guilá Naquitz cave in Oaxaca (Flannery, 1986), have been dated to between 8,000 and 10,000 years ago. The seeds from these squash are larger than those from wild varieties. This implies selection for rapid germination. The stems are larger than those from wild varieties, suggesting the fruits were also larger. These squashes were probably not used as food, but as containers to carry water.

The origin and evolution of maize has been the subject of often heated debate. First, there is no true consensus on the plant's wild ancestor. Some experts, especially Paul Mangelsdorf (1986), have argued that maize had a now extinct wild ancestor. Others, led by George Beadle (1980), contend a wild grass called teosinte is the ancestor. Teosinte is very close to maize genetically and hybridizes readily with maize, but morphologically it looks quite different. The debate is not over, but currently the evidence seems to be leaning toward teosinte. Harlan (1995) provides a good summary of the argument.

MacNeish's Tehuacan Valley survey (1964) described

the cultural context of the shift to maize-based agriculture (see Vignette). The Tehuacan sequence emphasizes three patterns. First, the phases immediately preceding domestication were characterized by broad-spectrum foraging. Second, sedentism precedes dependency on domesticates. MacNeish's archaeological sequence, coupled with dietary reconstructions based on carbon isotopes in ancient skeletons by Farnsworth and colleagues (1985), suggests a long period of sedentism before agriculture. Finally, the process of adoption was a long one. Foraged resources remained important components long after domesticated maize, beans, and squash entered the diet.

Eastern North America. Unlike those in many parts of the world, the primary domesticates in eastern North America, including sunflower, marsh elder, sumpweed, goosefoot, and pigweed, were not grasses. They were all producers of oil-rich seeds. The initial domestication of these seed producers took place in eastern North America between about 4,000 and 3,000 years ago. Domesticated sumpweed (*Iva annua*), recognized by increases in achene size, is found at Napoleon Hollow, Illinois, by about 4,000 years ago. Larger seeds, because of their greater energetic reserves, grow more quickly and provide more calories per seed to human consumers. Goosefoot (*Chenopodium berlandieri*) appears around 3,500 years ago at the Cloudsplitter rock shelter, Kentucky. Domesticated *Chenopodium* is recognized by a thinning of the seed testa or seed coat. Thin testa reflects selection for faster germination. Maize arrived from Mesoamerica much later, approximately 800 years ago.

History of Explanation. Many societies have stories to explain the beginning of agriculture. Harlan (1995) presents a series of origin myths from various agrarian societies. Typically, a divine being (Isis in Egypt; Demeter in Greece; a child of the Sun, the first Inca, in the Andes; Oannes in Mesopotamia) brings agriculture to the local, impoverished savages along with a system of laws, religious practices, and the arts of "civilized" life. Harlan summarizes the common elements of these myths as follows: (1) prefarming peoples led a primitive, wild, uncivilized, lawless, graceless, and brutish existence; (2) they did not farm because of ignorance or lack of intelligence; (3) divine intervention was required to bring agriculture, as well as the other elements of civilization, to humans; and (4) the agricultural descendants of these people recognize their general superiority to foragers.

Until recently, archaeologists' explanations were not so different. It was considered self-evident that agriculture was a superior adaptation. Hunters and gatherers lived their lives at the edge of starvation; because the food quest took so much time and effort, they lacked the energy to develop any trappings of civilized society.

GETTING THE MAIZE DATES RIGHT

In the early 1960s, archaeologist Richard S. MacNeish led a large, interdisciplinary project to uncover the earliest evidence of maize domestication. The project focused on a series of dry caves in the Tehuacan Valley of Mexico. The researchers excavated at Tehuacan not because they believed maize was domesticated precisely there, but because it was within the right geographic range and the caves offered excellent preservation conditions for organic materials.

The project was a success. MacNeish and his colleagues discovered tiny (< 2 inches long) maize cobs in stratified deposits in Coxcatlán Cave. Though small, the cobs clearly were domesticated maize. Because the cobs, as the world's oldest maize, were too important to be destroyed for radiocarbon dating, radiocarbon dates were produced from other organic remains associated with the same archaeological stratum. These dates led to the somewhat controversial conclusion that the earliest known domesticated maize cobs were about 7,000 years old.

Twenty-five years later, following important advances in dating methods, the maize cobs themselves were dated. The dates were produced using a method called accelerator mass spectronomy, which is a form of radiocarbon dating that requires much smaller samples than conventional radiocarbon dating. These direct dates were significantly younger than the original 7,000 years, but at 4,300 years old, the maize from Coxcatlán is still the world's oldest maize.

—RICHARD PAINE

Though superior, agriculture was such a complex concept it could not have been invented more than a few times and possibly only once. After agriculture was "invented" in one location, it would quickly spread as far as conditions allowed. For example, Robert Braidwood (1960) proposed that the invention of agriculture was a matter of cultural readiness and the right environmental conditions. As human technology and knowledge of the environment became more complex, humans exploiting environments that contained wild ancestors of agricultural staples, such as the "hilly flanks" of the Fertile Crescent, would eventually recognize the potential of these resources. It was only a matter of time before the first cultures would exploit that potential through cultivation and, eventually, domestication.

Beginning in the 1960s, studies of contemporary hunters and gatherers changed the picture of foraging as a lifestyle. The 1966 "Man the Hunter" symposium (Lee and DeVore, 1968), in particular, showed that foragers need only a few hours a day to obtain sufficient food. This appeared to be true even in marginal environments such as the Kalahari desert. Marshall Sahlins went so far as to call foragers the "original affluent society."

This new view of foragers changed the commonly accepted perspective on agriculture. Why would people choose an adaptation that required so much extra work? Climate change and population pressure have both been nominated as causes. Mark N. Cohen's *The Food Crisis in Prehistory* (1977) remains a powerful indicator of this change in perspective. Cohen argued that the only clear advantage of agriculture was its ability to support larger populations on less land. Given higher workloads, the only reason Cohen believed people would choose an agricultural adaptation would be population pressure. Using archaeological data from coastal Peru, he constructed a scenario in which population growth forced people to rely increasingly on less preferred foods until planting finally became the viable option.

Others made similar arguments. Lewis Binford (1968) argued that successful Near Eastern plant collectors would have encountered food shortages as their expanding populations spilled into areas marginally suited to wild grasses, and would have been forced to plant to increase food supplies. Similarly, Kent Flannery (1969) argued the "broad spectrum" strategies that preceded agriculture may have been a response to population growth in marginal environments. These strategies tend to emphasize high-cost resources like wild grasses.

David Rindos (1984) argues that domestication results from common interactions between humans and plants, which cause changes in plants without conscious human selection. People alter the local environment through their very presence. They disturb the soil, they concentrate nutrient-rich refuse, and they clear spaces for settlement. Rich disturbed environments around human settlement benefit plants that germinate and grow quickly. These plants tend to have seeds with thin testae or husks, which makes them easier to process, or larger seeds. When endosperms are larger, seeds have greater caloric reserves for faster growth; they also provide greater caloric returns for people. Rindos emphasizes that this, like other evolutionary trajectories, was not a directed process.

There is no consensus explanation today. Harlan contends that there is "no need to seek one single model to explain the origins of agriculture" (1995, p. 173). However, there are a number of clear commonalities, starting with the worldwide timing of the shift. Participants in a symposium at the School of American Research (see Price and Gebauer, 1995) identified a series of factors considered crucial to understanding the adoption of agriculture. These were, in order of suggested importance, (1) available protodomesticates, (2) human sedentism, (3) higher population density, (4) resource abundance, (5) geographic and/or social constraints, (6) processing and harvesting technology, (7) food storage technology, and (8) wealth accumulation. They suggested that the stages of domestication described by Rindos (1984) were likely as "old as humankind" but were not the appropriate concept of domestication for archaeologists (p. 17). They argued, appropriately, that assessments of the nature and causes of the transition to agriculture require consideration of the relative costs and returns of both foraged resources and those staples undergoing domestication. To quote David Rindos: "People, like any other animal, will not choose an obviously inconvenient, difficult, or inefficient subsistence strategy" (1984, p. 86). Population pressure is notably absent from the list of factors influencing the adoption of agriculture.

[*See also* Cultural Evolution, *article on* Cultural Transmission; Disease, *article on* Demography and Human Disease; Geochronology.]

BIBLIOGRAPHY

Bar-Yosef, Ofer. "The Natufian Culture in the Levant, Threshold to the Origins of Agriculture." *Evolutionary Anthropology* 6.5 (1998): 159–177.

Beadle, George. "The Ancestry of Corn." *Scientific American* 242 (1980): 112–119.

Binford, Lewis. "Post-Pleistocene Adaptations." In *New Perspectives in Archaeology*, edited by S. R. Binford and L. R. Binford, pp. 313–341. Chicago, 1968.

Braidwood, Robert. "The Agricultural Revolution." *Scientific American* 203 (1960): 130–141.

Chang, Kwang-Chih. *The Archaeology of Ancient China*. New Haven, 1986.

Childe, V. Gordon. *New Light on the Most Ancient Near East*. New York, 1953.

Cohen, Mark N. *The Food Crisis in Prehistory*. New Haven, 1977.

Cowan, C. Wesley, and Patty Jo Watson, eds. *The Origins of Agriculture: An International Perspective*. Washington, D.C., 1992. Excellent overview of the archaeological sequences, by region.

Diamond, Jared, and Scott F. Schafer. "How to Tame a Wild Plant." *Discover* 15.9 (1994): 100–107. This is a very good outline of the domestication process, straightforward and readable.

Farnsworth, Edward G., James E. Brady, Michael J. DeNiro, and Richard S. MacNeish. "A Re-Evaluation of the Isotopic and Archaeological Reconstruction of Diet in the Tehuacan Valley." *American Antiquity* 50 (1985): 102–116.

Flannery, Kent. "Origins and Ecological Effects of Early Domestication in Iran and the Near East." In *The Domestication and Exploitation of Plants and Animals*, edited by P. J. Ucko and G. W. Dimbleby, pp. 73–100. Chicago, 1969.

Flannery, Kent V., ed. *Guilá Naquitz: Archaic Foraging and Early Agriculture in Oaxaca, Mexico*. New York, 1986.

Gebauer, Anne Birgitte, and T. Douglas Price, eds. *Transitions to Agriculture in Prehistory*. Monographs in World Prehistory No. 4. Madison, Wis., 1992.

Harlan, Jack R. *The Living Fields: Our Agricultural Heritage*. Cambridge, 1995. Very complete overview of agriculture worldwide. Harlan is a plant geneticist and brings a different perspective to the topic. See especially the discussion of maize origins.

Harris, David R., ed. *The Origins and Spread of Agriculture and Pastoralism in Eurasia.* Washington, D.C., 1996.

Higham, Charles. "The Transition to Rice Cultivation in Southeast Asia." In Price and Gebauer 1995.

Lee, Richard B., and Irven DeVore, eds. *Man the Hunter.* New York, 1968.

MacNeish, Richard S. "Ancient Mesoamerican Civilization." *Science* 143 (1964): 531–537.

Mangelsdorf, Paul. "The Origin of Corn." *Scientific American* 255 (1986): 72–78.

Normille, Dennis. "Yangtze Seen as Earliest Rice Site." *Science* 275 (1997): 309.

Price, T. Douglas, and Anne Birgitte Gebauer, eds. *Last Hunters, First Farmers: New Perspectives on the Prehistoric Transition to Agriculture.* Santa Fe, N.M., 1995. Brings together current data and attempts to define state of knowledge of field. Highly recommended.

Rindos, David. *The Origins of Agriculture: An Evolutionary Perspective.* Orlando, Fla., 1984. Presents Rindos's coevolution view of domestication. An important work written for the professional audience.

Smith, Bruce. *The Emergence of Agriculture.* New York, 1995. Well presented and written for a wider audience than many other summary volumes.

Zohary, Daniel, and Maria Hopf. *Domestication of Plants in the Old World.* Oxford, 1994. Detailed discussion of domestication process; includes many less-used plants not considered here.

— RICHARD PAINE

FIGURE 1. By 4,000 Years Ago, Agriculture Became the Main Basis for Human Subsistence Over Most of Eurasia. This change brought about the clearance of forests, the cultivation of the soil, the maintenance of field systems, and the domestication of plants and animals. © Tim Hauf Photography/Visuals Unlimited.

Spread of Agriculture

As one of the more recent developments in the long course of human evolution, the transition from hunting and gathering to food production marked a significant turning point in the human way of life. The transformation in subsistence was accompanied by major changes across a broad front that ranged from new forms of habitation to demography and from social organization to the geographic distribution of human genes. In many cases, the first appearance of agriculture in a given area was not the result of its local origin in that place—that is, the domestication of wild plants and animals in situ— but was the consequence of the spread of agriculture from somewhere else. In short, food production first reached many places as an economic practice that was introduced from the outside.

Archaeological excavations at such sites as Abu Hureyra in Syria show that the earliest communities to engage in food production go back some ten thousand years. [*See* Agriculture, *article on* Origins of Agriculture.] It is also known that such early agricultural sites had a rather limited spatial distribution. Indeed, the vast majority of human populations at that time still lived exclusively on the basis of foraging. By about four thousand years ago, the situation had completely changed. Agriculture now constituted the main basis for human subsistence over most of Eurasia, and it was well established in many areas of sub-Saharan Africa and the New

World. Thus, over a span of only about 250 generations, most of the human populations had already completed what is sometimes called the Neolithic transition.

In contrast to the active interest taken in the origins of agriculture, it is fair to say that the study of the spread of agriculture was still out of fashion as late as the mid-1980s. This arose in part from an understandable reaction against the use and abuse of diffusionism by a previous generation of scholars, such as the prehistorian V. Gordon Childe. However, by the mid-1990s, growing interest developed, as seen in the contributions to *The Origins and Spread of Agriculture and Pastoralism in Eurasia* (1996), edited by David R. Harris, an environmental archaeologist at the University of London. The debate today centers on whether the spread of agriculture is best explained by cultural diffusion or by demic diffusion (see Explaining the Spread of Agriculture below).

Major improvements in archaeological recovery and the dating of sites may have acted as a catalyst for studies of how agriculture spread. It was only in the 1970s that the systematic recovery of plant remains and animal bones at archaeological sites became a standard feature of excavations in most parts of the world. In particular, the introduction of new flotation methods at this time facilitated the recovery of seeds. Dating by means of the radiocarbon method first came into use in the years just after World War II. The second revolution came in the mid-1980s with the introduction of a new method called accelerator mass spectrometry (AMS), which made possible the direct dating of individual seeds and bones.

The first attempt to measure the rate of spread of early farming in Europe drew upon the newly available radiocarbon dates for early farming sites in different parts of Europe (Ammerman and Cavalli-Sforza, 1971). These revealed a clear trend running from southeast to northwest across the continent. The dates showed that it took about 2,500 years—the equivalent of roughly 100 human generations—for early farming to travel from Greece to Scandinavia. The average rate of spread turned out to be about 1 kilometer per year. What was striking was the slowness of this rate—one that implied movement over a distance of only about 25 kilometers per generation (taking twenty-five years to be the length of a human generation at the time). The initial analysis also indicated that regional rates for the western Mediterranean and for central Europe could run twice as high as the average value for Europe as a whole. In a subsequent study, based on a different method of analysis and using a larger set of carbon-dated sites, the same overall rate was obtained for the spread of agriculture across Europe (Ammerman and Cavalli-Sforza, 1984). More recent AMS dates support this basic result, but they also document more variation in the rate of dispersal at the regional level. In particular, there is good evidence for the expansion's slowing down as it reached the northwestern periphery of Europe, where conditions at high latitudes were less favorable for the practice of early farming. In the Middle East, there is also the opportunity to track the spread of the early cultivation of wheat and barley in the opposite direction at such sites as Jeitun in Turkmenistan. Over the next twenty years, the systematic application of AMS dating should make it possible to trace in much greater detail the pattern of spread within Europe as well as in other parts of the world.

Explaining the Spread of Agriculture. Two processes to explain the spread of early farming in Europe are commonly invoked. The first of these involves cultural diffusion, by which cereals and farming techniques are passed from one local group to the next without the geographic displacement of the groups. This is the preferred mode of explanation for those who belong to the indigenist school of thought on the Neolithic transition. The second is what we have called demic diffusion, in which the spread arises from the movement of farmers. These two explanations are not necessarily mutually exclusive; indeed, the real issue may be to evaluate their relative importance in the case of a given region of Europe. However, they warrant being clearly distinguished at a conceptual level, because they give rise to different expectations when it comes to human population genetics and the distribution of languges around the globe, two areas of research interest today. Under the cultural mode of diffusion, the essential point is that those who are already farmers and pass on seeds and farming methods and those who receive them (and now change their way of subsistence) both continue to maintain their respective places in the landscape. In contrast, the demic model explains the spread of early farming by the movement of people over the landscape. Although demic diffusion itself can take place in several different forms—ranging from long-distance colonization to short-distance relocations of the kind envisioned in the wave of advance model (see discussion below)—the main point here is that early farmers, by relocating their settlements, brought about the Neolithic transition in Europe.

One of the early advocates of cultural diffusion was Marek Zvelebil (1986), an archaeologist at the University of Sheffield, who put forward the availability model for the transition to early agriculture in Europe. It is based on the idea that the late foragers in a given area initially adopted one or two elements of the farming package selectively and only gradually shifted to a major emphasis on food production. Another feature of the model was the emphasis placed on continuity both in settlement patterns and in levels of local population density over the transition. In other words, this model takes the position that late foragers (the final Mesolithic population) and first farmers (the early Neolithic population) lived in essentially the same way. It is not uncommon in the archaeological literature of the 1970s and 1980s to find claims for continuity between the Mesolithic and the Neolithic in various parts of Europe. However, except in northern Europe, few of these claims have turned out to be well substantiated upon closer examination. The evaluation of a claim for continuity requires comparing the final foragers in a given place with the first farmers in the same place in terms of their diets, settlement patterns, lithic (stone tool) technologies, and burial practices. At the present time, the best evidence for cultural diffusion in Europe is found in the case of Denmark, where the shift to the new subsistence economy, however, did not take place selectively and gradually but instead as a full package over a relatively short span of time.

The other main position, demic diffusion, was first put forward in 1973 (Ammerman and Cavalli-Sforza, 1973). There are three main elements to the formulation of the Neolithic transition in Europe. The first is the concept of demic diffusion itself. The second element, the wave of advance model, draws upon the work of Cambridge geneticist Ronald A. Fisher, who made it possible to model, in formal terms, the slow and continuous dispersal of a biological population that is experiencing growth. [See Fisher, Ronald Aylmer.] The third element deals with the interactions that populations of farmers and foragers can have with one another in a frontier situation (i.e., where both lifestyles may coexist in the same area).

In 1937, Fisher was able to show mathematically that, if an increase in population numbers coincides with a local migratory activity, a wave of population expansion will set in and progress outward at a steady radial rate. In Fisher's original formulation, the rate of advance of the wave front, r, is expressed by the equation

$$r = 2\sqrt{ma},$$

where m is the local migratory activity (the mean square distance of movement per unit time) and a is the initial growth rate. In our own treatment of the model, population growth is considered to take place according to a logistic form (i.e., an S-shaped curve where growth is most active at the wave's leading edge and slows down over time as local population density approaches an upper limit or carrying capacity). The other component, local migratory activity, is perhaps best understood as the summary of the frequency distribution of the distances that are moved whenever the first farmers relocated their settlements. The wave of advance model made it possible to bring local population growth, site relocation, and the rate of spread of agriculture together in a quantitative framework for the first time. In turn, this allowed researchers to explore in greater depth the hypothesis of demic diffusion.

In retrospect, the wave of advance is best seen as a first-generation model, a conceptual tool at the general level, that helped in reformulating the whole question of the Neolithic transition in Europe. In more recent studies, other researchers have gone on to develop second-generation models that treat site relocation in a more complex way. If one takes a closer look at the situation on the ground, one often finds that the expansion was less continuous in space than the general model predicts. Clusters of early farming sites are separated by open spaces on the map. Some places were preferred by early farmers over others. In Greece, for example, a demic model has been put forward in which alluvial plains with arable soils and good sources of water were preferred (Van Andel and Runnels, 1995). For certain parts of Portugal, a demic model that involves maritime colonization has recently been proposed (Zilhão, 1993). The idea here is that groups of early farmers, moving in small boats, may have relocated over even longer distances in search of new areas that had no late foragers. In the countries of southern Europe, such as Greece, Italy, and Portugal, support for the demic hypothesis is seen both in terms of the higher levels of population density observed among early farmers and the lack of continuity in settlement patterns, if one excludes cave sites (which do not represent a typical form of Neolithic settlement), over the transition. Indeed, one of the striking features of the situation in these three countries is the presence of whole areas that seem to show no evidence for habitation by final foragers whatsoever, then witness a sudden change with the arrival of early farm-

ers, who commonly make their first appearance in a given area with a full package of domesticated plants and animals.

The demic hypothesis for the spread of agriculture also carries with it clear implications for the evolution of genes and their patterns of geographic distribution in human populations. If agriculture spread solely by means of cultural diffusion, it should leave the previous gene distribution essentially intact. In contrast, a complete demic diffusion should lead to the replacement of the human populations that were there before—and consequently their genes—by those arriving from the place where agriculture originated. A mixture of cultural-demic diffusion should generate a gradient of values, or a cline, using a biological term, along the pathway of the expansion. In the case of Europe, for example, one would expect the gene frequencies of the original farmers to decrease proportionally as one moves from the Middle East across Europe. The statistical analysis of synthetic gene maps, which allows one to examine many different genetic markers all at the same time, reveals that the latter is just what is found in the case of populations living in Europe today. The full investigation of the classical genetic markers in human population around the world took many years to complete. In 1994, Cavalli-Sforza and colleagues published an important synthesis of the history and geography of human genes that includes populations throughout the world. In the case of European populations, the first principal component of the analysis of ninety-five classical polymorphisms, which accounts for 28 percent of the overall genetic variation, is linked with the spread of agriculture in Europe. What this means is that the Neolithic transition was a major event that left its lasting imprint on the genetic structure of populations in Europe. This result is now confirmed by the more recent analysis of an independent body of genetic data, the genetic markers of the Y chromosome. This chromosome is passed only along the male line of inheritance and thus offers distinct advantages in tracing gene genealogies more directly. Specifically, four haplotypes (EU4, Eu9, Eu10, and Eu11) represent the male contribution of the demic diffusion of farmers from the Middle East to Europe. Together they constitute about 22 percent of the European Y chromosomes. Moreover, the analysis of the Y chromosome now makes it possible to go even further; by comparing the cline obtained for populations in southern Europe with the one for populations in central and northern Europe, the genetic evidence reveals that demic diffusion played a more active role in the former case than it did in the latter (Semino et al., 2000). This new result is in agreement with current views held by archaeologists on the relative importance of demic diffusion in these two parts of Europe.

[*See also* Cultural Evolution, *article on* Cultural Transmission; Disease, *article on* Demography and Human Disease.]

BIBLIOGRAPHY

Ammerman, A. J., and P. Biagi, eds. *The Widening Harvest*. Boston, 2002. Proceedings of the 1998 conference "The Neolithic Transition in Europe: Looking Back, Looking Forward."

Ammerman, A. J., and L. L. Cavalli-Sforza. "Measuring the Rate of Spread of Early Farming in Europe." *Man* 6 (1971): 674–688.

Ammerman, A. J., and L. L. Cavalli-Sforza. "A Population Model for the Diffusion of Early Farming in Europe." In *The Explanation of Culture Change*. London, 1973.

Ammerman, A. J., and L. L. Cavalli-Sforza. *The Neolithic Transition and the Genetics of Populations in Europe*. Princeton, 1984. The classic interdisciplinary study setting out the demic position for the spread of early farming in Europe.

Cavalli-Sforza, L. L., P. Menozzi, and A. Piazza. *The History and Geography of Human Genes*. Princeton, 1994.

Fort, J., and V. Mendez. "Time-Delay Theory and the Neolithic Transition in Europe." *Physical Review Letters* 82.4 (1999): 867–870. Refines the mathematical formulation of the wave of advance model.

Harris, D. R., ed. *The Origins and Spread of Agriculture and Pastoralism in Eurasia*. Washington, D.C., 1994. A broad review of the archaeological evidence for different regions.

Price, T. D., ed. *Europe's First Farmers*. Cambridge, 2000. Proceedings of a meeting held in 1996; the treatments of the archaeological evidence for southern Europe and of human genetics are weak.

Renfrew, C. *Archaeology and Language: The Puzzle of Indo-European Origins*. London, 1987.

Semino, O., G. Passarino, P. J. Oefner, A. A. Lin, S. Arbuzova, L. E. Beckman, G. De Benedictis, P. Francalacci, A. Kouvatsi, S. Limborska, M. Marcikiae, A. Mika, B. Mika, D. Primorac, A. S. Santstachiara-Benerecetti, L. L. Cavalli-Sforza, and P. A. Underhill. "The Genetic Legacy of Paleolithic *Homo sapiens sapiens* in Extant Europeans: A Y Chromosome Perspective." *Science* 290 (2000): 1155–1159. A landmark synthesis of work on the genetic markers of the nonrecombining Y chromosome.

Van Andel, T. H., and C. N. Runnels. "The Earliest Farmers in Europe." *Antiquity* 69 (1995): 481–500. Based on their research in Greece, the authors present a second-generation model of demic diffusion.

Zilhão, J. "The Spread of Agro-Pastoral Economics Across Europe: A View from the Far West." *Journal of Mediterranean Archaeology* 6 (1993): 5–63.

Zvelebil, M., ed. *Hunters in Transition: Mesolithic Societies of Temperate Eurasia and the Transition to Farming*. Cambridge, 1986. A volume that represents the position of the indigenist school of thought on the spread of agriculture.

— ALBERT J. AMMERMAN

AIDS. *See* Acquired Immune Deficiency Syndrome.

ALARM CALLS

Alarm calls are highly conspicuous vocalizations given by animals when they encounter a predator. In some bird species, alarm calls are tonal, high-frequency signals that are difficult for predators to locate. The alarm calls of many mammals, however, are more easy to place, by both conspecifics and predators. This poses a challenge for those interested in the evolution of alarm calls, because signalers who alert others to danger also call predators' attention to themselves, thereby potentially increasing their vulnerability.

What selective advantage, if any, does an individual gain by giving an alarm call? Several studies suggest that alarm calling is adaptive in part because it alerts predators that they have been detected and causes them to cease hunting. In the Tai Forest of the Ivory Coast, Diana monkeys (*Cercopithecus diana*) are preyed upon by four different predators. Leopards (*Panthera pardus*) and crowned eagles (*Stephanoetus coronatus*) are "surprise" hunters that take their prey by means of a sudden attack. When monkeys encounter a leopard or an eagle, they give loud, persistent alarm calls. Two other predators, chimpanzees (*Pan troglodytes*) and humans with guns, are "pursuit" hunters that take prey after tracking them through the forest. When the monkeys encounter chimpanzees or humans, they give no alarm calls, but instead fall silent and retreat to the upper branches of the canopy. The monkeys' different responses to surprise and pursuit hunters suggest that one function of alarm calls is to communicate to a surprise hunter that it has been detected and that its hunt is unlikely to succeed. Supporting this view, one study of a radio-collared leopard found that a hunting leopard did, indeed, leave the area when monkeys began alarm calling to it (Figure 1).

If the only function of alarm calling were to communicate with the predator, prey species would need only one loud, conspicuous vocalization to alert predators of their discovery, and they would give this call whenever they encountered a predator. Two observations, however, suggest that alarm calls also serve the function of alerting nearby conspecifics of the proximity of a predator, and, in some cases, of a specific predator type. First, animals vary the rate at which they give alarm calls depending on their audience. In some cases, data support the hypothesis that alarm calling has evolved to reduce the caller's risk; in other cases, calling has apparently evolved to warn offspring or collateral kin. Second, many birds and mammals give acoustically different alarm calls to different classes of predator. Each alarm call type elicits a different response from animals nearby. Current theory suggests that acoustically different alarm calls evolve whenever animals confront different predator species whose hunting strategies require different modes of escape.

Birds typically have two acoustically distinct alarms, one for terrestrial predators, such as cats, that attack on the ground and one for aerial predators, such as hawks,

FIGURE 1. Leopards Commonly Take Their Kills into Trees to Prevent Them from Being Taken Away By Lions or Hyenas. Photo courtesy of Gustl Anzenberger.

that attack from the air. Terrestrial predator alarm calls elicit flight into the trees; aerial predator alarms elicit looking up into the air followed by flight into bushes or trees. Some avian species give a third, acoustically distinct alarm call to snakes. This vocalization elicits vigilance, approach, and often mobbing.

Ground squirrels (*Spermophilus* spp.) have two distinct alarms, a "whistle" given to eagles and hawks and a "chatter-chat" given to snakes, terrestrial mammalian predators, and aggressive members of their own species. Although these calls appear at first to be linked to a particular predator class, they are in fact more closely linked to the level of urgency, or imminent danger, that a predator poses. California and Belding's ground squirrels, for example, give whistle alarms when a predator arrives suddenly and there is little time to escape. Most

sudden attacks come from raptors, but occasionally squirrels are surprised by a terrestrial mammal. When this occurs, the mammal elicits whistles. Similarly, chatter-chat alarms are typically given to predators that have been spotted at a distance. Typically, such predators are mammalian carnivores, but it is not unusual for the squirrels to give chatter-chat alarms to a distant hawk.

Vervet monkeys (Figure 2) (see Vignette) and Diana monkeys give acoustically distinct alarm calls to leopards, eagles, and snakes. In both species, different calls indicate the presence of a different predator regardless of the immediacy of its threat. Listeners respond in qualitatively different ways to these different alarm calls. In the Kalahari Desert of southern Africa, suricates (*Suricata suricatta*), members of the mongoose family, give acoustically different alarm calls that encode informa-

VERVET MONKEY ALARM CALLS

East African vervet monkeys give acoustically different alarm calls in response to a variety of different predator types, including mammalian carnivores, eagles, and snakes. Each alarm call elicits a different adaptive response from nearby listeners. For example, alarm calls to carnivores cause other vervets to run into trees, where they are safe from a leopard's attack. In contrast, alarm calls to eagles cause listeners to look up or run into bushes, and alarm calls to snakes cause others to approach and mob the snake. Alarm calls appear to designate particular classes of predators rather than particular escape responses. For example, whereas eagle alarm calls cause vervets to run into bushes or look up when they are foraging on the ground, the same alarm calls cause vervets in trees to run down to the ground. Infant vervets require experience before they learn to associate particular alarm calls with specific predator species and before they acquire adultlike escape responses.

—ROBERT M. SEYFARTH AND DOROTHY L. CHENEY

tion not only about predator class but also about the urgency of the threat. Listeners appear to take into account both predator class and level of urgency in their responses, reacting more strongly, for example, to "urgent" eagle alarm calls than to less urgent ones.

Although the alarm calls of some primate species, including vervets and Diana monkeys, are acoustically very different from one another, the alarm calls of other primate species such as baboons (*Papio cynocephalus ursinus*) and Barbary macaques (*Macaca sylvanus*) are acoustically graded. Nevertheless, listeners appear to distinguish among different alarm call types despite their acoustic similarity.

In all of these species, one function of alarm calls may be to communicate to predators that they have been detected. For example, alarm calls given to snakes by birds, suricates, and monkeys consistently elicit a mobbing response. Mobbing draws animals' attention to the predator and eliminates the element of surprise. However, the acoustically distinct alarm calls found in many species indicate that a second function is to inform conspecifics about the nature and immediacy of danger, or about the most adaptive escape response. Although such information is widely broadcast, data suggest that alarm calling has evolved primarily because it informs specific individuals, namely, offspring, collateral kin, or potential mates.

In Belding's ground squirrels, females live in groups with offspring and other close kin, whereas males rarely, if ever, interact with close relatives. When giving alarm calls to terrestrial predators, callers incur an increased risk of being taken, but calling seems to have evolved because it functions to warn offspring and collateral kin. Females are much more likely than males to give alarm calls, and females with close relatives nearby are more likely to alarm than those without. In black-tailed prairie dogs (*Cynomys ludovicianus*), which have a similar social organization, males and females alarm call at low rates when no close genetic relatives are near and at significantly higher rates when either offspring or nondescendant collateral kin are present. In another ground squirrel species, *S. tereticaudus*, males give alarm calls at high rates before dispersal, when close kin are nearby, but typically remain silent after dispersal.

Not all ground squirrel alarm calls have evolved to warn kin. When a Belding's ground squirrel gives an alarm call to an aerial predator, it decreases the risk of being taken and gives alarm calls at the same rate regardless of the presence or absence of kin. Presumably, these calls have evolved primarily to communicate to the predator that the signaler has seen it.

FIGURE 2. Two Young Male Vervets Grooming.
Photo courtesy of Dorothy Cheney.

The presence and composition of an audience also influences the alarm calls of other species. Experiments with captive vervet monkeys found that females were more likely to give alarm calls to a threatening stimulus when they were accompanied by their offspring than when they were accompanied by an unrelated juvenile of the same age and sex. Captive male vervet monkeys and jungle fowl (*Gallus gallus*) alarm call at higher rates in the presence of a female than in the presence of another male and at higher rates in the presence of a conspecific female than in the presence of a female from another species. In the wild, lone vervet monkeys seldom give alarm calls when they detect a predator. Similarly, downy woodpeckers (*Picoides pubescens*) give no alarm calls if they are alone, if they are the only member of their species foraging in a mixed-species flock, or if the only other woodpecker present is a member of the same sex. If a member of the opposite sex is nearby, however, they call at high rates.

As with other sorts of communication, the evolution of alarm calling can only be understood by considering separately the consequences for signalers and receivers. From the signaler's perspective, producing an alarm call may involve a cost, which is presumably overcome by the beneficial consequence of driving the predator away or warning close kin. However, kin benefit only if the alarm call carries information that is specific enough to allow them to take appropriate action. Natural selection therefore favors signalers who give acoustically different alarm calls to predators with different hunting strategies. From the receiver's perspective, natural selection favors individuals who can learn rapidly to associate a specific call with a particular mode of escape.

[*See also* Genetic Markers; Kin Recognition; Kin Selection; Social Evolution.]

BIBLIOGRAPHY

Cheney, D. L., and R. M. Seyfarth. *How Monkeys See the World*. Chicago, 1990. Summarizes data on vervet monkey alarm calls and reviews many other studies of alarm calling in animals.

Fischer, J., K. Hammerschmidt, R. M. Seyfarth, and D. L. Cheney. "Acoustic Features of Female Chacma Baboon Barks." *Ethology* 107 (2001): 33–54.

Hoogland, J. L. "Nepotism and Alarm Calling in the Black-tailed Prairie Dog, *Cynomys ludovicianus*." *Animal Behaviour* 31 (1983): 472–479.

Manser, M. B. "The Acoustic Structure of Suricate Alarm Calls Varies with Predator Type and the Level of Response Urgency." *Proceedings of the Royal Society, Series B*, in press.

Manser, M. B., M. B. Bell, and L. B. Fletcher. "The Information Receivers Extract from Alarm Calls in Suricates." *Proceedings of the Royal Society, Series B*, in press.

Marler, P. "Characteristics of Some Alarm Calls." *Nature* 176 (1955): 6–8. Argues that the alarm calls of many bird species are difficult for predators to localize.

Owings, D. H., and R. A. Virginia. "Alarm Calls of California Ground Squirrels." *Zeitschrift für Tierpsychologie* 46 (1978): 58–70.

Ryan, M. J. *The Tungara Frog*. Chicago, 1985. Demonstrates that calling behavior and the acoustic structure of alarm calls is the result of selection pressures that favor calling as a means of attracting mates but act against calling because it attracts predatory bats.

Sherman, P. W. "Nepotism and the Evolution of Alarm Calls." *Science* 197 (1977): 1246–1253. Demonstrates that alarm calling to terrestrial predators is costly and that females call most often when in the presence of descendant and collateral kin.

Sherman, P. W. "Alarm Calls of Belding's Ground Squirrels to Aerial Predators: Nepotism or Self-Preservation?" *Behavioral Ecology and Sociobiology* 17 (1985): 313–323. Provides evidence that alarm calls given to terrestrial predators increase the caller's vulnerability and function to warn nearby kin, whereas alarm calls given to aerial predators do not increase the caller's vulnerability but function to increase the caller's chances of escape.

Zuberbuhler, K., D. Jenny, and R. Bshary. "The Predator Deterrence Function of Primate Alarm Calls." *Ethology* 105 (1999): 477–490. Leopards leave the area when monkeys begin giving alarms to them.

Zuberbuhler, K., R. Noe, and R. M. Seyfarth. "Diana Monkey Long-distance Calls: Messages for Conspecifics and Predators." *Animal Behaviour* 53 (1997): 589–604. Monkeys give alarm calls to surprise hunters but not to pursuit hunters. Argues that calls have evolved in part to communicate to predators.

Zuberbuhler, K., D. L. Cheney, and R. M. Seyfarth. "Conceptual Semantics in a Nonhuman Primate." *Journal of Comparative Psychology* 113 (1999): 33–42.

— Robert M. Seyfarth and Dorothy L. Cheney

ALGAE

What are algae? Such a simple question, but there is no simple answer. Algae are important, ubiquitous, and diverse; they are also problematic, confusing, and polysemous. The term *algae* (singular *alga*) usually includes the prokaryotic blue-green algae (Figure 1) but according to some experts, these are prokaryotic bacteria, that is, cyanobacteria. The eukaryotic algae include a wide diversity of microscopic to macroscopic organisms. There are several different evolutionary lineages, but these lineages are often more closely related to protozoa, animals, or plants than they are to each other. Furthermore, scientists may disagree when naming an "alga" and a "protozoon." To a phycologist (one who studies algae), the photosynthetic *Euglena* is an alga, and it follows that the closely related, but nonphotosynthetic, genus *Peronema* is also an alga. To a protozoologist, the argument is reversed. In this case, the definition of *algae* hinges on the word *photosynthesis*. This is a functional, not evolutionary or phylogenetic, definition; no one considers the algae to be a monophyletic group (i.e., a single group of related organisms that share a common ancestor). Indeed, one current evolutionary view is that algae arose several times independently during the history of living organisms. Consequently, the algae are several groups of often unrelated organisms that are united for

FIGURE 1. Stromatolites.
Colonies of blue-green algae, the oldest form to exist.
Hamlin Pool, W. Australia. © Fred Bavendam/Peter Arnold, Inc.

reasons of photosynthesis, size (generally small), and habitat (generally aquatic).

Photosynthesis is often defined as "the process by which green plants convert carbon dioxide and water into sugar and oxygen." However, an estimated 40–50 percent of global photosynthesis is carried out by algae that are not plants and are often not green. Indeed, it is surprising to most people that every other oxygen molecule they breathe comes from an alga. Nor do most remember that over 70 percent of the earth's surface is covered by oceans. In the oceans, algae form nearly the entire base of the food chain because they are the only significant source of photosynthesis—there are no grasslands or forests in the oceans. Algae are also the original organic carbon source for crude oil and natural gas. Ancient plants formed the coal beds, but ancient algae formed petroleum deposits. Although algae are not often associated with petroleum, many people find that the term *algae* conjures an image of primitive organisms. They sometimes consider the algae to be early evolutionary "dead ends," and discovering the role of algae in forming petroleum deposits amplifies this "primitive" view. People are often further surprised to learn that some algae, such as the diatoms, are believed to have first evolved during the Jurassic period, approximately 100 million years after the first appearance of mammals. Some other algal groups appear late in the fossil record as well, but many others appear to have first evolved in more ancient times. Some would argue that microalgae make poor fossils and have an incomplete and therefore unreliable fossil record, and this argument cannot be denied totally. However, bacteria, for example, have organic walls and have fossilized for over three billion years, and diatoms, with siliceous cell walls, provide outstanding fossil deposits, sometimes hundreds of meters thick. Therefore, we have some confidence in stating that some algal groups first evolved after the appearance of mammals or vascular plants.

In terms of number of individuals, algae dominate the oceans. There are approximately one million cells per milliliter, so there are about 3.6×10^{25} cells in the oceans at any one moment. Assuming the average cell diameter is 2 μm, then laying these cells end to end would produce a chain of cells that extended from the earth to the moon and back 9.5×10^{10} times.

So, who are these marvelous "algae" that provide half of our oxygen, all of our petroleum, fill the oceans, span the geologic past, and yet are so difficult to define? Unraveling the evolutionary history for algae is even more complex than providing a definition. In the following paragraphs, the various groups of algae are described.

Prokaryotic Algae. The prokaryotic algae (blue-green algae, cyanobacteria, and chloroxybacteria) are the simplest and most ancient group. Fossil blue-green algae are found in the Apex Basalt of Western Australia that is dated at 3.5 billion years old. Living cells carry out oxygen-releasing photosynthesis, and many can convert atmospheric nitrogen to ammonia (via nitrogen fixation). The diverse species occur in every imaginable habitat—in water and ice, on soil, plants, and animals, as symbionts in other organisms (e.g., lichens), even inside rocks. Species found in hot springs can grow at temperatures as high as 72°C. The prokaryotic algae were once considered to consist of two groups, the Cyanophyceae and the Prochlorophyceae. However, molecular systematic studies indicate they belong to a single group, the class Cyanophyceae. Chlorophyll *a* and phycobilipigments (phycocyanin and phycoerythrin) are the most common light-harvesting pigments, and the carbohydrate storage product is glycogen (α-linked glucose sugars). Mitochondria, chloroplasts, and flagella are completely absent. A number of species produce toxins that affect human health, especially in drinking waters that are derived from lakes and reservoirs.

Eukaryotic Algae. The eukaryotic algae, although forming a single group in the ecological sense, appear to be an artificial group in the evolutionary sense. That is, phylogenetic trees do not place all the algae in a single group. However, some caution should be exercised. Molecular systematics studies frequently fail to recover any well-supported evolutionary relationship for protists, and there is a difference between failing to show relationships and documenting no relationship. At this writing, it appears that some evolutionary relationships among the groups are now beginning to be recovered in new phylogenetic analyses. Other scientists believe that rapid evolutionary radiations have occurred (e.g., the Cambrian explosion) that for some reason fail to leave an evolutionary "signal." Therefore, evolutionary relationships of protistan groups may remain unknown. [*See* Cambrian Explosion.] Furthermore, some scientists suggest that lateral gene transfer and endosymbiosis have occurred extensively during geologic time, causing a nonhierarchical distribution of genes (and the traits they code) among today's living organisms. If extensive lateral gene transfer has occurred, then it may be difficult (or impossible) to recover a typical Darwinian evolutionary tree for the eukaryotic algae.

Glaucophytes. The class Glaucophyceae (Glaucocystophyceae) consists of algae that have a nonphytosynthetic host cell that contains blue-green algal-like bodies called cyanelles. Each cyanelle is capable of carrying out photosynthesis, and each is surrounded by a thin cell wall, which has a chemical composition similar to those of Gram-negative bacteria. However, the cyanelle has many fewer genes than a blue-green alga, and the cyanelle cannot grow and reproduce outside the host cell. Approximately 90 percent of the genes that regulate the cyanelle are located in the nucleus, not in the cyanelle genome. Nevertheless, the cyanelle has many more genes than a typical chloroplast, leading some scientists to suggest that the glaucophyte cyanelle is an example of how the chloroplast may have evolved from a symbiotic event involving a prokaryote. The question of an endosymbiotic origin of chloroplasts is far from resolved, however. Some scientists argue for a single primary origin of plastids, whereas others argue for multiple primary origins of plastids, and still others against any endosymbiotic origin. The matter is further complicated because most scientists agree that secondary endosymbiotic events gave rise to plastids in at least some groups of algae. Unlike some other controversies, the origin of plastids may be resolved in the near future. The carbohydrate storage product is starch, and the mitochondria have flattened cristae. Their closest relatives are perhaps the red algae.

Chlorarachniophytes. The class Chlorarachniophyceae is a small group of photosynthetic organisms that are most closely related to certain nonphotosynthetic amoebae. The word *Chlorarachnion* means "green spider," referring to the green color of the chloroplasts and the radiating spiderlike pseudopodia. Their nonphotosynthetic amoeboid relatives were historically named the Rhizopoda, but in recent years the term *Cercozoa* has gained some popularity. Chloroplasts apparently arose via a secondary endosymbiotic event (i.e., an amoeba engulfed a eukaryotic green alga), and a remnant of the green algal nucleus, called a nucleomorph, is still retained. Chlorophyll *a* and *b* are the light-harvesting pigments, the mitochondria have tubular cristae, the carbohydrate storage product is unknown, and zoospores (as well as a single flagellate species) have a single flagellum.

Euglenophytes. The class Euglenophyceae contains common freshwater algae, with a few marine representatives. Protozoologists refer to this group as the Euglenozoa. Nearly all members are flagellates. The common *Euglena* is often the first microorganism that students observe using a microscope. Numerous discoid green chloroplasts are found in the photosynthetic species, and chlorophyll *a* and *b* are the light-harvesting pigments. Chloroplast genomes are found in some nonphotosynthetic species but not in others. From an evolutionary perspective, this and other data suggest that (1) the first euglenoids were nonphotosynthetic, (2) chloroplasts were obtained by a symbiotic event and gave rise to the photosynthetic forms, and (3) some photosynthetic forms lost the ability to photosynthesize but still retain some of the chloroplast genes from their ancestors. Their mitochondria have discoid cristae, the carbohydrate storage product is paramylon (β-linked glucose sugars), and swimming cells have one or two flagella (one species has four flagella). The closest evolutionary relatives to the photosynthetic euglenophytes are nonphotosynthetic flagellate protozoa, including organisms such as *Trypanosoma*, which causes the tropical disease sleeping sickness.

Red Algae. The red algae are a large monophyletic group that includes many seaweeds but also some single-celled organisms. Most red algae occur in marine environments, but a substantial number also occur in freshwater streams. Even though red algae are almost always found in water (a few grow on soils), flagellate cells are entirely absent. This remarkable absence of swimming cells has puzzled scientists for two centuries. Chloroplasts show no evidence of a secondary endosymbiotic origin. Chlorophyll *a* and phycobilipigments are the common chloroplast pigments, and the common carbohydrate storage product is a special type of starch (floridean starch). The mitochondria have flattened cristae. The cell walls often contain special polysaccharides that are commercially valuable. Agar and agarose are one class of polysaccharides that are important for biological research (as well as certain prepared foods), and the carrageens are a second polysaccharide group that are used widely in prepared foods. Red algae are also eaten directly, and nori (*Porphyra*) is the most important, with annual sales of several billion dollars. The life cycle of many red algae is complicated and contains three life stages (gametophyte, carposporophyte, and tetrasporophyte generations).

Cryptophytes. The Cryptophyceae are primarily photosynthetic flagellates. Both ultrastructural and molecular data show that their chloroplasts evolved from a secondary symbiotic event; that is, a nonphotosynthetic cryptophyte engulfed a eukaryotic alga (probably a red alga), and that alga has become the chloroplast. A nucleomorph is present. The chloroplast pigments are chlorophylls *a* and *c* as well as phycobilipigments. The mitochondria have flattened cristae, and the carbohydrate storage product is starch. Cells bear two flagella,

RED TIDES

A red tide is formed by a large number of microalgal cells that accumulate near the surface of water bodies, frequently causing a red discoloration of the water; hence the colloquial name "red tide." High concentrations of algal cells do not always produce red tides; they may be brown, golden, green, aquamarine, or even purple in color. It is commonly, but erroneously, assumed that the coloration is due to the color of the algal cells; that is, red tides are caused by algae that are red in color. The colors of most phytoplankton are hues of green or brown, whereas blooms of *Noctiluca*, a colorless dinoflagellate, cause some of the brightest red tides.

What, then, causes the coloring? Ocean color is primarily a function of absorption and backscattering of light, and different particles (e.g., phytoplankton cells) absorb and backscatter photons differently. Enhanced absorption of blue and yellow photons occurs with increased algal concentrations, reducing the probability that those colors will be reflected. When many different types of particles are present (e.g., several species of varying size), their backscattering characteristics tend to cancel one another. Conversely, a monotypic particle size, which occurs during a bloom of a single species, creates a unique backscattering spectrum that may cause a specific color (e.g., red or brown) to appear. Furthermore, water itself strongly absorbs red light, and densely accumulated cells at the very upper surface may appear red due to the reduced role water plays in photon absorption; the same dense accumulation at a depth of 20 meters (66 feet) may go unnoticed by the human eye. Blooms of toxic algae often produce red tides that are harmful to humans or aquatic life, and they give red tides a more sinister connotation.

Finally, the definition of red tide varies with languages. In the English language, a red tide is restricted to blooms that produce a reddish coloration of the water, but in the Japanese language, a red tide describes any algal bloom without regard for color.

—ROBERT A. ANDERSEN

and either one or both are adorned with bipartite hairs. The cells have a complex covering called a periplast that consists of organic plates that overlap in a manner similar to roof shingles. They occur in both freshwater and marine environments.

Haptophytes. The division Haptophyta (classes Prymnesiophyceae and Pavlovophyceae) is a group of predominantly marine flagellates that were once placed in the chromophytes but are now believed to represent a separate evolutionary lineage. The characteristic fea-

ture is the haptonema, a threadlike projection that superficially resembles a flagellum. There are several known functions for the haptonema, including attachment and particle feeding. Most members are photosynthetic, and the chloroplasts have apparently arisen by the engulfment of a eukaryotic alga that eventually became a chloroplast; no nucleomorph is present. The chloroplast pigments include chlorophylls *a* and *c* as well as the carotenoids fucoxanthin and diatoxanthin (or their derivatives). Their mitochondria have tubular cristae. Flagellate cells have two flagella that lack tubular hairs, and both flagella are attached near the anterior. Swimming is accomplished by at least two methods, one with the flagella projecting forward, the other with the cell inverted so the flagella push the posterior of the cell through the water. The carbohydrate storage product is formed from β-linked glucose residues, resembling paramylon and chrysolaminaran. Many of the flagellate cells are covered by organic scales, and a group known as the coccolithophores are covered coccoliths formed from calcium carbonate crystals deposited onto the organic scales. Chalk cliffs, such as the white cliffs of Dover, are primarily fossilized deposits of coccoliths. The haptophytes are apparently a somewhat ancient group, perhaps first evolving in the Carboniferous period (360–286 million years ago). Just as the end of the Cretaceous period brought the extinction of the dinosaurs and the rapid radiation of flowering plants, bony fishes, and mammals, the end of the Cretaceous period (c.65 million years ago) drastically reduced the number of coccolithophores, and they were replaced with many other algal groups, especially the diatoms. Unlike most microorganism groups, there are more fossil species than living species in the Haptophyta. A number of the marine flagellates produce toxins that kill fish.

Dinoflagellates. The dinoflagellates are, as their name implies, primarily flagellate cells. They occur in many marine and freshwater environments. Approximately half of the species are photosynthetic. The nonphotosynthetic species feed on diatoms, ciliates, and other microscopic organisms, and a substantial number (approximately 10 percent) are mixotrophic (i.e., capable of both photosynthesis and feeding on other organisms). The earliest dinoflagellates were probably nonphotosynthetic, and chloroplasts have apparently arisen two or more times as an engulfed eukaryotic alga became a symbiotic organism and then over time a chloroplast. No nucleomorph is present. Remarkably, photosynthetic dinoflagellates themselves have become symbionts (zooxanthellae) in corals where inorganic nutrients from the corals and photosynthetically produced organic molecules of the dinoflagellate are exchanged. The chloroplast pigments vary depending on the chloroplast type. Most chloroplasts have chlorophyll *a* and *c*, some have carotenoids (peridinin or fucoxanthin), and some have phycobilipigments. Their mitochondria have tubular cristae, and cells typically have two flagella. One flagellum extends posteriorly and pushes the cell forward; the second flagellum encircles the cell, providing thrust and rotation. Because many dinoflagellate cells are large, they have a complex cytoskeleton of microtubules and other structural molecules. Although most dinoflagellates are flagellate, some are coccoid, filamentous, or even amoeboid. The storage product is usually described as starch, but additional research may reveal other compounds. Many dinoflagellates are naked (i.e., they lack a cell covering), and many others have cellulosic plates arranged around the cell periphery. A few dinoflagellates produce toxins that are harmful to humans and other animals (i.e., paralytic shellfish poisoning and diarrheic shellfish poisoning). *Pfiesteria piscicida* was once claimed to kill fish and harm humans, but convincing scientific evidence has not been found. The dinoflagellates are most closely related to the ciliates and the apicomplexans, and these three groups are collectively referred to as the alveolates. The alveolates may be related to the stramenopiles (which include the chromophyte algae) and the haptophytes, but currently the scientific data supporting this hypothesis are weak.

Chromophytes. The chromophytes (Heterokonta, or heterokont algae, golden algae, and stramenochromes), are a very large and diverse group of algae. They have chloroplasts that apparently arose by a secondary endosymbiotic event, but no nucleomorph is present. The chloroplast pigments include chlorophylls *a* and *c* (several different types) and carotenoids such as fucoxanthin, diatoxanthin, and violaxanthin (specific carotenoid composition varies among the classes). Their mitochondria have tubular cristae. Flagellate cells have one or two flagella in most cases, and one flagellum is adorned with tripartite hairs. The carbohydrate storage product is chrysolaminaran (or related compound) formed by β-linked glucose residues. This product is very small (approximately 25–35 glucose units) compared to colloidal starch, for example; consequently, to avoid osmolarity affects on the cell, the chrysolaminaran is maintained in a vacuole. Cells are covered by a wide variety of materials, and in some cases the cells are naked. Filamentous and coccoid forms often have organic cell walls, diatoms have cell walls composed of (siliceous) opaline glass (like windows), synurophytes have biradial silica scales, and a few chrysophytes and dictyochophytes have organic scales.

The chromophytes include the diatoms (division Bacillariophyta), and in terms of species numbers, they are the "insects" of the algal world. Approximately 100,000 fossil and living diatom species have been described, and some researchers speculate that a million or more species exist. The class Bolidophyceae contains oceanic flagellates that are closely related to diatoms. The class

Pelagophyceae are marine algae occurring in both the open oceans and the coastal waters, including flagellate, coccoid, sarcinoid, and filamentous forms. The class Dictyochophyceae are primarily flagellates that occur in freshwater and marine environments, but some open ocean representatives are amoeboid. The pelagophytes and dictyochophytes are close relatives. The diatoms, pelagophytes, and dictyochophytes may form a subgroup within the chromophytes, but currently the supporting data are weak.

The brown seaweeds (class Phaeophyceae) range in size from microscopic filaments to the giant kelps that may reach 100 meters (328 feet) in length. The yellow-green algae (class Xanthophyceae) are primarily coccoid and filamentous forms, most occurring in freshwater. The class Phaeothamniophyceae consists primarily of freshwater filamentous and coccoid forms. The phaeophytes, xanthophytes, and phaeothamniophytes are closely related to each other, and they form a subgroup within the chromophytes.

The classes Chrysophyceae and Synurophyceae are predominantly freshwater flagellates, although the Chrysophyceae includes amoeboid, capsoid, and coccoid forms as well. The class Eustigmatophyceae occurs in both marine and freshwater environments and consist primarily of coccoid single cells. The class Raphidophyceae consists only of flagellates, both marine and freshwater. The chrysophytes and synurophytes are very closely related, and the eustigmatophytes are frequently considered their sister group. The evolutionary relationship of the raphidophytes is less clear, and in some analyses they appear to be related to the other three classes, but in other analyses they appear to be related to the phaeophytes, xanthophytes, and phaeothamniophytes.

Ultrastructural, biochemical, and molecular data strongly support the monophyletic nature of the classes listed above. That is, each class is distinct (monophyletic), and members within each class share a number of specific characteristics. Paradoxically, the evolutionary relationships among these classes are less clear, and no consensus exists in many instances. Nevertheless, all chromophytes share features that unite them in a clear and monophyletic group. As a group, the chromophytes are most closely related to the aquatic fungi (Oomycetes and Hyphochytridiomycetes) and certain protozoa (Bicosoecids, Labyrinthulids, and Thraustochytrids). Collectively, the chromophytes and these aquatic fungi and protozoa are labeled the stramenopiles. The stramenopiles, in turn, may be related to the alveolates (dinoflagellates, ciliates, and apicomplexans), haptophytes, and cryptophytes. However, the supporting data are weak, at best.

Green Algae. The green algae are a large and diverse group of predominantly freshwater and soil organisms. Most green algae are microscopic, including *Ostreacoc-*

cus, which is believed to be the smallest known eukaryotic cell. There are two evolutionary lineages that diverged long ago. The chlorophytes form one group, and the second group, the charophytes, gave rise to the plants. Thus, in an evolutionary sense, these two classes cannot be considered a monophyletic group unless mosses, ferns, and flowering plants are considered to be algae—a terrific idea, according to most phycologists, but something most botanists are unwilling to concede. This charophyte/plant group is now frequently termed the Streptophyta, and the second lineage is termed the Chlorophyta. Within the Streptophyta, the algal members include the stonewarts (e.g., *Chara, Nitella*), the desmids, *Spirogyra*, and its relatives, as well as organisms such as *Coleochaete* and *Klebshormidium*. Over five thousand desmid species have been described, making them the most abundant subgroup of green algae. The Chlorophyta lineage is also diverse. It is particularly rich in its diversity of flagellates, including the well-known *Chlamydomonas* and *Volvox*. The Chlorophyta has been divided into several taxonomic classes; for example, Linda Graham and Lee Wilcox (2000) recognize four classes, while Van den Hoek and colleagues (1995) recognize nine classes. The definition of and means for delineating these classes remain controversial, and further work will undoubtedly be required before a consensus classification is achieved.

All green algae have a chloroplast that apparently arose by a primary endosymbiotic event. The photosynthetic pigments are similar to those of plants, chlorophylls *a* and *b*, as well as some carotenoids. Their mitochondria have flattened cristae, and their carbohydrate storage product is starch. The flagella (typically two) of swimming cells and the related cytoskeletal structures are variable, providing one basis for separating the Chlorophyta into several classes.

[*See also* Diatoms.]

BIBLIOGRAPHY

Andersen, R. A. "The Biodiversity of Eukaryotic Algae." *Biodiversity and Conservation* 1 (1992): 267–292. General article on algal biodiversity and its importance.

Anderson, D. M. "Red Tides." *Scientific American* 271 (1994): 62–68. General article on toxic algae—their occurrence and causes.

Berger, S., and M. J. Kaever. *Dasycladales: An Illustrated Monograph of a Fascinating Algal Order*. Stuttgart, 1992. The Dasycladales are a small group of marine algae in terms of species numbers, but they are ancient (650 million years old), complex, and relatively large in size. The numerous illustrations are superb.

Bhattacharya, D., ed. *Origins of Algae and Their Plastids*. Vienna, 1997. Technical source that thoroughly explores chloroplast origin and evolution in the eukaryotic algae.

Canter-Lund, H., and J. W. G. Lund. *Freshwater Algae: Their Microscopic World Explored*. Bristol, England, 1995. This book is filled with beautiful color photographs taken through a light microscope. Some aspects are outdated.

Foster, K. W., and R. D. Smyth. "Light Antennas in Phototactic Algae." *Microbiological Reviews* 44 (1980): 572–630. Many flagellate algae can sense light and swim toward or away from it. Special structures, termed eyespots, are the subcellular basis for light detection. The article contains technical information on the physics of light and the biochemistry and cellular ultrastructure of eyespots.

Garbary, D. J., and M. J. Wynne, eds. *Prominent Phycologists of the Twentieth Century*. Hantsport, Nova Scotia, 1996. Each chapter describes the life and work of famous phycologists.

Graham, L. E., and L. E. Wilcox. *Algae*. Upper Saddle River, N.J., 2000. The most recent and thorough textbook on algae. Technical background may be required for some topics; extensive references to the primary literature.

Jeffrey, S. W., R. F. C. Mantoura, and S. W. Wright, eds. *Phytoplankton Pigments in Oceanography*. Paris, 1997. The most thorough book available on chloroplast pigments (chlorophylls, carotenoids, and phycobilipigments).

Leadbeater, B. S. C., and J. C. Green, eds. *The Flagellates: Unity, Diversity and Evolution*. London, 2000. This is a technical book, but it includes chapters on the history of flagellates, mechanisms of flagellar propulsion, cellular and molecular aspects of flagella and associated cytoskeletons, ecology, and evolution.

Lee, R. E. *Phycology*. 3d ed. Cambridge, 1999. This general textbook on algae is less comprehensive than Graham and Wilcox (2000), but nonspecialists may find it easier to understand.

Lobban, C. S., and M. J. Wynne, eds. *The Biology of Seaweeds*. Oxford, 1982. Provides many examples of seaweed life cycles as well as cellular and biochemical information relating to seaweeds.

Margulis, L., J. O. Corliss, M. Melkonian, and D. J. Chapman, eds. *Handbook of the Protoctista*. Boston, 1990. A through and technical compilation of protists (algae, protozoa, some fungi). It is not organized with respect to evolution, and it has some odd word usage (e.g., *undulapodium* for *flagellum*).

Menzel, D., ed. *The Cytoskeleton of the Algae*. Boca Raton, Fla., 1992. Provides a review for all eukaryotic algae.

Reisser, W., ed. *Algae and Symbioses*. Bristol, England, 1992. Comprehensive but technical examination of algal symbioses, including corals, lichens, and sponges.

Round, F. E., R. M. Crawford, and D. G. Mann. *The Diatoms: Biology and Morphology of the Genera*. Cambridge, 1990. Provides a synopsis of diatoms, including hundreds of outstanding photographs.

Stewart, W. D. P., ed. *Algal Physiology and Biochemistry*. Berkeley, 1974. Although dated, this book still represents the most comprehensive text on these topics.

Taylor, F. J. R., ed. *The Biology of Dinoflagellates*. Oxford, 1987. Addresses all aspects of dinoflagellates, from ecology to toxicity.

Van den Hoek, et al. *Algae: An Introduction to Phycology*. Cambridge, 1995.

Wetherbee, R., R. A. Andersen, and J. Pickett-Heaps, eds. *Protistan Cell Surfaces*. Vienna, 1994. Technical but well-illustrated book describing organic, siliceous and calcareous scales, cell walls, and other cell coverings of eukaryotic algae, protozoa, and certain fungi, especially as these relate to evolution.

Whitton, B. A., and M. Potts, eds. *The Ecology of Cyanobacteria*. Dordrecht, Netherlands.

Winter, A., and W. G. Siesser, eds. *Coccolithophores*. Cambridge, 1994. Features numerous photos and line drawings as well as descriptions of the organisms and the process of scale formation.

Wiessner, W., E. Schnepf, and R. C. Starr, eds. *Algae, Environment and Human Affairs*. Bristol, England, 1995. Addresses many current issues such as pollution, toxic algae, and economic importance.

— ROBERT A. ANDERSEN

ALLOPARENTAL CARE

Alloparental care is parentlike behavior that is typically performed by individuals that are not the parent of the recipient. The term, coined by E. O. Wilson in 1975, was intended to include the workers of social insects, such as honeybees and other eusocial species. *Helping behavior* is virtually synonymous and is the term more often used for alloparental care in birds, mammals, and fish. Helping is similar to alloparenting but with the additional stipulation that it be done in company with the parents, thus excluding brood parasitism (as in cuckoos, cowbirds, and cuckoo bees), slave making (ants), and brood capture or brood amalgamation (as in some ducks).

In practice, the term is best reserved for behaviors that parents do only when they have young and that are directed only at specific recipients. Typically, a bird is called a helper if it feeds young that are not its own in company with one or more parents of the young (Figure 1). In contrast, helpers and nonhelpers may give alarm calls that inadvertently benefit young that are not offspring of the caller (especially in colonies), but alarm calling is not considered to be sufficient grounds for identifying helping behavior because it occurs in nonreproductive contexts and is not directed exclusively at particular recipients. The terms *helper* and *alloparent* are conventionally not applied to the individual members of clonal colonies, such as tunicates, even though individuals in these colonies may have role differentiation and may sometimes act for the good of the colony rather than their own good.

Importance to Evolutionary Biology: Departure from Classical Fitness Theory. Helping or alloparenting poses a problem for evolutionary biologists. In the classical view, individuals that are selected are those that leave the most mature offspring to the next generation. Thus, traits that confer more mature offspring on

---

FIGURE 1. Alloparental Care.
Helpers feeding young and defending the nest against a snake.   Jerram L. Brown. *Helping and Communal Breeding in Birds: Ecology and Evolution*. Copyright © 1987 by Princeton University Press. Reprinted by permission of Princeton University Press. Drawing by Takashi Taniguchi.

THE RED-COCKADED WOODPECKER

© Volker Steger/
Peter Arnold, Inc.

Pioneer settlers in southeastern United States in the nineteenth century first encountered extensive pine forests. These pines have thick, insulating bark that protects them from wildfires. Unlike most woodpeckers, which create cavities in dead trees for their nests, the red-cockaded woodpecker (*Picoides borealis*) uses living pine trees that are at least sixty years old and able to withstand the formerly frequent fires. At this age, the pines begin to be affected by a fungus at their core. The woodpeckers use this rotten core for their nests, after first breaking through the living outer layers. The woodpeckers make many holes below their nest that exude thick, gummy pine resin that tends to discourage tree-climbing snakes, a major predator. Breaking through is not easy, and it appeared that the availability of holes in living trees was limiting breeding opportunities, causing newly mature birds to delay breeding and stay with their parents, where they often turned to feeding the young of the parents. When new holes were drilled through the outer bark, many of the previously nonbreeding helpers became breeders, thus demonstrating the importance of ecological factors in delaying breeding.

—JERRAM L. BROWN

parents are selected. [*See* Natural Selection.] Helping does not appear to have been selected by this mechanism because helpers may not be parents at all and often appear to receive no obvious benefit in terms of own offspring for their efforts. Helping would instead increase the numbers of offspring of other parents. Indeed, many workers in social insects and helpers in some birds appear to sacrifice their entire classical or direct fitness (measured in descendant kin), leaving instead only an increment to their nondescendant kin, such as offspring of their parents, siblings, or half-siblings. Because helping does not appear to conform to the classical view, it has forced biologists to rethink one of the most important concepts of natural selection, namely, the idea of fitness.

Additionally, because helping is potentially a form of altruism (loss of classical fitness by the donor, gain by the recipient), it became of great interest to researchers. For this reason, the study of worker behavior in social insects and helping behavior in birds and other vertebrates came under scrutiny.

Is Helping Altruism? Altruism involves an act that is costly to the donor but beneficial to the recipient. The evolutionary theorist William Hamilton showed formally that apparently altruistic acts could actually benefit donors if directed at close kin. The donor indirectly increases the fitness of genes he shares with his kin by helping those kin. In the social insects, self-sacrifice in defense of the group does appear to satisfy Hamilton's concept of altruism (loss of direct fitness by donor coupled with gain by recipient). For birds and mammals, however, this is not an easy question to answer. In some cases, the helpers appear to be benefiting mainly their direct fitness by improving their chances for reproduction in the future. In other cases, such as in bee-eaters and certain individual pied kingfishers, careful accounting of costs and benefits in the field has shown that some helpers are acting altruistically; that is, they appear to benefit from indirect kin selection.

Are Recipients Kin? If recipients of help are usually close kin, Hamilton's theory would be supported. In most social insects and birds, most individual alloparents are related to their recipients, but even in these species a small minority of the alloparents are not related. In some ants, birds, primates, and fish, the relationship between alloparents is reciprocal and the alloparent and recipient are not related.

Do Alloparents Augment Reproductive Success of Recipients in a Measurable Way? For alloparental care to support Hamilton's interpretation, it is necessary to show that the reproductive success of recipients is augmented by the help received. In most species of birds, a correlation between breeding success of the recipient parents and the aid given by helpers was positive; however, such correlations might not represent cause and effect if both breeding success and amount of help were correlated with other variables, such as parental experience or territory quality. Experiments in which these variables were controlled, however, have shown that alloparental aid does cause increased reproductive success.

Do Helpers Augment Their Own Future Reproductive Success? Although helpers appear to benefit recipients, the main selective pressure favoring helping could be that it augments future reproduction by the helper. This explanation will not work for sterile workers of social insects, but it could work in some birds. To consider this question carefully, one must separate the benefits of delayed breeding and delayed dispersal from the benefits of helping per se. This cannot be done in group territorial birds, but it can be done in colonial birds that lack feeding territories (and therefore lack the ecological constraints associated with group territoriality). Thus, some nonbreeding male pied kingfishers (*Ceryle torquata*) appear to feed nestlings mainly as a way to establish social relations with the mother as a future mate (selfish), whereas others are more dedicated to the young (altruists). The selfish ones were usually unrelated and did not work hard; the altruists were usually closely related and did work hard. The selfish ones survived better and were more successful in mating the following year than the altruists. Both results were consistent with inclusive fitness theory. Here, then, some alloparental behavior is revealed as an attempt to curry favor with females for future matings.

Phylogenetic Constraints: Evolutionary Convergence. Helping is found in animals that are relatively simple, such as insects, and in those that are relatively complex in their behavior, such as primates and birds. All ants and termites have workers (alloparents), as well as many bees and wasps, an occasional beetle, aphid, thrips, and weevils. In fish, a few mouth-breeding cichlid varieties may have helpers. Many mammals have helpers. The behavior is common in marmosets, tamarins, and other primates, jackals, wolves, hyenas, dwarf mongooses, and pine voles. It is especially well developed in the naked mole rat. The phylogenetic pattern of occurrence of helping suggests that helping has evolved independently in these groups. Even in birds, helping appears to have evolved independently in many orders and families. Within families or genera, however, species with helpers are commonly closely related.

Geographic Occurrence. In birds, helping behavior is least common in cold climates with strong seasons while quite common in regions with reduced seasonality where more species can persist through the nonbreeding season on their breeding territory. Thus, helping is common in the tropics, Australia, and South Africa. Species such as flycatchers, whose diet forces them to migrate, rarely have helpers, whereas many babblers, corvids, and shrikes are able to find other kinds of food in winter, thus allowing them to stay in their territories. Staying in the home territory allows the offspring to stay with their parents in a territory that is familiar to them, whereas migration tends to disrupt family togetherness.

Occurrence of Helping in Various Kinds of Social Systems. Most animals with alloparents live in relatively permanent, close-knit social groups in which the same individuals are always in company with each other. Most avian species with helpers live in groups that defend group territories in which they live all year, with perhaps a brief absence during particularly difficult weather. Some species (about 14 percent) are colonial without feeding or breeding territories (beyond their nest hole), but some colonial species (bee-eaters) maintain feeding territories year-round.

The territorial groups may be divided into two types. In the most common type, only a single female breeds (singular breeding, as in the Florida scrub-jay, *Aphelocoma coerulescens*). Less commonly, the groups are composed of two or more breeding females, their mates, and some nonbreeding birds (plural breeding, as in the Mexican jay, *Aphelocoma ultramarina*). Singular breeding groups are typically smaller, consisting of a breeding female, one or more mates, and some younger nonbreeders. In plural breeding groups, the females may lay jointly in the same nest (anis) or in separate nests (certain jays).

In most species with alloparents or helpers, the majority of helpers consists of the young who are almost of breeding age but who have not yet left their parents. Thus, in attempting to analyze the possible causes of helping, attention has centered on the ecological conditions that constrain dispersal.

Dispersal and Ecological Constraints. Skeptics of the applicability of Hamilton's theory to alloparental care tended to view such behavior as self-serving in the long run even though it might appear altruistic in the short run. Confusion on this point was generated because some helpers delay breeding and dispersal, suggesting that the nonbreeding state was simply a step on the road to acquiring breeding status, rather than an act of altruism. Thus, such delays were widely regarded as self-serving. Alloparental behavior (as opposed to delayed breeding) is not so easily interpreted as self-serving. Nonbreeders do not have to help, and some do not.

Nevertheless, most workers agree that delayed dispersal sets the stage for selection to favor helping by keeping the nonbreeders together with their parents and siblings. The primary ecological factor that constrains dispersal is thought to be the difficulty that inexperienced birds have in obtaining control of a breeding territory or other critical resource. When most territories that are reasonably productive are already held by long-lived birds, it appears to be more productive for aspiring reproductives to use their home as a base from which to reconnoiter nearby possibilities and finally to claim a breeding position promptly as soon as an opening occurs.

It is difficult to show conclusively that limitation of territories with a good expectation of breeding success causes young birds to delay breeding. Usually only correlations are available. A threatened species of bird, the red-cockaded woodpecker (*Picoides borealis*), provides one of the best examples of an ecological limitation (see Vignette).

Parent–Offspring Relations. Hamilton's inclusive fitness theory provides a new perspective on parent–offspring relations, emphasizing parent–offspring conflict. In some situations, it is possible for offspring to serve their own interest rather than that of their mother. For example, worker ants are able to bias the sex ratio toward their predicted optimum as opposed to the optimum predicted for their mother even while helping the mother to rear her young.

Studies in the vertebrates have been suggestive but not so conclusive. For example, dominant-breeding bee-eaters harass subordinates in their group; these subordinates often give up their own breeding, then help to feed the young of the harassing male, who is often their father. In mammals, clear-cut examples of young gaining a goal that conflicts with that of their mother are missing.

Another perspective is that in helper systems, the tolerance of the parents for the continued presence of their offspring actually aids the young to become established as breeders (parental facilitation). Such tolerance has been reported to facilitate survival of the offspring in jays and may be associated with territorial bequeathal in squirrels.

[See also Eusociality, article on Eusociality in Mammals; Group Living; Naked Mole-Rats; Parental Care; Reciprocal Altruism.]

BIBLIOGRAPHY

Brown, J. L. *Helping and Communal Breeding in Birds: Ecology and Evolution.* Princeton, 1987.
Choe, J. C., and E. J. Crespi eds. *The Evolution of Social Behaviour in Insects and Arachnids.* Cambridge, 1997.
Dugatkin, L. A. *Cooperation among Animals: An Evolutionary Perspective.* New York, 1997.
Solomon, N. G., and J. A. French eds. *Cooperative Breeding in Mammals.* New York, 1997.
Taborsky, M. "Sneakers, Satellites, and Helpers: Parasitic and Cooperative Behavior in Fish Reproduction." *Advanced Studies in Behavior* 23 (1994): 1–100.

— JERRAM L. BROWN

ALTRUISM

Altruism is a form of cooperation (i.e., an act that enhances the fitness of others), but it is distinguished from mutualistic behavior in which the fitness of both the actor and the recipient is enhanced (e.g., the act of cleaning parasites from a host by an organism that feeds on those parasites). Altruistic acts decrease the direct fitness of the actor while increasing the direct fitness of a recipient, although sometimes they also refer to developmental decisions (e.g., the development of sterile castes among eusocial organisms that help others within the colony to reproduce).

Two questions about altruism have arisen. First, what criteria should be used to classify a behavior as altruism? Second, what selective mechanisms can cause altruism to evolve? The evolution of altruism is theoretically troublesome because selection would seem to favor free riders who take the benefits bestowed on them by altruists but who are not altruistic themselves and so do not incur any costs.

Classification. It is an inherent part of social interaction that one individual's action often induces a response by another. Some think it important to consider not only the fitness effects of the act on the actor but also how the responses of others affect the actor, before deciding to classify the act as altruism. However, this blurs the fact that an action and a response are distinct behavioral events. When one individual shares food with another, for example, the act is altruistic because the donor incurs some cost that benefits the recipient. If the recipient returns the favor at some later time, each discrete act is still altruistic, although the overall exchange appears to be mutualistic.

An act (considered separately from responses) is altruistic if it imposes costs on the actor while benefiting another. If the outcome of a decision on the actor cannot be determined until both parties have made their choices (e.g., as seen in the Prisoner's Dilemma game), an altruistic act must decrease the actor's fitness relative to what it would have been if the actor had acted otherwise, while holding the partner's choice constant.

By this principle, cooperating in the Prisoner's Dilemma is clearly altruistic. [*For details on the Prisoner's Dilemma, see* Cooperation.] Given that one player cooperates, the other player receives the payoff of $b - c$ by cooperating and b by defecting. Similarly, if the first player defects, the second player receives the payoff of $-c$ by cooperating and 0 by defecting. In both situations, player 2's option of cooperating is altruistic because it yields a lower payoff than the option of defecting.

Conversely, consider a situation in which two individuals must decide the level of effort they will put out in a cooperative hunting venture. Assume that B is putting out 50 percent effort and A is currently putting out 25 percent effort. If both A and B get a higher return when A increases his effort, then A's action is mutualistic. This is often likely to be the case when hunting game because the probability of success is synergistically higher when hunters coordinate their efforts. The synergistic effect also explains why some mutualisms are resistant to the free rider problem—free riding on the part of A reduces his overall rate of return, and vice versa.

Evolutionary Mechanisms. There are three theoretical classes of selective mechanisms that can cause altruism to evolve. The concept of a vehicle is intimately tied to them. Selection usually does not act on genes directly, but on the vehicles in which they are housed. A vehicle is a package in which genes are trapped so that they share a common fate when subjected to a selective event. For example, a human being, a bird, or any individual organism is a vehicle because if an allele enhances an organism's reproductive success, all the genes within that individual will share the same enhanced chance of being passed on to the next generation. Chromosomes, organelles, cells, groups, and other structures below and above the level of the individual can also be vehicles. Altruism is usually defined in reference to its effects on individuals.

Kin selection. An act increases the direct fitness of an actor if it enhances the propagation of the genes housed within the actor. An act increases the indirect fitness of an actor if it enhances the propagation of identical copies of genes housed within close relatives of the actor. The impact of an act on the actor's inclusive fitness refers to both the direct and the indirect fitness effects of the act on the actor.

With kin-selected altruism, an altruistic allele propagates itself by trading off the direct fitness costs it imposes on itself and the individual who bears it against the indirect fitness benefits on identical copies housed within related individuals. Thus, kin-selected altruism is costly to the direct fitness of the actor, beneficial to the relative (thereby increasing the actor's indirect fitness), and results in a net gain to the actor's inclusive fitness. Hamilton's rule specifies the condition favoring the tradeoff: an allele for altruism can evolve if the benefit to the recipient (b) multiplied by the average probability that the donor and recipient share the same altruistic allele by virtue of common descent (r) is greater than the cost to the donor (c) (i.e., $rb - c > 0$). The essence of the tradeoff is embodied in J. B. S. Haldane's famous quip that he would not sacrifice his life for that of his brother (who on average would share only half of his genes), but he would "for two brothers or eight cousins." [*See* Haldane, John Burdon Sanderson.]

Kin-selected altruism is responsible for the evolution of sterility and self-sacrificing behavior in eusocial insects. Parents can also manipulate their offspring into being altruistic. For example, male white-fronted bee-eaters, an avian species, often disrupt the breeding attempts of their male offspring. The enhanced difficulty of raising their own offspring makes it more profitable for male offspring to make the tradeoff embodied by Hamilton's rule and help raise their fathers' offspring instead. However, male offspring don't disrupt the breeding attempts of their fathers. This could be because the father (due to greater experience) can fledge more offspring than the son, thereby making it more costly for the son to disrupt the breeding attempts of the father than for the father to disrupt the breeding attempts of the son.

For kin-selected altruism to evolve, individuals must interact with kin, and they must be able to distinguish kin from nonkin. Because organisms cannot directly assess genetic similarity, they must use proximate cues of relatedness (e.g., physical proximity and chemical cues).

Compensatory mechanisms. Although including the effects of responses poses problems for the classi-fication of altruistic behavior, considering how others respond to an altruist can be crucial for understanding its evolution. Altruism can evolve if altruists receive compensatory responses from others.

Direct reciprocity. Direct reciprocity concerns mechanisms in which altruists receive compensation from their recipients. For example, contingent altruists continue to behave altruistically toward their partners provided their partners continue to behave altruistically toward them. When contingent altruists interact, they compensate each other through the reciprocal exchange of altruistic acts. Simultaneously, hermaphroditic fish produce both eggs and sperm. They do not fertilize their own eggs; rather, they swap eggs with other fish. There is often an incentive to cheat (i.e., to accept the eggs of another fish but without donating any) because it is cheaper to fertilize the eggs of another fish than to have one's eggs fertilized by that fish. Thus, the decision to donate eggs is often altruistic. The system of reciprocal altruism is maintained because fish remember individuals with whom they have swapped eggs in the past and donate fewer eggs to those who donated less on past encounters.

For contingent altruism to evolve, individuals must associate together for a long enough time that they can frequently exchange roles as potential altruist and recipient. Thus, contingent altruism is expected to be most prevalent among long-lived organisms that live in stable social groups. Moreover, contingent altruists must be able to detect cheaters (nonreciprocators) so that future aid may be withheld from them. The design features of cheating detection have been most extensively studied in humans.

Indirect reciprocity. Indirect reciprocity concerns mechanisms in which altruists receive compensation from third parties. In order for indirect reciprocity to evolve, third parties must have an incentive to compensate the altruist. One possibility is that the altruistic act enhances the actor's attractiveness as a social partner because it provides the third party with reliable information about the actor's phenotypic quality, skill, resources, mental state, or strategy. The altruist then receives compensation through enhanced social opportunities provided by third-party observers. Studies have shown that people are more likely to help someone in need when they know that observers are present or that observers are able to monitor their actions. This strongly suggests that acts of human altruism at least sometimes serve a signaling function.

Selection at other levels. Altruism can also evolve if there is a selective force at a higher level (such as a group) that overwhelms the costly effects at the level of the individual. Group-selected altruism is theoretically plausible, but the stringent evidentiary criteria that must be satisfied to demonstrate it suggest that it occurs relatively infrequently. Perhaps the most difficult requirement is to show that groups of individuals constitute vehicles. For example, individuals within human social groups are rarely, if ever, sufficiently trapped together so that they share a common fate when subjected to a selective event. More often, individuals within human social groups experience different selective events, or when they are exposed to the same selective event, they experience different outcomes.

Group-selected altruism nevertheless may have played an important role in some evolutionary phenomena. For example, mitochondria are thought to be the symbiotic descendants of bacteria that invaded host cells long ago. Because mitochondria have their own DNA and are often inherited from the maternal line, they have no interest in the sexual production of males. Selection acting on groups of mitochondria (trapped within the cell) may favor suppression of this interest if there are detrimental consequences to the cell (and all within it) by interfering with sexual reproduction. Moreover, selection acting on groups of pathogens (trapped within their hosts) may favor nonvirulence because of the detrimental effects of virulence on host (and group) survivorship. Nonvirulence may constitute altruism to the extent that less virulent pathogens use host resources for reproduction at a slower rate than more virulent pathogens.

[*See also* Kin Selection; Reciprocal Altruism; Social Evolution.]

BIBLIOGRAPHY

Alcock, J. *Animal Behavior: An Evolutionary Approach*, 5th ed. Sunderland, Mass., 1993. Discusses the free rider problem and sterility and self-sacrificing behavior in the eusocial insects.

Alexander, R. D. *The Biology of Moral Systems.* Hawthorne, N.Y., 1987. Discusses the concepts of direct and indirect reciprocity.

Andrews, P. W. "The Psychology of Social Chess and the Evolution of Attribution Mechanisms." *Evolution and Human Behavior* 22 (2001): 11–29. Discusses evidence that human altruistic acts function as signals and how they may provide reliable information about the altruist's social intentions.

Boyd, R. "Is the Repeated Prisoner's Dilemma a Good Model of Reciprocal Altruism?" *Ethology and Sociobiology* 9 (1987): 211–222. Discusses how cooperation in the prisoner's dilemma is altruistic.

Cosmides, L., and J. Tooby. "Cognitive Adaptations for Social Exchange." In *The Adapted Mind: Evolutionary Psychology and the Generation of Culture*, edited by J. H. Barkow, L. Cosmides, and J. Tooby, pp. 163–228. Oxford and New York, 1992. Reviews evidence suggesting that people have cognitive adaptations for detecting cheaters.

Dugatkin, L. A. *Cooperation among Animals.* Oxford and New York, 1998. Broad theoretical and empirical treatment of cooperation and altruism among animals.

Emlen, S. T., and P. H. Wrege. "Parent Offspring Conflict and the Recruitment of Helpers Among Bee-eaters." *Nature* 356 (1992): 331–333.

Keller, E. F., and E. A. Lloyd. *Keywords in Evolutionary Biology.* Cambridge, Mass., 1994. Contains articles on the concepts of altruism and group selection.

Keller, L. *Levels of Selection in Evolution.* Princeton, N.J., 2000. Contains articles on the concepts of vehicle and group selection.

Palmer, C. T., B. E. Frederickson, and C. F. Tilley. "Categories and Gatherings: Group Selection and the Mythology of Cultural Anthropology." *Evolution and Human Behavior* 18 (1997): 291–308. Important source for the concept of a vehicle.

— PAUL W. ANDREWS

AMPHIBIANS

Amphibians are named for their two-phased life consisting of larva and adult. Typically the larva is aquatic and metamorphoses into a terrestrial adult. In a descriptive sense, amphibians bridge the gap between fishes, which are fully aquatic, and amniotes (reptiles, birds, mammals), which have completely escaped a watery environment and have abandoned metamorphosis. However, amphibians are not in any sense trapped in an evolutionary cul-de-sac. In fact, they exhibit a far greater diversity of life modes than do amniotes.

Amphibians have generally not been regarded with much favor. An oft-cited paraphrasing from Linnaeus suggests that they are loathsome, slimy creatures and that the Creator saw fit not to make many of them. In fact, the number of living amphibians, about 4,800, exceeds that of our own lineage, mammals.

Each type of living amphibian—frog, salamander, and caecilian—is distinctive. Frogs are all squat, four-legged creatures with generally large mouths and eyes and elongate hindlimbs used for saltation. There is no tail (the meaning of Anura), because the caudal vertebrae have coalesced into a bony strut. About 90 percent of the living amphibian species are frogs; they rely mostly on visual and auditory cues. Salamanders are more typical-looking tetrapods, all with a tail (hence, Caudata) and most with four legs. Some are elongate and have reduced limbs and girdles; these are usually completely aquatic or fossorial species. In general, they rely more on olfactory cues. Living caecilians are all limbless and elongate. Grooved rings (annuli) encircle the body, evoking the image of an earthworm. All caecilians have reduced eyes (*caecus* is Latin for "blind"), although they are not blind. Just below the eye is a unique protrusible tentacle used for olfaction. The tail is essentially absent.

A possible fourth group of amphibians is the Albanerpetontidae, known only as fossils from the Jurassic to the Miocene. This group closely resembles salamanders in skull shape and in the primitive tetrapod features of a generalized body shape, four limbs, and a tail. Their relationship to other amphibians is not clear.

Features of Modern Amphibians. It is useful to distinguish the monophyletic "modern" amphibians (246 million years ago–present) from their Paleozoic relatives that gave rise to all modern tetrapods including amniotes. Modern amphibians include frogs, salamanders, and caecilians, and their Mesozoic (245–65 million years ago) and Cenozoic (65 million years ago–present) extinct relatives (including albanerpetontids), all of which are readily identifiable as belonging to this group.

In contrast, their Paleozoic relatives include the groups traditionally termed the Labyrinthodontia and Lepospondyli. Labyrinthodonts, including the earliest four-legged vertebrates, ranged from the Upper Devonian (375 million years ago) through the Permian (end of the Paleozoic, 290 million years ago), with numbers declining into the Triassic and one small lineage persisting into the Cretaceous. Lepospondyls range from the Lower Carboniferous (240 million years ago) to the base of the Upper Permian (250 million years ago). Labyrinthodonts are a paraphyletic group and also gave rise to amniotes. Lepospondyls are characterized by their vertebral morphology; the monophyly of the group is unclear.

Several features set living amphibians apart from other vertebrates. Living amphibians have teeth that are pedicellate and bicuspid. Pedicellate teeth have a zone of reduced mineralization between the crown and the base (pedicel). In fossils the crowns are often broken off here, leaving a cylindrical base with an open top. Pedicellate teeth are also found in a few temnospondyl labyrinthodonts believed to be closely related to modern amphibians.

Living amphibians use a buccal force-pump mechanism for breathing. Air is forced back into the lungs by positive pressure from the mouth cavity. Amphibians have distinctive short ribs that do not form a complete rib cage as in amniotes. In contrast, amniotes use aspiration to fill the lungs, in which the rib cage and/or diaphragm creates negative pressure in the thorax.

In addition to the stapes-basilar papilla sensory system of tetrapods, modern living amphibians have a second acoustic pathway, the opercular-amphibian papilla system. This system is sensitive to lower frequency vibrations than is the stapes-basilar papilla pathway. The operculum is also connected to the shoulder girdle by way of a modified levator scapulae muscle, the opercularis. This muscle transmits vibrations from the ground through the forelimb and shoulder girdle and opercularis muscle to the inner ear.

The skin is a significant respiratory organ; it is supplied by cutaneous branches of the ductus arteriosus (the presence of these has not been determined in caecilians). The skin has a stratum corneum like other tetrapods, although it is thinner than that of amniotes. However, living amphibians retain the primitive feature of granular and mucous glands. Granular glands secrete poisons of varying toxicity. Mucous glands keep the skin moist; the mucous allows the dissipation of heat, as well as the loss of water through the skin. Distinctive fat bodies associated with gonads and both the fat bodies and gonads develop from the germinal epithelium, a pattern unique among tetrapods.

Living amphibians are remarkable because many of the major features they share that distinguish them from other living tetrapods are the result of paedomorphosis. Paedomorphosis results from a change in the timing of development; specifically, a species becomes sexually mature (adult) at an earlier stage of development than its immediate ancestor. As a result, the adult of amphibians resembles the juvenile (or larval) stage of Paleozoic relatives. A secondary result of paedomorphosis is miniaturization; because living amphibians mature at an earlier age, they are typically much smaller than the Paleozoic forms.

Specifically, living amphibians share the paedomorphic absence or reduction of several skull bones. On the dorsal skull, the jugals, postorbitals, postparietals, supratemporals, intertemporals, and tabulars are absent. On the palate, the pterygoid, ectopterygoid, and palatines are reduced or absent so as to produce a large space, the interpterygoid vacuity, below the eye sockets. Albanerpetontids lack most of the same dorsal skull bones as living amphibians. However, they retain a jugal and the teeth are neither pedicellate nor bicuspid.

Relationships to Paleozoic Taxa. The exact relationship of modern amphibians to extinct Paleozoic forms is not clear. The favored family of hypotheses (Figure 1) posits that frogs, salamanders, and caecilians are monophyletic, and that this clade is nested within the dissorophoid temnospondyls (temnospondyls are a group of labyrinthodonts including Edopoidea, Trime-

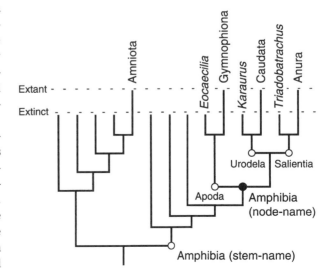

FIGURE 1. Relationships of the Major Groups of Amphibia. The open circle indicates stem-name and the closed circle a node-name.   After Cannatella and Hillis, 1993.

rorhachoidea, Eryopoidea, Stereospondyli, and Dissorophoidea). A more recent variant of the monophyly hypothesis is that modern amphibians are nested within the lepospondyls, particularly within the Microsauria. Because temnospondyls are distantly related to amphibians under this second hypothesis, the derived similarities between them and dissorophoid temnospondyls are convergent.

A very different hypothesis claims polyphyly of living amphibians, with caecilians derived from goniorhynchid microsaurs, and salamanders and frogs from temnospondyls. The polyphyly hypothesis gained some strength with the discovery of the fossil *Eocaecilia* (see below), which possessed a suite of characters intermediate between goniorhynchid microsaurs and living caecilians.

The Name Amphibia. In the *Systema Naturae* of Carolus Linnaeus, Amphibia was one of six major groups of animals (the others being mammals, birds, fish, insects, and mollusca) and it included frogs, salamanders, and caecilians. However, his Amphibia also included reptiles and some fish lacking typical scales. Later, as early fossil tetrapods were uncovered, these were also relegated to "Amphibia" because of their presumed ancestral position to other tetrapods. The great German biologist Ernst Haeckel in 1866 divided Amphibia (also known as Batrachia) into Lissamphibia for salamanders and frogs, and Phractamphibia for caecilians and fossil labyrinthodonts. "Liss–" refers to the naked skin of frogs and salamanders, and "phract–" to the skin armor of dermal scales present in early tetrapods and (in a reduced form) in caecilians. In 1901 Hans Gadow moved the caecilians from the Phractamphibia to the Lissamphibia.

For most of the twentieth century, the name Amphibia was used for tetrapods that were not reptiles, or their presumed derivatives, birds and mammals. Thus, the earliest tetrapods (labyrinthodonts from the Devonian) were included in Amphibia as were the Lepospondyli. This concept of Amphibia appeared in almost all comparative anatomy and paleontology texts, largely due to the influence of the great paleontologist Alfred S. Romer. The modern amphibians were believed to be polyphyletic, derived from different "amphibian" lineages: frogs from Labyrinthodontia, and salamanders and caecilians from Lepospondyli. In the 1960s Thomas Parsons and Ernest Williams adduced evidence for the monophyly of modern amphibians and resurrected Gadow's usage of Lissamphibia. However, Lissamphibia is used mostly among specialists to distinguish the modern and extinct forms. For most biologists, frogs, salamanders, and caecilians are simply amphibians.

Among contemporary biologists the name Amphibia refers to a monophyletic group according to either a phylogenetic node-based or stem-based definition. In the node-based usage, the name Amphibia includes the last ancestor node common to the living forms frogs, salamanders, and caecilians. Therefore, Amphibia includes this ancestral node and all its descendants, which are the modern forms. In the stem-based usage, the name Amphibia is defined as the branch that not only contains the living frogs, caecilians, and salamanders, and all other taxa (extinct in this case) more closely related to these than to amniotes (Figure 1). In other words, the stem-based name Amphibia includes all taxa along the stem leading to modern amphibians; this includes either the nonlissamphibian temnospondyls, the lepospondyls, or both groups, depending on which hypothesis one accepts.

Node-based and stem-based names have their proper advantages. In this case, however, the use of a stem-based definition for a name in general parlance has an undesirable effect, because generalizations about the biology of modern amphibians may be wrongly extended to temnospondyls and/or lepospondyls. Most of these groups bear little resemblance to the living forms, and their biology was presumably very different. For example, under a stem-based definition of Amphibia, the statement "all amphibians have mucous glands" would be interpreted to mean that lepospondyls had mucous glands, an inference for which there is no evidence. In contrast, under the node-based definition of Amphibia, one can infer that extinct frogs, salamanders, and caecilians had mucous glands, while making clear that this inference is not overextended to include extinct nonamphibian temnospondyls and lepospondyls.

Although paleontologists generally appreciate the semantic distinction between Amphibia and Lissamphibia, the vast majority of biologists use Amphibia to mean frogs, salamanders, and caecilians; this node-based usage is advocated here. Other names, such as Batrachomorpha, are available for the stem leading to modern amphibians.

Interrelationships of Amphibia. Two alternative phylogenetic trees have been considered for the Amphibia. One, based primarily on nonmolecular data, allies frogs and salamanders, with caecilians as the odd one out. In the other, analyses of DNA sequence data have slightly favored salamanders and caecilians as closest relatives. However, this is an area of active research and these findings should be regarded as preliminary; the first of these trees is shown in Figure 1. The extinct albanerpetontids (not shown for simplicity) have been considered to be either nested within salamanders or the sister group of Batrachia (and thus within Amphibia), or possibly the sister group of Amphibia.

Caecilian Diversity. The node-based name for modern caecilians is Gymnophiona, meaning a "naked snake." Caecilians include 160 extant species, restricted to the tropical regions of America, Africa (excluding Madagascar), and Asia. They are grouped into several families

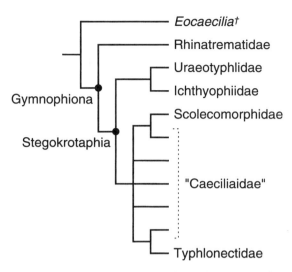

FIGURE 2. Relationships Among the Major Groups of Gymnophiona.
After Wilkinson and Nussbaum, 1996; and Wilkinson, 1997.

(Figure 2 and Table 1). The "Caeciliaidae" are paraphyletic with respect to Scolecomorphidae and Typhlonectidae.

Most caecilians are 0.3 to 0.5 meter long, although one species is as large as 1.5 meters. Although all caecilians are elongate, some are more elongate than others; the number of vertebrae ranges from 86 to 205. Caecilians are almost unique among amphibians in having an intromittent (copulatory) organ, the phallodeum, and internal fertilization occurs during copulation (the two species of the frog *Ascaphus* also have an intromittent organ).

Living caecilians have reduced eyes with small orbits, and scolecomorphids have eyes that are covered by the skull bones. Compared to other amphibians, the skulls of caecilians are highly ossified and many bones are fused. The resulting bullet-shaped crania are used for digging and compacting the soil. Most caecilians are oviparous, with free-living larvae. Viviparous (live birth) species occur in a few families. In some viviparous species the embryos derive nutrition from the lining of the oviduct. Most caecilians are fossorial, but the Typhlonectidae are aquatic and have laterally compressed bodies for swimming. Because of their habits, caecilians are rarely seen in nature. A dedicated searcher might find them by digging, and occasionally individuals are found on the surface after a heavy rain in tropical regions.

Although living caecilians are limbless, the earliest fossil caecilian had legs! *Eocaecilia micropodia* is a form from the Jurassic (144–213 million years ago); the body is elongate and has small but well-developed limbs. Like living caecilians, it has pedicellate teeth, but most importantly it has a groove in the edge of the eye socket that appears indicate the presence of the tentacle that is shared with all living caecilians. This suggests that *Eocaecilia* is the sister group of all other caecilians. The stem-based name for the clade of *Eocaecilia* + Gymnophiona is Apoda. Fossil caecilians are also known from the Upper Cretaceous and Tertiary, but only from vertebrae.

Salamander Diversity. The node-based name for salamanders is Caudata. There are about 415 species of living salamanders, arranged into 10 families (Figure 3 and Table 1). Historically, salamanders are a primarily Holarctic group of the north temperate regions; one clade has diversified in the neotropics. The largest salamanders are those of the Cryptobranchidae; adult *Andrias* may reach 1.5 meters in total length. The smallest are in *Thorius* (Plethodontidae), which may have an adult total length as small as 30 millimeters.

Of the 415 species of salamanders, about 270 are in the Plethodontidae. This clade is also most diverse morphologically and is the only lineage with a tropical radiation, extending through Central America into South America. Remarkably, this entire group lacks lungs and uses primarily cutaneous respiration. The release of the hyoid musculoskeleton from the constraints of buccal force-pump breathing has apparently permitted the evolution of a diverse array of mechanisms of tongue protrusion for prey capture.

Salamandrids are also diverse in morphology and life history, although not as speciose as plethodontids. Whereas most plethodontids are cryptic, many species of salamandrids are brightly colored with highly effective poison glands to deter predators. At least two species are viviparous.

Several salamanders are elongate and have reduced limbs. Some are larger aquatic forms such as Sirenidae, Proteidae, and Amphiumidae. Fully aquatic salamanders typically retain gill slits, and in addition some have external gills resembling crimson tufts of feathers. Elongate terrestrial salamanders typically have reduced limbs and digits and occupy a semifossorial niche in leaf litter or burrows. Sirens (Sirenidae) have completely lost the hindlimbs and girdles.

Most of the major groups of salamanders have internal fertilization, which is accomplished by way of a spermatophore. The male deposits a spermatophore either in water or on land, depending on the group. The female retrieves the spermatophore with her cloaca. The sperm may be retained live for months or even years in a cloacal pocket called the spermatheca. Fertilized eggs are deposited and develop either directly, in which case a small salamander hatches, or indirectly, in which a larval salamander emerges, and later metamorphoses.

Karaurus sharovi, the oldest salamander, is a fully articulated Middle Jurassic fossil from Kazakhstan. The stem-based name for the clade of *Karaurus* + Caudata is Urodela ("with a tail").

TABLE 1. Geographical Distribution of the Major Extant Groups of Amphibia

Taxon	Distribution
GYMNOPHIONA	
Rhinatrematidae	Northern South America
Ichthyophiidae	India, Sri Lanka, and Southeast Asia
Uraeotyphlidae	South India
Scolecomorphidae	Africa
"Caeciliaidae"	Mexico, Central and South America; Africa, Seychelles, India, and Southeast Asia
Typhlonectidae	South America
CAUDATA	
Hynobiidae	Continental Asia to Japan
Sirenidae	Eastern United States and adjacent Mexico
Cryptobranchidae	China, Japan, and eastern United States
Ambystomatidae	North America
Rhyacotritonidae	Northwest United States
Dicamptodontidae	Western United States and adjacent Canada
Salamandridae	Eastern and western North America, Europe and adjacent western Asia, northwest Africa, eastern Asia
Proteidae	Eastern United States and Canada, and Adriatic coast of Europe
Amphiumidae	Southeast United States
Plethodontidae	North and Central America, and northern South America; Italy and adjacent France, Sardinia
ANURA	
Ascaphus	Northwest United States and adjacent Canada
Leiopelma	New Zealand
Bombinatoridae	Europe and eastern continental Asia, Borneo and nearby Philippine islands
Discoglossidae	Europe and north Africa
Pipidae	South America and adjacent Panama, sub-Saharan Africa
Rhinophrynidae	Central America, Mexico, and south Texas
Pelobatidae	North America, Europe, and western Asia
Pelodytidae	Western Europe and western Asia
Megophryidae	South Asia to Southeast Asia
Heleophryne	Southern Africa
Myobatrachinae	Australia and New Guinea
Limnodynastinae	Australia and New Guinea
"Leptodactylidae"	South America, Central America, Mexico, and southern United States
Bufonidae	All continents, including Southeast Asia, except Australia and Antarctica
Centrolenidae	Mexico, Central America, and South America
Dendrobatidae	Northern South America, southeast Brazil, and Central America
Sooglossidae	Seychelles
Hylidae	The Americas, Europe and adjacent Asia, north Africa, east continental Asia, Japan, New Guinea, and Australia
Pseudidae	South America
Rhinoderma	Southern South America
Allophryne	Northern South America
Brachycephalidae	Atlantic forests of southeast Brazil
Microhylidae	Southern United States, Mexico, Central America, South America, sub-Saharan Africa, Madagascar, south continental Asia, Southeast Asia, New Guinea, and northeast Australia
"Ranidae"	All continents, including only northern South America and northeast Australia
Arthroleptidae	Sub-Saharan Africa
Hyperoliidae	Sub-Saharan Africa, Madagascar, and the Seychelles
Hemisus	Sub-Saharan Africa
Mantellinae	Madagascar
Rhacophoridae	Sub-Saharan Africa, Madagascar, south continental Asia, Southeast Asia, and Japan

Frog Diversity. *Triadobatrachus massinoti* is known from a single fossil from the Lower Triassic of Madagascar. It has been called a proanuran to indicate its position to frogs, and retains many ancestral features, such as 14 presacral vertebrae (living frogs have nine or fewer) and lack of fusion of the radius and ulna and also of the tibia and fibula (living frogs have fused elements, the radioulna and tibiofibula). The stem-

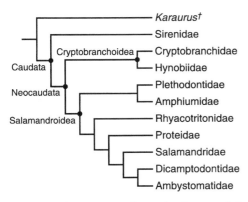

FIGURE 3. Relationships Among the Major Groups of Caudata. After Larson and Dimmick, 1993.

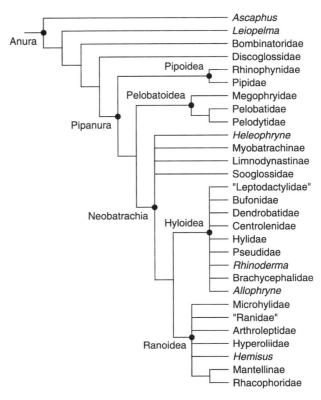

FIGURE 4. Relationships Among the Major Groups of Anura. After Ford and Cannatella (1993) and modified based on work in progress by the author.

based name for the clade of *Triadobatrachus* + Anura is Salientia.

The earliest forms considered as complete frogs are from the Middle Jurassic of Argentina; these are *Notobatrachus* and *Vieraella*. *Prosalirus* from the Lower Jurassic of Arizona and *Czatkobatrachus* from the Early Triassic of Poland are clearly frogs but more fragmentary. All of these, however, have skeletal features that indicate that the distinctive saltatory locomotion of frogs had evolved by this time.

Living frogs include about 4,200 species arranged in 25–30 families, depending on the authority (Figure 4 and Table 1). The smallest species is only 10mm in adult body length. A group of plesiomorphic lineages includes *Ascaphus*, *Leiopelma*, Bombinatoridae, and Discoglossidae; these have been called discoglossoids and are paraphyletic with respect to other frogs, the Pipanura. *Ascaphus* and *Leiopelma* are viewed as primitive, narrowly distributed relicts (*Ascaphus* in the Pacific Northwest of the United States and *Leiopelma* on only a few islands in New Zealand) of a once more widely distributed Mesozoic frog fauna.

The Pipanura comprise the Pipoidea, Pelobatoidea, and Neobatrachia. The Pipoidea and Pelobatoidea are regarded as "intermediate" lineages. Pipoidea includes the Rhinophrynidae, Pipidae, and the extinct Palaeobatrachidae (not shown in Figure 4). The Pelobatoidea include Megophryidae, Pelobatidae, and Pelodytidae. Pelobatoids and pipoids have a relatively large number of fossil forms in the Cretaceous and Tertiary.

The Neobatrachia consist of the "advanced" frogs and include 95 percent of living species. Except for the Plio-Pleistocene they are not well represented in the fossil record. There are two large clades within Neobatrachia, the Hyloidea (probably monophyletic) and Ranoidea. The Hyloidea is primarily a New World clade and the Ranoidea an Old World group, although the hyloids have

significant radiations in the Australopapuan region as do the "Ranidae" in the New World. Hyloids include Hylidae, "Leptodactylidae," Bufonidae, Centrolenidae, Pseudidae, Dendrobatidae, Brachycephalidae, Myobatrachinae, and Limnodynastinae. These last two are usually grouped in the "Myobatrachidae," although there is evidence that the Sooglossidae is most closely related to Myobatrachinae. Some odd taxa have been placed in distinct families because they are not clearly related to a particular group: *Allophryne* and *Rhinoderma*. *Heleophryne* is often placed in its own family, Heleophrynidae, and may be the most basal of the Neobatrachia. Ranoidea includes Microhylidae, "Ranidae," Arthroleptidae, *Hemisus*, Hyperoliidae Mantellinae, and Rhacophoridae.

Frogs have a dazzling array of evolutionary novelties associated with reproduction. Their diverse vocal signals of the males are used for mate advertisement and territorial displays. Parental care is highly developed in many lineages, including brooding of developing larvae on a bare back, in pouches on the back of females, in the vocal sacs of males, and in the stomach of females. Some females in some unrelated lineages of hylids and dendrobatids raise their tadpoles in the watery confines of a bromeliad axil and supply their own unfertilized eggs as food. Whereas amniotes escaped from the watery environment only once in their evolution of the amniote egg, frogs have done so many times; direct development with associated terrestrial eggs, in which the tadpole stage is bypassed in favor of straightforward development to a froglet, has evolved at least twenty times.

Although some frogs have escaped an aquatic existence, most have embraced it, taking the biphasic life to an extreme. Frog tadpoles are highly specialized to exploit their transitory and often unpredictable larval niche. They consist mostly of a feeding apparatus in the head and locomotor mechanism in the tail. The feeding apparatus is a highly efficient pump that filters minuscule organic food particles from the water. Tadpoles do not reproduce; there are no neotenic forms. They live their lives eating until it is time to make a quick and awkward metamorphosis to a froglet.

[*See also* Vertebrates.]

BIBLIOGRAPHY

Cannatella, D. C., and D. M. Hillis. "Amphibian Phylogeny: Phylogenetic Analysis of Morphology and Molecules." *Herpetological Monographs* 7 (1993): 1–7.

Carroll, R. L. "The Lissamphibian Enigma." In *Amphibian Biology. Volume 4*, edited by H. Heatwole and R. L. Carroll, pp. 1270–1273. Chipping Norton: Surrey Beatty and Sons, 2000. An overview of the problem of monophyly of Amphibia.

de Queiroz, K., and J. Gauthier. "Phylogenetic Taxonomy." *Annual Review of Ecology and Systematics* 23 (1992): 449–480. A general treatise of the principles of phylogenetic taxonomy, including node names and stem names.

Duellman, W. E. "Amphibian Species of the World: Additions and Corrections." *University of Kansas Museum, Natural History Special Publications* 21 (1993): 1–372. An update of Frost (1985).

Duellman, W. E., ed. *Patterns of Distribution of Amphibians: A Global Perspective*. Baltimore, 1999.

Duellman, W. E., and L. Trueb. *Biology of Amphibians*. New York, 1986. The most in-depth treatment of amphibian biology.

Ford, L. S., and D. C. Cannatella. "The Major Clades of Frogs." *Herpetological Monographs* 7 (1993): 94–117. A review of the evidence supporting the clades of frogs.

Frost, D. R., ed. *Amphibian Species of the World: A Taxonomic and Geographic Reference*. Lawrence, Kans., 1985. The standard for amphibian taxonomy.

Glaw, F., and J. Köhler. "Amphibian Species Diversity Exceeds that of Mammals." *Herpetological Review* 29 (1998): 11–12.

Heatwole, H., and R. L. Carroll, eds. *Amphibian Biology, Paleontology* vol. 4, *The Evolutionary History of Amphibians*. Chipping Norton: Surrey Beatty and Sons, 2000. An excellent series of review papers on fossil amphibians, labyrinthodonts, and lepospondyls.

Larson, A., and W. W. Dimmick. "Phylogenetic Relationships of the Salamander Families: An Analysis of Congruence Among Morphological and Molecular Characters." *Herpetological Monographs* 7 (1993): 77–93.

Laurin, M., and R. R. Reisz. "A New Perspective on Tetrapod Phylogeny." In *Amniote Origins. Completing the Transition to Land*, edited by S. S. Sumida and K. L. M. Martin, pp. 9–59. San Diego, 1997. A new hypothesis for the relationships of Amphibia.

McDiarmid, R. W., and R. Altig, eds. *Tadpoles: The Biology of Anuran Larvae*. Chicago, 1999. The most comprehensive treatment of frog larvae.

McGowan, G., and S. E. Evans. "Albanerpetontid Amphibians from the Cretaceous of Spain." *Nature* 373 (1995): 143–145.

Parsons, T. S., and E. E. Williams. "The Relationships of the Modern Amphibia: A Re-Examination." *Quarterly Review of Biology* 38 (1963): 26–53. A classic paper.

Pough, F. H., R. M. Andrews, J. E. Cadle, M. L. Crump, A. A. Savitzky, and K. D. Wells. *Herpetology*. Upper Saddle River, N.J., 2001. An excellent textbook, including a chapter on amphibian relationships.

Sanchíz, B. *Salientia. Encyclopedia of Paleoherpetology. Volume 4*, P. Wellnhofer (ed.). Munich, 1998. The most complete treatment of frog fossils.

Trueb, L., and R. Cloutier. "A Phylogenetic Investigation of the Inter- and Intrarelationships of the Lissamphibia (Amphibia: *Temnospondyli*)." In *Origins of the Higher Groups of Tetrapods: Controversy and Consensus*, edited by H.-P. Schultze and L. Trueb, pp. 233–313. Ithaca, 1991.

Wilkinson, M. "Characters, Congruence, and Quality: A Study of Neuroanatomical and Traditional Data in Caecilian Phylogeny." *Biological Reviews* 72 (1997): 423–470.

Wilkinson, M., and R. A. Nussbaum. "On the Phylogenetic Position of the Uraeotyphlidae (Amphibia: *Gymnophiona*)." *Copeia* 1996 (1996): 550–562.

Zug, G. R., L. J. Vitt, and J. P. Caldwell. *Herpetology: An Introductory Biology of Amphibians and Reptiles*. San Diego, 2001. Another excellent textbook, with consideration of alternative hypotheses of relationships.

ONLINE RESOURCES

"Amphibian Species of the World." http://research.amnh.org/herpetology/amphibia/index.html. An up-to-date online version of Frost (1985). Maintained and updated by Darrel Frost and the Division of Herpetology of the American Museum of Natural History.

"Amphibia Web." http://elib.cs.berkeley.edu/aw/index.html. A site inspired by global amphibian declines and spearheaded by David Wake. Access to information on amphibian biology and conservation, organized by species account.

"The Tree of Life: Salientia." http://phylogeny.arizona.edu/tree/eukaryotes/animals/chordata/salientia/salientia.html. A tree-based collection of Web pages to frogs, organized by higher taxa.

— DAVID CANNATELLA

ANCIENT DNA

The first recovery of DNA from preserved biological remains dates to 1984, when Russ Higuchi and his colleagues isolated mitochondrial DNA from a 140-year-old skin of a quagga (*Equus quagga*), an extinct species from southern Africa. These researchers were able to demonstrate by sequence analysis that the quagga had been more closely related to the zebra than to horses. This finding was quickly followed by Svante Pääbo's publication, in 1985, of a nuclear DNA sequence obtained from a 2,400-year-old mummy of an Egyptian boy.

The cloning methods used in both projects were complex and slow, and they were made significantly more difficult by the damaged nature of the ancient DNA. This type of research therefore seemed destined to become more of a biological curiosity than a significant field of study. However, the rapid development of polymerase chain reaction (PCR) technologies during the latter half of the 1980s led inevitably to its application in ancient DNA. The results were striking: during 1989, researchers recovered and characterized DNA from a 7,000-year-old human brain preserved in a peat bog and from museum skins of the extinct marsupial "wolf," *Thylacinus cynocephalus*. The subsequent discovery of DNA within preserved bones, teeth, and plant remains increased the potential for the application of this technology. This initial phase of research also gave rise to several claims of DNA recovery from extremely old specimens, such as chloroplast DNA from a 17- to 20-million-year-old magnolia leaf and a 38-million-year-old spider preserved in amber. Though interesting, these results severely contradict estimates of DNA survival based on theoretical rates of chemical degradation; more important, they have been impossible to replicate when more stringent conditions were employed during DNA recovery. For these reasons, they are not considered further in this article.

Ancient DNA has now become a tool used by workers in a number of fields, including molecular systematics, conservation genetics, archaeology, forensic science, and palaeontology. Even though a tool does not constitute a field of study in its own right, there are a number of features that distinguish ancient DNA work. Paradoxically, age is not one of these; although there is a generally accepted upper limit of around 120,000 years for ancient DNA recovery, characteristically "ancient" DNA has been recovered from samples as young as four years old. Nor, for the most part, is the type of material a significant issue; although bones, teeth, and dried tissues (generally from museum-held specimens) are still the bulk of ancient DNA sources, it has also been recovered from materials such as archaeological seeds, museum feathers, paleofeces, and bloodstains on historical metal tools and fabrics. The three factors that seem to be common to most ancient DNA projects are more technical in nature: the small quantity of DNA and large degree of damage to it; the presence of substances that inhibit the action of the polymerase enzyme used in PCR; and concern about the presence of exogenous DNA (contamination) that could be preferentially recovered.

Damage and Loss of DNA from Buried and Archival Materials. Ancient DNA is usually recovered in extremely small amounts. Exactly how much has been difficult to quantify because the majority of DNA extracted from ancient material is likely to be of modern microbial origin, but quantitative PCR methods suggest a figure on the order of thousands of copies of mitochondrial DNA sequences 100 base pairs in length per gram of bone. In order to improve the chance of recovering DNA from ancient material, an obvious step is to target DNA sequences that are present in high copy numbers. The mitochondrial genome is present in the order of tens to thousands of copies per cell in mammals, and it is primarily for this reason that it has been the molecule of choice for ancient DNA studies. However, the repeated recovery of nuclear DNA sequences suggest that mitochondrial DNA may not have this monopoly much longer.

The problem of the low amounts of DNA in ancient material stems from the post-mortem damage incurred by the molecule. Chief among the damaging processes are chain scission events that lead to the fragmentation of the DNA into small oligomers or even individual bases. This type of damage is probably initiated immediately after death by the host organism's own nucleases, during the period known as autolysis. The enzymatic decay generally continues until microbial utilization of the available resources is complete. Further modification of the DNA is by chemical means. Hydrolysis reactions lead to both deamination of bases and cleavage of the bond between base and phosphate-sugar backbone. This loss of bases leads to a weakening of the chain, which then can be broken easily. Oxidation, induced by the effects of ionizing radiation, leads to base modifications and to the formation of inter- and intramolecular cross-links, both of which lead to further strand breakage.

Estimates of the extent of fragmentation vary, but the majority of DNA strands in buried bone appear to be around 100–200 base pairs in length, and under 100 base pairs in plant seeds. Museum samples and bones from sites with exceptional preservation conditions (such as permafrost or recent cave sites) may contain molecules over 1,000 base pairs in length. It should be noted that, even if strand length is long, the presence of chemically modified or missing bases may lead to the failure of PCR amplification, or the affected sites may be incorrectly copied during strand extension, leading to errors in the sequence subsequently obtained.

Inhibition. The presence of a substance or substances that inhibit the action of the thermostable polymerase used in the PCR has been noted from the outset of ancient DNA amplification. The inhibitory effect is found in extracts from a variety of sources, and it seems likely that multiple compounds are responsible. Suggestions as to the source of the effect include metal ions and tannins, haem and haematin, and the Maillard products of reducing sugars. It seems likely that one of the major contributors to inhibition of amplification from buried bone samples is the presence of humic and fulvic acids—soluble, heterogenous, highly phenolic breakdown products of soils that are known to be potent in-

NEANDERTHAL DNA: ANCESTORS OR JUST RELATIVES?

Central to debates about the origins of modern humans is the question of the relationship between *Homo sapiens* and *Homo neandertalis*. Those who subscribe to the theory that modern humans evolved from archaic forms throughout the Old World over the last 1 to 2 million years (the multiregional model) believe that Neanderthals made a significant genetic contribution to modern humans, particularly those in Europe and Western Asia. Conversely, those who believed that modern humans evolved from archaic forms exclusively in Africa before emerging from Africa between 60 and 150 thousand years ago (the recent African origin theory) believe that Neanderthals contributed little or no genetic material to modern humans.

In 1997 Matthias Krings and colleagues obtained mitochondrial (mt) DNA control region sequence from the humerus of the Neanderthal-type specimen, found in 1856 near Düsseldorf, Germany. They used a range of methods to support the authenticity of their results, including being able to show that the mtDNA sequence they obtained was not present in modern humans (thus excluding the possibility of contamination by nuclear-mitochondrial sequences [numts]). The Neanderthal mtDNA sequence was found to lie outside the range of variation found in modern human mtDNA sequences. Furthermore, estimates of mtDNA common ancestry dates based on phylogenetic analysis were found to be four times greater for the human/Neanderthal split than for the deepest human/human split. Since this study was published a number of other groups have found similar results. Thus, ancient DNA from Neanderthals has provided strong support for the recent African origin theory.

—MARK THOMAS

hibitors of PCR. Various approaches have been employed to overcome inhibition of PCR, including the use of high concentrations of polymerase, addition of enzyme-stabilizing, inhibitor-binding, or DNA liberating agents in the reaction, and highly stringent DNA extraction methods.

Contamination. PCR is an extremely sensitive technique. Ancient DNA is present in small quantities and can be chemically modified; therefore, PCR preferentially amplifies any modern DNA in the extract to which the PCR primers can anneal. When the target DNA is of nonhuman origin, the most likely source for this modern DNA is PCR products of previous amplifications. In the case of broadly targeted primers, or where microbial DNA is the intended target, soil microbes may be amplified. The situation in which human DNA is the intended target is considerably more complex because contamination of human origin may also arise during excavation or post-excavation handling. Further problems arise through the presence of "look-alike" damaged modern contaminant DNA—chimeric sequences generated through the linking of two distinct strands by the polymerase enzyme during PCR, and the presence of nuclear encoded copies of mitochondrial-like sequences (abbreviated as numts). Such sequences occur at high copy number in the nuclear genome of some taxa. This situation is the most likely explanation for the source of proposed dinosaur sequences obtained from an 80-million-year-old partially fossilized bone sample recovered from a coal mine in eastern Utah. These sequences have subsequently been demonstrated to show a high degree of homology to human numts, suggesting that modern human DNA is their most likely source.

Applications Relevant to Evolution. The complexities of working with ancient DNA have limited the wide-scale application of the technology until very recently, and a large number of scholarly publications have focused on methodological and technological issues, reviews, suggestions for potential work, and small-scale applications that serve as a "proof of concept." In terms of evolutionarily relevant work conducted using ancient DNA, scientific publications can be divided into two groups: those dealing with the taxonomy of extinct taxa, and those comparing the genetic profiles of past and present populations to reveal aspects of population history.

Systematics and taxonomy. Phylogenetic placements have been made by using ancient DNA technologies for a number of extinct taxa. As well as the thylacine and quagga already noted, some mitochondrial sequence data now exist for the ground sloth *Mylodon darwinii*, the cave bear *Ursus spelaea*, the mammoth *Mammuthus primigenius*, Neanderthals *Homo Neanderthalis*, and a number of other taxa. Perhaps the most extensive study has been made on moas, an order of New Zealand ratite birds that became extinct during the past millennium. The entire mitochondrial genome sequences have been recovered for representatives of two species and compared with equivalent data from other ratite birds. Because this group is flightless, and representative species are present in all the southern continents, molecular clock estimates of their divergence times can be used to help date the breakup of Gondwanaland and other tectonic movements.

Population genetics. The potential to sample past populations and compare them with modern populations is extremely attractive for evolutionary biologists because it would allow the direct testing of population genetics theories and better understanding of the impact of bottlenecks and population expansions. However, the technical difficulties of working with ancient DNA have made it difficult to obtain sufficiently large sample sizes with which to work. In addition, most sets of samples tend to derive from a range of dates—from tens of years in the case of museum material up to thousands of years for paleontological samples. This renders meaningless the notion of a single contemporary source population. Where sufficient suitable samples are obtainable, analyses of historically recorded bottleneck events have been possible; an analysis of the northern elephant seal (*Mirounga angustirostris*) provides one example. A more qualitative approach has been taken when analyzing a small number of brown bear (*Ursus arctos*) specimens from Alaska and the Yukon, where a change in the dominant mitochondrial haplotype was observed to occur subsequent to the last glacial maximum.

[*See also* Archaeological Inference; Hominid Evolution, *article on* Neanderthals; Modern *Homo sapiens*, *article on* Human Genealogical History.]

BIBLIOGRAPHY

Cooper, A., C. Lalueza-Fox, S. Anderson, A. Rambaut, J. Austin, and R. Ward. "Complete Mitochondrial Genome Sequences of Two Extinct Moas Clarify Ratite Evolution." *Nature* 409 (2001): 704–707.

Higuchi, R., B. Bowman, M. Freiberger, O. A. Ryder, and A. C. Wilson. "DNA-Sequences from the Quagga, an Extinct member of the Horse Family." *Nature* 312 (1984): 282–284.

Hofreiter, M., D. Serre, H. N. Poinar, M. Kuch, and S. Pääbo. "Ancient DNA." *Nature Reviews Genetics* 2 (2001): 353–359.

Krings, M., A. Stone, R. W. Schmitz, H. Krainitzki, M. Stoneking, and S. Pääbo. "Neandertal DNA Sequences and the Origin of Modern Humans." *Cell* 90 (1997): 19–30.

Thomas, M. G., E. Hagelberg, H. B. Jones, Z. H. Yang, and A. M. Lister. "Molecular and Morphological Evidence on the Phylogeny of the Elephantidae." *Proceedings of the Royal Society of London Series B* 267 (2000): 2493–2500.

— MARK THOMAS AND IAN BARNES

ANGIOSPERMS. *See* Plants.

ANIMALS

The Metazoa (equals animals) is one of the great kingdoms of life. Estimates of the current numbers of living animal species vary considerably, but several million seem likely to exist. Their complexity, beauty, and cultural significance have made them the topic of endless discussion from ancient times. In an evolutionary context, the recognition that humans are themselves animals has added a new dimension to our relationship to them, with the new subject of "sociobiology" hoping to understand human behaviour in terms of their animal predecessors and evolutionary history. Although animals in popular parlance are unsatisfactorily divided into "invertebrates" and "vertebrates," vertebrates form only a small component of total animal diversity, and as a result are only briefly covered here.

Animals: What Are They? Animals are eukaryotes that are heterotrophic and multicellular. In addition, they are usually considered to be mobile and sensing. They typically reproduce sexually, and have diploid cells, with the exception of their haploid gametes. In order for a group of cells to coexist in a single organism, there generally needs to be some sort of junction developed between them. Furthermore, truly multicellular organisms, as opposed to colonies, demonstrate differentiation: not all of the cells carry out the same functions. It is therefore clear that the origin of animals must have been a quite complex event, with several steps required before true multicellularity could be attained. Another particularly important characteristic feature of animals is development: the coordinated and usually permanent differentiation and proliferation of cells from a single-celled zygote into the complex adult animal. In recent years, the molecular basis of this remarkable process has begun to be understood.

Animals are so diverse that their natural unity has sometimes been doubted, although modern molecular and morphological studies have tended to reinforce a unique origin for them. The broad groupings of living animals, called phyla, are typically very distinct, and determining the relationships between them has thus proved to be rather difficult. Their distinctness has led to the controversial idea that each phylum is united by its own "body plan." About thirty or so phyla are recognized in the extant fauna, and, although broad groupings of some phyla are generally accepted, such as the arthropods, onychophorans and tardigrades, phyletic inter-relationships have been and continue to be a source of fierce controversy.

Animal Organization. Animals can be thought of in terms of their organization (Figure 1): sponges (with differentiated cells and tissues but no proper epithelia); the cnidarians (with tissues and epithelia); and the bilaterians (with the addition of organs and mesoderm). Of these, the sponges (phylum Porifera) are widely regarded as being the most primitive metazoans. Although they can have a couple of dozen different types of cells, the degree to which these are permanently differentiated is limited: some cells can de-differentiate and even re-differentiate into new types. Furthermore, unlike higher animals, they do not form true epithelia, so their bodies are not compartmentalized. Sponge cells are organized into functional units around inhalant channels, through which water is drawn by flagella action into the internal cavity of the sponge, and from which food particles are removed. Remarkably for organisms that ap-

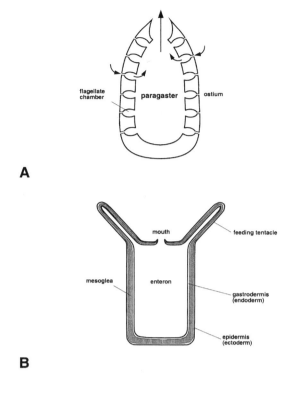

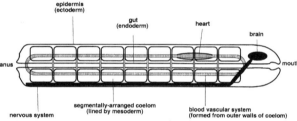

FIGURE 1. Three Ways of Organizing Animals.
(A) Typical sponge, forming a hollow body, into which feeding currents are drawn by beating cilia: water is then expelled dorsally; (B) hydroid cnidarian, epithelial-bound and bilayered, with a central digestive cavity; and (C) a segmented bilaterian, showing some typical features including the through-gut, blood vascular system, nervous system, and coelom. The derivation of some structures from endo-, ecto-, or mesoderm is indicated.   Courtesy of Graham E. Budd.

parently lack any sort of nervous system, a carnivorous sponge was recently described.

Cnidarians have definite tissues such as true epithelia, muscle, a nerve net, and reproductive tissues. Their cells are divided into two thin but definite layers, gastrodermis (lining the body cavity) and epidermis (forming the outer surface), and cnidarians are thus often referred to as having "diploblastic" (i.e., two-layered) organization. Between the two layers is a jellylike noncellular substance called mesoglea, although in some

groups the mesoglea contains cells, thereby somewhat weakening the classical "two layers" view of cnidarians. An enigmatic group of organisms with a somewhat similar organization, the ctenophores or sea-gooseberries, are also gelatinous in composition. Nevertheless, they are widely considered to possess a third tissue layer, the mesoderm, that generates cells lying between the outer epidermis and inner gastrodermis, a critical character that is considered to unite them with the bilaterians (see below). Another enigmatic group of organisms is the Placozoa, consisting only of the species *Trichoplax adhaerens*, a small, amoebalike organism with only four cell types. Although originally thought to be extremely basal in metazoans, it now seems more likely to lie close to the Cnidaria. Finally, the Mesozoa, consisting of the dicyemids and orthonectids, are also extremely simple in structure. Again once thought to be very basal, these taxa are now thought on both morphological and molecular grounds by some to lie within the bilaterians.

Bilaterians. By far the most numerous animals are the bilaterians, including many familiar animals such as vertebrates, molluscs and arthropods; but the grouping also contains many much less familiar phyla as well, which, although often poorly studied are often fascinating and challenging groups to investigate. Many of these poorly known forms are small, and are an important constituent of the "meiofauna" living between sediment grains. Several new bilaterians have been described recently, including loriciferans and cycliophorans. Bilaterians range in size from minute flatworms only 50 µm in diameter up to the blue whale, weighing some 150 tons and with a maximum length of over 30 m, a size range of some six orders of magnitude. Bilaterians all possess mesoderm, and are thus termed "triploblasts": the mesoderm gives rise to important tissues such as muscle, germ cells and the walls of the blood vascular system. The differentiated cells of bilaterians are organized into structures called organs that carry out specific tasks. As the name suggests, bilaterians are bilaterally symmetrical organisms (even starfish are developmentally bilateral), but they share several other important features too, such as usually possessing an anterior concentration of nervous tissue and a through-gut with mouth and anus. Some bilaterians (at least chordates, arthropods and annelids) also possess segmentation, the co-ordinated serial repetition of organ systems. Whether or not segmentation was present at the base of the bilaterians is a matter of considerable debate.

Bilaterian development. Early divisions of the zygote (fertilized egg) give rise to a solid or hollow ball of cells, the blastula (Figure 2). This ball then typically undergoes a distinctive rearrangement to form firstly the endoderm, that will go on to form the gut, and secondly, the mesoderm, the entire process being known as gastrulation: the remaining outer cell layer, the ectoderm,

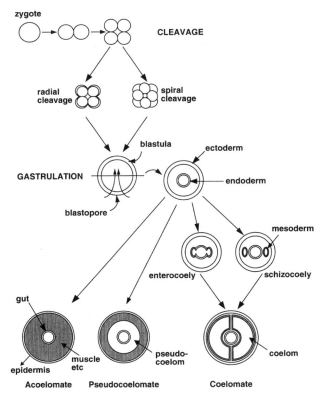

FIGURE 2. A Highly Schematic View of Bilaterian Development.
The fertilized egg (zygote) divides with spiral or radial cleavage to produce an often hollow ball of cells (blastula). During gastrulation, cells migrate into the interior through the blastopore that forms the mouth in protostomes; in deuterostomes, the mouth forms later. Mesoderm (marked in gray) then fully or partly fills the primary body cavity to form the acoelomate or pseudocoelomate condition; or may cavitate in one of two ways to produce a coelom. It should be stressed that this cartoon hardly does justice to the great variety at every stage that actually exists. Deuterostomes are classically meant to be distinguished by radial cleavage and enterocoely; protostomes by spiral cleavage and schizocoely.
Courtesy of Graham E. Budd.

goes on to form nerve and epidermal tissues. In many marine invertebrates, the resulting "gastrula" develops into a feeding larva that swims in the plankton for a few days before metamorphosing into the adult, often by settling on the sea bottom, a process called indirect development. Primarily terrestrial organisms (for example, reptiles, mammals, and birds), conversely, typically have direct development, wherein the gastrula develops smoothly into the adult without any intermediate larval stage; and human development is a good example of this. Life cycles and modes of early development have played an important role in theories about bilaterian relationships (see Vignette).

Like all animals, the bilaterians have several impor-tant functional problems to solve, including respiration, support, movement, nutrition, excretion, and reproduction. Clearly, the solutions to such problems vary greatly within such an enormous size range. The tiniest bilaterians, such as flatworms (platyhelminthes) and small rotifers rely on diffusion for most processes such as excretion and respiration and cilia for locomotion, but organisms greater in length than about 1 mm or so typically rely on specialized systems. One of the most important features that enables the development of specialized systems is a body cavity of some sort that can allow isolation of muscle movement in different organs; specialised digestive organs (the gut); storage of gametes; internal transport systems (e.g., the blood vascular and lymphatic systems) and development of an internal hydrostatic skeleton.

Body cavities and their origins have played an important role in speculations about bilaterian relationships. They fall into three main categories: the primary cavity that may form during early development inside a hollow blastula, and two sorts of secondary cavities: the gut, formed from endoderm, and a particularly important one, the so-called coelom. Classically, the coelom is a cavity that is entirely lined by mesoderm. Typically, only relatively large animals are coelomate. Bilaterians are sometimes classified according to the sort of body cavity they possess. Acoelomates at most possess only a gut, with the rest of the body being a rather solid, tissue-filled structure (e.g. flatworms, gnathostomulids); and coelomates (e.g., chordates and annelids) possess a coelom. A third category, pseudocoelomates (e.g., kinorhynchs, nematodes, and rotifers), retain the unlined primary body cavity into the adult. Arthropods possess a peculiar arrangement, in which the coelom breaks down early in development to merge with the primary body cavity, forming a "mixocoel." There have been many attempts to use body cavity type as a way of determining evolutionary relationships between bilaterians, but it is now universally considered that these types of organization do not fall completely into phylogenetically coherent groupings.

Models of Animal Relationships. Virtually all workers who have considered animals to have a single origin have placed the sponges basal in the group, with cnidarians the next most derived, as sister group to the bilaterians (ctenophores are sometimes also placed in this position) (Figure 3). However, relationships within bilaterians have proved much more controversial; and have depended very much on the model assumed for the morphology of the (hypothetical) basal bilaterian. One tradition has viewed the acoelomate flatworms as being basal, and representing the ancestral morphology of bilaterians, from which the advanced systems of coelomates were derived, perhaps more than once, via an intermediate stage represented by the pseudocoelomates.

PROTOSTOMES AND DEUTEROSTOMES

Early development is supposed to allow the identification of two large bilaterian groupings: the deuterostomes and the protostomes. In *protostomes*, the first small pore formed as cells migrate into the inside of the blastula during development to form the endoderm is meant to go on to form the mouth of the adult; whereas in *deuterostomes* this pore forms the anus, and the mouth is formed secondarily (hence the names, meaning mouth first and second, respectively).

Protostomes are also meant to be characterized by so-called spiral cleavage, referring to how the first divisions of the zygote take place, and deuterostomes by radial cleavage; and other features such as details of how early the identity of the cells is fixed and what sort of larva is formed are also meant to be congruent with these two important ones. Mode of formation of the coelom is also meant to be characteristic: with the protostome coelom forming by so-called schizocoely (the coelom forms by cavities appearing in clumps of mesodermal cells) and the deuterostome one by enterocoely (the coelom first forms by outpouchings from the gut-to-be).

Although some of these features are readily identifiable much doubt has recently been cast on their overall usefulness. This is partly because some organisms do not actually possess the developmental features they are meant to. Some protostomes, such as priapulans, have radial cleavage, while several deuterostomes are schizo-, not enterocoelous. Although the "cores" of deuterostomes (echinoderms, hemichordates, and chordates, including vertebrates) and protostomes (annelids, molluscs, echiurans, and sipunculans) thus seem fairly cohesive groups, overall these features seem too variable to be used as completely reliable guides to relationships on their own.

—GRAHAM E. BUDD

This view has seen ancestral bilaterians as resembling the solid larval stage of the cnidarians called the planula; so the origin of the bilaterians would therefore represent an example of heterochrony.

Conversely, another school of thought has seen the basal bilaterians as being a relatively large animal that already possessed a coelom and segments, and was derived directly from an adult similar to living cnidarians. In this case, important features of bilaterians are probably homologies, with the pseudocoelomates and acoelomates that typically lack them representing cases of independent loss. Other theories for the origins of bilaterians are inspired by coloniality and the importance of the planktonic feeding stages of many marine invertebrates.

Three important sources of information have recently been brought to bear on this vexed issue: increased understanding of the genetic basis of animal development, molecular systematics methods, and new fossil finds from the Cambrian. The first of these has indicated that the genetic basis of many important animal features such as segmentation and the blood vascular system appear to be highly conserved throughout the bilaterians. Some have taken this to imply that the structures these genes are now implicated in building were also present in the last common ancestor of bilaterians. This might in the extreme imply that the earliest bilaterians already had eyes, a heart, a coelom, segments, and even limbs of some sort. Other workers have interpreted these surprising conservations much more cautiously however, pointing out that genes can change their function through time, and thus that the function genes have now cannot necessarily be taken as an indicator of what their original role was. If so, then genes may have been repeatedly "co-opted" to carry out similar functions in different animals. Secondly, molecular systematics have markedly changed views of how the broad groupings of animals are related. While sponges and cnidarians are normally still considered basal, sponges often resolve as being paraphyletic, that is, living sponges do not form a natural group, but rather gave rise to all higher animals.

It is in the area of bilaterian relationships, however, that most change has occured. Traditionally, bilaterians have been divided into the protostomes and deuterostomes (see Vignette). However, the composition of the deuterostomes has been considerably narrowed, by removal of the so-called lophophorate animals (brachiopods, phoronids, and bryozoans) to the protostomes—a move that is not entirely surprising given the long-controversial nature of these phyla, which show a particularly puzzling mixture of deuterostome and protostome features. More controversial still, however, have been the rearrangements in the protostomes themselves. The arthropods have been removed from their traditional segmented relatives, the annelids, to form a completely separate grouping with the nematodes, priapulans, and some other pseudocoelomate phyla. This latter grouping can hardly be said to be characterized by important morphological features, but one they all share is the process of molting under the control of steroids called ecdysones; hence the name for it that has been coined, Ecdysozoa. Conversely, annelids are grouped together with brachiopods, molluscs, and the rest of the old protostomes in a poorly resolved group usually referred to as the Lophotrochozoa. In all these groups, the acoelomates and pseudocoelomates occur in rather a scattered way, giving some credence to the idea that they represent independent lines of secondary reduction and loss of features such as the coelom and segmentation. Nevertheless, the issue is far from resolved: some phyla, such as the chaetognaths, stubbornly refuse to be pinned

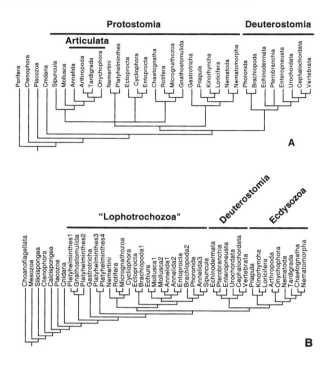

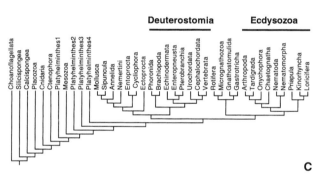

FIGURE 3. Three Recent Views of Metazoan Phylogeny. Diagram based on (A) morphology (Nielsen, 2001); (B) 18S rDNA (Peterson and Eernisse, 2001); and (C) a combination of morphology and 18S rDNA (Zrzavy et al., 1998). Note that in A, echiurans are included in annelids, and that acanthocephalans are included in Rotifera in all analyses. Enteropneusta + Pterobranchia = Hemichordata. Micrognathozoa have been added as sister group to Rotifera in all analyses. The presence of several flatworm groups appears to have a disturbing effect on the rest of the tree in the molecular analyses, and their exclusion leads to a fairly different tree shape.   Courtesy of Graham E. Budd.

ganisms that are now generally placed in a rather unsatisfactory Kingdom of their own, the Protoctista, i.e., eukaryotes excluding animals, plants, and fungi. Nowadays, animals are usually restricted to multicelluar heterotrophs, making the word synonymous to "Metazoa." Nevertheless, the relationships of some of the protoctists to the animals need to be considered. In particular, the choanoflagellates, single-celled flagellate heterotrophs, seem to share many important features with the animals. They resemble cells that were originally known only in sponges, but are now known from a variety of other phyla, the so-called collar cells, a similarity that has led many to believe that the earliest animals evolved from colonial, choanoflagellate-like protoctists. This view has received support from molecular systematics. The broader affinities of animals plus choanoflagellates are, however, more uncertain. Most studies support a relationship between plants, animals (and choanoflagellates) and fungi, but the order of relationships between the three is far from resolved. Recently, an animal plus fungi grouping has received some support, but the more traditional grouping of plants plus fungi also has its backers. It is probably best at present to regard these three important groups as forming an unresolved trichotomy.

History of the Animals. Fossils or trace fossils that can be unequivocally assigned to the animals do not appear in the fossil record until very close to the beginning of the Cambrian period, perhaps some 560 million years ago. These early beginnings, consisting of fossils that are usually considered to have cnidarian affinities, are quickly followed by a very substantial radiation of phyla, invariably referred to as the "Cambrian explosion." While the exact extent of the early assembly of the phyla has been tempered to some extent recently, there is little doubt that this period of time represents an important event in animal history. Nevertheless, the advent of molecular clock dating techniques has provided some controversial evidence that the bilaterians diverged well before this, even as early as 1.2 billion years ago or more. There is therefore a *prima facie* conflict between the molecules and the evidence of the fossil record, a problem that has led to some attempts to characterize early animals as being small and/or planktonic, and thus unlikely to be preserved in the fossil record. These attempts have proved to be far from satisfactory to some, and there have been efforts made to show that the events recorded in the fossil record close to the Cambrian boundary, broadly considered, represent the true history of bilaterian diversification.

The subsequent history of the animals is much clearer, because the advent of hard parts such as shells at this time makes the ensuing fossil record much more reliable. The resulting pattern of diversification is disputed, but one recent model is of a broadly exponential upwards curve, with this increase being repeatedly offset by the various mass extinctions. As well as patterns

down phylogenetically, while some recent molecular work has indicated that at least some flatworms may, after all, belong to the base of the bilaterians. Furthermore, in general, many morphologists would resist conclusions based only on molecules.

Broader Affinities of the Animals. For many years, "animals" were taken to include certain single-celled or-of diversity, the fossil record of animals has documented many important evolutionary events such as the rise of terrestrial faunas including the amphibians and reptiles, and the origins of many important, especially vertebrate, groups such as fish, birds, and mammals. Perhaps surprisingly, animal diversity is now more or less the highest it has ever been: whether this will be maintained in the face of human intervention, however, remains to be seen.

[*See also* Amphibians; Arthropods; Birds; Cnidarians; Dinosaurs; Echinoderms; Fishes; Insects; Mammals; Molluscs; Reptiles; Vertebrates.]

BIBLIOGRAPHY

Aguinaldo, A. M., A., J. M. Turbeville, L. S. Linford, M. C. Rivera, J. R. Garey, R. A. Raff, and J. A. Lake. "Evidence for a Clade of Nematodes, Arthropods and Other Moulting Animals." *Nature* 387 (1997): 489–493. Still controversial evidence for the "Ecdysozoa," that is far from being accepted by many morphologists.

Brusca, R. C., and G. J. Brusca. *Invertebrates*. Sinauer Associates, Inc. Sunderland, Mass., 1990. Standard "brick" textbook, it has good coverage centered on the "body plan" concept.

Budd, G. E., and S. Jensen. "A Critical Reappraisal of the Fossil Record of the Bilaterian Phyla." *Biological Reviews* 75 (2000): 253–295. A thorough review of the evidence for the origins of bilaterian phyla, including the biomechanical aspects to animal construction and the resulting constraints placed on early animals.

Carroll, S., J. F. Grenier, and S. D. Wetherbee. *From DNA to Diversity: Molecular Genetics and the Evolution of Animal Design*. Blackwell Science: Oxford, 2001. A beautiful and clear introduction to what is now known about the molecular basis for animal development, and its evolutionary significance.

Conway Morris, S. "The Question of Metazoan Monophyly and the Fossil Record." In W. E. G. Müller (ed.), *Molecular Evolution: Towards the Origin of the Metazoa*, pp. 1–11. Springer: Berlin, 1998. A recent discussion of the evidence for metazoan monophyly, concluding that metazoans are likely, but not quite definitely, monophyletic.

Dewel, R. A. "Colonial Origin for Eumetazoa: Major Morphological Transitions and the Origin of Bilaterian Complexity." *Journal of Morphology* 23 (2000): 35–74. A controversial but interesting theory that bilaterians arose out of a colonial grade of organization.

Hewzulla, D., M. C. Boulter, M. J. Benton, and J. M. Halley. "Evolutionary Patterns from Mass Originations and Mass Extinctions." *Philosophical Transactions of the Royal Society of London* B354 (1999): 463–469. A review of some models for animal (and plant) diversification through time, arguing that the essential pattern is of exponential increase through time, tempered by the catastrophic mass extinctions.

Kardong, K. V. *Vertebrates: Comparative Anatomy, Function, Evolution*. 3d ed. McGraw-Hill Higher Education: Columbus, Ohio, 2001. A very thorough comparative anatomy of the vertebrates, with an emphasis on function and evolution.

Löytynoja, A., and M. Milinkovitch. "Molecular Phylogenetic Analyses of the Mitochondrial ADP-ATP Carriers: The Plantae/Fungi/Metazoa Trichotomy Revisited." *Proceedings of the National Academy of Sciences of the USA* 98 (2001): 10202–10207. Technical discussion of the problem posed by plant/fungi/meta-

zoan relationships, concluding that plants and fungi may after all be sister groups, without being too dogmatic on the topic.

Nielsen, C. *Animal Evolution: Interrelationships of the Living Phyla*. 2d ed. Oxford, 2001. A splendidly concise and knowledgeable introduction to the animals, with an emphasis on their morphology and its phylogenetic significance. Molecular data is less well-covered. Includes coverage of the author's "trochaea" theory of animal origins.

Peterson, K. J., and D. J. Eernisse. "Animal Phylogeny and the Ancestry of Bilaterians: Inferences from Morphology and 18S rDNA Gene Sequences." *Evolution and Development* 3 (2001): 170–205.

Ruppert, E. E., and R. D. Barnes. *Invertebrate Biology*, 6th ed. Saunders College Publishing: Forth Worth, Texas, 1994. Out of the ordinary zoology textbook, it has an interesting emphasis on function and "compartmentation," the division of animals into functional subunits.

Snell, E. A., R. F. Furlong, and P. W. Holland. "Hsp70 Sequences Indicate that Choanoflagellates Are Closely Related to Animals." *Current Biology* 11 (2001): 967–970. A recent molecular study arguing for a choanoflagellate—animal sister-group relationship.

Vacelet, J., N. Bouryesnault, A. Fialamedioni, and C. R. Fisher. "A Methanotrophic Carnivorous Sponge." *Nature* 377 (1995): 296. The report of a remarkable life habit innovation within the most basal metazoans.

Wainright, P. O., G. Hinkle, M. I. Sogin, and S. K. Stickel. "Monophyletic Origins of the Metazoa—An Evolutionary Link with Fungi." *Science* 260 (1993): 340–342. A very influential paper arguing that fungi are the sister group to animals, based on ribosomal RNA. It also presents an analysis of broad-scale animal relationships, including the choanoflagellates.

Willmer, P. *Invertebrate Relationships: Patterns in Animal Evolution*. CUP: Cambridge, 1990. A book that, although highly controversial in its very polyphyletic views of animal evolution, nevertheless still contains much useful data, much of which is viewed critically.

Zrzavy, J., S. Mihulka, P. Kepka, A. Bezdek, and D. Tietz. "Phylogeny of the Metazoa Based on Morphological and a8S rDNA Evidence." *Cladistics* 14 (1998): 249–285. The first big combined morphology and molecules phylogeny of animals.

— GRAHAM E. BUDD

ANTIBIOTIC RESISTANCE

[*This entry comprises three articles:*

    Origins, Mechanisms, and Extent of Resistance
    Epidemiological Considerations
    Strategies for Managing Resistance

The first article provides a discussion of antibiotic resistance from an evolutionary perspective; the second article outlines the success that microorganisms have had in responding to change and how this has affected public health measures; the third article focuses on the strategies that have been proposed to slow or halt the spread of antibiotic resistance. For related discussions, see Bacteria and Archaea; Pesticide Resistance; and Resistance, Cost of.]

Origins, Mechanisms, and Extent of Resistance

From the clinical point of view, bacterial antibiotic resistance means inability to successfully treat with a given antibiotic an infection caused by a particular organism. From the microbiological point of view, it is essential to distinguish between natural or intrinsic resistance (in this case, all members of a given bacterial species or genus are resistant) and acquired resistance (the members of the species are normally susceptible, but some of them, the resistant ones, are not inhibited by the antibiotic). In general, the term *antibiotic resistance* should be reserved only for acquired resistance. In this case, it is appropriate to differentiate between low-level resistance, when the minimal concentration of antibiotic needed to inhibit the bacterial growth (also termed minimal inhibitory concentration, or MIC) is only slightly above that required to inhibit the susceptible organisms, and high-level resistance, where the MIC of resistant organisms greatly exceeds (frequently more than tenfold) the MIC of the susceptible ones. Many bacterial organisms with low-level resistance maintain a clinical susceptibility to the antibiotic (infections caused by these bacteria may be cured) because the pharmacokinetics of the drug may assure sufficient antibiotic concentration at the infection site to inhibit the infecting organism. Nevertheless, low-level antibiotic resistance should not be overlooked as an irrelevant phenomenon. From the viewpoint of resistance surveillance, the emergence of low-level antibiotic resistance should be considered an alarm signal, a hallmark of a possible evolutionary trend toward high-level, clinical resistance.

In most cases, antibiotics exert their selective pressure in a gradual way, both at the individual human or animal level (gradients of antibiotic concentrations are formed in the body) and at the population level (differences in antibiotic consumption between individuals) and in the free-environment (gradients of antibiotic concentrations surrounding antibiotic-producing microorganisms, or around inert soil particles to which antibiotics released in the environment adhere). Under gradual antibiotic pressure, and if the cost of resistance is tolerable, low-level resistant variants are selected and may evolve to high-level resistance. The distribution of fitness values over the space of resistance genotypes constitutes a resistance fitness landscape. Small hills or hillsides may correspond to low-level resistance mechanisms, and mountains correspond to very efficacious, high-level mechanisms. Evolution toward resistance is frequently a hill-climbing process; to reach a high peak, successive slopes must be climbed first.

Antibiotic resistance can sometimes be obtained as a result of a mutational event in housekeeping genes. Several types of resistance mutations are able to convert a susceptible organism into a resistant one. First, target gene mutations involve changes in a gene encoding the production of an antibiotic target, which modify the interaction with the drug so as to reduce the effect of the antibiotic. For instance, mutational changes in the sequence of some genes encoding transpeptidases-transglycosilases (PBPs), involved in synthesizing the bacterial cell wall, reduce the affinity of these enzymes for _-lactams and produce resistance to these agents in *Streptococcus pneumoniae* or *Neisseria meningitidis*. Similarly, mutations in genes encoding topoisomerase II or IV render many microorganisms resistant to fluoroquinolone antibiotics. Mutations in genes encoding ribosomal components (23S rRNA, or ribosomal proteins) may counter resistance, as is the case for *Helicobacter pylori* to clarithromycin (a macrolide) or *Escherichia coli* to streptomycin. Second, target access mutations prevent the antibiotic from reaching its cellular targets. For example, mutations in outer membrane proteins (OMPs) may reduce the uptake of antibiotic into the bacterial cell; that is frequently the case for _-lactams in *Klebsiella pneumoniae* and *Pseudomonas aeruginosa*. Third, target protective mutations, frequently in housekeeping genes, protect the target from the antibiotic. For example, mutations may increase the expression of preexisting chromosomally encoded antibiotic-detoxifying enzymes, such as chromosomal AmpC-type _-lactamases, or the expression of preexisting efflux pumps.

A major step in the evolution of mechanisms of resistance is the generation of resistance genes, encoding for proteins whose primary function is to resist to antibiotics. In some cases, these genes may originate from housekeeping genes that have been modified by mutation and recombination. Gene duplication is sometimes a primitive strategy to overcome the inhibitory action of an antibiotic. Eventually mutations in one of the copies may slightly increase resistance; after a number of successful modifications (perhaps sequentially selected by low antibiotic concentrations), the mutated copy may produce proteins that are much more effective in promoting resistance, but at the expense of losing part of its original physiological function. The product of the evolving resistance gene may compete with the ancestral product for the physiological substrate (reducing fitness), which results in new rounds of genetic modification to increase divergence (and finally total loss of the original function) from the ancestor gene. One example of such evolutionary trends may be the kinship and diversification of genes encoding bacterial PBPs (targets of _-lactams) and _-lactamases (detoxifying _-lactam antibiotics by hydrolisis). A second possible example is the evolution of the vaneomycin-resistance gene (*vanA*) product from bacterial dipeptide (D-Ala-D-Ala) ligases. A third one is the origin of aminoglycoside-modifying (inactivating) enzymes in sugar kinases or acetyl transferases. Indeed, the high diversity in resistance genes strongly suggests that many of them have evolved to their current function from housekeeping genes in many different evolutionary lineages. At least thirty different aminoglycoside-modifying genes and twenty different tetracycline-resistance genes are involved in resistance to a limited number of molecules in these antibiotic families, and at least some of the genes still retain functions unrelated to resistance. Multiple-sequence alignments also suggest different evolutionary origins for _-lactamases. In several instances, resistance genes may have evolved first in antibiotic-producing microorganisms. It is presumed that both antibiotic biosynthetic pathways and the mechanisms of resistance to avoid self-damage may result from a coevolutionary process. In fact, resistance can be viewed as a precondition for significant antibiotic production. The benefit to microbes associated with antibiotic production, probably preventing invasion of the same niche by sensitive competitors, may also have selected for those producer strains with greater resistance. Thus, the resistance genes may encode for specific systems to pump out the drug, to destroy the antibiotic warhead, or to protect the target structure.

The next evolutionary step is the construction of resistance operons, in which the expression or the function of the resistance gene is optimized, for example, by ensuring that transcription occurs only in the presence of the antibiotic challenge. Horizontal gene transfer may favor clustering genes in a single operon, as was suggested by Jeffrey Lawrence and John Roth. The production of gene clusters involved in resistance to multiple types of antibiotics ensures the survival of bacteria in environments (e.g., hospitals) where they are simultaneously exposed to a wide variety of antimicrobial agents. This impressive process of natural genetic engineering involves plasmids, transposons, and gene cassette–integron systems, and proceeds through successive site-specific recombination events.

The spread of resistance genes among bacterial strains and species is strongly facilitated by plasmids, transposable sequences, and probably temperate phages. Transmissible resistance determinants represent an excellent example of natural "genetic engineering" at work in evolution. Plasmids have coevolved with bacteria and, apart from their transfer functions, have specific maintenance systems to ensure self-perpetuation in bacterial populations. Exposed from the 1950s to an increasingly selective antibiotic environment, historical (preantibiotic) plasmids rapidly incorporated antibiotic resistance determinants. The study of preantibiotic plasmid collections strongly suggests that the emergence of resistance genes in such elements has largely occurred during the last decades.

To a certain extent, the acquisition of a resistance gene by a plasmid or transposable element helps to maintain these elements in antibiotic-polluted environments. In the presence of a novel drug or multiple drugs, plasmids encoding only a single type of resistance will probably be lost. In transmissible plasmids, however, there is always the possibility to transfer to another host that is resistant to the new drug, which may harbor another plasmid-determining resistance to this drug. Plasmids from natural populations of *E. coli* frequently show a mosaic structure, indicating high rates of recombination. The plasmid may also pick up resistance genes from the bacterial chromosome. No wonder that a multiple antibiotic environment has led plasmid evolution toward the acquisition of multiple antibiotic determinants in single replicative units, and even in the same gene cluster. Integrons are the primary systems for the capture of antibiotic resistance genes in Gram-negative bacteria. These elements are formed by gene cassettes (about forty types are known) located downstream of a recombinase-encoding conserved sequence that includes a strong promoter. The promoter is important both for expression of the gene antibiotic resistance cassettes and for regulating the frequency of recombination, as the cassettes may be integrated or deleted from their receptor integron structure. It is important to note that antibiotics may thus select not only for antibiotic resistance genes, but also (with them) mechanisms involved in gene mobility. Antibiotic pollution may thereby contribute to all types of bacterial phenotypic evolution, including virulence determinants (such as genes involved in mucosal colonization or toxin production) that are often located on accessory elements.

An interesting question is whether antibiotic selection, by facilitating interbacterial genetic exchange, could ameliorate the current ecological imbalance among bacterial species. That is, movement of resistance genes among bacterial taxa could help to maintain the normal integration of bacterial communities challenged by antibiotic pollution—a sort of "collective" adaptation.

Resistance may be widespread in nature, but the problems of resistance in medicine are frequently related with the emergence and spread of a relatively few successful resistant bacterial clones. As in any other strong selective process, the interaction of antibiotics and bacterial variants tends to reduce the bacterial diversity. Because of the elimination by antibiotics of potentially competing microorganisms, the resistant bacteria increase their number, which may facilitate dispersal and adaptation to novel environments, including host-to-host spread. Interestingly, in some cases, the same process that led to the acquisition of antibiotic resistance may result in other, unexpected advantages in bacterial fitness. For instance, the bleomycin-resistance gene contained in the transposon *Tn5*, and the tetracycline-resistant gene in the transposon *Tn10* confer improved survival and growth in *E. coli* in certain environments.

The advantages associated with resistance genes may be reduced by certain costs associated with the presence of these genes. That is, mutations that confer novel resistance phenotypes often have maladaptive pleiotropic effects, reducing competitiveness. Under antibiotic selection, however, the competitor organisms may be unable to take advantage of this reduction, and therefore the resistant bacteria genotypes have a chance to compensate by selection of modifiers that reduce the cost of resistance. This process of compensatory adaptation may even eliminate the biological cost of resistance. Following compensation, resistance may persist for a long time even in the absence of antibiotics. For instance, the KatG catalase-peroxidase activity is an important factor for the survival of *Mycobacterium tuberculosis* in both human and animal hosts. Mutations that eliminate this activity are the major cause of resistance to isoniazid. It might be predicted that *M. tuberculosis* isoniazid-resistant mutants would be less virulent than wild-type, isoniazid-susceptible strains, owing to the costs of maintaining resistance. However, when isoniazid-resistant isolates were analyzed, such reduced virulence was not found. It turned out that the resistant isolates had compensatory mutations in the gene encoding for the alkylperoxidase AphC, which can substitute for KatG for surviving inside the host, thereby avoiding the cost of resistance.

Recent observations suggest that bacteria may increase their mutation rates under antibiotic-mediated stress. This process is mediated by stress-responsive error-prone DNA polymerases V (*umuCD*) and IV (*dinB*), whose expression transiently increases the rate of bacterial mutation. Also, heritable hypermutable ("mutator") populations could be selected during antibiotic exposure. Mutator genotypes have mutation rates hundreds or even thousands of times above wild type, usually as the consequence of inherited defects in genes involved in the methyl-directed mismatch repair system (MMR). The consequence is that these strains have increased likelihood of finding mutational outcomes to survive antibiotic challenges. Among MMR-defective strains, the ability to recombine (homologous recombination) also increases, which may facilitate the acquisition of genes encoding antibiotic resistance from related organisms. Thus, antibiotics may select for hypermutable (and hyperrecombinable) organisms, accelerating the evolutionary rate.

Antibiotic resistance mechanisms can also emerge and evolve under the selective pressure of nonantibiotic challenges. These include heavy metals, toxic chemicals, organic solvents, household disinfectants, and non-

antibiotic drugs. Stuart Levy (1992) has shown that Mar mutants (overexpressing MarA) of *E. coli* are resistant not only to tetracycline, chloramphenicol, _-lactams, and fluoroquinolones, but also to oxidative stress agents (menadione and paraquat), to uncoupling agents (2,4-dinitrophenol), disinfectants (chloroxylenol and quaternary ammonium compounds), and other chemicals (such as cyclohexane). Antibiotics from human, veterinary, and agricultural origin are also released into the environment and may remain active for long periods of time. The key message here is that many environmental circumstances are potentially able to select antibiotic resistance.

Nevertheless, antibiotic consumption is the major driver of antibiotic resistance. Countries with the highest rates of antibiotic consumption per capita in the community are those where the highest resistance rates tend to occur. Antibiotic resistance in community-acquired bacterial pathogens is high in southern Europe and the United States (e.g., in *S. pneumoniae*, 30–40 percent are resistant to _-lactams and macrolides; in *E. coli*, 50 percent are resistant to ampicillin), as well as in the Far East (Korea and Japan have the highest rates of resistance in the world). In some cases, resistant bacterial clones that originated in these countries have subsequently invaded another country (e.g., a penicillin-resistant clone of *S. pneumoniae* from Spain to Iceland). In developing countries, antibiotic resistance is frequent among epidemic organisms, such as *Salmonella*, *Shigella*, and *Vibrio*.

Many of the life-threatening problems of antibiotic resistance emerge and evolve in hospitals, particularly in neonatal wards and intensive care units. Among them, the most worrisome organisms are found in some institutions at frequencies over 20 percent, including methicillin-resistant *Staphylococcus aureus*, frequently showing a multiresistant phenotype; vancomycin-resistant *Enterococcus*; *Klebsiella pneumoniae* and *E. coli* with extended-spectrum _-lactamases; and multidrug-resistant *Pseudomonas aeruginosa*, *Acinetobacter baumanii*, *Mycobacterium tuberculosis*, and *M. bovis*. In some cases, one or a few resistant clones may persist in the same hospital for years, suggesting frequent host-to-host transmission.

In conclusion, antibiotics constitute one of the most powerful forces acting on microbial communities. Antibiotic selective pressure in the human, animal, and external environments contributes to the increase of resistant mutants, to the selection of the most competitive among them, and to the dispersal of resistance genes.

[*See also* Bacteria and Archaea; Disease, *article on* Infectious Disease; Emerging and Re-Emerging Diseases; Mutation, *article on* Evolution of Mutation Rates; Pesticide Resistance; Plasmids; Resistance, Cost of.]

BIBLIOGRAPHY

Austin D. J., K. G. Kristinsson, and R. M. Anderson. "The Relationship between the Volume of Antimicrobial Consumption in Human Communities and the Frequency of Resistance." *Proceedings of the National Academy of Sciences USA* 96 (1999): 1152–1156. Data from epidemiological studies of consumption and resistance are used to develop mathematical models.

Baquero, F., and J. Blázquez. "Evolution of Antibiotic Resistance." *Trends In Ecology and Evolution* 12 (1997): 482–487. Reviews antibiotic resistance as a general evolutionary case study.

Baquero, F., and M. C. Negri. "Selective Compartments for Resistant Microorganisms in Antibiotic Gradients." *BioEssays* 19 (1997): 731–736. Examines the influence of variability in antibiotic concentrations on the selection of resistant variants.

Bjorkman J., I. Nagaev, O. G. Berg, D. Hughes, and D. I. Andersson. "Effects of Environment on Compensatory Mutations to Ameliorate Costs of Antibiotic Resistance." *Science* 25.287 (2000): 1479–1482. Examines how the environment, and not only antibiotics, determines the type of compensatory mutants that arise.

Courvalin, P. "Evasion of Antibiotic Action by Bacteria." *Journal of Antimicrobial Chemotherapy* 35 (1996): 855–869. Basic concepts related to the biochemistry, genetics, and evolution of antibiotic resistance in bacterial pathogens.

Davison, H. C., J. C. Low, and M. E. J. Woolhouse. "What Is Antibiotic Resistance and How Can We Measure It?" *Trends in Microbiology* 8 (2000): 554–558. Examines how to quantify antibiotic resistance.

De la Cruz, F., J. M. García-Lobo, and J. Davies. "Antibiotic Resistance: On How Bacterial Populations Respond to Simple Evolutionary Force." In *Bacterial Resistance to Antimicrobials: Mechanisms, Genetics, Medical Practice and Public Health,* edited by K. Lewis, A. Salyers, H. Taber, and R. Wax. 2001. About the origins and evolving trends of antibiotic resistance.

Lenski, R. E. "The Cost of Antibiotic Resistance—from the Perspective of a Bacterium." *Ciba Foundation Symposia* 207 (1997): 131–140. Reviews biological costs (reductions in fitness) of antibiotic resistance and compensation for the costs.

Lenski R. E., S. C. Simpson, and T. T. Nguyen. "Genetic Analysis of a Plasmid-Encoded, Host Genotype–Specific Enhancement of Bacterial Fitness." *Journal of Bacteriology* 176 (1994): 3140–3147. An important demonstration that genes encoding resistance to antibiotics may sometimes increase the fitness of microorganisms by other means than antibiotic resistance.

Levin, B. R., and C. T. Bergstrom. "Bacteria Are Different: Observations, Interpretations, Speculations, and Opinions about the Mechanisms of Adaptive Evolution in Prokaryotes." *Proceedings of the National Academy of Sciences USA* 97 (2000): 6981–6985. A review contrasting bacterial evolution with that of other organisms.

Levy, S. B. *The Antibiotic Paradox: How Miracle Drugs Are Destroying the Miracle.* New York, 1992. An easy-to-read book to introduce the general public to the problem of antibiotic resistance.

Lord Soulsby of Swaffham Prior and R. Wilbur, eds. *Antimicrobial Resistance.* International Congress and Symposium Series 247. London, 2001. A collection of short articles on various problems of antibiotic resistance, from genetics to sociology.

Martínez, J. L., and F. Baquero. "Mutation Frequencies and Antibiotic Resistance." *Antimicrobial Agents and Chemotherapy* 44 (2000): 1771–1777. A review of factors influencing antibiotic resistance mutation rates.

Massova, I., and S. Mobashery. "Kinship and Diversification of Bacterial Penicillin-binding Proteins and lactamases." *Antimicrobial Agents and Chemotherapy* 42 (1998): 1–7. An excellent analysis of the evolutionary origins of important resistance genes.

Oliver, A., R. Cantón, P. Campo, F. Baquero, and J. Blázquez. "High Frequency of Hypermutable *Pseudomonas aeruginosa* in Cystic Fibrosis Lung Infection." *Science* 288 (2000): 1251–1254. Shows the relationship between hypermutation and antibiotic resistance in the clinical setting.

Rowe-Magnus, D. A., and D. Mazel. "Resistance Gene Capture." *Current Opinions in Microbiology* 2 (1999): 283–488. A review on integrons as elements involved in gene capture and dissemination in Gram-negative bacteria.

Taddei, F., M. Radman, J. Maynard-Smith, B. Toupance, P. H. Gouyon, and B. Godelle. "Role of Mutator Alleles in Adaptive Evolution." *Nature* 387 (1997): 700–702. A key paper on the role of hypermutation in bacterial adaptation.

Walsh, C. "Molecular Mechanisms that Confer Antibacterial Drug Resistance." *Nature* 406 (2000): 775–781. A review on the molecular targets of classic and novel antimicrobial agents, along with the corresponding mechanisms of resistance.

World Health Organization. *Overcoming Antibiotic Resistance.* World Health Organization Report on Infectious Diseases, 2000. The WHO considers antibiotic resistance a major threat for public health.

— FERNANDO BAQUERO

Epidemiological Considerations

Microorganisms have been extremely resourceful in responding to change. In the distant past, some microorganisms evolved the ability to produce antibiotics that would inhibit or kill potential competitors, and the competitor microorganisms evolved the ability to resist their effects. Thus, a major discovery of the twentieth century, the use of antibiotics to treat infections, simply provided another selective pressure to challenge these organisms. In retrospect, it was not surprising that the introduction of new antimicrobial agents, from penicillin in the 1940s through fluoroquinolones in the 1980s, was often followed by the rapid evolution of resistant strains of bacteria. Yet in recent decades, the problem of antimicrobial resistance has become even more acute. The combination of increasing frequency of resistance in a variety of human pathogens (Figure 1) coupled with decreasing numbers of new antimicrobial agents has challenged the ability to treat once-susceptible infections. Antibiotic resistance has thus become a major public health issue. This article will examine the changes in society, technology, and the microbes themselves that are contributing to the emergence of antimicrobial resistance as an important public health problem.

Public Health Aspects of Antimicrobial Resistance. Two major elements, antibiotic use and disease transmission, influence the emergence of antimicrobial resistance and its rise as a public health problem. The first element, antimicrobial use, provides the selective

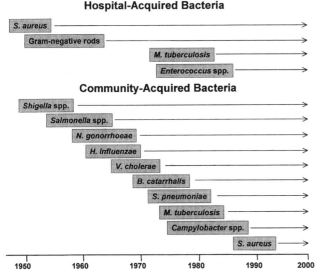

Hospital-Acquired Bacteria

Community-Acquired Bacteria

FIGURE 1. Hospital-Acquired versus Community-Acquired Bacteria. Emergence of clinically important antimicrobial resistance in the twentieth century among bacteria acquired in the hospital and in the community (*Staphylococcus aureus, Mycobacterium tuberculosis, Enterococcus* species, *Shigella* species, *Neisseria gonorrhoeae, Haemophilus influenzae, Vibrio cholerae, Branhamella catarrhalis, Streptococcus pneumoniae, Campylobacter* species).   Drawing by Mitchell L. Cohen.

pressure that leads to the emergence and encourages persistence of drug-resistant genes and strains. Since the discovery of antibiotics, this selective pressure has become intense. Each year in the United States, over 145 million courses of antibiotics are prescribed to outpatients; over 190 million daily-defined doses are given to persons in hospitals; and over 16 million pounds of antimicrobial agents are fed to animals to prevent or treat disease or to promote growth. Even the most appropriate use of antibiotics provides selective pressure; unfortunately, much of the current use is not appropriate. In humans, antimicrobial agents are often used when they will not be effective, such as to treat viral illnesses (e.g., upper respiratory tract infections or bronchitis), or they are used at excessive dosages or for too long, such as inappropriate regimens to prevent postoperative infections. Similar inappropriate use occurs in the treatment and prevention of infections in animals, and here the problem is compounded by the extensive use of antibiotics to make animals grow faster. Whether in the community, in the hospital, or on the farm, antimicrobial use provides the selective pressure for the emergence and persistence of resistant bacteria.

The second element of antimicrobial resistance becoming a public health problem is disease transmission, which causes resistant organisms to persist and spread by infecting other hosts or environments. An emerging

resistant bacterial strain in one individual, human or animal, may cause a serious or life-threatening infection in that individual. When the resistant bacteria are transmitted to others, however, the problem becomes one of public health. Epidemics from plague to cholera were well-recognized outcomes of disease transmission. For much of the twentieth century, there was optimism concerning the control of both sporadic and epidemic infectious diseases. This optimism was in part because of societal and technologic advances that were reducing transmission of infectious diseases; these included better housing and nutrition, safer food and water, improved hygiene and sanitation, and antibiotics and immunizations. Unfortunately, other societal and technologic changes were leading to new, reemergent, and drug-resistant infectious diseases. Since the introduction of antibiotics, a series of drug-resistant bacteria have caused new endemic and epidemic infections in hospitals, communities, and on the farm in both the developed and developing worlds.

Factors Influencing the Emergence of Antibiotic Resistance. The factors influencing the emergence of drug resistance are often the same ones that are contributing to the emergence of other infectious diseases. In a 1992 report entitled *Emerging Infections: Microbial Threats to Health in the United States*, the Institute of Medicine (IOM) identified six factors that were influencing emergence: changes in human demographics and behavior, changes in industry and technology, changes in the environment and land use, international travel and commerce, microbial adaptation and change, and breakdown in public health measures. The emergence of a drug-resistant bacterial strain is often the result of one or more of these factors leading to increased antibiotic use or increased disease transmission. Often, resistance emerges after these factors cause an increase in drug-susceptible infections, leading in turn to an increase in antibiotic treatment, then to an increase in drug-resistant infections. If an effective alternative antibiotic is available, the cycle of infection, treatment, and resistance can begin again.

The factors influencing the epidemiology of drug resistance may be examined in different settings, such as the developed and developing worlds, or on the farm, in the hospital, and in the community. Yet these different geographic areas or ecosystems are often linked, and resistance that emerges in one setting may spread and affect another. A drug-resistant *Salmonella* strain that emerges on the farm can spread through the food chain to the community and can also be introduced by food or ill patients into the hospital. In a similar fashion, the problems of developed and developing worlds are not always exclusive. Resistant shigellosis, tuberculosis, and gonorrhea strains have been introduced from the developing world into the developed world by travelers and immigrants.

To varying degrees, each of the factors identified by the IOM report is influencing antibiotic resistance. One important factor has been the changes that have occurred in human demographics. Aging, immunosuppressive therapy, underlying diseases, and malnutrition enhance susceptibility and lead to more infections, antibiotic use, and transmission of drug-resistant bacteria. Increasingly susceptible populations in intensive care units, for example, have been a major contributor to the emergence of vancomycin-resistant enterococcal infections. Also, in the United States, the increased susceptibility from HIV infection that occurred in the early 1990s was an important cofactor in the emergence of drug-resistant tuberculosis. Societal changes such as larger numbers of single-parent and two-income families increased the use of out-of-home day care and contributed to the emergence of drug-resistant *Streptococcus pneumoniae* (DRSP). The movement of immigrants and refugees has had an impact on diseases such as tuberculosis. Many of these immigrants and refugees are coming from parts of the world in which tuberculosis and drug-resistant tuberculosis are common. Currently, over one-third of new cases of tuberculosis in the United States are occurring among the foreign-born. These changes are affecting population susceptibility, social structure, and population movements and are increasing the transmission of a number of infectious diseases. Susceptibility to infections is increasing in both the developed and developing worlds.

Changing behaviors are also influencing emergence. Risky behaviors such as unprotected sex have resulted in increases in sexually transmitted diseases, again leading to more antibiotic use and subsequent resistance in disease such as gonorrhea and chancroid. Behaviors involving less clear risk, such as the choice and use of antibiotics, are also influencing resistance. Although inappropriate antibiotic use is becoming apparent as a contributor to resistance, even appropriate use can often have an unexpected impact. Such has been the case with the emphasis on the development and use of broad-spectrum antibiotics. Broad-spectrum antibiotics are by definition effective against a wide range of bacteria and appeal to both the pharmaceutical industry and to physicians. The appeal to the pharmaceutical industry is a broader market. The appeal to physicians, who are often using antibiotics empirically, is that a single broad-spectrum drug will more likely be effective. The impact of using such drugs, however, is that selective pressure is not confined to one or a few bacteria, as would occur with a narrow-spectrum drug such as penicillin, but the selective pressure is exerted against a much wider range of organisms. The extensive use of broad-spectrum cephalosporins in the 1980s in both the hospital setting and the community at large likely contributed to the emergence of vancomycin-resistant *Enterococcus* (VRE) and DRSP. In the case of VRE, enterococci are intrinsically resistant to cephalosporins; thus, use of the drugs killed off competitors and gave the enterococci an ecological advantage in settings where vancomycin was increasingly being used for the treatment of methicillin-resistant *Staphylococcus aureus* infections.

Changes in industry and technology particularly have influenced the resistance of foodborne pathogens such as *Salmonella* and *Campylobacter*. Practices associated with the intensive animal production and large-scale slaughter and processing that began in the 1950s have contributed to increases in the incidence of salmonellosis and the emergence of campylobacteriosis. Animal production was often coupled with antibiotic use for growth promotion, disease treatment, or prophylaxis. Thus, not only was salmonellosis increasing but so too was resistance, doubling between 1979 and 1994 to a point where over a quarter of all salmonellae were resistant to one or more antibiotics. The 1990s have witnessed similar events for *Campylobacter* as approval of fluoroquinolones for use in poultry has contributed to an increase in fluoroquinolone resistance from 0 to 13 percent between 1990 and 1998.

The impact of change in international travel and commerce is most obvious. Individuals and products can be moved between widely separated parts of the world within days or hours. Foods are increasingly important in international commerce. The potential impact of such commerce on resistance was demonstrated in a 1995 *Salmonella* outbreak associated with alfalfa sprouts in the United States and Finland in which the drug-resistant strain was eventually traced to contaminated seeds shipped to both countries from the Netherlands. Foreign travel has also increased dramatically within the last several decades, and people may introduce drug-resistant strains to new geographic areas. This is the case with the increase in fluoroquinolone-resistant *Campylobacter*, which is associated in part with acquisition during foreign travel. Other resistant bacteria, including *Shigella*, *Streptococcus pneumoniae*, and *Neisseria gonorrhoeae*, can also be acquired during travel and may be transmitted subsequently from person to person and become established in new geographic areas.

A final important factor in the emergence of resistance has been the breakdown in public health measures. This breakdown ranges from the loss of prevention and control programs in some areas to failures in basic hygiene and sanitation in others. In the United States, the emergence of drug-resistant tuberculosis followed the discontinuation in the 1980s of directly observed treatment programs as health departments moved scarce resources to address issues such as AIDS and chronic diseases. Thus, when tuberculosis reemerged in certain areas as a result of the HIV epidemic and immigration, the programs to ensure appropriate treatment no longer existed. Multidrug-resistant tuberculosis quickly became an important public health problem. Breakdowns in public health measures are not restricted to the developed world. Since the late 1970s, an epidemic of multidrug-resistant *Shigella dysenteriae* infections has plagued central and southern Africa, causing hundreds of thousands of deaths. This epidemic has persisted, in part, because of the public health breakdowns that have accompanied the wars, natural disasters, and population movements in this area.

Understanding the factors that are influencing the emergence of drug resistance can identify opportunities for intervention. The control and prevention of resistance require efforts to reduce the selective pressure of antibiotic use and the transmission of both susceptible and resistant bacteria in the developed and developing worlds. Efforts to improve appropriate antibiotic use in humans and animals and to reduce the incidence of infectious diseases will reduce the selective pressure that promotes the evolution of resistance. At the same time, improvements in hygiene, sanitation, infection control, and vaccine development and use will address the transmission of resistant bacteria. The continued development of new drugs and new classes of drugs is critical to treat infections that are rapidly becoming untreatable with currently available drugs. Nevertheless, unless efforts are made to address antibiotic use and disease transmission, the introduction of new drugs will only continue the cycle of infection, treatment, and resistance.

[*See also* Bacteria; Disease, *article on* Infectious Disease; Emerging and Re-Emerging Diseases; Plasmids.]

BIBLIOGRAPHY

Bloom, B. R., and C. J. L. Murray. "Tuberculosis: Commentary on a Reemergent Killer." *Science* 257 (1992): 1055–1064. A nice analysis of the factors influencing the reemergence of tuberculosis.

Centers for Disease Control and Prevention. *Preventing Emerging Infectious Diseases: A Strategy for the Twenty-first Century.* Atlanta, 1998. This report outlines strategies for addressing emerging infectious diseases, including those that are resistant to antimicrobial agents.

Ciba Foundation. *Antibiotic Resistance: Origins, Evolution, Selection and Spread.* Chichester, England, 1997. Examines issues surrounding resistance from the microbe to the human.

Dixon, D. M., et al. "Fungal Infections: A Growing Threat." *Public Health Reports* 111 (1996): 226–235. Examines factors influencing the emergence of fungal infections and emphasizes the changes in population susceptibility that are also influencing antimicrobial resistance.

Garrett, L. *Betrayal of Trust: The Collapse of Global Public Health.* New York, 2000. Examines the evolution and current state of public health and the array of public health challenges, including antimicrobial resistance.

Golub, E. S. *The Limits of Medicine.* New York, 1994. An interesting perspective on the evolution of medicine and public health.

Institute of Medicine. *Emerging Infections: Microbial Threats to Health in the United States.* Washington, D.C., 1992.

Institute of Medicine. *Antimicrobial Resistance: Issues and Options.* Washington, D.C., 1998. Report of a workshop examining the problem and potential interventions.

Levy, S. B., and R. V. Miller. *Gene Transfer in the Environment.* New York, 1989. A nice discussion of genetic processes influencing the emergence of resistance.

Mahon, B. E., et al. "An International Outbreak of *Salmonella* Infections Caused by Alfalfa Sprouts Grown from Contaminated Seeds." *Journal of Infectious Diseases* 175 (1997): 876–882. Demonstrates the international spread of drug-resistant organisms by a food commodity.

— MITCHELL L. COHEN

Strategies for Managing Resistance

Modern physicians have more than a hundred different antibiotics at their disposal for the treatment of bacterial infections. Almost without exception, however, these compounds belong to classes of drugs introduced into clinical use before 1970. Since the 1970s, pharmaceutical chemists have "evolved" new antibiotics by tinkering with the chemical structure of existing compounds (Figure 1). This stepwise modification of antibiotic structure has facilitated the rapid evolution of bacterial resistance mechanisms. Often when a new drug is introduced, genes conferring resistance to this drug already exist, or, if they do not, they evolve rapidly by small changes in existing resistance mechanisms.

The dearth of novel classes of antibiotics has coincided with substantial increases in the prevalence of antimicrobial resistance in nearly every important bacterial pathogen. The resulting need to maintain the efficacy of existing drugs has focused attention on strategies to limit the appearance and spread of resistance.

Appropriate Strategies Depend on Pathogen Genetics, Epidemiology. The appropriate design of such strategies depends on the genetic mechanisms by which bacteria become resistant, as well as on the selective mechanisms by which antibiotic use increases the frequency of resistant organisms in treated individuals and the population. Both kinds of mechanisms vary widely among different pathogens, but two broad groups may be defined.

The first group includes those infectious agents that are obligate pathogens—causing disease in humans is a necessary part of the transmission cycle—and for which antibiotic resistance can arise readily during treatment of an individual patient, usually by one or a few nucleotide substitutions in the pathogen's genome. The best exemplar of this class is tuberculosis (TB). (We do not consider viral infections here, but human immunodeficiency virus and other viral infections are similar to TB in this respect.) For such infections, preventing the emergence of resistance during treatment of individual

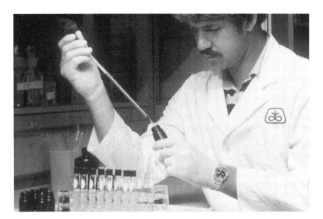

FIGURE 1. Since the 1970s, Pharmaceutical Chemists Have "Evolved" New Antibiotics by Tinkering with the Chemical Structure of Existing Compounds.
Visuals Unlimited.

hosts is central to the effort to control resistance at the population level. The World Health Organization promotes directly observed therapy of tuberculosis (DOTS)—a treatment regimen in which patients are prescribed multiple drugs and health care workers watch them take every dose of medication to ensure adherence to the regimen—as the centerpiece of programs to control TB and to minimize the spread of drug-resistant infections.

The scientific basis for regimens like DOTS was articulated as early as 1913, by Paul Ehrlich, the inventor of Salvarsan, one of the first effective antibacterial drugs. To maximize the chance of success in treating an individual patient, he advocated that treatment should "hit hard and hit early," and that patients should be treated with a combination of different antimicrobial agents. As understanding of the genetic basis of antimicrobial resistance has grown, it has become clear that for infections like TB, these principles—now phrased as early treatment, maintenance of adequate doses for long enough to eradicate infecting bacterial populations, and combination therapy—not only maximize treatment success but also reduce the chances that resistant organisms will emerge in treated patients. Early treatment and adequate dosing minimize the size of the bacterial population, thereby reducing the probability that resistance mutations will occur. Combination therapy helps to ensure that even when such mutations do occur, any bacterium resistant to one of the drugs will still be subject to killing by the other drugs. The basis for combination therapy can be understood quantitatively. In tuberculosis, mutants resistant to any individual drug will constitute between one in 10^9 and one in 10^6 bacteria. Because resistance to each of the major antituberculosis

drugs is by a different mechanism, the probability that an individual bacterium will be resistant to two drugs is the product of these probabilities, or about one in 10^{18} to 10^{12} (for three drugs, one in 10^{27} to one in 10^{18}). The size of the mycobacterial population in a typical infection is such that, at the start of treatment, mutants resistant to a single drug will usually be present; those resistant to two drugs may be present in rare cases; and those resistant to three or more drugs will almost never be present.

For infections like TB, measures to minimize the transmission of resistant infections when they do arise are necessary supplements to preventing the emergence of resistance in the first place. For a second group of bacterial pathogens, preventing transmission of resistant organisms assumes a more central role, because of differences in the epidemiology and genetics of resistance in these pathogens. Many of the bacterial pathogens of greatest concern at present—for example, *Streptococcus pneumoniae* (pneumococcus), *Staphylococcus aureus*, *Enterococcus faecium*, and Gram-negative bacilli responsible for many hospital-acquired infections—differ in two important respects from TB. First, these organisms typically are members of the commensal flora of the upper respiratory tract, gut, or skin. As a result, antibiotics exert selection for resistance not only when these organisms are themselves the targets of therapy (as, for example, when *S. aureus* from the skin enters the bloodstream through a wound and causes a blood infection), but also when an individual is treated for an infection caused by another organism and these commensals are exposed to antibiotics as "innocent bystanders." Second, in most cases, antibiotic resistance in these organisms cannot develop by a single point mutation, as it does in TB. Instead, development of resistance requires the acquisition of a new allele, gene, or set of genes by horizontal transfer of DNA from another organism. In these pathogens, preventing emergence of resistance during treatment takes second place to preventing transmission of resistant organisms, both because emergence of resistance is rare and because the success of treatment in curing a patient's infection is often decoupled from the selective effect of treatment on commensal organisms.

Interventions to reduce transmission of resistant bacteria are particularly important in hospitals. Such measures include use of disposable gowns and gloves by health care personnel to avoid contamination, procedures to screen patients for resistant bacteria and "cohort" or isolate those who carry them from the rest of the population, and frequent hand washing. Controlling transmission of resistant organisms in the community (outside of hospitals) is more difficult, but basic hygiene measures in other settings where resistant organisms are present, such as day care centers and nursing homes, serve similar purposes. In an effort to control the spread of penicillin-resistant *Streptococcus pneumoniae*, Malmöhus County in southern Sweden instituted a policy in 1995 of excluding children carrying this resistant organism from group day care (Ekdahl, 1998).

A simple strategy for controlling antimicrobial resistance in these pathogens is to reduce use of the selecting antibiotics or, more generally, to encourage use of antibiotics in ways that minimize the selective pressure for resistance. Such strategies, variously referred to as "judicious" or "prudent" use of antibiotics, have been widely promoted both within hospitals and for outpatient prescribers. Although the basic principle of avoiding unnecessary use of antibiotics (e.g., for treatment of viral infections) is undoubtedly useful in the effort to control resistance, the principles of prudent antibiotic use for treatment of many bacterial infections are poorly defined. Rules governing the licensing of antibiotics give little or no consideration to resistance, so there is often little information to aid physicians in defining the optimal choice of drug or dosing regimen to minimize resistance while maintaining efficacy in curing the patient.

Further complicating the implementation of "judicious use" policies is a conflict between the real or perceived interests of the individual patient and the interests of society. For a patient with an infection of the upper respiratory tract or a mild diarrheal infection, which may or may not be bacterial, antibiotic use offers the possibility of reducing the duration of symptoms. The increased risk of selecting resistant organisms that may later cause infection in that same individual may be small or, in some cases, even nonexistent. From the perspective of the community as a whole, however, each individual use of antibiotics provides additional selective pressure for resistance. In effect, this problem may represent a kind of "tragedy of the commons," whereby there is a conflict between the costs and benefits to an individual and to the community as a whole. Other strategies that have been proposed and tested for controlling resistance include deliberate diversification or cycling of antimicrobial drug classes and the use of "narrow spectrum" agents that are active against only a limited set of bacteria, in an effort to minimize the intensity of selection for resistance to any single drug or drug class.

A final class of strategies for controlling resistance relies on reducing individuals' exposure to resistant pathogens. Recently, a number of countries have initiated regulations to ban or restrict the use of antibiotics for growth promotion and treatment of food animals, in an effort to reduce human exposure to foodborne resistant pathogens. Vaccines can in some cases also be tools in the fight against resistance. In 2000, a new vaccine against the pneumococcus was introduced. This pathogen is highly polymorphic, with some ninety immunologically distinct serotypes; the vaccine includes only seven serotypes, which are especially common causes of severe disease. As it happens, these serotypes are also the ones with by far the highest levels of antibiotic resistance, so the selective effects of the vaccine against these serotypes will also indirectly select against resistant organisms. Furthermore, by preventing a small but significant fraction of childhood ear infections, the vaccine is expected to reduce the largest single source of demand for antibiotics in human medicine.

Forces Tending to Reduce Resistance in the Absence of Selection. The premise of most strategies for managing resistance is that if selective pressure in favor of resistance is reduced, then levels of resistance will decline. This will occur only if there is some force, selective or otherwise, acting to increase the prevalence of antibiotic-susceptible bacteria. There are two major candidates for such forces. In some cases, resistance to antibiotics carries a "fitness cost": resistant organisms grow more slowly or colonize hosts less efficiently than their susceptible counterparts. The physiological basis for this fitness cost depends on the mechanism of resistance, but may take the form of slower gene transcription or translation, changes in molecular transport across the cell membrane, or reduction in the activity of an enzyme important for cellular metabolism. Experimental studies have shown that many (though not all) mechanisms of resistance, including both point mutations and plasmid acquisition, when introduced into a bacterial strain, place that strain at a growth disadvantage in vitro or in animal infections compared to an otherwise isogenic, drug-susceptible strain. However, following passage of these resistant strains through several hundred bacterial generations in the absence of antibiotic selection, the fitness disadvantage has in most cases been reduced or even eliminated by the appearance of one or more compensatory mutations in the bacterial chromosome. The bacteria achieve this without losing their resistance to antibiotics. Investigations into the importance of such compensatory evolution in "natural" (clinical) isolates are just beginning.

A second mechanism that operates to reduce the frequency of resistant organisms, particularly in settings such as hospitals, is the import of susceptible bacteria by newly admitted patients. Many resistant organisms are more common in hospitals and other settings with high levels of antibiotic use and transmission than in the community at large. Because many of the pathogens of clinical importance in these settings are members of the normal, commensal microflora, the discharge of patients and the admission of new patients whose commensal flora is more susceptible may serve to "dilute" the prevalence of resistant organisms in the hospital. In hospitals, where the average length of stay is on the order of a week, mathematical models have suggested this phenomenon may be responsible for the rapid declines in resistance sometimes observed following interventions to curtail antibiotic use, pathogen transmission, or both. Such declines, due to dilution effects, can occur even in the absence of a fitness cost of resistance. Dilution of resistance by susceptible bacteria is comparable to the deliberate establishment of pesticide-free "refugia" in an effort to control pesticide resistance. Similar effects for antibiotic resistance may be present in other semiclosed populations with high levels of transmission and antibiotic use, such as day care centers and nursing homes.

Evaluation of Strategies to Control Resistance. Evaluation of strategies to control resistance has been impeded by the expense and complexity of testing a given strategy in even a small number of hospitals or populations. Especially in hospitals, many interventions have been undertaken in response to an identified problem with a resistant organism or group of organisms, so appropriate controls are not available to evaluate the success of the intervention. Because of such problems, it remains difficult to state with certainty which interventions will be most effective in which circumstances.

For community-acquired infections, there are few systematic studies of the success of efforts to reduce resistance, and those that do investigate these effects suffer from lack of control and replication. In the few studies that do exist, the effectiveness of such interventions has been difficult to interpret. For example, use of macrolides (a class of antibiotics) in Finland was reduced by almost 50 percent between 1988 and 1995. Published data show that resistance in group A streptococci to macrolides in Finland rose from 5 percent in 1988–1989 to a peak of 19 percent in 1993, then declined to about 9 percent in 1996 (Seppala et al., 1997). Although this is often interpreted as a successful effort to control resistance, the temporal trends do not show any clear relationship between the level of antibiotic use in a given year and the level of resistance in that year. One hypothesis to explain this and similar findings is that there is a time lag between reductions in use and declines in resistance. Mathematical models suggest that such a time lag would be expected for community-acquired infections if the fitness cost of resistance was quite low.

Many bacterial pathogens are resistant to more than one class of drugs. This can occur either because the same bacteria acquire multiple chromosomal genes or mutations conferring multiple drug resistance or because a strain acquires a plasmid carrying multiple resistance genes. In these cases, efforts to control resistance by reducing one antibiotic may be stymied by continuing selection by other, unrelated drugs.

Few novel antibiotic classes are expected in clinical practice in the near future. As a result, efforts to control resistance by promoting appropriate dosing of antibiotics and adherence to treatment regimens and by discouraging unnecessary antibiotic use will remain critical to preserving antibiotic effectiveness. Theoretical and empirical understanding of the precise impact of these efforts is still at an early stage.

[*See also* Acquired Immune Deficiency Syndrome, *article on* Dynamics of HIV Transmission and Infection; Bacteria and Archaea; Disease, *article on* Infectious Disease; Emerging and Re-Emerging Diseases; Pesticide Resistance; Plasmids; Resistance, Cost of; Vaccination.]

BIBLIOGRAPHY

Andersson, D. I., and B. R. Levin. "The Biological Cost of Antibiotic Resistance." *Current Opinion in Microbiology* 2 (1999): 489–493. A highly readable summary of evidence about the effects of antibiotic resistance on the fitness of bacteria.

Austin, D. J., and R. M. Anderson. "Studies of Antibiotic Resistance within the Patient, Hospitals and the Community Using Simple Mathematical Models." *Philosophical Transactions of the Royal Society of London B: Biological Sciences* 354 (1999): 721–738. An overview of several different approaches to mathematical modeling of antibiotic use and resistance.

Ehrlich, P. "Address in Pathology on Chemotherapeutics: Scientific Principles, Methods, and Results." 1913. *The Lancet* (1913, August 5): 445–451. A classic statement of principles of antimicrobial treatment.

Ekdahl, K., H. B. Hansson, S. Molstad, M. Soderstrom, M. Walder, and K. Persson. "Limiting the Spread of Penicillin-Resistant *Streptococcus pneumoniae*: Experiences from the South Swedish Pneumococcal Intervention Project." *Microbial Drug Resistance* 4 (1998): 99–105.

Kollef, M. H., and V. J. Fraser. "Antibiotic Resistance in the Intensive Care Unit." *Annals of Internal Medicine* 134 (2001): 298–314. A comprehensive review of measures taken to curb antibiotic resistance in hospitals (most such interventions have taken place in intensive care units).

Lipsitch, M. "Measuring and Interpreting Associations between Antibiotic Use and Penicillin Resistance in *Streptococcus pneumoniae*." *Clinical Infectious Diseases* 32 (2001): 1044–1054. Using a case study, describes methodological approaches to separating the effect of drug treatment on resistance in treated individuals and in populations.

Lipsitch, M. "The Rise and Fall of Antimicrobial Resistance." *Trends in Microbiology* 9 (2001): 438–444. Reviews evidence that addresses the question of what determines the rate of evolution of antimicrobial resistance in pathogen populations.

Schrag, S. J., B. Beall, and S. F. Dowell. "Limiting the Spread of Resistant Pneumococci: Biological and Epidemiologic Evidence for the Effectiveness of Alternative Interventions." *Clinical Microbiology Reviews* 13 (2000): 588–601. An excellent review, informed by evolutionary thinking, of current knowledge about the effectiveness of interventions to control resistance in *Streptococcus pneumoniae*.

Seppala, H., T. Klaukka, J. Vuopio-Varkila, A. Muotiala, H. Helenius, K. Lager, and P. Huovinen. "The Effect of Changes in the Consumption of Macrolide Antibiotics on Erythromycin Resistance in Group A Streptococci in Finland." *New England Journal of Medicine* 337 (1997): 441–446.

— MARC LIPSITCH

APOPTOSIS

Apoptosis, programmed or physiological cell death, is an active self-killing process of cells that regulates cell number and tissue architecture, in both embryonic and adult animals. It is a genetically controlled and evolutionary conserved process that comprises two distinct and sequential processes: the death of cells and their subsequent removal by phagocytes (see reviews in Heemels, 2000). In humans, many viral, immune, and degenerative diseases and some tumors are associated with impaired or excessive apoptosis.

The idea that cell death constitutes a physiological cell behavior characteristic of multicellular organisms was advanced in the 1950s on the basis of the observation of massive degenerative processes in the course of normal embryonic development and larval metamorphosis (Glücksmann, 1951). Since then, the study of cell death has received increased attention. Information accumulated in the last decade revealed the existence of a suicide program within the cell that can be activated or repressed by a variety of physiological and pathological stimuli. This suicide program constitutes the molecular basis of apoptosis and allows us to distinguish this active form of cell death from necrosis, which is a passive phenomenon consisting of the disruption of cell integrity caused by external agents. The extraordinary biological interest of apoptosis is because it provides a major explanation of how living organisms sculpt and maintain their forms and structures during development and evolution and because of the emerging possibility of manipulating the death program for therapeutic purposes in clinical medicine.

Identification of Apoptosis. Apoptosis was first defined by Kerr et al. (1972) on the basis of the morphological appearance of cells in sections of fixed tissues, including nuclear and cytoplasmic condensation followed by cellular fragmentation (Figure 1A and B). However, different morphologies of active cell death might be present in different animals or tissues (see Bursch et al., 2000). Currently, the so-called TUNEL assay based on the detection of cells with fragmented DNA (Figure 1C) and the use of Annexin V to detect changes in the plasma membrane of apoptotic cells are two widely used methods for the diagnosis of apoptotic cells. In living embryos, apoptosis can be detected by vital staining, but this technique stains mainly the phagocytic cells responsible for removing cell death corpses (Figure 1D and E). Biochemically, the detection of DNA fragmen-

C. ELEGANS

The nematode *Caenorhabditis elegans* (*C. elegans*) constitutes a paradigmatic model for the study of the cellular and genetic mechanisms that regulate embryonic development. The small size of the worm and the ease with which embryonic and adult development can be observed have allowed precise tracing of the fate of all cells from the single-cell egg to the adult worm, facilitating the identification and characterization of molecular pathways involved in cell proliferation and migration, lineage specification and cell death. During development of the adult hermaphrodite worm, 131 of the 1,090 somatic cells die by apoptosis. Genetic studies in mutant worms with an altered pattern in cell death have identified a complex machinery of factors required for the execution and regulation of apoptosis and for the clearance and degradation of apoptotic bodies. At the present time, fifteen different factors have been identified (see Liu and Hengartner, 1999).

These factors have been divided into four groups based on the order of their activity during the process of programmed cell death: (1) the *Ces-1* and *Ces-2* (cell death–deficient) genes are involved in the decision making for cell death and regulate apoptosis in specific neurons; (2) the *Ced-3*, *Ced-4*, *Ced-9*, and *Egl-1* genes participate in the execution of apoptosis in all the cells; (3) the *Ced-1*, *Ced-2*, *Ced-5*, *Ced-6*, *Ced-7*, *Ced-10*, and *Ced-12* genes coordinate the engulfment of dying cells by phagocytic cells; and (4) the *Nuc-1* gene is involved in the phagocytic degradation of dead cells.

—JUAN M. HURLE AND RAMON MERINO

tation by electrophoresis and the characterization of caspase activation (discussed in the following) are good approaches to demonstrate apoptotic processes.

Genetic Regulation of Apoptosis. The suicide genetic program of apoptosis appears to be complex and to include genes involved in the onset of cell death and genes protecting cells from entering the death program. A major advance in the knowledge of the molecular basis of apoptosis has been the study of cell death in the nematode *Caenorhabditis elegans* (see Liu and Hengartner, 1999). Initial approaches revealed the occurrence of cell death in the course of development of this worm. These studies were followed by the isolation of mutants with alterations in the pattern of cell death and by the subsequent identification of several genes responsible for their control. Four genes, termed cell death genes (*Ced*) *Ced-4*, *Ced-3*, and *Ced-9* and *Egl-1*

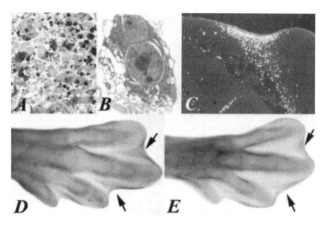

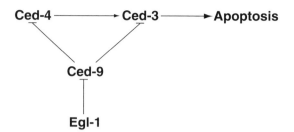

FIGURE 2. Schematic Representation of the Genetic Regulation of Apoptosis in the Nematode *Caenorhabditis elegans*. Juan M. Hurle.

FIGURE 1. Apoptosis. Identification of apoptosis in tissue sections (A) and in transmission electron microscopy (B). Note that the condensation of the nuclei and cytoplasm is the most significant alteration of the cells. A histological section (C) of an interdigital region of the developing chick leg bud (see D, labeled by TUNEL). Note that most interdigital cells are positive for this assay, which indicates that the DNA is fragmented in cells. The feet of a chick (D) and a duck (E) embryo at equivalent stages of development vital stained with neutral red to show the pattern of interdigital cell death. Whereas most interdigital cells in the chick are eliminated by apoptosis, only the most distal interdigital mesoderm undergoes apoptosis in the duck. Arrows indicate the interdigits to compare the intensity of apoptosis in both species.   Juan M. Hurle.

(egg-laying defective), are the major players accounting for the survival or death of cells during the embryonic development of the worms. *Ced-4* exerts a pro-apoptotic effect by activating the protease *Ced-3*, which is directly responsible for the induction of cell death. Both *Ced-4* and *Ced-3* activities are inhibited by *Ced-9*, which in turn is negatively regulated by *Egl-1* (Figure 2). One additional major advance in the knowledge of the regulatory mechanisms of apoptosis has been the discovery that this basic cell death machinery operating in *C. elegans* is highly conserved throughout the animal kingdom (see Aravind et al., 2001). Gene products homologous to the *C. elegans Ced*-genes have been discovered in vertebrates and invertebrates. However, a marked expansion in the number of members of each individual *C. elegans* component is observed in species such as mammals (Figure 3). The vertebrate homologous of *Ced-4* is *Apaf-1* (apoptotic protease-activating factor), which is the prototype of a still growing family of pro-apoptotic factors (see Strasser et al., 2000). At least 14 *Ced-3*-like proteases, called caspases, have been identified at present in mammals and, like *Ced-3*, are the direct effectors of cell death in these animals. In addition, the *Bcl-2* family of apoptotic regulators (B cell lymphocyte leukemia gene) is the mammalian counterpart of *Ced-9* in *C. ele-*

gans (see Gross et al., 1999). All members of this family possess three or four conserved motifs essential for their activity known as *Bcl-2* homology domains (BH1 to BH4). The *Bcl-2* family evolved primarily through gene duplication, giving rise to many members, that, like *Ced-9*, inhibit cell death (e.g., *Bcl-2*, *Bcl-xL*, *Bcl-w*, *Mcl-1*, and *A1*) and also to members with pro-apoptotic activity (i.e., *Bax*, *Bak*, or *Bcl-xs*). Such a functional diversification of *Bcl-2* family members might have been established at the level of coelomates. An additional group of apoptotic regulators in mammals are homologues of *Egl-1*. These members (*Bad*, *Bid*, *Bim*, *Bik*, and *Hrk*) share with each other the BH3 domain, which is also present in the *Bcl-2* family. Like *Egl-1*, the BH3-only members are potent inducers of cell death by repressing the anti-apoptotic activity of *Bcl-2* members. Factors with homology to these cell death regulators have also been discovered in *Drosophila melanogaster* (see Meier et al., 2000).

The apoptotic program is triggered by multiple signals such as growth factor deprivation, hormones, intracellular stress such as DNA damage, and gamma irradiation, among others. In the course of evolution, vertebrates developed additional mechanisms for the transmission of the external cell death stimuli to the executors of apoptosis described above. A family of cell surface receptors, the tumor necrosis factor (TNF) receptor family, is responsible for transduction of cell death signals after binding with specific ligands belonging to the TNF family (see Strasser et al., 2000). The death signal mediated by these receptors is transduced through interactions with intracellular adaptor proteins resulting in caspase activation (see Figure 3). These interactions are mediated by specific functional cell death regulatory domains shared by the receptors, the adaptors, and the caspases, and also by other apoptotic regulators, such as *Apaf-1/Ced-4*. Different types of these regulatory domains have been identified (death domains, death effector domains, caspase recruitment domains, and pyrin domains). The specificity of these domains for the proteins implicated in the control of

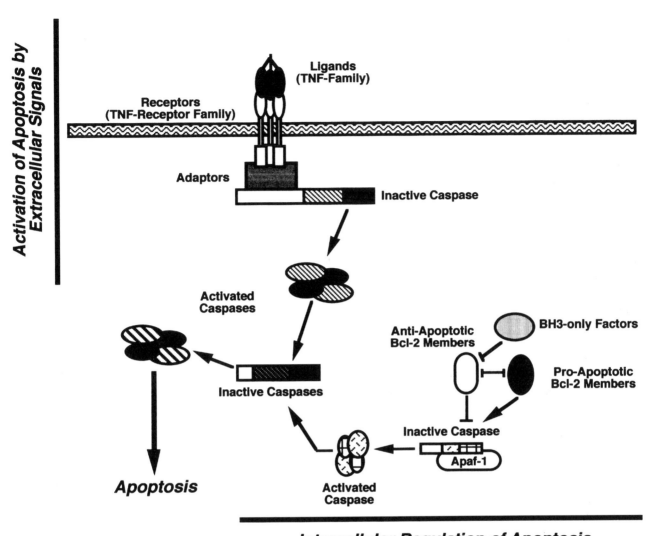

FIGURE 3. Genetic Regulation of Apoptosis in Vertebrates.
The upper part of the figure represents the signal transduction pathway mediated by specialized cell death membrane receptors, belonging to the TNF-receptor family, resulting in caspase activation and apoptosis. In the lower part of the figure is represented the basic intracellular elements that regulate cell death. These elements are the vertebrate counterparts of apoptotic regulators in *C. elegans* (see Figure 2) and control cell death mediated by intracellular and extracellular signals. Juan M. Hurle.

apoptosis is currently used as a tool to identify new factors of the apoptosis molecular machinery.

Physiological Significance of Apoptosis. In an early theoretical article, Umansky (1982) proposed that in evolution, cell death genes were activated at the early stages of the formation of multicellular organisms to eliminate damaged or abnormally functioning cells that presented a real or potential danger for the survival of the whole organism. This hypothesis has received considerable support with advances in the knowledge of the molecular basis of cell death. As mentioned above, the presence of damaged or single-stranded DNA consti-

tutes one of the mechanisms activating the cell death program. Conversely, many viruses carry genes encoding proteins that block the cell death program, suggestive of an evolutionary specialization of these viruses to inactivate the possibility of an antiviral defensive strategy of the multicellular organisms based on destruction of the cells infected by the virus.

In the course of evolution, organisms would have found a number of uses for the cell death program. In fact, there are many examples of the participation of cell death in physiologic processes. As mentioned above, during embryonic development and larval metamorphosis, the formation of many parts of the body involves the elimination by cell death of large cell populations, or even of all the cellular components of an organ rudiment (reviewed by Sanders and Wride, 1995). The formation of the digits during vertebrate limb development provides a paradigmatic example of this sculpturing role of apoptosis during development. In amniota, digits develop as chondrogenic rays, which differentiate into hand or feet primordia. Initially, the digits are joined by undifferentiated mesodermal tissue, but this interdigital tissue is next removed by massive apoptotic processes (formerly called interdigital necrotic areas). The extent of interdigital cell death correlates closely with the free digits (chick, lizard, or human) or webbed digits (duck, tortoise, or seal), phenotypes characteristic of the different species (refer to Figure 1D and E). In birds, at the molecular level, the diversification of the free digit/webbed digit phenotypes might be caused by minor quantitative differences in the local production of secreted factors (bone morphogenetic proteins and fibroblast growth factors) that regulate the interdigital apoptotic process (see Gañan et al., 1998).

The development of the structural connections of the nervous system in vertebrates constitutes another remarkable example of morphogenesis mediated by apoptosis. In developing neural nuclei and ganglia, neuroblasts are produced in excess, their survival requiring specific neurotrophic factors (nerve growth factor, neurotrophin-3 and -4, and brain-derived neurotrophic factor) delivery by their target tissues. Apoptosis eliminates neurons that fail to connect with their corresponding target.

In addition to these examples, in which cell death occurs at precise temporal and spatial patterns, apoptosis accounts for cell turnover and for the controlled elimination of cells associated with changes in the functional activity of organs or tissues. The elimination of lymphocytes during receptor repertoire selection and the differentiation of red blood cells or the atrophy of tissues linked to endocrine modifications are examples that illustrate the widespread participation of cell death in differentiation and tissue homeostasis.

[See also C. elegans; Key Innovations.]

BIBLIOGRAPHY

Aravind, L., V. M. Dixit, and E. V. Koonin. "Apoptotic Molecular Machinery: Vastly Increased Complexity in Vertebrates Revealed by Genome Comparisons." Science 291 (2001): 1279–1284. Focuses on the evolution of the apoptotic molecular machinery.

Bursch, W., A. Ellinger, C. Gerner, U. Frohwein, and R. Schulte-Hermann. "Programmed Cell Death (PCD): Apoptosis, Autophagic PCD, or Others?" Annals of New York Academy of Sciences 926 (2000): 1–12. Reports different types of self-killing cell processes.

Gañan, Y., D. Macias, R. D. Basco, R. Merino, and J. M. Hurle. "Mor-

phological Diversity of the Avian Foot Is Related with the Pattern of msx Gene Expression in the Developing Autopod." Developmental Biology 196 (1998): 33–41. Experimental analysis of the control of the different morphology of the digits in chick and duck embryos.

Glücksmann, A. "Cell Death in Normal Development." Biological Reviews 26 (1951): 59–86. This is one of the first reviews in which cell death is considered as a physiological process.

Gross, A., J. M. McDonnell, and S. J. Korsmeyer. "Bcl-2 Family Members and the Mitochondria in Apoptosis." Genes and Development 13 (1999): 1899–1911. A detailed review of the structure and function of the Bcl-2 members in the control of apoptosis.

Heemels, M. T., ed. "Nature Insight: Apoptosis." Nature 407 (2000): 769–816. A series of highly interesting reviews covering all aspects of apoptosis.

Kerr, J. F., A. H. Wyllie, and A. R. Currie. "Apoptosis: A Basic Biological Phenomenon with Wide-ranging Implications in Tissue Kinetics." British Journal of Cancer 26 (1972): 239–257. In this article, the term apoptosis is proposed for the first time, to define an active form of cell death.

Liu, Q. A., and M. O. Hengartner. "The Molecular Mechanism of Programmed Cell Death in C. elegans." Annals of the New York Academy of Sciences 887 (1999): 92–104. A detailed review of the regulation of apoptosis in C. elegans.

Meier, P., A. Finch, and G. Evan. "Apoptosis in Development." Nature 12 (2000): 796–801.

Sanders, E. J., and M. A. Wride. "Programmed Cell Death in Development." International Review of Cytology 163 (1995): 105–173. Detailed account of developmental processes involving apoptosis.

Strasser, A., L. O'Connor, and V. M. Dixit. "Apoptosis signaling." Annual Review of Biochemistry 69 (2000): 217–245.

Umansky, S. R. "The Genetic Program of Cell Death. Hypothesis and Some Applications: Transformation, Carcinogenesis Aging." Journal of Theoretical Biology 97 (1982): 591–602. This review proposes for the first time the potential functions and the biological significance of programmed cell death.

— JUAN M. HURLE AND RAMON MERINO

ARCHAEA. See Bacteria and Archaea.

ARCHAEOLOGICAL INFERENCE

The earliest archaeological sites on earth, dating to around the same time as the earliest fossils of the genus *Homo*, are simple patches of stone tools, intermixed sometimes with broken bones of animals. These sites testify to the lives of early hominids, but what exactly are they telling us? Were early human ancestors managing to scavenge an occasional marrow bone from lion or sabertooth kills and retreating to bash them open in a safe refuge? Or were hominid males regularly hunting large prey, carrying meaty parts to share with their mates and offspring at a home base? These and other possible scenarios for the formation of these ancient sites continue to be debated by researchers. The controversy exists because only some vestiges of hominid behavior have been preserved for more than a million and a half years. Because of this scant evidence and our awareness that many other processes contribute to forming and transforming ancient sites, methods and standards of archaeological inference are important research components.

Archaeology shares with paleoanthropology, paleontology, historical geology, and astronomy the study of unseen processes at a great remove in time, using the traces those processes left behind. Unlike scientists studying shorter term processes, historical scientists cannot repeat experiments or directly monitor relationships to test hypotheses about what created the evidence they have. Like specialists in criminal forensics, historical scientists must use indirect evidence to reconstruct past agents, processes, and events. The lack of directly observable causes for their evidence requires that investigators specify how it reflects past processes and events, and why they believe it does; in other words, they must build discipline-specific "casebooks" of relevant evidence. Historical sciences also share general frameworks for making and assessing inferences, that is, a distinct methodology.

This entry reviews archaeological inference in Anglophone archaeology. Because regional traditions still strongly influence archaeological practices, no claim can be made for universal consensus on archaeological inference. However, English-language research accounts for the preponderance of early paleolithic studies, where a wider consensus on methodology is emerging among researchers of diverse nationalities.

Analogy in Archaeology. Despite ongoing debate about whether archaeology is a humanity, a social science, or an evolutionary science, a surprising degree of unanimity exists on some points of method, based on a widespread acknowledgment that archaeological inference parallels that in other historical sciences. Historical sciences share the use of analogies and uniformitarian assumptions in making inferences about past agents, processes, and events. They also differ from the nonhistorical sciences in the degree to which analogy permeates the research process.

Analogy usually plays an exploratory role in scientific research, permitting researchers to reconceptualize an unfamiliar phenomenon and evoke new information about it, or to extend a theory by asserting that one phenomenon resembles another, admittedly nonidentical, one.

Historical scientists continue to use analogies in every phase of their research. This research approach depends on the assumption that those processes have remained uniform in their operation and outcomes over long spans of time. Simpson (1970) called this assumption "methodological uniformitarianism."

Following controversies about the proper use of anal-

ogy in the 1960s and 1970s, most archaeological theorists agreed that archaeological inference is fundamentally and inescapably analogical.

Theory that identifies processes responsible for archaeological evidence has been called "interpretive theory" or, more widely, "middle range theory" (see Binford, 1981). It depends on the assumption that investigating contemporary processes and their products will shed light on analogous processes in the remote past. Archaeologists also use analogical reasoning to formulate and select among hypotheses, against a largely unarticulated backdrop of analogies from modern cases. Although this less explicit use of analogy in problem definition and hypothesis formation seems to lack rigor, it has been discussed as a powerful research tool in other sciences (see Wylie, 1985).

Actualism, a term imported from a parallel use in paleontology, denotes research on causal relations in the present to understand evidence from the ancient past. It encompasses experimental archaeology, ethnoarchaeology, taphonomy, and other research on the "source side" of analogues for prehistoric evidence. Monitoring both processes and their consequences, actualistic research has built up an impressive casebook of "signatures" of various agents in archaeological materials. Lithic tool edge use and bone modification are particularly well-studied and productive areas in such agent-identification research. (See Vignette on The Case of the "Killer Ape.")

Agent-identification studies "work" because one can trace a clear causal path from a process to the evidence it produces in present-day situations. In most cases, the causal relations are straightforward: a specified process is "necessary and sufficient condition" for its outcome. Historical scientists use such strong cause–effect relations to infer from prehistoric evidence that an analogous, albeit unobservable, cause operated in ancient times. According to philosophers of science, inferences from such "relational analogies" are strongly warranted, because they involve clear cause–effect relations (see Vignette on the site at Olduvai Gorge).

Ultimately, however, investigators are less interested in whether hominids used stone tools to bash or cut bones than they are in the circumstances of those actions. Was it a daily activity, indicating reliance on meat, or a very rare event of the kind that left much more archaeology than the usual daily feeding activities? In an evolutionary paradigm, we expect that archaeologically visible behaviors emerged as part of new adaptations, implying some causal relationship between the broader adaptive context and the evidence at hand. The problem is how to move systematically from the evidence to these higher-order "target" issues.

In an apparent paradox, good actualistic identification of various agents' "signatures" on archaeological materials did not appreciably curb controversies over

THE CASE OF THE "KILLER APE"

The case of the "killer ape" is an excellent example of applying actualistic research to a research problem in human evolution. In the 1950s, Raymond Dart, who named the Taung *Australopithecus* find in 1929, vigorously argued that australopithecine and other animal bones from cave breccias of the Transvaal region testified to violent deaths at the hands of other australopithecines. Dart contended that the pattern of predominantly jaw and limb bones in the deposits reflected australopithecine selectivity in choosing weapons of the hunt. Moreover, he found on bones of both hominids and other animals what appeared to be marks of violence, including double depressed fractures in some australopithecine crania. Dart's thesis that the roots of modern human warfare and crime lay in our remote "killer ape" ancestors were popularized by playwright Robert Ardey in *African Genesis*. In the 1970s, C. K. Brain, of the Transvaal Museum, began modern experiments to assess rates of survival of skeletal parts in various contexts, including leopard lairs and other carnivore kills, as well as patterns of damage to bones of carnivore prey animals. Brain's careful studies, synthesized in *The Hunters or the Hunted? An Introduction to South African Cave Taphonomy* (1981), showed that hominid action was not the only means of producing the patterns of damage and bone element representation Dart had cited as "proof" of australopithecine hunting violence. In fact, the double depressed fractures to the cranium of one such hominid were more likely the result of a leopard bite. Thus, research on modern cases widened the view of "necessary and sufficient" causes of archaeological evidence.

—DIANE GIFFORD-GONZALEZ

the hominid behaviors and adaptive strategies that produced prehistoric assemblages.

Defining Levels of Cause and Effect. In "Bones Are Not Enough: Analogues, Knowledge, and Interpretive Strategies in Zooarchaeology," Diane Gifford-Gonzalez (1991) tried to address the linkage problem by proposing a nested hierarchy of causal terms to clarify archaeology's analytic targets. An *actor* is what we normally think of as a causal agent, a hominid wielding a hammerstone to crack a bone open, or a hyena chewing on it. However, the actual bone modification results from a more specific, physical *causal process*, mediated by an *effector* of the modification: the hammerstone in the hominid hand, the tooth in the hyena jaw.

These distinctions enable a better understanding of some problems of "equifinality" (disparate agents producing similar evidence) in archaeology. For example,

two *actors*, a hominid butchering with a sharp stone flake and a hoofed animal trampling a bone against an angular pebble, both produce cutlike effects on bone, because the *causal process* and the *effector* are the same. Here, the actor cannot be inferred from the trace alone. Archaeologists have had to seek other evidence through further actualistic research that, when considered with such marks, creates a higher level of confidence about which actor was involved.

Archaeologists' ultimate inferential targets include two levels of "context." *Behavioral context* comprises integrated patterns of behavior such as "scavenging," or "storage," within which archaeological evidence was produced. *Ecological context* refers to the environment and ecosystem in which the actors existed. Evolutionary accounts of the emergence of new hominid behaviors require this level of context as an important aspect of "selective context."

Different archaeologists can agree on actor, effector, and causal process of prehistoric evidence but disagree on its behavioral and ecological contexts. Some archaeologists have assumed that rigor at the agent-identification level of researches would inevitably reveal an unequivocal view of behavioral context. This is not the case. Inferences about these higher order contexts must stem from understanding archaeological evidence itself in the behavioral and environmental contexts.

Until recently, few empirical studies of the relationship between lower order events and processes, such as cut-mark infliction, to the behavioral and ecological contexts in which they occurred were attempted. Three approaches to research on paleolithic archaeology and human ecology have emerged since the 1980s that seek to bridge these gaps.

One approach studies animal ecology and animal remains in contemporary landscapes to bridge to prehistoric landscape context and investigate ancient hominid behavior. The most ambitious project of this sort, that of Blumenschine, Peters, and associates at Olduvai, has devised a predictive model of the regional Plio-Pleistocene landscape, as a backdrop for investigating early hominid land use. Landforms, vegetational, and animal communities are reconstructed in detail, as are their geological signatures. Hominids are "mapped on" to resources according to general principles drawn from primate and other animal ecology, generating hypotheses about land use to test with archaeological evidence. Modeling of hominid social organization and extractive strategies is less developed, for the research takes a more inductive approach to seeing what patterns emerge from the archaeological data. Its strength is its ability to predict patterning in the location and nature of archaeological evidence relative to geographic and biotic features.

Behavioral ecological research on modern people

who hunt and gather for a living is a second approach to understanding earlier hominid behavior and ecological context. The goal is discerning uniformitarian relations between archaeologically visible consequences (e.g., site location, animal carcass part discard and transport decisions, food debris) and behavioral variation that can be explained with adaptationist models. Strengths of this work are its theoretical links to evolutionary ecology and discovery of hitherto unknown patterns in age- and gender-specific labor and food allocation. Problems include lack of unanimity about the explanations for behavioral patterns, including especially the role of hunting; implications of differences between modern foragers and earlier hominids in tool-mediated foraging behavior and social relations; and, in some cases, unclear links between behavioral contexts and their potential archaeological signatures.

A third approach that aims for the same goals is Lewis Binford's massive *Constructing Frames of Reference* (2001), which addresses global regularities in hunter-gatherer ecology and their material effects.

Noting weaknesses and strengths does not imply any of these tactics are wrong. These are long-term and serious efforts to elucidate the link between evidence and target contexts of paleoanthropological interest. However, even though progress is being made working with higher-order contexts, problems of archaeological inference remain. This is so because biological systems do not operate in simple cause–effect patterns, and the nature of archaeological inference necessarily alters as it addresses the behavioral and ecological contexts.

Causality and Historical Inference in Biological Systems. Writings of biologist Ernst Mayr, especially *The Growth of Biological Thought* (1982), elucidate the problem of inferring ancient behavioral patterns or ecological context.

Mayr notes that causation is multivariate and probabilistic at higher levels of biological systems (community, ecosystem). The strength of relational analogies lies in their clear "if A, then B" chain of causation in observable contemporary contexts, which strongly warrants inference of a single analogous cause from an analogous prehistoric effect. Probabilistic causation, on the other hand, depends on the interaction of multiple variables, each of which could display a range of possible values as they interact. Outcomes of ecosystemic interactions are in fact often predictable, but they are better described statistically, as is typical of most statements about community ecology. Probabilistic causation is an open and somewhat contentious area of philosophy of science but is also ubiquitous in "real world" situations. For example, discussions of lung cancer and other diseases necessarily implicate probabilistic causal models. Considering debates about disease etiology, we can intuitively grasp the lessening of certainty involved.

OLDUVAI GORGE, TANZANIA

The FLK *"Zinj"* site at Olduvai Gorge, where the *"Zinjanthropus"* robust australopithecine cranium was found by Mary Leakey in 1958, also yielded thousands of well-preserved animal bones. This sample displays distinctive bone element frequencies that depart from that of the skeleton. They also bear many stone tool cut marks, hammerstone impact marks, and stone anvil scars, which were discerned by pioneering, actualistically informed research in the 1970s and 1980s. The evidence has been the basis for multiple analyses in the 1980s and 1990s. Bunn and Kroll (1986) inferred from the assemblage that ancient hominids were predaceous carnivores, carrying limbs of their prey back to home bases, whereas Binford (1988) concluded the evidence indicated that hominids scavenged carnivore kills. Other archaeologists have subsequently offered opinions on the implications of cut-mark patterns, hominid transport versus in-place destruction in determining bone frequencies. The controversy could partly result from researchers' varied methods of estimating bone frequencies, but more stems from lack of well-warranted linkages between the primary evidence and higher-order generalizations about behavior.

—DIANE GIFFORD-GONZALEZ

Probabilistic Causation and Archaeological Inference. Archaeologists using modern behavioral and ecological systems as analogues for ancient contexts are thus on very different footing than they were when working with causal processes, effectors, and actors. Because the relation of cause to effect on this level is probabilistic rather than universally necessary and sufficient, inferences of unseen prehistoric causes are less secure, requiring a new approach to uniformitarianism.

Because "explanation," or inference of causal context, at this level necessarily has some degree of uncertainty attached, researchers have sought to reduce it. Lyman (1987) calls for a "forensic approach" to the indeterminacy of the evidence: the more independently derived inferences point to the same circumstance, the more likely it is to be causal. Each line of evidence used should possess uniformitarian qualities, whether deterministic or probabilistic in operation, that, when juxtaposed, indicate the most likely explanation for the evidence (see also Binford, 2001; Gifford-Gonzalez, 1991). Independent lines of archaeological evidence noted in the literature include bone modification patterns, lithic technology, site organization, and environmental context.

This form of archaeological inference requires better coordination of data among specialists and creativity in constructing integrative research frameworks. Multidisciplinary team research on human evolution anticipates this process. What has been lacking, perhaps, is the perception that simple reductionist accounts of what caused past hominid behaviors will not emerge from archaeological and paleoanthropological data sets, and that we will have to live with some level of uncertainty about the ancient behaviors, relations, and ecosystems we want to study. However, compared to earlier logical leaps from lower order relational analogies to behavioral and ecological contexts, we can expect that an informed integrative approach will produce more realistic inferences about ancient human behavior and adaptation.

[*See also* Geochronology; Hominid Evolution, *article on* Early *Homo*; Modern *Homo sapiens, article on* Human Genealogical History; Paleolithic Technology.]

BIBLIOGRAPHY

Achinstein, P. "Models, Analogies, and Theories." *Philosophy of Science* 31 (1964): 328–350.

Binford, L. R. *Bones: Ancient Men and Modern Myths.* New York, 1981. A clear statement of the role of uniformitarianism, actualistic research, and "middle range theory."

Binford, L. R. "Fact and Fiction about the *Zinjanthropus* Floor: Data, Arguments and Interpretations." *Current Anthropology* 29 (1988): 123–135.

Binford, L. R. *Constructing Frames of Reference: An Analytical Method for Archaeological Theory Building Using Hunter-Gatherer and Environmental Data Sets.* Berkeley, 2001. Extends earlier work on uniformitarian resources for archaeology.

Blumenschine, R. J., and C. Peters. "Archaeological Predictions for Hominid Land Use in the Paleo-Olduvai Basin, Tanzania, during Lowermost Bed II Times." *Journal of Human Evolution* 34 (1998): 565–607. A holistic model of land use.

Bunn, H. T., and E. M. Kroll. "Systematic Butchery by Plio-Pleistocene Hominids at Olduvai Gorge, Tanzania." *Current Anthropology* 27 (1986): 431–452. An interpretation of cut marks and other modifications to Olduvai fauna.

Carloye, J. C. "An Interpretation of Scientific Models Involving Analogies." *Philosophy of Science* 38 (1971): 562–569.

Gifford, D. P. "Taphonomy and Paleoecology: A Critical Review of Archaeology's Sister Disciplines." *Advances in Archaeological Method and Theory* 4 (1981): 365–438. Actualism and uniformitarianism in archaeology discussed comparatively.

Gifford-Gonzalez, D. "Bones Are Not Enough: Analogues, Knowledge, and Interpretive Strategies in Zooarchaelogy." *Journal of Anthropological Archaeology* 10 (1991): 215–254.

Gould, R. A., and P. J. Watson. "A Dialogue on the Meaning and Use of Analogy in Ethnoarchaeological Reasoning." *Journal of Anthropological Archaeology* 1 (1982): 335–381. Defines the poles of opinion at that time; Watson's view is now more prevalent.

Hesse, M. *Models and Analogies in Science.* Notre Dame, Ind., 1966. A classic source on analogy in science.

Hodder, I. *The Present Past: An Introduction to Anthropology for Archaeologists.* Cambridge, 1982. Though disagreeing on other theoretical points, agrees with Binford about power and necessity of relational analogies.

Kaplan, H., et al. "A Theory of Human Life History Evolution: Diet, Intelligence, and Longevity." *Evolutionary Anthropology* 9 (2000): 156–185. Behavioral ecologists disagree with O'Connell et al. (1999).

Leatherdale, W. H. *The Role of Analogy, Model, and Metaphor in Science.* Amsterdam, 1974. Another overview of analogy in science.

Lyman. "Arachaeofaunas and Butchery Studies: A Taphonomic Perspective." *Advances in Archaeological Method and Theory* 10 (1987): 249–337.

Mayr, E. *The Growth of Biological Thought.* Cambridge, Mass., 1982.

O'Connell, J. F., et al. "Reanalysis of Large Mammal Body Part Transport among the Hadza." *Journal of Archaeological Science* 17 (1990): 301–316. Behavioral ecology of modern hunter-gatherers as a source of analogues of archaeological interest.

O'Connell, J. F., et al. "Grandmothering and the Evolution of *Homo erectus.*" *Journal of Human Evolution* 36 (1999): 461–485. Reflects unanticipated findings of long-term study.

Salmon, W. C. "Probabilistic Causality." *Pacific Philosophical Quarterly* 61 (1980): 50–74.

Simpson, G. G. "Uniformitarianism: An Inquiry into Principle, Theory, and Method in Geohistory and Biohistory." In *Essays in Evolution and Genetics in Honor of Theodosius Dobzhansky,* edited by Max K. Hecht and William C. Hecht, 1970. The classic source on methodological uniformitarianism by a major evolutionary theorist.

Weitzenfeld, J. S. "Valid Reasoning by Analogy." *Philosophy of Science* 51 (1984): 137–149.

Wylie, A. "The Reaction against Analogy." *Advances in Archaeological Method and Theory* 8 (1985): 63–111. A comprehensive literature review and philosophical analysis of the role of analogy in archaeology.

ONLINE RESOURCES

Hull, D. 1997. "Probabilistic Causation." *Stanford Encyclopedia of Philosophy.* http://plato.stanford.edu/entires/causation-probabilistic/. Reviews history, logic, and main philosophical positions on this form of causation, with links to related topics.

—Diane Gifford-Gonzalez

ART

[*This entry comprises two articles. The introductory article is a wide-ranging overview of art in human prehistory that discusses its possible function in social communication; the companion article considers its possible adaptive function. For a related discussion, see* Cultural Evolution.]

An Overview

Art is considered one of the distinctive attributes of *Homo sapiens.* It consists of deliberate communication through visual forms. The earliest forms of art include perforated shells, rock paintings, and sculptures in bone or ivory. Art and language provide parallel evidence for the evolution of cognitive and social skills, but this does not necessitate the use of a linguistic model to analyze visual communication, nor does it preclude the possibility that art expresses aesthetic values.

Art is an aspect of human culture. The evolutionary psychologists Cosmides, Tooby, and Barkow (1992) argue that the structure of culture derives directly from the structure of the mind. Boyd and Richerson give more weight to the shared character of cultural traits. Culture consists of "patterns . . . that are characteristic of groups of people rather than individuals" (Boyd and Richerson, 1985). Sociocultural anthropologists consider culture to be a structured system that emerges from and is constantly renegotiated through social interaction, rather than an assemblage of traits in the individual mind. The meaning of elements is defined by their place in the system. Current social theory, expressed in the work of Bourdieu and Giddens, would agree with Rousseau (Rousseau, 1963) that the cognitive structure of a cultural system and its expression in performance are mutually dependent and reinforce each other.

This has important implications for the evolution of culture, including art. Kauffman (1993) and Conway Morris (1998) argue the environment to which organisms adapt is transformed by interaction with other organisms. Art functions as a medium of communication in social interaction. Odling-Smee and colleagues' (1996) concept of niche construction contends that wherever organisms modify their environment, they will modify the selection pressures to which subsequent generations of the same population are subjected. The art objects produced by each generation enhance the environment experienced by their offspring.

The origins of art and language lie in the gulf that separates modern human and modern chimpanzee cultures. The oldest evidence for the use of ocher comes from the Howieson's Poort industry of South Africa, between 50,000 and 75,000 years before present (bp). Unfortunately, there is no indication of what it was used for. Even if ocher were used to color artifacts or the body, it would not necessarily constitute a visual language in the modern sense, because it is no more complex than a nonhuman call system. Many species, including nonhuman primates, use a call system, in which single cries signify "predator," "aggression," and so on.

Early art and language have no modern parallels. There are, however, analogs of art and decoration in the animal world that may help us understand the evolution of creative expression. Male bowerbirds build large, elaborate structures decorated with flowers, fruits, leaves, bits of fungi, butterfly wings, and feathers of other birds. These structures are not nests, which are built by the females, but displays that attract females and influence their decision to mate with particular males. Geoffrey Miller (2000) has suggested that serial monogamy and female choice in Pleistocene human populations provided selection pressure for analogous creative displays by human males, giving rise to art and dance. The evolutionary parallel between bowerbird displays and human art is, however, likely to be as partial—if not as provocative—as the parallels between bird song and human speech.

Aitchison (1996) favors a "bonfire" model for the origin of language, where a few small sparks are around for a long time, then the whole suddenly catches fire. The archaeological evidence suggests something similar may have occurred with art, although unequivocal evidence for art is (with the exception of the Châtelperronian) confined to modern humans. Kuhn et al. (2001) have recently published evidence for the more or less simultaneous appearance of personal ornament in West Asia, Europe, and Africa between 41,000 and 43,000 bp. This, as they point out, argues strongly against correlating the physical appearance of ornaments such as shell necklaces with a specific event in the cognitive evolution of a single population, but rather for the culmination of longer term evolutionary processes.

The oldest secure dates for rock art come from the paintings in the French cave of Chauvet, where paintings of two rhino and a bison have been dated to around 30,000 bp. The Upper Paleolithic cave art of France and Spain spans a continuous period from 30,000 to 12,000 bp. Remarkably, the distinctive animal style is already in full flower at Chauvet. Fallen slabs bearing paintings excavated at Apollo 11 shelter in southern Namibia have been dated to between 19,000 and 26,000 bp. The geometric rock art of southern Australia may date from 30,000 bp or earlier. If so, this would be the oldest continuously practiced art tradition, persisting in the recent rock art of central Australia and contemporary commercial Aboriginal art. Calcite covering a rock painting at Piauí, Brazil, has recently been reported to date from 30,000 to 40,000 bp. Fallen rock fragments bearing possible traces of paint in the Serra da Capivara Brazilian National Park have been dated to 29,860 ± 650 bp. Consistent early dates for South American rock art occur at around 10,000 to 11,000 bp. Most southern African rock art dates to the last few hundred years.

Another way of reconstructing art's original function is to discover what it is good at. Art is better than language at providing spatial information, such as how to tie knots or find routes to destinations. Art motifs cannot transform a proposition into a negative or a question, but they can be combined into compositions (narratives) or composite figures (visual metaphors). Art objects (necklaces, statuettes) can be worn to express status or exchanged to create and sustain a network of social relationships. Visual motifs can be more rapidly recognized than words (trademarks, road signs). In small-scale societies, art can function as a boundary marker. The structure of a spoken statement unfolds through time, but the structure of a pictorial composition is spatial and can be appreciated simultaneously. Narrative compositions in art are indisputably unique to modern humans.

Aitchison (1996) suggests two models for the origin of language. These can also be envisaged as possible origins for art as a cultural system. The arbitrariness of sounds in spoken language seems to benefit from, indeed to depend on, tightly structured cognitive oppositions. This implies that language began as a clear but simple structure consisting of a limited number of opposed signs. Although the sounds of speech are arbitrary, pictures resemble, or are naturally associated with, what they stand for. The French semiologist Mounin (1970) pointed out that an illustration of a refrigerator loaded with bottles and cartons succinctly conveys its capacity, but pictures of any number of alternative products associated with refrigerators would express the same message. The iconicity of art may have facilitated Aitchison's second model of origins. Here, everyone is babbling away about all kinds of things, but there is very little mutual comprehension.

No doubt human culture had been practiced for some time, perhaps a long time, before cultural behavior became sufficiently formalized and engrained in material artifacts to leave a recognizable trace. This does not justify accepting all fragmentary hints of expressive material as the beginnings of art. Early examples of apparently decorative or iconic artifacts may have been the result of idle, idiosyncratic play. The Makapansgat cobble, with natural depressions resembling two eyes and a mouth, may have been picked up by an australopithecine who spotted the resemblance and carried it to the site, where it was deposited some three million years ago. The ability to perceive suggestive forms is merely a precondition for the use of art. Three ostrich eggshell "beads" from an Acheulean site in Libya and five from two sites in India are not in themselves evidence of a cultural system of ornamentation, even if they were artificially perforated.

D'Errico and Nowell (2000) accept evidence for personal ornaments and decorated artifacts from Châtelperronian sites in western Europe, such as Roc de Combe and Arcy-sur-Cure as art, along with perforated and ochered shells associated with 100,000-year-old burials of anatomically modern humans at Qafzeh. They also rightly points out there is no ethnographic model for the initial development of symbolic communication in humans. We cannot assume Neanderthal necklaces relied on an expressive system of modern human complexity. Our knowledge of the use of ornaments in living human societies does not allow us to deduce a similar complexity in the significance of premodern human ornaments.

For Mithen (1998), the diagnostic feature of modern human cognition is its ability to link cognitive domains. This is an essential aspect of modeling reality, so as to explain and predict. Art is particularly important as a source of evidence for metaphors or analogical thought, because animal-headed humans appear in the earliest European Paleolithic art, c. 30,000 years old: the lion-headed human statuette of Hohlenstein-Stadel and the bison-man painting from Chauvet cave. Mellars (1998), on the other hand, points out that language would need the capacity to refer to things and actions distant in time and space from the speaker before it could sustain social networks on the scale found among modern hunter-gatherers. He regards Upper Paleolithic cave art (undoubtedly drawn from memory) as the best evidence for the cognitive skills that modern languages make possible.

Bourdieu (1990) and Derrida (1976) argue that, even in language, practice lies somewhere between the extremes of tight system and random babble. If the same is true of art, the crucial archaeological evidence for a cultural system will be for regular patterning in performance. It then becomes important to ask what social processes are tending to pull performances toward mutual intelligibility or agreement about a system and what processes tend to undermine consensus.

It is ironic that the best early evidence for art comes from western Europe, far in time and space from the early spread of modern humans in the tropics. This may be due to the exceptional conditions for preservation in deep limestone caves, rather than the shallow sandstone shelters typical of the tropics. The apparent "cultural explosion" in Europe may partly have been a response to specific ecological conditions, the arrival of modern humans approximately midway through the last glacial period, or, possibly, the encounter with Europe's existing Neanderthal inhabitants.

The long time span of hunter-gatherer art in Europe and Australia provides good evidence for continuity and change. Although horse and bison predominate in later European caves, in the early cave of Chauvet rhino, felines and mammoth together contribute more than 60 percent, and a similar preponderance is found in other Aurignacian sites. Work by French archaeologists shows that until the Middle Magdalenian, the frequency of animal species in art either side of the Pyrenees differed, but afterwards converged, implying greater cultural contacts. Deer and humans, on the other hand, are far more frequent in portable art, which also contains scenes of a kind unknown in cave art apart from the Lascaux "pit scene."

In Arnhem Land, northern Australia, the subject matter of rock art has been shown to record ecological changes consequent on the postglacial rise in sea level, as well as changes in Aboriginal material culture. The earliest art consists entirely of land animals. Although none has been directly dated, this art is inferred to date to the glacial maximum, when New Guinea and Australia formed a single landmass. The following "Estuarine" period is dated to between 6,000 and 1,200 bp when sea levels were rising. Fish appear, together with evidence of new mythological subjects such as the Rainbow Serpent. With the flooding of estuaries by black soil washed down rivers during the annual monsoon, the development of rich seasonal lagoons is recorded in the "X-ray" art of the last one to two thousand years. Changes in the distribution of motifs between sites suggest clan totemism may also have appeared in the last few thousand years, possibly in response to changing ecological conditions.

[*See also* Cultural Evolution, *article on* Cultural Transmission; Human Evolution, *articles on* History of Ideas *and* Morphological and Physiological Adaptations; Modern *Homo sapiens, articles on* Human Genealogical History *and* Neanderthal–Modern Human Divergence.]

BIBLIOGRAPHY

Aitchison, J. *The Seeds of Speech: Language Origin and Evolution.* Cambridge, 1996. A good review of current theories, influenced by Aitchison's own work on creoles.
Bednarik, R. G. "The Role of Pleistocene Beads in Documenting Hominid Cognition." *Rock Art Research* 14 (1997): 27–41.
Bourdieu, P. *The Logic of Practice.* Translated by R. Nice. Stanford, 1990. Complex but highly influential theoretical work.
Boyd, R., and P. J. Richerson. *Culture and the Evolutionary Process.* Chicago, 1985.
Clottes, J., ed. *La Grotte Chauvet: l'art des origines.* Paris, 2001.
Conway Morris, S. *The Crucible of Creation: The Burgess Shale and the Rise of Animals.* Oxford, 1998. A coevolutionary approach to explaining the Cambrian "explosion" of new life forms.
Cosmides, L., J. Tooby, and J. Barkow. "Introduction: Evolutionary Psychology and Conceptual Integration." In *The Adapted Mind: Evolutionary Psychology and the Generation of Culture,* edited by H. Barkow, L. Cosmides, and J. Tooby, pp. 4–136. Oxford, 1992.
D'Errico, F., and A. Nowell. "A New Look at the Berekhat Ram Figurine: Implications for the Origins of Symbolism." *Cambridge Archaeological Journal* 10 (2000): 123–167.
Derrida, J. *Of Grammatology.* Translated by G. C. Spivak. Baltimore, 1976. Not for beginners! A reformulation and development of Wittgenstein's later approach to language.
Gell, A. *Art and Agency.* Oxford, 1998. Rejects a linguistic model for the analysis of art. Argues art objects function as agents in the construction of social relationships.
Giddens, A. *The Constitution of Society.* Cambridge, 1984. Together with Bourdieu, Giddens lays the foundations for current social theory.
Hockett, C., and R. Ascher. "The Human Revolution." *Current Anthropology* 5 (1964): 35–147. The classic statement of the concept of "call systems."
Kauffman, S. *The Origins of Order: Self-Organisation and Selection in Evolution.* Oxford, 1993.
Klein, R. G. "Anatomy, Behaviour, and Modern Human Origins." *Journal of World Prehistory* 9 (1995): 167–198.
Kuhn, S., M. C. Stiner, D. S. Reese, and E. Güleç. "Ornaments of the Earliest Upper Paleolithic: New Insights from the Levant." *Proceedings of the National Academy of Sciences* 98 (2001): 7641–7646.
Layton, R. *Australian Rock Art: A New Synthesis.* Cambridge, 1992.
Layton, R. *Anthropology and History in Franche Comté: A Critique of Social Theory.* Oxford, 2001. Not a book about art, but it contains summaries of the theories of Bourdieu, Derrida, and Giddens, and a discussion of ways in which social and evolutionary theories might be reconciled.
Mellars, P. "Neanderthals, Modern Humans and the Archaeological Evidence for Language." In *The Origin and Diversification of Language,* edited by N. G. Jablonski and L. Aiello, pp. 89–115. Memoirs of the California Academy of Science. San Francisco, 1998.
Miller, G. *The Mating Mind.* New York, 2000.
Mithen, S. "A Creative Explosion? Theory of Mind, Language and the Disembodied Mind of the Upper Palaeolithic." In *Creativity in Human Evolution and Prehistory,* edited by S. Mithen, pp. 165–191. London, 1998.
Mounin, G. *Introduction à la sémiologie.* Paris, 1970. A contemporary of the famous French semiologist R. Barthes, but less well known in the English-speaking world.
Nobbs, M., and R. Dorn. "New Surface Exposure Ages for Petroglyphs from the Olary Province, South Australia." *Archaeology in Australia* 28 (1993): 18–39.
Odling-Smee, J., K. Laland, and M. Feldman. "Niche Construction." *American Naturalist* 147 (1996): 641–648.
Rousseau, J. J., ed. *The Social Contract and Discourses.* London, 1963.
Rowe, M. W. "Dating of Rock Paintings." *International Newsletter on Rock Art* 29 (2001): 5–13. A review of current dating techniques.
Sauvet, G., and S. Sauvet. "Fonction sémiologique de l'art pariétal animalier franco-cantabrique." *Bulletin de la Société Préhistorique Française* 76 (1979): 340–354.
Strecker, M., and P. Bahn, eds. *Dating the Earliest Known Rock Art.* Oxford, 1999. A series of papers reporting recently obtained dates for rock art in South America; also a paper by Bahn on the Makapansgat cobble.
Wendt, W. E. "'Art Mobilier' aus der Apollo-11 Grotte in Südwest-Afrika." *Acta Praehistorica et Archaeologica* 5 (1974): 1–42. A report was published in English in *South African Archaeological Bulletin* 31 (1976): 5–11.

— ROBERT LAYTON

An Adaptive Function?

Human artifacts plausibly labeled as art date to around 300,000 years ago, indicating that humans have produced art throughout their history. Traditionally, the capacity to produce art has been seen as an outgrowth of other human capacities, rather than a trait that has evolved for that purpose. However it is intriguing, given art's antiquity and universality, to speculate that the capacity to produce and understand art has its own adaptive function. This requires a plausible account of how art gave individuals or groups higher Darwinian fitness.

The thesis suggested here is that art is a pre-linguistic form of communication, a way of transmitting information reliably, and may itself have been an early form of writing. The ability to represent objects and ideas in the form of physical images and artifacts may have existed prior to language, as a form of symbolic communication. Such abilities could have been advantageous to their bearers or to the groups in which they lived. It is possible to consider that linguistic abilities themselves evolved from artistic capabilities, although this must remain a matter of speculation.

Three accounts for the origin of art are implicit in this summary. On the traditional view, art is merely a by-product of other evolution, and art itself did not evolve to perform any particular function (although this would not preclude it acquiring a function such as communication). The evolution of the brain is linked to other functions, such as logic and language, as these perform important functions to the individual. Individuals who can communicate and think clearly have higher rates of survival and reproduction. Artistic capability is simply an emergent feature of this kind of brain.

An alternative view is that artistic capability confers a "selective" advantage to individuals. Individuals who possessed "genes for art" would have had greater fitness than those who did not. Possibly artists were better communicators, or by virtue of producing art an individual enjoyed higher reproductive success by virtue of being more attractive and entertaining. In this scenario the evolution of the ability to create art would have to be matched by the ability to read art. This may suggest a group selection view.

On the group selection view, groups of individuals who had an artist in their ranks performed better than groups who did not have artists among them. This does not require the artist to behave altruistically for the good of the group. Rather, groups possessing artists were better at their tasks (e.g., hunting, trading, and gathering). Perhaps they could communicate important features of their environment to each other. Group selection models show that the genes for artistic ability need not spread through the population to fixation (i.e., everyone has them), that is not everyone needs to be an artist. All that is required is that the average fitness of individuals is higher in groups possessing artists.

Artistic capability using even the most primitive signs such as marks on trees or animal skins or rocks could improve cooperation and "planning," could serve to communicate or reinforce local customs or beliefs, or could act as a form of group defense to indicate the whereabouts of a hostile tribe and their numbers. One example of North American Indian picture writing has ten marks, preceding a canoe, preceding a fish—the fish is "read" as an identifier of a particular tribe, the Passamaquoddy. Thus ten Passamaquoddy Indians in a canoe. Whether this kind of communication is art or simply a form of written language is irrelevant as in the view advanced here art is a form of written language.

It is not possible to distinguish among these three scenarios with the available data. However, art's capacity to communicate, provoke, and possibly to compel suggests that it might have granted a group selected advantage. This view emerges from the recognition that a work of art has a dual existence. Once completed it is an object, with an existence independent of its creator. But the secondary existence is that, when it works, i.e., is re-created in the mind of the viewer, it communicates that which moved the artist—the *erlebnis*. To put it another way the lines, the colors, and to some extent the materials are there to be "read" by the viewer who then (ideally) re-creates what the artist intended. The viewer of an art work can take part in this process of re-creation and not be a passive onlooker, as might perhaps describe someone contemplating a natural object such as a stone, for which there is no need to re-create or imagine a creator's lived experience. Art is intended to communicate from artist to reader, the artist's beliefs and emotions.

A Brief History of Art and Language. It is acknowledged today that in ancient toolmaking many artifacts, as much as a million years old, appear as incisions or markings going beyond what was technically required of the tool. But while the pebble face of Makapansgat, South Africa, has been dated to Australopithecus (2–3 million years ago) it is the *Venus of Berekhat Ram*, Israel, which is accepted as the oldest known carving in the world and has been dated as 242,000–800,000 years before present (there is much known cave art dating to 30,000 bp).

When humans evolved language is still a matter of speculation, although it is probably reasonable to suppose that language is at least as old as *Homo sapiens*. This could place it back to perhaps 300,000 years or more conventionally to around 100,000 years, coinciding roughly with the time that modern humans spread out of Africa and then around the world. Certainly known art predates the oldest characterized language, Nostratic, a Proto Indo-European language inferred to exist some 12,000 years ago, and the oldest form of known writing, Sumerian (cuneiform), which is between 6,000 and 5,000 years before present.

But What Did Art Communicate? As suggested, art may have communicated information about many features of the environment. Much cave art is devoted to animals. Equally, though, there is a persistent tendency for art to communicate themes of sex, fertility, and reproduction. This is interesting in its own right because indicators of mating ability and fertility are among those most commonly subject to what is known as sexual selection, or selection for traits that confer mating or reproductive success. Implicit, of course, in the view that animals respond to sexually selected displays—such as the exaggerated tail of a peacock—is that animals have some sort of aesthetic sense or preference. Could artistic appreciation have its roots in these very ancient biological systems?

Two works spanning some 23,000 years—both with their sexual overtones—communicate important biological subjects. The extraordinary *Venus of Laussel* (Figure 1), artist unknown, is a conceptual work of art dating from about 23,000 years ago. The details of the hands are unimportant—the left hand is shriveled compared to the voluminous breasts and the fingers striated (similar to the magnificent wood statues of the Dogon tribes). But the hand lies over her stomach—a possible reference to communication of an unborn child. If the head is in relief facing the horn then we have a work in which the body is seen from the front and in profile. Her head is of a size equivalent to her breasts indicating the importance of the function of breasts (nourishment) and/or the lack of importance of the head. The neck is straight, and her right hand is at an impossible angle being parallel to the body almost tilted back. This "distortion" is a sign of good art "in the modern sense" and a conceptual piece, not a mere reproduction of a woman. Her gaze is down into the horn, which is curved and of equal size to her breasts. This curvature of the horn is

FIGURE 2. Les Demoiselles D'Avignon. Paris (June–July 1907). Oil on canvas, 8′ × 7′8″ (243.9 × 233.7 cm). The Museum of Modern Art, New York. Acquired through the Lillie P. Bliss Bequest. Photograph © 2002 The Museum of Modern Art, New York.

FIGURE 1. Venus of Laussel. 27,000–22,000 BP; 54 × 36 × 15.5 cm; Musée d'Aquitaine, Bordeaux Inv 61.3.1. Photo: J. M. Arnaud.

repeated, echoed by her hair and her left thigh. An echo is usually an indication of an important stress. The horn if not a drinking vessel (her leaning down) could be that of power—an Orpheus-like instrument that controls the world of animals—a key instrument for hunters. One can imagine its use by her—she holds it, or it on her face symbolizes a powerful rhino-like creature. The symbolic content of the Venus is thus food, nourishment, reproduction, and if we look at the hips of the woman—the powerful concept of "genetically" sound childbearing hips. *The Venus of Laussel* is the work of an artist communicating womanhood.

Picasso's *Les Demoiselles D'Avignon* (Figure 2) portrays five figures (Dame 1 to Dame 5 from left to right) that spell his more complex concept of womanhood. The emphasis is on the object of desire rather than procreation, although Dame 5 has the most ample childbearing hips and enormous round breasts. Details of hands are unimportant—only two have striated visible fingers. Dame 1 is the enigmatic "type"—we cannot see if she is walking. Is it her left or right foot in front? Is it her hand above her head? Dame 2 is the most giving of women, the giving concept, her breasts are round, her head hangs sideways Christ-like, her right leg moves across to protect herself as does the sheet cover—or of course the sheet may suggest her showing her leg. There is good symmetry in this figure with its possible evolutionary references. Dame 3 is more threatening with one pointed and one round breast, both arms up, hair up. Dame 4 has a protective element, a concern for the living, very differently displayed than the unborn child of the *Venus of Laussel*. But it is Dame 5 that mainly catches our attention. Here in 1907 Picasso has painted "liberated" woman equal to man—she is a thinker, her left arm rests on her knee and chin, a classic thinker position and its thickness reads that she is formidably powerful in this field. While she has her back to us, she has twisted her head around not unlike the owl—the symbol of wisdom. She will not be taken for granted, yet her features with the large breasts, broad hips are the most childbearing and conventionally attractive.

Both works have a mystery and beauty. Both are acts of artistic creation separated by 23,000 years and deriving from vastly different cultures—one pre-agricultural, the other technological. And yet in spite of this, they explore similar themes with a similar "language." Just as we can appreciate the *Venus of Laussel*, one can imagine prehistoric people grasping the meanings of Picasso's work—concepts of womanhood. Picasso's work at the very least, then, may suggest ways in which art exists independently of the linguistic milieu of the artist.

The many common elements in these two works of art show an ability to communicate universally and a wealth of information. The type of information also gives us a sociological insight as to what was seen as important at the time. In both cases a tribe or person who has access to these artworks was able to see the key elements were not the size of the hands or facial characteristics but concepts.

Did art precede language and evolve as an advantageous form of communication? Art is old, every culture has it, and it communicates universals as well as packages of information reliably and efficiently. It does not require the anatomical specializations of language, and was reasonably well developed in Neanderthals. A hunter in the pre-linguistic and pre-writing age may depict information graphically and to someone who has never been on a hunt this would be valuable and informative. Fertility images may have served to reinforce group norms and beliefs. One might argue that memes communicated artistically shift survival from being based not simply on the biologically fittest but on the fittest who work within the language of art (create, interpret, read, re-create), and those who best make use of the information it transmits.

BIBLIOGRAPHY

Bahn, P. G., *Prehistoric Art*. Cambridge, 1998. A comprehensive book of illustrations and overall descriptions of prehistoric art.
Barber, C. L., *The Story of Language*. London, 1967. An excellent catch-all book from the perspective of research in 1967.
Cavalli-Sforza, L. L., and F. Cavalli-Sforza. *The Great Human Diasporas: The History of Diversity and Evolution*. New York, 1995. A readable account of research on human migrations around the world following the evolution of modern humans.
Kahnweiler, D.-H. *Confessions Esthétiques*. Paris, 1963. A deep but accessible thinker and writer on modern art and aesthetics, indispensable for any evaluations of Cubism.
Lewin, R. *Human Evolution—An Illustrated Introduction*. Cambridge, Mass., 1993. A solid introductory book on evolution.
Lommel, A. *Prehistoric and Primitive Man*. London, 1966. Well illustrated and excellently thought out and researched ethnographic orientated book.
Pinker, S. *The Language Instinct*. London, 1995. A key book for research on language.

— ORDE LEVINSON

ARTHROPODS

Arthropods (hexapods, myriapods, crustaceans, and chelicerates) are triploblastic Metazoa characterized by a segmented, hardened, chitinous cuticular exoskeleton and paired, jointed appendages. This exoskeleton is composed of a series of dorsal, ventral, and lateral plates that periodically undergo molting (ecdysis). Primitively, arthropods share a compound eye with a subunit structure that is unique within the animal kingdom. Arthropods are the most diverse creatures on Earth, with the number of known species approaching one million, and perhaps ten times as many left to discover. Arthropods are found on all continents, in the deepest oceans, and on the highest mountains. They can be extremely small (< 1 mm, mites and parasitic wasps) to rather large (> 4 m Japanese spider crabs). They are herbivores, predators, and parasites, solitary and intensely social. Not only are they hugely diverse, but they also occur in amazing numbers, constituting the great majority of animal biomass.

The geological history of arthropods extends over 525 million years (to the Lower Cambrian) with now extinct lineages of great diversity (e.g., trilobites). This history has undergone several dramatic rounds of extinction and diversification, most prominently in the Paleozoic era near the end of the Ordovician period and at the Permian-Triassic boundary. The Cambrian and Ordovician fossil record of arthropods is exclusively marine, but terrestrial forms (including arachnids, millipedes, and centipedes) appear from the Upper Silurian, more than 400 million years ago.

Today, there are four main lineages of arthropods: Hexapoda (insects), Myriapoda (centipedes, millipedes, and relatives), Crustacea (shrimps, crabs, lobsters, crayfish, barnacles, etc.), and Chelicerata (sea spiders, horseshoe crabs, and arachnids). There are several extinct groups including trilobites, marrellomorphs, anomalocaridids, and euthycarcinoids, which may well be equal in stature to those we know today.

Hexapoda. The insects are by far the most diverse known arthropod group (but mites might come close), with hundreds of thousands of species known to science. Hexapods are characterized by possession of three body tagma (head, thorax, abdomen), the second of which possesses three limb-bearing segments. Insecta comprise most of the diversity within the Hexapoda, insects being those hexapods with an antenna developed as a flagellum without muscles between segments. The hexapod head (like that of crustaceans and myriapods) has a large, generally robust mandible used for food maceration, a single pair of sensory antennae, and both compound and simple eyes. There are thirty commonly recognized hexapod "orders" further organized into several higher groups: Entognatha (those with internal mouthparts)—Protura, Diplura, and Collembola (springtails); Archaeognatha (bristletails); Zygentoma (silverfish); Ephemerida (mayflies), Odonata (damselflies and dragonflies); orthopteroids—Plecoptera (stoneflies), Embiidina (web spinners), Dermaptera (earwigs), Grylloblattaria (ice insects), Phasmida (walking sticks), Orthoptera (crickets, grasshoppers), Zoraptera, Isoptera (termites), Mantodea (praying mantises), Blattaria (roaches); hemipteroids—Hemiptera (true bugs and hoppers), Thysanoptera (thrips), Pscoptera, Pthiraptera (lice); and the Holometabola—Coleoptera (beetles), Neuroptera (lacewings, dobsonflies, snakeflies), Hymenoptera (bees, ants, and wasps), Trichoptera, Lepidoptera (moths and butterflies), Siphonaptera (fleas), Mecoptera (snow fleas), Strepsiptera and Diptera (flies). Basal hexapods (Protura, Collembola, Diplura, Archaeognatha, and Zygentoma) are wingless, whereas the more derived insect orders generally possess two pairs of wings. The Neoptera (Pterygota—winged insects except for the "paleopteran" ephemerids and odonates) possess wing hinge structures that allow their wings to be folded back over their abdomen. Those insects with complex development, Holometabola, are the most diverse, with beetles leading the way with over 300,000 recognized species. Insects are found all over the world in terrestrial and freshwater habitats, and many have economic importance as pests, or medical interest for causing or carrying disease. There is an extensive fossil record of insects from the Devonian *Rhyniella*, through other Paleozoic and Mesozoic deposits, to the dramatic and beautiful amber-preserved insects from Lebanon, the Baltic, and the Dominican Republic.

Myriapoda. The centipedes, millipedes, symphylans, and pauropods are multilegged soil-adapted creatures. Generally without compound eyes (except for scutigeromorph centipedes), but possessing a single pair of sensory antennae, the myriapods are most easily recognized by their large numbers of legs and the trunk not being differentiated into distinct regions (tagmata). Almost all postcephalic segments bear a single (centipedes, pauropods, symphylans) or double pair of legs (millipedes) numbering into the hundreds in some taxa. These ar-

thropods are generally small ($< 5–10$ cm), but there are several dramatically larger examples (*Scolopendra gigantea*) at 30 centimeters or more. There are four main lineages of myriapods: Diplopoda (millipedes), Chilopoda (centipedes), Pauropoda, and Symphyla. The basic division among myriapods lies between the Chilopoda, which have the genital opening at the posterior end of the body, and the other three lineages, grouped as Progoneata on the basis of the genital opening being located anteriorly on the trunk, behind the second pair of legs. The millipedes are the most diverse group, with approximately 10,000 species. The chilopods are the other diverse group (approximately 2,800 known species). Pauropods and symphylans are less speciose, with a few hundred described taxa. In general, myriapods are soil creatures feeding on detritus, with the centipedes exclusively predatory and possessing a modified fang and the ability to deliver toxins to their prey. It is probable but not universally agreed that the myriapods share a single common ancestor. The movement of the head endoskeleton, structure and musculature of the mandible, and most DNA sequence evidence support the single origin of Myriapoda, but several hypotheses place myriapod lineages with hexapods. There are few well-preserved myriapod fossils, but the extant chilopod order Scutigeromorpha and the diplopod group Chilognatha both have fossil representatives from the Late Silurian. The extinct group Arthropleurida, thought to be members of the Diplopoda, may have reached 2 meters in length.

Crustacea. Crustaceans are perhaps the most morphologically diverse group of arthropods (over 30,000 species known), with huge variation in numbers and morphology of appendages, body organization (tagmosis), mode of development, and size (< 1 mm to > 4 m). These creatures are generally characterized by having two pair of antennae (first and second), biramous (branched) appendages, and a specialized swimming larval stage (nauplius). They usually possess both simple ("naupliar") and compound eyes (the latter frequently stalked). Like myriapods and hexapods, crustaceans possess strongly sclerotized mandibles that are distinguished by frequently having a segmented palp. The Crustacea are generally marine, with several freshwater and terrestrial groups (e.g., some isopods, the wood lice). Crustacean phylogeny is an area of active debate, with the status of some long-recognized groups under discussion. Currently, several higher groups are recognized, with their interrelationships (and even interdigitiation) unclear: Remipedia (twelve species; *Speleonectes*, *Lasionectes*, and three other genera), Cephalocarida (few species; *Hutchinsoniella* and three other genera), Branchiopoda (1,000 species; fairy shrimp, water fleas, tadpole shrimp, clam shrimp), Maxillopoda (10,000 species; copepods, barnacles, ostracods, fish lice), and Malacostraca (20,000 species; mantis shrimp, crayfish, lobsters, crabs, isopods, amphipods). Many of the debates on crustacean relationships center on the position of the recently discovered remipedes as either the most basal lineage resembling, in some respects, the first Crustacea or a more derived position having little to do with crustacean origins. The fossil group Phosphatocopina are probably the earliest Crustacea or the closest relatives of the extant Crustacea, first occurring in the Lower Cambrian in England, and being known from fine preservational quality (notably in the three-dimensional Orsten Cambrian fauna).

Chelicerata. The sea spiders, horseshoe crabs, and arachnids are characterized by division of body segments into two tagmata: prosoma and opisthosoma (generally), and the first leg-bearing head segment being modified into chelifores or chelicerae. With the exception of horseshoe crabs (the American *Limulus* and the Asian *Carcinoscorpius* and *Tachypleus*), chelicerates do not possess compound eyes, and none have antennae. Horseshoe crabs and arachnids have one pair of median eyes, whereas sea spiders have a second pair. Of the three main divisions of chelicerates (Pycnogonida—sea spiders [1,000 species], Xiphosura—horseshoe crabs [four species], and Arachnida—spiders, scorpions, etc. [60,000 species]), the sea spiders and horseshoe crabs are marine and the remainder terrestrial, with the exception of some groups of mites. Many groups of Acari (mites and ticks) are parasites of plants and animals, both vertebrates and invertebrates, and being ecto- and endoparasitic, mostly of respiratory organs. The arachnids are the most diverse component of the Chelicerata, with the Acari and Araneae (spiders) constituting the vast majority of taxa. Other arachnid groups include Opiliones (harvestmen, daddy longlegs), Scorpiones (scorpions), Solifugae (sun, camel, or wind spiders), Pseudoscorpiones ("false" scorpions), Ricinulei, Palpigradi (micro-whip scorpions), Amblypygi (tailless whip scorpions or whip spiders), and Uropygi (vinegaroons). The Paleozoic eurypterids are an aquatic (mostly brackish water) group, generally considered to be the closest relatives of Arachnida, though some workers consider them especially related to scorpions. The largest eurypterids are 1.8 meters long, among the largest arthropods ever. The sea spiders graze on corals, anemones, or seaweeds, and vary in size from quite small (< 1 cm) to almost a meter in leg span. Horseshoe crabs and arachnids are almost entirely predatory, with spiders the dominant arthropod predators in many environments. Horseshoe crabs scavenge and prey on small animals in seaweeds, and like the Opiliones, they digest their food internally. Most arachnids, however, digest food extraorally, ingesting their prey in the form of digested fluids.

Close Relatives. The closest relatives of the arthropods are the enigmatic water bears (Tardigrada) and velvet worms (Onychophora). All of these animals share paired appendages and a chitinous cuticle. There are approximately 800 species of tardigrades that live in marine, freshwater, and terrestrial habitats. Terrestrial tardigrades are mostly found on mosses and bryophytes and may occur in huge densities (hundreds of thousands to millions per square meter). Tardigrades are small (between 150 and 1,000 microns), have a round mouth, four pairs of legs, the last one being terminal, and, like arthropods and a few other taxa, grow by molting. Terrestrial tardigrades live in extreme environments supporting desiccation or freezing by entering into cryptobiosis. Tardigrades have been experimentally subjected to temperatures between $-272°C$ and $+340°C$, or between 160,000 psi to pure vacuum, excessive concentrations of gases, and radiation, and returned to active life. The cryptobiotic stage has been recorded to last over 100 years, and in this stage they can be dispersed by wind. The Onychophora are a group of exclusively terrestrial, predatory creatures that live in humid temperate (mostly Southern Hemisphere) and tropical forests. The velvet worms are characterized by a soft body with pairs of "lobopod" walking limbs, a pair of annulated antennae, jaws, and oral ("slime") papillae. About 150 extant species have been named, but there were many more types including marine "armored" or plated lobopods in the Early Paleozoic. Onychophorans and arthropods share a dorsal heart with segmental openings called ostiae, and a unique structure of the nephridia, the excretory organs. Lack of these organs in tardigrades may be due to miniaturization. It is thought that the Tardigrada are the sister taxon (closest relative) of the Arthropoda and the Onychophora the next closest relative.

More Distant Relatives. It has been long thought that there was an evolutionary progression from wormlike creatures, to lobopodous forms like Onychophora, to modern arthropods. This was expressed in the "Articulata" hypothesis that linked annelid worms (polychaetes and oligochaetes, including leeches) to the Onychophora and Arthropoda. Recent work, especially from DNA sequences, has largely replaced this view, allying arthropods, tardigrades, and onychophorans with other molting creatures such as the nematodes, kinorhynchs, and priapulids in the Ecdysozoa (after ecdysis or molting); and uniting the annelids with molluscs, nemerteans, sipunculans, and entoprocts in the Trochozoa (or Lophotrochozoa of some authors).

Extinct Lineages. No doubt there are more extinct than extant lineages of arthropods. More likely than not, most will remain unknown to science, but several major groups we do know about have a great effect on our notions of higher level relationships among the arthropods (living and extinct). Trilobites are the best known group of extinct arthropods. First known from the Lower Cambrian, trilobites had huge radiations in the

Paleozoic. Trilobites were an exclusively marine group (4,000 species described) characterized by two longitudinal furrows dividing the body into three lobes (hence the name). The body segments are organized into three tagma (cephalon, thorax, pygidium). Trilobites possessed compound eyes, a single pair of antennae, and had biramous appendages. All postantennal appendages in trilobites are basically similar in structure. Trilobites are closely related to the Chelicerata, together with numerous other extinct lineages constituting the group Arachnata. Anomalocaridids are a group of large (up to 2 m), predatory Cambrian arthropod relatives. With unmineralized, but sclerotized cuticle, they were known initially only by their raptorial feeding/grasping appendages, which were anterior to a circular mouth that was surrounded by a ring of plates. Their phylogenetic affinities are uncertain, but most recent work places them in the stem group of the Arthropoda, probably more closely related to extant arthropods than are tardigrades. Marrellomorphs are a clade known from the Burgess Shale (Middle Cambrian, Canada) and Hunsrück Slate (Lower Devonian, Germany) that possess two pairs of antenniform limbs and two pairs of long spines that curve back over the body. *Marrella* is the most abundant arthropod in the Burgess Shale fauna. Euthycarcinoids are a somewhat enigmatic group from the Lower Silurian to the Middle Triassic with potential affinities with myriapods or crustaceans. They possessed a single pair of antennae and numerous pairs of uniramous legs. Lopodian taxa were largely unknown until recent soft-part preserved specimens (mainly from China and from the Burgess Shale) were found. The marine lobopodians are thought to be related to living terrestrial Onychophora or Tardigrada, and possessed elaborate spines and armored plates. The "Orsten" fauna of Sweden contains amazingly well-preserved three-dimensional Upper Cambrian fossils, most importantly of basal crustacean taxa. Several of these forms (e.g., *Martinssonia*) are important to understanding the origins and relationships of the Crustacea. Among the most productive Paleozoic fossil deposits are the Burgess Shale, Chengjiang, Orsten, Rhynie Chert, Gilboa, and Mazon Creek deposits.

Arthropod Interrelationships. The question of arthropod relationships has been and is still unsettled. Of the living taxa (Chelicerata, Crustacea, Myriapoda, Hexapoda), it seems clear that those groups that possess mandibles (robust, sclerotized, chewing mouthparts), Crustacea, Myriapoda, and Hexapoda, share a unique common ancestor. The biting edge of mandibles is formed by the same segment, the coxa, of the same limb (third limb-bearing segment in Crustacea). Within this group things become less clear. There are two main competing hypotheses: Tracheata (myriapods and insects) and Tetraconata or Pancrustacea (crustaceans and insects). The Tracheata hypothesis is supported by

anatomical evidence, notably the similar tentorial head endoskeleton, an absence of limbs on the head segment (intercalary segment) innervated by the third brain ganglia, and similar respiratory and excretory organs. Molecular sequence data and an alternative set of anatomical features, notably eye structure and neurogenesis, support the Tetraconata.

One of the major ecological questions surrounding arthropod evolution is the question of the number of invasions of land, or at least the number of transitions between marine and terrestrial environments. The origin of the Arthropoda is undoubtedly marine, but the Onychophora, Arachnida, and several Malacostraca have independently invaded land. If the group Tracheata is upheld, there was one more invasion by the common ancestor of myriapods and insects. If, however, crustaceans and insects are the closest relatives, two additional transitions are implied. Several other issues remain unresolved as well, especially regarding the placement of the Pycnogonida, which could well constitute the sister group to all the remaining extant arthropods, and the relationships of extinct lineages both to living taxa and the overall scheme of arthropod history.

[*See also* Animals; Insects.]

BIBLIOGRAPHY

Edgecombe, G. D. *Arthropod Fossils and Phylogeny*. New York, 1998.

Fortey, R. A., and R. H. Thomas. *Arthropod Relationships*. London, 1998.

Giribet, G., D. L. Distel, M. Polz, W. Sterrer, and W. C. Wheeler. "Triploblastic Relationships with Emphasis on the Acoelomates, and the Position of Gnathostomulida, Cycliophora, Platyhelminthes, and Chaetognatha: A Combined Approach of 18S rDNA and Morphology." *Systematic Biology* 49 (2000): 539–562.

Giribet, G., G. D. Edgecombe, and W. C. Wheeler. "Arthropod Phylogeny Based on Eight Molecular Loci and Morphology." *Nature* 413 (2001): 157–161.

Melic, A. *Evolución y filogenia de Arthropoda*. Sociedad Entomológica Aragonesa, 1999.

Wheeler, W. C., and C. Y. Hayashi. "The Phylogeny of the Extant Chelicerate Orders." *Cladistics* 24 (1998): 173–192.

Wheeler, W. C., M. F. Whiting, J. C. Carpenter, and Q. D. Wheeler. "The Phylogeny of the Insect Orders." *Cladistics* 12 (2001): 1–57.

—WARD WHEELER, GONZALO GIRIBET, AND
GREGORY D. EDGECOMBE

ARTIFICIAL LIFE

Evolution in nature is characterized by many levels of hierarchy, including the gene, the individual, the population, and the ecosystem. Intensive study can be made within any of these levels of organization, or others, but the interactions across these levels of hierarchy are also important in determining the patterns of evolution. For example, genetic interactions affect individual behaviors, but so do populational interactions. Selection at the level of the individual affects gene frequencies, while at the same time affecting ecosystems by modifying factors such as food webs. It is often difficult to garner sufficient empirical evidence to study these interactions carefully. The sample size is small, the conditions for each experiment are subject to temporal and geographic vagaries, and each observation may require considerable effort.

In the late 1980s, an alternative method was proposed to investigate complex ecosystems, interactions between individuals, and even the effects of relatively simple genetic systems. The prescription was to simulate evolutionary processes on a computer—that is, to program routines of variation and selection and create an "artificial life" within the machine, where the conditions, factors, and parameters of the artificial system could be studied, replicated at will, and executed in fast time. The hope was to study "life as it could be," rather than "life as it is."

Of particular interest within this field of artificial life were studies that sought to put in place the essential elements of an artificial ecosystem and let its constituents evolve without specifying explicit performance criteria. There had been a long history of using so-called evolutionary algorithms for optimization (e.g., Bremermann, 1962; Fogel et al., 1966; Kaufman, 1967), but in these cases, an explicit payoff function was employed to guide the selection process and eliminate inappropriate individuals in a simulated population.

For instance, a task might be given to solve a system of linear equations, $Ax = b$. Candidate vectors are created at random and stored in a population. Each is assigned a figure of merit, here the inverse of "fitness," based on the squared error norm $\| Ax - b \|$. New candidate vectors x can be created based on the success or failure of other trial solutions in the population, with selection serving to eliminate those with the most error. Although this can be an effective means for optimization, here the payoff function is provided explicitly by the programmer before the evolution is begun.

In contrast, fitness in natural systems can be viewed as an emergent property of the interaction between individuals in light of their particular needs for survival and reproduction. These are not prescribed a priori. Rather than seeking to use evolutionary algorithms for optimization, the artificial-life approach sought to create simulations in which fitness would similarly emerge from the interaction of elements in the population, acting within the constraints of the "physics" of their artificial world. Perhaps the best-known research in this area is Tom Ray's simulation called *Tierra* (Ray, 1992), but an earlier simulation (much earlier, in fact) deserves first mention.

Michael Conrad and Howard Pattee (Conrad and Pattee, 1970) offered one of the earliest simulations of a hierarchical artificial ecosystem, using computers at Stanford University. A population of cell-like individual organisms, each of which relied on evolvable rules of behavior, was subjected to a strict materials conservation law that induced competition for survival. The organisms were capable of mutual cooperation, as well as of executing biological strategies that included genetic recombination and the modification of the expression of their genome. No fitness criteria were introduced explicitly as part of the program. Instead, the simulation was viewed as an ecosystem in which genetic, individual, and populational interactions would occur and behavior patterns would emerge.

Whereas Conrad and Pattee's simulation required competition from these cell-like structures for "energy" placed in an artificial environment, Ray's simulation required competition between computer programs for CPU cycles offered from a main central processor. Ray's program itself evolved programs in assembly language. It was first seeded with a hand-coded program capable of replicating itself into another area of RAM (random access memory). Upon starting the simulation, this program self-replicated, and its progeny in turn self-replicated, and so forth, ultimately to the point of filling the available environmental arena.

Each program competed for available CPU cycles, with the rule that generally shorter programs were able to replicate more quickly than longer ones. Mutation was imposed on the programs in two forms: (1) errors could occur to the program instructions during reproduction; and (2) at a background rate, random bits in the RAM were altered. This provided a source of random variation in the behavior of the replicating assembly-language programs.

Upon creation, programs were placed in a queue, and once the capacity of the RAM reached a specified threshold, akin to a carrying capacity, the programs were subject to having their memory deallocated (the equivalent of being selected out of the competition).

Ray demonstrated a repeatable effect of the emergence of "parasitic" programs. These programs were sets of instructions that could not copy themselves in the absence of a "host" program, but instead relied on part of a host's own code. Ray handcrafted his original self-replicating program in three parts. The first part checked the length of the program to determine the amount of space required to replicate. The second part used a programming loop to repeatedly call the third part, which accomplished the actual replication. The parasitic programs evolved to do away with the third part entirely, instead relying on calling another host program's replication loop to copy itself. In this fashion, the parasitic code not only took advantage of the host but actually replicated faster than its host—the parasites used fewer lines of code than their hosts.

Ray observed a second level of emergence, wherein so-called hyper-parasites emerged that could take advantage of the prior parasites. These new programs evolved a modified form of the replication loop, which instead of returning control to the parasite program that called the loop, giving the parasite a free chance to replicate, returned control to the hyper-parasite's own reproduction loop, which then took over and copied the instructions of the hyper-parasite. Any call from a parasite to this new hyper-parasite resulted in more copies of the hyper-parasite, rather than more copies of the parasite.

The sequence of two levels of such emergent properties, starting only with a handcrafted set of instructions for self-replicating and a simple set of rules for imposing mutation and selection based on available CPU cycles, was both surprising and encouraging. Unfortunately, the sequence of emergence appeared to halt at that point: no further examples of higher-order effects (i.e., hyper-hyper-parasites) or alternative directions were observed. Ray (1992) offered some possible plausible programs that might evolve to demonstrate symbiosis, where neither program could replicate in the absence of the other, but there was no evidence that such programs were ever evolved in the *Tierra* system.

Subsequent efforts have been aimed at implementing *Tierra* on a network, even the Internet, and observing the effects of much larger simulations. The evolving programs in these "NetTierra" experiments would rely on spare CPU cycles from unused machines that are connected to the Internet. A hypothesized emergent behavior would be that of being nocturnal: more activity would be expected from the programs in the early morning hours, when computers are left without a user. Another behavior might be that of migration: programs would move about the network seeking those machines located in the current early hours of the day.

The ultimate goal of the *Tierra* effort, and other similar efforts, is to understand the most basic requirements for open-ended evolution. To date, the emergence of new properties in these systems has always fizzled out, to put it plainly. It is hoped, however, that a basic understanding of life, in all its potential forms, could eventually be grasped through the study of artificial ecosystems with many similarities to Ray's *Tierra* and the earlier efforts of Conrad and Pattee. Such an understanding, and such a simulation, remains for future work.

[*See also* Natural Selection.]

BIBLIOGRAPHY

Bremermann, H. J. "Optimization through Evolution and Recombination." In *Self-Organizing Systems-1962*, edited by M. C. Yovits, G. T. Jacobi, and G. D. Goldstein, pp. 93–106, Washington D.C., 1962.

Conrad, M., and H. H. Pattee. "Evolution Experiments with an Artificial Ecosystem." *Journal of Theoretical Biology* 28 (1970): 393–409.

Fogel, L. J., et al. *Artificial Intelligence through Simulated Evolution.* New York, 1966.

Kaufman, H. "An Experimental Investigation of Process Identification by Competitive Evolution." *IEEE Transactions on Systems Science and Cybernetics*, SSC-3 (1967): 11–16.

Ray, T. S. "An Approach to the Synthesis of Life." In *Artificial Life II*, edited by C. G. Langton, C. Taylor, J. D. Farmer, and S. Rasmussen, pp. 371–408. Reading, Mass., 1992.

— DAVID B. FOGEL

ARTIFICIAL SELECTION

The term "artificial selection" is generally applied to situations in which humans deliberately mold the features of another species by choosing those individuals with the characters they value most as parents of the next generation. This is, arguably, one of the most successful enterprises in which humans have ever engaged. Through repeated application of this simple process, our ancestors were able to "domesticate" a wide variety of plant species, transforming them into more stable and higher-yielding crops that transformed human societies and led ultimately to modern civilization. The success of artificial selection also provided the critical clue to Charles Darwin in his search for a mechanism underlying natural evolutionary change, work that resulted in his theory of evolution by natural selection.

Shaping Ancient Agriculture. Essentially all present-day crops have been extensively reshaped from their wild ancestors by the repeated application of artificial selection. For example, the major cereals (rice, wheat, maize, sorghum, and millet) were all developed from wild grasses by artificial selection between 7,000 and 12,000 years ago. These developments occurred independently in Asia (rice), Africa (sorghum and millet), the Middle East (wheat), and the Americas (maize; see Vignette). All involved the conversion of small-seeded plants with loosely attached seeds into large-seeded plants whose seeds remain on them longer. This transition effectively domesticated these lines, making their survival dependent on farmers sowing their seed. Similarly dramatic transformations were achieved by artificial selection in the development of various fruit and vegetable crops and in breeding animals to improve milk, meat, and egg production.

Artificial selection has also been used to shape the aesthetic features of organisms that surround us in everyday life. Many ornamental plants are the result of selection for attractive features. Likewise, the diversity in domesticated pets, such as breeds of dogs, cats, rabbits,

TEOSINTE INTO MAIZE

One of the most striking examples of the power of artificial selection is the transition of the wild grass teosinte into maize by early plant breeders in the Central Highlands of Mexico between 6,500 and 8,000 years ago. There are radical differences in both the plants and especially in the ear architectures of teosinte and maize. These differences are so dramatic that teosinte was originally placed in a separate genus (*Euchlaena*) than maize (*Zea*), and many early plant breeders favored other hypotheses for the origins of maize. The Mexican agronomist José Segura made the first experimental teosinte–maize hybrids in the late 1800s, demonstrating a much closer connection between the two than was originally thought. The Nobel-prize-winning American geneticist George Beadle had a lifelong interest in the origins of maize; he was instrumental in promoting the view that maize is the result of an artificially selected line derived from teosinte. Beadle found that about one in 500 of the F_2 maize–teosinte hybrids was phenotypically very similar to one of the original parents. This frequency of parental recovery is consistent with the hypothesis that about 5 major loci (or tight linkage groups) account for much of the maize/teosinte difference. Recent QTL mapping work by John Doebley at the University of Wisconsin has shown that substitution of just a few maize alleles for their teosinte counterparts indeed recovers much of the change in phenotype between the two lines.

—BRUCE WALSH

FIGURE 1. Products of Artificial Selection: A Cultivated (left) and a Wild Strawberry.
© Wally Eberhart/Visuals Unlimited.

erwise highest-performing individuals. Such background natural selection can severely limit or even halt response to artificial selection.

The difference between "natural" and "artificial" selection can be difficult to define. For example, there is strong natural selection for both insecticide and antibiotic resistance as a result of extensive human usage of these chemicals and drugs to control harmful insects and bacteria. Although there is no intent by those using these agents to increase resistance, their application results in a change in the environment such that individuals with greater resistance leave more offspring than do less resistant individuals. A related issue is the fact that laboratory strains and zoo animals are subjected to selection pressure to optimize their performance in artificial environments. This can lead to the accumulation of traits and genotypes that are less than optimal if the organisms are introduced back into the natural environment. Reducing the effects of such inadvertent artificial selection to laboratory conditions is thus of great concern to conservation biologists.

Quantifying the Amount of Artificial Selection. Usually the goal of artificial selection on a trait is to move the population mean toward a more desirable value. However, selection changes the entire population distribution of the trait, and hence one can select for changes in the variance (variability of the trait across individuals) as well—for example, selecting for more uniform seed set time.

One standard measure of the amount of selection applied to change the mean is given by the *directional selection differential* S,

$$S = \bar{z}^* - \bar{z} \qquad (1)$$

which is the difference between the trait mean in the selected parents ($\bar{z}^*$) and the trait mean in the population before selection ($\bar{z}$). A positive S indicates selection to increase the trait value, while a negative S indicates selection to decrease it.

Equation (1) assumes that each selected parent contributes an equal number of offspring. If the selected

or horses, is to a large extent the result of selection for features that particular breeders regarded as desirable.

Natural versus Artificial Selection. Darwin, noting the enormous success of artificial selection in generating a seemingly endless array of variation, suggested that a related process—natural selection—occurs in nature. Under natural selection, the conscious choice of individuals by a breeder is replaced by disproportionate survival and reproduction of individuals whose traits help them cope better with their environment. Natural selection chooses those individuals with the highest fitness (those leaving the most offspring), changing trait values to improve fitness. Artificial selection is, thus, just a special case of natural selection in which the breeder's choice of desirable trait values alters the intrinsic fitness of an individual (Figure 1). Note that even when artificial selection is being practiced, natural selection also is still operating. Particular traits that may be highly desirable to the breeder may come with an increased cost from natural selection, such as reduced viability in the oth-

parents disproportionately contribute offspring, then S is weighted by offspring number. If parent k has trait value z_k and leaves o_k offspring, then $\bar{z}*$ in equation (1) is replaced by the trait value weighted by offspring number:

$$\bar{z}* = \sum_k z_k \frac{o_k}{\sum_k o_k} \qquad (2)$$

where Σ(sigma) denotes "the sum of." Thus, for example, Σo_k is the total number of offspring produced by the selected parents. Both equations (1) and (2) are special cases of the *Robertson–Price identity*, which expresses the directional selection differential as the covariance (like a correlation) between relative fitness (w) and the trait value (z):

$$S = \text{Cov}(w,z) \qquad (3)$$

If W represents the number of offspring an individual leaves (its fitness), then its relative fitness is given by $w = W/\bar{W}$, where $\bar{W}$ is the mean fitness in the population (the average number of offspring left by a random individual). The Robertson–Price identity formalizes the connection between fitness and character value, with selection increasing the mean of a character when the character is positively correlated with fitness, and decreasing the mean when the character is negatively correlated with fitness.

In the literature on breeding, S is often replaced by the *selection intensity i* (also known as the *standardized selection differential*), the selection differential divided by the standard deviation of the trait,

$$i = \frac{S}{\sqrt{\text{Var}(z)}} \qquad (4)$$

where $\text{Var}(z)$ is the phenotypic variance of the trait. Breeders typically express selection in terms of the fraction p of the population selected each generation. When the distribution of the trait in the population is normal, p uniquely defines the selection intensity, independent of the mean or variance of the trait. For example, if one selects the uppermost 5 percent, then (for a large population) the expected value of i is 2.06. This translates into an S value of $2.06 * \text{Var}(z)^{1/2}$, showing that for the same fraction chosen, S increases with the character variance, while the selection intensity is unchanged by this scale effect from the variance.

Response to Artificial Selection. It is important to distinguish between artificial selection (choosing parents to improve the value of a desired character) and the actual response to selection, the improvement (if any) in the trait in the next generation. The former represents the within-generation change from selection, while the latter is the between-generation change from selection.

No matter how much artificial selection is applied, there will be a response in the next generation only if at least some of the gain in the selected parents can be passed on to their offspring. Define the response to selection by

$$R = \bar{z}' - \bar{z} \qquad (5)$$

where z' is the trait mean in the offspring. The amount of heritable variation determines how large a fraction of the within-generation selection (S) is translated into a between-generation response (R), with the predicted response having the form

$$R = b S \qquad (6)$$

When offspring forming the next generation result from a sexual cross between selected parents, $b = h^2$, the *heritability*, or more formally, the *narrow-sense heritability*. When offspring are genetically identical to their parent (as would occur if they are produced by cloning the parent or by selfing a completely inbred parent), $b = H^2$, the *broad-sense heritability*. Both these heritabilities are measures of the fraction of the trait variation that is due to genetic differences among individuals. In the absence of the appropriate genetic variation, there is no tendency for offspring to resemble their parents (the appropriate heritability is zero), and there is no response to selection.

To define the broad- and narrow-sense heritabilities, a slight digression into the field of quantitative genetics is required. The observed trait value z in any individual can be thought of as consisting of both a genetic G and an environment E component, so that $z = G + E$. The genetic (or genotypic) value G is the mean character value that one would observe if the individual is cloned and replicated over a number of environments. The population variation in G defines the broad-sense heritability, with

$$H^2 = \frac{\text{Var}(G)}{\text{Var}(z)} \qquad (7)$$

H^2 is thus is the fraction of the phenotypic variation in the trait that is attributable to differences in the genetic values of individuals. If most of the variation is attributable to genetic differences, then most of the selection advance in parents is passed onto their clones.

When offspring are produced by sexual reproduction, parents do not pass their entire genotypes on to their offspring. Rather, each passes on a gamete, and hence the genetic value G of an individual is not an exact predictor of the genetic contribution passed on to its offspring. This is formalized by the notion of a *breeding* (or *additive-genetic*) *value*, A. The average value in the offspring ($\bar{z}_o$) from two parents equals the mean breeding values of the two parents, so that the difference between the mean offspring from two parents and the population mean ($\bar{z}$) is just the average of the parental breeding values:

$$\bar{z}_o - \bar{z} = (A_{\text{father}} + A_{\text{mother}})/2 \qquad (8)$$

By definition, the mean breeding value for a randomly chosen individual is taken to be zero. Note that this, in conjunction with equation (8), implies that one could estimate the breeding value of an individual by crossing it at random with a number of other individuals from the population (indeed, this is routinely done in animal breeding). Suppose a male is mated to a large number of random females. From equation (8), its mean breeding value is estimated as

$$A_{\text{father}} = 2\,(\bar{z}_o - \bar{z}) \qquad (9)$$

Just as the phenotypic value of an individual can be decomposed into a genetic (G) and environment (E) component, its genetic value can be further decomposed into a breeding value A and residual I component, $G = A + I$. It is only the fraction A that is (on average) passed onto the offspring. The narrow-sense heritability is defined as the fraction of the total trait variation attributable to variation in the breeding values of individuals:

$$h^2 = \frac{\text{Var}(A)}{\text{Var}(z)} \qquad (10)$$

To predict the response to selection, one thus needs an estimate of h^2 (for sexually produced offspring) or H^2 (when offspring are identical to their parents). A large number of experimental designs for doing this have been developed by quantitative geneticists. Perhaps the simplest is to plot the mean value of offspring against the mean character value of both parents (a mid-parent–offspring regression). The expected slope of the best-fitting line through these data is h^2.

Improving the Efficiency of Artificial Selection. Various strategies are used by plant and animal breeders to improve the response to selection. Simple *individual selection* (also refereed to as *mass selection*) is the choosing of individuals based solely on their individual phenotypes, and this was the practice probably used by early plant and animal breeders. However, relatives of an individual also provide clues to their suitability for selection. Animal breeders use extensive pedigree information to improve the estimate of an individual's breeding value, and hence to improve the response to selection. For some species, such as dairy cattle, the pedigree information may involve a million or more individuals. By using an appropriately weighted index of the values of an individual and its relatives, the coefficient b in the response to selection equation can be significantly increased above its value (h^2) under individual selection. Likewise, when selecting among clones, modern plant breeders use multiregional trials, in which clones are grown in a number of different regions or countries and over several years before the best lines are chosen. Such replication of clones in different environments (sampling both regional and year-to-year variation) results in a much larger broad-sense heritability (H^2) value, and hence a greater response to selection, than the H^2 value based on individual selection (using only a single individual for each clone).

Response after Multiple Generations of Selection. Both Var(G) and Var(A) are functions of the distribution of genotypes within a population. Selection changes this distribution both by changing allele frequencies at individual loci and by generating linkage disequilibrium (creating correlations between alleles at different loci in the gametes of selected parents). As selection changes the initial values of these variances, the heritabilities, and hence the response to selection, also change. After sufficient time selection tends to erode these variances; in particular, it tends to drive Var(A) toward zero as favorable alleles are fixed and unfavorable alleles lost. Although this is partly offset by mutations generating new variation, and to a lesser extent by recombination breaking up groups of tightly linked loci under selection (and hence freeing up new variation for selection to act on), the net result is that the rate of response generally declines as selection proceeds. Populations can reach what appears to be a *selection limit* where there is little apparent further response to selection. Such limits can result from a lack of suitable genetic variation for additional response, requiring the introduction of new variation, either by waiting for new mutations to occur or by introducing individuals from other lines. Often an apparent limit occurs even when there is still significant additive variance in the character under artificial selection. In such cases, the lack of response often results from artificial selection being countered by natural selection.

Genetic drift, the random sampling of alleles, also tends to reduce both Var(A) and Var(G) as alleles drift toward loss or fixation; thus, population size is a very important component in the response to artificial selection. Population size is set by the number of parents selected, so intense selection (choosing only a small fraction of parents) results in an increased value of S but a smaller h^2 than expected under less intense selection. Thus, strong selection gives a larger short-term response but a smaller total long-term response, as it tends to erode genetic variation much more quickly because of increased genetic drift. The British geneticist Alan Robertson showed that the optimal amount of selection to maximize the long-term gain is to select half the population in each generation. Unfortunately, many commercial breeding situations are more governed by short-term than by long-term response, so they use strategies that are less than optimal for maximizing the ultimate response to selection.

Correlated Responses to Selection. Even when artificial selection is strictly applied to a single character, changes in a number of other characters are often observed. Such *correlated responses* can result in improvement of one character at the expense of undesirable changes in other characters. In the extreme, improvement of a character may result in a decrease in overall fitness, eventually leading to a selection limit and no further response. In theory, one can use estimates of genetic covariances between characters to predict the nature of short-term correlated response. The response (R_2) in character 2 given selection (S) only on character 1 follows from a modification of equation (6):

$$R_2 = \frac{\text{Cov }(A_1, A_2)}{\text{Var }(z_1)} S \qquad (11)$$

where $\text{Cov}(A_1, A_2)$ is the covariance (within an individual) in breeding values between traits one and two, and $\text{Var}(z_1)$ is the phenotypic variance of trait 1. In practice, correlated changes are quite unpredictable, especially after a few generations of selection.

There are two biological mechanisms that can generate a correlated response. First, the character experiencing the correlated response may share part of a developmental pathway with the character under selection. Second, the character may be influenced by genes that are linked to genes for the trait under selection. The latter phenomenon is referred to as *genetic hitchhiking*: selection on a particular locus results in a region of DNA surrounding the selected site increasing in frequency, the size of the region being determined by the amount of recombination with the homologous chromosome. In regions of low recombination, such hitch-hiking effects can extend over large chromosomal regions and can result in potential correlated responses in a number of traits. With our present ability to uncover an essentially endless amount of DNA polymorphism, breeders are attempting to exploit hitchhiking to increase selection response. This is done by selecting for marker alleles at these random polymorphisms that have been shown to be linked to regions that also contain loci influencing the trait of interest. Selecting on the appropriate marker alleles results in an increase in the frequency of desirable alleles at these linked genes. Such *marker-assisted selection* can significantly increase the response over individual selection because the breeder can apply stronger selection to individual loci.

Are There Genetic Differences in the Responses to Artificial and Natural Selection? Artificial selection produces much sharper selection for a particular character than would be expected under natural selection. This occurs because selection in nature acts on a broad constellation of traits, diffusing the amount of direct selection on any particular component. Consequently, the amount of selection acting on a trait locus has been perceived as being stronger under artificial selection. Given that alleles having large effects on a character often have deleterious side effects on a host of other characters, it may require strong selection to overcome these effects and contribute to the selection response. If selection to increase major alleles is not sufficiently strong, it may not overcome deleterious side effects; in such cases, major alleles would not make a significant contribution to selection response.

It is widely recognized that major genes are important in the response of natural selection to dramatic human modifications of the environment, such as the introduction of pesticides; however, it has often been claimed that such cases are exceptions rather than the rule, and that most of the selection response to natural selection arises from the total contribution of a large number of genes, each of small effect (the *infinitesimal model*). A critical evaluation of the available data shows both that adaptation in natural populations is often caused by a few major genes, and that selection response to artificial selection is often consistent with the infinitesimal model. At present, there is little evidence for a significant difference in the genetic basis of response under artificial and natural selection.

BIBLIOGRAPHY

Doebley, J. "George Beadle's Other Hypothesis: One-Gene, One Trait." *Genetics* 158 (2001): 487–493. A very readable account of the great debate on the competing hypotheses for the origin of maize.

Falconer, D. S., and T. F. C. Mackay. *Introduction to Quantitative Genetics*. 4th ed. Harlow, UK, 1996. The classic introductory text on quantitative genetics and selection response.

Harlan, J. R. *Crops and Man*. 2d ed. Madison, Wis., 1992. An excellent review on the domestication of wild plants into our current crops.

Lynch, M., and B. Walsh. *Genetics and Analysis of Quantitative Traits*. Sunderland, Mass., 1998. A comprehensive treatment of advanced topics in quantitative genetics.

Orr, H. A. "The Genetics of Species Differences." *Trends in Ecology and Evolution* 16 (2001): 343–350. A good review of evidence for major vs. polygenes in the response to natural selection.

Walsh, B., and M. Lynch. *Evolution and Selection of Quantitative Traits*. Sunderland, Mass. The companion volume to Lynch and Walsh (1988), this is a thorough treatment of natural and artificial selection as applied to quantitative characters. Chapters are available on the Web at http://nitro.biosci.arizona.edu/zbook/volume_2/vol2.html

— BRUCE WALSH

ASSORTATIVE MATING

Assortative mating may be defined as mating based on similarity or difference between the characteristics of the partners. Positive assortative mating occurs if both members of a pair are similar, so that, for example, large individuals tend to pair with large partners and small individuals with small ones. Negative assortative mating, sometimes called disassortative mating, occurs if a trait is negatively correlated within the pair; for example, smaller males are paired with larger females and vice versa.

Although mate preferences are involved in assortative mating, it is important to realize that it is distinct from sexual selection and that the two processes can occur separately or together. Sexual selection occurs when individuals have differential reproduction arising from their ability to compete for mates, but assortative mating may occur even if individuals all have equivalent reproductive success; if all green males mate with green females and blue males with blue females, but neither green nor blue individuals have higher fitness, sexual selection is not acting. Conversely, if both green and blue females prefer green males, assortative mating is not apparent but sexual selection is operating because the green males have higher reproductive success than the blue ones. [See articles on Sexual Selection.] It is also worth noting that individuals of similar phenotype may be paired as an indirect effect of mate choice occurring using other criteria than the trait on which partners are similar. If high-quality individuals are able to attract other high quality individuals as mates, leaving the low-quality individuals to choose among themselves, it may appear as though an individual chooses a partner on the basis of its own phenotype, when in fact, given the opportunity, everyone would prefer to have a high quality mate. Similarly, if sizes of males and females co-vary over time and pairs are sampled over a long period, it may appear as if large and small individuals are selecting each other, but the available pool of mates is merely changing over time. Finally, assortative mating may occur if the act of courtship or copulation is difficult when partners differ substantially in size; constraints of this type have been suggested for frogs and certain kinds of arthropods, particularly aquatic species such as water striders in which one sex carries the other. [See articles on Mate Choice.]

Assortative mating has many interesting implications for evolution. Because mating of like types within a population will subdivide the gene pool, assortative mating can facilitate divergence and hence lead to speciation. Several models of sympatric speciation, the divergence and eventual reproductive isolation of populations living in the same place, incorporate assortative mating in their pathways to complete separation. It can also help maintain genetic and phenotypic variability within a group. Theoretical models suggest that populations can move more easily from one adaptive fitness peak to another if assortative mating exists, because it allows the mean fitness of a population to increase gradually. Assortative mating may also moderate the rate or intensity of sexual selection by influencing the variance

in reproductive success of particular types of individuals. In addition, ecological specialization, in which members of a species eat a particular type of food or congregate in a particular habitat, may interact with assortative mating. Specialization in recently diverged taxa may often be due to characters that also produce assortative mating.

Inbreeding, or the tendency for relatives to mate with each other, is also distinct from assortative mating because inbreeding occurs with respect to all loci, whereas individuals may mate assortatively on the basis of only a few aspects of the genotype. Nevertheless, inbreeding may occur when individuals use characteristics of their own genotype in mate choice, which is a form of assortative mating. Although offspring of close relatives are more likely to express deleterious recessive alleles and hence may have lower fitness than outbred offspring, a certain amount of inbreeding has been suggested to be beneficial, because it may keep advantageous gene combinations together and allow animals to be adapted to local conditions. This balance between inbreeding and outbreeding, sometimes called optimal outbreeding, may be achieved if animals exhibit positive assortative mating using some traits but negative assortative mating using other traits.

Both invertebrates and vertebrates, including humans, exhibit assortative mating. In many cases, positive size assortative mating occurs; for example, in the isopod *Asellus aquaticus*, the amphipod *Gammarus pulex*, and several species of insects including *Plecia nearctica* flies, large males are generally found coupled with large females. This pattern may occur because larger males are both more successful in intrasexual competition and able to choose highly fecund larger females. In two species of soldier beetles (*Chauliognathus basalis* and *C. deceptus*), however, little evidence of size assortative mating was found except in locations where the two coexisted; where only one species occurred, mate choice was apparent instead. Arctic charr (*Salvelinus alpinus*), a fish occupying postglacial lakes of northern Europe, exhibit from one to four sympatric morphs, with mating generally occurring within morphs.

Animals may also mate assortatively using behavioral characteristics. In many species of songbirds, females prefer to mate with males singing the same dialect of the song that their fathers sang. This type of pairing may occur through a process called sexual imprinting, in which young animals learn a set of characteristics during a critical period, then use those characteristics as criteria when finding a mate after they mature. Although sexual imprinting involves learning, it can have important genetic repercussions, because hybridization is discouraged and separation of populations may occur.

Among humans, positive assortative mating has been found for a variety of traits, including physical attributes such as height and relative weight as well as psychological ones such as a tendency toward antisocial behavior. Environmental factors such as education level and socioeconomic status also seem to covary in members of couples, although in many societies a tendency for women to "marry up" is seen, so that women are more likely than men to have partners who are better educated and from a higher social class than they are themselves. Interestingly, couples who are living together are at least sometimes found to be more different than married couples. Sociologists and psychologists point out that such covariation of nongenetic characteristics may obscure attempts to study heritability or the relative contributions of genes and the environment to the development of personality traits or such conditions as obesity.

Although positive assortative mating is more common, some fascinating examples of negative assortative mating occur. In the North American white-throated sparrow (*Zonotrichia albicollis*), two morphs are found: a white-striped one, with brighter crown stripes, brighter yellow eye stripes, and less striping in the throat patch; and a tan-striped morph, which is also smaller and less aggressive. Both sexes exhibit the dimorphism, which is caused by a chromosomal inversion, with white-striped birds heterozygous and tan-striped birds homozygous. Up to 98 percent of the birds in a population mate with an individual of the opposite morph. When birds of both sexes and both morphs were presented with a choice between partners of each morph, both tan-striped and white-striped females preferred tan-striped males, at least so long as the males were able to interact with the females. If the females only observed their potential mates, they exhibited no preference. Both types of males, in contrast, preferred white-striped females, but only when they did not interact with the females. Anne Houtman and J. Bruce Falls (1994) suggested that the negative assortative mating is maintained via the higher competitive ability of white-striped females, which are able to pair with the preferred tan-striped males.

One of the most interesting examples of negative assortative mating has been found in mice and, at least potentially, in humans, and is linked to selection for resistance to disease. The major histocompatibility complex (MHC) is a group of genes found in one form or another in all vertebrates. Its original function is still questionable, but it is involved in self-nonself recognition, perhaps originally important in maintaining cohesion among colonial ancestors of vertebrates similar to the modern ascidians. In modern-day vertebrates, MHC genes are part of the immune system, involved in presenting foreign objects such as virus particles to the cells that will eventually recognize and attack the invader. The complex is extraordinarily variable, with a higher degree of polymorphism than anywhere else in the genome, and biologists have long speculated about the function of this variation. One possibility is that greater degrees of heterozygosity enable an individual to recognize more types of pathogens, and associations between some MHC types and a few diseases, such as malaria in humans or Marek's disease in poultry, have been found. Other attempts to link disease resistance and MHC have not been as successful, but it is generally agreed that the high levels of polymorphism are adaptive.

In the 1970s, a group of researchers at the Sloan Kettering Cancer Institute noticed that in several strains of mice that had been inbred to differ only at the MHC, mating was more likely to occur if the two mice were from different strains than if they were identical at the MHC. It has now been established that male mice show the preference more consistently than females, and that the preferences also occur in wild house mice, which have been studied under seminatural conditions by Wayne Potts and his colleagues at the University of Utah. At least three adaptive hypotheses have been proposed for MHC-dependent mating preferences. First, MHC-disassortative mating preferences produce MHC-heterozygous offspring that may have enhanced immune competence, as suggested above. Although this hypothesis is not supported by tests of single parasites, MHC heterozygotes may be resistant to multiple parasites. Second, MHC-dependent mating preferences may allow hosts to escape immune recognition by rapidly evolving parasites if rare, less easily recognized, alleles are continually being provided. Finally, MHC-dependent mating preferences may also function to avoid inbreeding, which is consistent with other evidence that MHC genes play a role in kin recognition by allowing parents and offspring, and perhaps other relatives, to recognize each other.

The mice appear to use odor cues to detect MHC type. Humans also seem to be able to distinguish between individuals of different MHC type by odor, and at least under some circumstances to prefer either certain types or individuals that differ from them the most at this genetic region. In one experiment, college students were asked to refrain from using perfumes or eating strongly flavored food such as garlic, and the male students were asked to wear a T-shirt for two consecutive nights. The next day, each female student was asked to rate the odors of six T-shirts. They scored male body odors as more pleasant when they differed from the men in their MHC than when they were more similar. This difference in odor assessment was reversed when the women rating the odors were taking oral contraceptives. Furthermore, the odors of MHC-dissimilar men reminded the test women more often of their own actual or former mates than do the odors of MHC-similar men.

This test and others like it have been criticized for not having a large enough sample and because they may not have controlled for all of the possible confounding variables. At the very least, however, such studies suggest that the MHC—and assortative mating—may be important in human behavior as well as that of other mammals.

[*See also* Speciation.]

BIBLIOGRAPHY

Bateson, P., ed. *Mate Choice*. Cambridge, 1983. Written for a scholarly audience, but not necessarily a specialized one. Contains several articles relevant to assortative mating, including one on mate choice in snow geese.

Burley, N. "The Meaning of Assortative Mating." *Ethology and Sociobiology* 4 (1983): 191–203. Although written for a specialized scientific audience, this article is easy to understand and painstakingly sets out some of the limitations of our ability to detect assortative mating, as well as some examples.

Crespi, B. J. "Causes of Assortative Mating in Insects." *Animal Behaviour* 38 (1989): 980–1000. A technical article that is well written and explains some of the hypotheses about why insects mate assortatively.

Hedrick, P., and V. Loeschke. "MHC and Mate Selection in Humans?" *Trends in Ecology and Evolution* 11 (1996): 24–25. This letter is critical of the work by Wedekind et al. (1995), and is followed by another criticism and a reply by the authors of the original article.

Houtman, A. M., and J. B. Falls. "Negative Assortative Mating in the White-throated Sparrow, *Zonotrichia albicollis*: The Role of Mate Choice and Intra-sexual Competition." *Animal Behaviour* 48 (1994): 337–383. A fascinating description of the mating system of birds, with some clearly explained experiments examining the basis for the observed preferences.

Howard, D. J., and S. H. Berlocher, eds. *Endless Forms: Species and Speciation*. Oxford, 1998. A somewhat technical collection of articles, but several of the contributions briefly discuss the role of assortative mating in species formation.

Wedekind C., and S. Furi. "Body Odour Preferences in Men and Women: Do They Aim for Specific MHC Combinations or Simply Heterozygosity?" *Proceedings of the Royal Society of London B* 264 (1997): 1471–1479. Another consideration of the role that the MHC may play in odor preferences, which concludes that humans do not seem to converge on particular MHC types.

Wedekind, C., T. Seebeck, F. Bettens, and A. J. Paepke. "MHC-dependent Mate Preferences in Humans." *Proceedings of the Royal Society of London B* 260 (1995): 245–249. One of the first articles to suggest that humans use the MHC in mate choice; the source of some controversy, as the letters by Hedrick and Loeschke (1996) and others indicate.

— MARLENE ZUK

ATAVISMS

Atavisms are the reappearances in individual members of a species of characters once possessed by all the members of an ancestor. Atavisms are old structures that appear anew, not newly evolved ones. Well-known examples include tails in humans; extra toes in horses; and elements of the hindlimb skeleton in whales, snakes, and legless lizards. Atavistic muscles also are known and especially studied in birds. Whales and snakes evolved from limbed ancestors. Horses have only a single toe but evolved from ancestors with five toes, the number of toes having been reduced progressively during evolution. The loss of such complex structures as legs and tails does not mean that the ability to make these structures has also been lost. Atavisms tell us that the information used to produce limbs, tails, and toes in the ancestors has not been lost but remains in the genome and in the processes of embryonic development. Indeed, for an atavism to appear, the ability to produce the structure (or behavior, for atavistic behaviors are known) must have been retained, even though the structure (or behavior) does not form. This was well known to Charles Darwin:

> I have stated that the most probable hypothesis to account for the reappearance of very ancient characters, is—that there is a tendency in the young of each successive generation to produce the long-lost character, and that this tendency, for unknown causes, sometimes prevails. (Darwin, 1859, p. 201)

Identification of atavisms requires an extensive knowledge of morphology and a well-founded phylogenetic tree, as demonstrated in the recognition that, of eight arrangements of wrist and ankle bones in a single population of the salamander *Taricha granulosa*, two are atavistic.

In phyletic or taxic atavisms—as seen in species such as Hawaiian *Drosophila* and cichlid fishes of South Africa, whose evolution has been characterized by the frequent reemergence of ancestral characters—all members of the species share the atavistic character that was not present continuously in ancestral lineages.

Atavisms can appear spontaneously in natural populations, following artificial breeding and selection, or can be evoked experimentally. Atavistic skeletal elements in whale hindlimbs are an example of an atavism seen in nature.

During their evolution from limbed terrestrial animals that possessed both fore- and hindlimbs, whales retained their forelimbs as flippers but lost their hindlimbs, which either are missing completely or are represented as vestigial pelvic bones; a Middle Eocene "whale" with hindlimbs is appropriately named *Protocetus atavus*. Early in development, whale embryos possess hindlimb buds, but these limb buds do not persist into later development. The occasional appearance of hindlimb bones (1 in 5,000 individuals in sperm whales) results from the continuation of hindlimb bud development beyond when it would normally cease. What is the underlying developmental mechanism?

Wingless mutant chicken embryos lack wings as adults but as embryos possess wing buds. The demise of the limb buds is driven by a mutation. We assume that a mutation allows hindlimb buds to persist longer than is usual in whale embryos showing the atavism. The likely mechanism is a delay in the onset of programmed cell death (apoptosis) that removes the limb buds in limbless vertebrates. The limb buds then develop for long enough that skeletal elements can form.

Atavisms represent genetic change, as demonstrated clearly through recognition of mutations and in breeding experiments on laboratory animals. Excessive facial and upper body hair in humans (hypertrichosis), which can be so extensive as to completely obscure the skin, results from a mutation and is inherited. Guinea pigs have three toes: In only four or five generations of breeding from a single guinea pig with a poorly developed and nonfunctional fourth toe that arose after a spontaneous mutation, all individuals had a fully functional extra toe. Importantly, for the digit to be atavistic, it must arise as an additional digit, not by subdivision of an existing digit; in guinea pigs, it is a digit 4, not a subdivided digit 3.

Can we therefore speak of atavistic genes, that is, of a class or classes of genes more likely than others to produce atavisms when mutated or misexpressed? atavistic features have been identified following over- or underexpression of *Hox* (homeobox) genes. Expression of *Hox* genes out of position during mouse embryonic development can be associated with the development of structures that are out of place, and those structures can resemble ancestral features. The disruption in mice of the gene *Hoxd-13* affects the timing of limb development, resulting in an increase in the number of wrist bones to numbers seen in ancestors. Mutations in other *Hox* genes in mice result in the formation of extra cartilages in positions in which reptiles have normal skeletal elements, although whether such extra cartilages are truly atavistic remains in question.

Other developmental mechanisms can produce atavisms. One is prolongation of the growth of a structure that normally becomes rudimentary or vestigial. As already noted, modern horses have one functional toe on each leg but are descended from horses with three or four toes on each limb. These ancestors, in turn, arose from animals with five toes on each leg. Embryos of modern horses develop the rudiments for three toes, but the growth of the central toe normally outstrips that of the lateral toes, which are left behind as splint bones. Only the central toe contacts the ground and is functional. In three-toed horses—of which Julius Caesar's horse, and Bucephalos, the war horse of Alexander the Great, are examples—we hypothesize that a mutation

allows the primordia of the side toes to continue growing for a longer period, resulting in toes of similar lengths. Associated changes in the ankle, muscles, ligaments, and tendons leads to the formation of a functionally three-toed horse. Functioning atavistic toes speak volumes to the adaptability of embryonic processes in producing evolutionary change.

Atavisms also can be induced experimentally, as in the production of teeth in birds, of ancestral patterns in the musculoskeletal system of birds' legs, and of ancestral bristle patterns in *Drosophila*. Atavisms therefore demonstrate the evolutionary conservation of genetic potential and of developmental mechanisms.

[*See also* Apoptosis; Homeobox; Vestigial Organs and Structures.]

BIBLIOGRAPHY

Darwin, C. *The Origin of Species by Means of Natural Selection.* London, 1859.

Dollé, P., A. Dierich, M. LeMeur, T. Schimmang, B. Schuhbaur, P. Chambon, and D. Duboule. 1993. "Disruption of the *Hoxd-13* Gene Induces Localized Heterochrony Leading to Mice with Neotenic Limbs." *Cell* 75 (1993): 431–441.

Figuera, L. E., M. Pandolfo, P. W. Dunne, J. M. Cantú, and P. I. Patel. "Mapping of the Congenital Generalized Hypertrichosis Locus to Chromosome Xq24–q27.1." *Nature Genetics* 10 (1995): 202–207.

Gould, S. J. "Hen's Teeth and Horse's Toes." *Natural History* 89.2 (1980): 24–28.

Hall, B. K. "Developmental Mechanisms Underlying the Formation of Atavisms." *Biological Reviews of the Cambridge Philosophical Society* 59 (1984): 89–124.

Hall, B. K. "Atavisms and Atavistic Mutations." *Nature Genetics* 10 (1995): 126–127.

Hall, B. K. *Evolutionary Developmental Biology.* 2d ed. Dordrecht, Netherlands, 1999.

Kessel, M., and P. Gruss. "Murine Developmental Control Genes." *Science* 249 (1990): 374–379.

Raikow, R. J., A. H. Bledsoe, B. A. Myers, and C. J. Welsh. "Individual Variation in Avian Muscles and Its Significance for the Reconstruction of Phylogeny." *Systematic Zoology* 39 (1990): 362–370.

Shubin, N. H., D. B. Wake, and A. J. Crawford. "Morphological Variation in the Limbs of *Taricha granulosa* (Caudata: Salamandridae): Evolutionary and Phylogenetic Implications." *Evolution* 49 (1995): 874–884.

Stiassny, M. L. J. "Atavisms, Phylogenetic Character Reversals, and the Origin of Evolutionary Novelties." *Netherlands Journal of Zoology* 42 (1992): 260–276.

— BRIAN K. HALL

AUSTRALOPITHECINES. *See* Hominid Evolution, *article on Ardipithecus* and *Australopithecines*.

B

BACTERIA AND ARCHAEA

Although they are generally much smaller than can be distinguished by the human eye, microbes are almost everywhere: from rocks two miles below the earth's surface, to the boiling waters spewing from the geysers in Yellowstone Park, to the air we breathe. Microbes were initially considered a mere curiosity when Antonie van Leeuwenhoek turned his self-made magnifying lenses on samples of lake water in the late 1600s and described a myriad of "wee animalcules" that inhabited this microscopic world. It was not until the late 1800s that the extraordinary power of microbes began to be appreciated when Robert Koch established the germ theory of disease, proving beyond a doubt that microbes were the agents of infectious diseases. Unfortunately, as a result of the attention that has been given to the small percentage of microbes that cause disease, many of us fail to realize the crucial role that microbes have played and continue to play in making the earth habitable for other forms of life. Without the metabolic activity of microbes, plant and animal life on earth could simply not exist!

The goal of this article is to provide a framework for understanding some of the basic aspects of microbial life and address a general question about microbes, "How can beings so small have such a dramatic impact on the composition of our planet's soils, water and atmosphere, and influence the evolution of other forms of life?" In this article, *microbe* refers to all single-celled organisms, although the sections below focus on bacteria and archaea. Additional information on eukaryotic microbes, including the fungi and protists, can be found in other entries in this encyclopedia.

Who's There? The challenges of studying organisms that are 1,000 times smaller than the period at the end of this sentence are great (Figure 1). Consider the task of conducting a biological inventory of all species present in a hardwood forest (i.e., asking the question "Who's there?"). You might imagine a group of ornithologists armed with binoculars, nets and tape recorders heading out to conduct an inventory of the birds that inhabit a wood lot, but how would one begin to inventory the microbes present in this ecosystem? Each plant and animal in a forest has literally millions of microbes associated with them, while every gram of soil contains on the order of ten billion microbes and lake water typically contains one million microbes per milliliter. To complete a survey, one would need to determine which microbial species are present and then estimate the number of the

microbes that belong to each species. With microbes, that leads to a difficult problem: How can we even determine if two microbes are members of the same species? This problem is exacerbated by the lack of diversity in the shapes assumed by microbes—even though we can count the number of microbes present in a sample, there is generally insufficient diversity in sizes and shapes to provide meaningful grouping of microbes.

In the late 1960s, Carl Woese at the University of Illinois began exploring the diversity of the microbial world by comparing the sequence of bases in a gene that codes for ribosomal RNA (rRNA). rRNAs are an essential component of ribosomes, and all forms of cellular life on earth have ribosomes as part of the cellular machinery that translates messenger RNA into protein. Building on earlier studies by Emile Zuckerkandl and Linus Pauling, Woese reasoned that organisms of the same species would have nearly identical rRNA sequences, whereas the rRNAs from more distantly related organisms would be less similar in the sequence of bases. The differences between rRNAs result largely from point mutations, and so the likelihood of mutations being present increases with the length of time that has passed since species diverged from a common ancestor. This simple yet elegant reasoning yielded extraordinary results.

Rather than representing a group of closely related organisms, microbes were found to be extraordinarily diverse and to exist in three independent lines of evolutionary descent, now known as the Bacteria, the Ar-

FIGURE 1. Bacteria and Archaea.
A magnification of various bacteria on a kitchen scrub pad.
© David Scharf/Peter Arnold, Inc.

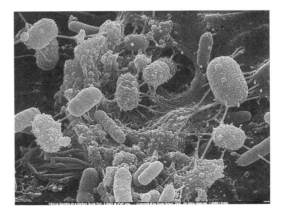

ORIGIN OF MITOCHONDRIA

Phylogenetic analysis of gene sequences has confirmed earlier suspicions that mitochondria and chloroplasts arose by endosymbiosis—the process by which one cell engulfs another, followed by a period of evolution in which both members develop a mutually dependent lifestyle. Both mitochondria and chloroplasts contain their own genome, on which reside genes including those encoding rRNAs. When the sequence of these rRNA genes were compared to those from other organisms, it was recognized that chloroplasts were most closely related to cyanobacteria, whereas mitochondria shared ancestry with a group of bacteria known as the Proteobacteria (Figure 2). This remains perhaps the most convincing evidence that distant ancestors of modern-day plants engulfed cyanobacteria that subsequently evolved into modern-day chloroplasts (although they still retain rRNA encoding genes that reveal their origin). Similarly, this phylogenetic analysis of mitochondrial rRNA suggests that an early eukaryote engulfed a bacterium that was the precursor to the modern-day mitochondrion. Humans and other animals therefore have two distinct single-celled ancestors: the one giving rise to our mitochondria as well as the one(s) giving rise to our nuclear genome.

The endosymbiotic origin of these organelles is supported by the comparison of other gene sequences and represents perhaps the best-documented and most important cases of lateral transfer of DNA—that is, the transfer of DNA between different species of organisms. Lateral gene transfer has tremendous potential to influence the evolution of life, and we are just beginning to appreciate the magnitude of this process.

—THOMAS M. SCHMIDT

chaea and the Eukarya (Figure 2). This viewpoint differed radically from the conventional wisdom of the day that described microbes as members of either the Monera (microbes with no membrane-bound nucleus) or Protoctista (eukaryotic microbes), and implied that the majority of biological diversity resided in plants and animals.

Microbial Lifestyles. The essence of life, survival and reproduction, is revealed clearly at the microbial level. And it is in the extensive evolutionary diversity of microbes that we are most likely to uncover the variety of ways in which the basic demands of life are met. Before considering some of the diverse lifestyles of microbes, it is worth noting that bacteria and archaea generally reproduce by binary fission—one cell divides into two. Although DNA can be exchanged among microbes

(roughly equivalent to sexual recombination), it is not required for reproduction. The more fundamental requirements are the capacity to extract energy from the environment and a means to obtain the basic building blocks of a cell, including water, carbon, nitrogen, phosphorus, sulfur, potassium, a variety of metals, and sometimes vitamins.

The uptake of resources from the environment and their conversion into biomolecules requires energy. Microbes harvest energy from their environment through a series of oxidation-reduction reactions. It is important to remember that for every oxidation reaction, in which electrons are removed from a substrate, there must be a compensatory reduction reaction that consumes the released electrons. As electrons are passed from higher to lower energy states (measured as redox potential) the energy released is ultimately captured in the form of adenosine triphosphate (ATP). For instance, human mitochondria oxidize glucose to carbon dioxide and reduce oxygen to water. The energy liberated from this oxidation-reduction reaction is captured in the form of

FIGURE 2. Representation of Evolutionary Relatedness. This representation is known as a phylogenetic tree and is based on a comparison of ribosomal RNA gene sequences. This analysis reveals that microbes are not constrained to any single taxonomic group, but rather comprise the majority of biological diversity in each of the three primary lines of evolutionary descent: Bacteria, Archaea, and Eukarya. Drawing by Thomas M. Schmidt.

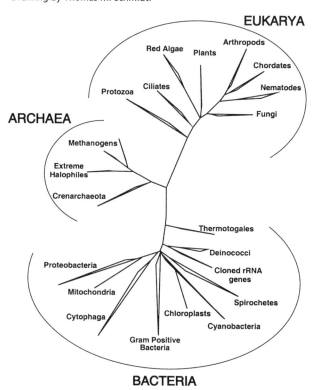

ATP and distributed throughout the cell. Knowing that mitochondria have evolved from bacteria (see Vignette on the Origin of Mitochondria), it is not surprising to find that many bacteria gain energy by the same basic mechanism as mitochondria. However, in addition to the oxidation of organic compounds like glucose (represented simply as CH_2O in Figure 3), there is a broad spectrum of inorganic compounds that can be oxidized, and these oxidation reactions can be tied not only to the reduction of oxygen, but to the reduction of other inorganic and organic compounds (Figure 3).

It has been suggested that the total number of bacterial and archaeal microbes on earth is approximately 10^{30}, that is 1 with thirty 0s! Even if this estimate is off by several orders of magnitude, the net effect of this astounding number of microbes deriving their existence from coupling the oxidation-reduction reactions is incredibly important for the cycling of elements on earth. Global cycles, including those of carbon, nitrogen, and sulfur, are driven primarily by the metabolic activity of microbes, and it is these activities that support plant and animal life on the planet.

Most of what we know about microbes and their contribution to global cycles is derived from studies of microbes that have been isolated from all other organisms and grown in "pure cultures." Growing a microbe in a pure culture permits detailed study of its nutrient requirements, energy-generating mechanisms, growth characteristics, and genetics of the organism, but it is not easy to extrapolate that information to predict how a microbe behaves in its natural environment. It is difficult to reproduce the natural habitat of a microbe in the laboratory, in part because microbes frequently function as part of a microbial community. The interactions among microbes in nature are undoubtedly im-

MICROBIAL RESPIRATION AND PHOTOSYNTHESIS

Specific enzymes in microbes catalyze the oxidation (removal of electrons) from organic or inorganic compounds. Every oxidation reaction is coupled to a reduction reaction; the arrows in Figure 3 indicate examples of such coupled reactions. The amount of energy available from coupling these reactions increases with the difference between the redox potential of the oxidation and reduction reaction (i.e., steeper slopes = greater potential energy). Energy released from the coupled oxidation-reduction reactions is used primarily to move protons across a microbe's cytoplasmic membrane. The accumulation of protons on the outside of the membrane represents potential energy that can then be harvested to:

1. synthesize ATP;
2. transport molecules (resources or waste products) across the cytoplasmic membrane, or;
3. move a microbe by turning flagella.

The oxidation of organic compounds, represented as CH_2O in this figure, can be coupled to oxygen (C; aerobic respiration) or to sulfur (A) or nitrogen (B) containing compounds in the environment (anaerobic respiration). Photosynthetic microbes also catalyze oxidation-reduction reactions, and use light energy to drive energy-consuming reactions that have an upward slope in this figure (E and F). These reactions result in the fixation of carbon dioxide to organic compounds that are used on biosynthesis. Like plants, cyanobacteria release oxygen in their light-driven metabolism (E), whereas other microbes release no oxygen during their metabolism but contribute important links in other global cycles including the sulfur cycle (F).

—Thomas M. Schmidt

FIGURE 3. Microbial Respiration and Photosynthesis. Drawing by Thomas M. Schmidt.

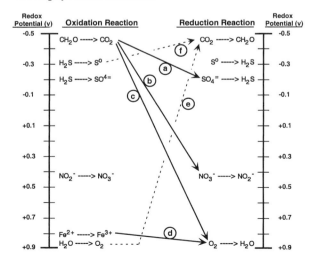

portant in understanding the behavior of each member of the community.

However, there is a far more serious difficulty arising from the heavy reliance on studies of pure cultures. Microbiologists estimate that a mere 1 percent of the microbes that are visible microscopically have been grown in pure culture. Studying these uncultivated microbes remains a tremendous challenge to microbiologists, and one that is being aided by the capacity to clone and analyze portions of their genomes. By studying the genomes of uncultivated microbes, it becomes possible to infer both their metabolic capacity and their evolutionary relatedness to other microbes.

Genome Sequences. In May 1995, a landmark in science was achieved with the announcement that the complete DNA sequence of a free-living organism had been determined. The sequenced genome was that of the bacterium *Haemophilus influenzae*. This organism was chosen for the sequencing project because the size of its chromosome (1.8 million base pairs) is fairly typical among bacteria and the percentage of its DNA that is made up of guanine and cytosine bases (G + C content = 38 percent) is close to that of the human genome. The rapid completion of this genome project, and at lower than expected costs, opened the floodgates for additional genome projects. There were nearly 200 microbial genome projects completed or under way as of late 2001 (http://www.tigr.org/tdb/mdb/mdbcomplete.html).

Just as ribosomal RNA sequences have been used to explore the evolutionary relatedness of microbes, the comparison of other gene sequences can provide clues not only about the origin of a gene but also about its potential function. The field of comparative genomics is filled with efforts to compare gene sequences and deduce the likely structure and function of the resulting proteins. From the collection of proteins present in an organism's genome, one can move forward to inferring biochemical pathways and ultimately describe the network of metabolic pathways that define that organism.

It is even possible to obtain large segments of the genome from the 99 percent of microbes that have not yet been grown in the laboratory. This approach is based on the purification of DNA from a community of microbes and the cloning of large fragments of that DNA into an artificial bacterial chromosome. If the large fragment of DNA encodes for a ribosomal RNA, that gene sequence can be used to link the rest of the cloned DNA fragment with organisms previously identified by rRNA sequence analysis (Figure 2). In a dramatic example of the potential of this approach, Edward DeLong and colleagues discovered that some of the uncultivated bacteria that are abundant in the earth's oceans use an unusual type of photosynthesis (previously known to exist only in the extremely halophilic Archaea) to drive their metabolism. These organisms have not yet been cultivated, but a crucial bit of information about their lifestyle and potential contribution to global mineral cycles has been revealed through analysis of a portion of their genome. By coupling advances from molecular genetic approaches with imaginative and persistent efforts to cultivate microbes and study microbial communities, we stand at the brink of tremendous advances in our understanding of the essential roles of microbes in the evolution and maintenance of the earth's biosphere.

Some Practical Applications. Understanding microbes and their interactions with ecosystems is far from an esoteric pursuit. For instance, since the advent of agriculture and industry, the human race has inadvertently undertaken a global ecological experiment whose outcome is far from certain. Extensive regions of land have been converted to agricultural use, leading to massive changes in soil nutrient cycles and runoff into coastal waters, while the burning of fossil fuels has released greenhouse gases with the resulting global warming. In both instances, the metabolism of microbes may either dampen or exacerbate potential problems. Microbes in soil influence the exchange of greenhouse gases between the atmosphere and soil, and the cycling of nutrients in soil and groundwater. Understanding their physiology and ecology can only help in predicting future trends and developing potential solutions.

On a more personal note, we experience benefits of research on diverse microbes in the antibiotics we use and those therapeutic agents under development, the thermally stable enzymes that are added to our laundry detergents to remove stains, and even the enzyme used to amplify DNA for use as evidence in paternity and criminal cases. Pathogens that resist cultivation are being diagnosed through an analysis of their gene sequences, and the strategies used by pathogenic bacteria to capitalize on their hosts are being revealed. There are still numerous discoveries to be made in the microbial world, and an understanding of the evolution and ecology of the diverse microbial world will surely stimulate these discoveries.

[*See also* Antibiotic Resistance, *article on* Origins, Mechanisms, and Extent of Resistance; Cellular Organelles; Clonal Structure, *article on* Population Structure and Clonality of Bacteria; Disease, *article on* Infectious Disease; Emerging and Re-Emerging Diseases; Plasmids.]

BIBLIOGRAPHY

Béjà, O., L. Aravind, E. V. Koonin, M. T. Suzuki, A. Hadd, L. P. Nguyen, S. B. Jovanovich, C. M. Gates, R. A. Feldman, J. L. Spudich, E. N. Spudich, and E. F. DeLong. "Bacterial Rhodopsin: Evidence for a New Type of Phototrophy in the Sea." *Science* 289 (2000): 1902–1906. The use of clone libraries to explore the microbial world is described in this article, along with the discovery of a new type of photosynthesis in the oceans.

Doolittle, W. F. "Phylogenetic Classification and the Universal Tree." *Science* 284 (1999): 2124–2128. The potential influence of lateral gene transfer on establishing the evolutionary relationships amongst microbes is addressed, along with a thoughtful view of the evolution of life.

Fraser, C. M., J. A. Eisen, and S. L. Salzberg. "Microbial Genome Sequencing." *Nature* 406 (2000): 799–803. This article provides an overview of the methods currently available for determination and analysis of the sequence of microbial genomes.

Madigan, M. T., J. M. Martinko, and J. Parker. *Brock Biology of Microorganisms.* 9th ed. Upper Saddle River, N.J., 2000. This excellent and up-to-date textbook reviews key discoveries in the history of microbiology and provides an encyclopedic coverage of the physiological diversity in the microbial world.

Maidak, B. L., J. R. Cole, T. G. Lilburn, C. T. Parker, Jr., P. R. Saxman, R. J. Farris, G. M. Garrity, G. J. Olsen, T. M. Schmidt, and J. M. Tiedje. "The RDP-II (Ribosomal Database Project)." *Nucleic Acids Research* 29 (2001): 173–174. This article describes the Ribosomal Database project in which RNA gene sequences that are used routinely to compute the relatedness of microbes are stored.

Olsen, G. J., C. R. Woese, and R. Overbeek. "The Winds of (Evolutionary) Change: Breathing New Life into Microbiology." *Journal of Bacteriology* 176 (1994): 1–6.

Pace, N. R. "A Molecular View of Microbial Diversity and the Biosphere." *Science* 276 (1997): 734–740.

Relman, D. A., J. S. Loutit, T. M. Schmidt, S. Falkow, and L. S. Tompkins. "An Approach to the Identification of Uncultured Pathogens: The Causative Agent of Bacillary Angiomatosis." *New England Journal of Medicine* 323 (1990): 1573–1580. This article reports the first use of ribosomal RNA sequences to identify a human pathogen that had not been grown in culture.

Staley, J. T., and A.-L. Reeysenbach, eds. *Biodiversity of Microbial Life: Foundation of Earth's Biosphere*. New York, 2002. This book includes chapters written by some of today's best microbiologist, and provides a current view of the physiological and evolutionary diversity of microbial life on earth.

Whitman, W. B., D. C. Coleman, and W. J. Wiebe. "Prokaryotes: The Unseen Majority." *Proceedings of the National Academy of Sciences* 95 (1998): 6578–6583. Using estimates of bacterial and archaeal abundance from numerous habitats, the authors approximate the total number of prokaryotic microbes on planet Earth.

Woese, Carl R. "Interpreting the Universal Phylogenetic Tree." *Proceedings of the National Academy of Sciences* 97 (2000): 8392–8396. This article, by the founder of microbial phylogenetics, addresses current issues related to the ribosomal RNA-based tree of life, including the potential influence of lateral gene transfer.

— Thomas M. Schmidt

BASIC REPRODUCTIVE RATE (R_0)

The basic reproductive rate, often designated R_0, is an important quantity that arises in two distinct areas of population, biology and evolution. In the context of infectious disease, R_0 represents the number of secondary infections that can be expected to arise from each primary infection. In the context of life history theory, R_0 is the average number of progeny produced over the life of an individual. The basic reproductive rate is therefore fundamentally the same in both contexts, as it reflects the number of secondary events—infections or offspring—produced by each primary event. In both contexts, R_0 also serves as a simple, summary quantity that is used to understand the evolution of the multiple underlying parameters that govern life histories and host–pathogen interactions.

To a first approximation, the quantity R_0 can be understood as follows. If $R_0 < 1$, and if it remains there indefinitely, then a population will become extinct because each generation of infection or reproduction leaves fewer descendants than there were progenitors. In contrast, a population will increase in number when $R_0 > 1$, although eventually some aspect of the environment must limit population growth. Such limitations may occur when a population depletes its resources or, in the case of an infectious disease, when the susceptible hosts have been depleted by prior infection, including recovery with immunity as well as death. When a population is at dynamic equilibrium (i.e., it is exactly replacing itself), then $R_0 = 1$.

The interpretation of the inequalities described above apply in most cases, but not in all. In particular, there can be short-term deviations from these relationships when the population in question has an age distribution and that distribution changes over time. In an age-structured population that includes disproportionate numbers of individuals of reproductive age, the population may in fact increase over the short term (births exceeding deaths), even though $R_0 < 1$. In fact, this situation presently exists in the human population in many developed countries, including the United States. In such cases, the average female will produce fewer than two children (less than one daughter per mother), but the population size is nonetheless increasing because the age distribution is overrepresented by women of reproductive age. In the absence of further changes in age-specific survival and reproduction, the age distribution will eventually stabilize and the basic reproductive rate, R_0, should then indicate whether a population is growing or declining in number (excluding the effects of migration).

In life history evolution, it might seem that natural selection should invariably increase R_0, with degradation of the environment providing the only limit to its increase. Such is the case in a population with nonoverlapping generations (e.g., annual plants). However, the situation is more complicated in species with overlapping generations. In such a species, if the population is expanding, then selection may favor mutations that increase the production of offspring earlier in a parent's life, even at the expense of fewer offspring over the parent's lifetime. The emphasis on early reproduction occurs because the mutant offspring will also reproduce at an earlier age, compounding the investment in earlier reproduction faster than a corresponding investment made later in life.

The quantity R_0 has played an important role in mathematical models of host–parasite (and host–pathogen) interactions, including efforts to better understand the evolution of parasite virulence. R_0 should increase with the transmission rate of a parasite, as well as with the length of the period before an infected host either dies or recovers. In certain parasites, however, there is a tradeoff between these two components. For example, a mutation that increases parasite growth within an individual host may increase its transmission rate while also increasing host mortality, thus shortening the window for transmission. Depending on the precise quantitative relationship between the transmission rate and infectious duration, a parasite of intermediate virulence may be favored by natural selection. In such cases, maximizing the quantity R_0 is used to solve for the parasite's optimal virulence, provided that certain other biological assumptions are met.

[*See also* Disease, *article on* Infectious Disease; Life History Theory, *article on* Human Life Histories; Population Dynamics; Reproductive Value; Transmission Dynamics.]

BIBLIOGRAPHY

Anderson, R. M., and R. M. May. *Infectious Diseases of Humans.* Oxford, 1991. A mathematical treatment of infectious disease by two of the leading contributors to this field.
Bulmer, M. *Theoretical Evolutionary Ecology.* Sunderland, Mass., 1994. Includes models of both age-structured populations and host–parasite interactions.
Caswell, H. *Matrix Population Models.* Sunderland, Mass., 2000. A thorough treatment of the mathematics of age-structured populations.
Lenski, R. E., and R. M. May. "The Evolution of Virulence in Parasites and Pathogens: Reconciliation between Two Competing Hypotheses." *Journal of Theoretical Biology* 169 (1994): 253–265. A mathematical analysis of the feedback between ecological and evolutionary dynamics in host–parasite interactions.
Wilson, E. O., and W. H. Bossert. *A Primer of Population Biology.* Sunderland, Mass., 1971. A clear and concise treatment of basic population models in ecology and evolution.

— RICHARD E. LENSKI

BAUPLÄNE. *See* Body Plans.

BIODIVERSITY. *See* Species Diversity.

BIOGEOGRAPHY

[*This entry comprises three articles:*

Island Biogeography
Vicariance Biogeography
Human Influences on Biogeography

The first article discusses the particulars of island biogeography; the second article provides an overview of possible future directions of the field of vicariance biogeography; the third article serves as an account of human effects on the biogeography of other species and the cascade of consequences. For a related discussion, see Primates, *article on* Primate Biogeography.]

Island Biogeography

Biogeography is the study of the distribution of life across space and how, through time, it has changed. The great complexity of the earth's ecosystems and the vagaries of past events render neat experimental approaches to biogeography difficult to attain. Islands, being discrete, internally quantifiable, numerous, and varied entities, provide us with a suite of "natural experiments." Scientists can select sets of islands that allow research problems to be isolated, enabling theories of general applicability to be developed and tested. Under this "natural laboratory" umbrella, a number of distinctive "island biogeographical" traditions have developed, focused to varying degrees on studies on real islands (i.e., isolated land masses set in the ocean). They span a broad continuum and have exchanged ideas with the fields of evolution, ecology, and conservation biology (Whittaker, 1998).

Evolutionary Island Biogeography. Biotic interchange between oceanic islands and other landmasses is generally limited. Remote oceanic islands typically have few species, but are rich in endemics, species restricted to one or a group of islands. Paleo-endemics are old lineages that long ago reached an island and have persisted while dying out from their original mainland areas. Neo-endemics, by contrast, are species that have developed in situ, diverging on the island from their ancestral forms. Evolutionary change can take the form of the continuation of a single lineage, in which the progenitor form becomes extinct. This is termed anagenesis. Or it may involve the splitting of lineages, termed cladogenesis where the progenitor is partitioned into two lines and becomes extinct in its original form. The term anacladogenesis is used when the progenitor survives essentially unchanged alongside the derived species. On the Juan Fernandez islands, over 600 kilometers west of Chile, the three models apply respectively to 71 percent, 5 percent, and 24 percent of the endemic plant species (Stuessy et al., 1990). It should be noted that as on continents, lineages may converge or cross via hybridization, but the extent to which hybridization has led directly to speciation is hard to quantify (Grant, 1998).

Studies of island populations have revealed much about mechanisms of evolutionary change such as founder effects, genetic drift, character displacement, natural selection, and sexual selection. Classic "model organisms" featured in such studies include the *Anolis* lizards of the Caribbean and the Hawaiian Drospholids (fruit flies). The founder effect (described by Ernst Mayr) is based on the observation that in general an immigrant population to a remote island will establish by means of a small founding population. This founding population contains only a subset of the genetic variation in the source population and it subsequently likely receives no further infusion. This creates a bias in the genetic base of the island population, on which other evolutionary processes then operate. Genetic variability can be increased by mutation and resorting after establishment. One form of change thought important under sustained conditions of low population size (and thus in the evolution of some island lineages) is genetic drift, the chance alteration of allele frequencies from one generation to the next. In addition to such nonselective drift, there may be distinctive "selectional" features of the novel island environment. The outcome of these phenomena can, it is thought, be a rapid shift to a new, coadapted combination of alleles, hence contributing to reproductive isolation from the original parent population.

Island species can be distinctive in a number of ways, sometimes displaying giganticism (large forms of normally small creatures, e.g., small mammals) or nanism (dwarf forms of normally large creatures, e.g., large mammals such as elephant or mammoth), loss of flight (birds and insects), loss of bright coloration and defensive behavior (birds), and loss of dispersability (plants). These shifts can be understood in terms of the differences in the selective environment on islands lacking the array of competitor and predator species found in mainland environments (Whittaker, 1998). On Wrangel island, some 200 kilometers from north-east Siberia, woolly mammoth were stranded on separation of the island from the mainland approximately 12,000 years ago. The climatic regime maintained the favored steppe vegetation of the mammoth, which survived on the island to between 7,000 and 4,000 bp. Elsewhere, through climate change and hunting pressures, the mammoth became extinct by 9,500 bp. Not only did the island population thus attain relictual status, but during the period between 12,000 and 7,000 bp it appeared (on the basis of fossil teeth) to shrink in body size by at least 30 percent (Vartanyan et al., 1993).

As exemplified by the mammoth, some island evolution occurs with little or no radiation, as island forms change away from the colonizing phenotype along a singular pathway (i.e., anagenesis). This may in time produce a distinct island species. A second important model is that termed the taxon cycle, first formulated for Melanesian island ant populations by E. O. Wilson in papers published in 1959 and 1961. The taxon cycle is best envisaged within the context of an island archipelago, driven by a series of colonization events of taxonomically and/or ecologically related forms. Immigrant species undergo niche shifts, which are in part driven by competitive interactions with later arrivals. Through time the earlier colonists lose mobility, decrease in distributional range, and may eventually be driven to extinction. Evidence for the taxon cycle is largely, but not exclusively distributional, and the model awaits confirmation or amendment by application of modern genetic analyses (see Whittaker, 1998).

The most spectacular evolutionary patterns, the radiation of monophyletic lineages, are to be found on the most isolated of large, oceanic islands. Although some radiations have been postulated to be essentially nonadaptive (featuring "drift" in within-island isolates rather than clear niche changes), most radiations involve clear alterations of niche and may thus be termed "adaptive." Examples of the former appear to be of mostly land snails, whereas the latter include the best-known plant, bird, and insect examples from Hawaii, the Galapagos, and Macaronesia (see entry by Grant). The term *archipelago speciation* is sometimes given to the great radiations of species across an archipelago: in such cases the separation (allopatry) of populations on different islands often appears crucial to the pattern that emerges. However, radiations can also occur sympatrically, that is, where species co-occur on the same islands (e.g., radiations of *Aeonium* on Tenerife), often aided by finer scale separation within a large and heterogeneous island.

When viewed in relation to a simple framework of area versus island or archipelago isolation, it becomes apparent that different evolutionary models have relevance to different geographical circumstances (*Global Ecology and Biogeography*, 2000). Thus, for example, the taxon cycle may be an appropriate model to test in a not too remote archipelago, whereas adaptive radiation is more typical of the most isolated, high-island archipelagos. As effective dispersal range varies between (and within) taxa, different evolutionary patterns may emerge within the same archipelago by comparison of different types of organisms.

Ecological Island Biogeography. Many natural scientists associate a single theme with "island biogeography": the explanation, modeling, and prediction of species richness and turnover and how these properties vary with factors such as island area and isolation. The dominant theory in this field has been the equilibrium or dynamic model of Robert H. MacArthur and Edward O. Wilson. First published as a zoogeographic theory in 1963, their ideas were developed more fully in their 1967 monograph *The Theory of Island Biogeography*. At its core, their theory is remarkably simple. It posits that the number of species on an island is a dynamic product of the opposition of rates of species immigration (afforced on remote islands by in situ speciation) and rates of species loss. The authors further argued that immigration rates would tend to decline predictably with isolation, and that extinction rates would vary in relation to the resource base of the island, best represented as a first approximation by island area. Extinction rate is suggested to depend on average life span of the taxon, the richness of the taxon, and the mean size of the species populations, these being necessarily related to island area.

Thus, species richness should be predictable as a function of area and isolation of the islands, and island biotas should on the whole be in a dynamic equilibrium, featuring predictable rates of turnover. MacArthur and Wilson (1967) presented this dynamic model as a component of a larger body of evolutionary ecological theory, including for instance the taxon cycle model, but it is the simple species richness/turnover model that most refer to as "the" theory of island biogeography. It was offered as a model for all islands, large or small, near or far, and it was this hoped-for generality, combined with the apparent testability of the model, that provided much of the attraction of the thesis.

The MacArthur–Wilson theory provided a plausible theory that might account for the species richness-area relationships demonstrated by earlier ecologists. For a given taxon and region, a fairly simple relationship often exists, expressed by the power function (or Arrhenius) model: $S = CA^z$ where S = species number, A = area of island, and where C and z = estimable parameters (constants). By logging species number and area, the species-area curve typically becomes a straight line, of the form $\log S = z \log A + \log C$, enabling the parameters C and z to be determined using simple linear regression. In this equation, z describes the slope of the log-log relationship and $\log C$ determines its intercept.

MacArthur and Wilson (1967) showed that the slope of the log-log plot appeared to be steeper for islands; thus, any reduction in island area lowers the diversity more than a similar reduction of sample area from a contiguous mainland habitat. Their dynamic model provided an explanation consistent with this and other empirical patterns concerning island species number. It thus provided a premise for testing hypotheses by means of mathematical analyses of species richness and turnover patterns. Since the publication of their theory, numerous studies have been undertaken (including manipulative experimental studies), many of which are consistent with their ideas, but some of which are inconsistent (Lomolino and Weiser, 2001). Subsidiary and alternative hypotheses have also been published that could account for the species-area effects found empirically. The key feature of the MacArthur–Wilson theory that sets it aside from the competing hypotheses is the occurrence of predictable rates of turnover at equilibrium: here the test results have been somewhat equivocal. It appears that in the real world, many islands are in a nonequilibrium condition much of the time: this may apply particularly to remote archipelagos where immigration rates and evolutionary response times are insufficiently rapid to maintain the form of dynamic equilibrium postulated by the theory. For recent reviews and developments, see Brown and Lomolino (1998), Whittaker (1998), and *Global Ecology and Biogeography* (2000).

An interesting and ambitious development of the MacArthur–Wilson theory is provided by Hubbell's (2001) "unified neutral theory of biodiversity and biogeography," which appears to forge the links between population abundance distributions and species richness more convincingly than MacArthur and Wilson. Hubbell's theory is based on a new distribution of relative species abundance called the zero-sum multinomial, which exhibits negative skewness to a degree he suggests is determined by island size and immigration rate. Hubbell's theory, like MacArthur and Wilson's, is founded on a number of simplifying assumptions, which may ultimately limit its applicability, but it seems likely that it will provide a considerable stimulus to the study of diversity constraints and patterns. While Hubbell's theory has things to say about islands it is not premised on the island–mainland dichotomy in the way the MacArthur–Wilson model is: it is thus illustrative of the way island biogeography feeds into general theory rather than being an island biogeographical theory itself.

The MacArthur–Wilson model is principally concerned with species numbers, rather than with compositional patterns, but island biotas are not simply random draws from the available regional species pools. Instead, some species types are found more frequently and others less frequently than expected by chance. For instance, remote islands lack terrestrial mammals and other taxa that have a poor trans-oceanic dispersal capability. Such distinctions can be made at a much finer taxonomic resolution, such that for instance, within the guild of fruit-eating pigeons, or flycatchers, some species combinations appear to occur together preferentially, whilst others are rare or cannot be found. Such observations led Jared Diamond to propose several "assembly rules" in a seminal study on birds of the New Guinea region published in the mid 1970s. He interpreted these empirical patterns in terms of the outcomes of inter-specific competition, contrasting distributional origins (including the occurrence of single-island endemics), habitat requirements, successional development on islands recovering from volcanic disturbance, and so forth. Diamond's island assembly theory thus provided a dynamic, ecological-evolutionary interpretation for island biogeographical patterns, building considerably on the MacArthur and Wilson model. It has proven to be a controversial body of work, because the statistical tools to demonstrate the existence of non-random patterns have been under continual criticism and development, and because present (and especially past) competitive effects are generally hard to isolate from other factors that might cause distributional patterns of this form.

Conservation Biology. The idea that remnant fragments of intact habitats, surrounded by altered or converted landscapes, might be considered from an island biogeographic perspective has led conservationists to lean heavily on island theory in designing or predicting the outcome of particular reserve network configurations. The application of island biogeographical ideas and tools to conservation biology has centered on the MacArthur–Wilson theory. The theory seemed to provide a basis for predicting the extinction of species that would follow from large-scale habitat conversion, as this results in the reduction of contiguous area and the increased isolation of remaining "habitat island" patches. Early contributions argued that theory favored fewer large areas over many small patches of the same total area, round rather than elongated reserves, and well-connected rather than very isolated reserves. Other authors have argued that the theory is inadequate to provide such specific guidance and to predict extinction accurately. This and other "island biogeographical" approaches to conservation biology are reviewed by Whittaker (1998).

Concluding Remarks: Future Directions. The island biogeographical themes discussed in this article are all the subject of continuing research. The distinction between the different headings is largely for convenience, being rather arbitrary in nature, as to some degree is that between "island" and "other" biogeographical theory. The development of novel techniques of genetic analysis provides tremendous opportunities to test theories concerning the taxon cycle and so-called adaptive radiations, and the relative importance of allopatric and sympatric phases, of multiple-colonization events, and of back-crossing, etc. in the development of the biogeographical patterns (e.g., Grant, 1998). In island ecology, the MacArthur–Wilson theory had a crucial role in that it was a dynamic theory, attempting to capture a fundamental process and its associated mechanisms, and thus offering the promise of extracting laws of general validity. It is now apparent that it is an insufficient basis on its own, but it remains an important part of biogeographical theory. New statistical tools provide the possibility of more powerful analyses of outstanding problems, while the development of nonequilibrium theories and the inclusion of scale explicitly within theory seem particularly promising (Whittaker, 1998; Hubbell, 2001). The application of island biogeographical tools and constructs has a part to play in solving the problems of conservation biology, although in practice an armory of approaches is required to solve the practical issues of species conservation in the twenty-first century.

BIBLIOGRAPHY

Brown, J. H., and M. V. Lomolino. *Biogeography*, 2d ed. Sunderland, Mass., 1998. An excellent textbook including an insightful review of island biogeography.

Damuth, J. "Cope's Rule, the Island Rule and the Scaling of Mammalian Population Density." *Nature* 365 (1993): 748–50.

Global Ecology and Biogeography, 2000, volume 9(1), pp. 1–92. A multiauthored special issue reviewing progress in island biogeography theory.

Grant, P. R., ed. *Evolution on Islands*. Oxford, 1998. Useful case studies of island evolution.

Hubbell, S. P. *The Unified Neutral Theory of Biodiversity and Biogeography*. Princeton, 2001. A monograph building on some of the foundations of MacArthur and Wilson's theory by connecting species-abundance distributions to diversity patterns in a novel way.

Lomolino, M. V., and M. D. Weiser. "Towards a More General Species-Area Relationship: Diversity on All Islands, Great and Small." *Journal of Biogeography* 28 (2001): 431–445.

MacArthur, R. H., and E. O. Wilson. *The Theory of Island Biogeography*. Princeton, 1967. Reprinted in 2001, with a new preface by E. O. Wilson.

Stuessy, T. F., D. J. Crawford, and C. Marticorena. "Patterns of Phylogeny in the Endemic Vascular Flora of the Juan Fernandez Islands, Chile." *Systematic Botany* 15 (1990): 338–46.

Vartanyan, S. L., V. E. Garutt, and A. V. Sher. "Holocene Dwarf Mammoths from Wrangel Island in the Siberian Arctic." *Nature* 362 (1993): 337–340.

Whittaker, R. J. *Island Biogeography: Ecology, Evolution, and Conservation*. Oxford, 1998. An accessible text reviewing all aspects of island biogeography (including the application of island theories to conservation biology). Requires minimal technical or mathematical knowledge of the reader.

— ROBERT J. WHITTAKER

Vicariance Biogeography

Biogeographers attempt to decipher the origins and distributions of populations, species, and biotas at different geographic and temporal scales. The field has attracted scientists from natural history, systematics, geology, evolutionary biology, and ecology.

Two major controversies dominate current biogeographical studies. The first is the issue of organismal dispersal versus prior ancestral existence. Do species exist in an area because they dispersed into the area and diverged, or because their most recent common ancestor existed there and diverged into the descendant species? The second controversy involves the interface between ecological (proximate) and historical (ultimate) explanations for the origins of taxa and their distributions. Thus, a dichotomy has developed between ecological and historical biogeography, where the former focused more on the processes involved in maintaining current distributions and the latter focused on the history of how species evolved in the areas that they currently occupy.

Vicariance biogeography has emanated from decades of dialogue in the history of biogeography as the predominate theory and method necessary to enhance our understanding of current and ancestral distributions of taxa. A comparison of the processes and predictions of vicariance and dispersal biogeography (including the field more recently called "phylogeography") is provided in Figure 1.

Vicariance biogeography evolved out of phylogenetic systematics (sensu Hennig 1966) and the legacy of historical biogeography of both panbiogeography (Croizat, 1964) and centers-of-origin, dispersal biogeography (Darlington, 1957). In its simplest form, vicariance biogeography is aimed at discovering and testing shared, ancestor-descendant relationships of organisms occupying common areas of endemism ranging in scale from

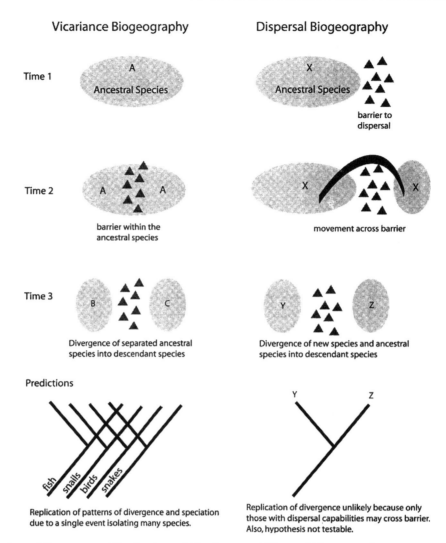

FIGURE 1. A Comparison of Processes and Predications of Vicariance and Dispersal Biogeography. Replicated speciation events across many clades in vicariance biogeography can be explained by one isolating event; replicated patterns in dispersal biogeography require individual explanations for each clade.   Courtesy of Richard Mayden.

caves and springs to continents and oceans (Nelson and Rosen, 1981; Wiley, 1980, 1988). This method of inquiry is built on the thesis that biodiversity evolves in concert with historical events of the earth, and that species occupying areas of earth subjected to incidents terminating or significantly impairing gene flow should respond in like manners through divergence and speciation. As a consequence, one can predict that there will be repeated, predictable, and testable patterns of speciation and divergence. Early on, vicariance biogeography targeted areas of endemism for investigation; more recent studies have not necessarily focused on areas of endemism but are more holistic in geographic coverage and cognizant of species with widespread distributions.

As a heuristic example of vicariance biogeography, imagine three areas of endemism (areas I, II, III), each

containing related species (clades 1-2-3, 4-5-6, 7-8-9) and individual species not part of clades shared across the areas (species 10 and 11) (Figure 2). In this example, without using the phylogenetic relationships of the species but knowing that the species are parts of "groups," one can identify shared biogeographic tracts and may hypothesize that the species endemic to each of the areas dispersed and diverged individually and independently. Such a conclusion requires multiple explanations for each of the dispersal events to each of the areas and is largely untestable, given that the dispersal occurred at some time in the past under unknown conditions. Furthermore, one may ask the question how multiple groups of organisms with differing dispersal capabilities were able to colonize these areas. Alternatively, the phylogenetic relationships of taxa provide essential evolu-

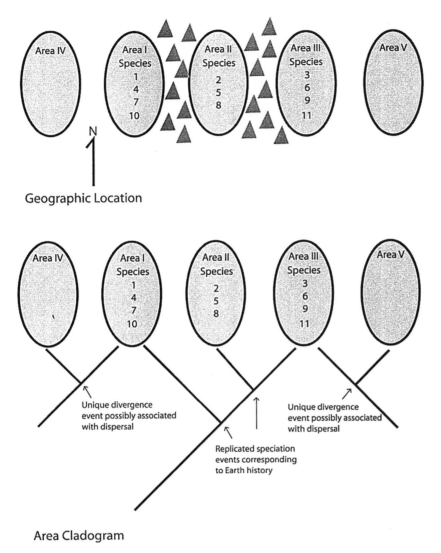

Geographic Location

Area Cladogram

FIGURE 2. Simple Example of an Area Cladogram Derived from Relationships of Species Inhabiting the Areas of Endemism. The replicated pattern is consistent with the disruption of the areas by river drainages separating terrestrial habitats. Some species, however, possess dispersal capabilities to overcome barriers that are part of communities because of dispersal.   Courtesy of Richard Mayden.

tionary information to enhance our understanding of history and refute many alternative explanations. By incorporating these data in biogeographic studies one can begin to ask questions related to the occurrence of shared ancestral species, dispersal, and divergence. In this example, the sister-group relationships among the three clades for the three areas are identical. One may hypothesize that areas I, II, and III share a common ancestral species for each of the clades and two extrinsic events in the areas led to the divergence patterns observed. Thus, in the once contiguous area ancestral species would have existed throughout (see Figures 1 and 2). Subsequent extrinsic events isolated a western region (continental drift, glaciation, river basins) from a combined central-eastern region. Where dispersal ca-

pabilities of taxa prevented migration or dispersal to other areas the biota in the western region I diverged independently of the ancestral species in the eastern-central region. Subsequent to this event, the eastern-central region and biota experienced an isolating event disrupting dispersal and gene flow that ultimately led to the existence of replicated patterns of sister-group relationships.

Thus, much of the evolution of the biodiversity of these three clades can be explained by the existence of widespread ancestral species that simply underwent two sequential phases of disrupted gene flow that ultimately led to replicated patterns of sister group relationships across the areas. However, some diversity cannot be explained so easily through the predictions of vicariance biogeography. Species 10 and 11 do not have their sister species in the three areas of endemism. Rather, these species exist in one or more of the areas of endemism, but their closest relatives lie outside the identified areas of endemism. In vicariance biogeography, the first-order explanation of replicated divergence in isolation accounts for the majority of diversity, and one can predict sister-group relationships based on knowledge of species relationships in the biotas and earth history. However, for species that are alien to such explanations researchers have typically invoked dispersal such as species 10 and 11 into communities I and III, respectively, from other communities as second-order explanations. This is viewed as a more parsimonious interpretation because it involves only seven historical explanations (three widespread ancestors, two extrinsic isolating events, and two invasions by dispersal) versus at least eleven historical explanations of dispersal to each of the areas of endemism (not counting many other assumptions and the fact that none of them are testable). Thus, vicariance biogeography incorporates the phylogenetic relationships of taxa, seeks first-level explanations for common patterns, provides testable hypotheses of area relationships based on species relationships, and does not exclude the possibility of dispersal in its holistic interpretation of the evolution of biodiversity.

In recent years as more phylogenies of organisms and biological inventories have become available, the field of vicariance biogeography has evolved to include more sophisticated and varied analyses to address previously intractable questions. As the number of areas and taxa biogeographers typically examined increased, it soon became clear that investigating the number of possible alternative relationships that can exist between three or more areas was analogous to recovering phylogenetic relationships of taxa using characters. As such, phylogenetic algorithms became a part of vicariance studies and researchers began to employ codings of areas of taxa based on "known" histories provided by the phylogenetic relationships of the taxa found in the areas. In such a case, the coding of a three-species monophyletic group wherein dichotomous relationships were available (Figure 2) would include two "character transformations" (0-1) for the species relationships and three "transformations" (0-1), each indicating the presence or absence of the individual descendant species in the particular areas. In vicariance biogeography, the sister-species relationships, or the existence of shared-ancestral species, serve as the analogy to the synapomorphy in organismal phylogenies. Many studies have incorporated these analytical advances in recovering hypotheses of area relationships in biogeographic studies, or the coevolution of diversity and earth, as well as coevolutionary studies between clades.

One example of biologically diverse areas with replicated patterns of speciation and divergence is the evolution of the aquatic biotas of the Central Highlands in eastern North America. The four highland areas (eastern, Ozark, Ouachita, and unglaciated highlands) are characterized by clear, cool, high-gradient rivers and faunas adapted to these types of environmental conditions. Separating these areas are lowland habitats less favorable for the existence of the endemic highland species that have resulted from historical events ranging from the inundation of parts of central North America by the Cretaceous sea to the destruction of intervening ancestral highland areas by multiple Pleistocene glacial advances. Previous hypotheses for the explanation of these diverse biotas invoked dispersal between the areas, in no particular pattern, with subsequent divergence and colonization of northern faunas displaced south by advancing glacial fronts, and posited that the aquatic diversity of these highland areas evolved during and following Pleistocene glaciation. Employing the theoretical and general methodological principles of vicariance biogeography and the analytical tools available for phylogeny reconstruction, Mayden (1988) examined a number of clades of species and individual species in an effort to test the Pleistocene hypothesis for the origin of these biotas. The areas in these analyses were river basins and the characters used to decipher historical relationships were independently resolved species relationships of clades of fishes. The resulting area cladograms depicted river system (area) relationships consistent with pre-Pleistocene drainage patterns, refuting the hypothesis of a Pleistocene origin of the diversity. Sister-species relationships more clearly reflected the geological and hydrological patterns existing prior to the Pleistocene than those existing afterwards. Thus, the first-order explanation for the origin of the diversity in the Central Highlands was one of consistently repeated vicariant events occurring over a landscape no longer in existence today, not the dispersal of species between areas.

Vicariance biogeography holds great promise for not only increasing our skills at interpreting the coevolution of the earth and its organisms by grounding many studies in a meaningful context for comparative evolutionary biology. Through concerted efforts to examine the evolution of biotas from different areas, researchers are more likely to identify particular, meaningful species relationships to target particular evolutionary questions. For example, if a researcher were concerned with the impact of competition or keystone species in community formation, it would be useful to know if the species in question is one inherited from ancestral communities or one that dispersed into the area in question.

The future of vicariance biogeography also has some potential pitfalls. The analytical side of historical biogeography is a growing discipline in need of considerably more attention to the underlying assumptions and the development of useful methods unique to the field. The analogy between area relationships and species relationships, as assumed in the implementation of phylogenetic algorithms, is not complete. More dialogue is needed for the development of analytical tools necessary for the discovery of relationships between biological and geological information, the treatment of different types of data, and the development of appropriate analytical tools. One assumption of phylogenetic analyses is that the characters of a transformation series coded in a matrix are homologous and that the taxa being examined possess traits through descent with modification that were inherited from previous ancestral species. This is also an assumption of analytical vicariance analyses. However, it must be acknowledged that areas likely contain diversity prior to the existence of a particular replicated pattern (or tract) and there are very likely multiple tracts represented in areas that date from different time periods. Furthermore, dispersal is a fact of nature. Without dispersal, ancestral species could not exist over expansive areas to respond to vicariant events.

FIGURE 1. Human Influences on Biogeography. Throughout history, humans have contributed to extinction by the use of such methods as forest clearing to manage agricultural areas. © Mark Newman/Visuals Unlimited.

BIBLIOGRAPHY

Brooks, D. R. "Historical Ecology: A New Approach to Studying the Evolution of Ecological Associations." *Annals of the Missouri Botanical Garden* 72 (1985): 660–680.

Brooks, D. R. "Parsimony Analysis in Historical Biogeography and Coevolution: Methodological and Theoretical Update." *Systematic Zoology* 39 (1990): 14–30.

Brundin, L. "Transantarctic Relationships and Their Significance as Evidenced by Midges." *Kungliga Svenska Vetenskapsakademiens Handlinger* (Series 4) 11 (1966): 1–472.

Croizat, L. *Space, Time, Form: The Biological Synthesis.* Caracas, 1964.

Hennig, W. *Phylogenetic Systematics.* Urbana, Ill., 1966.

Mayden, R. L. "Vicariance Biogeography, Parsimony, and Evolution in North American Freshwater Fishes." *Systematic Zoology* 37 (1988): 331–357.

Nelson, G., and D. E. Rosen. *Vicariance Biogeography: A Critique.* New York, 1981.

Wiley, E. O. "Phylogenetic Systematics and Vicariance Biogeography." *Systematic Botany* 5 (1980): 194–220.

Wiley, E. O. *Phylogenetics: The Theory and Practice of Phylogenetic Systematics.* New York, 1981.

Wiley, E. O. "Vicariance Biogeography." *Annual Review of Ecology and Systematics* 19 (1988): 513–542.

— RICHARD MAYDEN

Human Influences on Biogeography

The transition from the Middle to Upper Paleolithic period some forty thousand years ago signaled a "cultural takeoff" in human behavior, including more sophisticated tools of stone and bone, fiber and weaving industries, broader social organization, long-distance resource extraction, and cognitive changes expressed in prepared burials and art. Ample evidence demonstrates that humans significantly influenced and exploited their environment long before even the earliest civilizations, let alone the industrial age. Recognition of this challenges our beliefs that prehistoric peoples were either intrinsic environmentalists or simply too few and too primitive to be anything but environmentally benign.

Humans influenced biogeography by contributing to extinctions, through the use of fire and forest clearing to manage landscapes, from selective hunting and gathering, and perhaps through the spread of microorganisms. Humans colonized all but the harshest terrestrial ecosystems on the planet by about fifteen thousand years ago, and through the simple fact of making a living were participants in world biogeography. By about ten thousand years ago in some regions and later in others, human populations were large enough to produce additional effects, including urbanization, extensive deforestation, soil erosion, resource extraction, and waste disposal.

Despite these long-term trends of human influences on the environment, foragers and small-scale farmers vary in their expressions of what we today would judge to be either exploitive or environmentalist behaviors. Circumstances occasionally existed that produced behaviors consistent with modern values of conservation, for example, seasonal rounds, pastoralism, dispersed fields, and slash-and-burn horticulture. These events are short-lived on archaeological time scales, and the behavior is the product of evolutionary circumstance, rather than anything intrinsic to particular cultural evolutionary stages, or cultural types.

Human influences are not simply limited to "other" species. Early hominids may have contributed to the ex-

ELK IN YELLOWSTONE

From the 1920s to the 1960s, the U.S. National Park Service restrained elk numbers in Yellowstone National Park by trapping and culling. Since then, elk populations have soared to perhaps 100,000 animals, and evidence is mounting that they are damaging riparian and aspen habitats, in turn harming bird, small mammal, and bear populations. Prehistoric human hunters may have been a "keystone predator," taking ungulates such as elk whenever they were encountered, and thus unintentionally holding them to low numbers. Large animals worldwide rank high in forager diets because they produce more food for less effort than smaller animals, nuts, roots, seeds, and plants requiring extensive processing. The archaeological record of the western United States supports the argument that elk numbers were low in prehistoric times. Pre-Columbian hunting would have produced a mosaic of large animal abundances, low in some areas, high in others. In 1803–1806, the explorers Meriwether Lewis and George Rogers Clark observed superabundant game whenever they were in tribal buffer areas or war zones, and less game in areas of human presence. The elimination of pre-Columbian human hunters produced a historic ecosystem very different from the past. This knowledge warrants the inclusion of humans in our understanding of the world's ecosystems over many millennia, not just in recent centuries.

—STEVEN R. SIMMS

tinction of their relatives via niche absorption and competitive exclusion. Four million years ago, East Africa harbored two genera of hominids, *Paranthropus* and *Australopithecus* and by two and a half million years ago, there were from three to five Australopithecine species. Between two and a half and one million years ago, the last Australopithecines overlapped with the earliest members of the genus *Homo*. Multiple hominid species persisted to at least 300,000 years ago. Evidence of a *Homo erectus*/archaic *Homo sapiens* fossil from Java only 130,000 years ago raises the possibility of pockets of recent coexistence among hominid species. If the European Neanderthals were a distinct species, then their extinction by thirty thousand years ago is a dramatic statement on the strength of even the early human presence.

Our earliest ancestors' influence on nonhuman biota can be inferred from fossil and other evidence. Many large mammal genera went extinct about the time that the genus *Homo* arrived in southern Africa about two million years ago. This was followed by extinctions in

Southwest Asia and Europe, about one million years ago, and much later in Australia and the Americas. Given the sample sizes and changing chronology, the spikes in extinction events corresponding with human arrival may simply be chance. It is noteworthy that between five and a half and a half million years ago, African carnivore extinctions exceeded herbivore extinctions. During this period, technology and social organization may have increased hominid hunting ability, raising the possibility that hominids accelerated extinctions among carnivores.

The evidence for a human role in extinctions in the Late Pleistocene is stronger, but controversial. A decades-old, polarized debate continues over whether the extinction of thirty-five North American genera during the past twenty thousand years was caused by climate or human colonization. The extinctions were primarily, but not exclusively, of megafauna (e.g., mammoth, ground sloth, gyptodonts, horses, camelids), and some familiar animals survived (e.g., deer, elk, pronghom). Proponents of human agency argue that despite repeated advances of ice during the Pleistocene period, extinctions are most numerous during the last glaciation, the only one co-temporaneous with humans. Proponents of climatic causes emphasize the complexity of the problem, show that direct archaeological evidence for hunting megafauna is sparse, note that some extinctions occurred among birds and small mammals unlikely to have been significantly hunted, and emphasize the lack of precise chronological correlation between humans and particular extinction events.

Since the early twentieth century, archaeological finds have pointed to the "Clovis first" model of a rapid wave of human advance less than thirteen thousand years ago from Siberia across the Beringian continent to America. New lines of evidence, improved chronologies, and a more hemispheric view have led to a sea change in thinking. Humans were present before thirteen thousand years ago. Colonization along the now-inundated Pacific coast and perhaps other routes seems more probable. The emerging archaeological, linguistic, and genetic picture is of multiple colonization events, various human adaptive strategies, and an initial period of low use until rapid population increase after about thirteen thousand years ago.

When humans are a "keystone predator" (see Elk in Yellowstone Vignette), human selectivity of large game can have cascading effects on the ecosystem. Study of the growth rings of Late Pleistocene mammoth tusks indicates that they did not usually die under conditions of nutritional stress, an indicator of a climatic cause. Furthermore, female tusks show birth spacing consistent with living elephant populations under hunting pressure, not the longer birth spacing typical of elephants under nutritional stress. Human impact may also have been indirect with human colonists serving as vectors for microbes and disease into American animal populations with no prior human contact.

Human agency is likely significant to Late Pleistocene extinctions in the Americas, but the human role is short of the "blitzkrieg" metaphor often portrayed. It seems obvious that both climate and humans are relevant, warranting research into the effects produced by some human adaptive strategies in some ecosystem states that select for particular evolutionary outcomes and trends.

In Australia, the extinction of over half of the large mammals appears correlated to human colonization between forty thousand and sixty thousand years ago, a period of relatively stable climate. Australian extinctions are poorly dated, however, and the human colonization of the continent keeps inching toward older dates. Research focuses on regional and short-term expressions of climate, and the role of human-induced fire regimes combined with human predation. Australian research draws attention to the influence of generalist hunters in contrast to the presumption of many human-caused extinction studies that hunters must be specialists in order to have a significant impact. Studies of foragers show that generalist hunters will have disproportionate impacts on large animals that rank high in return rate for their search, killing, and processing because such prey will be taken upon encounter even if they are rare.

An extremely early South African case further illustrates the benefit of an ecological perspective. Reductions in the individual sizes of tortoises and shellfish some fifty thousand years ago reflect collection pressure from rising human populations. At the same time, humans developed the technology to include birds, fish, and dangerous prey such as buffalo in their diet. Despite broadening the diet, the impacts on tortoise and shellfish populations continued, suggesting that human mismanagement is a more accurate description than catastrophic overkill.

Islands were among the last places colonized by humans. If the blitzkrieg metaphor does not work for the Americas and Australia, it does seem to describe the extinction and extirpation of dozens of species of ground-dwelling and seabirds in New Zealand, Hawaii, Easter Island, Samoa, and many other islands of Polynesia and Melanesia. Islands document the additional impacts of fire, land clearing, and the introduction of nonnative species. Humans brought "cultural landscapes" to Polynesia over the past two thousand years, including domestic pig, dogs, and chickens and crops such as taro, yams, and bananas. Incidental introductions include rats, geckos, skinks, insects, microbes, and a host of weeds. Colonists had clear intentions for the development of uninhabited islands. Stratigraphic records show frequent burning for land clearing, and paleobotanical records show large-scale deforestation for wood extraction to the point of

THE WILDERNESS MYTH

Some observations by early European colonists to America seemed to substantiate Western culture's idea of wilderness. The superabundance of game, areas of dense forest, and few natives astounded early observers. Often the same observers also described deforested areas, great fields of corn, few animals, and many natives, but these did not capture the European mind. A great deal of archaeological, paleoecological, and ethnohistoric evidence shows that the Americas were occupied by tens of millions of people, that virtually all regions held anthropogenic ecosystems, and that the continent's biota and appearance were shaped by millennia of human presence. A dozen diseases introduced repeatedly in the sixteenth and seventeenth centuries decimated Native American populations, spreading along chains of concentrated habitation in the eastern and southwestern United States, Mexico, and South America, and along the Pacific coast. Long isolation from the urban areas of Eurasia left Native Americans with no resistance to diseases of density such as smallpox and measles. Thus, America was not a wilderness, nor was it "virgin." It just seemed that way to colonists who by the time of the westward movement saw only the rebound of a "widowed" landscape devoid of up to 90 percent of its previous inhabitants.

—STEVEN R. SIMMS

catastrophic impacts, such as the abandonment of Easter Island. By 1650 CE in Hawaii, population may have reached 300,000, and virtually all of the lowland habitats were anthropogenic.

Humans colonized Madagascar only about two thousand years ago. Rather than initiating change, humans altered the vegetation and fire patterns of the previous forty thousand years. The infrequent, intense fire pattern became one of less intense but pervasive burning. All of Madagascar's megafauna and many smaller animals became extinct within the span of human occupation. Like many islands colonized late in prehistory, Madagascar felt the full brunt of technologically sophisticated but nevertheless preindustrial humans.

The Mediterranean was not immune, and human influence preceded the well-known civilizations of the region. The hippos, elephants, and giant deer that populated Cyprus, Sicily, and Sardinia were exterminated in the Late Pleistocene epoch, repopulated by swimming ungulates only to be exterminated by human recolonization. Until the advent of farming on these islands after about six thousand years ago, the lack of game maintained strikingly low human populations.

The evolution of farming, first in Southwest Asia, China, and Southeast Asia, and later in Mexico and the Andes, caused human population to rise exponentially. Combined with the harness of animal power sources and irrigation, human impacts became more diverse, more extensive, and more intensive. Ain Ghazal, Jordan, was a typical Neolithic (early farming period) site, where a cascade of impacts is documented between 9,250 and 7,500 years ago. Land was cleared for farming, vegetation was altered by goat pasturage, and timber harvests provided construction materials and fuel. Local deforestation grew from an estimated 70 hectares to over 250 hectares during the occupation. Large quantities of wood were needed to process lime into stucco plaster, but deforestation caused stucco to become so expensive that construction methods changed. Neolithic communities like Ain Ghazal were common in the region, indicating the magnitude of cumulative, local impacts. Similar effects are associated with the spread of agriculture across the Mediterranean into Europe (5,000–7,000 years ago), where deforestation intensified over the earlier, Mesolithic pattern of human burning to manage acorn-producing oak forests.

In the arid American Southwest, large agricultural populations caused deforestation and soil erosion, as well as depletion of large game, by 1000 CE in places like Chaco Canyon, New Mexico. In the eastern United States, native species such as lambsquarter and sumpweed were domesticated by about four thousand years ago, altering the vegetation patterns in the river valleys of the Midwest and Southeast. Native exploitation of mast (nuts, fruits, berries) was so pervasive in the eastern United States that small mammal and passenger pigeons were held in check. When European colonization decimated Native Americans, the populations of animals dependent on nuts skyrocketed, and species such as the passenger pigeon became a "pestilence" in accounts from the eighteenth and nineteenth centuries. Prehistoric burning by Native Americans was likely pervasive and created a more open continent and a mosaic of habitats. Such examples are increasingly common and show the systemic effects of human participation in ecosystems regardless of whether ancient humans saw themselves as ecologically noble, had certain conservation practices, or considered the natural world sacred.

By the times of the Babylonians, Greeks, and Romans, the habitats of Southwest Asia were largely anthropogenic, indigenous animals (such as lions) were largely extinct, and deforestation created the open landscapes that characterize the region today. Humans produced a "moving mosaic": as centers of human activity shifted, habitats would regenerate until the next pulse of human activity. Pollen and zooarchaeological records are nevertheless clear that as forests regenerated and plants and animals returned, they were dominated by species capable of competing in anthropogenic situations. The modern scale of human impact is exponentially greater than only a few centuries ago, but this does not mitigate the significance of our deeply ancient role in shaping world biogeography.

[*See also* Agriculture, *article on* Origins of Agriculture; Disease, *article on* Demography and Human Disease.]

BIBLIOGRAPHY

Baleé, W., ed. *Advances in Historical Ecology*. New York, 1998. Up-to-date papers showing the scientific and sociopolitical complexities associated with understanding human interaction with the environment.

Bottema, S., G. Entjes-Nieborg, and W. Van Zeist. *Man's Role in the Shaping of the Eastern Mediterranean Landscape*. Rotterdam, 1990. Twenty-eight papers on the botany, zoology, and geomorphology of human–environment interactions in the region over the past ten thousand years.

Crosby, A. W. *Germs, Seeds and Animals: Studies in Ecological History*. London, 1994. Recent book by the eminent biogeographer focusing on disease and demography.

Denevan, W. "The Pristine Myth: The Landscape of the Americas in 1492." *Annals of the Association of American Geographers* 82 (1992): 369–385. Evidence for the proposition that America was a humanized landscape upon "discovery" by Europeans. Excellent introduction.

Kay, C. E., and R. T. Simmons, eds. *Wilderness and Political Ecology: Aboriginal Influences and the Original State of Nature*. Salt Lake City, 2002. Collection of empirical studies of human impacts, popular myths, and their implications for land policy and politics. Polemic, but eye-opening.

Klein, R. G. "The Impact of Early People on the Environment: The Case of Large Mammal Extinctions." In *Human Impact on the Environment: Ancient Roots, Current Challenges*, edited by J. E. Jacobsen and John Firor. Boulder, 1992. Brief overview of the evidence for the influence of very early humans and the difficulty of addressing the problem.

Krech, S. III. *The Ecological Indian: Myth and History*. New York. Clearly written, compassionate evaluation of pre-Columbian influence on the natural world regardless of people's intelligence, knowledge, intentions, or ideology.

Martin, P. S., and R. G. Klein, eds. *Quaternary Extinctions: A Prehistoric Revolution*. Tucson, 1984. A classic collected works on this complex subject.

Pyne, S. J. *World Fire: The Culture of Fire on Earth*. New York, 1995. Highly readable overview of the significance of human-caused fire from prehistory to the present.

Redman, C. L. *Human Impact on Ancient Environments*. Tucson, 1999. A brief but excellent introduction and reference work not limited to biogeography.

— STEVEN R. SIMMS

BIOINFORMATICS

Bioinformatics may be taken to mean the application of information science to biology. This definition is too broad for our purpose, and we restrict the meaning to the study of the processing of information in biological systems. However, we will not consider neurobiology. Our concerns are molecular and cell biology and genetics, and the use of computers to document, analyze, and model biological processes studied in these disciplines. Genetic inheritance and evolution underpin the whole of biology, and there is a remarkable uniformity of molecules and processes across all life forms with, of course, remarkable exceptions. Molecules, cells, individuals, and populations are often well defined, whereas organelles and organs are more subjectively defined. In humans there are more than 300 cell types that need to be characterized at the molecular level.

The major biological processes that are documented by bioinformatics are DNA replication and repair, DNA transcription to RNA, RNA translation to protein, and post-translational modification of proteins. Large databases of genomic DNA, RNA transcripts, and protein sequences and structures (EMBL Data Library, GenBank, SwissProt, PDB) form the first building blocks and have been accumulating over the past twenty years, with a remarkable expansion in the past five owing to the industrialization of DNA sequencing and the will to determine entire genome sequences. We now have complete genome sequences of numerous viruses and bacteria, yeast, fruit fly, nematode worm, thale cress (a *Brassica*), and rice. The first draft of the human genome has been published, and correct ordering of the fragments and identification of the genes is in progress. The complete genomes of human, mouse, and Pacific puffer fish will be characterized within a few years. This information will form a systematic index for functional genomics and proteomics, which in turn will feed into biotechnology and medicine.

The comparative method combined with evolutionary theory has enlightened biology for the past two centuries. The comparative method is equally important in bioinformatics and is used at all levels. DNA and protein sequence comparisons are the first steps in understanding a newly determined sequence. Gene expression patterns are used to try to understand differences between healthy and diseased tissue. Protein structures are compared to shed light on similarities and differences in protein function. Metabolic or signaling pathways are compared to detect differences in the ways that organisms function.

Chromosome maps represent the natural way of organizing the arrangement of genes. Genetic linkage maps were the first such maps to be constructed and differ from other chromosome maps because they can be constructed only for two or more polymorphic loci, and they give a statistical estimate of distance. Radiation hybrid maps are constructed by breaking the chromosomes with radiation and rescuing random samples of the fragments by their incorporation into rodent cells. A panel of about 100 such cell lines is characterized using markers of known position. The panel can then be used to map further markers that need not be polymorphic. Overlapping clone maps are constructed by breaking DNA at random and selecting fragments of a certain size range to be incorporated into a cloning vector. To sequence the human genome, a bacterial artificial chromosome library with inserts of about 150 kilobytes (kb) was used. Each clone was fingerprinted by cutting with a restriction enzyme and sizing the fragments. Overlapping clones are assembled to give a tiling path, from which clones for sequencing can be selected. Cytogenetic maps are constructed by attaching fluorescent dyes to DNA markers and then visualizing the chromosomal position to which the markers bind.

Genotype and phenotype have to be related by genetic mapping in order to define the chromosomal position. This should then lead to the identification of the gene and the mutations responsible for the phenotype. Single gene disorders are rare in the human population, cystic fibrosis being among the most frequent. Disorders caused by multiple gene variants are common and include diabetes, asthma, and propensity to heart disease.

Sequence comparison is one of the most common tasks in bioinformatics and has been refined to a considerable degree. Large collections of DNA and protein sequences can be searched rapidly by programs such as BLAST and FASTA. The most sensitive methods use a complete probabilistic description of known sequence families, called a Hidden Markov Model (HMM). One database of protein families discovered in this way is known as PFAM. Sequence data are often used to build phylogenetic trees of organisms, to study deviations from an evolutionary model (such as probabilities of transitions of the four nucleotides, or the Point Accepted Mutations [PAM] table for proteins), or to detect selective constraints on sequences, such as the conservation of an active site. We must not estimate all three from the same data.

RNA and protein molecules are frequently employed to estimate phylogenies of surprisingly distantly related organisms, with apparent success. These molecules are under considerable selective pressure to perform their function and have no brief to record the events of history. It is of considerable interest to attempt to reconstruct phylogeny from the gene order data that are now becoming available. Complete genome sequencing, though highly desirable, is not necessary; sufficient gene order data to determine the relationships of the mammalian families might be obtained by radiation hybrid mapping. Gene order is also, no doubt, under selective constraints, but these are less important than in the case of sequences. We believe that about 70 breaks are necessary to convert the gene order of a mouse into that of a human. There are undoubtedly local rearrangements, and a more complex picture will finally emerge.

Gene structure and function are crucial to our understanding of biology, but it is very hard to predict gene structure from sequence. Knowledge of RNA transcripts obtained by sequencing cDNA libraries is very helpful in understanding gene structure. It is likely that each human gene produces, on average, three alternatively spliced RNA products. Prediction of coding sequence, introns, and promoters is highly unreliable for human DNA. Comparative genomics for related species such as human, mouse, chicken, and Pacific puffer fish is very informative about gene structure.

Gene expression can now be measured or compared for different conditions for tens of thousands of genes simultaneously by using synthetic oligonucleotides or PCR products from cDNAs arrayed on glass. The objective of these experiments is to detect genes that are co-regulated in order to shed light on gene function. It is also crucial to know which protein products are produced, including their post-translational modifications. Protein separation and identification techniques involving gel electrophoresis and mass spectroscopy have reached a level of sophistication that makes this possible.

Protein structure and function lie at the heart of understanding biological processes. The protein world is large but finite. It requires about 6,000 proteins to make a living yeast cell function, and it is believed that there are fewer than 1,000 unique protein structures. Humans probably have several hundred thousand proteins that are involved at some stage of the life cycle. The number of genes is considerably less than this. The evolutionary conservation of life processes makes the study of protein evolution a worthwhile and fascinating endeavor. Globular proteins that can be crystallized have their structure determined by X-ray crystallography. The coordinates are stored in the Protein Structure Databank (PDB). Membrane proteins cannot usually be crystallized, but the transmembrane regions can be predicted and ligand binding studied. Determining the structure of a protein such as an enzyme is generally insufficient to provide a clue to how it might work. The basic building blocks of protein secondary structure are alpha helices and beta sheets. An individual protein may contain regions of one or both of these elements. Protein structure comparison is a necessary step in understanding protein function and evolution. There are two major databases of protein structure comparisons, called SCOP and CATH. Secondary structure prediction from protein sequence using multiple sequence data for comparisons can reach almost 80 percent accuracy using a program such as PHD. Structural genomics projects are under way to attempt to determine the structure of all the proteins of an organism such as a bacterium, and eventually to understand the entirety of its biology.

Protein interaction networks are the keys to cellular functions and are being extensively studied at the present time. The evolution of such networks will form a fascinating part of biology. Metabolic pathways involving enzymes have been studied and understood for many years. We are now able to study other pathways such as signaling. The yeast-two hybrid methodology enables high throughput screening of pair-wise protein interacts. These interactions give clues to how protein interaction networks are constructed. There are databases of protein interactions, such as KEG.

The central dogma of molecular biology was that DNA makes RNA makes protein. The reality is that each is involved in the processes of making the others, and there is feedback among them. The holy grail of bioinformatics was the hope that sequence predicts structure predicts function. The reality is that we need to determine sequence, structure, and function and try to relate them in order to gain understanding.

BIBLIOGRAPHY

Attwood, T. K., and D. J. Parry-Smith. *Introduction to Bioinformatics*. New York, 1999. An introduction suitable for the postgraduate.

Balding, D. J., M. J. Bishop, and C. Cannings, eds. *Handbook of Statistical Genetics*. New York, 2001. Includes bioinformatics, population genetics, evolutionary genetics, genetic epidemiology, animal and plant genetics, and applications.

Doolittle, R. F., ed. *Computer Methods for Macromolecular Sequence Analysis* (Methods in Enzymology, 266). Academic Press, 1996. A comprehensive account of techniques.

ONLINE RESOURCES

EMBnet http://www.embnet.org/. A foundation of bioinformatics centers comprising thirty-one national nodes and nine organizations.

"European Bioinformatics Institute." http://www.ebi.ac.uk/. Compiles the DNA sequence database EMBL Data Library and the protein sequence database SwissProt as well as other databases, including ENSEMBL.

"IUBio Archive." ftp://ftp.bio.indiana.edu. Provides bioinformatics programs to download and run on the user's own computer.

"National Center for Biotechnology Information." http://www.ncbi.nlm.nih.gov/. Compiles the DNA sequence database GenBank and other databases and provides indexing via Entrez and Medline.

"The Bioinformatics Resource." http://www.hgmp.mrc.ac.uk/CCP11/. Comprehensive listing of bioinformatics courses, research groups, meetings, etc. for the UK with plans to extend it internationally.

"The Genome Web." http://www.hgmp.mrc.ac.uk/GenomeWeb/. Comprehensive index to on-line resources for genomics and proteomics.

— M. J. Bishop

BIOLOGICAL WARFARE

Anyone that has ever grasped a rose by the stem or stepped on a sea urchin has had first-hand experience of direct mechanical defenses. In some cases specific behaviors may increase the efficacy of the mechanical defenses. When confronted, the porcupine *Erethizon dorsatum* turns its back towards the source of danger, raises its quills and lashes out with its tail. The quills are barbed and easily detached, so contact could leave the enemy with a number of spines lodged in its body. Defense may also involve the production of auditory signals, such as the hissing of cats or the Madagascar hissing cockroach, *Gromphadorhina portentosa*. Visual cues, like the broken-wing display of killdeer adults help reduce predation by leading potential predators away from the nest. Auditory, physical and visual signals are also commonly used to deter both intraspecific and interspecific competitors, as exemplified by the many interactions observed when lions, hyenas and vultures occur around a freshly killed carcass.

Chemical Defenses. Perhaps one of the most spectacular and well known forms of chemical defenses are those deployed by animals such as skunks (members of the Family Mustellidae), and bombardier beetles (members of the family Carabidae): they spray their enemies with repellents from a distance. The skunks produce a very pungent repellant from anal glands which, depending on the species, may contain thiols (such as *E*-2-buten-1-thiol) and/or thioacetate derivatives of the thiols (such as *S*-(*E*-2-butenyl thioacetate). Not only does the mixture have disagreeable odor at high concentrations, these compounds may induce vomiting and lacyrmation ("crying"). Bombardier beetles produce a spay of hot toxic chemicals, reminiscent of the middle age practice of pouring boiling oil from castle ramparts on the enemy below! Hydroquinones and hydrogen peroxide are produced by specialized secretory cells and accumulate in a reservoir. When needed, these compounds are introduced into a thick-walled reaction chamber where enzymes (peroxidases and catalases) break down the hydrogen peroxide and catalyze the oxidation of the quinones to p-quinones. The reactions produce sufficient heat to raise the temperature of the p-quinones to boiling point and the pressure of the gases released results in the expulsion of the hot spray from the abdominal tip. The beetle is able to orient the jet, which in some species may be pulsed while in others it may be a expelled as a continuous stream.

Other organisms may produce chemical irritants that work upon contact. The leaves of the stinging nettle, *Urtica dioica*, are covered with trichomes that contain a complex mixture including acids, quinones and neurotransmitters. When broken the trichomes release the chemical mixture and may induce a painful dermatological reaction. Similarly, damaged leaves of poison ivy, *Rhus radicans*, release the oleoresin urushiol (containing catechols and other phenolic resins) that binds with skin proteins and causes considerable physical discomfort. Similar mechanisms are found in a number of animals such as some caterpillars (e.g., the puss moth *Me-

galopyge opercularis* and the saddleback moth, *Sibine stimulae*), and jellyfish. Upon contact defense compounds (often polypeptides, proteins and neurotoxins) are released from urticating hairs or nematocysts and may inflict considerable pain to potential enemies.

Some chemical defenses must be tasted or ingested to be effective, based on a very wide array of compounds that include alkaloids, cardiac glycosides, cyanogenic glycosides, phenols, tannins and terpenes. At high concentrations they may be lethal while at lesser doses they affect behavioral and/or physiological processes. Behavioral changes induced by chemical defenses have obvious gains. Chemical defenses that deter or reduce feeding by herbivores will result in lower defoliation and thus increase the probability of plant survival. Similarly, alkaloids in the skin of certain amphibia that induce yawning behavior (dyskinesias) in the predator facilitate the escape of prey. Physiological changes may negatively affect growth, survivorship and/or subsequent adult reproductive success providing potentially long term benefits through lower population densities of natural enemies in subsequent generations. The same class of compounds may serve as the basis of defense in both plants and animals. For example, the deadly nightshade, *Atropa belladona*, mandrake, *Mandragora officiarum*, and henbane, *Hyoscyamus niger*, have chemical defenses based on the tropane alkaloids atropine, hyoscyamine and scopolamine. Alkaloid defenses are also found on ladybird beetles (precoccinelline, myrrhine and hippodamine), while poison-dart frogs in the family *Dendrobatidae* and several species of birds in the genus *Pitohui* rely on the related batrachotoxins.

Some plants also produce hormones, or closely related hormone mimics, of invertebrate and vertebrate herbivores, such as juvenile hormone and ecdysone, two major hormones implicated in many biological processes during different stages in the insect life cycle and the mammalian hormone estrogen. Ingestion of these hormones, or closely related mimics, may disrupt normal physiological processes of herbivores. A variation on this theme is the production of precocenes, compounds that inhibit the normal production of juvenile hormone in insect herbivores, resulting in precocious molting and high levels of sterility in surviving adults.

Given the role in chemical defense against attack by natural enemies it is not surprising to find that these compounds are not generally distributed uniformly within an organism. Higher concentrations are often found in more important tissues, such as the reproductive organs, and within tissues at sites that ensure maximal efficacy when the organism is attacked. For example, in common groundsel, *Senecio vulgaris*, the concentration of pyrrolizidine alkaloids is about 5 times higher in flowers than the stem and the outer cells of the stem have about 10 times more than internal ones.

GENERALIZED SCHEME OF INDUCED PLANT DEFENSES

When a plant is attacked the endogenous wound signal cascade is initiated by elicitors produced by infecting pathogens (e.g., coronatine) or herbivores (e.g., volicitin) (1.). The elicitor stimulates different biochemical pathways, such as the octadenoid or jasmonate pathway leading to the production of jasmonic and epi-jasmonic acid from linolenic acid (2). In the case of the jasmonate pathway one may see the accumulation of endogenous jasmonic acid in the tissues near the wound and then somewhat later in different parts of the plant, especially at the site of synthesis of the defense compounds. Products such as jasmonic acid then trigger the up-regulation of genes governing the subsequent synthesis of species-specific secondary defense compounds at different sites in the plant (3). In addition there may be a marked change in the profile of the green leaf volatiles. The actual composition of volatiles emitted following damage (see 4 for examples) will depend on the plant species and also the natural enemy inflicting the damage. Furthermore, the profile of emitted volatiles may vary with time following defoliation and also show marked diel periodicity. The production of green leaf volatiles following attack by natural enemies may modulate a number of different interactions (indicated by 5, 6, and 7). The first is the induction of chemical defenses in undamaged parts of the plant being attacked. It has also been postulated that these messages may induce increased chemical defenses in neighboring conspecifics and although there is some experimental evidence to support the hypothesis the subject is still very much open to debate. A second relates to host location by natural enemies for there is now clear evidence that in certain plant/insect interactions volatiles from herbivore damaged plants serve as cues for foraging parasitoids and predators while those emitted by artificially damaged ones do not. It should be noted while there is a selective advantage for natural enemies to exploit these signals to locate and attack the herbivore, this "attraction of bodyguards" is a secondary advantage to the plant, above and beyond the increased levels of induced chemical defenses. The third possible interaction relates to information for the herbivore about plant quality. In many cases an attacked plant would be a poor host, in quantity due to competition from herbivores already present and in quality due the increased levels of induced defenses. The ability to detect and avoid attacked hosts could benefit the herbivore but this avoidance behavior may also benefit the plant by reducing the probability of subsequent exploitation by natural enemies.

—JEREMY N. MCNEIL

Similarly, homobatrachotoxin in *Pitohui* birds is concentrated in the skin, feathers and breast muscle. While extracts of these tissues induced convulsions and death in mice extracts from internal organs, such as heart or liver, did not.

Chemical defenses may be constitutive, that is to say present at all times in the organism's tissues. However, defense may be induced, only being produced in response to an attack by natural enemies and thus reduces the cost of continuously maintaining high levels of defense compounds. Our understanding of induced plant defenses is increasing at an astonishing rate and a generalized scheme is presented in Box 1. Induced defenses are also found in the animal kingdom. Daphnia may occur in two morphs, with and without spines, and at low densities of natural enemies the spined morph is quite rare. However, the production of spines is in response to chemical cues from insect predators so as predator densities rise so does the frequency of the spined morph. This change in shape makes it much more difficult for the predator to manipulate the prey and facilitates escape following initial capture. Induced defenses of animals to chemical cues from predators are not just limited to morphological characteristics for there is accumulating evidence of significant changes in life history traits, such as developmental rates that result in temporal asynchrony between predator and prey.

The source of defense compounds is varied and in many cases synthesis is *de novo*. However, some herbivores incorporate the chemical defenses produced by the host through direct sequestration or following some modification, into their own defense system. The Monarch butterfly, *Daneus plexippus*, the milk weed bug, *Oncopletus faciatus* and the milkweed aphid, *Aphis neri*, specialist herbivores of milkweed, *Asclepias* spp acquire cardenolides from their host plant when feeding and deploy them to defend against their own natural enemies. All three of these insect species are good examples of a phenomenon termed aposomatic coloration. They are all brightly colored and represent visual cues indicating that the animal is not palatable and/or is toxic. These warning colors are the basis for some mimicry systems, where palatable species have evolved very similar patterns to unpalatable sympatric heterospecific species and are thus avoided by potential predators. The exploitation of plant chemical defenses may also extend over several trophic levels. The predatory ladybird beetle, *Coccinella septempunctata* uses pyrrolizodine alkaloids, originating from the host plant, *Senecio inaequidins*, which it obtains by feeding on the aphid herbivore *Aphis jacobaeae*.

A more indirect way of using plant defenses is behavioral sequestration. Decorator crabs cover their carapace with pieces of seaweed, postulated to provide camouflage against natural enemies. However, some species preferentially select seaweed that is chemically defended against herbivory by fish. Crabs decorated with chemically defended seaweed suffer lower levels of predation than conspecifics decorated with palatable seaweed. There are some extreme examples of an undefended species exploiting another chemically defended species to reduce predation, such as the amphipod, *Hyperiella dilatata* "kidnapping" a species of sea butterfly. The amphipod which, when undefended, is highly susceptible to fish predation significantly reduces the probability of being preyed upon when it places the mollusk, which contains the feeding deterrent pteroenone, on its back.

Certain animals may use chemical messages to attract "bodyguards" when confronted with danger from natural enemies. Larvae of the lycaenid butterfly, *Polyommatus icarus* provide the ant *Lasius flavus* a nutrient reward in exchange for protection. When attacked, the caterpillar everts its tentacles and from specialized glands produces droplets of liquid onto its body surface. The tentacles are the source of the chemical recruitment message while the droplets are the nutrient reward for the bodyguards that respond. This is not an all or none process, as caterpillars modulate both tentacle extrusion and the production of droplets as a function of bodyguard density. At low ant densities the rates of tentacle extrusion and droplet production following an attack are significantly higher than to a similar stimulus when ant densities are high. However, not all lycaenid-ant interactions are mutualistic relationships. In some species the ant, attracted by a chemical signal produced by the caterpillar, transports the larva back to the nest, where it is protected from natural enemies. However, once in the nest the caterpillar usurps much of the food brought back by workers and may actively feed on the ant brood.

There may also be deception at the higher trophic levels associated with a mutualistic interaction. Ants recognize the aphids they protect by their cuticular chemical profile and will eject or kill organisms with different chemical signatures. To circumvent this certain predators adopt a "wolf in sheep's clothing" approach. Wooly aphids secrete waxy strands over much of their body surface and certain predators actively cover themselves with these fibers. By presenting an "I am an aphid" chemical they are able to deceive the ant bodyguards and gain access to their aphid prey. Some specialist enemies of aphids and social insects have evolved more direct counter measures. They actually possess cuticular chemical profiles very similar to those of their prey and this chemical mimicry allows them to exploit resources without detection.

Chemical Defense against Competitors. Competition for resources is of considerable importance in plant community ecology, and some species have evolved a chemical means of resolving this problem. This is referred to as allelopathy, where plants release allelochemicals into the soil (e.g., juglone from the walnut tree, *Juglans regia* or sorgoleone in sorghum, *Sorghum bicolor*) that negatively affect seed germination and/or growth of potential competitors through disruption of normal physiological and biochemical processes.

Animals, like plants must compete for essential resources and an array of chemical signaling systems have evolved to reduce intraspecific and interspecific competition. In most cases these compounds act to modify behaviors rather than through direct physiological effects. Females of several fly such as the apple maggot, *Rhagoletis pomonella* (Tephritidae) or the alfalfa blotch leafminer *Agromyza frontella* (Agromyzidae) leave a marking pheromone on the surface of the site after laying an egg. These chemical messages, N (15 (β-glucopyranosyl)-oxy-8 hydroxypalmitoyl)taurine for the cherry maggot, *R. cerasi*, reduce oviposition at the site and consequently the probability that female's progeny is subjected to larval competition. Bark beetles (Scolytidae) have evolved very elaborate chemical communication systems, that not only help overcome plant host defenses but also to modulate both intra- and interspecific competition. The first individuals locating a suitable host tree emit an aggregation pheromone to attract conspecific adults so that, by force of numbers, they may overcome the host's defenses. Frequently this pheromone also acts as a repellent for other sympatric species and thereby decreasing the incidence of interspecific competition. Once numbers reach a certain density on a given host the beetles no longer release an aggregation pheromone but rather produce an anti-aggregation pheromone to ensure that the available resources are not overexploited through intraspecific competition.

Territorial marking with chemical signals is also found in many vertebrates, from lizards to antelope. For example, the Canadian beaver, *Castor canadensis* produces a very complex secretion called castoreum that is laid down at different points within the territory. The optimal size of a territory will be a trade off between the availability of resources within the area and the costs of maintaining effective marking. Therefore, the frequency of marking by residents may be density dependent, being much higher when the potential of competition for resources is greatest.

Conclusion. Within the context of "chemical warfare" it is necessary to move back from the narrow vision of chemicals for defense, for the same chemical may serve different roles within a given ecological community. For example, crucifers are well defended against many generalist herbivores using compounds such as the glucosinolate sinigrin. However, this same chemical serves as an oviposition and feeding cue for females and larvae of specialist *Pieris* spp. cabbage butterflies. Similarly,

receptive females in many moth species release a sex pheromone to attract conspecific males. However, this same message may be exploited by parasitoids that lay eggs within those of the moth. In this context it would be classified as a kairomone, a message emitted by one species (the moth) is of benefit to the detecting species (the parasitoid). Similarly, certain spiders emit volatiles that mimic moth sex pheromones and effectively lure in males who respond to a message perceived as a potentially beneficial intraspecific cue, the receptive female. However, in this case the message would be considered as an allomone, a message that induces a response in the receiving species but to the definite advantage to the species emitting the signal.

BIBLIOGRAPHY

Eisner, T. and J. Meinwald. *Chemical Ecology: The Chemistry of Biotic Interactions.* Washington, D.C., 1995.

Johnston, R. E., D. Muller-Schwarze, and P. Sorensen, eds. *Advances in Chemical Signaling in Vertebrates.* Vol. 8. Plenum, N.Y., 1999.

McClintock, and B. J. Baker, eds. *Marine Chemical Ecology.* Boca Raton, Fla., 2001.

Rosenthal, G. A., and M. R. Berenbaum, eds. *Herbivores: Their Interactions with Secondary Plant Metabolites.* 2d ed. New York, 1991.

Spencer, K. C., ed. *Chemical Mediation of Coevolution.* New York, 1988.

ONLINE RESOURCES

Chemoecology. http://link.springer.de/link.service/journals/00049

Journal of Chemical Ecology. http://www.cas.usf.edu/JCE/jce.html

— JEREMY N. MCNEIL

BIRDS

Birds are unique among extant animals in having feathers which enable flight, provide insulation, and are used in visual communication. Modified feathers aid in swimming, sound production, protection via camouflage, water repellence, water transport, tactile sensation, hearing, and support of the body. Birds have distinctive bills, produce external eggs, and demonstrate complex parental and reproductive behaviors. Birds have evolved homeothermy (nearly constant body temperature) independently from mammals. Features shared with other reptiles, but not with mammals, include nucleated red blood cells, a single middle ear bone, and a single occipital condyle on the back of the skull. Adaptations for flight include fusion and reinforcement of lightweight bones and presence of a keeled sternum supporting flight muscles. Birds have highly developed color vision and use vocalization in social interactions, and some are able to detect and react to magnetism. Birds are the most species-rich group of terrestrial vertebrates, with over 9,000 species known.

Origin of Birds. Although it is widely agreed that the closest living relatives of birds are crocodilians, there is controversy over which extinct lineage of reptiles gave rise to birds. Birds originated among a group of reptiles known as archosaurs, which also includes crocodilians, pterosaurs, and dinosaurs. There are two primary hypotheses regarding phylogenetic placement of birds within archosaurs. The predominant view is of birds as arising from theropod dinosaurs. Theropods are bipedal, mostly predatory dinosaurs, including the ceratosaurs, allosaurs, and dromeosaurs, such as *Velociraptor*. There are two primary groups of dinosaurs, the Ornithischia and the Saurischia, and theropods are members of the latter. Shared derived characters linking theropods and birds include bipedality and unique ankle and wrist morphologies. An alternative, thecodont hypothesis posits that birds arose from small, quadrupedal, arboreal archosaurs predating divergence of dinosaurs from other archosaurs. Current phylogenetic analyses support the view that birds are theropod dinosaurs, though supporters of the thecodont hypothesis note that the initial coding of characters to be analyzed is both crucial and contested. Whether avian flight evolved "from the ground up" by means of running, jumping, and flapping, or "from the trees down" by means of tree climbing, gliding, and flapping (or some combination of these scenarios) is actively debated and cannot be resolved by phylogenetic arguments alone.

The oldest undisputed fossil bird is the crow-sized *Archaeopteryx lithographica*, known from seven specimens collected in south-central Germany, and dated to the late Jurassic, about 150 million years ago. *Archaeopteryx* is a good example of a transitional form. It has teeth, claws and tail vertebrae as typical of many reptiles but not birds, yet it also has a splendid set of feathers. Another putative Jurassic bird, similar in age to *Archaeopteryx*, is *Confuciusornis sanctus*. This form has unfused wrist bones and long fingers, similar to *Archaeopteryx*, though it lacks teeth, and only downy body feathers are associated with the specimen. The fossil *Protoavis texensis* from late Triassic rocks in west Texas, is about 225 million years old and similar in size to *Archaeopteryx*; however, its identification as a bird is controversial.

Multiple avian lineages have been described based on Cretaceous fossils, and phylogenetic analyses place them in the following order of increasing relatedness to extant birds. The first, Alvarezsauridae includes three flightless genera, *Mononykus*, *Alvarezsaurus* and *Patagonykus*, from the Late Cretaceous. *Mononykus olecranus* was a turkey-sized creature with extremely shortened forelimbs in which the hand bones had been fused and reduced to include a single, massive digit that may have been used for digging. Some doubt that *Mononykus* was a bird, though phylogenetic analyses place

it between *Archaeopteryx* and modern birds. A second lineage, Enantiornithes or "opposite birds," are characterized by adaptations for powered flight and an arboreal lifestyle, including a primitive keeled sternum, shortening of the tail and fusion of some vertebrae to form the pygostyle, a solid bony structure that supports the tail feathers. Enantiornithes likely represent the most diverse and widespread group of Mesozoic birds, ranging from toothed, sparrow-sized perching birds such as *Sinornis santensis* and *Cathayornis yandica*, only ten million years younger than *Archaeopteryx*, to the turkey vulture-sized *Enantiornis leali* having a wingspan at least 1 meter in length. Other fossil forms placed in the Enantiornithes by some paleontologists include *Iberomesornis*, *Concornis*, and *Noguerornis*. The bones of some enantiornithines appear to be nearly devoid of blood vessels, which are necessary for rapid and continuous growth, suggesting they may have grown slowly and seasonally, rather than continuously up to adulthood, as characteristic of true homeotherms including modern birds. A third lineage, Hesperornithiformes, is one of the best known Cretaceous bird groups, with some representatives surviving until about 70 million years ago. They were heavy-boned, flightless diving birds with long, toothed bills, approximately the size of modern loons. Their forelimbs were reduced to a single, slender bone, and they used their feet to propel themselves on the surface and below water in search of fish and other aquatic prey. A fourth lineage, the ternlike Ichthyornithiformes, also present until about 70 million years ago, had large keeled sternums and powerful wings making them capable of long-distance flight.

There has been much debate regarding the age of extant avian orders and whether only one or many of the lineages that arose during the Cretaceous survived the extinction events marking the end of that period 65 million years ago. Current fossil evidence indicates that three extant orders of birds were represented by Late Cretaceous or Early Paleocene forms. Anseriformes (waterfowl) are represented by *Presbyornis*, geographically widespread waterbirds having long legs, long necks and ducklike bills and skulls. The oldest fossils are partial specimens from Late Cretaceous rocks in Mongolia and off the coast of Antarctica. *Presbyornis* was highly colonial, feeding in great flocks, as indicated by thousands of fossil bones at single Eocene lake sites in Wyoming, together with an abundance of eggshells, suggesting the presence of large nesting colonies. The extant order Gaviiformes (loons) is represented by *Neogaeornis wetzeli* from Late Cretaceous rocks in Chile and Uzbekistan, also suggesting a broad geographic distribution before the end of the Cretaceous. Fragmentary material from Late Cretaceous or Early Paleocene strata in New Jersey appear very similar to modern Charadriiformes (shorebirds, gulls), and several fossil genera have been ascribed to that order, including *Graculavus*, *Telmatornis*, and *Paleotringa*. A fourth modern order, Procellariiformes (albatrosses, petrels), may be represented in Late Cretaceous rocks by a few fossil bones from Mongolia and New Zealand.

Based on these finds, some posit that only one or a few avian lineages survived the Cretaceous, with the earliest diverging extant orders of birds being only 70-65 million years old, and most being less than 55 million years old. This view focuses on the timing of appearance for recognizable "crown groups" in the fossil record. Paleontologists focusing on the phylogenetic divergences for "ghost lineages" (lineages apparent in a phylogenetic tree but lacking a fossil record) and molecular systematists using estimates of rates of molecular sequence evolution derive older ages for the divergences among extant orders, ranging from over 120 to 70 million years ago, and suggest that a number of avian orders survived the Cretaceous extinction. Distinguishing between stem (older) and crown (younger) groups when considering a lineage's age is necessary, as fossils representing stem groups may not look like their descendants in taxonomically recognized crown groups.

Primary Lineages of Modern Birds. The number and orders recognized for birds, and their constituent taxa, vary widely among authors. The following provides a brief summary of 24 commonly recognized orders largely following Peters et al. (1931–1987). Sibley and Monroe (1993) present an alternative classification.

Anseriformes. Anseriformes (waterfowl, screamers) are waterbirds characterized by webbed toes and broad bills, often with lamellae to aid in straining food items from water. They are found worldwide with the exception of Antarctica. Most are powerful flyers and many northern species are highly migratory. Screamers differ from waterfowl in having long mostly unwebbed toes, fowl-like bills and sharp spurs on their lower legs. One hundred sixty species are recognized in 51 genera and 2 families.

Apodiformes. Apodiformes (swifts, hummingbirds) are small, fast, acrobatic birds that feed on insects (swifts) or nectar (hummingbirds) while on the wing. Both swifts and hummingbirds have powerful flight muscles and small, weak feet. The breeding season for most hummingbirds is synchronized with the local flowering cycle. Apodiformes can be found on all continents except Antarctica. Four hundred four species are placed into 137 genera and 3 families.

Caprimulgiformes. Caprimulgiformes (frogmouths, nightjars, potoos) are crepuscular and nocturnal birds with long, pointed wings, cryptically-colored plumage, and short legs. They also have small bills but large heads with unusually wide mouths for catching insects in flight or small prey on the ground. They are found in tropical and temperate habitats worldwide. One hundred two species are placed in 24 genera and 5 families.

Charadriiformes. Charadriiformes (shorebirds or waders, gulls, auks) are found around the world in coastal and inland shore habitats. Typical waders are diminutive fast flyers with long legs, necks and bills for feeding in shallow waters. Gulls are stout, scavenging birds of marine and freshwater habitats, and auks are compact seabirds, some with large, colorful bills, adept at swimming underwater to catch fish. Two hundred ninety two species are recognized in 82 genera and 14 families.

Ciconiiformes. Ciconiiformes (herons, storks, ibises, spoonbills, New World vultures) are primarily large, fish-eating, wading birds with long necks and bills and broad wings, inhabiting temperate and tropical habitats around the world. Herons are the most speciose group within this assemblage. New World vultures are recent additions based on analyses indicating their similarities to Old World vultures (Falconiformes) are convergent. Flamingos are also controversial as members of this order. One hundred seventeen species are placed in 48 genera and 7 families.

Coliiformes. Coliiformes (mousebirds or colies) are small-bodied, long-tailed gray birds, found in bushland and forest edge habitats throughout Africa south of the Sahara. Six species are placed in two genera and one family.

Columbiformes. Columbiformes (pigeons, doves, sandgrouse) are mostly fast-flying, seed- and fruit-eating birds with long pointed wings found throughout the world in tropical and temperate habitats. Most pigeon and dove species live in tropical and subtropical regions, whereas sandgrouse (considered Charadriiformes by many) are restricted to Old World arid regions. Extinct species include the flightless dodo from Mauritius and the migratory and gregarious passenger pigeon of North America. Three hundred and twenty species are placed in forty-five genera and two families.

Coraciiformes. Coraciiformes (kingfishers, todies, motmots, bee-eaters, rollers, hornbills) are a diverse group of primarily cavity nesting birds with plumed crests, pointed bills, rounded wings and small feet with some fusing of the three forward toes (syndactyly). Kingfishers are found worldwide, with centers of abundance in Southeast Asia and tropical Africa. Todies are confined to the West Indies, as are motmots to tropical America. Other Coraciiformes are found in Old World regions. Two hundred and six species are recognized in forty-seven genera and nine families.

Cuculiformes. Cuculiformes is traditionally comprised of Cuculidae (cuckoos) and Musophagidae (turacos), however, they appear not to be sister taxa, and some place turacos in a separate order, Musophagiformes. Cuckoos are a diverse, cosmopolitan group of medium-sized, long-tailed birds, known for the parasitic habits of about half its constituent species, in which eggs are laid directly in the nests of other species. Turacos, restricted to Africa south of the Sahara, are noisy, gregarious, mostly nonmigratory birds of forest and savanna habitats. One hundred fifty species are recognized in forty genera and two families.

Falconiformes. Falconiformes (hawks, eagles, falcons) are diurnal birds of prey having large eyes, hooked beaks, and powerful feet with sharp talons for killing and eating prey. They vary from swift, bird-catching falcons to huge carrion-eating Old World vultures, and are found on all continents but Antarctica and in all habitat types from tropical forests to arctic tundra. About three hundred species are recognized in seventy-six genera and three families.

Galliformes. Galliformes (chickens, grouse, pheasants, ptarmigan, quail, guans, currasows, megapodes) are vegetarian ground birds with short, stout beaks; most have short, rounded wings and fly only short distances. The megapodes, or moundbuilders, are exceptional among birds in using decaying plant material rather than their bodies to warm and incubate their eggs. Two hundred and sixty-four species are placed in seventy-six genera and seven families.

Gaviiformes. Gaviiformes (divers or loons) are heavy-bodied, foot-propelled diving birds with three webbed toes. They are found on lakes and in coastal areas across the northern hemisphere. Five species are placed in one genus and one family.

Gruiformes. Gruiformes (cranes, rails, bustards, sunbittern, kagu, mesites) are a heterogeneous group of mostly wetland-inhabiting ground nesters found the world over. Their diverse morphologies and lifestyles make simple characterization difficult, as they range from small, poorly flying rails to the 4-foot-tall sandhill cranes of North America which may migrate thousands of miles annually. Two hundred and twenty species are recognized in 61 genera and 12 families.

Passeriformes. Passeriformes (songbirds or perching birds) comprise nearly 60 percent of all extant birds and are characterized by small size, terrestrial lifestyles, and a complex syrinx system (responsible for vocalization). The suboscine Passeriformes comprise about 20 percent of the order, are characterized by relatively simple syringes, and include the flycatchers, antbirds and woodcreepers from the New World and the broadbills, pittas, asities, and New Zealand wrens from the Old World. All other passeriforms are placed in the oscine group, characterized by more complex syringes, and including well known groups such as lyrebirds, bowerbirds, fairywrens, crows, thrushes, starlings, tits, swallows, babblers, warblers, sparrows, finches, and tanagers. Five thousand seven hundred and thirty-nine species are recognized in 1,168 genera and up to 74 families.

Pelecaniformes. Pelecaniformes (pelicans, frigate birds, boobies, tropic birds, cormorants) are large fish-eating water birds with four webbed toes (totipalmate feet) for swimming. They are found across the globe in tropical, temperate, and occasionally subarctic regions. All species are colonial to some extent and have a strong tendency to return to the same breeding areas and mates each year. Fifty-six species are placed in 7 genera and 5 families.

Piciformes. Piciformes (woodpeckers, jacamars, puffbirds, barbets, toucans, honeyguides) are a heterogeneous assemblage of tree dwelling, cavity-nesting birds with brightly colored plumage, specialized bills and zygodactyl feet (two toes pointing forward and two pointing backward). They are found worldwide except in Australia, Antarctica, and Greenland. Three hundred and seventy-eight species are placed into 65 genera and 6 families.

Podicipediformes. Podicipediformes (grebes) are foot-propelled diving birds with lobed toes and are widely distributed in aquatic habitats around the world. Courtship rituals can be complex, including, in some species, a dance with long bouts of face-to-face head shaking. Nineteen species are placed in six genera and one family.

Procellariiformes. Procellariiformes (albatrosses, shearwaters, petrels) have a worldwide distribution and are highly adapted to life at sea, feeding on zooplankton, squid and fish. They have long, external, tubular nostrils and an ability to store and regurgitate food-derived oil to feed their young. Ninety-two species are recognized in twenty-three genera and four families.

Psittaciformes. Psittaciformes (parrots) are mostly arboreal, strong-flying, nonmigratory, birds, with brilliantly colored plumage, hooked bills, and zygodactyl feet. They live mainly in the Southern Hemisphere and are most prevalent in tropical regions. Most are cavity nesters with mated pairs staying together for multiyear periods. Three hundred and thirty species are recognized in seventy-eight genera and two families.

Sphenesciformes. Sphenesciformes (penguins) are flightless, social seabirds inhabiting cooler waters of southern oceans, from the Galapagos Islands to Antarctica. They retain a deep keel on the sternum and nonfolding, flipperlike wings for underwater propulsion. Seventeen species are placed in six genera and one family.

Strigiformes. Strigiformes (owls) are nocturnal or crepuscular predators with hooked beaks, sharp talons and soft plumage. They are found in diverse habitats, from tropical woodlands to arctic tundra, and on all continents except Antarctica. Like Falconiformes, they are characterized by reverse size dimorphism, in which females are larger than males. One hundred and sixty-two species are recognized in twenty-four genera and two families.

Struthioniformes. Struthioniformes, often called

"ratites," are mostly large, flightless birds with a typical "Gondawanaland" distribution, ranging across Central and South America (rhea), Africa (ostrich), Australia (cassowary, emu), and New Zealand (kiwi). Many extinct species of elephant birds from Madagascar and moas from New Zealand (up to 12 feet tall) are known from Pleistocene and Holocene sites. The number of toes has been reduced to three in rheas and emu and two in the ostrich as an adaptation for running, compared to four toes on most other birds. Ten species are recognized in 5 genera and 4 families.

Tinamiformes. Tinamiformes (tinamous) are partridge-sized birds ranging from southern Mexico through South America, found in many habitats from rainforest to barren, high elevation sites in the Andes. They tend to be poor flyers and ground-nesters, relying on cryptic coloration and stealth to escape predators. As in many Struthioniformes, their sister taxon, some male tinamous may associate with several females at nesting, and some males do all or most incubation of the eggs. About 47 species are placed in 9 genera and 1 family.

Trogoniformes. Trogoniformes (trogons) are brightly colored, pantropical forest-dwellers and include the well-known resplendent quetzal. Thirty-seven species are recognized in one family and eight genera.

Phylogeny and Classification of Modern Orders. Most researchers recognize two primary groups of extant birds, Paleognathae, which includes ratites (Struthioniformes) and tinamous (Tinamiformes), and Neognathae, which includes all other birds. Based on current evidence, neognaths have heteromorphic sex chromosomes, with the W chromosome significantly smaller than the Z chromosome, whereas paleognaths have nearly homomorphic W and Z chromosomes. Galliformes and Anseriformes are supported as sister taxa by diverse data sets. They are usually placed as sister to all other neognaths, though this remains uncertain due in part to the difficulties in rooting the avian tree. Diverse data sets also support close relationships between Procellariiformes and Pelecaniformes, Gaviiformes and Sphenisciformes, between Coraciiformes and Trogoniformes, and between Apodiformes and Caprimulgiformes (Figure 1). There is less consensus on broader avian ordinal relationships. For example, morphological analyses suggest a close relationship among Passeriformes, Piciformes, Coraciiformes, and Trogoniformes, whereas DNA-DNA hybridization analyses posit Passeriformes as being sister to a large group including Gruiformes, Ciconiiformes and Falconiformes. Some recent analyses of avian orders using mitochondrial DNA sequence characters and multiple reptilian and mammalian outgroups, place passeriform taxa as basal to representative Galliformes, Anseriformes, Struthioniformes, and Falconiformes. However, conventional relationships with early diverging Struthioniformes are found with mitochondrial

DNAs when fewer and more closely related outgroup taxa are used. Monophyly for some of the orders, particularly Ciconiiformes, Gruiformes, Piciformes, and Pelecaniformes remain contentious. The most taxonomically comprehensive ordinal phylogeny based on molecular data is from DNA-DNA hybridization analyses. Though the analyses have limitations, an exciting insight gained is that one of the two main lineages of oscine songbirds recognized, the Corvida, appears to represent a single diverse radiation of Australian endemic forms. Various species of Corvida that look like warblers, flycatchers, creepers, and thrushes are more closely related to each other than any are to the European or New World look-alikes with which they had previously been placed. This parallels the Australian marsupials being more closely related to each other than any are to similar appearing placental mammals.

Classifications of birds following the traditional sequence of orders beginning, approximately, with Struthioniformes, Procellariiformes, Sphenisciformes, and Gaviiformes and ending with Piciformes and Passeriformes, as found in most field guides and checklists, are only weakly connected with phylogenetic hypotheses and tell as much about the history of ornithology as about the history of birds.

Flightlessness. Flightlessness or near-flightlessness and associated morphological change has evolved in hundreds of extant or fossil taxa including ratites (ostrich, elephant bird), gruiforms (weka, takahe, kagu), ducks (steamer ducks), hoatzin, cormorants (Galapagos cormorant), pigeons (dodo), grebes (Atitlan grebe), parrots (kakapo), penguins, auks (great auk), and songbirds (rufous scrub-bird). Flightlessness is a dramatic example of and argument for evolution that has received relatively little attention. Evolutionary theory, including natural selection, accurately predicts that flightlessness should be most common on islands or other habitats devoid of predators, and that loss of flight should be accompanied by relaxation of selection for flight adaptations, yielding a tendency toward larger, heavier birds, with reduction in flight muscle, feather strength, and keel depth of the sternum. Flight muscles account for over 35 percent of body weight in some species (e.g., doves) and their reduction may also be seen as an adaptation for metabolic energy conservation. Flightless morphology appears to arise by means of arrested development, known as neoteny. By simple change in the timing of development, the general avian embryonic traits of disproportionately large pelvis and hind legs, greatly reduced wings, and a sternum with little or no keel may be maintained.

Recent and Potential Extinctions. About 90 percent of the 105 documented extinctions in the last four hundred years have been on oceanic islands and due to human predation or introduction of predators by hu-

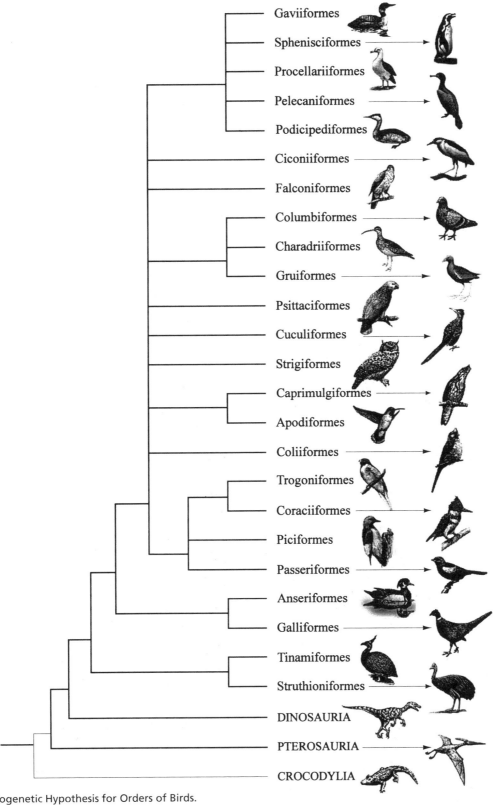

FIGURE 1. Phylogenetic Hypothesis for Orders of Birds.
Drawings by John Megahan.

mans. This is predictable given that island endemic species are more likely to have evolved without a cohort of competitors or predators, compared to species in continental habitats. Estimates place the number of currently endangered bird species at over 1,000, or more than 11 percent of the existing species diversity. This is certainly an underestimate, however, because as many as 50 percent of the world's bird species are also estimated to be declining in numbers or vulnerable to decline in light of continuing habitat loss. Future extinctions due to habitat destruction, international trade in wild birds, pesticides, and pollution will not be restricted to island forms.

Avian Genome Evolution. The genome in birds is both smaller in size and less variable across species than in other tetrapods. Mean DNA content for 165 avian species is 2.8 picograms per cell, compared to about 8.0 picograms per cell in mammals. Small cells have a greater rate of gas exchange per unit volume than large cells, due to their greater surface-to-volume ratio. Genome and cell size are strongly correlated in vertebrates. Thus, it seems the unique metabolic demands of flight in birds may have selected for reduced cell and genome size. Correspondingly, bats have smaller genomes and cell size compared to other mammals.

Birds tend to have slower rates of molecular sequence evolution relative to mammals of similar size, though the reasons for this are not clear. One hypothesis is that slightly higher body temperatures in birds yield a more stringent environment that reduces the fixation rate for naturally occurring mutations. Slower rates might also be due to higher rates of expression of antioxidant genes (such as superoxide dismutase [SOD] and catalase), which counter the mutagenic effects on DNA of oxygen free radicals, a normal byproduct of cellular respiration. This might also explain the tendency toward greater longevity in many avian species relative to mammals of similar body size. For example, captive reared pigeons often live twenty years compared to four years on average for captive reared rats, with pigeons showing significantly higher SOD expression levels compared to rats.

Birds have genetically determined sex as is found in most other vertebrates, although they have a ZW system in which females rather than males are heterogametic. Genetically determined sex is obviously an ancient trait within vertebrates; however, it is unclear whether or not it is independently derived in birds. All crocodilians and many turtles appear to have environmentally determined sex. Recent estimates of divergence time between sex-linked CHD-Z and CHD-W genes in birds suggest that some barriers to recombination for sex chromosomes in birds arose about 120 million years ago, prior to the radiation of most extant avian orders.

[*See also* Animals; Reptiles; Vertebrates.]

BIBLIOGRAPHY

Chatterjee, S. *The Rise of Birds*. Baltimore, 1997. An overview of avian evolution with discussion of controversial fossil *Protoavis texensis*.

Cracraft, J. "The Major Clades of Birds." In *The Phylogeny and Classification of the Tetrapods*, edited by M. J. Benton, pp. 333–355. Oxford, 1988.

Dingus, L., and Rowe, T. *The Mistaken Extinction: Dinosaur Evolution and the Origin of Birds*. New York, 1997. An overview of avian evolution, emphasizing birds as descendants of theropod dinosaurs.

Ellegren, H. "Evolution of the Avian Sex Chromosomes and Their Role in Sex Determination." *Trends in Ecol. and Evol.* 15 (2000): 188–192.

Feduccia, A. *The Origin and Evolution of Birds*. New Haven, 1996. An overview of avian evolution, emphasizing birds as arising from thecodont reptiles.

Gill, F. B. *Ornithology*, 2d ed. New York, 1995. A comprehensive reference for avian biology and evolution.

Groth, J. G., and G. F. Barrowclough. "Basal Divergences in Birds and the Phylogenetic Utility of the Nuclear RAG-1 Gene." *Mol. Phyl. Evol.* 12 (1999): 115–123.

Härlid, A., and U. Arnason. "Analyses of Mitochondrial DNA Nest Ratite Birds within the Neognathae: Supporting a Neotenous Origin of Ratite Morphological Characters." *Proceedings of the Royal Society, London B* 266 (1999): 305–309.

Hedges, S. B., P. H. Parker, C. G. Sibley, and S. Kumar. "Continental Breakup and the Ordinal Diversification of Birds and Mammals." *Nature* 381 (1996): 226–229.

Mindell, D. P., ed. *Avian Molecular Evolution and Systematics*. Academic Press, 1997. Treats utility and findings of molecular data sets for questions of avian evolution and phylogeny.

Mindell, D. P., M. D. Sorenson, D. E. Dimcheff, M. Hasegawa, J. C. Ast, and T. Yuri. "Interordinal Relationships of Birds and Other Reptiles Based on Whole Mitochondrial Genomes." *Systematic Biology* 48 (1999): 138–152.

Padian, K., and L. M. Chiappe. "The Origin and Early Evolution of Birds." *Biol. Rev.* 73 (1998): 1–42.

Payne, R. B. "Brood Parasitism in Birds: Strangers in the Nest." *BioScience* 48 (1998): 377–386. Reviews the evolution of brood parasitism behavior in birds, integrating behavioral and molecular genetic techniques.

Peters, J. L., E. Mayr, J. C. Greenway, Jr., et al. *Check-list of Birds of the World*, 16 vol. Cambridge, Mass., 1931–1987.

Ricklefs, R. E. "Density Dependence, Evolutionary Optimization, and the Diversification of Avian Life Histories." *Condor* 102 (2000): 9–22. A review of the diversity and evolution of avian life histories.

Sibley, C. G., and J. A. Ahlquist. *Phylogeny and Classification of Birds: A Study in Molecular Evolution*. New Haven, 1990. Comprehensive assessment of avian phylogeny based on DNA-DNA hybridization. Extensive literature reviews on history of avian classification.

Sibley, C. G., and B. L. Monroe, Jr. *Distribution and Taxonomy of Birds of the World*. New Haven, 1993. Summarizes for each of the world's 9,672 bird species, geographical distribution, preferred habitat, relationships, and classification based on the Sibley and Ahlquist's DNA-DNA hybridization based conclusions.

Starck, J. M., and R. E. Ricklefs, eds. *Avian Growth and Development: Evolution within the Altricial-Precocial Spectrum*. Oxford Ornithology Series, 8. Oxford, 1998. Presents an integrative perspective on avian development, ecology, and evolution.

— DAVID MINDELL

BIVALVES. *See* Molluscs.

BLOOD GROUP POLYMORPHISMS. *See* Genetic Polymorphism.

BODY PLANS

The body plan is one of the most useful, yet most elusive, concepts in biology. The equivalent in German, *bauplan* (plural *baupläne*), is also widely used. Although the word *bauplan* conjures images of nineteenth-century zoologists with immense beards, in fact in a biological sense it was coined only in 1945 (by the embryologist and philosopher J. H. Woodger). Even so, the concept of a biological plan, built on a common arrangement, has deeper roots that far predate Charles Darwin. This is because of the intuitive recognition of deep and pervasive similarities in organic architecture. Thus, animals as disparate as bats, toads, and snakes can be readily assigned to the vertebrate body plan and abalone, octopuses, and mussels to the mollusc body plan. From these examples, it is apparent that often a body plan is equated with a phylum, another term not entirely free of nebulosity.

Body Plans and Morphospace. Why, then, is the idea of the body plan potentially elusive? One difficulty, or arguably a virtue, is that the concept is clearly hierarchical. Consider, for example, flies. Technically, they are known as dipterans. To a reasonable first approximation, flies are pretty similar; hence, it is legitimate to talk about a dipteran body plan. But dipterans belong to the insects that in turn are arthropods. Consequently, it is equally valid to speak respectively of the insect and arthropod body plans. In essence, these identifications mirror the taxonomic hierarchy (order Diptera, class Insecta, phylum Arthropoda), but more importantly they reflect how animal life is tightly clustered around a series of particular morphological "solutions" that figuratively lie in a hyperspace of much larger anatomical possibilities. There is debate as to whether body plans (at whatever taxonomic level) are optimal constructions, or whether, in principle, innumerable alternatives might exist. The apparent distinctiveness of body plans also has a number of non-Darwinian connotations. First, any discussion of body plans often has an essentialist flavor reminiscent of the pre-Darwinian belief in archetypes. The archetype concept was vigorously espoused by Richard Owen, who saw life as effectively projections of a metaphysical realm where earthly vertebrates, for

example, corresponded to an "ideal" vertebrate form. [*See* Owen, Richard.] Even so, the definition of *body plan* is more protean than is sometimes imagined. One observation, which speaks against any essentialist concept, is the endless examples of redeployment of anatomical structures. A relatively familiar example is the transfer of certain lower jawbones in reptiles to the middle ear of mammals. A more arcane instance is the recruitment of an eye muscle in various fish, for example marlin, to form a heat organ that keeps the eye and brain warm. The above-mentioned clustering of morphotypes and the corresponding surrounding "empty zones" of morphospace beg the question of how body plans are actually assembled in an evolutionary context. It is scarcely surprising that there have been recurrent appeals to various mechanisms of macroevolution, which conveniently are no longer operative today.

Genetics and Phylotypes. Body plans serve to reveal a tension in biology, between ideas of unchanging form and evidence for continuing transformation. To many, the concept of a body plan epitomizes apparent stability of form, with the implication that evolution is hedged in by constraints. There is no doubt that certain factors must be significant. Functional interlocking, whereby a shift in an arrangement of musculature, for example, leads to a deleterious cascade of maladaptive changes, must act as a constraint. In a wider context, there are also physicochemical constraints, such as the Hagen-Poiseuille equation for fluid flow in pipes (which states that the rate of volume flow varies with the pipe's radius raised to the fourth power) or the surface-to-area relationship (whereby with increase of size volume grows at a much faster rate than area). These must contribute to limiting the range of evolutionary possibilities. Of the other factors imposing stability on body plans, and thereby restricting their potential for change, the constraints of development, notably during embryology, are also important (Hall, 1996). One reason is because a number of genes are now known either to play essential roles in the overall body plan or to initiate development of such major organs as the nervous system, eye, and heart. The epitomy of these are the homeobox (*Hox*) genes. In the favorite experimental organism the fruit fly (*Drosophila*), *Hox* genes effectively define the main divisions of the body. Their misexpressions explain how unusual redeployments come about, for example, where wings are duplicated or instead of an antenna a leg grows out of the head.

What was surprising to learn is that the body plan of mice and other vertebrates is controlled by effectively the same set of *Hox* genes as in *Drosophila*. These genetic controls on the body plan have been put in an evolutionary context via the so-called phylotypic stage. A widely used, and helpful, analogy is that of an hourglass (Figure 1). In this schema, the embryos of different spe-

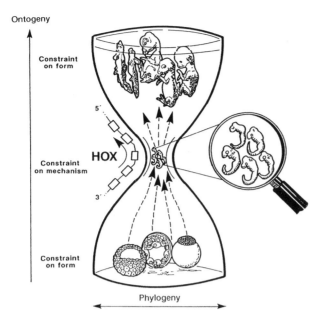

FIGURE 1. The Concept of the Phylotypic Stage. The "waist" of the hourglass indicates the point at which different types of embryos converge on a common form controlled by *Hox* gene expression prior to ongoing development that leads to the adult form.   Redrawn with permission from *Development,* Company of Biologists.

cies, which at an early stage often differ markedly from each other, are then channeled toward the phylotypic stage. In the analogy, this stage is represented by the "waist" of the hourglass, where the various embryos converge in form before expanding again into the very diverse range of adult morphologies. At first glance, this idea seems to have particular force with respect to the vertebrates. Here the various embryos—whether fish, bird, or mammal—are said to be almost indistinguishable at the phylotypic stage. Even so, some of the most famous illustrations of such embryos by Ernst Haeckel are, to put it bluntly, fudged. In fact, a careful examination of the phylotypic stage in the various vertebrates shows that the differences are more significant than is sometimes appreciated (Richardson et al., 1997).

Not only is the concept of a phylotypic stage somewhat shaky, but the connections between developmental constraints, homeotic genes, and body plans face some other difficulties. In the arthropods, for example, the experimental convenience of *Drosophila,* combined with exhaustive studies, has led to the pattern of *Hox* gene expression in the embryo receiving effectively an iconic status. Yet in other arthropods, such as the myriapods (e.g., centipedes), the same complement of *Hox* genes exists, but differentiation into body regions, so obvious in the fly, is limited (Averof, 1997). Moreover, the discovery that other genes essential for the development of such tissues as the heart and eye are shared

in widely disparate groups, to some extent undermines a genetically based concept of the body plan. Thus, when a gene such as *Pax-6* for eyes is transferred to an unrelated group, such as the cephalopods, it will still induce an eye and in certain circumstances even produce an ectopic monster, that is a body sprinkled with eyes. *Pax-6* "makes" eyes in flies and humans, but flies see us through their compound eyes, whereas we swat them using camera eyes. Genes such as *Pax-6* are obviously needed for the eye component of a body plan, but there must also be a whole series of gene networks, not only to ensure complete and successful development, but to organize the tissues into either the compound or the camera eye. To complicate matters, possession of a gene does not guarantee a particular component of a body plan. Thus, nematodes are equipped with *Pax-6,* but they lack eyes.

A further and more significant difficulty in equating genes and body plans is evidence for regular co-option of the former for other uses, either at later stages in ontogeny or where there has been a radical shift in the body plan. Echinoderms, in particular, show extensive evidence for such redeployment, consistent with the major reorganizations of their body plan from those of more primitive deuterostomes (Lowe and Wray, 1997). There is, moreover, a somewhat heretical view (Budd, 1999), suggesting that during evolution genes track body plans. In this scheme, once the latter are modified by the agencies of natural selection, only then are the genetic mechanisms reconfigured to optimize the developmental pathway. A suitable analogy would be to think of the switching gear for the lighting of a theater that is continuously reorganized to follow the plot. Even so, it would be unwise to assume that genetic pathways are always rigidly controlled. This is because some evidence suggests that effectively identical anatomical structures, such as the vulva of nematodes, can be assembled by different genetic pathways. This is not to say that all developmental systems that go to define body plans are labile. In arthropods, for example, there are recurrent correlations between genetic expression and patterns of limb development (Averof and Patel, 1997).

Phyla and the Cambrian "Explosion." Approximately thirty-five body plans are recognized among extant fauna, each supposedly discrete and disparate. Molecular biology, however, has shown that exact correlation between phyla and body plans is less obvious than might be thought. For example, the phylum known as echiurans has long been allied to the phylum of annelids (including the earthworm). Their body plans are markedly disparate; hence the phylum status awarded to Echiura and Annelida, respectively. Yet molecular evidence now indicates that the echiurans, far from being a distinct phylum, nest firmly within the annelids (McHugh, 1997). A similar case applies to the hemichordates, a primitive group of deuterostomes. They are divided into two major groups, the free-living acorn worms and the sessile rhabdopleurids, whose close relatives known as the graptolites flourished in the Paleozoic era. Despite highly disparate body plans, molecular evidence suggests that rather than being diphyletic, phylogenetically the rhabdopleurids nest within the acorn worms (Cameron et al., 2000). In both of these examples, major changes in the body plan may actually be achieved by small genetic changes. In conclusion, the overall effect of the new molecular phylogenies will, in general, downplay the importance of phyla.

A common conceit is that all, or nearly all, body plans appeared abruptly, perhaps over a few million years, as part of the Cambrian "explosion," about 540 million years ago. This has prompted suggestions that the mechanisms of evolution necessary for such a rapid and massive diversification stray beyond those invoked by neo-Darwinians. It is also supposed that in addition to the roster of familiar phyla that persist to the present day, there were at least as many extinct phyla. Stephen Jay Gould applies this view to bolster his argument that the history of life is dominated by contingent happenstance, with the principal conclusion that the evolution of humans is a fluke that is most unlikely to be repeated elsewhere. So too, one must suppose with any other group one cares to name.

Both ideas, of novel macroevolutionary mechanisms on the one hand and alternative histories and on the other, seem difficult to reconcile with what we know of evolution. Though far from perfect, the Cambrian fossil record, notably in the guise of Burgess Shale–type faunas, suggests that it is possible to plot the emergence of body plans in a steplike manner that makes coherent sense in terms of function and ecology (Budd and Jensen, 2000). Indeed, some experts suggest that the phyla per se took some forty million years to emerge from the onset of the Cambrian explosion. This argument relies on a cladistic framework (Figure 2). Here a distinction is made between the stem group, where there is a steplike acquisition of the characters that will define the phylum, and the crown group, where all characters are in place. Using this methodology, it is argued that most of the Cambrian fossils should be assigned to various stem groups. In principle, these stem groups can also accommodate the so-called extinct phyla, which, rather than representing excursions into novel regions of morphospace, are more likely to be nested in the overall scheme of animal phylogeny. Indeed, such supposedly bizarre body plans may play a key role in elucidating the early evolution of the various phyla. Furthermore, even if the earth were flung by some cataclysm back to a state analogous to the beginning of the Cambrian period, we have reason to believe that the new evolutionary outcomes would not be so different. This is because the ubiquity

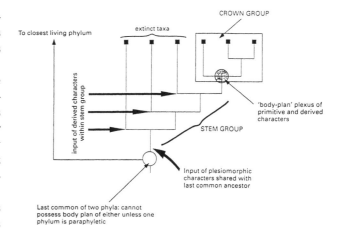

FIGURE 2. A Cladistic View of Bodyplan Assembly. The diagram shows the acquisition of character states in the stem group and the identity of a crown group, which is defined as the last common ancestor and all its living descendants. The crown group is equated with the phylum, the complete body plan.   Redrawn with permission from *Biological Reviews* 75 (2000): 135–142.

of evolutionary convergence make it probable that one biosphere and its body plans will look very much like another.

[*See also* Cambrian Explosion; Constraint; Homeobox; Metazoans; Multicellularity and Specialization; Phylotypic Stages.]

BIBLIOGRAPHY

Arthur, W. *The Origin of Animal Body Plans: A Study in Evolutionary Developmental Biology.* Cambridge, 1997.

Averof, M. "Arthropod Evolution: Same *Hox* Genes, Different Body Plans." *Current Biology* 7 (1997): R634–R636.

Averof, M., and N. H. Patel. "Crustacean Appendage Evolution Associated with Changes in *Hox* Gene Expression." *Nature* 388 (1997): 682–686.

Budd, G. E. "Does Evolution in Body Patterning Genes Drive Morphological Change—or Vice Versa?" *BioEssays* 21 (1999): 326–332.

Budd, G. E., and S. Jensen. "A Critical Reappraisal of the Fossil Record of the Bilateran Phyla." *Biological Reviews* 75 (2000): 253–295.

Cameron, C. B., J. R. Garey, and B. J. Swalla. "Evolution of the Chordate Body Plan: New Insights from Phylogenetic Analyses of Deuterostome Phyla." *Proceedings of the National Academy of Sciences USA* 97 (2000): 4469–4474.

Conway Morris, S. *The Crucible of Creation: The Burgess Shale and the Rise of Animals.* Oxford, 1998.

Gould, S. J. *Wonderful Life: The Burgess Shale and the Nature of History.* New York, 1989.

Hall, B. K. "*Baupläne*, Phylotypic Stages, and Constraint: Why There Are So Few Types of Animals." *Evolutionary Biology* 29 (1996): 215–261.

Hall, B. K. *Evolutionary Developmental Biology.* London, 1998.

Lowe, C. J., and G. A. Wray. "Radical Alterations in the Roles of

Homeobox Genes during Echinoderm Evolution." *Nature* 389 (1997): 718–721.

McHugh, D. "Molecular Evidence that Echiurans and Pogonophorans Are Derived Annelids." *Proceedings of the National Academy of Sciences USA* 94 (1997): 8006–8009.

Richardson, M. K., et al. "There Is No Highly Conserved Embryonic Stage in the Vertebrates: Implications for Current Theories of Evolution and Development." *Anatomy and Embryology* 196 (1997): 91–106.

Shubin, N., et al. "Fossils, Genes and the Evolution of Animal Limbs." *Nature* 388 (1997): 639–648.

Woodger, J. H. "On Biological Transformations." In *Essays on Growth and Form Presented to D'Arcy Wentworth Thompson*, edited by W. E. Le Gros Clark and P. B. Medawar, pp. 95–120. Oxford, 1945.

— SIMON CONWAY MORRIS

BRAIN SIZE EVOLUTION

The evolution of brain size in humans is best understood by referring to the neurological data and evolutionary histories of other mammals, including nonhuman primates. Compared to most mammals, primates have brains that are relatively large for their particular body sizes (i.e., they are "encephalized"), along with a generally high level of curiosity and intelligence. Among the more than 200 species of primates, *Homo sapiens* is by far the most encephalized. Humans, with an average cranial capacity of approximately 1,350 cm³ (cubic centimeters, which closely approximate brain mass in grams), have brains that are three times the size expected for monkeys or apes of equivalent body size. This remarkable statistic is the result of an evolutionary history that resulted in the ability for conscious thought and advanced communication processing that allowed our species to dominate the planet. Nevertheless, human brains are not the largest on Earth (whales and elephants have much bigger brains), and the simple ratio of brain size to body size (called relative brain size) in some of the smaller primates, such as the squirrel monkey, is every bit as large as that for humans. What, then, accounts for the high degree of encephalization in people? To approach this question, one must understand the scaling, physiological, and physical constraints that limit brain size in mammals, including humans, and then explore the nature of special features that evolved in the brains of certain species despite these constraints.

Constraints on Brain Size. Scaling factors (allometry) limit the manner in which subdivisions of growing organisms or organs increase in size relative to one another. For example, as humans mature, brain size generally does not keep up with the increase in body size. Consequently, babies have brains (and heads) that are absolutely smaller but relatively larger than those of adults. Allometry also applies at the species level, where relative brain size is usually greater in smaller-bodied

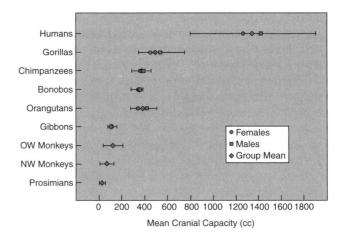

FIGURE 1. Mean Adult Cranial Capacities for Different Groups of Living Primates.
The great apes for which mean brain sizes of males are noticeably larger than those of females (orangutans, chimpanzees, and gorillas) are those in which males also have average body sizes that are larger. The same trend is found in humans. Although larger-bodied primates have greater absolute cranial capacities than smaller species, their relative brain sizes (the ratio of brain size to body size) are smaller because of allometric scaling.   Courtesy of Dean Falk.

(e.g., squirrel monkey) than larger-bodied (e.g., chimpanzee) species (Figure 1). This suggests that, as a general rule, bigger-bodied animals do not need a great deal more in the way of brains than do their smaller-bodied relatives. Thus, although gorillas are several times as large as chimpanzees and orangutans, their average cranial capacity of roughly 490 cm³ represents only about a 30 percent increase over the averages of approximately 375 cm³ for the other two apes (Figure 1). (The fact that humans have a mean brain size that is nearly three times that of much larger-bodied gorillas, and a mean relative brain size that equals that of much smaller-bodied squirrel monkeys, departs markedly from allometric expectations—testimony to how extremely encephalized humans are.)

To incorporate the brain–body size allometries, Encephalization Quotients (EQ) are often used instead of absolute brain size or brain/body ratios. EQs indicate the relationship between a particular brain size and the brain size expected of a model animal of the same body size given a particular allometric relationship. Departures from the model are assumed to measure changes other than those due solely to changes in body size. Even though the relative brain size (brain/body) of modern humans is similar to that of much smaller squirrel monkeys, our larger body size results in a substantially greater EQ.

In addition to limiting brain size/body size relationships, allometric scaling constrains the relative sizes of the different components that comprise mammalian brains in a highly predictable manner. Psychologists Barbara Finlay and Richard Darlington of Cornell University showed that, across 131 species of primates, bats, and insectivores, the relative sizes of 10 major subdivisions of their brains (e.g., cerebellum, neocortex, medulla) differed in a similar nonlinear manner that was highly predictable from the average absolute adult brain size for each species. These authors documented that the parts of the brain that increased most with increasing brain size are the result of longer periods of cell division during embryological development and therefore occur later in the development of organisms' brains (neurogenesis). Finlay and Darlington concluded that natural selection could favor an ability that depends on enlargement of one area of the brain (such as the inferior colliculus, which facilitates the ability to localize sounds) by extending the total duration of development for the whole brain, which would increase whole brain size and effect a suite of other behavioral capacities that depend on other enlarged areas that are swept along "on the coattails" of the targeted area. This hypothesis is in keeping with the previously unexplained fact that brain size in various groups of mammals increased independently during the past 65 million years.

Certain physiological constraints also limit brain size. Brains are metabolically expensive organs, both to grow and to sustain. It is estimated that approximately 60 percent of a newborn human's basal metabolic rate (BMR) is used for the brain's energetic requirements, and an adult brain utilizes a still considerable 20 percent of the BMR. Brains can therefore grow or evolve only to a size that can be supported by available energetic resources. Species with large brains accommodate the high energetic costs of growing them by extending the prenatal and postnatal periods of rapid brain growth, increasing the basal metabolism of the mother, and delaying weaning. Thus, according to the Maternal Energy Hypothesis of Robert Martin, it is the gestating and lactating metabolism of mothers that constrains fetal and neonatal brain size. The complementary Expensive Tissue Hypothesis of Leslie Aiello and colleagues suggests that, after weaning, primates are able to meet the energetic challenges of large brains because there has been an evolutionary tradeoff between an increase in size of the brain and a decrease in size of another metabolically expensive organ, the gut. Aiello notes that reduction of gut size in early hominids would have required a shift to a higher-quality, more digestible diet (e.g., roots, tubers, fish), along with the provisioning of weaned offspring. Aiello maintains that a shift to a high-protein diet would also have had a direct positive effect on the evolution of a relatively large brain. By focusing on the energetic needs of growth and maintenance of large brains after weaning, the Expensive-Tissue Hypothesis incorporates data

MEASURING THE SIZE OF HOMINID ENDOCASTS

Endocasts (or endocranial casts) are molds of the interiors of skulls that show the general sizes and shapes of the brains of early hominids (or other animals). With luck, endocasts may also reveal important details about the surface of the brain (cerebral cortex), including parts that are relatively enlarged. The volume of the braincase, or cranial capacity, is easy to measure from endocasts of fossil hominids; it approximates actual brain weight closely enough to be considered equivalent for all practical purposes. Thus, the average cranial capacity of 450 cm³ for South African australopithecines is taken to represent a brain mass of 450 grams. The volume of an endocast can be measured by submerging it in water and measuring the amount it displaces (water displacement method). Cranial capacity can also be measured directly from a whole skull by filling its braincase with a substance (such as mustard seed) and then pouring the substance into a graduated cylinder to determine its volume. Although these methods are tried and true, they do not work well on partial skulls or endocasts, and preparation of the latter may damage fragile fossils. These problems can be addressed by applying medical imaging techniques such as three-dimensional computed tomography (3D-CT) to the study of fossil braincases. With 3D-CT, we can scan a skull without damaging it and then use the data to generate, rotate, and measure a three-dimensional image of its endocast on a computer screen. Glenn Conroy and colleagues have validated the use of such "virtual endocasts" for estimating cranial capacities in hominids. With the application of medical imaging techniques, sample sizes of known cranial capacities for various species of fossil hominids should increase in the future.

—DEAN FALK

pertaining to variations in body composition, metabolic efficiency, energy allocation, and diet.

Just as reduction in gut size is hypothesized to have released a metabolic constraint that kept brain size in check during earlier hominid evolution, other mechanisms released evolutionary constraints on brain size related to the exquisitely heat-sensitive nature of this organ. Larger brains are associated with greater cooling needs than smaller brains. In humans who are overheated (hyperthermic), an elaborate network of cranial veins delivers cooled blood to the interior of the braincase, thereby keeping brain temperature within safe limits. In a sense, the human brain (like the engine of a car) has a radiator that prevents overheating. As shown by Wolfgang Zenker and Stefan Kubik, the tiny veins of the radiator riddle the bones of the skull and leave traces in the braincase. The fossil record of these traces suggests that the cranial radiator increased in complexity as brain size increased during the course of hominid evolution. According to this Radiator Hypothesis, the cranial vasculature of our australopithecine ancestors became modified in response to gravitational and thermal pressures associated with increasing bipedalism in hot, open habitats. The result was the beginning of a cranial radiator network of veins that could selectively cool the brain under conditions of intense exercise. More important, once in place, this system was itself modifiable and capable of keeping up with the increasing thermolytic needs of an evolving (enlarging) brain. Like the Expensive-Tissue Hypothesis, the Radiator Hypothesis is mechanistic; it suggests that the dramatic increase in brain size that occurred in *Homo* was facilitated (rather than directly caused) by the release of thermal constraints that previously kept brain size within ape ranges. The radiator network of veins is therefore seen as a prime releaser rather than a prime mover of human brain evolution.

In addition to energetic and thermal constraints, Michel Hofman has quantified certain physical limits for the human brain's internal connectivity, processing power, and potential for future size increase. Compared to those of other mammals, the brains of humans are disproportionately composed of neocortex. Hofman showed that the evolutionary process of neocorticalization in primates was due mainly to progressive expansion of the fibers that interconnect cortical neurons (white matter), rather than to the increase in the number of the neurons themselves (the brain's overlying gray matter). During evolution, the mass of white matter increased predictably and disproportionately with brain size; as a result, humans have a relative white matter volume of 34 percent, the highest value for any primate. Once a brain reaches a point where the bulk of its mass is in the form of connections, however, further increases would be unproductive because of declining neuronal integration and increased conduction time. Hofman mathematically modeled the size of the brain at which processing power would be optimized; he predicted that human brains would reach their maximum efficiency at a volume of 3,575 cm³ (i.e., two to three times that of modern humans). If so, this volume should represent an upper limit for future human brain evolution.

Selection for Neurological Adaptations. Because of the physiological, physical, and scaling constraints that limit brain size, it is difficult to disentangle the precise evolutionary pressures that resulted in special neurological adaptations (sometimes called "neurological reorganization") in different groups of mammals during the past 65 million years, with or without an associated increase in overall brain size. Although Finlay and Darlington hypothesized that selection for enlargement of

any one brain part is likely to result in a concerted enlargement of the whole brain, they noted that considerable ranges exist for the sizes of particular structures within a species, and that these leave room for additional, individual species-specific brain adaptations (enlargements of subareas) that are independent of overall brain size. Harry Jerison elucidated this idea with the Principle of Proper Mass: "The mass of neural tissue controlling a particular function is appropriate to the amount of information processing involved in performing the function." For example, Willem de Winter and Charles Oxnard analyzed brain proportions of primates, bats, and insectivores, and found that the relative proportions of different systems of functionally integrated brain components varied among the three groups in ways that mirrored their broad behavioral lifestyles and abilities. The authors also identified clusters of unrelated species within each of the three groups that occupied similar behavioral niches and had independently evolved similar brain proportions. Thus, although selection for specific adaptations appears to have resulted in increased overall brain size during the evolution of various mammals, there also seems to have been selection for integrated neurological components of special systems that was not merely related to overall brain size (e.g., for voluntary motor control in primates, integrating distance information in insectivores, and formation of spatiotemporal representations in bats).

In the case of hominid evolution, brain size increased dramatically during the roughly 3 million years for which a fossil record of cranial capacities is available (Figure 2). Known capacities in the earliest hominids, the australopithecines, were within the ranges of those of living great apes (e.g., australopithecines living in South Africa between 3.2 and 2.5 million years ago [mya] had an average capacity of about 450 cm³). By around 1.5 mya, cranial capacity had nearly doubled (to approximately 800–900 cm³) in some of the African *Homo* descendants of australopithecines. Cranial volume continued to increase dramatically as *Homo* evolved, reaching a maximum average of over 1,500 cm³ in classic Neanderthals (archaic *Homo sapiens*) that lived in Europe between 90,000 and 30,000 years ago. The dramatic increase in cranial capacity that occurred in early *Homo* is less marked when body size is taken into account. The early African specimen (KNM-WT 15000) shows that *Homo erectus* (*ergaster*) was much larger than the Australopithecines. Body size and shape in this taxon were similar to many living people although its cranial capacity was only two-thirds of the 1,350 cm³ mean for modern humans.

Since at least Darwin's time, scientists have wondered if one specific behavior was the target of natural selection that caused brain size to increase so dramatically in *Homo*. Given the highly speculative nature of attempting to identify a possible prime mover of brain

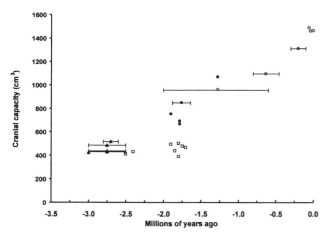

FIGURE 2. Cranial capacities plotted against time. Key to symbols; triangles, South African australopithecines (*Australopithecus africanus*) that may have been ancestral to *Homo*; squares, australopithecines from South and East African (*Paranthropus*) that became extinct without leaving descendants; circles, *Homo* (filled, individual specimens; open, group averages). Where appropriate, error bars indicate ranges of possible dates. Brain size has tripled in living people compared to the capacities of the australopithecines that lived from 3.0 to 2.5 million years ago. Absolute brain size is shown here. Because body size also increased with the appearance of Homo, the change in absolute brain size is less dramatic in the first widely successful members of our genus. The largest capacities shown are for Neanderthals that lived between 90,000 and 30,000 years ago.   Courtesy of Dean Falk.

size evolution, it is not surprising that numerous candidates have been proposed, including hunting, warfare, throwing, tool production, prolonged learning in infants, and language. Interestingly, Andrew Whiten and Richard Byrne suggest that the use of tactical deception to deprive others of desirable resources (so-called "Machiavellian" intelligence, in which seemingly altruistic acts are actually selfish strategies) may have been the prime mover of brain size evolution. Robin Dunbar concurs that social intelligence may have been a prime mover of brain evolution but suggests that the enlarged human brain evolved primarily to keep track of multiple social relationships in increasingly large groups. He notes that nonhuman primates maintain social cohesion by grooming each other's fur, and that both this activity and brain size increase across primates as group sizes enlarge. Because human groups are so large that social grooming cannot work as a bonding process, Dunbar proposes that language evolved to replace grooming during hominid evolution. Although it is extremely difficult to test prime mover hypotheses, it is worth noting that language is the one candidate that most clearly separates *modern* humans from other primates. It remains difficult, however, to pinpoint the appearance of language in hominid evolution.

[*See also* Primates, *articles on* Primate Classification and Phylogeny *and* Primate Societies and Social Life.]

BIBLIOGRAPHY

Aiello, L. C., and P. Wheeler. "The Expensive-Tissue Hypothesis." *Current Anthropology* 36 (1995): 199–221.

Conroy, G., et al. "Endocranial Capacity in an Early Hominid Cranium from Sterkfontein, South Africa." *Science* 280 (1998): 1730–1731.

de Winter, W., and C. E. Oxnard. "Evolutionary Radiations and Convergences in the Structural Organization of Mammalian Brains." *Nature* 409 (2001): 710–714.

Dunbar, R. I. M. "Coevolution of Neocortical Size, Group Size and Language in Humans" (target article and response to open peer commentary). *Behavioral and Brain Sciences* 16 (1993): 681–735.

Falk, D. "Brain Evolution in *Homo*: The "Radiator" Theory" (target article and response to open peer commentary). *Behavioral and Brain Sciences* 13 (1990): 333–381.

Falk, D. *Primate Diversity*. New York, 2000. Each of the fourteen chapters contains a Neural Note that provides information about primate brain evolution.

Falk, D., et. al. "Early Hominid Brain Evolution: A New Look at Old Endocasts." *Journal of Human Evolution* 38 (2000): 695–717.

Falk, D., and K. R. Gibson. *Evolutionary Anatomy of the Primate Cerebral Cortex*. Cambridge, 2001. Part I contains six chapters on the evolution of brain size.

Finlay, B. L., and R. B. Darlington. "Linked Regularities in the Development and Evolution of Mammalian Brains." *Science* 268 (1995): 1578–1584. Links embryological neurogenesis to independent evolution of overall brain size in different groups.

Gibson, K. R. "Cognition, Brain Size and the Extraction of Embedded Food Resources." In *Primate Ontogeny, Cognition and Social Behavior*, edited by J. Else and P. C. Lee. Cambridge, 1986.

Godfrey, L. R., et al. "Teeth, Brains, and Primate Life Histories." *American Journal of Physical Anthropology* 114 (2001): 192–214.

Hofman, M. Brain evolution in hominids: are we at the end of the road? In *Evolutionary Anatomy of the Primate Cerebral Cortex*, edited by D. Falk and K. R. Gibson. Cambridge, 2001.

Jerison, H. J. *Evolution of the Brain and Intelligence*. New York, 1973. The classic work on the evolution of brain size.

Martin, R. D. "Scaling of the Mammalian Brain: The Maternal Energy Hypothesis." *News in Physiological Sciences* 11 (1996): 149–156.

Milton, K. "Diet and Primate Brain Evolution." *Scientific American* 269 (August 1993): 86–93.

Whiten, A., and R. W. Byrne. *Machiavellian Intelligence II: Extensions and Evaluations*. New York, 1997.

Zenker, W., and S. Kubik. "Brain Cooling in Humans—Anatomical Considerations." *Anatomical Embryology* 193 (1996): 1–13. Provides a rare photograph of the human cranial "radiator" network of blood vessels that helps to regulate brain temperature.

— DEAN FALK

BUFFON, GEORGES-LOUIS LECLERC

Georges Louis Leclerc de Buffon (1707–1788), eighteenth-century French naturalist. Leclerc, later comte de

Buffon, was born in Burgundy on 7 September 1707. His early scientific interests focused on mathematics and the physical sciences, although he also conducted research on forestry and on other practical topics. Buffon belonged to a circle of French scientists who championed what was then the new Newtonian science—a mechanical vision of the world that held that matter in motion, operated on by natural forces such as gravity, could explain all that is observed in nature.

Buffon developed his interest in natural history after being appointed by Louis XV in 1739 as director of the Jardin du Roi (Royal Gardens and Natural History Collections) in Paris. In the eighteenth century, major collections often had printed catalogs, and Buffon set out to produce one for the Royal Collection. He was, however, ambitious, and he transformed the task into producing a comprehensive natural history of all nature's products: animals, plants, and minerals. Beginning in 1749 he published (along with a set of collaborators) the first three volumes of his famous *Histoire naturelle, générale et particulière* that ultimately comprised thirty-six volumes published over a period of fifty years. The *Histoire naturelle* proved an enormous success and became one of the most widely read books of the century. It was responsible for Buffon's reputation at the time as one of the four major figures of the French Enlightenment. The volumes were handsomely bound in leather, had hundreds of engravings, and were written in a beautiful style. Unfortunately, Buffon did not live long enough to complete his survey of all nature's products. At his death in 1788 he had published volumes on the history of the planet, the history of humans, and volumes on all the quadrupeds, birds, and minerals. A team of specialists later completed the entire survey.

Buffon believed that natural history suffered from a lack of sufficient empirical knowledge. He believed that it was useless to devise taxonomic systems, as his contemporary Carolus Linnaeus was doing, until naturalists had a deeper understanding of the facts of natural history. [*See* Linnaeus, Carolus.] His great work on natural history, therefore, consists largely of articles on individual animals and attempts to compile all known information: internal anatomy, external anatomy, life stages, behavior and breeding habits, geographical distribution, geographic variation, economic value, and a summary of what earlier naturalists had written on each animal. Buffon held that only once all information was compiled might it be possible to discern the order in nature.

Buffon also included a number of theoretical essays in his *Histoire naturelle* that were historically important for ideas on the evolution of life and on the concept of species. Of critical importance was Buffon's theory of generation. He held that each individual has an "internal molding force" responsible for its form and function.

This force worked on "organic molecules" that entered the body through the process of nutrition. Reproduction resulted when excess organic molecules (which traveled to the gonads and constituted a "seminal fluid") from the male and female mixed and produced a primitive embryo. The stronger of the two seminal fluids determined the form of the offspring.

Buffon's theory of generation held important implications for the concept of species. In his early writings, he had argued that a species should be understood as those animals that can successfully breed and produce offspring. Later, he drew out the significance of this breeding definition. Because the internal molding force was responsible for the internal and external form, Buffon came to see these forces as carrying the traits that characterize different species. His knowledge of geographical variation led him to conclude that the expression of the internal molding force could be influenced by the environment—the different races of horse, for example, being explained by the different environmental conditions in which these races originated.

Of greater significance, Buffon carried out a set of breeding experiments on his estate. In these he discovered that some individuals of closely related but different species produced some fertile hybrids: the horse and ass, and the dog and wolf. From these unexpected data, he concluded that morphologically similar animals were descendants of a common stock. He surveyed what he knew of all the then known quadrupeds (200) and decided that they could be grouped into thirty-eight natural families that shared common internal molding forces and that had differentiated in time. Buffon speculated that, in the early history of the globe, the various internal molding forces arose, and that as the surface of the earth changed, so to did the distribution of the descendants of these primitive stocks. To understand the diversity currently existing, then, naturalists needed to know the history of the earth as well as the the history of the distribution of life-forms.

Buffon did not believe that the environment could fundamentally alter the internal molding forces: frogs could not turn into horses. Change in time was limited. The descendants of primitive horses eventually became the different races of horse, as well as the closely related ass and zebra. Dogs, wolves, and foxes were all from a primitive wolf stock. Although Buffon was among the first naturalists who believed that to understand nature we need to consider change in time, he did not believe in the extensive evolution of life as Charles Darwin was to later describe. Buffon had a critical importance, nonetheless, by stressing change and by arguing that to understand contemporary life we need to understand its past history.

[*See also* Biogeography, *article on* Vicarance Biogeography; Classification; Species Concepts.]

BIBLIOGRAPHY

Buffon. Paris, 1952. A set of articles covering Buffon's life, work, and impact.

Buffon 88. Paris, 1992. A collection of articles celebrating the anniversary of Buffon's death and evaluating his historical importance.

Farber, P. *Finding Order in Nature: The Naturalist Tradition from Linnaeus to E. O. Wilson.* Baltimore, 2000. A general treatment of the naturalist tradition that discusses Buffon's legacy and the theory of evolution.

Fellows, O., and S. F. Milliken. *Buffon.* New York, 1972. A popular and accessible biography that describes his life and work.

Roger, J. *Buffon.* Ithaca, N. Y., 1997. A comprehensive biography by the leading authority on Buffon.

Sloan, P. "John Locke, John Ray, and the Problem of the Natural System." *Journal of the History of Biology* 5.1 (1972): 1–53. A discussion of Buffon's concept of species and its philosophical foundation.

— Paul Lawrence Farber

C

C. ELEGANS

The free-living nematode *Caenorhabditis elegans* (Figure 1) is an important model system for studying the principles of development and pattern formation and, more recently, the evolution of developmental processes. Besides the ease with which *C. elegans* can be cultured and manipulated in the laboratory, several features of nematodes make it possible to study the interactions among cells and tissues in multicellular organisms.

Sydney Brenner, a British molecular biologist, introduced *C. elegans* as a laboratory model organism because it feeds on bacteria such as *Escherichia coli* and can reproduce within three days. *C. elegans* exists as two sexes, self-fertilizing hermaphrodites and males, which are about 1 millimeter in length. In the laboratory, the life cycle consists of the egg, four juvenile stages (referred to as L1–L4), and the adult worm. In the wild, *C. elegans* and other nematodes can survive unfavorable environmental conditions by arresting in an alternative stage. For developmental biologists, *C. elegans* is of special interest because most nematodes display invariant cell lineages, that is, a similar pattern of cell divisions in all individuals of a given species. Also, adult *C. elegans* worms are anatomically and genetically simple. The adult hermaphrodite has only 959 somatic nuclei. Initially, 1,090 somatic cells are generated, 131 of which die by apoptosis (programmed cell death), a process that has been studied in detail in *C. elegans*.

Pioneering work by Sulston and Horvitz (1977) and Sulston and coworkers (1983) determined the complete cell lineage of *C. elegans;* that is, the fate of each cell during development was mapped. This allowed embryonic and postembryonic developmental processes to be studied at a cellular level. All cells have a numeric name; their lineage history is known, and their division pattern is predictable. How cells interact during development can be studied experimentally in *C. elegans*. Using laser microbeam irradiation, individual cells can be removed from the living animal without disturbing further development. Such cell ablation experiments revealed that cell–cell interactions are common during embryonic and postembryonic development. Thus, although the cell lineage is fixed, multiple interactions are required for normal development to occur, indicating that nematode development is not mosaic.

The cellular analysis has been complemented by genetic studies that have helped to identify the molecular

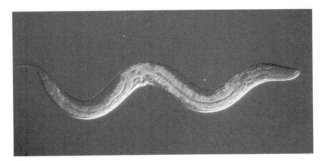

FIGURE 1. Adult *Caenorhabditis elegans*.
© Nathalie Pujol/Visuals Unlimited.

mechanisms controlling cell differentiation. Also, these studies have indicated how multipotent stem cells choose between alternative fates. On 11 December 1998, the genome sequence of *C. elegans* was reported as the first metazoan genome to be completely sequenced, thereby providing a new platform for investigating the biology of this organism.

A Particular Case Study: the Vulva. A process that has been studied in very great detail in *C. elegans* is the development of the vulva, the egg-laying structure of nematode females and hermaphrodites. The vulva is a derivative of the ventral epidermis and in *C. elegans* is made up of twenty-two cells (Figure 2A). These cells derive from three precursor cells called P(5–7).p, which divide within five hours to form the complete vulva. Although always the same three precursor cells form vulval tissue in wild-type animals, a group of six cells, P(3–8).p, is competent to participate in vulva formation. This is the vulva equivalence group. A signal from the gonadal anchor cell induces vulva formation in P(5–7).p. Genetic and molecular studies revealed that a highly conserved signal transduction process, involving an epidermal growth factor–like molecule, Ras, and other important oncogenes, is crucial for this inductive interaction (Figure 2B). Based on the involvement of oncogene homologues in *C. elegans* vulva development, genetic and molecular aspects of this developmental process have been investigated in great detail in recent years.

Evolution of Developmental Processes. Given the cellular resolution at which developmental processes can be analyzed and the genetic and molecular knowledge provided by *C. elegans*, nematodes are an excellent model system for studying the evolution of developmental processes. Many free-living species can be cultured

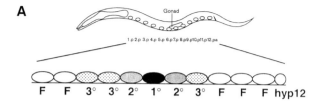

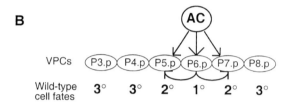

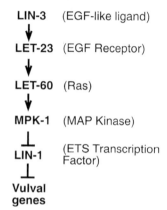

FIGURE 2. Schematic Summary of Ventral Epidermal Cell Fate Specification in *C. elegans*.

(A) In the first larval stage, the twelve ventral epidermal cells P(1–12).p are equally distributed in the region between the pharynx and the rectum. P(1,2,9–11).p in the anterior and posterior region fuse with the surrounding hypodermis (F, white ovals), whereas the VPCs P(3–8).p remain unfused. Later in development, P(3–8).p adopt one of three alternative fates as indicated by different symbols. P(5–7).p form vulval tissue. P6.p has the 1° fate (black oval), generates eight progeny and forms the central part of the vulva. P(5,7).p have a 2° fate (gray ovals), generate seven progeny each, and form the outer part of the vulva. P(3,4,8).p have a 3° fate (speckled ovals) and remain epidermal. The hierarchical cell fate designation indicates that cells with a lower fate (i.e., 3° cells) can replace cells with a higher fate (i.e., 2° or 1° cells) after cell ablation. (B) Proper development is ensured by signaling between the anchor cell (AC) and the vulval precursor cells and among the vulval precursor cells. Vulval induction by the AC is mediated by a conserved EGF/Ras/MAPK signaling pathway: *LIN-3* encodes an EGF-like ligand, which is expressed in the AC and signals underlying VPCs; *LET-23* encodes an EGF-receptor tyrosine kinase; *LET-60* encodes a Ras molecule acting as a molecular switch; *MPK-1/ SUR-1* encodes a MAP kinase and *LIN-1* encodes an ETS-domain transcription factor which functions to inhibit vulva formation. In P(5–7).p, AC signaling inactivates LIN-1, which then leads to vulva formation. *Development* 125, p. 3867, Company of Biologists Ltd.

in the laboratory, and cell lineage and cell ablation studies can be performed in a way similar to *C. elegans*. Vulva development has been compared among more than fifty species of seven different nematode families, and multiple differences in vulva formation have been observed. In general, the cells forming the vulva are homologous between all studied species. However, the cell–cell interactions required for the formation of a proper vulva differ between species, indicating important evolutionary changes. For example, the position of vulva formation within the worm body varies between species. *C. elegans* and most other free-living species form their vulvae in the central body region, whereas others form it in the posterior body region (e.g., *Mesorhabditis* sp. and *Teratorhabditis palmarum*). In *T. palmarum*, no inductive signal from the anchor cell is required for vulva formation, indicating that the mechanisms underlying cell fate specification are flexible and can be modified during evolution.

One species that has been studied in great detail at the genetic and molecular level is *Pristionchus pacificus*. This species diverged from *C. elegans* approximately 200 million years ago, and various differences at the cellular and molecular level have since evolved. Comparing homologous genes between *P. pacificus* and *C. elegans* indicates that genes can be recruited to serve completely new functions in a new regulatory linkage (co-option), they can change their molecular specificity while remaining in the original (homologous) developmental program, and at the same time, they can retain other functions. Similar studies have also been initiated in a third nematode species, *Oscheius* CEW1, providing a unique platform for studying the genetic basis of the evolution of developmental processes.

Nematode Phylogeny. Nematodes inhabit all ecosystems and are highly prolific. Besides free-living species, animals and plant parasites are of major importance. For example, the causal agents for elephantiasis and trichinosis, or river blindness, are the nematodes *Wuchereria bancrofti*, *Trichinella spiralis*, and *Onchocerca volvulus*, respectively. Using small subunit ribosomal DNA sequences from a wide range of nematodes, Baxter and coworkers (1998) provided a phylogenetic framework for the phylum Nematoda. This study suggests that animal and plant parasitism arose independently several times.

[*See also* Apoptosis; Cell Lineage.]

BIBLIOGRAPHY

The *C. elegans* Sequencing Consortium. "Genome Sequence of the Nematode *C. elegans*: A platform for investigating biology." *Science* 282 (1998): 2012–2018.

Eizinger, A., B. Jungblut, and R. J. Sommer. "Evolutionary Change in the Functional Specificity of Genes." *Trends in Genetics* 15 (1999): 197–202.

Epstein, H. F., and D. C. Shakes. *Methods in Cell Biology*, vol. 48,

Caenorhabditis elegans: *Modern biological analysis of an organism.* San Diego, Calif., 1995.

Malakhov, V. V. *Nematodes.* London and Washington, D.C., 1994.

Riddle, D. L., T. Blumenthal, B. J. Meyer, and J. R. Priess. *C. elegans II.* Cold Spring Harbor, N.Y., 1988.

Sommer, R. J. "Evolution and Development—the Nematode Vulva as a Case Study." *BioEssays* 19 (1997): 225–231.

Sommer, R. J. "Evolution of Nematode Development." *Current Opinion in Genetics and Development* 10 (2000): 443–448.

Wood, B. W. *The Nematode* Caenorhabditis elegans. Cold Spring Harbor, N.Y., 1988.

— RALF J. SOMMER

CAIN, ARTHUR JAMES

English zoologist and evolutionary biologist, ultimately Professor of Zoology at the University of Liverpool.

During the 1950s Arthur James Cain (1921–1999) became a leader in the new field of ecological genetics, carrying out observations and experiments on natural populations, trying to detect and measure natural selection. He studied genetic polymorphism in the colors and patterns on the shell of the European land snail *Cepaea nemoralis.* This polymorphism had become widely accepted as a text-book example of neutral variation. The differences between the morphs had no known function, and their frequencies seemed to vary haphazardly from place to place. The French zoologist Maxime Lamotte had shown that, if a high mutation rate was assumed, the distributions of gene frequencies agreed well with the expectations of random genetic drift.

The differences between the morphs were striking (brown, pink, or yellow shells with or without varying numbers of dark bands), and Cain doubted that they were neutral. With P. M. Sheppard he examined the distributions of the variants in different habitats, chosen to be clearly distinct in their vegetation. There were strong associations between the frequencies of morphs and the backgrounds on which the snails were found. In beech woods, on the dark uniform backgrounds of leaf litter, there were high frequencies of brown, pink, and unbanded shells. On the paler and more heterogeneous backgrounds of grasslands there were significantly higher frequencies of yellow and banded shells. Cain and Sheppard showed by experiment that thrushes ate more of the snails that stood out visibly against their background, demonstrating strong natural selection on the variation.

These observations convinced Cain that it is unwise to assume selective neutrality until there has been a serious study of relations between the variation and the ecology. The null hypothesis of neutrality must be tested rigorously before it can be regarded as an acceptable explanation. Cain was moved to write *The Perfection of Animals*, in which he argued trenchantly for the power and delicacy of natural selection. Not only must we understand the basic molecular, biochemical, and physiological processes that allow organisms to exist, but we must also discover the reasons for their diversity. He denied the distinction between "adaptive" and "ancestral" characters, and maintained that the differences between major groups, such as phyla, have persisted because they are soundly functional, not mere products of history. The important constraints upon form and function are those of physics, chemistry, and the nature of the environments in which organisms live. Assertions that characters are non-adaptive, Cain believed, are too often just results of our ignorance about their functions and workings. He went on to give many examples of the precision with which organisms have adjusted to their environmental circumstances. *The perfection of Animals* is one of the clearest and most cogent general statements of the "selectionist" position, and of scepticism about the roles of selective neutrality and historical contingency in evolution.

Cain also carried out important studies on the geography of gene frequencies, speciation, taxonomy, and the history of science.

[*See also* Mass Extinctions.]

BIBLIOGRAPHY

Cain, A. J. *Animal Species and Their Evolution.* London, 1954. An excellent short book on speciation, reprinted in 1993 by the Princeton University Press.

Cain, A. J. "The Perfection of Animals." In *Viewpoints in Biology 3* edited by J. D. Carthy and C. L. Duddington. London, 1964. Reprinted in 1989 in the *Biological Journal of the Linnean Society* 36: 1–29.

Cain, A. J. "Ecology and Ecogenetics of Terrestrial Molluscan Populations." In *The mollusca.* Volume 6. Ecology, edited by W. D. Russell-Hunter. London, 1983. A robust expression of his views about polymorphism.

Cain, A. J., and P. M. Sheppard. "Natural Selection in *Cepaea.*" *Genetics* 39 (1954): 89–116. The classic paper on polymorphism in snails.

— BRYAN C. CLARKE

CAMBRIAN EXPLOSION

The Cambrian explosion refers to the observation that an immense number and diversity of fossils appear suddenly beginning about 545 million years ago. It is a central theme in evolutionary biology, serving to integrate information from the fossil record and developmental and molecular biology, as well as the possible influence of extrinsic factors such as changes in atmospheric composition (especially oxygen) and major glaciations ("Snowball Earths"). It is paradoxical, therefore, that some workers argue that the Cambrian explosion never happened. By this, it is meant that the event was an "explosion" of fossils, representing the breaching of thresholds of preservation potential, specifically the appearance of robust skeletons and animals large enough to leave imprints in the sediment. This would be a dramatic change in the fossil record, but it would have little to do with evolution.

A related theme to this debate is whether most, if not all, of the metazoan (multicellur animals) body plans actually evolved in the Cambrian, and, if so, how rapidly. Alternatively, did the metazoans have a prior and protracted history, perhaps stretching back for hundreds of millions of years? If so, it was presumably cryptic and implies some sort of miniaturized fauna. The latter view has, at least to date, no direct support from the fossil record. If the animals were indeed tiny, then their chances of fossilization are greatly diminished. So such an absence may simply be negative evidence. The notion that animals do have a deeper history is derived from another department of biology, specifically molecular biology and the use of so-called molecular "clocks." In principle, the more distantly related the organism, the greater the difference in molecular sequence, say, the composition of the strings of amino acids that make up a protein. If the substitutions (e.g., at site X a glycine replaces a valine) occur at a more or less constant rate—for example, one on average every million years—then in principle the time of original divergence of the two living species can be estimated. Everyone knows that there are a variety of molecular "clocks," some running fast, others slow. So it is important to first choose the appropriate "clock." Only recently, however, has it been realized that many, maybe all, molecular "clocks" are also spasmodically erratic. The utility of molecular "clocks" is therefore in serious doubt, although it must be acknowledged that a number of independent studies agree that metazoans may have diverged long before the Cambrian. To the first approximation, the fossil record is probably a fair guide to understanding the Cambrian explosion.

Time, Time, Give Me Time. The timing of the Cambrian explosion has undergone significant revision. Radiometric dates, obtained by analyses of the U-Pb (uranium-lead) system in grains of zircon (zirconium silicate) that give precise dates with errors bars as small as a million years, have led to a new framework of events. The entire Cambrian period is now significantly shorter than once thought, with the base of the system (and thus the Vendian/Cambrian boundary) placed at 543 million years ago and the famous Burgess Shale being deposited at about 510 million years ago. Other key intervals are less precisely dated, but the last of the major glaciations, perhaps of global extent ("Snowball Earth"), terminated approximately 600 million years ago. Prior to that there is little direct evidence for metazoans. The geological interval between about 575 and 550 million years ago was marked by an important assemblage, known as the Ediacaran faunas. These fossils

may be central in understanding the origins of the Cambrian explosion.

The rock record yields paleontological information, which, it might be necessary to recall, includes not only the body fossil record (most obviously, but not exclusively, shells and other skeletal remains) but also remains of activities (trace fossils) and degradation products of their biochemistry (chemical fossils). It also gives insights into the positions of Cambrian continents and oceans (paleogeography), as well as some indication of the state of oceans (paleo-oceanography). Certain stable isotopes, notably the carbon 12/13 (δ^{13}C) and strontium 86/87 (δ^{87}Sr) systems, can be useful for stratigraphic correlations (via a technique known as chemostratigraphy). In the case of δ^{13}C, estimates of ocean productivity and, indirectly, levels of atmospheric oxygen are also potentially available. Some, or all, of the significant changes in the environment that have been registered across the Vendian-Cambrian boundary may have played a role in the evolutionary diversifications.

Skeletons. By far the most obvious manifestation of the Cambrian explosion is the geologically abrupt appearance of skeletal remains. When such faunas are described most often there is reference to a rather standard assemblage consisting of trilobites, brachiopods, molluscs, echinoderms (usually found as dispersed ossicles), and sponges (usually as isolated spicules). Prior to the Cambrian, the evidence for skeletal remains is very restricted. The principal exceptions occur only in the latest Neoproterozoic, in the form of rather simple calcareous tubes (*Cloudina*) and gobletlike organisms (*Namacalathus*). Both seem to have been constructed by cnidarian-grade (i.e., equivalent to the corals and sea anemones) animals. Nevertheless, for all intents and purposes, skeletons are an invention of the Cambrian.

The first shelly faunas that appear in the Cambrian pose a particular problem of interpretation. This is because many are patently the skeletal pieces of once larger animals, possessing complex skeletons that upon death and decay disintegrated into their myriad components. Reconstruction can be problematic, but discoveries from a number of Burgess Shale–type deposits have yielded intact specimens of certain groups. Another important component of these early skeletal faunas is a variety of tubes, but for the most part the nature of the original inhabitants is conjectural.

The phylogeny of the early metazoans is still controversial, but it seems that biomineralization evolved independently many times. Such convergent evolution is entirely unsurprising. In the majority of cases, the possession of hard parts can be linked to ecological necessity, especially a protective role. Evidence for attack includes small bore holes and larger bite marks. Curiously, in a census of bite marks inflicted on Cambrian trilobites, there was a striking preponderance for attack on the right-hand side of the animal. Even if a shell is intact, it did not necessarily escape engulfment. Such is apparent from the gut contents of priapulids, preserved in the Burgess Shale. The intestine may contain up to six hyoliths (an extinct group of probable molluscs), none of which shows evidence such as crushing or scratch marks for their gastric predicament.

Given the manifest contribution of skeletal hard parts to the fossil record, the origins and early evolution of biomineralization remain a focus of attention. Yet there are other important categories of body fossil. Principal among these are algal remains, referred to as acritarchs, that at one stage of their life cycle produces a tough, preservable cyst. Acritarchs, which may be related to the dinoflagellates, were photosynthetic and as such provide some insights into the evolution of the phytoplankton and the basis of primary productivity. The acritarchs show a very significant decline in diversity shortly after the major glaciations, remain depauperate during Ediacaran times, but then rebound dramatically in taxonomic richness during the Lower Cambrian. It has been suggested that this "explosion" in diversity, with the appearance of acritarchs possessing complex surface sculpture, might be due to increasing ecological pressure exerted by filter-feeding arthropods and other grazers. Geochemical evidence from across the Neoproterozoic-Cambrian interval also suggests that there was a significant shift in the way the organic carbon in the oceans was processed. Initially, the particles of organic material evidently descended as a sort of marine "snow." This slow descent enabled bacteria to degrade the organic material, and so leave a characteristic geochemical imprint on the carbon. In the Cambrian, however, the carbon was evidently descending through the water column much more rapidly, presumably because it was being packaged into fecal pellets produced by the newly evolved metazoan grazers. Such a process imposes a recognizably different geochemical signature.

Burrows and Tracks. In principle, there is no reason why animals could not have had a protracted pre-Cambrian history and only entered the fossil record with the "invention" of skeletons. Any idea that the Cambrian explosion was, however, simply an artifact engendered by the appearance of skeletons can be rejected by looking at the record of trackways and burrows. Trace fossils are a product of animal activity, such as strolling or burrowing, and as such have a behavioral dimension that by implication has a connection to the animal's neurology. Significantly trace fossils diversify rapidly in the Lower Cambrian and are an integral part of the Cambrian explosion.

With some highly contentious exceptions, no metazoan traces are known from sediments that predate the major glaciations. Accordingly, if any metazoans were present, they must have been minute, slithering around the sediment grains rather than pushing them aside to leave a trace. Interestingly, the first definitive traces, of Ediacaran age, are simple and consist of burrows restricted to single horizons and with little behavioral sophistication. This is not to say that the makers of these traces were brainless. Rather, these animals lived in a world that was only seeing the beginning of a sophisticated ecology, and hence reacted accordingly.

Burgess Shale–Type Faunas. Trace fossils are important, not only as a guide to behavioral repertoires but also as a proxy for the soft-bodied component of ancient faunas that normally never preserve because of the rapidity of postmortem decay. Scattered through the Lower and Middle Cambrian, however, there is a sequence of faunas (and to some extent floras) whose preservation was very far from typical. The cynosure of these is the Burgess Shale, of Middle Cambrian age. About forty similar deposits are known, of which Zhenjiang (Yunnan, China), Kaili (Guizhou, China), and Sirius Passet (Peary Land, Greenland) are preeminent. The extraordinary diversity of these faunas has greatly enhanced our understanding of the range and evolutionary importance of the Cambrian explosion. It has also been claimed that the overall occupation of morphospace, however defined, was significantly higher in the Cambrian and has subsequently shown a significant narrowing of what is termed disparity. This is a complex issue, but to the first approximation detailed study and comparison of disparity in at least the Cambrian and living arthropods and priapulids suggest little change in overall disparity.

Burgess Shale–type faunas provide three cardinal insights. First, they confirm the long-held supposition that the proportion of species with skeletal hard parts robust enough to fossilize in the normal circumstances is only a small fraction of the fauna, seldom more than 20 percent. Accordingly, a "typical" Cambrian fossil assemblage is a highly depauperate cross section of marine life. Even in the case of species with robust hard parts, the Burgess Shale–type faunas provide exceptional insights. As noted, dispersed sclerites are a common skeletal component. Until their recognition in Burgess Shale–type deposits, however, their reconstruction as an entire scleritome was necessarily conjectural. Outstanding in this regard are two examples. The first concerns the discovery of articulated halkieriids, from the Lower Cambrian of North Greenland. These not only confirmed the scleritome to have a chainmail-like array, but in addition sensationally revealed a large shell at either end, a completely unpredicted find. The second concerns a reticulate fossil known as *Microdictyon*, the biological affinities of which had been mysterious. To everyone's surprise, discovery of soft-bodied specimens from Zhenjiang showed these phosphatic disks to form "shoulder pads" above the legs of this primitive arthropod. Articulated trilobites also occur. In itself this is not unusual, but in Burgess Shale–type faunas the trilobites sometimes show exquisite preservation of appendages.

A second series of insights from the riches of the soft-bodied faunas concerns Cambrian ecology. Most notable has been the previously underappreciated importance of predation. This is most apparent from gut contents, but it can also be inferred by various fearsome grasping appendages and mouth parts. This new view of Cambrian ecology, which has dramatically overturned the previous view of a gentle world largely inhabited by somnolent deposit and suspension feeders, reinforces the general perception that the rapid rise of skeletons is linked principally to protection from predation. Additionally, it supports the view that the Cambrian explosion involved radical shifts in the ecologies, and such features as the invention and elaboration of nervous systems, eyes, and motility were driving forces in the diversification.

The third reason for the importance of Burgess Shale–type faunas, and arguably the one of widest biological significance, is the new insights provided into the early evolution of animal body plans. The evolutionary relationships of the metazoan phyla have been revolutionized by insights from molecular biology. A notable feature has been the radical reassignment of a number of phyla. Brachiopods, for example, long thought to be close to the basal deuterostomes (the superclade that encompasses the chordates and echinoderms), are now firmly embedded in another superclade, referred to as the lophotrochozoans. This latter group is a very disparate assemblage, including not only the brachiopods, but also the annelids, molluscs, and platyhelminthes. A similar set of surprises accompanied the recognition of the ecdysozoans. These are animals that shed their skeleton by a process known as ecdysis. Such a process, to accommodate growth, is characteristic of the arthropods (e.g., flies, scorpions, and shrimps), but until the revelations of molecular biology the existence of ecdysis in the nematodes (typified by the model organism, *Caenorhabditis elegans*) and the priapulids (abundant in the Cambrian) was thought to be unconnected. Now it appears to be otherwise: despite their very different appearances, these phyla are closely related.

The identifications of the ecdysozoans and lophotrochozoans are only two of a number of important rearrangements in metazoan phylogeny. Exciting as these new phylogenies are, they have, however, one significant, but inevitable, drawback. The molecular data are necessarily obtained from living organisms: for all intents and purposes, fossil DNA is a fiction. Living members of the major phyla are, however, very different from each other, and simply knowing they are closely related is of little help in determining how a given body plan actually evolved. Comparing, for example, two lophotrochozoans—say, a brachiopod and earthworm (anne-

lid)—and asking how each body plan emerged from a common ancestor is an exercise in futility. Hence the importance of the Cambrian fossil record, because this gives key insights into how certain phyla emerged and sets this process in the context of functional ecology, notably improved methods of locomotion, feeding, and defense. For example, the halkieriids and their relatives, the wiwaxiids, may provide important evidence into the diversification and differentiation of the lophotrochozoans. It seems likely that their precursor was a mollusclike creature, perhaps similar to the sluglike *Kimberella* of Ediacaran age. From such an animal it is possible via the halkieriids to trace the origin of both the annelids and the brachiopods. The former became active strollers on the seabed, whereas the brachiopods became sessile and bivalved.

The origins of the ecdysozoans may lie in a burrowing worm, effectively priapulidlike. The arthropods emerged as soft-bodied lobopodians, similar to the living velvet worms (*Peripatus*), using their lobopods to walk across the seafloor. Thereafter the arthropods evolved an increasingly complex body plan, effectively based on the invention of an articulated exoskeleton, a process that culminated in animals as disparate as the lobster and tarantula. Of course, we need to remember that future discoveries may throw doubt on the reality of such superclades as the Ecdysozoa. At the moment, evidence from quite a wide variety of molecules seems consistent with this phyletic group. Ironically, therefore, two of the most intensively studied animals in terms of molecular biology, fruit fly (*Drosophila*) and worm (*C. elegans*), are quite closely related.

At present, the early evolution of the deuterostomes is less clear, and the nature of the ancestral forms that would serve to connect the widely disparate chordates, echinoderms, and hemichordates has been largely conjectural. However, rather remarkable finds from the Zhenjiang biota, in the form of a group known as the vetulicolians, may now provide key evidence in as much as although superficially arthropodlike, these fossils are equipped with gill slits.

So What Really Happened? Imagine that we have access to a time machine, and we can visit three points in the geological past. The time dial is set at 800, 560, and 500 million years ago. Fortunately, the time machine not only has a minibar but also a complete molecular laboratory. First stop is at 800 million years ago. The planet is occupied by an immense diversity of microbial life, both prokaryotic and eukaryotic. Enthusiasts for the molecular "clock" have told us to collect as many metazoans as possible, confident that the divergence times estimated from the substitution rates in DNA and proteins point to a metazoan history that far precedes the Cambrian explosion. But what did these early metazoans look like? Much against her better judgment, the director of Tyme-Travel® has yielded to persistent lobbying and agreed to take two specialists, to collect, respectively, planktonic larvae and meiofauna. This is on the supposition that the earliest metazoans will be found either floating in the oceans or slithering around sediment grains. The director was, however, correct. Metazoans are found, but they have no significant similarity to either living larvae or meiofauna. Minute but multicellular, these early animals are convergent on the group of protistans known as the ciliates, which 800 million years ago are their main competitors. A further surprise is that a good part of the molecular architecture characteristic of more advanced metazoans is already present, but the circuitry of the gene networks is considerably less complex.

Now we advance to 560 million years ago. Microbial life is still abundant, but the seafloor is littered with large Ediacaran animals. Comparative anatomy and histology reveal the Ediacaran animals to represent the stem groups of all the principal divisions of metazoan life. A last stop, at 500 million years ago. The hatch opens, and cheers echo across the deserted landscape. The continents are still deserts, but life teems in the seas and oceans. Among the metazoans most of the principal body plans are now well established. The Cambrian explosion is over, but the director points to some hardy arthropods scuttling across the tidal flats. She reminds us that the story of evolution is by no means finished.

[See also Body Plans; Metazoans; Molluscs.]

BIBLIOGRAPHY

Arthur, W. *The Origin of Animal Body Plans: A Study in Evolutionary Developmental Biology.* Cambridge, 1997.
Ayala, F. J., et al. "Origin of the Metazoan Phyla: Molecular Clocks Confirm Paleontological Estimates." *Proceedings of the National Academy of Sciences USA* 95 (1998): 606–611.
Babcock, L. E. "Trilobite Malformations and the Fossil Record of Behavioral Asymmetry." *Journal of Paleontology* 67 (1993): 217–229.
Budd, G. E., and S. Jensen. "A Critical Reappraisal of the Fossil Record of the Bilaterian Phyla." *Biological Reviews* 75 (2000): 253–295.
Butterfield, N. J. "Plankton Ecology and the Proterozoic-Phanerozoic Transition." *Paleobiology* 23 (1997): 247–262.
Conway Morris, S. *The Crucible of Creation: The Burgess Shale and the Rise of Animals.* Oxford and New York, 1998.
Conway Morris, S. "The Cambrian 'Explosion': Slow-Fuse or Megatonnage?" *Proceedings of the National Academy of Sciences USA* 97 (2000): 4426–4429.
Conway Morris, S., and J. S. Peel. "Articulated Halkieriids from the Lower Cambrian of North Greenland and Their Role in Early Protostome Evolution." *Philosophical Transactions of the Royal Society of London B* 347 (1995): 305–358.
de Rosa, R., et al. "Hox Genes in Brachiopods and Priapulids and Protostome Evolution." *Nature* 399 (1999): 772–776.
Dewel, R. A. "Colonial Origin for Eumetazoa: Major Morphological Transitions and the Origin of Bilaterian Complexity." *Journal of Morphology* 243 (2000): 35–74.
Erwin, D. H. "The Origin of Bodyplans." *American Zoologist* 39 (1999): 617–629.
Grotzinger, J. P., et al. "Calcified Metazoans in Thrombolite-Stromatolite Reefs of the Terminal Proterozoic Nama Group, Namibia." *Paleobiology* 26 (2000): 334–359.
Hou, X.-g., et al. *The Chengjiang Fauna: Exceptionally Well-Preserved Animals from 530 Million Years Ago.* Kunming, China, 1999.
Knoll, A. H., and S. B. Carroll. "Early Animal Evolution: Emerging Views from Comparative Biology and Geology." *Science* 284 (1999): 2129–2137.
Logan, G. A., et al. "An Isotopic Biogeochemical Study of Neoproterozoic and Early Cambrian Sediments from the Centralian Superbasin, Australia." *Geochimicaet Cosmochimica Acta* 61 (1997): 5391–5409.
Lowenstam, H. A., and S. Weiner. *On Biomineralization.* New York, 1989.
Lynch, M. "The Age and Relationships of the Major Animal Phyla." *Evolution* 53 (1999): 319–325.
Martin, M. W., et al. "Age of Neoproterozoic Bilatarian [sic] Body and Trace Fossils, White Sea, Russia: Implications for Metazoan Evolution." *Science* 288 (2000): 841–845.
McHugh, D., and K. M. Halanych, eds. "Evolutionary Relationships of Metazoan Phyla: Advances, Problems, and Approaches." *American Zoologist* 38 (1998): 813–982.
Rodríguez-Trelles, F., et al. "Erratic Overdispersion of Three Molecular Clocks: GPDH, SOD, and XDH." *Proceedings of the National Academy of Sciences USA* 98 (2001): 11405–11410.
Shu, D.-G., et al. "Primitive Deuterostomes from the Chengjiang Lagerstätte (Lower Cambrian, China)." *Nature* 414 (2001): 419–424.
Valentine, J. W., et al. "Fossils, Molecules and Embryos: New Perspectives on the Cambrian Explosion." *Development* 126 (1999): 851–859.
Wills, M. A. "Cambrian and Recent Disparity: The Picture from Priapulids." *Paleobiology* 24 (1998): 177–199.
Xiao, S.-h., and A. H. Knoll. "Phosphatized Animal Embryos from the Neoproterozoic Doushantuo Formation at Weng'an, Guizhou, South China." *Journal of Paleontology* 74 (2000): 767–788.
Zhuravlev, A. Yu., and R. Riding. *The Ecology of the Cambrian Radiation.* New York, 2001.

— SIMON CONWAY MORRIS

CANALIZATION

Canalization is the property of developmental pathways to produce standard phenotypes despite mild environmental or genetic perturbations. The term was proposed by Conrad Hal Waddington (1940; 1942, p. 563) to describe the phenomenon that "developmental pathways . . . are adjusted so as to bring about one definite end-result regardless of minor variations in conditions during the course of the reaction." Developmental biologists and evolutionary biologists emphasize slightly different aspects of canalization in their definitions. The first sentence reflects the definition of Hall (1992) and points to the role of canalization as a developmental genetic mechanism to explain the constancy of phenotype. Similarly, Wilkins (1997, p. 257) emphasizes its developmental aspect when he defines canalization as "the stabilization of developmental pathways by multiple genetic factors within the genome, a form of genetic buffering." Gibson and Wagner (2000, p. 372) emphasize the evolutionary outcome of canalization as a reduction in variability, and they define canalization as "genetic buffering that has evolved under natural selection in order to stabilize the phenotype," although canalization may have components other than genetic.

Canalization allows mutations to accrue in the genotype without being expressed in the phenotype (and therefore without being immediately accessible to natural selection). Thus, in the short term, canalization limits the variability of the phenotype by promoting cryptic genetic variation. However, in the long term, canalization can act as a capacitor for phenotypic change because it allows mutant alleles to accumulate in a genome without their individual expression. Such genetic variability can be made manifest by changing the environmental conditions and can then be selected. The notion of canalization has been proposed several times under different names, including stabilizing selection (Schmalhausen, 1949), genetic homeostasis (Lerner, 1954) and universal pleiotropy (Wright, 1968).

Waddington (1942, p. 564) noted that canalization would limit variations in development such that "if wild animals of almost any species are collected, they will usually be found 'as like as peas in a pod.'" Indeed, canalization has been seen across the animal and plant kingdoms and has been invoked where the phenotypes of the wild-type organism have much less variance than phenotypes of mutants (see Eshel and Matessi, 1998; Rendel, 1967; Scharloo, 1991). The ability of developmental pathways to resist perturbations also has been demonstrated by computer models of phenotype production. Nijhout and Paulsen (1997) have shown that the phenotypic effect of variation at a single locus depends critically on the allelic values of other genes in the same pathway and on the frequency of those genes in the population. Moreover, they found that genetic background—the other genes in the genome—buffers pathways so that only a small fraction of the genes that affect the development of a particular trait can be identified in a single sampling. Von Dassow and colleagues (2000) have shown that highly evolved developmental pathways are robust entities that can regulate to produce the same phenotype even if the genotype varies within certain limits.

Hsp90 As an Agent of Canalization. The genetic mechanisms of canalization have recently become amenable to study. Two, in particular, have received attention in recent years: Hsp90 and functional redundancy. In 1999, Rutherford and Lindquist showed that a major agent responsible for buffering the phenotype was the "heat shock protein" Hsp90. Hsp90 is a protein that binds to a set of signal transduction molecules that are inherently unstable. Binding stabilizes their tertiary structure so that they can respond to the upstream signaling molecules. However, heat shock causes other proteins in the cell to become unstable, and Hsp90 is diverted from its normal function (of stabilizing the signal transduction proteins) to the more general function of stabilizing any of the cell's partially denatured peptides. Because Hsp90 is involved with stabilizing the structure of unstable proteins, Hsp90 might be involved in buffering developmental pathways against environmental contingencies that would destabilize proteins and against genetic mutations that might produce unstable proteins.

Evidence for the role of Hsp90 as a developmental buffer first came from mutations of *Hsp83*, the gene for Hsp90. Homozygous mutations of *Hsp83* are lethal in *Drosophila*. In their heterozygous state, these mutations increase the proportion of developmental abnormalities in the population into which they are introduced. In populations of *Drosophila* heterozygous for *Hsp83*, deformed eyes, bristle duplications, and abnormalities of legs and wings appeared. When different mutant alleles of *Hsp83* were brought together in the same flies, the incidence and severity of the abnormalities increased. The same abnormalities could be seen when a specific inhibitor of Hsp90 (geldanamycin) was added to the food of wild-type flies, whereas the types of defects differed between different stocks of flies.

The abnormalities did not show simple Mendelian inheritance, but were the outcome of interactions between several gene products. Selective breeding of flies with the abnormalities led, over a few generations, to populations where 80–90 percent of the progeny had the mutant phenotype. Moreover, these mutants did not keep the *Hsp83* mutation. In other words, once the mutation in *Hsp83* allowed the cryptic mutants to become expressed, selective matings could retain the abnormal phenotype even in the absence of abnormal Hsp90. Thus, Hsp90 is probably a major component of the buffering system that enables the canalization of development. Hsp90 might also be responsible for allowing mutations to accumulate but keeping them from being expressed until the environment changes. In other words, transient decreases in Hsp90 (resulting from its aiding stress-damaged proteins) would uncover preexisting genetic interactions that would produce morphological variations. Most of these morphological variations would probably be deleterious, but some might be selected for in the new environment. Canalization might thus be responsible for the long periods of stasis in the paleontological record of certain species, and the releasing of hidden morphological variation may be responsible for periods of radiation and morphological change.

Genetic Redundancy As an Agent of Canalization. One of the major discoveries of recent developmental biology has been the stability of phenotype even after the deletion of major developmentally important genes (Wilkins, 1997). In many instances, the loss of function of a particular gene is compensated for by the activation of another gene, sometimes from a different family than the one deleted. In other instances, there is already another protein in the cell whose activities are partially redundant to those of the protein encoded by the lost gene (Erickson, 1993; Wilkins, 1997). Nowak and colleagues (1997) have provided mathematical models to explain how redundancy can be selected for by natural selection and how redundancy can be made evolutionarily stable.

Canalization As a Link for Genetics, Evolution, and Development. Waddington's use of the term *canalization* to describe this limiting of phenotypic variability may have its origins in his interpretation of Alfred North Whitehead's *Process and Reality* (1929), a book used by several British embryologists seeking a philosophy of organization in which to ground their data (see Gilbert, 1991). Within his own theories of development and evolution, canalization had a central role. Canalization caused the formation of predictable trajectories of cell development, or chreodes; we would now call these developmental pathways. Such developmental pathways were organized into the "epigenetic landscape," wherein canalization increased as the pathways became more completely separated from each other. Genetic assimilation could occur when the canalized pathway of development was originally initiated by an external inducer. If, by mutation or by the chance assortment of different alleles, the same pathway could be initiated by an internal inducer, the same phenotype would be produced genetically as had been induced externally (Waddington, 1942, 1953). The Hsp90 studies mentioned earlier provide a mechanism for genetic assimilation as well as for canalization. Canalization thus provides an important link uniting genetics, development, and evolution.

[*See also* Phenotypic Plasticity; Phenotypic Stability.]

BIBLIOGRAPHY

Erickson, H. P. "Gene Knockouts of *c-src*, *TGFβ1*, and Tenascin Suggest Superfluous Nonfunctional Expression of Proteins." *Journal of Cell Biology* 120 (1993): 1079–1081.

Eshel, I., and C. Matessi. "Canalization, Genetic Assimilation, and Preadaptation: A Quantitative Genetic Model." *Genetics* 149 (1998): 2119–2133.

Gibson, G., and G. Wagner. "Canalization in Evolutionary Genetics: A Stabilizing Theory?" *BioEssays* 22 (2000): 372–380.

Gilbert, S. F. "Induction and the Origins of Developmental Genet-
ics." In *A Conceptual History of Modern Embryology*, edited by S. F. Gilbert, 181–206. New York, 1991.

Hall, B. K. "Waddington's Legacy in Development and Evolution." *American Zoologist* 32 (1992): 113–122.

Lerner, I. M. *Genetic Homeostasis*. Edinburgh, 1954.

Nijhout, H. F., and S. M. Paulsen. "Developmental Models and Polygenic Characters." *American Naturalist* 149 (1997): 394–405.

Nowak, M. A., M. C. Boerlijst, J. Cooke, and J. M. Smith. "Evolution of Genetic Redundancy." *Nature* 388 (1997): 167–171.

Rendel, J. M. *Canalization and Gene Control*. London, 1967.

Rutherford, S. L., and S. Lindquist. "Hsp90 as a Capacitor for Morphological Evolution." *Nature* 396 (1998): 336–342.

Scharloo, W. "Canalization: Developmental and Genetic Aspects." *Annual Review of Ecology and Systematics* 22 (1991): 65–93.

Schmalhausen, I. I. *Factors of Evolution: The Theory of Stabilizing Selection*. Philadelphia, 1949.

Von Dassow, G., E. Meir, E. M. Muro, and G. M. Odell. "A Segment Polarity Network Is a Robust Developmental Module." *Nature* 406 (2000): 188–192.

Waddington, C. H. "Genetic Control of Wing Development in *Drosophila*." *Journal of Genetics* 39 (1940): 75–139.

Waddington, C. H. "Canalization of Development and the Inheritance of Acquired Characters." *Nature* 150 (1942): 563–565.

Waddington, C. H. "Genetic Assimilation of an Acquired Character." *Evolution* 7 (1953): 118–126.

Whitehead, A. N. *Process and Reality*. Cambridge, 1929.

Wilkins, A. S. "Canalization: A Molecular Genetic Perspective." *BioEssays* 19 (1997): 257–262.

Wright, S. *Evolution and the Genetics of Populations*, vol. 1 Chicago, 1968.

— SCOTT F. GILBERT

CANCER

Only a small portion of human cancer cases are caused by familial cancer syndromes, but there is strong evidence that most cancers are influenced by genetic factors. The identification of genetic variants that increase or decrease an individual's risk would provide valuable information that could lead to strategies to avoid or prevent cancer, detect it earlier, or treat it more effectively. However, this availability of genetic profiles for cancer susceptibility raises important privacy and ethical issues that have implications for individuals and their families.

Familial Cancers. In some families, cancer is inherited as a genetic disease. The prototype example of this is the eye tumor retinoblastoma. In 1971, Dr. Alfred Knudson, then of the University of Texas, proposed that these individuals inherited a defective copy of a gene present in all of the cells in their body. If a mutation occurred in the other copy of the gene, in any of the individual's retinoblasts (precursor cells to the retina), then that cell could develop into a tumor. Because there are millions of retinoblasts, there is a high probability that at least one will develop a defect and become cancerous. Knudson correctly hypothesized that those individuals that did not have the familial form of retinoblastoma would have tumors in which the same gene was defective.

The gene for retinoblastoma, *RB*, was discovered in 1986 and shown to code for a protein that plays a critical role in the regulation of the cell division cycle. The complete loss of this gene is a critical step in the development of many cancers. This class of genes is called tumor suppressors, and their identification has led to a greatly increased understanding of how the cancer process works in all cells. To date, there are over twenty such tumor suppressor genes known, and defects in them are responsible for familial cases of many different tumor types, including breast, colon, pancreas, kidney, and skin cancers. In most cases, these genes fit the same pattern as RB: the same gene that is responsible for the familial cancer is also mutated in the cells of the non-familial or sporadic tumor. For example, the *APC* gene (adenomatous polyposis coli) is responsible for one form of inherited colon cancer, and this gene is also mutated in nearly all colon tumor cells. The tumor suppressor genes for the most part are involved in the regulation of growth of cells or the recognition and repair of damage to DNA.

In many cases, the understanding of the pathway that the tumor suppressor gene fits in has provided insight into the disease and implicated other genes in the cancer process. For example, the gene for nevoid basal cell carcinoma syndrome is a gene identified originally in *Drosophila* as an embryonic development gene known as *patched*. The patched protein is the receptor for a morphogenic protein, hedgehog, and this binding initiates a pathway of gene regulation critical to cell differentiation. Several genes in this pathway are also tumor suppressors or oncogenes, further supporting a connection between differentiation and growth control.

The identification of the genes responsible for inherited cancer syndromes provides the possibility to screen potentially affected individuals in the population for the purpose of diagnosis or prenatal testing. Because many of the cancers occur in the later stages of life, individuals carrying a mutation can be counseled and put under increased medical surveillance or provided with preventive therapies. Some patients with inherited colon, breast, or ovarian cancer susceptibilities opt for prophylactic removal of these organs, reducing their risk of disease. For individuals at high risk of breast cancer, chemo-preventive drugs such as tamoxiphen (an estrogen inhibitor) have been shown to lower risk. Similarly, an inhibitor of the cyclooxygenase-2 enzyme, celecoxib, has been demonstrated to reduce the occurrence of colon polyps in individuals that inherited a defective *APC* gene.

Sporadic Cancers. Most cancers do not occur in individuals with an obvious family history of cancer and are therefore referred to as sporadic. However, genetic factors are still often believed to play an important role in an individual's cancer susceptibility. Such factors could include genes involved in the repair of damaged DNA, immune recognition of abnormal cells, and metabolism of mutagenic compounds. In addition, genetic variation in the prognosis and response to therapy may also be genetically controlled. Identification of the relevant genes and variants that influence cancer phenotypes is a considerable challenge to the field of cancer genetics.

The population-based approaches to identify cancer include case-control studies in which individuals with cancer are matched to individuals from the general population that do not have cancer. Typically, cases and controls are matched by age, race, sex, and other environmental factors thought to be relevant (e.g., smoking history for lung cancer). This approach has the advantage that the samples are usually easy to obtain for the common cancers and the subjects are likely to reflect the general population. However, improper matching of cases and controls can lead to spurious results. Positive results need to be replicated in additional samples, or via functional validation of the genetic variant. Prospective cohorts involve collecting samples from patients who are at risk for cancer, then comparing individuals who get cancer over time to those who do not. This method may be less susceptible to ascertainment biases.

A variant of the case-control test is the trio design. In this method, affected individuals and their parents are sampled. The genotypes of heterozygous parents (parents with different alleles or forms of the two copies of a gene at a given locus) are examined to determine which allele was transmitted to the affected individual and which allele was not. Because there is a 50 percent probability for any given allele to be transmitted, an excess over this represents a potential linkage association between the allele and the phenotype. This test is known as a transmission disequilibrium test, and several variations of the test have been described. The method has the disadvantage that the parents need to be studied; this is difficult for late onset diseases, such as many cancers.

Another method applied to identify the genes for complex diseases is sibling-pair analysis. This method involves the ascertainment of two or more siblings affected by the disease in question. For a gene in which both parents are heterozygous, there is a 25 percent chance that the siblings will share neither allele, a 50 percent chance of sharing one allele, and a 25 percent chance of sharing both alleles. The data set is analyzed for distortions in this expected ratio, and significant distortions indicate linkage between the marker and a gene that influences the trait. This method is not dependent on the mode of inheritance of the trait and is therefore nonparametric. Identifying and collecting siblings with a disease is harder than unrelated cases; however, this method is powerful for identifying regions of the genome that are involved in disease susceptibility.

In the direct gene analysis approach, a candidate gene is screened for variation in a sample of affected individuals. Any variants detected that are likely to alter function are examined in a larger set, by biological approaches, or both.

Examples of Genes Involved in Cancer Susceptibility.

APC. Adenomatous polyposis coli is a familial disorder that predisposes individuals to have large numbers of colon polyps. These polyps are a risk factor for colon cancer. The gene responsible for this disorder, *APC*, encodes a large protein involved in a signaling pathway that regulates cell–cell interactions and growth. Examination of the sequence of the gene in colon cancer patients led to the identification of a thiamin (T) to adenine (A) mutation that causes the replacement of an isoleucine for a lysine residue at position 1,307 of the gene. This alteration does not change the function of the protein, but instead changes a stretch of the DNA from AAATAAAA to AAAAAAAA, creating an eight-base pair stretch of a single nucleotide. Such mononucleotide stretches are known to be more prone to further mutation, and indeed colon tumors from these individuals have a high frequency of mutation at this site. The I1307K variation confers only a modest increase in colon cancer risk to the individual; however, the allele is rather common in the Ashkenazi population.

P53. The *P53* gene plays a central role in the control of cell division. The encoded protein is the target for some DNA tumor viruses such as human papilloma virus (HPV), and a variant of *P53* has been shown to interact with HPV less effectively in vitro. This variant is associated with cervical cancer in some studies but not others, and the allele is more frequently lost in cervical tumors.

BRCA. The *BRCA1* and *BRCA2* genes are mutated in a number of families with high incidence of breast or ovarian cancer. Three mutations in these genes (185delAG and 5382insC in *BRCA1*, 6174delT in *BRCA2*) are common in the Ashenazi population. Individuals with these alleles have about a 50 percent risk of developing breast cancer over their lifetime, and many of them choose prophylactic mastectomy or ovarectomy to reduce their risk.

Relevance for Health Care. As more variants are identified that influence an individual's risk to develop cancer, there will be more opportunity to provide genetic testing and intervene medically as appropriate. The interventions that are envisioned include prenatal testing, increased screening for cancer, chemoprevention to reduce the risk, surgical procedures to remove the at-risk tissue prophylactically, and avoidance of en-

vironmental factors that might increase risk further. Further research should lead to better detection and prevention of cancer.

Societal Implications. The identification of alleles that influence cancer risk creates issues of privacy that need to be addressed. It is widely agreed that individuals should not be subjected to discrimination in employment or insurance because of their genetic backgrounds. There will also be individuals who choose not to know whether they have a genetic risk factor, and this may complicate genetic testing of their family members. These issues have not been thoroughly addressed by society. They will require increased education of the public and heath care professionals on genetic issues, as well as an increased number of genetic conselors to fulfill an increasing need.

[*See also* Disease, *article on* Hereditary Disease; Linkage; Mutation: An Overview; Pedigree Analysis.]

BIBLIOGRAPHY

Dean, M. "Toward a Unified Model of Tumor Suppression: Lessons Learned from the Human Patched Gene." *Biochimica et Biophysica Acta* 1332 (1997): M43–M52.

Knudson, A. G. "Mutation and Cancer: Statistical Study of Retinoblastoma." *Proceedings of the National Academy of Sciences USA* 68 (1971): 820–823.

Laken, S. J., G. M. Petersen, S. B. Gruber, C. Oddoux, H. Ostrer, F. M. Giardiello, S. R. Hamilton, H. Hampel, A. Markowitz, D. Klimstra, et al. "Familial Colorectal Cancer in Ashkenazim Due to a Hypermutable Tract in APC." *Nature Genetics* 17 (1997): 79–83.

Steinbach, G., P. M. Lynch, R. K. Phillips, M. H. Wallace, E. Hawk, G. B. Gordon, N. Wakabayashi, B. Saunders, Y. Shen, T. Fujimura, et al. "The Effect of Celecoxib, a Cyclooxygenase-2 Inhibitor, in Familial Adenomatous Polyposis." *New England Journal of Medicine* 342.26 (2000): 1946–1952.

Storey, A., M. Thomas, A. Kalita, C. Harwood, D. Gardiol, F. Mantovani, J. Breuer, I. M. Leigh, G. Matlashewski, and L. Banks. "Role of a *p53* Polymorphism in the Development of Human Papillomavirus-associated Cancer." *Nature* 393.6682 (1998): 229–234.

Struewing, J. P., P. Hartge, S. Wacholder, S. M. Baker, M. Berlin, M. McAdams, M. M. Timmerman, L. C. Brody, and M. A. Tucker. "The Risk of Cancer Associated with Specific Mutations of *BRCA1* and *BRCA2* among Ashkenazi Jews." *New England Journal of Medicine* 336.20 (1997): 1401–1408.

Weber, B. L. "Update on Breast Cancer Susceptibility Genes." *Recent Results in Cancer Research* 152 (1998): 49–59.

Weinberg, R. A. "The Retinoblastoma Protein and Cell Cycle Control." *Cell* 81 (1995): 323.

— MICHAEL DEAN

CELL EVOLUTION

All living organisms are composed of cells, which constitute the basic structural and functional unit of the biological world. According to one definition, a cell is a small, usually microscopic mass of protoplasm surrounded by a semipermeable membrane. A cell is capable of carrying out all the fundamental functions of life, thereby forming the least structural unit of living matter able to exist independently. Although cells differ markedly in size, appearance, and physiological properties, they all share a number of defining characteristics, most importantly the presence of an outer lipid membrane (plasma membrane) that encloses the molecules and any specialized structures (organelles) found within the cell. This bounding membrane separates the cell constituents and the biochemical reactions they perform from the outside environment and (in the case of multicelled organisms) from adjacent cells or extracellular matrices.

Organisms that consist of a single cell, such as bacteria, protozoa, and many kinds of algae, constitute most of the species in the biological world, and roughly half of its biomass. However, because cells are almost always submicroscopic, the unicellular world is largely invisible to the unaided eye. Instead, we are most aware of multicellular organisms such as animals and plants. In these complex organisms, cells are differentiated into distinct organs and tissues that have specialized physiological roles and biochemical functions.

Using a simple microscope in the last half of the seventeenth century, the Dutch naturalist Anton van Leeuwenhoek first viewed the single-celled world, seeing what he termed *animalcules*. In 1665, Robert Hooke first applied the term *cells* to the microscopic structure in cork. In fact, what Hooke saw in dead cork tissue was a pattern generated by the remnants of the cellulose walls that had surrounded the plasma membrane of cells in living cork tissue.

Two centuries later, Theodor Schwann and Matthias Jakob Schleiden realized that cells possessed not only the property of life but also the ability to reproduce through cell division. Moreover, it became clear that cells could only come from other, preexisting cells. This concept (the cell theory) dealt a severe blow to vitalism, the prevailing dogma of the time, which held that no single part of an organism was alive.

Despite their incredible diversity of size and appearance, cells of whatever biological source are remarkably alike in basic organizational and molecular properties and in the metabolic pathways by which they survive, carry out their functions, grow, and divide. For example, the chemistry of all life-forms is based on amino acids, sugars, fatty acids, and nucleotides, and these components are all synthesized via similar biochemical pathways. Moreover, the way in which genetic information is stored, replicated, repaired, and expressed is essentially the same in all living organisms. This fundamental similarity is due to the fact that all cells that we know are descendants of a common evolutionary ancestor.

Origin of the Cell. Although there is little debate that the contemporary biological world traces its origin to a single cellular progenitor, how the ancestral cell originated, some 3.5 to 3.8 billion years ago, remains an issue of intense speculation (see, for example, recent essays by Ingber, 2000; Cavalier-Smith, 2001). The concept of structural hierarchies is an attractive one in considering how the transition from inanimate atoms and molecules to the first cells might have occurred. Self-assembly of matter in a hierarchical fashion (with the accumulation of stable subassemblies) would have greatly accelerated cellular evolution. By catalyzing essentially solid-state reactions that would otherwise be limited by diffusion, clay minerals may have been instrumental in helping to promote the synthesis of the first simple organic molecules (including amino acids and nucleotides) and the corresponding polymers (proteins and nucleic acids, respectively).

Once formed, oligopeptides and oligonucleotides spontaneously fold into three-dimensional structures that not only stabilize these molecules but also endow them with characteristic chemical reactivities and the potential to interact with other macromolecules (both nucleic acid and protein). The emergence of an "RNA world," in which RNA molecules not only played an informational role but also served a catalytic function, is thought by some to have been an important phase in the pathway to cellular life. The ability of RNA to support its self-replication, in concert with a catalytic ("ribozyme") capacity, would have allowed the accumulation of novel proteins displaying an increasingly wide range of potential functions. By virtue of their increased catalytic accuracy and efficiency, proteins could progressively modify and/or take over reactions initially carried out by RNA. DNA would have made its appearance at a later stage, after the enzymes required for its synthesis and that of its precursors emerged. As DNA assumed a primary role in information storage and as proteins became the major biological catalysts, the stage was set for the modern DNA-RNA-protein world, in which RNA acts principally as an intermediary between the other two classes of macromolecule.

Because a defining feature of any cell is its bounding lipid membrane, the emergence of this membrane must have been a central event in cell evolution. Initially, lipids might have been concentrated from the surrounding environment through association with hydrophobic (water-insoluble) patches in protein-nucleic acid complexes. In this way, primitive membrane scaffolds would have evolved. Simple phospholipids are amphipathic, one part of their structure being hydrophobic and another hydrophilic. The characteristics allow phospholipids to associate spontaneously into bilayers, such that their hydrophobic portions are clustered together as much as possible, away from water, while their hydrophilic parts are in contact with the surrounding aqueous medium. Recruitment of amphipathic proteins into the membrane scaffold would have generated a lipoprotein network, with a consequent increase in biochemical versatility, leading to the eventual development of cross-membrane transport systems for both small and large molecules.

Although it is often assumed that a spontaneous assembly of phospholipid molecules eventually served to enclose a self-replicating mixture of RNA and other molecules, the primordial arrangement may actually have been an "inside-out" cell (termed *obcell* by Cavalier-Smith), with self-replicating and catalytic nucleoprotein complexes arrayed on the outside of a membrane globule, rather than inside. In-folding of a single obcell or fusion of two cup-shaped obcells would generate a protocell in which membrane-imbedded nucleoprotein complexes now became positioned inside the cell, as well as on the exterior surface.

Whatever the membrane topology of the precursors to the first cell, enclosure of the processes of genetic information transfer and metabolism would have had a profoundly autocatalytic effect, both by increasing the potential for intermolecular interactions within the cell and by concentrating the substrates and products of various reactions. This condition would have been a prerequisite for evolution of the complex intermediary metabolism carried out by modern cells. Because lipid membranes are able to grow (by incorporation of additional phospholipids and proteins) and divide (as a result of their susceptibility to shear forces), they display a limited form of heredity. The emergence of a lipid membrane not only allowed the first cells to come into existence, but also made subsequent cell division, and therefore reproduction at a cellular level, possible.

From Progenote to Cenancestor. Nowadays, there is a tight coupling between genotype (the genetic information carried by the genome of an organism) and phenotype (the expression of that genetic information as recognizable traits). However, during the course of cell evolution, there was undoubtedly a phase when genotype and phenotype were not so precisely linked, because genetic information was more primitively organized and inefficiently (even inaccurately) expressed. The term *progenote* has been used to designate a quasi-cellular entity (or collection of such entities) representative of this error-prone phase in the evolution of genetic information transfer (see Woese, 1998). The progenote phase of cell evolution may have given rise to a number of distinct lineages, but only one of these survived as the common ancestor of contemporary organisms.

By comparing similarities and differences in cellular organization and metabolism in extant organisms, we infer that the last common ancestor of all life (variously termed "last universal common ancestor [LUCA]" or "ce-

nancestor") was a fairly advanced cell, already possessing much of the biochemistry and molecular biology we recognize today (see Doolittle, 1996). The cenancestor would have had a DNA genome, ribosomes, and a genetic code very similar to the standard one used by most organisms; it would have had the capacity to replicate, repair, and transcribe DNA, and to carry out template-directed protein synthesis (translation); and it would have had a number of other features characteristic of modern organisms, such as molecular chaperones (so-called heat-shock proteins). In short, the cenancestor was likely a primitive bacterial-type cell.

Two Cell Types and Three Domains of Life. Two fundamentally distinct types of organism, prokaryotes and eukaryotes, have been distinguished on the basis of both molecular analysis and cellular organization. Prokaryotes are single-celled organisms, although some species do associate as multicellular assemblages. Prokaryotic cells are usually a few (1–10) micrometers long, generally spherical or rod-shaped, and often have a tough peptidoglycan cell wall. The interior of a prokaryotic cell is typically featureless, lacking obvious subcellular structure, although its DNA may be visible under the electron microscope as a condensed structure (nucleoid) confined to a discrete portion of the cell interior. In contrast to the situation in eukaryotic cells, where the genetic material is segregated from the rest of the cell in a membrane-bound nucleus, prokaryotic cells lack such an organelle.

Prokaryotes, which divide in two by simple binary fission, are further distinguished by their rapid growth and ability to adapt quickly to changing environmental circumstances. Because the DNA in a prokaryotic cell is not separated from the rest of the cytoplasm, transcription and translation are able to be physically and temporally coupled, with translation often beginning before transcription of the messenger RNA template is complete. Such coupling underlies the metabolic versatility and adaptability typical of prokaryotes.

In 1977, Woese and Fox presented evidence for two primary divisions within the prokaryotes: eubacteria, or "true" bacteria (domain Bacteria) and archaebacteria (domain Archaea). Eubacteria comprise most of the typical bacteria with which biologists are generally familiar; archaebacteria, on the other hand, inhabit unusual (often extreme) environmental niches, such as concentrated salt brines, anaerobic muds, and acidic hot springs. The two prokaryotic domains were considered by Woese and Fox to be as distinct from one another as either was from eukaryotes (domain Eucarya). The division of extant organisms into three primary domains (Woese, Kandler, and Wheelis, 1990) is supported by various kinds of evidence, although a surprisingly large number of genes characteristic of the Archaea have been discovered recently in completely sequenced eubacterial genomes, and vice versa. Lateral gene transfer (LGT) during organismal evolution has been invoked to account for such mixed gene patterns (Doolittle, 1999).

Much debate continues around the issue of how the three primary domains are related to one another in evolution. It is now generally accepted that the domains Archaea and Eucarya are phylogenetic sister groups, rather than the two prokaryotic domains, Bacteria and Archaea. For example, in Archaea, information transfer systems (e.g., transcription) more closely resemble the same systems in Eucarya rather than in Bacteria, although the conclusions that can be drawn from such comparisons are somewhat tempered by the genome scrambling effect of LGT. Nevertheless, it appears that the domain Bacteria emerged first in evolution, and that Archaea and Eucarya diverged subsequently.

Structure and Origin of Eukaryotic Cells. Compared to prokaryotic cells, eukaryotic cells are generally much larger (5–100 μm) and are characterized by an abundance of subcellular, membrane-bound structures termed organelles. [*See* Cellular Organelles.] A basic feature of eukaryotic cells is the presence of a nucleus, which houses the eukaryotic genome and is the site of transcription. Ribosomal RNA (rRNA) synthesis and ribosome assembly take place on a discrete subnuclear structure called the nucleolus, made up of multiple copies of rRNA genes, usually organized as tandem arrays.

Other eukaryotic organelles include the endomembrane system, comprising the endoplasmic reticulum, the Golgi body, various types of intracellular transport vesicles (including lysosomes), and the plasma membrane. The endomembrane system functions in both inter- and intracellular trafficking of proteins and other material. Organelles such as the mitochondrion, the chloroplast and the peroxisome function in energy production and oxygen metabolism. The first two organelles are of special interest here because they are the only ones known to contain genetic information and a translation system, both of which testify to their endosymbiotic origin from eubacterial progenitors (see below). Other, specialized organelles have a restricted distribution within the eukaryotic lineage: vacuoles in plant cells, hydrogen-producing hydrogenosomes (thought to be derived mitochondria) in many anaerobic eukaryotes, and glycosomes in kinetoplastid protozoa.

Accumulating evidence suggests that the eukaryotic cell is an evolutionary chimera, a combination of more than one genome. The eukaryotic genome contains basically archaeal informational genes (genes involved in genetic information transfer and processing) and eubacterial operational genes (genes mediating intermediary metabolism). How this genetic chimerism came about is not clear, but various schemes have been proposed that involve fusion between an archaeal cell and a eubacter-

ial cell to produce the original proto-eukaryotic cell. Whether the eubacterial cell involved in this fusion event was also the direct progenitor of the mitochondrion (in which case the origin of the mitochondrion would have been coincident with the origin of the eukaryotic cell as a whole, see below), or whether the mitochondrion originated in a later, separate, endosymbiotic event, is also a matter of current debate (see Gray et al., 1999). Although the eubacterial ancestry of mitochondria seems firmly established (Gray, 1992), it increasingly appears that the ancestors of amitochondriate eukaryotes (eukaryotes that do not have mitochondria) did, in fact, possess these organelles: that is, that the amitochondriate condition does not reflect an early divergence of the ancestors of these organisms away from the main line of eukaryotic evolution, before the acquisition of mitochondria, but rather is a derived trait, reflecting secondary loss of mitochondria.

Endosymbiotic Origin of Mitochondria and Chloroplasts. The presence of a genome in mitochondria and chloroplasts, and the sequence information these genomes provide, offers compelling evidence that mitochondria and chloroplasts are the direct descendants of eubacteria that formed a symbiotic relationship with the rest of the eukaryotic cell, either (in the case of the mitochondrion) at the earliest stages of eukaryotic cell evolution or at a later time. [*See* Endosymbiont Theory.] The closest eubacterial relatives of mitochondria are found within the α-Proteobacteria (nonsulfur purple bacteria), particularly a group that contains such obligate intracellular parasites as *Rickettsia prowazekii*, the causative agent of epidemic louse-borne typhus.

Chloroplasts, on the other hand, are specifically affiliated with the Cyanobacteria (formerly known as blue-green algae), from which they derived their photosynthetic apparatus. The genomes of chloroplasts and particularly mitochondria are highly reduced compared to the genomes that their eubacterial progenitors must have possessed. Much of this reduction reflects evolutionary transfer of originally organellar genes to the nucleus. As a result of this evolutionary streamlining, only a few residual organellar genes remain—which are, nevertheless, essential for the formation of a functional organelle. Most of proteins that make up and function in the mitochondrion and chloroplast are now encoded in and expressed from the nuclear genome.

Endocytosis, Cytokinesis, Mitosis, and Meiosis: The Essence of Being Eukaryotic. Eukaryotic cells have the ability to bring material into their interior (endocytosis) and to export material to the outside (exocytosis). Endocytosis occurs when the cell membrane folds in on itself, eventually surrounding the material to be imported. Pinching off of the infolded membrane internalizes vesicles whose contents can then be released and utilized by the cell. Phagocytosis, a special form of endocytosis, allows the eukaryotic cell to take in very large particles or even entire cells (e.g., bacteria).

The emergence of a phagocytic eukaryotic host ("eating cell") was a necessary precondition for endosymbiosis: it permitted the uptake of intact bacteria, certain of which (according to classical endosymbiont theory) were not digested, but instead entered into an endosymbiotic relationship with the host, thereby becoming the progenitors of mitochondria and chloroplasts. Because phagocytosis can only occur in the absence of a rigid cell wall, and because the latter structure probably originated in the cenancestor, loss of the cell wall must have been another critical step in the evolution of the primordial eukaryotic cell.

An additional feature that distinguishes the cells of prokaryotes from those of eukaryotes is the presence in eukaryotes of a cytoskeleton composed of actin filaments and microtubules. These proteins are highly conserved within the eukaryotic domain, indicating that the cytoskeleton must have developed very early in eukaryotic cell evolution. Both proteins underlie eukaryotic cell movement: actin filaments enable individual eukaryotic cells to move about, and microtubules are the main structural and functional elements in cilia and flagella, whose whiplike action propels eukaryotic cells through the medium

Actin filaments and microtubules play an essential role in the internal cytoplasmic movements that occur in eukaryotic cells. Microtubules associate to form the mitotic spindle, part of the machinery that ensures that the replicated genetic information is apportioned equally between the two daughter cells during eukaryotic cell division. In the absence of microtubules, a eukaryotic cell could not carry out either mitosis (normal somatic cell division) or meiosis (the reductive cell division that leads to production of haploid cells): in other words, it could not reproduce.

Exons and Introns in Eukaryotic Gene Expression and Gene Evolution. A final unique feature of eukaryotic cells is the presence of intervening sequences of DNA (introns) that do not code for proteins and that are imbedded in and effectively disrupt the reading frame of protein-coding genes. During transcription, these introns are transcribed together with their flanking exons, which represent the informational portions of an intron-containing gene. The resulting pre-mRNA is then assembled into a specialized RNA processing structure, the spliceosome. This ribonucleoprotein complex precisely cuts out the intron sequences and splices together the flanking exons, thereby generating a translatable open reading frame. This process, known as splicing, must be highly accurate because it has to join exons together at exactly the right position in the nucleotide sequence and in the right order.

In complex multicellular eukaryotes, alternative splicing often occurs; in this case, certain exons may be skipped over during the splicing process, generating mature proteins with different sequences. Alternative splicing, which often occurs in a tissue-specific fashion, has the potential to generate a number of structurally and functionally distinct proteins from a single gene.

Since the discovery of spliceosomal introns, there has been much debate about their origin and their evolutionary role in creating new protein combinations through "exon shuffling." The "introns early" hypothesis proposes that spliceosomal-type introns were initially present in all three organismal domains and that, through recombination between introns, they played a key role in protein evolution in the earliest stages of cellular (or even precellular) life. The competing, "introns late" hypothesis argues that spliceosomal introns arose specifically within the eukaryotic lineage and rapidly invaded (perhaps as some sort of transposable element) genes that were initially devoid of introns. Although the accumulating evidence seems to favor the "introns late" hypothesis, it is clear that in certain classes of genes in complex eukaryotes, exon shuffling has indeed played a role in the creation of novel gene combinations that specify proteins having new functions.

[*See also* Cortical Inheritance; Multicellularity and Specialization; Prokaryotes and Eukaryotes.]

BIBLIOGRAPHY

Alberts, B., D. Bray, J. Lewis, M. Raff, K. Roberts, and J. D. Watson. *Molecular Biology of the Cell.* 3rd ed. 1994. Chapter 1 of this general undergraduate textbook provides a concise and lucid description of cells and cell evolution. Material in this text may be accessed online by links to abstracts available through the PubMed facility of the National Center for Biotechnology Information (NCBI; URL http://www.ncbi.nlm.nih.gov/entrez/query.fcgi?db=PubMed).

Cavalier-Smith, T. "Obcells as Proto-Organisms: Membrane Heredity, Lithophosphorylation, and the Origins of the Genetic Code, the First Cells, and Photosynthesis." *Journal of Molecular Evolution* 53 (2001): 555–595. An essay that attempts to provide a unified picture of the origin of living organisms in their genetic, bioenergetic, and structural aspects.

de Duve, C. "The Birth of Complex Cells." *Scientific American* 274 (1996): 38–45. An eminently readable and clearly illustrated discussion of the basic elements of classical endosymbiont theory.

Doolittle, W. F. "Some Aspects of the Biology of Cells and Their Possible Evolutionary Significance." In *Evolution of Microbial Life*, edited by D. M. Roberts, P. Sharp, G. Alderson, and M. Collins, pp. 1–21. Society for General Microbiology, 1996. A comprehensive look at cell origins in the context of genomic chimerism.

Doolittle, W. F. "Phylogenetic Classification and the Universal Tree." *Science* 284 (1999): 2124–2129. A provocative discussion of the role of lateral gene transfer in genome and organismal evolution and its implications for attempts to infer a tree of life.

Gray, M. W. "The Endosymbiont Hypothesis Revisited." *International Review of Cytology* 141 (1992): 233–357. A comprehensive compilation and critical review of molecular data relating to the endosymbiont theory.

Gray, M. W., G. Burger, and B. F. Lang. "Mitochondrial Evolution." *Science* 283 (1999): 1476–1481. This review comments on recent challenges to the traditional endosymbiont model of mitochondrial origin, summarizing alternative scenarios of eukaryotic cell evolution.

Ingber, D. E. "The Origin of Cellular Life." *BioEssays* 22 (2000): 1160–1170. An essay that presents a scenario for the origin of life based on analysis of biological architecture and mechanical design at the microstructural level.

Woese, C. "The Universal Ancestor." *Proceedings of the National Academy of Sciences USA* 95 (1998): 6854–6859. The author presents a "genetic annealing" model for the universal ancestor of all extant life, including a discussion of the progenote concept.

Woese, C. R., and G. E. Fox. "Phylogenetic Structure of the Prokaryotic Domain: The Primary Kingdoms." *Proceedings of the National Academy of Sciences USA* 74 (1977): 5088–5090. An rRNA-based phylogenetic analysis that presents the concept of three primary lines of biological descent. The authors introduce the archaebacteria as a second prokaryotic lineage, fundamentally distinct from both eubacteria and eukaryotes.

Woese C. R., O. Kandler, and M. L. Wheelis. "Towards a Natural System of Organisms: Proposal for the Domains Archaea, Bacteria, and Eucarya." *Proceedings of the National Academy of Sciences USA* 87 (1990): 4576–4579. The authors propose that a formal system of organisms be established in which a new taxon called a "domain" exists above the level of kingdom. Life on earth is seen as comprising three domains, each containing two or more kingdoms.

— MICHAEL W. GRAY

CELL LINEAGE

Fertilized embryos in some animal groups, particularly nematodes, rotifers, and ascidians, undergo a pattern of cleavage that is determinate, with predictable cleavage numbers and planes. Each cleavage generates two daughter blastomeres, and determinacy of cleavage means that an individual blastomere has a predictable, defined pattern of mitotic descent, or cell lineage. In other groups the early lineage of a cell may not be known, but its later mitotic activity can become determinate, and so it is also legitimate to refer to a cell lineage. Such is the case for lineages in insect embryos, which lack conventional early cleavages, but which give rise to stem cells which generate neural progeny by determinate cell lineages. Likewise, individual structures in a postembryonic organism, such as the organules of the insect epidermis, may have a defined cell lineage during their late development, and to judge from the fossil record, some of these have been conserved. Postmitotic cells that arise from a rigid pattern of cell lineage occupy the position within the embryo conferred upon them by that lineage—through the number and geometry of the preceding cleavages—and can also assume a predictable fate.

The relationship between cell lineage and cell fate can, in principle, be one of two types: typological (also known as histotypical) or topographical. In the former, cell lineage may serve to partition maternally derived determinants in the embryo to particular blastomeres, so that the lineage of a blastomere *per se* can determine the fate of its progeny. Alternatively, a topographical lineage may simply organize morphological space, independent of determining cell fate, allocating cells to a particular place in the embryo, with determination of each cell's fate then occurring via interactions with its neighbors. In both cases, cell lineage has a causative role in the developmental fate of cells. Embryos which suffer a change in cell division generate a change in cell fate. These are two extremes: real embryos are a pastiche of mixed lineages, the exact blend of the two mechanisms possibly reflecting an evolutionary trade-off between maximizing the spatial distribution of cytoplasmic determinants and minimizing the need for differentially committed embryonic cells to migrate. In either case, changes in lineage are one pathway underlying evolutionary change, but the relationship between lineage and evolutionary progression is not at all straightforward, any more than it is for the underlying relationship between lineage and developmental fate and morphogenesis.

The term cell lineage, or simply lineage, is often used loosely, to refer to cells that adopt a particular fate (for example, "neural crest lineage"), or that express a particular phentoypic character (for example, "Na⁺ conductance channel"), but whose pattern of mitotic descent may not be known. In this sense, the germ line is an immortal lineage that is transferred without change in fate from one generation to the next. Cell lineage should not be confused with cell fate, however, which is usually independently controlled. To claim that a pattern of embryonic cell lineage is truly fixed requires sound methological validation, something not always available even in canonical species. Descriptive reports of cell lineage tend to homogenize the patterns of cleavage for an embryo, insofar as they derive from observations on many embryos. True four-dimensional lineage-tracing in individual embryos of the nematode *Caenorhabditis* now reports variations in a lineage previously thought wholly determinate. Such variations in cell lineage, or in their correlation with cell fate, provide an important source of variation upon which evolutionary selection may act.

The relationship between cell lineage and developmental fate has been evaluated in various groups, especially nematodes, but also in insects, ascidians, echinoderms, and vertebrates; valuable work also comes from gastropod molluscs and hirudinean annelids. Nematodes and molluscs, in particular, allow studies of the genetic programming of precision in cell lineages and of the evolution of its flexibility. Stochastic specification of cell fate may evolve to generate a fixed cell lineage; specification by cell interactions may evolve into autonomous specification. Our perspective on the influence that cell lineage exerts as a developmental force in evolutionary change is to some extent shaped by whether one looks at rules or exceptions in these cases.

Homology in fate can be shared by cells not necessarily enjoying a conserved lineage in related groups. In malacostracan Crustacea, for example, some cell lineages, such as those that generate four pairs of mesoteloblasts, are highly conserved; others, such as the primary ring of 19 ectoteloblasts, have been subject to phylogenetic change. The early differentiation of neurons in crustacean and insect embryos shows clear homologies in cell fate, yet the early cleavage of the embryos differs totally, and the intermediate stem cells producing these neurons, the neuroblasts, are not clearly homologues. On the other hand, the fate of an identified cell lineage need not be invariant even in closely related forms. Changes in life-cycle, such as those accompanying direct-development in sea urchins from the ancestral life-cycle with a pluteus larva, indicate that so-called gene expression territories, which partition the cells of a sea urchin embryo into the functional units of a pluteus larva, arise in quite different lineages in direct- and indirect-developing urchins. Thus cell lineage is not congruent with cell fate. Developmental modules, a characteristic feature of embryonic development and essential to its evolution, have evolved in two ways in direct developing sea urchins, either by changes in the expression of regulatory genes within individual gene expression territories, or by the elimination of entire modules.

Fixed cell lineages have arisen more than once yet have been conserved. Comparative cell lineage studies in nematodes reveal that a gradual shift has occurred from a nondetermined, more variable pattern of embryonic development to a faster, more highly determined one. In the deuterostomes, several features of cleavage in echinoderms are probably over 550 million years old. In an extreme case of conservatism among their sister group, the protostomes, embryonic cleavage patterns in different groups with trochophore larvae, which undergo spiral cleavage, are remarkably conserved during evolutionary change. So, too, is cell fate, which is conserved for identified blastomere quadrants A,B,C, and D not only in annelids, nemerteans, and molluscs, but also in polyclad turbellarian flatworms.

Cell lineages can also change, sometimes rapidly. In an extreme case, the spiral cleavage inherited for so long by ancestral molluscs, is completely abandoned in cephalopods. Even among more conservative spiralians, there is variation in the relative contributions of various micromere quartets to the formation of particular larval structures in different embryos. Mollusc embryos exhibit differences in their early cleavages that are not found in other groups. In a more detailed, nonspiralian case, comparative studies on vulva development in nematodes reveal evolutionary modification at different levels: the numbers of vulva precursor cells differ among species; the cell lineage of two of three particular lineages generated by these precursor cells in *Caenorhabditis* are altered, whereas a third lineage is conserved; variation can occur in this third lineage with respect to the number of precursor cells adopting that fate and the number of progeny so formed. Such evolutionary changes in developmental mechanisms specifying cell fates can occur without obvious morphological change, and correlate instead with evolution of cell fate: whether to divide, migrate, differentiate, or die.

Some evolutionary changes in cell lineage are the product of simple timing. Within gastropod molluscs, a heterochronic shift in the formation of the mesentoblast corresponds with the successive emergence of different groups from ancestral forms. In deuterostomes, early cleavage patterns in urochordate, cephalochordate, and chordate embryos closely conserve early cell lineages. Fate maps at the 8-cell stage are also strikingly similar. Cell lineage reveals an evolutionary shift, however, in the number of cell cycles required to attain equivalent morphogenetic stages. For example ascidian gastrulation is attained after 6–7 cleavages, while amphioxus takes 9–10 cleavages, and amphibian embryos take 12–15 cleavages. A distant urochordate relative, the larvacean embryo, gastrulates yet one cleavage earlier than ascidians. Thus the number of blastomeres sharing an early lineage and assuming a similar fate progressively increases in the chordate evolutionary series, with concomitant increase in cell number.

[*See also C. elegans*; Cell Evolution; Regulatory Genes.]

BIBLIOGRAPHY

Boyer, B. C., J. J. Henry, and M. Q. Martindale. "The Cell Lineage of a Polyclad Turbellarian Embryo Reveals Close Similarity to Coelomate Spiralians." *Developmental Biology* 204 (1998): 111–123. Contemporary methods to trace cell lineage in a neglected embryo reveal a fate map similar to those of annelid and mollusc embryos, thus confirming that polyclad turbellarians are a basal group of spiralians.

Dohle, W., and G. Scholtz. "How Far Does Cell Lineage Influence Cell Fate Specification in Crustacean Embryos?" *Seminars in Cell Develop. Biology* 8 (1997): 379–390. A summary of the comparative early cleavage patterns in Crustacea and the extent of their conservation in different groups.

Schnabel, R., H. Hutter, D. Moerman, and H. Schnabel. "Assessing Normal Embryogenesis in *Caenorhabditis elegans* Using a 4D Microscope: Variability of Development and Regional Specification." *Developmental Biology* 184 (1997): 234–265. The first complete attempt at 4D microscopy, reveals hitherto unsuspected variations in cell lineage in *Caenorhabditis*.

Sommer, R. J. "Evolutionary Changes of Developmental Mechanisms in the Absence of Cell Lineage Alterations During Vulva Formation in the Diplogastridae (Nematoda)." *Development* 124 (1997): 243–251. A cell lineage analysis of vulva formation in seven species of nematodes.

Stent, G. S. "Developmental Cell Lineage." *International Journal of Developmental Biology* 42 (1998): 237–241. A short, general review of cell lineage and its causative role in development.

Sternberg, P. W., and M. A. Felix. "Evolution of Cell Lineage." *Current Opinion in Genetics and Development* 7 (1997): 543–550. A good overview and useful current review of this topic.

van den Biggelaar, J. A. M., and G. Haszprunar. "Cleavage Patterns and Mesentoblast Formation in the Gastropoda: An Evolutionary Perspective." *Evolution* 50 (1996): 1520–1540. Documents that cleavage patterns predict evolutionary rank in gastropod embryos.

Wray, G. A. "The evolution of cell lineage in echinoderms." *American Zoologist* 34 (1994): 353–363. A summary of evolutionary changes in cell lineage in echinoderms, indicating the role played by adaptation within the context of developmental constraints.

Wray, G. A., and R. A. Raff 1989. "Evolutionary Modification of Cell Lineage in the Direct-Developing Sea Urchin *Heliocidaris erythrogramma*." *Developmental Biology* 132 (1989): 458–470. Documents the radical changes that have occurred in the embryonic development of a direct-developing sea urchin.

— IAN A. MEINERTZHAGEN

CELL-TYPE NUMBER AND COMPLEXITY

Multicellular organisms contain numbers of differentiated cell types, distinct in morphology and function, that form the basic building blocks of the body. Different types of cells were originally recognized and classed on the basis of their morphological distinctiveness under the microscope; such types are termed cell morphotypes. Most cell types show a range of morphological variation, and when the range is great, the limits of the type become somewhat subjective. Furthermore, cells with similar morphologies may exhibit functional differences, especially during development, when some cells can induce particular differentiations in adjoining tissues, whereas other cells of similar morphotype cannot. Such observations indicate that different genes have been expressed in cells that appear to be similar but that must differ on the molecular level. Nevertheless, many cell types are quite distinctive and discrete, and there is an obvious relationship between the number of cell morphotypes and the complexity of an organism's body plan.

A useful measure of the complexity of an object is given by its minimal description. Under this definition, complexity increases with the number of parts of an object, and with disorder among the parts, for increases in these features require longer descriptions. Multicellular organisms are organized as hierarchies, with cells forming tissues, tissues forming organs, organs forming organ systems, and all forming the individual organism. The entities on each of the hierachical levels are distinctive parts of the organism; simple organisms do not have as many parts, or even levels, as more complex ones. If the cell is accepted as the basic constructional unit, it can be argued that differences between cells that do not affect cell morphology are not consequential to measuring morphological complexity. A thorough assessment of an organism's anatomical complexity should take account of the number of cell morphotypes and also the pattern of their associations and positions in tissues, organs, and so forth, but descriptions at such detail are not available. Although cell morphotype number estimates the number of body parts at the cellular level, it does not reflect the number of parts at higher hierarchical levels, so that the difference in complexity suggested by counting cell morphotypes must be a minimal estimate of the actual difference. Thus, cell type numbers generally provide an ordinal index but not a scalar estimate of the actual magnitudes of morphological complexities. The approximate number of cell morphotypes is known for representatives of many major animal (metazoan) taxa and can be used as an index of their relative morphological complexities. The simplest free-living metazoans are members of the phylum Placozoa and have only four somatic (nonreproductive) cell types. Humans must be among the more complex metazoans, and they have about 210 somatic cell types other than neurons. Nerve cell types are so difficult to deal with that they are lumped in many cell-type estimates; in humans, there may be as many types of neurons as there are other types of cells.

Production of Cell Types. The ability to produce differentiated cell types lies at the root of the multicellular condition. Starting from a single egg (ovum), cells are proliferated, at first by cleavage, and later by cell growth and division; these processes are usually set in motion at fertilization. The oocyte is not isotropic, that is, its molecular contents vary from place to place, so that as cleavage proceeds, the new cells (blastomeres) contain different maternal gene products, some of which are factors that regulate gene expression in the nuclei of the blastomeres. Different portions of the nuclear genes are thus expressed in different blastomeres, producing different cell phenotypes. Furthermore, some of these different cell types influence the identity of still other cells by communicating via molecules (ligands) that activate signaling pathways in the receptor cells to mediate the repression or expression of genes in their nuclei. Thus, there is a relay of regulatory signals among cells that produces differential patterns of gene expression. Many of the resulting cell phenotypic differences can be observed in cell morphologies and serve to identify morphotypes. The genes in a given cell lineage are expressed in an orderly fashion that produces a series of descendant cells beginning with the founding blastomere, a generalized cell type that has more or less broad potentials, and proceeding to increasingly specific types among the proliferating cell offspring, culminating in the terminally differentiated cell morphotypes. The various cell types are commonly produced in places and numbers appropriate to the body plans of the organisms, although, in some cases, cells migrate from their points of origin to assemble in appropriate tissues, particularly in vertebrates.

Gene Expression. Because each cell type requires the expression of a different portion of the genome, one would expect more complex organisms, such as advanced metazoans, to have more genes than simpler forms, in order to encode the many cell phenotypes that they require. In a general way, this expectation is met. The yeast *Saccharomyces* (Figure 1) is single-celled and belongs to the Fungii, which is related to the Metazoa; it has been sequenced and contains an estimated 6,340 genes. Invertebrate metazoans have up to 100 cell morphotypes, and those that have been sequenced contain between 13,600 and 27,350 genes. The human genome sequence revals that humans have about 35,000 genes. Thus, there is a rough correlation between complexity as measured by the number of different kinds of cells and the number of genes. However, within invertebrate metazoans, there is little correlation; for example, simple worms (nematodes) have significantly more genes than complex flies with four or five times as many cell morphotypes. Evidently flies organize the genetic specification of their many cell types differently than nematodes do of their few.

In animals, gene expression is regulated at many levels, providing the opportunity for producing any number of expression events from a given gene sequence. For example, transcription of a gene is inititiated at a promoter complex, a segment of DNA that lies just upstream from the transcribed gene. The signals to transcribe a gene are carried along molecular pathways that end by the binding of regulatory factors to the promoter complex. If a gene is involved in the development of more than one structure, a separate signal is sent for its expression in each structure—there are multiple binding sites for transcription factors on each promoter complex. Thus, in flies (and arthropods in general), the genes that mediate the development of a cell type that is found in many segments are expressed by different signaling events in each of those segments. This sort of compartmentalization of development permits the segments to become differentiated, for a separate subroutine of gene expression can be evolved for each different sort of segment, producing some segments with legs, or others with wings or antennae, for example. In such a system, the same genes may be used over and over again. Even genes that apparently contribute to terminal cell mor-

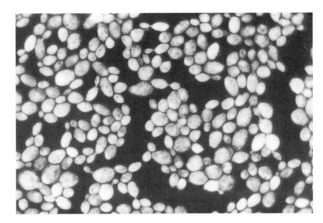

FIGURE 1. The Yeast *Saccharomyces cerevisiae* Has Been Sequenced and Contains an Estimated 6,340 Genes.
© John D. Cunningham/Visuals Unlimited Inc.

photypes may be expressed in more than one structure, or at different times within the same structure. Thus, complex organisms may not utilize many different genes, but may utilize the same genes in different combinations and at different times. The number of gene expression events in a complex organism will exceed those in a simple one, but the number of genes may not.

There are other ways that gene expression events may become multiplied without seeming to increase the number of genes. Gene transcripts may be processed in more than one way, so that different segments are discarded before translation, producing different proteins (thus, the same DNA sequence is used to make more than one gene product). Proteins with identical structures when synthesized may be involved in separate, distinctive functions through interactions with other gene products. Still other opportunities to expand the functional repertoire of the genome are presented at various steps along the signaling pathways. However, little is known about the relative contributions of the various possible modes of encoding complexity within metazoan genomes. When the diverse mechanics underlying gene expression events are better understood, the evolution of morphological complexity will surely be put on a firmer foundation.

Fossil Evidence. Fossil evidence indicates that the body plans of phyla first evolved over a half billion years ago and that many lineages in which those body plans evolved may have separated 100 million years earlier. Some phyla have remained at about their original complexity, others have become simplified, and others have evolved more complex lineages through time (arthropods evidently have more than doubled their cell morphotype numbers, whereas chordates evidently have more than quintupled theirs). Through the hundreds of millions of years of their separate histories, the animal phyla have taken different routes with regard to complexity, reflected not only in cell-type numbers but also in the patterns of expression within their genomes, which are amenable to historical interpretation. Relations between cell-type number, genome expression, and complexity in other multicellular groups, such as Fungii and green plants (Chlorobiota), should harbor important evolutionary information as well.

BIBLIOGRAPHY

Alberts, B., D. Bray, J. Lewis, M. Raff, K. Roberts, and J. Watson. *Molecular Biology of the Cell.* 2d ed. New York, 1989. Thorough account of cells, including discussions of cell types in animals (humans in particular) and plants.

Harrison, F. W., et al., eds. *Microscopic Anatomy of Invertebrates.* 15 vols. New York, 1991–1999. The most up-to-date, comprehensive account of the histology and cell morphotypes in all animal phyla.

Hinegardner, R., and J. Engleberg. "Biological Complexity." *Journal of Theoretical Biology* 104 (1983): 7–20.

McShea, D. W. "Complexity and Evolution: What Everybody Knows." *Biology and Philosophy* 6 (1991): 303–324. Clear discussion of the concept of complexity in biology.

Valentine, J. W. "Two Genomic Paths to the Evolution of Complexity in Body plans." *Paleobiology* 26 (2000): 513–519.

Valentine, J. W., A. G. Collins, and C. P. Meyer. "Morphological Complexity Increase in Metazoans." *Paleobiology* 20 (1994): 131–142. Traces rise in cell-type number through time.

— JAMES W. VALENTINE

CELLULAR ORGANELLES

One of the primary characteristics that distinguish prokaryotic and eukaryotic cells is the presence of organelles, complex, specialized, often compartmentalized structures within the cell. Although this is not a perfect generalization, all eukaryotes contain some complex membrane-bound compartments, although similar structures are very rare in prokaryotes. Defining an organelle is difficult; there is often only a vague distinction between an organelle and a particle (e.g., ribosomes or spliceosomes) or a diffuse complex (e.g., the cytoskeleton). This discussion will take a somewhat classical view of what constitutes an organelle. Similarly, there are a host of highly specialized organelles that are restricted to one or a handful of eukaryotic lineages. Although these may be intriguing and sophisticated structures, only organelles with widespread distribution or organelles arguably derived from them will be considered here.

Endosymbiosis and the Origin of Organelles. A penetrating and repetitive theme in any discussion of organelle evolution is endosymbiosis, or the idea that an organelle arose by the uptake and gradual specialization of one organism by another. An endosymbiotic origin was first postulated for chloroplasts in the late nine-

teenth century but was largely ignored. In the 1960s and 1970s, the discovery of organelle deoxyribonucleic acid (DNA) led to the wide acceptance of an endosymbiotic origin of mitochondria and plastids, which is now overwhelmingly supported by molecular and biochemical evidence. Endosymbiotic origins also have been postulated for many other eukaryotic organelles, and the weight of evidence in each of these cases is variable. For instance, interesting characteristics of peroxisomes and glycosomes might indeed be explained by an endosymbiotic origin. In the case of the flagella and endomembrane system (including the nucleus), no good evidence for an endosymbiotic origin is known, and these organelles appear to have arisen endogenously.

Nucleus and Nucleolus. The nucleus is the defining character of eukaryotes, and is accordingly found in all eukaryotes. It is a membrane system surrounding and containing eukaryotic genomes (with the exception of the mitochondrial and plastid genomes discussed below). The nuclear envelope is made up of two layers of lipid membrane, although topologically this is not two individual membranes but rather one membrane folded back on itself. The envelope is topologically continuous with the endoplasmic reticulum and could be considered an extension of the endomembrane system. The nuclear envelope is studded with nuclear pores, which are complex proteinaceous structures that allow transport of messenger ribonucleic acid (RNA), protein, or protein complexes. The nucleolus, a region of the nucleus that is dedicated to the transcription of ribosomal RNAs, is often a major component of the nucleus, clearly visible in light microscopy. No structure comparable to the nucleus exists in prokaryotes, although DNA is often sequestered to part of the cytoplasm called the nucleoid.

Because division of the nucleus is synonymous with division of the genome, nuclear division has been well characterized in many eukaryotes. In the best studied organisms (such as animals or plants), the nuclear envelope disintegrates prior to mitosis and reforms in the daughter cells (called open mitosis). In many fungi and protists, however, the nuclear envelope persists through mitosis and divides by binary fission (called closed mitosis).

The nucleus likely arose during the massive reorganization that took place at the origin of eukaryotes. As an extension of the endomembrane system, its origin is linked with the new capacity for membrane invagination, which is often considered to be the primary characteristic that drove the origin and specialization of eukaryotes. The origin of the nucleus has been specifically hypothesized to have arisen from the endoplasmic reticulum as a means to protect the genome from the shearing forces of the cytoskeleton. An endosymbiotic origin of the nucleus has also been proposed on several occasions, but these proposals usually rest on the interpretation of the nuclear envelope as a double membrane structure, when it is actually a single membrane folded on itself.

The Endomembrane System: Endoplasmic Reticulum and Golgi. The endomembrane system is a complex of membrane invaginations and vesicles responsible for a variety of functions in eukaryotes, including endocytosis, secretion, intracellular trafficking, ionic control, and protein glycosylation. Golgi dictyosomes, or lamella, are specialized endomembrane structures consisting of a series of stacked, flattened vesicles where sorting and glycosylation take place. All eukaryotes have a complex endomembrane system, although various elements of it are missing in some organisms and specialized elements have been developed in others. The origin of the endomembrane system is therefore thought to be tied to the origin of eukaryotes (see preceding discussion of nuclear origins).

Mitochondria. Mitochondria are double membrane-bound organelles that house oxidative phosphorylation, the primary pathway of energy production in aerobic eukaryotes, in addition to a variety of other metabolic activities such as fatty acid oxidation. Mitochondria consist of an outer membrane closely appressed to a highly invaginated inner membrane, the folds of which are called cristae and typically appear flat, tubular, or discoid (Figure 1). All known mitochondria contain a genome that encodes a small number of proteins and RNAs used in self-expression and oxidative phosphorylation. Mitochondrial genomes display an unusual number of deviations from the universal genetic code. Most mitochondrial proteins are encoded in the nucleus and are targeted posttranslationally to the mitochondrion. Targeting takes place through the recognition of an aminoterminal transit peptide by a protein import system that involves a number of proteins on the inner and outer membranes. Molecular evidence from the mitochondrial genome and from nucleus-encoded, mitochondrion-targeted proteins consistently shows that all mitochondria originated from a single endosymbiotic uptake of an alpha-proteobacterium. This is assumed to have taken place very early in eukaryotic evolution, because mitochondria are common to all eukaryotes except for a small number of protists and fungi that contain hydrogenosomes (see the following discussion) or appear to lack energy-producing organelles altogether. It was once thought that some of these amitochondriate protists may have evolved prior to the origin of mitochondria; however, it has now been shown that even these protists harbor mitochondrial genes in their nuclear genomes. This shows that the origin of mitochondria predates the divergence of all known eukaryotes, and amitochondriate eukaryotes either lost or degenerated their mitochondria.

Hydrogenosomes. Hydrogenosomes are membrane-

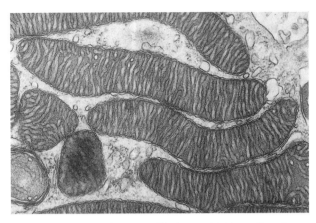

FIGURE 1. Mitochondria Magnified at 6,000 Times Showing the Cristae, or Folds of the Inner Membrane.
© Don W. Fawcett/ Visuals Unlimited, Inc.

bound metabolic compartments found only in parabasalia and certain members of heterolobosea, ciliates, chytrid fungi, and perhaps euglenozoa. Organisms with hydrogenosomes are anaerobic or microaerophilic and lack mitochondria. The hydrogenosome carries out a unique set of anaerobic metabolic reactions, releasing ATP through the conversion of pyruvate or malate to acetate, carbon dioxide, and hydrogen gas. Most hydrogenosomes do not have a genome (with one exception; see below).

Hydrogenosomes have been hypothesized to have originated either by the alteration of an existing organelle (e.g., mitochondria or peroxisomes) or by endosymbiosis with an anaerobic bacterium. In the 1990s it was discovered that nuclear genes encoding hydrogenosome-targeted proteins in the parabasalian protist *Trichomonas* were closely related to mitochondrial homologues, showing that parabasalian hydrogenosomes evolved from mitochondria. There is also evidence that hydrogenosomes in other lineages arose from mitochondria: chytrid hydrogenosomes contain structures resembling mitochondrial cristae, and in one ciliate hydrogenosome there are remnants of a mitochondrial genome.

Plastids and Chloroplasts. These are membrane-bound organelles responsible for photosynthesis in plants and algae. The general term for such organelles is *plastids; chloroplasts* refers to green plastids (as opposed to red, brown, or colorless). In addition to photosynthesis, the biochemical role of plastids may include biosynthesis of amino acids, fatty acids, isoprenoids, and heme. Plastid structure varies between algal groups, but typically there is an outer membrane closely opposed to an inner membrane, both surrounding an internal cavity called the stroma. Within the stroma there is a network of membrane structures called thylakoids, often stacked, flattened vesicles. All known plastids contain a genome that encodes a small number of proteins and RNAs related to gene expression, protein import, and photosynthesis. Most plastid proteins are encoded in the nucleus and targeted to the plastid posttranslationally using a system similar to that of mitochondria.

Plastids ultimately all arose from a single endosymbiotic uptake of a cyanobacterium, but the modern distribution of plastids in eukaryotes can only be explained by two processes, primary endosymbiosis and secondary endosymbiosis. Primary endosymbiotic plastids are those found in lineages that descended directly from the cell that engulfed the cyanobacterium, and their plastids are surrounded by two membranes (these are found in glaucocystophyte, red and green algae, and plants). At various times in evolution, one of these primary plastid-containing algae was itself engulfed by another eukaryote, and the endosymbiont alga was reduced so that typically all that remains is its plastid surrounded by one or two extra membranes. Secondary plastids are found in heterokonts, haptophytes, dinoflagellates, apicomplexa, euglenozoa, cryptomonads, and chlorarachniophytes (in cryptomonads and chlorarachniophytes, the endosymbionts also retain a relic algal nucleus). This process represents a transfer of extremely complex traits and sets of genetic information between distantly related organisms. It also accounts for most of the diversity of primary producers.

Peroxisomes. Peroxisomes are single membrane-bound organelles found in most eukaryotes; they are also called microbodies (a term that includes peroxisomes, glycosomes, glyoxisomes, and sometimes hydrogenosomes). Peroxisomes are responsible for different biochemical pathways in different organisms, including long-chain fatty acid degradation, purine metabolism, and hydrogen peroxide degradation. Peroxisome biogenesis is not well understood: some evidence suggests that peroxisomes may be derived from endoplasmic reticulum or from existing peroxisomes. Plant glyoxisomes are storage organelles that are most likely specialized peroxisomes.

The peroxisome is argued to have originated endogenously, by endosymbiosis of an unknown bacterium, or to have arisen from an existing organelle. No peroxisome contains a genome, but many nucleus-encoded peroxisomal proteins have been characterized. These employ a novel targeting system and show various phylogenetic affinities, suggesting that peroxisomal enzymes are derived from various sources.

Glycosomes. Glycosomes are single membrane-bound organelles found only in kinetoplastid protozoa (e.g., *Trypanosoma*). In these organisms, glycolysis is compartmentalized in the glycosome; in other eukaryotes, glycolysis takes place in the cytoplasm. The origin of glycosomes remains highly uncertain, but it has been suggested that glycosomes may be highly derived peroxisomes.

Flagella, Cilia, and Basal Bodies. Eukaryotic flagella and cilia (essentially the same structure—cilia are shorter than flagella) are long whiplike appendages anchored at one end and used in locomotion and feeding. Prokaryotes also have flagella, but they are structurally distinct and evolutionarily unrelated to the eukaryotic organelle. Prokaryotic flagella are stiff helical tubes that are anchored to a rotary motor and propel the cell much like a ship's propeller. Eukaryotic flagella are primarily composed of microtubules, cytoskeletal tubes assembled from alpha-tubulin and beta-tubulin proteins. The shaft of a flagellum consists of nine microtubule doublets forming a ring around two single microtubules, a configuration commonly called "9 + 2." These microtubules slide against one another using molecular motors to produce a beating or undulation of the shaft, which propels the cell or moves liquid over the surface of the cell.

At the base of each flagellar shaft is a basal body, which is a short "9 + 0" microtubule configuration (nine triplets and no central microtubules). Basal bodies are usually found in pairs or multiples of two. The basal bodies are microtubule organizing centers that direct the polymerization of tubulin monomers in the assembly of flagella and mitotic spindles (where they are often referred to as centrioles). Basal bodies do not divide, but during cell division a pair of basal bodies will move apart and two new basal bodes will assemble, each associated with one of the parent basal bodies. A debate over the presence of DNA in basal bodies took place from the 1960s to the 1980s, but current evidence suggests that they do not contain any genetic information.

Flagellar and basal body structure is remarkably conserved and common to most eukaryotic lineages, although in some eukaryotes, flagella are restricted to a portion of the life cycle or absent altogether. Because flagella are composed of proteins and structures common throughout the cytoskeleton, it is likely that they arose as an extension of the cytoskeleton very early in eukaryotic evolution. It has been proposed that flagella arose from a symbiotic association with spirochete bacteria, but all current evidence suggests otherwise.

[*See also* Cell Evolution.]

BIBLIOGRAPHY

Cavalier-Smith, T. "The Origin of Eukaryote and Archaebacterial Cells." *Annals of the New York Academy of Science* 503 (1987): 17–54. A theoretical scheme for the origin of eukaryotes, including ideas on the origin of organelles endogenously or by endosymbiosis. Of particular note is the discussion on the origin of the nucleus and endomembrane system.

Graham, L. E., and L. W. Wilcox. *Algae.* Upper Saddle River, N.J., 2000. A comprehensive and current text on all algal groups, with considerable attention given to plastids and their evolutionary histories.

Gray, M. W. "Evolution of Organellar Genomes." *Current Opinion in Genetics and Development* 9 (1999): 678–687. This and the following review provide an excellent current synopsis of the endosymbiotic origin and subsequent evolution of mitochondria (and to a lesser extent plastids), focusing on molecular biology and genomics.

Lang, B. F., M. W. Gray, and G. Burger. "Mitochondrial Genome Evolution and the Origin of Eukaryotes." *Annual Reviews in Genetics* 33 (1999): 351–397.

Margulis, L. *Symbiosis in Cell Evolution.* San Francisco, 1981. A detailed argument for the endosymbiotic origin of mitochondria, plastids, and flagella. This is a work of historical and scientific interest, but the reader should be aware that the endosymbiotic origin of flagella is not supported by the evidence and is not widely accepted.

McFadden, G. I., and P. R. Gilson. "Something Borrowed, Something Green: Lateral Transfer of Chloroplasts by Secondary Endosymbiosis." *Trends in Ecology & Evolution.* 10 (1995): 12–17. A well-illustrated review of the primary evidence for secondary endosymbiosis and plastid origins by researchers involved in making many of the most important discoveries.

Rizzotti, M. *Early Evolution: From the Appearance of the First Cell to the First Modern Organisms.* Basel, 2000. A well-illustrated general account of eukaryotic origins and evolution, with a heavy emphasis on organelles. This is a good source of general information on all organelles, but some of the conclusions (e.g., the multiple origins of primary plastids) are somewhat out of date.

Schenk, H. E. A., R. G. Herrmann, K. W. Jeon, N. E. Müller, and W. Schwemmler, eds. *Eukaryotism and Symbiosis: Intertaxonic Combinations versus Symbiotic Adaptation.* Berlin, 1997. The fifth in a series of books that contain several seminal papers in endosymbiosis research. The first four volumes are entitled "Endocytobiology I–IV." Offers a fascinating history of the concept of endosymbiosis in organelle origins.

— Patrick Keeling

CELLULAR SELECTION. *See* Developmental Selection.

CEPHALOCHORDATES. *See* Vertebrates.

CHARACTER DISPLACEMENT

Character displacement is trait evolution resulting from competition between species for resources. Appreciation of the role of character displacement in evolution goes back to Darwin (1859), who was convinced that food supplies in nature were limited and that natural selection should favor divergence between closely related species overlapping in diet and habitat. The idea was simple, but convincing evidence for character displacement has built only in the past two decades. Virtually all the evidence is observational, but experimental study of the process has begun. We cannot yet claim to understand character displacement fully, but doubts about its importance in divergence have been laid to rest.

Observations from Nature. David Lack's (1947) study of the Galapagos ground finches provided the first compelling case. The small and medium ground finches, *Geospiza fuliginosa* and *G. fortis*, have different beak sizes on islands where they occur together ("sympatry"), but each species has an intermediate beak size when it occurs alone ("allopatry"). Brown and Wilson (1956) presented other examples from animals, and called the phenomenon "ecological character displacement." These two studies had a profound influence on the development of ecology and evolutionary biology, and competition between species came to be regarded as a driving force in the divergence of species.

Other examples have since been discovered. Exaggerated divergence in sympatry, whereby species occurring together are more divergent than species occurring alone, remains the most common pattern described from nature. A second pattern has also been attributed to character displacement: species exploiting similar types of resources are sometimes evenly spaced apart in body size, beak size, or other traits involved in resource use: For example, mean canine tooth diameter of Israel's three species of small wild cats (*Felis*) forms a regular series, with a nearly constant separation of 0.6–0.8 mm, when males and females are included separately: 4.7, 5.4, 6.2, 6.9, 7.6, and 8.4 mm (Dayan et al., 1990). Such regularity cannot be accounted for by chance, and divergence stemming from resource competition provides a ready explanation.

The quality of the observational evidence for character displacement has been much debated. The problem is that alternative explanations for patterns are difficult to rule out and, indeed, a few once-promising cases later turned out to be caused by other mechanisms. Exaggerated divergence of beak length of two species of Asian nuthatches *Sitta* is better explained by parallel responses of species to an east-west gradient in natural selection rather than by character displacement (Grant, 1975). Exaggerated divergence in body size of two species of *Hydrobia* mud snails where their distributions overlap (Fenchel, 1975) probably results from a nongenetic developmental shift induced by competition (Grudemo and Johannesson, 1999), not from the evolutionary process of character displacement.

However, such findings are the exceptions. More frequently, tests of alternative hypotheses have found support for character displacement (Schluter, 2000). A few cases, including Lack's seminal example from Darwin's finches (Grant, 1986), have withstood tests against all the most likely alternatives. Further testing continues to be a priority, but the list of examples of character displacement is burgeoning—seventy-five, at last count—with many now having strong support.

Case Study Involving Experiments. Careful observations have brought us close to an understanding of character displacement in nature, but further progress will require experiments, as well. Experiments will allow us to test predictions of theory that observations alone might not be able to address. For example, character displacement predicts that adding a competing species to an environment should alter natural selection pressures on species already present, that divergence should be favored, and that changes in selection should be frequency dependent (that is, they should depend on the phenotype of the competitor). Some progress in testing these predictions has been made in a study of freshwater three-spined sticklebacks.

The sticklebacks exhibit exaggerated divergence in sympatry, suggesting a role for character displacement in their divergence (Schluter and McPhail, 1992). Small coastal lakes in southwestern British Columbia contain one or two species. When two species are present, one of them (the "limnetic") is small, slender, and feeds mainly on zooplankton, in the main water body, whereas the other (the "benthic") is larger, more robust, and feeds on bottom-dwelling invertebrates. Species occurring alone in similar lakes are intermediate. Experiments in ponds have shown that stickleback species compete for resources, and that reduced competition has accompanied divergence in morphology (Pritchard and Schluter, 2001).

Further experiments measured changes in relative fitness of phenotypes in an intermediate species when a second species was added. Each summer-long experiment measured selection but not evolutionary response. In the first experiment, the intermediate species was placed on both sides of two divided ponds, and a limnetic species was then added to one side. As predicted, the individuals within the target species most strongly impacted by the added competitor were those closest to it in morphology and diet, which generated natural selection in favor of the more benthic like phenotypes (Schluter, 1994). The target of the second experiment was again the intermediate species. This time, limnetics were added to one side of each divided pond and benthics were added to the other side. Selection on the target species depended on which competitor was added. In each case, the competitor most reduced the fitness of those phenotypes in the target closest to it in morphology and ecology, confirming that competition favors divergence.

What Next? The strength of evidence for character displacement has grown tremendously over the past decade, and debates over its role in evolution have subsided. The evidence has also brought forth interesting patterns that raise new questions (Schluter, 2000). We don't understand why known cases of character displacement are usually asymmetric (one species changes more than the other) and always involve divergence between species that are closely related. Theory suggests that competition should sometimes lead to convergence (Abrams, 1996), and distantly related species also compete for food. It is also not understood why character displacement is found predominantly in species that prey upon animals (relatively few examples are known from herbivores, detritivores, and plants). Theory has suggested that other antagonistic interactions such as "competition for enemy free space" (Holt, 1977) might also lead to divergence, but there are no convincing cases. Such findings continue to challenge researchers hoping to understand the evolutionary consequences of species interactions.

BIBLIOGRAPHY

Abrams, P. A. "Evolution and the Consequences of Species Introductions and Deletions." *Ecology* 77 (1996): 1321–1328.

Brown, W. L., Jr., and E. O. Wilson. "Character Displacement." *Systematic Zoology* 5 (1956): 49–64.

Darwin, C. R. *On the Origin of Species by Means of Natural Selection.* London, 1859.

Dayan, T., D. Simberloff, E. Tchhernov, and Y. Yom-Tov. "Feline Canines: Community Wide Character Displacement among the Small Cats of Israel." *American Naturalist* 136 (1990): 39–60.

Fenchel, T. "Character Displacement and Co-existence in Mud Snails (Hydrobiidae)." *Oecologia* 20 (1975): 19–32.

Grant, P. R. "The Classic Case of Character Displacement." *Evolutionary Biology* 8 (1975): 237.

Grant, P. R. *Ecology and Evolution of Darwin's Finches.* Princeton, 1986.

Grudemo, J., and K. Johannesson. "Size of Mudsnails, *Hydrobia ulvae* (Pennant) and *H. ventrosa* (Montagu), in Allopatry and Sympatry: Conclusions from Field Distributions and Laboratory Growth Experiments." *Journal of Experimental Marine Biology and Ecology* 239 (1999): 167–181.

Holt, R. D. "Predation, Apparent Competition, and the Structure of Prey Communities." *Theoretical Population Biology* 12 (1977): 197–229.

Lack, D. *Darwin's Finches.* Cambridge, 1947.

Pritchard, J. R., and D. Schluter. "Declining Competition during Character Displacement: Summoning the Ghost of Competition Past." *Evolutionary Ecology Research* 3 (2001): 209–220.

Schluter, D. "Experimental Evidence That Competition Promotes Divergence in Adaptive Radiation." *Science* 266 (1994): 798–801.

Schluter, D. "Ecological Character Displacement in Adaptive Radiation." *American Naturalist,* supplement, 156, (2000): S4–S16.

Schluter, D., and J. D. McPhail. "Ecological Character Displacement and Speciation in Sticklebacks." *American Naturalist* 140 (1992): 85–108.

— DOLPH SCHLUTER

CHLOROPLASTS. *See* Cellular Organelles.

CHORDATES. *See* Vertebrates.

CHROMOSOMES

Chromosomes are the physical structures in which genetic material is stored within a cell and is passed between generations. Although discovered in 1882 by the German embryologist Walther Fleming, it was not until 1903 that Walter Sutton first proposed that these thread-like structures, visible in dividing cells, carry Gregor Mendel's factors of inheritance. It was then another fifty years before the phage experiments of Alfred Hershey and Martha Chase provided conclusive proof that it is the DNA within chromosomes that is critical for inheritance, one year before James Watson and Francis Crick's famous discovery of the double helix in 1953.

From an evolutionary point of view, a chromosome is any molecule of DNA or RNA that is replicated within organisms and passed between generations. However, in terms of structure and genetic composition, they vary enormously, both within and between organisms. Some viruses contain just a single RNA molecule encoding a handful of genes (a term first coined in 1909 for Mendel's factors). Bacteria usually have a single, circular chromosome that is often supercoiled and that has a single origin of replication. Eukaryotic chromosomes are typically much larger but have a much lower gene density. For example, the single chromosome of the bacterium *Escherichia coli* consists of 4.6 million DNA base pairs and contains 4,300 genes, whereas the X chromosome in humans contains over 150 million base pairs but encodes for only about 3,500 genes.

There are many other important differences between eukaryotes and prokaryotes in terms of chromosome organization. DNA in eukaryotic chromosomes is bound by proteins called histones to form structures called nucleosomes, and other types of proteins form structural scaffolds that organize DNA during replication. Eukaryotic chromosomes are separated from the cytoplasm in a structure called the nucleus, and each chromosome has multiple origins of replication. In addition, most chromosomes exist in two copies, which are paired during cell division. Despite the differences in appearance and composition, however, the function of chromosomes in all organisms is the same: the coordinated replication of genetic material and the even segregation of the products of replication into daughter cells (or particles in the case of viruses).

Origin of Chromosomes. The evolutionary transition from a primeval soup of independent and competing replicators to the physically linked, cooperative gene complexes we call chromosomes is one of the key stages in the origins of life. Although we can never be sure of the selective forces that gave rise to chromosomes, the scenario outlined by John Maynard Smith and Eörs Szathmáry in *The Major Transitions in Evolution* (1995) is the most plausible. They suggest that

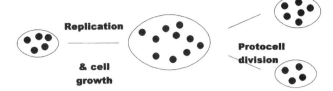

Figure 1. Protocell Structure of the Earliest Replicators. Replicators multiply within protocells, which grow and divide in response. Mutualistic interactions between replicators increase both the rate of replication and the rate of protocell growth.   Drawing by Gilean McVean.

prior to the origin of chromosomes, independent replicators (genes) were grouped within compartments, or protocells (Figure 1). Compartments favor the evolution of mutualistic interactions because they force each replicator's offspring to interact with the offspring of the other replicators within the protocell. Compartments can also restrict the spread of genetic parasites that benefit from the presence of mutualistically interacting replicators but give nothing in return. Parasites have a replication advantage over the mutualists, so they will rapidly come to dominate those cells in which they arise. But they are detrimental to the fitness of the protocell, so that selection between cells will favor those without parasites.

Under what circumstances might we expect variation in replicator content between protocells? One possibility is that bottlenecks occur during cell division, such that only a few replicators enter into each daughter cell. This results in stochastic differences between cells in replicator content, which can be sufficient to limit the spread of parasites, a model known as the stochastic corrector (Figure 2). However, although random segregation at cell division can generate cells that are free from parasites, it also has a high probability of generating cells that lack one or other of the mutualistically interacting replicators. It is this selective force that could have selected for the evolution of chromosomes. Physical linkage between replicators ensures balanced segregation of replicators during cell division, so even if chromosomes take longer to duplicate than unlinked ones (e.g., if they share a single origin of replication), they can be selectively favorable (Figure 3).

A remarkable parallel to this model can be seen in the modern-day segmented viruses, such as those of the influenza and Hanta viruses. These viruses have genomes that are split into several segments. For example, influenza A has eight segments, each coding for one or two genes. Prior to cell lysis, segments are packed into viral particles by a process that is at least partly random. Many particles contain incomplete segments that are the result of deletions, known as defective interfering par-

Parent cell   Offspring   Growth

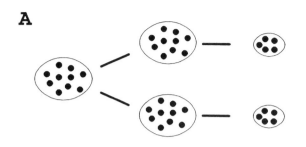

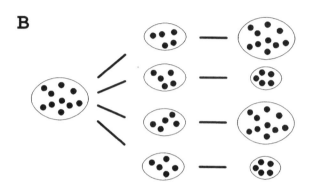

FIGURE 2. The Stochastic Corrector Model.
There are three types of replicator: blue and green, which interact mutualistically, and a red parasite, which replicates faster than either mutualist but reduces the rate of protocell growth. If large numbers of replicators are passed to each daughter cell at division (A), almost all cells will contain parasites and the growth rate, or fitness, of the offspring will be low. If only a few replicators pass to each daughter cell (B), some will be free from parasites and will consequently have a high growth rate.   Drawing by Gilean McVean.

ticles, which are parasitic as they reduce the infectivity of viral particles. Stochastic differences in the genetic content of viral particles and the presence of genomic parasites mirror key selective pressures of the stochastic corrector model. Why, then, do the viruses not evolve fused segments? The most likely answer is that fused segments cannot replicate fast enough to compete with unfused ones, despite their other advantages. Experimentally manipulating such viruses provides an excellent opportunity for testing models for the evolution of chromosomes and, more generally, genetic cooperation.

The Structure and Organization of Modern Chromosomes. Modern-day chromosomes are much more than the strings of genes generated by the stochastic corrector model. From an evolutionary viewpoint, perhaps the most important difference is that almost all chromosomes possess the ability to recombine with equivalent, or homologous, chromosomes. Recom-

bination shuffles genes and allows each gene an independent evolutionary fate. This is not true for the few chromosomes, such as the Y chromosome in mammals and the mitochondrial genome, that apparently do not recombine.

A second key development in the evolution of chromosomes was the strict association between chromosome replication and cell division. The vast majority of genetic material within prokaryotes and eukaryotes obeys the paradigm of a single round of replication per cell division, with exceptions being the intracellular genomes of plasmids and organelles, such as mitochondria and chloroplasts. The selective force responsible for the origin of the cell cycle may have been the elimination of parasitic replicators. Concerted replication ensures that no single gene can benefit by replicating faster. Likewise, a strict relationship of one replication to one cell

FIGURE 3. The Evolution of Chromosomes under the Stochastic Corrector Model.
In the absence of linkage between mutualistic replicators (blue and green), stochastic differences between daughter cells can generate offspring that are free from parasites (red), but that will also produce offspring that lack one or other mutualist (A). Linkage between mutualists ensures that both are inherited together (B), so cells containing chromosomes will have a higher average fitness (growth rate).   Drawing by Gilean McVean.

Parent cell   Offspring   Fitness

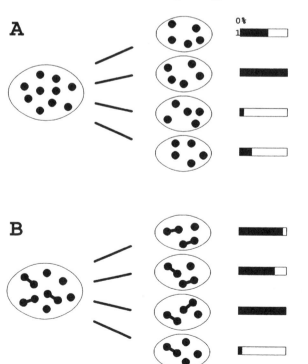

division removes any potential for competition between chromosomes within a cell. For this reason, the persistence of intracellular genomes, which replicate outside the cell cycle, is something of a paradox, because it allows the spread of fast-replicating parasites. For example, in the baker's yeast *Saccharomyces pombe* deletions of the mitochondrial genome that abolish oxidative phosphorylation and stunt growth are known as petite mutations and occur in up to 1 percent of the population each generation. In normal yeast, mitochondria are inherited from both parents; however, some petite mutations, called hypersuppressive, are dominantly inherited. It is thought that the mutant DNA found in mitochondrial structures can replicate faster than the complete molecules, and so displace the wild-type mitochondria, despite their deleterious effect on fitness.

Further developments in chromosome structure and organization evolved in the lineage leading to the eukaryotes. Of particular importance were the separation of genetic material from metabolic activity through the origin of the nucleus, the proliferation of proteins, such as histones, associated with the packaging of DNA in chromosomes, and the evolution of a cellular machinery devoted to ensuring correct segregation of chromosomes during cell division. The origins of syngamy (sexual reproduction through the fusion of gametes), diploidy, and the details of mitotic and meiotic division are beyond the scope of this article, but have been critical to the evolution of chromosome structure and organization in eukaryotes.

The Genetic Composition of Chromosomes. Chromosomes vary enormously in terms of genetic composition. At one extreme are the economical genomes of some bacteriophages such as ΦX174, in which every base either contributes to protein coding (in some cases several, overlapping proteins) or is involved in gene regulation or replication. At the other extreme are the Y chromosomes of mammals and many other animals, which may contain only a handful of genes embedded in millions of bases of nonfunctional DNA. There are even chromosomes known as supernumerary, or B, chromosomes, found in plants and animals, that appear to have no function at all.

The large variation between species in the extent of noncoding DNA has given rise to the C-value paradox. This is the observation that in higher eukaryotes there is no clear correlation between the DNA content of a genome and the number of genes encoded (or, more arbitrarily, organismic complexity). For example, tulips have ten times and some amphibians up to twenty times as much DNA as humans, but both probably have fewer genes. Even in humans, only about 5 percent of DNA in chromosomes actually encodes for genes or regulatory sequences. The rest of the genome is composed of non-coding DNA embedded within genes, known as introns, and different classes of repetitive element. Certain types of repetitive element, such as centromeric repeats and telomeres, play a structural role within the chromosome. The function of the vast majority of noncoding DNA is, however, unknown and is sporadically the subject of heated debate. Some authors claim that it represents the accumulation of selfish genes, which can only be eliminated if there is strong selection to reduce the time and energy required for DNA replication and cell division. Others maintain that noncoding DNA provides a structural function. Although the debate is far from resolved, there is evidence that many types of repeat can duplicate, either through active transposition (as in transposable elements) or by gene conversion, replication slippage, and unequal crossing over. In the absence of strong selection to streamline the genome, repetitive elements will tend to accumulate.

Chromosome Evolution and Speciation. Species vary widely in the number of chromosomes they possess, from a single pair in the ant *Myrmecia* (males have just a single chromosome) to 630 pairs in the fern *Ophioglossum*. Chromosome number can differ considerably even between species of the same genus. For example, the Atlantic salmon (*Salmo salar*) has twenty-nine pairs of chromosomes, whereas the sea trout (*Salmo trutta trutta*) has forty pairs. Because chromosomes segregate independently in meiosis, organisms with more chromosomes will tend to have more recombination per generation. However, the selective forces determining chromosome number remain obscure.

Chromosomes, like the DNA sequences they encode, can accumulate mutations, the most common forms being inversions, deletions, and duplications. In addition, chromosomes can fuse, split, or have parts of other chromosomes translocated onto them. For example, the majority of chromosomes in humans and chimpanzees are very similar, but chimpanzees have an extra pair, and several chromosomes show evidence of rearrangements. More dramatic changes in karyotype are complete genome duplications (autopolyploidy) or the combination of two genomes from different species (allopolyploidy), as is likely to have been important in the origin of the modern species of bread wheat, *Triticum aestivum*. DNA sequence–based approaches to identifying chromosomal regions with shared synteny (gene order) have been developed to probe ancient gene and genome duplications, such as those thought to have occurred in yeast and vertebrate evolution.

Chromosomal mutations that alter the genomic content are likely to have been important in evolution and speciation, by generating novel genes and, in the case of polyploidy, reproductively isolated organisms. However, the evolutionary significance of chromosomal mutations that have no effect on overall gene content, such as translocations and inversions, is unclear. It has been argued that inversions and translocations can lead to reproductive isolation and incipient speciation, because heterozygotes for the mutant and wild-type forms of chromosome may generate unbalanced gametes through recombination or unequal segregation during meiosis. Chromosomal races resulting from inversions or translocations at an appreciable frequency (the origin of which is aided by population subdivision) could represent the early stages of speciation. However, in many species, such as fruit flies and house mice, there appear to be mechanisms that prevent the production of unbalanced gametes, so heterozygotes for inversions and translocations do not suffer reduced fitness.

BIBLIOGRAPHY

Bush, G. L., S. M. Case, A. C. Wilson, and J. L. Patton. "Rapid Speciation and Chromosomal Evolution in Mammals." *Proceedings of the National Academy of Sciences USA* 74 (1977): 3942–3946. A comparative study showing a correlation between species richness and rates of chromosomal evolution across vertebrate genera. The data are suggestive of a causative correlation between chromosome changes and speciation, but far from conclusive.

Cavalier-Smith, T., ed. *The Evolution of Genome Size.* Chichester, England, 1985. A very detailed discussion of phylogenetic patterns regarding genome size. Chapter 4 presents an adaptationist explanation for the C-value paradox by the editor.

Doolittle, W. F., and C. Sapienza. "Selfish Genes, the Phenotype Paradigm and Genome Evolution." *Nature* 284.5757 (1980): 601–603. One of the most influential papers arguing in favor of selfish DNA.

Eigen, M., W. Gardiner, P. Schuster, and R. Winkler-Oswatitsch. "The Origin of Genetic Information." *Scientific American* 244.4 (1981): 78–94. A review article discussing many of the ideas of Manfred Eigen, perhaps the single most important theorist in the origins of life debate.

Maynard Smith, J., and E. Szathmáry. *The Major Transitions in Evolution.* Oxford, 1995. Chapters 4 and 7 provide a digested account of the stochastic corrector model and the origin of chromosomes. Chapter 8 discusses a related topic, the origin of mitosis.

Michod, R. E. "Population Biology of the First Replicators: On the Origin of the Genotype, Phenotype and Organism." *American Zoologist* 23.1 (1983): 5–14. The first model to suggest that stochastic differences between protocells may be the key to understanding some of the events in the early stages of the origin of life.

Orgel, L. E., and F. H. C. Crick. "Selfish DNA: The Ultimate Parasite." *Nature* 284.5757 (1980): 604–607. One of the most influential papers arguing in favor of selfish DNA.

Szathmáry, E., and L. Demeter. "Group Selection of Early Replicators and the Origin of Life." *Journal of Theoretical Biology* 169 (1987): 125–132.

Wolfe, K. H., and D. C. Shields. "Molecular Evidence for an Ancient Duplication of the Entire Yeast Genome." *Nature* 387 (1997): 708–713. An elegant use of DNA sequence comparisons to identify anciently duplicated regions of the yeast genome.

— GILEAN MCVEAN

CLASSIFICATION

Biological classification is the process of organizing knowledge about the natural world by arranging organisms into groups; these groups are called taxa (singular: taxon). The term *classification* is also used for the schemes that result from this process. Biological classifications are usually considered to have two major roles: (1) a representation of the natural order associated with present and past biodiversity, and (2) an information storage and retrieval system. Although these two roles might be considered "scientific" and "practical," respectively, achieving the first goal imparts heuristic value to the second so that the storage and retrieval system entails much more than simply pigeonholing.

Views on the underlying basis for the order seen in nature have changed markedly over time. For Aristotle it reflected the harmony of nature; for the nineteenth-century natural theologians it expressed the plan of a divine creator. With the widespread acceptance in the latter portion of the nineteenth century of Charles Darwin's theory of evolution through common descent, these metaphysical interpretations of the order in nature were replaced by a scientific one. In classifications that are based on evolutionary history, species in the same taxon are expected to share attributes that are absent in other taxa and reflect their common ancestry. Thus, evolutionary classifications should have heuristic and predictive value in comparative studies among organisms. Because of their scientific basis, these classifications can be tested using additional characters and the incorporation of additional taxa (e.g., newly discovered species).

Linnaean Classifications. Current methods of classification began with the work of the Swedish botanist Carolus Linnaeus (1707–1778). [*See* Linnaeus, Carolus.] The Linnaean system is a nested or inclusive hierarchy in which taxa at one level or rank in the hierarchy are grouped together in a taxon at a higher level in the hierarchy, and, in turn, taxa at that higher level are grouped together in a taxon at a still higher level. In other words, the units at the lowest (least inclusive) rank are aggregated into taxa at successively higher and higher ranks. This type of hierarchy differs from exclusive hierarchies such as military ranks (e.g., private, corporal, sergeant, etc.) where lower ranks are not subdivisions of higher ranks.

Linnaeus's classifications of organisms employed six ranked categories (from least inclusive to most inclusive): variety, species, genus, order, class, and kingdom. Thus, species are grouped into genera, genera are grouped into orders, and so on. Our knowledge about biological diversity has increased by several orders of magnitude since the middle of the eighteenth century and as a result modern classifications employ many ad-

TABLE 1. Linnaean Classification of Modern Humans

Kingdom: Animalia	animals
Phylum: Chordata	vertebrates and other animals with a notochord
Class: Mammalia	mammals
Order: Primates	monkeys, apes, humans
Family: Hominidae	humans
Genus: *Homo*	modern humans and closest fossil relatives
Species: *Homo sapiens*	modern species of humans

ditional taxonomic ranks in addition to those used by Linnaeus. The additional categories of phylum (zoology) or division (botany) and family are almost always used, and this augmented hierarchy (minus variety which is only used in botany) is often considered mandatory in classifications. In other words, a species is assigned to a supraspecfic taxon at each of these higher ranks (Table 1). Other categories are often included as well (e.g., tribe, cohort, etc.), and additional complexity can be introduced, as necessary, through the use of prefixes (e.g., orders can be divided into suborders, families can be grouped into superfamilies).

With the general acceptance of the view that the hierarchy so obvious to Linnaeus and others was generated by evolution (descent with modification), most taxonomists have attempted to reflect those evolutionary relationships in their classifications. However, the acceptance of evolution did not result in immediate changes to the classifications of organisms. Instead, evolution provided a new explanation for the order and groupings in existing Linnaean classifications. In part, this situation reflected the fact that the nested hierarchies in Linnaean classifications are consistent with an explanation that involves common ancestry and the divergence of related groups. Common ancestry provided an explanation for more inclusive taxa and divergence from that common ancestry provided a rationale for the subdivision of those more inclusive taxa into less inclusive taxa.

Regardless of the underlying rationale for classifications, the taxonomic hierarchy is inferred through some measure of the similarities and differences among organisms. Within an evolutionary context, the similarities are seen to reflect common ancestry whereas the differences are seen to reflect divergence from a common ancestry. These similarities and differences can be measured and weighed in different ways, leading to potential differences in the resulting classifications.

An emphasis on some measure of overall similarity among organisms in the recognition of taxa tends to result in taxa that reflect "general levels of organization or adaptation" or "ecological niche or role." Taxa recognized in this way are referred to as grades. The recog-

nition of grades has conceptual links with the preevolutionary idea of a "great chain of being"—the idea that biodiversity could be represented as a liner series that extends from the most complex species (*Homo sapiens*) down to the simplest organisms (bacteria, etc.). Within an evolutionary context this linear pattern is reflected in linear, progressive views of evolutionary change. Thus, fish evolved into amphibians, which evolved into reptiles, which in turn evolved into either birds or mammals. The division of this continuous linear sequence into discrete taxa is necessarily subjective; when these taxa are incorporated into the Linnaean hierarchy the result is an exclusive linear hierarchy of taxa, often at the same Linnaean rank (Class Osteichthyes, Class Amphibia, Class Reptilia, Class Aves; see Figure 1).

Subjective measures of the degree of divergence or morphological gaps between groups of species have traditionally also been used to recognize and rank taxa. For example, in addition to the grade concepts described above, Aves and Mammalia are ranked as classes along with Class Reptilia because the members of the former two taxa are considered to have diverged sufficiently both from their reptilian ancestors and within their own group to also be recognized at that rank. Thus, the range of variation within Mammalia and Aves is considered at least as large as that within Reptilia. Rank may also be

FIGURE 1. A Cladogram Showing Phylogenetic Relationships Among Selected Vertebrates.

The Linnaean taxa class Reptilia and class Aves are indicated along the top of the cladogram, and selected clades (monophyletic groups) are indicated to the right of the cladogram. Class Reptilia as traditionally circumscribed is a paraphyletic group because, although its members share a single common ancestry, some of the descendants of that ancestry (birds) are excluded. Clade Sauropsida is a monophyletic group that includes both living reptiles and birds. Aves is both a Linnaean class and a clade. Lissamphibians are living amphibians and squamates include lizards and snakes. Courtesy of Harold N. Bryant.

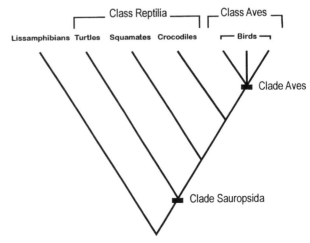

based on species diversity, with larger groups being assigned to higher ranks.

Many grade taxa are paraphyletic. The members of paraphyletic taxa share a single common ancestry but not all descendants of that ancestry are included because some descendants are considered to have evolved into a higher grade (Figure 1). Thus, Class Reptilia is paraphyletic because two descendant groups are excluded and placed in separate classes (Class Aves, Class Mammalia). Although Aves and Mammalia are thereby conceptualized as grades, they are the end points of the linear evolutionary progression and are also clades. Clades are monophyletic groups; these groups also share a single common ancestry and all descendants of that ancestry are included. Unlike grades, which are delimited subjectively, clades can also be discovered objectively using phylogenetic analysis.

In some instances, evolutionary classifications have also included polyphyletic groups in which the members of the taxon are derived from two or more separate ancestries. In other words, the members of the group do not share a single phylogenetic origin (Figure 2). For example, in the mid-twentieth century it was often argued that Mammalia included subgroups that had each evolved separately from different groups within Class Reptilia. A polyphyletic Mammalia was allowed in an evolutionary classification because the group was conceptualized as a grade. Recent phylogenetic analyses indicate, however, that Mammalia, as traditionally delimited, is a monophyletic group.

Most classifications constructed during the century following the acceptance of evolutionary theory included some mixture of grades and clades. Some taxonomists placed more emphasis on delimiting grade taxa and others placed more emphasis on recognizing clades. Concepts such as grade, degree of divergence, morphological gaps, and measures of diversity inject considerable subjectivity into classification. Subjectivity in the classification process was accepted in part because taxa were seen as artificial constructs and not a direct reflection of evolutionary history. The imprecise relationship between phylogenetic pattern and the classifications based on it led to the widespread view that the classification process involves as much art as science.

One attempt to make the inference of taxa more objective is the taxonomic approach termed phenetics. Phenetic methods measure overall similarity among taxa through the computational analysis of character/taxon matrices. Branching diagrams that cluster taxa based on their relative degree of similarity are the product of these analyses. No necessary connection to evolutionary history is implied. The approach was popular in the 1960s and 1970s but its use declined when it was shown that there was no objective way to choose among

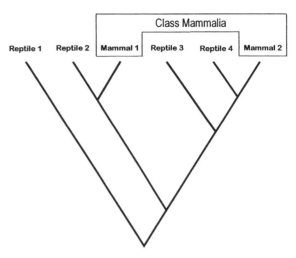

FIGURE 2. A Cladogram Showing Hypothetical Phylogenetic Relationships Among Four Reptiles and Two Mammals. Given this phylogenetic pattern, a class Mammalia consisting of the two mammals would be polyphyletic because the two mammals do not share a single common ancestry. In other words, the two mammals are more closely related to reptiles than they are to each other; the most recent common ancestor of the two mammals would have been a reptile. Courtesy of Harold N. Bryant.

alternative measures of "phenetic similarity" and because most systematists want their classifications to reflect evolutionary history.

The Trend Toward Phylogenetic Classification. Beginning in the middle of the twentieth century, there has been an increasing trend toward seeing taxa, not simply as abstract conceptual units that represented evolutionary relationships in some imprecise fashion, but as the actual historical entities produced by phylogeny. This conceptual change occurred initially to species. One of the major contributions of "The New Systematics" of the mid-twentieth century was the replacement of the view that species represented groups of similar organisms with one that saw species as interbreeding populations or population lineages. This change in the species concept led to the view that there was a fundamental difference between the species category and more inclusive taxonomic categories. Whereas species were seen as actual entities, genera and more inclusive taxa were still seen as subjective artificial groupings.

The first step toward conceptualizing supraspecific taxa as historical entities involved recognizing as taxa only those groups of species that share a single common ancestry. This view led to the increasing rejection of polyphyletic groups because they included species with separate phylogenetic origins. The second phase in this process was the increasing acceptance of the basic tenets of "phylogenetic systematics" (or cladistics), including the view that only complete units of exclusive common ancestry, monophyletic groups or clades, should be recognized as taxa. Clades are historical entities, not human constructs. Limiting taxa to clades results in the rejection of paraphyletic taxa. Paraphyletic groups are human constructs, rather than historical entities, because they are delimited subjectively based on the specific descendants that the taxonomist chooses to exclude.

The recognition of only clades as supraspecific taxa in classifications is inconsistent with various aspects of the Linnaean hierarchy and current conventions regarding its use in taxonomy. Cladistic analysis reveals a much more detailed taxonomic hierarchy than is accommodated using traditional Linnaean classifications. Attempts to represent the detailed topology of cladograms within the Linnaean hierarchy can lead to such a proliferation in the number of taxonomic categories that communication is impeded. This situation indicates that traditional Linnaean classifications with their limited number of ranks cannot adequately represent the nested relationships among monophyletic taxa. Proposed modifications to the Linnaean system, such as using the sequence of names of equal rank to convey phylogenetic relationships or the use of a special unranked category, the plesion, for fossil taxa, that would alleviate this perceived problem, have not been widely accepted.

Various rules and conventions associated with Linnaean classification help to perpetuate the recognition of paraphyletic taxa. Because of the binomial nature of species names, whereby each species must be referred to a genus, the genus category is mandatory in Linnaean classification. Genera in phylogenetic classifications should be monophyletic. However, the actual or potential ancestral species of a more inclusive taxon cannot be assigned to a monophyletic genus. The ancestor of the taxon is a member of that taxon but not any of its subgroups, and therefore will not have the characters that would allow it to be assigned to those subgroups. Linnaean conventions, however, dictate that it be assigned to one or more of those subgroups.

Some paraphyletic taxa are byproducts of other rank-based conventions of the Linnaean taxonomic system. If a taxonomist wishes to recognize and name a taxon at a particular rank in the Linnaean hierarchy (e.g., subfamily), convention states that other members of the more inclusive taxon (e.g., family) are also assigned to subfamilies. Because phylogenetic pattern rarely matches the horizontal arrangement of taxa produced by this convention, it is almost inevitable that one or more of the subfamilies so constructed will be paraphyletic, even if the taxonomist is consciously trying to avoid the creation of such groups. The difficulty in eliminating paraphyletic taxa from Linnaean classifications has led some to conclude that paraphyletic taxa are inevitable in this classification system.

Other problems have been identified with the use of

TRADITIONAL VERSUS PHYLOGENETIC CLASSIFICATION OF VERTEBRATES

Traditional classifications of vertebrates divided Subphylum Vertebrata into a number of classes, seven in the scheme presented in Table 2. Four of these classes (Agnatha, Osteichthyes, Amphibia, and Reptilia) are paraphyletic, at least when fossil taxa are included. The others are clades. The unranked phylogenetic classification consists only of clades and has a larger number of hierarchical levels. Hierarchical relationships are presented using relative indenting; taxa include all other taxa that are listed below it and are indented to its right. The cladistic status and membership of some taxa are altered; other taxa are eliminated or added. Clade Osteichthyes has a much increased membership because, unlike class Osteichthyes, it includes not only the bony fishes, but also their descendants, the tetrapods. Paraphyletic taxa such as Agnatha, Amphibia, and Reptilia are not included, whereas clades that were not recognized in traditional classifications (Tetrapoda, Amniota) are added. Clade Lissamphibia, which is limited to the extant amphibian radiation, is added. Reptilia could have been included as an alternative name for clade Sauropsida, but unlike class Reptilia it would not include the fossil "reptilian" outgroups to mammals and it would include clade Aves. Unlike the traditional scheme, clade Aves is deeply embedded in the hierarchy, reflecting its phylogenetic position within clade Vertebrata.

—HAROLD N. BRYANT

rank-based classifications. The use of ranks encourages the use of comparative studies that focus on a particular hierarchical level (e.g., species diversity within, or geographic range of, families), despite the fact that it is widely recognized that ranking is arbitrary and that rank-based comparisons across taxa have little or no biological meaning.

These problems with rank-based classifications have resulted in the proposal that ranks be abolished and that the hierarchical structure of phylogenetic classifications be conveyed in other ways. Hierarchical information can be presented through the use of indented lists (see Table 2) or an alphanumeric code associated with taxon names that indicates relative position in the taxonomic hierarchy. Hierarchical information can also be presented in accompanying cladograms.

Proponents of rank-based classification have argued that ranks are important for communication because they convey information regarding the diversity and the relative inclusiveness of taxa. For example, a taxon assigned to the category of class can usually, but not al-

TABLE 2. Comparison of Linnaean and Phylogenetic Classifications of Vertebrates

(1) Linnaean Classification:

Subphylum Vertebrata
  Class Agnatha
    Order Myxiniformes   hagfishes
    Order Petromyxoniformes   lampreys
  Class Chondrichthyes   sharks, etc.
  Class Osteichthyes
    Subclass Actinopterygii   ray-finned bony fishes
    Subclass Sarcopterygii
      Order Dipnoi   lungfishes
  Class Amphibia   frogs, salamanders, etc.
  Class Reptilia
    Order Chelonia   turtles
    Superorder Lepidosauria   lizards, snakes, etc.
    Order Crocodylia   crocodiles
  Class Aves   birds
  Class Mammalia   mammals

(2) Phylogenetic Classification:

Myxiniformes   hagfishes
Vertebrata
  Petromyzoniformes   lampreys
  Gnathostomata
    Chondrichthyes   sharks, etc.
    Osteichthyes
      Actinopterygii   ray-finned bony fishes
      Sarcopterygii
        Dipnoi   lungfishes
        Tetrapoda
          Lissamphibia   frogs, salamanders, etc.
          Amniota
            Sauropsida
              Chelonia   turtles
              Diapsida
                Lepidosauria   lizards, snakes
                Archosauria
                  Crocodylia   crocodiles
                  Aves   birds
              Mammalia   mammals

ways, be expected to be more speciose and morphologically diverse than one assigned to the rank of family. Information about relative inclusiveness includes the mutual exclusiveness of taxa assigned to the same category and the relative inclusiveness of taxa with names based on the same genus name (e.g., subfamily Felinae must be a subset of family Felidae). At issue is whether these the benefits outweigh the inability of the Linnaean system to translate explicit phylogenetic pattern into the form of a classification. The increasing, although still minority, answer of taxonomists and other biologists to that question is "no."

[*See also* Nomenclature.]

BIBLIOGRAPHY

de Queiroz, K. "Systematics and the Darwinian Revolution." *Philosophy of Science* 55 (1988): 238–259. Discussion of the failure of the Darwinian revolution to immediately initiate a similar revolution in taxonomy.

Hennig, W. *Phylogenetic Systematics.* Urbana, Ill., 1966. Seminal work on phylogenetic systematics (cladistics).

Huxley, J. S., ed. *The New Systematics.* Oxford, 1940. An important contribution to "the new systematics."

Linnaeus, C. *Species Plantarum.* Halmiae: Laurentii Salvii, 1753. Foundation of Linnaean classification of plants.

Linnaeus, C. *Systema Naturae per Regna Tria Naturae, Secundum Classes, Ordines, Genera, Species cum Characteribus, Differentiis, Synoymis, Locis.* 10th ed. Helmiae: Laurentii Salvii, 1758. Foundation of Linnaean classification of animals.

Mayr, E. "Cladistic Analysis or Cladistic Classification?" *Zeitschrift für Zoologishe Systematik und Evolutionsforschung* 12 (1974): 94–128. Critique of an exclusively phylogenetic approach to the recognition of supraspecific taxa.

Mayr, E. *The Growth of Biological Thought.* Cambridge, Mass., 1982. Detailed history of approaches to biological classification.

McKenna, M. C. "Towards a Phylogenetic Classification of the Mammalia." *Contributions in Primatology* 5 (1975): 21–46. Early phylogenetic classification demonstrating the proliferation of Linnaean ranks in such classifications.

Oosterbroek, P. "More Appropriate Definitions of Paraphyly and Polyphyly, with a Comment on the Farris 1974 Model." *Systematic Zoology* 36 (1987): 103–108. Discussion of the meaning of the terms paraphyly and polyphyly.

Panchen, A. L. *Classification, Evolution, and the Nature of Biology.* Cambridge, 1992. Historically based discussion of the relationship between classification and evolutionary theory.

Rieppel, O. *Fundamentals of Comparative Biology.* Basel, 1988. Discussion of the role of the investigation of order in nature on various aspects of comparative biology.

Simpson, G. G. *Principles of Animal Taxonomy.* New York, 1961. An example of the approach to classification during the middle of the twentieth century before the adoption of cladistic methods.

Sneath, P. H. A., and R. R. Sokal. *Numerical Taxonomy: The Principles and Practice of Numerical Classification.* San Francisco, 1973. An overview of phenetic approaches to classification.

Wiley, E. O. "An Annotated Linnean Hierarchy, with Comments on Natural Taxa and Competing Systems." *Systematic Zoology* 28 (1979): 308–337. Discussion of possible modifications to Linnaean classifications that would allow them to better reflect phylogenetic pattern.

— HAROLD N. BRYANT

CLONAL STRUCTURE

[*This entry comprises three articles:*

  An Overview
  Population Structure and Clonality of Bacteria
  Population Structure and Clonality of Protozoa

The first article provides a general overview of the genetic structure of populations and the processes that influence their structure; the second article describes the various parasexual processes that allow genetic exchange and recombination among lineages in bacteria;

the third article describes the various life cycles of parasitic protozoa, including sexual versus asexual reproduction and the potential for recombination. For related discussions, see Bacteria and Archaea; Linkage; *and* Protozoa.]

An Overview

In most animals and plants, reproduction is tied to the sexual exchange of genes between parents. Through meiosis and fertilization, each individual animal or plant is formed as a 50:50 mix of genes from its two parents. However, in the microbial world, many organisms can reproduce clonally for an indefinite number of generations: a single parent can reproduce without recombination with another individual, such that offspring are genetically identical to the parent (except for mutation). Recombination in bacteria further differs from recombination in animals and plants, in that only a small fraction of the genome is recombined: a short segment from a "donor" individual replaces the homologous segment in a "recipient" individual. Also in contrast to animals and plants, bacteria can recombine with organisms with which they are only distantly related.

Rates of Recombination in Nature. Microbiologists have recently investigated the rates at which microbes recombine in nature. One could, in principle, estimate recombination rates in nature from experiments in laboratory microcosms. However, the rate of recombination is sensitive to many environmental parameters, and so it would be difficult to determine a typical rate of recombination in nature from experiments in a limited set of environmental conditions. Instead, microbiologists have taken a "retrospective" approach, in which average recombination rates from the past are estimated from the pattern of genetic variation among contemporary organisms.

Recombination within populations. We first consider the rate at which recombination occurs within a population, as estimated from the genetic diversity among strains sampled from the environment. Here we assume that the strains sampled are members of the same population, with no intrinsic or geographical barriers to genetic exchange among the strains sampled. Also, all organisms are assumed to be ecologically interchangeable.

The rate at which entire genes are recombined can be assayed by measuring linkage disequilibrium (LD), the extent to which alleles at different gene loci are associated in a population. Strong LD results when the rate at which a gene is recombined is about the rate at which that gene is mutated (or less); LD disappears entirely when a gene's rate of recombination is much greater than mutation.

Linkage disequilibrium within microbial populations

has been measured through multilocus enzyme electrophoresis (MLEE), in which alleles at several enzyme-encoding genes are identified on the basis of the proteins' electrical charge and size. More recently, allelic variation has been assayed by sequencing regions from each of several genes.

Several dozen bacterial and microbial eukaryotic species have been assayed for LD, using either MLEE or sequencing data. Nearly all bacterial species have shown LD levels indicative of a history of rare recombination. For example, *Neisseria meningitidis* is fairly typical of many species with significant linkage disequilibrium, consistent with a recombination rate not much higher than that of mutation. The rarity of recombination is further evidenced by a related observation: a given multilocus genotype is typically seen in high frequency worldwide, and over many years. If recombination were frequent, individual multilocus genotypes would not show such longevity. These results suggest that the *N. meningitidis* lineages are very nearly clonal.

By comparison, two bacterial species have shown absolutely no linkage disequilibrium: *N. gonorrheae* and *Helicobacter pylori*. These results indicate a history of recombination at a rate higher than that of mutation, although one cannot tell from LD analysis alone how high the recombination rate is.

Among microbial eukaryotes, some species have also exhibited linkage disequilibrium levels consistent with rare recombination. For example, the protistan parasites *Trypanosoma cruzi* and *Leishmania peruviana* appear to be nearly clonal. Different populations of the malaria parasite *Plasmodium falciparum* show a range of recombination rates: some populations are nearly clonal, whereas others recombine frequently enough to eliminate LD.

Several methods estimate the rate at which portions of a gene are recombined within a population, using DNA sequence data. For example, an approach by J. Hey and D. Wakeley (1997) infers the rate of recombination from the extent to which different portions of the gene yield different phylogenetic relationships among strains. The rationale is that under clonality, all parts of the gene will be inherited together and thus yield the same phylogeny, but under recombination they will not. These methods have yielded low rates of recombination within bacterial species, generally within an order of magnitude of the mutation rate.

John Maynard Smith and colleagues (1993) have pointed out several weaknesses of retrospective approaches to estimating recombination rates within a population. Recombination rates may be artificially deflated if the strains sampled in a survey are actually members of different populations, especially if there are intrinsic barriers to recombination between members of the sample (e.g., DNA sequence divergence between different populations hinders their recombination) or if the strains sampled were obtained from different sites between which migration is rare. Recombination rates will also be deflated if strains are sampled from ecologically distinct populations, even if there are no barriers to genetic exchange between them. These potential problems can be addressed if strains are sampled from a single site and if care is taken not to include strains from long-divergent populations (i.e., strains from distinct DNA sequence clusters). The studies reported here have attempted to avoid these potential pitfalls.

Recombination between populations. One may estimate the rate at which recombination occurs between organisms that fall into discrete DNA sequence clusters, indicative of separate populations. In an approach developed by Brian Spratt (1999), strains are assigned to a sequence cluster on the basis of being identical for at least six of seven genes sequenced. Their method identifies DNA sequence substitutions arising from point mutations, and those arising from recombination with another sequence cluster. This yields estimates of the relative rates of mutation and between-cluster recombination.

This approach has so far been used to measure rates of recombination between sequence clusters in *N. meningitidis* and in *Streptococcus pneumoniae*. The rate at which a segment is involved in recombination with another cluster ranges between about four and ten times greater than the mutation rate for that segment.

Highly distinct sequence clusters (as identified in Spratt's approach) generally correspond to ecologically distinct populations. Therefore, this approach probably estimates the rate of recombination between ecologically distinct populations. As will be seen, predicting the evolutionary consequences of rare recombination requires knowing the rates at which a gene segment is involved in recombination both between and within populations.

Evolutionary Consequences of Rare Recombination.

Evolution of new adaptations. Although recombination is rare in many microbes, it is frequent enough to foster evolution of new adaptations. This is because recombination need act only once within a population to create a fortuitous combination of alleles, and microbial population sizes are enormous. In the case of bacteria, recombination has transferred adaptive alleles (e.g., conferring penicillin resistance) across widely divergent taxa.

Adaptive divergence between populations. In highly sexual organisms, recombination between nascent species is a major obstacle for speciation: nascent species cannot maintain their integrity unless the between-population recombination rate is reduced severely from the high within-population rate. However, recurrent recombination between rarely sexual populations does not hinder their divergence. Recombination within many microbial populations is already so low that, even if between- and within-population recombination were to occur at the same rate, recombination between populations could not disrupt the integrity of populations' adaptations. The evolution of sexual isolation is therefore not a necessary step in the origin of microbial diversity.

Purging of diversity within populations. In a rarely sexual population, each mutation conferring higher fitness sets in motion a round of natural selection that purges the population of its genetic diversity. This is because the genome that contains the mutation rapidly sweeps through the population, replacing the less fit variants. Under rare recombination, the entire genome of the adaptive mutant remains intact as it sweeps through the population, and so the population loses its genetic diversity at all loci. The intensity of this diversity-purging effect (called periodic selection or a selective sweep) is determined in part by the rate at which a gene segment is involved in recombination within the population. Given the recombination rates within bacterial populations, 99.9 percent or more of the diversity at each locus is expected to be purged with each periodic selection event.

Speciation. A microbial ecotype may be defined as a set of microbial strains using the same ecological niche, such that each adaptive mutant outcompetes to extinction all other strains from the same ecotype but does not drive to extinction strains from other ecotypes (Figure 1). Within an ecotype, genetic divergence is only temporary because periodic selection favoring the next adaptive mutant can quash all genetic variation within the ecotype. Divergence becomes permanent only if a new mutant (or recombinant) can escape the diversity-purging effects of selection within the ecotype. For example, if a mutant can metabolize a new resource, the new genotype and its descendants may form a separate ecotype, no longer susceptible to periodic selection events within the parental ecotype (i.e., they cannot be competitively extinguished by adaptive mutants from the parental ecotype). In summary, periodic selection fails to constrain the divergence between ecotypes; as discussed earlier, recombination would likewise fail to constrain interecotype divergence. Microbial ecotypes may represent evolutionary lineages that are irreversibly separate, a fundamental property shared by species throughout the tree of life.

Discovery of microbial ecotypes. Provided that recombination between ecotypes is sufficiently rare, the theory of T. Palys and colleagues (1997) predicts that each microbial ecotype should eventually be identifiable as a sequence cluster, where the average sequence divergence between ecotypes is far greater than the av-

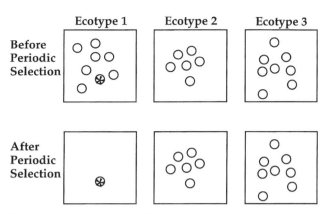

FIGURE 1. The Transience of Diversity within Ecotypes and the Permanence of Divergence among Ecotypes.

Each circle represents an individual organism, and the distance between circles represents the genetic divergence between organisms. The asterisk represents an adaptive mutation in one individual of ecotype 1. The adaptive mutant and its nearly clonal descendants replace the rest of the genetic diversity within the ecotype, owing to the rarity of recombination. Because ecotypes differ in the resources they use, an adaptive mutant from one ecotype (e.g., ecotype 1) does not outcompete the cells of other ecotypes. Once populations are divergent enough to escape one another's periodic selection events, the populations begin to diverge permanently.   F. M. Cohan. *ASM News*, 1996.

erage sequence divergence within them. Moreover, each ecotype should eventually be identifiable as a clade in a sequence-based phylogeny, because periodic selection repeatedly purges the sequence diversity within ecotypes but not the divergence between them. The predicted correspondence between ecologically distinct groups and sequence clusters has been corroborated by a survey of ecological and sequence diversity within dozens of bacterial taxa.

The correspondence between ecologically distinct populations and sequence clusters allows identification of unknown isolates from the environment. For example, multilocus sequence typing may allow rapid diagnosis of pathogens by sequence data over the Internet. Also, the correspondence between sequence clusters and ecotypes allows discovery of cryptic ecological diversity within a named species. Surveys of sequence diversity within bacterial species have revealed multiple sequence clusters that were later found to be ecologically distinct. It is likely that sequence-based approaches will reveal previously unknown ecological diversity within many rarely sexual microbial species.

[*See also* Bacteria and Archaea; Linkage; Protists.]

BIBLIOGRAPHY

Anderson, T. J., et al. "Microsatellite Markers Reveal a Spectrum of Population Structures in the Malaria Parasite *Plasmodium*

falciparum." *Molecular Biology and Evolution* 17 (2000): 1467–1482.

Cohan, F. M. "Bacterial Species and Speciation." *Systematic Biology* 50 (2001): 513–524. The consequences of rare but promiscuous recombination for bacterial evolution and speciation.

Feil, E. J., J. Maynard Smith, M. C. Enright, and B. G. Spratt. "Estimating Recombinational Parameters in *Streptococcus pneumoniae* from Multilocus Sequence Typing Data." *Genetics* 154 (2000): 1439–1450. A method for estimating the rate of recombination between sequence clusters within a named species.

Haubold, B., M. Travisano, P. B. Rainey, and R. R. Hudson. "Detecting Linkage Disequilibrium in Bacterial Populations." *Genetics* 150 (1998): 1341–1348. An improved method of detecting linkage disequilibrium, showing that bacterial recombination is rarer in some taxa than previously estimated.

Hey, J., and D. Wakeley. "A Coalescent Estimator of the Population Recombination Rate." *Genetics* 145 (1997): 833–846. A method for estimating the within-population rate of recombination within a gene segment.

Maynard Smith, J., N. H. Smith, M. O'Rourke, and B. G. Spratt. "How Clonal Are Bacteria?" *Proceedings of the National Academy of Sciences USA* 90 (1993): 4384–4388. Discussion of the assumptions underlying estimates of within-population recombination rates, based on linkage disequilibrium.

Palys, T., L. K. Nakamura, and F. M. Cohan. "Discovery and Classification of Ecological Diversity in the Bacterial World: The Role of DNA Sequence Data." *International Journal of Systematic Bacteriology* 47 (1997): 1145–1156. A theoretical rationale for using sequence data for discovering bacterial ecotypes: why ecotypes should correspond to sequence clusters based on any gene in the genome.

Spratt, B. G. "Multilocus Sequence Typing: Molecular Typing of Bacterial Pathogens in an Era of Rapid DNA Sequencing and the Internet." *Current Opinion in Microbiology* 2 (1999): 312–316. Discussion of multilocus sequence typing, which can predict the ecological qualities of an unknown strain based on the sequence cluster to which it belongs. This approach works because periodic selection purges nearly all the diversity within an ecotype, such that the members of an ecotype fall into clusters of nearly identical sequence types.

Suerbaum, S., J. Maynard Smith, K. Bapumia, G. Morelli, N. H. Smith, E. Kunstmann, I. Dyrek, and M. Achtman. "Free Recombination within *Helicobacter pylori*." *Proceedings of the National Academy of Sciences USA* 95 (1998): 12619–12624.

Tibayrenc, M., and F. J. Ayala. "Evolutionary Genetics of *Trypanosoma* and *Leishmania*." *Microbes and Infection* 1 (1999): 465–472. Documentation of the clonal nature of two major protistan parasites.

— FREDERICK M. COHAN

Population Structure and Clonality of Bacteria

Like the eukaryotes, the prokaryotes are capable of genetic recombination, but the characteristics of recombination are very different between these major groups. These differences have profound effects on the modes of adaptive evolution and speciation available to bacteria.

Mechanisms of Genetic Exchange. Bacteria can recombine by three parasexual mechanisms: transduction, conjugation, and transformation. These mechanisms introduce DNA unidirectionally, from a donor cell to a recipient cell, and the transferred DNA is generally a small fraction of the genome. These mechanisms bring DNA into the recipient cell in different ways, but in each case the donor DNA integrates into the recipient genome by homologous recombination (a process by which the introduced genetic material is incorporated into the recipient's genome at a site that contains the same or similar segment of DNA). None of these mechanisms is tied to reproduction; thus recombination is much rarer in bacteria than in most plants and animals, where recombination is obligately tied to reproduction.

Transduction is genetic exchange mediated by bacteriophages (bacterial viruses). During a viral infection, a genomic segment from the infected cell (the donor) becomes mistakenly packaged into the virus. When the virus later attaches to another cell (the recipient), the donor DNA is injected into the recipient cytoplasm, setting the stage for homologous recombination. Transduction can transfer a length of donor DNA about as large as the phage genome, up to several hundred kilobases.

In the process of conjugation, genetic exchange is mediated by plasmids, which are self-replicating circles of DNA infecting a bacterium. Recombination between bacteria becomes possible when DNA from a donor's genome becomes integrated into an infecting plasmid. The plasmid may then promote its own transfer to another cell, thus introducing donor DNA into a recipient. Plasmid conjugation can transfer larger segments of the genome than is possible in transduction. In both transduction and conjugation, the recipient cell plays a passive role in the uptake of donor DNA.

In contrast, the recipient plays an active role in the uptake of DNA in transformation. Through pathways coded by the recipient's genome, transformation introduces very small segments of free DNA from the environment, usually less than 1 percent of the genome, into the cell. Many bacterial species lack pathways for DNA uptake that are necessary for transformation.

Rarity of genetic exchange. Rates of recombination in nature have been estimated by "retrospective" approaches utilizing surveys of sequence or allozyme variation in natural populations. The rationale is that low recombination rates can be inferred when alleles at different loci show high degrees of association (i.e., linkage disequilibrium), or when different DNA segments yield congruent phylogenies. These retrospective approaches have shown that recombination in most bacterial species occurs at about the rate of mutation or somewhat higher. However, two species (*Neisseria gonorrheae* and *Helicobacter pylori*) have shown absolutely no linkage disequilibrium, consistent with recombination rates much higher than mutation. The high levels of recombination in these species are likely attributable to

their high "competence" for transformation. Nevertheless, high competence alone does not guarantee high rates of transformation in nature, as several naturally transforming species recombine in nature at a rate not much higher than mutation (e.g., *Bacillus* species).

Promiscuity of genetic exchange. While bacteria do not recombine frequently, they are not fussy about their choice of partners in genetic exchange. Bacteria can undergo homologous recombination with organisms that differ as much as 25 percent in DNA sequence. There are, nevertheless, some important constraints on genetic exchange between divergent bacteria. First, transduction and conjugation require donor and recipient to share vectors of recombination (i.e., the phage or plasmid). Also, donors and recipients must inhabit the same microenvironment. Finally, homologous recombination is limited by molecular constraints on integration of divergent DNA sequences. DNA repair systems tend to reverse integration when they detect nucleotide mismatches between recipient and donor; also, integration requires a 20–30 base-pair stretch of nearly perfectly matched DNA. The rate of recombination decays exponentially with donor-recipient sequence divergence, a relationship caused by the need for an exact match of DNA sequences.

Recombination in bacteria is not limited to the transfer of homologous segments. In heterologous recombination, bacteria can "capture" new gene loci from other organisms, sometimes from organisms that are extremely distantly related.

Adaptive Evolution. Although bacterial recombination is rare on a per-cell basis, it is frequent enough to bring about adaptations. Recombination need act only once within a population to bring about an adaptation, and population sizes are enormous. It is likely that recombination has often brought together adaptive combinations of alleles from within the same population, although no examples are known.

Alternatively, because recombination in bacteria is promiscuous, a bacterial population can import adaptive alleles from other species. For example, John Maynard Smith and coworkers have shown that antibiotic-resistant alleles have spread across species of *Neisseria*, replacing antibiotic-sensitive alleles by homologous recombination. Also, because bacteria can import new gene loci and whole operons from other species, they can acquire entirely new functions and invade new ecological niches almost instantaneously. It is not clear how much of bacterial adaptation within a particular population is attributable to acquiring new gene loci versus making changes in existing genes by mutation and recombination.

Acquisition of adaptations from other species is facilitated not just by the promiscuity of genetic exchange, but also by the small sizes of segments that are transferred. Because only a small fraction of a genome is transferred, a broadly adaptive gene (e.g., an antibiotic resistance gene) can be transferred without the cotransfer of other genes that might be narrowly adapted to the donor species.

Speciation. Because recombination in bacteria is so rare, it is insufficient to prevent adaptive divergence between sympatric (co-occurring) populations. That is, even if recombination between populations were to happen at the same rate as within populations, it is not sufficient to prevent divergence. Therefore, evolution of sexual isolation is not a necessary step in adaptive divergence between sympatric populations or in their speciation.

The origin of bacterial species is primarily a process of ecological divergence, as has been discussed by F. Cohan (2001). A bacterial "ecotype" can be defined as a population using a particular ecological niche, so that each adaptive mutant from the ecotype outcompetes to extinction all other strains from that ecotype; however, an adaptive mutant cannot extinguish strains from other ecotypes. Because of the rarity of recombination, each such adaptive mutant by spreading through the population quashes the genetic diversity within its ecotype, over all loci (a process called "periodic selection"). A new ecotype is created with a mutation or recombination event that thrusts a cell into a new ecological niche. If the nascent ecotype can escape periodic-selection events stemming from the parental ecotype, then the nascent and parental ecotypes have achieved a level of divergence consistent with species status: the ecotypes are ecologically distinct; they may diverge without constraint of recombination or periodic selection or any other force; diversity within each ecotype is limited by a force of cohesion (in this case, periodic selection); and each ecotype is a monophyletic group.

Bacteria should have a greater opportunity for speciation than is possible in the highly sexual animals and plants. First, speciation in animals and plants typically requires both reproductive and ecological divergence, but speciation in bacteria requires only ecological divergence. Second, animal and plant species are closed genetic systems, and so they must evolve all their adaptations on their own; bacteria, by contrast; can acquire adaptations from almost anywhere in the bacterial world. Third, heterologous gene transfer can give a recipient an entirely new metabolic function which allows it to utilize resources not available to the parental ecotype. Heterologous transfer can thus instantaneously produce a new ecotype. In addition, the extremely large population sizes of bacteria provide a bacterial ecotype with an enormous number of mutants and recombinants from which new ecotypes may be selected.

Discovery and diagnosis of bacterial ecotypes. Because periodic selection purges the genetic diversity

within ecotypes but not between them, each bacterial ecotype is expected to form a separate DNA sequence cluster for each gene in the genome. This correspondence between ecotypes and sequence clusters has led to the discovery of multiple ecologically distinct taxa within what was previously considered a single species. For example, strains from *Borrelia burgdorferi* (sensu lato) were found to fall into multiple sequence clusters, and these clusters were later found to correspond to ecologically distinct groups causing different diseases. Patterns of sequence clustering suggest that many named species contain multiple ecotypes, each with its own niche and its own periodic-selection events. Sequence-based approaches hold great promise for discovery of ecotypes, even within well-characterized species. Moreover, these approaches allow us to diagnose unknown strains into ecotypes that, without molecular study, would not have been suspected to exist.

[*See also* Bacteria and Archaea; Linkage.]

BIBLIOGRAPHY

Cohan, F. M. "Bacterial Species and Speciation." *Systematic Biology* 50 (2001): 513–524. The consequences of rare but promiscuous recombination for bacterial evolution and speciation.

Dubnau, D. "DNA Uptake in Bacteria." *Annual Review of Microbiology* 53 (1999): 217–244. Discussion of the mechanisms of transformation in a diversity of bacterial taxa.

Majewski, J. "Sexual Isolation in Bacteria." *FEMS Microbiology Letters* 199 (2001): 161–169. Discussion of the mechanisms of sexual isolation between bacterial species, as well as the forces of natural selection determining the level of promiscuity in bacterial genetic exchange.

Maynard Smith, J., C. G. Dowson, and B. G. Spratt. "Localized Sex in Bacteria." *Nature* 349 (1991): 29–31. Examples of adaptive transfer of antibiotic resistance alleles between bacterial species.

Ochman, H., J. G. Lawrence, and E. A. Groisman. "Lateral Gene Transfer and the Nature of Bacterial Innovation." *Nature* 405 (2000): 299–304. Documentation that large fractions of many genomes have been acquired from distantly related bacterial species, and that in many cases, the acquired genes have played a central role in adaptation to new ecological niches.

Sokurenko, E. V., D. L. Hasty, and D. E. Dykhuizen. "Pathoadaptive Mutations: Gene Loss and Variation in Bacterial Pathogens." *Trends in Microbiology* 7 (1999): 191–195. A counterpoint to the Ochman et al. paper; here a case is made for the importance of changes in existing genes, rather than acquisition of new genes, in bacterial adaptation.

— FREDERICK M. COHAN

Population Structure and Clonality of Protozoa

Parasitic protozoa are still among the main scourges of the human species. For example, the malaria epidemic is presently spreading rapidly as a consequence of antimalarial drug resistance of the parasite and insecticide resistance of the mosquito vector. Even in the industrial world, parasitic diseases constitute an important health problem, especially through the development of opportunistic infections associated with the AIDS epidemic.

Our knowledge of the mating systems of parasitic protozoa some twenty years ago led to some assumptions, but few data, on their population structure. Most parasitic protozoa were supposed to be strictly asexual. Malaria parasites, by contrast, were known to have an obligatory sexual phase at each transmission cycle and therefore were supposed to have a population structure similar to those of higher organisms, such as humans. In such organisms, the only obstacles to genetic exchange are time and geographical distance. The development of molecular tools and their population-genetic application to parasitic protozoa yielded totally unexpected results. This shows that logical assumptions made following the known biology of a species are not exempt from extensive verification when appropriate methods become available. The methods used to elucidate the population structure of parasitic protozoa and bacteria are based on the analysis of genetic recombination, the main consequence of sexuality. In sexual species, the genes located on different parts of a given chromosome or on different chromosomes are randomly reassorted, which means that every sexual individual has a unique genetic inheritance. The consequences of this random process in terms of gene distributions in natural populations can be predicted by simple statistical laws, similar to those that describe the distribution of playing cards that have been shuffled many times. When the genes are obviously not distributed at random in a population, as verified by appropriate statistics, this is taken as circumstantial evidence that sexual reproduction either does not occur or its effect has been strongly thwarted. In extreme cases, it is concluded that the population is clonal; that is, the offspring has exactly the same genetic makeup as its parent (obviously not observed in sexual species such as ours).

A Broad Spectrum of Mating Patterns and Population Structures. Using the general approach described above led to the "clonal theory of parasitic protozoa" proposed by Michel Tibayrenc and colleagues (1990), which states that clonal propagation is predominant in many medically important species. An important distinction has been subsequently made by John Maynard Smith and colleagues (1993) between sexual species that occasionally produce bouts of ephemeral clones (epidemic clonality) and species that undergo sustained clonal evolution on time scales of millions of years, with sexual recombination very rare or nonexisting. The present state of art confirms that the impact of clonal propagation in many parasitic protozoa is strong, although notable differences are observed among species. Species that undergo predominantly clonal evolu-

tion include American trypanosomes responsible for Chagas disease, pathogenic amoeba, the parasites responsible for the cosmopolitan diseases called leishmanioses, and even the parasite that causes toxoplasmosis, although this species is known to have a sexual cycle in the cat. The African trypanosomes responsible for sleeping sickness were once supposed to be strictly asexual, but successful mating experiments were made by L. Jenny and colleagues (1986). However, analysis of natural populations of African trypanosomes supports the clonal theory. According to A. Tait (2000), it seems that this clever parasite actually plays two games according to the ecological cycle involved. It is very clonal in its populations that infect humans but uses more sex (epidemic clonality model) in populations circulating in wild mammals. Finally, as shown by D. Walliker (2000), *Plasmodium falciparum*, the agent of the most malignant form of malaria, exhibits in Africa a population structure that indicates free exchange of genes, which is a logical consequence of its known biology (sex in the mosquito vector). Surprisingly, the pattern is different in South America, in which a picture of epidemic clonality is seen. This latter pattern could be explained by self-fertilization (if the two sexual cells have exactly the same genetic makeup, the offspring will be identical, in other words, will appear clonal).

Medical Implications of Mating Patterns and Population Structure. The molecular epidemiology approach is used to track emerging parasites and pathogens, and it is based on the assumption that genetic identity of a set of strains implies a common descent. Clonal species are therefore the most appropriate ones for such an epidemiological tracking purpose, with their multilocus genotypes that propagate unchanged for long periods of time (genetic photocopies). Moreover, when a species undergoes long-term clonal evolution, the fate of different clones is to accumulate more and more divergent mutations, including some in genes that govern relevant medical properties such as pathogenicity or resistance to drugs. One therefore expects a strong correlation between genetic divergence and biomedical differences in clonal species. In clonal species, the entire genotype that carries a mutation providing a selective advantage, such as resistance, will spread (successful genotypes). In a sexual species, only the beneficial mutations will spread (successful genes). In the first case, the "hitchhiking" effect will involve the whole genome, whereas in the second case, it will involve only the genes that are tightly linked to the selected gene. Finally, the predominance of clonal propagation in many parasites renders inappropriate the biological species concept, which is based on the presence or absence of free genetic exchange. Clonal species descriptions must therefore be based on criteria of phylogenetic divergence (phylogenetic species concept) or biomedical specificities (phenotypic species concept), or on a combination of the two.

The development of powerful technologies such as automated sequencing, DNA microarrays, and bioinformatics techniques, together with the various parasite genome projects sponsored by the World Health Organization, are bound to expand greatly our knowledge of the population genetics and evolution of parasitic protozoa. Such knowledge offers considerable promise for molecular epidemiology, taxonomy, and vaccine and drug development.

[*See also* Linkage; Malaria; Protists.]

BIBLIOGRAPHY

Jenni, L., S. Marti, J. Schweizer, B. Betschart, R. W. F. Le Page, J. M. Wells, A. Tait, P. Paindavoine, E. Pays, and M. Steinert. "Hybrid Formation between African Trypanosomes during Cyclical Transmission." *Nature* 322 (1986): 173–175.

Maynard Smith, J., N. H. Smith, M. O'Rourke, and B. G. Spratt. "How Clonal Are Bacteria?" *Proceedings of the National Academy of Sciences USA* 90 (1993): 4384–4388.

Tait, A. "Trypanosomiases." *Science* (2000): 258–282.

Tibayrenc, M., F. Kjellberg, and F. J. Ayala. "A Clonal Theory of Parasitic Protozoa: The Population Structure of *Entamoeba*, *Giardia*, *Leishmania*, *Naegleria*, *Plasmodium*, *Trichomonas* and *Trypanosoma*, and Its Medical and Taxonomical Consequences." *Proceedings of the National Academy of Sciences USA* 87 (1990): 2414–2418.

Walliker, D. "Malaria." *Science* (2000): 93–112.

— MICHEL TIBAYRENC

CLONING

The term *cloning* has two important meanings in contemporary biology. The discovery in the late 1960s and development in the 1970s of enzymes that can cut DNA at specific sites and other enzymes that can ligate the resulting pieces back together underlie the process of isolating a gene (or other DNA fragment) and placing it into a bacterial plasmid for further manipulation. This cloning of genes (molecular cloning) results in duplicates (constructs) that retain the same sequence as initial DNA, although for many purposes changes are often subsequently and purposefully introduced. These duplicates can be replicated manyfold by allowing the bacteria that hold them to divide repeatedly in culture. These techniques are the core of classical molecular biology.

Cloning whole animals (especially mammals) from both nonspecialized embryonic and somatic adult cells derived from specialized tissues has seen major progress in the last few years. Initially, this research flowed from attempts by developmental biologists to understand what is necessary and sufficient for development to proceed, specifically how the hereditary material changes in the hundreds of specialized cell types that make up an animal. The term *cloning* has been used to refer to production of an organism derived from specialized adult tissue and genetically identical to the donor. The term also has been applied to a range of techniques to produce animals that are genetically identical, from the splitting of embryos (generating twins) to the use of the nucleus from embryonic cells or from differentiated somatic tissues. The recent successes in cloning such as Dolly the Sheep (see Vignette) were driven by researchers whose goals are to replicate livestock and introduce genes that result in therapeutic proteins.

This work has an important ethical dimension. Techniques that range from cloning differentiated adult tissues, to in vitro twinning, to production of cloned stem cells, to in vitro fertilization of eggs with nonsperm cells have provoked a rich and varied ethical discussion that touches on a number of important questions regarding the use of human tissues and reproduction and has involved many governments and their agencies in drafting policies and legislation. Currently, there is broad consensus to restrict or ban human reproductive cloning, but much more heated debate surrounds the issue of cloning to produce human stem cells (somewhat specialized cells that can still give rise to several kinds of specialized cells) that could be used therapeutically, without any intention of producing a viable embryo for implantation.

Attempts in the 1950s and early 1960s at vertebrate cloning involved removing the diploid nucleus from embryonic frog cells and introducing it into an enucleated egg cell, an oocyte from which the nucleus had been removed. When nonembryonic frog cells were used to donate nuclei, the frogs survived only through the tadpole stage. During development in organisms with many different kinds of tissues, cells specialize in a process called differentiation. This process of differentiation involves a complex regulation of the gene transcription and metabolism. Although many adult tissues (e.g., brain, heart, lungs) are made up of cells whose nucleus contains a full complement of chromosomes, the cells and the DNA are in a state that makes them no longer totipotent, or able to be modified into any tissue. Early embryonic cells are not committed in this way; that is, they retain totipotency.

As the number of failed attempts at cloning from adult tissues mounted, a consensus arose that cloning from differentiated tissues was not possible. This conclusion was shaken in the late 1970s when a claim was made that mice had been cloned. This was truly remarkable, as it would have been the first example of cloning in mammals. Difficulties in replicating this feat caused many to question the result and seemed to confirm and

DOLLY AND THE MICE

In 1996, Ian Wilmut and his colleagues at the Roslin Institute in Scotland derived a viable embryo by cloning from a cell taken from adult mammary tissue. In February 1997, the first animal cloned from adult cells was born. The approach used had previously succeeded with embryonic donor cells. Unfertilized eggs were flushed from a sheep and the maternal nucleus removed by sucking it out with a fine needle. Mammary tissue from a six-year-old ewe had previously been cultured. Nutrients in the cell culture medium were reduced, causing the cells to go into the resting G0 part of the cell cycle, during which DNA transcription has stopped and the DNA is compactly folded into chromosomes in these quiescent cells. A donor cell and much larger recipient enucleated oocyte were abutted and fused using an electric current. This egg cell now contained the contents of the donor cell, including the nucleus containing the donor chromosomes. This newly constructed embryo was allowed to grow in culture for seven days, then transferred to a sheep at the appropriate stage of estrus. Only 10.5 percent of the 277 attempts resulted in growing embryos, and only one of the thirteen recipients became pregnant. Twenty-one weeks later, Dolly was born.

This procedure raised a series of questions: Was there something special about mammary tissue that made it more like embryonic stem cells than like other terminally differentiated adult tissue? Was there something critical about culturing the cells before using them to clone? Was the key to success forcing the cells into quiescence before using them? Was cell fusion the only way to move the nucleus into the egg? Did some other important factors come along with the nucleus? Could only females be cloned? Could transgenic animals be produced by introducing genes into the donor cells? Many of these questions have begun to be answered.

In 1997, Teruhiko Wakayama and his colleagues at the University of Hawaii successfully injected the nuclei of mouse cumulus cells into enucleated eggs. Cumulus cells are differentiated cells that coat the surface of the egg and are often found in the G0 stage of the cell cycle. This was accomplished without preparing donor cells in culture or fusing the donor and enucleated recipient. Once the nucleus was injected, the cells were cultured for an hour, after which cytochalasin B was added. This chemical inhibits the formation of a polar body, the second cell that forms before fertilization and takes half of the chromosomes of an egg cell in preparation for receiving chromosomes from a sperm cell. Wakayama and his colleagues then implanted the resulting embryos in surrogate mothers. On 3 October 1997, the first mouse clone was born. The team has since produced many more cloned mice, including clones of clones. In 1999, this group reported cloning males from adult tail-tip cells. The Roslin researchers, in 1997, went on to produce sheep with targeted changes in their genomes and have also since, with some modification, cloned pigs from cultured ovarian granulosa cells using the electrofusion approach.

Although there may be something important about the donor cells, clearly multiple cell types can be used. Electrofusion is not mandatory, and neither is culturing of the donor cells, although this may be a mechanism that can be used to place various cell types into an appropriate state. These techniques can be used to make transgenic animals and may become the preferred way for doing so. Both males and females can be cloned, but there are important questions surrounding how the inactivated X chromosomes in female donors fare in the process. Only one of the X chromosomes is active in mammals, and if cloning allows the inactivated one to be reactivated, the clone could then have access to different alleles than its mother. This and many other questions will be answered in the coming years.

—JEREMY CREIGHTON AHOUSE

retrench the impossibility of cloning. In the mid-1980s, sheep and later cattle were cloned from early embryonic cells. This led in the mid-1990s to the work of Ian Wilmut and Keith Campbell—first the cloning of a sheep from differentiated embryonic cells, then in 1996 cloning from adult cells (Dolly). This success has been followed by live births of a cloned mouse, cow, pig, and goat, and even attempts at cloning to recover endangered species by using eggs and surrogates from closely related breeds. There also have been viable clones produced that have had genetic modifications, as well as numerous successes with early embryonic twinning.

Although cloning experiments are often described in terms of the removal or manipulation of just the nucleus or chromosomes, it is important to note that mitochondria contain DNA and are present in the donor egg cell.

There are also many proteins and other cytosolic components that are moved along with the donor nucleus, while critical factors are left behind in the enucleated recipient egg cell. The cell fusion technique used to clone sheep combines the nucleus-containing donor cell and the enucleated egg, allowing all of the cell contents of the donor cell to be included. Natural monozygotic twins (often called identical twins) share maternal mitochondria and the same early cellular context, the same maternofetal environment, including exposure to nutrients, hormones, and other maternal factors (e.g., antibodies that are transported across the placenta to the neonate in primates). In these ways, twins share many more factors and conditions than clones would.

Many factors are involved in successfully removing genetic material from adult cells, generating a viable embryo, then getting those embryos, once implanted, carried to term and grown to adulthood. It is clear that not all adult cell types have been amenable, and despite a great deal of careful work, several species have not been successfully cloned (including cat, dog, chicken, rabbit, and monkey). The state of the DNA seems to be critical, both how the DNA is methylated (a process cells use to switch off some regions of DNA) and whether the cells are actively transcribing (cells were starved to quiescence for the Dolly experiment, and cells in the appropriate part of the cell cycle were used in generating mouse clones). The state of the enucleated recipient egg cell is also key. Developmental biologists are especially keen to understand the process of cellular commitment to a given tissue type and the loss of totipotency and whether it can be reversed. DNA does break down, and some cell types are irreversibly modified.

This topic is in tremendous flux as every month brings new examples of successful clones with slight variations in techniques and conditions. Still, the yield of adult animals from cloning is very low. Development requires the appropriate expression of proteins, not merely the presence of genes that could be translated, and many pre- and postnatal developmental problems are due to the misregulation of proteins. Success rates are below 2–3 percent even in those animals in which cloning from somatic tissues has been demonstrated. There have been recent results indicating that the genetic regulation of cloned animals is thrown off, suggesting both that development is tremendously resistant to variation in gene regulation and that cloning may often result in unhealthy offspring.

The current excitement around mammalian cloning has focused on the cloning of transgenic livestock that could produce therapeutic proteins, the duplication of rare farm animals with desirable features, and the therapeutic potential of cloned human stem cell tissue. Specific suggestions have ranged from producing ruminants that have the prion locus inactivated (and which thus could not transmit the various forms of encephalopathies such as scrapie and bovine spongiform encephalopathy), to goats that produce therapeutic proteins in their milk, to cloned human embryonic stem cells that are induced to proliferate without differentiation and then subsequently differentiated in the context of particular growth factors for subsequent therapeutic use without the risk of tissue rejection. However, the efficiency of cloning is still so low that it seems unlikely that perfectly tissue-matched cloned stem cells could be routinely produced at hospitals. Even if the focus of therapeutic transformation of cells from differentiated to totipotency moves away from cloning for ethical and practical reasons, however, cloning will have inspired and informed this pursuit. We can look forward to gaining much more understanding of the role of the nucleus, differentiation, and the factors in egg cells that allow development to proceed.

BIBLIOGRAPHY

Campbell, K. H. S., et al. "Sheep Cloned by Nuclear Transfer from a Cultured Cell Line." *Nature* 385 (1997): 810–813. The initial report of the first successful attempt to clone using donor cells derived from adult tissues.

Kitcher, P. *The Lives to Come: The Genetic Revolution and Human Possibilities.* New York, 1997. A good introduction to frame ethical questions in the face of biotechnological changes.

Kolata, G. *Clone: The Road to Dolly, and the Path Ahead.* New York, 1999. A journalistic review of the technologies, history, and ethical discussions that surround cloning. Includes a number of introductions to the scientists involved.

Lanza, Robert P., et al. "Cloning of an Endangered Species (*Bos gaurus*) Using Interspecies Nuclear Transfer." *Cloning* 2.2 (2000): 79–90. The description of the first attempt to clone an endangered animal.

McGee, G., ed. *The Human Cloning Debate.* 2d ed. Berkeley, 2000. Covers the ethical dimensions of human cloning.

McLaren, A. "Cloning: Pathways to a Pluripotent Future." *Science* 288 (2000): 1775–1780. A concise review of the issues and possibilities presented by cloning.

Pennisi, E., and G. Vogel. "Clones: A Hard Act to Follow." *Science* 288 (2000): 1722–1727. An overview of the difficulties involved in cloning mammals.

Wakayama, T., et al. "Full-term Development of Mice from Enucleated Oocytes Injected with Cumulus Cell Nuclei." *Nature* 394 (1998): 369–374. The initial report of mouse cloning using injection to introduce the nucleus into the recipient egg cell.

Wakayama, T. W., and R. Yanagimachi. "Cloning of Male Mice from Adult Tail-tip Cells." *Nature Genetics* 22.2 (1999): 127–128. The first example of cloning from male mammals.

— JEREMY CREIGHTON AHOUSE

CNIDARIANS

Cnidarians, such as jellyfish, anemones, corals, and hydroids, are a very diverse and an ancient group of simple metazoans (multicellular animals) found largely in marine and freshwater environments (with the exception of a few hydrozoans, like *Hydra*). Along with the distantly related Ctenophora (or comb jellies), cnidarians are the simplest animals with true differentiation of tissues. Although they were once grouped with ctenophores as the coelenterates, it is now clear that they represent an independent evolutionary origin.

The Cnidarian Body Plan. Cnidarians are among the oldest extant animals. Fossil deposits from the Precambrian era (such as the Ediacara deposits in Australia) contain numerous imprints of animals that are most likely ancient, extinct cnidarians. The cnidarian body is radially symmetric and takes the shape of either a polyp (like a hydra) or a medusa (like a jellyfish). Polyps are sessile (attached to the bottom) and are characterized by a stalk, or tube, topped with a mouth that is surrounded by tentacles (for prey capture). Medusae are usually free-swimming, generally bell-shaped, often with tentacles surrounding the margin of the bell, and a mouth on the underside (or subumbrellar) surface of the bell. Although some cnidarian life cycles consist only of a polyp or medusa stage, many cnidarians have both polyps and medusae in their life cycle. The gut, or gastrovascular cavity, is a simple sack that only opens at the mouth. The nervous system is arranged as a nerve net with no centralized processesing centers. Oxygen is taken in, and wastes are excreted, by diffusion through the tissue layers. Cnidarians can be hermaphrodites or have separate sexes.

Cnidarians are diploblasts, having only two tissue layers, called the epidermis and gastrodermis, which are separated by a layer of gelatinous material called the mesoglea. The gastrodermis lines the inside of the animal and contains cells that function in digestion and absorption of nutrients. Additionally, many cnidarians also house symbiotic algae in their gastrodermal cells. The mesoglea is a jellylike substance sandwiched between the epidermis and gastrodermis and helps give cnidarians their shape. The epidermis forms the outside of the animal and contains the cnidarian-specific cnidocyte cells. Cnidocytes contain an organelle, called a cnidae, that produces an eversible, threadlike structure that is discharged from the cell in response to mechanical and chemical stimuli (such as contact with a prey item). Cnidocytes function mostly in feeding and defense. The most common cnidocytes, the stinging nematocysts, are found in all major groups of cnidarians. The eversible threads of nematocysts have barbed tips and contain toxins that disable their prey.

Cnidarians are predators, catching zooplankton and other small marine animals with their tentacles. In addition to ingesting prey, many cnidarians are symbiotically associated with populations of single-celled marine algae, which they house in their tissues. These algae transfer photosythetically derived, energy-rich molecules to the host, in return for access to nitrogen, phosphorus, and other nutrients derived from host metabolism. The energy transferred to the host contributes to daily maintenance, growth, and reproduction. Generally the algae are either *Chlorella*-like green algae or dinoflagellates (and sometimes both), and are housed either in the gastroderm or mesoglea of the host. The population sizes of the algae are very large, with over a million algal cells found under a 1 square centimeter area of host tissue. Corals and other tropical, symbiotic cnidarians mostly harbor dinoflagellates (loosely termed zooxanthellae) in the genus *Symbiodinium*, though several other dinoflagellate groups contain a few symbiotic members. Recent studies of dinoflagellate phylogeny indicate that the *Symbiodinium*-like algae belong to an ancient, and diverse, lineage (possibly Cambrian age or older), suggesting that algal-cnidarian symbioses have been around for quite some time.

Characteristics of the Four Cnidarian Classes. Cnidarians are grouped into four classes: Hydrozoa, Scyphozoan, Cubozoa, and Anthozoa. Hydrozoans are generally small, mostly colonial animals that usually exist as polyps. They live attached to rocks, sea grasses, and other solid objects, where they feed on small crustaceans and other zooplankton. There are around 2,700 described species of hydrozoans, and most species have a life cycle with both polyp and medusa stages, although many species have lost the medusoid stage by a shift of developmental timing (paedomorphosis *Hydra* is one of these exceptions, having only a polyp stage).

Hydrozoans start life as small, oblong swimming planula larvae. Larvae settle on solid surfaces and undergo metamorphosis into the polyp form. Polyps grow and, in colonial species, asexually divide to make colonies with many connected polyps. Often these colonial polyps are polymorphic, with some dedicated to feeding or defense (gastrozoids and dactylizoids), whereas others are involved in reproduction (gonozoids). After reaching reproductive size, medusae develop from either the body wall of the polyp or directly from gonozoids. Medusae then develop gametes (eggs and sperm) that are released into the water for external fertilization.

Most hydrozoans are small and are rarely seen by the casual observer. Siphonophores, however, are large hydrozoans that consist of a float or swimming bell that supports a "fishing net" of polyp-covered feeding tentacles. Most siphonophorans live beneath the surface, far from shore. However, siphonophores like *Physalia* spp., or the man-of-wars, float along the surface supported by a large, brightly colored float and may be found near shore or washed up on beaches. Although they superficially resemble medusae, it is thought that they are actually "floating colonies" and, like many sessile hydrozoan colonies, show extensive polymorphism in their zooids.

Scyphozoans, or jellyfish, are found throughout the oceans of the world and feed mostly on small planktonic animals (zooplankton) and sometimes fish. There are about 200 described species of scyphozoans, and, like most hydrozoans, they have three separate life stages: a sexually produced planula stage, a strictly asexual polyp stage (usually not colonial), and a dominant, sexual, medusa stage. Scyphozoan medusae have a thick, elastic mesoglea and range in size from a few centimeters to over a meter wide. Feeding is done using tentacles arranged around the margin of the bell or with oral arms that develop out of the center of the subumbrellar surface. Several jellyfish species also harbor symbiotic dinoflagellates. One such jellyfish, *Mastigias* sp., is found in an isolated, saline lake in the Palau islands. This species vertically migrates every day, spending nights in deep, nutrient-rich water and days in well-lit, near-surface water. This daily movement exposes the algae to both the nutrients they require for growth and the high light levels necessary for photosynthesis.

Jellyfish are not normally considered "nuisance" species, but several jellyfish species have become so numerous that they may be altering the ecosystems in which they are found. Moon jellies (*Aurelia aurita*) and the recently introduced Australian spotted jellyfish (*Phyllorhiza punctata*) have reached unprecedented population sizes off the coast of Louisiana, in the United States, aided by the numerous oil rigs on which the polyp stage can grow and the rich phytoplankton blooms that are sustained by the unnaturally high levels of nutrients arriving from the Mississippi. By consuming large numbers of fish eggs and larvae, as well as competing with juvenile fish for zooplankton prey, these jellyfish may pose a serious threat to important commercial and sport fisheries.

Cubozoans, or box jellyfish, are another class of cnidarians in which the medusa is the dominant life stage. Medusae of cubozoans are small, rectangular boxes (1–10 cm in length), with four nematocyst covered tentacles, or bundles of tentacles, at the subumbrellar corners of the box. Cubozoans feed on fish and crustaceans, and are fast and agile swimmers. Unlike any other cnidarians, cubozoans have numerous well-developed eyes along the margin of the bell. These eyes are remarkably complex, having a cellular cornea, a lens, pigment cells, and a photoreceptive, enervated retina. It appears the eyes are used for feeding and during reproduction. There are only about twenty described species of box jellyfish, but the complete life cycle is known from only a few species. The life cycle is very similar to scyphozoans, with one intriguing exception: box jellies mate. In species so far examined, sperm transfer occurs when males and females form mating pairs. Developing eggs are either brooded until the planula larvae emerge or are broadcast into the plankton. Cubozoans are well known for being highly toxic, with some species capable of killing a human.

Anthozoa has the greatest species diversity of the four cnidarian classes, with over 6,000 described species. Anthozoans also have the greatest level of morphological diversity, encompassing sea anemones, sea pansies, corallimorphs, sea whips and fans, and soft and hard corals. The polyp is the dominant life stage and shows considerably greater complexity than hydrozoan and scyphozoan polyps. Unlike other cnidarians, the anthozoan gut is partially subdivided by thin, folded layers of tissue (mesenteries) that are vertically arranged around the inside of the gut. Also unique to the Anthozoa is the complete absence of a medusa stage: planula develop into a reproductively capable polyp. Planula development can occur either within the gastrovascular cavity (in brooding species) or in the external milieu (broadcasting spawning species).

Stony corals, perhaps the most famous cnidarians, are anthozoans in the order Scleractinia. Corals can be either colonial or solitary and secrete an external, calcium carbonate skeleton that provides protection and support to the overlaying polyp(s). In reef-building or hermatypic corals, the skeletons can grow to considerable size and provide the foundation upon which tropical coral reefs are built. Coral reefs are the largest structures built by biological processes, with the Australian Great Barrier, at over 6,000 kilometers long, being the largest biological structure in the world. Coral reefs are home to an incredible diversity of marine life, rivaling tropical rainforests in species diversity. With one known exception, all hermatypic coral species harbor symbiotic dinoflagellates, and it is this symbiotic relationship that promotes the incredible success of reef corals. Reef corals are primarily restricted to nutrient-poor, shallow tropical oceans. The presence of the symbiotic algae provides corals with a reliable source of food that would otherwise be difficult to obtain in these relatively barren seas. In addition, symbiotic dinoflagellates also enhance coral skeleton growth, allowing the corals to grow at considerably faster rates than they could without them. Therefore, the presence of the algal symbionts is largely responsible for the awesome diversity of organisms found in modern coral reefs.

Coral colonies may lose some or all of their algal symbionts if stressed. This process is termed bleaching because the loss of algal symbionts allows the underlying, white skeleton to become visible. Given the importance of algal symbionts to coral health, it is no surprise that losing those symbionts can lead to reduced growth and reproduction and to increased mortality. The most common causes of bleaching are high seawater temperatures and increased light levels, which appear to damage the symbionts' photosynthetic machinery (photosystem 2). The combination of high light and temperature occurs most frequently during summer doldrums, when seas are calm and water clarity is maximal. During periods of very high seawater temperatures ($> 30°C$), mass bleaching, where large fractions of the corals present on a reef may bleach, can occur over large geographic areas. During the El Niño of 1987, a mass bleaching event on several coral reefs along the east coast of Panama and Costa Rica resulted in almost 100 percent of the corals becoming bleached. Over 90 percent of the corals found on these reefs died, drastically changing the ecosystem. Mass bleaching events have increased in frequency and intensity over the last two decades and have become a major concern to reef biologists and conservationists. The cause of this increase has been linked to elevated seawater temperatures resulting from global warming. Given current global climate models, seawater temperatures are expected to continue increasing, and this may threaten the vary existence of coral reefs. However, our understanding of the causes and consequences of bleaching is still scant, and thus limits our ability to make strong predictions. For example, it is hypothesized that corals that can harbor more than one species of symbiont may be more stress resistant and may recover from bleaching more quickly. The reasoning is that different species of symbionts may be more resistant to stress than others, so that bleaching corals that contain, or can acquire, these resistant algae will do better. In the Caribbean, corals *Montastrea annularis* and *M. faveolata*, colonies that live in shallow water, have a different complement of symbiont species than colonies that live in deep water. Furthermore, during a bleaching event, the symbiont population will shift toward algae that are found in shallow water, presumably because they are better adapted to the higher light and temperature levels typically found in shallow waters. However, the extent to which algal symbiont variation can reduce the effect of bleaching is currently unknown and is being actively investigated by biologists around the world.

Evolutionary Relationships of the Four Cnidarian Classes. The relationships of the four cnidarian classes have been debated for nearly 150 years. Much of this debate has centered on "which came first," the medusae, or the polyp. Because of the orientation of their mouth and tentacles—polyps superficially appear to be inverted medusae—it has been proposed that they are homologous structures. In the Cubozoa, the polyp actually metamorphose directly into a swimming medusa. Because some Cnidarian taxa have both polyps and medusae, and others have one or the other, cnidarian relationships can shed light on the evolutionary origin of the curious alternation of generations seen in many extant cnidarians.

This debate was started by Ernst Haeckel, who argued that if "ontogeny recapitulates phylogeny," because the polyp comes first in the developmental sequence, it must represent the primitive condition, placing the medusa-less anthozoa at the base of the Cnidaria. Although Haeckel's credo is not universally applicable, molecular phylogenetic methods strongly support his hypothesis that the Anthozoa is the sister taxon to the remaining cnidarian classes. This is supported not only by phylogenies of DNA sequences, but by the observation that only the Anthozoa have the primitive condition of possessing circular mitochondrial DNA. The three classes with medusae (Scyphozoa, Cubozoa, and Hydrozoa) all have linear mitochondrial DNA, the only instance in the animal kingdom, suggesting that the "medusozoa" form a monophyletic group (Bridge et al., 1992).

[*See also* Animals.]

BIBLIOGRAPHY

Arai, M. N. *A Functional Biology of Scyphozoa.* New York, 1997. The most comprehensive book on the biology of scyphozoans recently published, with an extensive bibliography that appears to contain all jellyfish references published up to 1996.

Bridge, D., C. W. Cunningham, B. Schierwater, R. DeSalle, and L. W. Buss. "Mitochondrial DNA Structure and Cnidarian Phylogeny." *Proceedings of the National Academy of Sciences USA* 89 (1992): 8750–8753.

Brusca, R. C., and G. J. Brusca. *Invertebrates.* Sunderland, Mass., 1990. Chapter 8 introduces the Cnidaria and is organized around functional systems, largely ignoring classification until the last few pages. This approach is somewhat confusing, but the chapter is well worth a read for its interesting and illuminating examples.

Rowan, R., N. Knowlton, A. Baker, and J. Jara. "Landscape Ecology of Algal Symbionts Creates Variation in Episodes of Coral Bleaching." *Nature* 388 (1997): 265–269.

Rowan, R., and D. A. Powers. "Ribosomal RNA Sequences and the Diversity of Symbiotic Dinoflagellates (Zooxanthellae)." *Proceedings of the National Academy of Sciences USA* 89 (1992): 3639–3643. One of the first papers to examine the phylogenetic relationships among symbiotic dinoflagellates. This paper, as well as others by Rowan, revolutionized the study of zooxanthellae symbioses.

Rupert, E. E., and R. D. Barnes. *Invertebrate Biology.* Fort Worth, Tex., 1994. The fourth chapter of the book is a very lucid introduction to the cnidarians. The chapter introduces basic cnidarian biology, then proceeds to examine each of the major cnidarian groups in detail.

Veron, J. E. N. *Coral of the World.* Australian Institute of Marine Science and CRR Qld Pty Ltd., Townsville, Australia, 2000. An absolutely incredible book that covers all of the scleractinian corals known to the author, and probably to science. The book contains an excellent introductory chapter that covers coral biology, the coral geological history, and coral reef structure, as well as several chapters at the end of the book on biogeography, coral speciation, and a binomial key to the genera and species. The book is well illustrated, with each species description accompanied by beautiful color photographs, close-up photographs of skeletal features, and range maps; truly a phenomenal book.

— TOM WILCOX

COALESCENT THEORY AND MODELING

The coalescent is a probability model for the tree underlying a sample of homologous DNA sequences drawn from a population. It is designed to model within-species (genealogical) trees, and not between-species (phylogenetic) trees, for which different modeling assumptions are usually appropriate.

In some applications of coalescent models, the genealogical tree is the focus of interest—for example, in human history and in some aspects of conservation genetics. In the majority of applications, however, the genealogy is not of primary interest but must be accounted for so that we can investigate effectively the unknowns of interest. These include evolutionary parameters such as mutation or recombination rates, or demographic parameters such as historic population sizes or migration rates, or the locations of genes involved in a phenotype of interest, such as a human disease.

The usefulness of coalescent models stems from the fact that they focus on the observed sample, tracing the ancestral lineages of the sequences in the sample back in time to their Most Recent Common Ancestor (MRCA). This focus leads to important gains in computational efficiency, particularly for simulation, over methods that model the entire population forward in time. Moreover, the focus on samples is well suited to statistical inference: it leads to likelihood-based statistical methods, giving statistical efficiency and facilitating the quantitative comparison of models. Coalescent models are mathematically tractable, largely because of the separation of mutation from the demographic processes. Finally, coalescent models can approximate a wide class of neutral population genetics models, as well as some nonneutral models.

The standard coalescent model is based on the following assumptions: neutrality and panmixia; a population size that is large and constant; and recombination that is absent or negligible. These assumptions can all be weakened to some extent, and at some computational cost. A wide class of mutation models can be incorporated into the coalescent; although we speak here of the observed data as "sequences," this is intended to incorporate, for example, microsatellite and Single Nucleotide Polymorphism (SNP) data, or even haplotypes composed of a variety of data types.

The Standard Coalescent. Consider a haploid population of size N, evolving according to the Wright–Fisher model, which assigns probability $1/N$ to the event that two given sequences in any generation are descended from a common ancestor in the previous generation. It follows that the probability that the two sequences "coalesce" in their MRCA k or more generations ago is

$$P(\text{coalescence time} \geq k \text{ generations ago}) = \left(1 - \frac{1}{N}\right)^{k-1}$$

This formula specifies the geometric probability distribution with parameter $1/N$, which has mean and standard deviation (sd) both equal to N. Unless N is very small, this distribution is well approximated by the ex-

ponential distribution with rate parameter $1/N$ (written Exp($1/N$)), which also has mean = sd = N. Since mean coalescence times, and many other quantities of interest, scale linearly with the population size N, it is convenient to express time in units of N generations. Under this time scaling, the coalescence time T_2 of two sequences has approximately the Exp(1) distribution, which has mean = sd = 1 and

$$P(T_2 > t) = \exp(-t) \qquad (1)$$

Now consider tracing the genealogy of $n = 3$ sequences. There are initially three possibilities for the first pair of sequences to coalesce, and hence the (scaled) time to the first coalescence event has approximately the Exp(3) distribution (mean = sd = 1/3). Tracing it further back in time, there are now two lineages, and so we are in the same situation as for the $n = 2$ case. Thus T_3, the overall coalescence time of the sample, is the sum of independent Exp(1) and Exp(3) random variables. Note that coalescence events occur at distinct times, but the time difference between two consecutive events may be arbitrarily small.

These arguments generalize readily to arbitrary sample size n, provided that $n \ll N$. The time during which there are exactly j remaining lineages ($2 \leq j \leq n$) ancestral to the sample has approximately the Exp($j(j - 1)/2$) distribution. Expectations and variances of T_n and U_n, the total branch length of the tree, are given in Table 1.

Mutations in the standard coalescent occur along the branches of the genealogical tree independently and at constant rate $\theta/2$, where

$$\theta = 2N\mu$$

and μ is the mutation rate per sequence per generation. Therefore, conditional on the value of U_n, the number of mutations occurring in the ancestry of the sample, since its MRCA, has the Poisson distribution with mean $\theta U_n/2$.

Although the coalescent allows great flexibility in the choice of mutation model, attention is often restricted to a few standard models, both for comparability of results and for computational efficiency. For DNA sequence data, the infinitely-many-sites model is often adopted, under which each mutation occurs at a distinct site in the sequence.

Insights from the Standard Coalescent. Although mathematically straightforward and based on simple assumptions, the standard coalescent model has powerful implications that give important insights into the nature of genetic samples, and that are likely to hold at least approximately even when the modeling assumptions are not strictly valid.

One such insight is that the distribution of a genealogy under the standard coalescent model is dominated

TABLE 1. Expectations and Variances of T_n and U_n

Mean and variance of T_n, the time since the MRCA of a sample of size n, and U_n, the total branch length of the tree, under the standard coalescent model (1 time unit = N generations).

	Formula	$n = 2$	$n = 3$	$n = 10$	$n \uparrow \infty$
E (T_n)	$2(1 - 1/n)$	1	1·33	1·8	2
V (T_n)	$\sum_{j=2}^{n} 8/j^2 - 4(1 - 1/n)^2$	1	1·11	1·16	1·16
E (U_n)	$\sum_{j=2}^{n} 2/(j - 1)$	2	3	5·66	$1 + 2\log_e(n)$
V (U_n)	$\sum_{j=2}^{n} 4/(j - 1)^2$	4	5	6·16	6·58

by the time during which the sample has only two ancestral lineages. Over 50 percent of $E(T_n)$, and 85 percent of $V(T_n)$, can be attributed to this period. Consequently, samples may exhibit clustering into two or more groups, and resulting frequency distributions are often bimodal. This observation is important because researchers may be tempted to interpret clustered or bimodal data sets in terms of, say, population structure or selection, whereas the coalescent model shows that they can arise under neutrality and panmixia.

Another insight follows from the fact that U_n increases logarithmically with n. For many purposes, U_n corresponds to the amount of information in the sample, and the logarithmic increase in information here contrasts with the linear increase in standard statistical settings. For example, the most efficient estimators of the mutation parameter θ have variances that decline only like $1/\log(n)$, and some recently popular estimators have variances that do not even converge to zero as n increases. It follows that there is often a diminishing benefit from increasing the sample size, which has implications for the design of genetic surveys.

Note further the high variance of T_n relative to its mean. The same applies to U_n, except when n is extremely large. Distinct loci may display very different levels of allelic variability, not because of differences in mutation rate or selection pressure, but simply because the genealogies at the two loci are of very different heights. This phenomenon may confound naive attempts to investigate the heterogeneity of mutation and/or recombination rates. It does, however, suggest methods for detecting recombination events under an assumption of constant mutation rate.

The probability that the MRCA of the sample is the same as the MRCA of the entire population is $(n - 1)/(n + 1)$, which reaches 1/2 for a sample size as small as three. However, in the case that the two MRCAs are distinct, the time difference between them is expected to be very large.

Generalizations.

Beyond Wright–Fisher and haploidy. The range of applicability of the standard coalescent can be extended greatly by the simple device of a linear rescaling of time. Instead of interpreting each coalescent time unit as N generations, where N is the (census) population size, we write N_e for the time-scale factor, called the *effective population size*. The definition of N_e depends on the system to be modeled.

Under the Wright–Fisher model, the family sizes (i.e., the numbers of copies of each sequence transmitted to the next generation) have jointly the multinomial distribution, each with mean one. Constant population size requires the mean family sizes to be one, but beyond this they need only be *exchangeable*, which means that the joint distribution is unchanged under reordering of the families. When σ^2, the variance of the family size, is high, most sequences have no copies in the next generation, while a few have many copies. This makes it more likely that two sequences will have the same parent and hence speeds up unscaled coalescence events. The standard coalescent model still applies, provided that each unit of coalescent time is interpreted as $N_e = N/\sigma^2$ generations. In the Wright–Fisher case, $\sigma^2 = (1 - 1/N) \approx 1$.

The assumption of discrete, nonoverlapping generations in the derivation of the standard coalescent model is not essential but is retained here for ease of presentation. The assumption of haploidy is also dispensable. (In nonhaploid settings, we use n and N for numbers of sequences, not individuals.) Two sequences in a diploid, nonselfing organism cannot be descended from the same sequence in the previous generation. However, under random mating there are no restrictions on earlier generations, and a single generation is negligible in the coalescent time scaling. If the population is dimorphic, with sex ratio $p/(1 - p)$, the time-scaling constant is $N_e = 4p(1 - p)N$. The smaller the value of p, the faster coalescences occur (in unscaled time), because half the sequences in the population come from the minority sex and are thus more likely to have a common parent. Hermaphrodites with selfing proportion F can also be accommodated under the standard coalescent, with $N_e = (1 - F/2)N$.

Although powerful, the standard coalescent with linear time-scaling is not universally valid for modeling genealogies in natural populations, as illustrated by the situations described below. Similarly, the notion of effective population size is not universally meaningful.

Variable population size. Suppose that the population size $N_e t$ generations ago was $N_e \lambda(t)$, where N_e denotes the current effective population size, so that $\lambda(0) = 1$. The (scaled) coalescence time distribution for a pair of sequences is now given by

$$P(T_2 > t) = \exp\left(-\int_0^t \frac{ds}{\lambda(s)}\right) \qquad (2)$$

When the population size is large—that is, $\lambda(t)$ is large—the integral in equation (2) increases only slowly with t, corresponding to the fact that coalescences rarely occur (because two sequences are unlikely to have the same parent if the population size is large). Notice that the integral simplifies to just t in the standard coalescent case that $\lambda(t) \equiv 1$, thus recovering equation (1).

Exponential growth or decline forward in time at rate r per generation corresponds to $\lambda(t) = \exp(-Rt)$, where $R = N_e r$. This is unlikely to provide a good model for many natural populations, but it offers a simple one-parameter family for investigating the robustness of inferences about changes in population size, a family that includes a wide range of distributions for the shape of the genealogical tree.

When $R > 0$ (exponential growth), there are fewer recent coalescences than under the standard coalescent with the same value of U_n. Consequently, there tend to be more mutations that are represented only once in a sample. In the extreme case of very large R, all coalescence events occur at about the same time. Moreover, T_n is approximately constant over realizations of the model.

When $R < 0$ (exponential decline), there are relatively many recent coalescences; consequently, samples display even more clustering than under the standard coalescent. In fact, each coalescence event has positive probability that it does not occur in finite time:

$$P(T_2 > t) = \exp((1 - \exp(Rt))/R) \to_{t\uparrow\infty} \exp(1/R) \neq 0$$

Although infinite coalescence times are not strictly realistic, they can be interpreted in terms of sequences separated by a very large number mutations.

Population subdivision. A simple model of population subdivision is the so-called (finite) "island model," in which N_i individuals reside in subpopulation i, $i = 1, 2, \ldots, K$, and each has (independently) probability m_{ij} of being descended from a parent in subpopulation j. See Herbots (1997) for details of the "structured coalescent" approximation to the genealogy of samples drawn from an island model.

Properties of genealogies under the structured coalescent depend on the m_{ij} and the N_i. In the simplest setting of $N_{ij} \equiv N/K$ and $m_{ij} \equiv m$, for all i and $j \neq i$, the

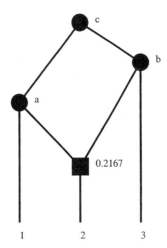

FIGURE 1. Illustration of an Ancestral Recombination Graph (ARG) with $n = 3$ DNA Sequences.
Coalescence events are marked with circles, the single recombination event is marked with a square and labeled with the breakpoint. Tracing lineages backwards in time (up the page), leftmost 21.67 percent of sites on sequence 2 (rounded to the nearest site) take the left fork at the square and coalesce with the corresponding sites on sequence 1 at a. The remaining sites on sequence 2 take the right fork and coalesce with sequence 3 at b. The MRCA of all three sequences occurs at c. Mutations are not shown. Any mutation occurring on the lineage from c to a will affect observed sequence 1; it will also affect sequence 2 if its location on the sequence is to the left of the breakpoint. Mutations occurring on the lineage from a to the recombination and left of the breakpoint will affect sequence 2; if to the right of the breakpoint they will have no effect on the observed sequences.   Drawing by David Balding.

value of $E(T_2)$ for sequences drawn from the same subpopulation is the same as under the standard coalescent. However, this should not be misinterpreted as implying that subdivision can be ignored: genealogies are very different under the two models. Under the structured coalescent, both short and long genealogies are more numerous than under the standard coalescent. Indeed, $V(T_n)$ becomes large without bound as m becomes small: with small probability one or both of the sequences has migrated since their MRCA, in which case T_2 tends to be extremely large because sequences in different subpopulations cannot coalesce until further migration events bring them to the same subpopulation.

Recombination. If data are obtained at several unlinked loci, then the standard coalescent model can be applied independently at each locus. For loci that are neither unlinked nor completely linked—for example, haplotypes of closely spaced SNP loci or long autosomal DNA sequences—a natural assumption is that the standard coalescent applies at each locus (or site), and that recombination events occur independently at a rate that is constant over time but may vary along the chromosome.

The Ancestral Recombination Graph (ARG; see Figure 1) is a device for representing on a single diagram the genealogical trees at every locus under this model. The leaves of the graph correspond to homologous chromosome segments that include all the observed loci, with two of them forming its endpoints, labeled arbitrarily "left" and "right." Lineages are traced back in time, with each pair coalescing at unit rate, as in the standard coalescent. Moreover, mutations occur independently on each lineage at a rate that is constant over time but may vary across loci. The novel feature of the ARG is that branching events also occur, at rate $\rho/2$ along each lineage, where $\rho = 2Nr$ and r is the recombination rate per segment per generation. Each branching of lineages is interpreted as a recombination event that occurs at a break point whose location along the segment is chosen with probabilities in proportion to the recombination rate. Lineages of loci to the "left" of the break point are regarded as following the left fork of the branch, while loci to the "right" take the right fork (Figure 1). Since $n \ll N$, recombinations are assumed not to occur between two chromosome segments both ancestral to the sample.

Recombination events increase the number of lineages, but since the overall coalescence rate is quadratic in the number of extant lineages, whereas the recombination rate is linear, the ARG eventually reaches a single lineage, the overall MRCA. Note that individual loci may reach their MRCA before the overall MRCA of the segment is reached. (See Griffiths and Marjoram, 1997, for further details of the ARG and its properties.)

Selection. The derivation of the standard coalescent outlined above relies crucially on neutrality, which allows demography to be modeled separately from mutation. However, to date there have been two successful approaches to introducing models of selection into the coalescent framework. The Ancestral Selection Graph (ASG) resembles the ARG in that branching points arise in the ancestral lineages. However, one of the forks at each branch will eventually be deleted, but only after mutations have been laid down on all the branches, so that branches bearing deleterious alleles may be preferentially deleted (see Neuhauser, 2001). Some models of selection may be reformulated as models of population subdivision, in which allelic states correspond to subpopulations, and mutation between states corresponds to migration (see Nordborg, 2001).

Statistical Inference.

Exact likelihood methods. As a simple illustration of likelihood-based methods, consider a sample of $n = 2$ DNA sequences under the standard coalescent with the infinitely-many-sites mutation model. All of the information in the data is captured in S, the number of sites

at which the sequences differ. Before observing its value, the coalescent model specifies an Exp(1) distribution for T_2. After S is observed, T_2 has the gamma distribution with

$$E(T_2 | S) = \frac{1 + S}{1 + \theta} \quad \text{and} \quad V(T_2 | S) = \frac{1 + S}{(1 + \theta)^2}$$

This may be compared with pre-data values $E(T_2) = V(T_2) = 1$. Given the mutation parameter θ, we have $E(S) = E(\theta U_2/2) = \theta$. As might be anticipated, if we observe $S < \theta$, then $E(T_2 | S) < E(T_2)$, and vice versa. Further, $V(T_2 | S) < V(T_2)$ unless S turns out to be more than twice its expected value.

As this example illustrates, the coalescent model can be thought of as specifying a *prior* distribution for the genealogical tree, which after data are observed can be updated to specify a *posterior* distribution. Inferences about the tree are best presented in terms of this posterior distribution. For example, a method-of-moments estimator of T_2 that ignores the coalescent prior has uniformly worse mean square error than the posterior median (see Tavaré et al., 1997).

For inferences about θ, the tree parameters need to be integrated out with respect to their coalescent prior distribution in order to obtain $L(\theta)$, the marginal likelihood for θ. Continuing the $n = 2$ example:

$$L(\theta) = P(S | \theta) = \int_0^\infty P(S | T_2, \theta) \pi(T_2) dT_2$$

Here π denotes the probability density function (pdf) specified by the coalescent model, that is, the Exp(1) pdf. Here, $P(S | T_2, \theta)$ is the Poisson (θT_2) probability mass function, and the integration over T_2 can be performed exactly to obtain

$$L(\theta) = \frac{\theta^S}{(1 + \theta)^{S+1}}$$

Maximizing $L(\theta)$, we obtain the MLE $\hat{\theta} = S$.

When $n > 2$, a marginal likelihood for θ can be evaluated exactly in the special case that no variation is observed in the sample, and it is assumed that these data reflect no mutations in the underlying genealogy. Then $L(\theta)$ is the product over j of the probability that no mutation occurs when the genealogy has exactly j lineages:

$$L(\theta) = P(S = 0 | \theta) = \prod_{j=2}^n \frac{j - 1}{j - 1 + 2\theta}$$

The MLE is $\hat{\theta} = 0$, which is implausible *a priori*.

The likelihood $L(\theta)$ can be multiplied by a prior pdf for θ based on any background information about N_e and μ—for example, an independent estimate of μ. A posterior distribution for θ can then be obtained via Bayes Theorem, as well as marginal posterior distributions for N_e and μ. Note that the data bear directly only on the compound parameter θ, so that although inferences about θ may be robust to the prior assumptions given sufficient data, inferences about N_e and μ are always sensitive to the prior.

Approximate likelihood methods. In most settings of interest, exact calculation of $L(\theta)$ is intractable and it is necessary to resort to stochastic approximation methods. Writing π for the coalescent prior pdf, and G for the genealogy (including branch lengths and topology but not mutations, so that G does not depend on θ), we can write:

$$L(\theta) = P(\text{data}|\theta) = \int P(\text{data}|G, \theta)\pi(G)dG$$

A simple approach is to replace this integral by the approximation

$$L(\theta) \approx \frac{1}{m} \sum_{i=1}^{m} P(\text{data}|G_i, \theta)$$

where $G_1, G_2, \ldots, G_m$ are independent trees drawn from π. Unfortunately, unless n is small, this simple approach is usually very inefficient, because trees consistent with the data arise only rarely, so that $P(\text{data}|G_i, \theta)$ will be negligible for all but a few values of i.

It may be possible to improve matters if, instead of drawing the G_i from π, we choose some other pdf π' that is concentrated on trees supported by the data. We then compensate for this choice by reweighting terms in the sum:

$$L(\theta) \approx \frac{1}{m} \sum_{i=1}^{m} P(\text{data}|G_i, \theta) \frac{\pi(G_i)}{\pi'(G_i)}$$

This is known as *importance sampling*; it is potentially a very powerful method, but its success depends on careful choice of π'. The theoretical best choice is the posterior distribution of G given the data:

$$\pi'(G) = P(G|\text{data}, \theta) = \frac{P(\text{data}|G, \theta)\pi(G)}{L(\theta)}$$

This, however, requires knowledge of $L(\theta)$, whose computation is the problem we are trying to solve. Nevertheless, useful choices for π' are sometimes available; *see* Stephens and Donnelly (2000).

Another approach is offered by a class of methods known as *Markov Chain Monte Carlo* (MCMC). In these methods, $G_1, G_2, \ldots, G_m$ is generated via a random walk in the space of possible trees, constructed such that each G_i is approximately a draw from $P(G|\text{data}, \theta)$. (The G_i are not usually independent, but their correlation can be made arbitrarily weak by retaining only G_{ki}, for some suitably large k.) Since MCMC methods generate approximate random samples from the posterior distribution for G, they lead immediately to approximate inferences about T_n or other aspects of the genealogical tree. Moreover, by specifying a prior distribution for θ and including it in the MCMC algorithm, one can draw inferences about all values of θ from a single MCMC run, thus avoiding separate runs to evaluate $L(\theta)$ at different values of θ. For a survey of importance sampling, MCMC, and related methods for drawing inferences about genetic samples under coalescent models, see Stephens (2001).

Nonlikelihood methods. Although likelihood methods are more statistically efficient than alternative methods—for example, those based on summary statistics—they remain computationally intensive. For some nonlikelihood methods of inference based on coalescent models, see Fu and Li (1999), and also the rejection-sampling approach of Pritchard et al. (1999).

[*See also* Molecular Evolution; Molecular Systematics; *articles on* Phylogenetic Inference.]

BIBLIOGRAPHY

Donnelly, P., and S. Tavaré. "Coalescents and Genealogical Structure under Neutrality." *Annual Review of Genetics* 29 (1995): 410–421.

Fu, Y-X., and W-H. Li. "Coalescing into the 21st Century: An Overview and Prospects of Coalescent Theory." *Theoretical Population Biology* 56 (1999): 1–10.

Griffiths, R. C., and P. Marjoram. "An Ancestral Recombination Graph." In "Progress in Population Genetics and Human Evolution," edited by P. Donnelly and S. Tavaré. *IMA Volumes in Mathematics and Its Applications* 87 (1997): 257–270.

Herbots, H. "The Structure Coalescent." In "Progress in population genetics and human evolution." *IMA Volumes in Mathematics and Its Applications* 87 (1997): 231–255.

Hudson, R. "Gene Genealogies and the Coalescent Process." In *Oxford Surveys in Evolutionary Biology*, edited by D. Futuyma and J. Antonovics, vol. 7, pp. 1–44. Oxford, 1991. A seminal paper introducing coalescent concepts to the biological literature, and still a useful introduction.

Kingman, J. F. C. "The Coalescent." *Stochastic Processes and Their Applications* 13 (1982): 235–248. A landmark paper setting out the mathematical foundations of coalescent theory.

Neuhauser, C. "Mathematical Models in Population Genetics." In *Handbook of Statistical Genetics*, edited by D. J. Balding et al., pp. 153–178. 2001.

Nordborg, M. "Coalescent Theory." In *Handbook of Statistical Genetics*, pp. 179–212. Chichester, 2001.

Pritchard, J. K., M. T. Seielstad, A. Perez-Lezaun, and M. W. Feldman. "Population Growth of Human Y Chromosomes: A Study of Y Chromosome Microsatellites." *Molecular Biology and Evolution* 16 (1999): 1791–1798.

Stephens, M. "Inference under the Coalescent." In *Handbook of Statistical Genetics*, pp. 213–238. Chichester, 2001.

Stephens, M., and P. Donnelly. "Inference in Molecular Population Genetics" (with discussion). *Journal of the Royal Statistical Society* B 62 (2000): 605–635.

Tavaré, S., D. J. Balding, R. C. Griffiths, and P. Donnelly. "Inferring Coalescence Times from DNA Sequence Data." *Genetics* 145 (1997): 505–518.

— DAVID BALDING

CODON USAGE BIAS

Living organisms use 20 amino acids to construct their protein molecules. These amino acids are coded for by

TABLE 1. The "Universal" Genetic Code

Codon	AA	Codon	AA	Codon	AA	Codon	AA
TTT	Phe	TCT	Ser	TAT	Tyr	TGT	Cys
TTC	Phe	TCC	Ser	TAC	Tyr	TGC	Cys
TTA	Leu	TCA	Ser	TAA	Stop	TGA	Stop
TTG	Leu	TCG	Ser	TAG	Stop	TGG	Trp
CTT	Leu	CCT	Pro	CAT	His	CGT	Arg
CTC	Leu	CCC	Pro	CAC	His	CGC	Arg
CTA	Leu	CCA	Pro	CAA	Gln	CGA	Arg
CTG	Leu	CCG	Pro	CAG	Gln	CGG	Arg
ATT	Ile	ACT	Thr	AAT	Asn	AGT	Ser
ATC	Ile	ACC	Thr	AAC	Asn	AGC	Ser
ATA	Ile	ACA	Thr	AAA	Lys	AGA	Arg
ATG	Met	ACG	Thr	AAG	Lys	AGG	Arg
GTT	Val	GCT	Ala	GAT	Asp	GGT	Gly
GTC	Val	GCC	Ala	GAC	Asp	GGC	Gly
GTA	Val	GCA	Ala	GAA	Glu	GGA	Gly
GTG	Val	GCG	Ala	GAG	Glu	GGG	Gly

Abbreviations: Codon: A, T, C and G are the four nucleotides adenine, thymine, cytosine and guanine. AA (amino acids): Phe—phenylalanine, Leu—leucine, Ile—isoleucine, Met—methionine, Ser—serine, Pro—proline, Thr—threonine, Ala—alanine, Tyr—tyrosine, His—histidine, Gln—glutamine, Asn—asparagine, Lys—lysine, Asp—aspartic acid, Glu—glutamic acid, Cys—cysteine, Trp—tryptophan, Arg—arginine, Gly—glycine.

64 codons of three nucleotides (e.g., the sequence TTT codes for phenylalanine). Almost all organisms use the same genetic code, which is referred to as the "universal code" (Table 1). Of the 20 amino acids, 18 are encoded by more than one codon. However, these synonymous codons (different codons that code for same amino acid) are not used uniformly; in almost every organism that has been studied, there are distinct preferences for particular codons. For example, although there are six codons that code for leucine in the universal genetic code, almost half of all leucine codons in the bacterium *Escherichia coli* are CUG (Table 2). This pattern is referred to as "synonymous codon bias" or "synonymous codon usage bias."

Patterns of synonymous codon bias differ both between organisms and between genes within an organism. For example, while TTT is preferentially used to encode phenylalanine in *E. coli*, TTC preferred in humans (Table 2); and within humans, the gene alpha-globin almost exclusively uses TTC, whereas dystrophin has no strong preference for either TTT or TTC. Even though there is a great variety of patterns of synonymous codon bias, most organisms basically fit into one of two categories: in some organisms, synonymous codon bias arises from the effects of natural selection on translation (the transfer of genetic information from DNA to further steps in protein synthesis); in others, synonymous codon bias reflects patterns of base composition. There are also some organisms, such as the fruit fly *Drosophila melanogaster*, that show aspects of both patterns.

Translational Patterns. In a variety of organisms—including bacteria, yeasts, nematodes, plants, and insects—it is evident that synonymous codon bias is a consequence of natural selection. The evidence for this comes primarily from two sources: first, the codons used most frequently are those that bind the commonest tRNA (the molecule responsible for matching amino acids to codons during translation), or that bind the tRNA with optimal base pairing; second, synonymous codon bias is positively correlated with gene expression levels. Both of these points are illustrated in Table 3 with data from *E. coli*. For leucine, there is a clear correlation between the use of each of its codons and the concentration of the matching tRNA; this correlation is strong in highly expressed genes but is greatly reduced in genes that are not highly expressed. For phenylalanine, there is a single tRNA that translates both synonymous codons; the preferred codon is the one that forms natural Watson-Crick base pairs at all three codon-positions. As with leucine, the preference for the optimal phenylalanine codon is reduced in genes that are not highly expressed.

Population genetic data also support the role of selection in maintaining synonymous codon bias in *E. coli* and *D. melanogaster*, the two organisms that have been studied in this respect. If we divide codons into those that are preferred in highly expressed genes (e.g., TTC in *E. coli*), and those that are not (e.g., TTT), then we find more preferred to unpreferred synonymous mutations segregating in populations (e.g., TTC → TTT), than unpreferred to preferred mutations (e.g., TTT → TTC). Furthermore, unpreferred mutations segregate at lower frequencies in the population than do preferred mutations. To understand why this is expected under selection, let us imagine that the rate of mutation from preferred to unpreferred codons is equal to the mutation rate from unpreferred to preferred. Selection will increase the number of sites fixed (all members of the population have the same codon) for preferred codons and decrease the number fixed for unpreferred codons; therefore, there will be an excess of preferred to unpreferred mutations. However, each unpreferred mutation is deleterious, whereas each preferred mutation is advantageous; the new unpreferred mutations are therefore expected to segregate at lower frequencies than do the new preferred mutations.

The correlation among tRNA concentration, gene expression level, and synonymous codon use suggests that synonymous codon bias is a consequence of selection to optimize some aspect of translation. However, it is less clear what this is. There are at least three possibilities: selection could be acting to maximize the rate of elongation, to minimize the cost of proofreading, or to maximize the accuracy of translation. The quantitative relationship between tRNA concentration and codon us-

TABLE 2. Proportion of Leucine and Phenylalanine Codons Encoded By Their Synonymous Codons in Four Species

	Codon	Escherichia coli (bacterium)	Saccharomyces cerevisiae (yeast)	Drosophila melanogaster (fruit fly)	Homo sapiens (human)
Leucine codons	TTA	0.13	0.28	0.05	0.07
	TTG	0.13	0.29	0.17	0.12
	CTT	0.11	0.13	0.10	0.13
	CTC	0.10	0.06	0.16	0.20
	CTA	0.04	0.14	0.08	0.07
	CTG	0.49	0.10	0.44	0.41
Phenylalanine codons	TTT	0.57	0.59	0.34	0.44
	TTC	0.43	0.41	0.66	0.56

age in *E. coli* grown under different conditions fits the predictions of a model in which growth rate is determined by elongation rate. However, other data do not fit this model. In *E. coli*, the glutamic acid codon GAA is translated about 3.5 times faster than GAG, irrespective of context. Yet the frequency with which GAA is used only increases in frequency across genes of different expression levels, when the first base of the adjacent codon in the gene is G. This suggests that selection on the elongation rate is not the principle determinant of synonymous codon bias in *E. coli*. In *D. melanogaster*, codon usage bias is strongest in the codons that appear to encode the most important amino acid residues—that is, in the codons in which an error is likely to have the most deleterious effect. This suggests that translational accuracy may be the main determinant of synonymous codon bias, at least in *D. melanogaster*.

Although translational selection appears to be the major selective force acting on synonymous codon use in organisms like *E. coli*, there is evidence that other selective forces are at work as well. For example, synonymous codon bias is lower at the start and end of genes in *E. coli*. This appears to arise from constraints imposed by ribosome binding, although the specific mechanisms are not completely understood.

Compositional Patterns. In many organisms, including most vertebrates, synonymous codon bias does not appear to be associated with the optimization of translation: no correlation is observed between codon bias and the level of gene expression. Instead, synonymous codon bias reflects the underlying base composition of the genome, or the chromosomal segment in which the gene resides. The best-studied pattern is that found in mammals and birds, in which codon usage can be summarized largely in terms of the GC content: some genes tend to use G- and C-ending codons, while others use A- and T-ending codons. The differences between genes in codon usage can be great; in humans, GC content at the third codon-position (GC_3) varies from less

TABLE 3. Proportion of Leucine Codons Used in Highly and Lowly Expressed Genes in *E. coli* and tRNA Concentrations

Codon	Low	High	tRNA concentration*
TTA	0.21	0.01	0.10
TTG	0.15	0.01	0.25
CTT	0.12	0.02	0.30
CTC	0.11	0.03	0.30
CTA	0.05	0.01	minor
CTG	0.36	0.92	1.00

*Concentrations are given relative to the concentration of the tRNA which recognizes CTG, which is given the arbitrary concentration of one.

than 30 percent in some genes to greater than 90 percent in others. In mammals, codon usage bias is correlated with the GC content of the chromosomal region, or "isochore," in which the gene resides. GC_3 is also correlated with the GC content of introns, 5′ and 3′ flanking regions, and the first two codon-positions (GC_{12}). Synonymous codon bias reflects large-scale variation in GC content along mammalian chromosomes, and all sequences, both coding and noncoding, reflect this variation to some extent.

It is generally believed that synonymous codon bias and large-scale variation in base composition are principally caused by the same factor. It has been suggested that synonymous codon bias might arise from (i) mutation bias, (ii) natural selection, or (iii) biased gene conversion (BGC). BGC is a process associated with genetic recombination in which heterozygous individuals produce an excess of one allele in their gametes (e.g., in an individual heterozygous for C and T at some site more than 50 percent of the gametes contain C). Although there are several molecular mechanisms that might generate variation in the pattern of mutation bias across a genome, analyses of data on single nucleotide polymorphism suggest that mutation bias is not responsible for synonymous codon bias in mammals. If mutation bias were the cause of synonymous codon bias, we would expect as many A or T mutations segregating at sites that were ancestrally G or C, as G or C mutations segregating at sites that were ancestrally A or T; in fact, there are many more GC → AT than AT → GC mutations, particularly in genes with high GC_3.

The excess of GC → AT mutations is consistent with both natural selection and BGC favoring high GC content. Unfortunately, it has so far proved impossible to differentiate between these explanations. Under selection, we might expect the repetitive DNA elements *Alu* and *L1* to be found in different regions of the genome because, although they are thought to share the same integration pathway, they have different GC contents (~52 percent versus 37 percent). However, the distributions of young *Alu* and *L1* elements are very similar; both are preferentially found in regions rich in A + T. Two lines of evidence support the BGC hypothesis: first, there is a correlation between GC content and recombination rate across the mammalian genome; second, sequences that have stopped recombining either decline in GC content or have lower GC content than their recombining counterparts. However, not all the data are consistent with the BGC hypothesis. In particular, some genes on the nonrecombining part of the Y chromosome are relatively GC rich; because they do not recombine, they do not undergo BGC, and they should therefore not be GC rich. Furthermore, BGC is a sensitive process; if $4N_ew \ll 0.1$, where N_e is the effective population size and w is a measure of BGC, then BGC will not affect base composition; but if $4N_ew > 4$, then every site not under selection will be converted to G or C. However, although we expect N_e to differ between mammals, levels of synonymous codon bias are generally very similar.

Conclusions. In some organisms, such as *E. coli* and *D. melanogaster*, our understanding of synonymous codon evolution is at an advanced stage. Unfortunately, this is not the case for species such as our own, but it is anticipated that this will change rapidly over the next few years.

BIBLIOGRAPHY

Akashi, H., and A. Eyre-Walker. "Translational Selection and Molecular Evolution." *Current Opinion in Genetics and Development* 8 (1998): 688–693. A review of synonymous codon evolution in systems with translational selection.

Bernardi, G. "Isochores and the Evolutionary Genomics of Vertebrates." *Gene* 241 (2000): 3–17. An extensive review of isochores and their evolution.

Duret, L., and D. Mouchiroud. "Expression Pattern and Surprisingly, Gene Length Shape Codon Usage in *Caenorhabditis, Drosophila* and *Arabidopsis*." *Proceedings of the National Academy of Sciences USA* 96 (1999): 4482–4487.

Eyre-Walker, A., and L. D. Hurst. "The Evolution of Isochores." *Nature Reviews Genetics* 2 (2001): 549–555.

Ikemura, T. "Codon Usage and tRNA Content in Unicellular and Multicellular Organisms." *Molecular Biology and Evolution* 2 (1985): 13–34. A classic early paper linking synonymous codon use to tRNA concentrations and gene expression levels.

International Human Genome Sequencing Consortium. "Initial Sequencing and Analysis of the Human Genome." *Nature* 409 (2001): 860–921. Analysis of the draft human genome sequence, including analyses of composition and repetitive DNA element distribution.

Sharp, P. M., C. J. Burgess, A. T. Lloyd, and K. J. Mitchell. "Selective Use of Termination and Variation in Codon Choice." In *Transfer RNA in Protein Synthesis*, edited by D. L. Hatfield, B. J. Lee, and R. M. Pirtle. Boca Raton, 1992. Overview of patterns of codon usage in a broad variety of species.

Smit, A. F. A. "Interspersed Repeats and Other Mementos of Transposable Elements in Mammalian Genomes." *Current Opinion in Genetics and Development* 9 (1999): 657–663. A review of interspersed elements and their distribution with respect to composition in mammalian genomes.

— ADAM EYRE-WALKER

COEVOLUTION

Coevolution is the process of reciprocal evolution between interacting species driven by natural selection. Many of the most commonly cited examples of how natural selection adapts organisms to their environments are examples of coevolutionary adaptation. These include some of the remarkable reciprocal adaptations found in many plants and their pollinators, predators and their prey, and parasites and their hosts. By linking species together, coevolution has shaped the fundamental structure of life on earth, and it continues to reorganize the earth's biodiversity.

The coevolutionary process has been responsible for many of the most important events in the diversification of life. The eukaryotic cell, which is the basis of all multicellular life, including all plants and animals, is actually an ancient and now obligate coevolved symbiosis between two more ancient forms of life, one acting as the host and the other now acting as organelles called mitochondria. The mitochondria have become a fundamental part of the machinery of the cell, controlling respiration and energy production. Although the host cell and the mitochondria each have their own DNA, neither of these formerly independent species now has a genetic code complete enough to allow them to function as separate organisms. Through coevolution, the host and symbiont have become highly specialized to one another, each relying on the other for certain cellular functions and, in the process, losing some genes that allow them to function independently.

Similarly, plants are ancient coevolved symbioses between eukaryotic cells and yet another bacterial symbiont that have become cellular organelles called chloroplasts. The chloroplasts, which give plants their green color, are responsible for photosynthesis. All photosynthetic plants are therefore a complex coevolved symbiosis between three ancient species: the host, the mitochondria, and the chloroplasts.

Many other organisms rely on yet other obligate or facultative coevolved symbioses to survive and reproduce, and many of these involve fungi or bacteria. Lichens are coevolved symbioses between fungal and algal species. Most flowering plants rely on symbioses called mycorrhizae, which are complex associations between fungi and plant roots. Legumes and a few other plants form different kinds of coevolved root symbioses, called rhizobia, with bacteria rather than fungi. These rhizobia fix nitrogen, providing the plant with nutrition. The rhizobial relationship involves a complex interplay of many genes in both the plant and the bacteria, with genes of each species turning on and off in response to cues produced by the other species.

Animals, too, rely heavily on a wide range of coevolved symbioses. Hundreds of specialized symbioses have been described between animals and the symbionts they harbor. Nutrition in most animals relies on symbionts that break down complex chemical compounds or produce vitamins needed for growth. Aphids, for example, harbor coevolved bacteria in sixty to eighty special cells within their bodies. These bacteria produce compounds essential for aphid nutrition, and the bacteria cannot survive independently of their aphid hosts. The bacteria are transmitted from female aphids to their offspring through their eggs. [*See* Mutualism.]

How Coevolution Proceeds. How species coevolve depends to a large extent on the ways in which they interact with one another. Many mutualistic symbioses evolve toward tightly coevolved interactions between two species or a small group of species. In contrast, mutualistic coevolution between free-living species often evolves toward incorporation of a number of unrelated species into the interaction. This coevolutionary recruitment of species develops as additional species, not initially part of an interaction, evolve to exploit the interaction. The evolution of many relationships between pollinators and plants and fruits and fruit eaters are the most familiar examples. As plants have evolved flowers that are pollinated by animals and fruits that are dispersed by fruit-eating animals, other plant and animal species have evolved to exploit these interactions. Hence, many pollinator species and fruit-eating animals feed on a number of flowers or fruits, and many plants rely on a diversity of pollinators to spread pollen to neighboring flowers and a number of fruit-eating species to disperse seeds.

The result of the accumulation of new species into these mutualisms has been the development of pollination syndromes and fruiting syndromes, in which unrelated taxa of plants and animals converge on similar traits. Many unrelated plant families, for example, have converged on tubular red flowers that are pollinated by

COEVOLUTIONARY ALTERNATION IN EUROPEAN CUCKOOS AND THEIR HOSTS

European cuckoos (*Cuculus canorus*) lay their eggs in the nests of other bird species, and the unsuspecting host birds raise the young cuckoos at the expense of their own young. Some bird species can recognize cuckoo eggs, and they toss them from their nests. In response, the cuckoos have evolved a set of behaviors and morphologies that minimize the chance that the host bird will discover and recognize the cuckoo egg. Cuckoos approach nests cautiously when the host leaves for a few minutes, lay an egg in only a few seconds, and retreat rapidly. Different cuckoo individuals have eggs that mimic different host species, and these different egg forms are a genetic polymorphism within cuckoo populations. In Britain, for example, cuckoos have eggs that closely mimic three of their current major hosts. Although cuckoos also use dunnocks as a major host, there is no genetic form of the cuckoo that produces eggs that match dunnock eggs, and dunnocks do not reject cuckoo eggs. Dunnocks may be a new host for the cuckoos. Research has suggested that cuckoos alternate among a group of host bird species over time, shifting to less defended hosts as defenses build up in current hosts. Further evidence that cuckoos alternate among hosts over time is found in the observation that a number of British birds that are not current hosts of cuckoos eject experimental cuckoo eggs. These bird species may have been past hosts that have recently been abandoned by the cuckoos.

—JOHN N. THOMPSON

hummingbirds, and most hummingbird species are capable of extracting nectar from a wide range of flowers adapted to hummingbirds. In these cases, the reciprocal nature of coevolutionary change is harder to pinpoint to any particular two species, but the development of these mutualistic syndromes over evolutionary time is one of the most important organizing factors in biological communities.

Antagonistic coevolution between parasites and their hosts, predators and prey, and grazers and their victims favor very different forms of coevolution from that found in mutualistic interactions. These interactions favor either escalating coevolutionary arms races or other forms of coevolution in which the focus of selection on the host or prey is to escape the interaction, and the focus of selection on the enemy is to overcome those escape mechanisms or defenses. The result can be sophisticated arsenals of defense and counter defense mechanisms in interacting antagonists. Many of the defenses found in species are directed against parasites. Parasitism of other species has been such a successful strategy over evolutionary time that it has become one of the most common lifestyles on earth, and almost all species on earth are attacked by one or more parasite species.

As a result of the ubiquity of parasitism, defense genes against parasites are an important component of all organisms whose genetic structure has been studied in detail. More than ten thousand different chemical compounds not involved in primary metabolism are found among the quarter million plants on earth, and many of these compounds have been shown to be involved in reducing attack by pathogens and herbivores. The immune system of vertebrates and the induced defenses found in many invertebrates and plants are sophisticated defense systems that have evolved in response to relentless attack by a wide diversity of fast evolving enemies.

Competition between species for limited resources is also based on antagonistic coevolution, but neither species benefits from the interaction. Consequently, coevolution between competitors follows very different pathways from coevolution between parasites and hosts or predators and prey. Coevolving competitors generally evolve toward decreasing the interaction with the other species. Over evolutionary time, competing species with similar requirements diverge in their use of habitats, prey, or other resources through the process called competitive displacement, as natural selection favors individuals that do not compete with other species for locally limited resources. As a result, the effects of competitive coevolution are seen more in the lack of current interaction between the species than in the interaction itself (something that has been called "the ghost of competition past"). Through competitive displacement, then, locally coexisting populations of some species are often more different in morphology, feeding habits, or use of habitats than populations of the same species in places where the competing species do not co-occur. Examples include some large mammalian carnivores in Eurasia, lizard species in southwestern North America and in the Caribbean, snail species within some Pacific islands, and plant species competing for pollinators.

Both antagonistic and mutualistic coevolution can take yet different forms in species that interact with each other only indirectly through a third species. Many predators use visual or chemical cues to identify prey, for example, and prey species have evolved to exploit these cues. A number of species worldwide that harbor toxins against their predators have evolved visual or chemical signals that warn predators that they are distasteful. In places where prey species are attacked by the same predator species, the opportunity has arisen for prey to evolve the same signaling systems. If two or more prey species are all distasteful, they often converge on the same color pattern or other cue. Among the most visually impressive are unrelated butterfly species that have converged on the same wing colors and patterns. The local avian predators learn the color pattern early in life after sampling one or more of the butterflies and becoming ill. They avoid butterflies with this color pattern thereafter. The same butterfly species sometimes converge on different color patterns in different geographic areas. The crucial component to this form of coevolution, called Müllerian mimicry, is that the prey species all provide the same cue to the predator.

These signaling systems, however, also provide the opportunity for exploitation by other species that are not toxic, distasteful, or otherwise dangerous to predators. In many cases, natural selection has favored individuals that mimic the signals of the distasteful species, thereby fooling the predators. If the mimic is rare, it will have little or no effect on natural selection on the distasteful model. If, however, the mimic is abundant, then the predator would not learn to associate the color pattern with toxicity. Under these conditions, antagonistic coevolution develops between the model and the mimic, with selection favoring models that diverge from the general pattern and favoring mimics that track the evolutionary changes in the model. This form of mimicry, called Batesian mimicry, is similar in its coevolutionary dynamics to the kinds of dynamics found in predators and prey, even though the model and mimic never directly interact with one another.

The same distasteful species are sometimes involved simultaneously in Müllerian and Batesian mimicry complexes, occasionally linking together very distantly related organisms. Sea slugs, which are highly distasteful to many marine predators, are often brightly colored and form Müllerian mimicry complexes in temperate and tropical oceans. Some marine flatworms, which are highly palatable to many marine predators, have converged on these sea slug patterns, forming Batesian mimicry complexes atop the Müllerian complexes. [*See* Mimicry.]

Coevolution, Specialization, and Local Adaptation. The major result of the coevolutionary process is that most species become specialized to interact with only a small number of other species. Although there are millions of species on earth, most species interact with one to a few dozen other species. Many predators and prey species become tightly linked as prey species develop new defenses and predators evolve new ways of overcoming those defenses. Most predators feed on more than one prey species, but they become specialized to feed on a small group of prey taxa that share similar traits. Some hawk species, for example, are specialized to attack mostly small mammals, whereas others are specialized to attack small birds. Different populations of these species attack a small number of the earth's mammal or bird species. Some hawk species are even more specialized. The snail kite of the Everglades and Central America is a specialist on one genus of snails and has a hooked shaped beak that is highly effective at extracting the snails from their shells.

Extreme specialization to a single species is particularly common in intimate symbioses between species, whether parasitic or mutualistic. Many mutualistic symbioses rely on specialized traits in both the host and the symbiont; as the species continue to coevolve, they become increasingly specialized to each other. Coevolution of parasites and hosts favors similar, often even greater, degrees of specialization. Most parasites are capable of attacking only a single host species or a small group of closely related host species. Countering host defenses often requires highly specialized counterdefenses that are effective only against a single host. In turn, some host defenses are effective against only a single parasite species. Hence, partly through coevolution, the earth's biodiversity has become a complex network of highly specialized interactions.

This specialization driven by coevolution can lead to local adaptation in interactions between species. Through natural selection, local populations of one species can become adapted to the specific genetic makeup of the local population of another species. Local adaptation to an enemy or mutualist can evolve even in the absence of coevolutionary response, but coevolution can be a particularly powerful process shaping local adaptation of species to one another. As the populations of two or more species become more locally coadapted over time, they may lose their ability to interact effectively with different populations of the same species. Some parasites are effective at attacking only the local population of their host. For example, some trematode parasites in New Zealand lakes are better able to attack the host snail population in their local lake than they are at attacking snails of the same species from nearby lakes. In turn, the snails from different lakes differ in their ability to defend themselves against different populations of the trematodes.

The coevolutionary process in the snails and their trematodes has been studied in detail, and it illustrates that antagonistic coevolution need not always lead to escalating defenses and counterdefenses. What often matters for a defense to be successful is that it is different from the defenses used by most others in the population. The parasites will be adapted to the most common defenses. Thus, any host that harbors a rare form of defense will be at an advantage. Experimental studies have shown that natural selection in the New Zealand lakes favors trematodes that attack the locally most common genetic form of the snail. That, in turn, favors rare genetic forms of the snail, which increase in fre-

quency in the population. When a rare form of the snail becomes sufficiently common, natural selection on the parasites switches and favors parasites that are adapted to attack the newly common snail form. In this way, genetic forms of the parasite and the host oscillate in their relative frequency over time within the local populations through the process called frequency-dependent natural selection. The same defenses and counterdefenses may therefore be constantly recycled within the host and parasite populations as the different genetic forms change in frequency over time. [See Natural Selection: An Overview.]

There is an additional component to selection favoring rare or novel forms of hosts that may be important to the way in which coevolution has generally organized life on earth. Parasitism may favor the maintenance of sexual reproduction in host populations as a defense mechanism, because sexual reproduction produces offspring with novel genotypes. In the case of the New Zealand snails, for example, snail populations heavily attacked by parasites reproduce sexually, whereas in populations where parasites are rare, the snails reproduce asexually (i.e., producing all female clones). Experiments have shown that coevolution favors sexual reproduction in the snails, because sexual reproduction produces rare genetic forms to which the parasites are poorly adapted. Therefore, sexual snails are favored over asexual snails in populations under heavy attack. [See Sex, Evolution of.]

If the parasite or predator, however, is capable of attacking more than one local host or prey species, a more complicated coevolutionary process, called coevolutionary alternation, is possible. As the parasite species heavily attacks one host species, natural selection will favor higher defenses in that host population. The coevolutionary response of the parasite can be a shift to a poorly defended host species rather than a counterdefense against the highly defended host species. That is, natural selection favors those parasite individuals that attack the poorly defended host species rather than the highly defended host species. That, in turn, places selection on the new host species to increase defenses. In the meantime, as the parasite decreases its use of the initial host, defenses in that host population will tend to decrease, because natural selection will favor individuals that do not devote resources to unneeded defenses. Over thousands of years, then, this one parasite species could coevolve with several host species, alternating among them as their defenses increase or decrease over time. [See Vignette for an example of coevolutionary alternation between European cuckoos and several bird species that serve as hosts.]

The Geographic Mosaic of Coevolution. The local adaptation that may result from coevolution is a major part of the coevolutionary process. Most species are groups of populations that differ from each other in a number of genetic traits, including the genes that affect their interactions with other species. Consequently, coevolution of species generally is a complex geographic process, with the same species often coevolving in at least slightly different ways in different geographic areas. Much current research on coevolution is devoted to understanding how the geographic structure of species and species interactions shapes the overall direction of coevolution among taxa.

The geographic mosaic theory of coevolution argues that most interactions between species have a complex geographic structure that is built upon genetic differences among populations and local adaptation in interacting species. This geographic structure makes coevolution a highly dynamic process and may keep species coevolving for much longer than would be possible if they interacted only locally. The theory asserts that there are three components of the coevolutionary process that together shape the dynamics of species interactions. First, species exhibit selection mosaics such that the natural selection on an interaction favors different traits in different biological communities. For example, particular chemical compounds may be more effective as defenses against enemies in some communities but not in others. Similarly, some methods of overcoming host or prey defenses may be more effective in some communities than in others. The result can be a complex geographic mosaic of traits favored in interacting species, with few traits favored across all populations. Second, true coevolution may occur in only some geographic locations. These coevolutionary hot spots are those places where the two species place reciprocal selection on each other. In other places, one of the species may be rare or absent, or the interaction may be common but have little effect on one of the species. A parasite may be highly virulent in some environments but have little effect on host survival and reproduction in other environments. Third, as individuals move between populations and some populations become extinct, the geographic distribution of coevolved traits constantly changes. This trait remixing continues to shift the overall distribution of coevolved traits within and among populations.

Mathematical models of selection mosaics, coevolutionary hot spots, and trait remixing have shown that all three of these components of the geographic mosaic of coevolution can have major effects on how species coevolve. The models have suggested that geographically structured interactions may be more likely to persist over long periods of time than interactions that are locally isolated. Hence, excessive fragmentation of landscapes has the potential to change the long-term prospects for continued coevolution of many taxa. The models also suggest that selection mosaics can maintain a great deal of genetic diversity within and among populations of coevolving species. Perhaps most importantly, the models indicate that a small number of coevolutionary hot spots may have powerful effects on the overall direction of coevolutionary change across all populations. That is, two interacting species need not coevolve everywhere for coevolution to be important in shaping the direction of evolution in those species.

Together, empirical and mathematical studies of the geographic mosaic of coevolution have suggested that most species interactions may be highly dynamic as the specific traits favored by the interaction constantly change over time and across landscapes. Few coevolved traits may spread to all populations of a species, because local coadaptation will favor different traits in different places. Also, because the species are constantly evolving in response to each other and because genes are constantly moving across populations, many local interactions will often be slightly maladapted at any single moment in time. Defenses and counterdefenses will sometimes be mismatched due to time lags, as one species evolves in response to new mutations or recombinations in the other species.

The geographic structure of coevolution therefore may be responsible for maintaining complex geographic patterns in the relationships among species. A result of this geographic structure can be the creation of new taxa. In the Rocky Mountains of the United States and Canada, for example, red crossbills feed on the seeds of a number of conifer species. It is now known that red crossbills are actually a complex of very closely related species that are adapted to different conifer species. One of these cryptic species is a specialist on lodgepole pine (Figure 1). Throughout much of the mountainous area of northern Idaho and nearby regions, lodgepole pines are heavily attacked by red squirrels and have coevolved with those squirrels. The red crossbills are rare in comparison to the squirrels and appear not to put heavy selective pressure on the pines. Outside of the northern Rockies, however, the red squirrels do not co-occur with the pines. In these outlying populations, the pines have coevolved with red squirrels. These pine populations have cone shapes that have evolved as defenses against the crossbills, and the crossbills have bill morphologies that have evolved to overcome those defenses. These outlying crossbill populations are so different from the more centrally located populations in the Rockies, that they appear to be separate species. They differ not only in bill morphology but also in mating songs and a variety of other traits. It seems that coevolution between pines and crossbills in these peripheral populations has resulted in the creation of new species through the process called diversifying coevolution.

FIGURE 1. Coevolution.
Crossbills (left) and lodgepole pines (right) coevolve in some parts of the Rocky Mountains but not in other parts. As a consequence, both crossbills and lodgepoles differ geographically in their traits.   Left: © Paul J. Fusco/Photo Researchers, Inc.; right: © Inga Spence/Visuals Unlimited Inc.

Overall, the coevolutionary process permeates the history of life on earth, the continued development of biodiversity, the maintenance of genetic diversity, and the organization of biological communities. The more we learn about how species have adapted to their environments and to other species, the more it seems that much of evolution is actually about coevolution. Few species on earth are able to survive and reproduce without interacting with other species, and that in itself guarantees that coevolution will be an important part of almost every aspect of the ongoing adaptation and diversification of life.

BIBLIOGRAPHY

Benkman, C. W. "The Selection Mosaic and Diversifying Coevolution between Crossbills and Lodgepole Pines." *American Naturalist* 153 (Suppl., 1999): S75–91.

Brodie, E. D., III, and E. B. Brodie, Jr. "Predator–Prey Arms Races." *Bioscience* 49 (1999): 557–568. Discusses how selection on predators and prey can be asymmetric.

Burdon, J. J., and P. H. Thrall. "Spatial and Temporal Patterns in Coevolving Plant and Pathogen Associations." *American Naturalist* 153 (Suppl., 1999): S15–S33. Summarizes the complex structure of the genetics of coevolution between plants and pathogens across geographic landscapes.

Davies, N. B., and M. Brooke. "Coevolution of the Cuckoo and Its Hosts." *Scientific American* 264 (1991): 92–98. Outlines the defenses and counterdefenses associated with brood parasitism.

Gray, M. W., G. Burger, and B. F. Lang. "Mitochondrial Evolution." *Science* 283 (1999): 1476–1481. Discusses how ancient bacteria became an integral part of eukaryotic cells.

Kraaijeveld, A. R., J. J. M. van Alphen, and H. C. J. Godfray. "The Coevolution of Host Resistance and Parasitoid Virulence." *Parasitology* 116 (Suppl., 1998): S29–S45. Examines the complex patterns found in the adaptations of parasitic insects that lay their eggs in or on other insects and the defenses of the insect hosts.

Lively, C. M. "Host–Parasite Coevolution and Sex." *Bioscience* 46 (1996): 107–114. Explains how sexual reproduction in many species may be a result of coevolution between parasites and hosts.

Pellmyr, O., and J. Leebens-Mack. "Forty Million Years of Mutualism: Evidence for Eocene Origin of the Yucca–Yucca Moth Association." *Proceedings of the National Academy of Sciences USA* (1999): 96 9178–9183. Shows evidence that one of the most commonly cited examples of coevolution between free-living species has persisted for millions of years.

Thompson, J. N. *The Coevolutionary Process.* Chicago, 1994. Discusses the relationship between specialization and coevolution, the diversity of outcomes of the coevolutionary process, and the geographic mosaic theory of coevolution.

— JOHN N. THOMPSON

COMPARATIVE METHOD

The comparative method is one of evolutionary biology's most enduring approaches to testing hypotheses of adaptation. Evolutionary biologists use it to investigate how the characteristics of organisms, such as their size, shape, life histories, and behaviors, evolve together across species. This article describes the foundations of comparative methods, reviews their applications and interpretations, and describes new methods for analyzing comparative data.

One of the best known comparative relationships is

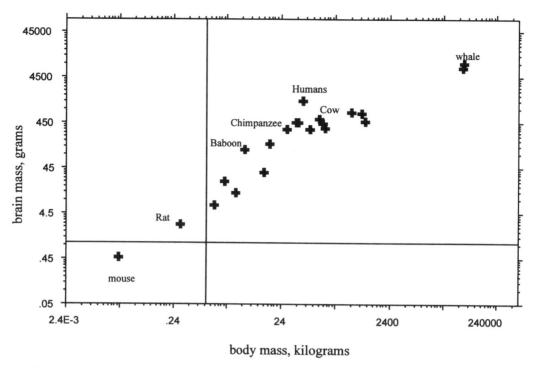

FIGURE 1. Brain Mass Versus Body Mass in Mammals, Shown on Logarithmic Scales.
The correlation between the two variables is 0.96 and the slope of the line relating brain mass to body mass is 0.59 (see text and Tables 1 and 3 for further discussion).   Drawing by Mark Pagel.

that between the size of an organism's body and the size of its brain (Figure 1): brain size increases steadily and predictably with body size in mammals, and this relationship holds within most other animal groups. Table 1 lists a number of other such trends and the species in which they have been found. These relationships reveal that as organisms evolve and grow larger, their life spans, ages at maturity, brain size, and basal metabolic rates increase, but their heart rates slow down.

The natural interpretation of comparative relationships is that they reveal adaptive trends. Organisms that conform to the trend are somehow more fit or suited for survival than those that do not. A larger animal may require a larger brain to process the increased number of neuronal signals coming from the organism's body and internal organs. Larger animals also tend to roam over wider areas, and a larger brain may be advantageous for remembering and mapping their territory. Small animals could have larger brains, but if the adaptive argument is correct, a large brain in a small animal would be a disadvantage. Brains are costly to produce and maintain (they must be supplied with blood and nutrients), and so animals are unlikely to evolve "spare capacity."

Alternatively, some commentators regard comparative trends as evidence not of adaptation but rather of some form of constraint on evolution. Animals with larger bodies may have larger brains simply because the

TABLE 1. Selected Comparative Relationships

The slope records how much the trait variable changes in response to a unit increase in body weight, when both variables are plotted on logarithmically transformed axes.

Trait	Slope of relationship to body size	Species in which measured
Brain mass	0.67–0.75	mammals
Age at maturity	0.25	birds, mammals
Life span	0.25	birds, mammals
Heart rate	−0.25	birds, mammals
Basal metabolic rate	0.75	wide range

genes that make larger bodies also make larger brains, an effect known as genetic pleiotropy. In this view, no adaptive explanation for the larger brains of larger-bodied species is required. Similarly, larger animals may have longer life spans because it takes longer to grow a larger body than a small one, given constraints on how rapidly cells can divide and replicate.

The tension between adaptive and nonadaptive explanations for observed comparative trends is what pushes investigators to make ever more precise tests of their hypotheses. Is the relationship between brain size and body size a nonadaptive trend arising from pleiotropy? A chimpanzee and a cow have brains similar in size despite the cow's considerably greater body size (Figure 1). This shows that it is possible to produce a large brain without having a large body, and raises the question of why large brains have evolved in primates but not in other mammals. One possibility is that primates have relatively sophisticated social structures, which perhaps require individuals to have a large brain. By investigating suitable alternatives, comparative studies can frequently test adaptive hypotheses and suggest new avenues of research.

The Problem of Independence. Over evolutionary time, species evolve and eventually give rise to daughter species that in turn give rise to more species. The collection of such ancestral-descendant relationships among a set of species is described by a phylogenetic tree. Phylogenetic trees are characterized by a series of branching points leading from the "root," or common ancestor of the species, up to the tips, or contemporary organisms. This hierarchical nature of biological evolution has important implications for how to interpret comparative data. In particular, when considering the support that a group of species provides for an adaptive relationship between two traits, it is essential to determine to what extent a species' value on a trait arises from its ancestry and to what extent it arises from adaptive evolution in that species.

Figure 2 shows a phylogeny of eight species, used to illustrate what is often called the problem of independence in comparative studies. Four of the species depicted in Figure 2A descend from a common ancestor with the character states {G1,G2} on two traits: G1 refers to the character state of trait 1, and G2 to the character state of trait 2. Four other species, all descendant from a different common ancestor, take the values {g1,g2}. Let

G1 signify the presence of some trait and g1 its absence; likewise for G2 and g2. The relationship counted across the eight species is perfect: G1 always pairs with G2 and g1 always pairs with g2. But the phylogeny reveals that the pairing {G1,G2} probably evolved only once from {g1,g2}. By comparision, the relationship {G1,G2} evolves repeatedly in Figure 2B: that is, on a number of independent occasions, when G1 evolves, so does G2.

The summary of the number of species with traits {G1,G2} and {g1,g2} is identical in the two figures. However, the result in Figure 2B compels us, in a way that is missing from Figure 2A, to seek evidence for a link between G1 and G2. For some reason, G1 and G2 in Figure 2A have evolved in the same phylogenetic branch, but quite possibly independently of each other. Their retention in the four contemporary species is no evidence at all that they *coevolve*, or that they are linked in some way, although each may be adaptive in its own right.

These considerations lead to the following important conclusion: the strength of evidence for an adaptive relationship in a comparative study derives not from knowing how many species come to inherit some set of traits, but in how many times the relationship between two traits has arisen independently. Much of the effort among researchers who design methods for comparative studies has been directed at this problem of independence and how to exploit phylogenies to reveal independent events of evolution.

Phylogenies and independent evolution. Several techniques have been proposed over the years to identify independent evolutionary events on phylogenetic trees, but a method known as independent contrasts, suggested by Joe Felsenstein in 1985, came to dominate in the 1990s. The method that Felsenstein devised avoids

FIGURE 2. A Phylogeny of Eight Species.
Shows (A) two sets of four species that have retained their ancestral values on each of two traits. Assume that the ancestor at the root of the tree was {g1,g2} and that the traits G1 and G2 evolved along the branch leading to the ancestor to the four species on the left of diagram A. These traits were then retained in the four descendants; (B) the evolution on four separate occasions of the pair of traits {G1,G2}. Panel B provides the greater evidence for correlated evolution between the two traits, despite both figures having the same overall number of species with trait combinations {G1,G2} and {g1,g2}. Drawing by Mark Pagel.

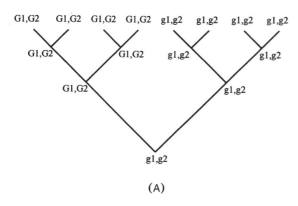

(A)

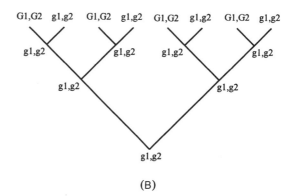

(B)

the problem of nonindependence among species by using the phylogeny to define a set of mutually independent pieces of information.

A "contrast" on a phylogeny is defined as a difference between two species, a species and an internal node (or ancestor) of the phylogeny, or between two nodes. Careful choice of the contrasts yields a set of difference scores such that each member of the set is statistically independent of each other member and which collectively account for the observed variability among species.

Figure 2 can be used again, this time to illustrate how the independent-contrast method works. Moving from left to right, species 1 and 2 share an immediate common ancestor. Assume that information is available on the brain size of these two species. Then the simple numerical difference in brain size between them identifies some amount of evolution that occurred since they branched from their common ancestor. Similarly, differences among the three other pairs of species identify amounts of evolution that arose since they diverged from their respective common ancestors and which are independent of the other pairs of species in the phylogeny.

Moving back one level on the tree, two additional differences can be calculated if it is possible to develop an estimate of the value at the corresponding internal nodes of the phylogeny. These differences also measure a unique portion of the evolutionary change on the tree. Moving back another level there is one more comparison, that between the nodes that represent the two descendants of the root or common ancestor of the tree.

A set of contrasts can be calculated separately for each of two or more traits. One might add body size information to that of brain size. Then the two sets of contrasts can be plotted and their correlation used to test the hypothesis of coevolution. Each pair of contrasts provides an independent test of the hypothesis of coevolution: if, for example, brain size and body size are suspected to covary, then each positive difference in body size between a pair of species (or nodes) should be accompanied by a positive difference in brain size. Further, larger differences in body size should be accompanied by larger differences in brain size.

Models of trait evolution. For the independent-contrast approach to produce a set of mutually independent data points, it is necessary to know how the traits evolved along the branches of the phylogeny. This sort of historical information is rarely known and so assumptions are made about the ways that traits evolve. The most commonly used model for continuously evolving traits is the constant-variance random-walk process (sometimes called Brownian motion). In this model, traits evolve each instant of time by some random

amount that might be positive (for example, the trait gets larger) or negative. The value of the trait at time $t + 1$ is whatever its value was at time t, plus the small random change.

The random walk is presumed to unfold independently at each instant of time and along each of the branches of the phylogeny. From this it is easy to appreciate that two species that have a recent common ancestor will tend to be more similar to each other than pairs of species with more distant ancestors: two bats will be more similar than a bat and a cow; two antelopes will be more similar than an antelope and a bear, and so on. In this way, the phylogeny specifies the expected similarity or lack of independence among species. Thus, closely related species will tend to have similar trait values, even under a random walk, because throughout most of their evolutionary history they were in fact the same ancestral species.

An alternative to the simple random walk is a biased random walk in which at each instant of time the trait changes by some random amount plus some constant. A biased random walk produces trends such that over time traits get larger or smaller (or faster or earlier, and so on). Other models of trait evolution allow the trait to vary randomly around some initial value, but large deviations from the starting point are biased statistically toward evolving back to the starting point. This is known as the Ornstein–Uhlenbeck process, and under it the trait behaves as if it were on an elastic band that is being stretched.

When real trait evolution violates the assumptions of the underlying model of trait evolution, comparative methods can produce poor estimates of correlations among traits. However, comparative methods tend to be quite robust to a range of realistic violations of the models, and statistical procedures are available to transform the phylogenetic branch lengths such that trait evolution conforms more closely to the presumed statistical model.

Incomplete phylogenetic information. The procedure for computing independent contrasts as described here makes use of a completely specified phylogeny— that is, one in which all species are known and in which at all nodes of the tree there are exactly two descendants. Actual phylogenies may fail to include some species or even entire groups of species, and the phylogeny may not be fully resolved. This can arise if, owing to insufficient information, it is impossible to specify the precise branching order of some groups of species. When this happens, species are, by convention, drawn on the phylogenetic tree as all emerging from one node.

A valuable feature of the independent-contrast technique is that it can be adapted to accommodate phylogenies that are not fully resolved. In effect, unresolved regions of the tree are reduced to contributing a single

Variance-Covariance matrix

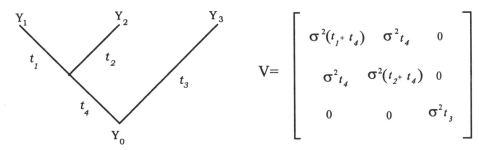

FIGURE 3. The Variances and Covariances Among Species (matrix V) Implied By the Phylogeny (left), Under the Assumption of a Constant-Variance (Brownian motion) Model of Evolution.
The symbols "Y" represent some trait that is measured on the three species. The value Y0 corresponds to the ancestral value at the root of the tree. The values inside parentheses in the matrix V are proportional to the implied variances and covariances, with σ2 the constant of proportionality.   Drawing by Mark Pagel.

piece of information. The effect on comparative methods of missing out on some of the species is generally not severe when the goal is to estimate correlations among traits.

New Comparative Methods for Continuously Evolving Traits. A new set of comparative methods based upon the statistical approach known as generalized least squares (GLS) is now replacing the independent-contrast techniques. The new methods, like their predecessors, identify independent instances of evolution in the phylogeny. They do so by using the phylogeny directly to adjust the statistical analysis to reflect the expected lack of independence among species. All of this is done without the mechanical procedure of forming contrasts or of calculating values at internal nodes: the data are simply analyzed across species.

For the simplest case of an unbiased random walk, the GLS and independent-contrast methods are mathematically equivalent. However, the GLS methods give greater flexibility for manipulating the phylogeny and the species data, and they retain information that is lost in the independent-contrast approach. For example, contrast methods cannot estimate the ancestral state at the root of the phylogeny.

Figure 3 shows a phylogeny of three species. It is reasonable to expect that species 1 and 2 in this phylogeny will be more similar to each other than either is to species 3. This is because species 1 and 2 share a more recent common ancestor. The generalized least squares model formalizes this intuition by applying a statistical model of evolution to the traits represented on a phylogeny.

The expected similarity between two species is estimated by their covariance (like a correlation). It is proportional to the lengths of the branches of the phylogeny that they have in common—the more of their ancestry

that is shared, the greater the expected similarity (larger covariance). The covariance between a species and itself will always be perfect and is called its variance. This simply reflects how much a species might vary were evolution rerun many times. The information on the expected variances and covariances among all pairs of species is represented in a matrix, usually denoted V (Figure 3). The species data points are analyzed as if they are independent, but the GLS method uses the information in the V matrix to gives less weight to species that are expected to be similar.

Testing the tempo, mode, and phylogenetic associations of trait evolution. Comparative methods, especially those using the GLS approach, can also test hypotheses about the nature of trait evolution, and determine whether the assumed constant-variance or Brownian motion model provides an adequate description of the data. These tests are made by estimating statistical parameters that are applied to the phylogeny. In this way it is possible to detect punctuational versus gradual modes of trait evolution, search for evidence of adaptive radiations, test whether the rate or tempo of trait evolution has been constant, and detect whether the phylogeny correctly predicts the patterns of covariance among species (Table 2). When violations of the assumption of the presumed models are detected, it is often possible to transform the phylogenetic branch lengths such that they conform to the Brownian motion model.

Application to mammalian brain-size evolution. The evolution of mammalian brain size illustrates the sorts of results that can emerge from analyzing comparative data. Theories of the evolution of mammalian brain size link brain volume to body mass either through body surface area or through basal metabolic rate. The surface-area hypothesis requires that the slope of the line relating brain volume to body mass is 2/3; the met-

TABLE 2. Testing Hypotheses about Trait Evolution and Adequacy of Assumed Models

Action	Test
Scale patterns of covariance implied by the phylogeny (Figure 3)	Degree of nonindependence among species
Scale individual branch lengths in tree	Punctuational versus gradual evolution
Scale total path lengths (root-to-tip distance) in tree	Constancy of rate of evolution; adaptive radiations (early and rapid evolution) versus temporally later evolution

abolic-rate hypothesis predicts that the slope of brain volume against body mass will be 3/4, owing to the relationships of basal metabolic rate to body and brain size. Empirically derived brain-body slopes generally lie somewhere in the 2/3 to 3/4 range.

The GLS model can be applied to the data of Figure 1, in combination with a phylogeny of the mammals. As a first step, these data can be shown to conform quite reasonably to the random-walk model of trait evolution. The GLS model estimates the slope relating brain mass to body mass for these data to be 0.59; this is lower than the expected 2/3 to 3/4, despite a very strong correlation between the two traits (Table 3). The 95 percent confidence intervals exclude 3/4, and nearly exclude 2/3, despite small sample sizes. The GLS model also estimates the ancestral mammalian brain and body sizes, here found to be a brain mass of 20 grams and a body mass of about 5 kilograms. By comparison to the brain and body size relationship, the slope of basal metabolic rate to body mass (Table 3) is near to the expected value (Table 1) of 0.75.

Correlated Evolution of Discrete Traits. The discussion thus far has emphasized continuously evolving traits. In many instances, investigators wish to test hypotheses about discrete traits. Discrete or categorical-variable methods are applied to data that fall into a small number (often two or three) of discrete classes or categories. Diet type, mating system, and the presence or absence of some feature or behavior are examples of this kind of data. The statistical Markov-transition model provides a natural framework for discrete traits.

The Markov-transition model is the statistical approach appropriate to describe the evolution of traits that adopt only a finite number of states. The simplest discrete trait is one that adopts only two states, corresponding perhaps to the presence or absence of some feature. The Markov model estimates the rates at which a trait or character makes transitions among its possible states as it evolves over time. These rates are sufficient

to calculate the most probable states at any node or tip of a phylogeny, given some starting point at the root. Because transition rates are typically small, closely related species tend to have similar trait values under a Markov model, whereas more distantly related species are not expected to be as similar.

To test for correlated evolution between two binary discrete characters, the Markov model compares the fit of two models to the data. In one the two traits are allowed to evolve independently; in the other they evolve in a correlated fashion. Evidence for a correlation is found if the latter model fits the data significantly better than the model of independent evolution.

For a trait that can take only two values (e.g., 0,1), two rates must be estimated, one for transition from "0" to "1," and the other for transitions from "1" to "0." Four parameters are required for two traits evolving independently (Figure 4). The model of correlated evolution considers the four possible states that two binary characters can jointly adopt (0,0; 0,1; 1,0; 1,1). It then allows one of the variables to change state in any branch of the tree, yielding eight possible transitions to be estimated (Figure 4). These can be shown to be sufficient to calculate the probability of any kind of change in any branch of the tree, and they can be used to chart the most probable course of evolution from the ancestral state to the contemporary derived state.

The evolution of lactose tolerance. The lactase enzyme confers an ability to digest milk. Human infants can digest it, but most adults cannot (widespread tolerance to lactose among some European groups is exceptional). A dominant view is that adult lactose-tolerance in humans is an adaptation to reduced exposure to the sun: both the sun and the lactase enzyme promote calcium absorption. The hypothesis predicts the lactose tolerance of some northerly-dwelling cultural groups such as the reindeer-herding Lapps, and the prevalence for lactose tolerance in some European groups.

An alternative to the latitude theory is that adult tolerance to lactose is advantageous in cultures that keep animals for milk. If milk forms a significant portion of the diet, selection pressures on adults to develop the ability to digest it could be strong. Thus, two binary traits are available (lactose tolerant/intolerant and dairy/nondairy subsistence) and the adaptive hypothesis is that they coevolve.

The Markov-process model applied to a phylogeny of human cultural groups shows that adult lactose-tolerance has arisen independently in humans up to three times in cultures that keep animals for milk, but never in nondairying cultures. This association is sufficiently strong to reject the model that the two traits have evolved independently in human groups, in favor of the view that they have coevolved. The Markov approach also reveals the probable course of the evolution of lac-

TABLE 3. Correlations and Slopes of Regression Lines of Brain-Volume and Basal Metabolic Rate against Body Size in Mammals

Ninety-five percent confidence intervals in brackets.

Variable	Number of species	Correlation	GLS regression slope, 95% CI	Theoretical expectation
Brain volume	23	0.96	0.59 (0.52–0.67)	0.67–0.75
Basal metabolic rate	15	0.95	0.72 (0.60–0.84)	0.75

tose tolerance: the ancestral condition of nondairying and no adult lactose-tolerance was first replaced by dairying perhaps as early as 6,000–8,000 years ago, which then favored the evolution of tolerance to lactose. The data do not support the alternative scenario that human groups with adult lactose-tolerance were more likely to adopt dairying subsistence practices.

Conclusion. Comparative methods provide a powerful set of techniques to investigate adaptive evolution. New statistical approaches are easy to use and make use of more of the information in the data than previous methods. Coupled with increasingly accurate and informative phylogenies based upon gene-sequence information, comparative methods can be used to test new hypotheses and to reexamine old ones. The avail-

ability of statistical parameters for scaling branch lengths, and the phylogeny itself usher in a new range of hypothesis tests for comparative data.

Future developments will allow comparative methods to be combined with techniques that allow the uncertainty in the phylogeny to be taken into account. Phylogenies are seldom known with certainty, and yet they are treated in comparative studies as if they are perfectly known. This is problematic because different phylogenies can give different answers to the comparative question. Procedures based upon a set of techniques known as Markov-Chain Monte Carlo (MCMC) make it possible to collect a random sample of phylogenetic trees for a given sample of species. That is, of the many possible different phylogenies that can be erected to describe a group of species, the MCMC methods identify a random sample of the most probable. The comparative relationship or parameter of interest—for example, a correlation between two traits or the ancestral state of a trait—can then be estimated in each tree in the sample, and the estimates combined by weighting them by the probability of the tree on which they were found. The result is a test of a comparative hypothesis that is not conditional upon any one phylogenetic tree.

[*See also* Adaptation; Brain Size Evolution; Constraints on Adaptation; Phylogenetic Inference, *article on* Methods; Punctuated Equilibrium.]

FIGURE 4. Independent Transitions Between Two Binary States in Two Traits (upper); Linked or Correlated Transitions in Two Binary Traits (lower).
Dashed lines are not calculated.   Drawing by Mark Pagel.

Independent transitions in two binary traits

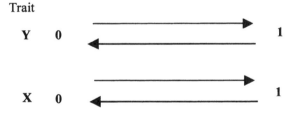

Linked transitions between two binary traits

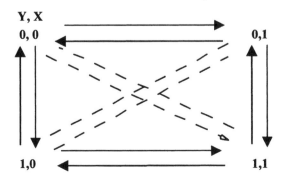

BIBLIOGRAPHY

Felsenstein, J. "Phylogenies and the Comparative Method." *American Naturalist* 125 (1985): 1–15. The article that introduced the method of independent contrasts.

Garland, T. Jr., and A. R. Ives. "Using the Past to Predict the Present: Confidence Intervals for Regression Equations in Phylogenetic Comparative Methods." *American Naturalist* 155 (2000): 346–364.

Harvey, P. H., and M. Pagel. *The Comparative Method in Evolutionary Biology.* Oxford, 1991. A general introduction to and overview of the field.

Holden, C., and R. Mace. "A Phylogenetic Analysis of the Evolution of Lactose Digestion." *Human Biology* 69 (1997): 605–628. Illustrates the application of comparative methods to human cultural evolution.

Huelsenbeck, J., B. Rannala, and J. P. Masly. "Accommodating Phylogenetic Uncertainty in Comparative Studies." *Science* 288 (2000): 2349–2350.

Larget, B., and D. L. Simon, "Markov Chain Monte Carlo Algorithms for the Bayesian Analysis of Phylogenetic Trees." *Molecular Biology and Evolution* 16 (1999): 750–759. An introduction to MCMC methods in a phylogenetic context.

Lutzoni, F., M. Pagel, and V. Reeb. "Major Fungal Lineages Are Derived from Lichen-Symbiotic Ancestors." *Nature* 411 (2001): 937–940. This paper demonstrates the application of Markov chain Monte-Carlo methods to account for phylogenetic uncertainty in comparative studies.

Martins, E. P., and T. H. Garland, "Phylogenetic Analyses of the Correlated Evolution of Continuous Characters: A Simulation Study." *Evolution* 45 (1991): 534–557. Useful information on how comparative methods fare when their assumptions are violated.

Martins, E. P., and T. F. Hansen. "Phylogenies and the Comparative Method: A General Approach to Incorporating Phylogenetic Information into the Analysis of Interspecific Data." *American Naturalist* 149 (1997): 646–667. A technical article on methods for continuously varying traits.

Pagel, M. "Detecting Correlated Evolution on Phylogenies: A General Method for the Comparative Analysis of Discrete Characters." *Proceedings of the Royal Society, B*, 255 (1994): 37–45. A technical article on methods for discrete traits.

Pagel, M. "Inferring the Historical Patterns of Biological Evolution." *Nature* 401 (1999): 877–884. Discusses how statistical models can be used to reconstruct historical events of evolution and test comparative hypotheses. Gives an introduction to the generalized least squares model.

Schluter, D., T. Price, A. Ø. Mooers, and D. Ludwig. "Likelihood of Ancestor States in Adaptive Radiation." *Evolution* 51 (1997): 1699–1711. Technical article on estimating ancestral states on phylogenies.

— MARK PAGEL

COMPLEXITY THEORY

A complex system is a group or organization that is made up of many interacting parts. Archetypal complex systems include the global climate, economies, ant colonies, and immune systems. In such systems, the individual parts—called components or agents—and the interactions between them often lead to large-scale behaviors that are not easily predicted from a knowledge of only the behavior of the individual agents. Such collective effects are called emergent behaviors. Examples of emergent behaviors include short- and long-term climate changes, price fluctuations in markets, foraging and building by ants, and the ability of immune systems to distinguish "self" from "other" and to protect the former and eradicate the latter.

Another important example of a complex system is an ecosystem. Depending on one's point of view, one may regard either individual organisms or entire species as being the agents from which an ecosystem is built. Interactions among these agents take a variety of forms. Much interest traditionally has focused on predator–prey and host–parasite interactions. These interactions are asymmetric, the two agents involved playing differ-

ent roles. There are also symmetric interactions, such as competition among agents for resources like food and space. Such competition may be among members of different species or among members of the same species. Other competition, such as competition for mates, is only among members of the same species. Symbiotic relationships between individuals or species are another form of symmetric interaction, in this case beneficial to both partners.

What is the emergent behavior of an ecosystem? There are many emergent behaviors, in fact. The very structure of an ecosystem is itself an emergent property. For example, the fact that there are many competing species rather than only a single one is a result of species interactions. Competition and cooperation between species make it advantageous for species to inhabit restricted "niches," feeding on specific resources, or living in particular environments. The many different forms of life seen today are as much the result of interactions between organisms as they are the result of the influence of the external physical environment. Animal and plant behaviors are also substantially the result of interactions. However, perhaps the classic emergent behavior of an ecosystem is evolution, and in fact the other behaviors can themselves be regarded as merely one aspect of evolution.

Evolution and Ecosystem Diversity. Evolution, the compounded result over long time periods of variation and selection, is responsible for every feature of ecosystem diversity that we see today. Conversely, ecosystem diversity is itself responsible for evolution. The selection pressures that make one variant of a species more successful, on average, than another come in large part from interactions with other individuals or species. Textbook examples of the effects of such selection pressures include the trees of the rain forest canopy, which have evolved to great height to reach the sunlight—the tallest tree will receive the light while shading others from it. Thus, some species of trees have become far taller than they should be for optimum structural soundness. The coevolution of predators and prey, such as cheetahs and antelope similarly can drive each to run faster, the one to catch its dinner and the other to avoid becoming dinner (Figure 1).

A principal result to which evolution gives rise is sophisticated organismal forms that are highly adapted to their particular niches. If we agree to call unexpected collective behaviors of complex systems emergent, then surely the evolution of current organismal forms is an extraordinary example of emergence. There are many people in the world today who refuse to believe that a "blind" process such as evolution could have crafted these forms, overwhelming evidence notwithstanding.

What can the study of complex systems contribute to evolutionary theory? There are at least two major ways in which it can help. The first is in the contribution of novel methods of mathematical and computational modeling that aid our understanding of emergent behaviors. The second is in the identification and elaboration of ideas from other complex systems that are relevant to ecologies and evolution.

Agent-Based Modeling. Agent-based computer modeling is one method relevant for evolutionary theory that has been developed in the complex systems research community. The term *agent-based modeling* (sometimes called individual-based modeling) refers to a collection of computational techniques in which individual agents and their interactions are explicitly simulated and emergent properties observed. This contrasts with more traditional differential-equation modeling methods in which much larger-scale properties of a system—population densities of species, densities of resources, and the like—are the atomic elements of the model, rather than individual agents. The goal of agent-based modeling is to design models that are sufficiently simple that the mechanisms of emergence can be understood and yet elaborate enough to show interesting behavior.

Genetic algorithms (GAs) are one class of agent-based modeling techniques that were designed to capture the essence of evolution and adaptation and yet be simple enough to be mathematically tractable. In GA methods, one studies the evolution of simple strings of symbols on a computer, or fragments of computer code, rather than attempting to simulate the behavior of real organisms. An early result from research on GAs was the mathematical characterization of adaptation as a near-optimal trade-off between exploitation of traits that have already been found to be useful and exploration for new useful traits. GA research has also led to mathematical characterizations of the roles of mutation, sexual recombination, diploidy, and other genetic processes and characteristics. In addition, GAs have a practical use as computational search and learning methods inspired by evolution. Textbooks on GAs discussing these various results include Goldberg (1989), Fogel (1995), Bäck (1996), and Mitchell (1996).

Artificial life simulations are another class of agent-based models, in which organisms and interactions are explicitly simulated. These models tend to include more complex interactions than do typical GAs and attempt to represent the conditions of the evolution of real organisms to a greater extent. Some well-known examples of artificial life simulations are T. S. Ray's (1992) tierra system, which demonstrated that increasingly efficient methods of self-reproduction can emerge in a simple ecological model; J. H. Holland's (1992, 1994, 1995) echo system, which attempted to demonstrate the emergence of multicellularity via cooperative and competitive interactions among agents; and M. A. Bedau and N. H. Packard's (1992) bugs model, in which the "evolutionary

FIGURE 1. Cheetahs Preying on Thomson's Gazelle.
© Fritz Polking/Peter Arnold, Inc.

activity" of the system—the rate at which the system generates novel adaptations—is quantified and measured. This measurement has also been applied to other simulations and evolutionary data. In each of these simulations, an emergent behavior of the system is identified and quantified, and proposals are made for identifying and quantifying similar behaviors in more realistic systems.

The simulation of macroevolutionary processes is a further class of agent-based modeling relevant to evolutionary theory. Macroevolutionary theory describes evolution at the level of higher taxa—species, genera, families, and so on—and concerns itself with such large-scale phenomena as species extinction and origination and long-term patterns of biodiversity. Probably the best-known example of a macroevolutionary model is P. Bak and K. Sneppen's (1993) coevolution model, which attempts to explain mass extinction as a result of species interactions. Other examples include M. E. J. Newman's (1997) extinction model, which views extinction as the result of environmental influences on species, and P. Sibani and colleagues (1994) "reset" model, which models evolution and extinction as a nonequilibrium process and makes predictions about patterns of change on very long time scales.

Concepts from Other Complex Systems. In addition to simulation and modeling methods, the other major contribution to evolutionary theory from complex systems research is the appropriation of concepts and results from other complex systems for the purposes of explaining evolution. A good example of this is the recent adoption of ideas from statistical physics in the field of evolution. One such idea is that of "energy landscape." Building on Sewall Wright's (1967) original proposal that evolution could be characterized as movement on a "fitness landscape," some complex systems researchers have modeled evolutionary dynamics as many-body dynamics on appropriate energy landscapes, similar to the physics concept of spin glasses. S. A. Kauffman (1993), for example, characterized evolutionary dynamics as adaptive walks on rugged fitness landscapes and correlated the statistics of these landscapes (in terms of quantities such as average numbers of local peaks, average distance between peaks, and correlations between fitnesses at fixed distances on the landscape) with the effectiveness of evolution on these landscapes. Other researchers have built on Motoo Kimura's (1983) idea of selective neutrality and applied statistical physics concepts such as percolation to characterize evolutionary dynamics on neutral networks. Others still have adapted concepts and methods from statistical mechanics and advanced statistics to describe population dynamics in simple evolutionary systems at a coarse-grained level. For example, E. van Nimwegen and colleagues (1997, 1999) have used such methods to demonstrate that metastable behavior in evolutionary systems can be the result of finite-population effects and can in some simple simulated cases be predicted in detail, and have proposed that these and related results may explain emergent behaviors seen in molecular evolution.

BIBLIOGRAPHY

Bäck, T. *Evolutionary Algorithms in Theory and Practice: Evolution Strategies, Evolutionary Programming, Genetic Algorithms.* Oxford and New York, 1996.

Bak, P., and K. Sneppen. "Punctuated Equilibrium and Criticality in a Simple Model of Evolution." *Physical Review Letters* 71 (1993): 4083–4087.

Bedau, M. A., and N. H. Packard. "Measurement of Evolutionary Activity, Teleology, and Life." In *Artificial Life*, edited by C. G. Langton, C. Taylor, J. D. Farmer, and S. Rasmussen, vol. 2, pp. 431–461. Reading, Mass., 1992.

Bedau, M. A., E. Snyder, and N. H. Packard. "A Classification of Long-term Evolutionary Dynamics." In *Artificial Life*, edited by C. Adami, R. Belew, H. Kitano, C. Taylor, vol. 6, pp. 189–198. Cambridge, Mass., 1998.

Fogel, D. B. *Evolutionary Computation: Toward a New Philosophy of Machine Intelligence.* New York, 1995.

Goldberg, D. E. *Genetic Algorithms in Search, Optimization, and Machine Learning.* Reading, Mass., 1989.

Holland, J. H. *Adaptation in Natural and Artificial Systems.* Cambridge, Mass., 1992.

Holland, J. H. "Echoing Emergence: Objectives, Rough Definitions, and Speculations for Echo-Class Models." In *Complexity: Metaphors, Models, and Reality*, edited by G. Cowan, D. Pines, and D. Melzner. Reading, Mass., 1994.

Holland, J. H. *Hidden Order: How Adaptation Builds Complexity.* Reading, Mass., 1995.

Kauffman, S. A. *The Origins of Order: Self-Organization and Selection in Evolution.* New York, 1993.

Kimura, M. *The Neutral Theory of Molecular Evolution.* Cambridge, 1983.

Mitchell, M. *An Introduction to Genetic Algorithms.* Cambridge, Mass., 1996.

Newman, M. E. J. "A Model of Mass Extinction." *Journal of Theoretical Biology* 189 (1997): 235–252.

Prugel-Bennett, A., and J. L. Shapiro. "An Analysis of Genetic Algorithms Using Statistical Mechanics." *Physical Review Letters* 72.9 (1994): 1305–1309.

Ray, T. S. "An Approach to the Synthesis of Life." In *Artificial Life*, edited by C. G. Langton, C. Taylor, J. D. Farmer, and S. Rasmussen, vol. 2, pp. 371–408. Reading, Mass., 1992.

Schuster, P., and W. Fontana. "Chance and Necessity in Evolution: Lessons from RNA." *Physica D* 133 (1999): 427–452.

Sibani, P., M. R. Schmidt, and P. Alstrom. "Evolution and Extinction Dynamics in Rugged Fitness Landscapes." *International Journal of Modern Physics B* 12 (1999): 361–391.

van Nimwegen, E., J. P. Crutchfield, and M. Huynen. "Neutral Evolution of Mutational Robustness." *Proceedings of the National Academy of Sciences U.S.A* 96 (1999): 9716–9720.

van Nimwegen, E., J. P. Crutchfield, and M. Mitchell. "Finite Populations Induce Metastability in Evolutionary Search." *Physics Letters A* 229.3 (1997): 144–150.

Wright, S. "Surfaces of Selective Value." *Proceedings of the National Academy of Sciences USA* 58 (1967): 165–179.

— MELANIE MITCHELL AND MARK NEWMAN

CONSCIOUSNESS

There is no consensus on the correct concept of consciousness, or even whether it is unitary construct. Hence, a discussion of its evolution is difficult and controversial. It may be possible, however, to determine which living taxa have specific capacities related to consciousness, and then make the best cladistic inference as to the timing of the appearance of these capacities. A number of types or features of consciousness are widely recognized. For example, many scholars accept a distinction between *experiences* and *thoughts*.

Experiences are usually acknowledged as having evolved earlier than thought and are often considered to be prerequisite for thought. To have an experience is, as Thomas Nagel (1984) put it, for there to be some way "that it is like" to be in that state. The capacity for experience may have evolved quite early, as minimal forms of this capacity may include the ability to experience pain or experience stimulation of sense receptors. But it is extremely difficult to determine which organisms have experiences of any specific type. For example, no simple behavioral criteria would reliably indicate the presence of pain. Some single-celled organisms will move away from (or be attracted by) certain types of stimuli, but the ability to discriminate among (behave differentially in response to) different stimuli does not guarantee a conscious experience. Planaria, for example, will reliably retreat from light sources, but most scholars would not wish to attribute to them the experience of pain.

In contrast to experience, conscious thought is more abstract and requires concepts. Highly flexible, context-dependent responses have been taken by some researchers as an indicator of the presence of thought. And at least some degree of flexibility of response is present in most organisms with full-blown perceptual organs. Certainly, behavioral evidence for categorization, problem solving, and "concept formation" has been accumulated in laboratory studies of various species of birds and mammals, and especially nonhuman primates. Whether the mental processes underlying such behaviors are "conscious thought," however, is controversial and difficult to establish.

The division between experience and thought is a psychologically important one, but it cuts across what many, if not most, would consider to be the essential feature of consciousness, namely reflective self-consciousness—what Thomas Natsoulas (1985) brands "Consciousness 4." Many scholars would happily grant consciousness of objects to those species with sensory systems that are capable of detecting and responding to those objects in the first place. They would further allow that these creatures are aware of something (and, in some cases, aware of a great deal) about their environment, and that some of these species may even possess concepts of such objects and think about them. Nonetheless, many of these same scholars would not grant that these organisms are conscious in the sense of being aware of their own mental states. This self-reflective component may be a crucial difference between humans and other animals. According to this view, humans—and only humans—not only perceive objects and feel hungry, but also know that they perceive and feel.

This form of self-knowledge is distinct from, though possibly dependent upon, knowledge of oneself as a distinct object in the environment (i.e., a minimal ability to explicitly represent and think about one's body as distinct from other objects). Implicit knowledge of one's body must be widespread among animals and would have evolved very early on—possibly as early as the evolution of full-blown sense organs. But this would not count as evidence for the type of reflective consciousness under discussion. Instead, Consciousness 4 is a type of *second-order mental state*—a representation of a mental state—and in this case, a representation of one's own mental state.

The evolution of reflective consciousness is controversial (see Gallup, 1998, and Povinelli, 1998, for a brief recap of the controversy). The central issues concern the nature of the self-concept in our closest living relatives, chimpanzees, and further, whether they are aware that they (or others) have mental states.

The most popular view is that chimpanzees in particular, and perhaps other species as well, possess a robust self-concept—representing to themselves their own bodies and mental states, as well as representing the mental states of others. Although part of this view is based upon experimental evidence, much of it appears to derive from the degree of similarity in the spontaneous social behavior of humans and other species. For example, many species, especially among social primates, practice forms of behavioral deception. Because we believe we are reasoning about the mental states of others when we deceive them, we assume that roughly the same thing happens in other species who practice deception.

However, both parts of the popular view may be false. With respect to the question of a self-concept, the most widely cited argument in favor of reflexive consciousness in chimpanzees comes from their well-demonstrated ability to recognize themselves in mirrors. Gallup (1998) and others have shown that chimpanzees and orangutans (but not other primates) use mirrors to explore their own bodies. Further, if their ears or eyebrow ridges are surreptitiously marked with a red mark, and they are then allowed to interact with a mirror, they will reach up and explore these marks on their own bodies (see, e.g., Gallup, 1998), whereas other nonhuman primates presented with the same task touch the mirror. Gallup has interpreted these behaviors as evidence of a robust self-concept in chimpanzees and orangutans (and its absence in other species). He has further suggested that chimpanzees and orangutans may therefore be capable of introspection, and as such, capable of using their own internal mental states to model or simulate the mental states of others.

Others have suggested alternatives to this interpretation of self-recognition in mirrors (see, e.g., Povinelli, 1998). First, passing the mirror test of self-recognition may indicate only that an organism appreciates the equivalence between what they see in the mirror and what is happening with their own body, not a recognition that they see *themselves* per se (see Povinelli, 1998). Although this would imply that chimpanzees and orangutans, unlike most other primates, may possess at least an on-line, explicit, and usable representation of the kinesthetic aspects of their bodies, it need not imply that they possess a reflective awareness of mental states (such as emotions, desires, or beliefs). One potential evolutionary account of why great apes and humans may possess a more explicit and integrated representation of their bodies than other primates has been proposed by Povinelli and Cant (1995). Their "clambering hypothesis" begins by noting that the common ancestor of the great apes and humans quadrupled in body size over a 10–20 million year period, but were forced to remain almost exclusively arboreal. Using the "clambering" behavior of modern orangutans as a rough model, they demonstrated that these two factors (body size and extreme arboreality) would have created severe difficulties for these animals in translocating their bodies across the gaps in the canopies, and therefore would have created strong selection pressures for a self-representational system dedicated to planning movements and their effects upon the environment. This evolutionary increase in body size—unprecedented in other highly arboreal primate groups—might have left a psychological imprint on the great ape/human lineage: a more explicit representation of the body and its movements.

Some researchers have attempted to assess more directly whether animals are aware of their own mental states. These "meta-memory" tasks generally involve conditional discrimination problems in which, on certain very difficult trials, the animals are given the option not to make a choice and instead advance to the next, possibly easier, trial (e.g., Smith, Shields, Washburn, and Allendoerfer, 1998; Hampton, 2001). Although intriguing, the amount and type of pretraining involved in such tasks make it unclear whether the responses involve conscious recollection and assessment (as opposed to implicit knowledge). Further, it seems quite possible that the animals are making assessments of their own behavior (as opposed to their mental states).

With respect to the question of whether chimpanzees or other species are aware of (or represent) the mental states of others, there is growing reason to be cautious. Consider something as seemingly simple as realizing that others have visual experiences—that is, understanding that they "see" things. Aspects of the behavior of chimpanzees, such as spontaneously following the gaze direction of others, would seem to suggest that they must appreciate that others "see." However, demonstrating that chimpanzees make this simple inference about a perceptual state (i.e., "seeing") has proven difficult, and further, there is a substantial body of carefully controlled, experimental evidence to suggest that they do not (see Povinelli, 2001). This stands in contrast to the ease with which this ability can be demonstrated in young children.

For some, this may seem biologically counter-intuitive. Why would similar behaviors in humans and other primates be attended by different kinds of awareness? For example, if humans follow the gaze of others in order to determine what they "see," then why should we suppose that gaze-following in chimpanzees is triggered by different representations? A model that we have labeled the "reinterpretation hypothesis" reconciles the experimental and naturalistic evidence by arguing that many of the spontaneous behaviors that humans and chimpanzees share in common, behaviors which seem to be driven by an awareness of the mental states of others (e.g., gaze following, "cooperation," "deceit," reconciliation after fights), originally evolved under the control of psychological systems that represent the behavior, but not the mental states of others (see Povinelli

and Giambrone, in Povinelli, 2001). The model assumes that before there were any organisms on earth capable of reasoning about the mental states of others, many species already possessed complex nervous systems that were capable of detecting the various statistical regularities in the behavior of others. As the central nervous system initially diversified and enlarged, natural selection acted to favor variants that could detect and process the important regularities in the world of a given species, and in highly social species that would have included the regularities in the behavior of conspecifics with whom they must live and mate. The reinterpretation hypothesis postulates that the human lineage uniquely evolved an additional ability, not present in other primate species, which allowed us to interpret these already existing behaviors as being caused by internal psychological states. This new system (probably intimately related to human language) may have forever altered our commonsense interpretation of our own behavior and the behavior of those around us.

But if the fabric of human social behavior initially evolved without the operation of second-order intentional states, then what did second-order mental states add to human behavior? Put another way, what causal role do our second-order mental states play in generating our behavior? We have argued that this role may be complex. First, it is possible that second-order mental states may simply be rapid, after-the-fact explanations of behaviors that were actually prompted by other "unconscious" psychological systems. In other cases, although a given second-order mental state may be generated before the execution of a behavior, it may still play no causal role in generating it (although this temporal pattern may create the cognitive illusion that it did play such a role). Finally, in still other instances, the second-order mental state may occur prior to the behavior *and* play a causal role in launching a behavior. On this view, one of the main initial functions of second-order mental states was to modulate already existing behaviors, giving humans added flexibility in the timing and context of their deployment. The complex causal role that second-order mental states may play in human behavior is illustrated in Figure 1.

Two concluding points are in order. First, we have deliberately made no appeal to neuroscience to help determine the evolutionary history of various aspects of consciousness. Although some claim support for early emergence of consciousness on the basis of gross anatomical similarities or particular kinds of cortical organizations between human brains and those of various other animals, at this point we are reluctant to make any judgments on the basis of neurological similarities or differences. As Todd Preuss (2000) has emphatically noted, neuroscience has vigorously pursued investigating neurological similarities between humans and pri-

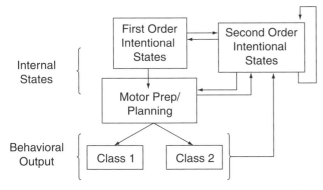

FIGURE 1. Complex Social Behaviors Shared by Humans and Chimpanzees.
The shared behaviors are directly generated by first-order intentional states, but in humans, second-order intentional states sometimes prompt the internal states that produce those behaviors. In many cases, humans may generate second-order intentional states after we have already begun (or completed) the very behaviors that our folk psychology uses them to explain. Second-order intentional states have initially evolved to simply play a role in regulating, organizing, and planning behaviors that were already present. The uniquely human aspects of this system are connected by the shaded arrows.   Drawing by authors.

mates, but has placed far less emphasis on understanding the neural differences between humans and other primates. Indeed, so few resources have been expended investigating the unique aspects of human neurology that Francis Crick has described the situation as "scandalous." The relatively few investigations of the differences between human brains and those of other primates have revealed very important dissimilarities, even areas where one would least expect to find such differences (e.g., primary visual cortex). Finally, from the opposite perspective, even if the human brain turned out to have numerous neural systems not found in our closest primate relatives, we know of no reasonable argument to support a claim that the same mental states, conscious or otherwise, must be underpinned by the same neurological states.

Finally, no evolutionary explanation of consciousness explains in any ultimate sense why experiences and *explicit* higher-order mental states (i.e., higher-order mental states that have some experiential feel to them) should exist as they do. The qualitative feel of mental states, as opposed to the information they bear, does not serve any obvious biological function. It could simply be an accident of evolutionary history on earth that there should be something "that it is like" to be in some mental states. But it could also be the case that internal states could not function as they do in humans without having a qualitative feel to them—although no one has, as of yet, offered any good reason for believing such a claim.

[*See also* Human Sociobiology and Behavior, *article on* Evolutionary Psychology; Primates, *article on* Primate Societies and Social Life.]

BIBLIOGRAPHY

Damasio, A. R. "Time-Locked Multiregional Activation: A System-Level Proposal for the Neural Substrates of Recall and Recognition." *Cognition*, 33 (1989): 25–62. Presents a systems-level theoretical framework for the understanding of memory and consciousness. Claims that only primate brains have the systematic connections that underlie consciousness.

Dretske, F. *Naturalizing the Mind*. Cambridge, Mass., 1997. There are natural representations which get their function through adaptation. The central theme is that consciousness states are information-bearing states available for reprocessing ("calibration"), and for use by the behavioral systems. Hence, they are adaptive. Also claims that introspection is displaced perception.

Ellis, R., and N. Newton, eds. *The Cauldron of Consciousness*. Philidelphia, 2000. Recent papers on consciousness with two themes: the relation of emotion and motivation to consciousness, and self-organizing systems; and a potpourri of other papers. Both themes are very current, and many of the papers are excellent, but uneven in quality.

Gallup, G. G., Jr. "Can Animals Empathize? Yes." *Scientific American Presents: Exploring Intelligence*, 9 (1998): 66, 68–71; Povinelli, Daniel. J. "Can Animals Empathize? Maybe Not." *Scientific American Presents: Exploring Intelligence*, 9 (1998): 66–75. A cordial, focused debate on the meaning of the ability of chimpanzees, orangutans, and human toddlers to recognize themselves in mirrors. Gallup maintains that it suggests that these organisms are capable of introspecting on their own mental states, and using this self-knowledge to model the experiences of others. Povinelli argues that the capacity suggests the presence of a rich kinesthetic self-concept, but does not imply an ability to reason about mental states.

Griffin, D. R. *Animal Minds*. Chicago, 1992. A very thoughtful, sophisticated, but easy to read, treatment of the subject, packed with examples of "intelligent" behavior in animals. Suffers from the defects of (1) treating any skepticism about the level of mentation in any animal as latent behaviorism, and (2) a mistakenly high level of confidence in reasoning by analogy to mental states in other creatures.

Humphrey, N. *A History of the Mind: Evolution and the Birth of Consciousness*. New York, 1992. Argues that consciousness began as sensation and that sensation began as outward directed response to stimulation. Suggests a neurological criterion for sensation as certain kind of sensory feedback loops that create a specious present.

Nagel, T. "What Is it Like to Be a Bat?" *Philosophical Review*, 83 (1984): 435–450. Claims that to be in a conscious state is for there to be something that it is like to be in that state. Argues that humans cannot know what it is like to be a bat, or any other creature with radically different sensory apparatus. The claim rests upon using "knowing what it is like" to mean something like "being able to imagine, as opposed to conceptually describing, what it is like or at least be able to recognize the experience as such if one were to have it."

Natsoulas, T. "Concepts of Consciousness." *Journal of Mind and Behavior*, 4 (1985): 333–356. An intelligent and informative discussion of the different meanings of the word *consciousness*. The labels he adopts for these, "Consciousness 1, Consciousness 2," etc., are frequently referenced in the literature on consciousness.

Nelkin, N. *Consciousness and the Origins of Thought*. New York, 1996. A very interesting philosophical piece informed by scientific theory and results. Argues against the traditional unity of consciousness, claiming that phenomenality (the qualitative feel of experiences), intentionality (the aboutness of conscious states), and apperception (awareness of one's own mental states) are separable and can be seen to be so in certain abnormal mental conditions such as blindsightedness.

Pinker, S. *How the Mind Works*. New York, 1997. A lively and informative exposition of a thoroughgoing form of the computational theory of mind. Concedes that the computational theory does not explain the enigmatic features of the mind, such as the "aboutness" of mental states and the qualitative feel of conscious states. Suggests that humans may simply not have been equipped by evolution to understand these features since there would be no obvious adaptive advantage to doing so.

Povinelli, D. J. *Folk Physics for Apes*. Oxford and New York, 2001. A detailed look at how the content of chimpanzees' awareness may differ from our own. Begins with a review of what is known about chimpanzees' understanding of mental states, but the bulk is dedicated to reporting more than two dozen new experiments examining chimpanzees' understanding of physical causality. Advances the thesis that chimpanzees (and other species) know much about the observable regularities of the social and physical world, but that humans alone have evolved psychological systems for redescribing these aspects of the world in terms of unobservable mental and physical phenomena such as emotions, intentions, beliefs, gravity, force, and mass.

Povinelli, D. J., and J. G. H. Cant. "Arboreal Clambering and the Evolution of Self-Conception." *Quarterly Review of Biology*, 70 (1995): 393–421. Uses research on locomotor behavior of modern orangutans to model for the selective forces which might have led to the emergence of an explicit kinesthetic body representation in the ancestor of the great apes and humans. Argues that this kinesthetic self-concept was the initial "platform" from which more elaborated forms of human self-awareness evolved.

Preuss, T. M. "Taking the Measure of Diversity: Comparative Alternatives to the Model-Animal Paradigm in Cortical Neuroscience." *Brain, Behavior and Evolution*, 55 (2000): 287–299. A "call-to-arms" for comparative neuroscience to investigate the differences in brain organization among living species, especially with respect to comparisons of humans and other primates. Suggests that the concept of diversity—so critical to other areas of evolutionary biology—has not been properly incorporated into the science of comparative neuroanatomy.

Smith, J. D., W. E. Shields, D. A. Wasburn, and K. R. Allendoerfer. *Journal of Experimental Psychology: General*, 127 (1998): 227–250; Hampton, Robert. R. "Rhesus Monkeys Know When They Remember." *Proceedings of the National Academy of Sciences*, 98 (2001): 5359–5362. Papers present the methods and results from experiments designed to assess the meta-memory capacity of rhesus monkeys, pigeons, and rats. Results show that after extensive training on the component parts of the tasks, some species can selectively elect not to respond on difficult trials (trials so difficult that they would be forced to respond at chance levels).

Tomasello, M., and J. Call. *Primate Cognition*. Oxford and New York, 1997. An exhaustive encyclopedia of studies of primate cognition. Although not directly dealing with questions of con-

sciousness, the first 5 chapters provide a thorough review of what is known about learning, concept formation, abstraction, and categorization in non-human primates.

— STEVE GIAMBRONE AND DANIEL J. POVINELLI

CONSENSUS TREES. *See* Phylogenetic Inference.

CONSTRAINT

Constraint implies a limitation to evolutionary change. Although very old, the concept has been revitalized in recent years as evolutionary biologists increasingly recognize that there are limits to adaptive evolution and that not all phenotypes are possible. Nonetheless, constraint remains a controversial topic and there is no consensus on exactly what it is or how to define it. This article first discusses some background concepts, then explores several classes of constraints operating on organisms.

In general, constraints are mechanisms or processes that inhibit the ability of the phenotype to evolve or bias its evolution along certain paths. The historical operation of constraints within a clade leads to evolutionary patterns, such as the unusually persistent stability of certain characters. Such patterns may initially suggest the action of constraint because they deviate from our expectation for how evolution would have proceeded in the *absence* of constraint. This expectation represents a "null model" for evolution against which the results of constraint can be compared. An alternative scenario is that the persistant stability of some characters reflects the action of natural selection, maintaining the character at some optimum.

Establishing a null model is a critical aspect of constraint theory; different null models lead to different notions of constraint. For molecular biologists interested in the evolution of DNA nucleotide sequences, for example, the null model of evolution is point substitution of nucleotides by random mutation. The expected pattern is random variation of nucleotide sequences among species. When a particular site or sequence is found to be invariable (it deviates from the null model), it is sometimes said to be "constrained," in this case meaning it is under stabilizing selection because a change in nucleotides would cause harmful functional effects in the proteins specified by the affected codon. In contrast, organismal biologists interested in phenotypic evolution most often use adaptive evolution by natural selection as their null model. The failure of the phenotype to evolve in response to selection is taken as evidence of constraint. Thus, we are led to nearly opposite conclusions about the nature of constraint: in the first case, selection is the mechanism of constraint; in the second, constraint is something that resists or opposes selection! These different conceptions follow directly upon the particular null model of evolution applied.

What is the best null model for evolutionary constraint? The answer depends on one's interests and perspective, but a good place to begin is the Darwinian theory of evolution by natural selection. As organisms evolve they change in a way that promotes survival and reproduction in the environments they inhabit, the process of adaptation. Adaptation refers to the "match" between an organism and its environment, with natural selection the mechanism that creates this match. Therefore, an appropriate null model for organismal evolution is *adaptive evolution by means of natural selection.*

Finally, constraint is appropriately applied only to *parts* of organisms, not to whole organisms or evolutionary lineages (clades). As such, only characters or groups of characters are constrained. When stating, for example, that birds are "constrained" because they have failed to evolve live birth, one means that the characters relating to reproduction are constrained in their evolution. Birds, as a lineage, are remarkably diverse and can hardly be thought of as constrained. Second, the relationship between constraint and selection is often unclear or contradictory. Sometimes constraint and selection are the same thing, and sometimes they are opposing forces. One must be explicit about the relationship between constraint and selection. It helps if we distinguish two, complementary components of natural selection.

Internal and External Components of Selection. "Adaptive evolution" in the sense used above refers to changes that enhance the match between an organism and its environment. Selection pressures of the environment itself drive this process, as emphasized by Darwin. However, another source of selection pressures may be equally important. These derive from the need for all parts within an organism to work together properly in a coordinated way. For example, if the upper and lower slicing teeth of a carnivorous mammal do not fit together, the animal cannot cut its food and will eventually die. Selection against this animal would occur in almost any environment. Likewise, a mutation that delays ossification of the jaws in an embryonic opossum would probably result in its death. The embryo depends on the early maturation of these parts for survival in the pouch while development of other systems continues. As for tooth function, this selection is mostly independent of the environment inhabited by the animal. Thus, it is called "internal selection" to distinguish it from the more typical type of environment-dependent "external selection." External selection promotes the match between environment and organism. Internal selection helps maintain functional integration within the organism. At any given time, a particular character is subject to internal and external components of selection, the balance be-

tween them determining the total strength and direction of selection acting on the trait.

A Definition of Constraint. Given the background and null model specified above, we can now define constraint as *a mechanism or process that limits or biases the evolutionary response of characters to external selection.* As such, constraints are intrinsic features of organisms that affect their ability to evolve in response to environmental selection pressures.

Types of Constraint. Evolutionary constraints can be categorized according to underlying process, hierarchical level and the life stage during which they act. The types of constraint identified here are widely used in the literature, though rarely defined. To ensure consistency, it is important to delineate them within a uniform context. Because we are most often concerned with the evolution and adaptation of adult phenotypes, all constraints are characterized from the perspective of their affect on external selection acting during the adult life stage.

Genetic constraint. Genetic constraints are of two sorts: *lack of genetic variation* and *genetic correlation.* Lack of genetic variation for a character limits its ability to evolve in response to selection. Except in rare cases, however, variation accrues fairly rapidly in a population (if only by mutation) and a lack of variation is unlikely to persist as a long-term constraint. Genetic correlation is a far better candidate for evolutionary constraint. The most important type of correlation in this context is *pleiotropy*—multiple phenotypic effects of a single gene. As such, selection on one character will indirectly affect the phenotype of one or more additional characters. If the effect of change in a correlated (unselected) character is harmful, adaptive evolution of the selected character will be slowed or prevented, depending on the strength of the genetic correlation. For example, the failure of terrestrial vertebrates to evolve more than five digits is probably the result of harmful pleiotropic effects of mutations that increase digit number. As such, selection might sometimes favor more than five digits, and we know that the developmental system can build such a limb (polydactyly appears sporadically in many animals), but deleterious "side-effects" of the genes specifying additional digits outweigh the benefit of more digits. Similar arguments have been made for the remarkable conservation of neck vertebral number (seven) in mammals—increases in vertebral number are pleiotropically linked to increased risk of cancer.

Developmental constraint. Most characters are physically generated during the developmental period of early ontogeny. Therefore, most phenotypic variation is introduced into a population at this time. Because adaptive evolution proceeds through selection acting upon (heritable) variation within a population, the failure to generate variation during development could be

FIGURE 1. Functional Constraint.
Changes to this green iguana's tongue surface would disrupt its adhesive mechanism and interfere with its ability to capture prey.   © Fred Bruemmer/Peter Arnold, Inc.

an important source of constraint—no variation, no opportunity for selection to "choose" among different phenotypes, no evolution. Therefore, developmental constraints result from limitations or biases in the production of character phenotypes caused by the dynamics of developmental processes. The resulting reduction in phenotypic variation will constrain the effects of external selection acting during later life stages.

Although conceptually straightforward, developmental constraints are difficult to identify at a mechanistic level. At least two processes might lead to similar outcomes: (1) Variations fail to arise in the first place. As such, there may be genetic variation for a character, but the developmental system "absorbs" this variation, or other perturbations, so that the underlying variation is not expressed phenotypically. Development in this case acts as a "buffer" ensuring consistency in the production of a character despite variation in the background information or developmental *milieu*. This process is referred to as *canalization* and it is the best candidate for "pure" developmental constraint. (2) Variations arise during development, but they disrupt subsequent development stages or otherwise interfere with survival of the embryo. The death of these embryos means that the variant characters will not enter the adult population, but in this case, the mechanism responsible for the reduction in variation is *internal selection*.

An example of developmental constraint is the failure of geophilomorph centipedes to evolve any species with an even number of body segments (the 1,000 species range from 29 to 191 segments, all odd numbers). Even-numbered individuals do not arise within a population, but whether because of canalization or internal selection (or conceivably, ontogenetic constraint; see below) is unknown. Another example of developmental con-

straint is the parallel evolution of similar wrist skeletal morphologies in salamanders—cartilages of the wrist repeatedly fuse in certain combinations and not in others. The fusions follow patterns of connectivity laid down during early development. Thus, selection has a restricted set of variants available in the population upon which it can act, leading to the biased evolution of certain morphologies in the clade.

Functional constraint. Functional constraint results either from *functional integration* or *functional trade-off*.

In functional integration, two or more characters operate in a coordinated way to produce an important functional outcome. If the phenotype of one character is changed by random mutation, the functionality of the whole system may be reduced or disrupted. The more tightly coordinated the characters are, the stronger the internal selection to keep them the same. Internal selection stemming from functional integration acts as a constraint when there are external selection pressures tending to modify one (or more) of the characters in the system. Internal selection on the characters that is strong enough (as in functionally critical systems) will override the external selection, so that the characters are unable to "adapt" to the local environment. The constraint applies only to the ability of particular characters to respond to external selection pressures. The remarkable conservation of the lingual prey capture system in iguanian lizards is an example of a putative functional constraint. In this case, internal selection has acted to maintain the system's phenotype, despite external selection pressures to modify for other functions some of its components, such as the tongue. Changes to the tongue surface, for example, would disrupt its adhesive mechanism and interfere with the ability to capture prey.

Functional trade-off results when a single character participates in more than one function (as opposed to participation of multiple characters in a single function). In this case, selection to enhance performance of the character in one function is overridden by strong selection on the same character stemming from another function. The failure of iguanian lizards to evolve an optimal vomeronasal chemosensory system, for example, seems to reflect the trade-off between the tongue's essential role in prey capture and its function as a chemical collector during tongue-flicking behavior. External selection on the tongue for enhancing chemoreception is constrained by the opposing internal selection of the feeding system—the tongue cannot perform both functions optimally.

Ontogenetic constraint. Any reduction in the variation of a population at an earlier life stage will constrain the effects of external selection acting at a later (adult) stage. Developmental constraints accomplish this by limiting the *introduction* of variation into a population. Ontogenetic constraints, in contrast, result from the reduction through the action of external selection of variation after it has been generated. For example, predation on juvenile vertebrates is severe and external selection for locomotor performance and escape behavior is therefore strong. Strong selection acting on juvenile stages can (1) lead to juvenile-specific adaptations and/ or (2) limit the range of phenotypes available to selection acting during the adult life-stage. This limited variation potentially constrains adaptive evolution of the adult phenotype, particularly if selection on adults acts in opposition to juvenile adaptation. The decoupling of juvenile and adult stages through metamorphosis (as in frogs and many insects) may be one way to escape ontogenetic constraints on adult phenotypes.

Constraint and Evolution. All organisms walk an evolutionary tightrope between the pressure to change and the pressure to remain the same—change in order to adapt to new environments, stay the same in order to preserve functional integration within the body. Character changes that would disrupt important developmental interactions or the functions of critical systems would hardly promote survival of the organism. Thus, the balance between evolvability and stability is an essential quality of living systems. If the environment is the driving force of evolutionary change, then constraints are the mechanisms that temper its effect on the phenotype so that the balance is maintained.

Although the evolutionary effects of constraints on character evolution are negative, their impact on the evolutionary success of organisms or lineages may not be. Constraints can serve as "regulators," ensuring that only appropriate character changes are admitted into a population. By modulating the kinds of variation available, constraints might sometimes facilitate adaptive evolution of the organism. Functional and developmental systems may evolve more rapidly if their characters are genetically correlated because selection on one character can produce immediate synergistic responses in related characters. Thus, the role of constraint in the history of life is much more than that of simple counterpoint to adaptation. Constraints remind us that not all things are possible and that organisms themselves set the stage for their own future evolution.

BIBLIOGRAPHY

Antonovics, J., and P. H. van Tienderen. "Ontoecogenophyloconstraints? The Chaos of Constraint Terminology." *Trends in Ecology and Evolution* 6 (1991): 166–168. The first study to emphasize the need for an explicit null model in constraint theory.
Arthur, W. *The Origin of Animal Body Plans. A Study in Evolutionary Developmental Biology.* Cambridge; 1997. An emphasis on developmental constraints and a lengthy discussion of internal selection.
Arthur, W., and M. Farrow. "The Pattern of Variation in Centipede

Segment Number As an Example of Developmental Constraint in Evolution." *Journal of Theoretical Biology* 200 (1999): 183–191. Excellent case study of developmental constraint.
Galis, F. "Why Do Almost All Mammals Have Seven Cervical Vertebrae? Developmental Constraints, *Hox* Genes, and Cancer." *Journal of Experimental Zoology (Molecular Development and Evolution)* 285 (1999): 19–26. Pleiotropy as constraint.
Gould, S. J. "A Developmental Constraint in *Cerion*, with Comments on the Definition and Interpretation of Constraint in Evolution." *Evolution* 43 (1989): 516–539.
Gould, S. J., and R. C. Lewontin. "The Spandrels of San Marco and the Panglossian Paradigm: A Critique of the Adaptationist Programme." *Proceedings of the Royal Society, London B* 205 (1979): 581–598. A seminal paper that sparked the modern interest in constraint.
Hall, B. K. *Evolutionary Developmental Biology. Second Edition.* London, 1998. Good overview of the various types of constraint.
Maynard Smith, J., R. Burian, S. Kauffman, P. Alberch, J. Campbell, B. Goodwin, R. Lande, D. Raup, and L. Wolpert. "Developmental Constraints and Evolution." *The Quarterly Review of Biology* 60 (1985): 265–287. The classic treatment of developmental constraint.
Raff, R. A. *The Shape of Life: Genes, Development, and the Evolution of Animal Form.* Chicago, 1996. Includes an overview.
Reeve, H. K., and P. W. Sherman. "Adaptation and the Goals of Evolutionary Research." *The Quarterly Journal of Biology* 68 (1993): 1–32. Includes a discussion of genetic and developmental constraints in relation to selection and adaptation.
Schlichting, C. D., and M. Pigliucci. *Phenotypic Evolution. A Reaction Norm Perspective.* Sunderland, Mass., 1998. Good overview that attempts to divorce constraint from selection.
Schwenk, K. "A Utilitarian Approach to Evolutionary Constraint." *Zoology* 98 (1995): 251–262. A first pass at clarifying the relationship between constraint and selection.
Schwenk, K., and G. P. Wagner. "Function and the Evolution of Phenotypic Stability: Connecting Pattern to Process." *American Zoologist* 41 (2001): 552–563. Considers the roles of functional interaction, internal selection, and functional trade-off in constraint.
Wagner, G. P. "The Significance of Developmental Constraints for Phenotypic Evolution by Natural Selection." In *Population Genetics and Evolution*, edited by G. de Jong, pp. 222–229. Berlin, 1988. Discusses possible adaptive benefits of constraint.
Wake, D. B. "Homoplasy: The Result of Natural Selection, or Evidence of Design Limitations?" *American Naturalist* 138 (1991): 543–567. Discusses developmental constraints and their role in generating parallel evolution.

— KURT SCHWENK

CONSTRAINTS ON ADAPTATION

Does natural selection lead to a "perfection of animals," as the indefatigable taxonomist Arthur Cain suggested in the 1960s? Or are there constraints on what natural selection can achieve, what Richard Dawkins has called "constraints on perfection"? Some commentators see natural selection as a mostly unfettered and creative force. Critics of this view emphasize the limitations on natural selection, seeing differences among organisms as reflecting largely nonselective and thus nonadaptive forces. One's stance on the matter perhaps reflects a worldview, as opinions are unusually divided and tend to rouse academics to passionate debate.

The colloquial meaning of *constraint* as applied to evolution is of something holding back or preventing mutation and natural selection from producing traits that might improve an organism's fitness; in the technical jargon, something stands in the way of the organism moving toward an adaptive optimum. Organisms must obey the laws of physics, and so trivial constraints on what natural selection can achieve are easy to list: a "Darwinian demon" that reproduces at an infinite rate and lives forever is impossible; an animal's mass must be positive; organisms cannot escape the effects of gravity or thermal regulation; and so on. In other instances, the same laws of physics may make some combinations of traits unlikely. Is it possible for a mammal to be as small as a bumblebee? Mammals are homeothermic, regulating their body temperatures within a narrow range. Homeothermy is problematic for small animals, owing to their having an unavoidably large surface area for their volume. But what if the same animal lived in a medium that reduced heat loss? To take another example, could a pig fly? Some calculations suggest that a flying animal the size of a pig would require an impossibly large sternum to support its flight muscles. Could a large mammal ever jump as far relative to its body size as a grasshopper can? The forces required to propel a heavy mammal such a distance would probably snap its bones. Even if it survived the takeoff, gravity would haunt it upon landing.

Other forms of constraints on adaptive evolution are less severe. Natural selection may be held back if the relevant genetic variation for a trait has not arisen. Without genetically distinct variants there is nothing on which selection can act. Animal breeders have been unsuccessful in repeated attempts to select for cows that produce only daughters—sons, by virtue of not making milk, being of less value. Hens have similarly frustrated chicken breeders by resisting attempts to get them to produce only daughters. Is there no genetic variation among cows or chickens in the likelihood of producing female offspring? Variation on which selection can act may be absent if the amount of time for novel genetic variants to arise has been short. Traits that are demonstrably advantageous may be absent simply because the relevant variation has not yet appeared. These "evolutionary time lags" may explain why moths fly toward artificial lights (a behavior that is adaptive in a cave may prove injurious in a home) or why hedgehogs curl up in balls at the approach of a car. Here the assumption is that if enough time elapses, natural selection will eventually produce the optimal trait.

Natural selection may be frustrated in its attempts to move a trait toward some optimum if the values of the trait intermediate between the current form and the optimal form have lower fitness. In these cases, relevant genetic variants may repeatedly arise but are selected against. Sewall Wright's (1977) metaphor of an adaptive landscape suggests that organisms may occupy locally adaptive peaks. Different peaks correspond to different combinations or values of traits, and the height of the peak is its fitness. An organism may be at the top of its peak, but higher peaks may rise nearby. Adaptive landscapes constrain natural selection's efficacy to reach the global optimum (highest peak). To move to a higher peak requires first descending into a valley, or becoming less fit; by definition, natural selection always produces fitter variants. The random fluctuations in gene frequency from one generation to the next known as genetic drift can propel organisms across valleys of low fitness, but these effects are usually rare or may take a long time.

Adaptive landscapes may be part of the explanation for why the vertebrate eye differs from the eyes of non-vertebrates and is characterized by what appears to be a design flaw. The light-sensitive retina of the vertebrate eye lies behind a layer of capillaries. Moreover, the optic nerve departs from the retina by passing through the front of the light-sensitive layer, creating the well-known blind spot in human vision. Eyes have evolved independently many times in the animal kingdom and not all of them show these flaws. In the giant squid, the retina lies in front of the capillaries that supply its blood. The phenomenon of "irreversible evolution" may arise when there is no way to escape from an adaptive optimum. The large mandibles of some parasitoid larvae may be an example. Parasitoids evolve these mandibles to kill other larvae. Theoretical models show that once they have evolved, all individuals must adopt them (or be killed). Large mandibles become the evolutionarily stable strategy—no other strategy can win against them—and there is no going back.

The term *allometry* is conventionally used to describe the quantitative relationships between various traits and body size. Features such as brain size, leg length, and length of the life span all change predictably with body size when measured across species. Natural selection acting on one trait changes others if the same genes affect more than one trait, a phenomenon known as pleiotropy. Pleiotropic effects in principle could slow the pace of adaptive evolution in one trait by dragging other traits away from selective optima. As with adaptive landscapes, relevant genetic variants arise, but pleiotropic effects mean that many get selected against. Some biologists suggest that allometric relationships are examples of evolutionary constraints that arise from pleiotropy. This need not be the case, however, as both traits may be responding to selective forces that just happen to be correlated: larger bodies may, for example, require larger brains.

Phylogenetic constraints, phylogenetic inertia, or historical constraints arise because the legacy of evolution means that natural selection always builds upon existing forms—it does not start from scratch like an engineer might. Constraints deriving from phylogeny or history, by narrowing the starting point, limit the directions in which adaptive evolution is likely to go. The vertebrate eye may be an example. Another is the near universality of seven neck vertebrae in mammals. Giraffes achieved their impressively long necks by elongating existing vertebrae rather than by creating new ones. Perhaps it is easier to integrate a fixed number of elongated vertebrae into the rest of the giraffe body plan than to introduce new ones.

Developmental constraints refer to the tendency for an organism to produce a biased set of alternatives for selection to act on. Maynard Smith and colleagues (1985) suggest that such biases issue from the "structure, character, composition or dynamics of the developmental system." Mammals are generally not known to produce oviparous or egg-laying variants (the exception being the monotreme mammals, such as the platypus) and this may constrain the potential for mammals to become egg laying. Conversely, birds do not produce viviparous (live-birth) variants. Developmental constraints arise from and are a special case of phylogenetic constraints.

Other, more specific, forces may constrain the direction or rate of adaptive evolution. In small populations, genetic drift frequently overpowers natural selection. If rates of mutation are low, the production of new variants will be limited. The strength of selection itself alters the rate of adaptive change.

What can we expect of natural selection? It is important to distinguish between factors that make some adaptive solutions impossible from those that merely slow the pace or bias the direction of evolution. The existence of constraints on evolution says little about their role in any particular case. As a scientific gambit, seeking adaptive explanations for traits may prove more profitable than merely assuming that a trait is constrained. Is there, for example, some advantage to seven neck vertebrae, or to having a retina that lies behind (is protected by?) a layer of capillaries? Natural selection has produced specialized and complex traits such as eyes, brains and nervous systems, echolocation, flight, eusocial behaviors, long-range navigation, immune systems, sexual reproduction, alternative metabolisms, and even specialized genetic systems such as haplodiploidy on many independent occasions throughout history. It is not to be written off lightly.

[See also Adaptation; Evolution; Fitness; Game Theory; Genetic Drift; Natural Selection, article on Natural Selection in Contemporary Human Society; Shifting Balance; Wright, Sewall.]

BIBLIOGRAPHY

Bull, J. J., and E. L. Charnov. "On Irreversible Evolution." *Evolution* 39 (1985): 1149–1155.

Cain, A. J. "The Perfection of Animals." In *Viewpoints in Biology*, 3 edited by J. D. Carthy and C. L. Duddington, pp. 36–63. London, 1964. Cain reports striking examples of adaptation in apparently neutral traits.

Dawkins, R. *The Extended Phenotype*. Oxford, 1982. A stimulating account of adaptive evolution. Chapter 3 is especially relevant to this essay.

Gould, S. J., and R. C. Lewontin. "The Spandrels of San Marco and the Panglossian Paradigm: A Critique of the Adaptationist Programme." *Proceedings of the Royal Society of London (B)* 205 (1979): 581–598. This article criticizes some forms of adaptationist thinking.

Lewontin, R. C. "Sociobiology as an Adaptationist Programme." *Behavioral Science* 24 (1979): 5–14.

Maynard Smith, J., R. Burian, S. Kauffman, P. Alberch, J. Campbell, B. Goodwin, R. Lande, D. Raup, and L. Wolpert. "Developmental Constraints and Evolution." *Quarterly Review of Biology* 60 (1985): 265–287.

Williams, G. C. *Natural Selection: Domains, Levels, and Challenges*. Oxford, 1992. Many of the ideas in this essay can be found in Chapter 6 of this overview of evolution and adaptation.

Wright, S. *Evolution and the Genetics of Populations*. Vol. 4, *Variability within and among Natural Populations*. Chicago, 1977.

— MARK PAGEL

CONTINGENCY. *See* Overview Essay *on* Macroevolution *at the beginning of Volume 1*; Adaptation; Microevolution.

CONVERGENT AND PARALLEL EVOLUTION

The study of adaptive changes is one of the main tasks of evolutionary biology. One approach to investigating the constraints on adaptive evolution is to study the occurrence of convergent features at any organizational level (e.g., community, species, physiological, or molecular). Convergent evolution can be defined as a process in which changes occur from different ancestral character states to the same descendant character state in independent evolutionary lineages. It can be distinguished from parallel evolution, in which the same ancestral character state evolves to the same descendant state in independent lineages (Figure 1). Typically, reference to convergent and parallel evolution is reserved for cases in which these changes are caused by adaptation, as opposed to mere similarity owing to chance. At the molecular level, this distinction becomes very important because of the limited number of possible character states for nucleotides (one for each of the four nucleotides

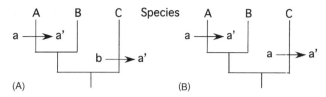

FIGURE 1. Convergent versus Parallel Evolution.
(A) Convergent evolution has occurred when two species or genes are identical in state but not identical by descent, and differ in their ancestral character state. (B) Parallel evolution has occurred when two species or genes are identical in state but not identical by descent, and share a common ancestral character state.   Drawing by Keith A. Crandall.

making up strands of DNA) and for amino acids (20 possible character states representing the 20 amino acids). The existence of convergent evolution in morphological, physiological, behavioral, and molecular changes indicates that there are significant constraints on the evolutionary process, and that these constraints are distributed across the hierarchy of organismal organization.

Detecting Convergent Evolution. Convergent evolution can seldom be observed directly, so it is typically inferred from reconstructing characters on phylogenies. Convergent changes are reconstructed as "homoplasious" changes on a well-supported phylogenetic tree; homoplasy refers to identical characters or character states that have arisen independently in distinct species. The reconstruction of ancestral character states in a phylogeny allows one to infer along which branches changes have occurred, as well as the probable ancestral character states. Using this information, it is in principle straightforward to classify changes as parallel or convergent, assuming one has independent evidence of adaptive significance. However, if the convergent characters are not few relative to the true characters of shared ancestry, they can overwhelm a phylogenetic analysis and cause the inference of incorrect phylogenetic relationships and, therefore, an incorrect interpretation of the characters as having evolved from common ancestry instead of by convergent evolution. Likewise, parallel evolution can be missed if the change in character state happens independently along sister lineages, because phylogeny reconstruction methods then tend to reconstruct the derived character as a common ancestor instead of two independently derived characters (Figure 2).

The demonstration of adaptive significance is not straightforward. At the molecular level, the standard approach is to compare the proportion of nucleotide substitutions in the gene sequence that change the amino acid (called "nonsynonymous substitutions") with the proportion of those that do not ("synonymous substitutions"). Higher proportions of coding changes are taken

as evidence of adaptive evolution. Sophisticated statistical approaches now allow the calculation of nonsynonymous and synonymous substitution rates along individual lineages, as well as for individual sites along a sequence, while incorporating more realistic models of molecular evolution. Using such approaches, selected sites along a DNA sequence can be identified and convergence can be inferred. One limitation to this approach is that it depends on the accurate reconstruction of ancestors at a given site. Alternatively, researchers have developed statistical tests for adaptive evolution that consider all sites simultaneously. Indeed, many of the most convincing cases of molecular adaptation have been argued on the basis of a high overall ratio of nonsynonymous sites to synonymous sites. Yang and Beilawski (2000) offer an excellent review of the many statistical approaches to detecting molecular adaptation. These approaches, however assume that the entire gene, or a large proportion of it, is under selection and thus tend to miss the effect of selection at a single site, or a few sites. Therefore, each approach has both advantages and disadvantages.

Different approaches are used to infer adaptation at the morphological level. The demonstration of adaptive characters has been a contentious field; adaptation has been invoked as an explanation for many features without rigorous data analysis and hypothesis testing. Now, however, the standard approach to demonstrating adaptation utilizes phylogenies and the comparative method (see Vignette). The comparative method allows one to test hypotheses of adaptive evolution in a statistically rigorous way within the context of a phylogeny. By reconstructing character states along a phylogeny, one can explicitly test for associations of various characters at the same branching point to infer adaptive significance.

Morphological Convergent Evolution. The classic examples of convergent evolution come from morphology. Some of the most intriguing examples come from cave-adapted organisms (Figure 3). Many such organisms share the common features of loss of pigmentation,

FIGURE 2. Parallel evolution can be undetectable if an identical character change occurs independently in two sister taxa. An ancestral state reconstruction in the absence of additional information will always infer a common ancestor of character a' instead of character a.   Drawing by Keith A. Crandall.

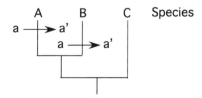

loss of eyes, elongation of sensory apparatus (e.g., antennae, sensory hairs), and reduced metabolic rates. Many other examples of convergent evolution at the morphological level exist, and any basic textbook in evolutionary biology provides many examples.

Physiological and Molecular Convergent Evolution. Perhaps the most famous non-morphological example of convergent evolution is the fermentation process in the foregut of two distinct and distantly related groups of mammals, a topic presented in a series of papers by Caro-Beth Stewart and colleagues. They studied this mode of digestion in both ruminants (e.g., cows) and colobine monkeys (e.g., langurs). In both cases, the fermentative foregut is accompanied by the recruitment of the enzyme lysozyme to be used in bacteriolysis. In both cases, the stomach lysozymes share physicochemical and catalytic properties as a result of convergent adaptation to functioning in stomach fluid. Indeed, the convergent evolution is evident down to the molecular level, where similar amino acid changes have evolved in both lineages independently. Not only did sequence similarity evolve in these two divergent lineages, but this was accompanied by an increase in the rate of evolution.

A well-established example of parallel changes at the molecular level comes from the evolution of drug resistance in HIV-1, the AIDS virus. Crandall and colleagues examined the molecular evolution of the protease gene in HIV-1 in eight patients on drug therapy that included protease inhibitors. In five of the patients, the virus evolved resistance to the drugs and the amount of virus in the bloodstream rebounded to high levels. The HIV from all five patients showed identical sequence evolution in terms of the amino acid changes at three different amino acid positions, and the majority showed identical changes at two more positions (Table 1).

Behavioral Convergent Evolution. Convergent evolution is not restricted to the evolution of molecules and morphologies. There are examples of convergent behaviors, many of which have associated convergent morphological changes. One example is the convergent evolution of courtship songs in cryptic (morphologically similar) species of green lacewing insects. Charles Henry and colleagues used DNA analysis to establish phylogenetic relationships among a group of lacewing species. They then identified the convergent evolution of courtship song in North American and Asian species, using an approach similar to that shown in the vignette.

Experimental Convergent Evolution. One of the most exciting areas of research in convergent evolution comes from the field of experimental viral evolution. Jim Bull and colleagues have exploited the bacteriophage (a virus that attacks bacteria) experimental evolutionary system to examine the extent and dynamics of convergent molecular changes. Replicate lineages were adapted to growth at high temperature on either of two hosts.

IDENTIFYING ADAPTATION USING PHYLOGENETICS

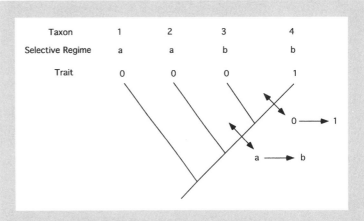

Baum and Larson have presented a phylogenetic framework for testing hypotheses of adaptation. Adaptation is the result of a character that provides current utility to the organism and has been generated historically through the action of natural selection for its current biological role. Alternative scenarios are possible, depending on the association of the selective regime (derived versus antecedent) and the relative utility (greater, lesser, or equal) (Table B.1). For an illustration of these concepts, examine Figure B.1, a phylogeny for four taxa (1–4) with the distribution of the character state for a trait (0,1) shown among the four taxa, as well as the selective regimes (a,b) for the four taxa. An adaptation, then, predicts—using taxon 4 as the focal taxon—that the utility of the derived trait (1) will exceed that of the ancestral trait (0) under the derived selective regime (b). Baum and Larson provide examples of this approach, using traits associated with climbing ability in salamanders.

TABLE B.1. Phylogenetic Definitions of Adaptation and Related Terms as Defined by Baum and Larson (1991)

Relative Utility	Selective Regime	
	Derived	*Antecedent*
Greater (aptation)	*Adaptation*	Exaptation
Less (disaptation)	Primary Disaptation	Secondary Disaptation
Equal (nonaptation)	Primary Nonaptation	Secondary Nonaptation

—KEITH A. CRANDALL

The researchers then documented the extent to which convergent evolutionary changes occurred during this adaptive evolutionary period. Amazingly, they found that more than half of the 119 observed nucleotide substi-

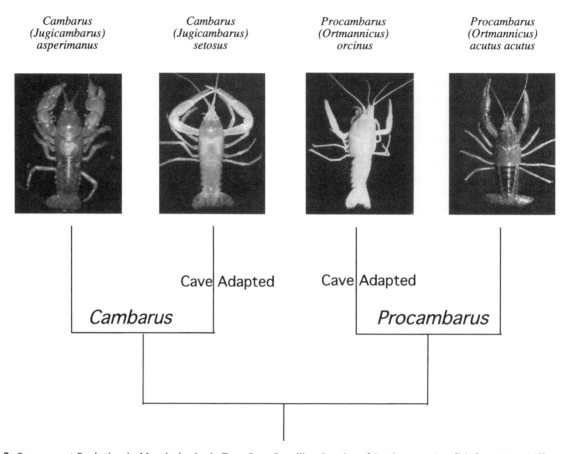

FIGURE 3. Convergent Evolution in Morphologies in Two Cave-Dwelling Species of Freshwater Crayfish from Two Different Genera (*Cambarus* and *Procambarus*).

Depicted are two pairs of species from each of the two genera. Each pair of surface-adapted and cave-adapted species shares a more recent common ancestor than do the two cave species, yet the morphologies associated with the adaptations to the cave environment are the same in each cave species. These include elongation of the pinchers, elongation of the antennae, loss of pigmentation, and loss of eyes.   Courtesy of Keith A. Crandall.

TABLE 1. Parallel Amino Acid Replacements in the Protease Gene of HIV-1 in Response to Drug Treatment in Patients Infected with HIV.

Patients were sampled at two different time points; 0—initial sampling at the start of drug therapy and some number of weeks later (59–75 weeks), as indicated for each patient.

Codon (nucleotides) Patient	10 (28–30)	54 (160–162)	71 (211–213)	82 (244–246)	90 (268–270)
P4-0	CTC	ATC	GCT	GTC	TTG
P4-75	ATC	GTC	GTT	GCC	ATG
P5-0	CTT	ATC	GCT	GTC	TTG
P5-59	ATT	GTC	GTT	GCC	ATG
P6-0	CTC	ATC	GCT	GTC	TTG
P6-71	ATC	GTC	GTT	GCC	ATG
P7-0	CTC	ATC	GTT	GTC	TTG
P7-60	ATC	No change	No change	GCC	ATG
P8-0	ATC	ATC	GCT	GTC	TTG
P8-72	GTC	GTC	No change	GCC	ATG

tutions were convergent. Some of these convergent changes were host-specific, and others were found in phages growing on both hosts. Attempts to reconstruct phylogenetic relationships using these terminal data failed to recover the true history of the lineages; thus, this experimental work suggests that convergent evolution under adaptive shift conditions can be extensive, and that the changes can obscure the true underlying evolutionary history.

BIBLIOGRAPHY

Baum, D. A., and A. Larson. "Adaptation Reviewed: A Phylogenetic Methodology for Studying Character Macroevolution." *Systematic Zoology* 40 (1991): 1–18.

Bull, J. J., M. R. Badgett, H. A. Wichman, J. P. Huelsenbeck, D. M. Hillis, A. Gulati, C. Ho, and I. J. Molineux. "Exceptional Convergent Evolution in a Virus." *Genetics* 147 (1997): 1497–1507.

Chang, B. S., and M. J. Donoghue. "Recreating Ancestral Proteins." *Trends in Ecology and Evolution* 15 (2000): 109–114.

Crandall, K. A., C. R. Kelsey, H. Imamichi, and N. P. Salzman. "Parallel Evolution of Drug Resistance in HIV: Failure of Nonsynonymous/Synonymous Substitution Rate Ratio To Detect Selection." *Molecular Biology and Evolution* 16 (1999): 372–382.

Doolittle, R. F. "Convergent Evolution: The Need To Be Explicit." *Trends in Biological Sciences* 19 (1994): 15–18.

Harvey, P. H., and M. D. Pagel. *The Comparative Method in Evolutionary Biology.* Oxford, 1991.

Henry, C. S., M. Lucía, M. Wells, and C. M. Simon. "Convergent Evolution of Courtship Songs among Cryptic Species of the *Carnea* group of green lacewings (Neuroptera: Chrysopidae: *Chrysoperla*)." *Evolution* 53 (1999): 1165–1179.

Messier, W., and C.-B. Stewart. "Episodic Adaptive Evolution of Primate Lysozymes." *Nature* 385 (1997): 151–154.

Sharp, P. M. "In Search of Molecular Darwinism." *Nature* 385 (1997): 111–112.

Stewart, C.-B. "Active Ancestral Molecules." *Nature* 374 (1995): 12–13.

Stewart, C.-B., J. W. Schilling, and A. C. Wilson. "Adaptive Evolution in the Stomach Lysozyme of Foregut Fermenters." *Nature* 330 (1987): 401–404.

Yang, Z., and J. P. Bielawski. "Statistical Methods for Detecting Molecular Adaptation." *Trends in Ecology and Evolution* 15 (2000): 496–502.

Zhang, J., and S. Kumar. "Detection of Convergent and Parallel Evolution at the Amino Acid Sequence Level." *Molecular Biology and Evolution* 14 (1997): 527–536.

— KEITH A. CRANDALL

COOPERATION

The evolution of cooperation poses an intriguing challenge to Darwinism. If life is a competitive "struggle for survival," a "war of each against all," how can natural selection ever lead to social traits that benefit other individuals? Charles Darwin was well aware of the problem. In fact, he termed one particular instance—castes of sterile workers in social insects—an "almost insuperable difficulty" for his theory.

In evolutionary biology, a social behavior is said to be cooperative if it benefits others. A cooperative act that is costly to the actor is termed altruistic. A cooperative act that is not costly but beneficial for the actor is said to be mutualistic. Costs and benefits are measured in terms of reproductive success, or Darwinian fitness. No intentions or emotions are implied in this context. A cow relieving its bowel performs a mutualistic act toward the grass. A bee stinging an intruder is acting altruistically toward its hive.

Many types of cooperative behavior have been documented, such as food sharing, cooperative hunting or foraging, cooperative defense of a common territory, and grooming. Studies in anthropology and primatology point to the overwhelming role of mutual help in early hominid societies. In ecology, mutualistic associations are increasingly seen as fundamental. Biologists find examples of cooperation at the level of cells, organelles, and even molecules.

As a rule, though, cooperative behavior offers opportunities for free riders, who profit from the help provided by others without returning it. It was only in the 1960s that this was generally recognized as a problem. Its most succinct formulation came via sociobiology: how can "altruistic" traits emerge in a world of "selfish genes"? A newly arisen gene (or gene complex) can spread only if more than the average number of copies are passed on to the next generation. What matters is that the descendants of the new gene increase their share in the gene pool. Whether these genes have a beneficial effect on the entire gene pool or not is irrelevant in this context.

Three major mechanisms are most commonly put forward to answer the question how a trait that curtails the success of its bearer can spread. They are based on kin selection, group selection, and reciprocal altruism.

Kin Selection. It is obvious that genes promoting the success of direct descendants will be favored by natural selection: they act to increase the proportion of their copies in the next generation. A genetic disposition to help close relatives will similarly help in propagating copies of itself. Indeed, the relatedness between two individuals measures the likelihood that these individuals share genes that are "identical by descent," that is, copies of some recent ancestor. In normal circumstances (excluding, for instance, inbreeding), the degree of relatedness between parent and child is 1/2, between siblings 1/2, between cousins 1/8, and so on. If a new mutation creates a gene causing its bearer to help siblings, and if that gene is passed on to an offspring, then the probability that it is also passed on to a sibling is 1/2 (whereas the probability that is has independently arisen in a nonrelated individual is almost negligible). Thus, genetic traits for nepotism (helping close relatives) channel their support toward other likely bearers of this trait and effectively promote their own spreading. This principle has been quantified in Hamilton's rule in the form

FIGURE 1. Selecting individual chickens with maximal egg production often leads to poultry with high stress-induced mortality.   © Sylvan H. Wittwer/Visuals Unlimited Inc.

$rb > c$, which means that a trait for helping relatives can spread as long as the relatedness (r) between donor and recipient times the benefit (b) to the recipient is larger than the cost (c) to the donor.

Kin selection is what helped Darwin overcome the "almost insuperable difficulty" posed by sterile worker ants. These workers support the closely related fertile males and females in their colony. The genes programming the sterile workers clearly cannot be passed on by the sterile workers themselves, but they can be transmitted (without being expressed) by their fertile relatives.

Kin selection explains cooperation within families. It is particularly effective among social insects like bees and termites, where—because reproduction is monopolized by a few females—the average degree of relatedness is extremely high.

Group Selection. Another mechanism leading to cooperation is group selection. Suppose that a population consists of many groups, which vary in their proportion of altruists, and that groups that contain more altruists produce more offspring. Within each group, an altruist will be handicapped, compared to a freeloader; but this disadvantage can be more than compensated by the greater success of the whole group, so that when averaged across all groups, the offspring of an altruist exceeds that of a nonaltruist. Group selection leading to the coexistence of altruistic and nonaltruistic types in the population can be shown to work if the progeny of all groups mix before forming new groups. While altruists fare worse than defectors within each group, groups with many altruists do better, have more progeny, and thus can "export" a higher percentage of altruists.

The term *group selection* has an unfavorable connotation. Before the 1960s, many behavioral traits used to be explained as having been selected because they

"benefit the group" (or the whole species). Examples included alarm calling, the widespread restraint in intraspecific fighting, the emergence of dominance hierarchies, and the maintenance of genetic diversity as a safeguard against environmental changes. This naive approach was discredited when theoreticians pointed out that a trait conferring an individual advantage will spread even if this implies, in the long run, a higher chance of the extinction of the whole population. It is only if the competition between groups (in the formation of new groups) is sufficiently high that it can override the effect of competition within groups. In this sense, group selection can be viewed as an individual-based selection for the ability to build successful groups.

A well-documented example of this is provided by artificial selection experiments operating on egg production. Selecting individual chickens with maximal egg production often leads to poultry with high stress-induced mortality; this changes when the breeding program operates, not on individuals, but on groups housed in separated cages (Figure 1). In that case, it is not the most productive hens, but the hens from the most productive groups, that are used as breeders for the next generation. This allows an increase in egg production without an increase in the mortality of chickens due to stress and aggression—in fact, it offers the means to select for egg production and adaptation to social group life at the same time.

Reciprocal Altruism. The third mechanism for cooperation is based on reciprocal altruism, a notion introduced by Robert Trivers in 1971. Here, the costs of helping are compensated when the recipient returns the help to the donor. The problem, of course, is that the recipient may fail to do so. More precisely, let us assume that two players meet twice, once as a potential donor, once as a recipient. In the role of the donor, they have to choose between helping or not. Being helped confers a benefit (b); helping implies a cost (c), assumed to be less than b.

This situation is an instance of the so-called Prisoner's Dilemma game, in which two players meet and have the choice between two strategies: to cooperate (play C) or to defect (play D) (Table 1). If both players cooperate, that is, help the other when they are in the

TABLE 1. Example of the Prisoner's Dilemma Game.
For the Prisoner's Dilemma game, one assumes T > R > P > S.

| | *If the Coplayer Plays* | |
	C	D
Payoff for playing C	R	S
Payoff for playing D	T	P

C = cooperate; D = defect; P = punishment; R = reward; S = sucker's payoff; T = temptation.

position of the donor, then their payoff is the reward (R) (in this instance, the difference $b - c$). If both players defect, they get no benefit and bear no cost, and their payoff (usually denoted P, for punishment) is 0. If one player cooperates and the other defects, the defector, who bears no costs, receives b points (the temptation, T, for unilateral defection), whereas the other player receives only the so-called sucker's payoff (S) for being exploited—which, in this case, is just $-c$, the cost of providing help.

More generally, one speaks of a prisoner's dilemma game whenever the elements of the payoff matrix (see Table 1) satisfy $T > R > P > S$. It is clear that a player is better off by playing D, no matter whether the other player opts for C or D. As a result, both players should play D. However, they would only receive the payoff (P), which is less than the reward (R) for mutual cooperation.

The Prisoner's Dilemma encapsulates in minimalistic form the message that in certain situations, the collective good (cooperation) and the individual interest (selfish defection) can be irreconcilable. What selfishness prescribes—namely, to defect—is economically counterproductive.

There are several ways out of this impasse. The simplest consists in repeating the Prisoner's Dilemma game between the same two players for an unspecified number of rounds, then summing up the payoff values obtained in each round. In this case, the moves of a player can influence the future decisions of the coplayer. There are infinitely many possible strategies (i.e., programs) for playing the repeated Prisoner's Dilemma game. If the probability of a further round is sufficiently high (specifically, higher than the cost-to-benefit ratio, c/b), then there exists no strategy that is the best reply to every possible strategy of the coplayer.

In the late 1970s, the political scientist Robert Axelrod introduced a new phase in the exploration of the Prisoner's Dilemma by setting up computer simulations where sundry strategies for the repeated Prisoner's Dilemma submitted by various experts were matched against each other in round-robin tournaments. The simplest of all the strategies that were submitted emerged as the winner: this was the strategy Tit for Tat, which prescribes cooperation in the first round, and from then on always repeats whatever the coplayer did in the previous round. This success was all the more surprising because Tit for Tat players can never do better than the coplayers they are matched with. In Axelrod's computer tournaments, the total sum of the payoffs against all coplayers was higher for Tit for Tat than for the rival strategies.

In his analysis, Axelrod (1984) highlighted a number of plausible reasons for the success of the Tit for Tat strategy. It was "nice" in never being the first to defect,

"provokable" by retaliating against every defection, "forgiving" by resuming cooperation after a cooperative move, and "robust" in the sense of offering an appropriate response against a wide variety of entrants (including other Tit for Tat players). This last aspect became particularly important when Axelrod followed the evolution of cooperation over many generations, by assuming that the players spawn offspring (who inherit the parental strategy) in proportion to their overall success. In such evolutionary chronicles, the frequencies of the strategies, and consequently their average payoff, vary with time. Significantly, although a single Tit for Tat player cannot invade a population of defectors (earning slightly less than defectors, by being exploited in the first round), a small cluster of Tit for Tat players can invade, and indeed take over: the slight disadvantage experienced against the majority of defectors is compensated by the much better performance in the few encounters with other Tit for Tat players.

Together with W. D. Hamilton, Axelrod (Axelrod and Hamilton, 1981) highlighted the impact of these findings for the theory of reciprocal altruism. [*See* Hamilton, William D.] Their seminal approach gave rise to a wide range of variants of the computer experiments. One topic of subsequent explorations was the effect of errors, both on the transmission of strategies from one generation to the next and on the strategic interaction between two players matched in a repeated Prisoner's Dilemma game. The first type of errors introduces mutants who, in the course of long computer simulations, can sample a large set of possible strategies. The second type of errors, reflecting the propensity for misimplementations or misperceptions in biologically realistic situations, proves particularly harmful in populations of Tit for Tat players. An erroneous defection gives rise to a chain of unilateral defections in the following rounds, each in retaliation to the previous one.

The corresponding evolutionary chronicles show that there is a distinct trend toward the emergence of a population that cooperates almost always. If $2P < T + S$ (in particular, if the benefit, b, is more than twice the cost, c), then the computer runs end often with the establishment of the Pavlov strategy. This strategy prescribes to cooperate in the first round, and from then on to cooperate if and only if the other player, in the previous round, has chosen the same move as oneself. Pavlov embodies a win-stay, lose-shift rule, and hence a rudimentary learning mechanism: it repeats a move if the payoff is large (R or T) and switches to the alternative move if the payoff is low (P or S). In particular, if a mistake occurs in the interaction of two Pavlov players, and one of them defects, then both will defect in the subsequent round and afterwards resume mutual cooperation. In this sense, the Pavlov strategy is more error-proof than Tit for Tat. On the other hand, a minority of

Pavlov players cannot invade a population of defectors. Such a population has first to be taken over by strong retaliatory strategies like Tit for Tat, and only then can Pavlov spread. (If $2P > T + S$, strategies based on the outcome of more than just the last round can emerge.)

These computer simulations are based on the assumption that the same two players interact for a sufficiently large number of rounds. Even if they interact at most once, however, cooperation can emerge. One such scenario is based on the notion of indirect reciprocation. If the players know each other by reputation (as is likely in small populations), then cooperative strategies can emerge—for example, the strategy of giving help only discriminatingly, namely, to those individuals who have given help to others more often than not. In this scenario, the altruistic act is returned, not by the recipient of that act, but by a third party. Someone who refuses to help eschews the cost of helping but is less likely to be helped, in turn, on future occasions. Interestingly, cooperation in such populations is mostly threatened by indiscriminate altruists; if those spread, defectors can invade without risking to be refused help. Cooperation based on such reputation effects requires a certain amount of cognitive abilities and is greatly strengthened by information transfer.

Another, less demanding scenario for cooperating in a one-shot Prisoner's Dilemma game relies on assortative interactions. So far, we have considered well-mixed populations of randomly meeting players. If players are territorial and interact only with their immediate neighbors, then a certain fraction of cooperators can stably subsist in the population. Theoretical models of spatially distributed populations show how waves of cooperators can spread.

This brief survey of theoretically possible explanations for altruistic behavior is necessarily incomplete. For example, it does not treat the extra difficulties faced by reciprocal altruism if the interactions occur not pairwise, but in larger groups, or the possibility that helping behavior is used to signal to potential mates the quality of the helper.

Advances in understanding how cooperation evolved are greatly hampered at this stage by the scarcity of precise empirical data. It is easy to find instances of cooperation in biological communities, but it is often very difficult to pinpoint their exact cause, especially because several of them—kin selection, reciprocal altruism, and so on—may well overlap.

This dearth of unequivocal data is caused by the inherent difficulties in measuring the fitness associated with behavioral traits in wildlife populations. For example, Sober and Wilson (1998) suggested that group selection may explain a spectacular form of self-sacrificing found among certain trematodes parasitizing ants.

In a group of some fifty such parasites ingested by an ant, one of them will migrate to the ant's brain. This causes the ant to climb on the tip of a grass blade, where it is likely to be eaten by a cow. The "brain worm" dies as a consequence, whereas the other parasites, who formed thick-walled cysts, survive to the next stage of their life cycle. The brain worm effectively sacrifices itself for the sake of the other group members residing within the ant. Despite the obvious appeal of this example, the exact parameters—genetic relatedness, average reproductive success of the trematodes, and so on—have not been measured yet.

A candidate for a Prisoner's Dilemma type of cooperation is predator inspection by sticklebacks. Frequently, pairs of sticklebacks cautiously approach a predatory fish, apparently in order to gain information on its motivational state. To cooperate within the stickleback pair means to approach the predator a few inches more. If both sticklebacks cooperate and approach jointly, their risk of being eaten is smaller than that of a single stickleback approaching the predator. With the help of mirror experiments, Manfred Milinski (1987) has shown that stickleback's tend to approach closer if their companions keep abreast with them.

A superficially similar situation occurs when two lionesses from the same pride join to investigate a perceived threat to their territory. It turned out that usually one of them defects by hanging back and letting the other risk being attacked. They do not reciprocate, as the Prisoner's Dilemma model would predict (Heinsohn and Packer, 1995). It could well be that if one lioness defects, the other lioness gains more by cooperating (i.e., by challenging the source of the threat) than by also defecting and thereby risking the loss of territory necessary for the survival of her pride. If this is indeed the case, then the game is no longer a Prisoner's Dilemma, and game theory predicts that the two lionesses should adopt distinct strategies.

Other examples of territoriality and group defense can be found in many groups of mammals and birds. Intruder mobbing and alarm calls are well-studied examples of cooperation. Usually, several possible explanations are still competing for such behavior. It seems that the degree of relatedness is often too low to provide an explanation via kin selection. In many species, grooming and nest helping offer similar challenges.

It may well be that the best data known so far on fitness payoffs have been obtained for certain phages—small chunks of RNA using the replication machinery of their bacterial host. If the multiplicity of infection is high, mutants that do not produce their full share of intracellular products spread. Such mutants act effectively as defectors, exploiting the contributions of the more cooperative phages within the bacterial cell. Thus, the tug-of-war between cooperation and exploitation works even on the level of of RNA molecules. It is likely to have been an issue from the origin of life onward.

BIBLIOGRAPHY

Axelrod, R. *The Evolution of Cooperation*. New York, 1984. A classic account on the early prisoner's dilemma tournaments.

Axelrod, R., and W. D. Hamilton. "The Evolution of Cooperation." *Science* 211 (1981): 1390–1396. A prize-winning, hugely influential paper using the repeated prisoner's dilemma game to explain reciprocal altruism.

Dawkins, R. *The Selfish Gene*. Oxford and New York, 1989. This new edition of the classic treatise on the "gene-centered" point of view is an ideal introduction to both kin selection and reciprocal altruism.

Dugatkin, L. A. *Cooperation among Animals: An Evolutionary Perspective*. Oxford and New York, 1997. An indispensable overview of many hundred studies on cooperation in a wide array of taxa.

Hamilton, W. D. "The Genetical Evolution of Social Behavior (Parts 1 and 2)." *Journal of Theoretical Biology* 7 (1964): 1–16, 17–52. A seminal, but highly intricate, paper introducing kin selection.

Heinsohn, R., and C. Packer. "Complex Cooperative Strategies in Group-Territorial African Lions." *Science* 296 (1995): 1260–1262.

Milinski, M. "Tit for Tat in Stickleback and the Evolution of Cooperation." *Nature* 325 (1987): 434–435. An elegant experiment on cooperation in predator inspection.

Nowak, M. A., and R. M. May. "Evolutionary Games and Spatial Chaos." *Nature* 359 (1992): 826–829. Introduces a spatial perspective to the study of the prisoner's dilemma.

Nowak, M. A., and K. Sigmund. "Win-Stay, Lose-Shift Outperforms Tit for Tat." *Nature* 364 (1993): 56–58. Highlights the Pavlov strategy.

Nowak, M. A., and K. Sigmund. "Evolution of Indirect Reciprocity by Image Scoring." *Nature* 393 (1998): 573–577. A minimalistic model for reciprocation through third parties.

Sober, E., and D. S. Wilson. *Unto Others: The Evolution and Psychology of Unselfish Behavior*. Cambridge, Mass., 1998. A thoughtful, if somewhat partisan, defense of group selection arguments.

Trivers, R. L. "The Evolution of Reciprocal Altruism." *Quarterly Review of Biology* 46 (1971): 35–57. A seminal article on reciprocal altruism; still an invaluable source of ideas.

Trivers, R. L. *Social Evolution*. Menlo Park, Calif., 1985. Includes a brilliant chapter on the evolution of cooperation, by the founder of the theory of reciprocal altruism.

Turner, P. E., and L. Chao. "Prisoner's Dilemma in an RNA Virus." *Nature* 398 (1999): 441–443. Contains probably the most reliable data on a "real-life" prisoner's dilemma game.

Zahavi, A., and A. Zahavi. *The Handicap Principle—A Missing Part of Darwin's Puzzle*. Oxford and New York, 1997.

— KARL SIGMUND

COPE'S RULE

Cope's rule is the observation that animal groups tend to evolve through time toward larger body size. Edward Drinker Cope (1840–1897) never formally stated this rule, but it is implicit in many of his writings. Cope is famous for his work on dinosaurs and fossil mammals, and his long rivalry with Othniel Charles Marsh (1831–1899) is well known. Cope was an evolutionist, and he wrote textbooks that promoted Darwinism. In these, he reconstructed phylogenies of reptiles and mammals, and in doing so, he noted how mean body size increased through time; classic examples include the evolution of the horses from the terrier-sized *Hyracotherium* of the Eocene epoch to the modern *Equus*, twenty times its size.

There has been debate about whether Cope's rule should be termed a "law." That is not possible because of the existence of exceptions to his observation, and also because calling it "Cope's law" would imply an innate force that drives animals to become bigger. Exceptions to Cope's observation include many cases of evolution to smaller size—for example, on islands (dwarf dinosaurs in the Romanian Cretaceous; dwarf Pleistocene elephants on Mediterranean islands), or in association with adaptations to cave life or burrowing. A rule is, according to Webster, "a generally prevailing condition," which is the true status of Cope's observation.

Why should body size generally increase within animal lineages? Most discussions have focused on the advantages of being big:

- improved ability to capture prey (e.g., in lions) or to escape predation (e.g., sauropod dinosaurs or elephants);
- greater reproductive success (sexual selection often favors heavier males);
- expanded size range of acceptable food (at greater height, for example);
- decreased annual mortality (as a result of expanded feeding opportunities);
- extended individual longevity (lifespan relates broadly to body size);
- better thermoregulation because of increased heat retention per unit volume (Bergmann's rule: body size of endotherms increases poleward).

The story is not so simple, however. There are disadvantages with large size; in particular, large animals require a great deal of food and hence can suffer when food is sparse. Elephants, for example, have to migrate huge distances merely to find enough fodder. In addition, large animals are rare, so, during crises, size is clearly exposed as a specialization, like an unusual dietary requirement, and a lineage of large animals is on the whole more exposed to extinction than is a related lineage of smaller animals.

Increase in size through time is just one of many possible evolutionary trends. The kind of trend expressed in Cope's rule could arise from natural selection, in which selection on individuals for larger size is sustained through lineages of many species and for millions of years; alternatively, it could arise from clade selection, or lineage sorting, in which larger species tend to survive preferentially over smaller species. Finally, it could be an artifact.

The third explanation views Cope's rule as a simple statistical consequence of the nature of clade diversification. Clades normally originate from comparatively unspecialized species that tend to be relatively small. As the clade diversifies, both larger and smaller species arise, although physiological and other constraints impose absolute lower and upper boundaries to the range of possible sizes for that particular taxon. If the ancestral species is relatively small—nearer to the lower size constraint than to the upper—then a random process of speciation and extinction will tend to result in an increase in mean size across the whole clade. Thus, Cope's rule may arise from the observation that ancestral species tend to be small, and that there are limits, in both directions, to the size of species within a clade.

Thus, Cope's rule is true—there are indeed genuine advantages to large body size—but there is no driving principle here. We tend to notice large animals and it may be that the classic examples of Cope's rule are more in the eye of the beholder than they are the results of evolution.

BIBLIOGRAPHY

Cope, E. D. *The Primary Factors of Organic Evolution*. Chicago, 1896. One of several publications in which Cope gave examples of the increase of body size in evolutionary lineages of mammals.

Damuth, J., and R. J. MacFadden, eds. *Body Size in Mammalian Paleobiology*. Cambridge, 1990. Case studies of body size patterns among mammals.

Jablonski, D. "Body-size Evolution in Cretaceous Molluscs and the Status of Cope's Rule." *Nature* 385 (1996): 250–252. Demonstration that Cope's rule does not apply as an overriding principle among mollusks.

Peters, R. H. *The Ecological Implications of Body Size*. Cambridge, 1993. The role of body size in microevolution and ecology.

Stanley, S. M. "An Explanation for Cope's Rule." *Evolution* 27 (1973): 1–26. An investigation of many examples of Cope's rule, and a demonstration that it is more a matter of evolving from small ancestors than specifically evolving to be big.

— MICHAEL J. BENTON

CORALS. *See* Cnidarians.

CORONARY ARTERY DISEASE

Coronary artery disease (CAD) is initiated by injuries to the endothelial lining of the coronary arteries. These injuries allow inflammatory cells to penetrate the lining, stimulate the proliferation of vascular smooth muscle cells, and lead to the deposition of lipids from low-density lipoprotein (LDL) particles. This results in an atherosclerotic plaque. The growing plaque restricts blood flow and changes the mechanical characteristics of the artery wall. These events facilitate plaque rupture, which in turn induces clotting and partial or total blockage of the flow of blood to some heart muscle cells. Depending on the extent and location of the blockages, symptoms range from mild effects to sudden death.

CAD accounts for about one-third of total human mortality in Western, developed societies, making it the most common cause of death. Both genetic and environmental factors contribute to this disease. How genes predisposing one to such a serious disease could have evolved to high frequencies will be addressed first. Given that genes do play a role in CAD predisposition, evolutionary approaches to detecting gene–CAD associations will be discussed next. Finally, the role of interactions among factors will be discussed from an evolutionary perspective, along with the clinical implications.

The Evolutionary Origins of CAD. The lateness in life in which CAD typically occurs and the environmental situation in which it is common (Western, developed societies) imply that it is unlikely that CAD itself has been the direct target of natural selection during human evolution. Rather, it is more likely that the genes that predispose one to CAD have effects on other traits that were subject to natural selection. The general phenomenon of a single gene having effects on multiple traits is called pleiotropy. The potential for pleiotropy's being involved in the evolution of the human predisposition to CAD is great because cholesterol, a lipid that is a major predictor of CAD risk, has several functions, including being a precursor for steroid hormones and modulating the structure and function of cell membranes.

The role of cholesterol as a precursor for steroid hormones suggests that the same genes that predispose humans to CAD may also play a role in reproduction. Indeed, studies in humans and other mammals indicate that genes affecting cholesterol metabolism also affect ovarian function, the menstrual cycle, and fertility. For example, the apolipoprotein (*ApoE*) locus in humans has three common polymorphic alleles, ε2, ε3, and ε4, that influence cholesterol levels and CAD incidence. These same alleles are associated with differences in male fertility. Thus, selection on reproductive traits could have altered the frequencies of alleles affecting CAD risk.

Lipids are also a critical component of cell membranes, including those of neurons. Many genes involved in CAD risk also have pleiotropic effects on brain development and functioning. For example, the alleles at the *ApoE* locus are associated with risk for Alzheimer's disease, variation in cognitive performance in healthy, middle-aged adults, differential neuronal growth in vitro and susceptibility to declines in cognitive performance after head trauma. Genetic variation at the *ApoE* locus also interacts with the dietary environment to influence the predisposition to fetal iodine deficiency disorder, the most common cause of preventable mental deficiency.

Human evolution has been characterized by the rapid and dramatic expansion of the brain and cognitive abilities. The development and maintenance of a large brain creates a high demand for cholesterol. Because the diet of early humans had much less cholesterol in it than the current diets of people living in developed countries, it is hypothesized that there was selection favoring genotypes that were "thrifty" in their absorption and production of cholesterol. When the life span increased and the diet become high in fat, the thrifty genotypes led to CAD as a pleiotropic consequence of the evolution of a large brain.

Evolutionary Analyses of Genetic Associations with CAD. The detection of genes that cause a predisposition to CAD is often based on association studies between genetic variation and variation in CAD and related phenotypes, such as cholesterol levels. Such associations arise during evolution. When a mutation influencing CAD first occurs on a chromosome, that chromosome also has a particular suite of genetic states at other genetically variable sites. Therefore, the original mutational event inevitably creates associations between itself and many other sites on the chromosome. This type of association is called linkage disequilibrium. Recombination will eventually break down disequilibrium, but disequilibrium can persist if there is little recombination between the sites. The linkage disequilibrium can also be enhanced by events in a population's history in which only a few chromosomes became the ancestors for most of the subsequent genetic variation. Many studies of disease association in humans are based on the evolutionary phenomenon of linkage disequilibrium and often seek out isolated human populations that have had an evolutionary history of small population size in the past.

One method of using disequilibrium to detect genes influencing CAD is to score a population for a large number of genetically variable marker sites scattered throughout the human genome. Any detected association between a marker with CAD is attributed not to the marker itself but to the chromosomal region located near the marker. For example, a scan of molecular markers found evidence for a gene or genes influencing elevated blood pressure (a risk factor for CAD) in a region on human chromosome 5. This region contains genes for the α_{1B} and β_2 adrenergic receptors, both of which participate in the control of vascular tone. Hence, this marker study suggests that these two loci are candidate genes for cardiovascular disease.

In many cases, candidate genes for influencing CAD can be identified directly from their known biochemical functions, as was the case for the *ApoE* locus. In the candidate locus approach, genetic variation is scored only within or close to the candidate locus. Recombination is often rare in such small genomic regions, making linkage disequilibrium strong. However, the strong disequilibrium makes it difficult to separate out functional variation from that merely associated through linkage disequilibrium. Moreover, recent genetic surveys based on DNA sequencing reveal so much variation even in a single candidate gene that many if not most of the sampled individuals are genetically unique, making it difficult to use standard statistics. Evolutionary biology provides solutions to disequilibrium and statistical sparseness. The simultaneous state of all the genetically variable sites in a candidate gene region on a particular chromosome defines a haplotype. The linkage disequilibrium among variable sites is incorporated into the haplotype states. When there has been no to little recombination, an evolutionary tree of the haplotypes can be estimated that reflects the history of the mutations that created the haplotypes. Pooling haplotypes into clades, or branches, on the haplotype tree overcomes the problem of statistical sparseness. Moreover, statistical power is enhanced by applying the evolutionary principle that the most meaningful comparisons are between clades that are evolutionary neighbors in the haplotype tree. Such evolutionary studies have detected associations between haplotypes and CAD-related traits at the *ApoAI-CIII-AIV* gene cluster and the apolipoprotein B (*ApoB*) region that were not detected by standard, nonevolutionary statistical procedures. Thus, evolutionary analyses can provide greater statistical power than nonevolutionary alternatives.

Genetic Architecture. Genetic architecture refers to the number of alleles and loci with their genomic positions that influence a trait. Genetic architecture also includes the patterns of interactions among different genes (epistasis), pleiotropy, and gene-by-environment interactions. Studies on candidate loci have revealed extensive pleiotropy (as mentioned above) and epistasis. For example, the $\epsilon4$ allele at *ApoE* is associated with elevated cholesterol, but more detailed studies revealed that the elevation of cholesterol actually arises from epistasis between $\epsilon4$ and the most common allele at the low-density lipoprotein receptor locus, a locus that codes for a protein to which the *ApoE* protein binds. Interactions are also found between genes and the environment, as mentioned previously for *ApoE* and iodine in the diet. An example related to CAD is provided by the *ApoB* gene and the response to a low-cholesterol diet. An evolutionary analysis of haplotypes at this locus revealed a haplotype clade that explained 22.3 percent of the variance in reduction of LDL cholesterol in response to a low-cholesterol diet. These and other studies reveal a genetic architecture of many loci with extensive pleiotropy, epistasis, and interactions between genes and environment. This complex genetic architecture has practical implications.

First, the impact of any given gene on CAD varies as a function of genotypes at other loci, age, gender, and environment. Context dependency undercuts the idea of a single "defective" gene for CAD. For example, the ε4 allele at the *ApoE* locus is associated with the highest average risk for CAD. However, the average association does not always translate well to the individual level—the level at which medical treatment is administered. For example, in the context of people who have high cholesterol levels, individuals bearing the "good" ε2 allele had a much higher CAD incidence than individuals bearing the ε4 allele. Indeed, just in the context of cholesterol levels, no single genotype at the *ApoE* locus is uniformly the best or worst with respect to CAD incidence. Thus, there is no obvious genetic cure for CAD.

Second, the existence of gene-by-environment interactions could allow existing treatments to be optimized for specific individuals. Perhaps the greatest use of evolutionary theory in understanding and treating CAD will be to foster an appreciation of human diversity and its clinical implication that treatment must focus on an individual with a disease and not on the disease of the individual.

[*See also* Overview Essay *on* Darwinian Medicine; Disease, *article on* Hereditary Disease; Linkage; Nutrition and Disease.]

BIBLIOGRAPHY

Flory, J. D., S. B. Manuck, R. E. Ferrell, C. M. Ryan, and M. F. Muldoon. "Memory Performance and the Apolipoprotein E Polymorphism in a Community Sample of Middle-aged Adults." *American Journal of Medical Genetics* 96 (2000): 707–711. Shows the effects of genetic variation at the *ApoE* locus on cognitive functioning in healthy adults.

Friedlander, Y., E. M. Berry, S. Eisenberg, Y. Stein, and E. Leitersdorf. "Plasma Lipids and Lipoproteins Response to a Dietary Challenge: Analysis of Four Candidate Genes." *Clinical Genetics* 47 (1995): 1–12. An evolutionary analysis of haplotype trees is used to identify a marker associated with the response to low-cholesterol diets.

Kardia, S. L. R., J. Stengard, and A. R. Templeton. "An Evolutionary Perspective on the Genetic Architecture of Susceptibility to Cardiovascular Disease." In *Evolution in Health and Disease*, edited by S. C. Stearns. Oxford, 1999. A review of some work on the genetic architecture of CAD and its evolutionary and clinical implications.

Krushkal, J., M. M. Xiong, R. Ferrell, C. F. Sing, S. T. Turner, and E. Boerwinkle. "Linkage and Association of Adrenergic and Dopamine Receptor Genes in the Distal Portion of the Long Arm of Chromosome 5 with Systolic Blood Pressure Variation." *Human Molecular Genetics* 7 (1998): 1379–1383.

Mahley, R. W., and S. C. Rall, Jr. "Apolipoprotein E: Far More Than a Lipid Transport Protein." *Annual Review of Genomics and Human Genetics* 1 (2000): 507–573. An extensive review of the many traits, including CAD, affected by the *ApoE* locus.

Mann, F. D. "Animal Fat and Cholesterol May Have Helped Primitive Man Evolve a Large Brain." *Perspectives in Biology and Medicine* 41 (1998): 417–425. Presents the "thrifty genotype" hypothesis for the evolution of CAD.

Pedersen, J. C., and K. Berg. "Interaction between Low-Density Lipoprotein Receptor (LDLR) and Apolipoprotein-E (*ApoE*) Alleles Contributes to Normal Variation in Lipid Level." *Clinical Genetics* 35 (1989): 331–337. One of the first papers to describe epistasis among genes involved in CAD risk.

Templeton, A. R. "The Complexity of the Genotype–Phenotype Relationship and the Limitations of Using Genetic "Markers" at the Individual Level." *Science in Context* 11 (1998): 373–389. A discussion of how interactions among genes and between genes and environments undermine the simplistic idea of a "gene for coronary artery disease."

Templeton, A. R. "Uses of Evolutionary Theory in the Human Genome Project." *Annual Review of Ecology and Systematics* 30 (1999): 23–49. A more detailed discussion of the issues raised in this article, along with other applications and types of diseases in humans.

Templeton, A. R. "Epistasis and Complex Traits." In *Epistasis and the Evolutionary Process*, edited by J. B. Wolf, I. E. D. Brodie, and M. J. Wade. Oxford, 2000. A review of the evidence for and the importance of epistasis in CAD and other traits.

Wang, H. Y., F. C. Zhang, J. J. Gao, J. B. Fan, P. Liu, Z. J. Zheng, H. Xi, Y. Sun, X. C. Gao, T. Z. Huang, Z. J. Ke, G. R. Guo, G. Y. Feng, G. Breen, D. St. Clair, and L. He. "Apolipoprotein E Is a Genetic Risk Factor for Fetal Iodine Deficiency Disorder in China." *Molecular Psychiatry* 5 (2000): 363–368. Shows how genetic variation at the *ApoE* locus interacts with the amount of iodine in the environment to predict risk for the most common form of preventable mental retardation.

— ALAN R. TEMPLETON

CORTICAL INHERITANCE

Cortical inheritance involves the serial transmission of cellular structural organization, through cell divisions, from parent cell to progeny by a non-Mendelian, nongenic (epigenetic or paragenetic) process. The phenomenon is also known as cytotaxis or directed assembly.

The cell cortex is a fixed, gel-like array of cytoplasm, containing cytoskeletal and membranous components, associated with the cell surface. It typically extends from the plasma membrane a few μmeters into the cytoplasm. The elaborate cortexes of ciliate protozoa (phylum Ciliophora) make them especially useful for study of the similar surface structures found in all eukaryotic cells. The cortexes of most ciliates contain an embedded series of repeated cortical units, each containing one or two basal bodies, and each unit having a distinctive anterior-posterior and left-right organization. These units are aligned into anterior-posterior ciliary rows; most of the cell surface is covered by these rows. The elaborate cortical array must be duplicated during every cell cycle to ensure that each daughter cell has a complete set of surface structures. During morphogenesis, the existing cortical pattern serves as a template for the formation of new structures. The template function of the existing cortex does more than just define the sites of assembly for new structures; it also communicates details of how and where the new structures are to be assembled. Templating is the basis for the nongenic nature of cortical inheritance. New cortical structures are assembled under the influence of existing structures in the cell; the source of some of this assembly information is not coded by the nucleic acid sequences of the cell.

Although there are many examples of cortical inheritance, the two most extensively analyzed are shown by *Paramecium* and *Tetrahymena:* doublet cells and cortical inversions. Doublet cells are essentially two normal cells fused back to back to make a twin with two oral apparatuses, twice the number of ciliary rows, and twice the number of other surface organelles. They are generated, not by mutagenesis, but by failure of conjugating cells to separate. Doublet cells are stable through many generations of asexual division. When mated with normal cells, there is a typical reciprocal transfer of Mendelian genes, but the progeny of the doublet exconjugants are consistently doublets, whereas the progeny of the normal cells are singlets. Even transfer of genetic markers in the fluid cytoplasm (mitochondria and endosymbiotic bacteria) does not alter the heritability of the doublet condition. The doublet trait is transmitted only through the structural progeny of the doublet cell.

Cortical inversions are produced by a structural alteration of the cell surface. They are not caused by a mutation. One or more ciliary rows are rotated 180° in the plane of the surface, producing a patch of cortex with reversed anterior-posterior orientation. Every structural and functional aspect of these rows is inverted—the cilia even beat "backwards," altering the cell's swimming behavior. During cell division, inverted ciliary rows produce more cortical units with the identical inverted orientation, ensuring the stability (for up to thousands of cell cycles) of the inverted region, unaffected by typical morphogenesis in adjacent, normal ciliary rows. Genetic crosses between cells with inversions and cells with normal cortexes show no transfer of the inversion, even though nuclear genes are reciprocally transferred between the partners. In both of these cases, a difference in cortical structure produces a significant difference in phenotype, but no evidence for nuclear or fluid cytoplasmic genetic changes specifying that difference can be found. The only identifiable source of heritable information to specify the altered cortical phenotype appears to be the existing structures of the cortex and their templating influence upon newly forming structures.

Phenomena similar to ciliate cortical inheritance are seen in many other organisms, including multicellular species. Individual cells show structural inheritance of components such as centrosomes, centrioles, and some endomembrane arrays. Patterning events before and after fertilization of the eggs of many organisms can be likened to similar processes in ciliates and identified with cortical inheritance. The apparent universality of cortical inheritance–like phenomena leads to the understanding that this process is not limited to the ciliates but applies to other living forms as well. The most spectacular demonstrations of something like cortical inheritance are those of the various prion systems. Prions are peculiar protein particles that have an "infectious" quality—the ability to induce a normal cellular protein (the prion precursor protein, coded by a nuclear gene) to alter its secondary/tertiary conformation and become a prion itself. Prions are the causative agents of several infectious neurodegenerative brain diseases and also of some non-Mendelian phenotypic conditions in yeast. The models for prion propagation all depend on a common basic mechanism: the prion protein provides structural information that promotes an irreversible alteration of the normal protein into the prion conformation. This process takes place at the protein level of organization, whereas the template of the ciliate cortex operates at the level of assembly of proteins into multimeric structures. The parallels of prion propagation to the structural templating shown by ciliates are nevertheless very clear. One could thus argue that cortical inheritance is one example of a general nongenic (epigenetic) structural information process present in all cells.

Cortical inheritance can play a role in evolutionary events, even though the long-term stability of cortical traits is lower than that of nucleic acid–encoded traits. Two possible aspects significant to evolution can be identified. One suggests that cortical inheritance can act as a developmental constraint (selective filter) against genetic changes that might be too abrupt. Because newly synthesized cortical proteins must be able to fit into the existing cortical complex assembly, major genetic changes in those proteins might not be compatible with the preexisting organization and thus would produce an aborted morphogenesis leading to lethality. The need for proper compatibility of new proteins as they are added to complex multiprotein machines such as ribosomes is already known, but in the cortex the consequence of the assembly is patterning, a crucial element of cellular organization in space. The other aspect suggests that cortical inheritance could produce non-Mendelian "hopeful monsters," new cortical organizations that could not be generated by gradual genetic changes, but could subsequently be maintained in a stable fashion by cooperative interactions between cortical processes and genetic expression. Possible examples of this kind of event might be the protozoan

Teutophrys, which has three food-capturing organelles and which looks as if it were derived from a triplet cell of its relative, *Dileptus*, which has only one such organelle. The other example is the symmetry-reversed (i.e., mirror-image) cells of *Tetrahymena*, which are stable in laboratory culture in spite of the abnormal, "backwards" assembly of the oral apparatus and some other cellular structures. Both of these aspects indicate that cortical inheritance could play a significant role in the evolution of an organism through interactions among genetic expression, phenotypic variation, and natural selection.

Even though it does not involve any nucleic acid–based information source, cortical inheritance does not invalidate the "central dogma" of molecular genetics. To the contrary, cortical inheritance depends on genetic expression for a source of all new component proteins. The elaborated formulation of the central dogma includes posttranslational events, and the information processes of cortical inheritance operate in those post-translational domains of gene expression. The phenomena of cortical inheritance (and related nongenic, epigenetic processes) remind us that the fundamental reproductive unit of life is not a nucleic acid molecule, but the remarkably versatile, intact, living cell.

[*See also* Cell Evolution; Maternal Cytoplasmic Control.]

BIBLIOGRAPHY

Aufderheide, K. J., T. C. Rotolo, and G. W. Grimes. "Analyses of Inverted Ciliary Rows in *Paramecium:* Combined Light and Electron Microscopic Observations." *European Journal of Protistology* 35 (1999): 81–91. Modern microscopic analyses of cortical inversions in paramecia.

Frankel, J. *Pattern Formation: Ciliate Studies and Models.* Oxford, 1989. A comprehensive overview of pattern formation and cortical inheritance.

Grimes, G. W., and K. J. Aufderheide. *Cellular Aspects of Pattern Formation: The Problem of Assembly.* Monographs in Developmental Biology, vol. 22. Basel, 1991. Summarizes cortical inheritance phenomena and the general implications of the phenomenon to eukaryotes.

Kirschner, M., J. Gerhart, and T. Mitchison. "Molecular 'vitalism.'" *Cell* 100 (2000): 79–88. A discussion of the general significance of phenomena such as cortical inheritance.

Locke, M. "Is There Somatic Inheritance of Intracellular Patterns?" *Journal of Cell Science* 96 (1990): 563–567. Discussion of cortical inheritance–like phenomena in many different metazoan species.

Nelsen, E. M., J. Frankel, and L. M. Jenkins. "Non-genic Inheritance of Cellular Handedness." *Development* 105 (1989): 447–456. Description of mirror-image *Tetrahymena* cells.

Prusiner, S. B., M. R. Scott, S. J. DeArmond, and F. E. Cohen. "Prion Protein Biology." *Cell* 93 (1998): 337–348. Review of prion phenomena, especially brain disease forms.

Serio, T. R., A. G. Cashikar, A. S. Kowal, G. J. Sawicki, J. J. Moslehi, L. Serpell, M. F. Arnsdorf, and S. L. Lindquist. "Nucleated Conformational Conversion and the Replication of Conformational Information by a Prion Determinant." *Science* 289 (2000): 1317–1321. Discussion and tests of various yeast prion propagation models.

— Karl J. Aufderheide

CREATIONISM

Creationism refers to the belief that some or all of the various forms of life on earth were brought into being by a creator. The emphasis in this article will be on those varieties of creationism that arise from interpretations of Genesis and of the Koran. These interpretations range from understanding the creation stories in religious texts to have a meaning that is literally true (e.g., direct creation of each "kind," or species, separately and almost simultaneously) to interpreting the texts as expressing important religious ideas in a metaphorical form. Proponents of the former have been the most prominent in raising issues of educational policy in the United States and elsewhere and in shaping the public debate regarding the validity of evolution generally.

A central philosophical question is how, if at all, creationism and evolution can accommodate each other. This question is captured by Michael Ruse's title, *Can a Darwinian Be a Christian?* (2001), but it could also be turned on its head: Can a religious person, a Jew, Christian, or Muslim, specifically, reasonably accept evolution? It will turn out that the answers given to the second of these two questions may depend on how strongly one believes that what evolutionary biologists call "macroevolution" has occurred. For purposes of this article, macroevolution can be defined as speciation, or the evolution of new species from potentially quite different ancestors. Macroevolution plays a central role in the creation–evolution debate because it directly challenges some creationist accounts of the origin of species.

Scientific Evidence for Evolution. It is not the purpose of this article to present a scientific case for the reality of evolution, including macroevolution. Rather, our interest is to ask what kinds of scientific evidence must a person who chooses not to believe in evolution either deny, interpret differently, or otherwise explain away.

Charles Darwin, in *On the Origin of Species* (1859), argued that multiple lines of evidence supported the conclusion that species had evolved from quite different ancestors (macroevolution) and that the primary mechanism for change had been evolution by natural selection. He also discussed why the difficulties then apparent should not be regarded as insurmountable. In the subsequent years, the relevant lines of evidence have deepened and diversified, the scientific difficulties have fallen one after the other, and mechanisms of inheritance and other causal processes have been clarified.

Some key facts of consensus that have emerged over the last century or so of research on large-scale geological and biological change and that are relevant to creationist views, listed by area of inquiry, are the following:

Geology
- The earth is old (roughly 4.5 billion years).
- The geological record formed incrementally over long periods of time.
- Plate tectonics and continental drift are clear.

Paleontology
- Transitional forms appear in sedimentary layers.
- Fossil sequences, beginning with only a few simple forms and adding diversity over eons, are commonplace.
- Vestigial structures are evidence of divergence from ancestral forms.

Genetics and Molecular Evolution
- Microevolution (evolution within a species) and macroevolution (speciation) are widely documented.
- Gene sequences reveal deep ancestral relationships among very disparate groups.
- DNA and RNA sequences are the core of inheritance (making evolutionary change both possible and physically inevitable, given limits to replication fidelity).
- Natural selection works incrementally.
- Reproductive isolation (incipient speciation) is enhanced both by simple isolation and by direct natural selection.
- Natural evolution of reproductive isolation leading to speciation can be observed in nature, especially in hybrid zones.

Scientific support for macroevolution is now comparable in strength to that for the planets' orbiting of the sun. One cannot actually observe the planets orbiting the sun. Indeed, some outer planets have not orbited even once since their discovery. One must infer that the planets orbit the sun. However, the evidence for this inference is so strong that planetary orbits are usually regarded as scientific fact. Evolution is seen by the scientific community to be a "fact" in precisely the same sense. Thus, if it is a fact that the planets orbit the sun, then large-scale evolution is at least as strong a fact. Conversely, if large-scale evolution (macroevolution) is regarded as a theory, then the idea that the planets orbit the sun also must be regarded as a theory, though as a theory with less support than that available for evolution.

There are, of course, deeper questions about both planetary motion and evolution that remain scientifically contentious. For example, the nature of gravity is far from clear, leading to searches for gravity waves and gravitons. Similarly, the sequences of phylogenetic diversification remain open for further study. These technical uncertainties are sometimes wrongly cited as suggestive of scientific uncertainty over the occurrence of evolution. However, they are not the cause of the public contention over evolution any more than confusion over the nature of gravity is a cause of public contention over planetary motion.

Can Creationism and Evolution Exist Side by Side? Advocates of any position that rejects macroevolution are faced with immense bodies of contrary information. Hence, the answers one might give to the question of whether creationism can accommodate evolution depend on the religious framework one adopts rather than on any differences in the scientific evidence. Several of the key creationist positions in this regard are discussed in this section.

Quick creation. The most publicly vociferous, politically active, and widely publicized creationists, at least in the United States, adopt the literalist view that God created the world in six days. Advocates of this position suggest that the scientific evidence for an old-earth, incremental geology and for macroevolution conflicts with "necessary inferences" from scripture. To accept the scientific conclusions would call religion into question. Perceived consequences of doing so include increasing immoral behavior, disrespect for or disbelief in God, and, consequently, damnation. Henry Morris (1972), a prominent quick creationist, asserted that the essence of evolution was transmitted by Satan to Adam in the Garden of Eden and that it is responsible for communism, racism, imperialism, militarism, pornography, promiscuity, and perversion.

The quick creationist analysis thus sees drastic consequences if one accepts evolution and it turns out to be false, and little or no payoff for accepting evolution, even if it is right. Given the perceived consequences, it would be irrational to accept either an old earth or evolution; hence, no amount of evidence, real or imagined, can be allowed to contradict the necessary inferences from the Bible. This stance depends on the assumption of catastrophic religious consequences and is essentially independent of the perceived strength of scientific evidence for evolution. Because no one should pursue damnation willfully, any disagreement with the quick creationist position must be based on a different view of the consequences that would arise from disbelief.

The core of one major theological disagreement with the quick creationist position had been articulated by the fourth century CE. Saint Augustine noted that

[u]sually, even a non-Christian knows something about the Earth, the heavens, and the other elements of the world.... It is a disgraceful and dangerous thing for an infidel to hear a Christian, presumably giving the meaning of Holy Scripture, talking nonsense on these topics.... If they find a

Christian mistaken in a field which they themselves know well and hear him maintaining his foolish opinions about our books, how are they going to believe those books in matters concerning the resurrection of the dead, the hope of eternal life and the kingdom of heaven . . . ?

Saint Augustine articulated a substantial risk to faith itself from denying strongly supported ideas about the natural world, even if they conflicted with Genesis. Martin Luther and John Calvin each also questioned the necessity of a literal interpretation of Genesis.

Less-literal creationisms. The increasing scientific evidence for evolution has, decade by decade, increased the relevance of the theological tradeoffs articulated by Saint Augustine. This, combined with what some creationists see as inconsistencies between the first two chapters of Genesis, has given rise to a variety of interpretations, each claiming to capture Genesis' real meaning. One suggestion is that the "days" of creation were each in fact millions of years long. Another possibility is that geological time spans millions of years, but new kinds, or species, were created only on a few widely separated days. Separate views are that organisms evolved gradually but were destroyed at the fall of Satan, then re-created in six days, or that organisms evolved gradually but that revelation occurred in six days.

Broad-based religious rejection of literalism. An Arkansas law required that a young earth, flood geology (the deluge) approach be taught in parallel with any public school presentation of evolution. The breadth of the theological rejection of this quick creationist position is evident in the list of plaintiffs who requested that the law be voided. These included "the resident Arkansas Bishops of the United Methodist, Episcopal, Roman Catholic, and African Methodist Episcopal Churches, the principal official of the Presbyterian Churches in Arkansas, [and] other United Methodist, Southern Baptist, and Presbyterian clergy . . ." (Judge William R. Overton, *McLean v. Arkansas Board of Education*, 529 F. Supp., 1255 E.D. Arkansas, 1982). Similar rejection is explicit in the many statements noting the lack of conflict between evolution and religion issued by various religious denominations and scientific groups.

The question of the literal truth of the creation narratives is not the only theological problem that has motivated alternate varieties of creationism. Of at least equal theological importance has been the problem of the existence of evil and of apparently unmerited suffering in a world created and overseen by an all-knowing, all-powerful, and benevolent God. A central part of the traditional Christian response has focused on the idea of original sin. We are all subject to suffering—even babies—because of the sins of Adam and Eve. Evolution, by weakening the belief in the reality of a literal Adam and Eve, makes this concept seem less plausible,

a problem that motivates, in part, some important additional, more metaphorical, versions of creationism.

Progressive creation. Progressive creationists accept an old earth, normal geology and microevolution (the evolution that occurs within a species). However, the appearance of major new forms in the fossil record without clear antecedents, a category taken to include humans, is attributed to new acts of creation by God. Creation is thus viewed as ongoing across geological time and progressive in the sense of creating both more complex forms and gradually greater diversity. By asserting the separate creation of humans, this approach preserves the literalness of Adam and Eve, and thus of original sin.

Gradual creation (theistic evolution). Many mainline Christian churches now accept an old earth, normal geology and the evolution of all organisms from very distant common ancestors. Evolution is seen as the way that God has created all organisms, including humans. The official Roman Catholic position is that there is no conflict between religion and the evolution of our bodies. However, our souls are individually created and did not evolve. This approach accommodates the key scientific facts while preserving the literalness of Adam and Eve, the first two of our ancestors into whom God chose to implant souls. It thus also preserves the literalness of original sin.

Finally, some other religious groups and philosophers simply accept evolution fully. This is easiest when science and religion are seen as dealing with distinct realms of knowledge, as in Stephen Jay Gould's *"Nonoverlapping Magisteria"* (1997), with wisdom arising from combining science (the basis for effective action) and religion (the key source of values). Such approaches advocate a variety of solutions to the problem of evil other than a literal Adam and Eve. For example, Adam and Eve can be taken as a metaphorical statement of an intrinsic human propensity to sin or to disobey God.

BIBLIOGRAPHY

Cain, D. "Let There Be Light." *Eternity* (May 1982): 30–31. Graphical contrast of fourteen "literal" interpretations of Genesis.

Dawkins, R. *The Blind Watchmaker*. New York, 1986. Argues that evolution makes atheism intellectually respectable (but see Ruse, 2001).

Gould, S. J. "Non-overlapping Magisteria." *Natural History* 106.2 (1997): 16–22, 60–62.

Gould, S. J. *Rocks of Ages: Science and Religion in the Fullness of Life*. New York, 1999. Good review of the controversy and an eloquent statement of the position that science and religion deal with nonoverlapping areas.

Kitcher, P. *Abusing Science: The Case against Creationism*. Cambridge, Mass., 1982. A philosopher of science succinctly evaluates creationist claims. Explains how science can select better claims from worse, even though usually we are forced to "believe where we cannot prove."

Matsumura, M. *Voices for Evolution*. Berkeley, 1995. (Also at http://natcenscied.org.) Official statements by religious and scientific organizations issued in support of evolution.

Miller, K. *Finding Darwin's God.* New York, 1999. A good explanation of the scientific problems with rejecting evolution and with intelligent design. Miller bases his own religious views on quantum uncertainty, in part.

Morris, H. *The Remarkable Birth of the Planet Earth.* Minneapolis, 1972. Espouses quick creationism.

National Academy of Sciences (USA). *Science and Creationism: A View from the National Academy of Sciences.* 2d ed. Washington, D.C., 1999. (Also at http://books.nap.edu/html/creationism/) The most prestigious scientific group in the United States affirms that evolution is a fact.

Nelson, C. E. "Creation, Evolution, or Both? A Multiple Model Approach." In *Science and Creation,* edited by R. W. Hanson. New York, 1986. Application of consequence analysis to creation and evolution trade-offs.

Nelson, C. E. "Effective Strategies for Teaching Evolution and Other Controversial Subjects." In *The Creation Controversy and the Science Classroom.* Arlington, Va., 2000.

Pennock, R. T. *Tower of Babel: The Evidence against the New Creationism.* Cambridge, Mass., 1999. Key philosophical critique of intelligent design.

Ruse, M. *Can a Darwinian Be a Christian?* Cambridge, 2001. Overview of the full range of theological approaches. Concludes that a Darwinian can without self-contradiction indeed be a Christian.

Saint Augustine. *On the Literal Meaning of Genesis.* Translated by J. H. Taylor. New York, 1982.

Scott, E. C. "Antievolutionism and Creationism in the United States." *Annual Review of Anthropology* 26 (1997): 263–289. Good overview.

Strahler, A. N. *Science and Earth History: The Evolution/Creation Controversy.* Buffalo, N.Y., 1999. Extensive review of claims that any scientific evidence supports a young earth, flood geology approach.

Trevathan, W., J. J. McKenna, and E. O. Smith, eds. *Evolutionary Medicine.* Oxford, 1999. How evolution matters in medicine.

ONLINE RESOURCES

Access Research Network. http://www.arn.org/. Articles supporting intelligent design. For critiques, see Pennock (1999), Ruse (2001), and the next two sites.

National Center for Science Education. http://natcenscied.org. Home page for this organization. Site provides key resources including essays on the varieties and history of creationism, "Seven Significant [American] Court Decisions," and "Evolution, Creationism and Science Education."

Talk.origins Archives. http://www.talkorigins.org. Best online source for overviews of evidence for evolution and analyses of creationist claims regarding science and evidence. Includes M. Isaak's "What Is Creationism" (2000), with its links to sites advocating most of the full range of American creationist positions, both Christian and non-Christian, and an introduction to creation myths generally.

— CRAIG E. NELSON

CRETACEOUS-TERTIARY EXTINCTION. *See* Mass Extinctions.

CRUSTACEANS. *See* Arthropods.

CRYPTIC FEMALE CHOICE

Charles Darwin distinguished two contexts in which males often compete for access to females: direct male–male battles and female choice. He apparently believed, perhaps because he shied away from thinking about intimate aspects of copulation, that the resulting sexual selection occurred only prior to copulation. He thus thought that a male's success in sexual competition could be measured in terms of his ability to obtain copulations. It is now clear that this view is incomplete, and that males also compete for access to the female's eggs after having achieved copulation. This competition is sometimes called sperm competition, but using Darwin's divisions, the postcopulatory equivalent of female choice is called cryptic female choice, and the postcopulatory equivalent of male–male battles is sperm competition in a strict sense. The word *cryptic* refers to the fact that any female choice among males that occurs after copulation has begun would be missed using the classic Darwinian criteria of success. The term *cryptic female choice* was first applied to biased oviposition in a scorpionfly in which females laid more eggs immediately after copulation with larger males and in the same year the idea played a central role in ideas regarding the evolution of animal genitalia.

In concrete terms, cryptic female choice can occur if a female's traits consistently bias the chances that particular conspecific males have of siring offspring when she copulates with more than one, so that males with traits that increase the probability of certain postcopulatory events are more likely to obtain fertilizations than others. Such female biases can be produced in many ways. At least twenty-one possible mechanisms have been described, including, for example, binding some but not other types of sperm with molecules in the zona pellucida of mammalian eggs, discarding the sperm of some males during or immediately after copulation, ovulating after some copulations but not others, and investing more in the offspring of some males than in those of others. In the end, the possible importance of cryptic female choice is related to the physical difficulty that a male has in guaranteeing that all, or even any, of the female's eggs will be fertilized by the sperm he deposits. Studies of sperm transfer in species with internal fertilization generally have shown that males very seldom deposit their sperm directly onto the female's eggs. Thus, insemination does not necessarily result in fertilization of a female's eggs. If, for instance, a female bird or mammal fails to ovulate within the time during which the male's sperm can survive within her reproductive tract following copulation, or if she fails to transport sperm to sites in her body where fertilization occurs, then that copulation will not result in offspring. If female failure to ovulate sometimes limits a male's reproduction and females mate with multiple males, any male trait that increases the chances that the female will ovulate can come under sexual selection by cryptic female choice. Similarly, copulation generally does not automatically result in transfer of sperm to the female. For example, male traits that induce the female not to exclude the male's genitalia from that portion of her genital tract where sperm have the best chances of surviving and fertilizing eggs or to refrain from ejecting his sperm from her body could also come under cryptic female choice. In sum, not all copulations are equally likely to produce offspring in many species, and male traits that increase the chances of fertilization can be subject to cryptic female choice.

Postcopulatory Choice. The potential importance of the female in influencing postcopulatory events is evident from basic morphology. Whereas precopulatory competition among males often occurs with relatively little direct female influence, postcopulatory competition is generally played out within the female's own body. Even small changes in her reproductive morphology and physiology can have consequences for a male's chances of fertilization. The extreme asymmetry between the tiny, delicate sperm and the hulking, complex female, with her extensive array of morphological, behavioral, and physiological capabilities, does not mean that males and their gametes are completely powerless in determining whether or not fertilization will occur; but it certainly seems to favor the possibility of female influence. In an analogy with human sporting events, the female's body constitutes the field on which males compete, and her behavior and physiology set the rules by which competitors must abide.

In addition to this basic asymmetry, there is reason to expect that females can gain from exercising postcopulatory choice, and that natural selection on many possible cryptic female choice mechanisms will tend to eventually lead to sexual selection. Take, for example, oviposition. In most species, natural selection on females will make it disadvantageous for a female to lay her eggs without first having mated. Females will thus tend to evolve the ability to use cues associated with copulation to trigger oviposition. Such stimuli will usually be produced by the male or his semen. Once the female has evolved to sense such stimuli and to respond by triggering oviposition, then male abilities to elicit this female response can come under sexual selection. If a female mates with more than one male, if her oviposition responses to males are not always complete (i.e., not all eggs are always laid immediately), and if some males are better able to elicit oviposition and thus obtain more offspring, then, other things being equal, those males better able to induce oviposition will tend to outreproduce others. Once such variant males are present, then selection can favor those females better able to

accentuate this bias in fertilization. For example, females with higher thresholds for triggering ovulation would tend to ovulate only after copulating with especially stimulating males and would be favored because they would produce male offspring better able to stimulate females in future generations. They would thereby exercise cryptic female choice in favor of such males. Changes in female thresholds, in turn, could set off a new round of evolution of male abilities to stimulate females.

Given the multitude of possible cryptic female choice mechanisms, the theoretical expectations that cryptic female choice can evolve readily, the frequent finding that females mate with multiple males in nature, and relatively strong evidence from a small number of direct studies (including tunicates, beetles, flies, frogs, birds, and water striders), there seems little doubt that cryptic female choice occurs in nature. However, the question whether or not it is widespread has been hotly debated. Because cryptic female choice was only recently carefully formulated and publicized, experiments testing directly for its occurrence have been performed with relatively few organisms. In addition, convincing direct tests are intrinsically difficult to perform, because there are so many different female processes that might be involved and that thus have to be checked, and because it is sometimes difficult to distinguish cryptic female choice from sperm competition. Thus, direct tests with particular species cannot yet resolve the issue of its general importance. The data with strongest general implications are more indirect. Three types in particular suggest that cryptic female choice may be very common indeed.

Genitalic Morphology. One of the most widespread trends in animal evolution is the tendency for the male genitalia of species with internal fertilization to evolve especially rapidly and divergently. The most likely explanation is that cryptic female choice is responsible. [*See* Genitalic Evolution.] An additional common phenomenon in females explained by this hypothesis involve, the long, tortuous ducts in the reproductive tracts of some groups that may screen access by males or their sperm to female sperm storage or fertilization sites. Data on genitalia are especially important because they are very abundant, and they speak rather strongly against the currently popular alternative hypothesis of male–female conflict and coevolutionary races between the sexes to explain events surrounding copulation and fertilization. The evidence against such races includes the frequent lack of interspecific differences among females that correspond to the differences among males, a strong trend in allometric scaling of genitalia in insects and spiders that is opposite to that predicted, and a general lack of female structures with mechanically appropriate designs for combating or repelling males. There are some

Figure 1. Male Jaguar "Love Biting" Female During Copulation.
If such male behavior patterns are indeed courtship, the likely conclusion is that the male is attempting to increase his chances of paternity.   © Tom McHugh/Photo Researchers, Inc.

female genitalic structures that mesh with species-specific male genitalia, but they are generally "selectively cooperative" structures, such as pits, slots, or grooves, that facilitate rather than oppose coupling with males. They are selective in that they facilitate coupling only with males possessing certain structures.

Male Courtship During or Following Copulation. A second, much less studied topic that nevertheless indicates that cryptic female choice may be widespread is male courtship that occurs during or following copulation. Such behavior is paradoxical under the usual interpretation that male courtship functions to induce the female to accept copulation—why should he continue to court after he is already copulating? Yet copulatory courtship is apparently very common. The only systematic search for copulatory courtship, using conservative criteria to define courtship behavior, found that it occurred in over 80 percent of a sample of 131 species of insects and spiders. There are also scattered reports of similar behavior in other groups, including nematodes, birds, scorpions, frogs, fish, reptiles, millipedes, mammals, molluscs, and crustaceans. If these male behavior patterns are indeed courtship (they include waving, rubbing the female, licking, squeezing rhythmically, kicking, tapping, jerking, rocking, biting, feeding, vibrating, singing, and shaking), the likely conclusion is that the male is attempting to induce additional, postintromission female cooperation that increases his chances of paternity (Figure 1). Studies of insects in three different orders have confirmed that copulatory courtship does indeed induce the female to favor the male's reproduction. If the 80 percent figure is anywhere nearly representative, then cryptic female choice would seem to be very common. Again, there are indications, though less definitive

than in the case of genitalia, that copulatory courtship is not the result of coevolutionary arms races between males and females. Male copulatory courtship behavior generally seems noncoercive and inappropriately designed to force the female to continue copulation or to perform other responses leading to fertilization. Indeed, the sites where many possible cryptic choice mechanisms occur are generally deep within the female's body, seemingly inaccessible to such male behavior.

Seminal Products. A third relatively widespread phenomenon also suggests, though less strongly, that cryptic female choice is common. Seminal products derived from male accessory glands frequently affect female reproductive processes. Experimental studies of these effects are almost completely limited to insects and ticks, but over seventy species have been studied in these two groups. With few exceptions, male seminal products produce one or more of the following effects: induce female to oviposit; induce female to ovulate or otherwise bring eggs to maturation; inhibit her from further mating; or, less frequently studied, induce her to transport his sperm. Such male products could evolve via cryptic female choice favoring those males best able to induce favorable female responses. It is less clear, however, whether these effects are best attributed to cryptic female choice or to male–female conflict. Some studies suggest, though not conclusively, that males may damage female reproductive interests via the effects of seminal products.

In summary, cryptic female choice and sperm competition extend the classic Darwinian context of sexual selection to include events that occur after copulation has begun. Cryptic female choice has been demonstrated in a number of species, and there are theoretical reasons to expect that it can evolve readily. Several indirect types of evidence suggest that it may be a widespread and important evolutionary phenomenon, but there are as yet only a few direct tests. Further tests, preferably in a variety of groups, will be needed to test the generality of its importance.

[*See also* Genitalic Evolution; Mate Choice; Parent–Offspring Conflict; Sexual Selection; Sperm Competition.]

BIBLIOGRAPHY

Cordero, C., and W. G. Eberhard. "Sexual Conflict and Female Choice." *Evolution.* Theoretical and empirical treatments of possible male–female conflict need reexamination because of flawed calculations of costs to females.

Darlington, M. B., D. W. Tallamy, and B. E. Powell. "Copulatory Courtship Signals Male Genetic Quality in Cucumber Beetles." More energetic copulatory courtship in a beetle induces the female to relax the walls of her reproductive tract and allow the male to deposit a spermatophore; male offspring of especially stimulating males are better stimulators.

Eberhard, W. G. *Sexual Selection and Animal Genitalia.* Cambridge, Mass., 1985. Proposes that male genitalia evolve rapidly and divergently due to sexual selection by cryptic female choice and critically evaluates this and other hypotheses.

Eberhard, W. G. "Evidence for Widespread Courtship during Copulation in 131 Species of Insects and Spiders, and Implications for Cryptic Female Choice." *Evolution* 48 (1994): 711–733. Apparent male courtship behavior occurred during copulation in 81 percent of 131 species of insects and spiders, suggesting that cryptic female choice is common.

Eberhard, W. G. *Female Control: Sexual Selection by Cryptic Female Choice.* Princeton, 1996. A summary of arguments and data indicating that cryptic female choice may be a major evolutionary phenomenon.

Eberhard, W. G. "Female Roles in Sperm Competition." In *Sperm Competition and Sexual Selection*, edited by T. Birkhead and A. P. Moller, pp. 91–116. New York, 1998. Summarizes arguments regarding cryptic female choice and discusses its possible relationships with male–female conflict.

Edvardsson, M., and G. Arnqvist. "Copulatory Courtship and Cryptic Female Choice in Red Flour Beetles *Tribolium castaneum.*" *Proceedings of the Royal Society of London B* 267 (2000): 559–563. Experimental manipulations showed that female perception of the rate of male copulatory courtship behavior in a beetle affected the male's fertilization success when the female mated with two different males.

Otronen, M., and M. Siva-Jothy. "The Effect of Postcopulatory Male Behaviour on Ejaculate Distribution within the Female Sperm Storage Organs of the Fly *Dryomyza anilis* (Diptera: Dryomyzidae)." *Behavioral Ecological Sociobiology* 29 (1991): 33–37. More postcopulatory genitalic tapping by the male increases the likelihood that his sperm will be used to fertilize the eggs that the female is about to lay.

Thornhill, R. "Cryptic Female Choice and Its Implications in the Scorpionfly *Harpobittacus nigriceps.*" 122 (1983): 765–788. Females laid more eggs immediately following copulations with larger males; also coined the term *cryptic female choice.*

Ward, P. "Cryptic Female Choice in the Yellow Dung Fly." *Evolution* 54 (2000): 1680–1686. Gives reasons for supposing that cryptic female choice occurs in this species when females shuffle sperm among their multiple storage organs and thus bias male chances of fertilization, a possibility that has been hotly debated.

—WILLIAM G. EBERHARD

CULTURAL EVOLUTION

[*This entry comprises two articles. The first article provides an introduction to the study of how cultures evolve, taking up the ideas of cultural inheritance in humans and in animals; the second article discusses methods by which ideas, behaviors, and practices might spread within cultures and across generations. For related discussions, see the* Overview Essay *on* Culture in Chimpanzees *at the beginning of Volume 1;* Art; *and* Human Sociobiology and Behavior.]

An Overview

Many researchers have noted analogies between the processes of biological evolution and cultural change.

For example, both genes and culture are informational entities that are differentially transmitted from one generation to the next. These similarities have led to the idea that culture evolves and prompted the development of mathematical models of cultural evolution.

The main scientific approach to the study of how culture evolves is a branch of theoretical population genetics known variously as cultural evolution, gene–culture coevolution, or dual inheritance theory. This intellectual tradition has nothing in common with the nineteenth-century cultural evolution schools, which, based on an erroneous progressive view of evolution, set out to model stages of societal progression (Morgan, 1877; Tylor, 1865). Rather, the population genetics approach regards culture as an evolving pool of ideas, beliefs, values, and knowledge that is learned and socially transmitted between individuals. Researchers focus on a single trait, such as a preference for drinking milk, or for sons over daughters, and employ a rigorous mathematical approach to describe how the cultural trait changes over time, sometimes coevolving with genetic variation. Where the cultural entity is a discrete trait, it has much in common with Richard Dawkins's (1976) idea of the meme, defined as a cultural analog of the gene.

Modern cultural evolution theory began in 1973 when geneticists Luca Cavalli-Sforza and Marcus Feldman of Stanford University published the first simple dynamic models of cultural transmission. Fueled by the ongoing sociobiology debate, an extensive series of theoretical papers culminated in the now classic text *Cultural Transmission and Evolution* (1981). Other mathematically minded researchers adopted Cavalli-Sforza and Feldman's approach, most notably anthropologists Rob Boyd and Peter Richerson, whose *Culture and the Evolutionary Process* (1985) introduced a variety of novel theoretical methods and stimulating ideas.

The emerging body of theory has developed in a variety of ways. One class of models is employed to explore how behavioral and personality traits are transmitted from parents to offspring. Other models address very general questions about the possible adaptive advantages of learning and culture. Others explore the processes of cultural change, as well as the nature of their interaction with genetics. More recently, these general methods have been applied to address specific cases, such as the evolution of the human sex ratio or of handedness.

Cultural Inheritance in Humans and Animals. For most social scientists "culture" is a given. The notion that much of the variation in the behavior of humans is brought about by their being exposed to divergent cultures is so widespread and intuitive that it is beyond dispute. Although it used to be fashionable to define culture as the interwoven complex of behavior, ideas, and artifacts that characterize a particular people (e.g., Tylor, 1871), among social scientists this view has been superseded by a more cognitive perspective that restricts culture to information stored in the brain. In contrast, biological approaches to culture (including those of most sociobiologists, human behavioral ecologists, and evolutionary psychologists) tend to regard the transmitted elements of culture as either exerting a comparatively trivial influence on human behavior or having limited influence that is strictly circumscribed by genes. Some of these biological approaches focus on the fitness related costs and benefits for behavioral strategies. Payoffs to individuals vary with their environment (including their social environment), and they can be frequency dependent. [*See* Frequency-Dependent Selection; Game Theory; Optimality Theory.]

For advocates of gene–culture coevolution (or dual inheritance theory), these biological perspectives underemphasize one critical factor, namely, socially transmitted culture. Too much culture changes too quickly to be feasibly explained by genetic variation, whereas the fact that different behavioral traditions can be found in similar environments would appear to render environmental explanations of behavior impotent a lot of the time. To give an example, Guglielmino et al. (1995) carried out an analysis of variation in cultural traits among 277 contemporary African societies and found that most traits examined correlated with cultural history rather than with ecology. Such findings suggest that most human behavioral traits are maintained in populations as distinct cultural traditions, rather than evoked by the natural environment.

Cultural evolution enthusiasts point to countless studies that have found that the attitudes of parents and offspring are rather similar. For example, a study of Stanford university students revealed that the religious and political attitudes of parents and offspring were strongly consistent (Cavalli-Sforza et al., 1982). Boyd and Richerson (1985) compiled the results of a variety of studies of the similarity of biological relatives that suggest that the correlations between parents and offspring for behavioral traits, such as religious and political views, attitudes, vocations, hobbies, and phobias, are quite high. These researchers also documented a large body of evidence that individuals learn from unrelated individuals. Although some portion of these correlations may be the result of genetic factors, the most obvious explanation for the patterns is that children learn social attitudes in the family.

Whether or not nonhuman animals can be said to have culture is a contentious issue that hangs on definitions of culture and the complexities of animal social behavior. It is well established that numerous animals have the ability to acquire from others skills, calls, and information concerning predators, mates, or resources. In some instances, this information transmission is suf-

FIGURE 1. Both Genes and Culture Are Informational Entities That Are Differentially Transmitted from One Generation to the Next.
© Nicole Duplaix/Peter Arnold, Inc.

ficient to propagate the diffusion of novel behavior patterns through animal populations and to maintain distinct behavioral traditions between different populations of the same species. However, animal traditions are rarely cumulative, and they tend to be ephemeral. Many researchers believe that an understanding of the ways in which social learning operates in animal populations will generate insights into the evolutionary roots of human culture.

Types of Cultural Selection. Cavalli-Sforza and Feldman define cultural selection as a process by which particular traits increase or decrease in frequency because of their differential probability of being adopted by other individuals through social learning. In contrast, natural selection can change cultural trait frequency through the differential survival of individuals expressing different types of cultural traits. For example, in developed countries, the trait of fertility control (e.g., via contraception) is at a clear disadvantage in natural selection, but it has spread by virtue of its advantage in cultural selection. Working at these two levels simultaneously helps us to understand how nonadaptive cultural traits could evolve. When they have sufficiently high cultural fitness, cultural traits, such as the use of contraception, can increase in frequency despite decreasing genetic fitness.

Boyd and Richerson (1985) split the general concept of cultural selection into subtypes. One such subtype is called guided variation, which refers to a process in which individuals acquire behavioral traits culturally (typically, although not necessarily, from their parents), then modify them on the basis of their personal experience. Here cultural variation is guided by individual experience, allowing behavioral traditions to gradually evolve toward the optimal behavior. Boyd and Richerson (1985) also considered biased cultural transmission, which occurs when, given a choice been two alternative behavior patterns, individuals are more likely to adopt one variant than another. Various types of biases exist, including genetic biases and frequency dependent biases that instigate conformity. The cultural processes in transmission models assume that humans are active decision makers who combine information available from both social sources and individual experience according to some mental algorithm.

BIBLIOGRAPHY

Boyd, R., and P. J. Richerson. *Culture and the Evolutionary Process.* Chicago, 1985.

Cavalli-Sforza, L. L., and M. W. Feldman. *Cultural Transmission and Evolution: A Quantitative Approach.* Princeton, 1981.

Cavalli-Sforza, L. L., M. W. Feldman, K. H. Chen, and S. M. Dornbusch. "Theory and Observation in Cultural Transmission." *Science* 218 (1982): 19–27.

Dawkins, R. *The Selfish Gene.* Oxford, 1976.

Guglielmino, C. R., C. Viganotti, B. Hewlett, and L. L. Cavalli-Sforza. "Cultural Variation in Africa: Role of Mechanism of Transmission and Adaptation." *Proceedings of the National Academy of Sciences USA* 92 (1995): 7585–7589.

Morgan, L. H. *Ancient Society, or Researches in the Lines of Human Progress from Savagery through Barbarism to Civilization.* New York, 1877.

Tylor, E. B. *Researches into the Early History of Mankind and the Development of Civilization.* London, 1865.

Tylor, E. B. *Primitive Culture.* London, 1871.

— KEVIN N. LALAND

Cultural Transmission

Culture is defined as those features of thought, speech, behavior, and technology that are learned and transmitted to other individuals. The process of transmission may involve active teaching and learning, imprinting, imitation, conditioning, or some combination of these. As a consequence of this transmission between individuals of the same or different populations, these aspects of culture are subject to a form of evolution over time. Just as in the biological sciences, the object of quantitative studies of cultural transmission is to understand the mechanism of transmission in order to predict rates of change of cultural variants and patterns of variation in the features under study. These features may be categorical—that is, discrete-valued—or they may vary continuously with the evolutionary dynamics determined by this classification.

Various names have been assigned to the units of cultural transmission since our designation of "traits" (Cavalli-Sforza and Feldman 1973). The term "meme" was used by Dawkins (1976) as a unit of information transmitted from one brain to another. To some commentators, "meme" has the connotation of imitation, and this may limit its descriptive utility. Lumsden and Wilson (1981) introduced the term "culturgen" as a trait processed through "genetically determined procedures that direct the assembly of the mind." This term places heavy weight on the biological basis of the process, which we regard as an unnecessary restriction.

Most discussions of cultural transmission and evolution focus on humans. It is possible to make analogous arguments for some animals, but the importance of transmission of traditions among animals for their evolution is strongly debated. Many animal behaviorists feel that, unlike humans, most animals do not intend to pass information to a conspecific. It is further claimed that only the powerful cognitive apparatus of humans permits the intentional accumulation of modifications to traits that may then be culturally transmitted (see Galef, 1992; Tomasello, 1999).

Cultural Transmission Modes. Using an analogy to epidemiology, Cavalli-Sforza and Feldman (1981) identified three modes of cultural transmission that are amenable to mathematical modeling and analysis: vertical, horizontal, and oblique.

Vertical transmission occurs from parents to offspring and may be quantified by a function that gives the probability that parents of specific types give rise to an offspring of their own or of another type. This is a clear extension of genetic transmission, which in large populations has the well-known property of preserving the variation in the trait. Thus, vertical transmission is conservative of variation and results in a slow evolutionary dynamic. By the same token, because it preserves existing variation in a population, it contributes strongly to the buildup of between-population variation.

Horizontal transmission takes place among peers within a given cohort (age group) in a population. It can be viewed in terms of the rate of contact between individuals and, given that contact occurs, the rate of conversion of one individual to take the trait value of the other. This process is expected to result in faster within-group evolution; and, if there is contact between groups, it could result in less between-group variation than is allowed by vertical transmission.

Oblique transmission passes from members of one generation to members of a later generation. Formal teaching is an example in which there can be rapid loss of variation as one individual transmits to many—a kind of cultural inbreeding. The result is rapid loss of variation within a population; depending on the amount of cultural migration between groups, there could be rapid buildup of between-group variation.

For some purposes, the distinction between oblique and horizontal transmission is blurred. In some situations, however, transmission across generations in a nonparental framework can be important; in these cases, the rate of evolution under oblique transmission—in the one-to-many model, for example—can be much faster than for the other modes (Cavalli-Sforza and Feldman, 1981).

Regardless of which mode of transmission is considered, there is in humans some species-specific (and therefore biological) predisposition that permits the sending and acquisition of cultural information. This predisposition is most likely to be manifest in early life stages, where it facilitates, for example, the acquisition of language and of sounds that are difficult to learn later in life. Part of this predisposition may come under the heading of imprinting, although, in humans, more active teaching and learning is involved than is usually connoted by the term "imprinting."

Cultural traits acquired early are likely to be long-lasting and, therefore, more likely to be transmitted later. In addition, it is likely that those features of culture that give the recipient an advantage in natural selection are the easiest to transmit and acquire. This forces us to consider the difference between cultural and natural selection.

Cultural Selection and Natural Selection. The parameters of the models of cultural transmission, whether vertical or horizontal, are quantitative summaries of complex psychological processes. These processes include details of the transmission mechanism—that is, how the information contained in the alternative forms of the cultural trait is sent; the reception of the information; the processing of the information; and the complex of influences that results in the receiver's accepting or rejecting the alternatives inherent in the original information (see Rogers, 1995). Insofar as the different forms of the trait differ in any of these processes, they will differ in their rates of transmission and will thus be subject to cultural selection. The result will be changes in the frequencies of the cultural variants within and between generations.

Some cultural traits, however, may affect the probability of survival and/or reproduction by their carriers. Insofar as the variants of the trait differ in these probabilities, there will be natural selection on the trait in the classical Darwinian sense. Over generations, this selection will lead to changes in the frequencies of the trait variants in the population. Note that cultural selection in favor of a behavioral or technological variant need not involve differences in mortality or fertility, but merely an increase in the fraction of the population practicing that variant (Campbell, 1965; Cavalli-Sforza and Feldman, 1981).

For some traits, the two types of selection may be in conflict. Celibacy in religious orders is an example in which there is a clear disadvantage under natural selection, but the structure of the society specifically, the system of primogeniture, or inheritance by the eldest son—led to cultural selection in favor of younger sons joining such religious orders. Other examples include the spread of contraception and the inception of the demographic transition, the spread of the use of dangerous drugs, and the practice of religious ceremonies that result in the acquisition of dangerous diseases, such as kuru among New Guinea highlanders. However, it is reasonable that human biological evolution has involved natural selection for predispositions that make it more likely for us to adopt cultural traits that are adaptive in the Darwinian sense (Cavalli-Sforza and Feldman, 1981; Boyd and Richerson, 1985). In this sense, one might say that, in the long run, the existence of culture is advantageous in the Darwinian sense.

Some authors have suggested that there has been natural selection for a specific class of cultural transmission propensities: those that result in conformity with the local majority (Boyd and Richerson, 1985). Under this scenario, preferences, values, or attitudes held by the majority would be easier to transmit and adopt than others. In terms of our quantitative model, this amounts to a frequency-dependent probability of transmission (rather than the constant one envisaged earlier), and one of its main effects will be to accelerate the evolution of the variant of a trait that is carried by the majority. Such frequency-dependent transmission of norms, values, or preferences has been included in recent developments of economic theory that incorporate cultural transmission (e.g., Bowles, 1998). One version of this group effect was treated by Cavalli-Sforza and Feldman (1973), who included in the transmission coefficient a component reflecting the group mean of the population. In the presence of such a group effect, variance of a trait within a group is restricted, while that between group means may grow without bound.

One area of debate involving both selection and chance phenomena involves the role of group selection. Here, the question is how a cultural trait—that apparently entails a disadvantage to the individual—for example, food-sharing—may be preserved at the group level. The most widely accepted version of the argument proposes that, even though the trait is individually disadvantageous within a group, the probability that a subgroup contributes to the whole set of groups increases with the frequency of that trait (see Uyenoyama, 1979). The greater the differences between the subpopulations, the stronger the effect of group selection is likely to be. Thus, frequency-dependent transmission of the conformist type, for example, may facilitate group selection (Boyd and Richerson, 1985). In general, however, restrictions on the amount of admixture between groups have led most scholars to regard group selection as of relatively minor evolutionary significance.

Cultural Drift. Since the 1930s and the pioneering work of Sewall Wright (1931), chance phenomena have been accepted as important in biological evolution, especially in small populations. Similar effects must be recognized in cultural evolution. A cultural innovation may have some intrinsic advantage over a competing variant but may be lost by chance while the number of its carriers is still too small to force its spread in the population. By the same token, in a small population, an innovation that does not represent the majority in some large neighboring population may be adopted and spread even though it has no intrinsic advantage—a kind of founder effect. Of course, the same result will occur if a custom or behavior is imposed by one or a few individuals with power over the remainder. As in biological evolution, removal of variation can occur by chance (cultural drift) or by restriction of cultural sources (cultural inbreeding).

Besides these direct analogies with biological stochastic phenomena, in the reproduction of cultural artifacts there is always the probability of random copy error. Manual copying of early manuscripts is an example. Errors may accumulate over time, with two possible effects. First, if the group remains intact, the statistical variation in the artifact broadens over time. But second, if the group fragments, one subpopulation may end up producing artifacts whose average properties differ significantly from those of the original combined population. All these cases fall under the rubric of cultural drift, which can be recognized as the analogue of Wright's random genetic drift.

In the discussion of chance in culture, it is important to recognize that innovation, which plays a role analogous to mutation in biology, may not be random in cultural traits. Biological mutations have no intentionality associated with them, but cultural innovations may well be the result of an attempt to improve an artifact or to suppress an idea. The stochasticity of cultural innovations may, therefore, be qualitatively different from that of genetic mutations.

Gene–Culture Coevolution. Rates of vertical cultural transmission are slow and probably comparable to those of genetic transmission, but horizontal transmission is expected to be much faster. Studies of human behavioral variation always invoke discussion of the roles of environment and genetics in the causes of this variation. In fact, what constitutes the environment of an individual may include the trait values possessed by his parents and peers that may be culturally transmitted. Complete description of such traits and their transmission therefore requires the simultaneous depiction of genetic and cultural contributors to the trait. We use the term "phenogenotype" for the joint description of an individual's genotype and phenotype.

The rules of genotypic transmission are understood from the science of genetics. The rules of cultural transmission are less precisely known, but in principle they can be represented by the vertical, horizontal, and oblique models described above. In combination, we have a dual transmission system, described first in mathematical terms by Feldman and Cavalli-Sforza (1976). The transmission of phenogenotypes should, in principle, include the possibility that the rate of transmission of the phenotypic variant depends on the genotypes of both transmitters and receivers.

Most of the mathematical results from complex transmission schemes are limited to the case in which only the genotype of the offspring affects the cultural transmission rates from the parents. Usually, if natural selection is included, it is assumed to act through differences in viability among the offspring cultural variants in the case of vertical transmission. At the population level, it is then natural to track the dynamics of the association between genotype and cultural trait in a manner analogous to linkage disequilibrium in standard evolutionary population genetics (Feldman and Zhivotovsky, 1992).

It is obvious that, when we consider phenogenotypes involving one diallelic gene and one dichotomous cultural trait, a full transmission system will involve a vast number of parameters, and we have little hope of estimating these from experiments with humans. The outcomes of the resulting evolutionary dynamics are correspondingly complicated, with multiple stable equilibria possible. Such complexities cannot be found in either simple one-gene models or one-cultural-trait models: they arise from the interactions between genotype and phenotype in producing the transmission.

As an example of this complexity, consider the case of an advantageous cultural variant that is transmitted at a higher rate to genetic heterozygotes than to homozygotes. It might seem obvious that this would be equivalent to classical heterozygote advantage; but this is not true unless the cultural transmission is absolutely perfect (Feldman and Cavalli-Sforza, 1976). Here, heterozygote advantage is modulated by the situation of dual transmission.

Statistical effects of dual transmission are extremely interesting in light of attempts by behavior geneticists to partition the causes of trait variation into genetic and environmental components (Plomin, 2001). All these attempts involve an analysis-of-variance framework that uses a linear model to partition variance into components called "genetic" and "environmental." They make no attempt to address the complexities of dual transmission.

Examples of Gene–Culture Coevolutionary Theory. One of the first cases studied under the joint-transmission paradigm described above was that of the evolution of adult lactose absorption (Aoki, 1986; Feldman and Cavalli-Sforza, 1989). The frequency of lactose absorbers in populations that have long practiced dairying often reaches over 90 percent, but it is typically less than 20 percent in nondairying groups. The ability to absorb lactose is determined by an autosomal dominant gene, with absorption dominant to malabsorption. Dual transmission analyses of the joint evolution of this gene and the custom of dairying have allowed for increased Darwinian fitness caused by the nutrition provided by milk as well as cultural transmission of milk use, which varies with the genotype at the lactose-absorbing locus. These analyses have demonstrated that the rate of vertical cultural transmission of milk use strongly affects the fate of the absorption allele. With strong vertical transmission, in as few as 300 generations a significant fitness advantage for absorbers can result in rapid increase of the absorption allele.

A second example of gene–culture coevolution, studied by Kumm and colleagues (1994), addressed how the widespread cultural bias against female children in South and East Asia can affect the evolution of sex-determining genes. These authors showed that the details of dual transmission strongly affect the evolution of the population sex ratio. If parents try to increase their number of sons without regard to the biologically fixed sex ratio among their children, and if they make up for lost daughters by having more children, a female-biased sex ratio will result. However, if parents with the bias against daughters attempt to achieve a desired sex ratio among their own children, and as a result have fewer offspring than unbiased parents, the initial sex ratio will evolve in a male-biased direction.

A third example derives from archaeological investigations of the origins of agriculture. Agriculture spread into geographical areas previously occupied by hunters and gatherers around 10,000 years ago. Ammerman and Cavalli-Sforza (1971) applied a model originated by R. A. Fisher (1937), for the spread of an advantageous gene, to interpret archaeological dating of the arrival of agriculture in various European locations. They estimated that the spread of farming to Europe from the Middle East occurred at a rate of about 1 kilometer per year. By studying geographical variation in gene frequencies over the same area, Ammerman and Cavalli-Sforza (1984) suggested that this spread was due to physical migration of the farmers rather than to the spread of the cultural idea of farming. Recent extensions of this analysis have involved more detailed models of demography and rates of cultural conversion of hunter-gatherers to farmers (Aoki and Shida, 1996), but the results are qualitatively similar.

[See also Human Sociobiology and Behavior, *article on* Behavioral Ecology.]

BIBLIOGRAPHY

Ammerman, A. J., and L. L. Cavalli-Sforza. "Measuring the Rate of Spread of Early Farming in Europe." *Man* 6 (1971): 674–688.

Ammerman, A. J., and L. L. Cavalli-Sforza. *The Neolithic Transition and the Genetics of Populations in Europe.* Princeton, 1984.

Aoki, K. "A Stochastic Model of Gene–Culture Coevolution Suggested by the 'Culture Historical Hypothesis' for the Evolution of Adult Lactose Absorption in Humans." *Proceedings of the National Academy of Sciences U.S.A.* 83 (1986): 2929–2933.

Aoki, K., and M. Shida. "Travelling Wave Solutions for the Spread of Farmers into a Region Occupied by Hunter-Gatherers." *Theoretical Population Biology* 50 (1996): 1–17.

Bowles, S. "Endogenous Preferences: The Cultural Consequences of Markets and Other Economic Institutions." *Journal of Economic Literature* 36 (1998): 75–111.

Boyd, R., and P. J. Richerson. *Culture and the Evolutionary Process.* Chicago, 1986.

Campbell, D. T. "Variation and Selective Retention in Sociocultural Evolution." In *Social Change in Developing Areas, a Reinterpretation of Evolutionary Theories,* edited by H. R. Barringer et al., pp. Cambridge, Mass., 1965.

Cavalli-Sforza, L. L., and M. W. Feldman. "Models for Cultural Inheritance. I. Group Mean and Within Group Variation." *Theoretical Population Biology* 4 (1973): 42–55.

Cavalli-Sforza, L. L., and M. W. Feldman. *Cultural Transmission and Evolution: A Quantitative Approach.* Princeton, 1981.

Dawkins, R. *The Selfish Gene.* New York, 1976.

Feldman, M. W., and L. L. Cavalli-Sforza. "Cultural and Biological Evolutionary Processes, Selection for a Trait under Complex Transmission." *Theoretical Population Biology* 9 (1976): 239–259.

Feldman, M. W. and L. L. Cavalli-Sforza. "On the Theory of Evolution under Genetic and Cultural Transmission with Application to the Lactose Absorption Problem." In *Mathematical Evolutionary Theory,* edited by M. W. Feldman, pp. 145–173. Princeton, 1989.

Feldman, M. W., and Lev A. Zhivotovsky. "Gene–Culture Coevolution: Toward a General Theory of Vertical Transmission." *Proceedings of the National Academy of Sciences U.S.A.* 89 (1992): 11,935–11,938.

Fisher, R. A. "The Wave of Advance of Advantageous Genes." *Annals of Eugenics,* London 7 (1937): 355–369.

Galef, B. G., Jr. "Weaning from Mother's Milk to Solid Foods: The Developmental Psychobiology of Self-selection of Foods by Rats." *Annals of the New York Academy of Sciences* 662 (1992): 37–52.

Hamilton, W. D. "The Genetical Evolution of Social Behaviour. I." *Journal of Theoretical Biology* 7 (1964): 1–16.

Kumm, J., K. N. Laland, and M. W. Feldman. "Gene–Culture Coevolution and Sex Ratios: The Effects of Infanticide, Sex-selective Abortion, Sex Selection, and Sex-biased Parental Investment on the Evolution of Sex Ratios." *Theoretical Population Biology* 46 (1994): 249–278.

Lumsden, C. J., and E. O. Wilson. *Genes, Mind, and Culture.* Cambridge, Mass., 1981.

Parker, G., and J. Maynard Smith. "Optimality Theory in Evolutionary Biology." *Nature* 348 (1990): 27–33.

Plomin, R. "Genetics and Behaviour." *The Psychologist* 14 (2001): 134–139.

Rogers, E. M. *Diffusion of Innovations.* 4th ed. New York, 1995.

Tomasello, M. "The Human Adaptation for Culture." *Annual Review of Anthropology* 28 (1999): 509–552.

Uyenoyama, M. K. "Evolution of Altruism under Group Selection in Large and Small Populations in Fluctuating Environments." *Theoretical Population Biology* 15 (1979): 58–85.

Wilson, E. O. *Sociobiology.* Cambridge, Mass., 1975.

Wright, S. "Statistical Methods in Biology." *Journal of the American Statistical Association* 26 (1931): 155–163.
                      — MARC FELDMAN AND L. LUCA CAVALLI-SFORZA

CULTURAL INHERITANCE. *See* Overview Essay *on* Culture in Chimpanzees *at the beginning of Volume 1.*

CULTURAL TRANSMISSION. *See* Cultural Evolution, *article on* Cultural Transmission.

CUVIER, GEORGES

Georges Cuvier (1769–1832), comparative anatomist and paleontologist, was born on 23 August 1769 in the Protestant town of Montbéliard, then part of the Duchy of Württemberg. Cuvier trained in Stuttgart for a career in administration. Because of limited opportunities, he took a position as the tutor for a family living in Normandy. The years he spent in Normandy kept him away from the storm of the French Revolution and gave him time to develop his intense interest in natural history, particularly the fauna of the seashore. His talent was soon recognized, and by 1795 he established himself in Paris at the Muséum national d'histoire naturelle and the Institut de France. Cuvier had a highly successful career in science, as well as in administration and education.

Cuvier is best remembered for his brilliant work in comparative anatomy. Building on the tradition in Paris established by Louis-Jean-Marie Daubenton (1716–1800) and Félix Vicq-d'Azyr (1748–1794), he extended comparative anatomy and developed it as a theoretical tool for understanding life. Cuvier used comparative anatomy to study living organisms as well as the remains of extinct animals. From his thousands of dissections, he believed he had discovered regularities that could be rigorously defined. Natural history, according to Cuvier, could be transformed from a largely descriptive subject to a scientific discipline that uncovered the laws of nature.

Cuvier held that an understanding of the animal body had to be grounded in relating structures to their functions (rather than approaching structure as a subject independent of function). He stands on one side of a long debate that occupied comparative anatomists for over a century: form versus function. Cuvier argued that animals should be considered "functional units," whose structure was determined by its relationship to its specific environment. By "functional unit," he meant that all of the individual's organs related functionally to one another and operated together. This intricate integration he called the "correlation of parts," and he believed it represented one of the most fundamental laws of comparative anatomy.

Cuvier also believed that a study of comparative anatomy could reveal the order in the living world. From his empirical studies, he concluded that all individuals of a species share a basic plan that can be precisely defined. So, too, could higher level groups—genera, families, and so on—be characterized by a basic plan, or "type." He further believed that he could organize all these types into a hierarchical system. Unlike many earlier classification systems, Cuvier believed that animals did not fall into a single, continuous system going from most simple to most perfect; rather, he argued that animals fell into four separate divisions without any intermediate links and without any hierarchy among the four divisions. He elaborated this system in his famous *Le Règne animal distribué d'après son organisation* (1817). Based on a vast empirical system, Cuvier's classification presented a more powerful tool than Carolus Linnaeus's earlier approach, and it became widely adopted in the scientific world.

Cuvier's use of comparative anatomy to study the fossil record helped revolutionize the study of the past. He claimed that because all the parts of an organism were organized into a functional unit determined by its environment, given a single important organ from a fossil he could "reconstruct" the rest of the animal and describe the environment in which it had lived. His many reconstructions caught the public's imagination and demonstrated the power of comparative anatomy.

From fossils excavated in the Paris basin, Cuvier sketched a picture of past life on earth. He showed that in the past distinct flora and fauna existed, and that the deeper one went from the surface, the more unlike the fossils were compared to contemporary forms. The distinctness of the geological strata especially struck Cuvier, and he speculated that the history of the planet had been characterized by sudden and violent change, a position called "catastrophism." He sketched his geological views in an essay republished in many editions, *Discours sur la théorie de la terre* (1812). Cuvier did not speculate in print or in public on the origin of the diversity of life or on the appearance of new forms. His belief in the complexity of the animal organism kept him from accepting the idea that animals could change in time. It is not surprising, then, that he strongly rejected the evolutionary theories proposed by his colleagues at the Muséum, Étienne Geoffroy Saint-Hilaire (1772–1844) and Jean Baptiste Lamarck. [*See* Lamarck, Jean Baptiste Pierre Antoine de Monet.]

Cuvier has an ambiguous relationship to evolutionary thought. During his life, he strongly opposed evolutionary theories, and his formulation of comparative anatomy provided the basis for later opposition to Charles Darwin's theory. Indeed, the critique from comparative anatomy presented one of the major hurdles evolutionary theorists had to contend with after 1859. Nonetheless, Cuvier's paleontology, by stressing the change in time in the fossil record, raised the question of the origin of species. By promoting a rigorous comparative anatomy, he helped to establish a discipline that ultimately rejected his static view and became evolutionary morphology.

[*See also* Body Plans; Classification; Species Concepts.]

BIBLIOGRAPHY

Appel, Toby. *The Cuvier–Geoffroy Debate.* Oxford, 1987. A discussion of Cuvier's antievolution argument.
Coleman, William. *Georges Cuvier, Zoologist: A Study in the History of Evolution Theory.* Cambridge, Mass., 1964. A general examination of Cuvier's thought.
Outram, Dorinda. *Georges Cuvier: Vocation, Science and Authority in Post-Revolutionary France.* Manchester, 1984. A reexamination of Cuvier's life, with an emphasis on the politics of science.
Smith, Jean Chandler. *Georges Cuvier: An Annotated Bibliography of His Published Works.* Washington, D.C., 1993. A complete description of all of Cuvier's published works.

— PAUL LAWRENCE FARBER

CYTOPLASMIC GENES

The main defining feature of eukaryotes is the division between the cellular nucleus, where genetic material is maintained and replicated, and cytoplasm, the repository of the cellular machinery. This division is not absolute, however. Genes are not found exclusively within the nucleus, they are also found in two types of vehicles within the cytoplasm. First, genes are found within mitochondria and chloroplasts, organelles derived from eubacteria and archaebacteria, respectively, which retain a small component of their ancestral genomes. Second, genes are present in a range of microorganisms that live inside cells, from bacteria to unicellular eukaryotes such as members of the Microspora. These intracellular symbionts are typically transmitted from parent to progeny via the cytoplasm in the same manner as mitochondria and chloroplasts. They differ from mitochondria and chloroplasts in having more extensive genomes, and in often in being "disposable," or not necessary for normal cellular function. They sometimes retain a low level of horizontal transmission from one host to another, despite being unable to replicate away from the host.

Within animals, cytoplasmic genes are (with very few exceptions) maternally inherited, passed from a mother to her progeny, but not from a father. The main reason for this is the unequal cytoplasmic content of eggs and sperm. Nearly all the cytoplasm in animals is inherited through the egg. In plants and protists (unicellular eukaryotes), there is a greater diversity of inheritance modes. Maternal inheritance of organelle DNA is most common in plants, but paternal inheritance is also known

(e.g., both chloroplasts and mitochondria in *Sequoia sempervirens*, the redwood tree), as is biparental inheritance (e.g., chloroplasts in *Pelargonium*, the geranium). It is notable that in some species, like the unicellular alga *Chlamydomonas*, mitochondria and chloroplasts are inherited uniparentally, but each from a different parent.

The observation that cytoplasmic genes generally have uniparental inheritance led Leda Cosmides and John Tooby (1981) to recognize that the rules of thinking we apply to nuclear genes do not straightforwardly hold true for cytoplasmic genes. Of primary importance, nuclear genes are inherited from both sexes and are selected to maximize the fitness of male and female carriers equally. In contrast, cytoplasmic genes are selected to maximize the fitness only of the sex through which they are inherited. Consider selection acting on maternally inherited mitochondria. Mutations in mitochondrial genes that increase female fitness will be selected for, but mutations that improve only male performance will not spread because the male does not transmit the mutation to future generations. In fact, mutations that reduce only male performance will not be removed by selection. More significant, mutations in maternally inherited genes that enhance the survival and production of female over male carriers (in species with separate sexes), and the production of female over male gametes (in hermaphrodites) will be favored.

The last point is exemplified by the cases of cytoplasmic male sterility in plants. In species such as thyme (*Thymus*) or sugar beet (*Beta*), some individual plants are sterile with respect to male function: they produce floral structures with normal ovule (female) development but aberrant anther (male) development. This trait is caused by certain mitochondrial genotypes. These mitochondria benefit from male sterility because the death of anther tissue (in which pollen is produced) results in greater resources being directed into female reproduction, or seed. Mitochondria are transmitted in ovules, but not in pollen. Thus, selection can favor traits in cytoplasmic genes that are deleterious to the interests of the majority of nuclear genes—a case of genomic conflict in which the mitochondria act as selfish genetic elements.

Interactions with Hosts. In line with the logic above, maternally inherited microorganisms are observed to enter into three types of interaction with their "host." These can be categorized according to their effects as beneficial symbiosis, sex ratio distortion, and cytoplasmic incompatibility.

Beneficial symbiosis. Researchers working with invertebrates have often observed that treatment of an individual with antibiotics that kill bacterial symbionts produces a drastic reduction in physiological performance, owing to the death of the cytoplasmic bacteria. These bacteria synthesize essential amino acids and vi-

EVOLUTIONARY GENETICS OF CYTOPLASMIC MALE STERILITY: GROUP SELECTION AND CYTO-NUCLEAR CONFLICT

Extinction is one possible outcome of invasion by a sex ratio distorter, but it is not an inevitable outcome. Study of the dynamics of cytoplasmic male sterility in natural populations has revealed two factors that can prevent fixation of sex ratio distorters (i.e., all individuals have the sex ratio distorter). The first is group selection against presence of the cytoplasmic male sterile genotype. For example, if the majority of pollination is by local plants, and if dispersal by seed is limited (i.e., most individual plants in a population are closely related), then patches with high frequency of cytoplasmic male sterility will suffer because insufficient pollen is produced locally to ensure full fertility. Under these conditions, seed production can decrease sufficiently to lead to local extinction—or, more specifically, to a higher rate of extinction than in patches without the cytoplasmic male sterile genotype. Selection at this level will maintain both mitochondria causing cytoplasmic male sterility and normal mitochondria in the population.

The second factor preventing cytoplasmic male sterile genotypes from becoming fixed is selection for nuclear genes that repress the action of the relevant mitochondria. When a cytoplasmic male sterility type spreads through a population, there is strong selection for nuclear genes that prevent the action of the effective mitochondria and restore male function. This occurs because when few individuals produce pollen, nuclear genes that increase pollen production are strongly favored. The presence of these repressor genes is typical of interactions between selfish genetic elements and their carrier genomes. Suppressors are also found in interactions between hosts and feminizing bacteria, and they are likely to be found for male-killing bacteria.

—Gregory Hurst

tamins that are absent or rare in the diet of their host. They pass from a female to her progeny through the egg and have lost the capacity for independent existence. A good example of such mutualistic symbiosis is the interaction between the bacterium *Buchnera aphidicola* and its aphid host: the bacteria supply essential amino acids that are in short supply in plant phloem, the aphid's diet. It is hard to escape the idea that these bacteria represent primitive organelles. The aphid dies following elimination of the bacteria. Phylogenetic studies reveal that the bacteria have been strictly vertically transmitted by the host for 200 million years. Furthermore, the genome of *Buchnera* is less than half the size of that of its free-living relatives; the bacterium has lost so many genetic functions catered for by its host that it cannot return to a free-living existence.

Sex ratio distortion. In arthropods, there are many records of individual females that produce offspring with female-biased sex ratios. Treatment of these females with antibiotics returns the sex ratio of their offspring to 1:1. There are three main categories of sex ratio distortion: parthenogenesis induction, feminization, and male-killing. All are alike in evolutionary logic, in that the distortion results in an increase in the production of daughters over sons. This is favored because cytoplasmic elements are maternally inherited, and selection therefore favors strains that increase the production of daughters.

Parthenogenesis induction occurs in several species of parasitoid wasp and thrips. In these species, males are haploid (having one set of chromosomes) and are produced parthenogenetically, whereas females are the diploid products of fertilization. When females are infected with *Wolbachia* bacteria, the haploid progeny (normally male) undergo chromosome doubling and therefore develop as parthenogenetically produced females.

In amphipod and isopod Crustacea, and in Lepidoptera, cytoplasmic bacteria and protists cause feminization. Individuals that would normally develop as males undergo female development. An example occurs in the pill woodlouse, in which there is a standard chromosomal sex determination system in uninfected individuals, but female sex determination, irrespective of sex chromosome constitution, when the mother is infected with *Wolbachia*.

The final manipulation observed is male-killing. A variety of insects carry one of a diverse assemblage of bacteria that kill male hosts during embryogenesis. This benefits the bacterium in females by reducing competition with their male siblings. In some cases, such as the ladybird beetle, male death directly increases resource availability to females, because the females consume the eggs containing their dead brothers.

A number of studies have been undertaken of the population dynamics of cytoplasmic genes that distort the sex ratio. These show that cytoplasmic genes can spread to high frequencies. In fact, they can potentially cause the extinction of the species that carries them by making all its individuals female. The reasons why extinction is not inevitable are best illustrated by study of cytoplasmic male sterility in plants (see Vignette on Evolutionary Genetics of Cytoplasmic Male Sterility). Selection on nuclear genes for resistance to the selfish phenotype, along with group-level selective effects associated with biased sex ratios, can stabilize infection within a population.

Cytoplasmic incompatibility. In another deleterious effect caused by inherited microorganisms, crosses between males infected with *Wolbachia* bacteria and females that are uninfected are incompatible, producing few or no viable offspring. The zygotes that result from their mating arrest during early embryogenesis owing to condensation of paternal chromatin. Treatment of the male partner with antibiotics restores viability to the cross. This trait was first observed in mosquitoes and has since been seen in a range of arthropods, including mites and woodlice.

The logic of cytoplasmic incompatibility can be understood in the context of structured populations. If we consider a population of 100 females and 100 males, where one female and one male are infected with *Wolbachia*, then the presence of the infected male within the population means that on average 1 percent of uninfected female offspring will die. Thus, the fitness of uninfected females will be lower than that of infected females, and the incompatibility trait will spread through the population. The advantage gained by the bacterium is indirect: it gains by lowering the fitness of uninfected hosts, rather than by directly increasing its own production and survivorship. Cytoplasmic incompatibility is bizarre at first sight, because *Wolbachia* causes its male host to have reduced reproductive success; however, because *Wolbachia* is inherited only through the female lineage, this is cost-free for the bacterium, whose fitness in male hosts is already zero.

Wolbachia that induce cytoplasmic incompatibility have been observed to spread rapidly through natural populations. In the early 1980s, a *Wolbachia* strain causing cytoplasmic incompatibility invaded a population of *Drosophila simulans* (fruit fly) from Riverside in southern California. Within 10 years, this strain had spread through most of California. The rapid spread of *Wolbachia* has caught the attention of applied biologists, who see the potential of attaching traits useful to humans (e.g., resistance to parasites vectored by mosquitoes) to *Wolbachia*, whose incompatibility phenotype can then spread the trait through an insect population. From a small initial release, the spread of *Wolbachia* through cytoplasmic incompatibility could deliver the trait to all the individuals within the population.

The interactions between *Wolbachia* and their hosts are further complicated by the observation that different strains of *Wolbachia* may be mutually incompatible. Infection of a male with one *Wolbachia* strain causes cytoplasmic incompatibility with a female infected with a different strain of the bacterium. This type of incompatibility between hosts infected with different strains is termed "bidirectional incompatibility." It has attracted the attention of workers interested in how speciation occurs. If different populations become infected with different strains of *Wolbachia*, those populations may become partially or completely reproductively isolated

WOLBACHIA AND SPECIATION?

The idea that *Wolbachia* may be important in producing reproductive isolation and speciation arose from observations of crosses between different populations of mosquitoes (*Culex pipiens*) and of jewel wasps (*Nasonia*). In both cases, very few hybrid individuals were produced. Hybrid inviability was caused by the presence of different strains of *Wolbachia* in the two populations, with concomitant bidirectional incompatibility.

These findings suggest that *Wolbachia* is potentially a widespread agent of speciation in insects. The issue remains contentious; however, the debate has changed from whether *Wolbachia* is important in isolation, to how often it is important. There is also a potential role for undirectional incompatibility in combination with other isolating factors. Even if *Wolbachia* alone does not cause complete isolation, the large effect of incompatibility it produces is likely to make it a highly significant force contributing to speciation, albeit in conjunction with other isolating barriers.

—GREGORY HURST

by virtue of the different infections; *Wolbachia* may, therefore, be important in speciation (see Vignette on *Walbachia* and Speciation).

Importance of Inherited Microorganisms in Host Evolution. Cytoplasmically inherited microorganisms cause a number of important alterations in host reproduction. Several scientists have asked whether they have an important role in altering host evolution, or whether they are just interesting curios of natural history. The answer to this rests on two questions. First, are these microorganisms important in host evolution in particular cases? Second, if they are sometimes important, are they common enough to be a general force in evolution?

Microorganisms that are beneficial to their host clearly fulfil the first criterion. To a large extent, the success of insects that feed on plant phloem or vertebrate blood is defined by their symbioses. Without their symbionts, they could not exploit these niches, and the niches would be unfilled. Symbionts are not omnipresent, however; the diet of predatory or parasitoid species contains the correct balance of nutrients for development and reproduction, so they rarely possess symbionts.

Microorganisms that distort the sex ratio are increasingly being viewed as important in arthropods. They may affect the evolution of host sex determination systems. In the pill woodlouse, for instance, the spread of *Wolbachia* that feminize the host has produced an alteration of the sex determination system from female heterogamety (XY females, XX males) to one in which the presence of *Wolbachia* controls sex determination (infected females and uninfected males). But how common are such cases? Sex ratio-distorting microorganisms occur in 10 to 15 percent of species in most arthropod groups. This suggests they may be an important force in evolution.

Wolbachia that induce cytoplasmic incompatibility may also be an important force in host speciation (see vignette 2). Such *Wolbachia* are probably present in around 10 to 20 percent of insect species, making them a potent force in the production of reproductive isolation, either in isolation or in combination with other factors.

BIBLIOGRAPHY

Bordenstein, S. R., F. P. O'Hara, and J. H. Werren. "*Wolbachia*-induced Bidirectional Incompatibility Precedes Other Hybrid Incompatibilities in *Nasonia*." *Nature* 409 (2001): 707–710. Evidence that *Wolbachia*-induced cytoplasmic incompatibility may contribute to speciation; a case is discussed in which there are no barriers isolating the species apart from presence of different *Wolbachia* strains.

Cosmides, L., and J. Tooby. "Cytoplasmic Inheritance and Intragenomic Conflict." *Journal of Theoretical Biology* 89 (1981): 83–129. This seminal work is the original source of inspiration for many workers in the field, and is still relevant today.

Hurst, G. D. D., and M. Schilthuizen. "Selfish Genetic Elements and Speciation." *Heredity* 80 (1998): 2–8. This article and Werren (1999) evaluate whether *Wolbachia*, and other selfish genetic elements, could be important in speciation.

Hurst, G. D. D., and J. H. Werren. "The Role of Selfish Genetic Elements in Eukaryote Evolution." *Nature Reviews Genetics* 2 (2001): 597–606. A recent account of selfish genes in evolution; deals with the diversity of cyto-nuclear conflicts, as well as the potential for them to induce speciation and evolution of sex determination systems.

Hurst, L. D., A. Atlan, and B. Bengtsson. "Genetic Conflicts." *Quarterly Review of Biology* 71 (1996): 317–364. An advanced review that defines genetic conflicts, and evaluates the role of genetic conflict in evolution, with interesting commentary on whether cytoplasmic genes may be important in driving the evolution of genetic systems, such as anisogamy and the presence of sexes.

O'Neill, S. L., A. A. Hoffmann, and J. H. Werren. *Influential Passengers: Inherited Microorganisms and Arthropod Reproduction.* Oxford, 1997. A recent advanced text with chapters written by specialists on the interaction between parasitic inherited microorganisms and their hosts.

Stouthamer, R., G. D. D. Hurst, and J. A. J. Breeuwer. "Detection of Sex Ratio Distorters in Natural Populations." In *Sex Ratio Handbook*, edited by I. C. W. Hardy. Cambridge, 2001. Examines how sex ratio distorters are detected, and reviews their incidence and importance.

Werren, J. H. "*Wolbachia* and Speciation." In *Endless Forms: Species and Speciation*, edited by D. Howard and S. Berlocher, pp. 245–260. Oxford and New York, 1999.

— GREGORY HURST

D

DARWIN, CHARLES

(1809–1882), English naturalist, the founder of modern evolutionary biology. Although Darwin was not the first scientist to advocate evolution or something like it, he was the first to present compelling arguments that evolution has in fact occurred and to show how it could be used as a basis for research. His theory of natural selection created a plausible mechanism and introduced a new way of thinking about nature. As a result, it could be argued that nobody has influenced modern intellectual culture more profoundly than he.

Born in Shrewsbury, England, on 12 February 1809, Darwin was well to do and well connected. Sent to study medicine at Edinburgh at the age of sixteen, he did not want to become a physician, and in 1827 he enrolled at Cambridge University to prepare for a career as a clergyman and part-time scientist. At both institutions he cultivated his interests in natural history with the encouragement of the faculty. He was invited to become an unofficial naturalist on board the HMS *Beagle* as it charted the coasts of South America.

The *Beagle* departed on 27 December 1831 and after circumnavigating the globe returned to England on 2 October 1836. Profoundly influenced by his reading of Charles Lyell's *Principles of Geology* (1830–1833) during the voyage, Darwin became one of Lyell's most able supporters. Communications from the field concerning his research on the elevation of South America created an immediate sensation. Later in the voyage he developed a theory of coral reefs that established his reputation as a geologist. Darwin's three books and many papers on geology occupied much of his time after his return to England. During the voyage, he also worked on marine biology, geographical distribution of species, and other aspects of natural history. He amassed a vast collection of specimens. Many of these were turned over to specialists, forming the basis for reports that he edited.

One result of having specialists examine the specimens was quite unexpected evidence for evolution. Birds from the Galapagos Islands that appeared to be mere varieties turned out to be different species. The implication was that they had differentiated locally. Likewise, some of the large fossil mammals that Darwin had collected were related to smaller species that were still extant. It makes sense that a talented young geologist would be the first scientist to appreciate the evolutionary implications of biogeography. Darwin had become accustomed to thinking in historical terms, and he was able to break with a tradition that focused mainly on physiology.

In March 1837, Darwin became convinced that evolution had in fact occurred. He began a concerted effort to understand its mechanism, speculating and reading about a wide range of topics. On 28 September he began to read Malthus's celebrated *Essay on the Principle of Population* (1798). Thomas Robert Malthus (1766–1834) was the world's first professor of economics. His essay dealt with what happens when human populations expand. For Darwin, the important point was that populations may grow when food is abundant, for example, after a plague, but the food supply does not increase as fast as the number of people does. Darwin reasoned that wherever there is such a "struggle for existence," some individuals will tend to reproduce more than others of their own species and pass on whatever gave them a competitive edge to their descendants. Recognizing the analogy with breeders of domesticated plants and animals "selecting" organisms of one kind over another, he referred to "natural selection" and "artificial selection" as variations of the same basic theme.

One result was a new kind of ecology, based on individualistic reproductive competition. More fundamental was the role that selection could play in biological thinking, which now became all the more historical. A biologist's task would now be to explain the living world in terms of how one organism had come to leave more offspring than another organism of the same species. There were profound philosophical implications as well. That kind of explanation was purely mechanistic, and one could dispense with older notions about the purposefulness of organic activities.

Although other scientists, including his grandfather, Erasmus Darwin (1731–1802), had speculated on the origin of new forms of life on earth, their thinking seems far less "evolutionary" than Darwin's. Many of them had spontaneous generation in mind rather than lineages being gradually transformed over many generations. Some treated "evolution" as if it were like the development of an embryo, with a definite goal. Others, notably Jean Baptiste de Lamarck (1744–1829), believed that there are unknown laws of nature that necessitate a change from simple to complex, with the human species the outcome.

Darwin considered Lamarck a bad scientist and maintained that he had learned nothing from his writings. He knew he had to do better. He gathered a large body of

FIGURE 1. Charles Darwin.
© Walt Anderson/Visuals Unlimited.

materials on evolution while he concentrated on his geological publications. A preliminary draft of a book on evolution by natural selection was ready by 1844. However, he put the draft aside and proceeded to publish some of his results on the zoology of the voyage, then unexpectedly spent about eight years, beginning in 1846, working on barnacles (subclass Cirripedia).

The treatise on barnacles, published in the 1850s, is a major evolutionary work, and later readers have had no difficulty understanding it as such. Well over a thousand pages long and therefore never widely read, it was a classic example of a systematic monograph. The work exemplifies Darwin's idea that classification should be strictly genealogical. With minor exceptions, his groups are branches of the phylogenetic tree (clades) that trace lineages of descent from a common ancestor. He used characters that change when a lineage splits in order to build an evolutionary genealogy, much as modern phylogeneticists do. As a result of having a good phylogeny, Darwin was able to show that some barnacles have evolved from a condition in which all individuals are simultaneous hermaphrodites to one in which there are dwarf males associated with females. This discovery was instrumental in his work on outcrossing and related phenomena, especially in plants, because it suggested that hermaphrodites do not always fertilize themselves.

In 1855, his work on barnacles finished, Darwin began to prepare his materials on evolution for publication. He also gathered original material on variation and conducted experiments on biogeography and ecology. Darwin explained natural selection to colleagues who were providing him with data, including the botanists

Joseph Hooker and Asa Gray. In a letter dated 1 May 1857, he informed Alfred Russel Wallace that he had a species theory. He explained his theory to the geologist Charles Lyell, who urged him to publish in order to maintain his priority.

Darwin intended to publish a very large book entitled *Natural Selection* and had most of a first draft written when, on 18 June 1858, a draft of a manuscript from Wallace arrived in which something like natural selection was elaborated. Lyell and Hooker arranged for a joint publication to be read before the Linnaean Society on 1 July. On 20 July, he began to write an "abstract." Entitled *On the Origin of Species by Means of Natural Selection, or the Preservation of Favoured Races in the Struggle for Life*, it was published on 24 November 1859. A total of six editions appeared in Darwin's lifetime, the last in 1872.

In *Origin of Species*, Darwin made a case for evolution in general as well as for natural selection. His basic stratagem was to show that a "host of facts," such as those of biogeography, embryology, and morphology, could readily be explained in terms of evolution and natural selection, but otherwise there was no scientific explanation for them. Arguments that seemed to offer difficulties he skillfully rebutted, often turning them on their heads. Remarkable adaptations such as the eye turned out not to be so difficult as even he had originally supposed. The lack of fossil evidence was answered by the imperfection of the record. Natural selection was rendered more plausible by the analogy of artificial selection. Darwin also appealed to the prospect for future scientific progress as a result of evolutionary thinking.

Darwin never published his comprehensive book on species. Instead, he brought out shorter books and papers that developed the implications of his theory. Many of these were on the reproductive anatomy and physiology of flowering plants. One unexpected result of his theory was that plants are highly adapted to ensure outcrossing. This led to another unexpected result, that inbreeding is highly deleterious. His book on orchids (1862) showed that such adaptations are quite elaborate, and also that they are not the sort of feature that one would expect had the flowers in question been designed by an intelligent creator. One outbreeding mechanism that he discovered was heterostyly, to which he devoted a book, *The Different Forms of Flowers on Plants of the Same Species* (1877). His experiments on the effects of inbreeding were presented in *The Effects of Cross and Self Fertilisation in the Vegetable Kingdom* (1876).

What had been a section of his species book was revised and published as *The Variation of Animals and Plants under Domestication* (1868). For Darwin, variation was crucial, because without it selection will not work, and he documented that there is a lot of it. He saw variation as an embryological matter, and he was interested in topics that are now discussed under such rubrics as pleiotropy and developmental constraint, which he treated as minor mechanisms of evolution. He also believed in the inheritance of acquired characteristics and the direct influence of the environment. He presented a "provisional hypothesis of pangenesis" toward the end of the book. It provided a "soft" theory of inheritance allowing for Lamarckian mechanisms, but, like modern genetics, it was a particulate theory.

Darwin recognized that natural and artificial selection are fundamentally the same mechanism, both depending on differential reproduction within populations, albeit differing with respect to what does the selecting. To these modes of selection he added a third, sexual selection, which played a crucial role in making the point that reproductive success is what really counts. Animals struggle, not just for existence, but for the opportunity to mate. Such reproductive adaptations as the antlers of deer and the brilliant plumage of birds, for example, exemplify how selection might produce features that are maladaptive from the point of view of the species but readily explicable in terms of one organism competing with another.

Sexual selection was a major topic in *The Descent of Man, and Selection in Relation to Sex* (1871). Also stressed in the same work was the evolution of morals, a phenomenon that required some refinements if it were to be explained in terms of individualistic competition. Darwin rejected the notion that individual species could be selected. However, he treated families, such as those of social insects, as "individuals" of a higher order that compete as units. For the human species, he invoked competition between larger units, such as tribes. The reason why he emphasized that is not obvious, nor is it obvious why he did not pay more attention to reciprocity. What we now call "kin selection" he recognized in the form of artificial selection based on "breeding from the same stock," on the assumption that relatives are apt to have the same inherited properties.

Darwin began to work on behavior and psychology from the outset of his evolutionary investigations. *Origin of Species* dealt at some length with the evolution of instincts, especially in social insects. In his book *The Expression of the Emotions in Man and Animals* (1872), he treated postures and movements from a phylogenetic and physiological point of view. Seeking to deal with the very earliest stages of behavioral evolution, he studied the activities of plants. The analogy of plants to animals is evident in his books *Climbing Plants* (1875) and *Insectivorous Plants* (1875). Darwin investigated the physiological basis of plant behavior experimentally and treated it in *The Power of Movement in Plants* (1880). Darwin is generally recognized as one of the greatest plant physiologists. His use of evolution as a guide to research in that area deserves broader appreciation.

His last book, on worms (1881), treated very simple animal behavior experimentally. He died 19 April 1882.

Darwin was a theorist, and he theorized about a vast range of subjects. There may be some merit in the suggestion by Ernst Mayr (1991) that Darwin had not one, but five theories, or any number that one might recognize. But such dissection should not detract from the organic unity of his accomplishment as a whole. Darwin was basically a synthesizer, with a unitary point of view. His separate contributions are not descriptions of unconnected fragments of the world. Rather, they illustrate the advantages of historical thinking in general, and his kind of evolutionary thinking in particular.

[*See also* Adaptation; Artificial Selection; Classification; Comparative Method; Darwin's Finches; Natural Selection.]

BIBLIOGRAPHY

Barlow, N. *The Autobiography of Charles Darwin, 1809–1882. With Original Omissions Restored.* New York, 1959. The expurgated version is available in F. Darwin (1887).

Barrett, P. H. *The Papers of Charles Darwin.* Chicago, 1977. A complete anthology of Darwin's publications other than books.

Barrett, P. H., P. J. Gautrey, S. Herbert, D. Kohn and S. Smith, eds. *Charles Darwin's Notebooks, 1836–1844: Geology, Transmutation of Species, Metaphysical Enquiries.* Ithaca, 1987. Transcriptions of Darwin's notes made as he created his theory.

Bowler, P. J. *Charles Darwin: The Man and His Influence.* Oxford, 1990. A solid and accessible biography recommended for the general reader.

Browne, J. 1995. *Charles Darwin: Voyaging: Volume One of a Biography.* New York, 1995–1999. A good read as well as a reliable source, unfortunately as yet incomplete.

Burkhardt, F., et al. *The Correspondence of Charles Darwin.* Cambridge, 1985–1999. This monumental work documents Darwin's life in minute detail but has been completed only up to the middle of the 1860s.

Darwin, F. *The Life and Letters of Charles Darwin, Including an Autobiographical Chapter.* London, 1887. Darwin's son Francis edited this and the next title, both of which are still very useful.

Darwin, F., and A. C. Seward. *More Letters of Charles Darwin: A Record of His Work in a Series of Hitherto Unpublished Letters.* London, 1903.

Ghiselin, M. T. *The Triumph of the Darwinian Method.* Berkeley, Calif., 1969. On the unity of Darwin's thinking.

Goldie, P., and M. T. Ghiselin, eds. *The Darwin CD ROM.* 2d. ed. San Francisco, 1997. Includes several of Darwin's books, including the barnacles treatise. Also a timeline, extensive bibliographies, and a biographical dictionary.

Kohn, D. *The Darwinian Heritage.* Princeton, 1985. A good sampler of recent scholarship.

Mayr, E. *One Long Argument: Charles Darwin and the Genesis of Modern Evolutionary Thought.* Cambridge, Mass., 1991. Written by a great evolutionary biologist, this brief biography is accessible to the general reader.

Richards, R. J. *Darwin and the Emergence of Evolutionary Theories of Mind and Behavior.* Chicago, 1987. An illuminating study of a largely neglected area of Darwin's contribution.

— MICHAEL T. GHISELIN

DARWINIAN MEDICINE. *See* Overview Essay *on* Darwinian Medicine *at the beginning of Volume 1.*

DARWIN'S FINCHES

Fourteen species of finches occupy the several islands of the Galapagos archipelago. A related species occurs on Cocos Island, 700 kilometers (1,120 miles) to the northwest. Their importance lies in illuminating the process of evolutionary diversification that occurs when a single species reaches an isolated and relatively unexploited environment (Grant, 1999). [*See* Galapagos Islands.]

According to the DNA evidence, the ancestral species, a relative of a modern group of seedeaters called grassquits, arrived from South America two million to three million years ago (Sato et al., 2001). They occupied the few islands in existence then and diversified on the different islands into two basic stocks of warbler finches (Petren et al., 1999). One stock gave rise to all other species, including the Cocos Island finch. Most of the ground finch species were formed in the last few hundred thousand years. Ideas about how all this happened go back to Darwin. [*See* Darwin, Charles.]

Charles Darwin, his assistants, and Captain FitzRoy collected several specimens of most of the species in 1835 on a visit to the Galapagos during the epic voyage of the *Beagle* around the world. The specimens became important evidence for the mutability of species when, back in England in 1837, Darwin learned from the taxonomist John Gould that all the specimens belonged to a single group of related species and not, as he had thought, to a few different families. He drew upon this information, their similarity in beak size and shape, and his recollections of them, when developing his evolutionary ideas. About one hundred years later, Percy Lowe called them *Darwin's finches* (Figure 1). Nevertheless, the evidence Darwin needed to explain how species form was not good enough to interpret the evolution of the finches because he failed to record the particular islands on which each species was collected. It was not until 1947 that David Lack was able to interpret their evolution.

The species on Galapagos comprise a group of ground finch species, a group of tree finch species and, least finchlike of all, warbler finches. Within the first two groups, there is little variation in plumage but striking variation in beaks. Because beaks are tools for gathering and dealing with foods, as graphically illustrated by Robert Bowman's analogy with pliers, Lack reasoned that each species exploits food in its own distinctive niche. To some extent, this can easily be seen, with tree finches feeding in the foliage of trees mainly at medium and high elevations on the Galapagos, ground finches feeding on

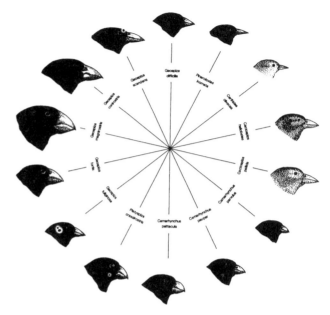

FIGURE 1. Adaptive Radiation of Fourteen Species of Darwin's Finches.

Grant, Peter. *Ecology and Evolution of Darwin's Finches.* Copyright © 1986 by PUP. Reprinted by permission of Princeton University Press.

seeds on the ground or shrubs, and warbler finches gleaning in the same way that real warblers do elsewhere. [*See* Speciation.]

To account for this diversity, Lack (1947) proposed that populations of the same species on separate islands diverged, becoming adapted to different food supplies as they progressed toward becoming two species. They were influenced either by the presence or by the absence of competitor species; their evolutionary directions were constrained or not. This part of the explanation of speciation was essentially the same as a sketchy version of Darwin's origin of species in general. Lack went further in considering what would happen when the two species eventually met on one of the islands. He suggested they would initially compete and some would interbreed, but there would be a period of divergence caused by natural selection that reduced competition and the chances of interbreeding. The result: two species formed from one and living in the same environment without breeding with each other. In support of the argument, Lack pointed to the strong geographical isolation of the Galapagos, the resulting scarcity of other continental species that might be competitors, and the large variety of islands on which evolutionary diversification could occur. The importance of isolation is highlighted by the fact that the solitary island of Cocos has only one species of finch.

Measurements have since confirmed the close association between beak size, food supply, and diet; demonstrated that islands differ in their food supplies; and shown that, in some instances, the beak size of a species on a particular island is more explicable in terms of both competitor species and food supply than by the food supply alone. Thus, Lack's emphasis on speciation being initiated on separate islands under the influence of competitors has gained support (Grant and Grant, 1997).

Long-term studies of individually marked ground finches on the islands of Daphne Major (Grant and Grant, 1998) and Genovesa (Grant and Grant, 1989) have revealed, surprisingly, two processes of evolution at work. The first is natural selection. It was first demonstrated on Daphne Major in 1977 with the finding that members of the medium ground finch (*Geospiza fortis*) population with large beaks survived a drought better than those with small beaks; as the food supply in general decreased, they were better equipped to deal with the large and hard seeds that remained. Beak size variation had been shown in the previous year to be highly heritable. Natural selection on heritable variation should lead to an evolutionary change in the next generation, and this in fact did occur, with offspring born in 1978 being larger on average than the preceding generation before the selection event occurred. Nearly ten years later, a selection event in the opposite direction occurred, this time with evolution in the direction of small beak size. Thus, here natural selection oscillates in direction (Grant, 1999).

The second process is hybridization. Rare but repeated hybridization was found to occur on Daphne between the medium ground finch and two other species, apparently as a result of young finches becoming imprinted on the song of another species. Unexpectedly, the hybrids survived as well, on average, as the parent species themselves; moreover, they had no difficulties in breeding. Over fifteen years a trickle of genes had passed from one hybridizing species to another, not enough to obliterate the differences between them, but enough to provide new genetic variation for the forces of selection to act on (Grant and Grant, 1998). Hybridization raises some unanswered questions, but one thing it shows clearly is that the process of speciation is not completed on separate islands and is continuing.

[*See also* Natural Selection.]

BIBLIOGRAPHY

Grant, B. R., and P. R. Grant. *Evolutionary Dynamics of a Natural Population: The Large Cactus Finch of the Galápagos.* Chicago, 1989.
Grant, B. R., and P. R. Grant. "Hybridization and Speciation in Darwin's Finches: The Role of Sexual Imprinting on a Culturally Transmitted Trait." In Endless Forms: Species and Speciation, edited by D. J. Howard and S. L. Berlocher, pp. 404–422. New York, 1998.
Grant, P. R. *Ecology and Evolution of Darwin's Finches*, 2d ed. Princeton, 1999.
Grant, P. R., and B. R. Grant. "Genetics and the Origin of Bird Species." *Proceedings of the National Academy of Science USA* 94 (1997): 7768–7775.
Lack, D. *Darwin's Finches.* Cambridge, 1947.
Petren, K., B. R. Grant, and P. R. Grant. "A Phylogeny of Darwin's Finches based on Microsatellite DNA Length Variation." *Proceedings of the Royal Society of London B* 266 (1999): 321–329.
Sato, A., H. Tichy, C. O'hUigin, P. R. Grant, B. R. Grant, and J. Klein. "On the Origin of Darwin's Finches." *Molecular Biology and Evolution* 18 (2001): 299–311.

— PETER GRANT AND ROSEMARY GRANT

DEMOGRAPHIC TRANSITION

The demographic transition refers to a phenomenon observed throughout the world, in which societies with high fertility and high mortality change to having lower fertility. This transition to having fewer children is generally (but not always) associated with increasing living standards and decreasing mortality risks and represents a drastic cultural change. Commonly, this transition begins among richer families, in which fertility declines earlier and often to lower levels than among the poorer families, erasing or even reversing previous positive correlations between wealth and number of offspring. Most societies in the developed world underwent a demographic transition in the nineteenth century, those in the developing world entered this phase much later, with Africa only now entering it. The transition usually occurs when populations are undergoing rapid population growth, which is frequently associated with the growth of some kind of impoverished underclass (such as landless peasants or industrial workers). Most human populations are actually still growing rapidly, but the rate of growth is now declining virtually everywhere. Populations after demographic transition have many fewer individuals in the lower age classes than those that are still experiencing high fertility and mortality, and their age pyramids are far less tapering (Figure 1).

The classic characterization of the demographic transition, first outlined in the 1940s by the American demographers F. W. Notestein and D. Kingsley, is of mortality declining first, followed by a decline in fertility. Although this typical characterization is common, the progress of this demographic and health transition is not exactly the same in each region. Examples exist of mortality declining contemporaneously with or even after a decline in fertility, and there are populations in which mortality has declined but no fertility decline has occurred at all. Yet there are sufficient similarities around the world to define a phenomenon in need of an explanation. An explanation for the demographic transition is of particular interest to evolutionary anthropologists, because the reduction in family size generally occurs at a time of improving living conditions. This presents a

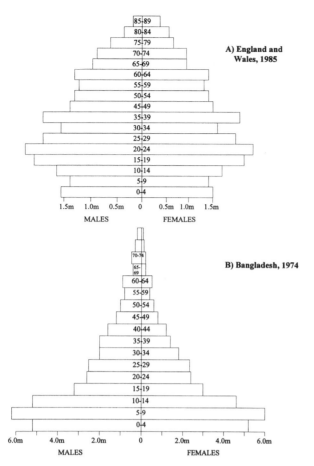

FIGURE 1. Population Pyramids.
Population pyramids from (A) England and Wales (1985), which had undergone demographic transition, and (B) Bangladesh (1974), which had not.

serious challenge to an adaptive view of human behavior and life history, which would predict the exact opposite—abundant resources should translate into higher fertility.

Evolutionary ecologists, however, are familiar with the notion that fertility might be reduced by some quantity/quality trade-off, based on the English ecologist David Lack's *Ecological Adaptations for Breeding in Birds* (1968). Human evolutionary ecologists have explored the possibility that very low fertility could indeed be adaptive in some circumstances. Models from evolutionary ecology share some common themes with economics, and in the area of fertility studies some common predictions emerge, although there are also key differences. Demographers have long argued that it is the economic costs and benefits of children that influence fertility. There are empirical examples of almost instant shifts in fertility occurring when the economic costs and benefits relevant to raising children change. Many de-

mographers argue that parents have large families when children can generate income for them. However, the logical consequence of this argument is that when children become a net cost, population fertility will decline to zero, which has never been observed.

Evolutionary anthropologists have argued that wealth actually always flows down the generations, even in traditional societies (i.e., children are always net costs) and that, in any case, fitness (as measured by long-term reproductive success) is the currency being maximized, not parental income. Economic demographers have also focused on the link between relative cohort size and fertility decline—the argument is that those individuals born in baby booms face stiff competition, particularly years later when they are trying to find jobs and build families; economic hardship relative to expectations then precipitates fertility decline. Social demographers have focused on lower fertility as a mechanism to maintain status in stratified societies. Others have highlighted the proximal effect of formal education, particularly of girls; even if it does nothing else, secondary and higher education at least deflects young women from motherhood during some of their more fertile years. All these notions touch on issues concerning parental investment.

In humans, parental investment can include anything from feeding to inherited wealth and education: anything that might enhance the long-term chances of gaining wealth or status. A state-dependent, dynamic model that optimizes reproductive success over two generations has been used to predict both optimal family size and the optimal amount of wealth to allocate to each child at the end of the parents' reproductive lives. Wealth inherited from parents was considered an important determinant of future reproductive success. When wealth is modeled as having the potential to generate more wealth, small families can be optimal if the costs of raising children to adulthood are high. The model shows that the higher the cost of raising children, the greater the optimal inheritance to allocate to each child. This makes the average family in such a population wealthier and smaller. In populations in which the costs of raising children are low, higher fertility and lower levels of wealth inheritance to each child are optimal, meaning the average family is poorer. This model offers insights into the paradox of the negative correlation between wealth and reproductive success that is commonly observed in societies that have experienced the demographic transition. Within each model population, wealth is positively associated with reproductive success, but across populations those model populations that had the poorest families also had the largest family sizes. Homogeneous populations, such as those from traditional cultures usually studied by anthropologists, frequently show a positive correlation between wealth and reproductive success, as predicted from evolutionary ecological theory. However, in the large, heterogeneous populations usually studied by demographers, this correlation is typically negative. If heterogeneous populations represent a group of subpopulations, each experiencing (or believing they experience) different costs and benefits associated with investing in children, then a decoupling or even a negative relationship between wealth and family size might be observed, potentially providing an adaptive explanation for the demographic transition.

Empirical evidence is mounting that many human populations are structured in this way. However, this does not explain ultimately why different levels of parental investment across populations have evolved. Formal population genetic models, where the quality of offspring is assumed to have long-term consequences for their competition against others, and hence for their future reproductive success, are needed. Differences between the sexes in the costs and benefits of varying levels of investment can be considerable. It is from models incorporating these variables that an understanding of the evolutionary basis of high parental investment might emerge, although as yet this branch of life history theory is relatively underdeveloped.

In addition to these economic and ecological approaches, other contributions to the puzzle of small family sizes come from evolutionary psychology. We are evolved to enjoy and hence seek out sex, and also to seek the resources that would help attract mates. Children would have been the inevitable consequences of these preferences in ancestral environments. We still have these preferences, and there is evidence that wealthy men in modern societies do have a higher number of sexual partners. But this does not necessarily translate into more offspring, owing to the use of contraception. However, this begs the question of why people choose to use contraception, and of why fertility started to decline in some countries long before any modern methods of contraception were available. Reproductive decision making is plastic in traditional and modern societies, and it is therefore interesting to examine how such an evolved psychology could have produced such a consistent, modern trend toward small family size.

Cultural transmission models may help explain some aspects of the historical trend. No simple biological, economic, or ecological correlate predicts the onset of the demographic transition across countries. Historical demographers, using language rather similar to the meme concept introduced by Richard Dawkins in *The Selfish Gene* (1976), tend to consider the demographic transitions as an idea that is spreading around the world. France experienced the first and probably best documented demographic transition starting in the nineteenth century. The prevalence of smaller families, presumably associated with higher parental investment in each child, spread from centers of sophistication over time. In remote areas, fertility decline came later, but once it started, it proceeded more rapidly. One of the best predictors of the onset of the trend was simply distance from these epicenters. Such findings describe rather than explain the demographic transition. But mathematical models of cultural evolution, in which units of culture are considered to evolve in a Darwinian way that maximizes their transmission rates (but without the need to enhance biological success), have been used to generate more formal explanations of this ideational change. Robert Boyd and Peter Richerson (1985) show that it is theoretically possible for a meme for low fertility to spread in a scenario where we copy successful individuals and success is defined through wealth. When child rearing competes directly with the ability to earn money, those with small families are more likely to be wealthy and thus are more likely to become role models.

Contributions from all these approaches add to our understanding. Even if a plausible, evolutionary model predicting the origin and maintenance of high parental investment combined with low fertility can be built, the existing empirical evidence from modern populations suggests that, at least over two generations, contemporary low fertility is not adaptive. In a survey of New Mexican men conducted in the early 1990s, it was found that those born into smaller families had greater educational achievements and earning power, but this did not translate into higher Darwinian fitness. These individuals lived in a period of economic and population expansion—extra parental investment might pay better fitness returns in a stable and constrained population, where competition would be more intense. However, there is actually evidence that familial correlations in fertility (assumed to be mainly environmental rather than genetic in origin) have increased since the demographic transition. Although much of this arises from population heterogeneity, much occurs at the individual level. Number of siblings is as good a predictor of fertility as is education level in some European countries.

It is possible that low fertility combined with high parental investment may yet prove a successful reproductive strategy over the very long term, but currently there is no evidence for that. Cultural change means that humans now have an unprecedented range of reproductive choices; thus, a diversity of reproductive preferences may be being expressed for the first time. If so, insufficient time has elapsed for evolution to act on those preferences in the current environment, which makes current fertility patterns essentially unpredictable.

[*See also* Life History Theory, *article on* Human Life Histories.]

BIBLIOGRAPHY

Borgerhoff Mulder, M. "The Demographic Transition: Are We Any Closer to an Evolutionary Explanation?" *Trends in Ecology and Evolution* 13 (1998): 266–270.

Boyd, R., and P. Richerson. *Culture and the Evolutionary Process.* Chicago, 1985.

Coale, A. I., and S. C. Watkins, eds. *The Decline of Fertility in Europe.* Princeton, 1986. Contains many historical accounts of fertility decline in Europe.

Dawkins, R. *The Selfish Gene.* Oxford, 1976.

Dunbar, R. I. M., ed. *Human Reproductive Decisions.* London, 1995. Examines adaptive studies human of demographic trends in general, including evolutionary approaches to the demographic transition.

Kaplan, H., J. Lancaster, J. A. Bock, and S. E. Johnson. "Does Observed Fertility Maximise Fitness among New Mexican Men? A Test of an Optimality Model and a New Theory of Parental Investment in the Embodied Capital of Offspring." *Human Nature* 6 (1995): 325–360.

Lack, D. *Ecological Adaptations for Breeding in Birds.* London, 1968.

Mace, R. "The Co-evolution of Human Fertility and Wealth Inheritance." *Philosophical Transactions of the Royal Society of London B* 353 (1998): 389–397. An optimality model in which grandparental reproductive success is maximized according to decisions on family size and inherited wealth.

— RUTH MACE

DEMOGRAPHY

Demography is the study of populations and the processes that shape them. Demographic concepts and techniques are important in evolutionary biology both for what they tell us about the ecological theater in which evolution is played out and, more specifically, as tools to study the evolution of life histories and senescence. Demographers are concerned with the size, distribution, and age structure of populations (statics) and the forces that cause these quantities to change (dynamics). In population biology, demographic ideas are used most extensively to study populations with overlapping generations and in which the probability of dying and giving birth depend on an organism's age.

Life Tables. Data on age-specific mortality and fertility are organized in a life table, which has a fundamental role in both the theory and practice of demography (including related fields such as human actuarial calculations). One form of the life table takes a cohort of individuals born at the same time and records the numbers remaining alive at regular time intervals until all are dead. From this table can be calculated quantities such as the probability of remaining alive to a certain age (traditionally called l_x, where x is age), the age-specific mortality rate, and the life expectancy of an individual of a certain age. Life tables also often include information about birth rates (traditionally called m_x, where x is again age; life tables with birth and death rates are colloquially called $l_x m_x$ life tables). An alternative to collecting longitudinal data about the fate of a single cohort is to measure the age-specific birth and death rates of individuals alive today to produce a current life table.

Geometric Growth. Before looking at the complexities of population growth in an age-structured population, consider the simplest case of a species with discrete generations in which each individual gives birth on average to λ offspring per year and then dies. We assume that there is no immigration or emigration, and that the birth rate is constant year by year. As Thomas Malthus first realized in the eighteenth century, the population will grow or decline depending on whether λ is greater or less than one, and the population t time intervals in the future will be $\lambda^t N_0$, where N_0 is the population density at time zero. Especially when we add greater biological detail, it is often useful to replace λ by its natural logarithm $r = \ln(\lambda)$, in which case the population at time t is $e^{rt} N_0$. Thus when r is less than zero the population declines exponentially, while when r is greater than zero it increases exponentially.

Age Structure and Stable Population Theory. The geometric population model tells us that if birth and death rates are fixed, the population will change by a factor λ (or e^r) each generation. Adding age structure complicates the picture. Now cohorts overlap and the probability of giving birth and dying depends on age. But even if you know from a life table the age-specific birth and death rates, it is not possible simply to write down a closed expression analogous to λ. Instead, as was first discovered by the Swiss mathematician Leonhard Euler and brought to prominence by the actuary A. J. Lotka, the logarithm of the population growth rate (r) is defined implicitly by the formula

$$ 1 = \sum_{x=\alpha}^{\beta} e^{-rx} \, l_x m_x, $$

where e is the exponential function, l_x denotes survival of the initial cohort to age x, and m_x denotes the number of female offspring born to a female age x. This expression, which has a crucial role in both population and evolutionary demography, is known as the Euler–Lotka equation.

So given that we know the age-specific fecundity and mortality schedules, we can calculate population growth rate, r, albeit only numerically on a computer. Predicting the consequences of current population parameters on future growth is called population projection, and it is used routinely in conservation biology and other areas of applied ecology. But the Euler–Lotka equation and its associated stable population theory is much more useful than that. It can be shown mathematically that irrespec-

TABLE 1. Parameters Derived from the Stable Euler–Lotka Population Model

The intrinsic rate of increase is the largest real root of the Euler–Lotka equation, computed using iterative (numerical) methods; it can be approximated using $r = \dfrac{\ln R_0}{T_p}$ where T_p is the mean age of parenthood.

Parameter/Model	Description	Notation	Formula
Net reproductive rate	Average lifetime number of offspring per female	R_0	$\sum_{x=\alpha}^{\beta} l_x m_x$
Intrinsic birth rate	Per capita births in a stable population	b	$1/\sum_{x=\alpha}^{\beta} e^{-rx} l_x$
Intrinsic death rate	Per capita death in a stable population	d	$b - r$
Finite rate of increase	Discrete (geometric) analog of intrinsic rate	λ	e^r
Doubling time	Time required for a stable population to double	DT	$\dfrac{\ln 2}{r}$
Mean generation time	Average time for population turnover	T	$\dfrac{\ln R_0}{r}$
Stable age distribution	Fraction of total population at age x	c_x	$\dfrac{e^{-rx} l_x}{\sum_{x=0}^{\omega} e^{-rx} l_x}$
Reproductive value	Contribution individual makes to future generation	v_x	$\dfrac{e^{-rx}}{l_x} \sum_{y=x}^{\omega} e^{-ry} l_x m_x$

tive of the initial population density and age structure, a population subject to constant birth and death parameters will assume a fixed growth rate and stable age structure determined only by these demographic parameters (a property called "ergodicity"). Moreover, the time it takes for the population to reach this stable form of growth (or decline) can also be calculated from the Euler–Lotka equation (technically, r is the largest real root of the equation and it is the other roots that provide this extra information). A variety of useful quantities employed by demographers and ecologists—for example, the mean generation time and the net reproductive rate—can also be derived from the Euler–Lotka equation, as is shown in Table 1. Finally, the contributions of each of the different age-specific birth and death terms to the overall growth rate of the population can be assessed. Typically, growth rate is more sensitive to proportional changes in fertility than mortality.

The Euler–Lotka equation applies to populations with strictly deterministic life-history parameters; that is, they are assumed not to vary from year to year. Moreover, populations are assumed to be large enough that we can ignore random effects such as the chance mortality of all individuals in the population at one time. Demographers have extended the basic theory to account

for both stochastic and cyclically varying birth and death rates, deriving estimators of population growth rates that are the stochastic equivalents of r, though these tend to be more complex and harder to measure (and a modified, weaker, concept of ergodicity applies).

The theory described here applies to single-sex age-structured populations, in which individuals age chronologically at the same rate. But many of the basic concepts apply to populations structured in other ways. In many cases it is important to include both sexes, especially if mating opportunities are limited or if males contribute to parental care and hence population growth rate. Botanists, in particular, frequently deal with size-structured populations in which transitions between size classes may not occur in each time unit, or individuals may "leapfrog" between classes. This complicates the calculations somewhat, but population growth rates and the other quantities listed in Table 1 can still be calculated.

Stable population theory is also important in evolution. Life history theory attempts to understand (among other things) how natural selection molds age-specific rates of investment in growth and reproduction. This poses the problem of the correct measure of fitness to use to distinguish different strategies, given that simply counting offspring does not work, as young produced early in life go on themselves to reproduce before young produced late in life. Sir Ronald Fisher showed that the solution of the Euler–Lotka equation where the parameters are strategy-specific is the appropriate measure to use. It is also possible to derive a measure of the "reproductive value" of individuals of different age, the contribution of the progeny produced over the rest of the organism's expected life. The study of how reproductive value declines with age has helped the understanding of the evolution of senescence and other aspects of life history.

The greatest limitation of stable population theory is that with few exceptions it assumes that demographic parameters are constant, and that they do not depend on population size. Population dynamics explores the importance of density dependence on population growth and persistence.

[*See also* Fitness; Life History Theory, *article on* Human Life Histories; Population Dynamics.]

BIBLIOGRAPHY

Carey, J. R. *Applied Demography for Biologists.* New York, 1993. A practical guide to life table and population methods.

Carey, J. R. "Insect Biodemography." *Annual Review of Entomology* 46 (2001): 79–110. A review of life table methods, mortality models, and factors favoring the evolution of extended life span in insects.

Caswell, H. *Matrix Population Models: Construction, Analysis, and Interpretation.* 2d ed. Sunderland, Mass., 2001. One of the best overviews of population modeling. Some basic mathematical background helpful but not required.

Charlesworth, B. *Evolution in Age-Structured Populations.* 2d ed. Cambridge, 1994.

Pressat, R. *The Dictionary of Demography.* New York, 1985.

Roff, D. A. *The Evolution of Life Histories: Theory and Analysis.* New York, 1992. Interesting and readable overview of life history analysis in evolutionary contexts.

Stearns, S. C. *The Evolution of Life Histories.* Oxford, 1992. Introduction to the evolution of life histories and discussion of both macroevolutionary framework and the microevolutionary conditions that determine how life histories evolve.

Tuljapurkar, S. "An Uncertain Life: Demography in Random Environments." *Theoretical Population Biology* 35 (1989): 227–294. State-of-science overview of stochastic demographic models and theory.

Wachter, K., and C. Finch, eds. *Between Zeus and the Salmon: The Biodemography of Longevity.* Washington, D.C., 1997. A collection of chapters from both biologists and demographers on aging and longevity.

— JAMES R. CAREY

DENSITY-DEPENDENT SELECTION

Density-dependent selection is a special type of natural selection in which survival and reproduction depend on the population density of other individuals of the same species. Density in this context can be thought of as the number of other individuals nearby. The action of density-dependent selection may be influenced by the varying density of neighbors during different life stages. For instance, insect survival from egg to adult may depend on the density of larvae, whereas the number of eggs laid by females may vary with adult density.

Development of Logistic Theory. Robert MacArthur (1962) was the first to develop the mathematical relationship between ecological theories of population growth and the effect of natural selection at different densities. MacArthur accomplished this with the aid of the logistic equation of population growth. This ecological model states that populations at low densities will grow exponentially at the intrinsic rate of population growth (r). The logistic model also assumes that, as a population becomes more crowded, the rate of growth declines. The population ceases to grow when it reaches its carrying capacity (K), which is the equilibrium population size.

Classical theories of natural selection measured fitness by calculating the intrinsic rate of growth (r) of a population. This parameter tends to be maximized by maximizing fertility and survival at low population density. MacArthur extended this idea by suggesting that, at high population density, the population size at carrying capacity (K) would be an appropriate measure of fitness. This theory has sometimes been referred to as r and K selection, drawing from the two parameters of the logistic equation. In 1971, Roughgarden generalized these ideas by suggesting that fitness may be equated with per capita rates of reproduction and population growth. An example of this theory is shown in Figure 1. At low population density, natural selection will favor the increase and ultimate fixation of the A_1 allele because the A_1 homozygotes have highest fitness. However, at high den-

FIGURE 1. Per-capita Growth Rates (fitness) for Three Genotypes. At low density, the A_1A_1 homozygote has the highest fitness, whereas at high density, the A_2A_2 homozygote has the highest fitness. The heterozygote has intermediate fitness at all densities.   Laurence D. Mueller.

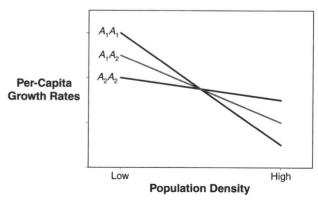

sity, natural selection would favor increases in the alternative A_2 allele. Most importantly, the outcome of evolution depends on the density of the environment.

Trade-offs at Low and High Densities. A key to this theory is the idea of trade-offs. As illustrated in Figure 1, the genotype that is best at low density has the lowest fitness at high density and vice versa. If these types of trade-offs did not exist, then there would be one best genotype for all environments. Although there are no first principles that can be invoked to prove that trade-offs exist, there are some simple arguments that suggest this is a reasonable assumption. Martin Cody (1966), first developed this idea in the context of life history evolution. Cody argued that all organisms must contend with limited amounts of time and energy. As soon as they devote more of their time and energy to, say, reproduction, they will have less time and energy for other activities, such as competing for food.

If the trade-off assumption is valid, those populations that have evolved to grow fastest at low density should do poorly at high density and vice versa. These ideas have been tested by maintaining populations of fruit flies (*Drosophila melanogaster*) at very low and very high densities. After eight generations of evolution, the population growth rates of the high- and low-density adapted populations differentiated, and the predicted trade-offs were observed (Figure 2).

These populations of fruit flies have been studied in more detail to determine which traits changed to cause the observed differences in population growth rates. At least three larval behavioral traits become differentiated between the low- and high-density populations. The high-density populations show elevated larval feeding rates compared to the low-density populations. In fruit flies, it is known that high feeding rates translate into increased competitive ability for limited food, which is certainly at a premium in crowded environments. However, larvae with high feeding rates show reduced survival at low density. Feeding rates then explain, at least in part, the trade-offs observed in Figure 2. When populations of fruit flies adapted to high larval densities are moved back to low densities, the flies' feeding rates rapidly evolve to a lower level, presumably as a consequence of the reduced survival of fast feeders at low density. Individuals from populations adapted to high density also move greater distances while foraging compared to individuals from populations adapted to low density. Finally, larvae from populations that have evolved at high densities are less likely to metamorphose into adults (pupate) on the surface of the food and tend to crawl farther from the food surface in search of a pupation site. This altered behavior also improves survival because larvae that pupate on the surface of the food in crowded cultures showed greatly elevated mortality rates.

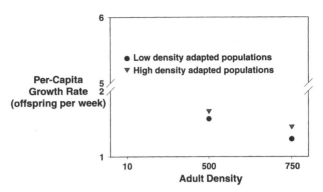

FIGURE 2. The Per-capita Growth Rates for Populations of Fruit Flies (*Drosophila melanogaster*) That Have Evolved at Either Very Low Density or Very High Density. The populations that had adapted to low densities for eight generations were tested at one low density (10 adults) and two high densities (500 and 7500 adults). The same tests were carried out simultaneously on the populations that had been maintained at very high densities.   Laurence D. Mueller.

Development of Verbal Theory. At the same time as MacArthur and Roughgarden were developing their quantitative theories of density-dependent natural selection, an extensive verbal theory of r and K selection was developed. A verbal theory is simply one in which the major assumptions and conclusions are argued in words without reliance on formal mathematics. Verbal theories are acceptable ways of developing ideas in biology. However, the logic supporting the conclusions of verbal theories is not always as obvious as it is with mathematical theories. The verbal theories of r and K selection suggested populations that evolved at high density, called K-selected, should be composed of individuals with increased competitive ability, larger body size, delayed reproduction, and repeated reproduction over many years, or iteroparity. Populations evolved at low density, or r-selected, under this theory would display the opposite set of characteristics: reduced competitive ability, small body size, high levels of reproduction early in life, and survival over fewer reproductive years.

Many of the logical flaws with the verbal theory of r and K selection quickly became obvious to many scientists. For instance, the mathematical theories developed by Roughgarden and MacArthur do not have adults of different ages; instead, all reproduction takes place at a single instant in time. The verbal theories mistakenly inferred that the evolution of high carrying capacities would lead to adults surviving and reproducing over many years. The demise of the verbal theory was signaled by a series of review papers by Steven Stearns in 1976 and 1977 that clearly revealed many of the flaws in the verbal theory.

Studies of Wild Populations. How important is density-dependent natural selection in wild populations?

This has been a difficult question to answer for several reasons. Much of the early empirical work focused on natural populations that were thought to have experienced different density environments. Any difference among these populations in characteristics like fertility and competitive ability was then attributed to density-dependent natural selection. The problem with these types of studies is that historical information on the past density conditions of populations was often anecdotal or incomplete. Likewise, because these natural populations were not under human control, it was often impossible to rule out other factors, such as predation and herbivory, that may have systematically differed between populations. Despite these problems, there are some well-studied natural populations where density-dependent selection is important.

Soay sheep in Scotland, for example, show phenotypic differences in coat color and horn type. Both of these characteristics are under single- or two-locus genetic control. Paul Moorcroft and his colleagues (1996) showed that females with dark coats and small twisted horns survived better at low densities than females with light coats and untwisted horns. However, at high population densities the advantage was reversed. Because population densities vary dramatically in the studied populations, density-dependent selection is probably important for the maintenance of the genetic polymorphisms in horn shape and coat color.

Populations that grow according to the logistic model are expected ultimately to reach an equilibrium size equal to the carrying capacity. However, depending on the characteristics of the population, the approach to this equilibrium can be gradual and smooth, or it may be oscillatory, with the population overshooting and undershooting the carrying capacity by decreasing amounts each generation, or some populations may never settle down to the equilibrium predicted by the logistic equation. These different scenarios reflect different kinds of population stability. Just as density-dependent natural selection may affect population growth rates, it may also affect the stability of population size. Population stability is an important area of biological research because the long-term persistence of populations, especially endangered species, can be affected by their tendency to fluctuate or stabilize. Theoretical work has come up with conflicting predictions. In some cases, density-dependent selection can lead to increased stability of a population, whereas in other cases, stability decreased. There has been one large experiment performed with fruit flies. In this study, fruit flies were placed in an environment that caused the population size to fluctuate. Despite evidence of substantial genetic change in these populations as they adapted to the successive environments, none of these changes appear to have affected the stability of the populations. In this case, the evolution of density-dependent traits did not affect the stability of population size.

[See also Demography; Fitness; Genetic Polymorphism; Life History Theory: An Overview; Population Genetics.]

BIBLIOGRAPHY

Cody, M. "A General Theory of Clutch Size." *Evolution* 20 (1966): 174–184.

MacArthur, R. H. "Some Generalized Theorems of Natural Selection." *Proceedings of the National Academy of Sciences USA* 48 (1962): 1893–1897. In this classic paper, MacArthur proposes that fitness be measured by the carrying capacity of the logistic equation at high densities.

MacArthur, R. H., and E. O. Wilson. *The Theory of Island Biogeography.* Princeton, N.J., 1967. The authors not only develop the ideas of *r* and *K* selection but consider many other interesting problems in ecology that influenced research in this area for the next twenty years.

Moorcroft, P. R., S. D. Albon, J. M. Pemberton, I. R. Stevenson, and T. H. Cluton-Brock. "Density-dependent Selection in a Fluctuating Ungulate Population." *Proceedings of the Royal Society of London B* 263 (1996): 31–38.

Mueller, L. D. "Theoretical and Empirical examination of Density-dependent Natural Selection." *Annual Review of Ecology and Systematics* 28 (1997): 269–288. This is a recent review of scientific research in this area.

Mueller, L. D., and F. J. Ayala. "Trade-off between *r*-selection and *K*-selection in *Drosophila* Populations." *Proceedings of the National Academy of Sciences USA* 78 (1981): 1303–1305. The authors demonstrate trade-off in population growth rates due to density-dependent natural selection.

Mueller, L. D., and A. Joshi. *Stability in Model Populations.* Princeton, N.J., 2000. Reviews some of the mathematical aspects of population stability analysis and discusses experimental research with fruit flies on the evolution of stability.

Roughgarden, J. "Density-dependent Natural Selection." *Ecology* 52 (1971): 453–468. Develops the basic theory and predictions of density-dependent natural selection in a straightforward, clear fashion.

Stearns, S. C. "Life History Tactics: A Review of the ideas." *Quarterly Review of Biology* 51 (1976): 3–47.

Stearns, S. C. "The Evolution of Life History Traits: A Critique of the Theory and Review of the Data." *Annual Review of Ecology and Systematics* 8 (1977): 145–171.

— LAURENCE D. MUELLER

DEUTEROSTOME. *See* Protostome-Deuterostome Origins.

DEVELOPMENT

[*This entry comprises four articles:*

An Overview
Evolution of Development
Developmental Stages, Processes, and Stability
Development and Ecology

The first article provides an overview of embryonic development; the second article focuses on how evolution and embryonic development have interacted over time; the third article surveys metazoan developmental stages; the fourth article focuses on how development is linked to and affected by an organism's ecology and environment. For related discussions, see Canalization; Germ Line and Soma; Life History Stages; Maternal Cytoplasmic Control; Phenotypic Plasticity; Phylotypic Stages; Protostome-Deuterostome Origins; Regulatory Genes; *and* Segmentation.]

An Overview

Embryonic development involves changes that occur in an individual between the time of zygote formation and the transition to the mature form (birth, hatching, or metamorphosis). It involves fertilization, cleavage of the large egg into many smaller cells, morphogenetic movements of these cells, and changes in their gene expression, resulting in their differentiation into various cell types that are coordinated with one another to form a mature organism. In direct development, observed in many plants, fish, reptiles, birds, and mammals, as well as some invertebrates, the mature form gradually and directly emerges from the fertilized egg or zygote by a process called epigenesis. Epigenesis (from the Greek *epi,* "upon," and *genesis,* "production") is the successive generation of new, more complex structures from preexisting, simpler ones. In contrast, the formation of the mature form may proceed through indirect development, whereby epigenesis generates a larval phase, which reaches the mature form through metamorphosis (a transition from one complex form to another), as in many invertebrates (e.g., caterpillar to butterfly) and amphibians (e.g., tadpole to frog). Embryonic development per se would end at this point; however, one should remember that significant maturational changes occur between birth (metamorphosis or hatching) and maturity, and others may continue until death. The study of developmental biology usually covers these events in addition to embryonic development, growth, and the formation of the germ cells (e.g., sperm and egg), which give rise to the next generation.

The History of Developmental Biology. Aristotle (384–322 BCE) was the first individual that we know of to leave written records of embryonic development. He observed that chick embryos changed from simple to more complex creatures (epigenesis) during the course of their maturation in the egg. He observed and explicitly described the changes in the heart and blood vessels, head and eye, and indicated that new structures were generated during the period of observation. His epigenetic view of development was not significantly improved upon for nearly two millennia. Improvements re-

quired two things: the addition of experimentation to observation, an advance in scientific process attributed to the Italian mathematician, astronomer, and physicist Galileo Galilei (1564–1642); and technological advances that allowed new ways of observing the structure and function of the embryo, such as the microscope and histological and molecular biology techniques.

In the early seventeenth century, scientists believed that embryonic development could begin with spontaneous generation (from fluid, mud, or meat). However, in 1651, the English physician William Harvey (1578–1657) concluded that all animals arise from eggs (*ex ovo omnia*). The most elusive evidence in support of this claim finally came in 1827, when the German zoologist Karl Ernst von Baer (1792–1876) first saw and described a mammalian egg. At that time, evolution and development were seen as similar and linear; researchers believed that more complex organisms evolved as additional development was layered on top of the adults of more primitive forms. The unicellular protists were likened to the egg, and simple invertebrates such as the hydra were likened to the gastrula stage embryo, when different cell layers form. At the later phenotypic stage, when human and other vertebrate embryos resemble one another and have a tail and apparent gill slits (pharyngeal arches), they were seen as comparable to an adult fish. Von Baer did not know about the evolutionary relationships between simple and complex organisms that would soon be put forward by Charles Darwin and Alfred Russel Wallace. [*See* Darwin, Charles; Wallace, Alfred Russel.] However, his observations and reasoning refuted the linear concept of evolution as it was known in his time. To counter this linear thinking, in 1828 Von Baer formulated four well-conceived laws that have provided a foundation of modern developmental biology:

- The general features common to a broad group of animals, such as vertebrates, appear earlier in development than the specific features of any of its subgroups, such as birds or mammals.
- Within an organism, the specific features (e.g., hands, wings, flippers, or paws) develop from the more general (e.g., limb buds).
- Embryos of a given species are divergent from, rather than similar to, the adult forms of simpler organisms.
- Therefore, an embryo of a more complex species may resemble an embryonic form of a simpler organism, but it will not resemble its adult form.

Evolution and development. A decade after von Baer published his laws, in Matthias Schleiden and Theodor Schwann proposed that all living things were made of cells. Twenty years later, the German pathologist Rudolf Virchow (1821–1902) proffered that all cells arise from other cells. These ideas were important to the study of development because they provided tangible structures (the cells themselves) that could be observed and assessed through experimentation. In 1856, same year as Virchow proposed his cell theory, Darwin and Wallace published their theories of evolution or descent with modification. Darwin drew several ideas for his evolutionary theories from von Baer. These included (1) a branching versus linear form of descent, (2) the idea that similar structures reveal common descent, and (3) the idea of using embryonic forms in classification. Because differences in adult form arise during development, von Baer's laws also led Darwin to see that the predictable similarities between the embryonic forms of different groups of organisms were strong support for his theory of evolution. Together, the ideas of von Baer, Schleiden and Schwann, and Darwin formed the foundation for our contemporary view that all organisms are related to each other through a cellular continuum, which can be drawn as a branched tree and which shows descent from shared ancestors back to some primordial cell. This important conclusion is called the principle of cellular continuity. August Weismann further refined this idea by proposing that the germ cells (sperm and egg) are segregated from the somatic cells of the body and that it is the germ-line lineage that is really immortal and continuous throughout the ages. [See Weismann, August Friedrich Leopold.]

As the ideas related to development and evolution (descent with modification) were resolving themselves in the late 1800s, two camps formed. The first of these included those who focused on descent through the unity that is the foundation of phylogeny, championed by the French naturalist Étienne Geoffroy de Saint-Hilaire (1772–1844) and by Richard Owen. [See Owen, Richard.] The strength of this focus comes from conserved similarities, conserved stages and processes that can be predicted with some success by applying von Baer's first law and Darwin's branching-tree concept of phylogeny. Typically, the more distantly related the organisms, the earlier in development their shared characters will manifest themselves. For example, the tetrapods (animals with four legs) branch from the vertebrates (animals with vertebrae), which branch from the chordates (animals with a notochord). As expected, the limb buds that form the legs of tetrapods develop later than do the somites, which form the vertebrae, and these develop later than does the notochord. Shared characteristics arising from inheritance from a common ancestor are called homologies. The second camp in the evolutionary debate included those who focused on modification through the diversity of specific traits that revealed adaptation to specific environments. The French comparative anatomist Georges Cuvier (1769–1832) and Charles Bell pioneered this view. With this focus, one can look at traits that are the result of distinct environmental pressures rather than homologous inheritance. If similar features in two organisms arose independently owing to the similar selective pressures of like environments, then these features are called analogous if they are only functionally similar, or homoplastic if they are both structurally and functionally similar. The mechanism resulting in homoplastic structures is called convergent evolution.

Genetics and development. In the early part of the twentieth century, it was believed that the only way to get from genotype (inherited information) to phenotype (embryonic or adult structure) was through development. Thus, the fields of genetics and embryology were perceived as one. At this time, the mechanisms for translating genotype into phenotype were unknown. In the mid-1920s, Hilde Mangold and Hans Spemann (1924) revealed that gene expression in one amphibian cell type could induce gene expression, and therefore development, in other amphibian cell types. Comparable inducing regions have since been identified in other vertebrates, as have some of the coordinated proteins involved as inducers, receptors, signal transduction molecules, and transcription factors. Spemann's evidence of induction revolutionized developmental biology by providing a zygotic mechanism for the epigenetic changes seen during development. Their work on induction won Spemann a Nobel Prize in 1935.

Thomas Hunt Morgan separated the fields of genetics and embryology in the 1920s: genetics informed about gene transmission; embryology related information about gene expression. As we learn more about both areas, however, the interesting questions in these fields seem to be converging again. Similarly, the fields of evolution and development are converging. For example, sometimes one cannot tell by observation alone whether a structure shared by evolutionarily diverse organisms is homologous or analogous. The single-lens eyes of vertebrates and cephalopods (octopus, squid) and the compound eyes of insects, though both involved with light perception, were thought to have arisen through convergent evolution because their common known ancestor lacked eyes and their structures were widely divergent. In the mid-1990s, however, a common gene, *Pax-6*, was identified that is expressed in all organisms with eyes, and which may provide a common foundation for eye formation. Thus, regardless of divergent anatomy, some now regard all eyes as homologous structures. Information such as this unites genetics, evolution, and development.

Molecular biology and development. The contemporary intersection of these disciplines has flourished because of advances in molecular biology, which have led to the discovery of many genes like *Pax-6* that are conserved across large phylogenetic distances and underlie the development of homologous structures or functions. Probably the most famous of these are the homeotic genes, a family of genes coding for transcription factors (genes that turn on and off other genes) and instigate changes in function and regional identity (pattern formation). Homeotic genes were first identified in the fruit fly (*Drosophila*). Their discovery has led to the unraveling of the mystery of the molecular nature of development in this complex invertebrate, and then, through investigations of homology, in many other organisms. In 1995, the Nobel Prize in medicine was given to Christiane Nuesslein-Volhardt, Eric Weischaus, and Ed Lewis for their work, beginning in the early 1970s, in revealing the network of gene expression that changes a fertilized fruit fly egg into a segmented fruit fly larva. However, their Nobel-worthy effort is not just about fruit fly epigenesis. Their analysis of gene networks in the fly has opened the doors to the present-day study of the molecular nature of development in many other organisms. Because many of the genes they characterized are involved in general processes, such as establishing orientation (head, trunk, tail) or body segmentation (seen in segmented insects or worms and human vertebrae and spinal nerves), they are conserved across large phylogenetic distances and are expressed in the embryos of many species, including humans. In fact, we have been able to use various model systems to their best effect in the study of contemporary developmental biology because of the similarities between diverse organisms and the conservation of the genes that generate them.

The twenty-first-century discovery, through the human genome project, that humans and other complex vertebrates have only 30,000 genes, rather than the 100,000 originally predicted, has led to the conclusion that many genes may be serving multiple purposes within the body. For example, one of the most extensive families of genes is called the bone morphogenetic factors (BMPs). Their first family member was discovered during a search for proteins inducing bone formation. In some tissues, however, they also regulate cell death (apoptosis), cell division, cell migration, cell differentiation, and pattern formation. Within this family, a single member in vertebrates, *BMP4*, plays a role in lens formation in the eye, induction of the skin, apoptosis of the tooth enamel knot, patterning of the dorsal central nervous system, migration of the hypaxial muscle cells from the somite, and induction of ventral cell fates. To be involved in many different functions, each protein is dependent on others, and different combinations of gene expression allow a moderate number of genes to produce significant complexity.

Contemporary developmental biology is an intersection between embryology, evolution, genetics, cell biology, and molecular biology. We are closer than ever to understanding the mechanisms that transform the apparently simple egg into the complex worm, fly, urchin, fish, frog, chick, mouse, or human.

GENE CONSERVATION IN DEVELOPMENT

Although new mutations probably occur in every individual, one would speculate that the function of some proteins are so fundamental that individuals with mutations affecting these genes' function would quickly be weeded out by natural selection. Thus, the functional aspects of these genes would change little over long periods of evolutionary time. As a corollary, scientists have proposed that genes that have remained conserved over long evolutionary periods must reflect homology of important function that has also been conserved through the same period. Such genes are used to provide evidence for homology between structures or functions in distantly related organisms. For example, scientists had been feuding since 1875 about whether the dorsal nerve cord of vertebrates is homologous to the ventral nerve cord of insects or whether they evolved independently. Recently scientists discovered that genes of the *orthodentical (odt)* family are expressed in the rostral nervous system of organisms from flies to humans. When fly embryos are produced in which both copies of the *odt* gene are mutated (homozygous mutants) a larva is generated lacking the front-end of its ventral nervous system. Embryos with such mutations can be rescued by over expressing fly *otd* in all cells of their bodies. Vertebrates, including humans, also have *odt* family members, called *OTX1 and 2*, which are expressed rostrally (forebrain and midbrain) in their dorsal nervous systems. These genes are so functionally conserved with the fly *otd* genes that fly *odt* homozygous mutants can be rescued by ubiquitously expressing the human *OTX2* gene in these fly embryos (Leuzinger et al., 1998). This and other recent evidence (reviewed by Arendt and Nübler-Jung, 1999) strongly supports an argument for the homology of the central nerve cord between flies and vertebrates.

Arendt, D., and K. Nübler-Jung. Comparison of early nerve cord development in insects and vertebrates. *Development* 126 (1999): 2309–2325.
Leuzinger, S., F. Hirth, D. Gerlich, D. Acampora, A. Simeone, W. J. Gehring, R. Finkelstein, K. Furukubo-Tokunaga, and H. Reichert. Equivalence of the fly *orthodenticle* gene and the human *OTX* genes in embryonic brain development of *Drosophila*. *Development* 125 (1998): 1703–1710.

—DIANA KAROL DARNELL

Developmental Processes and Mechanisms. Development begins with the egg, a large germ cell from the female that contains sufficient resources to sustain the embryo until it starts taking in nutrients. In addition to nutrients, cellular organelles, and zygotic genes, eggs contain maternally derived information molecules (RNA or protein) that can be used to regulate development of the early embryo until sufficient information molecules can be generated by the zygotic genome. Specific information molecules are often sequestered in a particular region of the egg, and these may be restricted to a subset of the cells as the early embryo divides. Because these molecules are from the mother, they are called maternal factors and the genes that produce them are called maternal effect genes. In the egg, maternal factors bind to the embryonic DNA in the nucleus and initiate (or suppress) gene transcription. Thus, they are also called transcription factors. Maternal transcription factors initiate the expression of zygotic genes in blastomeres (cells) of the early embryo, initiating a cascade of transcription-factor gene expression that will eventually led to zygotic control of differentiation. Before fertilization, these factors are inactive and the egg is quiescent. At fertilization, the entry of a single sperm into the egg activates metabolic and maternal signaling molecules to start embryonic development. At this point, the egg becomes a zygote.

Cleavage and morphogenesis. The first phase of zygotic development is called cleavage. During cleavage, the large fertilized zygote replicates its DNA many times and cleaves its cytoplasm to increase the nucleus to cytoplasm ratio within each cell. This segregates maternal determinants into specific cells, produces a larger number of sources for zygotic transcripts (the replicated nuclei), and allows zygotic transcription and translation to rapidly affect the behavior of the cells (smaller distances to travel, fewer inhibitors to overcome, less membrane to populate with receptors). Cleavage also generates a large number of (small) cells that can move relative to one another, increasing the number of specific inductive interactions that can take place. During cleavage of some embryos (e.g., insects and amphibians), the DNA is replicated so quickly that there is no time for transcription to take place. During this time, the embryos are dependent on maternal transcripts or proteins.

The maternal determinants (RNAs and proteins) localized within certain regions of an egg can cause the blastomeres (embryonic cells) derived from that region to become a particular type of cell. That is, the cell will express a particular fate, different from the fates of other blastomeres. The classic example involves the polar granules, maternal determinants laid down in the posterior end of many insect embryos. This RNA-containing material is segregated into the posterior cells, which later form the germ-line cells (sperm or eggs). Experiments in which the polar cytoplasm was removed or irradiated by ultraviolet light to destroy the activity of the polar granules yielded embryos without germ cells. This experiment shows that the polar granules are necessary for germ cell formation. When posterior (but not anterior) pole cytoplasm was transplanted back into the posterior end of

embryos without active polar granules, germ cell formation was rescued. This experiment shows that the polar granules are sufficient to promote germ cell formation. Being both necessary and sufficient are the main criteria used to support a cause-and-effect relationship between two items that are correlated. For this reason, polar granules are often referred to as germ cell determinants: they cause cells to have a germ cell fate.

In addition to specifying cell fates, maternal determinants are involved in setting up the general pattern of many embryos. About twenty maternal effect genes are involved in establishing the anterior-posterior body pattern of the fly larva. One gene, *bicoid*, produces RNA that is sequestered in the anterior pole of the egg. After fertilization, bicoid RNA is translated into a gradient of bicoid protein, which acts as a morphogen. Morphogens are transcription factors or other information molecules that can induce or suppress the expression of different genes depending on morphogen concentration. Thus, the zygotic nuclei that end up at the anterior tip of the egg are exposed to high bicoid protein concentration and express one set of genes, whereas those at the lower end of the gradient express another. A similar mechanism works from the posterior end of the embryo using a different maternal factor, nanos. The zygotic genes that are activated by these morphogens (gap genes) generate their own transcription factors, which in turn activate a third tier of transcription factors called the pair-rule genes. This cascade of localized factors activating other localized factors layered over simultaneous nuclear division, continues to generate a finer and finer pattern until the larva has identifiable, oriented segments, whose cells can begin to differentiate into the appropriate structures for that body region. *Drosophila* and many other organisms use maternal factors extensively during early embryogenesis and switch to zygotic gene expression at the midblastula transition, which is characterized by slower cell cycles (with larger gap phases when zygotic transcription can occur). At this time, embryos degrade their maternal RNAs. In contrast, other organisms, such as the mouse, begin expressing zygotic genes at the two-cell stage. Cleavage causes early embryos of these species (e.g., mammals and sea urchins) to form a dense cluster of cells, or morula, that expands to form a fluid-filled ball (blastocyst or blastula) before proceeding through morphogenesis, induction, determination, and differentiation.

Not all fates are directly affected by maternal factors. An alternative means of directing the fate of individual cells and tissues is through conditional specification. The conversion of one cell to a particular fate owing to signals from neighboring cells is called induction. With induction, one cell or group of cells, the inducer, sends a signal (often a secreted protein) into the extracellular milieu. Contact of this signal with an appropriate receptor on a neighboring cell (paracrine signaling) initiates a signal transduction event that results in changes in gene transcription in the responding cell. In this fashion, the fate of the responding cell can be altered. Therefore, its fate is conditional upon the induction by its neighbor. Although most embryos probably contain some maternal determinants that specify the early functions of the blastomeres, once zygotic genes are active, the dominant regulator of development is the interaction between populations of cells (induction) to choreograph the increasing complexity of the embryo (epigenesis).

Morphogenesis follows cleavage. During morphogenesis, cells and sheets of cells move relative to one another to acquire positions within the embryo appropriate to their fate. The first morphogenetic movements are those of gastrulation, the migration of cells out of the single layer on the surface of their blastocyst or blastula (the ectoderm or epidermal layer), then down into the interior of the embryo to form a middle (mesodermal) and deep (endodermal) layer of cells. These three originating layers (germ layers) are fated to form specific tissues within the body of the mature animal. Because the specificity of the layers is consistent across all animals, and because they occur very early in development, we infer that the formation of the germ layers was a very early branch point in the evolutionary history of animals.

As the three germ layers become appropriately positioned, their cells are still pluripotent, that is, able to form various tissues. Eventually, however, signaling events induce changes in their gene expression such that they become determined, sequentially limiting their potency, and finally they differentiate into cells of a specific type.

Determination and differentiation. Through the presence of maternal determinants, zygotic gene transcription, and signaling from neighbors, cells begin to differentiate. Grossly generalized, in vertebrates the ectoderm forms the skin and nervous system; the mesoderm forms the muscles, bones, reproductive organs, and supportive parts of many other internal organs; and the endoderm forms the gut tubing and digestive and respiratory organs. The causal mechanisms of determination and differentiation have been elucidated for some cell types. Striated muscle, of mesodermal origin, is a good example of the progression of epigenesis (general to specific) at the cellular level. Muscle cells begin as pluripotent mesodermal precursors that follow morphogenetic cues to form epithelial balls of cells, or somites, on either side of the neural tube. From these undifferentiated somites, the cartilage and bone (sclerotome), lower layer of the skin (dermotome), and muscle (myotome) progenitors will form. The myotome is induced on the dorsolateral and dorsomedial sides of the somite by signals from the neural ectoderm (probably by the proteins Wnt 1 and Wnt 3a dorsally and by Sonic Hedgehog ventrally), skin ectoderm (probably through Wnts

again) and lateral plate mesoderm (probably by bone morphogenetic protein 4 [BMP4]). These signaling molecules cause the myotome to become determined by inducing the expression of transcription factors (called MyoD and Myf5) that will turn on muscle-related genes. Any cell expressing the *MyoD* or *Myf5* gene will become a muscle cell and is called a myoblast. But a myoblast is still a far cry from a muscle. A myoblast is a single cell that does not express any functional muscle proteins, whereas a functional, striated muscle cell is a multinucleated fusion product of many cells that transcribes actin and myosin almost exclusively. To differentiate, the determined myoblast must divide extensively (regulated externally by fibroblast growth factors [FGFs]), stop dividing, and align with future fusion partners (regulated by cell surface molecules fibronectin, integrin, and cadherin/CAM), fuse into a syncytium with its neighbors (regulated by calcium and metaloproteinases called meltrins expressed in the muscle cells), and mature into a muscle fiber by massively upregulating the transcription of the contractile proteins actin and myosin.

Organogenesis and pattern formation. In order to function as an organism, not only must the cells differentiate properly (histogenesis), but they must become associated with appropriate other cell types within the body and in the appropriate location (organogenesis and pattern formation). [*See* Pattern Formation.] For example, the muscle cells not only must be able to contract but must be attached to tendons and bones so that their contraction produces movement, and to blood vessels and nerves so they can be fueled and controlled. How these functional and complex contacts between different cell types are arranged also appears to be dependent on intercellular signaling. For example, motor nerves from specific locations in the spinal cord send axonal processes out to connect with the muscle cells in the limbs. At their leading tip these axons have a mobile sheet of membrane called a growth cone that can crawl along the surface of other cells or their extracellular matrix. On the surface, this growth cone has receptors for specific information molecules (called chemotactic factors) in the environment, and concentration gradients of these molecules direct the growth cone's movement. Once chemotactic factors led the growth cone to the surface of the muscle, the nerve secretes agrin, a protein that causes acetylcholine receptors in the muscle cell membrane to aggregate under the growth cone. The neuromuscular junction, or synapse, becomes further specialized as a result of the secretion of extracellular matrix molecules, such as laminin by the neuron, and by folds that form in the muscle cell membrane to increase its surface area, presumably because of additional signaling between the cells. In this way, two cells that were originally separated can make a functional connection that is required for the survival of the organism.

Another area of pattern formation involves the position of structures or differentiated cells relative to the whole body. This area of pattern formation addresses why many starfish have five arms and humans have five fingers, how the stripes on a zebra are generated, and why the feathers on a bird are relatively evenly spaced. These questions of patterning have been addressed using mathematical modeling by Alan Turing (the mathematician who broke the Enigma code during World War II and one of the founders of computer science) and Hans Meinhardt. For some patterns, models support the hypothesis that there are inductive and inhibitory feedback loops (reaction-diffusion relationships) between signaling molecules of neighboring cells. With only minor modifications, these models can account for stripes, spots, and consistent fields, or they can produce the reiterated patterns of distinct size or duration found in nature.

Growth. Once histogenesis (the formation of different cell types), organogenesis (the formation of organs from those cell types), and pattern formation are complete, then the body continues to grow. In fact, for the human, all of the processes involved in morphogenesis, differentiation, and pattern formation occur during the first trimester of pregnancy. The vast majority take place in the first eight weeks of gestation, often before a woman knows she is pregnant. The remainder of the embryonic or fetal period is spent in growth. Growth for all organisms is the result of increased cell numbers. This is controlled intrinsically by growth factors (local regulators of cell division) and extrinsically by growth hormones (circulating regulators of cell division), both acting through receptors in the responding cells. Usually, different parts of an organism grow at different rates at different times. This is called allometric growth. Growth is coordinated for all tissues and for the body as a whole so that proportion is maintained. Were the number of cell divisions for your arms just one more than normal, you could tie your shoelaces while standing upright. But that does not happen. The number of cell divisions is elegantly regulated. If this elegant regulation breaks down, cells cease to be dependent on growth factors to induce their division and become cancerous. Basic research into the mechanisms of normal growth and development have helped in the understanding of this disease.

[*See also* Maternal Cytoplasmic Control; Regulatory Genes; Segmentation.]

BIBLIOGRAPHY

Carlson, Bruce M. *Human Embryology and Developmental Biology*. 2d ed. St. Louis, 1999. A clearly written and well-illustrated text on embryology, particularly the changes that occur during development rather than the mechanisms that regulate these changes.

Gilbert, Scott F. *Developmental Biology*. 6th ed. Sunderland, Mass., 2000. A comprehensive and exceptionally detailed text on developmental biology, with full-color illustrations and links to online resources at http://www.devbio.com (see below).

Halder, G., P. Callaerts, and W. J. Gehring, "Induction of Ectopic Eyes by Targeted Expression of the *eyeless* Gene in *Drosophila*." *Science* 267 (1995): 1788–1792. Provides evidence that *Pax-6* is sufficient to induce eye development in the fly.

Kalthoff, Klaus. *Analysis of Biological Development*. 2d ed. New York, 2001. An excellent and approachable introductory text of development, covering both principles and experimental evidence.

Moore, John A. *Science as a Way of Knowing*. Cambridge, Mass., 1993. A very readable book covering the history of science and its relationship to evolution, genetics, and developmental biology.

Slack, Jonathan. *Essential Developmental Biology*. Oxford. A concise coverage of modern developmental biology for undergraduates. Text assumes some knowledge of molecular and cellular biology.

Spemann, Hans, and Hilde Mangold. "Über induktion von Embryonalanlagen durch Implantation artfremder Organisatoren." *Roux Arch Entwicklungsmech Org* 100 (1924): 599–638. This seminal work changed the face of development worldwide. Tragically, Mangold died in 1924 at the age of twenty-six, and therefore could not be included in the Nobel Prize citation with Spemann, but her doctoral thesis is one of the few in biology to result in the award.

Wolpert, Lewis. *Principles of Development, Current Biology*. Oxford, 1998. This textbook is meant for both undergraduates and graduate students and emphasizes the key concepts associated with development, with a primary focus on vertebrates and *Drosophila*.

ONLINE RESOURCES

http://carol.wins.uva.nl/~roel/bauplan. "The Molecular Bauplan." An informative and well-referenced site maintained by Roeland Merks, of the Department of Biology, Utrecht University, Netherlands. Includes *Pax-6* information and more thorough coverage of the argument regarding homology versus analogy in eye development.

http://www.devbio.com. Contains abundant information on a wide range of developmental topics associated with Scott Gilbert's text, *Developmental Biology* (see above).

http://sdb.bio.purdue.edu. This site, supported by the Society for Developmental Biology, has links to numerous videos, books, and other Web sites.

— DIANA KAROL DARNELL

Evolution of Development

When we consider the million-odd known animal species, two impressions strike us. First, many animals are remarkably similar in form. Most are more or less bilaterally symmetrical and possess an anterior mouth and sense organs, a posterior anus, and a standard suite of organs (hearts, gonads, muscles, etc.). Paradoxically, we are also struck by the immense diversity of form. Some creatures have a symmetry that is radial, helical, or simply not apparent; others have tissues but no recognizable organs. Some animals weigh a microgram at adulthood, others weigh 150 metric tons; some animals have brains containing a 10^{11} highly organized neurons, others just a simple nerve network. Some animals lack mouths, others lack anuses; some have eyes on their genitalia.

Why are some creatures the same and others so different? Many researchers have sought the answer to this question by describing the way in which creatures develop from eggs into adults, that is, in their ontogenies. Classically, the unity and diversity of animal development have been sought in descriptions of the cellular events of ontogeny, a discipline known as comparative embryology. Today, unity and diversity are sought in the genetic programs that drive these cellular events, an enterprise known as evolutionary developmental biology, or simply evo-devo.

The modern search for unifying principles underlying the development of animals can be traced to the English Renaissance. In 1651, the English physician William Harvey, who had watched chicken embryos develop, asserted his famous principle: *ex ovo omnia*. To prove it, he sacrificed numerous does, but he failed to find their ova. It took a supreme anatomist, Karl Ernst von Baer, to find the mammalian ovum in 1827, which he did by dissecting a female dog in heat. The proof that all animals, even mammals, develop from eggs spoke profoundly of the unity of animal development. So, too, did a discovery made by a close friend and colleague of von Baer's, Christian Pander. Pander noticed that the blastoderm of a chicken embryo could be separated into three layers. Each of these "germ layers"—ectoderm, endoderm, and mesoderm, as they are now known— developed to form particular tissues and organs. What is more, as demonstrated in a host of exquisite nineteenth-century monographs on the embryology of minute marine creatures, the germ layers were found in nearly all animals. The immense diversity of animal morphology could be ordered and simplified by looking into the embryo. In his *Origin of Species* (1859), Charles Darwin offered this unity as one of the proofs of organic evolution.

The ontogeny of animals, however, also showed differences; von Baer, who studied a variety of vertebrates, noticed this and proposed that the earliest stages of ontogeny were most similar among species, the individual characteristics of reptiles, birds, and mammals appearing last in their development. Karl Ernst von Baer was no evolutionist, but others saw that embryology could be used to relate apparently diverse organs to each other, that is, to infer homology. If embryology could be used to infer homology, it could also be used to order phylogeny. In 1866, Alexander Kovalevsky found the tunicate tadpole larva and showed that it was astonishingly like a chordate. His collaborator, Elie Metchnikoff, used the tornaria larva of the acorn worm (*Balanoglossus*) to link vertebrates and echinoderms. The American embryologist E. B. Wilson showed that the early embryonic cell lineages of polychaete annelids and gastropod molluscs were identical; they, too, must be related. Wilson titled a paper *"Considerations on Cell Lineage and Ancestral Reminiscence"* (1898). Between the publication of Darwin's *Origin of Species* and the end of the nineteenth century, comparative embryologists gave an account of metazoan phylogeny that remained largely intact until the advent of molecular tools.

Comparative embryology also seemed to give a general account of how new morphological forms arise. Pre-Darwinian "transcendental morphologists" such as Étienne Serres and J. F. Meckel held that human ontogeny passes through successive stages of less "perfect" creatures such as fish, reptiles, and birds. It was an idea that was, at best, only weakly consistent with the embryological facts. Nevertheless, in the hands of Ernst Haeckel, a German proselyte of *Der Darwinismus*, or Darwinism, it became an evolutionary law. Brilliant, restless, and charismatic, Haeckel dominated continental evolutionary thinking until his death in 1919. Under the slogan "Ontogeny recapitulates phylogeny," his biogenetic law stated that evolution proceeds by the addition of novel structures at the end of ontogenies. It followed that an animal's embryology contained a complete record of its evolutionary history. All animals gastrulate, Haeckel observed; therefore, an adult creature much like a gastrula must once have existed. He called his invention the Gastraea; it was the first of a menagerie of hypothetical metazoan ancestors.

Many thought Haeckel was wrong—von Baer protested that the ontogenetic record revealed similar embryonic stages, not a chain of ancestral adults. Adam Sedgwick pointed out that embryos had features that no adult creature could ever have had. Others accused Haeckel of faking his facts, but to no avail: the biogenetic law swept all before it and became entrenched in popular thought before finally collapsing in the early twentieth century under the weight of its own exceptions. Today, terminal addition is recognized as just one of several ways in which changes in the timing of ontogeny can bring about morphological evolution. Another is paedomorphosis, in which ancestral juvenile features come to be expressed in the adults of descendants. Such "heterochronic" processes continue to arouse interest, albeit less among developmental biologists and geneticists than paleontologists.

With the rise of *Entwicklungsmechanik*, comparative embryology went into decline. A new breed of mechanism-minded biologists—Wilhelm Roux, Hans Driesch, and Spemann—were not interested in working out *Ahnengallerien* (ancestor portrait galleries); they wanted to know how animals were built. Evolutionists reciprocated the sentiment. Developmental biology was conspicuously absent from the evolutionary synthesis of the 1940s, which united evolutionary theory with genetics, ecology, and paleontology. In retrospect, this is understandable. The synthesis couched evolutionary theory in terms of gene frequencies, and in the 1940s the genetic study of development had scarcely begun.

The first *Drosophila* mutation was found by Thomas Hunt Morgan in 1910. Many more quickly followed. They were used to study the mechanics of inheritance. But in the 1960s, it became clear that morphological mutations had another use as well: they revealed, piece by piece, the genetic program that makes a fly. Among the most fascinating and intractable fly mutations were those that transformed appendages of one type into another, for example, an antenna into a leg. These were the homeotic mutations, from *homeosis*, a word coined by William Bateson in 1894 to describe malformations in which a structure of one type is transformed into another. Ed Lewis at the California Institute of Technology showed that many of these mutations disrupted a single gene cluster, the "bithorax complex." Molecular cloning of the bithorax complex in the 1980s showed that it encoded three transcription factors that shared a common DNA-binding motif, the homeobox. Another cluster of five homeotics, the Antennapedia complex, also encoded homeobox transcription factors. Collectively known as the homeotic complex (*Hom-C*) genes, their discovery brought evolution back into developmental biology.

The *Hom-C* genes are selector genes. They execute a combinatorial code that selects one identity rather than another for each region along the anterior-posterior axis of the embryo. Some *Hom-C* genes combine to specify the identities of head structures, others tail structures, others the parts in between. In the early 1980s, it became clear that many, perhaps all, animals had *Hom-C* genes. Mice had them, nematodes had them, and they seemed to work much the same way in each. Suddenly, evolutionary speculation became fashionable again. Molecular biologists, who proudly traced intellectual descent from Max Delbrück and Salvador Luria, began to cite eighteenth-century transcendental morphologists and wonder just what an enteropneust was. Perhaps the *Hom-C* genes were part of a universal program for building animal bodies. Perhaps possession of a *Hom-C* cluster was even the very essence of being an animal. There was talk of a new, molecular, comparative anatomy.

The *Hox* (homeobox) genes were not the only genes of interest. Christiane Nüsslein-Volhard and Eric Wieschaus's mutational screens had identified many genes that were needed for specifying the axes of the embryonic fly. They shared a Nobel Prize with Ed Lewis in 1995. Following Sydney Brenner's lead, hundreds of geneticists were also analyzing the nematode *Caenorhabditis elegans*. Vertebrate developmental biologists split their efforts between mice, chickens, *Xenopus*, and ze-

bra fish. As the molecular controls of embryogenesis, pattern formation, and morphogenesis were unraveled in each of these "model organisms," results began to converge, sometimes in surprising ways. Some of the genes that specified the dorsal-ventral axis in *Drosophila* appeared to have homologues in vertebrates that did the same thing, but inverted, such that dorsal genes in *Drosophila* were ventral in *Xenopus* and vice versa. The ideas of the Napoleonic anatomist Étienne Geoffroy Saint-Hilaire were revived. He had proposed that the anatomy of a lobster and a vertebrate could be homologized if you simply assumed that one was upside down relative to the other. Gene expression data now suggested he may have been right. Other, equally unexpected, homologies appeared. Walter Gehring showed that it was possible to grow eyes on the legs of a fruit fly by overexpressing *Pax-6*, a master regulator of eye development. Amazingly, mutations in the human *Pax-6* orthologue caused a failure of retinal development. Was the compound insect eye somehow homologous to the human eye? Most zoologists favored Ernst Mayr's view that eyes had evolved independently forty times in the animal kingdom. The guts of echinoderm and annelid larvae were shown to express the same set of regulatory genes. They, too, must be homologous. This undermined the raison d'être of the Protostomia and Deuterostomia, the great metazoan superphyla first proposed by Karl Grobben in 1914 and repeated in every zoology text since. But, then, molecular phylogeny based on sequences of 18S DNA found in ribosomal structures, had already rearranged the metazoan tree. The great verities of classical zoology were tumbling one by one.

In the 1990s, the Haeckelian game of hypothetical ancestors was taken up again with gusto but little consensus. Inevitably, the very meaning of homology came under scrutiny. Some experts cautioned that gene expression data could be misleading regarding organ-level homologies—genes, even entire signaling pathways, could be co-opted for new functions. The signaling pathways that specify the dimensions of a fruit fly's wings also specify the dimensions of a butterfly's wings—but are then reused to specify the butterfly's wing spots. The molecular basis of morphological diversity began to be studied. Ed Lewis had proposed in 1978 that changes in *Ultrabithorax* function underlay the evolution of the fruit fly's two wings from its four-winged ancestor. He was wrong, but many researcher thought that the *Hom-C* genes had to be important in morphological evolution. Previously obscure creatures, such as brine shrimp, onychophorans, amphioxus, and hydra, were brought back into the university laboratories from which they had been banished for fifty years. In 1996, the venerable German journal *Roux's Archives of Developmental Biology* changed its name to *Development, Genes, and Evolution*. A new journal, *Evolution and Development*, was founded by Rudy Raff in 1999. Evo-devo, if not exactly mature, had at least arrived at an unruly adolescence.

Is a new evolutionary synthesis at hand? Some think so. Developmental pathways, they argue, are not only shaped by evolution, but determine what is and what is not possible. Van Valen's famous aphorism, "Evolution is the control of development by ecology," should be turned on its head. Few population geneticists are convinced. The mechanics of development are fascinating, they reply, but do not dethrone natural selection as evolution's prime mover. Yet a modest new synthesis has emerged, one not so much of new principles, but rather of new possibilities. Ecological geneticists need not rest with estimating heritabilities; they can hope to clone the genes that cause phenotypic variation. Paleontologists cannot merely revel in stratigraphy; they must stay au courant with gene expression patterns as well. This is just part of a more profound revolution in the rest of biology. The deep homologies of animal development mean that worms, flies, chickens, mice, and humans are no longer studied in isolation from each other. Now every molecular geneticist is an evolutionary, or at least a comparative, biologist. In 2001, the complete sequence of the human genome was published. We proved to have only 30,000-odd genes against the 19,000 of *C. elegans*, a soil-dwelling worm. *Homo sapiens* was humbled. In a sense, it was the completion of a project that began in 1859.

[*See also* Germ Line and Soma.]

BIBLIOGRAPHY

Arthur, W. *The Origin of Animal Body Plans: A Study in Evolutionary Developmental Biology.* Cambridge, 1997. Provocative account of metazoan evolution from a population genetics perspective.

Bowler, P. J. *Life's Splendid Drama: Evolutionary Biology and the Reconstruction of Life's Ancestry, 1860–1940.* Chicago, 1996. Academic history of phylogenetic studies, with a fine chapter on comparative embryology.

Carroll, S. B., J. K. Grenier, and S. D. Weatherbee. *From DNA to Diversity: Molecular Genetics and the Evolution of Animal Design.* Oxford, 2001. The first true evo-devo textbook.

Gerhart, J., and M. Kirschner. *Cells, Embryos and Evolution.* Oxford, 1997. Two eminent cell biologists give an idiosyncratic account of the evolution of animal development.

Gilbert, S. F. *Developmental Biology.* 6th ed. Sunderland, Mass., 2000. Undergraduate developmental biology text, with a separate chapter on developmental mechanisms of evolutionary change. Essential.

Gilbert, S. F., and A. M. Raunio. *Embryology: Constructing the Organism.* Sunderland, Mass., 1997. Textbook for a revived comparative embryology, with contributions by most experts in the field.

Gould, S. J. *Ontogeny and Phylogeny.* Cambridge, Mass., 1977. The rise and fall of recapitulationism, along with an influential reformulation of heterochrony.

Hall, B. K. *Evolutionary Developmental Biology.* 2d ed. New York, 1998. A monographic survey of the field, particularly strong on historical controversies and vertebrate development.

Lawrence, P. A. *The Making of a Fly: The Genetics of Animal Design.* Oxford, 1992. The logic of *Drosophila* developmental genetics. Dated, dense, but beautifully written, with vignettes of the important fly geneticists.

Raff, R. A., and T. C. Kaufman. *Embryos, Genes, and Evolution: The Developmental-Genetic Basis of Evolutionary Change.* New York, 1983. The book that heralded the birth of evo-devo. Dated but still instructive.

— ARMAND M. LEROI

Developmental Stages, Processes, and Stability

Although there are numerous minor variations, the fundamental patterns of metazoan development have been remarkably well preserved throughout phylogeny. Other than cases of asexual reproduction by budding or regeneration after fission, metazoans (multicellular animals) reproduce by development of a new individual from an egg. In the vast majority of cases, the stimulus for development from an egg is fertilization by a spermatozoon, but examples of parthogenetic development (development from an egg in the absence of fertilization) are known in a number of animal groups.

The essence of animal reproduction is the joining together of gametes (sex cells) produced by females and males of the same species. The number of chromosomes in the gametes is reduced by half through the process of meiosis as the gametes develop in both the male and female body. The female gametes (eggs) undergo meiosis and mature in the female gonad, typically called an ovary. The male gametes, called spermatozoa, undergo meiosis and mature in the male gonad (testis).

Fertilization is defined as the process by which an egg is penetrated by a spermatozoon with the result that the genetic material in both the egg and the sperm cell is fused, restoring the full complement of genetic material, and embryonic development begins. Two principal strategies of fertilization have evolved across a large number of phylogenetic groups. External fertilization is characteristic of most aquatic or marine animals. The female lays very large numbers of eggs, which are then fertilized by spermatozoa that are emitted by the male or several males. External fertilization is a very inefficient process. Typically, only a small number of the eggs laid are fertilized, and an even smaller percentage of the fertilized eggs survive into adulthood. Terrestrial animals and a small number of aquatic species practice internal fertilization, in which the male must copulate with the female to bring the two types of gametes together. Internal fertilization is a much more efficient process than external fertilization, and far fewer eggs need to be produced to maintain the species.

For both types of fertilization, a thin noncellular

membrane, commonly called the zona pellucida, both protects the egg and, through specific receptor molecules intrinsic to the zona, ensures species specificity. With some exceptions, only spermatozoa of the same species are able to bind to the zona pellucida. Through the aid of trypsinlike enzymes, called acrosins, the spermatozoa make their way through the zona and are able to make contact with the egg. The first spermatozoon that makes contact with the surface of the egg penetrates the cell membrane of the egg. This releases a dual mechanism, called the fast and slow blocks to polyspermy (fertilization of the egg by more than one spermatozoon), that blocks entry of other spermatozoa into the egg. The fast block to polyspermy consists of a rapid (within seconds) change in the electrical charge of the surface membrane of the egg that prevents other spermatozoa that have penetrated the zona pellucida from fusing with the egg's surface. The second block to polyspermy, called the slow block (usually occurring within a minute), is mediated by the release of calcium ions from the egg and ultimately renders the zona pellucida impenetrable to other sperm cells.

Once the spermatozoon has made its way into the egg, its nucleus swells, and its chromosomal material combines with that of the egg. The process of fertilization is then complete, and the fertilized egg is called a zygote. Almost immediately the zygote enters a period of cleavage, during which it becomes subdivided from a single cell to a typically spherical mass consisting of many hundreds to thousands of cells, called blastomeres.

The characteristic patterns of cleavage define major taxonomic groups. In the protostomes, consisting of flatworms, roundworms, arthropods, annelids, and molluscs, cleavage typically occurs in a spiral pattern, in which the blastomeres are oriented at an angle to the principal axis of the embryo. In the majority of deuterostomes, consisting of the echinoderms and chordates, cleavage is of the radial variety, meaning that the blastomeres are oriented parallel to one another and to the main axis of the embryo. A second significant difference between cleaving protostome and deuterostome embryos is their response to the damage or removal of blastomeres. Cleavage stages of protostome embryos are characterized by mosaic properties. This means that if an early embryo is subdivided into individual or groups of cells, the remaining cells will develop into only those portions of the body that they would normally have formed, resulting in an incomplete embryo. Deuterostomes, in contrast, exhibit regulative properties, meaning that if an early embryo is subdivided, the remaining parts reorganize into complete embryos. Regulation is a fundamental mechanism by which identical twinning can occur.

After a number of cleavage divisions, the embryo in

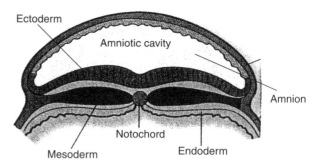

FIGURE 1. Cross-Section of an Early Mammalian Embryo, Showing the Location of the Three Germ Layers and the Notochord.
Above the embryo itself are the amnion and amniotic cavity, which will ultimately completely encircle the embryo.   From *Human Embryology and Developmental Biology*, 2d ed., Mosby, 1999.

species that possess little yolk becomes arranged into a hollow ball of cells, which is called a blastula. In embryos possessing large amounts of yolk, the blastula consists of a disk of cells situated on top of a large mass of yolk. Topographically, a blastula is defined by an axis that runs from an animal pole consisting of smaller cells to a vegetal pole, where the cells typically are larger because of a greater yolk content.

The blastula phase is shortly followed by one in which major morphological and molecular changes take place. This phase is called the gastrula, and at the morphological level, it is manifest by specific cellular movements that ultimately lead to the formation of three sheets of cells, called germ layers (Figure 1). Organisms with three germ layers are called triploblasts. Underlying these changes is a series of molecular signal calling in different areas of the embryo, leading to differential gene expression by various groups of cells in the embryo. These molecular changes determine not only patterns of cellular movements, but later specific structures that these cells will form.

The cellular movements that initiate gastrulation result in the formation of an indentation, called the blastopore, which marks the site at which migrating cells move into the interior of the embryo. In protostomes, the blastopore forms the mouth (*protostome* 'first mouth'), whereas in deuterostomes (*deuterostome* 'second mouth'), the blastopore typically forms the anus, with the mouth opening up later in development. In most spherical blastulae, the cells that indent or involute (migrate inward) through the blastopore form an interior sac that will ultimately comprise the lining of the gut. This lining represents the innermost germ layer, called the endoderm. The outer surface of the embryo constitutes the outer germ layer, or ectoderm. A sheet of cells between the endoderm and ectoderm forms the third germ layer, the mesoderm. Each of these germ layers gives rise to specific tissues as development proceeds.

Animals possessing large amounts of yolk, such as birds and reptiles, have adopted a different topographical form of gastrulation and germ layer formation. The essence of this mode of gastrulation is the formation of three flat layers of cells situated above a massive base of yolk. Instead of a blastopore, such embryos form a primitive streak through which cells migrate to form first the lowermost endodermal germ layer (sometimes called the hypoblast) and next the mesoderm, which is situated between the endoderm and the overlying ectoderm, which is often called the epiblast. Mammalian embryos, although not possessing significant amounts of yolk, nevertheless follow the pattern of their reptilian ancestors and form a gastrulation system similar to that of reptiles and birds.

One of the principal morphological manifestations of gastrulation in chordates is the formation of a condensation or rod of mesoderm, called chordamesoderm or the notochord, along the craniocaudal axis of the embryo. This structure, which is an evolutionarily conserved characteristic of all chordates, is a key to all subsequent development. Shortly after, or even during its formation, the notochord sends molecular signals to the overlying ectoderm, causing the ectoderm above it to thicken into a structure called the neural plate. The neural plate represents the primordium of the central nervous system. The signaling by the notochord to bring about this change is often called primary induction. (Induction is a process by which a signal from one tissue, called the inducer, brings about a change in the responding tissue that would not have occurred in the absence of the inductive signal. Inductive interactions are fundamental to the formation of most organs in the body.)

Shortly after its establishment, the neural plate undergoes a longitudinal folding into a cylindrical structure, called the neural tube. This process is called neurulation, and the embryo at this stage is often referred to as a neurula. As the neural tube is forming, two other events occur that are critical in shaping the overall form of the embryo. The first is the sequential formation of pairs of blocklike mesodermal condensations alongside the neural tube. These condensations, called somites, represent the principal morphological basis for segmentation of the vertebrate body, and their formation is the morphological manifestation of a strictly ordered expression of a family of genes called homeobox, or *Hox*, genes. *Hox* genes or their homologues are widely distributed throughout the animal kingdom, and there are striking parallels between these genes and their effects in both the invertebrates and vertebrates.

Another process of great developmental importance is the emigration from the neural tube of a population of cells, called the neural crest. Neural crest cells migrate throughout the body and form a wide variety of derivatives ranging from pigment cells and components of the peripheral nervous system, to bones of the head and connective tissue elements of the head, many glands, and parts of the heart and great vessels. It has been postulated that the cranial neural crest represents the evolutionary basis for the formation of the vertebrate face and feeding apparatus.

As the above changes are taking place, the body of the embryo is undergoing folding into a basically cylindrical structure, with an innermost tube (gut) lined by endoderm, an outer covering of ectoderm (epidermis of the skin), and a middle layer of mesoderm that will form muscle, skeletal elements, and the connective tissue stroma of many structures. In vertebrates, the nervous system, an ectodermal cylinder, becomes located on the dorsal side of the embryo, whereas in bilaterally symmetrical invertebrates, the nervous system is most prominently represented by a ventral nerve cord. After the cylindrical body has taken shape, the appearance of small buds of mesoderm covered by ectoderm indicates the formation of the appendages.

During this early period, corresponding to weeks 4–6 in human embryonic development, almost all organs in the body begin to form. Formation of almost any organ is the result of complex molecular signal calling, leading to one or more inductive interactions that produce the primordial elements of the organ in question. The first cells of an organ primordium are typically of an appearance common to most embryonic cells—stellate cells called mesenchyme or cuboidal or columnar epithelial cells. As the organ develops, these cells take on more specific shapes and functions through a process defined as differentiation.

In vertebrates, in particular, early organ primordia are often quite resistant to external perturbations. If some cells are damaged or removed, others take their place, and through its regulative properties, the organ forms normally. The more advanced the state of development, the less perfectly an organ primordium is able to restore its morphological and functional continuity following damage. It is a general rule in developing systems that in early periods more cells are capable of forming a structure than are needed. This accounts in part for the regulative ability of early embryos. As the embryo becomes older, its component cells become channeled into specific fates through a process commonly called determination. Differentiation represents the process by which already determined cells express their intrinsic genetic information as specific structures or functional entities.

The final stages in the development of the definitive shape of the individual are lumped into a process called morphogenesis. Morphogenesis is not a single process, but final form is attained in a variety of different ways in different structures. In the limbs, for instance, the formation of specific skeletal elements, muscles, and the digits themselves is the result of a large number of molecular signaling interactions that determine not only the course of differentiation of a specific cell but also the shape of the tissue or element in whose formation that cell takes part. In other structures, including limbs, mechanical function also plays an important role in the development of final form. Even external factors can have profound influences on development. Temperature, for instance, can influence the number of vertebrae in fishes or the sex of reptiles. Such factors are often termed epigenetic, because their influence is superimposed upon the genetic instructions that guide the fundamental construction of the structure.

[See also Canalization; Phylotypic Stages; Protostome-Deuterostome Origins.]

BIBLIOGRAPHY

Carlson, B. M. *Patten's Foundations of Embryology*. 6th ed. New York, 1996. An introductory textbook covering the basics of vertebrate embryology.

Carlson, B. M. *Human Embryology and Developmental Biology*. 2d ed. Saint Louis, 1999. A contemporary integration of human embryology and developmental biology.

Gilbert, S. F. *Developmental Biology*. 5th ed. Sunderland, Mass., 1997. A comprehensive approach to developmental biology with considerable emphasis on evolution.

Gould, S. J. *Ontogeny and Phylogeny*. Cambridge, Mass., 1977. One of the first modern books to take a combined look at evolution and development.

Hall, B. K. *Evolutionary Developmental Biology*. London, 1996. A contemporary integration of evolution and development from the perspective of a vertebrate-oriented biologist.

Kalthoff, K. *Analysis of Biological Development*. New York, 1996. An introductory developmental biology textbook with a strong treatment of invertebrates and molecular aspects of development.

Raff, R. A. *The Shape of Life: Genes, Development, and the Evolution of Animal Form*. Chicago, 1996. A contemporary integration of evolution and development from the perspective of an invertebrate-oriented biologist.

Torrey, T. W., and A. Feduccia. *Morphogenesis of the Vertebrates*. 3d ed. New York, 1971. A classic text that combines vertebrate embryology and comparative anatomy.

Wolpert, L., ed. *Principles of Development*. Oxford, 1998. An introductory text with a chapter on evolution and development.

— Bruce M. Carlson

Development and Ecology

The developmental sequence leading from a single-celled fertilized egg to a complex, multicellular individual is a most remarkable transformation. Considering the complexity of even the simplest developmental systems, it is clear that developmental processes are, to a large degree, buffered from external perturbations. Otherwise, environmental "bumps" might result in serious developmental mistakes. It is thus not surprising that for several centuries developmental biologists have focused on the most stable developmental systems and on developmental processes that minimize morphological variation among individuals in controlled laboratory situations. Indeed, until recently, morphological variation was thought to arise from embarrassing developmental defects.

Biologists are now recognizing the importance of phenotypic plasticity—the remarkable flexibility of many organisms to develop into different forms that will suit them best in the environment they will encounter. This perspective makes sense when we consider an organism developing in its natural environment. It may face an uncertain range of environmental conditions. For example, it may be a hot, dry summer or a cool, wet fall; there may happen to be few or many predators; or the habitat may be crowded or sparsely populated. In the face of such environmental variability, it is unlikely that one phenotype will do equally well in all possible environments. Rather, different phenotypes may be better suited to different conditions. In contrast to the earlier views, biologists are now appreciating that developmental processes can be remarkably responsive to an organism's environment, ecology, and life history. Van Valen (1973) summarized this: "A plausible argument could be made that evolution is the control of development by ecology."

Phenotypic variation can be produced by a variety of different developmental and genetic mechanisms, which have been split into the following categories. Genetic polymorphisms occur when phenotypic variation in a population arises from genetic differences between individuals. At the other end of the spectrum, polyphenism refers to different phenotypes that arise in response to different environments experienced by developing individuals. Between these two extremes, an individual's developmental trajectory may depend on both environmental conditions and its genotype. Such interactive relations are described by developmental reaction norms. Phenotypic plasticity appears to be nearly ubiquitous, occurring in all phyla of animals and plants. A few fascinating examples are described here.

Castes in Social Insects. Many social insects, such as ants and termites, show extreme phenotypic plasticity (Figure 1A). Female ant larvae are initially pluripotent—they are able to develop into any of the different forms, from small, medium, or large workers, soldiers, and queens. Such caste systems allow the efficient division of labor within a colony. The "decisions" that determine a particular individual's developmental pathway are influenced by the quality and quantity of the larval food, as well as by chemical cues and pheromones that reflect the social conditions in the ant nest. For example, if there are many soldiers, chemicals produced by the soldiers will suppress the development of more soldiers.

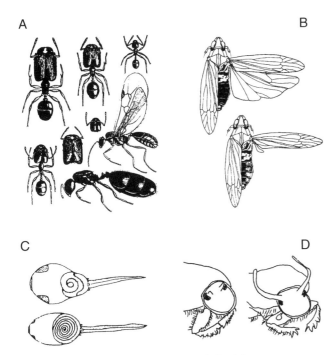

FIGURE 1. Examples of Developmental Plasticity.
(A) Morphological variation of different castes in the ant *Pheidole Kingi*. The male is the winged individual—all others are females. The queen is in the lower right; the other females range from the smaller, minor workers, through the larger major workers. (After Wheeler, 1910.) (B) Wing polyphenism in the planthopper *Prokelisia marginata,* typical of many insects. The nymphs develop into either the winged form (upper), which is a proficient flier and disperser, or the flightless form (bottom). (From Denno et al., 1985.) (C) The carnivorous morph (upper) and omnivorous morph (lower) of tadpoles of the New Mexico spadefoot toad (*Scaphiopus multiplicatus*). (From Pfennig, 1992.) (D) Hornless (left) and horned (right) phenotypes of male dung beetles *Onthophagus taurus.*   Sketched from photographs in Emlen and Nijhout, 1999.

This negative feedback loop creates a self-regulatory developmental system that ensures that appropriate numbers of different caste members are produced.

Wing Polyphenism. In many species of insects, individuals are able to develop into long-winged, flying forms or short-winged or wingless, flightless forms. For example, the planthopper (*Prokelisia marginata*) (Figure 1B) lives in estuaries along the Atlantic coast of the United States, where it feeds on the sap of marsh grass. Nymphs overwinter in stands of marsh grass along the coast. During the early spring, the density of planthoppers is relatively low, and the nutritional quality of the sap of the marsh grass is high. Life is good, and there is little reason to move; a relatively high proportion of individuals (especially females) develop into the flightless morph. However, densities of planthoppers increase rapidly, which results in severe overcrowding and a de-

crease in the nutritional quality of the sap. Meanwhile, patches of high-quality marsh grass develop inland along streams that flow into the estuaries. A higher proportion of individuals develop into the winged form and are able to migrate inland to the better food patches. This developmental flexibility represents an adaptive trade-off: when local resources are good, flightless individuals have saved energy that would have been spent on developing wings and flight muscles, and they can shunt these savings into higher reproductive rates; development of the wings costs more, but it allows individuals to follow the seasonal food supplies. The two factors that greatly influence planthopper fitness—nutritional quality of the food and crowding—are developmental triggers: high food quality suppresses wing development; crowding during the nymphal stage induces wing development.

Phenotypic Plasticity in Amphibian Larvae. Most frogs, toads, and salamanders live in water as tadpoles and on land as adults. The crucial breeding habitats can be very unpredictable. For example some desert spadefoot toads (*Scaphiopus multiplicatus*) breed in wetlands that form during monsoon rains. Depending on the rainfall, these can range from small, ephemeral puddles to large, longer-lasting ponds. The density of food, conspecific tadpoles, and predators are also variable and unpredictable. Mexican spadefoot tadpoles can develop into two different, environmentally induced forms (Figure 1C): carnivorous morphs eat fairy shrimp, have short intestines, large mouths and jaw musculature, and exhibit fast growth rates; omnivorous morphs eat detritus, have long intestines, relatively small mouths, and jaw musculature, and exhibit relatively slow growth rates. The development of the carnivorous morph is induced facultatively when shrimp are present, but tadpoles can revert to the omnivore morph when little prey is available. Because of intense competition among tadpoles for food, the performance of a particular morph also depends on the frequency of that morph in the pond. This flexible and reversible developmental program maximizes an individual's chances of successfully metamorphosing in the face of extremely unpredictable larval environments.

Horned Beetles. The males of many species of beetles have "horns" that are used in fights with other males while competing for females. The dung beetle (*Onthophagus taurus*) has been especially well studied. Beetles fly to fresh dung, where females dig tunnels under the dung, form round balls of dung, and roll them down into underground chambers, then lay an egg on each brood ball. The larva consumes the brood ball and metamorphoses in the underground chamber. Interestingly, horns do not come in all sizes, but rather represent an all-or-nothing developmental decision: larval males develop into either short-horned or long-horned forms

(Figure 1D). This developmental plasticity arises because of how these weapons are used by males in sexual competition to mate with females. Large-horned males attempt to guard a female's tunnel, and they use their horns to fight with other intruding males. Small-horned males are unable to compete successfully in these fights, but they use another "sneaking" strategy to successfully reproduce: they dig side tunnels that join the female's tunnel complex underneath the guarding male. Males with intermediate horns would not be able to compete effectively in either strategy, because they would lose in direct fights with males with longer horns, and they would be less mobile and slower than short-horned males in the sneaking strategy.

The developmental commitment to horn morphology is made during the larval stage. The growth rate and size of a larva depend on the nutritional quality and size of its brood ball, and activation of horn development occurs only if the larva has surpassed a certain size during development. In addition, the development of horns is costly and necessitates trade-offs with other adjacent, developing body parts. For example, beetles that develop long horns on their heads develop small eyes, because the developing horn structures take away resources from the adjacent developing eyes. The developmental reaction norms relating body size and horn size can be quickly shifted by selection. These developmental and genetic constraints associated with weapon production influence the patterns of evolution and diversification in beetles.

These examples illustrate some of the remarkable flexibility of developmental processes that can give rise to adaptive phenotypic plasticity. Developmental programs can be sensitive to a variety of environmental and ecological conditions, such as temperature, humidity, photoperiod, food quality and abundance, predators, and density of conspecifics. Understanding the causes and consequences of phenotypic variation is one of the most exciting fields in biology, requiring an integration of all levels of biological inquiry, including molecular and cellular levels of development, ecology, behavior, life history, patterns of speciation, and evolution. Some of the outstanding questions that remain are How do developmental events and the timing of those events influence phenotypic plasticity? In species that use environmental cues as induction triggers, how are those cues detected and monitored? How and why are those cues originally "captured" and integrated into developmental systems? When and why is phenotypic variation favored? What conditions favor genetic polymorphisms versus environmentally induced polyphenisms? How does phenotypic plasticity and the development machinery underlying it influence speciation and evolution? How do developmental constraints influence the possible range of evolutionary outcomes? These issues will be at the forefront of development, ecology, and evolution for some time to come.

[See also Life History Stages; Phenotypic Plasticity.]

BIBLIOGRAPHY

The July/August 1989 issue of *BioScience* (Vol. 39, No. 7) is devoted to issues of phenotypic variation, development, and ecology and contains many fine articles.

Denno, R. F., L. W. Douglass, and D. Jacobs. "Crowding and Host Plant Nutrition: Environmental Determinants of Wing-Form in *Prokelisia marginata*." *Ecology* 66 (1985): 1588–1596.

Emlen, D. J. "Alternative Reproductive Tactics and Male-Dimorphism in the Horned Beetle *Onthophagus acuminatus* (Coleoptera: Scarabaeidae)." *Behavioral Ecology and Sociobiology* 41 (1997): 335–341.

Emlen, D. J. "Integrating Development with Evolution: A Case Study with Beetle Horns." *BioScience* 50.5 (2000): 403–418.

Emlen, D. J. "Costs and the Diversification of Exaggerated Animal Structures." *Science* 291 (2001): 1534–1536.

Emlen, D. J., and H. F. Nijhout. "Hormonal Control of Male Horn Length Dimorphism in the Dung Beetle, *Onthophagus taurus* (Coleoptera: Scarabaeidae)." *Journal of Insect Physiology* 45 (1999): 45–53.

Greene, E. "Phenotypic Variation in Larval Development and Evolution: Polymorphism, Polyphenism, and Developmental Reaction Norms." In *The Origin and Evolution of Larval Forms* edited by M. Wake and B. Hall, pp. 379–410. New York, 1999.

Pfennig, D. W. "Polyphenism in Spadefoot Toad Tadpoles as a Locally Adjusted Evolutionarily Stable Strategy." *Evolution* 46 (1992): 1408–1420.

Van Valen, L. "Festchrift." *Science* 180 (1973): 488.

Wheeler, W. M. *Ants: Their Structure, Development and Behavior.* New York, 1910.

— ERICK P. GREENE

DEVELOPMENTAL SELECTION

Compare a tree with a neighboring building. There are profound differences between these complex structures. The arrangement of the various subunits of the building is repeated and predictable. In a tree, in contrast, there is a marked random element in the details of the individual branches, even though the overall form is functional and typical of the species and the local environment. If part of a tree is hidden, there is no way of reconstructing the precise details of its minor branches. This is common to most trees, but it is not a necessary, unavoidable characteristic: members of the genus *Araucaria*, for example, have a predictable, repeatable, form.

These examples point to two alternative ways a complex structure can be specified. A building is formed by a strict plan in which each stage of construction or development is the necessary consequence of previous events and immediate conditions. All the details of the final structure are specified and, in a sense, preformed. The orderly aspects of the whole can be no more predictable than those of its components. It is generally taken for granted that biological development is specified in some comparable program. This might well be due to characteristics of intuitive human thought, which rejects the uncertainty implied by random processes, assuming that a plan, a planner, or a *"watchmaker"* underlies all complex structures. Yet the form of trees and many other complex biological structures suggests that precise specification of all stages of development could not be the whole story. Instead, the overall form of a tree is specified more strictly than its components. This need not be mysterious: a thermostat is a simple device that selects a state without specifying the path by which it is reached and maintained. An ant's nest is another, more complex example: there is a large random element in the behavior of individual ants, yet the activity of the nest as a whole is functional and predictable. Could biological development be based on similar principles?

Mechanisms of Tree Morphogenesis and Their Implications. A reconstruction of tree development is a prerequisite for understanding the specification of its form. This reconstruction is relatively easy in trees and other plants, because they develop throughout their lives and are composed of large organs of all possible ages. Buds are always present in large numbers; responses to pruning and other damage demonstrate that they and small branches have developmental potentials that generally remain unrealized. Furthermore, most branches that do start growing eventually become dormant, and many are eventually shed. This is seen by comparisons of the degree of branching in young and mature regions and by the debris under most trees.

Experimental work with trees is necessarily limited, but evidence from herbaceous plants suggests the following outline. The various branches and buds (potential branches) interact and compete (Sachs et al., 1993). A developing branch must therefore be the source of signals (one of them being the phytohormone auxin) that inform the rest of the system of its presence, size, developmental rate, and light environment (Sachs, 1991). The branches that are the strongest producers of such signals inhibit their competitors and increase their own role in the canopy. The "success" of a branch, its continued development and contribution to the form of the tree, depends, at least partially, on its being the first to occupy a space that can be occupied by branches coming from different directions.

A decision in the competition between the developmental alternatives must be revised following the continued development of the tree or its neighbors, which necessarily change the role of any given branch. Damage to a developing branch is another common reason for the need for a review of branch relations. Yet the position of an unperturbed branch is fairly stable, and deterioration of a branch, when it occurs, is gradual. It follows that the relations between branches must depend not only on present conditions but also on past conditions. Such conditions include the stability of vascular supplies and the dependence of hormone production on the size of the branch.

It follows that genetic information does not specify the details of individual branch development. Genetics does, however, specify the formation and responses to the signals that mediate the competition between branches, and these parameters can determine the overall relations between branches, a balanced state that is reached in varied ways. There is an element of chance as to which branches belong to the limited number that is selected from among the excess that initially began to develop.

A tree was used to explain a general concept, and it was chosen because it is readily observed. The development of most trees is an example of developmental or "epigenetic" selection (Sachs, 1988), in which biological form is repeatedly "discovered" by a selection among processes that have a random component. Developmental selection concerns the relations between the genotype and the phenotype; it should therefore be important for understanding evolution.

Biological Evidence for Developmental Selection. *Developmental reversals.* The most direct evidence for developmental selection in trees is the nonrandom shedding of branches. There are only a few other examples of such reversals of development. The best known are neuronal connections. The number of neurons that connect to a muscle greatly exceeds the requirements for mature function. Most of these connections deteriorate, and only those that carry sufficient impulses survive. Function, rather than a strict plan, is thus used as a criterion for selecting a reliable though variable mature structure. This process, known as selective stabilization (Changeux, 1986) or neural Darwinism (Edelman, 1987), accounts for the stochastic element of neuronal connections, which differ even between identical twins. It is supported by experimental evidence; for example, paralysis leads to an increase rather than a decrease of neuronal connections.

At a subcellular level, microtubules, essential for cell function and divisions, assemble in a random, exploratory mode. Only those that find the "right" objects, such as chromosomes during mitosis, are stabilized. The rest depolymerize, and their building blocks are reused to initiate new microtubules (Gerhart and Kirschner, 1997; Kirschner and Mitchison, 1986). This allows for reliable structures and for regeneration following wounds. Another example is the development of stomata on the surface of some leaves. As many as half the stomatal initials fail to mature, and their cells become regular epidermal cells. As a result, the pattern of the mature stomata can be more spaced, or orderly, than that of all the stomata that are initiated (Kagan and Sachs, 1991). Failure of maturation presumably depends on developmental signals that indicate the presence and distances to other developing stomata.

The evidence of variable development. The reversal of developmental processes may be rare, possibly because of the cost of starting several developmental pathways, of which only a few succeed. Trees, however, show how the combination of randomness and selection produces the same functional structure in varied ways, with the final mature form specified more strictly than the details. In fact, variable details are a common feature of mature biological structures. Thus, blood vessels of vertebrates and trachea of insects display an "exploratory" rather than a precisely determined behavior (Gerhardt and Kirschner, 1997). The details of leaf venation provide another example (Sachs, 1991).

The most common evidence for developmental variability is found in embryonic cell lineages. The behavior of cells during the generation of vertebrate limbs, for example, is exploratory rather than genetically determined (Gerhardt and Kirschner, 1997). It is in chimeras, organisms composed of cells of varied genetic structure, that lineage variability is most readily seen. Where stem cells differ in pigment-forming abilities, their lineage is apparent in mature organs. Common examples are the garden chimeras of many species in which some of the tissues are albino, unable to synthesize chlorophyll. The variable tissues' origins are apparent even when different halves of a leaf are compared (Sachs, 1991). Chimeric mice, made by experimental implantation into an embryo of a cell with a distinct genetic constitution, have variable patches of colored fur. The contributions of embryonic cells are limited to "developmental compartments," but within compartments the products of a stem cell have no predictable, repeated form. Variable lineages are the rule in vertebrates and "higher" plants. Relatively predictable cell lineages are found in simple organisms, such as nematodes and mosses. Developmental variation is not characteristic of "primitive" systems.

Genetic Specification of the Outcome of Variable Development. A strict program, as the term is commonly understood, is not the only way in which development is specified. Instead, development reaches a predictable outcome by continuous modulation, with functional form produced by selection from varied, redundant paths (Sachs, 1994). Variable events are actually used to "find" the specified form, and they might even be essential in preventing the system from following an unfavorable path.

This means that the "correct" outcome can be recognized and selected. In trees, for example, form can be specified by a balanced feedback of developmental signals, or phytohormones (Sachs, 1991). At the tissue level, leaf vein patterning can be caused by similar principles: a positive feedback between cell polarization and the transport of the signal that induces this polarization results in a gradual canalization of differentiation. The resulting pattern is variable yet free of functional mistakes (Berleth and Sachs, 2001; Sachs, 1991). This can be generalized: the formation and consumption of developmental signals could be coupled to the size and development of the alternative structures. The effects of the signals are programmed so that whenever the system is unbalanced, whenever one factor is present in excess, the developmental processes that would tend to redress the situation are enhanced. Developmental changes cease when the system reaches an expected size and a balanced state. This paradigm is supported by regeneration phenomena, which are evidence for excess developmental potentials, for interactions between the parts of an organism, and for the ability of development to proceed from varied states in accordance with the state of these interactions. Furthermore, the loss of controls of the formation and sensitivity to such developmental signals is known to be a major cause of tumor development in both animals and plants.

This brings us to the question of the specification of the phenotypic form by the genotype. This is the question of how one-dimensional molecular information can specify the temporal and three spatial dimensions of macroscopic biological form. Although it is certainly a central biological question, its conceptual basis is rarely discussed, and available paradigms have serious drawbacks (Nijhout, 1990). The preceding discussion suggests two types of genetic specification, both essential rather than mutually exclusive. The first is the specification of programs in which the expression of a gene depends on previous gene expression, neighboring tissues, and environmental conditions. Research on this basis has had great success, but it might be ignoring important processes. The second type of specification causes an excess of competing developmental pathways governed partly by genetics and partly by chance, and a program to select only those pathways that lead to a specified form (Changeux, 1986; Edelman, 1987). The role of this alternative is indicated by the characteristic redundancy of genetic elements required for development (Cooke et al., 1997).

One consequence of developmental selection is the possibility of reaching a functional phenotype in varied, redundant ways. This is wasteful, perhaps, but it makes the system robust, correcting many of the "mistakes" that must occur during the complex processes of development. Such corrections could be expected to apply to the effects of many mutations. This could allow accumulation of genetic variability that is not expressed, or is expressed only under extreme, unusual conditions. This hidden variability could be the basis for unexpected, abrupt, phenotypic changes (Agur and Kerszberg, 1987).

[*See also* Pattern Formation.]

BIBLIOGRAPHY

Agur, Z., and M. Kerszberg. "The Emergence of Phenotypic Novelties through Progressive Genetic Change." *American Naturalist* 129 (1987): 862–875. A simple model demonstrating how novelties could arise from accumulated genetic change that have low penetrance.

Berleth, T., and T. Sachs. "Plant Morphogenesis: Long-distance Coordination and Local Patterning." *Current Opinion in Plant Biology* 4 (2001): 57–62. A short review of the available facts and theory of developmental signals that could integrate events at all levels and throughout a plant.

Changeux, J.-P. *Neuronal Man: The Biology of Mind.* Translated by L. Garey. Oxford, 1986. Chapter 7 reviews the facts and concept of the selective stabilization of neuronal connections. Illustrations and clear discussion make this a useful introduction to the topic.

Cooke, J., et al. "Evolutionary Origins and Maintenance of Redundant Gene Expression during Metazoan Development." *Trends in Genetics* 13 (1997): 360–364. A short review of the evidence that the genetic information that is essential for development tends to be redundant (the same is not true for housekeeping genes). Includes a discussion of ways in which this redundancy could be accounted for.

Dawkins, R. *The Blind Watchmaker.* New York, 1986. This clear exposition emphasizes the evolution of complex novel structures. This book (and others by the same author) considers the neglected conceptual problem of the relation between genetic information and its outcome, the functional phenotype.

Edelman, G. M. *Neural Darwinism—The Theory of Neuronal Group Selection.* New York, 1987. As in Changeux's book (which is much easier for a nonspecialist), the concept of developmental selection is presented in relation to the generation of neuronal connections. The evidence for such development and its consequences for evolution are considered. The emphasis is on brain function rather than the development of biological form.

Gerhart, J., and M. Kirschner. *Cells, Embryos, and Evolution.* Oxford, 1997. A general text of biological development (as in most other texts of this subject, it ignores plants). Unlike other texts, it includes (Chapter 4) a review of the exploratory behavior of biological systems. It also emphasizes the relation between animal development and evolution, including the concept of evolvability.

Kagan, M., and T. Sachs. "Development of Immature Stomata: Epigenetic Selection of a Spacing Pattern." *Developmental Biology* 146 (1991): 100–105. The spacing of stomata on leaves is one of the simplest biological patterns. Here a case is described in which measurements of the degree of order show that much of it appears during, rather than preceding, the development of stomata.

Kirschner, M., and J. Gerhart. "Evolvability." *Proceedings of the National Academy of Sciences USA* 95 (1998): 8420–8427. A review of ways in which development could both allow for and restrict evolution.

Kirschner, M., and T. Mitchison. "Beyond Self-Assembly: From Microtubules to Morphogenesis." *Cell* 45 (1986): 329–342. A review of the microtubule assembly, important both because of their essential roles and as an example of a bridge between macromolecular and cell organization.

Meins, F., Jr., and M. Seldran. "Pseudodirected Variation in the Requirement of Cultured Plant Cells for Cell-Division Factors." *Development* 120 (1994): 1163–1168. Demonstrates an ignored possibility, that an inherited variation that arises at random but in relatively high frequencies can be combined with internal selection between cells. This could result in predictable tissue differentiation.

Nijhout, H. F. "Metaphors and the Role of Genes in Development." *BioEssays* 12 (1990): 441–445. A discussion of the difficult and neglected problem of the relations between genetic information and the development of complex forms. Includes the difficult conceptual questions of the meaning of genes being causes of development and the need for new paradigms.

Raff, R. A. *The Shape of Life.* Chicago, 1996. Reviews the developmental basis of evolution, though not the point of view developed here. (As usual, it ignores plants, which must be, and always were, most of the living biomass.)

Sachs, T. "Epigenetic Selection: An Alternative Mechanism of Pattern Formation." *Journal of Theoretical Biology* 134 (1988): 547–559. A short, nonmathematical consideration of the possibility that development can resemble Darwinian processes.

Sachs, T. *Pattern Formation in Plant Tissues.* Cambridge, 1991. A review of various developmental processes that generate plant form. This is a personal rather than an accepted view, emphasizing developmental selection.

Sachs, T. "Variable Development as a Basis for Robust Pattern Formation." *Journal of Theoretical Biology* 170 (1994): 423–425. A short summary of the evidence for the fact, commonly ignored, that there is a large variable element to biological development.

Sachs, T., et al. "Plants as Competing Populations of Redundant Organs." *Plant Cell and Environment* 16 (1993): 765–770. Develops the concept that plants can be both individuals and systems of competing organs.

— TSVI SACHS

DIATOMS

The earliest mention of diatoms in a significant work on evolution was in Charles Darwin's *Origin of Species* (1859). Darwin wrote that "[f]ew objects are more beautiful than the minute siliceous cases of the diatomaceae: were these created that they might be examined and admired under the higher powers of the microscope?" Diatoms are indeed beautiful. Once a common object for observation by Victorian microscopy hobbyists, diatoms are now known to help create a world hospitable to human life. Consequently, knowledge of diatoms is important to several applied sciences. Living and fossil diatoms also interest basic scientists because of their unusual and beautiful shapes, their ecological functions, and the opportunities they present to unlock fundamental questions about life on earth.

Diatoms belong to a group of organisms called stramenopiles, which include organisms from giant kelp to some single-celled organisms formerly classified as fungi.

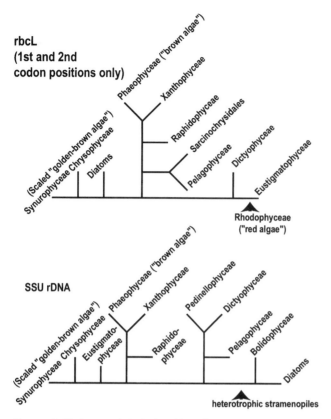

FIGURE 1. Phylogenetic Relationships of Major Autotrophic Protistan Stramenopiles (Heterokonta) Based on Ribulose-1,5-Bisphosphate Carboxylase/Oxygenase (rbcL) Data and Small Subunit Ribosomal DNA (SSU rDNA) Data.
Terminal taxa are reduced to class and redrawn from Daugbjerg and Andersen (1997) and Guillou et al. (1999), respectively. The trees are drawn as unrooted networks to facilitate comparison. Roots of each phylogenetic tree is indicated by arrows. The major result is that there is little similarity between the results of these two major studies.
Drawing by Edward Theriot.

The photosynthetic stramenopiles, including diatoms, are now thought to be a single branch on the tree of life and together are called Heterokonta or chromophytes. [*See* Algae.] The term *heterokont algae* is also used, but the reader should be aware that algae is an ecological grouping of nonvascular photosynthetic organisms that typically live in water. Organisms called algae occur in many places in the tree of life; there is no unique evolutionary group called algae. Morphological data and different kinds of DNA sequence data support different hypotheses about the closest relatives of diatoms among the Heterokonta (Figure 1). The situation will probably remain contentious for several years as much morphological and molecular data remain to be gathered and many microscopic organisms are still unknown to science.

Although there is uncertainty about the closest rela-

tives of diatoms, there has been little doubt that diatoms form an independent evolutionary branch since their discovery over 100 years ago. Many organisms use silica in their exterior covering, but usually only as scales (e.g., the synurophyte *Mallomonas*). Only the diatom vegetative cell wall (frustule) has strongly differentiated plates (valves) at poles of the frustule. Each half of a frustule is composed of a valve and a set of associated cingula (Figure 2). Each cingulum is a strip of silica wrapped around the cell. The end of each cingulum usually at least partly overlaps the other, and so the shape it describes is somewhat flexible. The bands and valves may each be ornamented, the valves typically more so (Figure 3). The valve of the half-frustule overlapping the other is called the epivalve. The valve belonging to the other is called the hypovalve. Data on structure of chloroplasts, the mitotic spindle, photosynthetic pigments, and genetic data all continue to support the notion that diatoms form a unique evolutionary branch.

Diatoms increase in number by cell division. As the diploid nucleus undergoes standard mitotic division, new valves and cingula are formed. Each new cell keeps half the original frustule (valve and associated cingula)

FIGURE 2. Generalized Life Cycle of the Diatoms.
Drawing by Edward Theriot.

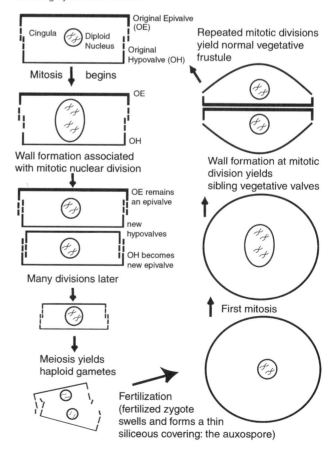

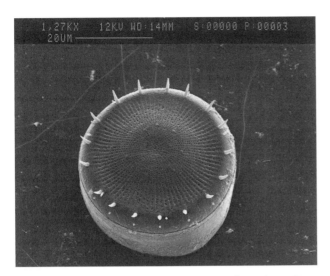

Figure 3. A Scanning Electron Micrograph of *Stephanodiscus yellowstonensis* Theriot and Stoermer.
The area of the frustule occupied by the smooth and unornamented cingula is marked by two arrows. One of the two valves is visible and is seen to be ornamented by various kinds of projections and pores. Some diatoms have even more ornate valves and some have extremely ornate cingula. This is a more typical diatom.   Courtesy of Edward Theriot.

and forms a new half-frustule for itself inside the old half (Figure 2). The old epivalve is still an epivalve, but the old hypovalve is now an epivalve and will remain so in all future cell divisions. The shape of the new cell is thus thought to be inherited in part by the physical constraints of the old valves. However, the new cell is not always a slightly smaller rendition of the old because cingula are somewhat flexible. Some species seem to reduce their cell length as they undergo vegetative size reduction, but cell width increases. Some cell lineages have been kept in culture for many generations without changing in size. Others sometimes have abrupt changes in size. Thus, this important part of basic diatom biology remains unexplained.

Size is regenerated when the diatom undergoes sexual reproduction. Individual cells on the small end of the size range undergo meiotic divisions to form haploid gametes, at least sometimes after receiving some sort of environmental cue (e.g., osmotic shock). The gametes fuse to create a diploid zygote, which swells and then develops a siliceous coating called an auxospore. The auxospore of most species is often shaped like a basketball or football. With each mitotic nuclear division of the auxospore, a typical vegetative valve is formed. Once a complete frustule is formed, the cycle begins anew (Figure 2).

Diatoms traditionally are classified into two groups: (1) the centric diatoms, Centrales, which typically are round, and (2) the pennate diatoms, Pennales, which typically are long and narrow. However, some pennate diatoms also have the primitive characteristics associated with centric diatoms, and much evidence supports an evolutionary grade from centrics to pennates. All diatoms are diploid in vegetative phase and undergo meiosis to form gametes. All centric diatoms are oogamous. In meiosis, some vegetative cells become eggs and others become meiospores (flagellated gametes). One pennate genus, *Rhabdonema*, is oogamous, but all other pennates have gametes that are completely or nearly identical (isogamous). All centric diatoms have small discoid chloroplasts. Some pennate diatoms also have small discoid chloroplasts, although most have one or a few platelike chloroplasts. The latter also have a raphe, a pair of longitudinal slits in the primary rib. Recent molecular studies continue to support the notion that pennate diatoms evolved from centric diatoms; thus, the higher level classification of diatoms is in need of revision.

Far too little is known about diatoms to be sure exactly how many species there are. A count is complicated by the fact that many authors have used different kinds of species concepts, some explicit but many not, and that many descriptions are based solely on relatively crude observations in the light microscope alone. Many modern species concepts do not distinguish subspecific taxa, but consider the species in some way to be the smallest evolutionary and/or identifiable unit, roughly corresponding to all of the various named taxonomic units diatomists typically use (species, subspecies, variety, and form). There are about 100,000 such names in records at the diatom herbarium of the Academy of Natural Sciences of Philadelphia. Even with various applications of species concepts, based on the fact that we are still describing 600 to 700 new names per year, and are just beginning to use more sensitive tools such as statistical analysis of shape and genetic data, it would not be surprising if the number of actual fossil and living diatom species was far greater than 100,000.

The great species diversity of diatoms is related to apparent adaptations to the wide variety of habitats in the aquatic environment. All photosynthetic diatoms require dissolved inorganic nutrients, including carbon dioxide, nitrogen, phosphorus, silicon, and a variety of metals and vitamins. Concentrations of these compounds vary from water body to water body and from season to season. A few heterotrophic diatoms require complex organic molecules from the environment as an energy source. Some diatoms are suspended in the water column for much or all of their life history, making up part of the ecological assemblage known as phytoplankton. Other diatoms are attached to various substrates, as part of the ecological assemblage known as periphyton. With increasing numbers of diatoms in living culture, the study of resource physiology of individual diatom

YELLOWSTONE LAKE DIATOMS

Yellowstone Lake, in Yellowstone National Park, and diatoms together have yielded an extraordinary record of the evolution of a new diatom species. Yellowstone Lake was completely covered with glaciers until about 14,000 years ago. As the lake began to thaw, it was invaded by several diatoms common to temperate North America today, including *Stephanodiscus niagarae* Ehrenberg. However, unlike other lakes across the continent, including lakes right next door, Yellowstone Lake spawned a new species, called *S. yellowstonensis* Theriot and Stoermer. Fossil samples taken from the sediments of Yellowstone Lake at intervals as small as forty years have now revealed that *S. niagarae* continuously and gradually evolved into *S. yellowstonensis* in less than 4,000 years. Field observations and culture experiments are showing that *S. yellowstonensis* is better physiologically at taking up dissolved nitrogen than is *S. niagarae* and that Yellowstone Lake is lower in dissolved nitrogen than are nearby lakes. Thus, part of the evolution of *S. yellowstonensis* seems to be adaptation to a relatively low nitrogen environment. Another unusual aspect of the study is that comparative analysis of morphology (phylogenetic analysis) had predicted that the ancestor to *S. yellowstonensis* would have the morphology of *S. niagarae*.

—EDWARD THERIOT

species is adding to our understanding of the true environmental optima of diatoms and also helps explain the adaptations of diatoms to new environments (see Vignette).

Diatoms are among the most important photosynthetic organisms on earth. They produce at least 20–25 percent of the net free oxygen produced by plants. Even in highly productive salt-marsh systems, where grasses form the most abundant visible part of the food web, diatoms account for as much as half of the carbon flowing through the system. This is possible because of their rapid reproductive rates and high food quality.

The complex relationship between diatom distribution and the environment makes diatoms useful to other fields of science. Diatoms are used to measure effects of eutrophication, acid rain, and natural variation in climate. Diatoms enter the lungs and bloodstream of drowning victims and are used to determine where an individual drowned. Diatom frustules preserve well enough to form thick deposits on the bottoms of oceans and lakes, called diatomites. Diatomites are mined, and the diatomaceous earth is used in abrasives, filters, pest control (as a physical barrier to insects), and animal feeds (as an adsorbent), and as a dulling agent in enamel paints. Diatoms are also used as guide fossils in geology and the mineral industry, particularly in the exploration for oil.

In summary, diatoms form a crucial link in the biosphere between the mineral and biological worlds. Their varieties of form and function link them to various pursuits of both practical and intellectual interest. A well-supported hypothesis of their closest relatives still remains elusive. Although the general outlines of the phylogenetic tree of diatoms is well corroborated (centrics giving rise to araphid pennates, which gave rise to raphid pennates), details are still not well understood, and the classification has yet to catch up with what is now known about relationships between diatoms. Despite over a hundred years of study, there is still much to learn about even the most basic aspects of diatom biology and of the importance of diatoms to the world.

[*See also* Paleontology.]

BIBLIOGRAPHY

Charles, D. F., R. W. Battarbee, I. Renberg, H. van Dam, and J. P. Smol. "Paleoecological Analysis of Lake Acidification Trends in North America and Europe Using Diatoms and Chrysophytes." In *Biological Monitoring of Freshwater Ecosystems*, edited by S. A. Norton, S. E. Lindberg, and A. Page, pp. 233–293. Boca Raton, Fla., 1989. An excellent example of the use of diatoms in gauging the effects of acid rain.

Daugbjerg, N., and R. A. Andersen. "A Molecular Phylogeny of the Heterokont Algae Based on Analyses of Chloroplast-encoded rbcL Sequence Data." *Journal of Phycology* 33 (1997): 1031–1041. A good recent example of molecular data supporting one version of diatom relationships in which diatoms are considered closely related to siliceous scaled algae (chrysophytes and synurophytes).

Fritz, S. C., S. Juggins, R. W. Battarbee, and D. R. Engstrom. "Reconstruction of Past Changes in Salinity and Climate Using a Diatom-based Transfer Function." *Nature* 352 (1991): 52–54. An excellent example of the use of diatoms to infer changes in climate using statistical models.

Guillou, L., M.-J. Chrotiennot-Dinet, L. K. Medlin, H. Claustre, S. Loiseaux-de Goer, and D. Vaulot. "*Bolidomonas*: A New Genus with Two Species Belonging to a New Algal Class, the Bolidophyceae (Heterokonta)." *Journal of Phycology* 35 (1999): 368–381. A good recent example of molecular data that contradict the traditional concept that diatoms are closely related to siliceous scaled algae such as chrysophytes and synurophytes.

Hoagland, K. E., S. C. Roemer, and J. R. Rosowski. "Colonization and Community Structure of Two Periphyton Assemblages, with Emphasis on the Diatoms (Bacillariophyceae)." *American Journal of Botany* 69 (1982): 188–213. A classic reference on the three-dimensional nature of diatom periphyton communities.

Kilham, S. S., E. C. Theriot, and S. C. Fritz. "Linking Planktonic Diatoms and Climate Change Using Resource Theory in the Large Lakes of the Yellowstone Ecosystem." *Limnology and Oceanography* 41 (1996): 1052–1062. An excellent example of using diatoms to infer climate change on the basis of experimental and field inferences about diatom competitive ability and physiological ecology.

Leipe, D. D., P. O. Wainright, H. J. Gunderson, D. Porter, D. J. Patterson, F. Valois, S. Himmerich, and M. L. Sogin. "The Stramenopiles from a Molecular Perspective: 16S-like rRNA Sequences from *Labyrinthuloides minuta* and *Cafeteria roenbegensis.*" *Phycologia* 33 (1994): 369–377. Molecular support for the placement of diatoms within stramenopiles.

Medlin, L. K., D. M. Williams, and P. A. Sims. "The Evolution of the Diatoms (Bacillariophyta): 1. Origin of the Group and Assessment of the Monophyly of Its Major Divisions." *European Journal of Phycology* 28 (1993): 261–275. A review of morphological and molecular data relevant to the internal classification of diatoms.

Patterson, D. J. "Stramenopiles: Chromophytes from a Protistan Perspective." In *The Chromophyte Algae: Problems and Perspectives*, edited by J. C. Green, B. S. C. Leadbeater, and W. L. Diver, pp. 357–379. Oxford, 1989. The early description of the stramenopiles.

Round, F. E., R. M. Crawford, and D. G. Mann. *The Diatoms, Biology and Morphology of the Genera.* Cambridge, 1990. Necessary for anyone with a serious interest in diatoms. Much information about diatom biology.

Stoermer, E. F., J. A. Wolin, C. L. Schelske, and D. J. Conley. "Siliceous Microfossil Succession in Lake Michigan." *Limnology and Oceanography* 35 (1990): 959–967. A review of the effects of phosphorus addition to the Great Lakes, especially Lake Michigan, as inferred from diatoms in lake sediments.

Sullivan, M. J., and C. A. Moncreiff. "Primary Production of Edaphic Algal Communities in a Mississippi Salt Marsh." *Journal of Phycology* 24 (1988): 49–58. An excellent example of the incredible importance of diatoms in the food web. A demonstration that diatoms' importance to the food web far exceeds their observed biomass.

Theriot, E. "Clusters, Species Concepts and Morphological Evolution of Diatoms." *Systematic Biology* 41 (1992): 141–157. The only example of speciation in diatoms from both cladistic and paleontological perspectives. A demonstration of extremely rapid morphological evolution.

Werner, D., ed. *The Biology of Diatoms.* Botanical Monographs, vol. 13. Berkeley, 1977. An older, but still valuable, text on the general biology of diatoms.

— EDWARD THERIOT

DINOSAURS

Dinosauria was first recognized as a group and named in 1842 by the great British anatomist Sir Richard Owen. The root, *deinos*, is commonly translated as "terrible," but in his original publication Owen chose the translation "fearfully great." To Owen, the name *dinosaur* referred to fearfully great saurian reptiles, known only from the fossilized bones of huge extinct animals, unlike anything alive today. But since that time, perhaps no other name in the history of biology has changed so much in its meaning.

Fueled by science fiction writers and Hollywood, the public still commonly embraces Owen's concept of dinosaurs as huge extinct behemoths. The name has even become an adjective used in many different languages to malign anything or anyone that is hopelessly out of date, maladapted, unintelligent, slow, inertia-bound, or obsolete. But in the wake of Charles Darwin's theory of evolution, scientists have come to recognize that ancestry, rather than outward appearance or popular conception, is what underlies biologically meaningful names. Darwin's theory was published in 1859, seventeen years after Owen coined the word *dinosaur*, and ever since scientists have been reexamining the meaning of names and the compositions of groups that were recognized through pre-Darwinian eyes. Owen's concept of Dinosauria has changed more profoundly than that of almost any other lineage.

From a modern evolutionary point of view, anything born to a dinosaur is a dinosaur, whether it is large or small, slow or fast, extinct or alive today. This simple idea is controversial when we consider the ancestry of birds. A steadily growing body of evidence indicates that birds are the living descendants of extinct Mesozoic dinosaurs. If this is true, then birds are dinosaurs, only some dinosaurs were fearfully great, and only some dinosaurs—the big ones—went extinct about sixty-five million years ago. From a modern evolutionary perspective, only some dinosaurs fit the stereotype, and the rest (the birds) are among the most active and intelligent creatures alive today.

The First Dinosaurs. The oldest known dinosaur fossils are *Herrerasaurus* and *Eoraptor*, which are from the Late Triassic period (roughly 230 million years ago) of Argentina. Slightly younger relatives of *Herrerasaurus* have been found in Texas and Arizona. Even these oldest known fossils are specialized in various ways, which suggests that the ancestral dinosaur species must have lived still earlier, in the Middle or Early Triassic period. But even if Dinosauria originated at the very beginning of the Triassic period, as far back as 245 million years ago, the appearance of this popular icon came only after 90 percent of the history of life had passed.

Although fossils from the very early or ancestral dinosaur species have not yet been found, *Eoraptor*, *Herrerasaurus*, and other Triassic fossils are sufficiently close to offer an image of what that ancestor probably looked like. It was rather small, weighing about the same as an adult human, with a long, narrow head and a pointed snout. Its jaws were lined with sharp serrated teeth, and its mouth extended from ear to ear. It had large eyes and probably had excellent color vision. It was habitually bipedal—meaning that it walked on its hind legs—with its knees turned in against the body, giving it an erect posture in marked contrast to the sprawling gait of other contemporary reptiles. It was probably one of the fastest and most agile creatures alive at that time. Because the head was held high off the ground on a long, flexible neck, its senses could be very rapidly directed over wide fields to quickly locate and track potential prey. Its hands were freed from their

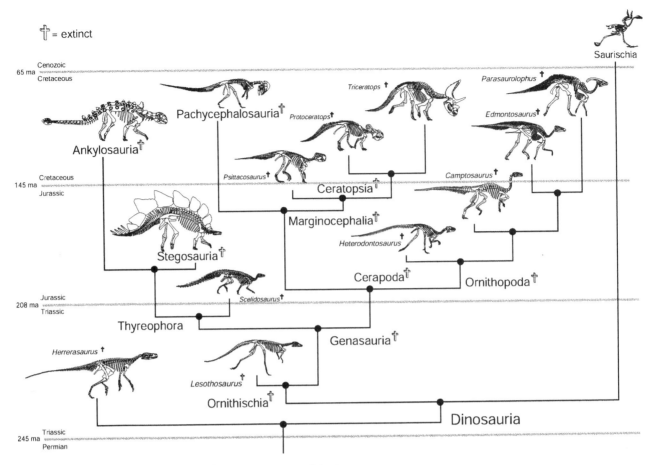

FIGURE 1. Phylogeny of Ornithischian Dinosaurs.   Courtesy of Tim Rowe.

role in locomotion and were modified for grasping. It was probably a deft predator that pursued a range of small prey items, from insects to animals larger than itself. From such an ancestor, we can trace a tremendous diversity of descendants, all of which belong in one of two major dinosaurian lineages.

Ornithischians. One of the two principal dinosaurian lineages is Ornithischia, which was named in reference to what early paleontologists thought to be a birdlike pelvis. Ornithischians are distinguished by teeth and jaws designed for tearing and grinding plants; their ribs and pelvis housed an enlarged digestive tract to accommodate large volumes of vegetation. A new bone, the predentary, appeared at the front of the ornithischian lower jaw, probably for cropping vegetation and stripping foliage off woody branches. In all but the oldest ornithischians, a horny beak rimmed the front of the mouth, while the cheek teeth are inset from the margins of the mouth and were probably covered by fleshy cheeks that assisted chewing (Figure 1).

Ornithischia includes highly divergent lineages, but all were herbivores. Ornithopoda, whose name means "birdlike foot," is distinguished by elaborate dental mod-

ifications for grinding vegetation. Early members were small, about 1 to 2 meters (3–6 feet) in length, but some of the forms during the Cretaceous period (145–65 million years ago) reached nearly 20 meters (66 feet) length and as adults weighed several tons. Despite their suggestive name, ornithopod dinosaurs are not the ancestors of birds, and they became progressively less birdlike as time went on. The sister lineage of Ornithopoda is Marginocephalia, the "margin-headed" ornithischians, and it includes two distinctive groups. Pachycephalosaurs are the "thick-headed" ornithischians. They evolved high domes of thickened bones over the top of the brain. Mechanical analyses suggest that pachycephalosaurs used their heads as battering rams, probably in battles for mates and territory. The ceratopsians are the horned ornithischians. Early ceratopsians were relatively small, like the early ornithopods and pachycephalosaurs, but the later forms were were rhino-sized animals that reverted to locomotion on all fours—they became quadrupedal—to support their great bulk. With increased size, a shelf at the back of the head expanded into a fan-shaped sheet of bone known as a frill, reaching lengths of almost 2 meters (6 feet) in some species.

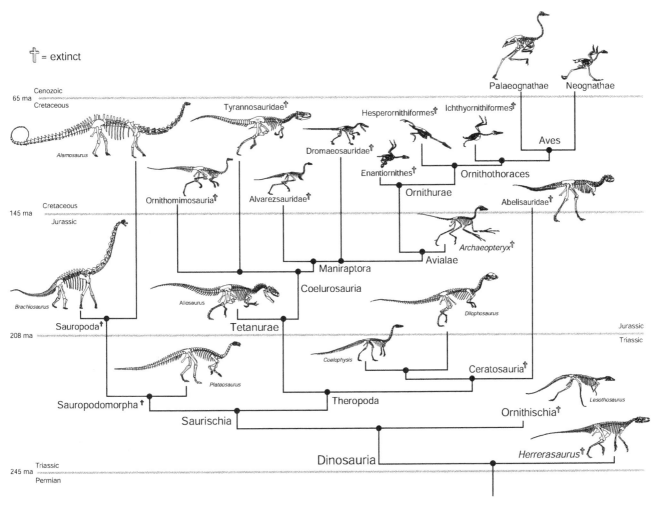

FIGURE 2. Phylogeny of Saurischian Dinosaurs.   Courtesy of Tim Rowe.

The third major group of ornithischian dinosaurs is Thyreophora, whose name means "shield bearer" in reference to their body armor. There are several thyreophoran lineages. The ankylosaurs became fully armored, and some became almost completely covered with a mosaic of protective bony scutes, almost like the shell of a turtle. They were squat, lumbering quadrupeds that approached 2 tons in weight. Stegosaurs, the plated dinosaurs, also evolved to elephantine size and became quadrupedal to support their weight. They have a distinctive paired row of spikes or plates of bone that rose along either side of the backbone. Their function may have been more for display than for protection or thermoregulation, although all three functions have been proposed. Despite suggestive names and some points of resemblance with birds, all of the ornithischians became extinct at the end of the Cretaceous period, about sixty-five million years ago.

Saurischians. Although paleontologists first noticed resemblances between birds and ornithischians, it is within the other major dinosaurian lineage, Saurischia, that we trace the ancestry of birds today. Saurischias, the so-called lizard-hipped dinosaurs, exhibit an extensive hierarchy of features that are found in living birds (Figure 2). The most ancient saurischian fossils are from Late Triassic rocks, and they have comparatively long necks and a distinctive hand in which the second or index finger is the longest. Birds, too, have very long necks, and although their hand is modified to support the wing feathers, this same basic pattern can be seen.

The earliest saurischians were 3 to 5 meters (10–16 feet) in length and weighed no more than a few hundred kilograms. Like other early dinosaurs, Triassic saurischians were habitually bipedal, and two major lineages had originated before the Triassic ended. At their first appearance, the sauropodomorphs were comparatively small bipedal animals that were probably omnivorous. This lineage survived until the very end of the Cretaceous period, a span of 155 million years, and over the course of its long history evolved to great size and produced the largest land animals ever. To sustain such mass, the later species reverted to walking on all fours. In the process, they evolved unbelievably long necks and tails and tiny heads. Giants like *Supersaurus* may have reached 40 meters (131 feet) in length. The limbs became columnar and elephantine to support a bulk estimated in the very largest species to range between 50 and 100 tons, at least ten times the weight of an adult male African elephant. These giants were herbivorous, with blunt spatula-shaped teeth that were probably used to strip foliage from stems. But there is nothing birdlike in the history of this group, and, like the ornithischians, it became extinct at the end of the Cretaceous period.

Theropods. All the other dinosaurs mentioned up to this point had histories that carried them in nonavian directions. But through the fossil record of theropod dinosaurs, we can trace a hierarchy of interested resemblances that are increasingly birdlike. Most of the unique features of basal theropods were to enhance the predatory lifestyle inherited from the ancestral dinosaur, but many persist in birds today. Theropods had a kinetic or flexible lower jaw that protected their lightly built jaws from being broken by struggling prey and possibly enabled them to swallow food items larger than their heads. A highly mobile ball-in-socket joint between the head and neck amplified the effectiveness of the long, mobile neck inherited from their saurischian ancestors. In addition to their sharp teeth and claws, their hands had greater leverage and strength for snagging and raking flesh, with strongly curved claws on each of three grasping fingers. The fourth finger was reduced, and the fifth finger is generally absent altogether. Like birds today, they had thin-walled, hollow bones forming a strong tubular skeleton. Also, their collarbones were fused together on the midline to form the wishbone or furcula, one of the most distinctive of avian characteristics. Throughout their history, theropods were obligate bipeds, their pelvis and hindlimbs modified to withstand the entire burden of swift, forceful running, jumping, and landing.

At least two theropod lineages had arisen by the end of the Triassic period. Ceratosaurs are the best-known theropods of the Triassic and Early Jurassic world. Although evolving several interesting parallels with birds in the way their skeletons grow, they disappeared from the northern hemisphere before the Cretaceous began and from the southern hemisphere by its end. However, their sister lineage, Tetanurae, has experienced a different fate. Early tetanurines are more birdlike than other contemporary dinosaurs in virtually all parts of their skeleton. Most importantly, the arms and hands were long and the hand consisted of only three adult fingers, the fourth being no more than an embryonic vestige. Together, these changes extended the reach and power

of the arms, setting the evolutionary stage for the evolution of a powerful flight stroke.

Some of the basal tetanurines that lived during the Jurassic and Cretaceous periods were giant predators. *Charchrodontosaurus* and *Giganotosaurus* exceeded even *Tyrannosaurus rex* in size. But the trail of anatomical clues to avian ancestry leads away from these giants and into Coelurosauria, a tetanurine theropod lineage that includes both large and some very small dinosaurs. Coelurosaurs are distinguished by some rather subtle characteristics. For example, the bones of the foot are tightly packed, representing a step toward the reinforced foot that enables birds to withstand the force of landing. The major coelurosaur lineages include the tyrannosaurids, ornithomimosaurs, and maniraptorans. The tyrannosaurids were giant predatory dinosaurs, the biggest of which approached 8 tons. They lived during the Cretaceous period and evolved huge heads, and their forelimbs became dwarfed in size in a very unbirdlike way. But avian resemblances are clear in Ornithomimosauria, the "ostrich mimics." These tetanurines were lightly built, with long hindlimbs designed for running and very long arms with slender hands. Some even lost their teeth and instead had a birdlike beak. The ornithomimosaur brain is very large and birdlike, approaching the relative size of the brain in modern flightless birds such as the ostrich. The eyes of ornithomimosaurs were also huge, suggesting further improvement in vision. Primitive members of the lineage were about 3 meters (10 feet) in length, but some later species range up to 6 meters (20 feet) long and were 2 meters (6 feet) at the hip. Ornithomimosaurs are indeed very birdlike, but they died out at the end of the Cretaceous period, along with the tyrannosaurids.

Feathered Dinosaurs. Early members of the maniraptoran lineage are even more birdlike than the other tetanurines. The most compelling evidence is from recent discoveries in China that include complete skeletons of primitive maniraptorans that preserve feathers. Even the earliest maniraptorans are small compared to virtually all other dinosaurs, and even the oldest members of the lineage have feathers. Maniraptorans have relatively longer arms and hands, and details of the hand and wrist structure closely resemble that seen in young birds. The tail is mobile at its base, but the rest is stiff, providing a rigid support for the tail feathers.

How feathers originated has been a persistent question in evolutionary biology. It was long argued that feathers, hollow bones, and the furcula arose as flight adaptations. But modern phylogenies suggest that hollow bones and a furcula are among the oldest features in theropods, and that feathers originated long before theropods learned to fly. The origin of feathers was probably tied to temperature regulation and may indicate the beginning of warm-bloodedness. Judging from living maniraptorans, feathers may also have arisen in connection with courtship displays.

The arms in basal maniraptorans already had most of the skeletal modifications that later proved essential for powered flight. Their elongated arms approached or exceeded the length of the hindlimb and were powerfully muscled. In the wrist is the semilunate carpal, a bone that helps constrain movement of the long hand to a horizontal plane. All these features probably originated as adaptations for hunting, grabbing, and holding struggling prey. The pelvis and tail are also more birdlike than in other dinosaurs. Much of the musculature that once attached to the tail has shifted to the pelvis, and the center of gravity is shifted forward, toward the arms. Birds carried this trend further by suspending the center of gravity between the wings during flight.

Maniraptora is a highly diverse lineage that includes several extinct side branches. The weird alvarezsaurids had reduced forelimbs and only a single large digit in their hands. The dromaeosaurs, including the well-known *Deinonychus* and *Velociraptor*, were distributed over much of the Cretaceous world. The oviraptorids included small, sleek cursorial bipeds, ranging up to perhaps 5 meters (16 feet) in length. Bizarre crests and pneumatized outgrowths of the head are highly distinctive of the lineage. There is evidence of feathers in all of the species. Several skeletons of *Oviraptor* have also been found preserved sitting on a nest of eggs in a brooding posture exactly like that of many modern birds. Like the appearance of feathers, parental brooding, so characteristic of modern birds, also predated the origin of flight. Some or all of the Mesozoic maniraptorans may have been warm-blooded, like their living descendants.

Flying Dinosaurs. The flying maniraptorans are members of Avialae. The oldest known fossil for this group is *Archaeopteryx*, long recognized as the world's oldest bird. Dating back about 150 million years, it is still the oldest known flying dinosaur. Its skeleton is only slightly modified over that in other maniraptorans, but several specimens are known in which flight feathers are preserved along the arms. These pigeon-sized dinosaurs have a pointed snout with teeth that are reduced in size and number, foreshadowing the origin of the avian beak. Their brain is expanded in characteristic avian fashion. The arms and hands are longer than the hindlimbs, providing a broad wing, and corresponding modifications in the hindlimb produced a more solid structure to withstand the forces generated in landing. The big toe or hallux has also moved onto the back of the foot, affording a degree of grasping capability.

Another persistent question in the evolution of birds involves the origin of flight. Earlier authors argued that flight must have arisen "from the trees down," with parachuting and gliding as intermediary stages. But the evidence that birds are on the maniraptoran branch of the dinosaurian tree suggests that flight evolved "from the ground up," because all the close relatives of birds were small ground-dwelling runners.

The Mesozoic history of flying dinosaurs has only recently come into focus as new discoveries filled what was long an almost compete gap in the fossil record of birds. The lineage christened Ornithurae includes all birds more closely related to living species than to *Archaeopteryx*. Ornithurines were capable of more powerful and sustained flight than *Archaeopteryx*. Their breastbone or sternum has a central keel to anchor more massive flight muscles. Complementary changes occurred in bones of the wrist and hand, which fused into a rigid skeleton to support the primary flight feathers of the wing.

Among the early orinthurines is a lineage that includes birds with further enhanced flight maneuverability known as ornithothoracians. Their shoulders have a robust strut to increase leverage of the flight muscles, and the wings are equipped with a "bastard wing" or alula, a modification of the thumb that prevents stalling during low-speed flight. Also, the vertebrae in the rear half of the tail are fused to form the pygostyle, the solid bony structure that supports the tail feathers as they are fanned and rapidly reoriented during flight to control lift and direction precisely. The oldest ornithothoracine fossil, which was found in rocks dating to the Early Cretaceous period (about 120 million years ago), was a finch-sized bird with the powerful bounding type of flight typical of sparrows and finches. An important extinct group of ornithothoracians is the enantiornithine lineage, which became diverse and widespread in the Cretaceous period. Early enantiornithines were small, sparrow-sized birds that were strong flyers; by the end of the Cretaceous period, there were turkey-sized forms with wingspans of more than 1 meter (3 feet), stilt-legged waders, and powerful runners.

Carinate birds have advanced flight still further. Their sternum has a deeper keel for larger flight muscles, and their trunk is shortened to provide a rigid armature for the larger flight muscles and for absorbing greater forces during landing. Several Cretaceous carinate lineages are known, including several whose fossils are found in marine rocks and mark the first exploitation of an aquatic habitat by dinosaurs. One marine lineage known as Ichthyornithiformes includes several powerful flyers, like modern pelicans or frigate birds, but they were not especially close relatives of either. Hesperornithiformes is another Cretaceous lineage that lived in oceans of the northern hemisphere. They were entirely flightless, with dwarfed forelimbs, and probably used their feet for propulsion under water while swimming across the surface like modern penguins and loons. Both of these lineages originated in the Early Cretaceous period and disappeared about 5 million years before the end of the Cretaceous.

Aves is the dinosaur lineage that comprises the last common ancestor to living bird species. Aves is distinguished by a combination of features including the replacement of teeth by a horny beak or bill and a greatly enlarged brain. The features summarized above are but a small sample of a huge body of evidence placing birds within the hierarchy of relationships among carinate, ornithothoracine, avialan, maniraptoran, tetanuran, theropod, and saurischian dinosaurs. Aves originated during the Middle or Early Cretaceous period, and it is this branch of the dinosaurian family tree that survived the terminal Cretaceous event (see Dinosaur Extinction). From Cretaceous onwards, avian dinosaurs speciated at a tremendous rate as they occupied the mountain and woodland areas of the world, nearly all of the world's islands, and all of the world's oceans and waterways.

Dinosaur Extinction. Much attention has been paid to the extinction of some of the dinosaurs in the terminal Cretaceous event. All of the ornithischians and sauropodomorph dinosaurs died out about sixty-five million years ago, along with several theropod groups such as the tyrannosaurids and enantiornithines. Between 100 and 150 species represented by fossils are known to have disappeared at that time. Their extinction was probably tied to environmental disruption caused by unprecedented volcanic activity that was ongoing when a large asteroid hit the earth. Because they have a poor fossil record, we do not know how birds were affected by the end of the Cretaceous period, except to say that avian dinosaurs survived to diversify into greater numbers of species than any other terrestrial vertebrate lineage.

At the height of dinosaurian diversity, just a few thousand years ago, Aves included between 12,000 and 20,000 species. A far greater episode of extinction than the terminal Cretaceous event began in the Pleistocene epoch (1.64 million–10,000 years ago). A first wave of extinction occurred with the environmental disruption of the ice ages. A far greater wave of extinction occurred as humans learned to sail and occupied the islands of the world, where overkill and habitat destruction led to the extinction of thousands of species. Yet another great wave of dinosaur extinction is under way today, as human overpopulation and resource exploitation have led to loss of habitat of the continental birds.

[See also Animals; Birds; Mass Extinctions; Reptiles.]

BIBLIOGRAPHY

Dingus, L., and T. Rowe. *The Mistaken Extinction—Dinosaur Evolution and the Origin of Birds.* New York, 1997.

Farlow, J. O., and M. Brett-Surman. *The Complete Dinosaur.* Bloomington, Ind., 1997.

Gauthier, J. A. "Saurischian Monophyly and the Origin of Birds." *The Origin of Birds and the Evolution of Flight*, edited by K. Padian, Memoir 8, pp. 1–47. San Francisco, 1986.

Padian, K., and P. E. Currie, eds. *Encyclopedia of Dinosaurs.* New York, 1997.

Rowe, T., K. Kishi, J. Merck, Jr., and M. Colbert. *The Age of Dinosaurs.* 3d ed. CD-ROM. New York, 1998.

Weishampel, D., H. H. Osmolska, and P. Dodson, eds. *The Dinosauria.* Los Angeles, 1990.

ONLINE RESOURCES

http://www.ctlab.geo.utexas.edu/dmg/index.html. This site, sponsored by the Digital Morphology Group, features three-dimensional imagery and ancillary information on the skulls of many birds and their extinct dinosaurian relatives.

— TIMOTHY ROWE

DIRECT DEVELOPMENT. *See* Life History Stages.

DIRECTED PROTEIN EVOLUTION

In the late 1980s, scientists began to develop techniques to alter protein function that do not require extensive knowledge of either protein structure or chemistry. They took notice of the power that Darwinian evolution has in nature to create diverse proteins and began to mimic it in the laboratory. The laboratory process was referred to as directed evolution because, instead of natural selection choosing which proteins are fit to be the parents of the next generation, scientists choose the "best" proteins and direct evolution experiments toward specific predefined goals. In this way, directed evolution is analogous to the artificial selection practiced by animal and plant breeders, but, instead of an entire organism, only proteins are selected and evolved.

The ability to evolve proteins in the laboratory is the result of recombinant DNA technology, which allows scientists to obtain a gene from one organism, manipulate it, then put it into another organism, which now produces a new protein. Proteins are long chains made up of amino acid building blocks, and each protein is encoded by a specific gene. A gene is a sequence of DNA, which is a long chain of nucleotides that serves as a recipe for building a protein. The nucleotides are read in "words" of three, and each three-letter word encodes one of twenty different amino acids. As the cellular machinery reads the recipe, it builds the protein one amino acid at a time.

Each protein has a specific recipe; changing one word, or even one letter, changes the recipe and thus changes the protein. The process of changing the words and letters in a gene is called mutation, and mutation provides the variation in the natural world that allows evolution to happen. Once the variation has been produced, individual organisms are different from each other. These differences can affect their success at competing for resources and, ultimately, their ability to grow and reproduce. If a mutation makes it more likely for an organism to reproduce successfully, natural selection may eventually spread that mutation throughout the population. In this way, organisms (and their proteins) evolve and become better adapted to their environment.

Directed evolution mimics natural evolution by alternating replication (with mutation) and selection. First, scientists isolate the gene that encodes an interesting protein. Next, they mutate the gene; this is done in such a way that thousands upon thousands of copies of the gene are produced, each with usually only a few mutations. The collection of mutant genes is called a library. Scientists then put the library into cells—often bacteria—and allow the cells to make the mutant proteins. The proteins are then tested to see which ones are best at performing the desired job. The genes for the best proteins can be collected and mutated again, and the whole process can be repeated as many times as required to obtain the desired function.

Creating Mutant Genes. Two sets of techniques are widely used to create the library of mutant genes; they may be used separately or combined, depending on the experiment. The most common techniques involve versions of error-prone polymerase chain reaction (PCR), a tool that allows many copies of a DNA molecule to be made. By varying the temperature, salt concentrations, and concentrations of nucleotides during PCR, scientists can control the level of single nucleotide (point) mutation. Usually the level of mutation is restricted to 0.1–0.5 percent so that beneficial mutations are not lost in a sea of deleterious mutations. Using PCR, millions of copies of an original gene, each with a few mutations, can be made in just a few hours.

The second set of techniques for creating a library, typically called DNA shuffling, introduces a version of sexual reproduction into the process. DNA shuffling creates "chimeric genes" by allowing recombination events to occur in the test tube. For example, scientists can use genes from several different organisms, cut them into many pieces of different lengths, then let the pieces join back together again to form full-length genes. The new genes are chimeric, meaning they contain bits and pieces from the different parent genes. The parent genes may have been different at only a few or at hundreds of nucleotides; therefore, the new chimeric genes contain combinations of sequences that were not present in the original parents. These chimeric genes in turn give rise to chimeric proteins.

After the library of mutant proteins has been created, their function can be evaluated either by selection or by high-throughput screening. Selection requires the protein to function for the cells to survive; thus, all non-functional proteins are eliminated from the analysis. For example, scientists may want a protein to break down a certain sugar. If bacteria were grown with only that particular sugar as food, then only those bacteria containing a mutant gene that encodes a functional protein would be able to grow. Bacteria whose genes have been mutated so that the proteins they encode cannot use the sugar will die. Because most mutations are harmful and destroy the protein's function, using selection can save scientists time they would otherwise spend analyzing dead, uninteresting proteins.

Selection gives scientists a simple yes or no answer to the question of whether a protein is still functional after mutation. Using selection, however, is not always convenient or even possible because it requires that the protein perform a function vital to the survival of the organism. Many useful and interesting functions are not vital. An alternative is screening for protein function, in which scientists evaluate how well all the mutant proteins do their job. Methods for screening protein function are plentiful, but to be useful for directed evolution studies, they must be high throughput. In other words, scientists must be able to screen hundreds or thousands of mutant proteins (ideally without purification) in a short period. To use the same example as above, if scientists wanted a particular protein to break down a specific sugar, they could mix that sugar with cells producing the protein. If the protein does its job, it produces a colored product. The amount of color produced over time can be measured and gives scientists a way to quantify how well the evolved protein breaks down the sugar.

Choosing or developing a method for either selecting or screening for the desired protein function is probably the biggest challenge for scientists performing directed evolution experiments. Once the method has been chosen, however, the directed evolution process itself is simple. In each generation, scientists create a library of mutants, analyze the library to find the best protein, mutate it to make the next generation, and continue until they create a protein with an appropriate level of the desired function.

Example of Directed Protein Evolution. One directed evolution success story has been the generation of a protease (a protein that breaks down other proteins) that can work at both high and low temperatures. The whole set of experiments began as an effort to develop a protein additive for laundry detergents that would be effective no matter what the water temperature. This may seem like a trivial use for directed evolution, but about 30 percent of industrial proteins currently produced worldwide are for use in detergents, and together they represent a multimillion-dollar market.

In addition, developing an "all-temperature" protein seemed like a difficult chemical problem to solve. For years, industrial biochemists searched the natural world in vain for proteins that could catalyze reactions efficiently at low temperatures but still retain their folded structures (and activity) at high temperatures. The biochemists' failure led many to speculate that stability at high temperature was incompatible with high activity at low temperature. Their reasoning was that a protein flexible enough to work at low temperature would fall apart at high temperature, whereas a protein rigid enough to work at high temperature would lock up at low temperature. Directed evolution, however, demonstrated that the two characteristics are not necessarily mutually exclusive. Experiments on the cold-loving protease subtilisin S41 showed that just thirteen amino acid substitutions increase the protein's stability more than 1,000 times over the parent at high temperature; most importantly, the activity of the protein at low temperature is increased threefold over the parent.

Two different evolutionary arguments can be invoked to explain why proteins that are both stable at high temperature and very active at low temperature are not commonly found in nature. The first argument proposes that because cold-loving proteins evolve inside cold-loving organisms, they are under no selective pressure to be stable at high temperature. Therefore, cold-loving proteins may have lost high-temperature stability as a result of random genetic drift. Alternatively, the second argument proposes that protein instability may actually be advantageous to the organism's fitness. Extremely stable proteins may resist the natural degradation process and accumulate, interfering with the cell's normal physiology. This would lead natural selection to select against extremely stable proteins.

Improving on Nature. Directed evolution can also be used to evolve proteins that perform new functions, and perform them better than their naturally evolved counterparts. An excellent example of this is the laboratory evolution of the dual protein indole-3-glycerolphosphate synthase (IGPS) and phosphoribosylanthranylate isomerase (PRAI). In nature, these two proteins are physically linked together, but they catalyze different chemical reactions. The synthase links two molecules together (synthesizes), whereas the isomerase literally flips a molecule into its mirror image. Targeting potentially important regions of the protein for random mutagenesis allowed scientists to switch the activity of IGPS to that of its partner, PRAI. In fact, the laboratory-evolved PRAI is six times better than the natural PRAI protein at flipping molecules. Using directed evolution, scientists turned one type of function into another type and improved it in the process.

The road scientists took to produce the PRAI protein is not the same as natural evolution would have, or even could have, taken. During several generations, the evolving protein was almost nonfunctional; scientists were able to coax the protein through these low points by including various chemicals necessary for survival. Nature would not have been able to coax the protein this way, and the organism housing it would have died. Natural selection could not have solved the problem the same way as the scientists. Directed evolution thus allows researcher to discover novel solutions for generating protein function.

As is clear from the examples given here, directed evolution by several rounds of random mutagenesis can be quite successful, but it can take many months to complete. This is because the mutation level is kept low (one to five mutations per gene) so that the library needed to screen the genes is manageable. Because the level of mutation is low, however, scientists are unlikely to find a sequence that drastically increases the protein's performance. Instead, they must patiently accumulate the necessary mutations over several generations (just as nature does).

Benefits of Directed Evolution. Recombination may be able to speed up directed evolution because it allows scientists to introduce more variation in gene sequence per generation. The parents in a recombination experiment may be different at many hundreds of nucleotides over the length of the gene. The result is that a recombined library contains sequences that differ at more positions than an error-prone PCR library. Even though the recombined sequences contain many more "mutations" than those in an error-prone library, the chimeric proteins they produce are likely to fold into their correct three-dimensional shapes because both parents have similar structures. Within a recombined library, there may be substantially "better" proteins, and recombined libraries have the possibility of containing members that have improved multiple functions. For example, DNA shuffling of twenty-six fragments from various subtilisin genes created chimeric proteins that had improved four properties relative to the parent proteins. It is possible that recombination will allow scientists to create desired proteins faster.

The real power of directed evolution stems from the fact that researchers are able to examine sequences in the laboratory that nature may not yet have explored. Because there are twenty possible amino acids at each position, the number of sequences of N amino acids in length is phenomenally huge (20^N). A typical protein has hundreds or even thousands of amino acids. Even in four billion years, nature has not had time to test even a fraction of the possible protein sequences via natural selection.

Scientists also are able to explore new sequences in the laboratory because they can ignore biological function to access new protein functions. In nature, sequences that prevent the host organism from successfully reproducing do not make it into the next generation. In the laboratory, scientists design the selection or screening method to reflect the function they are interested in, regardless of whether that function is necessary for cell survival in nature, then allow only those proteins that best perform the desired function to proceed. Because the protein is taken out of its normal environment and characteristics that may be important to fitness in nature are disregarded, directed evolution is able to work outside the boundaries that restrict natural evolution. Using directed evolution, scientists can tap into the potential hidden in every protein, investigating any function that is chemically or physically possible. They are limited only by their ingenuity in deciding where to explore and in designing selection or screening methods.

[*See also* Artificial Selection; Experimental Evolution, *article on* A Long-term Study with *E. coli*; Mutation: An Overview.]

BIBLIOGRAPHY

Altamirano, M. M., J. M. Blackburn, C. Aguayo, and A. R. Fersht. "Directed Evolution of New Catalytic Activity Using the α/β-Barrel Scaffold." *Nature* 403 (2000): 617–622.

Arnold, F. H., ed. *Evolutionary Protein Design*, vol. 55, *Advances in Protein Chemistry*. Edited by F. M. Richards, D. S. Eisenberg, and P. S. Kim. San Diego, Calif., 2001. With chapters that cover everything from the types of questions scientists are addressing using directed evolution to the new technology that is being developed to analyze evolved libraries of proteins, this is the most complete work on directed evolution currently available.

Arnold, F. H., and P. L. Wintrode. "Enzymes, Directed Evolution." In *Encyclopedia of Bioprocess Technology: Fermentation, Biocatalysis, and Bioseparation*, edited by M. C. Flickinger and S. W. Drew, pp. 971–987. New York, 1999. Review focusing on the methods and theory of directed evolution and its industrial applications.

The Arnold Group's Homepage, http://www.che.caltech.edu/groups/fha/. This web site provides information on directed evolution methods and applications, as well as a list of relevant publications and patents.

Joyce, G. F. "Directed Molecular Evolution." *Scientific American* (December 1992): 90–97. Written for the nonscientist, this article describes the directed evolution of RNAs just as the field was beginning to take off.

Maynard Smith, J. *The Theory of Evolution*. Cambridge, 1993. To learn how natural evolution works and how humans have adapted the process into artificial selection, you couldn't ask for a better teacher than John Maynard Smith.

Postgate, J. *The Outer Reaches of Life*. Cambridge, 1974. A delightful book that describes some of the proteins that allow life to exist at extremes of temperature, salinity, and pressure.

Sambrook, J., E. F. Fritsch, and T. Maniatis. *Molecular Cloning: A Laboratory Manual*. Cold Spring Harbor, N.Y., 1989. An indispensable guide to molecular biology for the experimental scientist, containing protocols for cloning, polymerase chain reaction, and bacterial transformations, as well as recipes for growth media and buffers.

Tagliaferro, L., and M. V. Bloom. *The Complete Idiot's Guide to Decoding Your Genes*. New York, 1999. This book is a good first step in learning about molecular biology and the laboratory

techniques used in directed evolution. It is much more accessible than a textbook, and it is funny.

Wilson, E. O. *The Diversity of Life*. New York, 1992. Describes what evolution is and how it works.

Zhao, H., J. C. Moore, A. A. Volkov, and F. H. Arnold. "Methods for Optimizing Industrial Enzymes by Directed Evolution." In *Manual of Industrial Microbiology and Biotechnology*, edited by A. L. Demain and J. E. Davies. 2d ed., 597–604. Washington, D.C., 1999. Step-by-step instructions for generating directed evolution libraries by error-prone PCR, DNA shuffling, and StEP recombination, as well as various selection and screening methods.

— KIMBERLY M. MAYER AND FRANCES H. ARNOLD

DISEASE

[*This entry comprises three articles:*

        Hereditary Disease
        Infectious Disease
        Demography and Human Disease

The first article provides a general overview of the nature of hereditary disease; the second entry contrasts infectious and hereditary disease from an evolutionary perspective; the final entry discusses how travel and trading patterns affect human diseases. For related discussions, see the Overview Essay *on* Darwinian Medicine *at the beginning of Volume 1;* Acquired Immune Deficiency Syndrome; Antibiotic Resistance; Coevolution; Emerging and Re-Emerging Diseases; Frequency Dependent Selection; Heterozygote Advantage, *article on* Sickle-Cell Anemia and Thalassemia; Immune System; Influenza; Malaria; Myxomatosis; Nutrition and Disease; Plagues and Epidemics; Red Queen Hypothesis; Resistance, Cost of; Sex, Evolution of; Transmission Dynamics; Vaccination; *and* Virulence.]

Hereditary Disease

After suffering from a particularly bad cold, one may wonder why natural selection, in its exquisite ability to render organisms adapted to their environment, has failed to produce an immune system that is capable of keeping away such annoying viral insults. In any population, a diseased state may arise from a wide variety of causes, and the prevalence of disease likewise involves many factors. This article limits its attention to diseases that have a genetic component—hereditary diseases—however, it soon becomes apparent that infectious agents may play a key role in hereditary as well as infectious diseases. Any natural population will have a distribution of fitnesses among its individuals, and those unfortunate enough to be in the lower tail of this distribution are generally, in some sense, diseased. But we will focus our attention on more specific diseased states, where a particular physiological system is incapable of dealing with the prevailing environmental conditions. Such conditions nearly always depend on the environment in some way and are triggered by exposure to a novel condition (including an infectious agent). Sometimes the infectious agent itself has no direct effect on health, but it triggers an autoimmune response (such as Reiter's disease). We shall call this a hereditary disease because only individuals with particular genotypes get Reiter's disease after infection. In other cases the infectious agent itself causes disease, and these may sometimes be considered hereditary if most individuals are fully immune to the pathogen, but a genetic lesion renders a subset of the population susceptible.

Returning to the question of disease prevalence, one obvious reason why diseases do not disappear altogether is that deleterious mutations continue to occur, so that at best the population is in a state of mutation-selection balance. The expected disease prevalence then depends on the rate of mutation and the severity of the disease, but in order to obtain a mathematical expression for the expected prevalence, one needs first to consider the genetic basis for the disease. Many disorders are transmitted as a single Mendelian gene, but more often chronic diseases exhibit complex, multigenic transmission. Because the methods and theory for these two cases (which are ends of a continuum) are so different, they will be considered separately.

Single-Gene Disorders. Human populations host a large number of genetic disorders that segregate like a normal Mendelian gene, and these have been relatively easy to map and eventually characterize at the molecular level. The prevalence of different single-gene disorders varies over orders of magnitude, and our understanding of the reasons for current allele frequencies is only partial. Simple population genetic models (Hartl and Clark, 1997) predict that recessive disorders will have an equilibrium allele frequency in the population of

$$\sqrt{\frac{\mu}{s}},$$

where μ is the mutation rate from a normal allele to the defective allele, and s is the selection coefficient (such that the AA, Aa, and aa genotypes have fitnesses 1, 1, 1 − s). Thus, the equilibrium gene frequency is higher when mutation rate to the disease-causing allele are high or selection against it is low. Similarly, if the disease is dominant, the equilibrium allele frequency for the defective allele is expected to be μ/s. Although the equilibrium allele frequency for the recessive case is often orders of magnitude higher than that for the dominant disease, the number of disease cases at equilibrium is in fact quite similar because the diseased state requires only one copy in the latter case. Given the enormous range in

prevalence of genetic disorders, one might suspect that they vary widely either in mutation rates or in selective coefficients. But even for lethal disorders, where $s = 1$, the frequency varies far more than the mutation rate. In these cases, it appears likely that the selection coefficients have not always been what they are today, and strong founder effects may also result in large changes in frequency. Now let us examine the details of a few particular single-gene disorders.

Sickle-cell anemia. The predominant adult form of hemoglobin has an oxygen-binding heme ring with two alpha subunits and two beta subunits. In 1957 V. Ingram found that the sickle or S-allele form of β globin has a different structure from the normal, or A, form. We now know that the sixth amino acid in the $β^s$ allele replaces the normal glutamic acid with a valine in this 146 amino acid protein. The result is a protein that tends to aggregate under conditions of hypoxia (low levels of oxygen in the blood), and results in a characteristic sickle shape to the red blood cells. In malarial regions of Africa, the incidence of the $β^s$ allele approaches 20 percent, much higher than expected under a mutation-selection balance unless the selection actually favors $β^s$. There is now quite direct evidence that heterozygotes for $β^s$ have reduced mortality by malaria, so that despite the deleterious consequences of $β^s$ homozygosity, the positive effects in heterozygotes drive the allele frequency up. Most other changes in the amino acid sequence of globins are deleterious. This can be inferred from the fact that there are over 700 structural variants of globins that have been cataloged, and all but HbS, HbC, and HbE are extremely rare. A striking thing about $β^s$ globin is that, despite the serious difficulties that some homozygotes experience, there are some homozygous individuals, having two copies of the same $β^s$ allele, who are virtually free of any symptoms. Observation of these individuals clearly establishes that even in the face of mutations of major effects, the genomic context of the alleles can make a large difference to the expression of the trait.

Thalassemia. Just as sickle-cell anemia tends to be clustered in regions of high malarial infection, so too are the thalassemias, diseases characterized by an imbalance in the relative amounts of α- and β-globin. The most common thalassemia is α-thalassemia, generally caused by a deficiency of copies of the α-globin gene. Normally we have two copies of α-globin on both copies of chromosome 16, for a total of four α-globin copies. There are individuals with from one to four copies (zero copies of α-globin is fatal before birth). Abnormalities in β-chain synthesis also occur, resulting in β-thalassemia. The phenotypes of individuals with β-thalassemia are highly variable, ranging from profound, life-threatening anemia to virtually normal, even though they may have the same globin alleles. The genetic causes for the variability in expression are also diverse, including factors that directly affect globin expression and other genes with quite indirect transacting modifiers. Over 200 variants that affect β-globin levels have been identified, and they exhibit local geographic distributions (Weatherall, 2001).

Glucose-6-phosphate dehydrogenase deficiency. Along with sickle cell and thalassemia, another genetic disorder that is elevated in frequency in regions where malaria is common is G6PD deficiency. This X-linked gene encodes the lead enzyme in the pentose phosphate shunt, which generates NADPH for many cellular oxidation-reduction or redox reactions. The precise mechanism whereby this change results in elevated resistance to malaria is not known. Complete loss of G6PD activity is fatal, and the various deficiency alleles are low activity variants. Populations in sub-Saharan Africa attain malarial resistance with the A and A⁻ alleles, whereas in the Mediterranean region, the low-activity G6PD allele has a different amino acid substitution, and is called Med. In Southeast Asia, the Canton and Mahidol alleles confer low activity. To make inferences about the evolutionary history of G6PD, Tishkoff and colleagues (2001) scored three linked microsatellites in individuals containing A/A⁻ and Med G6PD deficiency mutations as well as in nondeficient individuals from Africa, the Middle East, the Mediterranean, Europe, and Papua New Guinea. The G6PD A⁻ and Med alleles are widespread, but they occur on distinct haplotype backgrounds and have independent origins. The associated haplotypes have surprisingly low microsatellite variation and exhibit strong linkage disequilibrium. Population genetic models indicate that the A⁻ allele originated within the past 6,000 to 11,000 years and the Med allele arose within the past 1,600 to 3,000 years. The data are consistent with a rapid expansion in frequency, consistent with these alleles being favorable in malarial regions. The recency of G6PD deficiencies is also consistent with the inferred recency (within the past 10,000 years) of malaria as a common human pathogen.

Cystic fibrosis. The most prevalent genetic disorder in individuals of European descent is cystic fibrosis (CF), caused by mutations in the gene encoding the cystic fibrosis transmembrane conductance regulator (CFTR). More than 500 distinct defective alleles in CFTR have been characterized, but unlike phenylketonuria (see below), where the alleles are somewhat evenly distributed, the ΔF508 allele attains a frequency of 60 percent of the defective alleles in many populations. Why should one allele be so common? Many investigators assumed that there must have been a past selective advantage for CF, and the possibility that CF heterozygotes had elevated resistance to cholera remains a possible, but controversial, explanation (Gabriel et al., 1994). To learn more about the past history of CF alleles, Mateu and colleagues (2001) analyzed normal alleles and haplotype variation in four microsatellites and two single-

nucleotide polymorphisms (SNPs) in CFTR. They examined eighteen worldwide population samples, comprising a total of 1,944 human chromosomes and using gorilla as an outgroup. The CF allelic haplotype determined microsatellite diversity to a greater extent than did population of origin, suggesting that haplotype differences predate population splitting. Attempts to date the origin of the CF ΔF508 allele based on flanking variation suggest relatively recent origination, but the range of admissible ages is very large (Slatkin and Rannala, 1997).

Phenylketonuria. Another single-gene disorder that shows a wide range of phenotypic manifestations is phenylketonuria (PKU). More than 400 alleles of phenylalanine hydroxylase (PAH) have been identified that result in low PAH activity and can cause PKU. Low activity of PAH can result in severe mental retardation and many metabolic disturbances, including low melanin synthesis and resulting light pigmentation. Low phenylalanine diet at very early ages can make an enormous difference to the health of the patient. The many PAH alleles are sufficiently even in frequency that most cases of PKU are in fact heterozygotes for two noncomplementing alleles. Although PKU is listed as a simple autosomal recessive in human genetics textbooks, the range of phenotypes of individuals with the same PAH genotype suggests that there are important modifier genes as well (Scriver and Waters, 1999).

Animal models. Although the above discussion is restricted to human populations, the same principles apply to nonhuman animals. In fact, humans share more than 300 single-gene disorders with the mouse, and over 200 have been identified in the domestic dog. No doubt environmental factors, particularly pathogens, may impact genetic variation in natural populations in much the same way as in human populations.

Multigenic Disorders. Many common chronic diseases tend to aggregate in families, but they do not segregate like an ordinary Mendelian gene. Statistical tests show that the aggregation is significantly greater than expected by chance, and classical quantitative genetic methods may demonstrate that the diseases are heritable. We mean this in the formal sense that there is nonzero additive genetic variance, and this is typically manifested by familial resemblance (relatives of affected individuals have an elevated risk for the disease). From this point, the search for genes underlying these chronic diseases has not been easy.

Cancers. Cancer has been a particularly ellusive disease in part because in reality it is a collection of many distinct diseases. Some forms of cancer show no familial aggregation, but others aggregate strongly, suggesting a genetic basis. Many tumors are characterized by somatic deletions and aneuploidies, establishing at least a genetic basis for the somatic manifestation of the cancer. In some cases, there are variants of genes segregating in populations that confer widely differing risks of cancer. *BRCA1* and *BRCA2* are prime examples. *BRCA1* was found by laborious screening of some sixty-five genes in the region of chromosome 17 identified by a linkage study that ascertained multiply affected families (see Welsch and King, 2001, for review). Among families with early onset and multiple cases of breast cancer, *BRCA1* mutations account for about half the cases, and *BRCA2* mutations account for about 30 percent. Although this is very impressive, only about 5 to 10 percent of breast cancers appear to be heritable, and *BRCA1* and *BRCA2* account for only a negligible portion of sporadic cases. Furthermore, one would think that clones of the genes and identification of many mutations that result in cancer would allow investigators to determine means of preventing neoplasia. Although we have learned that both genes contribute to homologous recombination and DNA repair, to embryonic proliferation, and to transcriptional regulation, we still do not know exactly why these mutations lead to cancer.

Diabetes. Some form of diabetes affects nearly 4 percent of the adult population, and it occurs in a variety of forms with quite distinct etiology. Insulin-dependent diabetes mellitus (IDDM) generally appears in young patients and is characterized by destruction of the pancreatic beta cells, which produce insulin. Non-insulin-dependent diabetes mellitus (NIDDM) has a concordance rate in identical twins that approaches 100 percent, but still we do not fully understand its genetic basis. A recent attempt to identify genes underlying variation in NIDDM risk illustrates the difficulty in identifying candidate genes based on presumed function, and shows that evolution can create unexpected associations between genes and diseases. Positional cloning of a diabetes-susceptibility gene resulted in identification of calpain-10 (*CAPN10*), a ubiquitously expressed cysteine protease (Horikawa et al., 2000). Although the mechanism is not totally clear, insulin-mediated glucose turnover may be impaired, or insulin resistance may result from low expression of *CAPN10*, causing increased susceptibility to type 2 diabetes mellitus. Examination of sequences of diabetics revealed that the low-expressing alleles are associated with an intronic single nucleotide polymorphism (SNP) that is not even associated with a splice junction, nor translated because it is in the intron.

Cardiovascular disease. Cardiovascular disease, like cancer, actually consists of a collection of many diverse disorders spanning several distinct physiological systems. Defects in blood-clotting mechanisms can result in thrombosis, and resulting clots may starve the heart of oxygen and nutrients. Many disorders in lipid metabolism and transport may result in elevated cholesterol in the blood, with associated increased risk of ath-

erosclerosis. Diabetes, itself a complex of metabolic causes, often is associated with heart disease. In addition, elevated blood pressure results in arterial stress, and is also associated with risk of cardiac arrest. Given this broad set of physiological causes, it should not be surprising that a wide array of genes may be associated with heart disease.

Several genes involved in lipid metabolism and transport have allelic forms that result in a large change in function, including the low-density lipoprotein (LDL) receptor and lipoprotein lipase. These large-effect alleles often result in single-gene disorders. For example, null alleles of the LDL receptor result in familial hypercholesterolemia. Observation of these large-effect alleles raises the question of whether alleles of smaller effect have correspondingly smaller (but still nonzero) elevated risk. Studies of LPL (Clark et al., 1998) and apolipoprotein E (Fullerton et al., 2000) demonstrate the complexity of variation found even within single candidate genes. In *LPL* there were eighty-eight SNPs falling into eighty-eight distinct haplotypes (in a sample of seventy-one people), and in *ApoE* there were twenty-three SNPs and thirty-one haplotypes (in a sample of ninety-six people). Similar levels of complexity have been seen in allelic variation in β globin (Harding et al., 1997) and many other genes (Halushka et al., 1999). With this level of complexity, one fully expects that, even after finding a gene relevant to a complex disease, it may have multiple allelic forms with varying degrees of effect on the physiological phenotypes. This will make molecular diagnosis a challenging problem.

Long-term prospects for chronic hereditary diseases. Vast resources are being poured into efforts to identify genes that underlie chronic complex diseases. In addition to those listed above, asthma, osteoporosis, Alzheimer's disease, and schizophrenia are just a few on the list of disorders that exhibit familial clustering but are clearly not transmitted like a single gene. It is possible that some of these are like breast cancer, where some cases are strongly familial and may even have a fairly simple genetic basis, and other cases are truly sporadic. It is also possible that the disorders are the end result of a very large number of allelic effects at many genes, which add up to push one over a threshold of ill health. Even more likely is that many diseases will be caused by anywhere from two to twenty genes in relevant physiological pathways, and that these allelic effects interact both with each other and with the environment. These interactions imply that any single state of any single gene will not be very good at predicting disease risk. It also means that it will be much harder to get to the point of understanding disease well enough to predict or reduce risk from genotypic information. The fact that progress in understanding many of the complex diseases has been so slow suggests that these diseases truly are complex. In evolutionary terms, this is not surprising, as natural selection would have a much harder time reducing the incidence of disease caused by multiple interacting genes than it would for single-gene traits.

[*See also* Cancer; Coronary Artery Disease; Eugenics; Heterozygote Advantage, *article on* Sickle-Cell Anemia and Thalassemia; Modern *Homo sapiens, article on* Human Genealogical History; Pedigree Analysis; Senescence.]

BIBLIOGRAPHY

Clark, A. G., K. M. Weiss, D. A. Nickerson, S. L. Taylor, A. Buchanan, J. Stengard, V. Salomaa, E. Vartiainen, M. Perola, E. Boerwinkle, and C. F. Sing. "Haplotype Structure and Population Genetic Inferences from Nucleotide-Sequence Variation in Human Lipoprotein Lipase." *American Journal of Human Genetics* 63 (1998): 595–612.

Fullerton, S. M., A. G. Clark, K. M. Weiss, D. A. Nickerson, S. L. Taylor, J. Stengård, V. Salomaa, E. Vartiainen, M. Perola, E. Boerwinkle, and C. F. Sing. "Apolipoprotein E Variation at the Sequence Haplotype Level: Implications for the Origin and Maintenance of a Major Human Polymorphism." *American Journal of Human Genetics* 67 (2000): 881–900.

Gabriel, S. E., K. N. Brigman, B. H. Koller, R. C. Boucher, and M. J. Stutts. "Cystic Fibrosis Heterozygote Resistance to Cholera Toxin in the Cystic Fibrosis Mouse Model." *Science* 266 (1994): 107–109.

Halushka, M. K., J. B. Fan, K. Bentley, L. Hsie, N. Shen, A. Weder, R. Cooper, R. Lipshutz, and A. Chakravarti. "Patterns of Single-Nucleotide Polymorphisms in Candidate Genes for Blood-Pressure Homeostasis." *Nature Genetics* 22 (1999): 239–247.

Harding, R. M., S. M. Fullerton, R. C. Griffiths, J. Bond, M. J. Cox, J. A. Schneider, D. S. Moulin, and J. B. Clegg. "Archaic African and Asian Lineages in the Genetic Ancestry of Modern Humans." *American Journal of Human Genetics* 60 (1997): 772–789.

Hartl, D. L., and A. G. Clark. *Principles of Population Genetics.* Sunderland, Mass., 1997.

Horikawa, Y., N. Oda, N. J. Cox, X. Li, M. Orho-Melander, M. Hara, Y. Hinokio, T. H. Lindner, H. Mashima, P. E. Schwarz, L. del Bosque-Plata, Y. Horikawa, Y. Oda, I. Yoshiuchi, S. Colilla, K. S. Polonsky, S. Wei, P. Concannon, N. Iwasaki, J. Schulze, L. J. Baier, C. Bogardus, L. Groop, E. Boerwinkle, C. L. Hanis, and G. I. Bell. "Genetic Variation in the Gene Encoding Calpain-10 Is Associated with Type 2 Diabetes Mellitus." *Nature Genetics* 26 (2000): 163–175.

Mateu, E., F. Calafell, O. Lao, B. Bonne-Tamir, I. R. Kidd, A. Pakstis, K. K. Kidd, and J. Bertranpetit. "Worldwide Genetic Analysis of the CFTR Region." *American Journal of Human Genetics* 68 (2001): 103–117.

Scriver, C. R., and P. J. Waters. "Monogenic Traits Are Not Simple: Lessons from Phenylketonuria." *Trends in Genetics* 15 (1999): 267–272.

Slatkin, M., and B. Rannala. "Estimating the Age of Alleles by Use of Intraallelic Variability." *American Journal of Human Genetics* 60 (1997): 447–458.

Tishkoff, S. A., R. Varkony, N. Cahinhinan, S. Abbes, G. Argyropolous, G. Destro-Bisol, A. Drousiotou, T. Jenkins, G. Lefranc, J. Loiselet, A. Piro, M. Stoneking, A. Tagarelli, G. Tagarelli, E. Touma, S. M. Williams, and A. G. Clark. "Microsatellite/RFLP Haplotype Analysis of G6PD A- and Med Deficiency Mutations: Implications for the Origins of Malarial Resistance in Humans." *Science*, 2001.

Weatherall, D. J. "Phenotype-Genotype Relationships in Monogenic Disease: Lessons from the Thalassaemias." *Nature Reviews. Genetics* 2 (2001): 245–255.

Welcsh, P. L., and M. C. King. "*BRCA1* and *BRCA2* and the Genetics of Breast and Ovarian Cancer." *Human Molecular Genetics* 10 (2001): 705–713.

— ANDREW G. CLARK

Infectious Disease

Disease is any permanent or persistent condition that has a negative effect on an individual. In a human context, we think of diseases as causing discomfort, incapacity, illness, and even death. In agriculture, disease may lower yield or reduce the market value of a product. For an evolutionary biologist, the effect of a disease is quantified in terms of the reduced reproductive success or fitness of the host. Diseases can be caused by infectious or noninfectious agents. Examples of noninfectious diseases include genetic disorders and environmentally caused conditions such as vitamin deficiency. Infectious diseases are caused by "agents" that grow and reproduce on a living host and that are transferred from diseased to healthy individuals. Most commonly we think of diseases as being caused by viruses, bacteria, and other parasites.

Hereditary diseases are conceptually quite different from infectious diseases in that they are inherited, typically in a Mendelian fashion. Unless genetic diseases have some advantageous effect in the heterozygous condition (e.g., sickle-cell anemia), they can only be maintained by recurrent mutation. In practice, however, it may be difficult to distinguish genetic from infectious causation of disease. Biologists often speak of the "disease triangle"—the idea that for disease to be evident, one simultaneously needs a susceptible host genotype, an infectious agent, and an appropriate environment. Where these factors are not well known, or if there is a long delay between infection and disease expression, determining whether the disease is "genetic" or "infectious" is particularly difficult. Ulcers were thought to be the result of anxiety and increased stomach acidity; genetic factors were also known to predispose people to getting ulcers. Quite recently, it was shown that nearly all ulcers are associated with infection by the bacterium *Helicobacter pylori*. Environment, genetics, and pathogens interact as causative agents of disease.

By far, the majority of organisms have a parasitic mode of life. Peter Price (1980), an entomologist from Northern Arizona University, showed that there were nearly ten times as many parasitic as predatory species of insect. Even groups that are predominantly free-living (such as the crustaceans, molluscs, and flowering plants) have parasitic representatives. Parasites are often highly specialized, and nearly every host species can be attacked by many species of parasite. The better studied a species is, the more parasites it seems to have. Thus, tomatoes can be attacked by over a hundred different species of fungus, and humans have over thirty types of sexually transmitted disease. For an evolutionary biologist, to understand the evolution of pathogens and parasites is to understand the evolution of the majority of the planet's biodiversity.

The famous British cytologist Cyril Darlington in his book *The Evolution of Genetic Systems* (1958), first pointed out the parallels between transmission of disease among individuals and transmission of information by biological and cultural inheritance. Systems of DNA transfer between individuals in prokaryotes often involve infectious-type processes (transmissible plasmids, transduction, transformation). Semiautonomous genetic elements such as plasmids and transposons can be viewed as intracellular and genomic parasites. Infectious processes have been important in the symbiotic origin of mitochondria and chloroplasts in the eukaryotes. Uniparental inheritance may have evolved primarily as a mechanism to prevent the spread of defective but overreplicating organelles and other cytoplasmic parasites. Understanding infectious disease is therefore fundamental to understanding the evolution of genetic systems and perhaps life itself.

Infectious disease transmission may be either from parent to offspring (vertical) or between unrelated individuals in the population (horizontal). Vertical transmission is usually via the maternal parent, either in the cytoplasm, by adhesion or entry of pathogens into the egg or seed, or during and soon after birth. Horizontal transmission may be by direct sexual or nonsexual contact or by indirect contact with contaminated items (fomites). Diseases may also be transmitted aerially, by water, by infectious stages in the soil, or by vectors such as mosquitoes and ticks. Transmission of disease in hospitals (nosocomial infections) by health care workers is of particular concern. Transmission rate will increase with increasing density of the population if it results in more contacts or increased proximity between individuals. This makes disease an important factor in regulating the size of host populations. However, if the number of contacts is limited (as in many sexually transmitted diseases), then disease transmission may depend largely on the fraction of individuals that are diseased.

Many diseases (especially those caused by rust fungi and helminths) are characterized by complex life cycles involving alternate transmission between hosts of phylogenetically divergent taxa. For example, the life cycle stages of the liver fluke alternate between a vertebrate host, a snail, and sometimes a fish. Some rusts live alternatively on angiosperms and gymnosperms. Phylogenetic studies have shown that the evolutionarily more ancient host is not always ancestral, and that these complex life cycles may involve relatively recent host shifts.

Virulence. Some pathogens have a drastic effect on host survival, whereas others may sterilize their hosts. The latter process is sometimes termed "parasitic castration." A major issue of interest to evolutionary biologists has been understanding the enormous variation in the virulence of different pathogens. The conventional wisdom in the agricultural and medical fields has been that pathogens will evolve ever decreasing virulence so as not to endanger their hosts (or their host populations). This idea has been generally discredited as an overarching generalization. Not only does this conventional wisdom often rest on a group selection argument, but pathogens may evolve decreased virulence to increase their own fitness rather than that of their hosts. Often the greater the duration of an infection, the greater will be the period of disease transmission; thus, shortening a host's life span will decrease the fitness of the pathogen. The evolution of reduced virulence has been directly observed in the case of the myxomatosis virus, which was introduced into Australia because its high lethality made it an effective biological control agent of pest rabbits. In a decade or so after its introduction in the early 1950s, the virus evolved to an intermediate level of virulence which has been maintained ever since.

Transmission mode is critical to the evolution of virulence. Vertical transmission favors reduced virulence because survival and reproduction of the host is essential for transmission to its offspring. Rickettsial infections of rodents that are vertically transmitted are less virulent than those that are horizontally transmitted. Sexual transmission also favors reduced virulence with respect to mortality because such diseases can only be transmitted by adults during each breeding season. Sexual transmission may favor sterility if lesions on the sex organs increase transmission, or if it increases the frequency of matings (in ungulates, brucellosis caused by various species of *Brucella* results in fetal abortions that hasten the onset of estrus). Vector-based, waterborne, and soilborne transmissions favor reduced virulence of the pathogen in the vector, but increased virulence in the host. Transmission is less dependent on an active host, and biting rates may be higher on incapacitated hosts. A major generalization is that increased transmission opportunity results in the evolution of increased pathogen virulence. It has been argued by the evolutionary biologist Paul Ewald in his book *Evolution of Infectious Diseases* (1994) that there may have been strong selection for highly virulent viral strains during the 1918 worldwide influenza epidemic. The crowded conditions of soldiers in the trenches during World War I and the continual replacement of sick individuals with new recruits greatly increased the transmission opportunities for the virus. Conversely, mechanisms that reduce transmission opportunity may favor less virulent strains. Although the evidence is circumstantial, the reduction in the virulence of syphilis following its introduction into Europe may have been an evolutionary response to a reduction in sexual activity and the rise of puritanism (i.e., cultural changes in humans).

New host–pathogen associations are often thought to be more virulent than long-established ones, but there is no direct evidence for this generality. New virulent diseases attract our attention much more than new benign diseases. The high virulence of introduced diseases may also occur because of a lack of immunological response. Rapid pathogen growth and therefore high virulence may be important to evade induced host defenses or to outcompete other pathogen strains within a host. Virulence may be an accidental byproduct of "dead-end" evolution of the pathogen within the host. During an infection, pathogen variants arise that are able to infect critical tissues and organs, but these variants may not be transmitted. Acquired immune deficiency syndrome (AIDS) dementia, for example, is of no advantage to the human immunodeficiency virus (HIV) in that it cannot be transmitted from brain tissues.

Coevolution. Host–pathogen coevolution has resulted in a wide array of infection mechanisms in the pathogen and resistance factors in the host. Disease symptoms are not only the result of resources of the host being preempted by the pathogen, but can also be the result of pathogen manipulation of the host to increase pathogen transmission, or of the host mounting inducible defenses against the pathogen. Because these various adaptations involve fitness trade-offs with other traits, genetic variation in host–pathogen interactions is commonplace. Bacteria have evolved resistance to their viruses (the bacteriophages) by degrading alien DNA using restriction enzymes. These enzymes cut DNA at short highly specific base pair sequences. The bacterium protects its own DNA by methylation of bases in these sequences, and its enzymes cut the corresponding region of the phage DNA. Some phage in turn have mechanisms to overcome these restriction enzymes. The DNA of T2 phage. Which infects the bacterium *Escherichia coli*, contains the base hydromethylcytosine in place of normal cytosine. It thereby avoids recognition by the restriction enzymes. (The great variety of restriction enzymes incidentally, has proved to be an invaluable tool in molecular biology for the controlled manipulation of DNA.)

Host–pathogen coevolution has also been responsible for the presence of so-called gene-for-gene systems in plants and their pathogens. Such systems were first identified by a plant breeder in cultivated flax (*Linum usitatissimum*) and its rust disease caused by the fungus *Melampsora lini*. The host plant is resistant because an "elicitor" molecule present in the pathogen stimulates a hypersensitive response in infected plant cells. These cells self-destruct and so curtail the further growth of the pathogen. Resistance can be overcome by the pathogen mutations resulting in the loss of the elicitor. This serves to establish an "arms race," with the result that for most resistance genes in the host, there are usually corresponding genes in the pathogen that can overcome the resistance. Jeremy Burdon and A. M. Jarosz (1991) have shown that gene-for-gene systems are present in wild populations of flax in Australia and are not just the product of evolution in an agricultural context. Gene-for-gene systems have also been identified in insect pests (e.g., the Hessian fly on wheat).

There are many cases of genetic variation in resistance to infectious disease in humans. The best known example is the sickle-cell gene, which confers resistance to malaria. Specific blood groups have been associated with resistance to the smallpox virus. A small percentage of people are largely resistant to HIV as the result of a genetically altered receptor protein on the T cells. In all vertebrates, there is also a high degree of polymorphism in the major histocompatibility (MHC) genes involved in antigen recognition. The evolution of the MHC genes is very rapid, and they are highly polymorphic in human and animal populations. One of the "side effects" of this variation is tissue graft rejection; the tissue from a genetically unlike individual is misidentified by the MHC system as an invading "pathogen."

Intracellular and genomic parasites may be plasmids, bacteria or overreplicating organelles in the cytoplasm, or they may be transposable elements that invade DNA directly. Many arthropods contain *Wolbachia*-like bacteria, which are maternally inherited and have a variety of effects (sex-ratio distortion, sterility) on their hosts. Correspondingly, the hosts have genes that minimize these negative effects. For example, the pill bug (*Armadillidium vulgare*) has a nuclear gene that minimizes the effects of *Wolbachia*. Transposable elements are regions of DNA that replicate themselves into "uninfected" genomes and are transmitted through the germ line. Over 50 percent of the corn genome is made up of transposons. In the human genome, there are about one million copies of retrotransposon-like elements representing 25 percent of the total DNA. Some of these are still actively transposing, and some mutations to hemophilia and muscular dystrophy have been shown to result from the insertion of a transposable element into the functional gene.

The evolution of antibiotic resistance in medically important pathogens has been of enormous concern. The ability of pathogens to evolve in response to therapeutic agents is probably in part the result of their having enormous genetic flexibility to overcome the arsenal of a host's defenses. For example, several studies show that penicillin-resistant *Neisseria gonorrhoeae* (the bacterium causing gonorrhea) now occur in most clinical isolates, although for many years all strains were sensitive. Some pathogens are notable in that they have failed to evolve resistance to drugs. For example, penicillin has been used to treat syphilis for fifty years, but there is no recorded case of a resistant strain. The evolution of resistance to vaccines also appears to have been rare, even though they have been used extensively throughout the world. However, vaccine-escape mutants have been reported in the case of the hepatitis B virus. Because immunization activates whole families of antibodies, evolution to such multiple challenge may be less likely than evolution of resistance to a single antibiotic.

Over the past twenty-five years, the evolutionary mechanisms responsible for the maintenance of genetic recombination and sex have been widely debated. One of the favored explanations for the maintenance of sex is the need for continuous adaptation to ever-changing pathogen pressures. The susceptibility of genetically uniform crops to pathogens is well known, and in nature asexual forms of otherwise sexual species tend to be more common on the periphery of a species range, where they are likely to have escaped pathogen pressure. Moreover, all vertebrate hosts and some pathogens have somatic recombination mechanisms that act to increase their genetic variation. The immune system generates variants by gene rearrangements, and trypanosomes generate antigen diversity by insertion of different DNA sequences into the coding region for surface proteins. More direct evidence comes from natural populations, where there are both sexual and asexual forms of the same species. In the Central American fish *Poeciliopsis* and in the freshwater New Zealand snail *Potamopyrgus*, sexual forms are more abundant where there are more parasites. Experimental studies on sweet vernal grass (*Anthoxanthum odoratum*) have shown that the incidence of barley yellow dwarf virus is more common when the parents are surrounded by their own, rather than genetically unrelated, progeny.

Disease has also been implicated as a driving factor in sexual selection and the evolution of exaggerated secondary sexual traits. Secondary sexual characteristics may act as indicators of disease status, and females may choose particular males because they have fewer parasites or are genetically more resistant to those parasites. Strong evidence for this has come from the work of Anders Møller (1994) on barn swallows (*Hirundo rustica*). Males with longer tails have fewer ectoparasites and are preferred as mates by females.

Emerging Diseases. Newly emerging diseases are novel host–pathogen combinations that often involve evolutionary changes in the host and parasite. The species richness of many pathogen taxa argues that host shifts are often accompanied or followed by forces that lead to high rates of divergence and speciation. This has been well documented for the apple-maggot fly (*Rhagoletis*), which has undergone host shifts in the northern United States from the native hawthorn species to apples and cherries introduced for fruit production. Entry into a new host is accompanied by genetic changes in the pathogen. Experiments involving "passaging" a disease on a new host by successive infection or inoculation show that these changes may take only a handful of generations, and typically lead to decreased performance on the original host. Such rapid pathogen specialization to new hosts may explain the high species diversity of parasites.

When closely related pathogen species are found on closely related hosts, it is of interest to know if those pathogens evolved through cospeciation or host shifts. Highly coordinate or matching phylogenies argue for long-term associations and very rare host shifts. The endosymbiotic bacterium *Buchnera* shows a completely coordinate phylogeny with its aphid hosts, indicating long-term vertical transmission; whereas another endosymbiotic bacterium of insects, *Wolbachia*, shows intermittent cross-species transmission. The causative processes involved in these shifts can be determined by looking for common features of either the host or the pathogen that are associated with such shifts. Host shifts in the chrysomelid beetle genus *Blepharisma* on its plant hosts in the genus *Bursera* were more correlated with host secondary chemistry than with their overall phylogenetic similarity.

When a new disease appears, it is very important to know the nature of the causative agent so that therapeutic measures can be undertaken. Comparison of the pathogen's DNA with sequences of known organisms in existing databases is a complementary approach to simply studying symptoms or serological reactions. Moreover, it is important to identify the source or reservoir of the disease so that measures can be taken to prevent recurrences. For example, the clustered occurrence of hemorrhagic pneumonia in the western United States in 1993 was shown to be due to a virus similar to that isolated during the Korean War in Hantaan province (hence the common name, hanta virus). Natural populations of deer mice (*Peromyscus maniculatus*) were shown to harbor a virus with an almost identical DNA sequence. DNA-based methods have also established that HIV in humans arose independently several times from primate populations.

The use of phylogenetic methods in tracing the spread of diseases within and among populations is now well established and is termed *molecular epidemiology*. These methods have been used to identify sources of HIV infections. They are also being used to track and identify drug-resistant strains of tuberculosis. Many pathogens undergo evolutionary change while inside a particular host, and phylogenetic analyses of strains sampled from a host at different times have been used to characterize changes in the HIV virus during the course of an AIDS infection. Sequence analysis of the hemagglutinins genes of influenza viruses have shown that certain codons are under strong selection. The presence of changes in these codons can help predict which influenza strains are likely to give rise to descendant lineages. This illustrates the power of evolutionary analysis in pinpointing genetic factors in the emergence of new diseases, as well as its potential in monitoring programs.

Species Extinctions. Disease has been suggested as a factor in several of the major extinctions (e.g., the Cretaceous-Tertiary extinction of the dinosaurs and the mammalian extinctions in North America in the Pleistocene). However, the evidence for this is circumstantial at best, and it is not known whether disease has ever caused a major extinction event. On the other hand, because so many pathogens and parasites are highly host specific, the extinction of a particular host is almost certainly accompanied by the extinction of perhaps dozens of its pathogens. Such parasite extinctions usually go unnoticed in present counts of extinction rates, and parasites rarely appear on lists of endangered species. Highly host-specific diseases are unlikely to lead to direct extinction of their host populations, as the decline in the host abundance will lead to a decline in the transmission rate. Generalist pathogens are much more likely to lead to extinction because they may have a reservoir in a much more common species. The effect of one species on another via a shared parasite or predator has been termed *indirect competition*. Thus, the black-footed ferret (*Mustela nigripes*) was endangered by transmission of canine distemper virus from domestic and other zoo animals. Many birds in Hawaii have gone extinct because of the introduction of bird malaria (*Plasmodium* spp.) and the mosquito vectors (*Culex* spp.). In present-day populations, introduced species of birds act as disease reservoirs that restrict many native Hawaiian birds to high-elevation regions that are free of mosquitoes.

[*See also* Overview Essay *on* Darwinian Medicine; Acquired Immune Deficiency Syndrome, *article on* Origins and Phylogeny of HIV; Antibiotic Resistance, *articles on* Origins, Mechanisms, and Extent of Resistance *and* Strategies for Managing Resistance; Coevolution; Emerging and Re-Emerging Diseases; Frequency Dependent Selection; Heterozygote Advantage, *article on* Sickle-Cell Anemia and Thalassemia; Immune System, *article on* Microbial Countermeasures to Evade the Immune System; Influenza; Malaria; Myxomatosis; Plagues and Epidemics; Red Queen Hypothesis; Resistance, Cost of; Sex, Evolution of; Transmission Dynamics; Vaccination; Virulence.]

BIBLIOGRAPHY

GENERAL BOOKS ON DISEASE EVOLUTION
Crandall, K. A., ed. *The Evolution of HIV.* Baltimore, 1999.
Ewald, P. W. *Evolution of Infectious Diseases.* Oxford, 1994.
Ewald, P. W. *Plague Time: How Stealth Infections Cause Cancers, Heart Disease, and Other Deadly Ailments.* New York, 2000.
Fritz, R. S., and Simms, E. L., eds. *Plant Resistance to Herbivores and Pathogens: Ecology, Evolution, and Genetics.* Chicago, 1992.
Price, P. W. *Evolutionary Biology of Parasites.* Princeton, 1980.
Wills, C. *Yellow Fever, Black Goddess: The Coevolution of People and Plagues.* Reading, Mass., 1996.

OTHER READINGS
Muller, A. P. *Sexual Selection and the Barn Swallow.* Oxford, 1994. Oxford. Chapter 9 is an excellent account of theoretical issues and the experimental evidence for parasite-mediated sexual selection.
Stearns, S., ed. *Evolution in Health and Disease.* Oxford, 1999. Has several chapters on the evolutionary biology of pathogens, including the evolution of virulence, resistance to drug and vaccine therapy, and use of phylogenetic methods.

PAPERS
Becerra, J. X., and D. L. Venable. "Macroevolution of Insect–Plant Associations: The Relevance of Host Biogeography to Host Affiliation." *Proceedings of the National Academy of Sciences USA* 96 (1999): 12626–12631.
Bull, J. J. "Virulence." *Evolution* 48 (1994): 1423–1437.
Burdon, J. J., and A. M. Jarosz. "Host–Pathogen Interactions in Natural Populations of *Linum marginale* and *Melampsora lini*: 1. Patterns of Resistance and Racial Variation in a Large Host Population." *Evolution* 45 (1991): 205–217.
Bush, R. M., C. A. Bender, K. Subbarao, N. J. Cox, and W. M. Fitch. "Predicting the Evolution of Human Influenza A." *Science* 286 (1999): 1921–1925.
Ebert, D. "Experimental Evolution of Parasites." *Science* 282 (1998): 1432–1435.
Hughes, A. L., and M. Nei. "Pattern of Nucleotide Substitution at Major Histocompatability Complex Class I Loci Reveals Overdominant Selection." *Nature* 335 (1988): 167–170.
Kelley, S. E. "Viral Pathogens and the Advantage of Sex in the Perennial Grass *Anthoxanthum odoratum*." *Philosophical Transactions of the Royal Society of London B* 346 (1994): 295–302.
Lenski, R. E., and B. R. Levin. "Constraints on the Coevolution of Bacteria and Virulent Phage: A Model, Some Experiments, and Predictions for Natural Communities." *American Naturalist* 125 (1985): 585–602.
Lively, C. M., C. Craddock, and R. C. Vrijenhoek. "Red Queen Hypothesis Supported by Parasitism in Sexual and Clonal Fish." *Nature* 344 (1990): 864–866.
Moran, N., and P. Baumann. "Phylogenetics of Cytoplasmically Inherited microorganisms of Arthropods." *Trends in Ecology and Evolution* 9 (1994): 15–20.
Samson, M., F. Libert, B. J. Doranz et al. "Resistance to HIV-1 Infection in Caucasian Individuals Bearing Mutant Alleles of the

CCR-5 Chemokine Receptor Gene." *Nature* 382 (1996): 722–725.

Sassaman, D. M., B. A. Dombroski, J. V. Moran, et al. "Many Human L1 Elements Are Capable of Retrotransposition." *Nature Genetics* 16 (1997): 37–43.

Van Riper, C., S. G. van Riper, M. L. Goff, and M. Laird. "The Epizootiology and Ecological Significance of Malaria in Hawaiian Birds." *Ecological Monographs* 56 (1986): 327–344.

— JANIS ANTONOVICS

Demography and Human Disease

Several demographic trends of the past ten thousand years have had dramatic consequences for the spread of human diseases. First, agriculture, coupled with animal husbandry, largely replaced foraging as a subsistence strategy. Second, human populations grew substantially. Third, people became more sedentary. Finally, populations became more interconnected through trade, warfare, and population movements. Each of these trends continues today.

In *Guns, Germs, and Steel* (1997), Jared Diamond presents a revealing discussion of infectious disease from the microbe's point of view. Infection is an evolutionary battle between the microorganism, which derives its nutrition from the host's living body, and the host, which tries to stop it. The microbe's mission is to spread its offspring to new hosts. Many of the symptoms of infectious diseases are microorganisms' modifications of human bodies to facilitate this spread. Several disease microorganisms, including those that cause influenza, pertussis, and the common cold, use the host to broadcast them through coughing or sneezing. Wet sores associated with smallpox and venereal syphilis spread microbes through direct or indirect body contact. Violent bouts of diarrhea associated with fecal-oral diseases, such as amoebic dysentery and cholera, transport them back into the water supply, where they can be ingested by new hosts.

Epidemic diseases, such as smallpox and measles, have several common characteristics. They attack populations in waves, afflicting large percentages of the population over a brief period, then disappear for extended periods. They pass quickly from host to host, exposing entire populations in a short time. They tend to be acute, but of short duration; hosts either successfully fight the disease or succumb to it quickly. Survivors develop antibodies that provide long-term immunity to the disease. In many cases, including measles and smallpox, immunity lasts for the rest of the survivor's life. These characteristics mean the microorganisms that cause epidemic diseases require frequent human contacts for transmission, as well as a large population to maintain the pathogen lineage.

Other infectious diseases spend extended periods outside human hosts. Malaria and bubonic plague are transmitted by insect vectors: malaria by mosquitoes, and bubonic plague by fleas that infest wild rodents. Plague may have been endemic in black rat colonies of Asia since the thirteenth century. Schistosomiasis completes its life cycle in freshwater snails. Other parasites, such as the bacteria that cause botulism and tetanus, can survive for long periods in soils. Mark Cohen, in *Health and the Rise of Civilization* (1989), provides a good summary of disease vectors.

Domesticated Animals. Many epidemic diseases that affect contemporary or historical populations originated in domesticated animals. Smallpox and measles appear to be descended from diseases of cattle. Influenza probably came from a disease of pigs or chickens. William McNeill (1976) argues that most, if not all, distinctive infectious diseases of civilized society came from animal herds. He reports that humans share fifty diseases with cattle, forty-six with sheep and goats, forty-two with pigs, and twenty-six with poultry (some of these diseases afflict several domesticated species). Diamond (1997) argues that the main reason lethal crowd diseases were relatively absent in the Americas before European contact was the paucity of domesticated animals.

Agriculture. Agriculture fundamentally changes how humans interact with their environment. Farmers homogenize the landscape. They clear it of undesired species, which compete for nutrients and sunlight, and populate it with desired species. Although this simplification of the natural ecosystem may have eliminated, or retarded, some disease vectors, it benefited others.

Malaria and schistosomiasis are frequently cited examples of how human alteration of the landscape for agriculture promoted the spread of disease. Malaria, especially falciparum malaria, is one of history's great killers. Anopheles mosquito larvae thrive in standing water with a mix of light and shade. Bodies of stagnant water created by farming communities create precisely these environments. The snails that serve as the vector for the waterborne life cycle stage of schistosomiasis require narrow environmental conditions. They survive best in water that is nonturbid but rich in dissolved solids, dissolved oxygen, and weeds. These conditions are well met in irrigation ditches and shallow, artificial ponds. Schistosomiasis is thus widespread among irrigation and rice paddy farmers.

Population Growth. Population size and density affect the transmission of epidemic diseases. Epidemic diseases such as measles that leave victims immune to future infection must have a constant supply of new hosts to survive. Measles require something on the order of five thousand to forty thousand new hosts per year. Contemporary nations such as Guatemala that have high fertility rates may sustain birthrates close to forty per thousand population, per year. At that rate, measles would need a sustaining population of between 125,000 and one million people. McNeill (1976) argues that parasites that cause epidemic diseases are probably of relatively recent origin, on an evolutionary scale.

Sedentism. When humans stay in one place for an extended period, refuse can build up, increasing disease transmission from animals attracted to easily obtained food from human garbage. Flies, feeding on human garbage, spread fecal-oral diseases. Rats carry bubonic plague, as well as various hemorrhagic fevers. Canids (dogs, coyotes) carry rabies, and cats carry toxoplasmosis.

Sedentism leads to the accumulation of human waste. This promotes the spread of fecal-oral transmitted diseases, including diarrheal diseases caused by a variety of bacterial and protozoan species. Diarrheal diseases are among the leading causes of childhood death in the developing world today and were probably among the major killers in preindustrial cities and towns. Onchocerciasis, or river blindness, is an infection of simulium worms carried by flies. Dense human settlements fertilize rivers, making them better breeding grounds for flies, thus spreading the disease. Such disease cycles tend to be broken by mobility.

Communication. Increased levels of communication between groups expand host pools for epidemic diseases. McNeill (1976) sees the recent history of epidemic disease as the direct result of a series of expansions in human communication: the linking of China, India, and the Mediterranean world in the early Christian era, the spread of the Mongol empire in the thirteenth century, and the beginning of European seaborne exploration in the fifteenth century.

McNeill (1976) argues that the major civilized regions of the Old World had evolved their own distinct infectious disease regimes by around 500 BCE. By that point, Babylonian, Egyptian, and Chinese written records show that epidemic disease was well established. The linking of China, India, and the Mediterranean resulted from internal developments in all three regions. The Yangtze valley was not fully developed agriculturally and was not integrated into the Chinese social system until the end of the Han dynasty (221 CE). Greater India arose, through the efforts of traders and missionaries, between 100 BCE and 500 CE, creating a larger system that spread into Southeast Asia and to Indonesia. Extensive trade networks developed around the Mediterranean in the first millennium BCE as economically differentiated regions were linked by efficient (and therefore cheap) seaborne transportation. Roman imperialism and the expansion of trade networks effectively unified the Mediterranean world, including much of the Middle East and portions of Europe north of the Alps. The establishment of regular caravan trade along the overland Silk Road and corresponding sea routes, which linked these three host pools, was facilitated by the growth of these stable large-scale political entities. Small, densely settled political entities quickly arose along the way.

New epidemic diseases in China and the Mediterranean followed. China experienced a severe epidemic in 161–162 CE, which killed 30–40 percent of adult men. Four years later, the Antonine plagues (165–180 CE) devastated the Roman empire. The plague of Justinian (542 CE), probably bubonic plague, was spread by merchant ships from Egypt throughout the Mediterranean world. The earliest probable description of bubonic plague in China dates to 610 CE.

Warfare has acted as a vector of communication between populations, yielding dramatic consequences for disease. In *History of the Peloponnesian War*, Thucydides' firsthand description of the plague of Athens (430 BCE) presents a harrowing picture of an epidemic in wartime. Roman legions brought the Antonine plague back from campaigns in the Near East. Chinese armies carried the "barbarian pox" to China from their attacks on Nan-yang (317 CE).

The Mongol expansion may have been the catalyst for the fourteenth-century bubonic plague epidemic, the Black Death, in Eurasia. The plague originated in central Asia and reached China, which was under Mongol control, in 1331. It killed an estimated twenty-five million people in Asia in the fifteen years before it arrived in Constantinople in 1347. The plague arrived in Europe with a Mongol army that laid siege to the Crimean trade city of Caffa in 1346. From Caffa, it spread by ship through the Mediterranean and to western and northern Europe. McNeill (1976) estimates that approximately one-third of Europe's total population died of plague between 1346 and 1350. Other estimates hover between 20 and 45 percent, with locally higher losses. *Pasturella pestis* is spread by rodents (black rats). McNeill believes the Mongols facilitated its spread two ways: first, by moving trans-Asia transportation routes north, into the steppe, where a new host population of rodents became infected; second, by increasing the speed of movement across Asia.

McNeill suggests that as disease pools became larger and more connected, the period between successive exposures shortened. This meant larger proportions of persons with effective immunities, and therefore more limited impacts from later waves. Epidemic diseases such as measles became diseases of childhood. Short intervals between waves also created significant reproductive advantages for resistant individuals. This selective pressure would have increased overall population resistance to common parasites. This becomes abundantly clear during the period of initial contact between Europeans and New World peoples, brought about by European exploration in the fifteenth century. For example, in Veracruz, devastating outbreaks of smallpox and other Old World diseases followed the incursion in 1519 by Hernán Cortés and his forces. The population of central Mexico eventually collapsed, falling as much as 90 percent by the early seventeenth century (Cook and Borah, 1963). Virgin soil epidemics had similar effects among the indigenous populations of the Andes, the Mississippi valley, and other parts of the New World.

The expansion of host pools to its logical conclusion is evident by the Spanish influenza epidemic of 1917–1919. The worldwide death toll in just two years was estimated to be twenty million, with estimates of infection as high as five hundred million. The disease affected virtually every nation on earth.

[*See also* Overview Essay *on* Darwinian Medicine.]

BIBLIOGRAPHY

Cohen, M. *Health and the Rise of Civilization.* New Haven, 1989. Interesting study of the impact of cultural factors on disease loads. Pays particular attention to differences between foragers and agriculturalists.

Cook, S., and W. Borah. "The Aboriginal Population of Central Mexico." In *Ibero-Americana*, vol. 45. Berkeley, 1963.

DeSalle, R. *Epidemic! The World of Infectious Disease.* New York, 1999. Very readable introduction to epidemiology.

Diamond, J. *Guns, Germs, and Steel.* New York, 1997. Interesting model of the worldwide history of political domination, based on ecological factors such as disease history. Well-written accounts of disease transmission.

McNeill, W. *Plagues and Peoples.* Garden Ciry, N.Y., 1976. Highly influential history of disease and its impact on human history. Very readable.

Scott, S., and C. J. Duncan. *Human Demography and Disease.* Cambridge, 1998. More mathematical than the other resources listed here.

Thucydides. *History of the Peloponnesian War.* Translated by Rex Warner. New York, 1954. Includes graphic firsthand account of the Great Plague of Athens (430 BCE).

— RICHARD PAINE

DNA AND RNA

Deoxyribonucleic acid (DNA) and ribonucleic acid (RNA) are the principal units of heredity in all living things. DNA and RNA molecules are long linear strings, or polymers, of simpler molecules called nucleotides. Nucleic acids play a fundamental role in the storage and delivery of biological information. DNA is an exceptionally stable molecule, and fragments of intact DNA are routinely extracted from samples that are thousands of years old. RNA molecules are less stable, but they can fold into three-dimensional structures with catalytic properties. RNA molecules are distinguished by their functions within the cell; the principal types are messenger RNA (mRNA), transfer RNA (tRNA), and ribosomal RNA (rRNA).

The nucleotides that comprise nucleic acids consist of any of several nitrogen-containing *bases* linked to a 5-carbon pentose sugar that carries a phosphate group. The base component of a nucleotide can be either a pyrimidine—cytosine (C), thymine (T), or uracil (U)—or a purine, guanine (G) or adenine (A). The pentose sugar component can be either ribose or deoxyribose. Nucleotides can link together in a chain through the formation of phosphodiester bonds between the 5′ phosphate group on the sugar residue of one nucleotide and the 3′ hydroxyl group of the next. DNA and RNA are distinguished by the kind of 5-carbon sugar- and nitrogen-containing bases carried by their nucleotide subunits. The nucleotide subunits in DNA have deoxyribose as their pentose sugar component and either A, T, G, or C as a base. RNA consists of nucleotides with a ribose sugar, and containing any of the bases A, U, G, or C. Within each of these types of nucleic acid, the nucleotide subunits can be linked together in any order. The terminal nucleotide at one end of the polynucleotide chain has a free 5′ group; the terminal nucleotide at the other end has a free 3′ group. Nucleic acids can be represented in written form by the sequence of bases they contain (e.g., ATAGGTCT). It is conventional to write nucleic acid sequences in the 5′–3′ direction—that is, with the 5′ terminus on the left and the 3′ terminus on the right.

The central dogma of modern biology states that information flows through biological systems unidirectionally, from DNA through RNA to protein. Within a cell, the DNA genome is the repository of genetic information. The information contained in DNA is expressed indirectly via RNA. DNA is copied into mRNA by the process of *transcription*. The mRNA acts as an intermediary to protein synthesis, delivering genetic information to the locations in the cytoplasm where protein synthesis occurs. *Translation* of mRNA directs the assembly of the polypeptides (proteins) that determine the cell's chemical and physical properties. When a cell divides, the process of *replication* duplicates the information contained within the DNA genome.

As the concept of inheritance was established in the decades following the publication of Darwin's *Origin of Species*, the challenge to biologists became to identify the hereditary material and determine how it represented hereditary information. Initially it was assumed that molecules with diverse and complex structures, such as proteins, would form the basis of inheritance. However, a series of experiments in the early twentieth century clearly demonstrated the role of DNA as the repository of genetic information.

The earliest investigations centered on the phenomenon of *transformation* in the bacterium *Pneumococcus*. The presence or absence of a surface polysaccharide in this bacterium determines its pathogenicity. The smooth (S) strain, in which the polysaccharide is present, causes a fatal pneumonia in mice, while the rough

(R) strain, which lacks the polysaccharide, is avirulent. In 1928, Griffith demonstrated that while neither an injection of avirulent R bacteria nor an injection of heat-killed S bacteria was sufficient to cause death in laboratory mice, an injection that contained a mixture of the two was. Furthermore, virulent S type bacteria could be recovered from mice infected in this way. It was apparent from this that some factor retained in the heat-killed S bacteria contained the instructions for making the crucial polysaccharide, and that it had the capacity to *transform* the live R bacteria into the virulent S strain. The next step was to identify the biochemical factor responsible—the "transforming principle." In 1944, it was purified by Avery and colleagues and demonstrated chemically to be DNA.

The work of Hershey and Chase later reinforced the evidence that DNA could carry genetic information, by demonstrating it in a different system. Bacteriophage T2 is a virus that infects the bacterium *Escherichia coli*. T2 reproduces by subverting the machinery of an infected cell and directing to it to manufacture progeny phage. Electron microscope images had shown that the phage infected a bacterial cell by attaching to its exterior and injecting an unknown material. Hershey and Chase postulated that the material was the phage genetic information. To determine whether the material consisted of protein or DNA, they labeled viral proteins with a radioactive isotope of sulfur and the viral DNA with a radioactive isotope of phosphorus. After allowing labeled viruses to infect bacteria, they observed that the majority of radioactively labeled phosphorus entered the bacterium and was incorporated into progeny phage, whereas almost all of the radioactive sulfur remained outside. This demonstrated that the hereditary information of phage T2 was contained in DNA, reinforcing the general conclusion that the genetic material is DNA, regardless of whether the organism is viral or bacterial.

Modern biology has established that DNA is the genetic material in all known organisms. The only exception occurs in some viruses that use RNA instead. Proof that DNA was the genetic material in eukaryotic systems was based on experiments in eukaryotic cell culture in which transfer of DNA was shown to transfer specific properties from the cells of one species to the cells of another. Experiments of this type demonstrated not only that DNA is the genetic material in eukaryotes, but also that it can be transferred between different species and yet remain functional.

By the 1950s, the thrust of research had shifted toward explaining how DNA carries genetic information. The observation that bases are present in different amounts in the DNA of different species led to the idea that the sequence of bases is the form in which information is carried. The challenge to biologists was to work out the structure of the DNA molecule, and to explain how the sequence of bases encoded information. Francis Crick and James Watson were the first to describe the structure of DNA. Their model was partly based on X-ray diffraction photographs, taken by Maurice Wilkins and Rosalind Franklin, that showed great consistency and symmetry in the structure of DNA and gave important clues about its dimensions. Also key to Watson and Crick's model was Edwin Chargaff's 1949 observation that, in the DNA of any given type of cell, the amount of adenine approximately equals the amount of thymine, and the amount of cytosine approximately equals the amount of guanine (Chargraff's rule).

The model of DNA structure Watson and Crick developed is known as the *double helix*. In it, two polynucleotide chains associate by hydrogen bonding between the nitrogenous bases. The bases bond specifically, G with C and A with T, in reactions described as base pairing. The paired bases are said to be *complementary*, and the specificity of base pairing explains why the amounts of G and C, and of A and T, are approximately the same in DNA. The sugar phosphate backbones of the two chains are on the outside. They are twisted into a helical structure and carry the negative charges on the phosphate groups. The bases lie on the inside, perpendicular to the axis of the helix. The double helix structure resembles a spiral staircase, with the base pairs forming the steps. The model requires the two strands to run in opposite directions. Looking along the helix, therefore, one strand lies in the 5′ to 3′ direction, while its partner runs 3′ to 5′. Watson and Crick noted that the variable sequence of nucleotides running along the backbone of the double helix could provide a means of representing hereditary information in a simple, enciphered form—the genetic code. Furthermore, the specificity of base pairing meant that the sequence of bases contained within one nucleic acid could be copied or read by another, providing a simple means to transmit information. This elegant and straightforward principle underlies all heredity and evolution.

DNA in genomes is complexed with protein and occurs in a tightly packaged form. The bacterial genome forms a dense nucleoid, about 20 percent protein by mass, but precise details of the interactions of the protein and DNA are not known. Eukaryotic DNA is contained in nuclear structures called *chromosomes*, in which a single unbroken double-stranded DNA molecule is complexed with proteins called *histones* and with other non-histone proteins. The complex of DNA and protein is called *chromatin*. The basic structural subunit of chromatin, called a *nucleosome*, consists of approximately 200 base pairs (bp) of DNA and nine histone proteins.

Whenever a cell divides, the genome must be replicated to provide each daughter cell with a complete set of biological instructions. During replication, the DNA double helix duplex is locally "unzipped," exposing the individual strands in the helix. The specificity of base pairing means that each of the separated parental strands can act as a template for the assembly of a complementary daughter strand. An A in the parental strand directs incorporation of a T in the daughter, a parental G specifies a daughter C, and so on. This model of DNA replication predicts that when a DNA molecule is duplicated, the resulting daughter duplexes will contain one strand from the parental molecule, and one newly synthesized daughter strand; it is therefore called *semiconservative replication.*

Mathew Meselson and Franklin Stahl confirmed experimentally that DNA replicates semiconservatively by following the replication of radioactively labeled DNA through three generations of growth of *E. coli*. The first generation of bacteria was grown in a medium containing a heavy isotope, which was incorporated into the DNA. The second and third generations were grown in a medium containing normal "light" isotopes, so that any newly synthesized DNA would be labeled with the light isotope. DNA from each of the bacterial generations was extracted, and its density measured by centrifugation. The DNA of the first generation consisted of a duplex of two heavy strands and had high density. After one generation of growth in the light medium, however, the duplex DNA was hybrid in density, indicating that it contained one "heavy" and one "light" strand. In the third generation, each strand gained a light partner, so that half of the duplex remained hybrid while half was entirely light, and DNA of two different densities was seen. The individual strands of these duplexes were all entirely heavy or entirely light.

The synthesis of nucleic acids is catalysed by specific enzymes, which recognize a polynucleotide template and direct the synthesis of a new polynucleotide chain through the catalytic addition of nucleotide subunits according to the logic of base pairing. The enzymes are named according to the type of chain that is synthesized: DNA polymerases synthesize DNA, and RNA polymerases synthesize RNA. An enzyme that can synthesize a new DNA strand on a template strand is called a *DNA polymerase*. During replication, an enzyme called a helicase separates the paired strands of the parental DNA duplex. The local unwinding of the parental duplex forms a replication "bubble." Within the bubble, the individual strands of the parental template are exposed and can be accessed by a DNA polymerase. As replication proceeds, the parental template is unwound at one end and rewound at the other as the daughter duplex is formed. The replication bubble thus moves along the parental duplex, and only a small part of the DNA loses its duplex structure at any moment.

The two strands of a DNA duplex are oriented in opposite directions. As the replication bubble proceeds along the parental duplex, it moves in the $3'-5'$ direction relative to one strand, and $5'-3'$ relative to the other. However, DNA polymerases can extend polynucleotide chains only by adding nucleotides to the free $3'$ end of a growing chain. This complicates the process of replication. At the leading edge of the replication bubble, DNA polymerase encounters one template strand in the $5'$ to $3'$ direction, and it can therefore synthesize a new polynucleotide chain continuously as the parental duplex is unwound. The DNA synthesized in this way is called the *leading strand*. Synthesis of the opposite, *lagging strand* requires that the polymerase synthesizes new template in the direction opposite to which it is encountered. Consequently, only a restricted length of daughter polynucleotide can be synthesized at any one time, and the daughter strand has to be assembled from a series of short fragments. This mode of synthesis, continuous on one strand and discontinuous on the other, is called *semidiscontinuous replication*; it is thought to be common to prokaryotic and eukaryotic systems. The 1000–2000 base fragments that are formed during the discontinuous replication of the lagging strand are called *Okazaki fragments* after the scientist who discovered them.

DNA polymerases cannot actually initiate the synthesis of a deoxyribonucleotide chain; they can only add nucleotides to a chain that is already correctly paired with its complementary strand. Consequently, replication requires an RNA primer, synthesized by an enzyme called a *primase*, to provide free $3'$-OH ends to initiate the DNA chains on both the leading and the lagging strands. The leading strand requires only one such initiation event, but each of the fragments that make up the lagging strand requires a separate priming event for its synthesis. The DNA polymerase has a $5'-3'$ exonuclease activity that removes the RNA primer while simultaneously replacing it with a DNA sequence. Once the RNA primer has been removed and replaced, the adjacent Okazaki fragments must be linked together. The $3'$-OH end of the one fragment is adjacent to the $5'$ phosphate end of the previous fragment. An enzyme called *DNA ligase* covalently links adjacent fragments.

It is crucial that genetic information is reproduced accurately. When replication occurs, pairing errors between incoming nucleotides and those in the template occur with a frequency of approximately one every 10,000 base pairs. In bacteria these errors are corrected by the DNA polymerase itself, which has "proofreading" activity. In eukaryotes the basis of proofreading is unknown; it may be a function of the eukaryotic DNA polymerase, or of other enzymes involved in replication.

The DNA inherited by an organism leads to specific traits by directing the synthesis of certain proteins. The physician Archibald Garrod proposed the relationship between genes and proteins in 1909. He postulated that genes dictate phenotype through specifying enzymes that catalyse specific chemical processes in the cell. The first experimental evidence to provide strong support for this hypothesis came from an experiment devised by George Beadle and Edward Tatum. Beadle and Tatum began by isolating mutant strains of the red bread mold *Neurospora crassa*. By manipulating the supply of compounds in the mold's nutrient medium and seeing which allowed mutant strains to grow and which did not, Beadle and Tatum were able to deduce the sequence of biochemical reactions in cells that make necessary compounds like amino acids. Beadle and Tatum distinguished three classes of mutant, each of which, they concluded, was blocked at a different stage in the pathway that synthesizes arginine. They hypothesized that the inability of each mutant to manufacture a single enzyme arose from a defect in a single gene. This became known as the "one gene–one enzyme hypothesis."

Both DNA and protein are composed of a linear sequence of subunits. Proteins are composed of linear sequences of amino acids (polypeptides); DNA consists of a linear sequence of nucleotides. Eventually, analysis of the proteins made by mutant genes demonstrated that the nucleotides in DNA are arranged in an order corresponding to the amino acids in the protein that they specify, confirming Beadle and Tatum's hypothesis. However, it is more accurately stated as one gene–one polypeptide, because since not all proteins are enzymes, and some proteins are composed of multiple polypeptide subunits, each derived from a separate gene.

The genetic code that specifies the flow of information from gene to protein was deciphered in the early 1960s. It is based on triplets of nucelotides called *codons*. Since there are 4 distinct bases in each type of nucleic acid, there are 64 possible triplet combinations of nucleotides (4^3). These 64 codons correspond to 20 amino acids and three "stop signals" called amber, opal, and ochre. A gene includes a series of codons that is read sequentially and translated into a polypeptide chain. In principle, a nucleotide sequence can be translating in any one of three frames depending on where in the nucleotide sequence the decoding process begins. In almost every case, however, only one of these reading frames will produce a functional protein.

A variety of processes can cause alterations in the sequence of genomic DNA, and such alterations are called *mutations*. Mutations in coding sequences can have corresponding effects on the polypeptides they encode. In particular, mutations that involve insertion or deletion of nucleotides can have devastating effects on gene products. Since there are no punctuation signals except at the beginning and end of the RNA message, the reading frame of an RNA transcript is set at the initiation of the translation process and is maintained thereafter. Consequently, an insertion or deletion of bases that does not occur as a multiple of three will cause the entire reading frame to be shifted and will entirely alter the meaning of the sequence thereafter.

DNA is not directly translated into protein; it is first transcribed into mRNA, which retains all the information in the DNA sequence from which it was copied (but possesses a U where a T is found in the DNA template). Only one strand of a DNA duplex is translated into a messenger RNA. The DNA strand from which the RNA molecule is copied is called the *template strand* or *antisense strand*. The opposite strand, which bears a sequence corresponding to the mRNA sequence, is called the *coding strand* or *sense strand*. Transcription is initiated when an RNA polymerase binds to a specific sequence of DNA called a *promoter*. A promoter includes the initiation site where transcription actually begins, and also dozens of nucleotides "upstream" from the initiation site. As RNA polymerase moves along the DNA template, one turn of the double helix is untwisted at a time, separating the strands and exposing about ten DNA bases for pairing with RNA nucleotides. The polymerase adds nucelotides to the $3'$ end of the growing RNA molecule, proceeding until it reaches a specific termination sequence, whereupon it releases the completed RNA transcript. In prokaryotes, RNA transcripts are translated directly after their release from the template, but in eukaryotes, the products of transcription are processed before they leave the nucleus as mRNA molecules.

The codons in mRNA do not directly recognize the amino acids they specify; they use tRNAs, which act as adaptors, recognizing both the amino acid and its specific codons. There is at least one tRNA (but usually more) for each amino acid. The tRNAs fold to produce a compact L-shaped tertiary structure. An amino acid is covalently linked to one end of the tRNA structure, and at the other is an anticodon corresponding to that amino acid. A tRNA is named by using the three-letter abbreviation for the amino acid as a superscript (e.g., $tRNA_{Ala}$ is the tRNA for alanine). During translation, the assembly of polypeptides is directed by specific base pairing between triplet codons on mRNA and the corresponding tRNA anticodons. The specific binding of tRNAs to the mRNA transcript directs the sequence of amino acids during polypeptide synthesis.

The mechanics of ordering tRNA molecules onto mRNA transcripts are complicated and require coordination by cytoplasmic structures called *ribosomes*. Ribosomes are compact particles consisting of ribosomal RNA complexed with an array of different proteins. Each ribosome is a protein-synthesizing machine on which the tRNA molecules position themselves so as to read the genetic message encoded in a mRNA transcript. The ribosome first finds a specific start site on the mRNA that sets the reading frame and determines the amino terminal of the protein. Then, as the ribosome moves along the mRNA molecule, it translates the nucleotide sequence into an amino acid sequence one codon at a time, using tRNA molecules to add amino acids to the growing end of the polypeptide chain. Translation stops when the ribosome reaches one of the three stop codons. When a ribosome reaches the end of a message, both it and the freshly made carboxyl end of the protein are released from the 3′ end of the RNA molecule into the cytoplasm.

The genetic code that determines the translation process is, with a few minor exceptions, the same in organisms as diverse as bacteria, plants, and animals. Presuming that there is no a priori reason why a particular codon sequence should specify a particular amino acid, this fact strongly supports the idea that all life has a single origin.

[*See also* Genetic Code; *articles on* Origin of Life.]

BIBLIOGRAPHY

Avery, O. T., C. M. Macleod, and M. McCarty. "Studies on the Chemical Nature of the Substance Inducing Transformation of *Pneumonococcal* Types." *Journal of Experimental Medicine* 98 (1944): 451–460.

Crick, F. H. C., L. Barnett, S. Benner, and R. J. Watts-Tobin. "General Nature of the Genetic Code for Proteins." *Nature* 192 (1961): 1227–1232.

Griffith, F. "The Significance of *Pneumonococcal* Types." *Journal of Hygiene* 27 (1928): 113–159.

Hershey, A. D., and M. Chase. "Independent Functions of Viral Protein and Nucleic Acid in Growth of Bacteriophage." *Journal of Genetic Physiology* 36 (1952): 39–56.

Meselson and F. W. Stahl. "The Replication of DNA in *E. coli.*" *Proceedings of the National Academy of Sciences USA* 44 (1958): 671–682.

Watson, J. D., and F. H. C. Crick. "A structure for DNA." *Nature* 171 (1953): 737–738.

Watson, J. D., and F. H. C. Crick. "Genetic implications of the structure of DNA." *Nature* 171 (1953): 964–967.

Wilkins, M. F. H., A. R., Stokes, and H. R. Wilson. "Molecular Structure of DNA." *Nature* 171 (1953): 738–740.

— ROBERT GIFFORD

DNA SEQUENCE EVOLUTION. *See* Molecular Evolution.

DOBZHANSKY, THEODOSIUS

Theodosius Dobzhansky (1900–1975) was a key author of the synthetic theory of evolution, which embodies a complex array of biological knowledge centered around Charles Darwin's theory of evolution by natural selection couched in genetic terms. The epithet *synthetic* alludes to the artful combination of Darwin's natural selection with Mendelian genetics, but also to the incorporation of relevant knowledge from other biological disciplines. In the 1920s and 1930s, R. A. Fisher, J. B. S. Haldane, Sewall Wright, and others had developed mathematical accounts of natural selection as a genetic process. Dobzhansky's *Genetics and the Origin of Species*, first published in 1937, refashioned their formulations in language that biologists could understand, dressed the equations with natural history and experimental population genetics, and extended the synthesis to speciation and other cardinal problems.

Genetics and the Origin of Species had an enormous impact on naturalists and experimental biologists, who rapidly embraced the new understanding of the evolutionary process as one of genetic change in populations. Interest in evolutionary studies was greatly stimulated, and contributions to the theory followed, extending the synthesis of genetics and natural selection to other biological fields, such as zoology (notably, Ernst Mayr and Julian Huxley), paleontology (George G. Simpson), and botany (G. Ledyard Stebbins). By 1950, acceptance of Darwin's theory of evolution by natural selection was universal among biologists, and the synthetic theory had become widely adopted.

The line of thought of *Genetics and the Origin of Species* is surprisingly modern—in part, no doubt, because it established the pattern that successive evolutionary investigations and treatises largely would follow. The book starts with a consideration of organic diversity and discontinuity. Successively, it deals with mutation as the origin of hereditary variation, the role of chromosomal rearrangements, variation in natural populations, natural selection, the origin of species by polyploidy, the origin of species through gradual development of reproductive isolation, physiological and genetic differences between species, and the concept of species as natural units. The book's organization was largely preserved in the second (1941) and third (1951) editions, and in *Genetics of the Evolutionary Process* (1970), a book that Dobzhansky thought of as the fourth edition of the earlier one, but had changed too much for publication under the same title.

Dobzhansky extended the synthesis of Mendelism and Darwinism to the understanding of human nature in *Mankind Evolving* (1962), an unsurpassed synthesis of genetics, evolutionary theory, anthropology, and sociology. Dobzhansky expounded that human nature has two dimensions: the biological, which humans share with the rest of life, and the cultural, which is exclusive to humans. These two dimensions result from two interconnected processes, biological evolution and cultural evolution.

Dobzhansky was a prolific scientist, who published nearly 600 titles, a majority of which report results of experimental research, but there are numerous works of synthesis and theory (including more than a dozen books), essays on humanism and philosophy, and others. The incredibly numerous and diversified published works of Dobzhansky are unified by the theme of biological evolution.

Dobzhansky was an engaging teacher and a successful educator of scientists. Throughout his academic career, he trained more than thirty doctoral students and an even greater number of postdoctoral and visiting associates, many of them from foreign countries. Some of the most distinguished geneticists and evolutionists in the United States and abroad are his former students. Dobzhansky spent long periods of time in foreign academic institutions and was largely responsible for the establishment or development of genetics and evolutionary biology in various countries, notably Brazil, Chile, and Egypt.

Dobzhansky gave generously of his time to other scientists, particularly to students and young researchers, but he avoided administrative posts, alleging, perhaps correctly, that he had neither the temperament nor the ability for management. Most certainly, he preferred to dedicate his working time to teaching, research, and writing, rather than to administration.

Dobzhansky was born on 25 January 1900 in Nemirov, a small town 200 kilometers (124 miles) southeast of Kiev in the Ukraine. He graduated from the University of Kiev in 1921 and taught at the Polytechnic Institute in Kiev until 1924, when he became an assistant to Yuri Filipchenko, head of the new Department of Genetics at the University of Leningrad. Filipchenko had started a *Drosophila* genetics laboratory, where Dobzhansky was encouraged to investigate the pleiotropic effects of genes. In 1927, Dobzhansky obtained a fellowship from the International Education Board (Rockefeller Foundation) to work with Thomas Hunt Morgan at Columbia University. In the summer of 1928, he followed Morgan to the California Institute of Technology, where Dobzhansky was appointed assistant professor of genetics in 1929, and professor of genetics in 1936. In 1940, he returned to New York as professor of zoology at Columbia University, where he remained until 1962, when he became professor at the Rockefeller Institute (renamed Rockefeller University in 1965), also in New York City. In September 1971, he moved to the Department of Genetics at the University of California, Davis, where he was adjunct professor until his death in 1975.

[*See also* Speciation.]

BIBLIOGRAPHY

Dobzhansky, T. *Genetics and the Origin of Species.* 3d ed. New York, 1951.
Dobzhansky, T. *Mankind Evolving: The Evolution of the Human Species.* New Haven, 1962.
Dobzhansky, T. *Genetics of the Evolutionary Process.* New York, 1970.
Levine, L., ed. *Genetics of Natural Populations: The Continuing Importance of Theodosius Dobzhansky.* New York, 1995. A collection of essays on Dobzhansky's contributions to evolutionary biology, including both biographical and scientific issues.
Lewontin, R. C., et al., eds. *Dobzhansky's Genetics of Natural Populations,* I–XLIII. New York. Reprints of Dobzhansky's highly influential series of research papers on the genetics of fruit flies (*Drosophila*) in natural populations.

— FRANCISCO J. AYALA

DOLLO'S LAW

Dollo's law, named for the Belgian paleontologist Louis Dollo (1857–1931), is concerned with the irreversibility of evolution. As originally formulated by Dollo (1893), it is the principle that organs or complex structures cannot return to a condition seen in an ancestor (see Gould, 1970, for a review). The anatomist and zoologist Hans Gadow independently came up with the same principle in 1893. Nowadays, we believe that features that resemble ancestral features can reappear in descendant species but not reevolve; the genetic and developmental basis of the ancestral feature was retained in the descendant species.

An embryological implication of Dollo's law is that traces of ancestral features should be evident in descendants. Atavisms, the reappearance of ancestral states such as limbs in snakes or hind limb skeletal elements in whales (Hall, 1984), and phylogenetic character reversals, as frequently seen in fish (Stiassny, 1992) or in viviparity in lizards (Lee and Shine, 1998), demonstrate that the embryological capability to form some structures is retained.

Dollo's law was tested by Marshall et al. (1994) against knowledge of rates of degradation of genetic information. If unused genetic information is lost rapidly, as is the conventional wisdom, then a structure could not reappear exactly as it had been in an ancestor. Marshall and colleagues found that there is a significant chance that silenced genes or "lost" developmental programs can be reactivated if the time scale of loss was 500,000 to six million years ago, but not if more than ten million years ago. Gene or developmental programs are retained (they may serve other functions) and potentially could be involved again in the original function. Such knowledge of gene function and developmental processes provides the mechanistic basis for Dollo's law, a law that links development and evolution.

[*See also* Atavisms.]

BIBLIOGRAPHY

Dollo, L. "Les lois de l'évolution." *Bulletin of Belgian Society for Geology, Palaeontology and Hydrology* 8 (1893): 164–166.
Gould, S. J. "Dollo on Dollo's Law: Irreversibility and the Status of Evolutionary Laws." *Journal of the History of Biology* 3 (1970): 189–212.
Hall, B. K. "Developmental Mechanisms Underlying the Formation

of Atavisms." *Biological Reviews of the Cambridge Philosophical Society* 59 (1989): 89–124.

Lee, M. S. Y., and R. Shine. "Reptilian Viviparity and Dollo's Law." *Evolution* 52 (1998): 1441–1450.

Marshall, C. R., E. C. Raff, and R. A. Raff. "Dollo's Law and the Death and Resurrection of Genes." *Proceedings of the National Academy of Sciences USA* 91 (1994): 12283–12287.

Stiassny, M. L. J. "Atavisms, Phylogenetic Character Reversals, and the Origin of Evolutionary Novelties." *Netherlands Journal of Zoology* 42 (1992): 260–276.

— Brian K. Hall

DOSAGE COMPENSATION

In many organisms the X and Y sex chromosomes are highly differentiated. Typically the X carries many thousands of genes whereas the Y chromosome is largely composed of non-coding repeated sequences with just a few genes. For example, the human Y is as long as the X but only carries *Sry* which initiates male development, several male fertility factors which are absent from the X (e.g., *DAZ, ZFY*) and a few house-keeping genes that have similar though not identical X-linked homologs (e.g., *RPS4Y, DBY*).

Originally the X and Y were homologous chromosomes and the Y contained as many genes as the X. The Y has lost its genetic information owing to the absence of recombination between X and Y, resulting in the accumulation of deleterious recessive mutations on the Y. This is starkly shown by the pseudo-autosomal region, a small region of the Y which still recombines with the X. This region still possesses many active genes that are expressed from both the X and Y chromosomes. A second force leading to differentiation of X and Y has been the selection of male-specific functions on the Y. Over half the active genes on the human Y are specifically expressed in the male testis, and they all lack X-linked homologs. Some, maybe most of these genes arose by translocation of autosomal genes to the Y (e.g., *DAZ, RBM*). Alternatively, the X-linked copies may have been eliminated as Y-linked genes became specialized for males, but as yet such Y-linked genes have not been identified. Similar patterns of X-Y differentiation are seen in many species (e.g., worms, insects, snakes).

As Y-linked genes were lost there was a need for compensatory changes as most genes have the same optimal level of expression in both sexes. The imbalance between single X male and double X female leads to selection for increased X gene activity in males. But in turn this causes a double increase in XX females, which is disadvantageous. The way out of this dilemma is the evolution of dosage compensation that equalizes the dose in males and females. This can be achieved in many ways: random inactivation of one X in females, silencing of the paternally inherited X in females, down-regulation of both female Xs, or up-regulation of the single male X. The multitude of solutions reflects the rapid evolutionary turn-over of sex determining mechanisms which has caused multiple origins of sex chromosomes. We will now consider in greater detail these mechanisms and the evolutionary forces that have promoted dosage compensation.

Random X Inactivation. In placental mammals, dosage compensation is achieved by inactivation of one of the two X chromosomes in females. This is controlled by *Xic*, a multigenic region on the X chromosome. The main gene causing inactivation is *Xist* (X inactive-specific transcript). In females, *Xist* produces an RNA transcript that spreads along and coats the inactive X, whereas *Xist* is silenced on the active X. Deletion experiments show that *Xist* is essential for inactivation and only works *in cis*, which means that *Xist* action is limited to the inactive chromosome and has no *trans* effect on the active X or any other chromosome. *Xist* appears to work by attracting a complex of proteins to the X chromosome that cause histone deacetylation and DNA methylation. These changes to the structure of the chromosome cause it to become tightly condensed, blocking access to the DNA sequence and inactivating the chromosome (a similar form of inactivation occurs in genomic imprinting).

In somatic tissues, X inactivation is random. In XX females, the X derived from the mother is as likely to be inactivated as the X derived from the father. These two Xs initially differ in their histone acetylation and DNA methylation patterns, but the process leading to random X inactivation is preceded by genome-wide acetylation and de-methylation. This equalizes the epigenetic marks on the two X chromosomes, and so makes either equally likely to be chosen for inactivation. After these changes, a choice/counting process is activated that leads to all but one X being inactivated (i.e., in a XXX female, two Xs are inactivated). This process is not well understood but is thought to involve the blocking of sequences that turn-off *Xist*. If a limited amount of blocker is produced, and the blocker is self recruiting (i.e., only one blocking site ever attracts sufficient blocker), then only one copy of *Xist* will ever be turned off, and hence only one X will ever be active. This hypothesis remains to be validated and the nature of the presumed blocker substances is unknown (e.g., transcription factors, antisense RNA).

The process of X inactivation affects the whole of the X chromosome which appears visibly condensed (called a Barr body). However, there is good evidence that this dosage compensation evolved in a piece-meal fashion, following the accumulation of deleterious mutations at Y-linked homologs. The alternate hypothesis that full X inactivation evolving first, followed by loss of Y copies, is not supported as there are no known genes that show X inactivation that still have active Y-linked copies. The reverse pattern is seen: recent estimates suggest that around 15 percent of the genes on the human X chromosome escape X inactivation in females. Many of these have active Y-linked copies. But some do not, which suggests that the process of recruiting X inactivation signals has yet to take place.

A particularly interesting study of the association of X inactivation with the presence of Y-linked homologs was carried out by Jegalian and Page (1998). They compared three genes that escape X inactivation in humans (*ZFX*, *RSP4X*, and *SMCX*). In humans, they all have Y-linked homologs. But in other mammals (18 species compared), these genes had often gained X-inactivation. For example, *Zfx* showed the same human pattern of no X inactivation and a Y-linked homolog in most groups. But in the mouse and three other myomorph rodents, *Zfx* was subject to X inactivation and lacked a Y homolog. Each gene showed the same co-occurrence of X inactivation and the loss of the Y copy but in different mammalian genera. These results support the hypothesis that when genes are lost from the Y they also soon become X inactivated. In addition, the gain of X inactivation appears to occur on a gene-by-gene basis.

X Imprinting. In addition, mammals achieve X inactivation by genomic imprinting. This is a much simpler procedure in which the paternal X is exclusively inactivated. As the paternal X is always inherited by female offspring, its inactivation equalizes dose in males and females. The "imprinting" of the paternal X occurs during spermatogenesis. This is detectable by the 2–4 cell embryo stage, as the *Xist* gene of paternal X is visibly more active than *Xist* on the maternal Xs (in female or male embryos). This difference leads to the build up of histone deacetylation and DNA methylation, and exclusive inactivation of the paternal X very early in embryonic life. These differences are erased in cells destined to become the somatic tissues, which then undergo random X inactivation. However, in cells that form the extra-embryonic membranes (i.e., the placenta), imprinting X inactivation is the sole means of dosage compensation.

Imprinted paternal X inactivation seems in many ways a better dosage compensation mechanism. It does not involve counting or choice, it does not require resetting of maternal and paternal X chromosomes, and it occurs very early in development. In marsupials (e.g., kangaroos), it is the only mechanism of dosage compensation. Paternal X inactivation is a viable mechanism for achieving dosage compensation, so why did placental mammals also evolve random X inactivation? The big advantage of random inactivation is that both genes are expressed. This is likely to be beneficial for the same

reason that heterozygotes often have higher fitness (due to the masking of mildly deleterious alleles). In females, somatic tissues are chimeric with some cells expressing paternal X genes and others expressing maternal X genes. This has been shown to lower the effect of deleterious X-linked genes.

We also need to explain why placental mammals retained paternal X inactivation in the extra-embryonic tissues. Paternal X inactivation allows dosage compensation early in development. For random X inactivation to create a chimeric pattern of paternal and maternal inactivation in each tissue, it must occur later in development when a sufficient number of cells have been produced. If random inactivation occurred early, blocks of cells derived from the same precursor would all have the same X inactivated. This explains why the extra-embryonic membranes do not show random X inactivation. These membranes develop early in development as they are needed to initiate and establish the fetal-maternal link. This cellular differentiation occurs when there are very few cells in the embryo, so random X inactivation would bring few benefits.

Up or Down Regulation. In other species, dosage compensation is achieved by up or down regulation in one of the two sexes, rather than by inactivation. In *Drosophila*, the single male X is up-regulated to transcribe at the same rate as the two female Xs. The decision to carry out dosage compensation is signaled by the gene *msl2*. In females, *msl2* translation is blocked by the sex determination peptide Sxl that is only present in females. In males, protein produced by *msl2* and several other genes (*msl-1*, *msl-3*, *mle* and *mof*) forms a complex with two noncoding RNAs produced by the X-linked *roX1* and *roX2* genes. This complex initially forms close to the *roX* genes and then spreads to multiple sites along the single male X. The complex alters the chromosome structure by causing histone acetylation, and thereby opens up the single male X to hypertranscription.

In the worm *Caenorhabditis elegans*, dosage compensation is achieved in the opposite manner, through down-regulation of transcription in XX hermaphrodites. Males in this species are XO and have a single X. In a similar way to *Drosophila*, dosage compensation is tied to the sex determination pathway. In hermaphrodites, three *sdc* genes are active, and their protein products form a complex with *dpy* and *mix-1* gene products that localizes specifically to the X chromosomes. This results in down-regulation of both female Xs. The mechanism is not known, but proteins similar to those produced by the *sdc* genes are known to be important in chromosome packaging in other species. In males, *sdc* gene transcription is blocked by the sex determination gene *xol-1* that is only active in males.

Birds and Butterflies. Although there are similarities between mammals, *Drosophila* and worms, the genes, signaling systems and mechanisms involved are different. This reveals that these species have independent evolutionary origins of their dosage compensation mechanisms. This deduction is supported by the nature of their sex chromosomes, which also are thought to have independent origins. A good example of this is the insect *Sciara ocellaris*. This is a member of the same order of insects as *Drosophila* (the Diptera). *Sciara* has XX females and XO males and dosage compensation is achieved as in *Drosophila* by hyper-transcription of the male's single X. The key sex determination (*Sxl*) and dosage compensation genes (*msl*, *mle* and *mof*) of *Drosophila* can be identified in *Sciara* but are not involved in either of these processes. This example highlights the probable existence of several other systems of dosage compensation in non-model organism with differentiated sex chromosomes.

For a long time, biologists have wanted to know about dosage compensation in female heterogametic species, like birds and butterflies, which have ZW females and ZZ males. In both of these groups it has been confidently stated that there is no dosage compensation, and that differential expression of Z-linked genes is a normal part of sex differences. This raises the question of whether female heterogamety somehow precludes the evolution of dosage compensation.

However, these judgments are based on data from just one or two genes owing to the lack of good genome maps to define Z-linked genes in birds and butterflies. Recent advances have started to rectify this situation. A study by Heather McQueen and colleagues (2001) showed that six out of nine Z-linked genes (which lack W-linked homologs) had similar levels of gene expression in female and male chick embryos. The only gene to show a clearly higher expression in male embryos was *ScII*. However, this gene may be involved in sex determination or dosage compensation, as its ortholog in *C. elegans* (*mix-1*) acts to down regulate the two Xs in females. As yet there is no good evidence for butterflies, but it seems safer to assume that, just as in other species, dosage compensation evolves under female heterogamety to accompany the genetic erosion of W-linked genes.

Conclusion. The evolution of a genetically eroded Y chromosome creates dosage differences between male and females. This selects for dosage compensation. Our knowledge of dosage compensation in several model organisms reveals that equalization of dose is achieved in many different ways. It can occur through up-regulation in the heterogametic sex or down-regulation in the homogametic sex. Changes in expression are achieved by random inactivation, imprinting, hyper- and hypo-transcription. This diversity reflects the multiple origin of differentiated sex chromosomes, and the diversity of potential ways in which dosage can be adjusted.

[*See also* Genomic Imprinting; Heterozygote Advantage, *article on* Sickle-Cell Anemia and Thalassemia; Sex Chromosomes; Sex Determination.]

BIBLIOGRAPHY

Avner, P., and E. Heard. "X Chromosome Inactivation: Counting, Choice and Initiation." *Nature Reviews Genetics* 2 (2001): 59–67. An extensive review of the mechanisms underlying dosage compensation in mammals.

Jegalian, K., and D. C. Page. "A Proposed Path by Which Genes Common to Mammalian X and Y Chromosome Evolve to Become X Inactivated." *Nature* 394 (1998): 776–780. Comparison of three genes that show variable patterns of X inactivation and Y-linked homologs across most of the major mammalian groups.

Lahn, B. T., and D. C. Page. "Functional Coherence of the Human Y Chromosome." *Science* 278 (1997): 675–680. A recent paper with details of genes linked to the Y chromosome.

Marin, I., M. L. Siegal, and B. S. Baker. "The Evolution of Dosage Compensation Mechanisms." *BioEssays* 22 (2000): 1106–1114. A good general account of evolutionary forces.

McQueen, H. A., D. McBride, G. Miele, A. P. Bird, and M. Clinton. "Dosage Compensation in Birds." *Current Biology* 11 (2001): 253–257. The first clear demonstration of dosage compensation in a female heterogametic species (the chicken).

Ruiz, M. F., M. R. Esteban, C. Doñoro, C. Goday, and L. Sánchez. "Evolution of Dosage Compensation in Diptera." *Genetics* 156 (2000): 1853–1865. An excellent paper that shows that the dosage compensation genes of *Drosophila* are not active in producing dosage compensation in *Sciara*.

— ANDREW POMIANKOWSKI

E

ECHINODERMS

The Echinodermata is a group of invertebrate animals that include starfish, sea urchins, crinoids, brittle stars, and sea cucumbers. Echinoderms have a long and rich fossil history, consisting of many extinct and less familiar animals, such as blastoids and carpoids. Most living echinoderms have a characteristic fivefold organization of the body. This unique configuration is readily apparent in starfish, which are the most familiar members of the group. Despite their peculiar anatomy, echinoderms are evolutionarily closely related to the Chordata, a group that contains the vertebrates (including ourselves). All echinoderms live in the sea, where they are often quite abundant and ecologically important animals.

Anatomy. Echinoderms are among the most distinctive of all animal groups. They are characterized by five anatomical features.

1. The most apparent feature is the overall radial body organization of adult echinoderms. Fivefold (star-shaped) radial symmetry is most common, but many other kinds of body organization are known from living and fossil echinoderms. Nearly all of the organs within an echinoderm are repeated in multiples of five: a typical sea urchin, for example, contains five gonads, five radial nerves, and so forth.

2. Echinoderms possess a mineralized skeleton composed of many—typically hundreds—of individual ossicles that interlock to form a tough "shell." Unlike the true shell of a mollusc, the skeleton of an echinoderm is an internal structure, lying just under the epidermis, or skin. This skeleton is composed of calcite, a mineral that contains calcium carbonate. Individual ossicles are constructed as a three-dimensional meshwork of tiny spaces, rather like a sponge. This arrangement provides strength with a minimum of materials. Unlike vertebrates, the skeleton of an echinoderm continues to grow throughout life.

3. All echinoderms contain a special organ system called the water vascular system. The most prominent components of the water vascular system are podia (also called tube feet). These are small, fleshy organs shaped like fingers that lie in twin columns called ambulacra along the underside of starfish arms; in other groups of echinoderms, the ambulacra are distributed along arms (as in crinoids and brittle stars) or wrapped around the body (as in sea urchins). Podia are multifunctional structures: they are used to capture food, they are used to move about the seafloor by the nonsessile species, they contain most of the sensory neurons, and they are an important site of gas exchange. Podia contain muscles for contraction but rely on hydraulic pressure for extension. A system of internal vessels, valves, and muscular bulbs provides the water used to power the movement of podia.

4. The nervous system of echinoderms lacks any kind of brain. Instead, it is composed of a central nerve ring that surrounds the mouth, five radial nerves that run under the ambulacra, and an extensive network of sensory nerves within the skin. Echinoderms lack eyes, but their sensory neurons are capable of sensing light levels and are attuned to motion and chemical stimuli.

5. A final diagnostic feature of echinoderms is their unique connective tissue. Echinoderms possess a special form of collagen, a tough protein that makes up the bulk of ligaments in our own bodies and holds our bones together. In echinoderms, ligaments containing collagen hold the many skeletal ossicles together. Unlike our ligaments, however, these can be transformed from stiff to flexible as needs dictate. For instance, a crinoid can move its arms into the current and begin capturing suspended food particles, then stiffen its ligaments so that no muscular force is needed to hold that position; if the current shifts, the crinoid can relax its ligaments, reposition its arms using muscles, and lock them in place again by stiffening its ligaments. This unique ability has been exploited by echinoderms in many ways and is one of the keys to their ecological and evolutionary success.

Diversity. The phylum Echinodermata contains five living groups called classes: the Crinoidea (crinoids, or sea lilies, and feather stars), the Asteroidea (starfish), the Ophiuroidea (brittle stars and basket stars), the Echinoidea (sea urchins, sand dollars, and sea biscuits), and the Holothuroidea (sea cucumbers). These five living groups together contain approximately 6,000 species. The mineralized skeleton of echinoderms has left a rich fossil record consisting of more than 15,000 additional species. Although echinoderms are not the largest of the animal phyla in terms of numbers of species (arthropods, nematodes, chordates, and molluscs are all considerably larger), they are among the most diverse in terms of anatomy. Many variations on basic body symmetry are known. Some starfish and brittle stars have six, seven, nine, or even more arms. In most of these cases, five arms develop first, and the "extra" arms are added later. In two living groups, the sea cucumbers and the "irregular" sea urchins (consisting of sand dollars and sea biscuits), the fivefold radial body organization is secondarily arranged into a bilateral symmetry. The earliest members of the phylum did not have radial symmetry at all. For instance, the helicoplacoids were shaped like cigars, with three ambulacra wound around helically, and the stylophorans were oddly shaped animals that lacked any kind of body symmetry. The basic arrangement of even the "standard" fivefold symmetrical echinoderms differs in many ways. Some are attached to the seafloor by stalks with their mouth facing upward (crinoids), others lack stalks and move about on the seafloor with their mouth facing downward (starfish and sea urchins), and still others move about on their sides with their mouths facing forward (sea cucumbers). Thus, the basic orientation of the body differs among groups of echinoderms. In addition, many echinoderms have lost various organs. Some starfish and brittle stars lack an anus and must regurgitate their food when digestion is complete. Far from being a disadvantage, this has allowed starfishes to exploit a new method of feeding: they pry open clams and insert their stomach, which then digests the soft tissues of the hapless mollusc; this would not be possible with a stomach anchored to an anus.

Evolutionary History. The earliest echinoderm fossils come from the middle of the Cambrian period, approximately 520 million years ago. These early forms did not resemble any living groups of echinoderms. Although most lacked the radial body symmetry characteristic of modern groups, their skeleton and traces of the water vascular system unambiguously mark them as echinoderms. Echinoderms radiated rapidly during the Cambrian and subsequent Ordovician periods and came to dominate marine ecosystems. Crinoids, in particular, were spectacularly successful. By the end of the Ordovician, early representatives of the living groups had already evolved. Living alongside them were many other groups that did not survive, such as the blastoids and carpoids. A huge mass extinction about 250 million years ago, probably the largest in the earth's history, marked the end of the Paleozoic era. Only a few species of echinoderms survived this ecological meltdown, most of them probably ecological "generalists." These tough species went on to found evolutionary radiations during the Mesozoic era that generated considerable anatomical diversity and large numbers of new species. These evolutionary radiations eventually produced the five living groups as we know them today.

Ecology. Echinoderms are exclusively marine creatures; only a few can survive in brackish water, and none is known to live in freshwater. Within the marine realm, however, echinoderms occupy nearly every habitat, from the intertidal to the deepest depths, and from the equator to the poles. Within many marine ecosystems, echinoderms are a numerically and ecologically important component. Particularly in deeper water, echinoderms

are sometimes the most abundant kind of animal present. Echinoderms display a broad range of lifestyles: some are mobile on the seafloor (most sea urchins and starfish), others sessile (attached to the seafloor; many crinoids); some burrow within the sediment (many brittle stars and sea urchins), and a few swim (some crinoids and sea cucumbers). Echinoderms obtain their food using a broad range of methods: many sea urchins are herbivores, scraping algae that encrust rocks or feeding on larger algae; some brittle stars are active predators that are capable of capturing squid and fish; a large number of crinoids and brittle stars are suspension feeders, capturing plankton and detritus suspended in the water using their extended "arms"; finally, many sea urchins, sea cucumbers, and brittle stars are scavengers or detritus feeders. Nearly all echinoderms have complex life histories, consisting of an early larval stage that feeds in the plankton, followed by metamorphosis into an adult that lives on the seafloor. An echinoderm larva may drift tens or even hundreds of kilometers on ocean currents before metamorphosis. Two groups of echinoderms form commercially important fisheries: sea cucumber muscle is consumed in Asia, and sea urchin roe (whole ovaries) is a traditional item in both Asia and the Mediterranean region. Because the demand for these creatures is substantial, some species of shallow-water echinoderms have been overfished and are threatened over parts of their range.

[See also Animals.]

BIBLIOGRAPHY

Hess, H., et al. *Fossil Crinoids*. Cambridge, 1999. An informative and beautifully illustrated book on crinoid evolution.
Hyman, L. H. *The Invertebrates: Echinodermata*. New York, 1955. The classic comprehensive treatment of echinoderms.
Nichols, D. *Echinoderms*. London, 1962. A somewhat dated but still very useful general introduction to echinoderms.
Paul, C. R. C., and A. B. Smith. *Echinoderm Phylogeny and Evolutionary Biology*. Oxford, 1988. A collection of original articles covering various aspects of echinoderm evolution.
Smith, A. B. *Echinoid Palaeobiology*. London, 1986. An excellent book about sea urchins, with an emphasis on functional biology and paleontology.

— GREG WRAY

EEA. *See* Environment of Evolutionary Adaptedness.

EMERGING AND RE-EMERGING DISEASES

AIDS, Ebola, and hemolytic uremic syndrome (caused by certain strains of the bacterium *Escherichia coli*) are among the notable infectious diseases identified in the last few decades. Others, such as influenza, reappear periodically to cause major epidemics or even pandemics (epidemics that affect the entire world). The black death of the Middle Ages, tuberculosis in the late nineteenth century, the influenza pandemic of 1918–1919 (with an estimated 20 million deaths worldwide), and the AIDS pandemic of today, are all diseases that have had significant public health impact. (Humans have at times attempted to exploit the impact of infectious diseases through biological warfare or bioterrorism.) With increasing globalization, all the world is now vulnerable to an infection arising from anywhere.

Emerging infectious diseases are those that have newly appeared in the human population or are rapidly increasing in prevalence (the number of cases) or geographic range. As with the Ebola outbreaks periodically reported, emerging infectious diseases often seem mysterious and dramatic, but in fact specific factors responsible for emergence can be identified (Table 1). If we consider emergence as a two-step process—introduction and establishment/dissemination—these factors act to precipitate or promote one or both of these steps.

In considering infections that have emerged to date, it is striking how many already existed in nature and simply gained access to new host populations. This often happens as a result of changed ecological or environmental conditions that place humans in contact with previously inaccessible pathogens or the natural hosts that carry them. One important effect of evolution is therefore the generation of the vast biodiversity of pathogens now found throughout the world. The term *viral traffic* (or, more generally, *microbial traffic*) was coined to represent processes involving the access, introduction, or dissemination of existing pathogens to new host populations. In a sense, humans sample the vast microbial biodiversity in nature. Although this can occur anywhere, areas of high biodiversity would seem to deserve special attention.

The numerous historical examples of infections originating as zoonoses (infections transmissible from animals to humans) suggest that the "zoonotic pool"—introductions of pathogens from other species—is an important source of emerging pathogens or their precursors, some of which might become successful given the right conditions. HIV is a likely example. Although the original ancestors of the most prevalent HIV-1 strains are not known with certainty, the best current evidence suggests that HIV-1 is a zoonotic introduction, possibly from chimpanzees. HIV-2, another lentivirus that also causes AIDS, was identified in a man in rural Liberia whose HIV-2 strain closely resembled viruses from the sooty mangabey monkey, the presumed reservoir of a virus closely ancestral to HIV-2. This suggests that zoonotic introductions of viruses such as HIV may well occur periodically in isolated populations. In the case of HIV-1, social changes in the last half of the twentieth century (such as migration to cities) that allowed the virus to reach a larger population after introduction were key to its success. Other social changes that allowed the transmission of the virus to new individuals despite its relatively low natural transmissibility (although the long duration of infectivity also allowed the virus many opportunities to be transmitted) also contributed to its spread.

Periodic discoveries of "new" zoonoses suggest that the zoonotic pool is quite deep. Consider hantavirus pulmonary syndrome. In summer 1993, patients in the Four Corners region of the southwestern United States were admitted to local hospitals with fever and acute respiratory distress; about 60 percent subsequently died. Both serology and detection of genetic sequences by polymerase chain reaction (PCR) provided evidence for a previously unrecognized hantavirus as the cause of the outbreak. Samples from *Peromyscus maniculatus* (the deer mouse) matched the virus identified in the patients. The virus probably has long been present in mouse populations, rarely but occasionally getting opportunities to infect people. Unusual climatic conditions in the months preceding the outbreak led to a greatly increased rodent population, and thus greater opportunities for people to come in contact with infected rodents (and, hence, with the virus). Case finding and testing identified a number of sporadic cases (including some that occurred years before the outbreak but were not recognized at the time), and numerous related but distinct hantaviruses have been found throughout the Americas, suggesting that these viruses were probably widespread in nature for some considerable time.

Natural events were responsible for the 1993 outbreak of hantavirus pulmonary syndrome. But because people are major agents of ecological change worldwide, the precipitating environmental changes are often brought about by human activities, such as agriculture or irrigation. Hantaan virus (the prototype hantavirus, causing Korean hemorrhagic fever) is a natural infection of the field mouse *Apodemus agrarius*, for which rice fields are a favored habitat. Infected mice shed virus in urine and other secretions, and people may become infected by coming in contact with the virus left behind. Cases of disease (estimated at 100,000–200,000 annually) are increasing in concert with rice cultivation. Similarly, the unrelated Junin virus (the cause of Argentine hemorrhagic fever) is a virus of the rodent *Calomys musculinus*. Human cases increased in proportion with the expansion of maize agriculture. There are many other microbes with similar life histories. The emergence of Lyme disease (caused by the bacterium *Borrelia burgdorferi*) in the United States can largely be attributed to the fact that people like to have forests surrounding their homes, and thereby inadvertently interpose themselves in the ecology of the forest organisms (which for *Borrelia burgdorferi* normally involves a rodent reservoir, in North America often *Peromyscus*, and a tick that

TABLE 1. Factors in Infectious Disease Emergence*

Factor	Examples of Specific Factors	Examples of Diseases
Ecological Changes (including those due to economic development and land use)	Agriculture; dams, changes in water ecosystems; deforestation/reforestation; flood/drought; famine; climate changes	Schistosomiasis (dams); Rift Valley fever (dams, irrigation); Argentine hemorrhagic fever (agriculture); Hantaan (Korean hemorrhagic fever) (agriculture); hantavirus pulmonary syndrome, southwestern United States, 1993 (weather anomalies)
Human Demographics, Behavior	Societal events: population growth and migration (movement from rural areas to cities); war or civil conflict; economic impoverishment; urban decay; factors in human behavior such as sexual behavior (including urban prostitution and "sex-for-drugs"); intravenous drug use; diet; outdoor recreation; use of child care facilities (high-density settings)	Introduction of HIV; spread of dengue; spread of HIV and other sexually transmitted diseases
International Travel and Commerce	Worldwide movement of goods and people; air travel	"Airport" malaria; dissemination of mosquito vectors such as *Aedes albopictus* (Asian tiger mosquito); dissemination of dengue; ratborne hantaviruses; introduction of cholera into South America; dissemination of 0139 (non-01) cholera organism
Technology and Industry	Food production: globalization of food supplies; changes in food processing and packaging. Health care: new medical devices; organ or tissue transplantation; drugs causing immunosuppression; widespread use of antibiotics	Food processing: hemolytic uremic syndrome (*E. coli* contamination of hamburger meat), bovine spongiform encephalopathy. Health care: Transfusion-associated hepatitis (hepatitis B, C), opportunistic infections in immunosuppressed patients, Creutzfeldt-Jakob disease from contaminated batches of human growth hormone (medical technology)
Microbial Adaptation and Change	Microbial evolution, response to selection in environment	Changes in virulence and toxin production; development of drug resistance (antimicrobial resistant bacteria, chloroquine resistant malaria); "antigenic drift" in influenza virus
Breakdown in Public Health Measures	Curtailment or reduction in prevention programs; lack of, or inadequate, sanitation and vector control measures	Resurgence of tuberculosis in United States; cholera in refugee camps in Africa; resurgence of diphtheria in former Soviet republics

*Categories are not mutually exclusive; several factors may contribute to emergence of a disease.

From S. S. Morse, (1995).

feeds on mice and deer at different stages in its life cycle and can transmit the infection when it feeds). Infections transmitted by mosquitoes, which include malaria, dengue, yellow fever, West Nile, Rift Valley fever, and many others, are often stimulated by dams or irrigation projects, because the responsible mosquitoes breed in water.

Although we commonly think of infectious diseases as causing outbreaks, there has recently been increasing recognition that infections can also be associated with chronic diseases. It has long been known that hepatitis B virus is responsible for many cases of liver cancer worldwide. More recently, pioneering work by Marshall implicated the bacterium *Helicobacter* in gastric ulcers and cancer, and Chang and Moore identified a novel herpesvirus (now known as human herpesvirus 8) as the likely cause of Kaposi's sarcoma.

The role of evolution is perhaps more complex. Evolution is generating the biodiversity of actual and potential members of the zoonotic pool. The evolution of the host-pathogen interaction is also key. Most zoonotic infections, at least those identified as such, are more severe in their accidental hosts than in their natural hosts, largely due to differences in host response. Lentiviruses (the viruses in primates that most likely gave rise to HIV-1) are mostly asymptomatic in their primate hosts. In addition, microbes are constantly evolving. Selection for drug resistance, driven by the wide and sometimes inappropriate use of antimicrobials in a variety of applications, is a sobering demonstration of the power of natural selection. It is thought that inadequate antimicrobial treatment led to the evolution of multidrug-resistant tuberculosis (MDR TB), on multiple occasions. MDR TB could now spread rapidly by the same mechanisms as ordinary tuberculosis. Pathogens can also acquire antibiotic resistance genes from other, often nonpathogenic, species in the environment, in many cases through horizontal gene transfer. The recent discoveries of "pathogenicity islands" (clusters of genes in bacterial pathogens that confer the ability to infect and cause disease) also suggest the importance of horizontal gene transfer. On occasion, the evolution of a new variant may result in a new expression of disease. A possible example is Brazilan purpuric fever in 1990, associated with a newly emerged clonal variant of *Hemophilus influenzae*, biogroup *aegyptius*.

Many viruses show a high mutation rate and can rapidly evolve to yield new variants. Despite this, they also show apparent stability over relatively long periods of time, indicating that there are factors stabilizing viral phenotype and even the viral genome, probably natural selection. Influenza is probably the best-known example of viral emergence arising from the evolution of new variants. The familiar annual epidemics of influenza are caused by mutation in the hemagglutinin (H) (a key surface protein) in a virus strain already circulating in the population. The mutation allows the new variant to escape immediate recognition by host antibodies. This process, which is called antigenic drift, is probably driven (at least in part) by immune selection. However, periodically, influenza A undergoes a major antigenic shift in the H protein, and a pandemic may result. The evolution of pandemic strains involves a reassortment of viral genes from different influenza strains (rather than accumulation of point mutations as in antigenic drift). It appears that waterfowl (in which the virus is in evolutionary stasis) are the natural hosts for influenza. Humans appear relatively resistant to avian influenza, but pigs may be more easily infected with both avian

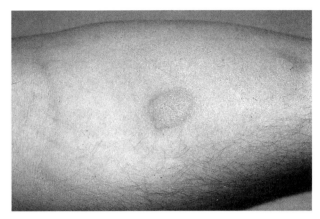

FIGURE 1. A Positive TB Skin Test.
© Ken Greer/Visuals Unlimited.

and mammalian influenza viruses. Scholtissek and Naylor have noted that the major influenza pandemics have historically originated in southern China, which also traditionally practices integrated pig-duck husbandry. They suggested that this practice may facilitate the development of new influenza "reassortants" by making it easier for pigs that are already infected with a mammalian influenza virus to become simultaneously infected with an avian influenza virus. Animal husbandry may therefore significantly influence, and even possibly accelerate, the evolution of influenza.

Once introduced, the success of the pathogen in a new population depends on its establishment and dissemination within the population. Many zoonotic introductions are highly virulent and not readily transmissible from person to person, preventing their establishment. Both the evolutionary potential of the pathogen and chance will play a role in whether the infection will be able to establish itself. An analytic framework was developed by Anderson and May, based on the basic reproductive rate (net transmission potential) of the introduced pathogen. Ewald (1994) made the interesting additional suggestion (still controversial to some) that selecting for greater transmissibility should favor greater virulence.

Human intervention and social changes, in addition to providing opportunities for introduction of pathogens, also provide increasing opportunities for dissemination. Ebola is usually introduced into humans by contact with its (still unknown) natural host in the forest, but most of the cases occur subsequently in the hospital through contaminated injection equipment. A number of factors have led to the increase in tuberculosis worldwide: HIV infection increases susceptibility to tuberculosis, while high-density settings (such as day care centers, homeless shelters, and prisons) enhance transmission. Human migration from rural areas to cities, especially in areas with a high degree of biodiversity, can introduce remote pathogens to a larger population. HIV has been the best-known beneficiary, but many other diseases stand to benefit. After its move from a rural area into an initial city, HIV-1 spread along highways to other regional cities, then later by long-distance routes, including air travel, to progressively more distant places. Rapid air travel affords many pathogens rich opportunities for globalization.

The globalization and industrialization of the food supply and other goods also offer pathways for microbial traffic. The strains of *Escherichia coli* that cause hemolytic uremic syndrome were probably once limited to a few relatively isolated populations of cattle, but have spread as cattle are collected into large central processing facilities. Bovine spongiform encephalopathy (BSE), also known as mad cow disease, appeared in Britain within the last few years as a probable interspecies transfer of scrapie from sheep to cattle. Widespread use of animal byproducts as feed supplements, coupled with changes in rendering processes that allowed the scrapie agent in sheep byproducts to contaminate the feed, may have been responsible.

Re-emerging diseases are those that were previously decreasing but are now rapidly increasing again. Usually these are diseases that were once controlled but are now staging a comeback owing to breakdowns in public health or control measures. The resurgence of diphtheria in the former Soviet Union in the 1990s (as immunization programs lapsed due to lack of resources) is a recent example. Re-emerging diseases should be a reminder that complacency can be as great a danger as many of the infectious diseases themselves.

Ecological changes and major demographic changes (such as population migrations) often precipitate emergence; these "signals" for microbial traffic should be seen as warning signs for increased public health surveillance. There may be many potential emerging infections in the biodiversity of nature and the pathways of evolution. Therefore, to truly understand and anticipate emerging infections requires understanding them in their ecological and evolutionary context.

[*See also* Acquired Immune Deficiency Syndrome, *articles on* Origins and Phylogeny of HIV *and* Dynamics of HIV Transmission and Infection; *articles on* Antibiotic Resistance; Disease, *article on* Infectious Disease; *articles on* Immune System; Plagues and Epidemics; Transmission Dynamics; Vaccination; Virulence.]

BIBLIOGRAPHY

Ewald, Paul W. *Evolution of Infectious Disease.* New York and Oxford, 1994. Innovative and thought-provoking consideration of how virulence evolves.

Haggett, Peter. *The Geographical Structure of Epidemics.* Oxford, 2000.

Hahn, B. H., G. M. Shaw, K. M. De Cock, and P. M. Sharp. "AIDS As a Zoonosis: Scientific and Public Health Implications." *Science* 287 (2000): 607–614.

Krause, Richard M., ed. *Emerging Infections.* San Diego, 1998. A recent multiauthor volume containing reviews on a variety of infections.

Lederberg, J., R. E. Shope, and S. C. Oaks, Jr., eds. *Emerging Infections: Microbial Threats to Health in the United States.* Washington, D.C., 1992. Particularly useful for policy implications.

Morse, Stephen S., ed. *Emerging Viruses.* New York and Oxford, 1993. Contains overviews by many of the leading scientists in the field.

Morse, Stephen S. *The Evolutionary Biology of Viruses.* New York, 1994. Special emphasis on the relationship between genetic variation and natural selection and on the evolutionary constraints imposed by the interaction of the virus with its host.

Morse, S. S. "Factors in the Emergence of Infectious Diseases." *Emerging Infectious Diseases* 1 (1995): 7–15 (January 1995). A general review, with references, covering the material in this article in more detail. Also available online (see Online Resources, below).

Persing, D. H., and F. G. Prendergast. "Infection, Immunity, and Cancer." *Archives of Pathology and Laboratory Medicine* 123 (1999): 1015–1022. A general review on infectious agents in chronic diseases such as cancer.

Webster, Robert G. "Influenza: An Emerging Microbial Pathogen." In *Emerging Infections,* edited by R. M. Krause. San Diego, 1998.

ONLINE RESOURCES

"Emerging Infectious Diseases." http://www.cdc.gov/ncidod/eid. Journal published by the Centers for Disease Control and Prevention (CDC), available in both online and print formats. Reviews and articles on many of the subjects touched on here can be found in its pages.

"ProMED-mail." http://www.promedmail.org. Begun as an offshoot of ProMED (the Program for Monitoring Emerging Diseases), and now a program of the International Society for Infectious Diseases. International network for reporting and discussion of infectious disease outbreaks and emerging disease events.

— STEPHEN S. MORSE

EMOTIONS AND SELF-KNOWLEDGE

Around the world, people recognize that humans experience transient alterations of subjective states in response to information, that these alterations are often (with varying degrees of voluntary control) outwardly expressed, that they motivate action, and that they involve a change in one's perception of oneself, one's surrounding, and one's goals. English speakers label these experiences *emotions,* and similar terms occur in many languages. Because the constituents of this category vary across languages, some anthropologists argue that emotions are primarily cultural in nature and are incommensurate across cultures. However, critics of this position note that, although labels and connotations differ across cultures, many emotional expressions (particularly those involving the face) are easily recognized by outsiders, and outsiders learn to map local labels onto their own emotional lexicon. This suggests that, whereas cultures may determine the meaning of events that elicit emotions, and whereas cultures may glorify, disparage, ignore, or combine particular aspects of the emotion repertoire, the core aspects of human emotions are species-typical, and hence are likely to be the products of selection.

In *The Expression of the Emotions in Man and Animals* (1872), Charles Darwin presented evidence that some emotional expressions are recognizable across cultures and appear spontaneously in children. Darwin used the similarities between these expressions and those seen in animals to argue for common descent. This perspective defied the Western anthropocentric tradition, but Darwin conformed to that tradition in viewing emotions' effects on reasoning as detrimental, describing human emotions as vestiges of earlier evolutionary stages in which the intellect was of less importance. Although the argument for phylogenetic continuity plays a role in modern explanations of emotions, Darwin's vestigialism has largely been replaced by adaptionism.

Contemporary investigations of emotion relevant to an evolutionary perspective take one of four forms. First, psychologists focus on documenting and categorizing the proximate components of emotion, emotional expression, and emotions' influence on cognition. Questions of adaptive utility motivate these investigations in a broad sense, but these issues often do not directly shape hypotheses tested. Second, employing perspectives taken from game theory, scholars explore the utility of particular emotions with regard to strategic dilemmas. Third, evolutionary psychologists seek to explain the influences of emotions on attention, memory, motivation, and other functions in terms of the recurrent challenges that have confronted humans. Fourth, comparative psychologists and primatologists seek to document similarities and differences in emotional responses and states between humans and other primates and within and among other primate species.

Language provides a focus for inquiry into the cross-cultural variation in human emotional expression. Variation in the nuances of emotion names shows that this domain is subject to considerable elaboration. However, despite such elaboration, even unschooled observers intuit that emotion terms are often thematically linked; that is, terms such as *terror, fear*, and *anxiety* all revolve around a single core experience. Furthermore, these core experiences are often associated with readily recognizable facial expressions. This has led many psychologists to propose the existence of discrete "basic" human emotions; it is thought that the human panoply of emotions is generated from these unitary, elementary constituents through processes of combination and/or fine cultural discrimination. Although there is no con-

sensus as to the number of basic emotions, commonly proposed members include happiness, surprise, fear, sadness, anger, disgust, contempt, shame, and guilt.

In contrast to the above perspective, a minority of psychologists argue that unified basic emotions are illusory; rather, any given emotional experience is described as occupying a position along such spectra as hedonic valence, degree of arousal, and tendency to approach versus avoid. Similarly, facial expressions are thought to rarely involve the full complements of muscular activation claimed to be indexical of particular basic emotions, and are instead composed of a variety of component expressive elements that vary depending on the details of context, degree of arousal, and tendency to approach versus avoid.

Proponents of both the discrete and the dimensional views make reference to the adaptive utility of the proposed psychological features. For example, both schools recognize the benefits of coping with imminent threats. However, whereas discrete emotion theorists argue that a unitary psychological mechanism, fear, directs attention to the threat, motivates the avoidance of harm, and focuses memory and problem-solving abilities on the task of coping with the threat, dimensional theorists point out that both the behavioral and physiological responses to threat, including signaling of intention, depend on the costs and benefits of different courses of action, suggesting the existence of a complex and flexible problem-solving system rather than a distinct psychological mechanism that is activated only in response to threats. The divide between discrete and dimensional emotion theorists thus partially mirrors the debate between evolutionary psychologists and sociobiologists/behavioral ecologists. [*See articles on* Human Sociobiology and Behavior.]

Most game theoretic investigations of emotion assume that distinct emotions exist. Robert Trivers (1971) argued that iterated exchanges create selective pressure for psychological mechanisms that lead the individual to avoid imperiling long-term benefits in favor of smaller short-term gains. Aggression in response to having been cheated is adaptive because it decreases the likelihood of future defections against one, particularly if it is disproportionate relative to the transgression. Because disproportionate response is costly, an auxiliary mechanism is necessary to impel aggression, and outrage serves this purpose. Similarly, because gratitude in response to generosity motivates reciprocation, actors who feel gratitude avoid the short-term gain to be reaped by defection and obtain the long-term gains provided by cooperation. Using more detailed models, Jack Hirschleifer (1987) demonstrated that the advantages of anger and gratitude are not limited to reciprocally altruistic exchanges, but pertain in any situation of potential cooperation.

Following Trivers, Robert Frank (1988) argued that the preference for short-term gains is a consequence of the tendency, common to all vertebrates, to discount the future. To this, Alan P. Fiske (in press) added a list of biases in human cognition, including propensities to overestimate one's own abilities, to ignore base rates, to overvalue immediate gains relative to probabilistic costs, and to both overestimate the probabilities of conjunctive events and underestimate the probabilities of disjunctive events. These biases undermine individuals' abilities to establish and maintain advantageous long-term relationships by causing the benefits of defection to loom large relative to the costs. However, because emotions such as love and guilt constitute immediate rewards and punishments, they change the subjective costs and benefits of present action in ways that countermand the temptation for short-term gain, committing an individual to a course of action that yields long-term benefits. In addition, by leading individuals to act in ways that are currently costly, these emotions generate advertisements of likely future behavior, thereby shaping the responses of others. Furthermore, via two pathways, emotions may shape behavior in an adaptive fashion even when future interactions will not involve the same opponents or partners. First, to the extent that an individual's inclination to act in a particular fashion is affected by past behavior, habits, and so on, and is expressed in a manner that others can detect, then it pays to adhere to a pattern consistent with a preference for long-term gain (by showing gratitude, reacting aggressively to transgressions, etc.). Second, to the extent that others know of an individual's past actions, his or her behavior in one interaction can profitably shape future interactions with others.

Many authors propose that guilt evolved to dissuade individuals from harming beneficial relationships or to motivate them to repair damage done to such relationships. One difficulty with these explanations is that, even granted the existence of cognitive biases, selection should favor the ability to surreptitiously cheat, yet guilt motivates reparation even when the transgression is undiscovered. This raises the possibility that guilt, although premised on emotions such as regret and sympathy, may be a product of cultural rather than biological evolution. The latter explanation accounts for the absence of universal facial or postural expressions of guilt, signals that would be expected if the capacity to feel guilt had been selected as a result of the advantages of reassuring others of one's good intentions.

Evolutionary theorists have noted that romantic love functions to commit individuals to a long-term cooperative mateship (at least temporarily) and signal that commitment to the partner, two valuable functions given human infant altriciality. Romantic love exhibits a distinct chronology, with an initial period of obsessive ideation eventually giving way to a less intrusive form of attachment. The early phase may both strongly dissuade defection during a period of scrutiny and motivate energetic signaling of commitment; once a cooperative enterprise has been established, sunk costs may reduce the necessary intensity on both counts.

Love is closely tied to jealousy. Evolutionary psychologists examine emotions such as jealousy in light of recurrent adaptive challenges. For example, David Buss (2000) notes that a principal hazard for men is misdirected investment due to cuckoldry, and a principal hazard for women is cessation of male provisioning due to abandonment. Consistent with these observations, Buss finds that men are more disturbed by the prospect of a mate's sexual infidelity, whereas women are more disturbed by the prospect of a mate's emotional attachment to rival women. Others counter that men, like males of other species, face competition over paternities. Because this gives rise to sexual jealousy in other primate species where there is no male provisioning, male jealousy should not be seen as evidence of a long history of paternal investment in humans.

Some evolutionary psychologists argue that emotions may be more diverse than either traditional psychologists or Western folk models suggest. For example, rather than discussing a single emotion, fear, that is a response to imminent danger, evolutionary psychologists predict the existence of multiple types of fear, each associated with such distinct classes of threats as predators, rival conspecifics, social exclusion, and snakes and spiders; further subdivisions may occur on the basis of context. In *Evolutionary Psychology and the Emotions* (cited in Lewis and Haviland-Jones), Leda Cosmides and John Tooby argued that each emotion is a superordinate program that serves to govern the activities of, and interactions between, specialized programs concerning a wide variety of psychological and physiological mechanisms. Activation of a given emotion, for example, fear of predators in the twilight, directs attention to relevant information (potential signs of predators), sharpens sensory modalities relevant to that information (visual, auditory, etc.), cues patterns of interpretation (ambiguous shadows look like predators, etc.), readies relevant motor patterns (flight or fight), reassigns priority among goals (escape vs. feeding, mating, playing chess, etc.), searches relevant memory categories (information about predators), and so on. This perspective provides a solution to the dimensional theorists' complaint that physiological and behavioral responses to a given stimulus vary with the physical and social context, because fear of predators when escape is possible will be qualitatively different from fear of predators when escape is impossible. Similarly, dimensional theorists' complaints regarding the disjunction between emotions and expressions, as well as the lack of uniformity of the latter, can be addressed with the observation that, because some adaptive challenges are best met with a communication of intent whereas others are better served by the absence thereof, only some emotions involve an expressive component. At the same time, similarities among many of the members of a class of emotions, or among the most memorable or commonly experienced member of that class, can explain the apparent existence of basic emotions.

Human emotions are clearly derived from a psychological foundation shared by social mammals. Primates in general are unusually social. [See Primates: Primate Societies and Social Life.] Other primate species share our admiration of successful individuals including a desire for proximity and a willingness to provide client services to obtain it. Humans rely on socially transmitted information in coping with physical and social environments to an especially profound extent. Our desire for close observation and imitation of successful others leads us to adopt ideas and practices of probably local utility. At a larger scale, conformity to cultural values, beliefs, and practices makes behavior predictable and allows for the advent of complex coordination and cooperation; shame and pride motivate an assessment of prevailing norms, an awareness of the presence of observers, and conformity to pervasive expectations when under observation, Conversely, contempt and moral outrage motivate publicizing the actions of nonconformists, excluding them from cooperative endeavors, and inflicting costs upon them. A richer understanding of the evolution of contempt and moral outrage is needed given that punishment plays a key role in maintaining cooperation.

Finally, although considerations of kin selection and reciprocal altruism indicate that many animals should experience emotions in a corporate fashion (i.e., harm to kin or allies is experienced as harm to self), evidence is accumulating that social primates link their identities to group membership in ways not directly explicable in terms of kin selection. We humans identify with groups that can be both enormously large and also diverse in composition, suggesting that the benefits of coordination and cooperation favored the evolution of highly developed human corporate sensibilities.

[See also Consciousness; Cooperation; Human Sociobiology and Behavior, article on Evolutionary Psychology; Prisoner's Dilemma Games; Reciprocal Altruism.]

BIBLIOGRAPHY

Buss, D. M. *The Dangerous Passion: Why Jealousy Is as Necessary as Love and Sex*. New York, 2000. An evolutionary psychological approach to jealousy.

Darwin, C. *The Expression of the Emotions in Man and Animals*. London, 1872.

Ekman, P. "An Argument for Basic Emotions." *Cognition and Emotion* 6.3–4 (1992): 169–200. A point-by-point enumeration of the case for basic emotions.

Ekman, P., and W. V. Friesen. "Constants across Cultures in the Face and Emotion." *Journal of Personality and Social Psychology* 17.2 (1971): 124–129. A classic study of universality in the facial expression of emotion.

Fessler, D. M. T. "Toward an Understanding of the Universality of Second Order Emotions." In *Biocultural Approaches to the Emotions*, edited by A. L. Hinton, pp. 75–116. New York, 1999. A cross-cultural examination of the evolution of shame and pride.

Fiske, A. P. "A Proxy Theory of Emotions: Motives that Represent the Value of Social Relationships." In *Relational Models Theory: Advances and Prospects*, edited by N. Haslam. Mahwah, N.J., in press. A recent blend of game theoretic and psychological approaches, arguing that pervasive biases in reasoning create a need for mechanisms (in the form of emotions) that keep actors from damaging valuable social relationships.

Frank, R. H. *Passions within Reason: The Strategic Role of the Emotions*. New York, 1988. A classic, extensive game theoretic approach to emotions.

Fridlund, A. J. *Human Facial Expression: An Evolutionary View*. San Diego, Calif., 1994. A presentation of the dimensional view of emotional expression.

Haidt, J. "The Moral Emotions." In *Handbook of Affective Sciences*, edited by R. J. Davidson. Oxford, in press. An overview of research on moral emotions and their effects on social behavior, frequently using an evolutionary perspective.

Henrich, J., and F. J. Gil-White. "The Evolution of Prestige: Freely Conferred Status as a Mechanism for Enhancing the Benefits of Cultural Transmission." *Evolution and Human Behavior*. Addresses the question of the evolution of psychological mechanisms, including admiration, designed to facilitate the acquisition of socially transmitted information.

Hirschleifer, J. "On Emotions as Guarantors of Threats and Promises." In *The Latest on the Best: Essays on Evolution and Optimality*, edited by J. Dupre et al., pp. 307–326. Cambridge, Mass., 1987. A game theoretic account of emotions in cooperative interactions.

Izard, C. E. *The Psychology of Emotions*. New York, 1991. A thorough introduction to the discrete emotion perspective.

Lewis, M., and J. M. Haviland-Jones, eds. *Handbook of Emotions*.

Nesse, R. M. "Evolutionary Explanations of Emotions." *Human Nature* 1.3 (1990): 261–289. An introduction to the evolutionary psychological approach to emotions, including discussions of fear and psychopathology.

Trivers, R. L. "The Evolution of Reciprocal Altruism." *Quarterly Review of Biology* 46.1 (1971): 35–57. A seminal game theoretic analysis of dyadic interaction, including evolutionary explanations of a number of social emotions.

Turner, T. J., and A. Ortony. "Basic Emotions: Can Conflicting Criteria Converge?" *Psychological Review* 99.3 (1992): 566–571. A brief criticism of the notion of "basic" emotions, introducing the dimensional perspective as an alternative.

— Daniel M. T. Fessler

ENDOSYMBIONT THEORY

Eukaryotes (animals, plants, fungi, protozoa, and algae) comprise organisms whose genome is normally contained within a membrane-bound compartment called the nucleus. A further defining feature of eukaryotic cells is the presence of other membrane-bound organelles that are morphologically and functionally distinct. The endosymbiont theory proposes that certain of these organelles, notably mitochondria and chloroplasts, originated as bacterial symbionts within a nucleus-containing host cell (xenogenous origin), rather than being formed from within the cell itself (autogenous origin). The beginnings of this theory can be traced to various scientific writings of the late nineteenth century, notably those of the German histologist Richard Altmann concerning "bioblasts" (renamed mitochondria in 1897 by the histologist A. Benda) and those of the German botanist Andreas Schimper on "chromatophores" (i.e., chloroplasts). The earliest and most fully elaborated version of the endosymbiont theory (specifically as it relates to chloroplasts) is undoubtedly that published in 1905 by the Russian botanist Constantin Mereschowsky.

In 1924, Ivan Wallin, an American biologist, enthusiastically declared that "mitochondria are, in reality, bacterial organisms, symbiotically combined with the tissues of higher organisms." However, this view and that espoused by Mereschowsky did not find favor among Wallin's contemporaries. For example, E. B. Wilson, a noted American cell biologist, dismissed Mereschowsky's endosymbiotic proposals as "entertaining fantasy" and "flights of imagination." In referring to Wallin's ideas about the origin of mitochondria, Wilson was no less skeptical: "To many, no doubt, such speculations may appear too fantastic for present mention in polite biological society." Wilson did allow that it was "within the range of possibility that [these speculations] may some day call for serious consideration"—although he hardly gave the impression in his writings that he believed this would happen. Books by Lynn Margulis (1993) and Jan Sapp (1994) provide detailed and engaging historical perspectives on the early debates about endosymbiont theory.

The endosymbiont hypothesis of organelle origins was largely forgotten or ignored during the ensuing three decades, and was only resurrected and reinvigorated after the discovery in the 1960s that mitochondria and chloroplasts contain unique species of DNA. In her influential book *Origin of Eukaryotic Cells* (1970), Lynn Margulis pointed out a number of structural and biochemical features shared by organelles and bacteria, arguing that these similarities could reflect an origin of the organelles through a series of bacterial endosymbioses (hence the serial endosymbiosis theory). During the 1970s, the accumulating data increasingly favored an endosymbiotic origin of chloroplasts; however, data supporting a similar origin of mitochondria remained controversial, prompting vigorous debates between proponents and opponents of a xenogenous origin of mitochondria. We now recognize that much of the contention at the time was attributable to the fact that most of the relevant data were from animals and fungi, whose mitochondrial genomes are highly derived compared to that of their bacterial ancestor, as are their mitochon-

drial transcription and translation systems. As plant mitochondrial DNAs (mtDNAs) began to be investigated, the tide turned strongly in favor of a bacterial origin of mitochondria. Plant mitochondrial genomes are much more slowly evolving than their animal and fungal counterparts, so their gene sequences and gene expression patterns more closely resemble those of their bacterial progenitor. Certain unicellular eukaryotes (protists) have mitochondrial genomes that in gene content and gene organization are even more ancestral (minimally derived) than those of plants.

Molecular Evidence Supporting the Endosymbiont Hypothesis. Early suggestions that mitochondria and chloroplasts might derive in evolution from bacterial symbionts were based on morphological and physiological similarities between bacteria and organelles and were put forth before the roles of these organelles in the cell had even been established. The discovery of organelle DNA suggested that genes encoded by mitochondrial and chloroplast genomes might provide clues to their origin. This prediction was borne out by the sequencing of organelle genes, initially ribosomal RNA (rRNA) genes and later protein-coding ones. These organelle gene sequences provided the first compelling molecular evidence of the endosymbiotic origin of mitochondria and chloroplasts, pointing to an ancestry of these organelles from within specific phyla of bacteria: α-Proteobacteria (originally called nonsulfur purple bacteria) in the case of mitochondria and Cyanobacteria (formerly blue-green algae) in the case of chloroplasts. Subsequent analysis of the sequences of proteins encoded by organelle DNA strongly confirmed these conclusions. These data also showed clearly that mitochondria and chloroplasts originated from within the eubacteria (domain Bacteria), not from within a newly discovered second prokaryotic group, the archaebacteria (domain Archaea).

Additional evidence supporting the endosymbiotic, eubacterial origin of organelles has come from complete sequencing of organelle DNAs, which has revealed the suite of genes these DNAs contain and the way in which these genes are organized and expressed. Again, the resemblance of chloroplast genomes to eubacterial (specifically cyanobacterial) genomes is quite striking. For example, chloroplast DNA encodes a eubacterialike RNA polymerase, and both photosynthetic and ribosomal protein genes are often arrayed in operonlike clusters that mimic their cyanobacterial counterparts. In the case of mitochondria, there is a great deal of variability in the extent to which these genomes still resemble their eubacterial ancestors, with evident vestiges of eubacterial ancestry primarily restricted to plant and especially some protist mitochondrial genomes. Nicholas Gillham's book, *Organelle Genes and Genomes* (1994), is an excellent source of information about mitochondrial and chloroplast DNAs and their evolutionary origins.

Origins of Chloroplasts and Mitochondria. Chloroplasts contain chlorophyll *a* as their primary photosynthetic pigment but differ in their content of secondary pigments: that is, chlorophyll *b* in green algae and plants; chlorophyll *c* in golden algae, brown algae, diatoms, and dinoflagellates; and phycobiliproteins (phycobilins) in red algae. In the 1970s, this distinction led some investigators to propose that chloroplasts had multiple origins from similarly pigmented eubacteria. Because the photosynthetic apparatus of cyanobacteria features chlorophyll *a* + phycobilins, these bacteria were assumed to be the specific ancestors of the chloroplasts of red algae. Eubacterial groups containing either chlorophylls *a* + *b* or chlorophylls *a* + *c* were then sought as the specific progenitors of similarly pigmented chloroplasts. However, molecular evidence continues to support the idea that the three types of chloroplast are descendants of a common eubacterial ancestor (i.e., that chloroplasts are monophyletic, not polyphyletic), and that the chloroplast ancestor arose from within the phylum Cyanobacteria.

What the molecular data also show is that, whereas the chloroplasts of red and (most) green algae are of primary origin (i.e., they derive directly from an endosymbiosis with a free-living eubacterium), the *a* + *c*–type algae appear to have acquired their chloroplasts secondarily, through endosymbiosis with another eukaryotic alga rather than directly from a eubacterial symbiont. In most *a* + *c*–type algae, all that remains of the algal symbiont is its chloroplast, surrounded by one or two additional surrounding membranes (a vestige of the engulfment process). However, in the cryptomonad algae, this reduction process has not proceeded to completion, and a residual nucleus, called a nucleomorph, remains. Comparison of rRNA genes encoded in the nucleomorph and the main nuclear genome shows convincingly that they had separate evolutionary origins, and that the symbiont in cryptomonads was a red alga. Current data indicate that, whereas chloroplasts per se had a single origin, subsequent diversification of the eukaryotic algae involved at least two and probably more secondary endosymbiosis events, the exact number remaining controversial (Figure 1).

Molecular data also support a monophyletic origin of mitochondria, despite enormous differences in the size and coding capacity of characterized mitochondrial genomes. In rRNA- or protein-sequence-based phylogenies, mitochondria cluster as a monophyletic group to the exclusion of their closest bacterial relatives, the rickettsial group of α-Proteobacteria. Moreover, the organization of ribosomal protein genes in plant and some protist mtDNAs is closely similar to that in corresponding eubacterial operons, except that the mitochondrial gene

(A) PRIMARY ENDOSYMBIOSIS
(prokaryote + eukaryote = eukaryote)

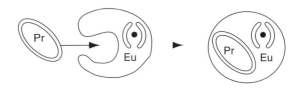

(B) SECONDARY ENDOSYMBIOSIS
(eukaryote + eukaryote = eukaryote)

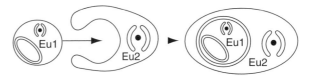

FIGURE 1. Distinction Between Primary and Secondary Endosymbioses.
Pr, Prokaryote; Eu, eukaryote.   Adapted from the work of G. I. McFadden, *Journal of Cell Science* 95 (1990): 303–308, with permission.

clusters share characteristic deletions not seen in the corresponding eubacterial operons. The simplest explanation of this pattern is that these deletions occurred only in the mitochondrial lineage and very early on, before the diversification of the main eukaryotic groups.

Reductive Evolution of Organelle Genomes. The most gene-rich chloroplast and mitochondrial genomes are at least an order of magnitude smaller than those of their closest living eubacterial relatives, and there is considerable variation in gene content among chloroplast and (especially) among mitochondrial DNAs. From the currently available data, we are able to infer that (1) organelle genomes have sustained massive gene loss since their origin from the genomes of eubacterial symbionts, (2) this gene loss has occurred to different extents in different eukaryotic lines, (3) individual genes have been lost from organelle DNA on more than one occasion (i.e., the same gene has disappeared from the organelle genome repeatedly and independently in different lineages), and (4) gene loss from organelle DNAs is an ongoing evolutionary process.

Many of the protein-coding genes that were once resident in organelle DNAs have been transferred to the nuclear genome, from which they are now expressed; in these cases, the protein products are targeted to and imported into the organelle. However, many (perhaps most) nuclear proteins that are transported to the organelles appear to have been newly elaborated or re-cruited after the acquisition of the organelles (i.e., their origin cannot be traced back to the eubacterial progen-itor of the organelle in question). In plants and some protists, nucleus-encoded tRNAs are also imported into mitochondria to supplement tRNAs encoded by the mtDNA.

Symbiotic Origins of Other Organelles. The serial endosymbiosis theory proposes that, in addition to mi-tochondria and chloroplasts, other eukaryotic organ-elles, such as peroxisomes and flagella, had a symbiotic origin. To date, no substantive evidence supports this thesis. Apart from mitochondria and chloroplasts, no other organelles appear to contain DNA; hence, it is not possible in these cases to use the same criteria of origin that have proven so compelling in the case of mitochon-dria and chloroplasts.

A few unicellular eukaryotes, mostly living in anaer-obic environments, lack mitochondria entirely. Consid-erable debate has centered on the question of whether these protists (collectively termed Archezoa by T. Cava-lier-Smith) diverged away from the main eukaryotic line prior to the acquisition of mitochondria (i.e., never had mitochondria), or whether they are secondarily amito-chondriate, having lost mitochondria at some point in their evolution. A phagocytic, primitively amitochon-driate eukaryote is just the sort of host cell envisaged by the endosymbiont hypothesis, as typically formu-lated.

Recent studies have identified genes encoding dis-tinctly "mitochondrial" proteins, such as the chaperonin Hsp60, in the genomes of protists that supposedly never had mitochondria. Moreover, in several cases it has been shown that these proteins are targeted to the hydroge-nosome, an organelle that generates ATP anaerobically, liberating hydrogen as a byproduct. These observations have led to the proposal that hydrogenosomes are de-rived in evolution from mitochondria. Certain protists, such as *Giardia lamblia*, lack hydrogenosomes or any other apparent membrane-bound energy-transducing or-ganelles. It remains possible that these amitochondriate eukaryotes never did possess mitochondria (i.e., are primitively amitochondriate).

Origin of Eukaryotes versus Origin of Organ-elles. According to classical endosymbiont theory, mi-tochondria and chloroplasts originated subsequent to the origin of the eukaryotic cell itself; that is, a cell al-ready having the defining characteristics of a eukaryote (nucleus, endomembrane system, cytoskeleton) was the host that engulfed the eubacterial progenitors of the or-ganelles. Although few dispute this scenario in the case of chloroplasts, some recent models propose that the origin of the eukaryotic cell was coincident in time with the origin of the mitochondrion. In this scenario, a fu-sion between a proteobacterium and an archaebacter-ium provided the genetic elements that formed the basis of both the nuclear compartment and the mitochon-drion. Although there is increasing evidence for a chi-meric origin of the nuclear genome, with some nuclear genes having a eubacterial and others an archaebacterial ancestry, current data do not exclude the traditional endosymbiotic model of mitochondrial origin. What is clear is that the origin of mitochondria, if not coincident with that of the eukaryotic cell as a whole (estimated to be some 800 million to 1.5 billion years ago), must have occurred very shortly thereafter on an evolutionary timescale.

What were the selective pressures driving the origin of mitochondria and chloroplasts? In the case of mito-chondria, it has been argued that the mitochondrion pro-vided the host with a more efficient means of energy production, in the form of ATP. However, an important consideration is that the bacterial symbiont initially would not have had the capacity to import ADP from and export ATP to the cytoplasm and could not do so until it subsequently acquired a membrane-localized ADP/ATP transporter system. For this reason, it has been suggested that the initial function of mitochondria was to protect against the deleterious effects of a grow-ing oxygen atmosphere, by converting potentially toxic oxygen to harmless water through the action of the elec-tron transport chain. Only later did the mitochondrion become the primary supplier of ATP to the aerobically growing eukaryotic cell.

In the case of chloroplasts, the acquisition of photo-synthesis, either through primary or secondary endosym-biosis, would have allowed new phototrophic combina-tions to diversify rapidly. As a result, their descendants could occupy a myriad of ecological niches that would not have been possible otherwise.

[*See also* Cellular Organelles.]

BIBLIOGRAPHY

Andersson, S. G. E., and C. G. Kurland. "Origins of Mitochondria and Hydrogenosomes." *Current Opinion in Microbiology* 2 (1999): 535–541. A critical discussion of alternative models for the origin of mitochondria and hydrogenosomes. The authors contend that phylogenetic reconstructions favor the simplest version of the endosymbiont theory over more recent models that derive the nuclear compartment and mitochondrion at the same time.

Andersson, S. G. E., A. Zomorodipour, J. O. Andersson, T. Sicheritz-Pontén, U. C. M. Alsmark, R. M. Podowski, A. K. Näslund, A.-S. Eriksson, H. H. Winkler, and C. G. Kurland. "The Genome Se-quence of *Rickettsia prowazekii* and the Origin of Mitochon-dria." *Nature* 396 (1998): 133–140. A landmark article reporting the complete sequence of the genome of the closest known bacterial relative of mitochondria.

Cavalier-Smith, T. "Eukaryotes with No Mitochondria." *Nature* 326 (1987): 332–333.

de Duve, C. "The Birth of Complex Cells." *Scientific American* 274 (1996): 38–45. An eminently readable and clearly illustrated dis-cussion of the basic elements of classical endosymbiont theory.

Delwiche, C. F. "Tracing the Thread of Plastid Diversity through the Tapestry of Life." *American Naturalist* 154 (1999): S164–S177. A comprehensive treatment of the role of primary and secondary endosymbiosis in chloroplast evolution.

Delwiche, C. F., and J. D. Palmer. "The Origin of Plastids and Their Spread Via Secondary Symbiosis." *Plant Systematics and Evolution* (Supplement) 11 (1999): 53–86.

Douglas, S. E. "Plastid Evolution: Origins, Diversity, Trends." *Current Opinion in Genetics and Development* 8 (1998): 655–661. A concise review that focuses on data from complete bacterial and chloroplast genome sequences in relation to chloroplast evolution.

Gillham, N. *Organelle Genes and Genomes.* Oxford, 1994.

Gray, M. W. "The Endosymbiont Hypothesis Revisited." *International Review of Cytology* 141 (1992): 233–357. A comprehensive compilation and critical review of molecular data relating to the endosymbiont theory. This review updates that of Gray and Doolittle (1982).

Gray, M. W. "Evolution of Organellar Genomes." *Current Opinion in Genetics and Development* 9 (1999): 678–686. A review focusing particularly on recent organelle gene and genome sequence data as they relate to the evolution of mitochondrial and chloroplast DNAs.

Gray, M. W., and W. F. Doolittle. "Has the Endosymbiont Hypothesis Been Proven?" *Microbiological Reviews* 46 (1982): 1–42. The first review of the endosymbiont theory to delineate the molecular criteria by which this theory might be proven. The review compiles and critically comments on relevant data that had accumulated to that point.

Gray, M. W., G. Burger, and B. F. Lang. "Mitochondrial Evolution." *Science* 283 (1999): 1476–1481. A focused review of mitochondrial genome diversity and of the role of mitochondrial genomics in revealing the origin and evolution of mitochondria. This review also comments on recent challenges to the traditional endosymbiont model of mitochondrial origin.

Kurland, C. G., and S. G. E. Andersson. "Origin and Evolution of the Mitochondrial Proteome." *Microbiological and Molecular Biological Reviews* 64 (2000): 786–820. An incisive analysis of the phylogenetic origins of all mitochondrial proteins that can be identified in the complete sequence of the yeast nuclear and mitochondrial genomes. The authors conclude that the endosymbiont theory for the origin of mitochondria "requires substantial modification" in the sense that proteins that have clearly descended from the α-proteobacterial ancestor of mitochondria represent only a minor fraction of yeast mitochondrial proteins.

Lang, B. F., G. Burger, C. J. O'Kelly, R. Cedergren, G. B. Golding, C. Lemieux, D. Sankoff, M. Turmel, and M. W. Gray. "An Ancestral Mitochondrial DNA Resembling a Eubacterial Genome in Miniature." *Nature* 387 (1997): 493–497. Reports the complete sequence of the most gene-rich and bacterialike mitochondrial genome characterized to date. The data reported served to underpin a subsequent confirmation of the rickettsial/mitochondrial connection reported in detail by Andersson et al. (1988).

Lang, B. F., M. W. Gray, and G. Burger. "Mitochondrial Genome Evolution and the Origin of Eukaryotes." *Annual Review of Genetics* 33 (1999): 351–397.

Margulis, L. *Origin of Eukaryotic Cells.* New Haven, Conn., 1970.

Margulis, L. *Symbosis in Cell Evolution.* San Francisco, 1993.

Martin, W., and K. V. Kowallik. "Annotated English Translation of Mereschowsky's 1905 Paper 'Über Natur und Ursprung der Chromatophoren im Pflanzenreiche.'" *European Journal of Phycology* 34 (1999): 287–295. A most informative translation of and commentary on an early and now-classic endosymbiosis paper.

Palmer, J. D. "A Single Birth of All Plastids?" *Nature* 405 (2000): 32–69. A succinct discussion of the question of whether chloroplasts are monophyletic in origin.

Sapp, J. *Evolution by Association.* Oxford, 1994.

Wallin, I. E. "On the Nature of Mitochondria. VII. The Independent Growth of Mitochondria in Culture Media." *American Journal of Anatomy* 33 (1924): 147–173.

Wilson, E. B. *The Cell in Development and Heredity.* Macmillan, 1925.

ONLINE RESOURCES

"The Organelle Genome Megasequencing Program (OGMP)." megasun.bch.umontreal.ca/ogmpproj.html. Maintained at the University of Montreal. Provides information on an interdisciplinary collaboration that predominantly involves research groups in eastern Canada, and which investigates the molecular evolution of mitochondria, chloroplasts, and bacteria. The focus of this collaborative effort is on organelle phylogeny, especially in protists. A centralized sequencing and bioinformatics facility at the University of Montreal serves as the major research hub of the OGMP.

"The Mitochondrial Database Project (GOBASE)." megasun.bch.umontreal.ca/gobase/gobase.html. Maintained at the University of Montreal. GOBASE is a taxonomically broad organelle genome database that organizes and integrates diverse data related to organelles. The current version focuses on the mitochondrial subset of such data.

— MICHAEL W. GRAY

ENGINEERING APPLICATIONS

The use of algorithms inspired by Charles Darwin's theory of evolution has become widespread in many scientific and engineering disciplines (Fogel, 1998). Perhaps the best-known class of these is the genetic algorithm, but others are also growing in popularity, including genetic programming, evolutionary programming, and evolution strategies. Although these different algorithm types were largely independently conceived, it is in keeping with their evolutionary inspiration that there is now considerable cross-fertilization, leading to a blurring of the traditional classifications. Nowadays, most researchers prefer the umbrella term *evolutionary computation* or *evolutionary algorithms* to embrace all types of algorithms inspired by evolutionary theory (Baeck et al., 1997).

A general outline of an evolutionary algorithm is shown in Figure 1. The algorithm starts with an initial population of trial solutions, often generated at random. A fitness score is calculated for each population member, depending on how closely it approaches the desired solution—this is measured by the fitness function. To generate new solutions, parents are selected, with the more fit population members being more likely to be chosen. Combining or mutating the parent solutions gives rise to offspring, which are in turn assessed for their fitness. These new solutions compete with the existing solutions, and the fittest survive to become the parents of the next generation. As this process continues

A REAL-LIFE EXAMPLE

Researchers at Roche (Basle, Switzerland) led by Gisbert Schneider have developed an evolution strategy called TOPAS for indirect drug design. TOPAS works by assembling molecular fragments and is guided by a fitness function that measures similarity to a template molecule. The aim of the program is to generate molecules that are similar enough to the template so that they exhibit the same type of biological activity, yet different enough to perhaps escape problems of toxicity or a competitor's patent.

A published example (Schneider et al., 2000; see Figure 2) illustrates the use of TOPAS to design a novel potassium channel-blocking agent aimed at treating various heart diseases. The initial template molecule (1) was supplied to the program, which after 100 generations arrived at the novel molecule (2). This was found to be somewhat less active than the template, but only a small change was required to produce (3)—a molecule with comparable activity to the template but belonging to a different structural class. Such a molecule would form a good starting point in the search for a drug.

—DAVID E. CLARK

over a number of generations, better and better solutions evolve, guided by the fitness function, until an optimal (or near-optimal) one is attained.

The attractive feature of these algorithms to the scientific and engineering communities is that they have proven extremely effective at solving optimization problems, that is, locating the best (or close to the best) solution when many other possible solutions exist (Coley, 1999). Optimization problems are encountered in nearly all areas of science and engineering; thus, applications of evolutionary algorithms can be found in fields as diverse as civil engineering, computer science, astronomy, chemistry and chemical engineering, and bioinformatics. Applications in computer-aided drug design have been among the most successful.

Applications in Computer-aided Drug Design. A simple way of viewing computer-aided drug design is to use the analogy of a key fitting a lock. The lock represents some biochemical target, and the key is the drug molecule seeking either to lock or unlock the target, depending on the requirements of the disease being treated. Sometimes, drug designers have detailed information on the structure of the lock and its keyhole; in these cases, structure-based drug design techniques can be employed to craft prospective keys to fit the lock exactly. Often, however, the designers are not so fortu-

nate, and instead must rely on drawing analogies from keys that have been tried before, including those that have not worked. This situation is sometimes called indirect drug design.

Many evolutionary algorithms have been developed for computer-aided drug design applications. In general, these have proven very successful, both in the structure-based and indirect drug design scenarios (Clark, 2000). In some cases, evolutionary algorithms have been successful enough to be incorporated in commercial software products.

Docking. One of the fundamental tasks in structure-based drug design is attempting to fit a prospective drug molecule into the active site of the biochemical target, a procedure known as docking. This can be done manually by a designer sitting at a graphics workstation, but this is both time-consuming and subject to the biases of the person in question. In the last twenty years, fast computational methods for docking have been developed that permit the rapid and objective assessment of thousands of molecules per day.

A docking algorithm comprises two components—a search algorithm to explore the different shapes (conformations) that a drug molecule can adopt and a scoring function to assess how well these fit the active site. With this in mind, it is possible to design an evolutionary algorithm to find the optimal conformation guided by a fitness function based on the principles of molecular recognition; indeed, docking has been one of the most fruitful application areas for evolutionary algorithms in computer-aided drug design (Morris et al., 2000). Researchers have explored different types of evolutionary algorithms (particularly genetic algorithms, but evolutionary programming has also been successful) and a host of fitness function variants in the search for more

---

FIGURE 1. An Outline of a Simple Evolutionary Algorithm. Drawing by David E. Clark.

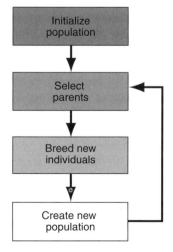

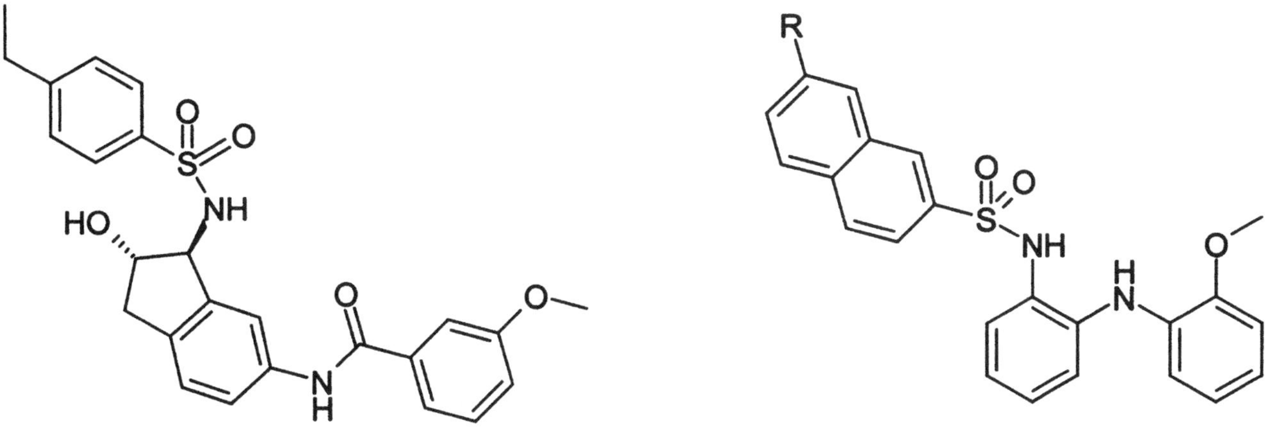

1 (IC50 = 110 nM)

2 (R = OMe, IC50 = 7340 nM)

3 (R = H, IC50 = 470 nM)

FIGURE 2. Engineering Applications.
Chemical structures and potassium channel blocking activities of a template molecule (1), a molecule designed by TOPAS (2), and a simple modification of it (3). Note that lower figures indicate greater biological potencies. After Schneider, G. et al. *Angew. Chem. Int. Ed.* 39 (2000): 4130–4133.

reliable and rapid docking methods. One of the most popular computer programs to emerge in recent years is GOLD (Genetic Optimization of Ligand Docking; Jones et al., 1997), which was developed at the University of Sheffield and is now distributed commercially by the Cambridge Structural Data Centre (Cambridge, England).

Quantitative structure-activity relationships. When no structural information about the biological target is available, one of the most important approaches for guiding drug design is the derivation of mathematical equations called quantitative structure-activity relationships (QSARs). These equations seek to relate the experimentally measured binding affinity of a compound (i.e., how well the key fits the lock) to various properties (or descriptors) that can be computed directly from the molecule's chemical structure. These descriptors may range from extremely simple quantities, such as the number of carbon atoms, to much more complex properties relating to the distribution of electronic charge in the molecule or its relative attraction for oily as opposed to watery environments (its lipophilicity). A typical QSAR will have the form

$$F(\text{activity}) = a \times \text{descriptor1} + b \times \text{descriptor2} \\ + c \times \text{descriptor3} \ldots$$

where $F(\text{activity})$ is some function (usually logarithmic) of the molecule's binding affinity; a,b, and c are coefficients of appropriate sign and magnitude; and descriptors 1–3 are selected molecular properties used to model the activity. The value of such an equation in drug design

is that, if properly constructed and validated, it can help to predict the binding affinity of proposed drug molecules, even before they are synthesized in a test tube. All that is needed is the molecule's chemical structure.

To construct a QSAR, three steps are required. First, a set of descriptors needs to be selected from all those that are available (this is called feature selection). Second, the coefficients for the equation relating the selected features to the binding affinity need to be determined using some statistical method. Third, the model must be validated.

It is in the feature selection step that evolutionary algorithms have proven to be very powerful (So, 2000). There are often tens or hundreds of possible descriptors to choose from, so to find the optimal subset manually would be virtually impossible. However, it is possible to construct a QSAR-generation program in the form of an evolutionary algorithm. An initial population is set up by selecting subsets of descriptors at random. Each member is then scored by generating and validating a model from the subset of descriptors it represents. The subsets of descriptors that give rise to the best models achieve the highest fitness scores and so are more likely to pass on their selected descriptors to future generations. Over a number of generations, the population of solutions should converge on a set of good QSAR models.

This kind of approach is embodied in the genetic function approximation (GFA) program that was developed by David Rogers (Rogers and Hopfinger, 1994) while at Molecular Simulations (now Accelrys) Inc. (San

Diego, Calif.). GFA is an integral part of the molecular modeling software marketed by Accelrys and has proven effective in several drug design situations.

[*See also* Genetic Algorithms.]

BIBLIOGRAPHY

Baeck, T., D. B. Fogel, and Z. Michalewicz, eds. *Handbook of Evolutionary Computation*. 1997.

Clark, D. E., ed. *Evolutionary Algorithms in Molecular Design*. Weinheim, Germany, 2000.

Coley, D. A. *Introduction to Genetic Algorithms for Scientists and Engineers*. 1999.

Fogel, D. B., ed. *Evolutionary Computation: The Fossil Record*. Piscataway, N.J., 1998.

Jones, G., P. Willett, R. C. Glen, A. R. Leach, and R. Taylor. "Development and Validation of a Genetic Algorithm for Flexible Docking." *Journal of Molecular Biology* 267 (1997): 727–748.

Morris, G. M., A. J. Olson, and D. M. Goodsell. "Protein-Ligand Docking." In *Evolutionary Algorithms in Molecular Design*, edited by D. E. Clark, pp. 31–48. Weinheim, Germany, 2000.

Rogers, D., and A. J. Hopfinger. "Application of Genetic Function Approximation to Quantitative Structure-Activity Relationships and Quantitative Structure-Property Relationships." *Journal of Chemical Information and Computer Sciences* 34 (1994): 854–866.

Schneider, G., O. Clement-Chomienne, L. Hilfiger, P. Schneider, S. Kirsch, H.-J. Böhm, and W. Neidhart. "Virtual Screening for Bioactive Molecules by Evolutionary de Novo Design." *Angewandte Chemie: International Edition* 39 (2000): 4130–4133.

So, S.-S. "Quantitative Structure-Activity Relationships." In *Evolutionary Algorithms in Molecular Design*, edited by D. E. Clark, pp. 71–97. Weinheim, Germany, 2000.

— DAVID E. CLARK

ENVIRONMENT OF EVOLUTIONARY ADAPTEDNESS

A central part of evolutionary theory is that there is a fit between the characteristics of any species and the environment in which it lives, and that this has been shaped over evolutionary history by the process of natural selection. Natural selection thus produces adaptation.

It follows from this process that for any particular species there is an environment to which it is best suited to live, and equally others where it cannot or can only do so with difficulty (i.e., zones of adaptation and maladaptation). Another outcome is that the traits of any organism are best understood in terms of the particular environment in which they evolved. Thus, for example, the thick fur of a polar bear and its color are explicable in the context of a cold, snowy landscape.

Evolutionary psychologists, concerned with the evolution of human cognitive characteristics, have extended and formalized this part of evolutionary theory with reference to what is known as the environment of evolutionary adaptedness, or EAA for short. The EAA refers to the evolutionary conditions under which human mental abilities evolved, and hence to the selective processes that shaped the human mind. The concept was developed by John Bowlby in 1969, but its use has flourished enormously in the past decade with the growth of evolutionary approaches to human psychology.

Evolutionary psychologists have argued that the human mind is not a tabula rasa on which culture can write anything, but is predisposed to certain cognitive abilities as a result of having been shaped by natural selection—hence the term *the adapted mind*. Thus, there are specific cognitive modules relating to mate choice, social relationships, technology, and spatial organization, each of which is an evolutionary product. Among the claims of evolutionary psychologists are that human brains are better at solving social problems than logical ones, and that males and females differ in their mental templates for preferred mates.

The reason why evolutionary psychology has developed the more formal notion of the EAA, rather than simply referring to the process of adaptation, lies in the particular problems of studying human evolution. If the human mind evolved under natural selection, it did so prior to the establishment of *Homo sapiens*, and over long periods of the past. However, few, if any, humans actually live under those conditions now. For example, at the time when the fossil evidence shows that the human brain was expanding (i.e., during the Pleistocene, between 2.0 and 0.2 million years ago), humans and their ancestors lived in very small scale societies, linked together by intimate webs of kinship, at low population densities, and with relatively little by way of technology. In contrast, the people who for the most part are studied by psychologists today are predominantly urban, live in very large scale societies, are surrounded by technology, are buffered from the direct effects of the environment, and rather than struggling to maintain a viable population size, are actively trying to keep reproductive levels low. There is thus a mismatch between the current environment of the human species, a product of only the last few thousand years, and that in which the human mind evolved. The EAA is designed to find a way around this problem. The adaptive nature of the human mind does not relate to current environments or current reproductive utility, but to the conditions of the past—it is those that form the "environment of evolutionary adaptedness." Any study that tries to explain the adaptive nature of the human mind—or, indeed, of any other human characteristic—must do so in terms of the actual environment in which the traits evolved.

The key question is therefore to define the human EAA and to use this as the framework against which human cognitive abilities can be assessed. This is a far from simple exercise. Given that it is in the past, it cannot be directly observed. The primary solution to this difficulty has been to assume that the best analogy for the human EAA is represented by the life of hunter-gatherers living today and observable ethnographically, or by inference from the archaeological record. This is not to say necessarily that surviving hunter-gatherers are relics from the world of the past in which humans evolved, but that they, rather than those living through agriculture, more closely approximate the living conditions of human Pleistocene ancestors. The primary traits on which evolutionary psychologists have focused are that they live in relatively small groups, in which face-to-face relationships are the norm, and in which a close degree of kin relatedness is probable; that there is a sexual division of labor and scheduling of activities, with consequences for energy budgets, social roles, and reproductive behavior; that a high level of parental investment, especially by females, is associated with a high risk of infant mortality; and that there is a relatively high risk of agonistic encounters with other groups.

However, the EEA as it has been used obviously comprises far more than the hunter-gatherer lifestyle. Many of the cognitive traits on which evolutionary psychologists have focused are not actually associated with unique human traits, but are general to the mammals or to primates. For example, one area of interest has been differences in cognition between men and women, and these have been associated with such adaptive problems as male paternal uncertainty and the high costs of reproduction in females. These, though, are not unique to humans, but are general mammalian characteristics, and it is likely that the basic differences were established in a much earlier EEA. Another example would be the importance of social relationships, which has given rise to the social intelligence hypothesis (i.e., that large brains are associated with high levels of social complexity); again, sociality, although highly extended in humans, is a trait found in all monkeys and apes.

The EEA is therefore not a particular environment or phase in evolution, but rather is an abstract concept that attempts to put together the range of selective pressures that have given rise to human characteristics. It is dubious, in the long term, whether it is a concept that is needed in addition to the general one of adaptation, but it does highlight the problem that the circumstances under which a trait may have evolved are not necessarily the same as those in which we may observe them today. For humans, the problem may be particularly intense because of the extraordinary scale of change that has occurred in the last few thousand years, a period of time that is far too short to account for major evolutionary changes. However, the issue of time lags between an environmental change and its impact on evolution is also important for evolutionary biology more generally,

for whereas humans may be an extreme case, they are certainly not unique in this respect.

BIBLIOGRAPHY

Foley, R. A. "The Adaptive Legacy of Human Evolution: A Search for the EAA." *Evolutionary Anthropology* 4 (1997): 194–203.

Irons, W. "Adaptively Relevant Environments Versus the Environment of Evolutionary Adaptedness." *Evolutionary Anthropology* 6 (1998): 194–204.

— ROBERT FOLEY

EPIDEMICS. *See* Plagues and Epidemics.

EPIDEMIOLOGY. *See* Disease, *article on* Infectious Disease; Plagues and Epidemics.

EPIGENESIS AND PREFORMATIONISM

One of the most persistent questions in the history of Western science has concerned the phenomena that surround animal generation. At least from the time of Aristotle in the fourth century B.C.E., natural philosophers have attempted to solve the problem of how to explain the process that regulates the orderly sequence of events associated with fertilization and development. Aristotle even dedicated a whole work, *De Generatione*, to the subject, describing animal generation as a process controlled by teleological causes guiding development from undifferentiated matter (male and female seminal material) to the designed adult form. This context served as the original formulation of epigenesis, or the gradual process in which the form of the organism emerged from a formless beginning. Quite literally, it stressed the new creation of a being with each new generative act.

Aristotle's treatment explained both heredity and development, since the new organism was the product of the combined seminal material of the male and female blended together in procreation to produce the new being. Additionally, it was internally consistent with his entire worldview, including his notion of causation, his understanding of the four elements in the natural world, and his principle of teleology in the biological world. Because of their completeness and consistency, Aristotle's ideas provided the framework for both inheritance and development well into the seventeenth century.

But the seventeenth century also marked the first wholesale attacks on the Aristotelian system, in terms of both a new philosophical alternative for natural philosophy—the mechanical philosophy—and a new appreciation to demonstrate mechanical principles empirically. The new philosophy, first articulated by the French philosopher René Descartes at the beginning of the seventeenth century, then receiving its most refined treatment in the work of the English mathematician and physician Isaac Newton at the century's end, called for an end to the qualitative treatment of the natural world inherent in the Aristotelian worldview. In its place was a new interpretation of the natural world featuring the existence of particulate matter constantly moving under the control of unseen but active forces. Natural philosophers who embraced the new worldview during the scientific revolution of the seventeenth century searched for mechanical explanations of natural phenomena, explanations that yielded to mathematical analysis, not philosophical debate.

As part of the new philosophy, investigators adopted the need to support their mechanical philosophy with direct observation and experimental manipulation of nature. New instruments emerged to accomplish these tasks, including the microscope. Now the reproductive and developmental processes could be observed directly for the first time, allowing naturalists to see for themselves what actually happened. Almost immediately, some observers noted problems with Aristotle's treatment of generation. Using the microscope, they observed that at no stage in the developmental process was there a period in which there was no undifferentiated material, as demanded by the epigenetic position. Instead, the microscope appeared to reveal the existence of preformed parts or even preformed wholes. According to these investigators, there was no real new generation, as Aristotle claimed, only the gradual unfolding of either preformed structures or preformed organisms; this was the idea of preformationism as an alternative to epigenesis.

Preformation continued to receive additional observational support in the seventeenth and the eighteenth centuries. The self-trained Dutch instrument makers Antonie van Leeuwenhoek and Nicolaas Hartsoeker, along with the cleric Malebranche, were among the microscopists at the end of the seventeenth century who saw and reported on preformed organisms. In addition to the observational evidence, it is equally important to note that the new mechanical philosophy embraced by many natural philosophers of the age practically demanded preformation. After all, how does one explain the sudden appearance of a new generation (epigenesis) with mechanical principles? Additionally, preformation fit better with theology, especially when the preformed germ was observed in the ovum; here was evidence that God created all of mankind at the original creation, explaining not only the preformation process but also the doctrine of original sin. Despite this compelling argument, there were disputes among preformations as to whether the preformed organism was encased within the female ovum (the ovist position) or within the male sperm (the animalculist or spermist position). Microscopical evidence provided evidence for each position,

as well as an abundant amount of observation of encased organisms throughout the animal world.

At the same time, some followers of Isaac Newton failed to see preformed structures and created a revised version of epigenesis by the early part of the eighteenth century. Designed and unseen mechanical forces, analogous to gravity and acting according to God's providence, were offered to explain the orderly process. This helped some natural philosophers to understand how Abraham Trembley's famous hydra, celebrated in the 1740s, could regenerate a whole new organism from a part sliced from an adult hydra. The hydra was a freshwater organism that regenerated lost parts, confounding the eighteenth-century naturalists. All agreed that it was unlikely that the hydra contained a preformed structure that "knew" it would be sliced and therefore would regenerate a new preformed structure. However, Newtonian forces that penetrated all matter could provide an explanatory framework for the regeneration of lost parts from undifferentiated material.

Thus, by the mid-eighteenth century, a full-scale debate emerged in the salons and scientific societies in Europe centered on whether preformation, in its myriad forms, or epigenesis explained animal generation and development. Georges Buffon, a naturalist of the eighteenth century, hired the English microscopist John Turberville Needham to investigate the problem in the Jardin du Roi in Paris. Buffon, already a strong proponent of Newtonian forces in the natural world, was prepared to accept similar forces for generation. Needham confirmed his assumption. Experimenting with the new miscroscope, Needham demonstrated that life only emerged from organic molecules under the guidance of the organizing *moule intérieur*, thereby updating epigenesis with a forcelike concept. But the abbé Spallanzani, working at the same time in Italy, repeated similar experiments, demonstrating to his satisfaction that preformed organisms emerged from the organic infusions they both used. His ideas were consistent with those of Charles Bonnet, who updated preformation by claiming that whole organisms were not preformed, but that preformed organized germs within the seminal material merged to form the organized parts observed in generation. Thus, the experimental demonstrations of the mid-eighteenth century actually took explanatory form under the influence of either preformation or epigenesis. The demonstrations themselves were not unequivocal, but the philosophical commitments of the natural philosophers were.

The debate continued throughout the eighteenth century, with little evidence that it ever received widespread resolution. In fact, both the *Encyclopédie* and the *Encyclopaedia Britannica*, the two major compendia of knowledge in France and England, contained articles on generation that stressed its mysterious nature and the futility of ever resolving its questions. Just as the microscope of the seventeenth century led to the creation of preformation alternatives to epigenesis, the microscope of the nineteenth century led to a reversal. One of the persistent problems behind microscopical observations until the nineteenth century was that the optics were of such poor quality that often the only witnesses to the microscopical observations were those who actually made the observation. For example, Needham was one of the eighteenth century's most esteemed microscopists simply because he could use his instrument. Chromatic and spherical aberrations within the devices wreaked havoc on observations, often preventing the untrained eye from seeing anything.

By the 1820s, the situation had changed dramatically. Improved instruments produced by the Zeiss Company in Germany (Leipzig) essentially eliminated the problems. Microscopists could duplicate the work of one another, demonstrate their observations to others, and replicate their own work. Furthermore, the new instruments had much greater magnification and resolution. Thus, when the Prussian naturalist Karl Ernst von Baer published his monumental work on animal generation in 1828, he was able to demonstrate with great microscopical acuity that the developing embryo emerged gradually from undifferentiated material in the female ovum. This work was continued by both Theodor Schwann and Matthias Schleiden in the 1830s, illustrating that all cell generation, including both plants and animals, emerged in a similar manner from previously undifferentiated material. By the mid-nineteenth century, it appeared as if epigenesis had once again become firmly established, supported by a vast scientific literature, and that preformation was an outmoded idea.

But preformation experienced a renaissance, this time in different garb. The German physician Rudolf Virchow, who also contributed to the modern notion of cell theory in 1858, proclaimed that *omnis cellula e cellula*, or all cells come from previously existing cells. Although this was not an overt restatement of preformation, it did indicate that there was some organized material that existed prior to the formation of a new organism. Subsequent microscopical studies, this time aided by yet more advanced instrumental techniques, uncovered the ever-present nature of chromatin matter and then chromosomes within the nucleus of all cells, even gametic cells prior to fertilization. Much of this work, primarily by German microscopists, seemed to indicate that chromosomal material spanned the gap between generations, serving as the "organized" connection between parent and offspring. By 1890, both the German morphologist August Weismann and the Dutch botanist Hugo de Vries suggested that particles located within the nucleus, perhaps as subunits of the chromatin material, represented the inheritable traits shared between gen-

eration. Thus, phoenixlike, preformation emerged anew by the end of the nineteenth century.

Similar to the events of the mid-eighteenth century, the new preformationism was sharply debated. Particularly at both the Stazione Zoologica in Naples, Italy, and the Marine Biological Laboratory in Woods Hole, Massachusetts, the two leading marine biology laboratories at the end of the nineteenth century, researchers attempted to frame experimental approaches to decide the issue and delivered long and polemical discussions debating the merits of preformation and epigenesis. The German embryologist Wilhelm Roux offered empirical studies of development that seemed to point in favor of preformation, emphasizing the mechanical and material nature of development; August Weismann wrote more on the continuity of the germinal material from generation to generation, again supporting a modern notion of preformation. At the same time, the German biologist Hans Driesch experimentally demonstrated why the developmental process had to be regulated by integrative processes that could not be immediately reduced to mechanical and materialistic notions. In the United States, C. O. Whitman, Edmund Beecher Wilson, and Thomas Hunt Morgan all provided laboratory studies and logical arguments in support of epigenesis. [*See* Morgan, T. H.] Wilson and Morgan were of particular importance; they were well aware of the work in Europe, had been influenced by the notion that the vitalistic notions of earlier biologists could probably be explained in terms of physical and chemical notions, and considered cells to be inherently organized entities. By the early twentieth century, therefore, the Americans rejected the completely materialistic and mechanistic interpretations of preformation that cited control of inheritance and variation in a specific structure of the cell (chromosome) and insisted that the phenomenon of development was epigenetic, caused by the integrity of the organized whole.

The first decade of the twentieth century ultimately brought another resolution to the debate when cytological evidence, from Wilson's laboratory, and experimental evidence, from Morgan's laboratory in the same building at Columbia University in New York City, demonstrated the validity of the chromosome theory of inheritance. This did not force either Wilson or Morgan, however, to accept the ideas of Roux or Weismann. Instead, Morgan changed the earlier interpretation of the action of the chromosome by separating the phenomena of inheritance from the phenomena associated with variation. From 1910 on, he considered inheritance to be associated with the chromosome, whereas variation was a product of the action of the cell in development. In other words, inheritance dealt with issues of the genotype, and variation was considered evidence of phenotypic phenomena. As a result, the long association (extending back to Aristotle at least) between inheritance and variation had been broken. In a sense, the modern tradition from Morgan stresses elements both of preformation and epigenesis. Chromosomal material, consisting of genes as operational units, is inherited by the offspring as discrete entities from parental stock. The developing organism is the product of epigenetic processes, controlled in part by the genetic material and in part by the organized matrix (cytoplasm) in which development occurs. It is interesting to note that modern genetics continues to emphasize the role of continuous units between generations, a legacy from the preformation tradition, whereas development biology (embryology) maintains that development is the gradual emergence of a new organism from undifferentiated material, illustrating its debt to epigenesis.

[*See also* Development, *article on* Developmental Stages, Processes, and Stability; Epigenetics; Recapitulation.]

BIBLIOGRAPHY

Benson, K. R. "Observation versus Philosophical Commitment in Eighteenth-century Ideas of Regeneration and Generation." In *A History of Regeneration Research*, edited by C. E. Dinsmore. Cambridge, 1991.

Bodemer, C. W. "Regeneration and the Decline of Preformation in Eighteenth-century Embryology." *Bulletin of the History of Medicine* 38 (1964): 20–31.

Gasking, E. *Investigations into Generation, 1651–1821.* Baltimore, 1967.

Maienschein, J. "T. H. Morgan's Regeneration, Epigenesis, and (W)holism." In *A History of Regeneration Research*, edited by C. E. Dinsmore. Cambridge, 1991.

Roe, S. A. *Matter, Life, and Generation: Eighteenth-century Embryology and the Haller-Wolff Debate.* Cambridge, 1981.

Roger, J. *Les sciences de la vie dans la pensée française du XVIIIe siècle.* Paris, 1963.

— KEITH R. BENSON

EPIGENETICS

Epigenetics is a term coined by the British embryologist and geneticist C. H. Waddington in 1947 "for the branch of biology which studies the causal interactions between genes and their products which bring the phenotype into being" (Waddington, 1975, p. 218). Although the term was little used during the first few decades of its existence, with the growth of molecular biology it began to be applied to studies of the control of gene activity during embryonic development and differentiation. Whereas the traditional genetic approach focused on variations that are caused by differences in genes, the epigenetic approach focused on phenotypic variation that is not coupled with genotypic variation. Epigenetics was concerned, on the one hand, with the problem that Waddington called canalization—how organisms of the same species, which often differ from each other genetically,

usually develop essentially the same phenotype—and, on the other hand, with how the cells within an individual, which are genetically identical, can differ markedly in structure and function. Brain cells, liver cells, and skin cells differ phenotypically, and their daughter cells inherit their phenotype, but the variation is epigenetic, not genetic. Similarly, the difference between a worker bee and a queen bee is epigenetic, not genetic, because whether a larva becomes a worker or a queen depends on the way it is fed.

By the end of the twentieth century, epigenetics had grown to become a widely recognized subdiscipline of biology. However, its growth was accompanied by a narrowing of its scope. Increasingly, epigenetics is now identified with cellular heredity, and it has become almost synonymous with epigenetic inheritance. One recent book about epigenetics defines it as "the study of mitotically or meiotically heritable variations in gene function that cannot be explained by changes in DNA sequence" (Russo et al., 1996, p. 667). Modern epigenetics is therefore primarily concerned with the mechanisms through which cells become committed to a particular form or function, and how that functional or structural state is then transmitted in cell lineages.

Mechanisms of Epigenetic Inheritance. The best understood epigenetic system is the methylation marking system. Cytosine (C), one of the four bases comprising the DNA molecule, sometimes carries a methyl group (C^m). This does not affect its coding properties, but the extent of cytosine methylation within and around a gene affects the probability that it will be active. Typically, genes in a highly methylated region of DNA are inactive, whereas genes in the same region may be transcribed if the methylation level is low. Developmental and environmental cues lead to changes in methylation, so the same gene may carry distinctly different methylation patterns (marks) in different cell types. These alternative patterns, called epialleles, can be transmitted to daughter cells, because methylation occurs symmetrically, usually in CG doublets in which C^mG on one strand is paired with GC^m on the other. Following DNA replication, the parental strand is methylated, but the daughter strand is not. However, maintenance methyltransferases recognize the asymmetrical, half-methylated sites, and methylate the cytosines of the new strand. Thus, by exploiting the semiconservative replication of DNA and the mirror symmetry of CG sites, the pattern of methylation can be reconstituted at every cell division.

In the early development of female eutherian mammals, one of the X chromosomes in each cell becomes transcriptionally inactive. This inactive state is transmitted to the cell's descendants. This inactive X, like inactive mobile genetic elements such as transposons and transcriptionally silent transgenes, is heavily methylated. Methylation also has a role in genomic imprinting: the activity of some genes depends on whether they were inherited from the male or female parent, and in many cases this differential activity is associated with differences in the methylation patterns of the maternally and paternally derived chromosomes.

Heritable methylation patterns are the best understood type of epigenetic marks, but there are others. Marks involving DNA-associated proteins that affect gene activity can also be transmitted in cell lineages, and are maintained and reconstituted following DNA replication. Differences in cell states can also be transmitted in other ways, such as through regulatory feedback loops. If a gene is turned on by a transitory external stimulus, and the gene's product acts as a positive regulator that maintains transcription of the gene, then, providing the concentration of gene product is high enough, gene activity is likely to be maintained following cell division. A very different type of epigenetic (cytoplasmic) inheritance is found in ciliated protozoa, where genetically identical individuals can have heritable variations in the pattern of cilia on their surface. Even experimentally altered patterns can be transmitted to daughter cells for hundreds of generations. The mechanisms underlying this are not understood, but structural inheritance is probably universal, because all cells seem to rely on preexisting structures when they produce new ones. Aberrant prions—infectious protein complexes that cause degenerative diseases such as bovine spongiform encephalopathy and scrapie—are thought to be formed through such structural templating.

Epigenetic Marks in Heredity. Epigenetic inheritance is important because development depends on induced epigenetic changes being transmitted to daughter cells. Cells "remember" their past, and new changes are built on past changes. However, epigenetic marks are usually reversible; they are sometimes erased, even in differentiated cells, which then produce a different cell type. Until recently, it was assumed that during sexual reproduction epigenetic marks from the previous generation are always erased before development gets under way. However, work on imprinting has shown that chromosomes retain marks associated with their parent of origin. Even more interesting are the cases in which epigenetic marks are transmitted through several generations. For example, it is now recognized that a well-studied heritable variation in mouse coat color is caused by an epigenetic modification, not a mutation. Similarly, the peloric form of toadflax, a mutant described by Carolus Linnaeus over 250 years ago, has turned out to be an epimutation—a difference in methylation pattern, not DNA sequence.

The extent of such transgenerational inheritance of epigenetic marks is at present unknown, and its significance is hotly debated. Does it mean that an environmental change affects not only the variations that are selected, but also the variations that are produced? Whatever the answer to this question, there is no doubt that Waddington was right to stress the importance of epigenetics in evolution. The refinement of epigenetic systems, which both respond to environmental cues and transmit the response to daughter cells, has been central to the evolutionary shaping of ontogenies.

[*See also* Epigenesis and Preformationism; Maternal Cytoplasmic Control.]

BIBLIOGRAPHY

Cubas, P., C. Vincent, and E. Coen. "An Epigenetic Mutation Responsible for Natural Variation in Floral Symmetry." *Nature* 401 (1999): 157–161. Reports the discovery that Linnaeus's peloric mutant is an epigenetic variation.

Grimes, G. W., and K. J. Aufderheide. *Cellular Aspects of Pattern Formation: The Problem of Assembly.* Basel, 1991. A thought-provoking account of structural inheritance in ciliates and other organisms.

Holliday, R. "Mechanisms for the Control of Gene Activity during Development." *Biological Reviews* 65 (1990): 431–471. Includes an account of methylation by one of the pioneers in the field.

Jablonka, E., and M. J. Lamb. *Epigenetic Inheritance and Evolution: The Lamarckian Dimension.* Oxford, 1995.

Jablonka, E., and M. J. Lamb. "Epigenetic Inheritance in Evolution." *Journal of Evolutionary Biology* 11 (1998): 159–183. This article is followed by thirteen others (185–260) that debate the views offered by Jablonka and Lamb.

Lyko, F., and R. Paro. "Chromosomal Elements Conferring Epigenetic Inheritance." *BioEssays* 21 (1999): 824–832.

Morgan, H. D., H. G. E. Sutherland, D. I. K. Martin, and E. Whitelaw. "Epigenetic Inheritance at the Agouti Locus in the Mouse." *Nature Genetics* 23 (1999): 314–318.

Prusiner, S. B. "Prions." *Proceedings of the National Academy of Sciences USA* 95 (1998): 13363–13383. A comprehensive review of the research on prions.

Russo, V. E. A., R. A. Martienssen, and A. D. Riggs eds. *Epigenetic Mechanisms of Gene Regulation.* Cold Spring Harbor, N.Y., 1996.

Waddington, C. H. *The Evolution of an Evolutionist.* Edinburgh, 1975. Reprints several of Waddington's early papers on epigenetics.

— Eva Jablonka and Marion J. Lamb

ESS. *See* Game Theory.

ETHOLOGY

Animals use their freedom to move and interact as one of the most important ways in which they are adapted to the conditions in which they live. These adaptations take many different forms, such as finding food, avoiding being eaten, finding a suitable place to live, attracting a mate, and caring for young. Each species has special requirements, and the same problem is often solved in different ways by different species. The biology of behavior, or ethology, as it is commonly called, is concerned about all aspects of the diverse ways in which animals engage their environments.

For many years, questions about adaptation were treated separately from questions about mechanism. The Nobel Prize–winning ethologist Niko Tinbergen refined these questions, separating both of them into two further problems. Issues to do with mechanism were to be divided into those connected with assembly of the adult and those to do with the workings of the fully developed behavioral process. Issues to do with why animals behave in the way they do were to be divided into the ways in which the process serves its current use and those by which the process acquired its present character. The evolution of behavior is about the second of these issues.

Charles Darwin's mechanism for evolutionary change consisted of three conditional steps. Each condition must have held if adaptation to the environment occurred in the course of biological evolution. First, variation must have existed. Second, some variants must have survived more readily than others. Third, the variation must have been inherited. In most cases, the source of the variation and the mechanism of inheritance are presumed to be genes. However, as animals became more complicated in the course of evolution, the mechanisms of inheritance were increasingly provided by social learning—most obviously in humans.

Behavior may be treated in the same way as any other aspect of an animal's biological makeup in the sense that behavior patterns often have a regularity and consistency that relate to the needs of the animal. Moreover, the behavior of one species often differs markedly from that of another. If behavior can be treated like anatomy, then it makes sense to look for similarities in closely related species. Konrad Lorenz believed that the use of stereotyped behavior patterns might provide a better way of constructing a taxonomy than the use of morphological characters. In 1973, he shared a Nobel Prize with Niko Tinbergen for their role in founding modern ethology and with Karl von Frisch, who, in wonderful field experiments, discovered the dance language of honeybees. Lorenz put into practice his ideas about the value of stereotyped behavior in uncovering the relatedness of different species by constructing a taxonomy of geese and ducks. Although modern taxonomists would not restrict themselves to one set of characteristics, Lorenz emphasized the point that behavior should not be treated differently from other characteristics. Some of the facial communication found in primates is remarkably conservative; for example, the so-called fear grin, given by subordinate animals to dominant ones, is found from primitive lemurs to chimpanzees, and it is not difficult to see how, in humans, it has come to be second-

arily associated with pleasure through the reduction of tension.

Tinbergen's distinction between current utility and evolution has not always seemed obvious. Surely, it was argued, if the function of a behavior pattern serves to enhance the survival of an animal and increase its reproductive success, then what is the difference from the evolutionary pressures that shaped the behavior in the first place? Biologists have been properly warned not to write evolutionary accounts in which the past is seen as leading purposefully toward the goal of the present blissful state of perfection. A clear distinction is necessarily and wisely drawn between the present-day utility (or function) of a biological process, structure, or behavior pattern and its historical, evolutionary origins. Darwin noted, for example, that whereas the bony plates of the mammalian skull allow the young mammal an easier passage through the mother's birth canal, these same plates are also present in the mammals' egg-laying reptilian ancestors. Their original biological function clearly must have been different from their current function.

The distinction between current function and historical evolution is all the more necessary because current adaptations may result from the experience of the individual during its lifetime. Muscles develop in response to the specific loads placed upon them during exercise. The detailed structure of the nervous system is sculpted by experience of use and disuse. Behavior, in particular, becomes adapted to local conditions during the course of an individual's development, whether through learning by trial and error or through copying others. These are all examples of adaptations that are acquired during the lifetime of the individual, and they are clearly distinct from adaptations that are inherited. Nevertheless, it was the insistence of Lorenz and Tinbergen that so much of behavior represents the product of biological evolution in past generations that brought the study of behavior into the renewed Darwinian synthesis that was being forged in the 1930s.

The interest in biological utility of behavior has led to many excellent studies of animals in natural conditions. A captive animal is usually too constrained by its artificial environment to provide a complete understanding of the functions of the great variety of activities that most animals are capable of performing. Studies in natural conditions have been an important part of ethology and played a major role in developing the distinctive and powerful methods for observing and measuring behavior. Field studies like those of John Crook on weaver birds and Jane Goodall on chimpanzees paved the way for the fields of behavioral ecology and primatology. Despite the emphasis on field work, it would be a mistake to represent ethologists as nonexperimental in their approach and merely concerned with description.

Tinbergen was a master of elegant field experiments, and the fine tradition he established has continued to the present day. Tape recordings of predators or conspecifics (such as offspring or potential mates) are played to free-living animals to discover how they respond. Dummies of different designs have similarly been used to gauge responsiveness to a particular shape or color, such as the pecking of gull chicks at different objects more or less resembling the bills of their parents. These and many other examples make the point that even mainstream ethology involves much more than mere observation. Moreover, a great many ethologists have devoted their professional lives to laboratory studies of the control and development of behavior. Indeed, some of the most striking ethological discoveries, such as imprinting and song learning in birds, have been made in artificial conditions and have markedly influenced how behavior in natural conditions has been interpreted.

When observed in hand-reared birds, the elaborate sequence involved in building a nest is not easily explained in terms of a series of learned actions, each triggered by a particular stimulus from the environment. Nonetheless, the readiness to consider what a particular behavior pattern might be for in the natural environment has been distinctive of the subject. When this approach was coupled with comparisons between animals, the easy assumption that all animals solve the same problem in the same way was quickly shown to be false. The comparative approach continues to be an important characteristic of broad ethology.

The Organization of Behavior. Two basic concepts of classical ethology were the sign stimulus and the fixed action pattern. The red belly of a male stickleback, a highly territorial freshwater fish, and the red breast of the equally territorial European robin are examples of sign stimuli. Sometimes the experimenter could construct something even more attractive than anything that occurs in the natural world by exaggerating the features detected by the animal. In his book *The Study of Instinct* (1951), Tinbergen illustrated an oyster catcher attempting to brood a much larger egg than its own in preference to the real thing. Stimuli that generated such bizarre behavior were labeled "supernormal." These concepts were productive in leading to the analysis of stimulus characteristics that selectively elicit particular bits of behavior. Fixed action patterns (or modal action patterns, as they are better called) provided useful units for description and comparison between species. Behavioral characteristics were used in taxonomy, and the zoological concern with evolution led to attempts to formulate principles for the derivation and ritualization of signal movements.

It is simply not possible to perform at the same time all the things that an individual must do if it is to stay alive and leave behind closely related descendants. The many requirements for behavioral action in an animal may be roughly classified as preservation, protection, and propagation. The many functional activities grouped under preservation and protection are necessary for survival, whereas propagation is necessary for genetic continuity from one generation to the next. A great many of these jobs are mutually exclusive. When a monitoring system tells the brain that blood sugars have fallen to low levels, for example, it makes sense that something somewhere suppresses all activities that might interfere with the business of seeking out food and raising the level of blood sugar.

When an alarm signal warns that a dangerous predator is approaching, feeding had better be suppressed in favor of bolting for safety lest the animal becomes the meal of another species. The need to express many different types of incompatible behavior requires well-defined internal rules of organization.

Organizing many different systems of behavior to maximize the chances of survival and reproductive success leads to large computational needs. In general, it is this pressure that has led to the evolution of ever more elaborate brains. At one time, animals were treated in the simplest possible way until good reasons were given for thinking otherwise. Nowadays, the tendency is to ask to what extent animals' cognitive abilities, perceptions, and self-awareness are similar to those of humans. This can lead to naive projection of human intentions and emotions into animals that are extremely unlikely to have them. Seemingly purposeful behavior may be expressed by negative feedback from the environment, just as it is in many human-made homeostatic devices such as those that regulate temperature in houses. Maintenance of humid environment by terrestrial isopod crustacea such as wood lice depends on a simple rule: stop walking when humidity is higher than the threshold value. Similarly, complicated optimizing and time-budgeting problems are often solved by simple rules of thumb that make time-saving shortcuts. Anticipation of the future can be generated by rather simple learning rules. Capacities to perform simple associations between neutral and biologically significant stimuli are known in many primitive invertebrates. Fruit flies associate certain odors with impending electric shock, and behave so as to avoid being shocked. Indeed, it must be said that humans do many complicated things entirely automatically without thinking about them.

The Origins of Adaptive Behavior. Because humans learn so much, it is easy to assume that if animals do clever things, then they must also learn what they do. However, much highly adaptive behavior develops without opportunities for practice or for copying other, more experienced animals. Indeed, a classic ethological concern was with the inborn character of much behavior, and the subject was strongly associated with the development of a theory of instinct. However, even the founders of the subject did not deny the importance of learning. On the contrary, they gave great prominence to developmental processes such as imprinting, for example, that specify what an animal treats as its mother or its mate and song learning that specifies the way a male bird sings a different dialect from another male of its own species.

Modern research has also eroded another belief of the classical ethologists that all members of the same species of the same age and sex will behave in the same way. The days are over when a field worker could confidently suppose that a good description of a species obtained from one habitat could be generalized to the same species in another set of environmental conditions. The variations in behavior within a species may, of course, reflect the pervasiveness of learning processes. However, some alternative modes of behavior are probably triggered rather than instructed by prevailing environmental conditions. In the gelada baboon, for instance, many adult males are much bigger than females. Once they have taken over a group of females, the males defend the females from the attentions of other males. Other males are the same size as females and sneak copulations when a larger male is not looking. The offsetting benefit for the smaller males is that they have much longer reproductive lives than the larger males. It seems likely that any male can go either way and that the particular way in which it develops depends on conditions. Examples such as this are leading to a growing interest in alternative tactics, their functional significance, and the nature of the developmental principles involved.

Konrad Lorenz saw adult behavior as involving the intercalation of separate and recognizable "learned" and "instinctive" elements. Few people share this view any longer, and the work by developmentally minded ethologists has been important in illustrating how the processes of development involve an interplay between internal and external factors.

After the early abortive attempts to classify behavior in terms of instincts, attention has increasingly focused on faculties or properties of behavior that bridge the conventional functional categories such as feeding, courtship, and caring for the young. Consequently, in modern ethological work, more and more emphasis is being placed on shared mechanisms of perception, storage of information, and control of output. As this happens, the interests of many ethologists are coinciding to a greater extent with what have been the traditional concerns of psychology.

In understanding behavioral development, it is important to understand that, in general, genes do not code for behavior in the sense that simple correspondence can be found between genes and behavior. Differences between individuals in some genes are often associated not only with variation in anatomy but also with variation in behavior. Such variation is used, for example, in the artificial selection of particular characteristics in domestic dogs. The general point about behavioral development is that a nervous system generating observable behavior is a product of both gene expression and the conditions prevailing as that individual matured. Furthermore, which genes are expressed depends on the state of the nervous system. Thus, external conditions may influence which genes are expressed. At least three different developmental processes lead to the adaptive complexity of behavior. Some do not depend on experience with the relevant environmental conditions, some are triggered by particular environmental conditions, and some involve specific rules for instruction by the environment. Any given behavior pattern may be affected by all three of these developmental processes.

The Rise of Sociobiology and Behavioral Ecology. In their time, the founding fathers of ethology were particularly successful, partly because they brought to behavioral biology a coherent theory of how behavior is organized and partly because they were interested in what behavior was for. Their functional approach marked them out as being distinct from the comparative psychologists. By the early 1970s, however, ethology seemed ripe for takeover. The much hoped for understanding of the links between behavior and underlying mechanisms was still fragmentary. Meanwhile, field studies relating behavior patterns to the social and ecological conditions in which they normally occur led to the enormous popularity and success of behavioral ecology, in which an understanding of mechanisms played very little part. A new subject called sociobiology moved into the available space, bringing to the study of behavior important concepts and methods from population biology, together with some all-embracing claims of its own. The pivotal moment for the growth of sociobiology was the publication of E. O. Wilson's book *Sociobiology* (1976). Imaginations were captured by the way the ideas from evolutionary biology were used. The appeal of evolutionary theory, in which sociobiology was embedded, was that it seemed once again to make a complicated subject manageable.

Individual animals interact with others, have relationships, and collectively form societies. Social behavior often seems to involve cooperation, and from Charles Darwin onwards, this aspect of behavior had continued to tease the theorists. If evolution depended on competition, how could cooperation have evolved? Three types of explanation were offered. The first benefited enormously from the thinking of William H. Hamilton. Hamilton pointed out that if parties are related, then the benefits of helping a cousin, say, are logically the same as helping a child, although quantitatively less effective in terms of gene propagation. This idea fostered the gene-centered approach expressed most vividly and accessibly in Richard Dawkins's book *The Selfish Gene* (1976). The second explanation was that both parties would benefit from the act of cooperation. Robert Trivers coined the term *reciprocal altruism*. The large predator fish opens its mouth to cleaner fish, which pick food remnants from the teeth of the predator without danger to themselves. The third explanation for cooperation is that the individual is part of a group of unrelated individuals that may survive better as an entity than as another group as a result of actions taken by the individuals within it. In the face of gene-centered theories, this explanation was widely thought to be implausible, but it may occur if individuals die out more rapidly than groups and immigration between groups is difficult.

The gene-centered approach to behavioral biology brought a new look to studies of communication where well-known instances of supposed transfer of information were reinterpreted in terms of selfish manipulation. It also led to a reexamination of apparent shared interests of both parents caring for the young or the mutually beneficial interactions of parent and offspring. Most adults of most species are able to produce more than one offspring during their lifetime. The characteristics of those offspring that are most successful in leaving descendants will tend to predominate in subsequent generations. Parents who sacrifice too much for one offspring will have fewer descendants. By the same token, offspring that do less to ensure their own survival than others will have fewer descendants. Such are the broad rules that shape any life span, but just how they look in detail will depend on the species and, within a species, on local conditions. In some species, parental care for the tiny progeny consists of nothing more than providing a small amount of yolk, enough to sustain the offspring until they can feed for themselves. Marine fish such as herring produce vast numbers of eggs and sperm, which fuse in the sea. Neither parent provides any care for the progeny. Most fertilized herring eggs consequently die at an early stage in their lives. Other fish produce far fewer fertilized eggs and care for them in a variety of ways, some keeping them like a mammal in their bodies until the young are born, others gathering the fertilized eggs into their mouths, where they are protected.

Of all animal groups, birds and mammals produce the smallest number of young and take the greatest care of those they do produce. Birds encase their fertilized eggs in a hard shell and eject them from their bodies, whereupon both sexes usually take turns in keeping the developing egg at body temperature until the chick hatches. While it is developing inside the egg, the embryonic chick is fed from the yolk, which at the outset is enor-

mous relative to the embryo. After the egg has hatched, both parents usually protect and bring food to the developing young. This has important consequences for the amount of care that is given by the parent and the time taken to become adult. The long period of development that is particularly characteristic of humans, but also true to a lesser extent of most other mammals, is made possible by the protection and the provisioning by the parent. As they prepare for their own eventual reproductive life, children meanwhile have to survive.

The traditional image of parenthood has been one of complete harmony between the mother and her unborn child. But evolutionary theory has cast doubt on this blissful picture. In sexually reproducing species, parents are not genetically identical to their offspring. Consequently, offspring may require more from parents than parents are prepared to give, creating the possibility of a conflict of long-term interests. Robert Trivers called this "parent–offspring conflict," a term that refers strictly to a conflict of reproductive interests, not a conflict in the sense of overt squabbling. The parent may sacrifice some of the needs of its current offspring for others that it has yet to produce; the offspring maximizes its own chances of survival. Parent and offspring "disagree" about how much the offspring should receive. The result of such evolutionary conflicts of interest is sometimes portrayed as an arms race, with escalating fetal manipulation of the mother being opposed by ever more sophisticated maternal countermeasures. However, limits must be encountered in the course of evolution. If the offspring is too aggressive in its demands, it will kill its maternal host and, of course, itself. Likewise, if the mother is too mean, her parasitic offspring will not thrive, and she might as well have not bred. Moreover, mutually beneficial communication often occurs between parent and offspring so that independence comes at a time that is beneficial to both parties.

Despite the invigorating debates about the function and evolution of social behavior, the impact of sociobiology on behavioral biology as a whole was that large chunks of the subject, which had been central concerns of ethology, were deemed to be irrelevant or uninteresting. After 1976, few students interested in whole animals wanted to work on the problems of how behavior develops or on how it is controlled. For many years, therefore, issues to do with mechanism were largely ignored. Eventually, sociobiology was deemed to have overreached itself. In the study of cognition, it was replaced by the derivative field of evolutionary psychology in the 1990s. In behavioral biology, the atrophied links between the "why" and the "how" questions were once again rebuilt.

Asking what something is for is never going to reveal directly the way in which that thing works. But the functional approach does help to distinguish between independent mechanisms controlling behavior and can lead fruitfully to the important controlling variables of each system. This is important in the design of experiments in which, inevitably, only a small number of independent variables are manipulated while the others are held constant or randomized. The experiment is a waste of time if important conditions that are going to be held constant are badly arranged. A functional approach can provide the knowledge that prevents expensive and time-consuming mistakes.

Those who worked on the most efficient ways to find food in a natural environment have appreciated that their work raised important issues about how behavior is controlled. As a result of the regained awareness, flourishing links have been formed most notably between the behavioral ecologists and the psychologists interested in the experimental analysis of learning. In behavioral development, too, functionally inspired approaches have played a useful role in making sense of what otherwise seemed a hopelessly confused area. Asking what might be the current use of behavior helps to distinguish juvenile specializations from emerging adult behavior and helps to explain the developmental scaffolding used in the assembly process. Functional assembly rules are important, for example, in determining when an animal gathers crucial information from its environment. With attention focused on the problem, attempts can be made to analyze the mechanisms. Here again, the optimal design approach frames and stimulates research on the processes of development.

The stream of ideas between "how" and "why" approaches flows both ways. Many of those who have focused on the behavioral ecology of animals are beginning to appreciate the need for knowledge of the mechanisms to address the functional and evolutionary questions in which they are most interested. This has happened notably in the studies of perceptual factors and learning processes influencing mate choice and their implication for associated evolutionary theories of sexual selection. It is also happening in areas of work generally lumped under the heading of "life history strategies," which raise important issues to do with conditional responses to environmental conditions. In general, these changes in thought are occurring because what animals actually do is being seen as important in stimulating (as well as constraining) ideas about the function and evolution. The mechanisms involved in the development and control of behavior may often feed back into evolutionary processes, as seems likely to have been the case with mate choice and with the active control of the social environment.

Although the barriers between the "why" and the "how" approaches have once again become more permeable, enormous strides have been made in neuroethology and in understanding the hormonal basis of be-

havior. Links between the physiology of metabolism and behavior are being made, as well as between behavioral state and the immune system. The relevance and value of molecular techniques are beginning to be realized. In general, studies of the development and control of behavior look very different in the new millennium from how they seemed twenty-five years ago. A systems approach is essential, and behavioral biologists are particularly well equipped to provide it. Some of the most interesting people studying the neural basis of behavior know only too well that the data they obtain are much the same as those obtained by a meteorologist in the middle of a hurricane working at ground level. They have realized that if you want a coherent sense of the whole system, you need the equivalent of a satellite picture. The people who study behavior provide the overview for those on the metaphorical ground. Moreover, the behavioral biologists have developed special techniques for the measurement of free-flowing, unconstrained behavior.

Active Role of Behavior in Evolution. Ethology has made a considerable impact on thinking about the dynamics of evolution. First, animals are able to modify their behavior in response to changed conditions, thereby allowing evolutionary change that otherwise would probably have been prevented by the death of the animals exposed to those conditions. Second, by their behavior, animals often expose themselves to new conditions that may reveal heritable variability, thereby opening up possibilities for evolutionary change.

Third, animals make active choices. It is widely appreciated now that the results of their choices have consequences for subsequent evolution. The role of mate choice in evolution was clearly recognized by Darwin in his principle of sexual selection, now the subject of much theoretical and empirical work. The choice of the habitat in which the animal lives can influence the subsequent course of evolution in a variety of ways. If, in the course of evolution, individuals developed a feature that would have been maladaptive in the environment in which their ancestors lived, they have the capacity to move to another environment where the feature works or, at least, is no longer costly. If, say, an animal became sightless because of the benefits of other evolutionary changes, it could move to a dark environment, such as a cave, where such a deficit is no longer a handicap.

Finally, by their behavior, animals also change the physical or the social conditions with which they and their descendants have to cope and thereby affect the subsequent course of evolution. By leaving an impact on their physical and social environment, organisms may affect the evolution of their own descendants, quite apart from changing the conditions for themselves. Some of the impact is subtle, such as when a plant sheds its leaves, which fall to the ground and change the characteristics of the soil in which its own roots and those of its descendants grow. Some of the impact is conspicuous and active, such as when beavers dam a river, flood a valley, and create a private lake for themselves. The effect of behavioral control on evolutionary change could be especially great when a major component of the environmental conditions with which animals have to cope is provided by their social environment. A similar type of positive feedback to that flowing from the effects of mate choice could operate in such circumstances. If individuals compete with each other within a social group and the outcome of the competition depends in part on each individual's capacity to predict what the other will do, the evolutionary outcome might easily acquire a ratchetlike property.

Conclusion. Ethology and closely related subjects differ from many other fields of biology in that considerable time and effort are needed for a discussion of conceptual issues. The range of material found in the literature on behavioral biology is enormous and reflects what is perhaps both a weakness and a strength. The weakness is that replications of experiments are unusual, and parametric variation of important conditions influencing behavior, as would be standard in experimental psychology, is not that common. The strength is that the new material constantly surprises and opens up new lines of research. Some tension is found between the view that variation in behavior is what matters most of all and the view that general principles must be uncovered. Of course, what this reflects is a difference in research goals. If they are pressed, most ethologists would agree that they were more interested in one goal rather than the other, even though they would disagree about which is the most important one. However, it would be too crude to suggest that the breadth of ethology has come about because an all-around approach to behavior helps each research worker to get more easily to where he or she had always intended to go. A great many ethologists simply take delight in the study of behavior in all its diversity.

The development and control of an individual's behavior can both be investigated at many levels—the whole animal, the systems physiology, and the molecular levels. Also, as Robert Hinde has repeatedly emphasized, behavior between individuals can be investigated in terms of interactions, relationships, and social structures. For these reasons, it is important to understand the differences between levels of organization as well as appreciate Tinbergen's distinction between the proximate questions of how behavior develops and how it is controlled once fully assembled.

Modern ethology abuts so many different disciplines that it defies simple definition in terms of a common problem or a shared literature. It overlaps extensively with the fields known as behavioral ecology and sociobiology. Moreover, those who call themselves ethologists are now to be found working alongside neurobiologists, social and developmental psychologists, anthropologists, and psychiatrists, among many others. Ethologists are accustomed to think in ways that reflect their experience with free-running systems that both influence and are affected by many things about them. These skills have enabled them to understand the dynamics of behavioral processes and are prized by those people with whom they collaborate. At one time, ethology looked as if it might succumb to its sister disciplines. However, it has reemerged as an important subject that will continue to play an important integrative role in the drive to understand what behavior patterns are for, how they evolved, how they developed, and how they are controlled.

[*See also* Adaptation; Altruism; Darwin, Charles; Human Sociobiology and Behavior, *articles on* Human Sociobiology, Evolutionary Psychology, *and* Behavioral Ecology; Social Evolution.]

BIBLIOGRAPHY

Alcock, J. *Animal Behavior.* 6th ed. 1998. A comprehensive introduction to behavioral biology.

Avital, E., and E. Jablonka. *Animal Traditions: Behavioural Inheritance in Evolution.* 2000. Describes the role of social learning in inheritance.

Bateson, P., and P. Martin. *Design for a Life: How Behaviour Develops.* 2000. Explains in an accessible way the intricacies of where behavior comes from.

Dawkins, M. *Through Our Eyes Only?* 1993. Excellent introduction to the problems of animal consciousness.

Dawkins, M. *Unravelling Animal Behaviour.* 2d ed. 1995. Useful guide for those attempting to understand the problems of ethology.

Dawkins, R. *The Selfish Gene.* Oxford, 1976.

Hauser, M. *Wild Minds: What Animals Really Think.* 2000. A readable introduction to animal cognition.

Laland, K. N., and G. R. Brown. *Sense and Nonsense: Evolutionary Perspectives on Human Behaviour.* 2001. Sane and reliable pilot through treacherous waters.

Manning, A., and M. Dawkins. *Introduction to Animal Behaviour.* 5th ed. 1998. A coauthored new edition of Manning's popular textbook.

Martin, P., and P. Bateson. *Measuring Behaviour.* 2d ed. 1993. A practical guide to what can be measured when studying behavior and how to do it.

McFarland, D. *Animal Behaviour.* 3d ed. 1999. A good general textbook, bringing together ethology and psychology.

Tinbergen, N. *The Study of Instinct.* 1951.

Wilson, E. O. *Sociobiology.* Cambridge, Mass., 1976.

— PATRICK BATESON

EUGENICS

Eugenics is based on the idea that the human gene pool (hence humanity) can be improved by selective mating. Almost since the rise of agriculture, people have realized

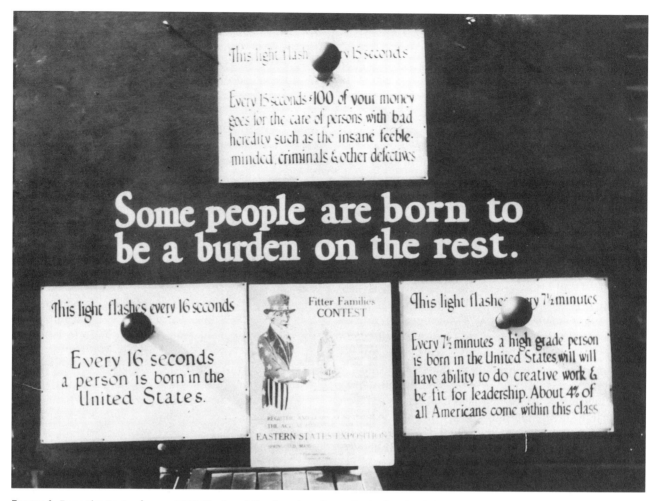

FIGURE 1. Eugenics Poster from a 1926 Display at the American Eugenics Society in Philadelphia.
This display used flashing lights to emphasize the potential impact on American prosperity if the reproduction of "inferior" persons was not controlled.   American Philosophical Society.

that careful breeding could enhance many useful features of domestic animals and plants. But systematic efforts to engender a wider racial improvement did not become widespread or popular until the late nineteenth century, when eugenics gained general acceptance among the educated middle classes. Socialists and reform liberals, as well as conservatives, came to believe that the worst kinds of people (as variously defined) were outbreeding the best. The eugenics movement reached its zenith in the twenty years prior to World War II, when countries around the globe implemented eugenic legislation and policies. The revelations of Nazi atrocities after the war and, most importantly, a change in social values in the 1950s and 1960s, however, lessened public support for eugenics, and many of the laws were repealed or fell into disuse. Instead, eugenic efforts were directed to genetic counseling and medical genetics. By the 1960s, eugenics had become a pejorative term, and the eugenics movement appeared dead. Today,

however, the advances of genetic medicine have revived concerns about eugenics. Many people fear that we have failed to learn the consequences of uncritically embracing genetic medicine. They worry that we will exaggerate the importance of genes to human disease and behavior, and hence again look to scientific fixes for what are ultimately social problems.

Charles Darwin's cousin, Francis Galton, coined the word *eugenics* in 1883, but he had been concerned about the course of human evolution for at least twenty years before that. This concern arose from the almost universal belief in the nineteenth century that human traits—most importantly, mental and behavioral characteristics—were highly heritable. Darwin's idea of natural selection was supported by the results of artificial selection on domestic animals, but his theory applied to people equally well, as Darwin himself showed in *The Descent of Man* (1871). Consequently, Galton urged the most intelligent and talented to have more children, and also to marry each other. Such a program has come to be called "positive eugenics" as opposed to the better known "negative eugenics," under which those with unwanted characteristics are discouraged or prevented from having children.

In many Western countries, eugenics became increasingly popular in the early twentieth century. The hereditarian beliefs of the previous century appeared to be confirmed by scientific studies of the inheritance of human (and other species') features. Almost every characteristic examined, whether in humans, fruit flies, or maize, turned out to have some genetic basis. Traits not directly heritable were held to be a consequence of others that were. In addition, traits that were considered undesirable were concentrated in races and social classes with apparently higher fertility. The inevitable deduction was made: unless something was done either to encourage the "better sort" of people to have more children or to reduce the number of children of the lesser type, humanity would degenerate and Western society would disintegrate under the weight of the undesirables. Because it is easier to prevent reproduction, the focus was on negative eugenics.

In practice, because it seemed to be the root cause of so many social problems (e.g., pauperism and alcoholism), the single trait that concerned eugenicists most was mental defect, or "feeblemindedness," as it was known. Paul Popenoe (1915), writing in the *Journal of Heredity*, which he also edited, quoted a standard, widely accepted definition: a feebleminded individual was one who was "incapable of performing his duties as a member of society in the position of life to which he is born."

Geneticists of the time all but universally believed feeblemindedness to be the result of genetic inheritance. The most influential studies were those conducted by the Eugenics Record Office, on Long Island, New York, especially those published by its head, Charles B. Davenport, and by Henry H. Goddard of New Jersey, who later introduced intelligence quotient (IQ) testing to the United States. Two of Goddard's books, *The Kallikak Family: A Study in the Heredity of Feeblemindedness* (1912) and *Feeblemindedness: Its Causes and Consequences* (1914), were particularly important, not just in arguing that feeblemindedness was hereditary, but in proposing that it resulted from being homozygous for a recessive allele. This Mendelian interpretation became the standard view for at least the next twenty years.

Estimates of the incidence of "feeblemindedness" varied greatly. Although a commonly used figure in the period 1910–1930 was 0.3 percent of the population of the United States, other surveys suggested that 2 percent of New York schoolchildren were feebleminded, as were an astounding 89 percent of black Americans drafted in World War I. Of even greater concern, however, were British and American studies of the fertility of the feebleminded, who appeared to have nearly twice the number of children as normal parents. Given the apparent hereditary nature of feeblemindedness, the long-term consequences were obvious and alarming.

The merits of various negative eugenic policies—chiefly sexual segregation and sterilization of the affected—were debated in both the scientific and popular press. Segregation for the years when individuals were fertile was mooted, but very expensive. Female sterilization was also expensive, although, with the advent of vasectomy, male sterilization was easy and affordable. Consequently, some thirty American states enacted sterilization laws (although not all survived court challenges on constitutional grounds), and over 60,000 people were eventually sterilized, about half of these in California. Similar legislation was passed in one Mexican state, two Canadian provinces, Japan, and several European countries, notably, of course, Nazi-controlled Germany, but also all of the Scandinavian nations. Swedish authorities sterilized 1,210 people on eugenic grounds in 1947, and just under 63,000 were sterilized in the period 1935–1975, although the recorded reasons shifted from primarily "eugenic" to "medical" in the 1950s. Opposition from organized labor and the Roman Catholic church, however, prevented the passage of sterilization laws in Britain.

One of the scientific debates over eugenics concerned the long-term results of eugenic selection and, in the first instance, relied explicitly on Goddard's Mendelian model. Selection, either eugenic or natural, against a recessive allele is not very effective when that allele is rare because, as the Hardy-Weinberg principle shows, most of those alleles are hidden from selection in heterozygotes. If the frequency of the allele is given by q, the proportion of those exhibiting the trait (i.e., those who are homozygous for the recessive allele, the only ones who may be selected against) is q^2, which is very small when q is small. When selection discriminates perfectly and all those affected are prevented from having children, then the frequency in the subsequent generation is given by $q' = q/(1 + q)$: after n generations the frequency of q is given by $q_n = (n + q^{-1})^{-1}$. This latter formula was implicitly used in 1917 by Cambridge geneticist Reginald C. Punnett to show that, for example, a reduction in incidence from 0.1 to 0.01 percent would take about sixty-nine generations, longer than the time since the fall of the Roman Empire.

This argument can often be found today in college genetics textbooks, where it is used to claim that eugenics could not have worked. But this interpretation is wrong on several grounds. First, feeblemindedness was not thought to be rare. Using Goddard's model, a single generation of eugenic selection would reduce the standard 0.30 percent estimate of its incidence to 0.27 per-

THE VIEWS OF EUGENICISTS

"Society must look upon germ-plasm as belonging to society and not solely to the individual who carries it."

—H. H. LAUGHLIN, 1914. *Eugenics Record Office Bulletin 10A.*

"From an ethical standpoint, so few people would now contend that two feeble-minded or epileptic persons have any 'right' to marry and perpetuate their kind, that it is hardly worth while to argue the point."

—P. POPENOE AND R. H. JOHNSON, 1918. *Applied Eugenics.* New York: Macmillan.

"When we know that men or women are not only the embodiments but also the bearers of hereditary taint and defect, we have no more right to allow them to reproduce than to allow a child with scarlet fever to be visited by all his school friends."

—J. S. HUXLEY, 1926. *Nature* 118: 844–846.

"To stop the propagation of the feebleminded, by thoroughly effective measures, is a procedure for the welfare of future generations that should be supported by all enlightened persons. Even though it may get rid of but a small proportion of the defective genes, every case saved is a gain, is worth while in itself."

—H. S. JENNINGS, 1930. *The Biological Basis of Human Nature.* London: Faber & Faber.

"All modern geneticists approve the segregation or sterilization of persons who are known to have serious hereditary defects, such as hereditary feeblemindedness, insanity, etc."

—E. G. CONKLIN, 1930. In E. V. Cowdry (ed.), *Human Biology and Racial Welfare.* New York: Hoeber.

"To state that reproductive selection against severe physical and mental abnormalities will reduce the number of affected from one generation to the next by only a few percent does not alter the fact that these few percent may mean tens of thousands of unfortunate individuals who, if never born, will be saved untold sorrow."

—C. STERN, 1949. *Principles of Human Genetics.* San Francisco: Freeman.

"There is certainly a need for eugenics. No one who has visited institutions for mental defectives and deaf mutes can doubt this need."

—J. B. S. HALDANE, 1965. In S. J. Geerts (ed.), *Genetics Today.* Vol. 2. Oxford: Pergamon.

—HAMISH G. SPENCER AND DIANE B. PAUL

cent, a 10 percent reduction. In 1924, R. A. Fisher, the leading statistician of the twentieth century and an ardent eugenicist, argued that such a reduction was not at all negligible. Moreover, Fisher pointed out that, if there were any degree of assortative mating, which seemed quite probable, a larger fraction of the population than expected under random mating is homozygous and the reduction would be even greater. Second, the argument was bypassed by eugenicists, who responded that at least some progress was better than none. Finally, these calculations were promoted not as a criticism of eugenics (as modern textbooks would have us believe), but as a call for eugenicists to discover methods to identify carriers. If heterozygotes too may be ascertained, selection is highly efficient.

A second sort of criticism of the science underlying eugenics concerned the mode of inheritance of feeblemindedness. Biometricians, such as Karl Pearson, the inventor of the well-known correlation coefficient and the director of the Galton Eugenics Laboratory at the University of London, and assistant director David Heron, disparaged the Mendelian model as well as the data on which it was based in a series of papers in 1913 and 1914. Their critiques were accurate and damning— even by the standards of the day, the quality of the data was poor, and there were numerous inconsistencies in the interpretation—but they had no effect at the time. Nevertheless, like the inefficacy of selection argument, these criticisms were not antieugenic; rather, they were intended to put eugenics on a proper, firm (i.e., biometric in the critics' view) footing. Indeed, in the same paper in which he showed that the Mendelian model would result in significant eugenic progress in a single generation, Fisher calculated that a biometric, polygenic model would also give rise to a substantial decrease in the incidence of feeblemindedness. Pearson, Heron, and almost every leading geneticist of the day supported eugenics, and few scientists (or members of the scientific laity, for that matter) had doubts that mental defect was hereditary. In the 1910s, work on the fruit fly (*Drosophila melanogaster*) in Thomas Hunt Morgan's Columbia University laboratory showed that variation in numerous characters was caused by "Mendelian factors" (i.e., genes), and Mendelian inheritance was a consequence of the way chromosomes behaved at meiosis. The flood of data implying that it was all in the genes swamped the biometricians' objections.

Davenport and Popenoe did note the illogic of a socially defined trait being inherited in a Mendelian fashion. This realization, however, did not detract from their hereditarian beliefs (or, in the case of Davenport, from his support of the Mendelian model). Pearson pointed out that correlation did not identify cause; therefore, one could not infer the hereditary nature of feeblemindedness solely from pedigrees. But before 1930, the only geteticist to make any unequivocally antieugenic statements was Morgan, who was an extremely private person politically, avoiding any publicly conspicuous criticism. Nevertheless, in the final chapter of the revised edition of *Evolution and Genetics* (1925), Morgan argued that the evidence for the Mendelian model for feeblemindedness was open to other interpretations, most especially an environmental effect of "demoralizing social conditions." Although he was never explicit on the point, all the mathematical models of the effects of eugenic selection, whether Mendelian or biometric, assumed there was no environmental effect. In addition, Morgan made the plainly political point that it was often not obvious what distinguished the normal and desirable from the abnormal and undesirable, and he pleaded for a "little more goodwill" than was shown by many eugenicists.

The conventional wisdom of modern college textbooks is that the advance of genetics led to the downfall of eugenics. In addition to the inefficacy of selection argument, they point out that variability may be intrinsically valuable, for example, at loci subject to heterozygote advantage. Indeed, during the 1930s, some geneticists did criticize sterilization laws and other eugenic policies. But most of these scientists, for example, J. B. S. Haldane (who, along with Fisher and Sewall Wright, founded the field of mathematical population genetics) and Julian Huxley (first director of the United Nations Educational, Scientific, and Cultural Organization and grandson of "Darwin's bulldog," Thomas Huxley), supported eugenics in principle; their opposition was restricted to then current practice. Often their motivation was political: as a Marxist, Haldane, for example, was concerned that eugenic policies were targeting the working class. Although college textbooks that were originally enthusiastic moderated their discussions in later editions, their authors remained fundamentally committed to eugenics. It was possible (indeed likely) in the 1930s, 1940s, and 1950s to recognize all the scientific flaws identified above yet remain a eugenicist.

While acknowledging the importance of the revelations of Nazi atrocities in the decline of the public's and scientists' support of eugenics, recent historiography highlights the change in social and political values of the latter part of the twentieth century. The increased emphasis on individual rights, most importantly patients' medical rights, the right to privacy, and reproductive autonomy (also a consequence of the rise of feminism), meant that the views of eugenicists (see Vignette) seemed outdated. Even geneticists who opposed eugenics often did so on ethical or political grounds. For instance, although he effectively ridiculed the shoddy science behind eugenics, Lionel Penrose, who became Galton Professor of Eugenics after World War II, opposed eugenics primarily because he believed a society should be judged by how well it looked after those who could not care for themselves.

This history, however, leaves us in a quandary: it implies eugenics was a political choice (albeit a currently disdained one). Certainly much of eugenics rested on poor science. But many of these errors have been eliminated or alleviated (at least in principle) by the new genetics: we can detect heterozygotes for most genes, and we are much more certain about the mode of inheritance of many characters. Much of today's argument about whether genetic medicine is eugenics in a new guise arises because of these scientific advances. The debate is so rancorous because, as in the past, it is fundamentally about ethical and political issues.

[*See also* Disease, *article on* Hereditary Disease; Fisher, Ronald Aylmer; Galton, Francis; Genetic Load; Haldane, John Burdon Sanderson; Heterozygote Advantage, *article on* Sickle-Cell Anemia and Thalassemia; Race, *article on* Population Genetic Perspectives.]

BIBLIOGRAPHY

Adams, M. B., ed. *The Wellborn Science: Eugenics in Germany, France, Brazil, and Russia.* New York, 1990. Analyzes the eugenics movements in four non-English-speaking countries, comparing them to the better studied examples of Britain and the United States.

Barker, D. "The Biology of Stupidity: Genetics, Eugenics and Mental Deficiency in the Inter-war Years." *British Journal for the History of Science* 22 (1989): 347–375. Documents the widespread and long-lived acceptance of the Mendelian model for the inheritance of feeblemindedness. Also discusses the pervasive faults in both data and interpretation, but fails to notice that such a critique had been made at the time by Heron and Pearson (see Spencer and Paul, 1998).

Broberg, G., and N. Roll-Hansen, eds. 1996. *Eugenics and the Welfare State: Sterilization Policy in Denmark, Sweden, Norway and Finland.* East Lansing, Mich., 1996. University Press. A collection of essays on the history of eugenics and the role of science in its implementation in Scandinavia.

Buchanan, A., D. W. Brock, N. Daniels, and D. Wikler. *From Chance to Choice: Genetics and Justice.* Cambridge, Mass., 2000. A sophisticated discussion by four distinguished philosophers of bioethical issues raised by advances in genetics.

Burleigh, M. *Death and Deliverance: "Euthanasia" in Germany c. 1900–1945.* Cambridge, 1994. An account of the eugenics movement in Weimar and Nazi Germany, especially its application in psychiatry.

Fisher, R. A. "The Elimination of Mental Defect." *Eugenics Review* 26 (1924): 114–116. The leading twentieth-century statistician argues that eugenic selection would initially effect a significant decrease in the incidence of feeblemindedness whatever its mode of inheritance.

Goddard, H. H. *The Kallikak Family: A Study in the Heredity of Feeblemindedness.* New York, 1912.

Goddard, H. H. *Feeblemindedness: Its Causes and Consequences.* New York, 1914.

Jones, G. *Social Hygiene in Twentieth Century Britain.* London, 1986. A study of the importance of hereditarian ideas in attempts to improve public health, including eugenics.

Kevles, D. J. *In the Name of Eugenics: Genetics and the Uses of*

Human Heredity. Berkeley and Los Angeles, 1985. A comprehensive history of the eugenics movement in Britain and the United States.

Mazumdar, P. M. H. *Eugenics, Human Genetics, and Human Failings: The Eugenics Society, Its Sources and Its Critics in Britain.* London, 1992. A study of the British eugenics movement that stresses the importance of the scientific interests, especially in mathematical statistics, shared by geneticists of the political right and left.

Morgan, T. H. *Evolution and Genetics.* Princeton, 1925.

Müller-Hill, B. *Murderous Science: Elimination by Scientific Selection of Jews, Gypsies, and Others, Germany 1933–1945.* Translated by G. R. Fraser. Oxford, 1988. A study of how the German scientific establishment became an integral part of the racial and eugenic policies of the Nazi government.

Paul, D. B. *Controlling Human Heredity: 1865 to the Present.* Amherst, N.Y., 1995. A brief history of eugenics, emphasizing its wide support among geneticists of the first half of the twentieth century. The author asks what lessons from this history can be applied to current ethical issues in genetic medicine.

Paul, D. B., and H. G. Spencer. "Did Eugenics Rest on an Elementary Mistake?" In *Thinking about Evolution: Historical, Philosophical and Political Perspectives*, edited by R. S. Singh, C. B. Krimbas, D. B. Paul and J. Beatty, pp. 103–118. New York, 2001. An examination of the textbook argument that a basic understanding of the Hardy-Weinberg principle shows that eugenics would never have worked.

Peel, R., ed. *Essays in the History of Eugenics: Proceedings of a Conference Organised by the Galton Institute, London, 1997.* London, 1998. An interesting collection of essays on specialized subjects including the development of biometry, women and feminism, "reform eugenics," and social class.

Poperoe, Paul. *Journal of Heredity* 6 (1915): 32–36.

Spencer, H. G., and D. B. Paul. "The Failure of a Scientific Critique: David Heron, Karl Pearson and Mendelian Eugenics." *British Journal for the History of Science* 31 (1998): 441–452. Examines the contemporary critiques of Goddard's Mendelian model for the inheritance of feeblemindedness and why they were unsuccessful.

Stepan, N. L. *"The Hour of Eugenics": Race, Gender, and Nation in Latin America.* Ithaca, N.Y., 1991. A history of the eugenics movement in Central and South America.

Turney, J. *Frankenstein's Footsteps: Science, Genetics, and Popular Culture.* New Haven, 1998. An exploration of the history of public concern about biological science, especially genetics.

Zenderland, L. *Measuring Minds: Henry Herbert Goddard and the Origins of American Intelligence Testing.* Cambridge, 1998. A history of the work of Goddard and how his IQ testing methods affected the very concept of intelligence.

— HAMISH G. SPENCER AND DIANE B. PAUL

EUKARYOTES. *See* Prokaryotes and Eukaryotes.

EUSOCIALITY

[*This entry comprises two articles. The first article focuses on the origin and evolution of eusociality in insects; the second article discusses eusociality in mammals. For related discussions, see* Altruism; Kin Recognition; Kin Selection; *and* Naked Mole-Rats.]

Eusocial Insects

Eusociality, or "true" sociality, refers to social systems in which there are two discrete types of individuals, those that reproduce at a relatively high level but engage in less work (reproductives), and those that reproduce less but work more (helpers). The main distinguishing feature of eusociality is the presence of these two "castes": reproductives specialized for producing offspring, and altruistic helpers specialized for tasks involved in rearing offspring, such as foraging for food, feeding the young, and building, maintaining, and defending the nest or other type of domicile. In most eusocial insects, the reproductives are female "queens," who are the mothers of the helpers, and the role of the helpers is to aid in the production of more helpers plus the next generation of reproductives, who leave the colony to mate and attempt to start families of their own. However, there is a tremendous diversity of forms among eusocial insects, varying in many aspects of their genetics, phenotypes, life cycles, and ecology.

Diversity of Eusocial Insects. Among insects, eusociality is found in about twelve thousand species of Hymenoptera (ants, bees, and wasps), six species of Thysanoptera (thrips), about two thousand species of Isoptera (termites), about forty species of Homoptera (aphids), and one species of Coleoptera (beetle). One of the most important features of these taxa, with regard to their social evolution, is their genetic systems, that is, the numbers of sets of chromosomes in females and males, and whether or not the females can produce females or males without sperm (parthenogenetically).

FIGURE 1. Queen Honeybee (*Apis mellifera*) Surrounded by Workers.

In most eusocial insects, the reproductives are female "queens," who are the mothers of the helpers. The role of the helpers is to aid in the production of more helpers plus the next generation of reproductives, who leave the colony to mate and attempt to start families of their own.
© R. Williamson/Visuals Unlimited.

The Isoptera, Homoptera, and Coleoptera are diploid in both sexes, although in aphids reproduction is by female-to-female parthenogenesis during most of the life cycle, whereby the offspring of a female are all her genetically identical clone. By contrast, Hymenoptera and Thysanoptera are all haplodiploid: the males are haploid, being produced parthenogenetically from unfertilized eggs, and the females are diploid, being produced from fertilized eggs.

The diversity of genetic systems is important to the evolution of eusociality because different genetic systems engender different levels of genetic relatedness (proportional sharing of genes identical by inheritance) between colony members. As discussed below, selection involving the fitness effects of interactions between various genetic relatives (kin selection) is critical to the origin and evolution of social systems.

Eusocial insects differ in a considerable number of traits associated with their colony composition and life cycles (Table 1). However, all eusocial insects inhabit domiciles of some kind, where the queen normally resides and offspring develop. Hymenoptera and termites dig or utilize burrows or cavities in soil, wood, or pithy stems, or construct nests from pulped wood, mud, masses of workers (in army and driver ants), or other materials. In Hymenoptera, workers must leave the nest to forage for animal or plant tissue, pollen, nectar, and water. In some termites, workers forage for living or dead plant tissue, whereas other termites simply consume their wooden nest from the inside; all termites utilize cellulases, derived from microrganisms in their guts or produced themselves, to digest their woody food sources. Eusocial thrips and aphids all induce galls (hollow, modified plant parts), within which they feed on living plant tissue by sucking cell contents or phloem; the eusocial beetle constructs tunnels in the heartwood of trees, within which it eats ambrosia fungus cultivated on the tunnel walls. The domiciles of eusocial insects, whether hives, burrows, open paper nests, or galls, all provide excellent conditions for the rearing of offspring: they are either relatively safe or relatively easily defensible from most threats, in many taxa food is available within the domicile, and in many insect taxa, as well as in naked mole-rats, the domicile can be extended and enlarged to provide a high-quality habitat for multiple generations.

Origins and Causes of Eusociality. Phylogenetic evidence indicates that eusociality has originated once in ants, several time in wasps, at least six times in bees, once or twice in thrips, at least twice in termites (with workers evolving once and soldiers evolving separately, also once), several or many times in aphids, and once in beetles. Arguments concerning the causes of the origins of eusociality that are based on relative numbers of origins in different taxa are problematic, because the num-

TABLE 1. Diversity of Colony Characteristics of Eusocial Insects

Characteristic	Diversity among taxa
Number of queens at colony initiation and afterwards	Some species have one queen, others have multiple queens; in some species, queen number varies over time and between colonies (Hymenoptera and some Isoptera)
Reproducting males present in colony or not	"Kings" in Isoptera and naked mole-rats
Tasks of helpers	Helpers forage, build, nurse, and defend (most Hymenoptera), only defend (aphids and thrips), or forage, build, and defend (Isoptera)
Morphology of helpers and reproductives	Reproductives are morphologically similar to helpers (some Hymenoptera, beetles) or profoundly different (some Hymenoptera, Isoptera, thrips, and aphids)
Colony size	Colonies may be small (ten or fewer adults, some Hymenoptera and beetles) or huge (millions of helpers, some ants and Isoptera), with all sizes in between represented
Colony reproduction	Colonies may reproduce via budding (with a group of queens and helpers leaving together, some Hymenoptera and Isoptera), by release of winged dispersers (most Hymenoptera and Isoptera, all thrips, aphids, and beetles), or by both methods (some Hymenoptera and Isoptera)

bers of lineages present in the past (when eusociality arose) cannot be ascertained, and because after eusocial groups have arisen, their presence should reduce the chances of related, noneusocial taxa evolving it later. As a result, the causes of the origins of eusociality are best sought by searching for similarities among eusocial taxa with regard to the aspects of their biology that appear relevant to eusociality, and by comparing eusocial taxa with their closest noneusocial evolutionary relatives to identify their differences.

Why and how does eusociality evolve in some lineages but not others? In virtually all taxa, eusociality has originated in the context of the nuclear family comprising a mother (sometimes also a father, as in termites) and her offspring. The likelihood that eusociality evolves in this context is expected to be a function of two main factors: (1) the genetic relatedness of incipient helpers to their own offspring versus the offspring of the queen (their brothers and sisters), and (2) the ecological and demographic costs and benefits to incipient helpers of leaving the nest and attempting independent reproduc-

TABLE 2. Genetic Relatedness in Haplodiploids and Diploids

These relatedness values correspond to the probability that a focal allele in one individual is also present in the other individual. Niece refers to sister's daughter, and nephew refers to sister's son. Relatedness values without parentheses are for the case of single mating by females, and values in parentheses refer to the case of multiple mating, at its limit (a large number of matings). Values in diploids are for autosomal genes.

HAPLODIPLOIDS

To:	Son	Daughter	Brother	Sister	Niece	Nephew
From:						
Female	0.5	0.5	0.25	0.75	0.375	0.375
	(0.5)	(0.5)	(0.25)	(0.25)	(0.125)	(0.125)
Male	0	1.0	0.5	0.5	0.25	0.25
	(0)	(1.0)	(0.5)	(0.5)	(0.25)	(0.25)

DIPLOIDS

To:	Son	Daughter	Brother	Sister	Niece	Nephew
From:						
Female	0.5	0.5	0.5	0.5	0.25	0.25
	(0.5)	(0.5)	(0.25)	(0.25)	(0.125)	(0.125)
Male	0.5	0.5	0.5	0.5	0.25	0.25
	(0.5)	(0.5)	(0.25)	(0.25)	(0.125)	(0.125)

tion versus staying and helping their mother to reproduce in their natal nest.

The role of genetic relatedness. Genetic relatedness depends on the genetic system exhibited by a given taxon (Table 2). In diploid taxa, if the mother has mated only once, offspring gain equal fitness from producing and rearing their own offspring (which are genetically related to them by one-half, sharing half of their genes) versus rearing an equal number of sisters and brothers (which are also related to them by one-half, on average). By contrast, if their mother has mated multiple times, offspring will be some mixture of full and half siblings (related to them by one-quarter), so they will be be more closely related to offspring than to siblings overall.

Under haplodiploidy and single mating by the mother, females are also related by one-half to their own offspring. However, under haplodiploidy, females are related by three-quarters to their full sisters, because full sisters share the same half of their genome from their haploid father, plus one-quarter of their genome shared via their mother. By contrast, females are related by only one-quarter to their brothers, because they are related to brothers only via their mother (because males have no fathers, being produced parthenogenetically). Because of this asymmetry in relatedness, incipient female helpers in haplodiploids may be expected to gain more fitness from rearing sisters (related to them by three-quarters) than from producing their own offspring (related to them by one-half). However, if incipient female helpers are also rearing brothers (related to them by one-quarter), then their average relatedness to siblings is still only one-half under an equal sex ratio (the aver-

age of three-quarters and one-quarter), so there is no genetic fitness advantage gained from staying and helping.

The only way that especially high full-sister relatedness in haplodiploids can favor the origin of eusociality is if, in some colonies or at some times, the sex ratios of reproductive offspring produced by mothers are biased toward females (such that helping is favored), whereas in other colonies or at other times, to balance the population-wide sex ratio, mothers produce sex ratios biased toward males (such that helping is not favored). Under such "split" sex ratios, eusociality evolves first in female-biased colonies, and later, it becomes fixed (i.e., all members of the population have the genetic trait) owing to the ecological benefits of cooperation and division of labor. Although full-sister relatednesses of three-quarter are common in eusocial insects, the degree to which split sex ratios have been instrumental in the origins of eusociality in Hymenoptera or Thysanoptera is not yet known.

Under a genetic system of clonality, all colony members are related genetically by one, such that a eusocial insect colony (with reproductives and helpers) is analogous to the cooperating, genetically identical cells of a metazoan body (with a germ line and a soma). In clonal insects, eusociality should be favored whenever clones gain higher fitness from producing some subset of individuals that engage in permanent helping behavior (e.g., by defending the colony, as in aphid species with soldiers) than they do from having all individuals become normal reproductives. There is, however, an important difference between cooperating insects and the cooperating cells in a metazoan body: in the insects, one clone may be infiltrated by another (e.g., reproductive aphids from one clone may invade another, to take advantage of the help provided by their soldiers). Under these circumstances, average relatedness between colony members falls below one, and the benefits to a clone of producing helpers are reduced.

The role of ecological benefits and costs. Whereas genetic relatedness determines the relative genetic benefits of helping to rear siblings or clonemates versus attempting to reproduce on one's own, the ecological benefits and costs of helping versus attempting independent reproduction determine the relative numbers of individuals that are produced by these two strategies.

The main benefits of helping behavior come about from increased colony fitness (colony survival and numbers of reproductives produced), arising from the advantages of increased group size and division of labor. Depending on the taxonomic group, helpers specialize in building, foraging, nursing young, defending the colony, or some combination of these tasks, whereas queens (and sometime kings) specialize in offspring production. Such division of labor provides benefits because each task is undertaken more efficiently by specialists, as in human societies. Increased group (colony) size may provide benefits to colonies in a number of ways: for example, more builders may result in faster domicile construction or extension, more foragers may lead to higher and more constant food intake, and more defenders may engender more effective defense. Some ecological circumstances, such as limited time windows for domicile building, high mortality rates of adults (which would lead to total loss of brood if an individual sought to reproduce alone), high vulnerability of colonies to starvation, or a high incidence of natural enemies seeking to usurp the domicile and brood, may thus favor the origin of helping and eusociality. Overall, the ecological conditions favoring helping appear to fall into two categories: (1) thrips, aphids, beetles, and most termites live in "factory fortresses," with food available within the domicile (the gall, burrow, or nest), juveniles that can feed themselves, and strong selection to defend their extremely valuable, resource-rich colony; and (2) ants, bees, and wasps live in more or less vulnerable domiciles that they must leave to acquire food for nourishing their helpless juveniles; foraging leads to high adult mortality rates, but the presence of multiple workers provides "life insurance" for the juveniles, because it is unlikely that all the adults will be killed.

The main cost of helping to helpers is that, because of their role in the society, they produce fewer offspring of their own (i.e., fewer daughters and sons), engaging instead in the rearing of sisters, brothers, and sometimes nieces and nephews. In addition, larger group sizes may lead to increased levels of parasitism or increased at-traction of other natural enemies, both of which could select against helping behavior.

The primary potential benefit of leaving the colony to attempt independent reproduction is that, if a successful colony results, the mother would likely have higher relative fitness than if she had remained at home as a helper. However, the odds of realizing such success appear to be extremely low, especially because success requires mating, finding or creating a domicile, and rearing a brood through to adulthood. The ecological conditions that can result in low relative fitness from independent reproduction are often termed ecological constraints, and strong constraints are expected to favor the origin of eusociality. The magnitude of such constraints in any given species depends on aspects of demography and timing as well as aspects of habitats, domiciles, and mating systems. For example, many individuals may become adult at a time of year (e.g., in midsummer in many temperate-zone bees and wasps) when success from independent reproduction is quite low relative to other times (e.g., in spring).

Integrating genetics and ecology. Evaluating the causes of eusociality requires joint consideration of genetic relatedness and ecological costs and benefits, with regard to the selective forces involved in the transition from independent reproduction to helping. In species of eusocial insects that are believed to be most similar to those in which eusociality has originated, average genetic relatedness is generally substantial (at least one-quarter, and usually one-half or above), and certain ecological circumstances, such as the factory fortress domicile and life history demographics that fit the life insurance model, appear to apply widely across eusocial taxa and thus may have convergently favored eusocial life in multiple lineages. However, integration of data on genetic relatedness with data on ecology to predict and explain the origins of eusociality has proven difficult for two reasons. First, quantitative information on the costs and benefits of helping versus dispersing is hard to obtain in natural field populations. Second, extant eusocial species need not closely resemble those in which eusociality originated, and we have no way of knowing the extent that they do resemble them. Moreover, in only a few cases have noneusocial species that are closely related phylogenetically to eusocial ones been studied in any detail, so it is difficult to tell just what genetic or ecological changes have precipitated the transition from a solitary life to a eusocial one. Future work will involve: development of testable models, such as skew models, that focus on events surrounding the origins of eusociality; inference of robust species-level phylogenies for additional eusocial taxa; and detailed study of eusocial forms, such as halictine bees, that exhibit considerable plasticity in the expression of eusocial behavior.

Conflict and Cooperation in Eusocial Colonies. Eusocial species are characterized by helper altruism, with helpers exhibiting reduced personal reproduction and aiding in the reproduction of the queen. Such cooperative behavior can be strongly selected for at the among-colony level, with more cooperative colonies surviving better and producing more dispersing reproductives at the end of the colony cycle. However, despite their altruistic and cooperative reputation, eusocial insects exhibit considerable conflict among individuals within colonies, primarily in circumstances where there are large differences in relative fitness between reproductives and helpers, and unequal relatedness of different colony members to reproductive female and male brood. Although within-colony conflict is selected against at the colony level (because it reduces colony reproduction), it can be selected for at the level of individuals, because individuals that "win" in the conflicts have higher relative fitness. Normally, these two opposing selective pressures balance one another in some way, resulting in complex mixtures of cooperation and conflict.

The expression and context of within-colony conflict depend in large part on the composition and nature of the colony itself. Eusocial species traditionally have been categorized into two main types, (1) primitively eusocial forms, in which queens and helpers do not differ in external morphology and colonies tend to be small, with a few to several dozen helpers, and (2) advanced eusocial forms, in which colonies are relatively large and queens differ discretely from helpers in morphology. The terms *primitive* and *advanced* have nothing to do with degree of adaptedness. Instead, they are meant to denote the differences between the two main forms of eusocial colony. They also imply an evolutionary sequence, whereby within any taxonomic group, primitively eusocial forms evolved first, and advanced eusocial forms evolved from them in one or more descendant lineages.

In primitively eusocial species, such as *Polistes* wasps and halictine bees, queens are normally aggressive toward workers, nudging them, monitoring their activities, and attacking workers who show signs of trying to lay eggs. Such obvious but constrained aggression appears to have two main selective bases. First, in such small-colony forms, any given worker has a reasonable chance of replacing the queen should she weaken or die, and queens normally have higher relative fitness than workers. As a result, workers are under selection to become more reproductive given any chance to do so, but queens are under selection to retain their reproductive dominance (via suppressing workers) so they can produce more offspring of higher relatedness to themselves. Second, workers may prefer to avoid risky tasks, such as foraging, to increase the odds that they can survive and reproduce to some extent. However, queens benefit from increased levels of work, so they are selected to suppress worker "laziness" via aggression and threats.

In advanced eusocial species, a worker has virtually no chance of supplanting the queen as the prime reproductive, because colonies are much larger, the two castes differ in morphology, with each more highly specialized for its respective role, and workers in these forms are normally unmated. However, workers are still under selection to reproduce to some degree, and in haplodiploid species, unmated workers can, and often do, produce males via parthenogenesis.

The presence and nature of within-colony conflict over male production in haplodiploid, advanced eusocial species depends on genetic relatednesses within the colony (Table 2). When queens have mated only once, workers are genetically related by one-half to their own sons, by three-eighths to nephews (sisters' sons), and by only one-quarter to brothers (mother's sons). Any given worker is thus under selection to produce males itself, but its genetic interests conflict with those of both other workers (who would each prefer to make the males themselves) and the queen (who is related to her own sons by one-half, but to workers' sons by only one-quarter). As a result, in this situation, we expect queen–worker conflict and worker–worker conflict over the production of males, which manifests itself in aggressive behavior, egg eating, and variation among colonies and among species in who produce the males. Such behavior is indeed found in many haplodiploid taxa with single mating by queens, such as stingless bees, some bumblebees, and some yellowjacket wasps. By contrast, if a queen has mated many times (as in honeybees), then workers can be more closely related to brothers (queen's sons, relatedness of one-quarter) than they are to nephews (other workers' sons, relatedness approaching one-eighth as the number of queen matings increases). Thus, although each worker in such colonies still wants to produce its own sons, it prefers brothers over nephews, so workers are selected to "police" each other's male production by eating any eggs laid by other workers, thus allowing the queen to make all of the males. Such worker policing has been well documented in honeybees. Finally, if there are substantial colony-level costs of engaging in worker policing (such as energy not spent in foraging and nursing), then workers may be selected to police themselves, that is, to fully suppress their own reproduction, and within-colony conflict over male production will no longer be expressed.

Sociality in Insects and Humans. Eusocial insects are among the most endlessly fascinating of all animals, in large part because their behavior resembles that of humans in so many ways: they live in complex, often very large groups with divisions of labor; some taxa, such as ants and termites, have profound ecological effects on their habitats; and their interactions involve spectacular examples of cooperation but also fierce conflicts. Indeed, for both social insects and humans, the interests of mutually dependent parties often diverge to a considerable degree, and the resolution of such conflicts can take myriad forms. Future studies of the evolution of eusociality in insects will lead to further insights into the nature of all societies, including our own.

[*See also* Altruism; Kin Recognition; Kin Selection; Naked Mole-Rats.]

BIBLIOGRAPHY

Abe, T., D. E. Bignell, and M. Higashi. *Termites: Evolution, Sociality, Symbioses, Ecology.* Dordrecht, Netherlands, 2000. Edited volume with up-to-date coverage of most aspects of termite biology.

Bourke, A. F. G. "Sociality and Kin Selection in Insects." In *Behavioural Ecology: An Evolutionary Approach,* edited by J. K. Krebs and N. B. Davies. Oxford, 1997. A comprehensive and authoritative review.

Bourke, A. F. G., and N. R. Franks. *Social Evolution in Ants.* Princeton, 1995. Excellent explanations of kin selection and sex ratio evolution, as well as thorough coverage of ant sociality.

Breed, M. D., and R. E. Page, Jr. *The Genetics of Social Evolution.* Boulder, 1989. Useful collection of papers focusing on role of genetics in social insect evolution.

Choe, J. C., and B. J. Crespi. *The Evolution of Social Behavior in Insects and Arachnids.* Cambridge, 1997. Comprehensive treatment of all social insect taxa, plus a review chapter.

Crozier, R. H., and P. Pamilo. *Evolution of Social Insect Colonies: Sex Allocation and Kin Selection.* Oxford, 1996. Excellent, albeit technically demanding, introduction to modeling in the context of social insect evolution.

Engels, W. *Social Insects: An Evolutionary Approach to Castes and Reproduction.* Berlin, 1990. In-depth treatment of the evolution of reproductive division of labor in various social insect groups.

Hermann, H. R. *Social Insects.* Vols. 1–4. New York, 1979. A bit dated, but a good treatment of the behavior of all social insect taxa, as well as some theory and ecology.

Keller, L. *Queen Number and Sociality in Insects.* Oxford, 1993. Excellent combination of theory and reviews of recent work, focusing on the explanation of variation in queen numbers among and within taxa.

Queller, D. C., and J. E. Strassmann. "Kin Selection and Social Insects." *BioScience* 48 (1998): 165–175. Outstanding recent review.

Ross, K. G., and R. W. Matthews. *The Social Biology of Wasps.* Ithaca, N.Y., (1991). Best, most thorough book on social wasps.

Michener, C. D. *The Social Behavior of the Bees.* Cambridge, Mass., 1974. Wonderful introduction to the natural history and diversity of bees.

Michener, C. D. *The Bees of the World.* Baltimore, 2000. The most comprehensive and authoritative work on all aspects of the biology of bees.

Thorne, B. L. "Evolution of Eusociality in Termites." *Annual Review of Ecology and Systematics* 28 (1997): 27–54. Excellent review of termite sociality.

Wilson, E. O. *The Insect Societies.* Cambridge, Mass., 1971. Dated with regard to application of theory and knowledge of different taxa, but still a useful introduction to social insects.

Wilson, E. O., and B. Holldobler. *The Ants.* Cambridge, Mass., 1990. Comprehensive coverage of all aspects of the biology of ants.

— BERNARD J. CRESPI

Eusociality in Mammals

Eusociality, a term coined by entomologists, means "true sociality." Three characteristics define eusocial societies: (1) overlapping generations (mother and offspring live together), (2) reproductive division of labor (i.e., only a few individuals bear offspring), and (3) alloparental care (nonreproductives assist in rearing the young of breeders).

The evolution of eusociality has puzzled biologists ever since Charles Darwin (1859) identified worker ants as presenting a "special difficulty" for his theory of evolution by natural selection. How, he wondered, could workers have evolved by gradual steps to be less and less reproductive? And how did the morphological and behavioral specializations of nonbreeders evolve if their bearers did not reproduce?

About a century later, W. D. Hamilton (1964) proposed a simple but elegant solution, now known as "kin selection." The idea is that individuals can reproduce not only by conceiving and rearing descendants but also by helping relatives to reproduce. The same reproductive benefits accrue to a helper from raising a sibling or an offspring of its own because both share the same proportion of their genes (one-half) with the helper. Morphological and behavioral specializations can be perfected among workers if they can help reproductive relatives sufficiently. This is because breeders carry and pass on the genetic specifications for worker traits, although they do not personally express them.

Until recently, eusociality was thought to occur only in ants, bees, wasps (order Hymenoptera), and termites (order Isoptera). Therefore, attempts to understand the evolution of eusociality focused on genetic, ecological, and phylogenetic factors that were unique to these insects. However, during the period 1980–2001, eusociality was reported in Japanese aphids (Homoptera), Australian weevils (Coleoptera) and thrips (Thysanoptera), as well as coral reef shrimp (Decapoda). Moreover, as detailed information has accumulated on the group structure and reproductive behaviors of social mammals and birds, it has become clear that there are close parallels with eusocial insects, especially those that live in equivalent-sized (small) colonies.

Traditionally, students of vertebrate social systems have labeled species that share characteristics 1–3 (above) as "cooperative breeders." Their theoretical and empirical research proceeded largely independently from the studies of entomologists interested in eusociality. Explanations for cooperative breeding have focused primarily on the extrinsic (ecological) factors that favored the formation of groups, such as predation and food distribution, whereas explanations for eusociality focused primarily on intrinsic factors that resulted in genetic benefits for alloparents (e.g., kin selection).

But are cooperative breeding and eusociality really so different, or are their convergences sufficient to justify attempting to unite these separate explanations? The following sections address this question, considering in turn each of the characteristics of eusociality.

Overlapping Generations. All cooperatively breeding mammals and birds, as well as eusocial insects, live in family groups. In vertebrates and many insects, these groups usually contain less than twenty adults, except for colonies of certain African mole rats, especially the naked mole-rat, whose colonies contain up to 300 individuals. [*See* Naked Mole-Rats.] The young stay at home when ecological conditions make dispersal especially difficult or group-living particularly beneficial. As a result, colonies typically are composed of two or more generations of reproductively mature, closely related individuals.

However, the length of time offspring remain at home varies. In black-tailed prairie dogs, African lions, and golden jackals, some offspring remain on their parents' territory for only one or two breeding seasons before leaving to start their own family. In wolves and dwarf mongooses, offspring may remain for four or five seasons (i.e., half their adult life). And in wild dogs and naked mole-rats, many individuals never disperse. When ecological constraints are severe enough to keep individuals at home for life, inbreeding can occur. In turn, inbreeding creates groups in which alloparents are extremely closely related to younger siblings and gain greater genetic rewards from assisting them.

Often natal philopatry is sex-biased; that is, there is a difference between males and females in which sex disperses and which sex remain on the natal territory. In prairie dogs, lions, lion tamarins, and pine voles, as in the Hymenoptera, it is the females that remain near home and the males that disperse at sexual maturity. Among wolves, wild dogs, and jackals, males are philopatric. In dwarf mongooses and naked and Damaraland mole rats, as in the Isoptera (termites), there is no sex bias in dispersal.

Alloparental Care. When ecological circumstances force the young to remain at home, their personal reproduction typically is restricted by larger, older, and more dominant individuals (usually their parents). Nonetheless, individuals can spread their genes by increasing the production and survival of kin, especially siblings. Alloparents also enjoy continued access to the safety of the natal territory, gain experience at parenting, and sometimes even inherit the established breeding site. The more philopatric sex is the more alloparental.

Alloparental care takes many forms. In primates, it involves grooming, carrying, and sheltering the young; in carnivores, it also involves transporting and regurgitating food; in prairie dogs and some voles, it involves allonursing; and in mongooses and mole rats, it involves babysitting, territorial defense, and group aggression against predators.

Mammalian (and avian) alloparents always are temporarily nonreproductive, but they are never physiologically sterile. Does this disqualify vertebrates from being "eusocial"? The answer is no, because sterility is not a prerequisite for eusociality. Indeed, workers in many eusocial insects are not sterile. In general, species in which helpers retain the ability to reproduce live in small colonies. Should a breeder die, each individual has a finite probability of filling the vacancy. Thus, it is to each individual's advantage to retain the ability to reproduce. Sterility is obligate only among eusocial insects that inhabit very large colonies.

Reproductive Skew. Vertebrate and invertebrate societies vary in the degree to which reproduction is evenly distributed among individuals, or concentrated into one or a few. Individual variation in lifetime reproductive success or "skew" results from social competition and suppression within groups and ecological factors that preclude dispersal and independent reproduction. The amount of skew within groups can be used as a yardstick to compare social systems: if every female reproduced, there would be no skew (i.e., 0.0), but if only one female reproduced, the skew would be 1.0. All cooperatively breeding vertebrates and eusocial invertebrates can be arrayed along this continuous scale of skew.

Behavioral and demographic information allows us to predict the relative positions of different mammalian societies along the skew continuum (Figure 1). At the lower end of the scale lie societies with multiple reproductive individuals per group as well as alloparental care by breeders, such as prairie dogs, some marmots, and lions. Societies in which alloparents do not reproduce while helping but have a reasonable chance of eventually dispersing and producing their own offspring lie a little farther up the scale (jackals, spotted hyenas, wolves, and pine voles). In the middle of the scale are societies with a single female breeder per group and in which opportunities for dispersal and independent breeding by alloparents are restricted—but possible—throughout life (golden lion tamarins, dwarf mongooses, and meerkats). Far to the right on the scale are societies with one female breeder and in which alloparents only rarely have opportunities to disperse or mate, such as painted hunting dogs, cotton-top tamarins, and naked and Damaraland mole rats.

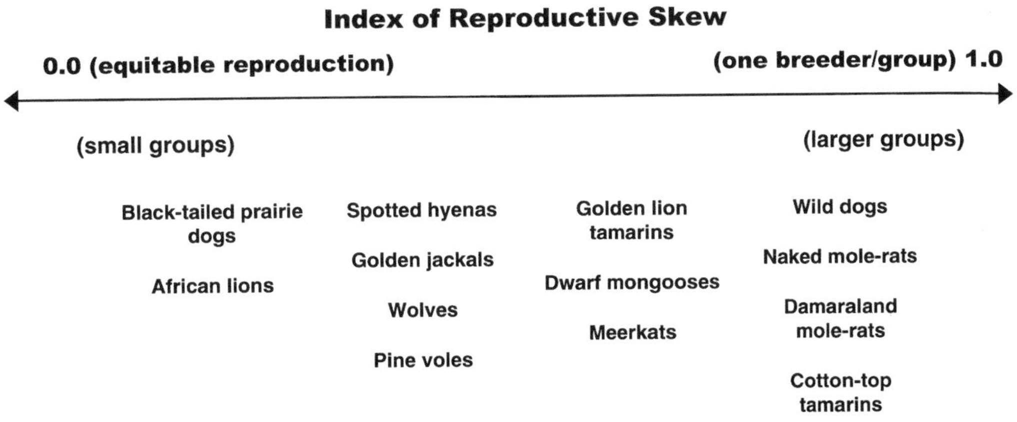

Index of Reproductive Skew

0.0 (equitable reproduction)   **(one breeder/group) 1.0**

←——→

(small groups)   (larger groups)

Black-tailed prairie dogs	Spotted hyenas	Golden lion tamarins	Wild dogs
African lions	Golden jackals	Dwarf mongooses	Naked mole-rats
	Wolves	Meerkats	Damaraland mole-rats
	Pine voles		Cotton-top tamarins

FIGURE 1. The Eusociality Continuum in Mammals.
Predicted locations of selected species of cooperatively breeding mammals along the eusociality continuum (a scale of the skew in lifetime reproductive success among members of social groups). When skews vary among groups or populations within species, that species should be represented as a segment of the continuum (rather than a single point) denoting the intraspecific range of skew values. In this figure, societies that probably have similar reproductive skews are grouped together. Drawing by Paul W. Sherman.

Societies that lie at different positions along the eusociality continuum differ not only in skew but also in group sizes. Those at the upper (right) end in Figure 1 live in larger groups. Of course, colonies of mammals never get as large as those of some invertebrates, probably because mammals are so much larger relative to the sizes of safe nesting sites that fewer mammals can pack into them. The evolution of specialized helper phenotypes (castes) and intragroup breeding conflict are apparently related to group size. Group size is itself a function of the ecological advantages of group living and natal philopatry and the disadvantages of attempting to disperse and reproduce independently. There are no permanent physical castes among vertebrate alloparents, although individuals often adopt different roles as they get older and larger. Because all alloparents are potential reproductives, the stage is set for competition and conflict whenever reproductive opportunities arise. Dominance interactions and physical aggression characterize most cooperatively breeding vertebrates.

These considerations imply that cooperative breeding and eusociality are not discrete phenomena, but rather form a continuum of fundamentally similar social systems whose main differences lie in group size and reproductive skew. The eusociality continuum unites all such societies under a single theoretical and terminological umbrella. It also highlights common ecological and genetic selective factors underlying social evolution and encourages intellectual cross-fertilization between entomologists, mammalogists, and ornithologists. Cooperatively breeding mammals and birds should be considered eusocial, just as eusocial insects clearly are cooperative breeders.

[*See also* Altruism; Kin Recognition; Kin Selection; Naked Mole-Rats.]

BIBLIOGRAPHY

Alexander, R. D., K. Noonan, and B. J. Crespi. "The Evolution of Eusociality." In *The Biology of the Naked Mole-Rat*, edited by P. W. Sherman, J. U. M. Jarvis, and R. D. Alexander, pp. 3–44. Princeton, 1991. A comprehensive treatment of eusocial evolution, particularly in insects.

Andersson, M. "The Evolution of Eusociality." *Annual Review of Ecology and Systematics* 15 (1984): 165–189. The initial attempt to draw attention to parallels between cooperative breeding in vertebrates and eusociality in invertebrates.

Bennett, N. C., and C. G. Faulkes. *African Mole-Rats: Ecology and Eusociality.* Cambridge, 2000. Discusses the ecology and phylogeny of sociality in the rodent family Bathyergidae.

Darwin, C. R. *On the Origin of Species* (1859). Facsimile-edition, Cambridge, Mass., 1967. The source.

Emlen, S. T. "Evolution of Cooperative Breeding in Birds and Mammals." In *Behavioural Ecology*, edited by J. R. Krebs and N. B. Davies, 3d ed., 301–337. Oxford, 1991. Illustrates ecological and behavioral convergences between cooperatively breeding birds and mammals.

Hamilton, W. D. "The Genetical Evolution of Social Behavior, 1 and 2." *Journal of Theoretical Biology* 7 (1964): 1–52. The seminal paper on kin selection theory.

Jarvis, J. U. M., and N. C. Bennett. "Eusociality Has Evolved Independently in Two Genera of Bathyergid Mole-Rats—but Occurs in No Other Subterranean Mammal." *Behavioral Ecology and Sociobiology* 33 (1993): 353–360. Describes similarities and differences between the two most social mole rats.

Jarvis, J. U. M., M. J. O'Riain, N. C. Bennett, and P. W. Sherman. "Mammalian Eusociality: A Family Affair." *Trends in Ecology and Evolution* 9 (1994): 47–51. Explains and tests the aridity–food distribution hypothesis to explain mole rat sociality.

Jennions, M. D., and D. W. Macdonald. "Cooperative Breeding in

Mammals." *Trends in Ecology and Evolution* 9 (1994): 89–93. Describes various forms of alloparental behavior in mammals.

Keller, L., and H. K. Reeve. "Partitioning of Reproduction in Animal Societies." *Trends in Ecology and Evolution* 9 (1994): 98–102. Explores the concept and uses of reproductive skew to characterize reproductive asymmetries in animal societies.

Mumme, R. L. "A Bird's Eye View of Mammalian Cooperative Breeding." In *Cooperative Breeding in Mammals*, edited by N. G. Solomon and J. A. French, 364–388. Cambridge, 1997. Explores parallels between cooperative breeding in birds and mammals.

Shellman-Reeve, J. S. "The Spectrum of Eusociality in Termites." In *The Evolution of Social Behavior in Insects and Arachnids*, edited by J. C. Choe and B. J. Crespi, pp. 52–93. Cambridge, 1998. Places various termite species at different points along the spectrum of eusociality.

Sherman, P. W., J. U. M. Jarvis, and R. D. Alexander, eds. *The Biology of the Naked Mole-Rat*. Princeton, 1991. A comprehensive monograph about the species that represents the extreme in mammalian eusociality.

Sherman, P. W., E. A. Lacey, H. K. Reeve, and L. Keller. "The Eusociality Continuum." *Behavioral Ecology* 6 (1995): 102–108. Proposes that eusociality is not a unitary phenomenon but rather a continuum of social systems whose main differences lie in group sizes and degrees of reproductive monopolization (skew).

Solomon, N. G., and J. A. French, eds. *Cooperative Breeding in Mammals*. Cambridge, 1997. An anthology highlighting mechanisms and functions of cooperative breeding and alloparental care in mammals.

— PAUL W. SHERMAN

EVOLUTION

Evolution is often used to describe the process by which ideas, designs or plans, and animate and inanimate objects change over time into different forms. Things that emerge, unwind, develop, spread out, or appear in succession or a series are all said to evolve. The term *evolution* is recorded as appearing in English for the first time in 1647. Charles Lyell used it in the early nineteenth century to describe changes to the earth's geology, and by the late nineteenth century Herbert Spencer applied the word *evolution* to change within human societies.

Applied to biology, evolution describes genetic changes that occur in organisms over time. It is distinct from maturational changes that occur during development or in organisms that have metamorphosis. In some contexts—most conspicuously when discussing "human evolution"—the term often acquires an implicit sense of progress such that later forms are assumed to be more complex and sophisticated than their primitive ancestors. In other contexts, "evolution" connotes a sense of adaptation, such that evolved forms are better suited to coping with the demands of their environment.

Evolution may have come to convey these senses of progressive biological adaptation from its origin in the Latin *evolvere*, meaning to unfold or unfurl. Those things that unfold or unfurl—such as flags—are already formed. Yet neither the notion of progress nor of adaptation is inherent to evolution, although the process of evolution can lead to both. Evolution may be simply defined as a change in the frequencies of genes in a population of individuals from one generation to the next. This uncomplicated definition can encompass apparent progress and instances of adaptation, but it can also encompass genetic change that arises from chance or random factors.

Inheritance in biological systems arises almost entirely from parents contributing genes to their offspring. A family in which both parents have brown eyes, for example, may have children who all have blue eyes. Genes for brown eyes are dominant over genes for blue eyes. Brown-eyed parents can produced blue-eyed children if both parents carry the recessive gene for blue eye color. All that is required is that the children receive the recessive gene from each parent. For this to happen in each of several offspring would be unusual but not out of the question.

If we view this family as a small population, we see that the frequency of the blue eye-color gene has increased, and the brown eye-color gene has disappeared, all in one generation. There has been evolution within this family, although it is not necessary to say that the children are better adapted. On a larger scale, this process of random gene-frequency change from one generation to the next is called genetic drift.

In other instances, the process of natural selection leads to individuals becoming better adapted. If running faster is useful for escaping predators, then natural selection will favor genes that make individuals run faster. The frequency of these genes will then increase in the population. Natural selection, therefore, implies evolution, although it should be clear from the eye-color example that evolution does not imply natural selection.

By similar reasoning, one can imagine that natural selection could bring about changes to organisms that human observers might describe as progressive. The path leading from ancestral hominid species to contemporary *Homo sapiens* is characterized by increases in brain size and creative intelligence. But "progress" of this sort is not inherent to the process of evolution. Neanderthals, for example, possessed brains even larger than those of contemporary modern humans, yet few would regard the Neanderthals as more sophisticated or intelligent than modern humans.

More broadly, animals can lose as well as acquire characteristics that most would regard as sophisticated or progressive. For instance, some cave-dwelling fish species lack eyes because their ancestors never evolved them. Other fish species that have adopted caves as their preferred habitat have evolved to an eyeless form; the eternal darkness of the environment that these fish inhabit makes eyes—sophisticated as they are—of no use.

Biological evolution, stripped to its barest essentials, is nothing more than the temporal changes in the genetic makeup of populations. It may come as a surprise to learn that, as widely used as the word *evolution* is today, Charles Darwin used it only once in his monumental work *The Origin of Species*. Perhaps he was anxious to avoid its more connotative meanings, because even when he did use it, he waited until the end of *The Origin*—in fact, until the very last word: "[W]hilst this planet has gone cycling on according to the fixed law of gravity, from so simple a beginning endless forms most beautiful and most wonderful have been, and are being evolved."

[*See also* Adaptation; Genes; Genetic Drift; Kimura, Motoo; Mendelian Genetics; Natural Selection, *article on* Natural Selection in Contemporary Human Society; Neutral Theory.]

BIBLIOGRAPHY

Bowler, P. "The Changing Meaning of 'Evolution.'" *Journal of the History of Ideas* 36 (1975): 95–114.

Kimura, M. *The Neutral Theory of Molecular Evolution.* Cambridge, 1983. One of the best introductions to evolution at the level of genes. Technical treatments follow a general expository introduction.

Mayr, E. *The Growth of Biological Thought: Diversity, Evolution, and Inheritance.* Cambridge, Mass., 1982. A general treatment of the history of evolutionary thought, along with an analysis of concepts of evolution.

Williams, G. C. *Adaptation and Natural Selection: A Critique of Some Current Evolutionary Thought.* Princeton, 1966. A landmark in the field. Nontechnical but without loss of depth or insight.

— MARK PAGEL

EVOLUTIONARILY STABLE STRATEGY. *See* Game Theory.

EVOLUTIONARY ALGORITHMS. *See* Genetic Algorithms.

EVOLUTIONARY COMPUTATION. *See* Genetic Algorithms.

EVOLUTIONARY ECONOMICS

The work of F. A. Hayek (1952, 1967, 1988) was a conceptual precursor of recent developments in the fields of evolutionary economics and psychology. He understood these important principles: (1) in *personal exchange*, the solidarity and sharing rules of the family, extended family, and tribe are instinctive, evolutionarily adaptive and efficient; (2) the emergence of *impersonal*

exchange over wide-ranging markets represented a form of cultural evolution, enabling gains from exchange and specialization for those groups adopting them relative to groups that did not; (3) the rules governing (1) are destructive if applied to (2), and vice versa; and (4) neither system evolved by conscious design—rather, individuals made choices reflecting circumstantial details that had broader implications of which they were largely unaware, achieving outcomes that were not their intent. Douglas North (1990) provides a historical analysis of the tradeoffs between personal and impersonal exchange and their effects on the success or failure of the economic growth of nation-states.

Evolutionary psychology and experimental economists have expanded and filled in many new details within this framework. Evolutionary psychologists (Barkow, Cosmides, and Tooby, 1992) argue that much of the conflict between personal exchange norms, on one hand, and modern institutions and practices like impersonal exchange, on the other, arises because our genetic resources evolved in our environment of evolutionary adaptedness (EEA). The EEA is represented by the 1- to 2-million-year-period in which the hominid line emerged and lived as hunter-gatherers under personal exchange regimes. These conditions were altered dramatically by the relatively rapid cultural and technological change occurring after the invention of agriculture 10,000 years ago, and the information age made possible by the invention of writing in the interval 9,000–5,000 years ago.

The human propensity to achieve cooperative outcomes in bilateral interactions is illustrated in the extensive form game experiments reported by McCabe, Rassenti, and Smith (1996). A truncated version of these experiments is illustrated in Figures 1 and 2. The first is a trust game in which offers to cooperate by Player 2 can lead either to a reciprocal cooperative response or to defection by Player 1; in the second trust/punishment game, if Player 2 defects, the choice passes back to Player 2, who can either accept defection or punish

FIGURE 1. One-Play Extensive Form Game Experiments. Trust game results for twenty-six pairs.   McCabe, Rassenti, and Smith (1996).

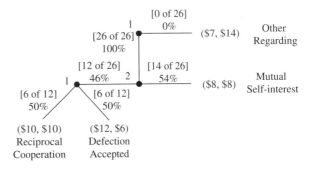

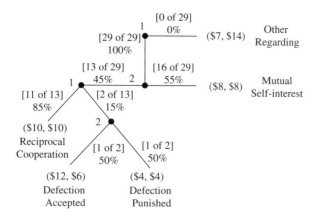

FIGURE 2. One-Play Extensive Form Game Experiments. Trust/Punish game results for twenty-nine pairs. McCabe, Rassenti, and Smith (1996).

Player 1 at a personal cost to Player 2. Player 1 begins the game (at the top) by deciding to move right (in which case the game ends immediately) or down. Players 1 and 2 then move sequentially until they reach a payoff— Player 1 receiving the left and 2 the right number of dollars. The games are played once between anonymously paired subjects; without a history or a future, strangers cannot build reputations. In these low-information environments, we hypothesize the following innate-experiential interactive solutions to this joint decision problem: (1) the altruistic outcome ($7, $14), in which Player 1 is assumed to have other-regarding utility for Player 2's payoff; (2) the no-risk equilibrium outcome ($8, $8), based on the assumption of mutually self-interested players; (3) the risky choice by Player 2 to offer to cooperate with Player 1; (4) a reciprocal cooperative response to this action by Player 1, making them both better off ($10, $10); or, alternatively, (5) the choice by Player 1 to defect in his own self-interest ($12, $6). In the punishment version of the trust game, if Player 1 chooses defection, then play returns to Player 2, who can self-interestedly accept Player 1's defection ($12, $6) or (6) choose to punish Player 1's defection, but at a personal cost to Player 2 ($4, $4).

The results from experiments with cash-motivated undergraduate subjects are shown in the figures. In 55 observations across the two data sets, we do not see one instance of the altruistic outcome, with a little more than half of our subjects playing the no-risk equilibrium. Of those who try to cooperate, the presence of a punishment option greatly increases the chances of cooperation. While 50 percent of players 1 defect (when given the chance) in the trust game, only 15 percent defect when there is a chance they will be punished.

Hirshleifer (1982) discusses the need to use evolutionary models to understand the human tendency to-

ward conflict and/or cooperation expressed these experimental results. More recently, it has been supposed that the human capacity to cooperate may have resulted from an evolved capacity for indirect reciprocity, through what Wedikind and Milinski (2000) call "image scoring," and McCabe et al. (2001) call "goodwill accounting."

[*See also* Human Sociobiology and Behavior, *article on* Evolutionary Psychology; Optimality Theory, *article on* Optimal Foraging.]

BIBLIOGRAPHY

Barkow, J., L. Cosmides, and J. Tooby. *The Adapted Mind*. Oxford, 1992.

Hayek, F. A. *The Sensory Order*. Chicago, 1952.

Hayek, F. A. *Studies in Philosophy, Politics, and Economics*. Chicago, 1967.

Hayek, F. A. *The Fatal Conceit*. Chicago, 1988.

Hirshleifer, J. "Evolutionary Models in Economics and Law: Cooperation versus Conflict Strategies." *Research in Law and Economics* 4 (1982): 1–60.

Hoffman, E., K. McCabe, and V. Smith. "Behavioral Foundations of Reciprocity: Experimental Economics and Evolutionary Psychology." *Economic Inquiry* 36 (1998): 335–352.

McCabe, K., S. Rassenti, and V. Smith. "Game Theory and Reciprocity in Some Extensive Form Experimental Games." *Proceedings of the National Academy of Sciences* 93 (1996): 13,421–13,428.

McCabe, K., S. Rassenti, and V. Smith, "Goodwill Accounting and the Process of Exchange." In *Bounded Rationality: The Adaptive Toolbox*, edited by G. Gigerenzer and R. Selten. Cambridge, Mass., 2001.

North D. C. *Institutions, Institutional Change and Economic Performance*. Cambridge, 1990.

Wedekind, C., and M. Milinski "Evolution of Indirect Reciprocity by Image Scoring." *Nature* 393 (1998) 573–577.

— KEVIN A. McCABE AND VERNON L. SMITH

EVOLUTIONARY EPISTEMOLOGY

Philosophy is commonly divided into metaphysics and epistemology. Metaphysics is the study of what there is in the world, and epistemology treats how we come to know what we know. Contrary to how these descriptions may sound, the goals are philosophical, not scientific. Metaphysics does not deal with such issues as the existence of atoms or prions but with more general questions such as the adequacy of a metaphysics of entities and their properties versus relations or fields. Epistemology does not concern our perceptual abilities or psychological regularities. The issue is the justification of our beliefs. In the early days of philosophy, the standards for justified belief were extremely high, so high that no one could justifiably claim to know anything, but in more recent epistemological works, these standards have been made more reasonable. Evolutionary epistemology is an instance of these more recent epistemological theories: it is an attempt to use the fact of evolution and natural selection to justify one's beliefs.

Although several early workers, including Karl Popper (1972), advocated evolutionary epistemology, it has received its most detailed exposition in the hands of Donald Campbell, the father of present-day evolutionary epistemology. Campbell (1974) began by treating evolutionary epistemology in a literal sense. Just as certain ordinary phenotypic traits, such as the structure of the vertebrate eye, are programmed into our genes, so are certain beliefs and behavioral responses. Given the nature of the evolutionary process, both sorts of characteristics come to "fit" the environment. Just as we have the right to expect our eyes to perceive correctly, we have the right to expect our behavior and beliefs to reflect the world with reasonable accuracy. However, this reasoning applies only to those entities and events that are the appropriate size and duration. We can see most multicellular organisms with our unaided senses, but we cannot see microscopic organisms or black holes with our unaided senses. On the literal interpretation, evolutionary epistemology applies to a relatively few characteristics and justifies only rough approximations. Evolutionary epistemology is literal if it depends on genes.

To increase the scope of evolutionary epistemology, its recent advocates replaced a literal interpretation of this philosophical research program with one that is metaphorical. The structure of selection processes is retained, but genes are replaced by such epistemological units as concepts, beliefs, or more generally, memes. On the literal interpretation, certain behavioral predispositions, such as the sucking instinct or fear of snakes, are in some sense "programmed" into our genes because they aided in the survival of our ancestors. On this literal interpretation, such things as behavioral predispositions and instincts are phenotypic traits. On the metaphorical usage, they are memes. Just as genes are selected in literal versions of evolutionary theory, memes are selected in this metaphorical version. If evolutionary theory can in some sense "justify" the characteristics for which they code, then evolutionary theory in its memetic forms should provide some metaphorical justification.

Next, Campbell moved from evolutionary epistemology as depending on a metaphor to a general analysis of selection, what he terms "selection theory." Initially, Campbell attempted to justify beliefs, including human beliefs, by reasoning analogously from gene-based biological evolution to meme-based conceptual evolution, but these are not the only examples of selection processes. The reaction of the immune system to antigens incorporates a selection process. So too is the development of the central nervous system. Individual operant learning has been characterized from its inception as a selection process. Social learning is an even better example of a selection process. Instead of treating selection in gene-based biological evolution as standard and all other uses as metaphorical, Campbell decided that what was really needed was a general analysis of selection. All the processes just listed and possibly more count equally as selection processes.

The whole point of evolutionary epistemology is to justify belief. On the literal interpretation, some characteristics receive at least some justification. This justification is only weakened by moving to a metaphorical interpretation. Finally, Campbell's general selection theory provides no warrant at all, not even metaphorical. From his earliest publications, Campbell alternated between these three versions of evolutionary epistemology. Calls for a general selection theory can be found in Campbell's earliest writings, but they came to prevail in his later writings, as Campbell reluctantly realized that such a general analysis of selection, as valuable as it might be, had lost the sort of epistemological warrant offered by early versions of evolutionary epistemology. Why should we believe the results of conceptual selection? Campbell's answer is in terms of social structures that encourage the production of true beliefs. The title of one of his posthumous papers, "From Evolutionary Epistemology via Selection Theory to a Sociology of Scientific Validity" (Campbell et al., 1997), traces Campbell's conceptual development. The rest of this article expands on Campbell's changing views about evolutionary epistemology.

Perfection and Nature versus Nurture. Although evolutionary biologists now recognize a half dozen or so mechanisms that play a role in biological evolution, natural selection remains the chief mechanism for organisms having the characteristics that they do and the extent to which these characteristics are adaptive. However, selection does not work perfectly. In the evolutionary process, good enough has to be good enough. In the nineteenth century, the human eye was the paradigm example of a structure that could not have been produced by natural selection because it is too complicated and too perfect. The human eye is complicated, but it is anything but perfect. Any halfway decent engineer could improve on the design. Beliefs or behavioral predispositions that arise through the process of natural selection may also be far from perfect. We are all aware of certain systematic perceptual errors to which human beings are prone.

The relative roles of genes and environments extend from certain traits that are as genetic as one can get to those that are largely a function of the environment. Suckling and fear of snakes are commonly presented as examples of traits that are largely genetic. Of course, just any old environment won't do, but the environment does not provide sequential instructions in the production of such traits. Any human baby that is to survive had better have sucking behavior built into its genes. With the presence of fairly unstructured releasing mechanisms, babies begin to suckle soon after birth. They need not be taught to suckle. This behavior arose a long time ago in our mammalian past. One can only imagine the number of early mammals who died before this instinct became lodged firmly in place.

The story for the fear of snakes is a bit more complicated. Monkeys are not automatically afraid of snakes, but they learn to fear them very easily and rapidly. Only a single instance of a snake climbing into a tree followed by the warning behavior of older monkeys is enough for younger monkeys to become afraid. They are not so ready to fear other new elements in their environment. At the other extreme are those responses that have to be learned by means of numerous structured and repeated interactions. Most human beings are capable of learning calculus, but it is in no way built into our genes. We must learn it generation after generation, and just a cue or two will not do.

Building "knowledge" into our genes is a long and arduous process. Building it into our central nervous system is also arduous but occurs much more quickly. Those behaviors that are consistently beneficial over numerous generations can come to be built into our genes. Those that occur much more sporadically can at most be built into our central nervous system—from scratch each generation. The former sort of occurrence is "evolutionary" in a literal sense. Genes are changed generation after generation by means of selective interactions with the environment. The process of building knowledge into our central nervous system is "evolutionary" in only a metaphorical sense. No changes in gene frequencies are responsible. It is a matter of learning, both individual and social.

Following Popper, Campbell argued that knowledge acquisition is not a matter of adhering to the right logic, including inductive logic, but trial and error or, in Popper's (1972) terms, conjectures and refutations. It is a function of blind variation and selective retention. According to Campbell, this process must exhibit three essential elements: a mechanism for introducing variations, consistent selection processes, and a mechanism for preserving and/or propagating variations. More generally, evolutionary epistemology must have three fundamental characteristics: it is evolutionary, it concerns the justification of our beliefs, and it is naturalistic (Campbell, 1974).

Blind Variation. No aspect of Campbell's analysis of evolutionary epistemology introduced more unnecessary confusion than the adjectives he used to modify "variation"—*blind, chance, random, unforesighted, nonprescient,* and so on. By referring to variations as blind, Campbell does not mean to imply that some variations can see and others not. Variations are not the sort of

thing that can literally see or be blind. He is using the word *blind* in a metaphorical sense. Nor is he using *chance* and *random* in technical senses, because few if any natural processes fulfill the requirements of these mathematical notions. In biological evolution, terms such as *chance* and *random* are used to characterize mutations. All that these adjectives are designed to deny is a correlation between the mutations that an organism might need and those it might get. An organism living in an environment in which rain is decreasing needs some way to survive this change. It is no more likely to get a mutation to accomplish this end than an organism living in an environment in which rainfall is increasing.

Campbell is concerned with variations in biological evolution, but he is even more concerned with variations in conceptual change. According to Campbell (1974: 422), variations are "produced without prior knowledge of which ones, if any, will furnish a selectworthy encounter." No one thinks that an amoeba can have prior knowledge of anything, let alone which variations will lead to selectworthy encounters, but people are in a much better position than amoebae. We are certainly not prescient. We cannot perceive the future in advance. But we can understand the natural world sufficiently well to allow us on occasion to predict the future so that we can anticipate it. In ordinary life as well as in science, we can make educated guesses about the way the world is. These guesses are far from infallible, but they are not the result of pure chance either. Initially, learning is a matter of variations and selective retention, but as organisms begin to accumulate knowledge of their environments, these decisions are no longer a simple matter of variations and selective retention. They are made in light of one's current beliefs. They structure later observations.

The remaining problem concerns intentionality. At times discoveries are made by "chance" in a significant sense, but few scientists are likely to urge their students to depend very heavily on chance events to further their careers. Scientists can intend to solve a particular problem. Their intending to solve this problem increases the likelihood that they will solve it but only slightly. Perhaps our intentions do not have the massive influence on the course of human events that we would like, but they do have some effects, and such effects are anything but mysterious. One thing is certain: scientists cannot solve a problem in advance of solving it. If Campbell included the word *blind* in his analysis of selection processes in order to exclude this misreading, it hardly seems worth the confusion that it has engendered.

Proliferation and Selective Retention. Selection is a good deal more straightforward. If an anaerobic amoeba slithers its way upward toward the top of the water in which it is living, it will die, whereas if it works its way down, it has a better chance of surviving to reproduce. After numerous generations, either this population of anaerobic amoebae goes extinct or the descendants of the original amoeba are strongly disposed to slither down rather than up. They may not "know" anything, but they are inclined to behave in ways consentient with past successes. In biological evolution, organisms pay for their mistakes with their lives. In different periods of the human species, what one might consider "scientists" also paid for their mistakes with their lives. So the story goes, ancient Egyptian potentates tied the engineer (or one of his sons) to the top of obelisks as they were raised into place, and early physicists who worked with radioactive material frequently died because of their ignorance of the effects of such material.

But as Popper (1972, p. 244) remarked, the development of cognitive mechanisms for error elimination now allows "our hypotheses to die in our stead." The result is evolution via selection but only in a metaphorical sense. Memes replace genes. Certain ideas proliferate at the expense of other ideas. In 1859, only a tiny percentage of the world's population believed that species evolve; nowadays, this number has increased to 5 or 10 percent, maybe more. But no one claims that any changes in gene frequencies are responsible for these changes in meme frequencies. At most, the memetic version of evolutionary epistemology is only a metaphor, and any warrant for our beliefs that it supplies is derivative of gene-based biological evolution.

Epistemological Justification. The chief objection to evolutionary epistemology is that, most charitably, it does not serve to provide much in the way of warrant for very many of our beliefs. Less charitably, it provides no warrant for any of our beliefs. "Why are squirrels so good at judging distances?" Because if they were not, they would have gone extinct long ago. Judging distances accurately has been too important to their survival and reproduction for their not being able to succeed enough of the time. Unfortunately, success does not guarantee truth anymore than truth can guarantee success. Too often it seems that those behaviors furthest removed from literal claims of knowledge receive the most warrant, and those that have the greatest claim to being considered knowledge receive the least warrant. We all know that holding false beliefs can be very effective. Those soldiers who are convinced that God is on their side are more likely to win armed conflicts than more reasonable soldiers. That a majority of the citizens in the most scientific society that has ever existed refuse to acknowledge that species evolve shows the force of false beliefs.

A general fear of snakes seems built into humans, implying something about the role of snakes in our evolutionary past. Being able to distinguish between dangerous snakes and their harmless relatives has not played much of a role in human evolution, at least not often enough for long enough. A belief that snakes are dangerous obtains some warrant from gene-based biological evolution. The ability to distinguish between dangerous and nondangerous snakes does not. At most, reasoning by analogy from gene-based biological evolution to conceptual change confers only a very weak warrant for only a small percentage of our beliefs, but Campbell following Popper is concerned primarily with science, and scientific beliefs are furthest removed from our evolutionary history. In fact, much of science has been devoted to counteracting our evolved beliefs, for example, that heavy objects fall faster than light objects, that all species have an eternal and immutable essence, or that space is Euclidean. Psychologists have gone to equal lengths to point out the numerous perceptual errors to which human beings are prone. Often the world is not the way we think it is.

But Campbell and Popper are aware of all the preceding. Even so, they think that construing the scientific process as a selection process provides some sort of warrant for scientific beliefs. They are a matter of conjectures and refutations. Scientists put forth hypotheses that may have very little warrant but receive increasing warrant as the scientific process proceeds. Those hypotheses that withstand serious attempts at refuting them are more likely to be true than those that have not. Science is so organized that hypotheses are subjected to testing and retesting, whether the authors of these hypotheses like it or not.

Naturalistic Epistemology. Traditional epistemologists take epistemology to be more than pure description but normative. It purports to tell us what knowledge must be like. Campbell's epistemology is naturalistic, eschewing much in the way of normative claims. On that ground alone, traditional epistemologists would reject his evolutionary epistemology as not being genuine epistemology at all. That Campbell's naturalistic grounds involve biological evolution would be at most beside the point.

Campbell argues that the search for anything like an ultimate justification of knowledge must be abandoned in favor of "hypothetical realism." We can come to know about the world in which we live through perception, reason, and especially science. However, this knowledge is always hypothetical and revisable. We can never know anything for certain, even geometry. As an uninterpreted calculus, geometry is neither true nor false. However, when physicists attempted to interpret geometry to be about such things as light rays, they discovered to their consternation that space is non-Euclidean; more accurately, relativity theory implied that physical space should not be Euclidean, and as it turned out, it is not. In fact, just about every principle that epistemologists have claimed to be a priori true turns out to be simply false. About the only normative element in Campbell's system is the recommendation that whenever one discovers a significant fit between a system and its environment, look for an explanation in terms of variation and selective retention.

Sociology of Scientific Validity. As the preceding discussion shows, Campbell, not to mention other evolutionary epistemologists, took evolutionary theory as a biological theory to form the basis of evolutionary epistemology. Until recently, Campbell has been interpreted as treating sociocultural evolution, in particular the evolution of science, as a system analogous to biological evolution. Because biological evolution has certain characteristics and implies certain things about organisms and their evolution, parallel claims about sociocultural evolution have some derivative warrant as well.

As might be expected, critics of evolutionary epistemology have attacked this analogy. Genes have certain characteristics; the basic units of sociocultural evolution lack these characteristics. Biological evolution is blind; sociocultural evolution is not. But from the very first Campbell had another interpretation of evolutionary epistemology in mind, and this alternative came to prevail as he developed his system of thought. From his earliest papers Campbell can be found calling for a general analysis of selection. Instead of reasoning by analogy from selection in gene-based biological evolution to other sorts of selection, examine all purported examples of change resulting from variation and selective retention to see if they all have something in common. For example, the reaction of the immune system to antigens exhibits an especially clear instance of variation and selective retention. The immune system produces millions of cells, each with its own affinity to some part of a possible invader. The overwhelming majority of these cells never meet up with an antigen that they are able to recognize as foreign, but when one does, it proceeds to multiply at a phenomenal pace until either the intruders are killed off or the host dies.

Like the central nervous system, the immune system allows organisms that reproduce very slowly to react to rapid changes in their environments. Operant learning also seems to contain a selective element, as does social learning and the development of the central nervous system itself. Just as philosophers attempt to produce a general analysis of functional systems, they might well turn their attention to a comparable general analysis of selection processes. In this analysis, gene-based biological evolution has no preferred status. Just because scientists discovered how gene-based biological evolution occurred before they had a comparable success in psychology or immunology gives it no pride of place. All selection systems are equally instances of selection. According to one set of authors, selection is "repeated cycles of replication, variation and environmental interaction so structured that environmental interaction causes

replication to be differential. The net effect is the evolution of the lineages produced by this process" (Hull et al., 2001, p. 53).

This definition is more general than other analyses in two ways. First, replication includes any transmission of information, regardless of its constitution. Genes make more genes, but knowledge claims can also give rise to revised knowledge claims. Second, the relation between those entities that function in replication and those that play a role in environmental interaction need not be limited to development. Genes code for their phenotypic traits. Human languages in their descriptive role refer to their subject matter. As I approach a barking dog, I might well say to myself, "Barking dogs do not bite." Very rapidly I would discover if this maxim is true. The relation between genes as replicators and the production of more inclusive entities is one of embryological development. The relation between memes and the testing of these memes is not.

As promising as this change of focus may or may not be, it has an implication that gave Campbell pause: working out the general structure of selection processes does not afford any epistemological warrant for any of the systems that turn out to count as selection processes. This realization forced Campbell to admit that whatever validity that science as a selection process may have stems from the social organization of science. In the early years of epistemology, the usual strategy was to ask what an isolated individual could come to learn about his or her surroundings. The usual answer was little if anything, and the usual example was Robinson Crusoe, but Crusoe became shipwrecked on his deserted island after he had learned a language, after he had been socialized. He was hardly starting with a clean slate. In recent years, philosophy has taken a social turn. We gain knowledge in consort with others, not in regal isolation, and the social genesis of knowledge is crucial. Language does not function primarily for description but for communication. Human languages could not have arisen by an isolated individual trying to describe his surroundings. Nor could it have arisen by numerous people trying to describe their surroundings. They must communicate with each other. Because people can communicate, they can describe.

Science has been so successful in describing the world in which we live because of the social structure it developed. First and foremost, scientists desire the high opinion of their peers, not an especially unusual phenomenon for human beings. In addition, they want to understand some part of the empirical world—the genetics of a species of fruit fly, a way for the nucleus of atoms to be split, and so on. A crucial element in this process is mutual use. No scientists can produce everything that they need to succeed in their line of research. They must use the work of other scientists, and mutual use is the coin of the realm. Scientists want credit for their contributions, also not an especially unusual desire for human beings. The chief way of conferring this credit is by using the work of other scientists. If I use your work and integrate it into my own, I must find it of value. But what if your work turns out to be mistaken? Then everyone who used your work will suffer, and you will be punished accordingly. How? By your fellow scientists being less likely to use any of your future work.

Students of science as well as scientists themselves tend to emphasize that scientists test each other's work. They not only test it, they test it seriously. In general scientists do not check each other's work before they use it for the simple reason that it would slow down their own research significantly, and in science the scientist who first makes a discovery public gets all the credit. Scientists reserve testing for those findings that threaten their own research, and who better to test a line of research than the person whose own work is being threatened by it? One reason why scientists do what they are supposed to do is that in general it is in their own best self-interest to do so.

All the preceding observations about science may not have been precisely what Campbell had in mind when he referred to the "sociology of scientific validity," but it is close. Campbell never abandoned his evolutionary epistemology, but he shifted the epistemological warrant for our beliefs from something being a selection process to its occurring in social groups that exhibit a particular sort of social organization, the sort of organization that promotes serious testing of hypotheses. Why science works as well as it does is that scientists have developed this sort of social organization to a greater extent than any other social institution has done. In fact, it defines science.

Conclusion. As Campbell sees the development of his philosophical research program, he never abandoned his quest for scientific validity. He proceeded from (1) evolutionary epistemology in a literal sense to (2) evolutionary epistemology in analogy to selection in gene-based biological evolution to (3) a general analysis of selection and from there to (4) the sort of social organization that encourages true belief. In a literal sense, evolutionary epistemology afforded warrant for only a very few fundamental ways of viewing the world. All other warrant was analogical. Campbell's selection theory placed all sorts of variation and selective retention on a par, but in the process he abandoned any claims for epistemological warrant from evolutionary theory in particular and selection theory in general. Instead, whatever epistemological warrant scientists have for their beliefs is grounded in the social structure of science which encourages them to set out hypotheses and test them seriously. That scientists come up with new hypotheses and expose them to serious criticism can be construed as a selection process, but even more fundamental is why scientists consistently behave in this way. Science owes whatever warrant it has to the reward system inherent in its social structure.

BIBLIOGRAPHY

Aunger, R., ed. *Darwinizing Culture: The Status of Memetics as a Science.* Oxford, 2000. Includes papers both supportive and highly critical of meme theory.

Blackmore, S. *The Meme Machine.* Oxford, 1999. Provides a popular exposition of memetics, the study of sociocultural change as a selection process.

Callebaut, W. *Taking the Naturalistic Turn, or, How Real Philosophy of Science Is Done.* Chicago, 1993. Argues that our understanding of science must rest on our knowledge of science and not on any a priori philosophical claims about knowledge.

Campbell, D. T. "Evolutionary Epistemology." In *The Philosophy of Karl Popper,* edited by P. A. Schilpp, pp. 413–463. LaSalle, Ill., 1974. In characterizing the evolutionary epistemology of Karl Popper, Campbell presents the classic exposition of this epistemological view.

Campbell, D. T., C. M. Heyes, and B. Frankel. "From Evolutionary Epistemology via Selection Theory to a Sociology of Scientific Validity." *Evolution and Cognition* 3 (1997): 5–38. This posthumous publication summarizes the course of Campbell's views on evolutionary epistemology.

Cziko, G. *Without Miracles: Universal Selection and the Second Darwinian Revolution.* Cambridge, Mass., 1995. Sets out the variety of systems that count as being instance of variation and selective retention.

Dawkins, R. *The Blind Watchmaker: Why the Evidence of Evolution Reveals a Universe without Design.* New York, 1986. While refuting creationism, Dawkins provides an excellent exposition of the relation between nature and nurture in development as well as the limitations of selection.

Durham, W. H. *Coevolution: Genes, Culture and Human Destiny.* Stanford, Calif., 1991. An evolutionary biologist sets out a theory of how biological evolution and sociocultural evolution proceed in consort.

Giere, R. N. *Science without Laws.* Chicago, 1999. Presents a naturalistic view of science.

Heyes, C., and D. L. Hull, eds. *Selection Theory and Social Construction: The Evolutionary Naturalistic Epistemology of Donald T. Campbell.* Albany, N.Y., 2001. Discusses the various parts of Campbell's evolutionary epistemology.

Hull, D. L. *Science as a Process: An Evolutionary Account of the Social and Conceptual Development of Science.* Chicago, 1988. Sets out a general analysis of selection and shows how it applies to science.

Hull, D. L., S. Glenn, and R. Langman. "A General Account of Selection: Biology, Immunology and Behavior." In *Science and Selection,* pp. 49–93. Cambridge, 2001. Sets out a general analysis of selection for the three most unproblematic examples of selection.

Plotkin, H. C. *Darwin Machines and the Nature of Knowledge.* Cambridge, Mass., 1994. Discusses evolutionary epistemology in a popular format.

Popper, K. R. *Objective Knowledge.* Oxford, 1972. Presents the connection between a nonstandard view of biological evolution and epistemological warrant.

Ruse, M. *Taking Darwin Seriously: A Naturalistic Approach to Philosophy.* Oxford, 1986. Presents systems of both evolutionary epistemology and evolutionary ethics as naturalistic theories.

— DAVID L. HULL

EVOLUTIONARY PSYCHOLOGY. *See* Human Sociobiology and Behavior, *article on* Evolutionary Psychology.

EXAPTATION

The term *exaptation* has been proposed to describe an organismal character that currently has a role in promoting fitness but that was not originally shaped by natural selection for that role. A "role in promoting fitness" means that the character interacts with the environment so as to enhance its bearers' chances of leaving viable offspring relative to other organisms that lack it. Exaptations are fit for their current role, hence *aptus*, but were not designed for it, and are thus not *ad aptus*, or shaped toward that fitness. They owe their fitness to features, or form, that arose for other historical reasons, and are therefore fit (*aptus*) by reason of (*ex*) their form, or *ex aptus*. The term *adaptation* is restricted to a character that was shaped by selection for the function in which it is currently enhancing fitness. Adaptations and exaptations form the two partially overlapping subsets of the set of aptations: all characters that currently increase the fitness of organisms (Table 1). Nonaptations are neutral characters: they are currently not subject to natural selection (Gould and Vrba, 1982).

A role of a character can remain the same for long periods. For example, the four-limbed morphology was an adaptation for the role of locomotion on a substrate in the earliest tetrapods and is still an adaptation for that role in numerous living terrestrial tetrapods. A new role may be added to, or replace, the original one. An example is the forelimb of contemporary moles. The forelimb retains the original adaptive locomotory function and also has the added exaptive role of digging. Its anatomical specializations, such as long strong claws, selected for more efficient making of burrows or changes to the anatomy of the forelimb itself to accommodate the requirements of digging are adaptations for digging. An example of an exaptive new role that replaced the old one is the use of the flipperlike wing of penguins for swimming that replaced the flight function of wings in antecedents. Here it could be said that all of the changes to the ancestral wing that have evolved for swimming are adaptations, but the wing structure itself is an exaptation in the penguin.

This classification of some characters as exaptations was proposed in the spirit of Charles Darwin's original intent, as illustrated by his example of skull sutures:

> The sutures in the skulls of young mammals have been advanced as a beautiful adaptation for aiding parturition [giving birth]; and no doubt they facilitate, or may be indispensable for this act; but as sutures occur in the skulls of young birds and reptiles, which have only to escape from a broken egg, we may infer that this structure has arisen from the laws of growth, and has been taken advantage of in the parturition of the higher animals. (in Gould and Vrba, 1982)

Darwin's meaning is clear: unfused sutures in newborn mammals cannot be labeled an adaptation because they were not built by selection for the role in which they are now selectively favored. Skull sutures may have originated as nonaptations, or as adaptations for growth of the skull, and later a new selection regime in placental mammals coopted them as an exaptation for yielding to stress during birth. George C. Williams in his classic 1966 book on adaptation agreed with Darwin: adaptation should be argued only when we can "attribute the origin and perfection of this design to a long period of selection for effectiveness in this particular role" (p. 6). He termed the operation of a useful character not built by selection for its current role a fortuitous "effect." The distinction is that characters, such as skull sutures, may be aptive in their current role, even if they did not originate as adaptations for that role.

Until the early 1980s there was no term in evolutionary biology for the status of characters such as unfused skull sutures in young mammals in relation to parturition or the forelimbs of moles with respect to digging.

TABLE 1. Classification of Organismal Characters that Enhance Fitness

Character and Current Use		*Historical Process*
Aptation A character that contributes to fitness	Adaptation	Natural selection shaped the character to function in its current role in promoting fitness.
	Exaptation	A character previously shaped by selection for a particular role (an adaptation) was later coopted by selection for a new and still current fitness role.
		A character originally not shaped by selection (a nonaptation) was later coopted by selection for its current role.

The term preadaptation is inappropriate. A preadaptation is an adaptation for a role earlier in time (e.g., early tetrapod forelimbs for terrestrial locomotion) that would in future be coopted by selection for a different role in a descendant lineage (digging in moles). This is similar to an exaptation except that exaptations need not have been adaptations in their original role—they may have been nonaptations. Preaptation does not apply to the coopted character in its new role, nor does it encompass the possibility that nonaptations may also be coopted by future selection regimes.

Exaptations may be more important in the history of life than commonly appreciated. There may be redundant genetic material, the products of which can be exapted into novel roles. Numerous examples of regulatory elements and genes with new roles have been identified. At the level of phenotypes, Darwin foresaw that "almost every part of each living being has probably served, in a lightly modified condition, for diverse purposes, and has acted in the living machinery of many ancient and distinct specific forms" (Jacob, 1983, p. 131).

Current methods for testing hypotheses of adaptation involve analysis of the phylogenetic context and comparisons of the locations in the phylogeny of the changes in characters, in selective regimes, and in functional capabilities.

[*See also* Adaptation; Evolution; Natural Selection.]

BIBLIOGRAPHY

Baum, D. A., and A. Larson. "Adaptation Reviewed: A Phylogenetic Methodology for Studying Character Macroevolution." *Systematic Zoology* 40 (1991): 1–18.

Brosius, J. "RNAs from All Categories Can Generate Retrosequences that May Be Exapted as Novel Genes or Regulatory Elements." *Gene* 238 (1999): 115–134. Although this is rather technical, it is interesting to see how many exaptive genomic elements have already been discovered.

Gould, S. J., and E. S. Vrba. "Exaptation—A Missing Term in the Science of Form." *Paleobiology* 8 (1982): 4–15.

Jacob, François. "Molecular Tinkering in Evolution." In *Evolution from Molecules to Men*, edited by D. S. Bendall, pp. 131–144. Cambridge, 1983.

Losos, J. B., and D. B. Miles. "Adaptation, Constraint, and the Comparative Method." In *Ecological Morphology*, edited by P. C. Wainwright and S. M. Reilly, pp. 61–98. Chicago, 1994.

Williams, G. C. *Adaptation and Natural Selection: A Critique of Some Current Evolutionary Thought*. Princeton, 1966. A seminal book that remains as thought provoking as ever.

— ELISABETH S. VRBA

EXONS. *See* Genes; Introns.

EXPERIMENTAL EVOLUTION

[*This entry comprises two articles. The first article provides a summary overview contrasting experimental and comparative/historical approaches for studying evolution; the second article provides a case study of an evolutionary experiment in which populations of E. coli have been propagated and monitored for 20,000 generations. For related discussions, see* Adaptation; Artificial Selection; Bacteria and Archaea; Comparative Method; Fitness; Genetic Drift; Natural Selection; *and* Senescence.]

An Overview

Experimental evolution is a scientific method in which populations of organisms are introduced into novel environments, and changes within those populations are then observed over many generations. In other words, it is a method for directly observing evolutionary change, including adaptation.

Experimental evolution is an alternative to the most common method for studying evolutionary adaptation, the comparative or historical approach. [*See* Comparative Method.] The latter examines characters in organisms from different natural environments and tries to understand the pattern of change that the descendants of a common ancestor might have followed in diversifying into their current forms. In essence, the comparative approach attempts to look backwards in time, given information about the current diversity of organisms. It necessarily has to make many assumptions about phylogenetic relationships among groups and the probable course of evolution.

In contrast, the experimental method creates an ancestral population and then watches, in real time, adaptation and diversification of descendant populations in one or more environments. It does not make assumptions about evolution, or constraints on it, but rather observes the adaptive solutions that appear within the evolving populations. The two approaches are complementary, each with its strengths and limitations. The comparative approach investigates evolution in populations of all kinds of organisms in the natural world, which is full of complexity and compromise. Experimental evolution is limited to strictly controlled conditions in the laboratory and is feasible for only certain types of organisms. However, this approach permits the application of the experimental scientific method to evolutionary studies, including control, replication, and repeatability. The comparative method attempts to understand events that are unique, at least in certain respects, which imposes limitations and assumptions on the application of statistical tests to comparative data. The two methods provide different insights into evolutionary adaptation to the environment.

An evolutionary experiment might proceed in the following way. A large population of organisms is created and placed in a controlled environment in the laboratory. It is important that a great number of individuals (certainly hundreds, preferably millions) be used, to limit the establishment of particular alleles (alternative forms of the same gene) by chance alone. [See Genetic Drift.] This population is then kept under these conditions for many generations, permitting adaptation to the general laboratory conditions. After this initial period, this population is then used as the ancestral population for the experiment itself. To that end, the ancestral population is randomly divided into experimental (selected) and control groups of populations, each of which is replicated ideally several times. The control populations are kept under the same conditions in which the ancestral population evolved; they serve to indicate the direction and amount of evolutionary change that can be attributed to further laboratory adaptation. The selected populations are placed in one or more novel environments, in which a single variable of interest (e.g., temperature, water or nutrient availability, breeding time, amount or duration of light exposure) is altered and all other conditions are maintained identical to those of the ancestral population. Both the selected and the control populations are then permitted to evolve, preferably for hundreds or even thousands of generations.

Types of Evolutionary Experiments. There are several forms that selection in an evolutionary experiment might take. The first is laboratory natural selection, in which replicated populations are exposed to a novel environment and changes within the populations over many generations are measured and analyzed, as described above. The experimenter provides the environment but does not otherwise directly select any character or choose individuals for differential breeding. The populations are left to their own devices to evolve solutions to the environmental challenge. The second variety is artificial truncation selection, in which only organisms possessing certain characters (or extreme character values) are permitted to reproduce the next generation. [See Artificial Selection.] This artificial selection is the familiar form used in animal and plant breeding, and it can sometimes quickly result in the creation of new types of organisms (e.g., breeds of dogs). However, by imposing a single desirable factor or set of factors, it may constrain the pathways along which evolution can proceed to solve a more general problem. The third approach is laboratory culling. In this design, a more extreme environmental condition (e.g., high temperature) is imposed every generation, and only a small proportion of the population is permitted to survive to reproduce the next generation. This type of selection has elements of the other two designs. It permits only a small fraction of the population to breed, in contrast to most laboratory natural selection protocols. However, it does not specify that survivors must possess certain characteristics, as does artificial truncation selection;

consequently, a diversity of pathways might be used in evolving solutions to the environmental challenge.

Though not absolutely necessary, it is very desirable that two basic measurements can be undertaken during an evolutionary experiment: ancestral comparison and relative fitness. For some types of organisms, the ancestral population can be preserved in a condition (typically frozen) from which it can be revived and compared directly to its descendants, both selected and control populations. Such comparisons permit a direct determination of the type and amount of evolutionary change that the descendants have undergone. In addition, preservation of samples of the selected populations during the course of the experiment allows later analysis of the time of appearance of novel adaptive traits. It is also highly useful to be able to measure the fitness of the descendant populations relative to their ancestor in both the novel and ancestral environments. [See Fitness.] Fitness can be measured by placing both the ancestral and a selected (or control) population in a common environment and measuring the number of offspring produced by each. A readily scored, but preferably neutral, genetic marker is often used to distinguish the ancestral and derived types. Differential reproduction serves as a quantitative measure of the adaptive improvement of an evolving population and is the single best measure of evolutionary adaptation.

Obviously, such evolutionary experiments cannot be undertaken on all kinds of organisms. Large organisms with long generation times are not feasible objects for such studies. Organisms that reproduce rapidly and can be grown easily in large numbers in the laboratory are the best subjects. Viruses or unicellular organisms such as bacteria, protists, and yeasts are ideal for experimental evolution, not only because of their short generation times and large population sizes in culture, but also because many are well characterized genetically. In such organisms, it may be possible to determine the exact genetic alterations that underlie adaptive change. Evolutionary experiments can also be undertaken on multicellular animals such as fruit flies and even mice, but these are more challenging because of the difficulty in maintaining large population sizes and many replicated populations.

Advantages of Experimental Evolution. Part of the utility of experimental evolution is the production of biological novelty, literally building a better mouse for some particular environmental condition. The creation of lineages of organisms with known evolutionary and environmental histories provides a valuable resource for future study. It may even be possible to analyze the genetic and functional bases underlying the observed adaptive changes. An issue of great interest in that regard is the degree of repeatability in evolution, that is, whether certain genetic changes would occur repeatedly in similar populations adapting to a novel environment or whether there are a large number of potential mechanisms of adaptive change. The replicated populations within an experimental group provide a means of examining the number of potential adaptive solutions to a common environmental condition and whether the same changes occur repeatedly.

Another advantage of this method is that it provides an experimental way to test general evolutionary assumptions and predictions. It is frequently assumed, for instance, that adaptive gains in a new environment are accompanied by a decline in function in other environments. In evolutionary experiments, it is possible to test for such "trade-offs" directly: a common assumption is thereby turned into a testable hypothesis. The importance of having replicated populations within each experimental group is that each population can be considered an independent observation in testing the hypothesis. Standard statistical methods can thus be applied in the analysis of the experiment, and the evolutionary hypothesis can be rejected or provisionally supported by the observations. Experimental evolution permits evolutionary studies to move beyond retrospective comparative analyses.

Review of Studies. There are many examples of experimental evolutionary studies, and only a few can be mentioned here; more are detailed in Bell (1997) and Bennett and Lenski (1999). In regard to testing evolutionary theory, a classic series of studies by Michael Rose and his coworkers (1990) examined predicted trade-offs between early reproduction and longevity. Using populations of fruit flies, and selecting some for early reproduction and preventing others from reproducing until later in life, they found that life span increased significantly in the latter group, as predicted by theory. Interestingly, and unanticipated, the longer-lived populations also evolved greater resistance to some types of environmental stress, including desiccation and starvation. When selection for delayed reproduction was removed, longevity decreased, as did stress resistance. These studies have greatly influenced thinking concerning aging and its evolutionary and physiological bases.

Another study of particular interest is that of J. Swallow and colleagues (1998), who selected populations of mice on the basis of their voluntary running behavior. Mice that ran the most in cage wheels were bred to similar mice, and these lineages were compared to controls that were bred at random. After ten generations, voluntary running behavior approximately doubled. Maximum oxygen consumption in the selected mice also increased at the same time. From a mechanistic viewpoint, it is now possible to examine which portions of the oxygen transporting and utilization systems responded to this selection and whether these changes were consistent across selected populations. These populations can also be used to test more general evolutionary theories concerning performance trade-offs, models for the evolution of endothermy, and constraints on organismal design.

In a study on the repeatability of evolution, Jim Bull and coworkers (1997) examined the adaptation of replicated lineages of a bacteriophage (virus that attacks bacteria) to a moderately high and stressful temperature, which greatly inhibited its growth rate. During adaptation, growth rates improved as much as four thousand fold. Because a bacteriophage has a small genome, it was possible to analyze the DNA of each of the different replicates; it was found that approximately half of the changes that occurred were identical in more than one of the lineages. These experiments indicate that the same evolutionary changes may appear repeatedly in a population exposed to a novel environmental stress.

In a final example, Mike Travisano and colleagues (1995) performed an experiment to examine the relative roles of adaptation, chance, and historical contingency in shaping phenotypic diversity among populations adapting to similar environments. Bacterial populations were permitted to evolve in two different environments, a novel sugar nutrient and low temperature. These populations initially differed among themselves in relative fitness and cell size. During evolution, relative fitness improved in both novel environments, and differences in fitness among populations were greatly decreased as a result of adaptation. The remaining differences attributable to the initial fitness condition (history) and chance divergence in fitness were very small. In contrast, chance and history were far more significant factors influencing the final diversity of cell size among the populations. These results indicate that historical condition and chance may have longer lasting influences on characters that do not strongly affect fitness, such as cell size, than on fitness itself.

[See also Adaptation; Natural Selection, *article on* Natural Selection in Contemporary Human Society; Senescence.]

BIBLIOGRAPHY

Bell, G. *Selection: The Mechanism of Evolution.* New York, 1997. Views selection as the primary force in evolution, with particular emphasis on experimental studies of evolution.

Bennett, A. F., and R. E. Lenski. "Experimental Evolution and Its Role in Evolutionary Physiology." *American Zoologist* 39 (1999): 346–362. A survey article on approaches and methods in experimental evolution.

Bull, J. J., et al. "Exceptional Convergent Evolution in a Virus." *Genetics* 147 (1997): 1497–1507.

Gould, S. J. *Wonderful Life: The Burgess Shale and the Nature of History.* New York, 1989. Discusses the very early fossil record and its relationship to the subsequent evolution of animals and speculates about the paths that evolution might take.

Harvey, P. H., and M. D. Pagel. *The Comparative Method in Evolutionary Biology.* Oxford, 1991. A textbook discussion of the theory and methods of the comparative/historical approach to evolutionary analysis.

Rose, M. R.. *Evolutionary Biology of Aging*. Oxford, 1991. An advanced book discussing evolutionary theories about aging and reviewing their experimental basis.

Rose, M. R., et al. "The Use of Selection to Probe Patterns of Pleiotropy in Fitness Characters." In *Insect Life Cycles: Genetics, Evolution and Co-Ordination*, edited by F. Gilbert. New York, 1990. Discusses designs for evolutionary experiments and the advantages of each.

Swallow, J. G., et al. "Effects of Voluntary Activity and Genetic Selection on Aerobic Capacity in House Mice (*Mus domesticus*)". *Journal of Applied Physiology* 84 (1998): 69–76.

Travisano, M., et al. "Experimental Tests of the Roles of Adaptation, Chance, and History in Evolution." *Science* 267 (1995): 87–90.

Wichman, H. A., et al. "Different Trajectories of Parallel Evolution during Viral Adaptation." *Science* 285 (1999): 422–424.

— ALBERT F. BENNETT

A Long-Term Study with *E. coli*

The paleontologist Stephen Jay Gould envisioned a thought experiment of "replaying life's tape" to explore the predictability, or repeatability, of evolution. Gould (1989) argued that evolution is not repeatable: "Any replay of the tape would lead evolution down a pathway radically different from the road actually taken . . . no finale can be specified at the start, and none would ever occur a second time in the same way, because any pathway proceeds through thousands of improbable stages." No one can replay evolution on the vast scale imagined by Gould. But on a much smaller scale, the following experiment examines the same issue by monitoring multiple populations, all founded from the same ancestor, as they evolve in identical environments for thousands of generations.

Using a homogeneous clone of the bacterium *Escherichia coli*, Richard Lenski started twelve populations in 1988. These populations have been propagated side by side for more than 20,000 cell generations in a simple, glucose-limited environment in the laboratory. This environment is a novel one for the bacteria, insofar as it differs in important respects from the conditions that have prevailed during their evolution in nature. Consequently, there is considerable scope for the bacteria to improve their functioning in the laboratory environment. Meanwhile, cells of the ancestral strain were stored in a nonevolving state (in a deep freezer), and samples from the evolving populations likewise have been stored at periodic intervals.

By simultaneously reviving ancestral and derived cells from the freezer, one can compare them directly to measure the evolutionary changes that occurred during the experiment. Changes in ecological, physiological, and genomic properties have been quantified in this manner.

Using a genetic marker that allows two strains to be distinguished when they are mixed, it is possible even to compete the derived cells against their ancestor, and thereby measure their relative fitness. In effect, this approach is comparable to resurrecting fossil hominids, such as Neanderthals—not merely their bones or even DNA, but actual beings—and comparing their performance capacities with those of modern humans.

Every day, the populations have been diluted 100-fold into fresh medium. The population size, after the cells have exhausted the daily supply of glucose, is about 3×10^8 individuals. The particular strain of *E. coli* used in this experiment lacks viruses and plasmids (which can promote intergenomic recombination in bacteria). Hence, these experimental populations are completely asexual, and spontaneous mutation provides the only source of genetic variation. The total mutation rate for *E. coli* is about 3×10^{-3} mutations per genome replicated (for DNA repair–proficient replication). This rate, coupled with the population size, implies that each population has had almost a million mutations every day. Mutation therefore generates abundant variation on which natural selection and genetic drift can act.

During 20,000 generations, the competitive fitness of the derived bacteria improved by about 75 percent on average, relative to their common ancestor (see Figure 1). The rate of improvement was much greater early in the experiment than later on; as the populations became better adapted to the experimental environment, further gains became progressively slower. Notice also that the populations evolved along roughly similar, though not identical, fitness trajectories. Thus, with respect to competitive performance, evolution appears to have been fairly repeatable.

The results in the preceding paragraph were based on competition assays under conditions identical to those that prevailed during the experimental evolution. Additional competitions were run in other environments, in which different resources (such as maltose) were substituted for glucose (the only sugar provided during the evolution experiment). The fitness of the derived lines relative to their ancestor tended to be lower in these "foreign" environments than in their "home" environment. The multiple derived lines were also more heterogeneous (diverse) in their response to foreign environments than in their performance in the glucose-limited environment in which they evolved. This diversity in foreign environments indicates that the twelve populations had diverged physiologically and genetically, despite their similarity in the selective environment.

The populations changed relative to their ancestor, and diverged phenotypically from one another, in other aspects as well. All twelve evolved lines produced larger individual cells than did the ancestor, but the extent of

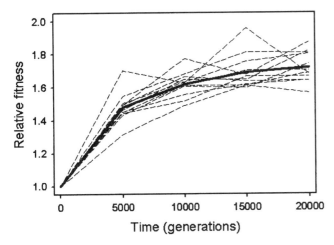

FIGURE 1. Improvement in Fitness of Twelve Evolving Populations, Measured Relative to Their Common Ancestor, During 20,000 Cell Generations.
The estimates of fitness were obtained by allowing each derived population to compete against a variant of the ancestor carrying a selectively neutral marker, such that the competitors could be readily distinguished. Relative fitness expresses the ratio of growth rates achieved by two populations during direct competition. The competitions were run under the same conditions in which the evolution experiment took place. The solid line is the mean of the twelve populations. Data from Cooper and Lenski, 2000.

the cell-size increase varied considerably. Also, the cells in a few lines (but not most) evolved from elongated rods to a more spherical shape. The functional significance of these morphological changes is not clear, because the variation among lines is not correlated with their competitiveness in the selective environment.

A striking change that occurred in several lines, but not all, affected the process of DNA repair. Most organisms, including bacteria, have systems used to repair their genomes. This repair process reduces the mutation rate below the level it would be without such systems. During the evolution experiment, several lines lost their DNA repair ability, with a corresponding increase in their mutation rates. The resulting "mutator" state is disadvantageous in one respect, because many of the extra mutations it causes elsewhere in the genome are deleterious. However, a mutator may spread by "hitchhiking" along with a beneficial mutation that it fortuitously causes.

These various phenotypic changes in the evolved lines are heritable. However, the precise genetic bases for all these changes are not yet fully known. A current focus of this research is to identify genomic changes in the evolving *E. coli* populations. Once a particular mutation is found in an evolved line, however, it cannot be automatically concluded that it was beneficial. Neutral and even slightly deleterious mutations may hitchhike

with beneficial mutations elsewhere in the genome, especially in asexual populations. Therefore, after a mutation has been found, it should be manipulated, for example, by restoring the ancestral allele to an evolved genotype, to measure its direct effects on fitness and other phenotypic traits of interest.

Millions of mutations occurred in each population, but the vast majority of them were eliminated by selection or drift. It is estimated that fewer than 100 mutations were actually substituted in a typical *E. coli* line during 20,000 generations (with somewhat more substitutions in the lines that became mutators). Finding these substitutions in a genome of five million base pairs is a challenge. Nonetheless, some mutations have been found, especially ones involving insertions and deletions (which are easier to find than point mutations that affect only a single base pair).

For example, an operon (set of coregulated genes) for ribose catabolism was deleted in all twelve lineages, although the extent of the deletion differed in each case. Genetic manipulations have confirmed that deletion of this operon is beneficial in the glucose-limited environment. However, the evolved cells have lost their ancestor's ability to grow on the sugar ribose, which would limit their success in an environment where ribose was an important resource. In contrast with these repeatable deletions of the ribose operon, some other mutations were apparently substituted in only a subset (or even only one) of the evolved lines.

In summary, this long-term experiment with *E. coli* shows that evolutionary change is both repeatable and idiosyncratic. All twelve populations have become more fit in glucose, produce larger cells than their ancestor, and have deleted genes used for ribose catabolism, indicating a degree of repeatability to evolution. On the other hand, the twelve lines have diverged from one another in their performance on maltose, cell shape, proficiency for DNA repair, and various genes that underlie these and other phenotypic differences.

[*See also* Adaptation; Bacteria and Archaea; Directed Protein Evolution; Fitness; Genetic Drift; Hitchhiking; Mutation, *article on* Evolution of Mutation Rates; Natural Selection.]

BIBLIOGRAPHY

Appenzeller, T. "Test Tube Evolution Catches Time in a Bottle." *Science* 284 (1999): 2108–2110. A brief overview of this and other evolution experiments with microorganisms.

Cooper, V. S., and R. E. Lenski. "The Population Genetics of Ecological Specialization in Evolving *E. coli* populations." *Nature* 407 (2000): 736–739. Examines the dynamic processes that led to the specialization of the long-term lines on glucose.

Cooper, V. S., D. Schneider, M. Blot, and R. E. Lenski. "Mechanisms Causing Rapid and Parallel Losses of Ribose Catabolism in Evolving Populations of *E. coli* B." *Journal of Bacteriology* 183

(2001): 2834–2841. Study reporting the parallel deletions of an operon in the long-term lines.

Gould, S. J. *Wonderful Life: The Burgess Shale and the Nature of History*. New York, 1989. A popular account of findings and questions from paleontology, including the idea of "replaying life's tape."

Lenski, R. E., and J. A. Mongold. "Cell Size, Shape, and Fitness in Evolving Populations of Bacteria." In *Scaling in Biology*, edited by J. H. Brown and G. B. West, pp. 221–235. Oxford, 2000. Examines the relationship between cell morphology and competitive fitness in the long-term *E. coli* lines.

Lenski, R. E., and M. Travisano. "Dynamics of Adaptation and Diversification: A 10,000-Generation Experiment with Bacterial Populations." *Proceedings of the National Academy of Sciences USA* 91 (1994): 6808–6814. An overview of the long-term study of evolving *E. coli*, drawing an analogy between time travel and the experimental approach to evolution.

Schneider, D., E. Duperchy, E. Coursange, R. E. Lenski, and M. Blot. "Long-term Experimental Evolution in *Escherichia coli*: 9. Characterization of Insertion Sequence-mediated Mutations and Rearrangements." *Genetics* 156 (2000): 477–488. Descriptions of some of the mutations that have been substituted in the evolving *E. coli* lines.

Sniegowski, P. D., P. J. Gerrish, and R. E. Lenski. "Evolution of High Mutation Rates in Experimental Populations of *Escherichia coli*." *Nature* 387 (1997): 703–705. Provides the evidence that DNA repair functions were lost in several of the long-term lines.

Travisano, M., and R. E. Lenski. "Long-term Experimental Evolution in *Escherichia coli*: 4. Targets of Selection and the Specificity of Adaptation." *Genetics* 143 (1996): 15–26. Contrasts the parallel fitness gains seen in the glucose environment with more divergent responses to maltose and other foreign environments.

— RICHARD E. LENSKI

EXTINCTION

Species do not persist indefinitely. The period from the origin to the extinction of a typical species is thought to be of the order of five million to ten million years, a fraction of a percent of the time that life has been present on the earth. This is likely to be something of an overestimate, however, because it is based largely on evidence from the fossil record, which is biased toward abundant and widespread species that may typically persist for longer. This average figure also implies a longer persistence than that which would be experienced by a high proportion of species, because the distribution of these times is skewed such that most species are relatively short-lived and only a few have relatively long persistence.

Against the broad sweep of the history of life, such comparatively brief persistence times mean that the vast majority of species that have at some point graced the earth are now extinct. The best working estimate is that there are about thirteen million extant species. They comprise at the very most a few percent of all the species that have ever existed, and likely much less. That is, perhaps 99 percent of all species that have ever existed are extinct.

Extinction results from interplay between features of the environment in which a species occurs (extrinsic factors) and the traits that the species exhibits (intrinsic factors).

Extrinsic Factors. Four broad classes of extrinsic factors can be recognized: habitat loss, degradation and fragmentation, predation and disease, competition, and extinction cascades.

Habitat loss, degradation, and fragmentation. Habitat change robs some species of the potential to maintain viable populations, resulting in their extinction, while often expanding the opportunities for others. It can be driven by many processes, ranging from the impacts of large meteors, to the cyclic variation in the parameters of the earth's orbit, to the economic forces underpinning many human activities. As such, habitat change may be a rare and unexpected event, or a regular occurrence with a long or a short return time.

Overall, habitat change has been the single most important factor that has led to the extinction of species. Thus, much of the change in species composition documented in the fossil record coincides with periods of habitat change. Similarly, the scale of habitat change that has been wrought by human activities in recent times is one of the main reasons that extinction rates in the near future are predicted to be three to four orders of magnitude greater than the background rates of the past, that is, ignoring mass extinctions. Thus, habitat loss and degradation constitute major causes of endangerment for 85 and 47 percent of threatened bird and mammal species, respectively.

The loss and degradation of habitat are often accompanied by its fragmentation, in which formerly more continuous areas are broken up into increasingly scattered patches. The environmental conditions of smaller patches tend to be different (in part because of their greater edge to core ratio). Moreover, smaller patches are less likely to be able to sustain populations of a species (individuals of which may be unable or unwilling to cross intervening areas) that are of sufficient size to be viable, resulting in local extinctions, and increasing the likelihood of overall extinction.

A significant time lag, sometimes of the order of several decades, may arise between reductions in habitat area and the extinction of species because populations can persist for some time after they have lost long-term viability (e.g., because their populations are now too small, or because other species, such as pollinators or dispersers, are no longer present). On the one hand, such an "extinction debt" (or body of "latent extinctions") means that the immediate effects of habitat destruction may markedly underestimate the overall consequences (often indicated by the numbers of highly threatened but extant species). On the other hand, it presents an opportunity for remedial conservation actions.

Predation and disease. Extinction of a species results when net mortality exceeds natality. One obvious way in which mortality can be increased is through increased predator pressure, and predation can indeed lead to extinction. Perhaps the best documented examples have concerned the introduction of exotic predators to lakes and islands. Thus, hundreds of species of fish, many endemic, from the lakes of the East African Rift Valley are extinct as a result of the intentional introduction of the Nile perch (*Lates niloticus*), a voracious predator (although other factors have also contributed). Likewise, the accidental introduction of the brown tree snake (*Boiga irregularis*) to the island of Guam around 1950 resulted, directly or indirectly, in the loss of perhaps nine species of native forest birds (some endemic to the island) and the reduction of most of the rest to small remnant populations. In both cases, the catholic tastes of the generalist predators involved is important, enabling them to maintain high abundances even when one of their prey species has been driven scarce.

It is probable that Guam, like many other Pacific islands, had already experienced the loss of a significant proportion of its native fauna much before the arrival of the brown tree snake, about the time of first human colonization; the Polynesians exterminated more than two thousand bird species from Pacific islands. Indeed, humans from early in their history have caused the extinction of many species. In some cases, this has occurred, at least in part, as a result of predation. Thus, the disappearance within the past forty thousand years of two-thirds or more of the species of large-bodied mammals from the Americas (including woolly mammoths, ground sloths, and glyptodonts), Australia (including diprotodonts and giant kangaroos), and Madagascar (including giant lemurs) appears to have approximately coincided with the establishment of human populations in these areas. All other things being equal, the energetic return to a consumer per unit of hunting effort will increase with the size of the prey, which must have made these species tempting targets. Of course, the harvesting activities of humans are not confined to the distant past, and they continue to imperil the persistence of species. Of the extant birds and mammals that are presently regarded as having a real risk of becoming extinct in the near future, 31 percent and 34 percent, respectively, are threatened by overexploitation.

An important influence on world history has been the diseases that invading peoples have transmitted to others. Because of the lack of immunity of the recipients, they have consequently been decimated and conquered. The potential of diseases to drive extinctions is clear.

Thus, for example, many of the birds of the Hawaiian islands that have gone extinct in the last two centuries, or are highly threatened, seem likely to have become so at least in part because of mortality resulting from the spread of avian pox and malaria by introduced mosquitoes.

Competition. The exclusion of a species by one or more others may occur not only by predation but also by interspecific competition. Such competition has been suggested as a cause of a number of large-scale changes in floral or faunal composition documented in the fossil record. Unfortunately, even for extant species, the roles of competition and habitat change are difficult to disentangle, although the two are not necessarily mutually incompatible; habitat change is likely to favor some species at the expense of others. This difficulty is exemplified by the debates that have raged over the reasons for the simultaneous reduction in the range in Britain of the native red squirrel (*Sciurus vulgaris*) and expansion in the range of the introduced gray squirrel (*Sciurus carolinensis*). The likelihood of a competitive interaction has long been entertained, but only recently has it come to be generally accepted that the red squirrel is a less efficient forager in deciduous woodland and anthropogenic habitats.

Extinction cascades. The extinction of one species may lead to the extinction of others. Indeed, this is inevitable where this species provides critical resources for others, such as specialist herbivores, parasites, or predators, or perhaps itself acts as a specialist pollinator or dispersal agent. Thus, for example, in New Zealand, the giant eagle (*Harpagornis moorei*) almost certainly preyed on moas (large, flightless, herbivorous birds), and its extinction most likely resulted when moas declined in numbers as a result of the hunting by the Maori that led to their demise. Likewise, the extinction of many species of birds and mammals may have been accompanied by the extinction of associated specialist lice and ticks.

More complex sets of interactions may also mean that one extinction is followed by others. This is evidenced by the dramatic, and often extensive, changes in floral and faunal composition that can result from changes in the abundance and occurrence of key species (e.g., large-bodied predators and herbivores). Some of these interactions may be subtle, perhaps helping to account for why, in many cases, it is impossible to say categorically why a particular species became extinct. Commonly, no single factor is found to provide a wholly convincing explanation, highlighting the enormous potential for complex relationships to occur among the different factors that can lead to the decline and ultimately to the loss of the last individual.

Intrinsic Factors. For a range of taxonomic groups, the extinction probabilities of species have been found

TABLE 1. Estimated Mean Duration (in Millions of Years) of Fossil Species

Marine	Duration (millions of years)	Nonmarine	Duration (millions of years)
Reef corals	25.0	Monocotyledonous plants	4.0
Bivalves	23.0	Horses	4.0
Benthic foraminferans	21.0	Dicotyledonous plants	3.0
Bryozoa	12.0	Freshwater fish	3.0
Gastropods	10.0	Birds	2.5
Planktic foraminiferans	10.0	Mammals	1.7
Echinoids	7.0	Primates	1.0
Crinoids	6.7	Insects	1.5

SOURCE: From compilation by M. L. McKinney. "Extinction Vulnerability and Selectivity: Combining Ecological and Paleontological Views." *Annual Review of Ecology and Systematics* 28 (1997): 495–516.

to be largely independent of time or of the period that has elapsed since their origin (ignoring the complications of mass extinctions and the present spate of extinctions). L. M. Van Valen (1973) explained this observation in terms of the Red Queen's hypothesis (based on a remark made by the character in Lewis Carroll's *Alice through the Looking Glass*). It states that the sum of the momentary absolute fitness of interacting species is constant. In other words, although species evolve constantly under the pressure of natural selection, there is no net gain or loss in overall fitness as a result because all species are evolving to counter one another's advances. [*See* Red Queen Hypothesis.]

Nevertheless, some groups tend to have characteristically higher rates of extinction than do others. Thus, there is substantial variation in the estimated periods for which, on average, species in different taxonomic groups persist (Table 1). Indeed, extinctions tend to be taxonomically clumped (parrots, pheasants, and primates are all disproportionately threatened at present), often disproportionately within species-poor groups, which may mean that more genetic diversity is lost than would be expected by chance. Such differences may reflect extrinsic factors. Thus, for example, in the fossil record, marine groups seem to have lower rates of extinction than terrestrial groups, which may reflect the greater buffering of marine systems to environmental change. However, the differences may also reflect intrinsic factors that make some species more vulnerable to extinction than others, with the relationship between the intrinsic characteristics of species and the likelihood of extinction depending fundamentally on the extrinsic factors that are posing the threat to continued persistence.

Abundance and range size. Perhaps the strongest correlates of the risk of extinction faced by a species are its abundance and range size. Rare and restricted species are, on average, at the greatest risk. Rarity may increase the risk of extinction for several reasons. First, at very low numbers, populations may suffer particularly

from demographic stochasticity, in which by chance all individuals may die during the same time interval, all may fail to breed, or sex ratios (the number of males to females) become highly skewed. Second, small populations may suffer from an increased risk of extinction because of an unavoidable increase in matings between close relatives. Such inbreeding reduces reproductive success in populations of species that naturally outbreed. I Saacheri and colleagues (1998) demonstrated for the Glanville fritillary butterfly (*Melitaea cinxia*) that larval survival, adult longevity, and egg-hatching rate were adversely affected by inbreeding, and that extinction risk increased significantly. Third, at low densities, Allee effects may also occur, in which population growth rates decline because the probability of individuals finding a mate becomes unduly small. Fourth, and perhaps often more importantly, small populations are more readily driven to extinction by more deterministic processes, such as habitat loss or exploitation.

A restricted distribution tends to increase the risk of extinction because there is a high likelihood of adverse conditions affecting the distribution as a whole (it is more likely that all areas will be in decline or undergo local extinction simultaneously). Moreover, the risks associated with abundance and range size tend to go hand in hand (conferring "double jeopardy"), because species with small range sizes typically have relatively small population sizes and low local densities compared with related species. This is one of the reasons that species restricted to small islands are especially vulnerable and why a high proportion of extinctions in recent times have concerned such species; wherever people have colonized isolated islands, extinctions have followed. These inevitably have rather small populations and range sizes compared with close relatives on the mainland.

This is not to say that abundant and widely distributed species cannot decline, perhaps to extinction, quite rapidly. When Europeans reached North America, and probably into the early 1800s, the passenger pigeon (*Ectopistes migratorius*) is estimated to have had a population of from three billion to five billion individuals (comparable to the current entire bird population of North America) and was distributed over most of eastern North America. Nonetheless, forest destruction and fragmentation, facilitated by high levels of exploitation and disturbance, resulted in dramatic declines such that the species was scarce in the 1880s and very rare by the 1890s. The last authentic recorded sightings occurred at the turn of the twentieth century, and the last individual died in captivity at the Cincinnati Zoo in 1914.

Specialization. The breadth of conditions or resources exploited by a species may influence the likelihood of its extinction, because the narrower this breadth, the less likely the species is to be able to persist in a changing world. This is especially so if there is heavy dependence on another particular species, as mentioned in the context of extinction cascades, but applies much more widely. Thus, for example, in the Paleozoic era, crinoids that had wider niche breadths (occupied more environments) persisted for longer than did those that had narrower niche breadths. Specialization is much more frequent than generalism, and rarity and restricted distribution are much more frequent than abundance and widespread distribution. This may serve in part to explain why such a high proportion of species have rather short persistence times.

Body size and reproductive strategy. Large-bodied species may have a disproportionate risk of extinction, particularly when faced with heavy exploitation (see above), because large size tends to be associated with low rates of population increase and with low population densities. For such species, even the loss of relatively small to moderate absolute numbers of individuals can soon outstrip their reproductive capacities, as witnessed, for example, by the impacts of the intentional exploitation of large-bodied marine mammals (e.g., whales), and of the accidental mortalities of albatrosses caused by longline fisheries.

Other facets of reproductive strategies may also influence the likelihood of a species becoming extinct. Thus, asexual populations tend to have higher extinction risks than sexual (because the former adapt more slowly to changing environmental conditions). Likewise, species that display forms of social facilitation, such as nesting in large aggregations to reduce the per capita effects of predation, may be very vulnerable when population sizes are small. This may have played a role in the catastrophic decline of the passenger pigeon.

Conclusion. While extinction ultimately befalls all species, the history of life has been characterized by an underlying tendency for rates of speciation to outstrip rates of extinction. This has resulted in overall progressive growth in global species richness, albeit overlain by periods of sometimes dramatic decrease and others of relative stability. The present period, however, is one in which rates of extinction are markedly outstripping those of speciation.

[*See also* Mass Extinctions.]

BIBLIOGRAPHY

BirdLife International. *Threatened Birds of the World.* Barcelona and Cambridge, 2000. BirdLife International. A detailed account of those species of birds at greatest risk of extinction.

Caughley, G., and A. Gunn. *Conservation Biology in Theory and Practice.* Oxford, 1996. Summarizes a number of case studies of species that have declined to, or close to, extinction.

Lawton, J. H., and R. M. May, eds. *Extinction Rates.* Oxford, 1995. Perhaps the most important collection of articles on the topic of extinction, particularly the estimation of extinction rates.

Mace, G. M., and A. Balmford. "Patterns and Processes in Contemporary Mammalian Extinction." In *Priorities for the Conservation of Mammalian Diversity: Has the Panda Had Its Day?,* edited by A. Entwistle and N. Dunstone, pp. 27–52. Cambridge, 2000.

MacPhee, R. D. E., ed. *Extinctions in Near Time: Causes, Contexts, and Consequences.* New York, 1999. Likely to become a classic text on extinctions over the last 100,000 years.

Pimm, S. L., G. J. Russell, J. L. Gittleman, and T. M. Brooks. "The Future of Biodiversity." *Science* 269 (1995): 347–350. Remains a useful overview.

Purvis, A., K. E. Jones, and G. M. Mace. "Extinction." *BioEssays* 22 (2000): 1123–1133.

Raup, D. M. "The Role of Extinction in Evolution." *Proceedings of the National Academy of Sciences USA* 91 (1994): 6758–6763.

Saacheri, I. "Inbreeding and Extinction in a Butterfly Metapopulation." *Nature* 392 (1998): 491–494.

Van Valen, L. M. "A New Evolutionary Law." *Evolutionary Theory* 1 (1973): 1–30.

— KEVIN J. GASTON

EXTRAPAIR COPULATIONS

Extrapair copulation (EPC) is a copulation between a female and a male other than her social partner. Many organisms form long-term pair-bonds for a reproductive cycle or longer. Although females copulate with their social partners, they may also copulate with additional males. In socially monogamous mating systems, but also to some extent in other mating systems, females may be constrained in their choice of mate because late-choosing females will not have access to the pool of males that have already been chosen. This constraint on choice of a sire can be circumvented by female engagement in EPCs. EPCs may be either solicited or resisted by females. The latter forced EPCs are taxonomically less widespread, and they may less frequently result in extrapair paternity.

EPCs are often difficult to observe because of the covert behavior of females seeking EPCs. This makes quantification of EPCs difficult, although their consequences in terms of fertilization can be assessed directly through genetic studies of paternity. The frequency of EPCs estimated as the proportion of females involved may reach 100 percent in some species of socially monogamous species of birds. The relationship between frequency of EPCs and extrapair paternity among species is weakly positive, although extrapair paternity often is much more common than expected from the number of EPCs. This suggests that EPC males may have superior fertilization ability owing to the quality of their ejaculates, the timing of EPCs, or the physiological response of females to EPCs.

EPCs have been recorded in most taxa investigated. Studies of extrapair paternity in more than 200 species of birds have shown variation in frequency from none to 68 percent of the offspring in different species, with a mean frequency of 14 percent. EPCs and extrapair paternity are common in primates. Covert female behavior in some species makes quantification difficult. For example, observations of EPCs with extragroup males in chimpanzees are very rare, but offspring are sired by such males. A behavioral study of humans suggests that perhaps as many as 5–10 percent of all copulations are EPCs, with a peak around the time of ovulation. Within-pair copulations in humans do not show any clear pattern during the menstrual cycle. Studies of extrapair paternity in humans have found 1–30 percent, with a median value of 9 percent, in thirteen different studies in western and traditional societies.

Males of most species attempt to copulate with additional females than their social partners, and such attempts are relatively indiscriminate with respect to females. In contrast, females often actively seek EPCs by soliciting copulations from males with a preferred phenotype or by differentially visiting the territories of such males. Whereas the EPC behavior of males often is overt, female behavior is covert. EPCs often lead to sperm transfer and fertilization, and extensive studies of paternity of several hundred species of animals have revealed that EPC males often fertilize a fraction or even a majority of the eggs of a single female. The function of EPCs for males is fertilization of partners in addition to their social mate(s).

Females often engage in EPCs without gaining direct material benefits from this behavior. However, several studies have demonstrated that the offspring of EPCs have enhanced viability due to superior genetic quality. A direct function of EPCs for females is insurance of fertilization in case of infertility or subfertility of the pair-bond male, perhaps by a male of superior genetic quality. Females of some species may also engage in EPCs to assess potential future mates, to gain material advantages in terms of access to the resources of neighboring breeding territories, or to benefit from the antipredator vigilance of more than a single male. Females of group-living species such as cooperatively breeding birds or multimale groups of mammals may also benefit from copulations with multiple males in terms of paternal care provided by

all males or the absence of infanticidal activity by males with whom a female has copulated.

EPCs and extrapair fertilizations have been shown to be more common when opportunities are frequent. EPCs are more common at high population densities and when neighboring pairs reproduce asynchronously. Females mated to less attractive partners may actively choose such ecological conditions for breeding to facilitate EPCs. Most extrapair fertilizations are between neighbors or individuals in the immediate neighborhood. Extrapair paternity and thus EPCs in birds are much less common in island than in mainland populations of the same species. This difference is associated with a lower degree of genetic variation among individuals on islands, apparently arising from founder effects and inbreeding. Even among mainland populations, those with higher levels of genetic variation have a larger frequency of extrapair paternity.

Males that are successful in acquiring EPCs and extrapair paternity generally differ in phenotype from males that are unsuccessful. For example, the expression of secondary sexual characteristics differs between these two groups of males. In birds, EPC males have brighter coloration, more extravagant plumage, larger song repertoires, or more intense song activity than males that lose paternity. Males engaged in EPCs are sometimes also found to be socially more dominant than the average male in the population.

Males may incur costs of EPCs in terms of lost paternity with their social mate, increased risk of sexually transmitted diseases, and the cost of time and energy of EPC searching behavior. With the exception of the first cost, these costs have not been shown to be important, although more studies are required to reach firm conclusions. Females may incur costs of EPCs in terms of mate loss, reduced paternal care, increased risk of sexually transmitted diseases, injury or death during EPCs, and the cost of time and energy of EPC searching behavior. Some studies have demonstrated that pair-bond males reduce the extent of their parental care when their female has engaged in EPCs. Females of species with forced EPCs run the risk of injury or death, for example, as commonly observed in waterfowl. There is little empirical evidence of costs in terms of sexually transmitted diseases in free-living animals.

Through their effect on fertilization, EPCs increase the variance in male fitness. For example, socially monogamous species generally have little variance in fitness when calculated from the number of offspring produced. When the paternity of the offspring is considered, the variance in male fitness may increase by a factor of twenty or more. Hence, the intensity of sexual selection measured as the standardized variance in fitness is increased considerably, explaining why males of many socially monogamous species have evolved extravagant secondary sexual characteristics. The variance in female fitness may also be increased by EPCs if this increases the quality of offspring, although this remains to be demonstrated.

EPCs increase the intensity of sexual selection acting on males, and males of species of birds with a high frequency of extrapair paternity generally have evolved more extravagant secondary sexual characteristics than males of species with little or no extrapair paternity.

[*See also* Cryptic Female Choice; Mate Choice; Sexual Selection, *article on* Bowerbirds.]

BIBLIOGRAPHY

Birkhead, T. R., and A. P. Møller, eds. *Sperm Competition and Sexual Selection*. London, 1998. A general overview of the current knowledge of the occurrence of extrapair copulations in animals and its functions and ecological and evolutionary consequences.

Møller, A. P., and P. Ninni. "Sperm Competition and Sexual Selection: A Meta-analysis of Paternity Studies of Birds." *Behavioral Ecology and Sociobiology* 43 (1998): 345–358. An analysis of factors accounting for intraspecific variation in paternity in birds.

— ANDERS PAPE MØLLER

F

FERNS. *See* Plants.

FISHER, RONALD AYLMER

Sir Ronald Fisher (1890–1962), the father of modern statistics, was for most of his working life a professor of genetics, first in London and then at Cambridge. He was born in London on 17 February 1890. At Harrow School he distinguished himself in mathematics despite being handicapped by poor eyesight, which prevented him from working by artificial light. His interest in natural history was reflected in the books chosen for special school prizes, culminating in his last year in the choice of the complete works of Charles Darwin.

Fisher entered Gonville and Caius College, Cambridge, as a scholar in 1909, graduating with a bachelor of arts degree in mathematics in 1912. While there, he instigated the formation of the Cambridge University Eugenics Society, through which he met Major Leonard Darwin, Charles's fourth son and president of the Eugenics Education Society of London, who was to become his mentor and friend. In 1919, he was appointed statistician to Rothamsted Experimental Station, and in 1933 he was elected to succeed Karl Pearson as Galton Professor of Eugenics (i.e., of Human Genetics, as it later became known) at University College, London. He moved to Cambridge in 1943 as Arthur Balfour Professor of Genetics. Retiring in 1957, he spent his last years in Adelaide, Australia, where he died on 29 July 1962.

Fisher made profound contributions to applied and theoretical statistics, to genetics, and to evolutionary theory. In his first term as an undergraduate at Cambridge, he had bought William Bateson's book *Mendel's Principles of Heredity*, with its translation of Gregor Mendel's paper. Before graduating, he had already remarked on the surprisingly good fit of Mendel's data, and in 1916, encouraged by Leonard Darwin, he completed the first paper on biometrical genetics and the analysis of variance, *The Correlation between Relatives on the Supposition of Mendelian Inheritance*, eventually published in 1918.

From his post of statistician at Rothamsted, Fisher made advances that revolutionized statistics, but his advances in evolutionary theory were just as epoch making. In a single publication in 1922 he proved that heterozygotic advantage at a locus with two different alleles gives rise to a stable gene-frequency equilibrium, introduced the first stochastic model into genetics (a branching process) to compute the chance of survival of individual genes, and initiated the study of gene-frequency distributions by introducing the chain-binomial model and analyzing it using a diffusion approximation. (This model is often called the Wright-Fisher model, but Sewall Wright's paper did not appear until 1931; Fisher's use of the diffusion approximation constitutes the foundation of stochastic diffusion theory.) In another 1922 paper he applied his new method of maximum likelihood to the estimation of linkage for the first time. Other contemporary papers dealt with variability in nature, the evolution of dominance, and mimicry, and in 1926 he started his long association with E. B. Ford, with whom he later measured the effect of natural selection in wild populations.

In 1930, *The Genetical Theory of Natural Selection* was published, containing a wealth of new evolutionary arguments, from the fundamental theorem of natural selection to novel ideas about sexual selection, kin selection, sympatric speciation, reproductive value, linked loci, and parental expenditure. More than any other work, *The Genetical Theory* established a firm basis for the modern view that evolution by natural selection is primarily a within-species phenomenon. It completed the evolutionary synthesis of Mendelian genetics and Darwinian evolution and opened the way for the neo-Darwinian revolution associated with the names of Ernst Mayr, Theodosius Dobzhansky, and George Gaylord Simpson.

Taking up his appointment at University College in 1933, Fisher's pace did not slacken. In 1935, he secured funds from the Rockefeller Foundation to establish the Blood-Group Serum Unit at the Galton Laboratory with the express purpose of initiating the construction of a linkage map for humans, for he had already seen the connection between "Linkage Studies and the Prognosis of Hereditary Ailments" (the title of his lecture to the International Congress on Life Assurance Medicine in that year). When he moved to Cambridge in 1943, he was reunited with his colleagues from the Blood-Group Unit, who had been evacuated there during the war. An immediate consequence was his brilliant solution of the *Rhesus* blood-group puzzle, involving three closely linked loci that between them explained the array of blood-group (serological) reactions that to everyone else had appeared chaotic.

Fisher was one of the great intellects of the twentieth century. In statistics his accomplishments rank with those of Karl Friedrich Gauss and Pierre Simon Laplace, and in biology he has been described as the greatest of Charles Darwin's successors.

[*See also* Eugenics; Fundamental Theorem of Natural Selection; Quantitative Genetics; Reproductive Value; Sex Ratios.]

BIBLIOGRAPHY

Box, J. F. *R. A. Fisher: The Life of a Scientist.* New York, 1978.
Edwards, A. W. F. "R. A. Fisher: Twice Professor of Genetics: London and Cambridge *or* 'A fairly well-known geneticist.'" *Biometrics* 46 (1990): 897–904.
Edwards, A. W. F. "The Genetical Theory of Natural Selection." *Genetics* 154 (2000): 1419–1426.

— A. W. F. EDWARDS

FISHES

Fossils tell us that fishes have lived and evolved on the earth for more than 500 million years. It is not surprising, then, to find almost 25,000 living species—more species than all other vertebrate groups combined—nor to learn that living fishes comprise a small fraction of the mostly extinct total diversity of fishes over time.

In terms of evolutionary history, organisms commonly referred to as fishes comprise an unnatural, incomplete group, so we must first carefully define our subject. Most modern scientists concur that named groups should include all descendants of the immediate common ancestor of the included organisms (or, in more technical terms, should be *monophyletic*). Most agree also that the most primitive living fishes are hagfishes (Myxiniformes), and that "primitive" fishes include lampreys, sharks, and rays. It is also commonly held that all of these share an ancestor with some protochordate invertebrate, probably a urochordate or cephalochordate. Moving up the evolutionary tree through the trunk that links hagfishes, lampreys, sharks, and rays together brings us to the divergence of two large branches. The Actinopterygii or ray-finned fishes is a very large (23,681 living species) branch including sturgeons, eels, minnows, catfishes, salmons, sardines, cods, basses, perches, tunas, flounders, and many more. The other branch, Sarcopterygii, includes not only fossil and still-living groups that everyone agrees are fishes, like coelacanths and lungfishes, but also all of the approximately 23,500 living tetrapods (amphibians, mammals, reptiles, and birds). Technically, therefore, if fishes are to include hagfishes, lampreys, sharks, and rays, and all of the ray-finned fishes, as it does in common usage, we humans, as well as all other tetrapods, should also be considered fishes because tetrapods share an ancestor with fishes and are nested within (and therefore part of) fishes. In other words, the term *fishes* refers to a *paraphyletic* group because it does not include all descendants of its common ancestor. This essay will therefore discuss the evolution of aquatic vertebrates with gills and fins commonly called fishes, or, more specifically, all descendants of an ancestor shared with the most primitive fishes, the Myxiniformes (hagfishes), except the tetrapods. It thus covers about 24,600 living species in 482 families and 57 orders (Nelson, 1994), as well as probably even greater, but poorly known extinct diversity.

The Earliest Fishes. Considerable debate surrounds relationships of the earliest fishes, now mostly extinct. The superclass Agnatha, or jawless fishes, into which all fossil jawless groups once were classified, is now generally recognized to be paraphyletic, and major lineages may have been independently derived from an acraniate ancestor, but this category name is used here for convenience. Each major lineage appears early in the fossil record (470–500 million years ago [mya]), characterized by lack of jaws derived from gill arches, lack of pelvic fins, and one or two semicircular canals in the inner ear. Fossil hagfishes (Myxini, 43 living species), though generally considered the most primitive Agnathans, enigmatically appear relatively late (300 mya). The living species are superficially eel-like, marine bottom scavengers up to 1 meter or more long; they lack dorsal fins and have degenerate eyes and a sucking mouth with teeth on the tongue, and 1–16 external gill openings. Hagfishes are famous for secreting copious, stringy mucus and knotting their long bodies. The only other living group of Agnathans, the distantly related lampreys (Petromyzontiformes, 41 extant species), are freshwater and anadromous (living at sea, spawning in freshwater), and also eel-like and superficially resembling hagfishes in many ways. These similarities are now recognized to be convergent, and the two are distantly related. Lampreys have one or two dorsal fins, well-developed eyes in adults, no barbels, teeth on an oral disc and on the tongue, and seven pairs of external gill openings. Some are parasitic.

Though it is by no means clear, lampreys may form the sister group to the jawed fishes (Gnathostomata); if true, between lampreys and hagfishes lie a number of extinct, diverse, and bizarre groups of other marine and freshwater jawless fishes extinct before 350 million years ago. Pteraspidimorphs were marine and freshwater, jawless fishes, which, in contrast to the naked hagfishes and lampreys, had heavy dermal armor or head plates covering much of their bodies, which sometimes exceeded a meter long. Smaller (10–20 cm), but also armored, though with dermal denticles instead of plates, some Theleodontiforms resembled modern minnows in body form. Appearing a bit later, during the upper Silurian (420 mya), the Cephalaspidomorphs diversified in at least three major lineages and 30 genera, many very speciose, before disappearing by the end of Devonian (355 mya). Like Pteraspidimorphs, many Cephalaspidomorphs had a broad head shield, but this included an

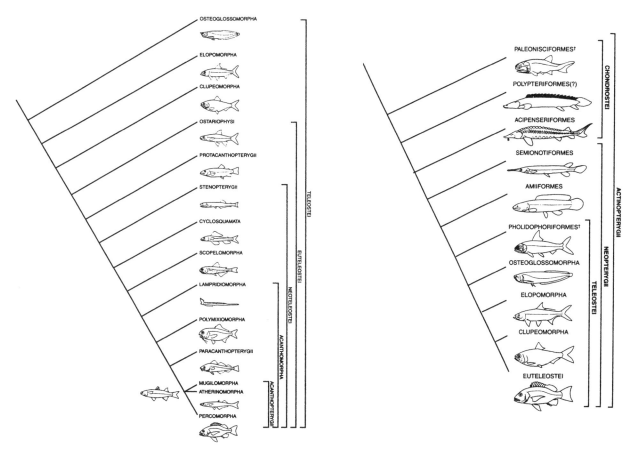

FIGURE 1. Hypothesized Relationships Among Major Groups of Fishes.
(† indicates extinct groups).   Helfman, Collette, and Facey. *The Diversity of Fishes*. Blackwell Science, 1997.

obvious organ that possibly had mechanoreceptive or electrogenerative or receptive functions. With abundant fossils, Cephalaspidomorphs are well studied, and details of cranial morphology clearly suggest a close relationship with the otherwise very different lampreys, making the 50-million-year gap between the two in the fossil record enigmatic. Cellular bone and an ossified endoskeleton first appeared in Cephalaspidomorphs, as did lateral, paired finlike appendages that undoubtedly increased maneuverability. Other Cephalaspidimorph lineages, the Anaspids and Galeaspids, roughly contemporaneous with the Cephalaspidiformes and Theleodonts, were also well armored. These groups appear to have been strictly marine during their early histories and later invaded freshwater.

Jaws, derived from gill arches, endochondral bone, paired limbs (fins), and three semicircular canals characterize all other vertebrates. Jaws were clearly a major innovation that allowed fishes to invade new habitats and utilize new food resources, but the relationship between known jawless and jawed groups remains unclear. Similarly puzzling is the fact that all major groups of jawed fishes, collectively the Gnathostomes, appear, all well advanced, almost simultaneously in the mid to late Silurian to early Devonian (400–425 mya).

Placoderms, generally considered the most primitive Gnathostomes, rapidly diversified in the Devonian but were relatively depauperate in Carboniferous, after which they disappeared, leaving no trace of direct descendants. Bony ornamented plates adorned the skull and anterior trunk of most, and they had a unique tooth structure and upper jaw attachment that allowed the upper jaw to be lifted. They appear first only in marine habitats and later invaded freshwater.

Chondrichthyes. Chondrichthyes (sharks, rays, and chimaeras or Holocephali—846 living species) appeared by the middle Devonian. Many major radiations of this group of primarily marine predators were extinct by the Triassic, and only two, Elasmobranchs and Holocephalans, have living representatives. Chondrichthyes is characterized by cartilaginous skeletons, jaw mechanisms that differ from bony fishes, placoid scales, serially replaced teeth, and multiple external gill slits in most, among many other characters. Though their cartilaginous skeletons left a poor fossil record, throughout their history a trend is seen toward improved feeding and locomotion. The world's largest fish, the whale shark, is an Elasmobranch, and some fossil species were far larger. Many Elasmobranchs are highly advanced midwater predators, including some that elevate their body temperature to improve swimming. Fertilization is internal, and advanced forms of embryonic nutrition provision, analogous to that of mammals, exist in many. All have direct development without a larval stage. Some have invaded freshwater. Holocephalans, or ratfishes or rabbitfishes, are relatively little studied marine oddities, clearly closely related to sharks and rays but differing primarily in the upper jaw being fused to the skull, and in having platelike, crushing teeth that grow continuously but are not replaced, as well as a very different body form. Most inhabit deep waters. The living 58 species are few in proportion to the far more diverse extinct forms.

Appearing in the early Silurian and persisting to the early Permian (430–250 mya), and thus living with Placoderms and early Elasmobranchs, as well as many jawless fishes at first, were the Acanthodii, or spiny sharks. The resemblance of this diverse group to sharks is superficial, and the two groups are distantly related. Acanthodians were mostly midwater feeders, some planktivorous, in both marine and fresh waters. Many characters support the hypothesis that this is the sister group to (sharing a common, immediate ancestor with) the bony fishes.

Osteichthyes. Fragmentary early Silurian fossils are believed to represent "bony fishes" (also called Osteichthyes), but two unambiguous major lineages are well represented in the mid-Devonian (390 mya): the now incredibly diverse Actinopterygii, and the Sarcopterygii. Living Sarcopterygians, outside of tetrapods, are few, but a famous "living fossil," a coelacanth, known previously only from Cretaceous (>65 mya) and earlier fossils (to Carboniferous, 355 mya), was discovered living off the coast of Southeast Africa in 1938, and a second extant species was discovered in Indonesia in 1997. Both are little changed from those in the fossil record of 200 mya, providing a rare and much publicized opportunity to study primitive relatives of the tetrapods. The Coelacanthamorpha form the sister of the remaining Sarcopterygians, with Dipnoi, the lungfishes, forming the sister of the extinct Osteolepimorpha (also called Rhipidistians). The Osteolepimorpha range from Devonian to late Paleozoic (410–250 mya). Abundant and excellent fossils clearly indicate it to be the sister group to tetrapods, sharing many traits with Labyrinthodont amphibians. Lungfishes, originally misidentified as amphibians, are now known from six living species in two families, both known from far back in the fossil record—Ceratodontidae as early as the early Triassic (240 mya). Fossil lungfishes include 10 families (>100 species) and many early marine forms, though extant species are all freshwater. Lungfishes are well studied as a result of their relationship to tetrapods, their extensive fossil record, and the availability of living material. All have lungs and are aerial breathers as adults, but the larvae have functional external gills.

Actinopterygii contains the vast majority of living fishes (23,681 species). The origin of this group is not clear, but fragmentary fossils appear in late Silurian and complete specimens in Mid- to Late Devonian, by which the group was diverse in both morphology and habitat (marine and freshwater). Actinopterygians share their common ancestor's characteristic of fins supported by rays derived from scales and moved by body musculature, whereas Sarcopterygians have fins with a bony central axis moved by musculature in the fin itself. The Actinopterygii diversified as the Agnathans, Acanthodians and Placoderms diminished, suggesting ecological interactions as a causal factor.

Actinopterygian evolution is characterized by trends of multiple innovations in the jaws that improved and diversified feeding abilities, and by changes in body and fin morphology that improved swimming. Geometry of the skull and jaw were altered to increase bite speed, strength, and dexterity, as well as to produce suction. Scales lightened and became more flexible. The caudal (tail) fin evolved from the heterocercal condition (unequal lobes) of primitive fishes to homocercal, with advanced underlying structural modifications to increase thrust. The skeleton lightened and sophisticated buoyancy control developed; these combined with advanced fin anatomy and repositioning to allow delicate and precise movements. Though these are general trends expressed throughout the group, there are notable exceptions to the trends. There are also cases of remarkable, extreme specialization in other anatomical and physiological systems, notably auditory and mechano- and electroreception, not to mention olfaction and vision, and even light production.

The extreme Actinopterygian diversity defies comprehensive discussion here, so only highlights from selected groups will be presented. The general references provide resources to help the reader more thoroughly explore this remarkable group.

Diverse by the time the group appears in the fossil record in the Mid- to Late Devonian through late Paleozoic, Paleonisciformes are an interesting but extinct group of primitive Actinopterygians. Considered sister to all other Actinopterygians, this group mirrors virtually all of the general evolutionary trends mentioned above as seen across all other Acanthopterygians. The range of body plans parallels that of modern teleosts. Though supporting evidence is controversial, some suggest that the living Polypteriformes (bichirs and reed fishes) and sturgeons (Acipenseriformes) may represent divergent, remnant Paleonisciformes.

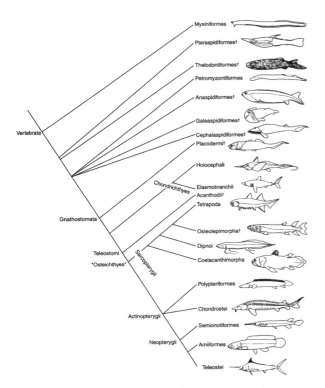

FIGURE 2. Hypothesized Relationships Among Major Groups of Living Teleosts.

Helfman, Collette, and Facey. *The Diversity of Fishes*. Blackwell Science, 1997.

Later radiations of Actinopterygians occurred in the Triassic (250–205 mya) and Jurassic (205–135 mya) and again in the Cretaceous (135–65 mya). Remnants of the five orders represented in the earlier two of these radiations persist today: the Semiontoform gars and Amiiform bowfin. Well-known fossil lineages, including other members of the morphologically diverse Semiontoformes, and three other less diverse orders are extinct. The more recent radiations of ray-finned fishes involved the teleosts, a monophyletic group represented by more than 23,637 living species (or more than 96 percent of all living fishes) that today occupy nearly every conceivable aquatic habitat on the earth. Teleosts first appear in the Triassic and are characterized by internal skeletal stiffening of the upper caudal fin lobe, a mobile premaxilla (paired front, upper jawbone), and unpaired basibranchial tooth plates, among other characters. Each of the three earliest teleost radiations is thought to have produced one of the three major monophyletic groups (superorders): the Osteoglossomorpha (bonytongues, mooneyes, mormyrids, and knifefishes), Elopomorpha (eels, tarpon, bonefish, and relatives) and Clupeomorpha (shads, herrings, sardines, and anchovies). The fourth major radiation, in the Cretaceous, capitalized on

locomotory advances and innovations in jaw mechanics to produce the majority of today's extremely diverse Euteleostean fishes (discussed below).

Osteoglossomorpha, more diverse in the fossil record than at present, is a freshwater group that is perhaps most interesting for the innovation of reception and production of complex patterns of weak electrical fields in the family Mormyridae. This African group developed complex communications and social systems based on varied pulses and patterns of electrical discharges; it diversified into about 90 percent of the extant 217 Osteoglossomorph species. Mormyrids are famous for their learning ability and a relative brain size comparable to that of humans.

Elopomorphs are remarkable for their unusual leptocephalous larva, a large, ribbon or leaf-shaped, relatively long-lived planktonic stage that reaches as much as 2 meters in length and looks nothing like the adult into which it metamorphoses. The eels are remarkably diverse, with many adept crevice dwellers and burrowers. The very few species that enter freshwater are catadromous, reproducing at sea and migrating great distances from rivers and small streams of North America and Europe, where they mature, to travel to reproductive areas at great depths (2,000 m) in the Sargasso Sea. Some Elopomorphs have diversified extensively in deep sea habitats, producing bizarre, light-producing predatory forms with huge mouths and diminutive bodies.

Clupeomorphs, with a unique connection between the swim bladder and inner ear that provides acute hearing and mechanoreception, include some of the world's most important commercial food fishes. Typically schooling, marine fishes (with some freshwater representatives) that strain plankton from midwater habitats, groups such as the sardines, anchovies, and herrings are directly exploited on a massive scale and also provide forage to support commercially important predators.

The Chondrostei, with its modern sturgeons and paddlefishes, probably arose in the Jurassic, since its earliest fossils are from Early Cretaceous. They are characterized by a secondarily derived cartilaginous skeleton, heterocercal tail (as in sharks), reduced scales, and spiral valve intestine, among other traits. Their fossil relatives, the Paleoniscoids, had bony skeletons. Few species remain, all restricted to large rivers of the Northern Hemisphere. Though some enter marine habitats, all depend on freshwater habitats for reproduction.

The very large, morphologically diverse, and successful Euteleostei, sister group to the Chondrostei, includes many fairly well defined and probably monophyletic major groups (superorders), but intergroup relationships, as well as the monophyly of the Euteleosts as a whole, remain controversial. Very general hypotheses of relationships among them will be discussed after basic accounts of salient features of the biology of each.

An early diversification of Euteleosts produced the highly successful Ostariophysi, a predominantly freshwater group containing the well-known minnows, suckers, catfishes, and characids and their relatives. Many traits uniquely characterize this large group (>6,500 species, or 27 percent of all living species), but those thought particularly important in their success are a gas bladder-ear connection that allows them to hear in a wide range of frequencies, and a fright pheromone (odor) and response repertoire, released upon injury of an individual that is often species-specific. The minnow family, technically Cyprinidae (in Cypriniformes), is the largest living family of freshwater fishes with 2,662 species. Catfishes (Siluriformes) and Characiformes are also diverse, with 2,405 and 1,343 species, respectively. These orders all have nearly global distributions, but Characiformes seem to replace Cypriniformes in South America. Gymnotiformes, clearly sharing an ancestor with catfishes, have independently developed keen electrosensory and electrogenerative capabilities and applications very similar to those of the Mormyrids, which they superficially resemble in body form.

The Protacanthopterygii, including Esociformes (pikes), Osmeriformes (smelts and diverse relatives), and Salmoniformes (salmons, trouts, and relatives), is generally considered one of the more primitive Euteleost lineages. Esociforms are a small freshwater group with circumarctic distribution. The 236 species in Osmeriformes are mostly marine, but best known are the 70 or so species that are anadromous (spawn in freshwater) or landlocked, including the Galaxioids of the Southern Hemisphere. Many bizarre Osmeriformes are important and diverse in deep-sea habitats. Salmoniformes includes the economically very important salmons and trouts and their relatives, charrs, whitefishes and ciscoes, all freshwater and anadromous fishes of the Northern Hemisphere.

Several exclusively marine, mostly highly specialized deep-sea groups are next in the phylogenetic tree of fishes. Stenopterygii consists of about 330 species in two marine orders. Most all are deep-sea (200–4,000 m) forms, many producing light. Some are the most abundant species in that little-known habitat, and they may be the most abundant and widely distributed vertebrates. Cyclosquamata (219 species) includes some of the most specialized and bizarre deep sea fishes, with tubular eyes, huge mouths, and greatly reduced skulls and body. Hermaphrodites are common in one family, introducing a trend toward sexual lability that continues through the higher teleosts. Scopelomorphs and all remaining teleosts share loss of the fifth pharyngeal tooth plate and associated musculature. The lanternfishes (Myctophiformes), with species-specific patterns of light-producing organs, and some migrating 200 vertical meters or more every day, comprise nearly all of this superorder.

True fin spines differentiate all remaining teleosts from those previously discussed, in which all fin rays are segmented though sometimes hardened into spine-like structures. This innovation characterizes the Acanthomorpha, members of which also share advances in skeletal and jaw anatomy that improve swimming and feeding abilities. Lampridomorphs and Polymixiomorphs are small groups that mix characters of Acanthomorphs and lower groups, causing their relationships to remain enigmatic. The Paracanthopterygians, literally a side branch (sister) of the Acanthopterygii (below), is relatively diverse in marine habitats, especially the deep sea, and has some freshwater representatives. In this group are the Gadiformes, or cods and hakes and their relatives, which support some of the world's most important fisheries, and a number of eel-like fishes that include the depth record holder (8,370 m). Also here are the well-lighted and highly vocal toadfishes (Batrachoidiformes) as well as unusual goosefishes, frogfishes, anglerfishes, and batfishes, with many bizarre, bottom-dwelling, and highly adapted deep-sea forms. The trend toward increased variations on sexuality is exemplified here by diminutive males in deep-sea anglerfishes that are obligate parasites on the females.

Remaining teleosts or Acanthomorphs belong to the Acanthopterygii, characterized by still more advanced modifications of the upper oral jaw to allow it to slide forward and achieve much greater, more rapid extension, and developments in the more posterior pharyngeal jaws that allow a greater diversity of foods. Presence of a pelvic fin spine also ties this group together. The connection between the gas bladder and gut is lost in these fishes (making them physoclistous as compared to physostomous in earlier groups), with concomitant repercussions on buoyancy control. Many other characters further align these fishes into a monophyletic assemblage, but again, details of relationships within this massive group remain controversial.

Recent treatments recognize a close relationship between the Mugilomorpha (mullets) and Atherinomorpha (silversides, rainbowfishes, topminnows, livebearers, pupfishes, and their relatives). The former contains about 45 common nearshore and catadromous detritivores (scavengers), some of considerable economic importance, and the latter contains highly successful, small marine and freshwater surface fishes. Some Atherinomorphs, especially those in Atheriniformes that bear live young, are well known for extremely specialized sexual dimorphism and variations in reproductive strategies, culminating in unisexual "species" that reproduce by mechanisms ranging from simple parthenogenesis to parasitism of males of other species. In the latter, the parasitized male's genome may (hybridogenesis) or may not (gynogenesis) be included in the offspring. In hybridogenesis, the male's contribution is dropped and a new one incorporated at each generation. Atherinomorpha also includes the flying fishes, with greatly expanded pectoral fins that allow them to glide great distances out of water, and pupfishes, famous for adaptation to extreme salinity and temperature in scarce desert waters. Others in the group have adapted to drought by developing desiccation-resistant eggs.

Finally, we come to the Percomorpha, where nearly half of the total diversity of living fishes resides. The group is characterized by forward displacement of the pelvic girdle and its attachment to the pectoral girdle, which is displaced upward, as well as other characters. Monophyly of this group was recently challenged, and an alternative hypothesis (Johnson and Patterson, 1993) is gaining support. The alternative proposes a monophyletic Percomorpha that consists of a new group, the Smegmamorpha, as sister to Perciformes + Tetraodontiformes + Pleuronectiformes + Scorpaeniformes (with no resolution of relationships among these). Smegmamorpha includes the Mugilomorphs and Atherinomorphs (from outside the classically defined Percomorpha), and three of the groups classically placed in the Percomorpha—Gasterosteiformes, Synbranchiformes, and Elassomatids. Once again, however, there is no resolution of relationships within Smegmamorpha. Stephanoberycids, Beryciformes, and Zeiformes of the classical Percomorpha thus fall outside the new Percomorpha as sequential sister groups to it.

Stephanoberyciformes and Beryciformes are, respectively, deep-sea and nocturnal groups, some with light production and other adaptations to total darkness. Some are abundant but cryptic nocturnal reef fishes with large eyes and red coloration. Zeiformes, another mostly deep-sea group, presents a number of confusing characters, so much debate surrounds its relationships. One treatment places it as sister to the Tetraodontiformes (puffer fishes and relatives).

Gasterosteiformes includes the highly divergent and well-known sea horses and sticklebacks, as well as lesser known elongate fishes with tubular snouts, like pipefishes, coronet fishes, trumpet fishes, and seamoths. Apparently closely related to them is the Synbranchiformes, with two families of air-breathing, burrowing swamp fishes with elongate, eel-like bodies.

The position of the 1,271 Scorpaeniformes (scorpionfishes, sculpins, searobins, lumpfishes, and snailfishes) has been enigmatic for some time. Monophyly is probably indicated by the characteristic spine below the eye, which gives them the common moniker "mail-cheeked" fishes. Most are nearshore marine bottom-dwellers, especially common in rocky, wave-swept habitats. Many have spines endowed with highly toxic venom, and they are often brightly colored. About 52 species are restricted to freshwater.

Perciformes, as treated here, is the largest of all vertebrate orders. Its 9,293 species dominate marine and many freshwater habitats. The group has diversified relatively recently in geologic time: confirmed Perciform fossils are all post-Cretaceous. It should come as no surprise that relationships of major groups within Perciformes are hotly debated. The classical composition of the group (Nelson, 1984) is followed here, recognizing that it almost certainly does not represent a monophyletic group. Much more study is required before a stable classification of this group is attained. Despite the importance of this group, space constraints force a very brief treatment that results in the exclusion of some families. Discussion is limited to selected suborders, which are likely mostly monophyletic, but we recognize that many may prove to be unnatural groupings.

The suborder Percoidei (2,860 species) includes Centropomidae (snooks), Moronidae (temperate basses), Perchichthyidae (temperate perches), Sciaenids (croakers) and Gerreids (mojarras), which include anadromous and landlocked freshwater species. Exclusively freshwater families include the economically important North American Centrarchidae, with 29 species of sunfishes and basses, and the Percidae (162 species). The diverse North American Percids called darters (>145 species) are well-studied, colorful occupants of streams and rivers that display complex territoriality, courtship, and parental guarding of eggs. The remaining groups are marine. The 450 Serranids, all with three cheek spines, range from small, colorful reef fishes to massive predators like the jewfish (400 kg). Many are sequential hermaphrodites—with individuals born with both ovaries and testes and functioning first as one sex and later in life as the other. Social cues determine the timing of the switch. Lutjanids (snappers, 125 species) are near-bottom predators important as sport and food fishes, as are Sciaenids (drums), named for their prolific gas-bladder-amplified sound production, especially during courtship. Butterfly fishes (Chaetodontidae, 114 species) and Angel fishes (Pomacanthidae) are brightly colored, conspicuous coral-reef fishes formerly placed in the same family. Chaetodontids have elongate snouts for snipping coral polyps from their surrounding matrix. Color patterns of angelfishes change during an individual's lifetime, with juveniles looking little like adults.

The genus *Elassoma*, the pygmy sunfishes of North America, was recently moved from Centrarchidae to its own new suborder, Elassomatidae, on the basis of morphological evidence that indicated it was clearly not a Centrarchid. Though its relationships remain unclear, reproductive behavior and other evidence support a possible relationship to Gasterosteiformes, and so it is now placed by some in the Smegmamorpha.

Labroidei is an extremely successful and interesting group characterized by unique pharyngeal morphology. As treated here, it consists of six families. Cichlidae is a famously diverse freshwater group (>1,300 species) popular with aquarists for bright colors and interesting behavior, including extensive parental care. The group is especially diverse in Africa, where large "species flocks" in the rift valley lakes diversified very rapidly and extensively from a common ancestor. It is not surprising that sex changes have recently been reported in Cichlids, since the closely related Labridae (>500 species, commonly called wrasses) are noted for this interesting behavior. Within single species of Labrids, many color patterns may exist, and individuals express them as a function of social context, which also triggers sex changes. Scarids (parrotfishes) are large, beautifully colorful coral-reef fishes, specialized for feeding on dead corals. Socially mediated sex change is also common here, as is changing color pattern as a function of age and social status.

Nototheniodei includes the ice fishes, which evolved an antifreeze that allows them to exploit supercooled brines under antarctic ice. Some lack red blood cells, making them nearly colorless.

Trachinoidei consists mostly of marine benthic fishes, many burrowing in sandy bottoms. Related to this, many have upward-oriented eyes that remain exposed while the body is buried. One, the stargazer, produces strong electrical discharges for defense and feeding.

Blennioidei is a diverse shallow marine group, often very colorful, with fascinating, complex courtship rituals and parental care. Most have odd filaments adorning their heads. Some (sabertooth blennies) mimic the color and shape of harmless cleaner wrasses to surprise unsuspecting victims by ripping off pieces of fin, some injecting a toxin. Gobioidei consists of small, mostly marine, benthic fishes, characterized by fusion of the pelvic fins into a sucking disk. Gobiidae, the largest family of marine fishes in the world, includes some of the tiniest fishes (mature at 8–10 mm), as well as some of the most beautifully colorful ones, with fascinating reproductive behaviors.

Acanthuroids include the surgeonfishes, with razor-sharp keels on the base of the tailfin that are thought to reduce turbulence.

Scomboidei includes the economically important tunas and billfishes and their relatives, barracudas and mackerels. These streamlined, open-water, marine, long-distance, fast swimmers are extraordinarily specialized. Three different ways to maintain elevated body temperatures have evolved independently in this predaceous group, surely improving their ability to outrace prey. They also demonstrate extreme streamlining, with fins that collapse into grooves in the body, eye covers, and other advances to reduce turbulence.

Anabantoids are mostly freshwater marsh occupants, with unique adaptations of the gills for breathing aerial oxygen in low-oxygen environments. They also have interesting parental care, many building nests of bubbles in which eggs are guarded.

Finally, we reach the crowning orders of fish phylogeny, Pleuronectiformes and Tetraodontiformes. Pleuronectiformes—the flounders, soles, and their relatives generally called flatfishes—are highly specialized marine bottom-dwellers (with a few invading freshwater) with fascinating life histories. Adults are asymmetrical and lie on their sides on the bottom, looking up through eyes that have rotated earlier in life to one side or the other. Larvae are normal, unremarkable, symmetrical fishes living in open water, but their heads are literally twisted as they grow. In some families, the left eye migrates (right-eyed) and in others the right eye moves (left-eyed). A few families have both left- and right-eyed groups or individuals. They are mostly sit-and-wait predators, often altering the coloration of their upper side to match the surrounding bottom.

Tetraodontiformes—the pufferfishes, triggerfishes, and ocean sunfishes and their relatives—are named for their characteristic pattern of four outer jaw teeth, which are actually not teeth but modified jawbones; some groups are characterized by fewer "teeth." Skull and skeletal bones are greatly reduced. Many are spiny or leathery-skinned, boxy fishes that swim by flipping their medial fins back and forth while hardly moving other fins. They specialize in eating sponges, corals, urchins, jellyfishes, and other well-protected organisms that other fishes cannot handle. Some are famous for tetraodontotoxin, one of the most potent toxins known; their meat, served as "fugu," is a delicacy requiring very careful preparation so as not to be lethal. The ocean sunfishes comprise an interesting family (Molidae) of three highly specialized huge, midwater species, rectangular in profile, which have largely cartilaginous skeletons and high water content. They lack a tail fin, but the dorsal and anal fins merge where it normally would be. Their two-meter bodies can attain weights of 1,000 kilograms or more, feeding primarily on jellyfishes as they float lazily in mid-ocean, sometimes lying on their sides at the surface on sunny days.

[See also Vertebrates.]

BIBLIOGRAPHY

Barlow, G. W. The Cichlid Fishes: Nature's Grand Experiment in Evolution. Cambridge, Mass., 2000. An enjoyable book on the evolution and behavior of this interesting and diverse family of rapidly evolving fishes. Packed with information, but written for a popular audience.

Bond, C. E. Biology of Fishes. Fort Worth, Tex., 1996. A classical ichthyology or fish biology text, written primarily for college students. Taxonomy and relationships hypotheses differ somewhat from this contribution.

Helfman, G. S., B. B. Collette, and D. E. Facey. The Diversity of Fishes. Malden, Mass., 1997. A recent ichthyology or fish biology textbook.

Johnson, G. D., and C. Patterson. "Percomorph Phylogeny: A Survey of Acanthomorphs and a New Proposal." *Bulletin of Marine Science* 52 (1993): 554–626. An important, relatively recent contribution in which the group Smegmamorpha is first proposed, along with several other major rearrangements of hypotheses of relationships of Percomorph fishes.

Johnson, G. D., and W. D. Anderson, eds. Proceedings of the symposium on phylogeny of Percomorpha, June 15–17, 1990, held in Charleston, South Carolina at the 70th annual meetings of the American Society of Ichthyologists and Herpetologists. *Bulletin of Marine Science* 52 (1993). A compilation of technical papers on systematics and evolution of a diversity of Percomorph and related groups.

Kocher, T. D., and C. A. Stepien. *Molecular Systematics of Fishes.* San Diego, 1997. A recent compilation of technical papers on the application of DNA studies to interpret evolutionary relationships of diverse groups of fishes.

Long, J. A. *The Rise of Fishes: 500 Million Years of Evolution.* Baltimore, 1995. A nicely illustrated book on extinct and fossil fishes, aimed at a popular audience.

Moyle, P. B., and J. J. Cech, Jr. *Fishes—An Introduction to Ichthyology.* Englewood Cliffs, N.J., 1988. Another classical ichthyology textbook. Taxonomy and relationships hypotheses differ somewhat from this contribution.

Nelson, J. S. *Fishes of the World.* New York, 1994. A thorough summary covering the entire diversity of fishes, with succinct summary information about each group at the family or higher level. Names and relationships differ slightly from those indicated in this contribution.

Paxton, J. R., and W. N. Eschmeyer. *Encyclopedia of Fishes.* San Diego, 1998. A beautifully illustrated compilation of 38 short, informative essays by authorities on their favorite selected groups (mostly orders) of fishes. Includes an introduction to fishes and essays on classification of fishes, fossil fishes, habitats and adaptation, behavior, and endangered species. The information is accurate, illustrations superb, and essays written for a lay audience.

— Dean A. Hendrickson

FITNESS

Evolution by natural selection requires individuals within a population to differ with respect to heritable traits associated with differences in fitness, that is, to differences in the genetic contributions of individuals to future generations. Neither Charles Darwin nor Alfred Russel Wallace used the term *fitness* in their original publication on natural selection (Darwin and Wallace, 1977); the catchphrase "survival of the fittest" was introduced later by the philosopher Herbert Spencer. The development of population genetics in the twentieth century, in which the rules of Mendelian inheritance are included in mathematical models of populations subject to various evolutionary forces, has allowed the formulation of precise definitions of fitness for a range of different genetic and ecological situations (Crow and Kimura, 1970; Hartl and Clark, 1997). Evolutionary ecologists have made use of these definitions to make predictions about the outcome of natural selection on many

traits of interest to naturalists (Bulmer, 1994; Frank, 1998). Studies of laboratory and natural populations have provided information on the relation between such traits and fitness (Endler, 1986). For these reasons, the concept of fitness plays a central role in modern evolutionary biology.

Determinants of Fitness in Simple Systems.

Viability as a component of fitness. To obtain a feeling for the modern use of the term *fitness*, it is convenient to start with the simplest type of situation, that of an organism that reproduces by discrete generations, such as an annually breeding plant whose seeds are produced at the end of the summer, germinate at the beginning of spring, and mature into flowering adult plants in the following summer. Suppose that there are two alternative forms of plant, lead-tolerant and lead-intolerant, respectively. The environment is contaminated with lead, which causes increased mortality of the intolerant plants compared with the tolerant plants (Jain and Bradshaw, 1966).

The formulas in Table 1 show that, if we started with 50,000 plants of each type (n_1 and n_2), and their respective survival probabilities (v_1 and v_2) were 0.004 and 0.002, the expected numbers of survivors would be 200 tolerant plants and 100 intolerant plants; that is, their frequencies would have changed from one-half to two-thirds and one-third, respectively. If the environment were less harsh, so that ten times as many of each type

TABLE 1. Viability Selection on Lead Tolerance in a Plant Population

Form of plant	Tolerant	Intolerant
Numbers of seeds	n_1 (50,000)	n_2 (50,000)
Frequencies among seeds	$n_1/(n_1 + n_2)$ (0.5)	$n_2/(n_1 + n_2)$ (0.5)
Probabilities of surviving to maturity	v_1 (0.04)	v_2 (0.02)
Numbers of survivors	n_1v_1 (2,000)	n_2v_2 (1,000)
Frequencies among survivors	$n_1v_1/(n_1v_1 + n_2v_2)$ (0.667)	$n_2v_2/(n_1v_1 + n_2v_2)$ (0.333)

Assume that tolerance is controlled by a single gene, with tolerance dominant to intolerance, and that the plant population is randomly mating. The new frequencies of the tolerant and intolerant forms among the seeds in the next generation can be shown to be 0.629 and 0.371, respectively; that is, the tolerant form increases in frequency over one complete generation by a factor of 1.25.

If tolerance is recessive to intolerance, the new frequencies of the tolerant and intolerant forms would be 0.728 and 0.272, respectively; that is, the tolerant form increases in frequency by a factor of 1.46.

survive but the relative survival probabilities are the same (survival probabilities of 0.04 and 0.02, as shown in Table 1), the numbers of survivors would be increased by a factor of 10, but the frequencies of the two types among the survivors would be the same.

Relative fitnesses and changes in frequencies. This simple example brings out two points. First, only the relative values of the survival probabilities v_1 and v_2 of the two forms are relevant to their final frequencies among the survivors. Second, the magnitude of the change in the frequencies of the two forms within a generation is determined by the magnitude of the difference in these relative values; for example, if v_1/v_2 were only 1.1 instead of 2.0, the frequency of tolerants among the survivors would be 52 percent instead of 67 percent.

To determine the new frequencies of the two forms among the next generation, which is necessary to describe the course of evolutionary change, we would also need to know their mode of inheritance, that is, which of the alleles is dominant and which is recessive (see note to Table 1). The relative fitness values can only tell us the direction of change across generations, not its rate (Crow and Kimura, 1970; Hartl and Clark, 1997). The details of the mode of inheritance can also be crucial in determining the outcome of evolution, for example, whether one type replaces the other, or whether both can coexist in a stable equilibrium.

Female fertility as a component of fitness. Survival is only one component of fitness. The lead-tolerant form might, for example, produce only three-quarters as much seed as the intolerant form, because it has an alteration in the transport system across the cell membrane that impairs cell function and so reduces fertility. If the plant in question is self-fertilizing, seed output is the only relevant aspect of reproductive success, and so the ratio of the net contributions (w) to the next generation of the tolerant and intolerant types is $w_1/w_2 = 0.75\, v_1/v_2$. This provides a measure of the net fitness of the tolerant type relative to the intolerant type.

Trade-offs of this kind between different components of fitness have been described in a large number of cases and are important for our understanding of the evolution of many important traits (Stearns, 1992). It is probably rare for there to be no cost to a change in the organism that confers a benefit in a new environment, such as the presence of metal pollution, and metal-tolerant plants are indeed at a fitness disadvantage in a metal-free environment (Jain and Bradshaw, 1966).

Male fertility as a component of fitness. If the population of plants in the example is completely outcrossing, so that pollen is distributed randomly to the ovules of members of the population, the relative contributions to the pool of pollen grains of the two forms must be included in the calculation of fitness, assuming that the two traits are controlled by nuclear genes for which one copy is inherited from the father and the other from the mother: contributions via male and female parents are given equal weight. In animals or plants with separate sexes, the relative male mating and fertilization successes of different types must similarly be included in calculations of fitnesses. These calculations can become very complicated, especially when the fertility of a mating depends on the genetic makeup of both partners (Crow and Kimura, 1970; Hartl and Clark, 1997).

More Complex Systems.

Effects of the genetic system. If a trait is controlled by a gene carried in a maternally transmitted cytoplasmic genome, such as the mitochondrion or chloroplast, transmission through male function is irrelevant, and fitness is determined solely by viability and female fertility. This brings out another important aspect of fitness: both the system of mating and the mode of inheritance of the genetic factors controlling the trait of interest must be included in the calculations. There can be radical differences between the outcomes of selection under different mating systems and modes of inheritance. For example, if lead tolerance had no effect on male fertility but reduced female fertility by 60 percent, a twofold advantage in survival probability would more than outweigh this fertility disadvantage with nuclear gene inheritance and random mating, but not with complete self-fertilization or strictly maternal inheritance. Conflicts of interest arise with respect to fitness between different components of the genome, parent and offspring, and mothers and fathers (Keller, 1999).

Gametic selection and segregation distortion. There are also cases in which genetic differences can directly affect transmission through the gametes. In flowering plants, many genes are expressed in pollen, and the success in fertilization of a pollen grain may be affected by the variant form (allele) of a particular gene that it carries (Walsh and Charlesworth, 1992). In animals, genes are not expressed in the sperm, but there are several cases of segregation distortion, where one allele (D_1) at a locus when heterozygous causes the destruction of gametes carrying the alternative allele (D_2). Provided that the fertility of $D_1 D_2$ males is affected less than linearly by the destruction of the D_2 sperm, D_1 will gain a transmission advantage and spread through the population, unless there is an opposing selective disadvantage at the level of individuals (Hartl and Clark, 1997).

Complex life cycles and frequency-dependent fitnesses. In many species, including humans, reproduction occurs repeatedly over several successive time periods, so that generations overlap, and we cannot apply the simple models just described. Methods for dealing with this problem have been developed, but there is no exact general formula for fitness (Charlesworth,

1994). A useful approach is to examine the rate of spread of a rare mutant gene introduced into the population; such a gene will spread only if its carriers have survival and fertility traits that would confer a higher rate of increase in numbers on a population that consists entirely of such individuals. The fitness of the mutants can be measured as the difference between their ultimate rate of increase in numbers and that of the other part of the population (Figure 1).

The method of determining the conditions for the spread of a rare mutant form is widely used in the context of evolutionarily stable strategy (ESS) theory, which determines the state of the population that resists invasion by any feasible alternative (Bulmer, 1994; Frank, 1998; Maynard Smith, 1982). [See Game Theory.] This greatly simplifies the study of selection when the relative fitnesses of different types of individuals are influenced by their frequencies in the population, as in the case of the sex ratio.

TABLE 2. Fitnesses That Vary Over Time

Generation	Fitness of type 1 (w_1)	Fitness of type 2 (w_2)	Ratio (w_2/w_1)
1	1.20	0.9	0.750
2	0.82	1.1	1.341
3	1.30	0.85	0.654
4	0.71	1.15	1.620
Mean	1.0075	1.000	1.091

Fluctuating selection pressures. In the real world, relative fitnesses may vary over both time and space. Much attention has been given to developing methods for determining the appropriate averaging procedures for predicting the course of selection when there is spatial and temporal variation (Crow and Kimura, 1970; Hartl and Clark, 1997). When fitnesses fluctuate between generations, for example, the product of relative fitnesses over time should be used to determine the outcome of selection (Table 2).

In Table 2, the fitnesses of two types are assumed to vary from generation in a regular cycle that repeats itself over four generations. If type 2 individuals are carriers of a rare mutant gene, and the numbers of type 1 and type 2 are n_1 and n_2 (with n_1 much bigger than n_2), respectively, the ratio of the number of type 2 to the number of type 1 in the next generation can be shown to be approximately $n_2 w_2/n_1 w_1$. Repeating this over all four generations, the ratio of the final numbers of types 2 and 1 over a complete cycle is the initial value of n_2/n_1 times the product of the successive values of w_2/w_1; that is, $0.750 \times 1.341 \times 0.654 \times 1.620 = 1.066$ in this example.

Type 2 is thus predicted to increase, even though the mean of its fitness values over the cycle of generations is smaller than that of type 1.

Indirect effects on fitness. Up to now, we have only discussed situations in which genes have direct effects on aspects of survival or reproduction. There are many important cases in which fitness differences result solely from the effects of genes on their mode of transmission. For example, in a species with separate sexes, a mutation that causes females to produce only daughters by asexual reproduction but has no other effect will have twice the female fertility of sexual females, and hence spreads rapidly through the population (Maynard Smith, 1978). [See Sex, Evolution of.]

The ratio of males to females in a sexual species may also influence fitness. With nuclear gene inheritance, each individual receives one gene from the mother and the other from the father. If there is an unequal number of male and female births, this means that there is a fitness advantage to individuals who produce more progeny of the sex that is in short supply, because individuals of this sex will have higher mating success (Fisher,

FIGURE 1. The Rate of Spread of a Rare Mutant Gene in a Population.
The filled circles indicate the changes in size of a human population over successive intervals of ten years, on a scale of natural logarithms. The open circles indicate the corresponding changes in the numbers of carriers of a rare mutant gene, present initially at a frequency of 1/1000. As is usual with overlapping generations, the population sizes fluctuate initially, but eventually settle down to constant rates of growth on the log scale, with the overall population growing at a rate of 0.0001 per year and the mutant at a rate of 0.0095. The difference in growth rates indicates a fitness advantage of $0.0095 - 0.001 = 0.0084$ to the mutant, in terms of growth rate per year. Human populations have a generation time of about 25 years, measured by the mean age of parents. The fitness advantage of the mutant is thus about $0.0084 \times 25 = 0.21$ on a per generation basis; this corresponds to the difference in relative fitnesses of the two types in the case of a simple discrete generation model.
Drawing by Brian Charlesworth.

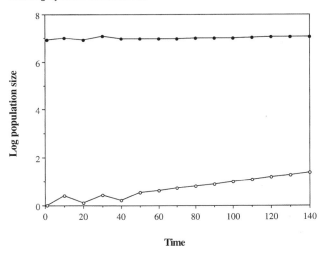

1999). This insight has stimulated the growth of elaborate models of the evolutionary modification of the sex ratio, which have been extensively tested against data (Bulmer, 1994; Frank, 1998). [*See* Sex Ratios.]

Genes that modify aspects of the genetic system, such as the frequency of genetic recombination or rate of inbreeding, may also gain fitness advantages, because they establish nonrandom associations with other genes that directly affect fitness (Barton and Charlesworth, 1998; Charlesworth and Charlesworth, 1998; Maynard Smith, 1978).

Higher-level fitness effects. Finally, fitness differences among groups of individuals may affect the course of evolution. The most clear-cut example of this is kin selection, in which an allele that causes individuals to sacrifice their own survival ability or fertility can gain an overall advantage because it increases the fitnesses of close relatives, who have a higher than average chance of carrying the allele (Fisher, 1999; Hamilton, 1964). [*See* Kin Selection.] Many aspects of the biology of social animals are thought to reflect the action of this form of selection, and methods for calculating the fitness effects of different behavioral traits in this context have been devised (Bulmer, 1994; Frank, 1998). If the species as a whole is divided into small groups that are partially isolated from each other, there is also an opportunity for differences in genetic makeup between groups to arise by random sampling effects, which may affect the contributions of different groups to the next generation. For example, a type that enjoys an advantage from superior competitive ability within a group (e.g., through aggressive behavior) may cause a reduction in the overall fitness of the group, and hence fail to spread through the whole population (Keller, 1999). [*See* Group Selection.]

The relative fitnesses of species in terms of their long-term rates of extinction or rate of species formation may also affect the distribution of traits among species. For example, asexual reproduction is comparatively rare among animals and plants, although it can easily be established within a species (Maynard Smith, 1978). This must in part reflect the fact that asexual species are likely to have higher rates of extinction than their sexual counterparts; for example, because they cannot evolve as fast in response to a changing environment (Barton and Charlesworth, 1998; Keller, 1999; Maynard Smith, 1978). The general importance of such higher-level fitness effects is controversial, because there are reasons to believe that they can usually be overridden in the short term by selection at lower levels (Fisher, 1999; Keller, 1999). [*See* Species Selection.]

BIBLIOGRAPHY

Barton, N. H., and B. Charlesworth. "Why Sex and Recombination?" *Science* 281 (1998): 1986–1990.

Bulmer, M. G. *Theoretical Evolutionary Ecology.* Sunderland, Mass., 1994.

Charlesworth, B. *Evolution in Age-structured Populations.* 2d ed. Cambridge, 1994.

Charlesworth, B., and D. Charlesworth. "Some Evolutionary Consequences of Deleterious Mutations." *Genetica* 102/103 (1998): 3–19.

Crow, J. F., and M. Kimura. *An Introduction to Population Genetics Theory.* New York, 1970.

Darwin, C. R., and A. R. Wallace. "On the Tendency of Species to Form Varieties; and on the Perpetuation of Varieties and Species by Natural Selection" (1858). Reprinted in *The Collected Papers of Charles Darwin*, edited by P. H. Barrett, vol. 2, pp. 3–18. Chicago, 1977.

Endler, J. A. *Natural Selection in the Wild.* Princeton, 1986.

Fisher, R. A. *The Genetical Theory of Natural Selection: A Complete Variorum Edition* (1930). Oxford, 1999.

Frank, S. A. *Foundations of Social Evolution.* Princeton, 1998.

Hamilton, W. D. "The Genetical Evolution of Social Behavior." *Journal of Theoretical Biology* 7 (1964): 1–16.

Hartl, D. L., and A. G. Clark. *Principles of Population Genetics.* 3d ed. Sunderland, Mass., 1997.

Jain, S. K., and A. D. Bradshaw. "Evolutionary Divergence among Adjacent Plant Populations: 1. The Evidence and Its Theoretical Analysis." *Heredity* 21 (1966): 407–441.

Keller, L. *Levels of Selection in Evolution.* Princeton, 1999.

Maynard Smith, J. *The Evolution of Sex.* Cambridge, 1978.

Maynard Smith, J. *Evolution and the Theory of Games.* Cambridge, 1982.

Stearns, S. C. *The Evolution of Life Histories.* Oxford, 1992.

Walsh, N. E., and D. Charlesworth. "Evolutionary Interpretations of Differences in Pollen-Tube Growth Rates." *Quarterly Review of Biology* 67 (1992): 19–37.

— Brian Charlesworth

FLUCTUATING ASYMMETRY

Asymmetries abound in nature. The human heart is normally displaced to the left, crabs have one large and one small claw, and flatfish are collapsed onto their left or right sides. These asymmetries are adaptive designs that reveal the originally symmetric structures from which they were derived.

More subtle asymmetries also exist. Swallow tail feathers, fruit fly wings, and human ears are all structures that on average are equally large on left and right sides. Yet take any individual and careful examination will reveal slight asymmetries. Most individuals are highly symmetric but others show quite obvious differences between left and right sides.

Perhaps surprisingly, this subtle form of asymmetry, known as fluctuating asymmetry (FA), has become an area of lively investigation as well as great contention. Interest has been galvanized by the claim that FA increases when organisms suffer a wide range of environmental and genetic stresses. In addition, many animals appear to assess asymmetries in order to avoid dangerous rivals and find mates with high fitness. This has led to the advocacy of FA as a cheap and reliable measure of fitness. But for every observation of higher asymmetry in stressed individuals and populations, there are counterexamples where FA and stress are unconnected. In this article we will investigate whether FA can be used as a reliable diagnostic tool of quality.

Measuring Asymmetry. As both sides of an organism are the product of the same genes, and environmental differences between the sides are likely to be minor, departures from symmetry are thought to arise from the accumulation of random errors in development. The ability of an individual to buffer its development against random disruptions is known as its developmental stability. If an individual has poor developmental stability, either because it is subject to genetic or environmental stress, it is likely to show a higher degree of asymmetry.

Many early workers concentrated on FA, which is a population measure of asymmetry. The typical frequency distribution of asymmetry (left side minus right side) measurements for a morphological trait follows a normal distribution (Figure 1A), and the mean asymmetry is expected to be zero. Once these properties have been checked, the absolute deviation between left and right measures is plotted (Figure 1B), and the mean absolute asymmetry is taken as the population measure of FA.

Most morphological traits studied demonstrate small degrees of FA, less than 1 percent of the absolute size of the trait. In the example in Figure 1, the mean wing length is almost 2 mm, yet the mean FA is only 0.022 mm. This is where the problems start. A pair of calipers may be good enough to correctly measure trait lengths, but considerably more precision is required to accurately measure the difference between the two sides, which have very similar lengths. One way of dealing with this scaling problem is to take repeated measures. Each side is measured two or three times. Then a set of statistical procedures can be followed to estimate the within-sides variance (i.e., the measurement error) and to remove this from the estimate of the between-sides variance (i.e., the difference between left and right), and so get a good estimate of the true fluctuating asymmetry.

However, serious errors can creep in even when these techniques are followed. As asymmetries are small relative to the object being measured, estimates often suffer from biases arising from inaccuracy in the measuring equipment or from the user's behavior. It is notable that these errors are often so small as to be irrelevant when measuring trait size, but become significant when measuring trait asymmetry. In addition, there is a more fundamental problem. Using the difference between left and right sides to estimate developmental stability is equivalent to using a sample size of two measurements to estimate the variance. Usually much larger samples are used to estimate the variance. So even at their best, asymmetry measurements can give only a

A)

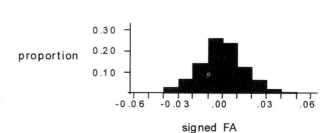

B)

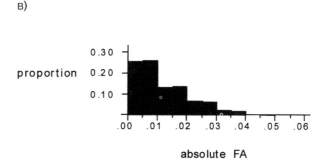

FIGURE 1. Signed and Absolute Fluctuating Asymmetry for Male Wing Length in the Stalk-Eyed Fly *Cyrtodiopsis dalmanni*. The distribution of asymmetry is shown for (A) signed fluctuating asymmetry (left side minus right side), and (B) absolute fluctuating asymmetry (sample size = 403). The distribution of signed asymmetry values follows a normal distribution with a mean that does not differ from zero. The mean wing length for this sample was 2.535 mm, and the mean fluctuating asymmetry was 0.013 mm. Andrew Pomiankowski.

hazy estimate of the underlying property of developmental stability, and this is the quantity that scientists would like to measure.

Asymmetry and Fitness. Despite these difficulties, studies of FA have burgeoned in the last decade. This has been largely attributable to the promise that asymmetry is a good measure of phenotypic stress. A vast range of associations between environmental factors and FA has been found, for example adverse temperatures, poor nutrition, a variety of pollutants, DDT, population density, parasite infection, and even high-intensity noises. It has also been reported that FA increases under genetic stresses like hybridization between species, inbreeding, chromosome duplications (e.g., as in Down's syndrome), and the introduction of new mutants. This positive picture has inspired several scientists to advocate the use of FA as a general bioindicator of individual and population health, with widespread applications in ecology, animal behavior, and conservation biology.

But does the evidence support this view? There are persistent worries that reporting bias has inflated the association between FA and stress, as only positive re-

HUMAN FACIAL ASYMMETRY

Do humans use facial symmetry as a cue in assessing potential mates? Females (or males) asked to assess male (or female) faces for attractiveness consistently favor those that are more symmetric. To show that this association arises purely through differences in asymmetry, further investigation has used images morphed to create symmetric faces. Digitized faces are either mirrored down the middle or warped so that the position of a number of landmarks coincides on both sides. In both cases, female and male raters given pairs of symmetric and normal images perceive the symmetric images as more attractive.

Doubts have been raised on a number of counts. The blending of the two sides tends to produce a more even image, with smoother skin tones. Undoubtedly such a property contributes to attractiveness as well as greater symmetry. However, when skin textures are held constant, preference for symmetry is still reported. Another worry concerns correlates of symmetry. Scheib and colleagues (1999) presented female raters with half images of male faces (just the left or right side) which had previously been assessed for asymmetry. Females still preferred low-asymmetry faces, even though the half images lacked any information about symmetry. This suggests that symmetry may not be a primary cue used in mate assessment, but may be correlated with other attractive features. In this study, females preferred males with relatively long faces and higher cheekbones. But it is unclear whether these "masculine" characters are generally associated with symmetry. Finally, just as in other species, it remains unclear whether human facial asymmetry correlates with properties such as health or fertility. Only weak and inconsistent associations have been reported between indicators of human asymmetry and health.

—ANDREW POMIANKOWSKI

sults get published (Simmons et al., 1999). In addition, most evidence comes from field observations in which many variables remain uncontrolled. To avoid these concerns, experiments have been carried out under uniform laboratory conditions with controlled manipulations of stress. A recent collation of experimental studies revealed a highly inconsistent pattern in which increases in FA were trait-, stress- and species-specific. A typical example is Roy and Stanton's (1999) experimental study of five separate stress regimes (boron, water, salt, nutrients, and light) applied to the wild mustard *Sinapis arvensis*. FA was measured in four traits (petal, leaf, fruit, and cotyledon). Although FA sometimes increased with stress, the trait that responded varied with each

stress, and there was no consistent pattern of change in FA within an individual. A similar inconsistent pattern has emerged from studies of genetic stress caused by inbreeding, with many carefully controlled experiments reporting no association between the level of inbreeding and the degree of FA. So, far from being a reliable general indicator, trait FA appears to react in unpredictable ways to environmental and genetic stresses.

A proposed solution to this variability is to study composite measures of FA in which several different trait FAs are measured and combined to produce a single FA index. The idea is that individual trait FAs carry weak signals that can be combined into a powerful measure of the underlying developmental stability. Composites have been used extensively in humans. For instance, data from ear, elbow, wrist, finger, ankle, and foot FA have been combined in an FA composite that correlates well with several aspects of male sexual behavior. However, many other studies have found that FA composites are no better, or are even worse, at predicting performance. The problem is that combinations increase noise as well as signal. So composites have proved to be no better than single-trait measures of FA. They are not a general solution to the weak power of FA to detect developmental stability.

Sex and Asymmetry. One promising way forward has been the analysis of asymmetry in sexual traits, such as the peacock's tail. As females use exaggerated sexual traits in their mate choice as indicators of male quality, several workers thought that sexual-trait FA might show a more consistent pattern reflecting environmental and genetic stress.

Evidence in favor of this idea has come from the surprising finding that females are able to assess male symmetry. This has been demonstrated using experimental manipulations of symmetry of male displays in birds (zebra finches, swallows), with the general finding that females prefer symmetric males. What is less clear is the importance of mate choice for symmetry under natural conditions. The experiments compared highly asymmetric with perfectly symmetric individuals, whereas in the wild, differences in symmetry will generally be very slight. It is not clear whether animals can directly distinguish such small differences. However, several studies of human face perception suggest that we can assess small facial asymmetries and use them in our assessment of attractiveness (see Vignette).

Why might females prefer to mate with symmetric males? Three hypotheses have been formulated suggesting that FA in sexual traits is special. First, sexual traits are highly exaggerated and therefore costly to produce. Hence development of these traits increases stress, which should be reflected in higher levels of FA. This predicts that the mean FA in sexual traits will be higher than in similar nonsexual traits. Second, sexual traits are predicted to show declining FA as they increase in size. This prediction is counterintuitive, as we expect scaling to cause asymmetry to increase with trait size. However, a negative relationship between asymmetry and size will occur if high-quality (i.e., low stress) individuals not only produce large sexual traits but also have low asymmetry. Finally, using a similar logic, sexual-trait FA is predicted to be more sensitive to stress than other traits.

The initial evidence supported all three of these predictions. Work on the swallow *Hirundo rustica* confirmed that the male's elongated tail feathers, which are used as a sexual display, had higher mean FA. In addition, FA was lower in individuals with longer tail length. The effect of stress was examined by experimentally altering mite infestations of nestlings. Adding mites caused a dramatic increase in tail feather FA, whereas removal of mites by spraying with pesticide reduced tail feather FA. Other nonsexual feather traits were unaffected by infection levels.

However, more recent work in a variety of species (dung beetles, stalk-eyed flies, and wild mustard) have failed to confirm any of the hypotheses. Higher mean FA turns out to be an artifact. Sexual traits tend to be large, so they tend to have high absolute FA. But sexual traits do not have relatively large mean FA compared with nonsexual traits when size is controlled for. The negative relationship between asymmetry and size is not found in other species. Their sexual traits either show no relationship or increasing asymmetry with size, patterns typical of nonsexual traits. Finally, the principal hypothesis that sexual-trait FA is more sensitive to stress has gained no further support. Stress often affects trait size but seems to have little consequence for FA. To sum up, sexual-trait FA appears to be similar to nonsexual-trait FA. The responses of FA to stress are once again inconsistent.

Conclusion. With the benefit of hindsight, the major claims about asymmetry and fitness seem naive. Enthusiasm got the better of sense. Although FA is cheap to measure, it is difficult to measure accurately, and it is a poor indicator of developmental stability. It may be that asymmetry is a good indicator in some traits, in some species, of some stresses. But as yet, there is no general theory to predict which traits respond to which stresses. The only clear pattern to emerge from the last decade of study is that FA is an inconsistent indicator of environmental and genetic stress.

[*See also* Male-Male Competition; Mate Choice, *articles on* Mate Choice in Plants *and* Human Mate Choice; Sexual Dimorphism; Sexual Selection, *article on* Bowerbirds.]

BIBLIOGRAPHY

Bjorksten, T., K. Fowler, and A. Pomiankowski. "What Does Sexual Trait FA Tell Us About Stress?" *Trends in Ecology and Evolu-tion* 15 (2000): 163–166. A critical review of experimental data showing that FA in sexual traits is not a good indicator of stress.

Leary, R. F., and F. W. Allendorf. "Fluctuating Asymmetry As an Indicator of Stress: Implications for Conservation Biology." *Trends in Ecology and Evolution* 4 (1989): 214–217. An early paper discussing the potential of FA as an indicator of stress.

Møller, A. P., and J. P. Swaddle. *Asymmetry, Developmental Stability and Evolution.* Oxford and New York, 1997. This book has become notorious for its advocacy of FA as a measure of fitness. For a critical review of the text see D. Houle, "High Enthusiasm and Low R-Squared," *Evolution* 52 (1998): 1872–1876.

Palmer, A. R., and C. Strobeck. "Fluctuating Asymmetry: Measurement, Analysis, Patterns." *Annual Review of Ecology and Systematics* 17 (1986): 391–431. Contains the standard techniques used for the analysis of FA.

Perrett, D. I., et al. "Symmetry and Human Facial Attraction." *Evolution and Human Behaviour* 20 (1999): 295–307. A recent study of how human facial symmetry influences attraction.

Roy, B. A., and M. L. Stanton. "Asymmetry of Wild Mustard, *Sinapsis arvensis* (Brassicaceae) in Response to Severe Physiological Stress." *Journal of Evolutionary Biology* 12 (1999): 440–449.

Scheib, J. E., S. W. Gangestad, and R. Thornhill. "Facial Attractiveness, Symmetry and Cues of Good Genes." *Proceedings of the Royal Society of London,* B, 266 (1999): 1913–1917.

Simmons L. W., J. L. Tomkins, J. S. Kotiaho, and J. Hunt. "Fluctuating Paradigm." *Proceedings of the Royal Society of London B,* 266 (1999): 593–595.

Swaddle, J. P., and I. C. Cuthill. "Preference for Symmetric Males by Female Zebra Finches." *Nature* 367 (1994): 165–166. The first clear experimental demonstration that females show mate preference for males with symmetric sexual traits.

Woods, R. E., C. Sgro, M. J. Hercus, and A. A. Hoffmann. "The Association between Fluctuating Asymmetry, Trait Variability, Trait Heritability and Stress: A Multiply Replicated Experiment on Combined Stresses in *Drosophila melanogaster*." *Evolution* 53 (1999): 493–505. A well-replicated study showing that stress has inconsistent effects on trait FA.

— ANDREW POMIANKOWSKI

FORAGING STRATEGIES. *See* Human Foraging Strategies; Optimality Theory, *article on* Optimal Foraging.

FORCED EVOLUTION. *See* Directed Protein Evolution.

FORENSIC DNA

Except for identical twins, no two people have the same genomic DNA sequences. Although pairs of people differ at about one position every one thousand bases (the human genome is about 3.2 billion base pairs long), far fewer than these three million positions need to be examined, in practice, to distinguish one person from another. The forensic uses of DNA rest on this power of discrimination and on two other features: the same DNA

DNA FROM A SNOWBALL

In October of 1994, a 32-year-old woman disappeared from her home in Richmond, Prince Edward Island, Canada. Three weeks later a man's leather jacket stained with blood was found in the woods 8 kilometers from her home, and in May of 1995 her body was uncovered from a shallow grave. DNA testing established a match between the victim and the blood on the jacket. How could the jacket be linked to the suspect, the woman's estranged common-law husband? The Royal Canadian Mounted Police found white domestic cat hairs in the jacket lining, and also found that the suspect lived with his parents and Snowball, a white American shorthair cat. Could DNA tie Snowball to the jacket, and therefore tie the suspect to the crime?

The Laboratory of Genomic Diversity of the National Cancer Institute, under the direction of Dr. Stephen O'Brien, is well-known for its feline studies and Dr. Marilyn Menotti-Raymond there has developed a panel of 400 simple tandem repeat (STR) genetic markers for the domestic cat. She also maintains a three-generation pedigree that can be used to establish genotyping protocols and establish linkage relationships for these markers. Upon being approached by the RCMP, Drs. Menotti-Raymond, O'Brien, and Victor David, extracted DNA from the root of one the hairs off the jacket and from a blood sample from Snowball. They found a 10-locus STR match between these two DNA samples, and the suspect was convicted of second-degree murder in 1996.

—BRUCE WEIR

is found in every tissue type, and relatives share a portion of their DNA. The first feature allows people to be associated with crimes on the basis of DNA extracted from biological samples, including stains of bodily fluids. The second feature allows the identification of remains or determination of parentage on the basis of profiles from samples and family members.

Currently, DNA sequencing as a means of identifying individuals is applied only to mitochondrial DNA. A region of about 600 base pairs in the highly variable region of the mitochondrion is sequenced. Nuclear DNA is characterized currently on the basis of small number of microsatellites, or short tandem repeats (STRs). In the United States, the Federal Bureau of Investigation uses a core set of 13 unlinked STR loci. The growing availability of data on single nucleotide polymorphisms (SNPs) and cheap SNP typing methods suggests that they will be used in the future. Earlier work used blood protein markers, especially in parentage analyses, and there was a short period during which minisatellites, or

variable number of tandem repeat (VNTR) loci, were used. Regardless of the nature of the genetic marker employed, the term "DNA profile'" is appropriate. There must be a reliable procedure for determining whether or not two profiles are the same. If no errors have been made, nonmatching profiles could not have originated from the same person. Ideally, matching profiles would imply a common origin, but courts of law currently look for some quantification of the evidence of a match, mainly to assess whether the match could be coincidental. The degree of profile similarity between relatives is a probabilistic quantity, and numerical calculations are routinely performed when relatives are involved. Those calculations may also need to take account of the possibility of mutation altering allelic states and lowering the degree of similarity.

Calculations for DNA profiles used for identification can be given only for the probabilities of these matches under specified hypotheses about the origins of those profiles. In the O. J. Simpson murder trial, for example, the prosecutor asserted that Simpson was the source of a series of blood spots found on the pavement next to the bodies of the two victims. Under that assertion, the probability of the evidence was 100 percent. Had the defense claimed that the spots came from some other person, the probability of the evidence would have involved the chance of a person having the same profile as Simpson at all the loci typed from those spots. That probability is very small. In fact, the defense also raised the possibility of errors in the handling of evidence. In a recent demonstration that the Cabernet Sauvignon grape variety had the varieties Cabernet Franc and Sauvignon Blanc as parents, various pairs of alternative hypotheses were considered: either those two varieties were the parents, or only one of the two was parental, or the two were more distantly related to Cabernet Sauvignon. Although juries and wine lovers would prefer to know the probabilities of guilt or pedigree given the DNA profiles, they can be told only the probabilities of the profiles, given alternative statements about profile origins. In a rational world, they would combine these probabilities with their prior beliefs to arrive at their conclusion.

The use of DNA profiles in criminal cases has been improved substantially by the possibility of amplifying tiny amounts of DNA from crime stains through the polymerase chain reaction (PCR). When only nanograms of DNA are needed, it is possible to link people to crimes on the basis of saliva residues on cigarette butts or postage stamps. The so-called Low Copy Number (LCN) typing being used by the Forensic Science Service in the United Kingdom needs only picograms of DNA; fingerprints or even nose prints may be used. The large number of mitochondrial (mt) genomes per cell and the relatively small size of the mt genome mean that mt profiles are valuable for typing old samples in which nuclear DNA may be degraded; they find great use in the analysis of hair and bone samples. Remains found at Ekaterinburg, Russia, thought to be those of the family of the last Russian tsar, were identified as being a family group because of shared mt profiles; then they were identified as Romanovs through comparison of the mt profiles to those of living maternal relatives of the family. The mt genome is maternally transmitted. By contrast, the Y chromosome is paternally transmitted, and this was used to link living male descendants of Thomas Jefferson with living male descendants of his slave Sally Hemming. DNA evidence, therefore, does not exclude the Ekaterinburg bones as being those of Tsar Nicholas and his family, and it does not exclude Thomas Jefferson from having fathered some, but not all, of Sally Hemming's sons.

DNA profiles are often used to link suspected perpetrators to crimes. In the simplest situation, the profiles are those of the suspect and those found on the evidence, although they can be from a victim and the evidence. There is increasing use of nonhuman DNA—for example, leaves from a specific Palo Verde tree matching leaves found in a suspect's pick-up truck, or hairs from a suspect's cat found on a jacket stained with a victim's blood. Profiles are also used to link crimes such as strings of burglaries or serial rapes. The typing of people at the time of arrest or conviction is allowing some of those people to be linked to other crimes, past or future. This requires the searching of DNA profile databases for matches to evidence profiles from crimes for which there is no suspect. The increasing miniaturization of PCR equipment will allow evidence to be typed at the scene of a crime, and the future identification of possible suspects within hours, or even minutes, of the crime's being committed.

Outside the world of forensic science, DNA profiles are now being used in the search for human disease genes and to predict adverse drug reactions. In the same vein, forensic scientists are searching for markers that allow physical characteristics to be described; for example, a "red-hair" marker is known. The implications of being able to describe a person's appearance, or present or future health status, on the basis of DNA extracted from a discarded facial tissue are profound. Privacy issues will need to be addressed.

[See also Genetic Markers.]

BIBLIOGRAPHY

Bowers, J. E., and C. P. Meredith. "The Parentage of a Classic Wine Grape, Cabernet Sauvignon." *Nature Genetics* 16 (1997):84–87.
Foster, E. A., M. A. Jobling, P. G. Taylor, P. Donnelly, P. de Kniffs, R. Mieremet, T. Zerjal, and C. Tyler-Smith. "Jefferson Fathered Slave's Last Child." *Nature* 396 (1998): 27–28.
Gill, P., P. L. Ivanov, C. Kimpton, P. Piercy, N. Benson, G. Tully, I. Evett, E. Hegelberg, and K. Sullivan. "Identification of the Re-
mains of the Romanov Family by DNA Analysis." *Nature Genetics* 6 (1994):130–135.
Menotti-Raymond, M. A., and S. J. O'Brien. "Pet Cat Hair Implicates Murder Suspect." *Science* 386 (1997): 774.
Weir, B. S. "DNA Statistics in the Simpson Matter." *Nature Genetics* 11 (1995): 365–368.
Yoon, C.-K. "Forensic Science—Botanical Witness for the Prosecution." *Science* 260 (1993): 894–895.

— BRUCE WEIR

FRANKLIN, ROSALIND

Rosalind E. Franklin (1920–1958) was a leading X-ray crystallographer of biological materials, including DNA. She discovered the "B" form of DNA, recognizing that two interchangeable forms of crystalline DNA existed (the other being the "A" form), and her unpublished diffraction data on the B form were used by James D. Watson and Francis Crick to construct their double-stranded helical model of DNA structure.

Raised in a prominent Jewish family in London, Franklin enrolled at Newnham College, Cambridge, for her undergraduate degree, continuing there for graduate work in physical chemistry. Franklin completed a Ph.D. in 1945, having taken time to do war-related work on the structure of coal. From 1947 to 1951, she investigated carbon and graphite in Paris as a researcher at the Laboratoire Central des Services Chimiques de l'État. There she was trained in X-ray crystallography with Jacques Méring. Upon returning to England in 1951 to continue crystallographic research, she shifted her focus to biology. Appointed as a Turner-Newall Fellow to John Randall's Biophysics Research Unit at King's College, London, she set up an X-ray diffraction apparatus to work on the structure of DNA. Another biophysicist at King's College, Maurice Wilkins, was also working on DNA structure. Franklin's independent work on the same subject caused friction between Franklin and Wilkins just as competition over DNA intensified beyond King's College.

In 1951, James Watson joined Sir Lawrence Bragg's Cavendish Laboratory to learn the techniques of X-ray crystallography and apply them to solve the structure of tobacco mosaic virus. Watson's real interest, however, was in the nature of the gene, and he and Bragg's senior graduate student, Francis Crick, began building structural models of the likely hereditary material, DNA. Their lack of detailed structural information (such as that available through X-ray crystallography) impaired their efforts to build an accurate model of DNA. At some point, Wilkins showed Watson one of Franklin's highly informative diffraction patterns for the B form of DNA, which she had set aside in order first to determine the structure of the A form. Because Bragg's and Randall's laboratories were both supported by the Medical Research Council, Watson and Crick also gained access to Franklin's unpublished results through an agency report. As Watson later explained, not only did her diffraction photograph offer compelling evidence for a helical structure, but it provided the crucial parameters for the model Watson and Crick built early in 1953. Even though Franklin and Wilkins each published papers presenting their data in the same issue of *Nature* as Watson and Crick's publication of their model, the papers from King's College were never as widely cited as that of their Cambridge rivals. Franklin's career was cut short by an early death (from cancer) in 1958. Four years later Watson, Crick, and Wilkins received the Nobel Prize in medicine or physiology for determining the structure of DNA. Watson's unflattering presentation of Franklin in his widely read chronicle of their competition (*The Double Helix*, 1968) fueled criticism that Franklin should have received more credit along with Watson and Crick for their double helical model, since it relied on her data (Sayre, 1975). Since then, authors have disagreed about whether Franklin would have reached an independent determination of the double-helical structure of DNA had Watson and Crick not used her data first (e.g., Klug, 1968; Klug, 1974; Judson, 1980).

In 1953, Franklin left King's College and the DNA project to join J. D. Bernal's X-ray crystallography unit at Birkbeck College. There she and her collaborators (including Kenneth C. Holmes, Aaron Klug, and John Finch) investigated the structure of tobacco mosaic virus, following early publications by Watson and Wilkins on this virus. Although Franklin confirmed Watson's observation that the tobacco mosaic virus rods were helical, she corrected other misconceptions with her more precise diffraction data. In particular, she showed that the virus rods were not solid but hollow, and that the nucleic acid component was embedded in the protein helix and was not in the center of the helix as previously thought. Her great success in this endeavor developed in part from her collaboration with two rival groups of biochemists working on the virus (led by Gerhard Schramm in Tübingen and Heinz Fraenkel-Conrat in Berkeley). Although Franklin's elegant work on tobacco mosaic virus drew praise from Watson, Crick, and the molecular biology community at the time, the growing importance of the structure of DNA has since eclipsed this recognition.

BIBLIOGRAPHY

Freeland Judson, H. "Reflections on the Historiography of Molecular Biology." *Minerva* 18 (1980): 369–421.

Klug, A. "Rosalind Franklin and the Discovery of the Structure of DNA." *Nature* 219 (1968): 808–810, 843–844.

Klug, A. "Rosalind Franklin and the Double Helix." *Nature* 248 (1974): 787–788.

Olby, R. "Rosalind Elsie Franklin." In *Dictionary of Scientific Biography*, edited by Charles Coulston Gillispie. New York, 1970.

This entry includes references to Franklin's important scientific publications.

Sayre, A. *Rosalind Franklin and DNA*. New York, 1975.

Watson, J. D. *The Double Helix: A Personal Account of the Discovery of the Structure of DNA: Text, Commentary, Reviews, Original Papers*. Edited by Gunther S. Stent. New York, 1980. This edition reprints Franklin's 1953 *Nature* paper as well as critical reviews of Watson's account.

— ANGELA N. H. CREAGER

FREQUENCY-DEPENDENT SELECTION

Models of short-term evolution usually assume that the selective values (fitnesses) of genotypes are constant from generation to generation. However, ecological studies of selective agents, such as predators, parasites, parasitoids, competitors, and symbionts, show that their effects often depend on the number of individuals in the affected population. If the agents discriminate between phenotypes, their action should depend on phenotypic frequencies. It might be supposed that abiotic selective factors (temperature, sunshine, precipitation, humidity, geology, soil structure, etc.) should be independent of numbers or frequencies, but biotic and abiotic factors often interact. For example, the ability of an Antarctic penguin to withstand cold can depend on the food available to support metabolic heating, which in turn may be influenced by the number of competitors. The majority of selective forces in nature probably change with the numbers of organisms and the frequencies of phenotypes.

The term *frequency-dependent selection* denotes the situation when the selective values of phenotypes depend on their frequencies. The dependence can be positive, so that the selective value rises with the frequency of the phenotype, or negative, so that it falls. The former (positive frequency-dependent selection, anti-apostatic selection, enstatic selection, or "majority advantage") tends to make gene frequencies unstable and to reduce genetic variation. The latter (negative frequency-dependent selection, apostatic selection, or "minority advantage") tends to maintain variety within the population.

An example of positive frequency-dependent selection is the tendency of predators to attack rare forms of prey that "stand out" in otherwise uniform flocks or schools. Another is that of phenotypes predisposed to mate among themselves. For example, some land snails have genetic variants in which the coil of the shell is reversed. This changes the position of the genital aperture, so that snails with opposite coils have difficulty in mating. A study of *Partula suturalis* on the Pacific island of Mooréa showed that rare sinistral snails from an area dominated by dextrals produced fewer young than those from an area where the two phenotypes were equally common.

There is a strong interest in negative frequency-dependent selection because it can help to explain the many genetic polymorphisms in characters that affect survival and reproduction (i.e., that are not selectively neutral). Examples of negative frequency dependence can be conveniently classified by the agents of selection, such as predators, prey, parasites, hosts, or competitors.

Selection by Predators and Prey. At the end of the nineteenth century, the English zoologist E. B. Poulton (1884) pointed out that a naive predator, learning to hunt for different varieties of a prey species, ought to attack the common varieties and overlook the rare ones. He argued that predators behave like people and need several encounters with a new object before they can search for it efficiently. Rare types of prey could gain an advantage if encounters were infrequent enough to hamper effective learning. The advantage would decline as the frequency of the type increased, thus favoring an equilibrium at which two or more types coexisted stably in the population.

Poulton's idea remained an interesting possibility until the 1940s, when the English zoologist E. J. Popham carried out experiments in which minnows preyed upon corixid water bugs. Popham used three varieties of the prey, differing in color, and offered them to the predator in varying proportions against a background of sand. When the varieties were offered at equal frequencies, the predators took proportionately fewer of the variety that most closely matched the background, and proportionately more of the most conspicuous one. When Popham altered the frequencies, the rarer varieties gained an advantage. Even the most conspicuous variety, if rare enough, was eaten proportionately less often than the most cryptic one. Thus, there was strong negative frequency-dependent selection.

Popham's observations have been confirmed and extended to include a wide range of predators. Many experiments have used wild birds feeding on artificial baits. When green and brown baits are exposed on short grass at moderate densities, the birds show negative frequency dependence. However, when the density exceeds about twenty or thirty per square meter, the selection becomes positively frequency dependent. Mice hunting by smell for baits flavored with peppermint or vanilla show a similar reversal.

Many species of animals and plants are polymorphic in their external colors and patterns. Often the alternative morphs are clearly distinct and seem to use visual "tricks" to emphasize their distinctness. It is very likely that predators are important in maintaining these polymorphisms.

Another source of negative frequency-dependent selection is Batesian mimicry, in which a palatable species, the mimic, gains a benefit by resembling a distasteful or dangerous species, the model. [*See* Mimicry.] This strat-

agem works best when the mimic is rarer than the model, because otherwise predators will come to associate the joint phenotype with a reward rather than a penalty. The constraint on the numbers of the mimic can be overcome if it is polymorphic, with each morph mimicking a different model. A famous example is the African swallowtail butterfly (*Papilio dardanus*), which has several morphs, so distinct from each other that they were originally classified as separate species. The resemblance between each morph and its model, always a distasteful species of butterfly, is striking and exact. In Batesian mimicry, the selective value of each morph declines as its numbers increase relative to its specific model.

Sometimes prey can exert frequency-dependent selection on their predators. A striking example is *Perissodus microlepis*, a scale-eating fish that inhabits Lake Tanganyika. It has two morphs differing in the orientation of the mouth. One faces to the left, the other to the right. The difference is the result of a single gene. *P. microlepis* attacks its prey from behind. "Right-mouthed" individuals snatch scales from the prey's left flank and "left-mouthed" ones from the right flank. Within a population, the frequencies of the two phenotypes fluctuate around 50 percent. Frequency-dependent selection occurs because the prey species finds it easier to avoid attack if the predominant threat comes from only one direction. When the frequencies of the two types of predator are unequal, the rarer morph has more success than the more common one.

Selection by Parasites and Hosts. The British geneticist J. B. S. Haldane first made clear the central importance of disease in evolution. He also pointed out that disease organisms can exert negative frequency-dependent selection on their hosts, acting to maintain immunological and biochemical polymorphisms. [*See* Haldane, John Burdon Sanderson.] A genetically uniform host is liable to have parasites that are finely adjusted to its biochemistry and physiology. Even a small biochemical change in the host may give it some degree of resistance to the parasite. If so, and if the change is inherited, the new allele will spread. However, as its frequency increases, the advantage will decline, because the parasite evolves adjustments to the new type of host. This frequency dependence can produce balanced polymorphisms in both host and parasite, but it does not necessarily do so. If the length of the parasite's life cycle resembles that of the host, the "arms race" between the two may generate successive transient polymorphisms; the first altering the host, the second altering the parasite so that it is better adapted to the new host, and so on. If parasite generations are much shorter than host generations, however, the likelihood of polymorphism is greater. For example, a bacterium might adjust within a single host generation. If the bacterium's adaptation to one host reduces its ability to infect others, it can generate a smooth frequency dependence rather than alternating bouts of directional selection.

Experiments by the Canadian plant breeder H. H. Flor investigated the relations between flax and flax rust, one of the fungal parasites that attack it. He found that, for each gene conditioning the reaction of the host, there was a specific gene conditioning the pathogenicity of the parasite. Such "gene for gene" relations have since been found in many interactions between plants and fungi, plants and insects, and plants and nematode worms. In principle, these interactions can produce negative frequency dependence on parasites as well as hosts.

In many animals, genes responsible for resistance to disease are polymorphic. They include genes for the Toll group of receptors, which condition responses to bacteria and viruses, and genes in the major histocompatibility complex (MHC), which condition vertebrate immune responses. Members of the human MHC complex, the HLA genes, include the most polymorphic loci known. There is a strong presumption that they are maintained by frequency-dependent selection, although there is also evidence of heterozygous advantage.

The American biologist R. T. Damian suggested that parasites in a vertebrate host should acquire, on their surfaces, molecules (antigens) mimicking those of the host. The host would then be unable to react against the parasites without attacking its own antigens. Vertebrates have mechanisms to minimize immunity against self, including the elimination of most self-reacting cells during the early stages of fetal development. The parasites would gain from the mimicry, exerting selection in favor of variant hosts with different antigens.

Human diseases associated with autoimmunity, such as rheumatoid arthritis, myasthenia gravis, and some forms of diabetes, may be precipitated by exposure to pathogens that mimic human antigens. In reacting to the parasites, the host may occasionally become immune to some of its own cells. There is currently a debate about the importance of molecular mimicry in autoimmune diseases, many of which are associated with particular alleles in the HLA complex.

Selection by Competitors. If two species compete for food or other resources, one does not inevitably eliminate the other. A necessary condition for stable coexistence is that the numbers of each species should be limited more by competitors of their own species than by competitors from the other. This condition can be satisfied if they exploit somewhat different components of the environment (ecological niches). The same is true of genotypes.

The American geneticist H. Levene studied differential selection in two or more niches, and found that polymorphism is maintained if the weighted harmonic mean selective value of the heterozygotes in the various niches

is higher than that of the homozygotes. Although Levene's model gives the conditions for polymorphism in terms of constant selective values within each niche, as far as the whole population is concerned the selection is negatively frequency dependent, because the numbers of organisms in each niche are assumed to be independently regulated after the selection has taken place. Without this regulation, there is polymorphism only with conventional heterozygous advantage.

Many experiments have investigated competition between genotypes of fruit flies. When tested for frequency-dependent selection, a high proportion gave evidence of it. The causes of this selection are unknown, but they are presumed to be competitive interactions of the kind mentioned above. Replicates in different laboratories have not always given the same results, perhaps because of variations in methods of culture.

Other interactions between members of one species can give rise to negative frequency dependence. Many plants are polymorphic for alleles that prevent self-fertilization. Pollen can fertilize an ovule only if the male and female parts of the respective flowers express different alleles. Consequently, a plant carrying rare alleles has a higher frequency of successful fertilizations than one carrying common alleles. The extreme case of self-incompatibility polymorphism is sex itself.

Mating preferences for rare phenotypes have been reported in fruit flies and other insects. The mechanisms are not clear, although there is evidence of olfactory and auditory cues. Rats and mice seem to prefer mating with individuals that differ at MHC loci, a choice apparently determined by scent.

Other forms of behavior may have a frequency-dependent component. For example, aggression may be individually advantageous when rare, but disadvantageous when common. [See Human Sociobiology and Behavior, *article on* Behavioral Ecology.]

Theoretical Models of Frequency-Dependent Selection. The first mathematical model of frequency-dependent selection was made by J. B. S. Haldane, and the theory was greatly extended and developed by the American geneticist Sewall Wright. The simplest model of frequency dependence in a diploid species assumes that there are two alleles, A and a, and two phenotypes, the dominant (AA or Aa) and the recessive (aa). If the frequency of A is p, and the frequency of a is q ($p + q = 1$), if mating is random, and if there are no other disturbing influences, the frequencies of phenotypes in zygotes will be $1 - q^2$ for the dominant and q^2 for the recessive. If the selective values of the two phenotypes are $1 - t(1 - q^2)$ and $1 - tq^2$, respectively, where t is a constant, the change in frequency per generation will be

$$\Delta q = tq(1 - q)(2q^2 - 1)/(a + t(2q^2 - 1))$$

and there will be an equilibrium ($\Delta q = 0$) when $2q^2 - 1 = 0$ (i.e., when $q = 0.7071$). Plotting Δq against q will show that the equilibrium is stable.

[*See also* Game Theory; Genetic Polymorphism; Red Queen Hypothesis; Sex Ratios.]

BIBLIOGRAPHY

Ayala, F. J., and C. A. Campbell. "Frequency-dependent Selection." *Annual Review of Ecology and Systematics* 5 (1976): 115–138. A useful review, emphasizing experiments on competition in *Drosophila*, but giving no treatment of frequency-dependent selection by parasites.

Bell, G. *Selection*. New York, 1997. A good general text on selection, reporting some of the more recent work on frequency dependence.

Clarke, B. "The Evolution of Genetic Diversity." *Proceedings of the Royal Society of London* B 205 (1979): 453–474. Emphasizes the importance of frequency-dependent selection in maintaining genetic variation.

Clarke, B., L. Partridge, and A. Robertson, eds. *Frequency-dependent Selection*. London, 1988. A symposium volume containing the most comprehensive discussions of the subject.

Fisher, R. A. "On Some Objections to Mimicry Theory: Statistical and Genetic." *Transactions of the Royal Entomological Society of London* 75 (1927): 269–278. The first treatment of frequency-dependent balance between polymorphic Batesian mimics. A copy of this paper, in pdf format, is available at http://www.library.adelaide.edu.au/digitised/fisher/.

Frank, S. A. "Polymorphism of Attack and Defense." *Trends in Ecology and Evolution* 15 (2000): 167–171. A review of frequency-dependent selection by parasitic elements, emphasizing those originating within a single organism. It discusses conflicts between mitochondria and nuclear genes in male sterility, the spread and containment of B chromosomes, and the dynamics of segregation distortion.

Poulton, E. B. "Notes upon, or Suggested by, the Colours, Markings, and Protective Attitudes of Certain Lepidopterous Larvae and Pupae, and of a Phytophagous Hymenopterous Larva." *Transactions of the Entomology Society of London* (1884): 27–60. A remarkably prescient paper, the first statement of frequency-dependent selection, and the first suggestion that predators could maintain polymorphisms in their prey. Relevant extracts were reprinted, with a commentary, in *Biological Journal of the Linnean Society* 23 (1984): 15–18.

Wright, S. *Evolution and the Genetics of Populations*, vol. 2, *The Theory of Gene Frequencies*. Chicago, 1969. Chapter 5 gives a good introduction to the mathematics of frequency-dependent selection.

— Bryan C. Clarke

FUNDAMENTAL THEOREM OF NATURAL SELECTION

Evolution feeds on the variability in fitness of the units under selection. Without such variability there can be no change. R. A. Fisher's "growth-rate theorem," which we might call his "Little Theorem," quantifies this: *If a population consists of independent groups each with its own growth-rate (the factor by which it grows in a given time interval), the change in the population growth-rate in this interval is equal to the variance in growth-rates amongst the groups, divided by the population growth-rate.*

Fisher applied the Little Theorem to the gene as the unit of selection and thereby discovered "The Fundamental Theorem of Natural Selection" (FTNS), but it may equally well be applied in demography or economics. When all the growth-rates are the same and their variance zero the population growth-rate remains unchanged. When there are differences, however, the fastest-growing units will gradually dominate, increasing the average growth-rate ("fitness") of the population. Eventually only the "fittest" remains and all the variability in growth-rates has been consumed. Fisher's Little Theorem gives the exact rate of change in the overall growth-rate as a result of this selection process. It is easily proved mathematically (see, for example, Edwards, 1994, where in addition the background to FTNS is described and its subsequent history fully discussed).

The problem that Fisher (1930) faced was how to adapt his Little Theorem (itself never explicitly published by him) to the genetical problem, because the "fitness" of a particular allele at a particular locus at a particular time is a transient concept, depending, as Fisher repeatedly stressed, on the entire environment in which it finds itself at that moment, including the "environment" generated by all the other genes.

FTNS, in modern terminology, is: *The rate of increase in the mean fitness of any population at any time ascribable to natural selection acting through changes in gene frequencies is exactly equal to its genic variance in fitness at that time.* Thus, a population in which there is a range of variation in fitnesses (for example, some low, some middle, some high), can evolve more rapidly than one in which all the fitnesses are the same.

Enunciated rather cryptically by Fisher in *The Genetical Theory of Natural Selection* in 1930, he considered FTNS of fundamental importance because it precisely describes that component of evolutionary change which may be said to arise from gene frequencies evolving, and it thus captures the "natural selection" component of evolution. This component forms only part of the total evolutionary change. For, as Fisher wrote in the opening sentence of *The Genetical Theory*, "Natural Selection is not Evolution."

Though it was long before it was appreciated as such, FTNS is in fact a straightforward application of Fisher's (1918) linear model to the effect of a gene on genotypic fitness. Considering in the first instance a single locus with two alleles, *G* and *g*, Fisher found the regression of genotypic fitness on the number of *G* alleles in the genotype, 0, 1, or 2. He called the variance removed by this regression the *genetic variance* because it was attributable to the genes (now usually the *additive genetic*, or the *genic*, variance). The slope of the regression line is a measure of the *average effect* on fitness of substituting one allele for the other, and the underlying model partitions the genotypic fitnesses into the linear or "regression" component made up of the average effects—or *breeding values*—of the individual genes, and a residual component representing all the non-linear effects such as dominance and epistacy (the "error" in linear-model terminology).

In the usual regression manner, the mean genotypic fitness is the same whether calculated from the actual values or the regression values. However, a change of genotype frequencies, such as is brought about by selection, will in general change these means differently, that based on the actual values by a total amount straightforwardly calculated but that based on the regression values by a *partial* amount attributable to the change in gene frequencies alone. It is only this *partial change*, predicted by the linear model, to which FTNS refers.

The change in gene frequencies itself will not, in general, be given by the average effect of the gene substitution, but by the *average excess* in fitness of one gene over the other, computed directly for each gene from the genotypic fitnesses by weighting them according to the numbers of the gene they contain. Using continuous-time mathematics Fisher (1930) showed that the genic variance was a function of both the average effect and the average excess and corresponded exactly to the partial increase in the mean fitness. Not until Ewens (1989) gave a simple proof, following a suggestion by Price (1972), was this clearly understood.

Fisher's achievement was, however, not recognized at the time by the readers of *The Genetical Theory*, and in 1941 he gave a further exposition of it. He reiterated that "the effect that is wanted is only that due to the change in the frequencies of the different possible genotypes . . . which the change in gene frequency may in fact bring about." He pointed out that it was possible to think of the overall change in the genotypic frequencies brought about by selection as consisting of two steps, the first being the change that would lead to the new gene frequencies conditional on genotype frequencies that would make the partial increase in mean fitness equal to the total increase, whilst the second would be the remaining change from these genotype frequencies to the final ones, the gene frequencies remaining constant. The first part is the partial change to which FTNS refers. Lessard (1997) gave a detailed analysis of this point of view.

Unfortunately many erroneous statements and interpretations of FTNS exist in the literature, the commonest being that it refers to the total, not the partial, change in the mean fitness, and therefore that it is true in only a very restricted class of models. However, in his original presentation of the theorem Fisher did go on to discuss the total change through an equation incorporating the effect of the deterioration of the environment. Subsequent discussion by Price (1972), Ewens (1989), and Frank (1997, 1998) has clarified the interpretation of the partition of the total change into the additive part owing to the changing gene frequencies alone and the residual part.

The exact form of the elementary calculus formula

$$d(uv) = v\,du + u\,dv$$

is

$$\begin{aligned} d(uv) &= (u + du)(v + dv) - uv \\ &= v\,du + u'\,dv \end{aligned}$$

where $u' = u + du$. If v symbolizes the average effect on fitness and u the gene frequency then uv is the mean fitness and its change $d(uv)$ over an interval in which the gene frequency changes from u to u' may thus be partioned into $v\,du$ referring to the part arising from the change in gene frequency, the average effect being held constant, and $u'\,dv$ referring to the change in the average effect weighted by the new gene frequency. FTNS refers only to the first part. The total change in mean fitness can thus be considered as arising partly from changes in gene frequencies and partly from changes in average effects.

In time the misunderstandings surrounding FTNS will be forgotten and it will be accepted as the original inspiration for a number of fundamental insights into the operation of natural selection.

[*See also* Fisher, Ronald Aylmer.]

BIBLIOGRAPHY

Edwards, A. W. F. The fundamental theorem of natural selection. *Biological Reviews* 69 (1994): 443–474.

Ewens, W. J. An interpretation and proof of the Fundamental Theorem of Natural Selection. *Theor. Pop. Biol.* 36 (1989): 167–180.

Fisher, R. A. The correlation between relatives on the supposition of Mendelian inheritance. *Transactions of the Royal Society of Edinburgh*, 52 (1918): 399–433.

Fisher, R. A. *The Genetical Theory of Natural Selection.* Oxford: Clarendon Press, 1930; 2nd ed: New York, Dover Publications, 1958; variorum ed: Oxford University Press, 1999.

Fisher, R. A. Average excess and average effect of a gene substitution. *Ann. Eugen.* 11 (1941): 53–63.

Frank, S. A. The Price Equation, Fisher's Fundamental Theorem, kin selection, and causal analysis. *Evolution* 51 (1997): 1712–1729.

Frank, S. A. *Foundations of Social Evolution.* Princeton University Press, 1998.

Lessard, S. Fisher's fundamental theorem of natural selection revisited. *Theor. Popul. Biol.* 52 (1997): 119–136.

Price, G. R. Fisher's 'fundamental theorem' made clear. *Ann. Hum. Genet.* 36 (1972): 129–140.

— A. W. F. Edwards

FUNGI

Fungi and funguslike organisms encompass an estimated 1.5 million species. However, fewer than 80,000 (5 percent) have been described, a reflection of the often microscopic fungal thallus (body) and ephemeral, often minute, spore-producing structures (sporocarps). Some fungi, however, develop large spore-producing structures such as mushrooms, galls of corn smut, spongelike masses of morels, and the dark fingerlike tissues of Xylarias. Microscopic fungi may be evident from their symptoms, including leaf spots, rotten wood, lung diseases, circular lesions of ringworm, and dead frogs and insects.

Some fungi have flagellated cells, a mark of their evolution in ancient aquatic environments. Later fungal diversifications, however, occurred on land at the time other groups invaded early terrestrial habitats (see Figure 1). Some representatives of terrestrial groups later reentered aquatic environments. Conversely, flagellated fungi and funguslike organisms moved to land within living plants and animals. Many species escaped drought by forming desiccation-resistant diaspores.

Fossils of fungi are difficult to distinguish from those of other filamentous microbes, and they are most clearly identified when they closely resemble extant fungi. The earliest such fossil is of a fungus that lived 460 million years ago, one that is similar to extant arbuscular mycorrhiza-forming fungi (Glomales). "Molecular clock" estimates of fungal age based on sequence divergence of macromolecules extend fungal origins even farther back in time, with a current estimate placing fungi on Earth 1,000 million years ago. Over the last decade, new phylogenetic and DNA methods have provided a basis for reinterpretation of other lines of evidence of fungal evolution.

Makeup of Fungi. Typical members of the monophyletic kingdom Fungi have a filamentous body (mycelium composed of individual threads called hyphae) with apical growth, heterotrophic carbon acquisition, absorptive nutrition with glycogen as a storage product, and sexual and asexual spores produced on a common mycelium. An easy separation of fungi and funguslike organisms is, however, not possible, because widespread convergence and rapid divergence confound our efforts based solely on the morphology and biochemistry.

Modern DNA analysis reveals that certain structural and biochemical characters do separate fungi from funguslike heterotrophs, but the degree of evolutionary difference cannot be judged solely by such characters. Members of kingdom Fungi are distinguished by a combination of characters including haploid thallus due to meiosis of the zygote nucleus; wall polymers including chitin and glucans; and ergosterol as a major component of cell membranes. Lysine synthesis via the alpha ami-

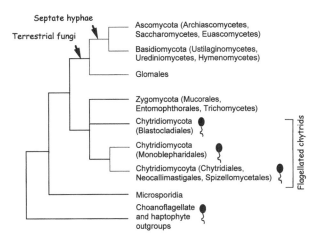

FIGURE 1. Members of the Kingdom Fungi Showing Microsporidia As a Basal Lineage.
There is a monophyletic grouping of the terrestrial fungi (Glomales, Ascomycota, Basidiomycota). Other traditional phyla may not be monophyletic. Groups bearing flagella indicated by symbol with tail.   Based on James et al., 2000; Keeling et al., 2000.

noadepic acid pathway contrasts with funguslike organisms that use the diaminopimelic acid pathway, as do plants. Compared with those of unrelated filamentous heterotrophs, fungal mitochondrial cristae are platelike rather than flattened, and the simpler Golgi bodies have a single cisternal element rather than stacked cisternae with differentiated proximal and distal faces. Only members of one phylum of fungi (Chytridiomycota) bear flagella, and their flagellar apparatuses were noted to be similar to those of choanoflagellates and metazoans before the advent of DNA analysis. Only flagellated fungi have centrioles associated with nuclear division; the nonflagellated fungi, however, have spindle pole bodies, apparently evolutionarily derived from centrioles, associated with nuclear division.

Phylogenetic Relationships and Classification of Fungi. Five major lineages are recognized as phylum-level taxa (Figure 1). Several divergent lineages have been removed from kingdom Fungi, including oomycetes and related heterotrophs (Hyphochytriales, Labyrinthulales, Thraustochytridiales) that have been classified as Straminipila (also known by the informal variants "stramenopiles" and "straminopiles" and as Chromista and Heterokonta). Their cells are biflagellate with anteriorly inserted flagella, one of which is smooth and the other covered with hairs. Other straminipiles include phototrophs with chlorophyll *a* and *c* (brown algae, chrysophytes, and diatoms). Other groups once placed among Fungi have been informally grouped as "slime molds," including plasmodial slime molds (Myxomycetes), cellular slime molds (Acrasiaceae and Dictyos-

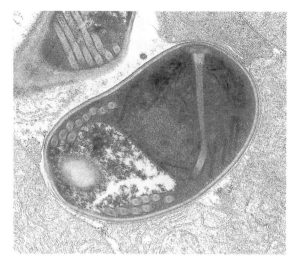

FIGURE 2. One of Eight Microsporidian Spores in the Cytoplasm of the Crustacean, Gammarus.
The thin enveloping membrane originally enclosing the spores has ruptured. The electron translucent polar filament is seen extending diagonally across the spore. Transmission electron micrograph, courtesy Earl Weidner.

teliaceae), and parasitic slime molds (Plasmodiophoraceae). Although these organisms are excluded from a monophyletic grouping of Fungi, they occupy the same habitats as fungi, have similar morphologies and symbiotic interactions, and continue to be studied by mycologists, who distinguish them from and compare them to true Fungi. In addition to the removal of these groups, molecular data have shown the Microsporidia to have affinity with Fungi; microsporidians likely comprise a sister group to, or perhaps are even members of, the kingdom Fungi. The placement of microsporidians, previously considered as basal protozoans, came about from phylogenetic analyses employing protein-coding genes in addition to the earlier use of nuclear-encoded ribosomal RNA genes. Such molecular analysis fill a critical need for phylogenetic studies, especially when organisms or genes have undergone rapid divergence.

Fungal Diversity.

Microsporidia (1,300 species). Molecular evidence has placed microsporidians within or as a sister group to the Fungi (Figure 1). The highly derived obligate parasites lack mitochondria and flagella and have a variety of animal hosts, including vertebrates (*Enterocytozoon*), but more than half are associated with insects (*Nosema*). The first formally described microsporidian was classified as a yeast, and they do share macroscopic features of yeast such as spore production. Spores arrive in an appropriate host gut where they are triggered to evert their unique polar filament for protoplast (sporoplasm) injection into the host (infective phase). The injected uninucleate or binucleate sporo-

plasm may be in direct contact with the host cytoplasm, where it undergoes nuclear and/or cytoplasmic divisions (proliferative phase). The onset of a third (sporogonic) phase is signaled by meiosis in the few microsporidians known to undergo meiosis and by thickened parasite cell membranes from accumulation of rough endoplasmic reticulum and secretions. Cytoplasmic divisions produce two to more than 100 walled spores (Figure 2) that are released from the host for reinfection.

Common ultrastructural and biochemical traits shared with fungi include intranuclear mitotic spindles, spindle pole bodies (microsporidian spindle plaques), paramural bodies, Golgi lacking stacked cisternae, and chitin in spore walls. Other fungal traits need to be investigated (e.g., presence of ergosterol, glycogen, and glucans in walls, and drug reactions).

Chytridiomycota (800 species). Somatic morphology is diverse, with some species having single-celled thalli and others developing extensive coenocytic (noncellular) mycelia. In asexual reproduction, posteriorly uniflagellated cells, hallmarks of shared ancestry with animals, are cleaved from the zoosporangium cytoplasm and then escape to settle on a substrate to undergo another round of the life cycle. Sexual reproduction is documented only in a few species. Phylogenetic studies delimit three lineages—a core group (Chytridiales, Neocallimastigales, Spizellomycetales), Monoblepharidales, and Blastocladiales—but it is not clear if these flagellated organisms constitute a monophyletic grouping, the traditional phylum Chytridiomycota, or paraphyletic groups at the base of the fungal tree (Figure 3).

The phylum includes cellulose-degrading anaerobes, exceptional for their multiflagellate condition with up to

FIGURE 3. Male Gametangia Develop Above Those of Female and Release Flagellated Gametes Earlier Than Those of the Female Shown Releasing Here.
Allomyces is one of few Chytrids with well documented sexual reproduction. Unlike most fungi it has alternating haploid and diploid generations in its life cycle. Light micrograph, M. Blackwell.

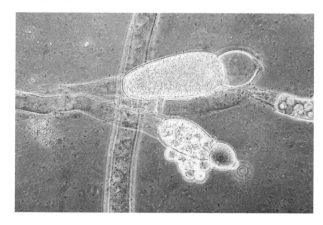

twelve flagella (Neocallimastigales), that proliferate in the gut of vertebrate herbivores. Chytrids garnered attention in the late 1990s as the agent of decimation of amphibian populations worldwide. Others parasitize aquatic invertebrate animals, plants, and fungi and often kill their hosts; some are plant pathogens that can occupy arid lands by living within the host. The majority of species, however, are saprobic, obtaining fixed carbon from dead bodies of plants and animals in aquatic habitats. Unusual features of some arthropod-parasitic chytrids are wall-less thalli and obligate use of two hosts (mosquitoes and copepods) in the life cycle (*Coelomomyces*); exceptional fungal life cycles with sporic meiosis, resulting in an alternation of haploid and diploid generations, occur in *Coelomomyces* and other Blastocladiales.

Zygomycota (1,100 species). Members of Zygomycota are largely terrestrial fungi, but a few are parasites of aquatic organisms (diatoms). They have been distinguished from chytrids by the absence of flagellated cells and presence of certain structures in sexual reproduction. The thallus typically is a well-developed coenocytic mycelium, although it is reduced and/or septate in several groups. Some species are referred to as dimorphic, and these are characterized by having a mycelial thallus under certain environmental conditions and a yeast stage under others. Hyphal fusions do not occur in zygomycetes.

Traditional zygomycetes comprise several monophyletic lineages characterized by distinct life histories. Members of Mucorales (*Mucor*, *Rhizopus*) produce abundant sporangiospores cleaved from the sporangial cytoplasm; when it occurs, sexual reproduction is usually accomplished by outcrossing with two mating types in each species. The functional gametes are nuclei of the fusing hyphalike gametangia (plasmogamy, the first step in fertilization). A zygosporangium results from plasmogamy, and a single zygospore is cleaved within. One or more nuclear fusions (karyogamy, the second step in fertilization) occur in the zygospore, and this cell is the functional zygote. The zygote is the only diploid stage in the life cycle, and is the site of meiosis. A second lineage (Entomophthorales) is characterized by reduced hyphal development, relatively undifferentiated asexual structures, and inbreeding (homothallic) sexual systems. These may be parasites of insects as their name implies; many more are saprobic or parasitic on soil invertebrate animals and even fern gametophytes. A third lineage (Zoopagales) has simple thalli but elaborate sexual reproductive structures. Some prey on invertebrate animals using hyphal traps "baited" with chemical attractants. Other fungal and animal parasites among the zygomycetes also parasitize small animals and fungi. Species of one of the most derived groups (Trichomycetes) live in the hind gut of arthropods, including crus-

taceans, aquatic insect larvae, and millipedes. Their life cycles are coordinated with those of the arthropods and may respond to host molt hormone by switching spore production from thin-walled spores to specialized thick-walled spores that better survive the external environment.

Glomales (200 species). With one exception, species in this phylum have obligate symbiotic relationships with the roots of approximately 80 percent of terrestrial plant species. The fungus-root structures formed by these fungi are called arbuscular mycorrhizae. *Geosiphon pyriforme*, placed by ribosomal RNA data within or at the base of the glomalean phylogenetic tree, is exceptional in its association with Cyanobacteria. Asexual reproduction is by spores that can be sieved from the surrounding soil. Sexual reproduction is not known. Traditionally the Glomales phylum was included among the zygomycetes, but molecular data place this monophyletic group basal to ascomycetes and basidiomycetes (Figure 1). The oldest fungal fossil assignable to a taxonomic group is a glomalean fungus collected in sediments dated 460 million years old.

Hyphae of arbuscular mycorrhizal fungi (AMF) grow between root cortical cells and penetrate the cell wall to cause invagination of the cell plasma membrane. Fungal hyphae produce highly branched structures, arbuscules with great surface areas in contact with invaginated plant membranes; this is the site of nutrient exchange between fungus and root. Glomalean fungi are efficient at phosphorus uptake and completely dependent on the plant as a source of carbohydrates. Certain AMF produce vesicles with storage function that act as nutrient sinks. AMF also produce antibiotics to prevent infection by root pathogens.

Basidiomycota (23,000 species). Members of the phylum Basidiomycota are well known because the group includes species with large, persistent sporocarps (agarics and polypores), as well as plant pathogens (rusts and smuts) with conspicuous symptoms. The diverse phylum (Figure 1) also contains microscopic forms. The thallus of most basidiomycetes is hyphal, and extensive mycelia can develop underground from a single spore, sometimes occupying >500 hectares. Hyphae are separated into compartments by internal walls, the septa, which are perforate. The wall around a septal pore may be slightly flared or thickened and donutlike, the so-called dolipore septum. Dolipore septa may have septal pore cap membranes acting as sieves to block nuclear passage. A few basidiomycetes grow as single-celled yeasts, either throughout their life cycles or at certain stages in their life cycles (e.g., plant parasitic smut fungi).

Basidiomycetes are characterized by basidiospore production in sexual reproduction. In hymenomycetes, spore-producing cells (basidia) are located in sporocarp

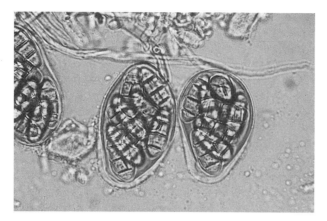

FIGURE 4. Two Asci Released from a Broken Ascocarp. Each ascus contains eight septate ascospores.   Light micrograph, M. Blackwell.

tissue (e.g., mushroom gills). Others (rust and smut fungi) lack sporocarps. The club-shaped basidia of hymenomycetes arise in gill tissue and are the site of karyogamy and meiosis. After meiosis of the zygote nucleus, each of the four haploid nuclei migrates into a stalk (sterigma) that develops at one end of a basidium and into basidiospores as they bud out and eventually separate from the basidium (Figure 7). Basidiospores are forcibly discharged from basidia and dispersed by air currents. On a suitable substrate, spores germinate to produce hyphae that form a primary mycelium, which may fuse with a second primary hypha soon after spore germination (plasmogamy), after which each hyphal compartment contains two nuclei. Hyphal fusion initiates a long-lived dikaryotic secondary mycelium. In some basidiomycetes a specialized hyphal branch, the clamp connection, forms at the apical side of each new septum and hooks back for passage of nuclei and maintenance of dikaryons in apical cells. Agarics have mating alleles that control nuclear fusion. Two mating systems are known. A bipolar system (*Heterobasidion*) has alleles at a single locus that control the fusions. The more complex tetrapolar system (*Schizophyllum*) has several closely linked alleles at each of two distinct unlinked loci, which direct the mating process by controlling events (e.g., membrane fusion, nuclear pairing and migration, and clamp formation). Outcrossing is the rule, and many basidiomycetes have multiple alleles, unique for fungi, at the mating loci, which dramatically increase the opportunity for outcrossing. Hyphal fusion (vegetative compatibility) is controlled by mating genes. Some produce asexual spores (mitospores or conidia) and other propugules (e.g., thick-walled resistant spores and hyphal fragments). Asexual reproduction usually precedes sexual reproduction, but some basidiomycetes reproduce solely by asexual means.

In addition to many plant pathogens (*Puccinia*, *Ustilago*, *Jola*), basidiomycetes are animal (*Cryptococcus*) and fungal pathogens (*Asterophora*). Many are saprobic and are efficient wood rotters (*Merulius*); a number of species form symbiotic associations with plants, often as ectomycorrhizae (*Boletus*, *Amanita*) or endophytes. Old World termites and New World ants cultivate basidiomycetes, and female wood wasps deposit them with their eggs for larval nutrition. Basidiomycetes are associated with insects for spore dispersal and may produce chemical attractants, as do stinkhorns. Visual cues attract pollinators to pseudoflowers formed by rust fungi on host leaves. Mushroom toxins affect mammalian cell division, autonomic and central nervous system responses, and gastrointestinal tract function. Notable among these are the deadly cyclopeptides (*Amanita*) and hallucinogenic psilocybin (*Psilocybe*) causing the sensation of minuteness of the sort experienced by the title character in *Alice in Wonderland*. Toxins protect basidiocarps against nematodes and other invertebrate animals. Other animals specialize on basidiocarps for feeding and rearing young, and scale insects receive protection from fungal thalli (*Septobasidium*) in exchange for the "sacrifice" of occasional scales to the fungi.

Ascomycota (33,000 species). Ascomycota is the largest phylum as measured by the number of described species and contains great diversity in morphology, habit, and life history. Three classes have been distinguished through DNA data (Figure 1). Thalli may be single cells (yeasts such as *Saccharomyces* and *Schizosaccharomyces*) or extensive mycelial systems (*Neurospora*). Interconvertible dimorphic forms are characteristic of mammalian pathogens growing as yeasts in their hosts and as mycelia on the outside. Hyphal compartments are divided by simple septa without dolipores or

FIGURE 5. Mite-Dispersed Ascospores Shown Near Mite Mouth Using Scanning Electron Microscope.
Light micrograph, Meredith Blackwell.

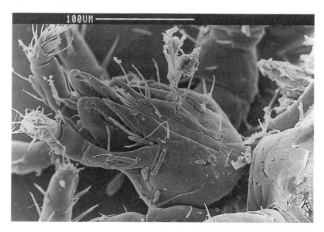

septal pore caps. The dikaryon is of short duration, immediately preceding karyogamy in a saclike cell (ascus). Asci (Figure 4) are produced within single cells, directly on hyphae, or within sporocarps. The short-lived zygote nucleus in the ascus undergoes meiosis, usually followed by mitosis, to produce eight haploid nuclei, but one to thousands of haploid nuclei may develop. Cytoplasm is cleaved to enclose nuclei, and walls develop to complete ascospore development; ascus pressure results in forcible ascospore discharge. Some ascospores (*Ophiostoma*) are specialized for arthropod dispersal (Figure 5). Sexual fusions of heterothallic ascomycetes are under the control of nonhomologous mating genes at a single locus. Hyphal fusion (vegetative compatibility) is controlled by genes other than those that control mating. Most ascomycetes reproduce asexually by production of mitospores (conidia). Conidia (Figure 6) and differentiated subtending hyphae are more complex than those of basidiomycetes. Some ascomycetes reproduce only asexually, a derived condition. Because fungal classification has been based on structures associated with sexual reproduction, asexual ascomycetes (and basidiomycetes) previously were grouped under artificial systems. Now that DNA data can determine the closest relatives of organisms, asexual fungi are classified as mitotic members of various taxa, not as a separate group of Fungi Imperfecti or Deuteromycetes as they once were.

Ascomycetes are saprobes and pathogens. They attack animals, including humans (*Pneumocystis*, *Histoplasma*), and excel as plant pathogens (*Diaporthe*), causing huge crop losses. Ascomycete interactions with photosynthetic organisms provide textbook examples of symbiosis. These include not only pathogenic interactions, but also lichens (with green algae and Cyanobacteria), symptomless leaf endophytic infections (*Balansia*) with virtually all land plants, and mycorrhizae

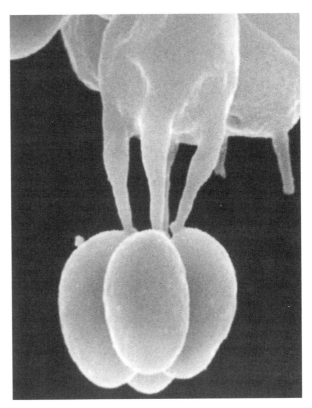

FIGURE 7. Basidiospores at the Tips of Sterigmata Issuing from the Basidium Apex.
Gravity ensures that basidia will be aimed for forcible spore release away from the sporocarp. Scanning electron micrograph, courtesy of Charles W. Mims.

FIGURE 6. Spiral-Shaped Multicelled Conidia of an Ascomycete Shown with High Magnification of a Light Microscope.
Light micrograph, Meredith Blackwell.

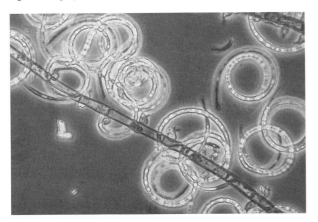

formed by a few ascomycetes. Ascomycetes have many associations with insects, including scolytid (bark) beetles that have specialized exoskeleton pockets (mycangia) to carry a fungal inoculum. The fungi may serve as a beetle food resource, but also benefit beetles by overcoming tree defenses. Diseases caused by tree pathogens may decimate forests (Dutch elm disease, chestnut blight). Yeasts are invaluable in production of fermented foods and drinks. Other yeasts and yeastlike fungi participate in detoxification and nitrogen recycling in insect guts. Ascomycetes produce secondary metabolites to interact with animals. Alkaloids from grass endophytes affect development of insects feeding upon them. A close endophyte relative, ergot, has alkaloids that cause hallucinations and excruciating symptoms of vasoconstriction. Sulfur compounds produced by some ascomycetes attract insects. False morels produce lethal toxins in their large sporocarps (ascocarps). Other ascocarps are valued as food (morels), and truffles may be costlier per gram than gold. At the right concentration, toxins synthesized by ascomycetes control bacterial infections and reduce high cholesterol levels. Mycotoxins (aflatoxin, tremorgenic toxin, and vomitoxin) cause hemorrhagic enteritis, suppression of immunity, convulsions, vomiting, liver injury, and other symptoms.

[*See also* Disease, *article on* Infectious Disease; Lichens.]

BIBLIOGRAPHY

Alexopoulos, C. J., C. W. Mims, and M. Blackwell. *Introductory Mycology.* 4th ed. New York, 1996. Advanced-undergraduate- and graduate-level text that is phylogenetically arranged and covers life cycles, biology, organismal relations, and the fossil record of fungi and funguslike organisms.

Baldauf, S. L., A. J. Roger, I. Wenk-Siefert, and W. F. Doolittle. "A Kingdom-Level Phylogeny of Eukaryotes Based on Combined Protein Data." *Science* 290 (2000): 972–977. Presents a phylogenetic tree and discussion of discrepancies based on analysis of several protein-coding genes versus ribosomal genes.

Blackwell, M. "Evolution—Terrestrial Life—Fungal from the Start?" *Science* 289 (2000): 1884–1885. A summary of fungal symbiotic associations with plants and, especially, animals that suggests where to look to discover new fungal fossils. Provides references to many earlier discussions of the topic.

Dix, N. J., and J. W. Webster. *Fungal Ecology.* London, 1995. Textbook introduction to fungal ecology, including discussions on field observations, experimental manipulations, and ecophysiology.

Griffin, D. *Fungal Physiology.* 2d ed. New York, 1993. Offers a recent basic review of fungal physiology.

Hawksworth, D. L., P. M. Kirk, B. C. Sutton, and D. N. Pegler. *Ainsworth and Bisby's Dictionary of the Fungi.* 8th ed. Wallingford, 1995. A single-volume illustrated text essential for information on fungi, providing a classification (although one that does not always adhere to principals of phylogenetic systematics), discussions of special topics such as nomenclature, literature, and genetic systems, and all described taxa to genus level.

Heckman, D. S., D. M. Geiser, B. R. Eidell, R. L. Stauffer, N. L. Kardos, and S. B. Hedges. "Molecular Evidence for the Early Colonization of Land by Fungi and Plants." *Science* 293 (2001): 1129–1133. Discusses the age of fungi based on estimates from extrapolated protein sequence divergence using calibration points from the fossil record.

James, T. Y., D. Porter, C. A. Leander, R. Vigalys, and J. E. Longcore. "Molecular Phylogenetics of the Chytridiomycota Supports the Utility of Ultrastructural Data in Chytrid Systematics." *Canadian Journal of Botany* 78 (2000): 336–350. Phylogenetic study that addresses evolution of the basal fungi and the monophyly of flagellated forms and problems of long branch attraction.

Keeling, P. J., M. A. Luker, and J. D. Palmer. "Evidence from Beta-tubulin That Microsporidia Evolved from Fungi." *Molecular Biology and Evolution* 17 (2000): 1–9. Discusses placement of microsporidians as fungal relatives rather than as basal amitochondriate protists.

Kwon-Chung, K. J., and J. E. Bennett. *Medical Mycology.* Philadelphia, 1992. Comprehensive text that discusses the major fungal pathogens of humans and provides information on treatment.

Kurtzman, C. P., and J. W. Fell, eds. *The Yeasts: A Taxonomic Study.* 4th ed. Amsterdam, 1998. Comprehensive coverage of the yeasts, a polyphyletic assemblage of economically and biologically important fungi, that arranges the species phylogenetically.

Lutzoni, F., M. Pagel, and V. Reeb. "Major Fungal Lineages Derived from Lichen-Symbiotic Ancestors." *Nature* 411 (2001): 937–940.

Margulis, L., J. O. Corliss, M. Melkonian, and D. J. Chapman, eds. *Handbook of Prototista.* Boston, 1990. Includes discussion of many groups of protistan-grade organisms, including funguslike groups.

McLaughlin, D. J., E. G. McLaughlin, and P. A. Lempke, eds. *Mycota.* Vol. 7, Parts A and B, *Systematics and Evolution.* Berlin, 2001. Multiauthored review of fungi and straminipiles to the level of family; essential as a starting point for future studies.

Suh, S.-O., H. Noda, and M. Blackwell. "Insect Symbiosis: Derivation of Yeast-like Endosymbionts within an Entomopathogenic Lineage." *Molecular Biology and Evolution* 18 (2001): 995–1000. Provides a review of insect-yeast associations.

Wittner, M., and L. M. Weis, eds. *The Microsporidia and Microsporidiosis.* Washington, D.C., 1999. Provides an essential introduction to the group of obligate pathogens of animals, especially of insects, through discussions of many facets of their biology, host distributions, and physiology.

— MEREDITH BLACKWELL

G

GALAPAGOS ISLANDS

Archipelagos provide some of the clearest evidence of adaptive radiation: the differentiation of a single ancestral species into many derived species, each adapted to a different way of life. A classical example is Darwin's finches, comprising fourteen closely related species living on several islands in the Galapagos archipelago. Here, well isolated and in the absence of all but a handful of land bird species, Darwin's finches have radiated on different islands into niches occupied on the South American continent by members of other families. Significantly, on the single island of Cocos, about 700 kilometers (1,120 miles) away, there is but one species of this group of finches. [*See* Darwin's Finches.]

The Galapagos comprise an archipelago of eighteen major volcanic islands and many more smaller ones in the Pacific Ocean, astride the equator and approximately 1,000 kilometers (1,600 miles) from the coast of continental Ecuador. Like the Hawaiian Islands, they were formed by volcanic activity at a hot spot, but unlike those islands, this has resulted not in a linear chain of successively older islands but in a two-dimensional array. The hot spot is under the western island of Fernandina. The oldest current island, San Cristóbal, is farthest from it in an east-southeasterly direction and is approximately four and a half million years old. Other islands dating back from ten million to twenty million years have sunk beneath the surface. Molecular data indicate that marine iguanas have been on the Galapagos for more than ten million years, much longer than the age of the current islands, whereas Darwin's finches in contrast have been present for an estimated two million to three million years.

Despite their equatorial location, the islands have a strongly seasonal climate, with a relatively dry and cool season and a hot and wet season. The length of the wet season and the amount of rain vary annually. El Niño years of abundant rainfall contrast with La Niña years of drought, each recurring at roughly four-year intervals. The consequences of climatic variation are (1) uplands are forested and generally moist, whereas lowlands are more sparsely vegetated and seasonally arid, and (2) conditions for terrestrial organisms oscillate between years of highly suitable conditions for breeding and years of stressful conditions that imperil survival.

The islands have never been connected to the mainland of South America, yet, despite their remoteness, their native animal and plant inhabitants show clear affinities to organisms on the South American continent. Charles Darwin was aware of this resemblance when he visited the Galapagos in 1835 as the naturalist on board the *Beagle* on its epic voyage around the world. He encountered tortoises, marine and land iguanas, finches, a few insects, and many plants, which he collected and described in his notes. He regretted such a short visit (four islands visited in five weeks of the dry season); nevertheless, in finding mockingbirds of different appearance on three of the islands, he made a crucial observation that contributed to the initial development of his evolutionary thinking. The mockingbirds showed him that species change, and the finches helped him to understand how and why they change.

— Peter Grant and Rosemary Grant

GALTON, FRANCIS

(1822–1911), polymathic man of science who wrote on many topics, including exploration, meteorology, anthropology, heredity, statistics, psychology, eugenics, and criminology. Charles Darwin's half-cousin, Galton spent a rather aimless youth, training in medicine and obtaining a mediocre mathematics degree from Cambridge University. His father's death in 1844 left him financially secure, which allowed him to indulge his tastes for adventure, traveling in the Middle East, exploring Africa, and investigating weather patterns. The publication of Darwin's *Origin of Species* transformed his intellectual life, removing any lingering religious sentiments and providing Galton with a more focused mission. He turned his attention to problems of human biology, including inheritance, psychology, anthropometrics, and development. His passion was to measure and quantify, in the service of which he also made important contributions to statistics.

Although Galton worked from the 1860s within a general evolutionary framework, his writings intersected evolutionary theory in several different areas. First, despite his admiration for Darwin, he was an early and severe critic of his cousin's theory of inheritance through the mechanism Darwin called pangenesis. Darwin postulated that small particles he called "gemules" were thrown off by the somatic cells and somehow influenced the reproductive cells, to be then passed on to offspring. Galton conducted a series of experiments with rabbits, transfusing them with blood and looking for evidence that Darwin's gemules influenced the characteristics of subsequent offspring. Finding that the young rabbits were like their parents rather than the rabbits whose blood had been transfused, he concluded that the heredity material, whatever it is, is confined to the reproductive elements, the egg and the sperm. He coined the word *stirp* to denote the totality of the germinal material transmitted from parents to offspring, a concept that had much similarity with the German biologist August Weismann's later notion of the germ plasm.

Second, this research confirmed Galton in his belief in "hard" heredity, that is, that inheritance is unaffected by characteristics acquired by the parents. One of the earliest characteristics whose inheritance he studied was what he variously called "talent," "genius," and "intelligence." In the paper "Hereditary Talent" and his first "Darwinian" book, *Hereditary Genius* (1869), Galton used pedigree studies to argue that these attributes run in families in patterns that cannot be explained by opportunity, education, wealth, or other environmental factors. As he later formulated it, nature is far more important than nurture in molding the attributes, physical as well as mental, of organisms. This was the basis of his theory of eugenics, a word he coined in 1883, whereby human society was to be improved by those with desirable attributes being encouraged to breed (positive eugenics), whereas those with undesirable traits (e.g., criminality, subnormal intelligence) were discouraged or prevented from breeding by segregation or sterilization (negative eugenics).

The continuity of the stirp from generation to generation meant that the individual's own stirp was an ancestral product, the parents contributing half, the grandparents between them a quarter, and so on. This "law of ancestral heredity" allowed Galton to explain how characteristics might reappear in an individual, having lain dormant for several generations. At the same time, Galton's own measurements and calculations showed that an individual possessing some unusual attribute, great height or high intelligence, for example, would have offspring tending to revert to the average for the species. As a motor of evolutionary change, he was forced to invoke the idea of the "sport," or "hopeful monster." He was thus an evolutionary saltationist, almost in spite of himself, his ordinary model of inherited generational relationships emphasizing stability. Evolutionary change thus occurred through specific leaps, or saltations.

The measurement of these relationships led Galton to develop his statistical ideas of regression and correlation, mathematical representations of the degrees of relatedness between individuals or groups. In the hands of his disciple and admirer Karl Pearson, Galton's concepts were developed into the modern discipline of statistics, which in turn formed the basis of biometry, the movement that attempted to explain evolution through mathematical methods, based on the assumption of continuous variation between generations. Galton also bequeathed the biometricians another important insight, that whole populations of organisms need to be investigated if evolution is to be understood. Pearson was the first holder of a chair in eugenics (now genetics) that Galton endowed at University College London. Between them, Galton and Pearson were instrumental in laying the groundwork for the switch from "type" (i.e., individual) to population thinking in evolutionary biology.

Galton was never an especially sophisticated mathematician, nor was his principal focus evolutionary theory per se. Rather, he studied human beings as products of evolution. In his consistent application of the tools of science to anthropology in its broadest sense, however, he worked comfortably within the Darwinian paradigm.

[*See also* Eugenics.]

BIBLIOGRAPHY

Bowler, P. J. *The Mendelian Revolution.* London, 1989. Contains an excellent brief discussion of Galton's ideas on heredity.

Forrest, D. W. *Francis Galton: The Life and Work of a Victorian Genius.* London, 1974. A sound biography, although now getting dated.

Galton, F. *Natural Inheritance.* London, 1889. Galton's mature exposition of his ideas on heredity.

Galton, F. *Memories of My Life.* London, 1908. Galton's autobiography, published in his old age.

Gillham, N. W. *A Life of Sir Francis Galton: From African Exploration to the Birth of Eugenics.* Oxford, 2001.

Keynes, M. *Sir Francis Galton, F.R.S.: The Legacy of His Ideas.* 1993. A useful collection of essays on many aspects of Galton's ideas and researches.

Pearson, K. *The Life, Letters and Labours of Francis Galton.* 3 vols. in 4 parts. Cambridge, 1914–1930. A massive biography, based on the Galton Archive, now at University College London. Very useful, although uncritical.

Porter, T. M. *The Rise of Statistical Thinking, 1820–1900.* Princeton, N.J., 1986. A good exposition of Galton's statistical ideas and their influence on Pearson.

— W. F. BYNUM

GAME THEORY

Evolutionary game theory is a way of thinking about evolution when the best thing for an individual to do depends on what other members of the population are doing—"best" here meaning that which maximizes the individual's fitness. For example, if most members of a population are timid and likely to run away if threatened, it may pay an individual to threaten; but if most members respond to a threat by attacking and if fights are expensive, threatening would not be a good tactic. Faced with such problems, the game theoretic approach is to look for an evolutionarily stable strategy (ESS). The word *strategy* need not refer to a behavior; it can be any aspect of the phenotype. The notion of evolutionary stability implies that, once the members of a population have the appropriate phenotype, evolutionary change will cease.

Verbally, strategy X is an ESS if, when almost all members of a population do X, the population cannot be invaded in evolutionary time by any alternative, mutant strategy, Y. This condition of noninvasibility will be satisfied provided that the benefit to an individual adopting X, interacting with another X, is greater than the benefit to any alternative strategy, Y, against a typical individual adopting X.

This verbal description of an ESS must now be formalized. First, we need to define *benefit*, or *payoff*, the term used in game theory. The payoff to an individual X, interacting with an individual Y, written $E(X,Y)$, means the change in the fitness of X resulting from an interaction with Y. Imagine a population consisting mainly of X individuals, with a small proportion (p) of Y individuals. Suppose that each individual interacts with a random opponent (it makes no difference if individuals interact with a series of random opponents). Then the fitnesses denoted by W of X and Y individuals are:

$$W(X) = W_0 + (1 - p) E(X,X) + pE(X,Y)$$
$$W(Y) = W_0 + (1 - p) E(Y,X) + pE(Y,Y), \qquad (1)$$

where W_0 is the fitness of all individuals before the interaction.

Provided that $W(X) > W(Y)$ for all possible alternative strategies Y, then a population of individuals adopting X cannot be invaded by any mutant. Remembering that p is small, that is, $p \simeq 0$, we can write down the following conditions for strategy X to be an ESS:

either
$$E(X,X) > E(Y,X)$$
or
$$E(X,X) = E(Y,X) \text{ and } E(X,Y) > E(Y,Y) \qquad (2)$$

This derivation of the conditions for X to be uninvadeable has made some tacit assumptions, which need to be spelled out. First, it assumes that individuals produce offspring identical to themselves: X produces X and Y produces Y. In other words, it is a model of an asexual population with perfect heritability. In practice, the model works well for a sexual population provided that there is additive heritability—that is, if offspring tend to resemble their parents. Second, the model assumes that individuals interact in pairs. It can, however, be applied whenever the fitness of an individual with phenotype X depends on the proportions of X and Y individuals in the population—that is, when fitnesses are frequency dependent. An example in which interactions are not pairwise—the evolution of the sex ratio—is discussed below.

The term *game theory* was borrowed from the classical *Theory of Games*, first developed by Von Neumann and Morgenstern to analyze how humans should behave in situations of conflict—a game of poker or a negotiation between employer and trade union, for example. The basic assumption was that a player should behave "rationally" so as to maximize his payoff, assuming that his opponent would also behave rationally. It is important to emphasize that evolutionary game theory makes no assumption of rationality: reason is replaced by fitness maximization. Thus, the notion of an ESS has been applied, not only to animal behavior, but to plant growth and even to the evolution of viruses. The best growth strategy for a plant, for example, depends on what others are doing, because a plant must not be overshadowed by its neighbors. The evolutionarily stable growth strategy for plants in a pure stand devotes far more energy to growth, and therefore less to reproduction, than would be optimal for the productivity of the population as a whole. Leaves are organs of competition as well as of photosynthesis.

The biological problem for which the idea of an ESS was invented was the evolution of ritualized behavior during animal fights. Why do animals often signal to one another during contests, instead of fighting in an escalated manner? The Hawk–Dove game, invented to analyze this problem, will be used to introduce the notion of a "payoff matrix" and to discuss the types of equilibrium that can arise.

Suppose that two animals are competing for a resource of value (V). Two alternative strategies are available to them:

Hawk (H) : escalate and continue until injured or opponent retreats.
Dove (D) : display, then retreat at once if opponent escalates.

If two opponents both escalate, it is assumed that one is injured and forced to retreat, and the other obtains the resource. Injury reduces fitness by a cost (C). These assumptions can be summarized in a payoff matrix:

	H	D
H	$(V - C)/2$	V
D	0	$V/2$

In this matrix, the entries are the payoffs to an individual adopting the strategy on the left, if its opponent adopts the strategy above. For example, the payoff to H when its opponent plays H is calculated as follows. Half the time H wins, receiving a payoff (V). Half the time H loses, suffering a cost (C). Thus, the expected payoff over many contents is

$$\frac{1}{2}V - \frac{1}{2}C = (V - C)/2$$

We now seek the ESS of this matrix. Suppose that $C > V$: if the chances of victory and injury are equal, an escalated fight is not worthwhile. For example, suppose that $C = 4$ and $V = 2$. The payoff matrix becomes:

	H	D
H	−1	+2
D	0	+1

It is easy to see that neither H nor D is an ESS: if everyone else plays H, it is better to play D, and vice versa. So what will happen? The stable solution is for an individual to adopt a "mixed" strategy—that is, sometimes to play H and sometimes D. Thus, consider the strategy M = (play H with probability P, and D with probability $1 − P$). If M is an ESS, it is common sense (and can be proved) that, when everyone is playing M, the payoff to an individual when, by chance, he plays H must equal his payoff when he plays D: if this were not so, it would pay him to abandon M, and always play the better pure strategy, H or D. A simple calculation shows that, for the chosen values of V and C, equality is achieved when P = 0.5, and the average payoff is also 0.5: 0.5 (−1) + 0.5 (2). Of course, this does not prove that M is an ESS, but only that, if there is an ESS, it must be M. However, checking condition 2 shows that M is indeed an ESS.

This ESS could be achieved in two ways. Either there could be a genetically uniform population of individuals adopting strategy M, or there could be a genetically polymorphic population of 50 percent H individuals and 50 percent D individuals. Formally, we have only shown that M is stable, but in practice it is usually the case that, if there is a mixed ESS, the corresponding polymorphic population is also stable. The principle that, if there is a mixed ESS, the payoffs when playing the alternative strategies must be equal is important when applying evolutionary game theory to specific cases. For example, it is sometimes the case in sexual species that there are two types of males—aggressive (H-strategy) males holding territories or harems, and "sneakers" (D-strategy). If the ESS is the mixed strategy, then the average lifetime fitness of the two types should be equal. The alternative is that all individuals should adopt a "conditional" strategy: if, for genetic or environmental reasons, a male is of high quality, adopt the aggressive strategy; otherwise, adopt the sneaker strategy. In the latter case, the fitness payoffs to the two strategies need not be equal.

Is the strategy M, in which individuals "choose" what to do randomly, really the best they can do? It seems that it is, but a subtle change in the "rules" makes an alternative strategy possible. What would rational humans do confronted with the Hawk–Dove game? Surely they would agree to toss a coin for the resource. By avoiding costly fights, the average payoff would be 1.0, instead of 0.5, which is the best that they can expect playing M. Animals cannot toss a coin. But if there is any asymmetry associated with each contest and known to both contestants, it can be used to settle it. The obvious

asymmetry is "ownership," or prior possession: for example, contests over a territory are often between an "owner" and an "intruder." This suggests a third strategy, "bourgeois," (B): play H if owner; play D if intruder. The payoff matrix now becomes:

	H	D	B
H	−1	+2	0.5
D	0	+1	0.5
B	−0.5	1.5	1.0

The "bourgeois" strategy is the only ESS of this matrix, having the highest average payoff against the other strategies. Note that this conclusion does not depend on any assumption that ownership makes an individual more capable of winning an escalated fight. The idea that ownership can be used to settle contests over an indivisible resource is often borne out. But complications often arise, in particular because more than one asymmetry may be associated with a contest (e.g., one contestant may be larger). There is a class of games known to theorists as consensus games. In such games, although the contestants have opposed interests, they also have a common interest in reaching a consensus without escalation. In such cases, animals may exchange signals to reveal an asymmetry (e.g., in size or need for the resource being contested), which is then used to settle the contest.

The Hawk–Dove–Bourgeois game is an example of a wide class of games, so-called asymmetric games: there is a difference, in this case of ownership, that is known with certainty to both contestants. Such games are common, for example, in interactions between a male and a female, or (as discussed below) between a signaller and a receiver of the signal. There is an important theorem about such games, derived by Selten: an asymmetric game cannot have a mixed ESS; both contestants must adopt a pure strategy.

The first use of game theoretic thinking in biology was R. A. Fisher's discussion of the evolution of the sex ratio. Why, he asked, is the ratio of males to females at birth usually 1:1? One might reply because meiosis produces equal numbers of X and Y sperm. But if it paid to produce a ratio different from 1:1, surely other mechanisms of sex determination would have evolved. Fisher's argument can be paraphrased as follows. An individual will be selected to maximize the number of his or her grandchildren. If there are more males than females in the population, it will pay to produce daughters, and vice versa. This follows from the fact that every offspring has one father and one mother, so whichever sex is rarer, that sex has, on average, more offspring. For example, if there is a surfeit of males, some will not have any reproductive success by virtue of not being able to find a mate. Therefore, the only "uninvadeable" sex ratio is

1:1. This argument has since been extended by Hamilton, Charnov, and others to more complex situations. It differs from the Hawk–Dove game in that game theoretic thinking is being applied to a situation with no "pairwise contests"; instead, each individual's fitness depends on the average behavior of the population.

An important question is whether there is always an ESS. If, as in the Hawk–Dove game, there are only two pure strategies (plus a set of mixed strategies in which these are played with different probabilities), the answer is yes. In a 2×2 matrix, there may be a mixed ESS, or a single pure ESS (e.g., example, if $V > C$, Hawk is the only ESS), or there may be two alternative ESSs. In the latter case, the population will evolve to one or the other, depending on the initial conditions. This possibility of two ESSs for the same game has an important implication. Given that a particular phenotype is an ESS (i.e., one of the two ESSs), it does not follow that the population will evolve that phenotype from all initial conditions: condition 2 guarantees "local" but not "global" stability.

If there are three or more alternative pure strategies, there may be no ESS. The simplest example is the children's game Rock–Scissors–Paper, in which R beats S, S beats P, and P beats R. The payoff matrix is:

	R	S	P
R	e	1	-1
S	-1	e	1
P	1	-1	e

If a population plays this game, the outcome depends on the (small) payoff (e) allocated to a draw. If e is negative (a draw has a small cost), the mixed strategy $M = \frac{1}{3}R, \frac{1}{3}S, \frac{1}{3}P$ in which each strategy is played randomly and equally often is an ESS. But if e is small and negative, there is no ESS: the population cycles endlessly, R—P—S—R—. Although first published to make a mathematical point—that games need not have an ESS—a real example has since been described. Male lizards are of three genetically different types. Males with orange throats hold a territory containing several females (harem holders). A population of harem-holding males can be invaded by "sneaker" males with green throats, which wait until an orange-throated male's back is turned and mate with one of his females. In this way, sneaker males come to replace the harem holders. But once the population consists mainly of green-throated sneaker males, it can be invaded by blue-throated males, each of which holds a territory large enough to contain one female (monogamous males), which they can guard successfully against sneakers, thereby ensuring, their reproductive success. And, of course, once blue-throated monogamous males are common, the original orange-throated males can invade by forming harems in which reproduc-

tive success is higher than with a single female, and the cycle returns to its starting position, poised to cycle again.

A field in which game theoretic models are playing and important role is in the evolution of animal signaling. The central problem is the reliability of signals: if lying would pay, why tell the truth? The simplest answer to this question is that it may be physically impossible to lie. For example, a funnel-web spider intruding onto a web signals its weight to the owner by vibrating the web. If the difference in weight between owner and intruder is greater than 10 percent the smaller spider retreats, regardless of ownership. A spider cannot "lie" about its weight: the smaller of two spiders can be converted into a winner by glueing a weight to its back. This is an example of a consensus game, as mentioned above; escalated fights between spiders are dangerous. But there are many cases in which lying signals are not ruled out by the laws of physics. Zahavi has suggested that honesty requires that signals be costly to make, so that only high-quality signallers can afford to make them. The proposal was controversial for some years, essentially because it was impossible to tell from a purely verbal model whether the "handicap principle" would work. Game theoretic models, pioneered by Grafen, have shown that an evolutionarily stable signalling system may require that signals be costly, but relatively less costly for high-quality individuals (or, in a different context, for individuals in greater need).

The form of such models is very different from the simple Hawk–Dove game. For example, in the context of sexual selection, one seeks a pair of strategies: for the male, an "advertising" strategy relating the cost of his signal to his quality, and for the female, a "response" strategy determining her likelihood of mating to the strength of the signal received. Both of these strategies are continuous (as opposed to the discrete Hawk or Dove strategies)—relating a variable output to a variable input (signal strength to quality: response likelihood to signal strength). Evolutionary stability requires a pair of strategies such that it would not pay a member of either sex to change, provided the other sex does not.

To sum up, evolutionary game theory is a powerful method of thinking about evolution when fitnesses are frequency dependent. Although most applications have been to animal behavior, the method is in principle applicable more widely, including to plants and microorganisms. But several limitations should be borne in mind. First, the method assumes that the traits analyzed are heritable but ignores genetic details. Second, an ESS may be only locally stable. If more than one ESS exists, their zones of attraction may have to be investigated numerically. Third, it is able only to analyze equilibria, and occasionally to suggest that no equilibrium is possible. Other methods are needed to analyze continuing evolutionary change.

[*See also* Frequency Dependent Selection; Mathematical Models; Prisoner's Dilemma Games; Social Evolution.]

BIBLIOGRAPHY

Maynard Smith, J. *Evolution and the Theory of Games*. Cambridge, 1982.

— JOHN MAYNARD SMITH

GENE–CULTURE COEVOLUTION. *See* Cultural Evolution, *article on* Cultural Transmission.

GENE DUPLICATION. *See* Gene Families.

GENE FAMILIES

Much of the genome in eukaryotes appears to consist of repeated sequences. Such sequences include noncoding DNA and also functional genes, as exemplified by the ubiquity of multigene families. Multigene families consist of groups of similar genes that have arisen by duplication from a common ancestral gene and that generally retain similar functions. One of the best-known examples is the globin gene family. Around 600 million years ago, a duplication from an ancestral globin gene was followed by divergence into genes encoding two types of functional protein: myoglobin, involved in oxygen storage in muscles, and hemoglobin, responsible for oxygen transport in the blood.

Multigene families have been shown to be important in all eukaryotes. For example, analysis of the whole genome sequence of the plant *Arabidopsis thaliana* showed that it contains about 25,000 protein-encoding genes, only about 4,000 of which are unique, the rest belonging to some 7,000 families. About half of these gene families have more than five members. Similarly, around 40 percent of the ~35,000 known or predicted proteins encoded by the human genome are encoded by genes belonging to gene families. A remarkable feature of some gene families is that even those with many thousands of members may show a high level of nucleotide sequence conservation (e.g., ribosomal RNA genes). However, the various copies generally differ in sequence and function. For example, the largest gene family in the worm *Caenorhabditis elegans* is that encoding the G-protein-coupled seven-transmembrane-domain receptors (about 800 members), which are involved in a large variety of signal transduction processes (e.g., vision, olfactory processes).

Pointing out the importance of gene duplications, Susumu Ohno's book *Evolution by Gene Duplication*

(1970) had a profound impact on evolutionary biology. It suggested that gene duplications are a necessary prerequisite for the evolution of new structures and functions. Ohno's central argument is that natural selection places a strong brake on the rate of evolutionary change. In the face of this constraint, duplication events could favor the evolution of greater complexity by providing additional raw genetic material. Gene duplication is therefore seen not only as important for understanding microevolution (small-scale changes within a species) but also as a major mechanism of macroevolutionary changes (changes at a level above the species) in eukaryotes.

Origin of Gene Duplications. Gene duplication can involve large or small regions of the genome. At one extreme, an entire genome can be duplicated. Several mechanisms can cause this to happen. Sometimes the paired chromosomes fail to divide during meiosis, producing gametes with a full diploid complement of chromosomes. Instead of two haploid gametes fusing to restore the normal diploid number of chromosomes, two diploid gametes occasionally fuse to produce an offspring with twice the number of normal chromosomes. This phenomenon is called autopolyploidy. A more common polyploidization event, allopolyploidy, involves hybridization between related species. If the analogous chromosomes of the two species are somewhat dissimilar, they may not pair properly, resulting in a doubling of the chromosome number without all division.

Genes may also be duplicated in a smaller scale by tandem duplication. Gene duplications are usually the consequence of rare accidents caused by the enzymes that usually ensure the normal recombination processes. In eukaryotes, an efficient enzymatic system joins the two ends of a broken DNA molecule together, so that duplications can also arise from the erratic rejoining of fragments of chromosomes that have somehow become broken in more than one place. When duplicated DNA sequences are joined head to tail, they are said to be tandemly repeated. Once a single tandem repeat appears, it can be further extended into a long series of duplications. This happens when unequal crossing over occurs between two duplicated sequences, giving rise to a single copy on one strand and three copies on the other (Figure 1). An important feature of tandem duplications is that the probability of unequal crossing over increases with the number of tandem copies, because a copy of the gene on one strand can pair with any of the copies on the other strand.

Finally, partial gene duplications may also occur via exon-shuffling. This mechanism increases genetic diversity by combining useful complementary domains from individual genes: recombination within introns may result in the independent assortment of exons, and repetitive sequences in the middle of introns may act as hot

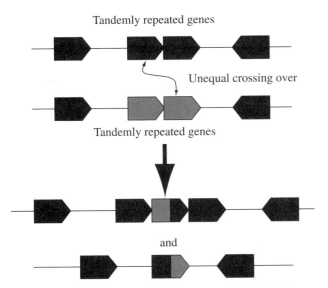

Tandemly repeated genes

Unequal crossing over

Tandemly repeated genes

and

FIGURE 1. Losses and Gains of Gene Copies as a Result of Unequal Crossing-Over in Family of Tandemly Repeated Genes.
This may be a general feature as long regions of homologous DNA sequence are favorable situations for genetic recombination.   After Alberts, B. et al. *Molecular Evolution of the Cell.* Garland.

spots for recombination, facilitating the reshuffling of exon sequences.

Selection and Fate of Duplication Events. Multigene families are ubiquitous; however, the development of multigene families from a single gene remains a little-studied aspect of genome evolution. The classical view of gene duplication imagined that most duplicated genes would be lost. Consider a rare duplication: because the duplicate has the same gene sequence as the original, it is unlikely that it will be favored by selection.

Although the new copy could spread through the population by genetic drift, there is little to stop from accumulating deleterious mutations. Indeed, as long as one of the genes of a duplicated pair retains its original function, the other is likely to accumulate degenerative mutations. Only on rare occasions in which a mutation causes a beneficial divergence in function will the duplicate gene be preserved. The classical view assumes that 99 percent of duplicate genes evolve into nonfunctional pseudogenes; however, empirical evidence suggests that a much larger fraction of duplicate gene copies is retained. The fraction of genes retained in duplicate has been estimated at ~8 percent in yeast over ~100 million years to more than ~70 percent in maize, *Xenopus*, and salmonids over ~11, ~30, and 25–100 million years, respectively. Such high levels of maintenance have led evolutionary biologists to reassess the selective forces acting on duplicate genes.

It is now thought much more likely that the invasion

and maintenance of duplicate genes is driven by some type of positive selection. One favorable case is when deleterious mutations are partially dominant—that is, when these mutations affect fitness even when present in a single copy. Such mutations frequently occur for multidomain proteins: if one copy undergoes a deleterious mutation in one domain, its product will still interact via the other, intact domains, destroying function even if the other copy is still functional. In such cases, duplicate copies will be positively selected by keeping multidomain proteins functional. Gene duplications may also be preserved if there is selection for a higher level of gene expression, as shown for the enzymes involved in insecticide resistance in numerous pest species. In yeast, genes displaying high concentrations of mRNA have also been shown to be more frequently retained as duplicates than are genes with low expression levels. Finally, the maintenance of duplicate genes may also be favored when gene copies diverge in expression pattern owing to mutations in promoter regions. Although the two copies retain the same function, they differ in the timing of expression and/or in the pattern of expression in different tissues. Direct evidence for such a scenario comes from examples where the shared expression pattern of duplicate genes coincides with the total expression pattern of the non-duplicated gene in closely related organisms.

Gene Duplication and/or Genome Duplication. The first line of evidence that genome duplication is important in evolution came from comparisons of karyotype and genome sizes reported by Ohno and colleagues in the early 1970s. However, complete genome sequencing has recently revealed that genome size in eukaryotes is dependent largely on the amount of noncoding DNA, rather than on the number of genes. One unanswered question subject to extensive debate is whether duplication of the entire genome or of numerous independent genes (or chromosome regions) is the driving force. The genome of the plant *Arabidopsis thaliana* clearly illustrates this issue. It contains 24 large duplicated segments, each 100kb in size or larger strongly suggesting that a tetraploidization event (a duplication of the entire genome) occurred about 100 million years ago. All but 17 percent of the genes are arranged in tandem arrays, indicating that single gene duplication is also a major phenomenon in this species.

In vertebrates, the most convincing evidence that genome duplications have played a role in evolution is provided by the structure and organization of the *Hox* gene clusters. The *Hox* genes encode a class of DNA-binding transcription factors that play a crucial role in the establishment of cellular and tissue identity during embryogenesis. The closest relative of vertebrates, *Amphioxus* (Cephalochordata), displays a single *Hox* gene cluster, whereas lampreys, representing jawless vertebrates, are endowed with two or three independent sets of *Hox* gene clusters. All jawed vertebrates examined to date, from fish to humans, are equipped with at least four separate clusters of *Hox* genes. The presence of four related sets of genes, on different chromosomes, is not confined to *Hox* genes: if a single gene is present in invertebrates, several (two, three, or four) paralogues are generally found in vertebrate.

During the 1990s, a large number of genes were cloned from *Amphioxus*, Ascidia, and nonmammalian vertebrates. The picture that has emerged from research into these genes is one of extensive gene duplication in the vertebrate lineage after divergence from cephalochordates (*Amphioxus*). It has been suggested that two separate phases of extensive gene duplication occurred early in vertebrate evolution (Figure 2). This is known as the one-to-four rule. One phase of duplication is thought to have occurred in the vertebrate lineage after the divergence of the cephalochordates, with the second phase of duplication occurring after the divergence of the extant jawless vertebrates (hagfish and lampreys), which probably dates back to the early Devonian period (395–345 million years ago). Peter Holland has suggested that gene duplication occurred by tetraploidization for the first event and by multiple tandem duplications for the second. However, the process (duplication of the entire genome or of individual genes or chromosome segments) and the timing of gene duplication in vertebrates is still a matter of debate, and further empirical data and analysis are required.

Importance of Duplications in Evolution. Polyploidy is widespread in plants, with an estimated frequency of 30–80 percent. Sally Otto and Jeannette Whitton found polyploidization to be an active and ongoing process in ferns, monocotyledonous plants, and herbaceous dicotyledonous plants, in which it is involved in about 7 percent of speciation events. In contrast, the available data suggest that polyploidization has played a relatively minor role in the ongoing evolution of woody dicotyledonous plants, gymnosperms, and animals. Although vertebrates probably underwent two genomic doubling events, polyploidization occurs only sporadically in animals and is restricted mostly to certain lineages, such as fish and amphibians. The presence or absence of sex chromosomes may be one of the key factors determining whether tetraploidy is possible, because sex chromosomes display abnormal segregation in tetraploids, thus limiting the prevalence of tetraploidy.

One of the most difficult questions to answer is whether gene duplication has important consequences for phenotypic evolution. Does the origin of new genes create opportunities for the evolution of novel characters, or could those characters have evolved through the differential use of pre-existing genes? Holland has suggested that the expansion of the *Hox* gene cluster by

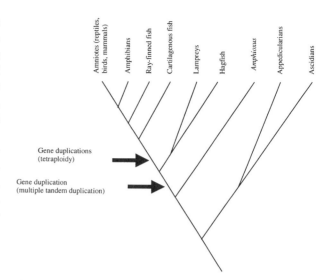

FIGURE 2. Probable Phylogeny of the Chordate Phylum, Showing the Relative Timings of Large-Scale Gene Duplications.
Holland, P. W. H. "Gene duplication: Past, present and future." *Seminars in Cell & Developmental Biology*, 10 (1999): 541–547.

tandem duplication, and the expansion of gene families in vertebrates, were important in the evolution of phenotypic complexity. However, gene duplications are also frequent in unicellular eukaryotes with no obvious increase in complexity, such as yeast, ciliates (*Paramecium*), and parabasalians (*Trichomonas*). Otto and Whitton also concluded that polyploidization results in phenotype shifts that may increase evolutionary diversification. There are indeed theoretical arguments suggesting that polyploidy should boost rates of adaptive evolution: polyploids have a greater chance of bearing new beneficial alleles and are better poised to evolve novel functions in duplicated gene families. However, these authors concluded that the role of polyploidy remains enigmatic because the evidence is elusive. Determining the effect of gene and genome duplication on the rate of adaptive evolution is certainly one of the major challenges in this field.

[*See also Hox* Genes.]

BIBLIOGRAPHY

Arabidopsis Genome Initiative. "Analysis of the Genome Sequence of the Flowering Plant *Arabidopsis thaliana*." *Nature* 408 (2000): 796–815.

Clark, A. G. "Invasion and Maintenance of a Gene Duplication." *Proceedings of the National Academy of Sciences USA* 91 (1994): 2950–2954.

Gibson, T. J., and J. Spring. "Genetic Redundancy in Vertebrates: Polyploidy and Persistence of Genes Encoding Multidomain Proteins." *Trends in Genetics* 14 (1998): 46–49.

Holland, P. W. H. "Gene Duplication: Past, Present and Future." *Seminars in Cell and Developmental Biology* 10 (1999): 541–547.

Li, W.-H., Z. Gu, H. Wang, and A. Nekrutenko. "Evolutionary Analyses of the Human Genome." *Nature* 409 (2001): 847–849.

Lynch, M., and A. Force. "The Probability of Duplicate Gene Preservation by Subfunctionalization." *Genetics* 154 (2000): 459–473.

Ohno, S. *Evolution by Gene Duplication*. New York, 1970.

Otto, S. P., and J. Whitton. "Polyploid Incidence and Evolution." *Annual Review of Genetics*. 34 (2000): 401–437.

Patthy, L. "Genome Evolution and the Evolution of Exon-shuffling—a Review." *Gene* 238 (1999): 103–114.

— Denis Bourguet and Herve Philippe

GENE REGULATORY NETWORKS

There are two extreme kinds of genetic regulatory networks, those governing metabolic control pathways and those involved in developmental regulatory circuits. In metabolic pathways, the phenotype of interest is the rate of production (or flux) of a product or an intermediate in the pathway. The activities of genes in metabolic pathways are expressed as activities of enzymes (the gene products), and genetic variation can be represented directly as variation in enzymatic activity. The quantitative analysis of the kinetic properties of metabolic systems is well developed and constitutes the fields of metabolic control theory and biochemical systems theory (Savageau, 1976; Cornish-Bowden and Cárdenas, 1990; Voit, 2000). In developmental regulatory circuits, by contrast, the phenotypes of interest are the steady-state levels of expression of gene products and the spatial patterns of that expression. Regulatory networks in development often consist entirely of interactions among regulatory genes—that is, genes that code for transcription factors that enhance or repress transcription of other genes. Genetic networks in development control the spatial and temporal patterns of gene expression that define the loci of subsequent differentiation or morphogenesis. Between these two extremes lie the majority of genetic regulatory networks that control the broad diversity of metabolic, physiological, and developmental processes. Typical genetic networks combine both regulatory and enzymatic processes. Their gene products are subject to various nongenetic physical and chemical processes (such as activation and inactivation) and interact with various structural features of a cell that ensure an appropriate spatial and temporal distribution and compartmentalization of their effects.

From an evolutionary perspective, it is of interest to understand the functional variation and diversification of genetic regulatory networks. Variation in the behavior or output of a genetic network can come about through variation in coding regions and regulatory regions of the participating genes. Variation in coding regions, unless it is functionally silent, leads to variation in the "activity" of gene products that is independent of their level of expression. Variation in regulatory regions, by contrast, can lead to variation in either the degree of transcription of a gene or variation in the spatial pattern of that transcription. Genetic variation in regulatory regions can therefore lead to localized variation in the level of expression of a gene product. It is believed that microevolutionary changes of gene activity in metabolic control systems occur primarily by selection for alternative alleles in the coding regions of genes, whereas evolution of developmental regulation occurs primarily by changing the level of expression of gene products, and must therefore be due to selection for alternative alleles in the regulatory region of a gene.

The regulatory regions of genes can be extremely complex. A regulatory region usually contains several enhancer and repressor sites for transcriptional regulation. These so-called cis-regulatory elements are short sequences (typically five to seven bases) where transcription factors bind. Many genes have cis-regulatory elements for several different kinds of activators and repressors, and their transcription can be enhanced or inhibited by several different combinations of activators and repressors. In development, transcription factors are often distributed in a graded fashion across tissues, and intersecting gradients produce locally unique combinations of activators and repressors that can result in spatially patterned transcription (Davidson, 2001; Gerhart and Kirshner, 1997). Regulatory interactions are often hierarchical. For example, the regulatory network that specifies segmentation and segment polarity in the developing embryo of *Drosophila melanogaster* (Figure 1) begins with a set of gene products deposited in the egg by the mother, which in turn activate a complex cascade of gene interactions. What is not captured in Figure 1 is that many of the gene products diffuse and become distributed in a graded fashion, so that only a subset of the possible interactions shown take place at any given location in the developing embryo (Bodnar, 1997; Carroll et al., 2001).

Evolution of regulation in a genetic network can occur by mutations that either remove or generate recognition sites for enhancer or repressor binding. The probability of such change is relatively high because these binding sites are small, and because there is some degeneracy in the recognition code (Carroll et al., 2001). The affinity by which transcription factors bind to regulatory sites is often low, and this leads to stochasticity in the activation and deactivation of transcription. It turns out that genes with stochastically unstable expression regulation exhibit a more predictable response to a gradient of enhancers than genes that have stable on/off switches. It appears that the combination of small regulatory sites, the low affinity of regulator binding to these sites, and the multiplicity enhancer and repressor sites for each gene, generate a particularly flexible and

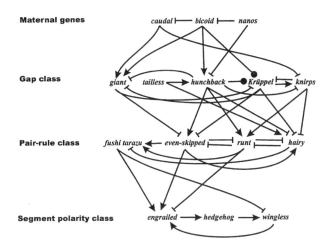

Maternal genes

Gap class

Pair-rule class

Segment polarity class

FIGURE 1. Genetic Regulatory Network for Segmental Patterning in the Early *Drosophila* Embryo.
Events lower in the hierarchy also occur at progressively later times in development. Arrows indicate activation; barred lines indicate repression. The two balled inputs to Krüppel indicate interactions that activate Krüppel when the transcription factors are at low concentrations but repress Krüppel at high concentrations.　Drawing by H. F. Nijhout.

adaptively responsive regulatory mechanism (Cook et al., 1998; Carroll et al., 2001).

An important consequence of the evolution of novel cis-regulatory sites is that genes can come under the control of new regulatory interactions. Thus an established genetic circuit can become co-opted to function in a novel context, and it is likely that such co-option, rather than mutations of structural genes, is a widespread basis of the origin of novelty in evolution. An example of co-option is the use of the regulatory network that leads to the patterned expression of the Distal-less gene. The expression of Distal-less in early embryonic development marks the sites where localized cell division that leads to the formation of appendages will occur, and this is believed to be the primitive function of this gene. Later in development, the expression of Distal-less defines many of the places where other projections of the body wall as diverse as the spines of sea urchins and the prolegs of caterpillars (neither of which is homologous to appendages) will grow out. Later yet, in the development of butterfly wing imaginal disks, the expression of Distal-less defines the cells that will become the organizing centers for the development of eyespots of the color pattern.

When the structure of a genetic network is known, it can be used to analyze and predict the sensitivity of the system to mutation and selection. This is done by writing the appropriate equations that describe the way each gene product is produced and destroyed, how gene products move from cell to cell (e.g., by diffu-

sion), how they are activated and inactivated, and the mechanism by which relevant gene products exert their phenotypic effects. A genetic network can usually be represented as a set of coupled differential equations. Most genetic regulatory networks produce a nonlinear relationship between variation in gene activity and variation in the phenotype. This is because the systems of equations that describe sequential reactions, gene-activation cascades, feedback regulation (positive or negative), and inhibition are nonlinear. Nonlinearity and nonadditivity of genetic effects on the phenotype have interesting consequences for the correlation between genetic variation and phenotypic variation. For instance, in the linear reaction network shown in Figure 2, the steady-state value of C (C*) is given by $C^* = k_1/k_4$ (assuming first-order kinetics). Thus variation in C^* is linearly related to k_1 and nonlinearly to k_4, which means that the *relative* effect of a particular amount of variation in k_4 will depend on the absolute value of k_1. Because k_2 and k_3 do not contribute to the value of C^*, variation in C^* is unrelated to variation in k_2 and k_3. The correlation between allelic variation in the genes that control k_2 and k_3 and variation in the trait will be zero, even though both genes are essential for the production of C. The only exception would be mutations that produce null alleles for k_2 or k_3, because these would effectively prevent C from achieving a nonzero steady state. Thus, even in a very simple network, the phenotype is not equally sensitive to genetic variation in all its determinants. This indicates that a system can be robust to variation in many of the genes that give rise to a trait even in the absence of special mechanisms that might convey robustness. Due to nonlinearities in the interactions, the properties of more complex networks, such as the one shown in Figure 1, are inherently relatively insensitive to quantitative variation in many of their constitutive genes (Bodnar, 1997; Von Dassow et al., 2000). In such inherently stable systems, evolutionary change is more likely to occur through quantal changes in regulation that change the connectivities of the network, than through gradualistic changes in the quantitative properties of the constitutive genes.

FIGURE 2. A Simple Linear Pathway in Which Four Genes Control the Synthesis and Destruction of a Series of Substrates.
The ks are rate constants.　Drawing by H. F. Nijhout.

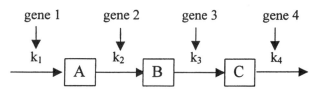

BIBLIOGRAPHY

Bodnar, J. W. "Programming the *Drosophila* Embryo." *Journal of Theoretical Biology* 188 (1997): 391–445.

Carroll, S. B., J. K. Grenier, and S. D. Weatherbee. *From DNA to Diversity: Molecular Genetics and the Evolution of Animal Design.* Malden, Mass., 2001.

Cook, D. L., A. N. Gerber, and S. J. Tapscott. "Modeling Stochastic Gene Expression: Implications for Haploinsufficiency." *Proceedings of the National Academy of Science* 95 (1998): 15641–15646.

Cornish-Bowden, A., and M. L. Cárdenas. *Control of Metabolic Processes.* NATO ASI Series A, vol. 191. New York, 1990.

Davidson, E. H. *Genomic Regulatory Systems.* San Diego, 2001.

Gerhart, J., and M. Kirschner. *Cells, Embryos, and Evolution.* Malden, Mass., 1997.

Savageau, M. A. *Biochemical Systems Analysis.* Reading, Mass., 1976.

Voit, E. O. *Computational Analysis of Biochemical Systems.* Cambridge, 2000.

Von Dassow, G., E. Meir, E. Munro, and G. M. Odell. "The Segment Polarity Network Is a Robust Developmental Module." *Nature* 406 (2000): 188–192.

— H. F. Nijhout

GENES

The Danish plant breeder Wilhelm Johannsen (1857–1927) first coined the term *gene* in 1909. However, the concept of the gene as the basis of inheritance dates back to Gregor Mendel's classic paper on peas in 1866. [*See* Mendel, Gregor.] This paper, "Experiments in Plant Hybridization," published in the proceedings of a local natural history society in Austria, was ignored until 1900, when it was rediscovered independently by Eric von Tschermak-Seysenegg, Hugo De Vries, and Carl Correns. Mendel saw that one "constant factor" (the gene) controlled one phenotypic character, such as flower color. He recognized that each individual must bear two versions of the factor controlling each character, one from the male and one from the female parent, and that these versions could be the same or different. Furthermore, he showed that one version of a factor could block or dominate the expression of another version in the phenotype and that the factors controlling different characters were inherited independently.

Alleles: Different Forms of a Gene. In his studies of pea plants, Mendel recognized that the heritable factors that controlled a specific trait could take more than one form. Johannsen termed the different forms of a gene *allelomorphs*, or *alleles*. Some genes are monomorphic, whereas others are highly polymorphic, existing as many different alleles. Alleles of a specific gene vary from one another in the DNA code of the gene. Thus, for example, the allele that causes sickle-cell anemia in humans is the result of a single nucleotide substitution (adenine replaced by guanine) in the second position of the sixth codon of the beta-globin gene. This causes glutamic acid rather than valine to be produced at this position of the beta-globin chain of individuals with this allele.

One Gene—One Enzyme. Mendel's ideas on segregation, independent assortment, and genetic dominance have been extended and refined over the last 100 years. In 1909, the physician Sir Archibald E. Garrod, in his book *Inborn Errors of Metabolism*, described studies of a variety of inherited human diseases. Garrod recognized that some recessive mutations caused defects in normal metabolic processes, concluding, in essence, that one gene mutation produces one metabolic block. George Beadle and Edward Tatum refined this proposal in 1941. Publishing results of work on the nutrient requirements of mutant strains of the bread mold *Neurospora crassa*, they concluded that each mutation led to the need for one specific growth factor to be added to minimal medium to allow growth. They thus deduced that each mutation resulted in the loss of activity of one enzymatic reaction, leading to the one gene–one enzyme concept.

The next refinement in the definition of the gene was the realization that enzymes can consist of several parts. For example, the oxygen-carrying hemoglobins in blood consist of four polypeptides, two alpha-globin chains and two beta-globin chains. Each chain is the product of a separate gene. Thus, the one gene–one enzyme concept became the one gene–one polypeptide chain concept that is now at the center of molecular genetics. The final refinement to date was the discovery that some genes do not code for polypeptides, but act simply as RNA molecules in a number of vital cellular processes.

Gene Structure. The idea of the gene as the fundamental unit of inheritance that could not be split into smaller units by recombination or mutation was rejected in 1940 following Clarence Oliver's report of recombination within the *lozenge* gene of *Drosophila*. This led to the acceptance that the gene must be made up of smaller subunits. Following James Watson and Francis Crick's discovery of the double helix structure of DNA in the 1950s, it was realized that genes are constructed out of sequences of nucleotide bases, which are the nondivisible units of genetic material.

The genetic material of organisms is arranged in chromosomes. Each gene has a specific location on a chromosome. This is called the locus for that specific gene. Chromosomes are elongate structures, leading to the view of genes as a linear string of beads.

Although this view is very simplistic, it accounts for the basic genetic structure in prokaryotes (e.g., bacteria), where genes are usually made up of sequences of nucleotide bases, all of which code for amino acids. This sequence of bases is flanked at one end, the 5′ end, by a short sequence of bases that regulate transcription of the coding sequence, including a start codon (UAG). The other end of the coding sequence, the 3′ end, terminates with a stop codon (e.g., UAA) and a short sequence that regulates the termination of transcription.

In eukaryotes, however, the coding sequence is often interrupted by noncoding sequences called introns. The complexity of many eukaryotic genes has led to two different terms, the *gene* and the *structural gene*. The gene includes the sequences controlling transcription, the exons or coding sequences, and any introns within the gene. The structural gene is just that part of the gene that is transcribed to produce an RNA product.

Transcription and Translation. The link between a gene and its specific expressed product (i.e., the polypeptide) is a two-step process. First, a complementary strand of RNA is made from one strand of DNA. Each of the four nucleotides in DNA—adenine, cytosine, guanine, and thymine—is replaced during this transcription by its RNA complement: uracil, guanine, cytosine, and adenine, respectively. This RNA product is known as the primary transcript. In prokaryotes, this transcript usually acts directly as the template for translation of the genetic code into amino acid sequences of polypeptide chains. However, in eukaryotes, most genes contain both exons (expressed sequences) and introns (intervening sequences). The intron sequences have thus to be excised from the primary transcript before translation into a polypeptide takes place. Intron sequences of the primary transcript are excised by splicing, involving another type of RNA called small nuclear RNA. The resulting RNA, called messenger RNA, migrates out of the nucleus and into the cytoplasm, where it is translated into protein on the ribosomes. Translation involves two further types of RNA, transfer RNA and ribosomal RNA. The various RNAs are also coded for by genes.

Transcription Factors and Promoters. Transcription of DNA into RNA is catalyzed by high molecular weight complex enzymes called RNA polymerases. The process of transcription is more complex in eukaryotes than in prokaryotes. Thus, whereas a single RNA polymerase is involved in all transcription in the bacterium *Escherichia coli*, eukaryotes contain three RNA polymerases. These three have different functions. RNA polymerase I is involved in the synthesis of most ribosomal RNAs. RNA polymerase II catalyzes the transcription of protein-coding nuclear genes, and RNA polymerase III aids the synthesis of transfer RNA molecules and one ribosomal RNA, 5S rRNA, which is not catalyzed by RNA polymerase I. Furthermore, although prokaryotic RNA polymerases can initiate transcription themselves, eukaryotic RNA polymerases require the assistance of other proteins, called transcription factors, to initiate the synthesis of an RNA chain. Transcription factors in turn have to bind to specific promoter regions in the DNA before the RNA polymerase can bind and commence transcription. These promoter regions are inte-

gral and essential elements within the coding structure of eukaryotic genes.

Different transcription factors and promoters are recognized and used by different RNA polymerases. The promoters used by RNA polymerase II, which transcribes most eukaryotic genes, are short nucleotide motifs of six to nine bases, located upstream from the point at which transcription starts. Four conserved motifs are associated with RNA polymerase II. The TATA box, which has a conserved sequence of seven nucleotides (TATAAAA), is positioned about thirty nucleotides upstream from the transcription start point and has an essential role in the determination of where transcription begins. The CAAT box (GGCCAATCT) is usually about eighty nucleotides upstream. The other two promoters, the GC box (GGGCGG) and the octamer box (ATTTGCAT), may be absent or present in one or more copies. These and the CAAT box are important in improving the efficiency of transcription.

William Bateson, who did much to champion Mendel's ideas on inheritance, and indeed coined the term *genetics* for the study of inheritance, from the Greek word meaning "to generate," wrote in 1900: "An exact determination of the laws of heredity will probably work more change in man's outlook on the world, and his power over nature, than any other advance in natural knowledge that can be foreseen." A century later, the truth of these prophetic words is self-evident. Our understanding of the structure and behavior of heritable material, in individuals and populations of individuals, impinges upon almost every area of biology and medicine. At the center of this understanding stands Mendel's "constant factor," the gene.

[*See also* DNA and RNA; Gene Families.]

BIBLIOGRAPHY

Bateson, W. *Mendel's Principles of Heredity.* Cambridge, 1909.
Beadle, G. W., and E. L. Tatum. "Genetic Control of Biochemical Reactions in *Neurospora.*" *Proceedings of the National Academy of Sciences USA* 27 (1941): 499.
Carlson, E. A. *The Gene: A Critical History.* 2d ed. Philadelphia, 1987.
Garrod, A. E. *Inborn Errors of Metabolism.* London, 1909.
Judson, H. F. *The Eighth Day of Creation: The Makers of the Revolution in Biology.* Cold Spring Harbor, N.Y., 1996.
Mendel, G. "Experiments in Plant Hybridization" (1866). In *The Origins of Genetics: A Mendel Source Book*, edited by C. Stern and E. Sherwood. New York, 1966.
Orel, V. *Gregor Mendel: The First Geneticist.* Oxford, 1996.
Sturtevant, A. H. *A History of Genetics*, New York, 1965.
Watson, J. D. *The Double Helix.* New York, 1968.

— MICHAEL E. N. MAJERUS

GENETIC ALGORITHMS

Genetic algorithms are computational methods whose design employs certain key concepts from the mechanisms of natural evolution. There are a number of such methods, the most prominent of which (apart from genetic algorithms) are evolutionary programming, evolution strategies, and genetic programming. The term *evolutionary computation* is now generally used to describe the field uniting researchers and practitioners of these methods, while an example algorithm from any of these categories is termed an *evolutionary algorithm*.

Evolutionary algorithms are mainly applied to *optimization* problems, which involve finding the best possible choice from a very large set (or "space") of candidates. In practice, the space is so large as to make exhaustive search of it unthinkable. Several branches of computer science and artificial intelligence are focused on how to search such spaces efficiently. For present purposes, to aid our appreciation of evolutionary algorithms, it will be instructive to contrast them with another class of algorithm called *local search*. In the following, both optimization and local search are further described before we focus our attention on evolutionary algorithms.

An exemplary optimization task is the Traveling Salesperson Problem, which involves a number of "cities" with known distances between them. A salesperson starts at a given city and needs to visit each of the others precisely once, then return to the city from which the tour began. The problem is to minimize the total distance traveled by the salesperson. In this case, it is not hard to see that any permutation of the cities corresponds to a candidate tour. Given such a permutation, the total length of the tour it encodes can easily be calculated. The problem is then to find a permutation for which this total length is as small as possible. Countless further examples exist in all areas of science, engineering, and industry. In general, optimization involves a set of structures, S (which may be permutations, aircraft wing designs, factory production schedules, and so forth), and a function that assigns a numeric value, called its *fitness*, to any individual s from S. Evolutionary algorithms, in common with other optimization methods, attempt efficiently to sample structures in S in search of one or more with optimal fitness. The most straightforward class of algorithm that regularly performs well on practical optimization tasks is local search, the generic structure of which is as follows:

1. Initialize—generate an initial solution c, and evaluate its fitness $f(c)$; call c the "current solution."
2. Generate and evaluate several solutions from the *neighborhood* of c.
3. If a termination criterion has been reached, then stop; otherwise, replace c with a suitably chosen one of the neighboring solutions generated in (2), and then return to (2).

The neighborhood of a candidate solution s is the set of all solutions that can be reached from s by a single ap-

EVOLUTIONARY COMPUTATION

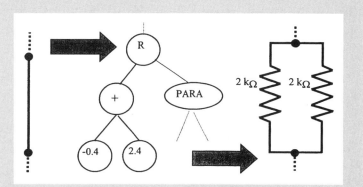

Evolutionary computation is being successfully used in electrical circuit design. A typical objective is to produce a *filter*, which may be required, for example, to remove all frequency components above 10 MHz from an input signal. Experts tend to modify standard designs, typically producing acceptable but nonideal results; some undesired frequencies will not be filtered, for example. Genetic programming has been used to evolve circuits whose performance surpasses that achieved by human experts. In fact, the process has rediscovered several previously patented "novel" circuits for filters, amplifiers, temperature sensors, and other types, as well as produced entirely novel designs.

The accompanying figure illustrates how a candidate circuit design is encoded as a tree-structured program. Part of a tree (center) is presented (by a higher-level node) with part of a circuit (in this case, the bare wire segment on the left), and makes modifications as follows. The label "R" specifies that a resistor be installed, with a resistance specified by its left sub-tree. The "PARA" node further specifies that the result be copied in parallel. Thus, the original bare wire has grown into the circuit segment on the right. Further operations may be specified when the nodes that spring from the "PARA" node are interpreted. The resulting complete circuit is then evaluated via a circuit simulation package.

—DAVID WOLFE CORNE

plication of a "small change" variation operator, also called a *mutation* operator. In our Traveling Salesperson example, a suitable mutation operator would be simply to swap the positions in the permutation of a randomly chosen pair of cities. Clearly, the "neighborhood" concept emphasizes *local* search. In seeking ever better solutions, local search methods employ a sensible tenet: *solutions that are similar in structure will generally be similar in fitness*. With a little thought, especially given that suitably good solutions tend to make up only a tiny fraction of the search space, this implies that it is best to search locally in the region of the best solutions found so far.

Local search methods therefore work via *exploitation* of the best candidate solutions attained so far. That is, the structures of such candidates are exploited constantly as a template for potentially better solutions. In contrast, there is usually little *exploration* in local search. Such emphasis on exploitation corresponds to a very high selection pressure strategy, with consequent well-known pitfalls. In particular, local search techniques are highly prone to become "trapped" at solutions that are locally optimal, with no means of escape toward better solutions that may exist elsewhere in the fitness landscape.

Evolutionary algorithms differ from local search in that they balance effort between exploitation and exploration in a way that turns out to be more effective in many applications. They are characterized by their use of ideas from evolution in support of this search process. In particular, they maintain a *population* of candidate solutions; they (usually) use a selection strategy biased toward the fittest in the population; and they involve operations that produce new candidate solutions from one or more old ones by processes analogous to natural mutation and/or recombination. The generic structure of an evolutionary algorithm is as follows:

1. Initialization—a population of candidate solutions is generated, and their fitnesses are evaluated.
2. Selection—a number of individuals in the population are selected to become "parents."
3. Variation—mutation and recombination operations are performed on the parents to produce a collection of children.
4. Population update—selected children will replace certain individuals in the population.
5. If a termination criterion has been reached, then stop; otherwise return to step 2.

Whereas in local search there is typically just one candidate to apply variation operators to, the presence of a population here forces us to select from many, and hence allows us to use mechanisms akin to natural selection. Further, we can also use multiparent variation operators, and hence indulge in an analogue of recombination. Naturally, how mutation and/or recombination may be implemented will depend on the problem at hand and the chosen encoding. Much thought enters these design choices. To illustrate a recombination operator, we will remain with our simple permutation encoding for the Traveling Salesperson Problem. If the letters *a–g* represent cities, let two parent candidate solutions be *abcdefg* and *cdagbfe*. We can recombine these to produce a valid child as follows. First, randomly choose some of the elements of the first parent; here is an example, with our random choices in boldface: *a**bcd**efg*. Next, reorder these elements in place, to match their order of appearance in

the second parent. This gives *a**dcgb**fe*, which is now the result of our recombination. In line with the goals usually sought when designing recombination operators, this combines structural aspects of two parent solutions to yield a child that attempts to exploit structural aspects from each parent, yet also begins to *explore* a region of the space quite nonlocal to either parent.

The use of recombination hence contributes exploratory activity, although the importance of recombination in evolutionary algorithms is much debated. The major contribution to exploration of the search space, in any evolutionary algorithm, arises from the existence of a population and the common use of a relatively low-pressure selection strategy. Rather than exploiting the single best region of solutions so far found, a well-designed evolutionary algorithm will maintain footholds in several promising regions at once. Most evolutionary algorithms will eventually converge to a point at which most individuals in the population are very similar, heralding a "commitment" to a certain area of the search space. However, suitable settings for population size and selection strategy will tend to postpone such commitment, with the result that solutions better than those obtained with simple local search methods will often be obtained, albeit sometimes at the price of elapsed time.

The strategy employed by all evolutionary algorithms, with or without recombination, can be seen as an *implicitly parallel* search for good "building blocks." A building block is a structural pattern that seems to correspond with better-than-average fitness when it is present in a candidate solution. For example, in the Traveling Salesperson problem, a certain ordered subtour of specific cities may be particularly compact in length. This subtour will tend to contribute strongly to the overall fitness of any tour in which it is a component; however, it may also occasionally appear in a solution whose overall fitness is poor, since other components are so ill-structured as to mask its benefits. Nevertheless, if we allow that the presence of good building blocks generally correlates with fitness, then we can view evolutionary algorithm search as an implicitly parallel sampling of building blocks. Each fitness calculation provides an estimate of the quality of each of the many potential building blocks in the candidate evaluated. Over time, with repeated, biased sampling of the same building blocks in different solutions, improved estimates begin to be reflected in the relative frequencies of different building blocks in the population as a whole. So, better building blocks—those that tend to contribute to fitness rather than detract from it—will increase their presence in the population over time. Further, when recombination is employed as a variation operator, the evolutionary algorithm increasingly samples *combinations* of building blocks, possibly discovering new ones as a result. There is, however, much debate over this analysis.

For example, the combination of high sample variance (with strong building blocks appearing in weak solutions and *vice versa*) and finite populations (with consequences analogous to genetic drift) makes it unwise to attribute an evolutionary algorithm's success largely to implicit parallelism.

The broad differences among the different families of evolutionary algorithms are as follows. Genetic algorithms use both mutation and recombination, and they are often associated with the use of binary string encodings, although such encoding is now rarely used unless it seems clearly applicable. Evolutionary programming and evolution strategies often eschew recombination, and they are characterized by their common application to problems that are naturally encoded as real-valued parameter vectors. Finally, genetic programming is characterized by the fact that the candidate solution structures are *computer programs*. Most often, tree-structured programs are used, with associated specialized mutation and recombination operators.

Evolutionary algorithms have achieved considerable success in applications and are now often used to address scheduling, timetabling, logistics, design, data analysis, and many other tasks in industry and science. In several cases, it is a moot point whether alternative optimization technologies could produce better results; however, the use of evolutionary algorithms in such cases can be seen as a consequence of their desirable "concept-to-result" properties. That is, it is relatively simple to set up and implement an evolutionary algorithm, and there are no formal requirements (such as differentiability or continuity) on the function to be optimized. Best results tend to arise from hybridization with application-specific algorithms and techniques at various stages. For example, well-known algorithms for shortest-path routing in networks provide good starting solutions for evolutionary algorithm approaches to more complex routing optimization problems. Another example is the use of evolutionary algorithms in scheduling manufacturing processes. Here, well-established and fast "dispatch heuristics" such as "do the most urgent task next" or "do the shortest task next" are commonly employed within each of the initialization, encoding, and variation operators of an evolutionary algorithm.

[*See also* Engineering Applications; Fitness; Genetic Drift; Mutation, *article on* Evolution of Mutation Rates; Natural Selection, *article on* Natural Selection in Contemporary Human Society; Sex, Evolution of.]

BIBLIOGRAPHY

Bäck, T. *Evolutionary Algorithms in Theory and Practice.* New York, 1996. A comprehensive coverage of evolutionary algorithms, focusing on evolution strategies. Rich on theoretical issues, and hence requiring confidence in mathematics.

Bentley, P. J., and D. W. Corne, eds. *Creative Evolutionary Systems.* San Francisco, 2001. An accessible introductory chapter is followed by a collection of contributions which show the employment of evolutionary algorithms in fields associated with creativity; in particular, there are sections on successfully exhibited art and published music produced using evolutionary algorithms.

Davis, L. *The Handbook of Genetic Algorithms.* New York, 1991. A comprehensive, practical introduction followed by contributed case-study chapters; prescient in its coverage of application domains, such as scheduling, telecommunications, and neural network design, that are now industrial and commercial success stories for the field.

Fogel, D. B. *Evolutionary Computation: The Fossil Record.* Piscataway, N.J., 1998. A unique annotated collection of early preprints, uncovering many examples of algorithmic ideas published several decades ago that have been reinvented by modern practitioners and researchers.

Fogel, D. B. *Evolutionary Computation: Towards a New Philosophy of Machine Intelligence.* 2d ed. Piscataway, N.J., 2000. A very readable and comprehensive overview of the field.

Goldberg, D. E. *Genetic Algorithms in Search, Optimization and Machine Learning.* New York, 1989. A classic and accessible text, focused primarily on genetic algorithms, although, in the light of modern experience, tending to overemphasize binary encodings.

Koza, J. *Genetic Programming: On the Programming of Computers by Means of Natural Selection.* Cambridge, Mass., 1992. The classic text on genetic programming; extremely clear at the cost of being rather wordy. Contains much that remains unaltered (in terms of algorithm design and standard test problems) in modern genetic programming.

Koza, J., D. Andre, F. H. Bennett, and M. Keane. *Genetic Programming III.* San Francisco, 1999. A useful recent update of Koza (1992), clearly describing many modern triumphs of genetic programming, including the circuit design application.

Michalewicz, Z. *Genetic Algorithms + Data Structures = Evolution Programs.* 3d ed. Berlin, 1996. Often, these days, the preferred undergraduate text, with accessible coverage of the broad field of evolutionary computation and many practical examples and illustrations.

ONLINE RESOURCES

"evoweb" http://evonet.dcs.napier.ac.uk/. Maintained at Napier University, Edinburgh, UK, this is a stylish and comprehensive resource with pointers to a great deal of introductory material, mature research material, as well as courses, careers, and up-to-date news in the field.

"The Genetic Algorithms Archive." http://www.aic.nrl.navy.mil/galist/. Set up primarily to maintain archive postings of the "GA-List", a periodical email discussion forum which began in 1985, various additions over the years have led to this becoming a favored and reliable general resource for evolutionary computation in general, with pointers to events, articles, research groups, theses, and more.

— DAVID WOLFE CORNE

GENETICALLY MODIFIED ORGANISMS

A genetically modified organism, or GMO, expresses a novel genotype as a consequence of biotechnological manipulation to itself or to one or more of its ancestors.

TABLE 1. Examples of Genetically Modified Organisms and Their Intended Utility

Host Organism	Introduced Gene	Intended Utility
Microbes		
Pseudomonas syringae, *P. fluorescens*	Deletion of ice-nucleating cell membrane protein	"Ice minus bacteria" reduce losses of strawberries and other crops to frost
P. putida	4-ethylbenzoate-degrading enzyme	Degrade benzene derivatives
P. fluorescens	Several genes for hydrocarbon degradation and light production	Detect and degrade polycyclic aromatic hydrocarbons
Plants		
Maize, cotton, potato	Toxin from bacterium *Bacillus thurigenensis*	Kills lepidopterans (butterflies and moths) eating plant tissues
Soybean, cotton, corn, canola, soybean, poplar	Glyphosate resistance	Ability to withstand application of herbicides
Rice	Provitamin A	Provides vitamin A precursor to address dietary deficiency
Rice	Soybean ferritin	Increases iron content of rice to reduce anemia
Strawberry	Flounder antifreeze polypeptide	Resistance to freezing
Tomato	Anti-sense polygalacturonase	Suppresses expression of enzyme softening the tomato, allowing on-vine ripening
Poplar	Modified lignin	Enhanced paper-making qualities
Animals		
Mouse	Virus-neutralizing monoclonal antibody	Model system for testing protection against viral encephalitis
Atlantic salmon	Atlantic salmon growth hormone	Accelerated growth rate, improved feed conversion efficiency
Pig	Insulin-like growth factor 1	Accelerated growth rate, improved feed conversion efficiency, leaner carcass composition
Pig	Phytase	Ability to utilize phytate in plant derived feeds, decreasing phosphorus in wastes
Goat	Human tissue plasminogen activator	Production of anti-clotting agent
Sheep	Human $\alpha 1$ anti-trypsin	Production of agent for treatment of asthma and emphysema
Pig	Human Factor VIII	Secrete blood clotting factor in milk, to be purified and administered to hemophiliacs

Use of the term GMO generally implies application of gene transfer to modify a targeted trait. Less frequently, the term may refer to other genetic modifications, such as chromosome set manipulation or cloning. Gene transfer involves introduction of an expression vector, a DNA construct containing a gene encoding a protein and regulatory elements controlling its expression, into a host. The methods used for gene transfer vary for microbial, plant, and animal hosts. In animals, the DNA construct is introduced into newly fertilized embryos in hope that, in some individuals, it will be integrated into a chromosome, incorporated into the germline, and then transmitted and expressed in their progeny. Individuals that have the DNA construct in their chromosomal DNA are termed transgenic; screening and propagation are needed to develop a transgenic line expressing and transmitting the desired phenotypic modification. Expression of introduced genes can result in dramatic impacts upon valued phenotypes, posing benefits to agriculture, medicine, bioremediation, and materials processing (Table 1). For example, transgenic farm animals are being developed for production of therapeutic proteins for human and veterinary medical uses; the new protein can be purified from the milk and manufactured into a pharmaceutical product.

Commercialization of GMOs has proven controversial for reasons including concerns about their possible ecological and evolutionary impacts. Ecological hazards include unintended effects upon nontarget organisms, colonization of ecosystems outside the area of application, alteration of population or community dynamics, and heightened predation or competition. Reproductively fertile GMOs could interbreed with natural populations; any genetic or evolutionary impacts would depend on the fitness of novel genotypes in the wild. When fitness is high relative to the wild type, there is the risk that a transferred gene could become permanently incorporated into natural populations. When fitness is low, maladaptive traits could be introduced into native populations, posing a hazard to the viability of the receiving population. Risks posed by a given GMO must be assessed on a case-by-case basis, considering not only the species and transferred gene, but also the particular transgenic line and receiving ecosystem at issue. This is because each transgenic line is different; expression of the transgene will depend not only on the structural gene transferred, but also on the regulatory element controlling expression of the transgene and the genomic site of integration. The receiving ecosystem provides the ecological context in which the fitness of transgenic genotypes will be expressed.

Ecological Effects. A number of laboratory studies have demonstrated ecological impact pathways. Many transgenic crop lines express a toxin gene from the bacterium *Bacillus thurigenensis* (*Bt*) as a means of protection from infestation by butterfly and moth larvae. While expression of the *Bt* transgene may pose fewer risks than application of chemical pesticides, ecological effects on nontarget organisms are not entirely eliminated. A prominent study showed that monarch butterfly caterpillars feeding on milkweed leaves dusted with pollen from corn expressing a *Bt* toxin transgene exhibited reduced viability and growth relative to a control group. Subsequent studies reached varying conclusions regarding the importance of *Bt* toxicity to nontarget organisms in the field.

Field studies have shown both direct and indirect ecological impacts of GMOs. Soybeans expressing a glyphosate resistance transgene will withstand application of the herbicide Roundup. Agricultural application of this herbicide will result directly in fields with dramatically fewer weeds, indirectly reducing the numbers of birds and other opportunistic feeders by reducing the supply of seeds produced by weeds. *Bt* toxin from transgenic plants is released into the soil from the roots of the growing plant, from pollen-fall during tasselling, and from breakdown of crop residues after harvest. The toxin remains active in the soil, where it binds rapidly to clays and humic acids, and persists for at least 234 days (the longest time studied) as determined by larvicidal bioassay. There is not yet indication of how soil communities might be affected by the *Bt* toxin. *Bt* toxin in the rhizosphere might improve control of insect pests, or might impose selection promoting the evolution of toxin-resistant insect pests. Nontarget insects and organisms in higher trophic levels could be affected by the toxin. Further field studies will be needed to investigate these possible impact pathways.

Genetic Effects. The possibility that an introduced gene might be transmitted from a fertile transgenic to a nontransgenic individual poses genetic effects to members of the same species and to members of closely related species.

Pollination of nongenetically modified (GM) plants by GM plants has resulted in transmission of the transgene in corn and rapeseed. Interbreeding of different lines of GM rapeseed, each resistant to a different herbicide, resulted in plants with multiple herbicide resistance.

Complicating the assessment of risk, the genetic effect of introgression of a transgene may vary with the genetic background of the receiving population. For example, different degrees of growth enhancement due to expression of a growth hormone gene construct were observed among wild and domesticated rainbow trout strains. Transgenic wild-strain rainbow trout retained the slender body morphology of the wild-type strain, but their final size at maturity was dramatically increased by transgenesis. The domesticated, selectively bred strain showed relatively little response to expression of the transgene. The greatest response to expression of the transgene was observed in hybrids of the wild and domesticated strains. The latter finding indicates the difficulty of predicting genetic risks posed by escape of transgenic salmon from culture facilities into wild populations.

One of the foremost concerns posed by GM plants is the possibility that transgenes might be transmitted from crops to wild populations of related species. Transmission of transgenes has been observed among cultivated and wild radishes, oilseed rape and weedy rape, and cultivated squash and wild gourd. Hybrid radishes and rape weeds were not as fit as their wild parents in the first generation, but seemed to regain reproductive fitness rapidly. Expression of a transgene in a wild plant could render it harder to kill or more resistant to insect pests, affecting its evolution.

Evolutionary Effects. The finding that genetically modified organisms may interbreed with wild populations raises the issue of possible evolutionary effects. Data useful for exploring possible evolutionary impact mechanisms and for quantifying associated risks were collected using a model species, Japanese medaka, expressing a growth hormone transgene. A deterministic model was developed to predict the frequency of the transgene in a receiving population and its effect on population demographics. The model led to development of the Trojan gene hypothesis, that is, that under certain conditions, the transgene could spread as a result of enhanced mating advantage enjoyed by larger individuals, but that the reduced viability of offspring eventually could cause local extinction of populations. A subsequent elaboration of the model used estimates of fitness parameters from a founder population to predict changes in transgene frequency. Aspects of an organism's life cycle were grouped into net fitness components regarding juvenile viability, adult viability, age at sexual maturity, female fecundity, male fertility, and mating advantage. For a wide range of parameter values, the model predicted that transgenes could spread through populations despite high juvenile viability costs if there were sufficiently high positive effects on other fitness components. Sensitivity analyses showed that transgene effects on age at sexual maturity should have the greatest effect on transgene frequency, followed by juvenile viability, mating advantage, female fecundity, and male fertility, with the least impact due to changes in adult viability.

Selective pressures imposed by genetically modified organisms may influence the evolution of other organisms. Transgenic crops expressing a *Bt* protein protecting the plant from insect pests impose selection upon targeted populations to develop resistance to the specific toxin. Insect resistance management strategies are implemented to minimize the likelihood that resistant populations of insects will develop and to assure the continuing effectiveness of *Bt* as a tool for pest control. Core elements of a resistance management plan are utilization of a variety of *Bt* genes, and maintenance of a refuge—including wild host plants, other crops, or non-*Bt* plantings of the crop in question—to ensure an adequate population of susceptible insects to mate with resistant individuals surviving exposure to the *Bt* crop. Farmers using *Bt* crops must sign written agreements to maintain such refuges, although the extent of compliance is unclear.

Empirical knowledge of hazards posed by environmental release of GMOs and their associated risks is limited. Applied ecologists have shown gene flow from GMOs to wild populations of the same or closely related species, although the implications of gene flow are poorly understood. It appears likely that some GMOs will pose ecological, genetic, or evolutionary hazards. Many critical experiments aimed at estimating fitness parameters in introgressed populations have yet to be conducted. In no case is there sufficient data to parameterize a fitness model and to predict whether a transgene will become permanently introgressed into a natural population. It is not yet possible to predict the consequences of introgression of a transgene on the evolution of a natural population or a species.

BIBLIOGRAPHY

Devlin, R. H., C. A. Biagi, T. Y. Yesaki, D. E. Smailus, and J. C. Byatt. "A Growth-Hormone Transgene Boosts the Size of Wild but Not Domesticated Trout." *Nature* 409 (2001): 781–782. Demonstrates that the effect of introgression of a transgene is difficult to predict. Technical background helpful.

Losey, J. E., L. S. Rayor, and M. E. Carter. "Transgenic Pollen Harms Monarch Larvae." *Nature* 399 (1999): 214. This straightforward demonstration of *Bt* toxicity to a nontarget organism sparked controversy and further impact assessment research.

Muir, W. M., and R. D. Howard. "Possible Ecological Risks of Transgenic Organism Release When Transgenes Affect Mating Success: Sexual Selection and the Trojan Gene Hypothesis." *Proceedings of the National Academy of Sciences USA* 96 (1999): 13853–13856. An important model showing that population viability can be affected by a tradeoff between high mating success for a large transgenic individual and poor viability of its offspring.

Muir, W. M., and R. D. Howard. "Fitness Components and Ecological Risks of Transgenic Release: A Model Using Japanese Medaka (*Oryzias latipes*)." *American Naturalist* 158 (2000): 1–16. A comprehensive model of six fitness parameters for transgenic organisms.

National Research Council. *Genetically Modified Pest-Protected Plants: Science and Regulation.* Washington, D.C., 2000. National Academy Press. Available online at http://www.nap.edu. A detailed evaluation of risks and benefits of plants genetically modified to resist pests, with recommendations for scientists and policy makers.

National Research Council. *Ecological Monitoring of Genetically Modified Crops.* Washington, D.C., 2001. National Academy Press. Available online at http://www.nap.edu. Brief report of a workshop addressing the challenges of ecological monitoring of genetically modified crops.

Royal Society of London, U.S. National Academy of Sciences, Brazilian Academy of Sciences, Chinese Academy of Sciences, Indian National Science Academy, Mexican Academy of Sciences, and Third World Academy of Sciences. *Transgenic Plants and World Agriculture.* Washington, D.C., 2000. National Academy Press. Available online at http://www.nap.edu. A brief report summarizing the potential of genetically modified plants for addressing world food needs, possible environmental effects, and recommended roles of government, private corporations, and research institutions.

Saxena, D., S. Flores, and G. Stotzky. "Transgenic Plants: Insecticidal Toxin in Root Exudates from *Bt* Corn." *Nature* 402 (1999): 480. Showed the release of *Bt* toxin to soil, its binding to soil particles, and continued insecticidal activity.

Tiedje, J. M., R. K. Colwell, Y. L. Grossman, R. E. Hodson, R. E. Lenski, R. N. Mack, and P. J. Regal. "The Planned Introduction of Genetically Engineered Organisms: Ecological Considerations and Recommendations." *Ecology* 70 (1989): 298–315. The earliest expression by ecologists of environmental concerns posed by environmental release of GMOs.

ONLINE RESOURCES:

"Information Systems for Biotechnology: A National Resource for Agbiotech Information." http://www.isb.vt.edu. Maintained at Information Systems for Biotechnology. Provides information resources to support the environmentally responsible use of agricultural biotechnology products. Documents and searchable databases pertaining to the development, testing, and regulatory review of genetically modified plants, animals, and microorganisms within the United States and abroad.

"Performance Standards for Safely Conducting Research with Genetically Modified Fish and Shellfish." http://www.isb.vt.edu/perfstands/psmain.cfm. Maintained by Information Systems for Biotechnology. With input from a wide range of aquatics sciences professionals, a U.S. Department of Agriculture (USDA)—sanctioned working group developed the Performance Standards as a tool for risk assessment and risk management for aquatic GMOs.

Scientists' Working Group on Biosafety. http://www.edmonds-institute.org. Maintained by the Edmonds Institute. A manual for assessing food safety and environmental hazards posed by aquatic and terrestrial GMOs, and management of any hazards identified.

— ERIC M. HALLERMAN

GENETIC CODE

The genetic code determines how a given sequence of DNA or RNA will be translated into protein: without it, genetic information would be meaningless. More important still, the genetic code defines the impact of mutations—the random changes to the sequence of a gene that provide the raw material for evolution—and thereby mediates evolutionary change. The elucidation of the genetic code in the fifteen years following James Watson and Francis Crick's discovery of the structure of DNA, therefore, ranks among the most fundamental discoveries in biology, and in 1968, the Nobel Prize for physiology or medicine was awarded to Robert Holley, Marshall Nirenberg, and Har Gobind Khorana for their work on the structure and function of the genetic code. Unraveling the code's evolutionary history, however, has so far proved less tractable. Despite much speculation, the origin and early evolution of the genetic code remain mysterious.

Elucidating the Code. Experiments carried out in the late 1950s showed that before a given gene is translated into protein, it is copied into an intermediate template molecule, known as messenger RNA (mRNA). It is this messenger that is translated, as dictated by the genetic code, into the string of amino acids that constitutes the corresponding protein. By the end of 1961, the general features of the code were understood. It was known that the sequence of the messenger is read from a fixed starting point in consecutive groups of three bases. In other words, the code is a triplet code: groups of three bases specify one amino acid, the building blocks of proteins. Each group of three is now known as a *codon*—a term coined by Sydney Brenner (one of Crick's closest colleagues). Because there are four possible bases in RNA (where DNA has adenine, guanine, cytosine, and thymine, RNA has adenine, guanine, cytosine, and uracil), there are $4^3 = 64$ possible codons. However, because the genetic code specifies only 21 entities—20 amino acids and a "stop" signal—the genetic code is technically "degenerate" or redundant. That is, there is not a one-to-one correspondence between codons and amino acids. Instead, most amino acids are specified by more than one codon.

Even before the general features of the code were clear, the mechanics of translation—the physical implementation of the genetic code—were starting to be worked out. In 1955, Crick predicted the existence of a class of small molecules that would bind to particular sequences of nucleic acid (at that time, nobody knew whether it would be RNA or DNA), each one carrying with it a particular amino acid. These small molecules, Crick supposed, would be loaded with their amino acid by special enzymes, with one enzyme for each of the 20 amino acids. Both the small molecules and the enzymes turn out to exist. The small molecules are now known as transfer RNAs (tRNAs). Each tRNA contains a region of three bases—termed the *anticodon*—that binds to mRNA with complementary base pairing: cytosine with guanine, adenine with uracil. The only difference is that the match between the third position of the codon and the first position of the anticodon is not always exact. Instead, it is sometimes allowed to wobble (see below). Each tRNA also has a region, distinct from the anticodon, that binds to a particular amino acid. The amino acid is in fact bound to the tRNA by an enzyme, an aminoacyl-tRNA synthetase, specific for that amino acid.

The development of cell-free systems of protein synthesis meant that the individual assignments of amino acids to codons could be solved. As soon as the assignments were known—a process that took five years—it was clear that the genetic code has a powerfully striking feature: the pattern of amino acid assignments to codons is not random (see Figure 1). Instead, if an amino acid is specified by more than one codon, the first two bases of the codons are usually the same. For example, the codons CCU, CCC, CCA, and CCG all specify the amino acid proline. This accounts for the wobbling: most tRNAs do not need to match the third position of the codon exactly. This pattern has important consequences for the effects that mutations have on proteins: mutations at the first and second positions of a codon almost always produce an amino acid substitution, but 66 percent of mutations at the third position of a codon produce no change.

Theories of the Code's Evolution. The obvious nonrandom nature of the genetic code has generated enormous speculation about the forces that could have led to its evolution. The speculation has been the more intense because, as far as we know, the same genetic code—called variously the "universal," "standard," or "canonical" genetic code—is used by all but a handful of extant organisms. The few exceptions are clearly derived from the universal code, differing from it only in the details of codon assignments. Thus, it is generally agreed that the genetic code had its origins in the RNA world—that is, a primitive genetic code arose early in the history of life—and reached its present form before the time of the last universal common ancestor. However, key questions remain unanswered. Most contentious of these is whether the genetic code is primarily a consequence of its early history, or whether it has been

2nd → 1st ↓	U	C	A	G	3rd ↓
U	Phe	Ser	Tyr	Cys	U
	Phe	Ser	Tyr	Cys	C
	Leu	Ser	Stop	Stop	A
	Leu	Ser	Stop	Trp	G
C	Leu	Pro	His	Arg	U
	Leu	Pro	His	Arg	C
	Leu	Pro	Gln	Arg	A
	Leu	Pro	Gln	Arg	G
A	Ile	Thr	Asn	Ser	U
	Ile	Thr	Asn	Ser	C
	Ile	Thr	Lys	Arg	A
	Met	Thr	Lys	Arg	G
G	Val	Ala	Asp	Gly	U
	Val	Ala	Asp	Gly	C
	Val	Ala	Glu	Gly	A
	Val	Ala	Glu	Gly	G

FIGURE 1. The Universal Genetic Code.
The bases at the first position of a codon are shown at the left, bases at the second position are shown across the top, and bases at the third position are shown at the right. Thus, the codon UUU is assigned to phenylalanine.   Courtesy of Olivia P. Judson.

shaped by natural selection and is, therefore, in some way adaptive—or even optimal.

Hypotheses to account for the genetic code fall into at least six basic types. The most famous is Crick's *frozen accident theory*. Crick suggested that the genetic code had, from simple beginnings, incorporated new amino acids until organisms were sufficiently complex that further changes were detrimental, and the code froze. Carl Woese, another doyen of molecular biology, advocated the *stereochemical theory* of the code's evolution. In its strongest form, this theory accounts for the particular assignments of amino acids to particular codons (or anticodons) as being necessary consequences of their chemical interactions—that is, the codons for, say, glycine, would have some powerful chemical attraction to glycine, and so on. Meanwhile, J. Tze-Fei Wong advanced the *coevolution hypothesis*. According to this idea, the universal genetic code is the final result of a series of gradual changes whereby amino acids most closely related to each other through their biosynthetic pathways were assigned to similar codons. In contrast, proponents of either of two theories—the *lethal mutation theory* and the *translation-error-minimiza-* tion theory—argue that the genetic code has evolved to minimize the occurrence and the impact of mutations, or the occurrence and impact of translation errors, respectively. (The distinction is subtle: translation errors affect the organism, whereas mutations affect descendants.) Finally, the *genetic flexibility hypothesis* supposes that the genetic code has evolved as a compromise between two opposing forces, robustness and mutability, where robustness is pressure for the genetic code to prevent amino acid substitutions in proteins, and mutability is pressure for the genetic code to allow proteins to undergo evolutionary change.

The frozen accident. In its strongest form, the frozen accident theory states that the particular assignment of amino acids to codons is arbitrary. The observation that similar amino acids have similar codons would be accounted for as being merely the most likely and the least disruptive method of amino acid incorporation. This view is now unpopular. Most authors believe that codon assignments reflect either chemical affinities from the code's murky beginnings, or the action of natural selection. In the broadest sense, however, the frozen accident theory must be correct: at some point, hundreds of millions of years ago, the code froze.

This conclusion is supported as follows. Codon reassignments are easy. Transfer RNA molecules are labile: a mutation to the anticodon can cause a tRNA bearing one amino acid to recognize a different codon, and in some cases, a single point mutation is enough to alter a tRNA molecule to bind to a different amino acid. Thus, mutations that cause codon reassignments probably arise frequently. However, codon reassignments are rare in nature. Just a score of alternative codes are known. Of these, most are found in the shrunken genomes of the mitochondria—the miniature organs that are found within eukaryotic cells and that are thought to be the remnants of ancient symbiotic bacteria. Only a few groups of organisms are presently known to use alternative codes. Therefore, unless the universal code is almost universally optimal (which some people believe, but which seems unlikely for reasons outlined below), or unless thousands of alternative codes remain undetected, we must assume, as Crick argued, that nowadays, further change is almost always lethal.

The early code. But what was the situation before the code froze? One obvious question arises. Was the genetic code originally based on codons of one nucleotide (with a maximum of four amino acids), which then expanded to two nucleotides (with a maximum of 16 amino acids), which then expanded to three nucleotides? This is unlikely, because each transition would have rendered all previous genetic messages meaningless. Moreover, although the maximum number of amino acids in a doublet code would be only four short of the number in the universal genetic code, a more realistic number of amino acids for such a code would be much smaller. Otherwise, all point mutations would cause an amino acid substitution, and proteins would be terribly unstable. But a smaller number of amino acids may not be enough to build sophisticated proteins. Thus, if a doublet code ever existed, it is unlikely to have been the direct ancestor of the universal genetic code. It is far more probable that the ancestral genetic code was a triplet code from the start—albeit one that encoded fewer amino acids.

How did the genetic code get from this primitive state to the code we see today? Both the stereochemical theory and the coevolution theory presuppose that the code's early history is preserved in its present shape, and that the code froze before codon reassignments obliterated the early pattern. Such preservation cannot presently be ruled out. Nonetheless, two factors suggest it is unlikely. First, there is the aforementioned ease with which tRNAs can change their allegiance. Second, some alternative codes—the yeast mitochondrial code (Figure 2), for example—contain several codon reassignments, suggesting that extensive reassignments would have been possible as the universal code was forming (see the Vignette).

To what extent, then, would the final codon assign-

MECHANISMS OF CODON REASSIGNMENT

Once a codon-reassigning mutation has occurred in a tRNA, the reassignment can become established through either of two mechanisms: codon capture, or the ambiguous intermediate. Codon capture involves several steps. First, a codon disappears from a genome by chance. When, through mutation, the codon reappears in the genome, it can then be captured by a tRNA loaded with a different amino acid. Codon capture is likely to occur in small genomes, especially those with a strong nucleotide bias, since under these circumstances codons are frequently lost. For example, one recent survey of 111 complete mitochondrial genomes found that 76 lacked one or more codons.

The ambiguous intermediate, in contrast, assumes that for a period of time a codon is read by two different tRNAs. Although this sounds as if it would be lethal for the organism, codon reassignments through an ambiguous intermediate have been engineered into both *Saccharomyces cerevisiae* and *Escherichia coli*, and both organisms survived in the laboratory.

—OLIVIA P. JUDSON

FIGURE 2. The Yeast Mitochondrial Genetic Code. Differences between the two presented in bold italics. Courtesy of Olivia P. Judson.

2nd → 1st ↓	U	C	A	G	3rd ↓
U	Phe	Ser	Tyr	Cys	U
	Phe	Ser	Tyr	Cys	C
	Leu	Ser	Stop	*Trp*	A
	Leu	Ser	Stop	Trp	G
C	*Thr*	Pro	His	Arg	U
	Thr	Pro	His	Arg	C
	Thr	Pro	Gln	Arg	A
	Thr	Pro	Gln	Arg	G
A	Ile	Thr	Asn	Ser	U
	Ile	Thr	Asn	Ser	C
	Met	Thr	Lys	Arg	A
	Met	Thr	Lys	Arg	G
G	Val	Ala	Asp	Gly	U
	Val	Ala	Asp	Gly	C
	Val	Ala	Glu	Gly	A
	Val	Ala	Glu	Gly	G

ments have been the result of natural selection? This is controversial. Fans of the lethal mutation theory and the translation-error-minimization theory argue that the assignments of amino acids to codons are the best available: that in the event of a mutation (or translation error) the relative codon assignments of amino acids either prevents amino acid substitutions or replaces one amino acid with another that is similar, and therefore least likely to be disruptive. Such claims must be viewed with caution, however. Testing these ideas is difficult. First, most models are highly sensitive to the underlying assumptions. Second, problems of classifying amino acids as similar or dissimilar are legendary. In 1965, the biochemists Emile Zuckerkandl and Linus Pauling complained that "the inadequacy of a priori views on conservatism and non-conservatism [of amino acid substitutions] is patent. Apparently chemists and protein molecules do not share the same opinions regarding the definition of the most prominent properties of [an amino acid] residue." Matters have not become easier. One recent study found that there are 43 different published hydrophobicity scales—scales measuring the extent to which an amino acid is soluble in water—and that although the results for some amino acids are consistent across the different scales, the results for others vary wildly, with proline found in as many as 16 different positions, and cysteine appearing as either the most hydrophobic amino acid or the least.

Moreover, a recent analysis of codon reassignments within mitochondria—and the first study of genetic codes to be based on molecular data rather than computer simulations or statistical analyses—suggests a different story. In the mitochondria, codon reassignments result in a better match to the amino acid requirements of proteins. That is, codons tend to be gained by amino acids that appear in proteins more than would be expected based on their representation in the genetic code before the reassignment. Conversely, codons tend to be lost by amino acids that appear in proteins less than would be expected based on their representation in the genetic code. To give an example, methionine appears in the universal genetic code once, yet, in the mitochondria, methionine is often used more frequently than one in every 61 amino acids. Acquiring a second codon for methionine means that the new genetic code better reflects the amino acid requirement of proteins and therefore reduces the burden of slightly deleterious mutations. Note that this idea, although superficially similar to the lethal mutation theory, is different in several important respects. For example, this idea makes no statement about the relative codon assignments of the amino acids, but instead predicts the number of codons that should be allotted to each amino acid.

If mitochondria are a good model for the processes that occurred while the universal code was forming, it would suggest the following picture of the code's evolution. In the first place, the striking redundancy of the third codon position would be a consequence of the early translation apparatus: matching two out of three requires less precision than matching three out of three. Indeed, one analysis of the effects of selection pressures on genetic codes suggests that selection for third base redundancy was by far the most important factor in shaping the universal genetic code. Second, the proportions of amino acids in the universal code may reflect those that most closely matched the amino acid requirements of proteins at the time the code froze. If this is correct, most modern organisms may be stuck with a code that is now suboptimal but that can no longer easily be changed.

In the continued absence of sufficient data, arguments will continue to rage. In years to come, however, the discovery of additional alternative codes, experimental attempts to recreate the amino acid and nucleic acid interactions of the RNA world, and the engineering of organisms with completely new genetic codes should shed light on how this essential, enigmatic mediator of evolution came to be.

[*See also articles on* Origin of Life.]

BIBLIOGRAPHY

Crick, F. H. C. "The Origin of the Genetic Code." *Journal of Molecular Biology* 38 (1968): 367–379. A description of the evolution of the genetic code, including Woese's stereochemical theory, and the paper in which Crick proposed the frozen accident.

Ellington, A. D., M. Khrapov, and C. A. Shaw. "The Scene of a Frozen Accident." *RNA* 6 (2000): 485–498. A lucid, if technical, account of the arguments against the notion that historical affinities of amino acids and codons are reflected in the code's present form.

Freeland, S. J., R. D. Knight, L. F. Landweber, and L. D. Hurst. "Early Fixation of an Optimal Genetic Code." *Molecular Biology and Evolution* 17 (2000): 511–518. The most recent paper to claim that the genetic code is the best possible genetic code.

Judson, H. F. *The Eighth Day of Creation: The Makers of the Revolution in Biology, Expanded Edition.* New York, 1996. An account of the discovery of the structure of DNA and the discoveries that followed, including the elucidation of the genetic code.

Judson, O. P., and D. Haydon. "The Genetic Code: What Is It Good For? An Analysis of the Effects of Selection Pressures on Genetic Codes." *Journal of Molecular Evolution* 49 (1999): 539–550. An analysis of claims that the universal genetic code is optimal.

Liu, D. R., and P. G. Schultz. "Progress Toward the Evolution of an Organism with an Expanded Genetic Code." *Proceedings of the National Academy of Sciences, USA* 96 (1999): 4780–4785. A highly technical paper describing the engineering of an organism with a brand new genetic code.

Osawa, S. *Evolution of the Genetic Code.* Oxford, 1995. A comprehensive, if dry, overview of the subject.

Swire, J., and A. Burt. "Mitochondrial Genetic Codes Evolve to Match Amino Acid Requirements." *Science* (submitted, 2001). The first analysis of patterns and mechanisms of codon reassignments based on molecular sequence data.

Wong, J. T.-F. "A Co-evolution Theory of the Genetic Code." *Proceedings of the National Academy of Sciences, USA* 72 (1975): 1909–1912. The first statement of the coevolution theory.

Zuckerkandl, E., and L. Pauling. "Evolutionary Divergence and Convergence in Proteins." In *Evolving Genes and Proteins*, edited by V. Bryson and H. J. Vogel. Academic Press, 1965.

— OLIVIA P. JUDSON

GENETIC DRIFT

Each genetic locus can exist in several versions (or alleles) that differ somewhat in their DNA sequences. Without selection, mutation, or migration, the relative frequencies of these alleles in a population are expected to remain the same from one generation to the next. However, because all populations are finite, allele frequencies can change by random chance, even when the frequency of alleles is expected to be the same on average. This change in allele frequency arising from the randomness of sampling alleles from one generation to create the next is called random genetic drift. Drift was first introduced into evolutionary biology by Sewall Wright, one of the founders of theoretical population genetics. As a result, drift was long called the "Sewall Wright effect." [*See* Wright, Sewall.]

Population Size. When a copy of a gene leaves more or fewer copies of itself than expected, this allows the potential for a change in the frequency of that allele in the population by genetic drift. The simplest case, when each descendant allele has an equal and independent probability of having any parent allele as its ancestor, is referred to as being an ideal population. In an ideal population, allele frequencies can change as a result of simple random sampling. When rolling a six-sided die, for example, one expects two ones out of twelve rolls but may by chance get anything from zero to twelve ones. The larger the number of rolls, however, the closer the frequency of any particular outcome—such as two ones in twelve rolls—should come to the expectation. Similarly, a genetic allele with a frequency of 0.4 in one generation is expected to be at a frequency of 0.4 in the next generation in the absence of selection, mutation, or migration; but the frequency of this allele could become anything between 0 and 1 owing to random sampling at alleles. The larger the population size, the closer the allele frequency in the next generation to its expectation; in smaller populations, allele frequencies can fluctuate widely. Figure 1 shows the change in allele frequency over generations in simulations of ideal populations of different sizes. A good experimental example was provided by Buri (1956), who followed the frequency of eye color alleles in *Drosophila* populations of different sizes.

In the ideal population, the variance in reproductive success (i.e., the variation in the number of surviving offspring) among diploid individuals turns out to be $2(1 - 1/N)$, where N is the number of individuals. If individuals vary more than this in their reproductive success, then the amount of change arising from genetic drift is expected to be greater than with an ideal population of the same size. Variance in reproductive success implies that some parents leave more offspring than others, and therefore a higher proportion of the individuals in the offspring generation have identical alleles. In the extreme case, where only one allele copy by chance gives rise to all alleles in the next generation, the population immediately becomes fixed for that allele and genetic drift will be at its maximum.

The changes in the amount of genetic drift expected with a nonideal population can be complicated. Fortunately, for most purposes, an effective population size can be defined to account for these deviations from the ideal assumptions. The effective size of a population is the size of an ideal population that has the same expected level of genetic drift. Effective size is often written as N_e. In a population that consistently has N individuals and variance in reproductive success V, the effective size is approximately $N_e = N/V$ for haploids and $(4N - 2)/(V + 2)$ for diploids. As V increases, the effective population size gets smaller. As the variance in reproductive success goes to 0, then N_e for haploids reaches a maximum of infinity. With haploids and $V =$

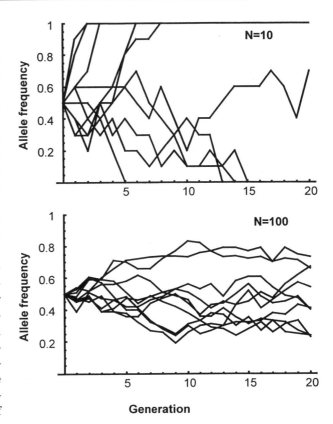

FIGURE 1. Genetic Drift in Ideal Haploid Populations with Either 10 or 100 Individuals. In these simulations, the populations start with two alleles, each at a frequency of 0.5. Each line on the graph represents the frequency of one of the alleles through time. Note that the populations diverge from one another as time progresses. This divergence is faster in the smaller populations. If an infinite number of populations were simulated, the mean allele frequency over all populations would not change, but each individual population would change in allele frequency.   Drawing by Michael C. Whitlock.

0, each allele copy is leaving exactly one copy of itself, and therefore the allele frequency is unchanged from one generation to another. (Thus, there is no genetic drift, and therefore the effective size is infinite.) With diploids, even when there is no variance in reproductive success among individuals, there can be variance in the success of the two alleles within individuals, so there can still be genetic drift. With diploids, the effective size reaches its maximum of about $2N$ if $V = 0$.

The variance in reproductive success can be increased by a variety of means. For example, in a given species, if there is a biased sex ratio, the effective size will be lower. (In the extreme, when one sex is very rare and the other is common, N_e is approximately equal to four times the number of the rare sex.)

Furthermore, generations with fewer individuals tend to cause a disproportionate amount of genetic drift. The cumulative drift over a series of generations is most affected by the generations of small population size. The total amount of drift in these temporally varying populations is similar to the amount expected in a population held constant at the harmonic mean of the effective sizes of the varying population. As an extreme example, a population that has billions of individuals, nine out of ten generations, but only two individuals for the tenth generation will have an effective size of only twenty, despite having an average size in the billions. Similarly, in founder events, when a large population is founded by a small number of individuals, there is an opportunity for much more genetic drift than normal.

As a result of these types of factors, the effective size of natural or lab populations tends to be much less than the census size. A review by Frankham (1995) suggests that the ratio of N_e to N is quite variable, but on average about 10 percent.

Drift Is Nonadaptive and Nondirectional. Although it causes the change of allele frequencies in populations and therefore is a process of evolution, drift in itself does not increase the adaptation of a population. Drift changes allele frequencies without regard to their adaptive properties and therefore is nonadaptive.

Furthermore, drift is as likely to increase allele frequencies as to decrease them and is therefore nondirectional. Moreover, drift is not correlated across generations; an increase in an allele frequency in one generation is as likely to be followed by a decrease in the next generation as another increase. Drift can by chance cause consecutive directional changes in allele frequency, but there is no correlation expected on average across generations.

Selection, in contrast, is by definition adaptive, and is quite likely to be directional. These properties of selection can allow us to distinguish selection from genetic drift in natural populations.

Direct Effects of Drift. Genetic drift has several direct effects. Because change arising from drift ultimately leads to allele frequencies approaching, then being fixed at, either 0 or 1, drift tends to cause a reduction in the genetic variance within populations to decrease. (On average, the genetic variance will decline by a fraction of $1/2N_e$ per generation for diploid populations.) Because the allele frequency in one population can change at random and independently of other populations, drift tends to cause the genetic variance among populations to increase. Also, because drift implies that some individuals share common ancestors, with more genetic drift the probability is greater that two allele copies are identical because of their common descent. This also implies that, on average, a small population will increase in homozygosity faster than a larger population.

Why Drift Matters. Drift affects several important evolutionary processes and patterns. First and foremost,

the study of microevolution is the study of allele frequency change; along with selection, mutation, and migration, drift is one of the basic mechanisms by which allele frequencies evolve. The eminent theoretical population geneticist Motoo Kimura argued that drift was the major cause of evolutionary change at the molecular level; he called this proposal the neutral theory of molecular evolution (Kimura, 1983). [See Kimura, Motoo.] Though contentious, it is clear that genetic drift plays an important role in at least the molecular evolution of most species.

Most of the important evolutionary aspects of genetic drift come from its interaction with natural selection. Because drift can allow allele frequencies to change in a way that deviates from the expected value under selection, it is possible that allele frequencies can change in a way opposite to that expected by selection alone. The effects of drift become more important as selection becomes weaker. Thus, the frequency of a deleterious allele can increase, even to fixation, because of genetic drift, and beneficial alleles can be lost. Small populations are more likely than large to fix deleterious alleles, and the mean fitness of small populations can decline over time because of these fixations, a process called mutational meltdown or drift load. Drift load may be the greatest genetic threat to small endangered populations (Lynch, 1996). There is good indirect evidence that many deleterious alleles have fixed in human populations as a result of drift (Eyre-Walker and Keightley, 1999).

If drift can temporarily overwhelm the effects of selection so that deterministic evolution can take the population to a different evolutionary stable state, it is possible that the evolutionary future of a population can be changed qualitatively. This is part of the so-called shifting balance theory, as proposed by Sewall Wright.

Furthermore, drift tends to cause the loss of genetic variance and can cause the loss of alleles entirely. Because genetic variability determines the future evolvability of a population and genetic variance determines the rate of future evolution, genetic drift can limit the future response to selection of a population. Again, this may be an important determinant of the survivorship probability of a small population.

Drift can cause species or populations within a species to differentiate genetically. These differences in allele frequencies can sometimes be used to estimate times of divergence or the degree of interconnectedness of species or populations.

[See also Metapopulation; Neutral Theory; Shifting Balance.]

BIBLIOGRAPHY

Buri, P. "Gene Frequency in Small Populations of Mutant *Drosophila*." *Evolution* 10 (1956): 367–402.

Crow, J. F., and M. Kimura. *An Introduction to Population Genetics Theory*. New York, 1970.

Eyre-Walker, A., and P. Keightley. "High Genomic Deleterious Mutation Rates in Hominids." *Nature* 397 (1999): 344–347.

Frankham, R. "Effective Population Size/Adult Population Size Ratios in Wildlife: A Review." *Genetical Research* 66 (1995): 95–107.

Hartl, D. L., and A. G. Clark. *Principles of Population Genetics*. Sunderland, Mass., 1997.

Kimura, M. *The Neutral Theory of Molecular Evolution*. Cambridge, 1983.

Lynch, M. "A Quantitative-Genetic Perspective on Conservation Issues." In *Conservation Genetics: Case Histories from Nature*, edited by J. C. Avise and J. L. Hamrick, pp. 471–501. New York, 1996.

— MICHAEL C. WHITLOCK

GENETIC LOAD

Genetic load is a way of describing how the members of biological populations are less well adapted than they could be. Genetic load has an exact definition in terms of fitness. *Fitness* is a technical term in evolutionary biology and has nothing to do with its everyday meaning of athletic fitness. The fitness of an individual (or, to be exact, of the individuals with a particular genotype) has two components: its relative chance of survival and its relative fecundity. Fecundity is the number of eggs laid, for females, and the number of eggs fertilized, for males. Fitness equals relative chance of survival multiplied by relative fecundity. To calculate the genetic load of a population, we first need to know the fitness of every genotype in the population. In practice, fitnesses are difficult to measure and genotypes can be difficult to identify. Genetic load is therefore more often a theoretical reasoning device than a practically measurable fact. The account of genetic load here is for a theoretical population.

We identify the genotype with the highest fitness and arbitrarily assign it a fitness of 1. We then calculate the mean fitness for all the genotypes in the population, relative to the figure of 1 for the best genotype. Mean fitness is usually symbolized as $\bar{w}$. For example, let us imagine a population with nine diploid individuals, three of whom have genotype AA, three have Aa, and three have aa. (A and a refer to alternative alleles at the same locus.) For simplicity, suppose they all have the same fecundity and differ only in their chance of surviving to reproduce. The absolute chance of survival for AA is 40 percent, for Aa 50 percent, and for aa 30 percent. We assign a fitness of 1 to the Aa genotype. Relative to that, the fitness of AA is 0.8 and of aa is 0.6. The mean fitness of the population is

$$\bar{w} = 1/3\,(0.8) + 1/3\,(1) + 1/3\,(0.6) = 0.8 \tag{1}$$

It is less than 1 because not everyone in the population has the best genotype. We can now define genetic load (symbolized by L) as

$$L = 1 - \bar{w} \tag{2}$$

In this imaginary example, $L = 0.2$. The genetic load of a population expresses the relative chance that an average member of the population will die without breeding because it has an inferior genotype. If every member of the population had the same genotype, $\bar{w} = 1$ and $L = 0$: the population has zero load. Load cannot be more than 1: when every individual in the population except one has a lethal genotype, load approaches 1 and the population is probably about to go extinct.

The formula for load is often expressed in a related form. If we use the absolute chances of survival (40, 50, and 30 percent in the example), we can define w_{opt} as the survival rate of the best genotype: it is 50 percent. We calculate mean fitness for the absolute survival rates ($\bar{w} = 40$ percent) with the absolute numbers

$$L = \frac{w_{opt} - \bar{w}}{w_{opt}} \tag{3}$$

which again comes out as 0.2 in the example. The fitness of an average member of the population is 20 percent lower than it could have been.

Genetic load can arise for a number of reasons; they are all cases in which the members of a population are less well adapted than they could be. [*See* Adaptation *for a related discussion of adaptive imperfection.*] The distinctive feature of the definition of genetic load is that it is relative to the best genotype that exists in a population. Sometimes biologists discuss the imperfection of adaptation relative to an imaginary but nonexistent genotype. If a biologist suggests that human beings would be better off with a pair of wings, he or she is not formally using the concept of genetic load. Evolutionary theorists have discussed three main kinds of genetic load.

Mutational Load. In every generation some new inferior genotypes arise by mutation. Consider the case in which the best genotype is AA, and A genes mutate to an inferior allele a at a rate m per generation. For simplicity, suppose that the inferior a allele is dominant. Aa heterozygotes will then have a lowered fitness; perhaps their chance of survival is reduced by 5–95 percent of the survivorship of AA homozygotes. The a allele is being selected against and will exist only at a low frequency, as it is created by mutation from A alleles. It can be calculated that the frequency of a dominant disadvantageous allele will be m/s, where s is the fractional reduction in fitness caused by the allele (5 percent in the example above). The more damaging a mutant allele is, the higher its value of s, and the lower its frequency will be in a population: natural selection is working against it more powerfully.

A surprising result is that the genetic load arising from mutation is approximately independent of the damage done by the mutant allele a. The mutation load is the chance that an average member of the population will fail to breed because it contains an inferior, mutant allele. For a mutant allele that causes a 5 percent reduction in survival, its frequency in the population will be one-fifth of the frequency of a mutant allele that reduces survival by 1 percent (if both have the same mutation rate m); but any one mutant bearer will be five times as likely to die if it has a copy of the more damaging mutation that reduces survival by 5 percent than if it has the less damaging mutation that reduces survival by 1 percent. The total chance of dying, for an average member of the population, is the same: it is less likely to have a damaging mutation, but if it has got it, it is more likely to die of it. The mutation load is independent of s.

To be exact, the chance that an average member of the population will die is $2m$ for dominant mutations and m for recessive mutations. (It is twice as high for dominant mutations because they are removed in heterozygotes that contain only one copy of the inferior gene. Recessive mutations are expressed only in aa homozygotes, and when one of them dies, it removes two copies of the bad gene.) In fact, the majority of damaging mutations are recessive and the mutation load on real populations probably approximately equals the mutation rate. For any one genetic locus, this will be low, because mutation rates per locus are one in a million or so. Some loci are exceptional: the gene that can mutate to cause achondroplastic dwarfism in human beings, for example, has a mutation rate of about 10^{-4} and causes a nontrivial genetic load. The total mutational load is given by mutations in the whole genome, and this may amount to a seriously high number, as we will see below. Only deleterious mutations, not all mutations, cause a mutational load.

Segregational Load. Segregational load arises when Mendelian segregation generates inferior genotypes. The most important case is when the heterozygote has the highest fitness (heterozygous advantage), for example, if the fitnesses are as follows:

genotype	AA	Aa	aa
fitness	0.9	1	0.9

In this case, it can be calculated that 25 percent of the offspring will be AA, 50 percent Aa, and 25 percent aa. The population has a load of 0.05: half the members of the population of each generation have a 10 percent higher chance of dying because they are homozygotes. Segregational load can also arise with certain types of frequency-dependent selection. Segregational load reflects an inefficiency of the Mendelian mechanism of inheritance. When the best genotype is a heterozygote, it is impossible for a sexual population to "breed true."

That is, even if natural selection eliminated all the homozygotes before the reproductive stage, the parents (all of whom are heterozygotes) could not give rise to a purely heterozygous set of offspring. When one AA parent breeds with another Aa parent, half their offspring are homozygotes. The problem does not arise with asexual reproduction, in which Mendelian segregation is lacking.

Substitutional Load. Substitutional load arises in a population in which natural selection is acting to substitute one allele for another. The fitnesses might be

genotype	AA	Aa	aa
fitness	0.9	0.95	1

Over time, selection will substitute the a gene for the A gene, but the complete process takes thousands of generations. Until the substitution is complete, the population has a load because some individuals contain the inferior genotypes. For example, during the nineteenth century in industrially polluted areas of Britain, a dark form of the peppered moth (*Biston betularia*) was substituted for a light form of the moth. If we had looked at the moth population during this time, some members of the population had lower fitness than others: the moth population had a substitutional load.

Substitutional load is often discussed in terms of a closely related concept called the cost of natural selection. The British biologist J. B. S. Haldane introduced the idea of a cost of selection in a paper in 1957 (by which time Haldane was taking up residence in India), and it has subsequently been incorporated in the general concept of genetic load. [*See* Haldane, John Burdon Sanderson]. Haldane formally defined the cost of natural selection as the ratio of

$$\frac{\text{chance that an individual dies because of selection}}{\text{chance that an individual survives}}$$

summed over all the generations it takes natural selection to substitute the superior allele. In one generation, mean fitness ($\bar{w}$) equals the chance an individual survives and load (L) equals the chance an individual dies because of selection. In one generation, cost equals $L/\bar{w}$. Haldane's "cost of selection" is another way of talking about genetic loads. In Haldane's language, natural selection imposes a "cost" on a population because when it acts, some members of the population will die prematurely or fail to breed. He argued that the cost places an upper limit on the rate of evolution because if natural selection is too strong, it will kill the whole population. For example, natural selection would be very strong if a few individuals had a superior AA genotype while most of the population had lethal Aa or aa genotypes. In theory, the frequency of the a gene could go up from a low value to 100 percent in one generation; but in this case, the population would probably be driven extinct in the process: almost everyone has to die without breeding to make the rapid evolution possible. This corresponds to a genetic load near the upper limit of 1.

Mutational, segregational, and substitutional loads are the kinds of genetic load that are most often discussed, but some other kinds of load have also been defined. For example, when mother and fetus have incompatible genotypes (such as with the Rhesus blood group), the mother may mount an immune reaction against the fetus. This can be expressed as an "incompatibility load." John Maynard Smith (1976) defined a "lag load" to express the way adaptation may be out of date because of environmental change. The lag load has been used in research on Van Valen's Red Queen hypothesis.

Hard and Soft Selection. Any population in which there is a genetic load is experiencing "genetic deaths": some individuals die without reproducing because of the action of natural selection. In human populations, this matters for moral and medical reasons. In evolutionary biology, loads matter because they may place an upper limit on certain properties of life. In a population with zero load, natural selection is not acting; none of the death in that population occurs for genetic reasons. If the load increases above 0, some genetic death does occur, but it poses no problem for the population as a whole; it is a normal condition of life, for natural selection to act. The problem comes when the load approaches 1. At this level, every individual in the population dies before breeding. The population is now unsustainable because the force of natural selection is too strong. If the mutation rate is so high that every offspring contains a new harmful mutation, the mutational load will reach 1. A life-form with so high a rate of mutation as this cannot exist. The need for a mutational load below 1 may set an upper limit on the complexity of life (if more complex life forms are coded for by more genes and mutation rates increase with gene numbers). Similar arguments for substitutional loads and segregational loads suggest that there is an upper limit on the rate of evolution and the amount of genetic variation in a population. In all these cases, the exact value of the upper limit depends on the way natural selection acts. In particular, we should distinguish "hard" from "soft" selection.

Some individuals die prematurely in every population for nongenetic, ecological reasons. Suppose, for example, that individuals in a population need a territory to obtain food to survive. If there are 100 territories and 1,000 offspring are produced each generation, then 900 will die for want of a territory. We can call these 900 deaths "background" ecological mortality. It happens every generation because more offspring are produced than can survive. It will happen if every individual is genetically identical and the genetic load is zero. Now consider some selective mortality, for example, that arising from mutation. Suppose that half the offspring (500 of them) contain new deleterious mutations. One extreme possibility is that the mutations have no effect on the ability to defend a territory; they may, for example, reduce fecundity to 0, sterilizing the individual. Then the genetic deaths that eliminate the mutations come on top of the background morality. Of the 1,000 offspring, 900 will fail to gain a territory (and 450 of them contain mutations). Of the 100 territory holders, 50 will be sterile because of mutation. The reproductive output of the population is halved. This extreme is called "hard" selection. The genetic load reduces the population's reproductive output.

The other extreme possibility is that the mutations reduce the ability of an individual to defend a territory. The 500 mutants will then be concentrated in the 900 unsuccessful individuals who fail to gain territories. The 100 territory holders will lack mutations and reproduce as normal. In this case, the high mutation load has had no effect on the population's reproductive output: this extreme is called "soft" selection. In reality, some genetic deaths may be substituted for background mortality and others will cause additional mortality, and selection will usually be some mix of hard and soft. The same number of genetic deaths are required whether selection is hard or soft (500, in the example), but more genetic death is possible when selection is soft. Thus, a load of 1 is an upper limit whether selection is hard or soft. But with hard selection, the realistic limit is below 1. If background ecological mortality eliminates half the population and selection is hard, then the upper limit on the load is about 0.5, whereas it can go up to 1 if selection is soft.

Genetic Loads and the Neutral Theory of Molecular Evolution. Arguments using genetic loads have been influential in two main areas of evolutionary biology. One concerned the neutral theory of molecular evolution, postulated in the late 1960s and 1970s. Early measurements of the rate of molecular evolution suggested it was much higher than had been foreseen. The Japanese geneticist Motoo Kimura (1983) argued that it was too high to be explained by natural selection. [See Kimura, Motoo.] He used Haldane's argument about the cost of natural selection. The observed rate of evolution would impose a cost (or substitutional load) that was too high to be possible. The amounts of genetic variation, when first measured at the molecular level in 1966, also were higher than expected. In this case, the puzzle was posed by segregational load. We saw above that one locus with a fitness difference of 10 percent and heterozygous advantage imposes a load of 5 percent. But the observations suggested that at least 10,000 gene loci are polymorphic in human populations (polymorphic means that more than one allele is present in the population). If one locus reduces average survival by 5 percent, a second would reduce it by a further 5 percent, a third by a further 5 percent, and so on, to the 10,000th power. With that many heterozygous loci, it hardly matters whether the load per locus is 5 percent or 0.05 percent: we have derived an impossibility. Natural selection could not maintain that much variation, without causing an impossible amount of death.

Kimura's conclusion was that most evolutionary change is not driven, and most variation is not maintained, by natural selection. It is caused instead by neutral genetic drift, that is, by genetic differences that have no effect on fitness. Neutral drift causes no genetic load. Kimura's argument was criticized in three ways. First, like Haldane, he assumed hard selection in his quantitative arguments. Natural selection can act more powerfully if it is soft. Second, natural selection may not act independently at each locus. For example, consider the problem of maintaining variation (a similar argument applies for the rate of evolution). The problem arose because selection at every locus reduced average fitness by 5 percent. But suppose instead that selection acted in some "threshold" manner, such that individuals who are heterozygous at 10,000 or more loci have high fitness and those with fewer than 10,000 heterozygotes have low fitness. It could be, for instance, that the individuals with the most heterozygotes win the territories and the individuals with the fewest heterozygotes do not. Natural selection now no longer acts against each homozygous locus independently. Much higher levels of genetic variation can be maintained without running up an impossible genetic load. Third, genetic variation theoretically can be maintained by frequency-dependent selection without imposing segregational load. Natural selection could maintain the observed levels of variation, provided it is not by heterozygous advantage.

The consensus of many evolutionary theorists is that Kimura's early argument was inconclusive. The neutral theory has been successful, but not because of its arguments about genetic loads. However, some theorists, such as G. C. Williams in his book *Natural Selection* (1992), argue that the problem was swept under the rug rather than solved. The problem is no longer the topic of active research, but remains to be solved.

The Mutational Meltdown. Through the 1980s and 1990s, arguments about genetic loads revived following the discovery that mutation rates are higher than had been foreseen in certain life-forms. Following the AIDS pandemic, RNA viruses have been extensively studied. RNA viruses have high mutation rates, of about 10^{-4} per nucleotide. Most RNA viruses are about 10^4 nucleotides long, making their total mutation rate about 1. The mutational load can be calculated to be $1 - e^U$, where U is the deleterious mutation rate. For $U = 1$, the load is about 0.7, a high figure. RNA viruses may be limited by their mutation rates: if the rate were much higher, the load would become unsustainably high. (Some biolo-

gists aim to exploit this, killing HIV by elevating its mutation rate.) When the mutation rate goes so high that natural selection is overwhelmed, what happens is that mutations accumulate in the population. The RNA message is then randomized over time, and the population soon goes extinct. This is an example of what Michael Lynch and colleagues (1999) have called a "mutational meltdown." The German chemist-turned-biologist Manfred Eigen (1992) has particularly argued that RNA viruses are limited by their mutation rates. He has also argued that mutational loads were the key hurdle that had to be overcome at the origin of life. [*See* Quasi-Species.]

A more controversial possibility is that complex life-forms, such as vertebrates and maybe many multicellular life-forms, have mutation rates near or even more than 1 (i.e., one mutation per genome per generation). Measurements of the harmful mutation rate in humans suggest that it is more than 1. This may imply that human beings have an unsustainable mutational load. However, with sexual reproduction, it is possible for the load to be reduced for any given mutation rate. Recombination can, if certain special conditions are met, combine multiple mutations in a single offspring. Natural selection then eliminates mutations in multiple form, and it requires less organismic death to cause any given amount of genetic death than in an asexual population. If selection acts nonindependently at many loci, it can work more powerfully and cause less of a load.

Genetic load is a general conceptual reasoning device, for thinking about imperfect adaptation, natural selection, and genetic death in biological populations. It has also been used to deduce upper limits on some properties of life. Currently, most evolutionary thinkers doubt that substitutional load limits the rate of evolution or that segregational load limits the amount of genetic variation. But there are skeptics who doubt whether the arguments have already been sewn up. Many theorists suspect that RNA viruses that reproduce asexually are limited by mutational loads. Whether mutational loads also limit complex life including human beings depends on an unsolved problem: the evolution of sexual reproduction.

[*See also* Coevolution; Red Queen Hypothesis; Sex, Evolution of.]

BIBLIOGRAPHY

Crow, J. F. "Mutation, Mean Fitness, and Genetic Load." *Oxford Surveys in Evolutionary Biology* 9 (1993): 3–42. Expository review chapter, particularly about mutation loads.

Eigen, M. *Steps towards Life.* Oxford, 1992. Popular book, including discussion of how entropic forces such as mutation had to be overcome during the evolution of life.

Haldane, J. B. S. "The Cost of Natural Selection." *Journal of Genetics* 55 (1957): 511–524. The original statement of costs of selection.

Kimura, M. *The Neutral Theory of Molecular Evolution.* Cambridge, 1983. Load arguments had become less important to the neutral theory by the time Kimura wrote this great book, but he includes an account of the arguments and back references.

Kondrashov, A. S. "Deleterious Mutations and the Evolution of Sexual Reproduction." *Nature* 336 (1988): 435–440. How sexual reproduction can reduce genetic loads.

Lewontin, R. C. *The Genetic Basis of Evolutionary Change.* New York, 1974. Includes an explanation of early arguments about loads and the neutral theory, and is written with a virtuoso degree of clarity and rigor.

Lynch, M., et al. "Perspective: Spontaneous Deleterious Mutation." *Evolution* 53 (1999): 645–663. Modern empirical research on the mutational meltdown.

Maynard Smith, J. "What Determines the Rate of Evolution?" *American Naturalist* 110 (1976): 331–338. Defining the lag load.

Ridley, M. *Mendel's Demon.* London, 2000. (Published as *The Cooperative Gene.* New York, 2001.) Popular book about mutation and meiotic-drive loads in the evolution of complex life, including explanation of Eigen and Kondrashov's ideas.

Wallace, B. "One Selectionist's Perspective." *Quarterly Review of Biology* 64 (1989): 127–145. Arguments why evolution may not be limited by loads.

Williams, G. C. *Natural Selection.* New York, 1992. Discusses whether the load problems identified by Kimura really have been solved.

— MARK RIDLEY

GENETIC MARKERS

Genetic markers are identifiable features of the DNA (or RNA) of organisms that vary among individuals. They are commonly used to identify individuals, to establish parentage, and to assess movement of individuals between populations.

Most markers in contemporary use are distinguished by their mobility under an electric gradient through a gel soaked in a buffer solution (gel electrophoresis). Markers have undergone several generations of development. Although initially most attention was paid to variation in proteins (allozymes), the markers in most common use today reflect variation among individuals in the amount of particular regions of noncoding DNA that repeat themselves.

The first revolution in technique came from the discovery by Alec Jeffreys and colleagues (1985) of variation in minisatellite DNA in humans. Minisatellites are noncoding regions of the genome where unusual repeat units (usually >10 bases in length) are repeated to give blocks of DNA ranging in size from about 500 bases to 30 kilobases. Because they are subject to very high rates of mutation and are present at many (hundreds to thousands) of loci, they can be used to generate a unique fingerprint of every individual in a population. Such DNA fingerprinting has proved valuable in forensic analysis and has been exploited in evolutionary studies. Minisatellites have the advantage that markers derived from human studies often also reveal variation in other species. In addition, because they integrate effects at several loci simultaneously, they are particularly good at measuring rare events such as the impact of mutations. For example, minisatellites revealed elevated mutation rates in humans after the nuclear accident at Chernobyl, Ukraine, in 1986.

However, most studies of population structure and mating systems now make use of microsatellite loci. Microsatellites (sometimes called simple tandem repeats) are short sequences (e.g., AAAG) that are repeated to give stretches of DNA containing as many as sixty copies of the original motif. They can be amplified via the polymerase chain reaction. In addition, because the sequences involved are quite short, amplification can sometimes be successful from DNA that has been partly degraded. Thus, typing is possible from very limited amounts of tissue from the target individual, so the extent of interference with the population can be minimized. For example, hairs collected from night nests and feces have been used as sources of microsatellite DNA in studies of chimpanzees, where sampling of blood and tissue would pose formidable ethical and logistic difficulties.

Microsatellite markers also exhibit allelism and codominance. This means that they can be associated with a particular region of a chromosome, and the expression of one allele does not usually hamper interpretation of the presence of the second allele on the homologous chromosome. This greatly facilitates interpretation and allows application of the mathematical theory that underlies population genetics. Microsatellites are also dispersed widely throughout the genome, and particularly when the motif is repeated many times, are subject to extreme mutation rates, generating high levels of variability. The primary constraint on the use of microsatellites has been that cross-species applicability is restricted, so there is a need to develop a set of markers for each species in turn. This difficulty has been alleviated by enrichment techniques that accelerate the process of isolating variable loci.

The applications of markers are legion. The article concentrates on four particularly common analyses in evolutionary biology.

Determining Parentage. In Shakespeare's *Merchant of Venice*, Launcelot has little trouble deceiving his father, Gobbo, leading to his famous lines: "Nay, indeed, if you had your eyes, you might fail of the knowing me: it is a wise father that knows his own child." The advent of genetic markers for the determination of paternity has revealed such skepticism true beyond initial belief. Uncertainty over paternity is understandably rife in societies where matings take place between casual acquaintances. However, first impressions from field observations suggest that confidence over paternity should be much greater in societies where females and males form permanent social relationships, such as the social monogamy so prevalent among birds, or in the polygyny commonly seen in mammals, where males seem to guard groups of females closely and effectively from the attentions of other males. Undermining this confidence, genetic markers have repeatedly shown that many species participate in matings outside the social bonds—so-called extrapair copulations—that seem to govern their societies. [*See* Extrapair Copulations.]

The first reliable genetic markers to reveal infidelity in socially monogamous birds came from color polymorphisms in geese. If two parents that are homozygous for a recessive color pattern produce offspring with a different dominant color pattern, we have to assume either a rare mutation or, more plausibly, that extrapair mating has occurred. Although some progress in unraveling extrapair mating was made with allozymes, dramatic advances occurred with the advent of minisatellite fingerprinting. It soon became evident that many socially monogamous bird species mate outside the pair-bond. Although such extrapair mating is often rare, in some socially monogamous species infidelity is the rule. Extrapair matings account for more than 70 percent of offspring in Australian fairy-wrens from the genus *Malurus*. Results from some societies that are not monogamous are equally surprising. For example, bull seals fighting each other on the beach to monopolize access to females is one of the most obvious manifestations of polygyny. Their vigorous defense of females that have come to shore led biologists to the obvious conclusion that the males that could best defend a harem would dominate paternity. However, paternity analysis using microsatellites in gray and fur seals reveals that many of the matings are achieved by males that never come to shore.

Although microsatellites afford many advantages, there are some difficulties that need to be considered. The first problem arises because, unlike allozyme and minisatellite methods, microsatellites have low transferrability between species, so the initial cost of development can be high. The second problem occurs when only one of the two alleles in the individual is amplified, often because of a mutation in the regions flanking the microstaellite. The presence of such null alleles means that the individual is falsely scored as a homozygote, confounding paternity assignment. The third problem is distinguishing among closely related individuals, as regardless of the variability of the locus, parents have only two alleles to transmit to their offspring. The final problem is that the advantages of high mutation rates are countered by the disadvantage that a parent can be incorrectly excluded because its offspring bear a mutant allele. Techniques are emerging that allow these difficulties to be treated statistically.

Determining Patterns of Relatedness. The extent of relatedness between individuals is predicted to influence the extent of cooperation and conflict between in-dividuals. However, because of uncertainty over mating patterns, it will not always be possible to guess patterns of relatedness on the basis of behavioral observations. In social insects, where cooperative behavior has developed to an unusual extent, the probability that females share genes by descent can vary from 0.75 to virtually zero, depending on patterns of dispersal and the mating behavior in the population. Thus, it is crucial to determine relatedness in order to understand the underlying causes of sociality. Are individuals social because of the benefits of cooperating with close kin, or are they social for more direct benefits?

The difficulties are illustrated with three unusual examples. Slime molds exist as individual free-living cells. When they are starved, however, they aggregate into a multicellular fruiting body (a slug) containing a stalk, which contributes nothing to subsequent generations, and fruiting cells, most of which become viable spores. If slugs were formed from a single clone, stalk cells may derive kin-associated benefits from supporting reproduction of their clone mates. However, microsatellite markers have been used to show that slugs are often chimeras (combinations) of many different clones. Some clones are cheating genotypes that are most likely to go on to produce spores at the expense of their chimeric partner, which contributes to the stalk. Distinguishing relatedness among aggregating cells would be impossible without recourse to genetic analysis, and genetic markers provide a convenient insight into this problem.

The second and third examples reflect the opposite pattern and show how behavior that seems competitive may actually have a cooperative basis. Many ducks are intraspecific brood parasites, laying eggs in the nests of other members of their own species. Such behavior seems hostile to the interests of the host and would be expected to be resisted. However, it has recently been shown that the genotype of the parasite can be estimated nondestructively by protein fingerprinting, a technique that involves taking a small sample of albumen from the eggs in the nest. In goldeneyes, eggs are usually dumped in the nests of close relatives, suggesting that the practice is a partly cooperative one. Another unusual type of kin association occurs in lekking birds, where males aggregate to display to females, which visit them exclusively for the purpose of copulation. Lekking males initially appear to be fiercely competitive at the display sites, but a number of studies have now shown that the birds that aggregate at leks are close relatives, suggesting that the relatives may benefit from communal display.

Determining Genetic Structure. There are often poor correlations between the phenotype and the underlying genotype. Understanding this structuring is of both practical and theoretical concern. For example, albatrosses are subject to appalling mortality when caught as bycatch by longline fisheries. These birds can forage thousands of kilometers from their natal site and show great plumage variability. Genetic markers that reliably distinguished between each breeding colony would identify the populations most at risk.

The best markers for studies of this sort show zero or limited recombination, and thus diverge consistently over time, so they can be used to estimate the age at which populations started to diverge from each other. Female lineages can be traced through variation in mitochondrial DNA, because mitochondria are not transmitted via sperm. By contrast, male lineages in mammals and many other organisms can be traced via the evolution of the Y chromosome, which is inevitably transmitted from father to son. In humans, for example, the Lemba are a South African ethnic group of Bantu appearance that claim Jewish descent. This claim was viewed with considerable skepticism, but it has been confirmed through the use of Y chromosome markers, which uncovered a high frequency of alleles associated with Jewish populations.

Indices of Selection. Most genetic markers are presumed to be selectively neutral, having no influence on the fitness of the individuals that bear them. However, because they are often in close physical association with areas under selection, alleles at marker loci can be used as indices of selection of the surrounding region of the chromosome. For example, wild Soay sheep on the Scottish island of St. Kilda undergo regular and dramatic population fluctuations, with populations building up to extreme numbers, then crashing once food supply has been exhausted. If alleles were sampled randomly, we would expect genetic variation to be depleted by the population crashes, but this is not the case. A variety of allozyme and microsatellite alleles fluctuate predictably in frequency with population density, suggesting that they are associated with genes with fitness effects that differ as the population cycle proceeds. In general, the efficacy of markers as indicators of selection cannot be anticipated, so a large number of alternative markers need to be sampled to provide reliable indicators of selection. This problem will disappear as the genome of a variety of species becomes precisely known, so that markers can be chosen because of their physical proximity on the chromosome to genetic loci of known biological interest. For example, a microsatellite locus known to be in the intron of an interferon gene in Soay sheep predicts resistance to parasitic nematodes.

[*See also* Cryptic Female Choice; Mate Choice, *articles on* Mate Choice in Plants *and* Human Mate Choice; Sexual Selection, *article on* Bowerbirds.]

BIBLIOGRAPHY

Andersson, M., and M. Ahlund. "Host–Parasite Relatedness Shown by Protein Fingerprinting in a Brood Parasitic Bird." *Proceedings of the National Academy of Sciences USA* 97 (2000): 13188–13193. Ducks are related to the birds they parasitize.

Avise, J. C. *Molecular Markers, Natural History and Evolution.* New York, 1994. An excellent overview.

Bancroft, D. R., et al. "Molecular Genetic Variation and Individual Survival during Population Crashes of an Unmanaged Ungulate Population." *Philosophical Transactions of the Royal Society of London B* 347 (1995): 263–273. Genetic markers help us visualize selection.

Coltman, D. W., et al. "A Microsatellite Polymorphism in the Gamma Interferon Gene Is Associated with Resistance to Gastrointestinal Nematodes in a Naturally Parasitized Population of Soay Sheep." *Parasitology* 122 (2001): 571–582. Markers are particularly useful if associated with genes of known function.

Constable, J. L., et al. "Noninvasive Paternity Assignment in Gombe Chimpanzees." *Molecular Ecology* 10 (2001): 1279–1300. Genetic variation can be determined from feces of wild chimpanzees.

Dubrova, Y. E., et al. "Human Minisatellite Mutation Rate after the Chernobyl Accident." *Nature* 380 (1996): 683–686. Genetic markers reveal DNA damage following exposure to radiation.

Gagneux P., et al. "Female Reproductive Strategies, Paternity and Community Structure in Wild West African Chimpanzees." *Animal Behaviour* 57 (1999): 19–32. Genetic variation can be determined from hairs of wild chimpanzees.

Gemmell, N. J., et al. "Low Reproductive Success in Territorial Male Antarctic Fur Seals (*Arctocephalus gazella*) Suggests the Existence of Alternative Mating Strategies." *Molecular Ecology* 10 (2001): 451–460. Most matings in seals are gained by males that do not come to shore.

Jeffreys, A. J., et al. "Individual-Specific 'Fingerprints' of Human DNA." *Nature* 316 (1985): 76–79. The classic paper that introduced DNA fingerprinting.

Marshall, T. C., et al. "Statistical Confidence for Likelihood-based Paternity Inference in Natural Populations." *Molecular Ecology* 7 (1998): 639–655. The confidence we can place in paternity analysis.

Mulder, R. A., et al. "Helpers Liberate Female Fairy-Wrens from Constraints on Extra-pair Mate Choice." *Proceedings of the Royal Society of London B* 255 (1994): 223–229. The least faithful socially monogamous species, a result unimaginable without markers.

Petrie, M., et al. "Peacocks Lek with Relatives Even in the Absence of Social and Environmental Cues." *Nature* 401 (1999): 155–157. Male peacocks prefer to display next to their close relatives.

Queller, D. C., and J. E. Strassmann. "Kin Selection and Social Insects." *BioScience* 48 (1998): 165–175. An excellent overview of cooperation and conflict in social insects.

Strassmann, J. E., et al. "Altruism and Social Cheating in the Social Amoeba *Dictyostelium discoideum.*" *Nature* 408 (2000): 965–967.

Thomas, M. G., et al. "Y Chromosomes Traveling South: The Cohen Modal Haplotype and the Origins of the Lemba—the 'black Jews of Southern Africa.'" *American Journal of Human Genetics* 66 (2000): 674–686. Y chromosomes indicate male lines of descent.

ONLINE RESOURCES

es.rice.edu/projects/Bios321/social.wasp.home.html. "Social Behaviour of Polistine Wasps." Maintained by Joan Strassman and Dave Queller's group at Rice University, this site provides clear introductions to the study of sociality in wasps, as well as the use of microsatellites to determine paternity and the extent of relatedness between individuals.

— ANDREW COCKBURN

GENETIC POLYMORPHISM

The term *genetic polymorphism* refers to the presence of multiple states for an inherited characteristic within a population. Variation can be found at multiple levels within populations, starting with phenotypic morphs or traits, progressing to chromosome variants, alternate forms of proteins coded by alleles at a specific genetic locus, and finally to nucleotide variation at the DNA level.

Phenotypic Polymorphism. There are many examples of morphological variation for discrete traits or characteristics. Most well-known examples that are inherited in a simple manner arise from a few alleles segregating at one or two genes. Gregor Mendel, among other geneticists, studied polymorphisms in flower color and seed shape in the sweet pea plant, and T. H. Morgan studied eye color and wing shape polymorphisms in the fruit fly *Drosophila*. The studies of these polymorphisms in the laboratory helped shape our basic understanding of genetic inheritance in the early part of the twentieth century. Phenotypic polymorphisms are also present in natural populations. For example, the shell of the land snail *Cepaea nemoralis* can vary in color (pink or yellow) and can be banded or unbanded. One locus controls shell color, with pink dominant to yellow, and another controls bandedness, with unbanded dominant to banded. Snail populations are also polymorphic for the height of the shell and the number of bands, but these characteristics have a more complex genetic basis. As the number of loci contributing to a trait increases, the discrete polymorphisms blend into continuous variation.

Chromosome Polymorphism. Variation in karyotype (the chromosome complement of a cell) can be observed within populations of all kinds of plants, invertebrates, vertebrates, and even mammals. This polymorphism may consist of variation for chromosome number and morphology. Extra (supernumerary) chromosomes, reciprocal translocations of fragments between chromosomes, and inversions of chromosome sections can all be found segregating. Chromosome polymorphisms are visualized by using microscopy to examine cells taken from rapidly dividing tissues or from cell culture. Supernumerary chromosomes (often called B chromosomes) have been studied extensively in grasshoppers and maize. In the grasshopper *Attractomorpha australis*, most individuals have eighteen autosomes, but individuals with up to six extra chromosomes have been observed. Inversions have been widely studied in *Drosophila* using the large polytene chromosomes found in the salivary glands of the larvae. These chromosomes have characteristic banding patterns, specific to particular areas of each chromosome. Samples from *Drosophila* populations often show subtle differences between individuals where the banding pattern in some parts of the chromosome has been reversed. These mark inverted segments that by chance have been excised from the chromosome and reinserted in the opposite orientation.

Molecular Polymorphism. One of the first types of molecular variation to be studied were immunological polymorphisms in the proteins on the surface of blood cells. These molecules are called antigens, and there are more than fifty different kinds known for human red blood cells. The best known is the ABO blood group system. The blood group phenotypes are recognized with antiserum, produced by injecting blood into a rabbit, which then makes antiserum against that blood type. The antiserum is collected from the rabbit. When mixed with a sample of a specific target blood from a person, the antiserum causes the blood sample to agglutinate. In the ABO system, there are three alleles, designated I^A, I^B, and I^O. People with genotypes $I^A I^A$ or $I^A I^O$ have blood type A, genotypes $I^A I^B$ have blood types AB, genotypes $I^B I^B$ or $I^B I^O$ have blood type B, and genotype $I^O I^O$ have blood type O. The frequencies of these alleles vary between different geographic and ethnic groups.

In the 1950s, scientists interested in the biochemical properties of proteins started to examine variation in metabolic enzymes and blood serum proteins. This was done using a technique called gel electrophoresis, which is used to find variation in macromolecules such as proteins or nucleic acids. The general principle is that samples of protein from homogenized tissue, or purified nucleic acids, are placed in a slot on a gel composed of starch, acrylamide, or agarose in an electricity-conducting buffer. Applying an electric current to the gel causes the molecules to move through the gel matrix at a rate proportional to their charge and size, sorting the molecules into charge or size classes. Most enzymes are run on starch gels. The position of the enzyme is visualized by applying a substrate specific to the enzyme together with a marker dye that precipitates when the enzyme-catalyzed reaction takes place. This produces a colored band on the gel, indicating the position of the specific enzyme.

Early investigators found many variants of individual proteins with different electrophoretic mobilities. These represent different allelic variants of the protein. The variation in electrophoretic mobility derives from amino acid substitutions that alter the relative charge of the protein. These different variants are termed allozymes. On an electrophoresis gel, an individual that is homozygous for the same allozyme allele at a specific locus will have a single band, whereas a heterozygote for two different alleles will show two separate bands appearing at different positions on the gel as a result of their differences in mobility. The first studies of allozyme variation at a population level were published in 1966, with classic papers by H. Harris looking at variation in humans and R. C. Lewontin and J. L. Hubby, who examined polymorphism in natural populations of the fruit fly *Drosophila pseudoobscura*. These studies showed that there was extensive variation at the molecular level in natural populations and inspired a whole new field of research studying patterns of molecular genetic variation in a range of species. This newly accessible source of genetic variation allowed a whole new approach to the study of evolution.

During the 1970s and 1980s, techniques were developed that allowed variation at the DNA level to be assayed. This included indirect methods such as restriction fragment length polymorphism (RFLP) analysis, which reveals sequence variation that creates or removes restriction enzyme sites (these enzymes, obtained from bacteria, recognize and cut specific DNA sequence motifs four to six base pairs long), and the actual reading of the sequence of DNA molecules. Both techniques use gel electrophoresis (on agarose gels for RFLPs and on polyacrylamide gels for DNA sequencing) to resolve the DNA fragments. The basic technique most commonly used for sequencing DNA today was invented by Fred Sanger in 1977 at Cambridge University.

The first study of the complete DNA sequence of a gene at a population level was conducted by the geneticist Marty Kreitman in 1983. He sequenced the DNA for two allozyme alleles of the enzyme alcohol-dehydrogenase (*Adh-f* and *Adh-s*) in 11 strains of *Drosophila melanogaster* to understand the actual DNA sequence variation underlying the polymorphism observed at the protein level. Kreitman found that the difference between the two allozyme alleles was the same single amino acid substitution in each case (threonine substituted for lysine at position 192 of the protein). He also found that the DNA sequences from the eleven strains were all different and that there were another forty-three nucleotide polymorphisms that did not cause amino acid changes as a result of the redundancy of the genetic code. This showed that genetic polymorphism is greatest at the DNA level and that much of this variation is not visible at other levels. It also illustrated that degrees of polymorphism vary according to the type of mutation or position in the DNA molecule. Often variation is greatest at sites not subject to strong selection, such as noncoding DNA sequences, or for mutations in coding sequences that do not cause amino acid substitutions.

Advances in the automation of sequencing technology now mean it is relatively easy to collect large DNA sequence data sets. However, for screening sequence variation at the population level, it is often quicker to fully sequence a subsample and then to use faster methods to screen for the polymorphisms identified. These single nucleotide polymorphisms (SNPs), which distinguish individual sequences, are now being used in disease mapping and evolutionary studies. They can be quickly assayed using gels, microarrays (DNA chips), and other novel methods currently being developed.

Another type of DNA variation comes from a class of markers called variable number tandem repeats (VNTRs). These are stretches of noncoding DNA consisting of DNA sequence motifs (repeat units) that are repeated many times in a continuous block or tandem array. These types of markers are usually highly polymorphic within populations for the number of repeat units in a specific array (equivalent to a discrete locus). This variation is assayed by using gel electrophoresis to measure the size of pieces of DNA carrying a specific repeat array, which will vary in size according to the number of times the sequence motif is repeated. The first VNTR markers to be widely used were minisatellites, using techniques pioneered by Alec Jeffreys and colleagues at the University of Leicester in the mid-1980s. Microsatellites, or short repeat arrays of ten to fifty units of two to six base pair motifs (e.g., polycytosine-adenine), are now the most commonly used VNTR markers because they are easily screened by the polymerase chain reaction and electrophoresis. The discovery of VNTR markers precipitated a revolution in the way that genetic markers were used. The huge amount of genetic variation accessible with the new marker class meant that for the first time individual identity and relationships between individuals could be easily resolved. (This generally was not possible with allozymes or RFLPs, as they lacked sufficient variation.) These markers are now widely used in paternity, forensic, evolutionary, and genetic mapping studies in many organisms.

Amount, Sources, and Maintenance of Genetic Polymorphism in Natural Populations. The amount of genetic variation at molecular markers is usually summarized as heterozygosity (the probability that two copies of a gene picked at random from a population will be different) for genic type markers (allozymes, microsatellites, etc.) and as sequence diversity (a measure of the number of pairwise nucleotide differences) for DNA sequences. For protein loci, the average heterozygosity is about 0.15 for insects and about 0.037 for large mammals. Among VNTR loci, heterozygosity can be 0.9 to 1.0. At the DNA level, the average sequence divergence is about 0.1 percent in humans and about 1 percent in *Drosophila*.

New genetic variation can arise in situ within a population through mutation and recombination, or it may be brought into the population by migrants after arising elsewhere. Different types of genetic loci have different mutation rates. For example, the average rate of point mutations in DNA sequences is about 10^{-8} per generation, whereas for microsatellites, it is around 10^{-5} per generation.

The fate of genetic polymorphisms in a population is determined by two main forces: genetic drift (random changes in allele frequency caused by chance variation in the transmission of alleles between generations) and selection. Selection is a powerful force that can maintain polymorphism despite drift or deleterious effects of alleles when combined in specific genotypes. Often this happens because selection favors different polymorphisms at different times, in different places and environments. The geneticist Theodosius Dobzhansky and colleagues studied many different chromosome inversions in the fruit fly *Drosophila pseudoobscura* from different locations around the United States. They found that the frequency of the inversions varied seasonally, geographically, and between different environments within the same area. These polymorphisms are maintained by different selective pressures in different places and at different times. Most individuals appear to be heterozygous for two different inversions, and heterozygotes survive better than homozygotes. Inversions suppress recombination of genes within the inversion. Dobzhansky concluded that the inversions contained groups of genes whose collective association conferred an advantage and that these "supergene" complexes were protected against recombination by the inversion. The chromosomal polymorphism in these populations is an example of balancing selection, in which the heterozygotes are the fittest genotype because they cope better with changing environments. In a similar way, selection can maintain polymorphisms when alleles confer different advantages at different times in the life cycle of an organism or in different sexes.

Selection can maintain a genetic polymorphism even if one of the alleles involved can have deleterious effects. For example, in humans, the homozygous sickle-cell hemoglobin condition causes severe anemia and is usually lethal in childhood. In malarial areas, normal hemoglobin homozygotes experience high rates of malaria infection, and many victims die in childhood. However, in malarial areas, heterozygotes for normal hemoglobin and sickle-cell hemoglobin have a selective advantage because of enhanced resistance to the malaria parasite, maintaining the sickle-cell allele in the population. This is an example of stabilizing selection. Outside of malarial areas, there is no advantage to being a sickle-cell heterozygote because there is mild anemia associated with this condition. This, combined with the consequences of being a sickle-cell homozygote, means that the sickle-cell allele is almost entirely absent in nonmalarial areas.

Frequency-dependent selection is another way in which genetic polymorphisms can be maintained in a population. This describes a situation in which a phenotype is selected against when common and selected for when rare. Batesian mimicry (mimicry of the color patterns of unpalatable species by a palatable species) in *Papilio* butterflies provides an example. When unpalatable models are rare and mimics are more common than nonmimics, mimics suffer more predation than nonmimics. When models are common and mimics are rare relative to nonmimics, the mimics experience lower predation than nonmimics. This can result in variation in the frequency of the mimics (and their color pattern genes) in relation to the frequency of the model species.

Most molecular polymorphisms are not actually affected by selection. This is termed neutral variation and applies to VNTRs and other noncoding regions of the genome. It can also apply to coding sequences for both mutations that do and do not cause amino acid substitutions. In the early days of population genetics, the paradigm was that all genetic variation was maintained by balancing selection (heterozygote advantage). However, as surveys of molecular polymorphisms developed, the large amount of variation discovered at protein loci became problematic for this theory. This was because even with very small fitness differences between alleles, the effect of selection on many thousands of loci would mean that most organisms should actually be extinct because of the high genetic load. Other arguments and data on rates of molecular evolution led Motoo Kimura, along with Jack King and Thomas Jukes, to propose in the late 1960s that the most genetic variation has no functional consequences for the organism. Initially, this theory of neutral molecular evolution was criticized as being non-Darwinian, but today it is mostly accepted.

Genetic polymorphism is the information that forms the basis of all evolutionary study. By comparing genetic polymorphism in different groups, biologists can investigate anything from the evolutionary relationships between taxa to the location of disease genes in the genome. Current important evolutionary questions for biologists include exactly how both overall genetic variation and specific polymorphisms at individual genes contribute to the ability of an organism to survive and reproduce.

[*See also* Neutral Theory.]

BIBLIOGRAPHY

Cavalli-Sforza, L. L., and W. F. Bodmer. *The Genetics of Human Populations*. San Francisco, 1971. Classic book with detailed descriptions of blood group polymorphisms in humans.

Griffiths, A. J. F., J. H., Miller, D. T. Suzuki, R. C. Lewontin, and W. M. Gelbart. *An Introduction to Genetic Analysis*. New York, 1996. An accessible introduction to many aspects of genetics.

Harris, H. "Enzyme Polymorphism in Man." *Proceedings of the Royal Society of London B* 164 (1966): 298–310. A report of the first study to examine enzyme polymorphisms in humans at a population level.

Hubby, J. L., and R. C. Lewontin. "A Molecular Approach to the Study of Genic Heterozygosity in Natural Populations: 1. The Number of Alleles at Different Loci in *Drosophila pseudoobscura*." *Genetics* 54 (1966): 577–594. One of the first studies to examine protein polymorphism in animal populations.

Jeffreys, A. J., V. Wilson, and S. L. Thein. "Hypervariable Minisatellite Regions in Human DNA." *Nature* 314 (1985): 278–281. One of the first papers about the use of VNTR markers in humans.

Kreitman, M. "Nucleotide Polymorphism at the Alcoholdehydrogenase Locus of *Drosophila melanogaster*." *Nature* 304 (1983): 412–417. This paper describes the first study of the DNA sequence of a gene at a population level.

Lewontin, R. C., and J. L. Hubby. "A Molecular Approach to the Study of Genic Heterozygosity in Natural Populations: 2. Amounts of Variation and Degree of Heterozygosity in Natural Populations of *Drosophila pseudoobscura*." *Genetics* 54 (1966): 595–609. One of the first studies to examine protein polymorphism in animal populations.

Maynard Smith, J. *Evolutionary Genetics*. New York, 1998. A good introduction to evolutionary biology with an excellent treatment of the underlying theory.

Ridley, M. *Evolution*. Cambridge, Mass., 1998. An introduction to evolutionary biology for undergraduate level students.

Slate, J., L. E. B. Kruuk, T. C. Marshall, J. M. Pemberton, and T. H. Clutton-Brock. "Inbreeding Depression Influences Lifetime Breeding Success in a Wild Population of Red Deer (*Cervus elaphus*)." *Proceedings of the Royal Society of London Series B* 267 (2000): 1657–1662. Shows how the amount of genetic variation an individual carries can influence its ability to reproduce.

Stephens, J. C., J. A. Schneider, D. A. Tanguay, J. Choi, T. Acharya, S. E. Stanley, R. H. Jiang, C. J. Messer, A. Chew, J. H. Han, J. C. Duan, J. L. Carr, M. S. Lee, B. Koshy, A. M. Kumar, G. Zhang, W. R. Newell, A. Windemuth, C. B. Xu, T. S. Kalbfleisch, S. L. Shaner, K. Arnold, V. Schulz, C. M. Drysdale, K. Nandabalan, R. S. Judson, G. Ruano, and G. F. Vovis. "Haplotype Variation and Linkage Disequilibrium in 313 Human Genes." *Science* 293 (2001): 489–493. Shows the modern approach to quantifying genetic variation in humans using single nucleotide polymorphisms.

Sunnucks, P. "Efficient Genetic Markers for Population Biology." *Trends in Ecology and Evolution* 15 (2000): 199–203. An accessible review which gives a good overview of some different techniques used to examine genetic variation in natural populations.

— SIMON J. GOODMAN

GENETIC REDUNDANCY

Darwin's theory of evolution and subsequently population genetics have stressed the role that natural selection plays in the fixation of beneficial mutations and the elimination of deleterious ones. By contrast, the neutral theory shows what level of genetic diversity we might expect when a mutant gene is selectively equivalent to a wild type. These mathematical theories provide the logical foundation on which our understanding of nature rests; they give us insight into how adaptations could have developed through the fixation of innumerable mu-

tations, each of small effect. Nonetheless, in genetic systems there is a great deal of diversity that is neither immediately beneficial nor always neutral yet has been preserved over long periods of time. Such genetic information is placed in a category of its own: genetically redundant information.

The term "redundancy" is somewhat problematic because it might lead one to assume that this information has no adaptive value in any circumstance. To appreciate the value of redundancy, however, one need only consider devices engineered by humans. Modern aircraft, for example, are controlled by sophisticated avionic circuits. Only a single system is required to fly the aircraft, and because each system is costly to the manufacturer, one might suppose that each airplane is equipped with only a single copy. In fact, most aircraft have four replicas of this circuitry, so that if one or more fails, there will be a backup to take its place. Redundancy ensures the robustness of the aircraft in the face of damage to the system. We can tell that a system is redundant precisely because damage leaves it functioning exactly as it did before the damage occurred. Therefore, the redundancy is not neutral, because it has a value when its duplicate is removed; nor is it immediately advantageous, because, in the absence of damage, duplicates can be removed without adverse consequences.

What is true of engineered devices often is also true of evolved devices. Thus, it comes as no surprise to find extensive redundancy among the genomes of living organisms. The best-characterized example of redundancy in genetic systems (apart from the redundancy of the genetic code itself) is polyploidy. In diploid genomes, for example, there are at least two copies of each autosomal allele. Diploidy has been explained as an adaptation to minimize the impact of deleterious recessive mutations: the second allele can mask the negative effects of mutation. A further characteristic of diploidy is haplo-insufficiency on the loss of an allele, or the reduction in fitness that results from that loss.

This article focuses on redundancy among independent genes rather than on polyploidy, since the former type of redundancy is harder to understand. Following the removal, or "knockout," of both alleles of a gene, there often is no detectable, or "scoreable," change in the phenotype or fitness of the organism. For example, knockout of the *Drosophila* genes *gooseberry* and *sloppy paired*, the mouse genes *tanascin* and *Hox*, the *Arabidopsis* gene *far*, or the yeast gene *myosin* do not result in a quantifiable change in phenotype under laboratory conditions. Therefore, these genes are thought to possess a degree of redundancy similar to that of avionics systems.

From the perspective of evolution, redundancy presents a series of fascinating questions. How does redundancy evolve in the first place? Once it has evolved, how is it maintained? Is redundancy simply a matter of preserving duplicates of a given gene, or does it involve the restructuring of whole genetic networks? What fraction of the genome is redundant, and how does this fraction relate to factors that influence the rate of mutation? What are the implications of large-scale redundancy for the rate of evolution of a species? Each of these questions is addressed in the sections that follow.

Evolution and Maintenance of Redundancy. The two principal theories describing the origin of redundancy are functional shift and gene duplication. Functional shift is the phenomenon in which two independent genes evolve toward some degree of overlap in their function, or toward a third, novel function. In contrast, gene duplication occurs when a single gene duplicates, resulting in the presence of two identical genes at different sites in the genome. For a population of individuals to possess redundant genes, the duplicate must spread to the point of fixation. This can happen even when there is a probability of both copies of the original gene being lost through mutation, as long as the rate of recurrent duplication is on the same order as the rate of mutation.

Once a degree of redundancy has arisen, it must be maintained. This presents a problem because random inactivation of one gene from a redundant pair will have minimal phenotypic effects, or even none. Several mechanisms for preserving redundancy have been proposed. These theories can be characterized by their focus on cumulative benefit, mutational error buffering, developmental error buffering, pleiotropy, or regulatory elements.

Cumulative benefit theories assume that having an extra active copy of a gene increases the quantity of the product (e.g., a particular protein) that the gene makes. If this increases fitness, genetic redundancy is easy to understand. Thus, each eukaryotic cell harbors multiple copies of the mitochondrial genome, which enables cells to metabolize efficiently, and multiple copies of tRNA and mRNA genes for efficient translation. Eliminating copies will reduce fitness; hence, redundancy is maintained by stabilizing selection acting on the ensemble of identical genes.

Mutational error buffering theories define genetic error buffering as any mechanism that reduces mutational load. Consider two genes that are otherwise identical but have slightly different mutation rates and slightly different efficiencies. The gene with the higher mutation rate will, on average, be eliminated from the population; that is, it becomes a pseudogene. If it is the less efficient gene that also possesses the lower rate of mutation, redundancy can be preserved, because loss of the more efficient gene owing to mutation will bring a relatively large drop in fitness. If the more efficient gene had the lower mutation rate, there would be no selective reason

to maintain redundancy, because loss of the less efficient gene would not bring a substantial loss of fitness. Thus, only by allowing a slight asymmetry in each gene's abilities to perform its respective functions might redundancy be preserved.

Under theories of developmental error buffering, it is observed that during the early growth and development of an organism, many different and strongly interlinked processes must occur. Sometimes random events or environmental variability disturb one or more of these processes. Developmental error buffering denotes mechanisms that can reduce the effects of such perturbations—mechanisms that, in effect, improve the stability of the phenotype. Duplicated or redundant pairs of genes can be maintained if one can take over the lost function when the other becomes defective owing to an event during development. This can occur only if the redundant member of the pair becomes defective at a lower rate than the other member of the pair does. In other words, a function that acts as a developmental buffer must have a low error rate and must be in support of a developmentally unstable function with a high mutation rate.

Some form of pleiotropy can also ensure the conservation of redundant function. The term "pleiotropy" refers to cases in which a single gene affects more than one function and therefore experiences selection in more than one context. Consider two genes with two independent functions. Assume that one gene (the pleiotropic gene) can, in addition to its primary function, also perform the function of the other gene, though less efficiently. Mutations to the pleiotropic gene can eliminate either its unique function, or both functions. Redundancy will be preserved whenever the rate of elimination of the pleiotropic function is lower than elimination of the unique function. In other words, when one gene's pleiotropic function is more robust than the other gene's unique function, the correlated function can be preserved.

Regulatory element theories note that there is redundancy not only among coding regions but also among regulatory elements associated with structural genes. For example, if each of two duplicated genes is accompanied by duplicates of subsets of regulatory elements (where each subset overlaps to some degree through shared elements), the redundant genes can be maintained by selection that acts through their unique patterns of expression and the shared regulatory element. The shared regulatory element controls the correlated function. If one assumes that the shared regulatory element is a smaller mutational target than the coding region, this will prolong the half-life of the redundant function. It is not sufficient, however, to prevent one or more shared elements from becoming silenced in the long term. To preserve redundancy indefinitely requires, as in the pleiotropic model, some asymmetry in mutation and/or in efficacy of the regulators.

Is Redundancy Simple Preservation or Large-scale Restructuring? Genetic redundancy can be the direct consequence of carrying duplicate copies of a gene, or the outcome of nonlinear interactions among many genes that are coordinately regulated. As in an ecosystem where predators are able to switch to other prey when their favored resources becomes limiting, so genes can modify their pattern of regulation when some target genes are removed or damaged. In this case, redundancy arises because each gene interacts with numerous others according to the abundance of all concerned. Redundancy is not a property of individual genes, but of a network.

In an effort to determine which type of redundancy is more prevalent in yeast, A. Wagner searched for correlations between fitness effects and genetic similarity in yeast knockout mutants. If gene duplication accounts for redundancy, then we expect that after knockout of a gene, the degree of fitness reduction will correlate positively with the genetic divergence between the target of the knockout and its duplicate. We expect genes that are more similar to compensate more effectively for the loss of their corresponding paired genes. This, however, is not what we find. There is no significant correlation between genetic divergence and the magnitude of the protective effect on a knockout. There are, however, many knockouts that have a minimal effect on fitness, even though the removed genes have a very weak similarity to other genes in the genome. This lends credibility to the hypothesis that redundancy emerges as a network effect.

Proportion of Redundancy and Relation to Factors Influencing Mutation Rates. What fraction of the genome is redundant, and how does this fraction relate to factors that influence the rate of mutation? These questions are largely unanswerable at present. In most organisms that have been sequenced so far, at least one-third of the genes in the genome belong to paralogous gene families—that is, they share homology through duplication. Assuming that redundancy exists wherever there is strong genetic similarity, this might lead us to expect very high levels of redundancy; however, we know from yeast studies that this is not always the case. Furthermore, selection for redundancy may have more to do with shoring up the genome against errors than with selection for redundancy in itself. This is likely because robustness in resistance to errors does not directly increase the reproductive fitness of the redundant genome, but it enables populations with high redundancy to outlast those with little redundancy.

Redundancy has been documented best among eukaryotes—organisms with large genomes and low mutation rates. Eukaryotes are thought to have undergone one or more whole-genome duplications during their evolutionary history. Mutation rates vary from a low of around 8.4×10^{-11} per base pair per replication in *Drosophila* to highs of 4.0×10^{-10} in *E. coli* and 2.4×10^{-8} in bacteriaphage lambda.

Interestingly, higher mutation rates (like those observed in viruses and bacteria) are not associated with higher levels of redundancy; if anything, the reverse appears to be true. Why this should be so is at present controversial. Many viruses and bacteria have "overlapping genes," stretches of DNA (or RNA) that encode more than one protein. This is the opposite of redundancy, because a single mutation is likely to affect two or more proteins. Through mechanisms that are not entirely understood, overlapping genes may actually make populations of viruses in the host more stable. Alternatively, viruses may have been selected for very small genomes and high replication rates, making redundancy disadvantageous.

Implications of Redundancy for Rates of Evolution. An important consequence of redundancy is the reduction of genetically encoded variance among phenotypes within a population. This occurs because redundancy tends to obscure the effects of mutations on phenotypes. We know from Fisher's fundamental theorem that selection is more effective when there is greater heritable variance among phenotypes, so we expect that redundancy will reduce the rate of adaptive evolution.

To understand this relationship, consider the consequences of complete redundancy. Adaptive evolution would come to a halt because no mutation would ever produce any change in phenotype, and the rate of fixation of mutations would then be determined exclusively by the rate of mutation, as is predicted by the neutral theory. There are circumstances, however, when redundancy may promote adaptive evolution—particularly in cases where local deviations from the wild type phenotype are deleterious. The effect of partial redundancy is to allow a sufficient quantity of genetic polymorphism to arise by neutral drift, so that new, nonredundant mutations can fall within the vicinity of the second fitness peak. Thus, the phenotype is able to "hop" from one fitness peak directly to another without having to go through intermediate phenotypes of lesser viability. Redundancy could provide for more effective exploration of genetic space. Neither of these two potential consequences of redundancy has yet been tested empirically.

BIBLIOGRAPHY

Krakauer, D. C., and M. A. Nowak. "Evolutionary Preservation of Redundant Duplicated Gene." *Seminars in Cell and Developmental Biology* 10 (1999): 555–559.

Wagner, A. "Robustness against Mutations in Genetic Networks of Yeast." *Nature Genetics* 24 (2000): 355–361.

— DAVID C. KRAKAUER

GENITALIC EVOLUTION

One of the most sweeping trends in animal evolution is for male genitalia and genitalic products such as spermatophores to diverge rapidly; they often differ even in closely related species. This pattern is extensively documented in the taxonomic literature on many different groups, including insects, spiders, scorpions, opilionids, mites, pseudoscorpions, nematodes, leeches, lizards, snakes, primates, rodents, pulmonate snails, and cephalopods. Taxonomists frequently have found that male genitalia have very complex forms that are especially useful in distinguishing closely related species. The often bizarre, exuberant diversity of male genitalic forms is exemplified by their extraordinary lengths. Male genitalia reach over several times the length of the male's entire body in some beetles, flies, and slugs. Although behavior during copulation is much less studied than morphology, some extraordinary activities nevertheless have been documented. These include removal of sperm of previous males from inside the female in damselflies and some other insects; perforation of the female's body wall to inject sperm directly into her (the sperm then move directly to her ovaries) in leeches and bedbugs; scraping holes in the female reproductive tract to allow other, nonsperm seminal products into her blood, where they induce oviposition or inhibit remating in some flies; apparently douching the female in some male sharks; pulling the lower part of the female's reproductive tract inside out from her body in some flies and katydids; and squeezing the female with complex, species-specific rhythms.

At first glance, the extremely complicated structure of male genitalia in many species seems to constitute a challenge to evolutionary theory: their complexity is far too great to be explained by natural selection favoring the ability to perform the relatively simple task of transferring sperm to the female. Several competing theories attempt to explain this problem. The oldest is the hypothesis of species isolation by "lock and key." This is the idea that females have evolved genitalia that are capable of mechanically excluding the male genitalia of other species. In so doing, they avoid losing their investment in expensive eggs from mistakenly mating with males of other species. Female genitalic "locks" are proposed to admit only the genitalia of conspecific males— the correct "keys." The lock-and-key hypothesis has failed numerous tests: females of many species lack any genitalic structure mechanically capable of impeding the entry of the genitalia of nonconspecific males, cross-specific pairings occur in some others despite interspecific differences in female genitalia, and there is divergence in genitalic structures even in species in which females are completely isolated from contact with males

of other closely related species (e.g., endemic species on depauperate oceanic islands and parasites that are limited to hosts on which no other related parasites occur). The magnitude of the tendency for male genitalia to diverge rapidly is so sweeping (literally millions of species are involved) that it is still possible that genitalia have evolved under lock-and-key selection in some groups. It is now clear, however, that lock and key is not a valid general explanation.

A second failed, but still occasionally cited, general hypothesis is Mayr's pleiotropy theory. This holds that the genes responsible for divergent genitalic structures have diverged because they also have pleiotropic effects on other, ecologically important traits that are under different selective regimes in different species. The most telling evidence against this hypothesis is the consistent correlation between the evolution of alternative male structures to transfer sperm (e.g., spermatophores, legs, pedipalps, and sternites) and the simultaneous rapid divergent evolution of these particular structures, even though structures associated with the male's primary genital opening remain relatively unaltered. Such consistent coincidences are very unlikely.

Two further hypotheses to explain genitalic complexity are more recent and less thoroughly tested. The sexual selection by female choice hypothesis proposes that females discriminate among conspecific males by biasing postcopulatory processes that affect the male's chances of fertilizing eggs on the basis of male genitalic form and behavior. Male genitalia are thought to perform internal courtship of the female during copulation, eliciting favorable responses such as ovulation, oviposition, sperm transport, sperm dumping, and inhibition of remating. The proposed advantage to the female of screening males this way is to obtain superior offspring, perhaps especially sons that are better able to induce similar female cooperation in subsequent generations. The second hypothesis, male–female conflict, proposes that selection on males and females to control events surrounding copulation leads to a coevolutionary arms race between male and female genitalia. The proposed advantages to females from resisting males involve natural selection rather than sexual selection, and include avoidance of damage from copulation, reduced ability to avoid predation during copulation, infection by venereally transmitted diseases, and interruption of oviposition of feeding.

These hypotheses make some similar predictions, such as a positive association between female re-mating frequency and genitalic complexity, that have been confirmed. Three lines of evidence argue against male–female conflict as a general explanation of genitalic complexity: intraspecific allometric relations between male genitalic size and overall body size differ sharply from relations typically shown by weapons such as horns in intraspecific conflicts; the prediction of reduced genitalic divergence in groups in which the male lures the female (rather than coercing her) appears not to be supported; and active rejection structures associated with female genitalia are almost completely absent in at least some groups. There are a few scattered direct observations that selection by female choice occurs. Also favoring the female choice hypothesis are other scattered observations of male genitalia being moved in ways during copulation that seem designed to stimulate females.

The female choice hypothesis is thus favored by the overall patterns in currently available data, but more by a process of elimination than by direct tests of its predictions. Given that rapid divergent evolution in genitalia is so extraordinarily widespread in nature, it will probably be some time before a truly general consensus is reached.

[*See also* Cryptic Female Choice; Male–Male Competition; *articles on* Mate Choice; Sexual Dimorphism; Sperm Competition.]

BIBLIOGRAPHY

Alexander, R. D., D. Marshall, and J. Cooley. "Evolutionary Perspectives on Insect Mating." In *The Evolution of Mating Systems in Insects and Arachnids*, edited by J. Choe and B. Crespi, pp. 4–31. Cambridge, 1997. Gives the first detailed presentation of Lloyd's argument that male genitalia are used as weapons to overcome female resistance to male control of the events associated with copulation, with a critical appraisal of the cryptic female choice hypothesis.

Arnqvist, G. "Comparative Evidence for the Evolution of Genitalia by Sexual Selection." *Nature* 393 (1998): 784–786. Confirms a prediction from sexual selection: more frequent remating by females is associated with more rapid divergent genitalic evolution in nineteen pairs of insect taxa that are relatively monandrous and polyandrous.

Arnqvist, G., and I. Danielsson. "Copulatory Courtship, Genital Morphology, and Male Fertilization Success in Water Striders." *Evolution* 53 (1999): 147–156. The form of a male water strider's genitalia is associated with his fertilization success when the female mates with two different males.

Eberhard, W. G. *Sexual Selection and Animal Genitalia*. Cambridge, Mass., 1985. Proposes that male genitalia evolve rapidly and divergently due to sexual selection by cryptic female choice and critically evaluates this and other hypotheses.

Eberhard, W. G. "Sexual Selection and Cryptic Female Choice in Insects and Arachnids." In *The Evolution of Mating Systems in Insects and Arachnids*, edited by J. Choe and B. Crespi, pp. 32–57. Cambridge, 1997. Summarizes arguments regarding cryptic female choice and critically discusses the alternative male–female conflict hypothesis.

Mayr, E. *Animal Species and Evolution*. Cambridge, Mass., 1963. Proposes that rapid divergent evolution of genitalia is the result of pleiotropic effects of genes that evolve due to their effects on other traits.

Rodriguez, V. "Relation of Flagellum Length to Reproductive Success in Male *Chelymorpha alternans* Boheman (Coleoptera: Chrysomelidae: Cassidinae)." *Coleopterists Bulletin* 49 (1995). Shows that males of a beetle that have especially long genitalic processes are more likely to father offspring with multiply mated females, possibly because the female is less likely to discard their sperm.

Shapiro, A. M., and A. H. Porter. "The Lock-and-Key Hypothesis: Evolutionary and Biosystematic Interpretation of Insect Genitalia." *Annual Review of Entomology* 34 (1989): 231–245. Reviews the repeated failures to confirm predictions derived from the lock-and-key hypothesis but argues that many studies are less than completely conclusive.

— WILLIAM G. EBERHARD

GENOME SEQUENCING PROJECTS

The genome is the repository of nucleotide sequences in an individual bacterium or in each cell of a eukaryotic organism. Viruses and some bacteria that are parasitic rely on host genes to complete their life cycle. Genome sequences are of DNA, except for some viruses that use RNA. Eukaryotes also contain mitochondrial genomes that are thought to be relics of symbiotic bacteria, and green plants in addition contain chloroplast genomes.

Genome size varies enormously, and prokaryotic genomes are smaller than eukaryotic genomes. Genome size in eukaryotes is not closely correlated with the number of genes. Instead, it is a function of the number and length of introns that are included in each gene sequence. An even more important determinant of genome size is the presence of repetitive sequence families. There are often longer repetitive families (LINEs) of around 1,000 nucleotides and shorter repetitive families (SINEs) of around 350 nucleotides. The copy number of some families can run to hundreds of thousands. As we explain below, the repetitive sequence families have been a considerable hindrance in determining the human genome sequence. A database of genome sizes is provided at http://www.cbs.dtu.dk/databases/DOGS/index.html.

The number of genes in a genome is a useful reference point. Gene identification is remarkably difficult, so the figures for the numbers of genes should be taken as indicative, not finalized. The human mitochondrial genome contains only 33 genes (2 rRNA, 20 tRNA, and 11 protein), and comparative studies with the mitochondria of other organisms indicate that genes from the mitochondrial genome have been variously incorporated into the nuclear genome. A typical bacterium such as *Escherichia coli* has 5,361 genes, and the yeast *Saccharomyces cerevisiae* has 6,183 genes. It is important to understand that genomes evolve by processes of gene duplication and even whole genome duplication. Thus, the comparison of the sequences within the yeast genome suggests that it has evolved by a number of whole genome duplications followed by subsequent selective loss of genes. Basic life processes are highly conserved across living organisms, so that bacterial and yeast gene homologues exist of genes known to be causative of human genetic diseases. A compilation of such genes from model organisms can be found at XREFdb (http://www.ncbi.nlm.nih.gov/XREFdb/).

Genes that evolve by divergence of species from a common ancestral species are called orthologues, whereas genes that have evolved by gene or genome duplication within an organism are called paralogues. Such assignments are based on sequence similarity and are inferences of evolutionary relationships that cannot be definitively proven. The order of genes in genomes can also help to substantiate these inferences. A compilation of orthologous genes is found at COGs (http://www.ncbi.nlm.nih.gov/COG/).

The number of genes in the *Caenorhabditis elegans* (nematode) genome is about 16,332, whereas in *Drosophila melanogaster* (fruit fly) it is about 13,601. This is somewhat surprising until one realizes that the number of genes is not necessarily the same as the number of transcripts and translated protein products. Alternatively spliced transcripts are rare in the nematode but common in the fruit fly. Human genes are also frequently alternatively spliced, so that the number of protein products is much greater than the estimate of the number of genes of about 30,000 would at first sight imply.

Genome sequencing of bacteria can be readily accomplished today by the "shotgun" methodology. This involves breaking the DNA of multiple copies of the genome into random overlapping fragments. If the entire sequence is unique, we can be guaranteed that when sufficient fragments have been sequenced, the whole genome sequence can be reassembled in the correct order. Problems arise when repetitive sequences are present, and it may be necessary to prepare genome maps to assist the assembly process.

The nematode genome-sequencing project was a model and precursor of the effort to sequence the human genome. The random genome fragments with an insert size of about 50 kb were sub-cloned into cosmids. The cosmids were fingerprinted by digestion with a restriction enzyme and the fragments sized. Overlapping cosmids share some of the same sized fragments, and a tiling path of overlaps could be built up for each chromosome. A huge amount of genetic linkage data and deletions could be related to the physical map. In difficult areas, it was necessary to bridge the cosmid contigs with Yeast Artificial Chromosome (YAC) clones that have a larger insert size of about 500 kb. All this information was compiled in a specially designed data repository called ACeDB. The best cosmids to sequence in order to complete the genome sequence were chosen by inspection of the database. Shotgun sequencing was used

to produce the majority of the sequence for each cosmid. In order to finish the sequencing of a cosmid, it may be necessary to adopt other strategies. Typically, extension of the sequence of unjoined contig ends could be made by PCR.

In 1988, when the human genome project was initiated, it was considered that many years of mapping would be required before sequencing could begin in earnest. At that time, hundreds of genes had been crudely mapped by linkage analysis or cytogenetic techniques, but it was realized that a dramatic increase in mapping efficiency was needed.

It had been discovered in the 1960s that human chromosomes could be stained with dyes, and that light and dark banding patterns could be used to identify chromosomes and to localize radioactive probes to chromosome regions. Use of fluorescent probes enhanced the methodology, and Fluorescent In Situ Hybridization (FISH) is still of value today. Producing fluorescently labeled PCR products from each individual chromosome permits very powerful chromosome painting methods. These paints can be used to study chromosomal abnormalities, including translocations. They can also be used to paint primate chromosomes to display breaks and joins that have occurred during evolution from a common ancestor.

Human genetic linkage analysis is very important because it is the only way of connecting a phenotype with an underlying genotype. In mice, second-generation backcross individuals can be bred to provide a very accurate linkage map. This has been done with two sibling species, *Mus musculus* and *M. spretus*, that are sufficiently different that every locus will be informative. The mouse is a very important model for understanding the human. For example, defects of development, including various forms of blindness and deafness, can be traced to homologous genes in the two species. This is cataloged in DHMHD (http://www.hgmp.mrc.ac.uk/DHMHD/dysmorph.html).

In human populations, we have to rely on existing individuals for ethical and practical reasons. The Centre d'Étude du Polymorphisme Humain (CEPH) has produced detailed linkage maps for humans by collecting family DNA. The CEPH families are three-generation families with large numbers of children. The markers initially used were Restriction Fragment Length Polymorphisms (RFLPs). More recently, it has been found that variation in the lengths of dinucleotide simple repeats (e.g., GC) are very informative. More than 6,000 such markers with wide coverage in terms of location have been mapped to the genome. It is now routine to perform a genome screen for a disease phenotype, with a good chance of localizing to a relatively narrow region.

Another approach to gene mapping was the use of somatic cell hybrids. Human cells are irradiated to break

TABLE 1. Completed genomes as of September 2001

Archaea

Aeropyrum pernix K1
Archaeoglobus fulgidus
Halobacterium sp. NRC-1
Methanobacterium thermoautotrophicum
Methanococcus jannaschii
Pyrococcus abyssi
Pyrococcus horikoshii
Sulfolobus solfataricus
Sulfolobus tokodaii
Thermoplasma acidophilum
Thermoplasma volcanium

Bacteria

Agrobacterium tumefaciens linear chromosome
Agrobacterium tumefaciens strain C58 circular
  chromosome
Aquifex aeolicus
Bacillus halodurans
Bacillus subtilis
Borrelia burgdorferi
Buchnera sp. APS
Campylobacter jejuni
Caulobacter crescentus
Chlamydia muridarum
Chlamydia pneumoniae AR39
Chlamydia pneumoniae CWL029
Chlamydia pneumoniae J138
Chlamydia trachomatis
Clostridium acetobutylicum ATCC824
Corynebacterium glutamicum
Deinococcus radiodurans chrI
Deinococcus radiodurans chrII
Escherichia coli K-12
Escherichia coli O157:H7 strain EDL933
Haemophilus influenzae
Helicobacter pylori strain 26695

Bacteria, continued

Helicobacter pylori strain J99
Lactococcus lactis subsp. *lactis* IL1403
Mesorhizobium loti strain MAFF303099
Mycobacterium leprae TN
Mycobacterium tuberculosis CDC1551
Mycobacterium tuberculosis H37Rv
Mycoplasma genitalium
Mycoplasma pneumoniae M129
Mycoplasma pulmonis strain UAB CTIP
Neisseria meningitidis serogroup A
Neisseria meningitidis serogroup B
Pasteurella multocida PM70
Pseudomonas aeruginosa
Rickettsia conorii Malish 7
Rickettsia prowazekii
Sinorhizobium meliloti 1021
Staphylococcus aureus strain Mu50
Staphylococcus aureus strain N315
Streptococcus pneumoniae R6
Streptococcus pneumoniae
Streptococcus pyogenes M1 GAS
Synechocystis sp. PCC 6803
Thermotoga maritima
Treponema pallidum
Ureaplasma urealyticum
Vibrio cholerae chrI
Vibrio cholerae chrII
Xylella fastidiosa
Yersinia pestis

Eukaryota

Caenorhabditis elegans
Drosophila melanogaster
Saccharomyces cerevisiae strain S288C
Homo sapiens
Arabidopsis thaliana

them, and the attempt is made to rescue a single human chromosome into a rodent cell. A panel of cell lines can be produced, one cell per chromosome, in order to permit the assignment of a gene to a chromosome. This was less successful than hoped because the rescued chromosome always had some deletions. Even worse, fragments of other chromosomes were incorporated into the cell.

Adaptation has led to a powerful method of gene mapping called radiation hybrid mapping. Instead of trying to be selective, researchers break the chromosomes into fragments, the size of which depends on the dose of radiation. Then, random fragments of chromosomes are incorporated into cells. A panel of 150 cells can give a very accurate map position. First, the panel must be characterized using known markers to describe the content of each cell. Once this has been done, new markers can readily be mapped using the panel. The map position is inferred by maximum likelihood, in a manner similar to genetic linkage. Unlike linkage, a single marker at a time can be mapped, and it need not be polymorphic.

Considerable effort was put into making human YAC libraries in the hope of making overlapping clone maps as was done for the nematode project. Unlike cosmids, it is hard to sequence the inserts of YACs. An even greater defect was the propensity of YAC clones to produce chimaeras, with a single insert being composed of DNA from different chromosomes. This is thought to be a result of recombination between very similar repetitive regions that exist in the human genome but not in the nematode. After almost ten years, YACs were abandoned in favor of Bacterial Artificial Chromosomes (BACs), which contain shorter inserts (150 kb) but are stable.

The draft human genome sequence that was published in 2001 has been produced by sequencing from ordered libraries of BAC clones. Mapping was essential to order the BAC clones into the "golden path" so that optimal clones can be selected for sequencing. As described for the nematode, each BAC insert is subjected to shotgun sequencing followed by a finishing stage. It is misleading to consider that the complete human genome sequence is available. Probably 95 percent of the sequence has been determined, but some remains unknown. Further, the sequence we have is not all correctly ordered, and a number of years' work remains before we can be fully satisfied with the result. ENSEMBL (http://www.ensembl.org) is one project that is attempting the automated annotation of the human genome.

We have up to now talked about the human genome sequence as though it were a single entity. This is highly misleading: each human individual (apart from identical twins) has a different genome sequence. The Human Genome Diversity Project is an effort by anthropologists, geneticists, physicians, linguists, and other scholars from around the world to document the genetic variation of the human species worldwide. This scientific endeavor is designed to collect information on human genome variation to help us understand the genetic makeup of all of humanity and not just some of its parts. The information will also be used to learn about human biological history and the biological relationships among different human groups, and it may be useful in understanding the causes and determining the treatment of particular human diseases.

The number of completed genomes in 11 Archaea, 50 Bacteria, and 5 Eukaryota and the list of species are given in Table 1.

Comparative genomics turns out to be a very important subject, for two reasons. First, gene order data can give evolutionary insights that may differ from those provided by nucleotide and amino-acid sequence data. Second, prediction of human gene structure in terms of intron and exon boundaries, start, stop, polyA site, promoters, and enhancers proves to be very difficult. Comparison of the mouse and puffer fish genomes with the human genome will be of enormous help in increasing our confidence in gene predictions. Even so, it may be necessary to characterize the proteins to make sure that we have understood correctly. Fortunately, this is becoming a realistic endeavor with advances in protein identification by mass spectroscopy. This is only the starting point in the vast effort to understand control of gene expression, protein interactions, and cellular pathways leading to a comprehensive description of function.

Complete genome sequencing, gene identification, and enumeration of gene order is the fullest method to produce data for genome comparison. Among vertebrates, human, mouse, and puffer fish genome sequences will certainly be finished. There are many species of interest for which it may be unrealistic to determine the entire genome sequence but for which detailed mapping may produce interesting results. This is particularly true for species of agricultural interest, including both animals and plants. Here, the objective is to identify quantitative trait loci (QTLs) that influence size and quality of a product.

Crop species often come from a family group of related species—for example, cereals from the Gramineae, potatoes, tomatoes, peppers, and aubergines from the Solanaceae, and cabbages, cauliflowers, rape, turnip, and the model *Arabidopsis* from the Brassicaceae. Having produced a map for one species, it then becomes very straightforward to map others. In cereals, rice appears to have a rather basic genome, with sorghum, millet, and maize being more complicated. Wheat is triploid, containing three distinct barleylike genomes.

Gene mapping is quite advanced in mammals, with projects for cattle, sheep, goats, pigs, horses, dogs, rats, and cats. In dog varieties, there are marked differences in size and levels of aggressive behavior that are of interest to map and may have human homologues. Mammalian evolution is of great interest, with a large part of the radiation having been established over a brief period. It will be fascinating to bring gene order data to bear on this problem as a complement to the existing phylogenetic inferences from sequence data.

— M. J. BISHOP

GENOME SIZE EVOLUTION

Because genes are made of DNA, the total amount of DNA in the genome (genome size) might be expected to be a rough measure of the total number of genes. Indeed, in prokaryotic genomes this expectation is generally met, with genes contributing most of the DNA. Surprisingly, however, in eukaryotic genomes, genes are often relegated to a minority status, with nongenic DNA constituting upwards of 90 percent of the total genome size in many cases.

Genome Size in Prokaryotes. Prokaryotic genomes vary over a ~twentyfold range, from ~0.6 million base pairs in bacterial pathogens and obligatory symbionts to 13 million base pairs in some cyanobacteria. Prokaryotes also display a corresponding twentyfold variation in the number of genes, with smaller genomes having fewer genes. Complete sequencing of a number of bacterial genomes has confirmed this—there is an almost perfect correlation between genome size and the total number of genes.

These results do not imply that bacterial genomes consist exclusively of genes. In fact, all studied bacteria do contain nongenic DNA, primarily in the form of transposable elements and insertions of bacteriophages. However, the nongenic fraction in bacteria is generally small in size and does not contribute significantly to genome size differences.

What are the evolutionary forces affecting gene number and genome size in prokaryotes? Prokaryotic gene content is in constant flux. Many prokaryotes readily pick up foreign DNA and incorporate it into their genomes. These segments of foreign DNA are preferentially retained if they contain useful genes. Genes can also be added to the genome by gene duplication. On the other side of the coin, genes are lost due to deletion, especially if they are no longer essential.

This perpetual gene turnover ensures that, at any given time, bacterial genomes primarily contain functionally important genes. Free-living bacteria, with a need for a full range of biochemical functions, tend to have larger genomes with more genes, whereas obligately parasitic or symbiotic bacteria have some of the smallest known genomes. In one particularly instructive case, some Rickettsia bacteria recently switched from a free-living to a parasitic lifestyle. As expected, much of the Rickettsia genome consists of partially deleted remnants of formerly useful genes (Andersson and Andersson, 2001). It appears that in bacteria, natural selection for informational content, combined with the mechanisms for gene acquisition (through gene duplications and horizontal gene transfer) and gene loss (through spontaneous deletion) are the key determinants of genome size.

The C-Value Paradox. In sharp contrast to prokaryotes, the story of eukaryotic genomes is not primarily about genes. Genome sizes in eukaryotes vary by ~200,000. This variation is both extreme and puzzling, because it does not correspond to our intuitions about organismic complexity. The largest genome size is found not in humans or other multicellular organisms, but in the single-celled *Amoeba dubia* (genome size of 690,000 million base pairs, approximately 230 times larger than the human genome).

Moreover, genome sizes of very similar (and therefore similarly complex) organisms often vary by ~100-fold. Why would fruit flies have 100 times less DNA than grasshoppers, or rice have 60-fold less DNA than lillies? The answer is not that these organisms differ greatly in the number of genes. Indeed, one of the surprises of the complete sequencing of several eukaryotic genomes has been the similarity in the number of genes between very distantly related multicellular organisms (13,600 in the fruit fly *Drosophila melanogaster*, 18,400 in the worm *Caernohabditis elegans*, 26,000 in the flowering plant *Arabidopsis thaliana*, and ~30,000 in humans). Even within this small and biased group (organisms were partly chosen for sequencing based on the compactness of their genomes), the ~2-fold variation in gene number coexists with a ~40-fold difference in genome size.

The observations that eukaryotic genome sizes are decoupled from organismic complexity and that some organisms have vastly more DNA than would be required for their informational needs are known collectively as the C-value paradox (Thomas, 1971), with C-value standing for a haploid genome size. Even though the C-value paradox is almost fifty years old, we are still far from solving it.

Genomic Content in Eukaryotes. On a broad phenomenological level, variation in genome size is attributable to noncoding, often repetitive, DNA. However, noncoding DNA is a very heterogeneous group including such distinct sequences as simple satellite repeats, multiple classes of transposable elements, introns, and insertions of viral and organelle DNA. We still do not know whether any one type of noncoding DNA is primarily responsible for genome size variation.

There are many instances in which one or another kind of noncoding DNA has been amplified (or deleted) to a great degree. For instance, 17 percent of the genome in the fly *Drosophila hydei* is composed of satellite repeats, whereas no satellite DNA has been detected in a closely related species *D. ezoana* (John and Miklos, 1988). A recent study demonstrated that ~50 percent of the maize genome is composed of recently amplified retrotransposable elements (SanMiguel et al., 1998). Other noncoding sequences in these organisms do not appear to change in size to a similar degree. Such studies demonstrate the dynamism and unpredictability of genome size evolution even between closely related organisms.

At the same time, there are some preliminary indications that on a deeper phylogenetic scale the amounts of different kinds of noncoding DNA tend to change together. For instance, across eukaryotes, intron size tends to correlate positively with genome size, albeit at a slower than linear rate (Vinogradov, 1999). Importantly, changes in intron length account for a small part in genome size changes, implying that other sequences change their size along with the introns. Changes in genome size have also been correlated with changes in the number of transposable elements, pseudogenes, amounts of simple repetitive DNA, and interenhancer distances. Such coordinate changes suggest the action of global evolutionary forces capable of affecting all or most sequences in the genome.

Phenotypic Correlates of Genome Size. One of the central observations about genome size is the large number of correlations between genome size and many cellular and organismic traits, such as cell and nucleus size, rates of basal metabolism, seed size, response of plants to CO_2, cell division rates in many taxa, and many others (for a fuller list see Cavalier-Smith, 1985, and Gregory and Hebert, 1999). Correlation between genome size and cell size is particularly phylogenetically robust. It is also very strong, positive, and essentially linear. These correlations suggest that genome size, independently of its genic content (the so-called organismal nucleotype), can be of adaptive significance.

Theories of Genome Size Evolution. Evolution of genome size must involve the interplay between length mutations, genetic drift, and natural selection. Focusing on noncoding DNA, we can easily identify many mechanisms of DNA addition, such as polyploidization, duplicative transposition of mobile elements, insertions of viral and organelle DNA, expansion of simple repeats through replication slippage, and many others. DNA can also be removed through various deletional mechanisms. Essentially, the ultimate fate of length mutations depends on their success in surviving the vagaries of random genetic drift and passing through the sieve of natural selection.

In principle, a change in the rate and pattern of any mutational process or a change in the strength, direction, or efficacy of natural selection could affect genome size. Not surprisingly, there are many competing theories that differ as to which forces are considered primary determinants in the evolution of genome size.

Adaptive theories give the ultimate importance to positive natural selection. Most adaptive theories emphasize phenotypic effects of the DNA, independent of its informational content. If the impact of genome size on phenotype is substantial, natural selection for the optimal phenotype would also determine the genome size best fit for any particular organism. In this view, the exact nature of mutational forces generating genome length variation is irrelevant, as long as there is enough variation for natural selection to act upon.

In sharp contrast, "junk" DNA theory proposes that much of the noncoding DNA consists of passively carried sequences of no functional importance to the organism (Ohno, 1972). The original proposal of the junk DNA concept was not specific about the exact scenarios of genome size evolution. Subsequent interpretations, however, suggested a model where insertions (especially expansions of "selfish" transposable elements [Doolittle and Sapienza, 1980; Orgel and Crick, 1980]) are prevalent in most genomes, generating large amounts of useless DNA. Junk DNA continues to expand until it becomes too burdensome, and any additional amounts of junk are eliminated by purifying natural selection. Although this scenario emphasizes the intrinsic expansionist tendencies of junk DNA sequences, in the end it is natural selection that sets the equilibrium genome size.

The "mutational" theory is similar to the junk DNA theory in assuming that much of the noncoding DNA in large genomes is useless and marginally maladaptive. However, mutational theory does not assume that junk DNA is destined to expand until checked by purifying natural selection. It assumes, instead, that the genome size in many organisms is evolving toward a value wherein the rates of mutational DNA loss and DNA gain are equal. In this theory, genomic rates of mutational addition and loss become larger as genome size grows, but at different rates. While the genome size is small, rates of addition predominate. As a result, the genome size grows until, at equilibrium, rates of DNA addition become equal to the rate of DNA removal. At larger genome sizes, DNA removal predominates and brings genome size back to the equilibrium value. Changes in genome size then correspond to changes in the mutation rates of DNA addition or subtraction.

Evidence. Most evidence for the role of natural selection comes from strong correlations between genome size and traits of clear selective importance. Unfortunately, phenotypic correlations with genome size are consistent with all theories of genome size evolution. It is true that they are most naturally explained and, in fact, predicted by the adaptive theories. They can also be explained, however, by the junk DNA and mutational theories. Because DNA content could have a direct impact on many phenotypic characters (for instance, genome duplication leads to an immediate increase in the cell size), a change in the amount of junk DNA through any means would necessarily affect many phenotypic characters.

The evidence for the junk DNA theories comes from the observation of poor sequence conservation of most noncoding DNA and from often profoundly fast expansions of some types of noncoding DNA. For instance, a recent study has demonstrated that the maize genome has grown twofold in the last 3 million years, primarily or exclusively through the expansion of retrotransposable elements (SanMiguel et al., 1998). Such experiments do demonstrate the strong expansive potential of at least some noncoding DNA. However, it remains unclear whether the rate of junk DNA expansion is indeed fast enough to overwhelm junk DNA removal either through natural selection or through spontaneous deletion and genetic drift. It is also unclear whether poor sequence conservation of noncoding DNA truly signifies its dispensability, or that instead its function does not depend on the precise sequence and therefore cannot easily be detected.

The best evidence for the mutational theory comes from the negative correlation between the rate of spontaneous DNA loss and genome size observed in some animals. For instance, *Drosophila*, with a relatively small and compact genome, loses DNA 10-fold faster than Laupala crickets (an 11-fold larger genome), 25-fold faster than humans (a 17-fold larger genome), and ~50-fold faster than Podisma grasshoppers (a 100-fold larger genome) (Petrov et al., 2000; Bensasson et al., 2001). The negative correlation between the rate of DNA loss and genome size is not predicted by either adaptive or junk DNA theories. According to the adaptive theory, DNA content is modulated in populations through directional or stabilizing natural selection acting on existing genome size variation. Conceivably, natural selection for or against genome size growth could act on the modifiers of mutational DNA loss. However, such selection is unlikely to be effective, since mutational DNA loss is very slow even in *Drosophila* (loss of ~1 base pair per generation) and affects genome size only over very long periods of time. In the junk DNA theory, rates of mutational DNA addition are always higher than the rates of DNA loss, leading to the increase of genome size until stemmed by selection. The equilibrium genome size is a function of the tolerance of organisms to the burden of useless DNA, not the rates of DNA addition or subtraction.

Although intriguing, these results are at best preliminary. The measurements of the spontaneous rate of DNA loss are few, limited mostly to insects, and thus cannot be extended broadly. Moreover, even if confirmed with future research, the mutational theory does not imply that genome size has no adaptive significance at present or that natural selection played no role in setting mutational trends toward DNA loss or gain.

Conclusion. Genome size evolution in eukaryotes is dominated by large-scale fluctuations in the amount of noncoding DNA. Given the often vast size, the varied molecular nature of noncoding DNA, and the multitude of both selective and mutational forces potentially affecting its amount, it is no surprise that genome size evolution remains enigmatic. Currently available evidence fails to distinguish among different theories of genome size evolution, although there is some hope that the modern tools of genomic analysis might change this by allowing more direct measurements of individual forces acting on genome size. In the end, it is likely that no one theory will be sufficient to explain the multiple features of genome size evolution.

[*See also* Gene Families; Genetic Polymorphism; Linkage; Repetitive DNA; Transposable Elements.]

BIBLIOGRAPHY

Andersson, J. O., and S. G. Andersson. "Pseudogenes, Junk DNA, and the Dynamics of Rickettsia Genomes." *Molecular Biology and Evolution* 18 (2001): 829–839. Instructive study of the progressive gene loss in a bacteria after the transition to an obligately parasityic lifestyle.
Bensasson, D., et al., "Genomic Gigantism: DNA Loss Is Slow in Mountain Grasshoppers." *Molecular Biology and Evolution* 18 (2001): 246–253. Further established the importance of DNA loss through small deletions, by extending it to the very large genomes.
Cavalier-Smith, T., ed. *The Evolution of Genome Size.* New York, 1985. To this day, the only published book devoted exclusively to genome size evolution. A key reference.
Doolittle, W. F., and C. Sapienza. "Selfish Genes, the Phenotype Paradigm and Genome Evolution." *Nature* 284 (1980): 601–603. Along with Orgel and Crick 1980, introduced the concept of selfish DNA.
Gregory, T. R., and P. D. Hebert "The Modulation of DNA Content: Proximate Causes and Ultimate Consequences." *Genome Re-*

search (1999): 317–324. An important recent review of genome size evolution with a focus on phenotypic effects of genome size.
John, B., and G. L. G. Miklos. *The Eukaryotic Genome in Development and Evolution.* London, 1988. A good summary of genome structure and function evolution.
Ohno, S. "So Much 'Junk' In Our Genomes." In *Evolution of Genetic Systems*, edited by H. H. Smith pp. 366–370. 1972. Introduced the concept of "junk" DNA.
Orgel, L. E., and F. H. C. Crick. "Selfish DNA: The Ultimate Parasite." *Nature* 284 (1980): 604–607. Along with Doolittle and Sapienza (1980), introduced the concept of selfish DNA. Suggested that genome size is established through the balance of DNA addition through the duplications of selfish DNA elements and purifying selection against such growth.
Petrov, D. A., et al. "Evidence for DNA Loss as a Determinant of Genome Size." *Science* 287 (2000): 1060–1062. This study demonstrated that rate of DNA loss through the imbalance of small deletions and insertions could be one of the causative factors affecting genome size.
SanMiguel, P., et al. "The Paleontology of Intergene Retrotransposons of Maize." *Natural Genetics* 20 (1998): 43–45. A seminal study of genome size evolution in plants. Demonstrated that eukaryotic genome size can sometimes grow quickly through rampant transposition of transposable elements.
Thomas, C. A. "The Genetic Organization of Chromosomes." *Annual Review of Genetics* 5 (1971): 237–256. Early review of genome evolution that coined the term *C-value paradox.*
Vinogradov, A. E. "Intron-Genome Size Relationship on a Large Evolutionary Scale." *Journal of Molecular Evolution* 49 (1999): 376–384. A key study demonstrating that intron size positively covaries with genome size across diverse eukaryotes.

— DMITRI A. PETROV

GENOMIC CONFLICT. *See* Selfish Gene.

GENOMIC IMPRINTING

Genomic imprinting is one of the most surprising and mysterious findings in evolutionary biology. Genes sometimes carry an imprint of their parental origin and have different expression patterns when they come from a mother or a father. For example, the copy of the mouse insulinlike growth factor *Igf2* inherited from the father is turned on, but the copy inherited from the mother is totally suppressed. These parental imprints are erased and reset each generation. Thus, male offspring will pass on active copies of *Igf2* to its children whereas female offspring will pass on inactive copies. The memory of whether these genes came from the grandmother or the grandfather is erased, and a new imprint reflecting the offspring's gender is reapplied during sperm or egg production. The mechanisms controlling imprinting are now well understood, but the evolution of this strange control of gene expression remains a matter of contention.

Imprinting was first discovered by nuclear transfer studies in mice carried out in the 1980s. Micromanipulation was used to move haploid female or male pro-

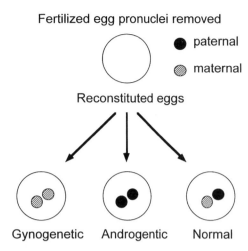

Fertilized egg pronuclei removed

● paternal

◍ maternal

Reconstituted eggs

Gynogenetic   Androgentic   Normal

Figure 1. Transplantation Experiments.
The maternal and paternal pronuclei were removed from fertilized mouse eggs, then replaced to create gynogenetic (2M:0P), androgenetic (0M:2P) and normal (1M:1P) diploid eggs. Only the normal eggs developed normally to term.
Courtesy of Andrew Pomiankowski.

nuclei between recently fertilized eggs (Figure 1). A normal fertilized egg contains one female and one male pronucleus (1M:1P). If the male pronucleus is removed and replaced with a female pronucleus, the resulting gynogenetic egg contains two female pronuclei (2M:0P). The reverse manipulation results in an androgenetic egg containing two male pronuclei (0M:2P). Gynogenetic and androgenetic eggs initiate development, implant normally, but then fail to thrive, and all eventually die: this despite the fact that both kinds of egg contain a full set of genes. In contrast, eggs reconstituted with the normal pattern of one female and one male pronucleus develop normally to term.

Imprinting is also evident at the level of chromosome regions. A dramatic example in humans is deletion of the q11–13 region of chromosome 15. Maternal inheritance (i.e., inheritance from the mother) of the deletion causes Angelman syndrome (hyperactivity, inappropriate laughter, repetitive movements), whereas paternal inheritance causes Prader-Willi syndrome (hypotonia, short stature, small gonads). These different clinical syndromes are not caused by the deletion of different genes. The same genes are missing in both cases, but these deletions have very different effects when inherited from the mother or the father.

These parent-of-origin effects were pinned down to particular genetic loci during the 1990s. At the last count, over forty genes have been shown to be imprinted in mice and humans. One of the best understood cases is the mouse *Igf2* gene. Denise Barlow and colleagues demonstrated imprinting by investigating a null allele of *Igf2* (i.e. a deletion of this gene). The null allele has no detectable phenotype when maternally inherited, but gives rise to mice that are considerably smaller than their littermates when paternally inherited. As the maternal copy of *Igf2* is silent, inheritance of the null allele from the mother has no effect. In contrast, the paternal copy of *Igf2* is usually active. Thus, paternal inheritance of the null allele causes a reduction in gene expression and in this case leads to a loss in body size.

Mechanism of Imprinting. To understand imprinting at a mechanistic level, we need to explain a number of key features. Imprinting involves generation of distinct maternal and paternal marks on the DNA that alter gene expression, maintenance of parental marks through somatic cell division, and erasure of marks during gametogenesis, along with the reapplication of novel marks to reflect current gender. Thus, imprints need to be permanent within a generation but reversible between generations.

Silencing of imprinted genes is achieved by changes in DNA structure, not by changes in DNA sequence. Most imprinted genes show differences between the maternal and paternal copies in histone deacetylation and DNA methylation. Histones are DNA-binding proteins that form nucleosomes, the basic structure of chromatin. When histones are deacetylated, the nucleosomes condense into a tightly bound structure. In addition, upstream of most imprinted genes are DNA sequences that are relatively rich in the dinucleotide CpG (cytosine followed by guanine). The cytosine base in these CpG dinucleotides can be altered by the addition of a methyl group (CH_3). Like histone deacetylation, methylation is associated with chromosomal condensation. These changes to the DNA structure are capable of causing differential gene expression in a number of ways. The main effect of condensation is to block access to DNA sequences that act as promoters (or repressors), thereby turning off (or on) gene transcription.

Once established, patterns of deacetylation and methylation are maintained through cell division. Most is known about the maintenance of methylation. When a DNA molecule is replicated, the two daughter chromosomes are methylated on one strand but not the other (Figure 2). A set of *Dnmt* methylation enzymes recognizes this hemimethylated state and reestablishes methylation on both strands of the DNA. Nonmethylated DNA is not recognized by the methylation enzymes and remains nonmethylated after replication. The net effect is that methylated and nonmethylated alleles can coexist in the same cell, and their states are stably maintained through cell divisions. This allows maternal and paternal alleles to retain their imprinting status throughout growth and development.

Imprinting is reset in the germ line of both sexes. Genome-wide demethylation occurs soon after embryonic establishment of the germ line. This removes the

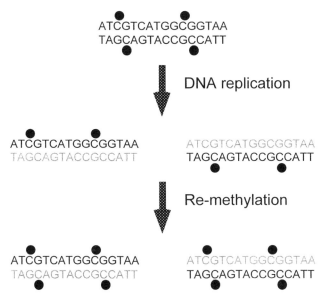

FIGURE 2. Replication of DNA Methylation.
Methylation of cytosine residues (black spots) in CpG dinucleotides in the DNA sequence. This occurs on the plus strand (top, running left to right) and the complementary minus strand (bottom, running right to left). After DNA replication, the new strand is not methylated. A set of enzymes recognize hemimethylated sites and remethylate the cytosine base of the new strand. This allows the pattern of methylated sites and remethylate the cytosine base of the new strand. This allows the pattern of methylation to be perpetuated through cell division. A = adenine. C = cytosine. G = guanine. T = thymine.   Courtesy of Andrew Pomiankowski.

preexisting maternal or paternal marks. As the germ cells start to differentiate into sperm and eggs, a process of de novo methylation begins. Unfortunately, little is known about the enzymes involved, how sites are identified for deacetylation and methylation, or how sex-specific imprints are established.

Conflict Hypothesis. The evolution of genomic imprinting at first appears paradoxical. There seems no obvious benefit to turning off gene expression from one gene, especially as this predisposes the individual to deleterious somatic mutations that are usually masked by the presence of the second copy. The lack of any straightforward benefit from imprinting has led to many ill thought out and poorly supported hypotheses. For instance, no mammal is known to develop asexually through parthenogenesis (i.e., from an unfertilized egg). This is ruled out by imprinting, as the paternal genome with its particular pattern of imprints is necessary for normal development. While the absence of parthenogenesis is a consequence of imprinting, it hardly seems a plausible reason for the evolution of around 100 imprinted genes in the mammalian genome.

The rejection of inappropriate hypotheses has be-

come easier since David Haig (2000) proposed a general mechanism for the evolution of imprinting. This is known as the conflict hypothesis and arises from a simple observation about relatedness. Haig's hypothesis has the virtue that it potentially can explain many of the phenomena associated with imprinting. However, as more data have been uncovered, it has become less clear whether conflict is the main selective force or just one of many.

The reason for conflict between maternal and paternal copies of the same gene within an individual arises from differences in relatedness. This is easily shown by considering the offspring of a single female (Figure 3). As a result of Mendelian inheritance, a maternally derived gene has a 50 percent chance of being present in other progeny of the same mother. This probability is constant and independent of the mating system (e.g., monogamous, polygamous). In contrast, the relatedness of paternally inherited genes depends on the degree of multiple mating. If the mother is monogamous (i.e., she mates only with a single male throughout her reproductive life), all her offspring will have a 50 percent chance of sharing identical paternal genes. But in most mammal species females mate with multiple males. Among siblings, paternally inherited genes are often from different

FIGURE 3. Relatedness Between Maternal and Paternal Genes in Offspring When a Female Mates with Two Males.
Offspring have a 50 percent probability of sharing maternal genes, but the probability is much lower for paternal genes, as these may originate from different fathers. This creates a relatedness asymmetry between maternal and paternal genes, which increases with the degree of multiple paternity. Courtesy of Andrew Pomiankowski.

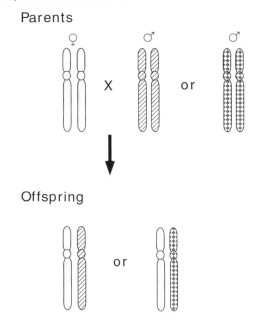

fathers, both within and between litters. Thus, average relatedness of paternally inherited genes is typically considerably less than 50 percent.

This difference in relatedness has a profound effect on mother–offspring interactions. Given that a mother is equally related to all her current and future offspring, she benefits from distributing her resources relatively evenly within a litter and retaining resources for the future. In offspring, maternally inherited genes are under similar selection, as they too are present in all current and future offspring. However, paternally inherited genes have much less interest in other offspring, which are far less likely to carry related genes. We thus expect paternally inherited genes to be far more demanding of resources from the mother, even if their own gain is to the detriment of others within the same brood. The paternally inherited genes also care less if the current offspring demand decreases the mother's survival chances as future offspring are unlikely to be fathered by the same male.

It is easiest to see the evolutionary consequence of selection on maternal and paternal genes in the extreme case in which all offspring have different fathers. Consider a gene coding for an embryonic growth factor. Mutants that increase paternal gene expression are favored by selection, as an embryo producing more growth factor receives more resources from the mother, and thus enjoys better survival chances. This reduces the number and survival of other offspring produced by the same mother. However, this does not affect selection on the mutant, as the paternal gene is not present in other progeny. Once increases in paternal gene expression are established, there is counterselection in favor of less demand from the maternal gene. Mutants that marginally reduce maternal gene expression are selected despite any loss in fitness because they benefit other offspring that have a 50 percent probability of carrying the mutant. These counteracting forces lead to a coevolutionary "arms race," in which increases in paternal copy demand are matched by decreases in maternal copy demand. The process continues until the maternal copy is silenced and growth factor is produced only by the paternally inherited gene.

Evidence for the Conflict Hypothesis. Some gross distortions of parental contributions to embryos fit the predictions of the conflict hypothesis. As mentioned above, gynogenetic (2M:0P) and androgenetic (0M:2P) embryos fail to develop, but for different reasons. Gynogenetic embryos abort because the placenta fails to develop properly, whereas androgenetic embryos abort because of disproportionately large growth of the placenta. These observations suggest that maternal genes favor low-growth demand and paternal genes favor high-growth demand.

The most stunning evidence in favor of the hypothesis comes from the mouse insulinlike growth factor *Igf2*. This embryonic growth factor is silent when maternally inherited and active when paternally inherited. However, its receptor gene, *Igf2r*, shows the reverse pattern, being active when maternally inherited and silent when paternally inherited. The *Igf2r* receptor binds to the *Igf2* protein, inactivates its signaling function, and leads to its degradation. Thus, the *Igf2r* gene acts as a growth repressor, and, as predicted by the conflict hypothesis, it is a maternally active and paternally inactive gene.

Data from a number of imprinted genes have accumulated during the last decade and now allow a more general test of the conflict hypothesis. The best data come from individuals with uniparental disomies (where both chromosomes are inherited from one parent) and mouse gene knockouts (genetically engineered microdeletions). Most maternal disomies are growth retarding, as predicted by the conflict hypothesis. However, nearly all paternal disomies are also growth retarding, only one being unambiguously growth enhancing. Thus, overall, the evidence from disomies does not support the conflict hypothesis. But it is difficult to put too much weight on this finding, as the effect of disomy per se is probably too great, and so obscures most differences between maternal and paternal origins of the disomy.

Knockouts of a number of imprinted genes show the predicted pattern of paternal knockout embryo growth suppression and maternal knockout embryo growth enhancement (e.g., *H19*, *Grf1*, and *Gnas*). However, not all data fit so nicely. Several genes have no or ambiguous effects (e.g., *Ins*, *Ins2*, and *Snrpn*), and some have the reverse effects on embryo growth (e.g., *Mash2*). In addition, some imprinted genes have effects that are hard to interpret as part of maternal-embryotic growth regulation. A number of imprinted genes are important in brain development. For example, *Peg1* and *Peg3* are both imprinted genes expressed in the brain that are paternally active. When the paternal copy is deleted, female behavior toward offspring is abnormal (male behavior is normal).

At first glance, these findings suggest that maternal-paternal conflict is not the only force governing the evolution of genomic imprinting. But knockouts are only a crude test of the theory, because they reduce gene expression to zero. It has been suggested that more minor changes in expression might still conform to the conflict hypothesis predictions. Although these are clever post hoc explanations, what is really needed is experiments, and these remain to be attempted. Another area where more data are needed is from species with lower or higher rates of multiple mating. At the moment, investigation has centered on mice and humans. Some imprinting differences have been identified (e.g., *Igf2r* is not imprinted in humans), but it is difficult to make sense of these. What is needed are comparisons between closely related species that differ in mating system.

X-linked Imprinting. The phenomenon of X-linked imprinting suggests an intriguing alternative to the conflict hypothesis. Two examples of imprinting are known from comparisons of individuals with a single X chromosome that is either maternally ($X^m O$) or paternally ($X^p O$) derived. These XO individuals develop as females. In mice, the imprinted X affects embryonic growth rate. $X^p O$ embryos are smaller than normal $X^m X^p$ females, which in turn are smaller than $X^m O$ embryos. This pattern of growth enhancement by the maternal X is contrary to the conflict hypothesis. In humans, XO females suffer from Turner's syndrome. In this case, $X^p O$ children have better social cognitive skills than $X^m O$ children. It is not obvious how this behavioral difference can be accounted for by the conflict hypothesis.

An alternative explanation for X-linked imprinting derives from the pattern of inheritance of the X chromosome (Figure 4). The maternal X is transmitted equally to female and male offspring, whereas the paternal X is only passed to female offspring. This inheritance asymmetry allows X-linked imprinted genes to have sex specific effects. In females, expression of X-linked genes is the average of X^m and X^p contributions (in female mammals, X-linked genes undergo dosage compensation). In contrast, male gene expression is just from the maternal copy. Thus, if the paternal copy is silenced by imprinting, it reduces female gene expression but has no effect on males, and if the maternal copy is silenced by imprinting, it predominantly reduces male gene expression. This allows us to make sense of the two examples of X-linked imprinting. In mice, the X-linked imprint is paternally silenced and leads to faster growth in male offspring (which carry the active maternal X). In humans, the X-linked imprint is maternally silenced and leads to better social cognitive skills in female offspring (who carry the active paternal X and the inactive maternal X). It is easy to imagine that these sex differences are the result of different selection pressures on the two sexes.

These findings are interesting, as they provide a concrete alternative explanation to the conflict hypothesis. Imprinting on the X appears to be the result of selection for sex-specific gene expression rather than conflict. The one caveat here is that the X-linked imprinted genes have not yet been mapped, so there may be some surprises when this happens. The sex-specific explanation cannot apply to imprinting on autosomes, as these are inherited equally by both sexes (except in some insects with haplodiploid patterns of inheritance). But the lack of evidence that conflict has shaped imprinted X-linked genes does suggest that conflict is not the only selective pressure and alternative forces may explain some autosomal imprints.

Conclusion. In this article, we have concentrated on

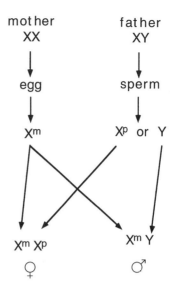

FIGURE 4. Sex Specific Inheritance of X-Linked Imprints. The maternal X^m is inherited by both sexes, whereas the paternal X^p is passed only to daughters. *Journal of Theoretical Biology.* Academic Press, 1999.

explaining genomic imprinting in mammals. This is because mammalian imprinting is well studied, and related vertebrates (birds, fish, and reptiles) appear to lack imprinting at the level of individual genes. However, imprinting is found in other groups. In particular, it has evolved and controls individual gene expression in the flowering plants, where it affects parental investment in seeds. The logic of the conflict hypothesis may apply here as well. There are also several examples of genome imprinting in insects, amphibia, and fish in which the whole maternal or paternal genomes are turned off or ejected from cells. The evolution and mechanistic basis of these phenomena are probably unrelated to individual gene imprinting in mammals and plants, but little is fully understood.

The next few years promise to be exciting for research in genomic imprinting. Many more imprinted genes will be characterized, and a variety of nonstandard species (i.e., not humans or mice) are starting to be investigated. This information will be important in gaining a better understanding the selective forces maintaining imprinting. It will then be possible to undertake better tests of the conflict hypothesis and other evolutionary explanations of imprinting.

[*See also* Dosage Compensation.]

BIBLIOGRAPHY

Haig, D. "The Kinship Theory of Genomic Imprinting." *Annual Review of Ecology and Systematics* 31 (2000): 9–32. A recent restatement of the conflict hypothesis.

Hurst, L. D., and G. T. McVean. "Do We Understand the Evolution of Genomic Imprinting." *Current Opinions in Genetics and*

Development 8 (1998): 701–708. A critical review of evidence from knockouts and uniparental disomies and whether they support the predictions of the conflict hypothesis.

Iwasa, Y., and A. Pomiankowski. "Sex-Specific X Chromosome Expression Caused by Genomic Imprinting." *Journal of Theoretical Biology* 197 (1999): 487–495. Summarizes the data on X-linked imprinting and proposes that it has evolved to permit sex-specific gene expression.

Moore, T., and D. Haig. "Genomic Imprinting in Mammalian Development: A Parental Tug of War." *Trends in Genetics* 7 (1991): 45–49. One of the earliest statements of the conflict hypothesis. It remains one of the clearest, though few imprinted genes were well characterized at the time.

Reik, W., and J. Walter. "Genomic Imprinting: Parental Influence on the Genome." *Nature Reviews Genetics* 2 (2000): 21–32. An excellent review concentrating on the mechanistic basis of imprinting.

Surani, M. A. H. "Evidences and Consequences of Differences between Maternal and Paternal Genomes during Embryogenesis in the Mouse." In *Experimental Approaches to Mammalian Embryonic Development*, edited by J. Rossant and R. Pedersen, pp. 401–435. Cambridge, 1987. A review of the experimental manipulations of maternal and paternal genomes, which provided the first clear evidence for genomic imprinting in mammals.

— ANDREW POMIANKOWSKI

GEOCHRONOLOGY

Geochronology is the scientific study of the ages of ancient geological and archaeological deposits. Geochronologists are geologists, physicists, or chemists who use a variety of scientific methods to analyze geological, biological, and artifactual materials to determine the time when they were formed or deposited. Geochronological measurements are based on three principles: first, that some physical process has been steadily affecting the material to be dated; second, that we can measure the extent to which the process has occurred, usually by measuring the amount of a product that has been generated; and third, that the process began just when the deposit was formed. A simple example is determining the age of a tree: most trees produce a distinct ring of dark wood each year. To determine the time since the tree began to grow, we count the number of tree rings (the product).

Geochronology is based mainly on processes of radioactive decay. Some naturally occurring atoms spontaneously decompose into other kinds of atoms, at a well-known rate determined by the laws of physics. If a geological sample was deposited at some time in the past with radioactive elements in it, we may be able to determine the time elapsed since deposition by measuring how much of the radioactive element is left, or by measuring how much of the product atoms (called daughters) have accumulated from the decay of their radioactive parent. It is essential that the rate of decay of the parent should be slow enough that there is still some parent decaying when the sample is collected.

Radioactive decay of atoms also results in the emission of radiation called *alpha, beta and gamma rays*. As these rays bombard (*irradiate*) the materials in geological deposits, they cause gradual changes in their atomic structure that can be detected in various ways. The amount of structural change is proportional to the length of time of irradiation and can be used to the date the material. As long as any prior changes in the atomic structure were erased before the deposit was formed, then only changes occurring since the formation of the deposit will be detected.

Methods of Chronological Dating. The following are the most widely used methods of geochronological dating.

Radiocarbon. Radioactive atoms of carbon, called carbon-14 or C-14, are constantly formed in the atmosphere by cosmic rays that bombard the earth from outer space. Every sample of living matter entraps a constant proportion of these atoms. That is, C-14 comprises a known proportion of the carbon atoms in an organism. As soon as the organism dies, the C-14 atoms begin to decay. Half of the C-14 atoms decay in 5,730 years (this is called the half-life); after another 5,730 years, only a quarter of the initial number are left, and so forth. By measuring the number of C-14 atoms remaining, we determine the time elapsed since the death of the organism. Samples up to 50,000 years old can be dated; older samples contain too little C-14 to measure. Because the amount of atmospheric C-14 in the past was not exactly constant, it is necessary to calibrate C-14 ages, by counting tree rings in C-14-dated trees dating up to about 12,000 years. Other correction methods are used to calibrate older dates. C-14 can be used to date charcoal, fossil wood, bones, and shells, with a precision of about 1 percent of the age.

Uranium series. Uranium is a radioactive element with a half-life of billions of years. As it decays, its atoms are successively transformed into atoms with much shorter half-lives, including thorium-230 and protactinium-231. The decay of these atoms can be used to date geological deposits that are sometimes associated with fossil humans and other organisms. Ancient humans lived in caves formed out of limestone. Stalagmites growing on the floors of caves may cover fossils and can be dated by measuring the increase in the ratio of thorium-230 (Th-230) to its parent uranium. Travertine (or tufa) formed by waters emerging from springs also can be dated. These ancient spring deposits may contain tools, faunal bones, and skeletal remains of hominids. Precision of U-series dates is about 1 percent of the age; the upper limit of dating is about 500,000 years, after which the amount of Th-230 reaches its maximum value. Bones, teeth, and shells also can be analyzed but tend to give less reliable dates.

Argon-potassium. Potassium (K) is a chemical element found in all rocks. A small fraction of all potassium is radioactive (K-40) and decays to atoms of argon (Ar-40). The ratio of Ar-40 to K-40 in a mineral is a measure of the time elapsed since the mineral was formed. Another kind of argon, Ar-39, is formed when K is irradiated in a nuclear reactor. This permits us to measure the age of minerals from the ratio Ar-40/Ar-39. Only minerals formed at the high temperatures of a volcanic eruption can be dated. Fortunately, many important fossil sites are capped by layers of volcanic ash, especially in East Africa. Dates for the evolution of *Australopithecus* and early species of *Homo* have been determined by this method. Ages with a precision of about 1 percent can be obtained down to a lower limit of about 100,000 years; there is no significant upper limit.

The next three methods are all based on measurement of changes in atomic structure caused when materials are irradiated by radioactivity from the small amounts of the radioactive elements uranium, thorium, and potassium found in ordinary sediment.

Thermoluminescence (TL). Flint, which was used by prehistoric humans for toolmaking, gives off a little flash of light when it is heated: this represents the release of atomic energy that has built up in the flint as it was naturally irradiated. Flint artifacts that have been heated in an ancient fireplace will start to build up a new store of energy. When an archaeological flint is heated in the laboratory, the brightness of the flash given off is a measure of the amount of radioactive energy stored in the flint while it was buried at a site. The TL age is calculated from the ratio of the amount of stored energy (brightness) to the rate at which it was being irradiated. TL-datable flints are found in many prehistoric sites and give ages up to a limit of about 500,000 years old, with a precision of 5–10 percent of the age.

Optical dating. Fossils are often found buried in layers of sand. When a sample of sand is exposed to a beam of green light, it gives off a flash of red light. Because white light contains all these wavelengths, the luminescent energy of the sand is bleached when it is exposed to sunlight for a few minutes. After bleached sand has been buried for thousands of years, the luminescent energy builds up again because of irradiation by uranium, thorium, and potassium in the sand. We determine the stored energy from the brightness of the red flash produced when it is hit by green light, and the age is determined from the ratio of the stored energy to the rate of irradiation. This method of optically stimulated luminescence (OSL) dating can give the age of sun-bleached sand up to about 200,000 years, with a precision of no better

THE ARRIVAL OF HUMANS IN AUSTRALIA

The continent of Australia was separated from the Asian landmass by open sea, even during the lowest sea levels of the last glacial period. The first humans to arrive must have been able to sail some sort of craft across the gap of 100 kilometers or more separating Indonesia from Australia. There is no archaeological evidence as to when early modern humans learned to sail, but we can at least ask, when did the earliest Australians make this trip? Thermoluminescence dates obtained on sediment found in artifact-bearing rock shelters in northeastern Australia indicate that humans arrived in northern Australia between 50 and 60 thousand years ago (Roberts et al., 1994). Although skeletal remains do not survive in the hot, humid climate of this region, it is assumed that the occupants of these sites were modern humans; no other hominid types have ever been found in Australia. A more direct demonstration of this timing was the chronometric study of the site at Lake Mungo, where a complete human skeleton was found (Thorne et al., 1999). Combined U-series and ESR dating of the skeleton gave a date of 62 thousand years; this was consistent with luminescence (OSL) dates of sediments thought to correlate with those at the discovery site. Nevertheless, some dispute has persisted about these dates. Recently, Roberts and colleagues (2001) showed that the extinction of a large number of larger vertebrate species, known to have been hunted by humans, occurred about 46,000 years ago and proposed that this event may have been tied to human predation.

— HENRY SCHWARCZ

than 5 percent. An example of the dating of humans in Australia is given in the accompanying Vignette.

Electron spin resonance (ESR). Radioactive bombardment knocks out electrons from their orbits around atomic nuclei and allows them to be lodged in energy "traps" in crystals. The number of these trapped electrons is counted in an ESR spectrometer and is a measure of the amount of radioactive bombardment of a sample since it was buried, assuming that it had no trapped electrons at that time. We can use this to date enamel from the teeth of large mammals in fossil sites, left as a result of hunting or scavenging by humans. The age is determined from the ratio of the number of trapped electrons to the rate of radioactive bombardment. Ages with a precision of 10 percent can be determined up to two million years or more.

Two other methods that require calibration by one of the methods already listed are the following.

Amino acid racemization. Amino acids found in proteins of seashells and eggshells are gradually transformed from their natural state at a rate that depends on the temperature where they are buried. Because the earth's temperature has varied through time, the method must be calibrated for each region.

Natural remnant magnetization. The earth's magnetic poles have periodically switched (north for south). Traces of these reversals are found in layers of sediment; the last reversal was 780,000 years ago. Times of all older reversals have been determined. Where a continuous series of reversals is present, they can be used to limit the age of fossil-bearing layers.

[*See also* Archaeological Inference; Geology; *articles on* Hominid Evolution; Molecular Clock.]

BIBLIOGRAPHY

Aitken, M. J. *Thermoluminescence Dating.* London, 1985. Still the best general reference on this topic.

Aitken, M. J. *Science-Based Dating in Archaeology.* London and New York, 1990. One of the best overviews of the subject.

Grun, R. "Electron Spin Resonance Dating." In *Chronometric Dating in Archaeology,* edited by R. E. Taylor and M. J. Aitken, pp. 217–260. New York, 1997. A short summary of the principal applications of ESR dating.

McDougall, I., and M. T. Harrison. *Geochronology and Thermochronology by the 40Ar/39Ar Method.* 2d ed. New York, 1999. The senior author is a world authority on application of argon dating to human evolution.

Roberts, R. G., et al. "New Ages for the Last Australian Megafauna: Continent-wide Extinction about 46,000 Years Ago." *Science* 292 (2001): 1888–1892.

Roberts, R. G., R. Jones, and A. Smith. "Beyond the Radiocarbon Barrier in Australian Prehistory." *Antiquity* 68 (1994): 611–616.

Schwarcz, H. P. "Uranium Series Dating." In *Chronometric Dating in Archaeology,* edited by R. E. Taylor and M. J. Aitken, pp. 159–182. New York, 1997. A brief review of this method.

Schwarcz, H. P., and W. J. Rink. "Dating Methods for Sediments of Caves and Rock Shelters." *Geoarchaeology* 16 (2001): 355–372. A general review of dating methods at fossil human sites.

Taylor, R. E. *Radiocarbon Dating: An Archaeological Perspective.* Orlando, Fla., 1987. A very accessible treatment of this subject with an archaeological orientation.

Thorne A., J. J. Simpson, M. McCulloch, L. Taylor, D. Curnoe, R. Grün, G. Mortimer, and N. A. Spooner. "Australia's Oldest Human Remains: Age of the Lake Mungo 3 Skeleton." *Journal of Human Evolution* 36 (1999): 591–612.

— HENRY SCHWARCZ

GEOGRAPHIC VARIATION

Populations of a species that are separated in space or time may vary in some characteristics; variation—for example, in sexual or ontogenetic characters—may be present even within a single population. Geographic variation is ubiquitous; only species with minute ranges are likely not to exhibit some facet of geographic variation. All facets of the phenotype and genotype can vary

geographically, including external morphology (e.g., size, shape, color), internal morphology (dentition, organ position), physiology (temperature tolerance, disease resistance), biochemistry (isozymes, blood proteins, venom), physical performance, reproductive mode (viviparity), nuclear DNA, and mtDNA markers.

Studies of geographic variation have aided the understanding of microevolutionary forces, speciation, species concepts, and the nature of species and biodiversity. Studies of geographic variation in the latter half of the twentieth century moved from a "taxonomic" approach of naming subspecies, through an approach that allowed an objective quantitative description of the pattern, to studies that integrate quantitative descriptions with molecular ecology, molecular phylogenetics, field experiments, and quantitative hypothesis testing. This allows a fuller understanding of the causative factors of geographic variation and its relation to speciation. This has been facilitated by the widespread availability of new techniques for investigation, such as DNA sequence variation, and conceptual advances regarding the role of historical factors in ecology.

Subspecies. During the 1960s, the conventional approach to geographic variation in a species was often "taxonomic" in that it centered on recognizing a taxonomic category (i.e., subspecies) that occupied a particular segment of the species's geographic range. Species were often identified by a trinomial: genus, species, and subspecies. For example, the Palaearctic grass snake, *N. natrix,* has the trinomial *N. n. corsa* applied to the Corsican populations. Species could therefore be monotypic, lacking subspecies, if they did not have obvious geographic variation; or they could be polytypic, with at least two subspecies, if they had obvious geographic variation.

As a generalization, subspecies were recognized on visually obvious single characters, without rigorous analysis or consideration of the noncategorical nature of the variation, or its cause. Often, a large number of trivial subspecies were recognized that were author-dependent and arbitrarily sectioned gradual clines into subspecific categories (Figure 1). Moreover, because the categories were often arbitrary, there was a tendency to use physiographic features such as rivers and mountains to delimit their range. This led to some circularity in the view that the interruption of gene flow by these features was a key factor in population differentiation, and that gene flow was a potent force in the cohesion of even polytypic species. Nowadays, much less emphasis is placed on naming subspecies, in part because of the advances described below, and in part because species concepts such as the phylogenetic species concept facilitate naming some populations, particularly allopatric forms, as full species rather than subspecies.

The Patterns of Geographic Variation. From the

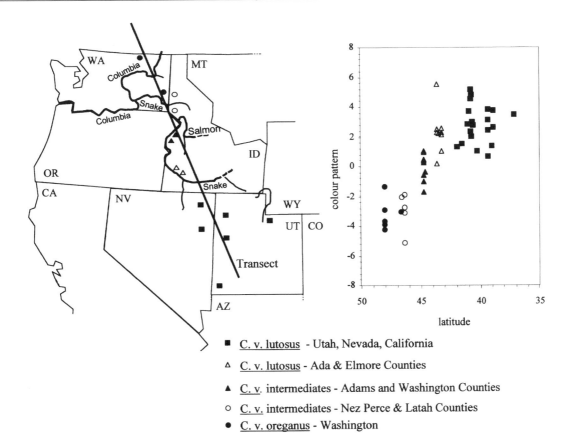

■ C. v. lutosus - Utah, Nevada, California

△ C. v. lutosus - Ada & Elmore Counties

▲ C. v. intermediates - Adams and Washington Counties

○ C. v. intermediates - Nez Perce & Latah Counties

● C. v. oreganus - Washington

FIGURE 1. Clinal Variation in the Color Pattern in the Western Rattlesnake, *Crotalus viridis*, in North America, from Washington State, South through Idaho, to Utah, and Nevada.
The graph illustrates the N–S clinal pattern between two presently recognized subspecies, *C. v. oreganus* and *C. v. lutosus*, when the component score representing color pattern is plotted against the latitudinal transect along the Snake River. The map illustrates the transect line, and the distribution of sampled populations. Data from C. Pook.

1970s onward, increasing interest has been paid to describing the nature of geographic variation. The use of multivariate statistical analysis to combine the information in several characters was a significant and long-lasting advance. Multivariate techniques as applied to geographic variation analysis have two basic conceptual stages: first, the construction of an actual, or implied, measure of (dis)similarity among geographic populations; and second, a summarization of this information to portray the relationships among the populations.

One simple way to summarize this information is to plot these taxonomic distances on a geographic network, joining adjacent populations on a map. Another way is to subject it to a hierarchical cluster analysis; however, this has the disadvantage that it imposes a categorical structure on a pattern that may be smoothly clinal. The most useful techniques are ordination techniques, such as principal component/coordinate analysis and canonical variate analysis. These techniques summarize variation in many characters in a reduced number of axes without assuming that the geographic vari-

ation is either clinal or categorical. They are particularly useful for generalizing across several characters in a character system. These methods are reviewed by Thorpe (1976) and Sokal (1983). If one wishes to assume both nature and cause, one can use hierarchical phylogenetic reconstruction methods. The appropriate quantitative methods reveal a variety of patterns of geographic variation on a larger, or smaller (microgeographic), scale.

Clinal variation is the gradual change over geographic space. Examples of this include the gradual increase in size of house sparrows northward in North America (Gould and Johnson, 1972), and the cline in color pattern of the western rattlesnake down the Snake River, USA (Figure 1).

Categorical variation may also be referred to as a *stepped cline*. Here one has distinct forms occupying distinct geographic areas with a more or less sharp transition or "hybrid" zone between them. Examples of this include the north–south variation in the color pattern of sexually mature male Tenerife lacertids (Thorpe and Richard, 2001), and the north–south Caribbean coastal

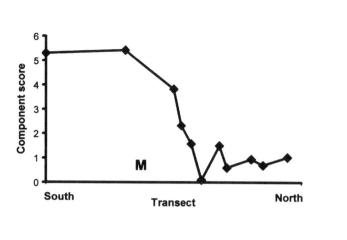

FIGURE 2. A Stepped Cline in the Component Score Representing Shape and Scalation Variation of the Anole (*Anolis oculatus*) along the Transect Down the Caribbean Coast of Dominica.
Unlike the geographic variation in the rest of the island (see Figure 3), here the changes in locality means are in a sharp step, are not associated with clear biotope variation, and are congruent between character systems. Both the color pattern (Figure 3) and mtDNA lineages (position of change marked as M) show patterns broadly congruent to the shape and scalation figured here. Moreover, recently it has been shown that there is reduced gene flow between these northern and southern Caribbean forms (Figure 3). Consequently, unlike the rest of the island, the primary cause of geographic variation here may be temporary vicariance due to volcanic activity.   Data from A. Malhotra and R. S. Thorpe.

forms of the Dominican anole (Figure 2). In some situations these categories may grade into, or be regarded as, full species. Much effort goes into studying the transitions or "hybrid" zones between these categories.

Mosaic, also called *crazy quilt*, variation refers to a mix of geographic patches of forms in what appears to be a haphazard fashion. North American gall aphids show such patterns (Sokal and Riska, 1981).

Causes of Geographical Variation

Genotype and phenotype. It is axiomatic for most studies of geographic variation that the characteristics that are being investigated are under some element of genetic control. With most phenotypic characteristics, such as morphology or performance, the observed character state is influenced by both genetic and environmental forces so that, under normal environmental conditions, a genotype can express itself as a range of phenotypes. This *phenotypic plasticity* is easily conceived if we consider height in humans, which can be influenced by inheritance from parents (genotype) and by food intake (environment). Almost all phenotypic characteristics have a degree of heritability. However, heritability cannot be readily measured in natural populations, and (since it is not 0 or 100 percent) it cannot indicate to what extent patterns of geographic variation arise from phenotypic plasticity.

Ecology, history, and gene flow. Given that the characteristics under study have some component of ge-

netic control, the pattern of variation may be influenced by historical factors, or by factors such as natural selection for current ecological conditions, or by a combination of both. Historical factors should be reflected in the phylogeny of the populations and can be referred to as *phylogenesis*; ecological factors (such as selection) can be referred to as *ecogenesis*. The differentiating effects of both ecogenesis and phylogenesis can be influenced by gene flow.

Methods of Investigation. Rigorously testing the causes of the geographic variation and assessing the relative importance of historical and ecological factors is currently one of the key emphases in studies of geographic variation. Several ways to do this are reviewed below.

Neutral markers. Some types of gene sequence data are thought to be selectively neutral and to change over a geological historical time scale. This type of data, analyzed using quantitative phylogenetic reconstruction methods, can reveal historical processes such as vicariance and dispersal; however, on their own they are not generally pertinent to studying ecogenesis. The widespread availability of neutral DNA markers has led to great expansion of this type of study and to the coining of the term *phylogeography* (Avise, 2000).

Simple repeat sequences of nuclear DNA called microsatellites are also thought to be largely selectively neutral. The allele frequency of these hypervariable loci varies at a time scale that makes them suitable for estimating contemporary gene flow between geographic populations. Under some conditions, microsatellites are also used phylogeographically, as in studies of humans, bears, lizards and bees. Although the recent accessibility of relatively neutral DNA markers has revolutionized aspects of geographic variation analysis, to study selection as a factor in geographic variation, one uses other characteristics such as morphology.

Introductions. The recent movements of humans from Europe to the New World have resulted in the introduction of many organisms that have rapidly expanded over large areas and have differentiated to show geographic variation. An example of this is the introduction of English sparrows into North America and New Zealand, and their subsequent differentiation. In these circumstances, ecogenetic rather than phylogenetic factors are probably responsible for the geographic variation.

Congruence. Several independent characters may show similar (congruent) patterns of geographic variation when vicariance influences a species. An example of this may be provided by the Palaearctic grass snake *N. natrix*, which shows an east–west pattern of geographic variation that is thought to be at least in part attributable to Plio-Pleistocene vicariance and subsequent contact. The grass snake's body dimensions, scalation, color pattern, internal morphology, dentition, and dermal sense organs show close congruence in geographic variation. On the other hand, if natural selection for current conditions is the primary cause of variation, one character or suite of characters may be adapted for variation in one ecological factor while another character or suite of characters is adapted to an ecological factor with a different pattern of geographic variation. In these circumstances, widespread congruence is unlikely, although sets of characters may be congruent because they are adapting to the same ecological factor, or one with a similar pattern of geographic variation. Examples of this come from island lizards. The Tenerife lacertid has altitudinal variation in body size that is in part incongruent with the variation in male color pattern (latitudinal biotopes) and shape (mosaic) (Thorpe and Richard, 2001). Moreover, strong selection can swamp the congruent effects of vicariance; this may also be seen in the island lizards, where selection appears to be strong enough to render at least some character systems incongruent with vicariance-induced DNA patterns and/or phylogenetic relationships (Thorpe et al., 1996).

Parallel patterns and trends. So-called ecogeographic rules have long been known for birds and mammals. For example, Bergmann's rule on body size and Allen's rule on relative protuberance size suggest that populations occupying colder climates (latitudinal or altitudinal) will have larger bodies and shorter protuber-ances to produce a volume/surface area ratio that allows heat retention (Mayr, 1963). The larger body size of house sparrows in northern North America is a well-known example (Gould and Johnston, 1972).

Such parallels are useful for understanding the causal factors in geographic variation and can be extended to other situations. For example, in island archipelagos with a prevailing wind direction (e.g., Canary Islands and Lesser Antilles), high-elevation islands may have parallel environmental zonation. Each island may have a separate congeneric lizard species with parallel geographic variation in concert with this zonation. Although there may be deep molecular phylogenetic divisions within an island species (Thorpe and Malhotra, 1996), one can exclude phylogenesis in favor of adaptation by natural selection when the ecological zonation is parallel, but not the molecular lineages. In the Canary Islands, both the Gran Canaria and Tenerife skinks show north–south variation in tail color (the tail being human-visual blue in the south) in concert with the latitudinal biotopes (Brown et al., 1991). It is thought that the blue tail attracts predators for sacrificial tail autotomy, and that is favored in the selection regime of the dry southern biotope. Similarly, the anoles in a series of high-elevation Lesser Antillean islands tend to be larger at higher altitudes, greener in the rainforest, and with fewer scales in wet areas, irrespective of deep molecular phylogenetic divisions within islands (Thorpe and Malhotra, 1996).

Field experiments. One convincing way to demonstrate that natural selection is responsible for a pattern of geographic variation is by field experiments. An experimental study of the color pattern of Trinidad guppies in response to predation levels showed, by manipulating predator pressure, that the balance between sexual selection and anti-predator crypsis (camouflage) was such that in low-predator regimes male guppies had more, larger, brighter markings (Endler, 1980). Another example is afforded by a large-scale enclosure experiment conducted on the Dominican anole (Malhotra and Thorpe, 1991; Thorpe and Malhotra, 1996). Within Dominica, this solitary anole (*Anolis oculatus*) generally shows intergrading ecotypes (Figure 3) associated with biotopes. Correlation evidence suggests natural selection as the cause for the geographic variation over most of the island, and a common garden experiment rejects phenotypic plasticity.

Correlations. Correlational studies are useful in elucidating the factors that may influence patterns of geographic variation. Many potential causative factors can be considered simultaneously by using appropriate statistical methods. Phylogenetic relationships may also be taken into account. By including molecular phylogeographic relationships and ecological factors in the same test, one can test the relative importance of historical

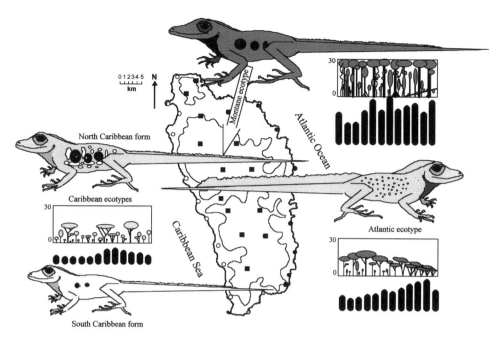

FIGURE 3. Ecotypes of *Anolis oculatus* on Dominica.
The habitat on Dominica, as on other high elevation Lesser Antillean islands, shows pronounced differentiation into intergrading biotopes. Hollow circles indicate seasonal Caribbean xeric woodland, solid circles indicate evergreen Atlantic littoral woodland, shaded circles indicate a transition between these former vegetation zones, and squares indicate evergreen rainforest. The 300 m contour is indicated. For the biotopes canopy height (in meters) is indicated diagrammatically, and relative monthly rainfall (Jan–Dec) is indicated by a histogram (the maximum monthly rainfall illustrated is c.850mm). Aspects of the size, shape and color pattern are indicated for the ecotypes associated with these biotopes. Geographic variation among biotopes tends to be smoothly clinal with some incongruence among characters that adapt to different facets of the environment. However, there may be parallel variation in some characters for the different anole species that inhabit the other islands in the archipelago with parallel biotopes. The associations between habitat and morphology and the parallels, together with common-garden and natural selection experiments, indicate that natural selection is responsible for this variation. In contrast, the geographic variation between the southern Caribbean and northern Caribbean form is not associated with habitat differences and is of a different nature (see Figure 2).   This figure is reproduced with kind permission of IIASA, Austria.

and current factors (Thorpe et al., 1996). For example, in the Malayan pit viper we can see that venom variation is more probably due to natural selection for diet than to phylogenetic relationships (Daltry et al., 1996); and in the Dominican anole (Malhotra and Thorpe, 2000) we can see that, though some morphological characteristics are associated with phylogeny, natural selection for ecological conditions influences a wide range of character systems (Figure 3).

One of the current interests in speciation is the relative importance of selection and vicariance in influencing gene flow among populations. Studies of geographic variation are central to this, and matrix correspondence analysis of geographic variation can be used to investigate the relative importance of different factors. For example, in the Tenerife lacertid these methods suggest that gene flow (measured by microsatellites) is primarily influenced by sexual selection via UV markings rather than by vicariance-induced phylogenetic lineages (Thorpe and Richard, 2001). However, much work remains to be done

in this area, and the factors that result in speciation rather than just geographic variation need to be more fully elucidated (Magurran, 1998). Hence, geographic variation analysis appears likely to remain at the center of the debate on speciation and microevolution.

BIBLIOGRAPHY

Avise, J. C. *Phylogeography: The History and Formation of Species.* Cambridge., Mass., 2000.

Brown, R. P., R. S. Thorpe, and M. Baez. "Parallel Within-Island Evolution in Lizards on Neighbouring Islands." *Nature* 352 (1991): 60–62.

Daltry, J., W. Wüster, and R. S. Thorpe. "Diet and Snake Venom Evolution." *Nature* 379 (1996): 537–540.

Endler, J. A. "Natural Selection on Color Patterns in *Poecilia reticulata.*" *Evolution* 34 (1980): 76–91.

Gould, S. J., and R. J. Johnston. "Geographic Variation." *Annual Review of Ecological Systematics* 3 (1972): 457–498.

Magurran, A. E. "Population Differentiation Without Speciation." *Philosophical Transactions of the Royal Society, London, B* 353 (1998): 275–286.

Malhotra, A., and R. S. Thorpe. "Experimental Detection of Rapid

Evolutionary Response in Natural Lizard Populations." *Nature* 353 (1991): 347–348.

Malhotra, A., and R. S. Thorpe. "The Dynamics of Natural Selection and Vicariance in the Dominican Anole: Patterns of Within-Island Molecular and Morphological Divergence." *Evolution* 54 (2000): 245–258.

Mayr, E. *Animal Species and Evolution.* Cambridge, Mass., 1963.

Sokal, R. R. "Analysing Geographic Variation in Geographic Space." In *Numerical Taxonomy*, edited by J Felsenstein, pp. 384–403. Berlin, 1983.

Sokal, R. R., and B. Riska. "Geographic Variation in *Pemphigus populitransversus* (Insecta: Aphididae)." *Biological Journal of The Linnean Society* 15 (1981): 201–233.

Thorpe, R. S. "Biometric Analysis of Geographic Variation and Racial Affinities." *Biological Review* 51 (1976): 407–452.

Thorpe, R. S., H. Black, and A. Malhotra. "Matrix Correspondence Tests on the DNA Phylogeny of the Tenerife Lacertid Elucidate Both Historical Causes and Morphological Adaptation." *Systematic Biology* 45 (1996): 335–343.

Thorpe, R. S. and A. Malhotra. "Molecular and Morphological Evolution Within Small Islands." *Philosophical Transactions of the Royal Society, London, B* 351 (1996): 815–822.

Thorpe, R. S., and M. Richard. "Evidence That Ultraviolet Markings Are Associated with Patterns of Molecular Gene Flow." *Proceedings of the National Academy of Sciences USA* 98 (2001): 3929–3934.

— ROGER S. THORPE

GEOLOGICAL PERIODS

Geologists have developed a standard geologic timescale that allows them to communicate consistently about events in the distant geologic past. The largest time divisions are eons, which are subdivided into eras, periods, epochs, and ages.

Origins of the Timescale. The geologic timescale (Figure 1) was developed in Europe, primarily between 1795 and 1840. It was not organized in a logical or systematic fashion from first principles, but grew in a haphazard way, with different parts of the timescale named at different times by different people. The names of the geologic periods follow no logical pattern, but reflect historical accidents of naming. However, it is such a fundamental tool that geologists find it the most useful way to communicate about the age of geological events. This system is much more practical than referring to such events by numerical dates (which are often poorly known and subject to change). The method is analogous to describing historical events by such landmarks as the monarchs of England. Although we can often refer to historical events by the precise year, terms such as *Elizabethan* and *Victorian* are often more useful to the historian in providing a time context and require only a knowledge of the sequence of kings and queens of England and the approximate years of their reign.

Before 1800, geologists viewed the rock record as a product of Noah's Flood, with Primary, or primitive, granites and metamorphic rocks (typically found in the cores of mountains) representing the rocks of the original created earth, tilted, layered Secondary rocks representing sedimentary deposits of the Flood itself, and horizontal, less consolidated "Tertiary" sediments above them as post-Flood deposits. Today, only the term *Tertiary* survives from this system, the term referring to rocks now dated between 65 and 2 million years ago.

Although scholars attempted to force the rock record into a biblical straitjacket, miners were more concerned with practical matters such as naming and describing distinct rock layers. All over Europe, the industrial revolution and the demand for coal had led to widespread mapping and mining of the beds, known as the "coal measures." In 1822 William Conybeare and William Phillips (following J. J. Omalius d'Halloy) formally renamed these rocks the Carboniferous, meaning "coal-bearing" in Latin. (In the United States, the early Carboniferous was named the Mississippian in 1870, and the late Carboniferous was called the Pennsylvanian in 1891; these terms are still widely used in the United States.) Likewise, the limestone of the white cliffs of Dover inspired their name for the Cretaceous period (from the Latin for "chalk"). Even earlier (1795), the famous explorer Alexander von Humboldt had coined the name *Jurassic* for the rocks exposed in the Jura Mountains of the southwestern Alps. In 1815 Friedrich von Alberti used the term *Triassic* for the threefold succession of rocks found over Germany and elsewhere in Europe.

In the 1830s Adam Sedgwick (first professor of geology at Cambridge University) and the gentleman geologist Roderick Impey Murchison did fieldwork in western England and Wales to decipher the base of the Secondary sequence. In 1834 Sedgwick named the lowest rocks the Cambrian (after the Roman name for Wales), and Murchison called the upper rocks the Silurian (after the Roman name for an ancient Welsh tribe, the Silures). However, they used different criteria for recognizing their systems. Murchison based the Silurian on distinctive fossils, so it could be recognized around the world, while Sedgwick used more parochial features of particular geological formations. Arguments arose about the boundary between their two systems, and soon Murchison's Silurian had swallowed up Sedgwick's Cambrian. The dispute was not settled until after they died, when Charles Lapworth recognized a distinctive interval between the Cambrian and Silurian in Scotland, which he called the Ordovician (after the Roman name for another Welsh tribe, the Ordovices) in 1879.

Before the Cambrian-Silurian feud destroyed their friendship, however, Murchison and Sedgwick collaborated on the naming of the rocks just beneath the Carboniferous. In most of the British Isles, they are river deposits known as the Old Red Sandstone, but in southwestern England (Devonshire and Cornwall), they

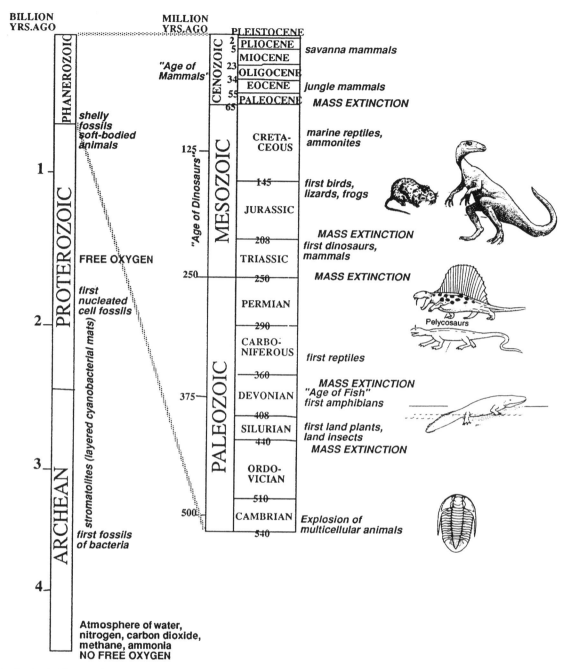

FIGURE 1. Geologic Time Scale.
Shown are the major events in earth and biologic history.   Donald Prothero.

change in character into marine beds that were long confused with the Carboniferous or Silurian. In 1840 Sedgwick and Murchison coined the term *Devonian* for rocks of this age, whether they were marine or nonmarine. In 1841 Murchison went to Russia at the invitation of the czar, where he studied the region's geology and confirmed much of what he had already discovered about European geology. In the Ural Mountains he rec-

ognized a sequence of rocks above the Carboniferous which he called the Permian (after the town of Perm in the Urals).

By the late 1830s, the character of these time divisions was becoming apparent, and large-scale patterns could be recognized. In 1838 Sedgwick proposed the Paleozoic era for the early part of the timescale (ending with the Permian), and in 1840 John Phillips used the term *Mesozoic era* for the Triassic-Jurassic-Cretaceous, and *Cenozoic era* for the post-Mesozoic. In 1833 Charles Lyell named the subdivisions (or epochs) of the Cenozoic and Tertiary as the Eocene, Miocene, and Pliocene; the Paleocene epoch was named by the paleobotanist W. P. Schimper in 1874, and the Oligocene epoch by H. E. von Beyrich in 1854.

Highlights of Geologic History. Some periods in the history of earth and life are particularly eventful. For example, the beginning of the Paleozoic era and Phanerozoic eon is the Cambrian period, which marked the first abundant fossils with shells. Likewise, the boundaries between the eras are marked by major mass extinctions. The boundary between the Paleozoic and Mesozoic eras (Permian and Triassic periods) was marked by the largest mass extinction in earth's history, when 95 percent of the marine species on earth vanished. The majority of typical seafloor invertebrates of the Paleozoic, such as brachiopods ("lampshells"), crinoids ("sea lilies"), bryozoans ("moss animals"), and horn corals vanished, and were replaced by modern shelly marine groups such as clams, snails, and sea urchins. The boundary between the Mesozoic and Cenozoic eras (Cretaceous and Tertiary periods) saw another major mass extinction, when the dinosaurs vanished from the land and many marine groups (such as the ammonites, distant relatives of the octopus, squid, and chambered nautilus) died out. When Phillips and Sedgwick first proposed the Paleozoic, Mesozoic, and Cenozoic eras, they did so with the full awareness that the typical fossil assemblage of each era was highly distinctive. Since that time, we have come to understand why these faunas look different, and we also know much more about the causes of these two major mass extinctions.

Geologic Time. The geologic timescale was created by geologists almost 200 years ago to describe the relative ages of geologic events, and has been internationally accepted ever since. The terms now in use, delineating the various eras, periods, and epochs, may not be logically arranged, since they grew by accident, but their meaning has remained consistent for over 100 years. Numerical age estimates of the duration of these time intervals have been known for less than a century, and sometimes are revised as newer and better dates become available. Thus, for geologists the relative timescale is the only practical means of describing events in the geologic past.

[*See also* Paleontology.]

BIBLIOGRAPHY

Cloud, P. *Oasis in Space: Earth History from the Beginning.* New York, 1988. A broad overview of earth history written for the general audience.

Hartmann, J., and R. Miller. *The History of the Earth: An Illus-trated Chronicle of an Evolving Planet.* New York, 1991. Beautifully illustrated book with paintings of the earth as it may have looked through the past 4 billion years.

Prothero, D. R., and R. H. Dott, Jr. *Evolution of the Earth.* 6th ed. New York, 2001. A college-level textbook on earth history.

Rodgers, J. J. W. *A History of the Earth.* Cambridge, 1994. A specialized review of the geologic and tectonic history of the earth.

Stanley, S. M. *Earth System History.* New York, 1999. A college-level textbook on earth history, emphasizing earth system interactions between the atmosphere, the biosphere, and the earth.

Windley, B. F. 1984. *The Evolving Continents.* 2d ed. New York, 1984. A college-level book focusing on the tectonic history of the continents, especially in the Precambrian.

— DONALD R. PROTHERO

GEOLOGY

Geological evidence relating to past life on the earth and the context in which it has evolved provides insights on four topics: geological time, past conditions, the history of life, and interactions between evolving life and the earth.

Geological Time. It is only over the last three centuries that researchers have realized the antiquity of the earth. Estimates have expanded six orders of magnitude from biblically derived figures of around six thousand years to today's radiometric value of 4.6 billion years. The unimaginable scale of geological time, which itself poses problems of interpretation for human psychology, has earned the graphic nickname "deep time."

Initial interpretations of the geological record by European intellectuals drew heavily upon the biblical concept of a universal deluge. By the end of the eighteenth century, however, such diluvial theories were giving way to the precepts of modern geology. Evidence for repeated cycles of sedimentation, deformation, uplift, and erosion caused the Scottish natural philosopher James Hutton (1726–1797), to remark "With respect to human observation, this world has neither a beginning nor an end." Also around this time, the recognition that distinctive suites of fossils appear in a consistent order in sedimentary sequences, allowing correlation of strata in different areas, launched the discipline of stratigraphy—the systematic ordering of geological formations. During the nineteenth century, stratigraphers established the relative geological timescale that is still used today (see Vignette).

Meanwhile, absolute geological time was variously estimated by extrapolation from inferred rates of erosion, sedimentation, accumulation of salt in the oceans, faunal turnover in the fossil record, and other such measures. Results varied widely, but often ran to millions of years or even hundreds of millions of years, just for the time elapsed since the beginning of the Paleozoic era.

THE GEOLOGICAL TIMESCALE

Geological time is divided into a hierarchy (eons, eras, periods) of formally named successive units of relative time. Absolute dating of their boundaries by radiometric methods has become increasingly refined, especially over the last few years, so different values may be encountered in older publications. The eons were named in the twentieth century. The Phanerozoic eon (from the Greek *phaneros*, "visible" + *zoos*, "living") refers to the abundant skeletal fossil record. The three preceding eons are sometimes referred to jointly as the Cryptozoic (from the Greek *kruptos*, "hidden"), or more informally as the Precambrian. The three eras of the Phanerozoic (Palaeozoic, Mesozoic, and Cenozoic) were introduced in 1838–1840, in recognition of three distinct phases in the fossil record (named from the Greek for, respectively, ancient (*palaios*), middle (*mesos*), and recent (*kainos*) life). Natural boundaries between these eras were provided by mass extinctions. The periods were named during the nineteenth century, largely by reference to the areas (or local tribes) where rocks of those ages were first mapped, although some acquired their names from distinctive rock types, while the Tertiary and Quaternary came from an older system of naming.

TABLE 1. Relative Geological Time

Eon	Era	Period	Millions of years ago (approximate)
Phanerozoic	Cenozoic	Quaternary	1.8
		Tertiary	65
	Mesozoic	Cretaceous	142
		Jurassic	206
		Triassic	248
	Paleozoic	Permian	290
		Carboniferous	354
		Devonian	417
		Silurian	443
		Ordovician	495
		Cambrian	545
Proterozoic			2500
Archaean			3800
Hadean			4600

—PETER W. SKELTON

The eminent Scottish physicist William Thomson (later Lord Kelvin; 1824–1907) temporarily caused a stir with his preferred estimate of a mere twenty to forty million years for the entire age of the earth, on the assumption that it had gradually cooled from a molten state. Such a short timescale was widely felt to pose a serious problem for Charles Darwin's theory of evolution by natural selection. The subsequent discovery of radioactivity undermined the calculation by pointing to a previously unknown source of heating. Moreover, it led to the development of powerful new dating methods, based on measurements of the ratios of unstable (radiogenic) isotopes and their decay products preserved in rocks. A variety of such radiometric methods have provided today's absolute timescale (see Vignette). Earth is now thought to have originated through accretion of material in the nascent solar system around 4.6 billion years ago, based on the dating of meteorites that represent residual matter from that initial phase.

To reconstruct events in the distant past, beyond human records, geologists had to develop a science of prehistory, with its own characteristic methods for generating and testing hypotheses. Hutton urged that geological phenomena should be interpreted only in terms of natural processes that can be seen to operate today—an approach encapsulated by one of his followers in the widely quoted aphorism: "The present is the key to the past." Hutton's ideas were developed further by his compatriot Charles Lyell (1797–1875) in his highly influential *Principles of Geology* (1830–1833). But Lyell's principle of uniformity, termed uniformitarianism by the English philosopher of science William Whewell (1794–1866), now embraced three distinct assertions. Besides explanation in terms only of known causes, he insisted on their gradual (or incremental) operation, as well as a long-term steady state in global conditions. The last two assertions were aimed particularly against speculative catastrophic explanations then being offered for such dramatic geological phenomena as mass extinctions, which Lyell regarded as unscientific vestiges of diluvial thinking. Yet his own steady state hypothesis flew in the face of the growing fossil evidence for faunal progression and changing climates, while critics pointed out that the short span of human history was unlikely to have sampled all kinds of catastrophe. Hence, only the first element of his uniformitarianism—in effect, conformity of explanation with the currently known laws of physics and chemistry—is generally adopted today by geologists.

Alongside the uniformitarian recipe for explanations, geologists developed a pragmatic approach to testing their hypotheses. Whereas a purported process or event in deep time is irretrievable for experimental manipulation, it can be expected to have left behind a variety of discrete clues. The more of these that are independently observed, the more robustly a hypothesized process or event is corroborated. Termed *consilience* (a "jumping together" of inductive inferences) by Whewell, this is essentially the method used in a legal trial. For example, the browsing (as opposed to grazing) habits of

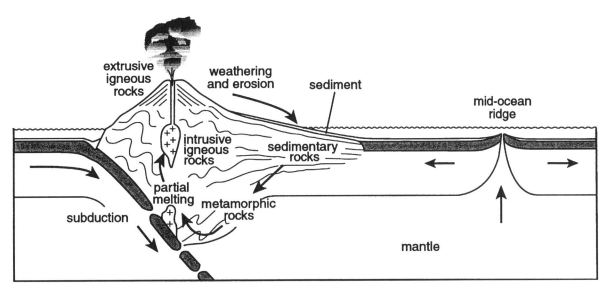

FIGURE 1. Diagrammatic Vertical Section Through the Earth's Lithosphere Showing the Basic Elements of Plate Tectonics and the Rock Cycle.

Diagram prepared by John Taylor, Department of Earth Sciences, Open University.

early horses was first inferred from the shapes of their fossilized teeth, but has since been corroborated by the discovery of wads of leaves in the stomach contents of exceptionally preserved examples from early Tertiary lake deposits in Germany.

Deep time often confounds expectations of what we should see in the geological record, because the rare becomes commonplace over such time spans and may have disproportionate effects compared to more familiar influences. For example, the geological record of deserts is often dominated by water-lain deposits, because flash floods, infrequent though they are, can carry much more sediment around than can wind. Rarer but even more dramatic events, such as major asteroid impacts, have to be allowed for in interpreting the geological record. Growing appreciation of this point in recent decades has led to a synthesis of the originally opposed concepts of uniformitarianism and catastrophism.

Past Conditions. The theory of plate tectonics, developed since the 1960s, reveals the earth's outer skin, or lithosphere, to be a mobile jigsaw of discrete rigid plates that move around as a result of convection in the hot mantle beneath. At constructive plate boundaries, magma welling up from the mantle is injected along midocean ridges, accreting onto the plates on either side, which steadily move away from one another like diverging conveyor belts (Figure 1, right). The increase in lithospheric area is compensated for at destructive plate boundaries, where the leading edge of one plate plunges beneath that of another (Figure 1, left). This process is termed *subduction*, and the zone along which it occurs, a *subduction zone*. Melting of the mantle above the de-

scending slab, due to release of water from the latter, generates magmas that rise through the margin of the overriding plate, forming a chain of volcanoes alongside the subduction zone (Figure 1, left). Meanwhile, the dense subducted slab sinks through the mantle, eventually melting much later. Elsewhere, conservative plate boundaries are where plates merely slide past one another, with neither construction nor destruction of the lithosphere. Most of the lithosphere is about 40 kilometers thick and is capped by a thin, mantle-derived rocky crust that is relatively dense, riding low on the ductile mantle beneath and forming the floor of the ocean basins. Continental lithosphere, by contrast, may reach 100 kilometers in thickness and includes a thick crust of relatively low density rock, which floats high on the mantle to form land over wide areas. The continental masses thus drift around (at about the pace that your fingernails grow), sometimes splitting apart and sometimes colliding (to form mountain chains), carried by the plates. Over the long term, it has been the steady accretion of lighter material that separates during subduction-related melting that has built up most of the continental crust.

Plate tectonics and radiation from the sun together drive the continual transformation of one rock type to another, in what is known as the rock cycle. Rocks at the earth's surface are weathered and eroded by air, water, and ice—the ultimate energy for these processes derived from solar heating—to yield sediments. When buried, compacted, and cemented, the latter form sedimentary rocks. These may either become reexposed to erosion or recrystallize in the solid state to form metamorphic rocks in response to heat and pressure at depth. Greater heating and pressure (with continental collision, for example) can initiate melting, to yield magmas that rise again toward the surface to form igneous rocks on cooling. At the surface, weathering and erosion repeat their work. Hence, over long periods, material is cycled between these three rock types (Figure 1, left).

This outline of the geological dynamics of the earth's surface has a twofold relevance to evolutionary science. First, the sedimentary part of the rock cycle entraps the remains of organisms as fossils, and thus provides our main archive for researching life's history. Second, reconstruction of past plate movements, hence changing arrangements of continents and oceans, is an essential aid to understanding the distributions of organisms today. A classic example is the presence of marsupial mammals in both Australia and South America, which can be explained as the legacy of the former conjunction of these continents with Antarctica until early Tertiary times.

Aspects of the environments in which ancient organisms lived can be gleaned from sedimentary rocks. The compositions and textures of the rocks, as well as their arrangement in characteristic successions, can be interpreted to reveal past conditions of deposition, using uniformitarian analogy. This is the business of sedimentology. The seaward advance of deltas, for example, creates successions that commence with muddy marine deposits and pass up to sandy shoreline and channel deposits together with overbank muds. When corresponding successions of strata are observed in ancient sedimentary formations, a similar mode of origin may be inferred. Ancient environments, such as river basins, coal swamps, and deserts, may accordingly be mapped out for given periods of geological time. Changes in inferred conditions, as well as climates, can to some extent be explained by reference to plate tectonics. For example, Britain's geological record for the Late Paleozoic era reveals a relatively hot, arid climate in the Devonian and earliest Carboniferous periods, followed by a more humid regime conducive to coal swamp formation later in the Carboniferous period, with a return to desert conditions in the Permian. This sequence of climatic changes reflects the latitudinal drift of the country across the equator, starting from the southern arid belt, then traversing the humid intertropical convergence zone, and finally passing into the northern arid belt. Geophysical evidence for this migration comes from measurements of the declinations of magnetic minerals found in certain rocks. In the original environment, these minerals tend to become aligned with the earth's magnetic field, thus increasingly tilted down toward the nearer pole, passing from low to high latitudes. When trapped in solidified lava or lithified sediment, for example, they impart to their host rock a faint residual magnetism (remanent magnetism), which can be measured with sensitive instruments, allowing the original latitude in which the rock formed to be inferred. Other aspects of geographical change that are revealed by sedimentary sequences, such as long-term rises and falls in sea level (over many millions of years), can also be ascribed to plate tectonics. Mid-ocean ridges widen with increased rates of plate divergence, and their total length increases as continental rifting gives rise to new ocean basins. Both effects cause volumetric expansion of the global ridge system, displacing ocean water and leading to worldwide rise in sea level. The opposite effects correspondingly produce falls in sea level. Coalescence of the continents to form a single supercontinent (Pangea) in the Late Permian period, for example, led to a large drop in sea level, with widespread continental emergence.

Plate tectonics cannot account for all past changes in conditions, however, and other factors need to be considered as well. For example, the sedimentary record shows that global climate patterns have undergone considerable fluctuation. The Quaternary world has been dominated by long periods of high- to mid-latitude glaciation, interrupted by relatively short warming (interglacial) phases—just the latest of which has been the entire setting for the short story of human civilization. Over the longer term, however, the "icehouse" regime of the Quaternary period proves to have been toward the colder end of a spectrum of climatic states. We have to go back to rocks of Late Carboniferous to Early Permian age (in the southern continents) to see evidence for similarly cold conditions again. Going back further, research in recent years has tentatively identified an earlier, considerably more intense icehouse period in the Late Proterozoic eon, when glaciation apparently extended even to low latitudes (the "snowball earth" theory). At the other end of the climatic spectrum is the "greenhouse" state, with high-latitude warmth and little or no polar ice. The Late Cretaceous record provides a particularly good example, showing ample fossil evidence for the growth of lush forests within just a few degrees of the poles. Although the changing dispositions of continents and oceans arising from plate tectonics can go some way to explaining these fluctuations, computer models of global climate suggest that other factors need to be called upon as well. In particular, model simulations yield a better fit with geological data when changes in atmospheric levels of greenhouse gases (e.g., carbon dioxide) are included. In the case of the Cretaceous period, gases emanating from widespread and prolonged oceanic volcanism in the Pacific—on a scale dwarfing anything seen today—was probably a major contributor to its greenhouse atmosphere. Much of this volcanism was sited within tectonic plates (rather than being limited to plate margins) and thus appears to reflect unusual mantle activity not normally involved in plate tectonics. Occasional massive outpourings of basaltic magma (flood basalts) on continental areas have likewise been implicated in atmospheric perturbations at other times.

How far back does the earth's geological record go? The ceaseless rock cycle has ensured that little evidence remains of the planet's early history. So far, the oldest rocks that have been radiometrically dated give an age of just over four billion years old. Slightly younger examples from East Greenland, dated at 3.8 billion years old, contain relics of limestones and sandstones as well as lava structures that are characteristically formed under water. Moreover, the ratio of carbon isotopes varies between the carbon in traces of organic compounds and that in the carbonate of the limestone (which is calcium carbonate). Such a differentiation typically arises in the presence of photosynthetic organisms, which preferentially take up carbon dioxide containing the lighter common isotope of carbon. Hence, not only liquid water, but probably also life, was already present on the earth by 3.8 billion years ago. How much of the ocean water was supplied by outgasing from the earth itself and how much was brought in by comets and other extraterrestrial debris are currently uncertain. The trail of the earth's earlier history can be taken up, surprisingly, on the moon. Rock samples collected by the Apollo missions in the late 1960s and early 1970s suggest that the moon formed from material flung into orbit when the young earth was struck by a small planet-sized body. Dating of the old lunar highlands indicates that an initial ocean of magma had solidified by about 4.5 billion years ago. In the absence of a subsequent rock cycle, the only disturbance thereafter came from the impacts of other solar system debris, of which the moon's surface thus kept a detailed diary. Dating of impact-melted rocks suggests that a particularly heavy bombardment occurred between about 3.85 and 3.95 billion years ago. There can be little doubt that the earth—a far larger target—would have suffered likewise. The scale and intensity of the bombardment would probably have repeatedly boiled off any liquid water, and extinguished any life that might have become established by then. Hence, to a first approximation, we can take 3.8 billion years as the likely time of permanent establishment of life on the earth. Although both plate tectonics and the rock cycle would have been in operation then, two major differences would have been the presence of much smaller continental masses and an oxygen-free atmosphere.

The History of Life. Fossils are the only direct clues to past life on the earth. They include not only the remains of organisms (body fossils) but also marks of their activities, such as burrows, footprints or droppings (trace fossils), and diagnostic chemical traces (chemofossils). Their formation is a chance affair, usually involving early burial and preservative mineral growth, although there are myriad ways in which fossilization can come about (which is the subject of taphonomy).

The fossil record is frequently accused of being "incomplete," with the implication of its being an unreliable witness to patterns of evolution. Charles Darwin, for example, expressed disappointment at the lack of then known examples of gradual evolutionary change, which he blamed on gaps in the record. However, two important qualifications need to be made to this complaint. First, the record would be better characterized as highly uneven, rather than uniformly incomplete. Over the last few decades, geologists have developed methods for estimating the proportion of time elapsed during the accumulation of sedimentary sequences that is physically represented by deposits, as opposed to hiatuses (resolution analysis). Results vary widely, but certain depositional settings (such as rift lakes, deep basins on the continental shelves, and shallower parts of the ocean floor) make favorable targets in the quest for sequences with good resolution. This is because reworking of sediment (i.e., deletion of record) by currents in such areas is relatively less frequent than it is in, say, river channels and shallow marine environments. Another aspect of unevenness is the wide variation in preservation potential of different kinds of organisms. Hardly surprisingly, skeletal hard parts stand a much better chance of becoming fossils than soft tissues (although even the latter may be fossilized in favorable circumstances). Understanding such taphonomic biases helps direct the search for well-resolved records of microevolution. Since the debate on punctuated equilibrium versus gradualism was launched in the 1970s, paleontologists have documented numerous examples. One of the more surprising and important findings from such studies is the frequency of stasis (long periods with no net change) seen in fossil lineages, often extending over several million years.

The second qualification to the charge of incompleteness is that "incomplete" does not mean wrong. Every fossil potentially tells us something. "Completeness" depends on the question to be answered. For example, a single confirmed fossil would suffice to answer the question Was there life in the Archaean? (the answer is yes, incidentally). A minimum temporal duration for a fossil taxon is established merely by the observed limits of its stratigraphical range; moreover, the frequency of gaps within that range can be used to estimate error bars on its limits. Thus, for groups with a good fossil record (many marine shelly organisms, for example), the stratigraphical distribution of taxa can, to a large extent, be taken at face value.

Perhaps the main historical contributions of paleontology to evolutionary science are the demonstration of extinctions and of the major patterns in the history of life. The first credible evidence for extinction was provided in 1796 by the brilliant French comparative anatomist Georges Cuvier, who demonstrated that a fossil elephant found in the Paris Basin could not be assigned to either of the two known living species. [*See* Cuvier, Georges.] During the next century, the documentation of droves of extinct taxa led initially to the notion of catastrophic revolutions in the history of life, mentioned earlier, but later, with increasing stratigraphical refinement, to the realization that extinctions were scattered throughout the record. In the latter part of the twentieth century, and especially since the presentation in the early 1980s of evidence for a catastrophic impact at the Cretaceous/Tertiary boundary, a continuum from piecemeal (background) extinctions to maxima of regionally or globally synchronous (mass) extinctions has been recognized. Mass extinctions have given us an important new perspective on large-scale turnover in the history of life, relative to the classical Darwinian view. No longer are biotic interactions such as competition and predation regarded as the predominant drivers of evolutionary replacements, as was usually supposed in the older literature. Instead, abrupt and widespread environmental perturbations are now seen as having played a major role by precipitating ecological collapse and hence mass extinction. According to this view, contingency plays a large part in the survival of taxa, and thus the phylogenetic roots of ensuing evolutionary radiations.

As far as the major patterns in the history of life are concerned, the main outlines of the Phanerozoic record were already apparent in the nineteenth century, but those of the aptly named Cryptozoic had to await the following century. For approximately the first half of its history on the earth, life seems to have consisted entirely of prokaryotes (bacteria and related organisms) the oldest possible eukaryote (nucleus-bearing) fossils are about 2.1 billion years old. Much later, between 1.2 billion and 1.0 billion years ago, a flurry of eukaryote diversification gave rise to most of the main unicellular and multicellular groups, although the first fossils of animals appear only in rocks dated to about 600 million years ago, shortly before the great ice age of the Late Proterozoic eon. The dawn of the Cambrian period was marked by a rapid proliferation of animals with skeletal hard parts (the Cambrian evolutionary explosion), thus commencing the rich fossil record of the Phanerozoic.

Fossils can also contribute valuable information to the reconstruction of phylogenies (cladistics). Any fossil showing traits diagnostic for a given clade (an entire branch of a phylogeny) can at least furnish a minimum date of branching from its sister group. Moreover, fossils of relatively early representatives of a clade may retain primitive character states that are missing in extant representatives. Such information can help to clarify the relationships of the clade in question to other taxa. For example, the skulls of some of the earliest fossil crocodiles, of Triassic to Early Jurassic age, retain an aperture in front of each eye socket (antorbital fenestra), which has been lost in extant crocodiles and alligators. This aperture is also found in dinosaurs and birds, with which they can thus be grouped in a larger clade (the archosaurs) that excludes, for example, lizards and turtles. Finally, where relatively continuous stratigraphical sampling is achievable, it may be possible to infer probable ancestor–descendant relationships—which are not detectable through cladistic analysis alone—from a direct literal reading of the record. A number of phylogenies for microfossil taxa (such as foraminifera) have been reconstructed on this basis using samples from cores of ocean floor sediments.

Interactions between Evolving Life and the Earth. The last few decades have seen the rapid development of earth systems science, which explores the interactions and feedbacks that exist between the solid earth, its oceans, atmosphere, and life. Insights on such dynamic relationships in the past can be gained by estimating long-term net fluxes between geological sources and sinks in biogeochemical cycles. In the case of the carbon cycle, for example, major geological sources include volcanic outgasing and weathering of carbonaceous deposits, whereas the main eventual sinks are buried organic carbon (e.g., coal and hydrocarbons) and carbonate rocks (chiefly limestone, much of which is formed from shelly debris).

Such an approach dramatically highlights the importance of life in modifying conditions on the earth, especially its atmospheric composition. The buildup of free oxygen, as a consequence of photosynthesis, has long been appreciated. However, no less remarkable has been the massive burial of carbon in the largely biotically mediated sinks noted above, with major consequences for climate (because of the drawdown of atmospheric carbon dioxide). The burial of vast amounts of peat in extensive Late Carboniferous delta systems, for example, was probably a critical factor in the shift of global climate to an icehouse regime at that time. The rise of land vegetation also profoundly altered the water cycle, and the rock cycle in turn, through the effects of recycled water (transpiration) on atmospheric humidity and of root systems on weathering and erosion.

The earth has thus experienced a very different history from that which it would have had without life. Nevertheless, major shifts in equilibrium states, especially between greenhouse and icehouse climatic regimes, which have entailed the demise of entire ecosystems such as polar forests, contradict any notion of planetary homeostasis (as proposed by the Gaia hypothesis, which asserts that life regulates conditions to suit itself). Instead, the picture that emerges is that of a complex, highly interactive system prone to chaotic shifts in states, with life evolutionarily tracking the changing conditions.

The Role of Geology in the Evolutionary Sciences. Geology thus gives shape and context to the history of life, charting its evolutionary radiations and extinctions, and identifying at least some of their possible causes. Through its insights into long-term geographical and climatic change, it also helps to explain many aspects of present-day distributions. At the microevolutionary level, the fossil record can offer—in favorable circumstances—informative case histories of phyletic patterns such as stasis, and punctuational or gradual change. It is also salutary to remember that probably at least 99 percent of all species that have ever lived are extinct and that fossils are the only witnesses that we have to their former existence.

[*See also* Extinction; Lyell, Charles; Mass Extinctions; Punctuated Equilibrium.]

BIBLIOGRAPHY

Briggs, D. E. G., and P. R. Crowther, eds. *Palaeobiology II.* Oxford, 2001. An encyclopedic compilation of essays by a large international authorship, covering a multitude of topics, including major events in the history of life, taphonomy, and fossil evidence relating to patterns of evolution and extinction.

Gradstein, F., and J. Ogg. "A Phanerozoic Time Scale." *Episodes* 19. 1–2 (1996). A widely used geological time scale, with a detailed compilation of absolute ages.

Hutton, J. *Abstract of a Dissertation . . . Concerning the System of the Earth.* (1785). Facsimile reprint, Edinburgh, 1987.

Press, F. and R. Siever. *Understanding Earth.* 3d ed. New York, 2000. A comprehensive standard text on physical geology, emphasizing earth systems and plate tectonics.

Rudwick, M. J. S. *The Meaning of Fossils: Episodes in the History of Palaeontology.* 2d ed. Chicago, 1972. An immensely readable account of the early history of paleontology, covering also the development of the geological timescale and the concepts of uniformitarianism and catastrophism.

Skelton, P. W., R. A. Spicer, and A. Rees. *Evolving Life and the Earth.* Milton Keynes, England, 1997. Reviews the history of life from an earth systems science perspective, focusing especially on the role of life in atmospheric and climatic change through time.

Smith, A. B. *Systematics and the Fossil Record: Documenting Evolutionary Patterns.* Oxford, 1994. A clear, didactic technical review of the use of fossil data for tackling evolutionary questions, approached from a cladistic perspective.

Smith, A. G., D. G. Smith, and B. M. Funnell. *Atlas of Mesozoic and Cenozoic Coastlines.* Cambridge, 1994. A recent compilation of paleogeographical maps. Pre-Mesozoic maps are more difficult to construct, relying exclusively on continental data because of the loss of ocean crust through subduction.

— PETER W. SKELTON

GERM LAYERS. *See* Development, *article on* Evolution of Development.

GERM LINE AND SOMA

The bodies of multicellular organisms such as plants, animals, fungi, and many protists are remarkable for their complexity and internal diversity. This diversity arises from fusion between sperm and egg in animals (or corresponding cells with different names in other kingdoms of life). This first cell or zygote undergoes a kaleidoscopic transformation into various cell types with different roles within the body. Arguably the most important of these roles is the production of sperm or eggs to produce the next generation of zygotes. However, in the most familiar animal groups (including humans), only a relatively few cells are able to make gametes. These cells (often called spermatogonia or oogonia) form a lineage called the germ line. These cells make up an exclusive fraternity: all other cells are restricted to nonreproductive fates in the body to form skin, blood, nerves, muscles, glands, or other structures collectively called the soma.

Membership in the germ line club can be determined very early in the embryos of multicellular organisms. In many cases, this lineage is determined even before the genes of the embryo itself are activated. All somatic cells produced after this stage will never have an opportunity to give rise to sperm or eggs. In this sense, only germ line cells are able to pass harmful mutations, new gene combinations, or rare beneficial genes to offspring. Whatever genetic changes are experienced by somatic cells, these changes are invisible to forces such as natural selection or genetic drift that shape the course of evolution. For this reason, the division between germ line and soma is centrally important to understanding the way that organic evolution works.

August Weismann and the History of Germ Line and Soma. What determines membership in the germ line? The history of this question has been analyzed by the Harvard University naturalist, systematist, and historian Ernst Mayr, and the following summary is largely drawn from his work. The early microscopical investigations of dividing cells identified the nucleus as the location of the genetic material and chromosomes as dividing along with the cell itself. This was enough information to identify the chromosomes as the heritable material passed from parents to offspring through gametes. The German biologist August Weismann combined these observations into a comprehensive theory of inheritance and development (which he continued to modify and expand throughout his lifetime). In 1885, he proposed in his book *Die Kontinuität des Keimplasmas als Grundlage einer Theorie der Vererbung* that the germ plasm is the unaltered genetic material passed from generation to generation in the form of the germ line cells and their descendant gametes. In Weismann's theory of inheritance, genetic material in somatic cells is dissected or partitioned into pieces, and the genetic particles that are passed to each cell determine the somatic form of those cells or its eventual descendants. Because each somatic cell contains only part of the genetic information present in gametes, these somatic cells cannot give rise to gametes themselves.

Although Weismann's theory of inheritance proved to be incorrect, his insight into the separation between germ line and soma proved to be nearly exactly true. He was correct that the germ plasm of gametes is entirely separate from the phenotype of the soma, and, as Mayr points out, Weismann was wrong only in the mechanism that separates the germ plasm from the soma. Weismann proposed that this mechanism depends on the early separation of germ line cells from the cells destined to become the soma. We now know that this mechanism depends on the one-way path of information from the nucleus to the cytoplasm of cells via messenger RNA.

The nature of the germ plasm has been investigated in a few organisms in detail, including nematodes, amphibians, and fruit flies. In the fruit fly embryo, all nuclei up to the thirteenth cell division are capable of giving rise to germ line cells, but only those at the most posterior part of the embryo do so because these nuclei encounter the pole plasm (or germ plasm) found only at the posterior pole of the embryo. This germ plasm consists of both messenger RNA and proteins that direct the fate of nuclei to form germ line cells. Some clever experiments by Karl Illmensee and Anthony Mahowald (1974) showed that injection of tiny volumes of this pole plasm into the anterior end of a fruit fly embryo could induce formation of germ cells from nuclei that would otherwise have become somatic cells.

The Consequences of Weismann's Germ Plasm Theory. The continuity of the germ plasm had several important consequences for evolutionary biology. The first concerned the debate over the inheritance of acquired characteristics advocated in the early nineteenth century by the French naturalist (and first evolutionist) Jean Baptiste de Lamarck. The second was the focus on individual organisms (rather than their component parts) as the units of natural selection.

Weismann formulated his germ plasm theory at a time (from the 1870s through the 1890s) when evolutionary biologists were establishing the validity and outline of evolutionary theory but had not yet discovered a mechanism for inheritance that could lead to evolutionary change under natural selection. Most (including Charles Darwin) believed in some form of inheritance of acquired modifications. Weismann opposed these views vigorously. He argued instead that acquired modifications are not heritable because the somatic cells do not contribute to the germ line and because the germ plasm (which forms the germ line cells) was segregated from the soma early in development. Thus, evolution could not occur through use or disuse of somatic parts (now called soft selection) but only through modifications acquired via the germ plasm of gametes (called hard selection). The logic of this argument was unassailable for organisms with an early determination of the germ line, and Weismann's views eventually won this debate after the rediscovery of Gregor Mendel's genetic research in 1900.

As a result of this emphasis on the germ plasm and hard inheritance, Weismann's ideas helped to focus evolutionary biologists on the importance of individual organisms. Because somatic cells could not contribute directly to the germ line, their evolutionary importance lay in their contribution to the function of the whole organism and its ability to obtain mates. The evolutionary fates of the diverse cells in the organism were intertwined, so only selection among whole organisms was relevant to understanding the course of evolution.

Was Weismann Right? Changes in somatic cells or tissues cannot heritably affect DNA and chromosomes. Thus, the germ plasm is indeed separate from the soma, and acquired traits cannot be inherited. However, many organisms do not separate their germ plasm into an early germ line isolated from somatic cells. In these organisms, some of the somatic cells (and their mutations) can contribute to the pool of gametes used in sexual reproduction. The Yale University evolutionary biologist Leo Buss (1987) wrote an excellent treatment of these issues, and much of the following discussion is drawn from his work.

Although Weismann focused on animals with a germ line, the germ line is not a universal trait of multicellular organisms. Buss pointed out that most higher taxonomic groups of protists (including algae and slime molds), all fungi, all plants, and a number of animal phyla have no germ line but develop germ cells from the soma. The examples include all flowering plants, in which specific somatic cells of the meristem can turn into pollen- or ova-producing cells of the flower, and sponges, in which flagellated feeding cells (called choanocytes) can lose the flagellum and transform into gametes. Many other colonial marine invertebrates (hydroids, tunicates, and bryozoans) also have no germ line. An isolated germ line is found only in some green algae and many animal phyla. Of the animal phyla, some (including various arthropods, annelid worms, molluscs, and vertebrates) have germ line cells determined early in embryogenesis so that there are limited opportunities for genetic variation to arise within an individual and be passed on to offspring. Thus, although Weismann and most other evolutionary biologists considered the germ line to be fundamental to understanding the mechanisms of evolution, the germ line is in fact an evolutionarily derived cell lineage whose appearance itself requires some explanation.

Buss's ideas changed the way evolutionary biologists think about development, inheritance, natural selection, and the nature of individual organisms. First, many multicellular organisms probably consist of many different genotypes in cells that can produce gametes. Thus, even though phenotypic modifications of the soma cannot be inherited, it is possible for genetic changes in somatic cells to be inherited if those somatic cells also transform into gametes.

Second, many organisms might experience competition among distinctive cell lines. If somatic mutations can be inherited, and if some of these mutations cause differences in the tendency to become germ line cells, then these genetically different cell lines may compete for access to the gonads and ability to produce offspring. This would amount to natural selection acting within a multicellular body. Buss (1982) himself showed the potential for such competition among cell lines in a slime mold. Mutants that advance their own reproductive interests even at the expense of the whole organism will tend to win out within a particular individual organism. Alternatively, mutants that allow one cell line to exclude some or all others from producing germ line cells would also succeed in such a competition.

Third, the development of multicellular embryos with germ lines can be viewed in a different light. It is tempting to look at the elaborate unfolding of cells and tissues within an embryo as a kind of cooperation among groups of cells striving together to build a functional adult organism capable of surviving and reproducing. However, competition among cell lines to become part of the germ line lends quite a different view of multicellular development. Rather than cooperation, the progressive specification of the fates of somatic cells can be seen instead as a kind of preemptive strike by germ line cells against the infiltration of the gonads by mutant somatic cell lines. The various molecular mechanisms that result in irreversible somatic fates for almost all cells in an embryo can be viewed as remnants of ancestral mechanisms that evolved initially as ways to manage or restrict cell lineage competition. Such mechanisms are similar to molecular and cellular mechanisms of the immune system that manage or restrict the proliferation of mutant cell lines that could form tumors. The nearly universal appearance of a one-cell stage (the zygote) in multicellular life cycles also can be interpreted as a mechanism for ensuring genetic uniformity among cells in the organism (and reducing the potential for conflict among genetically different cell lines).

Finally, Buss's ideas help to explain in part the taxonomic distribution of germ lines among major branches of the tree of life. Part of the explanation may depend on cell motility. Animal cells are mobile within the body, and mutant cell lines that can move to the gonad and exclude other lineages may put the remainder of the lineages at risk of extinction. Thus, in many animals, the presence of molecular mechanisms for policing the germ line might be especially advantageous. As a result, many animal groups have evolved early isolation of the germ line within the embryo. Plant cells, on the other hand, are constrained by the cell wall and do not move within the soma. As a result, a multicellular plant is in no danger of being taken over by a mutant cell line that gains access to the germ line of a single flower or stem. Thus, plants have no need of a segregated and genetically homogeneous germ line and have not evolved one.

The Evolution of Individuals. Evolutionary biologists recognize individuals as physiologically discrete bodies. If the germ line is separated early in development from the soma, then individuals are genetically discrete units as well, and natural selection operates entirely on those units. However, for most organisms, the germ line does not exist or is not separated early from the soma. In these cases, natural selection could act on several different kinds of units, including cell lines within bodies and bodies within populations. In such organisms, the meaning of an "individual" is less clear and could correspond to cell lines, bodies, or units of selection at higher or lower levels of organization (from genes to species).

[*See also* Development, *article on* Evolution of Development; Weismann, August Friedrich Leopold.]

BIBLIOGRAPHY

Buss, L. W. "Somatic Cell Parasitism and the Evolution of Somatic Tissue Compatibility." *Proceedings of the National Academy of Sciences USA* 79 (1982): 5337–5341.

Buss, L. W. *The Evolution of Individuality*. Princeton, 1987. The most important reference on this subject in the second half of the twentieth century. Includes several new ideas on the importance of conflict among replicating units and the resolution of this conflict as the basis for evolution at any hierarchical level of biological organization (from genes and cells to organisms or species).

Gilbert, S. F. *Developmental Biology*. 2d ed. Sunderland, Mass., 1988. This edition of Gilbert's text includes detailed descriptions (pp. 264–272) of the contents and function of germ plasm in insects and other animals (although the fifth edition, 1997, includes less detail on this subject).

Grosberg, R. K., and R. R. Strathmann. "One Cell, Two Cell, Red Cell, Blue Cell: The Persistence of a Unicellular Stage in Multicellular Life Histories." *Trends in Ecology and Evolution* 13 (3, 1998): 112–116. A review of the evolution of germ line and soma from the perspective of the single-cell zygote.

Illmensee, K., and A. Mahowald. "Transplantation of Posterior Polar Plasm in *Drosophila*. Induction of Germ Cells at the Anterior Pole of the Egg." *Proceedings of the National Academy of Sciences USA* 71 (1974): 1016–1020.

Mayr, E. *The Growth of Biological Thought: Diversity, Evolution, and Inheritance.* Cambridge, Mass., 1982. A general treatment of the history of ideas in evolutionary biology. Includes an extensive review of August Weismann's theory of the continuity of the germ plasm and the implications of this theory for mechanisms of inheritance, the targets of natural selection, and the meaning of multicellular development.

Michod, R. E. "Evolution of the Individual." *American Naturalist* 150 (Supplement, 1997): S5–S21. Though technical in nature, this is an important work because it provides a model for predicting some of the conditions under which rules of development should evolve so that cells in an organism will cooperate to protect the germ line (rather than compete to become germ line cells themselves).

Stoner, D. S., and I. L. Weissman. "Somatic and Germ Cell Parasitism in a Colonial Ascidian: Possible Role for a Polymorphic Allorecognition System." *Proceedings of the National Academy of Sciences USA* 93 (1996): 15254–15259. A dramatic demonstration of the potential for within-individual competition among cell lines in the race to form germ line cells and gametes, in this case within naturally forming chimeras of two different ascidian genotypes.

— MICHAEL HART

GLOBALIZATION

Globalization refers to economic, political, social, cultural, and demographic processes that transcend nation-state boundaries and to the ways in which diverse obstacles (physical, spatial, and cognitive) have been overcome successfully. Globalization may encompass the worldwide integration of trade and financial commerce, the internationalization of state power, and to the breakdown of communication barriers. The implications of such processes to evolution are diverse in that they involve important questions about human diversity as well as social and cultural change. Although the global will never replace the local, humans increasingly find themselves acting in a global arena, which makes the study of globalization important.

Just as there is no generally accepted definition of globalization, the extent and duration of globalization are disputed. Closely related to the concept of globalization is that of modernity, which theorizes that there has been a tendency over the past two hundred years for all societies to undergo improvement and rationality in the forms of production and institutions, and that there are centers and peripheries of development. Such Eurocentric notions were common during the late colonial period (1900–1960), when the assumed superiority of Western European ideas was sacrosanct. Modernist and global thinking thus involves an assumption that the world is composed of a core of civilized and a periphery of primitive societies. Immanuel Wallerstein argued that, since the fifteenth century, there has been a capitalist world system rather than a set of dyadic relationships between nation-states and their colonies, de-centering and erasing the boundedness of space and time. Ideas in his *Modern World-System* (1974) have been extended to include the rise and fall of global systems spanning thousands of years (Frank and Gills, 1993). Although the debates on the world system concept revolve around the scope and duration of such cycles, the very notion of a global system is frequently teleological and includes ideas of development and evolution. The extent to which humanity is entering a globalized world should be seen within a much wider and longer historical set of processes.

Whereas past global systems varied in their geographical scope, recent debate about the modern world system has focused on a number of issues, including the impact of flows of trade, capital, labor, and ideas across the globe. The main basis for social integration for several centuries has been the nation-state, which grew in conjunction with the development of overseas empires. Among the results of colonialism were long-distance mass migrations of up to 15 million Africans and many more Europeans. Indeed, European nation-states developed concurrently with an expansion of people and trade into global empires. From an early stage in colonial expansion, slave plantations were dependent on labor from one continent (Africa), which produced commodities in another (the Caribbean), which were in turn consumed in yet other continents (Europe and North America). A highly organized system of production involved the extension of European state power through military and bureaucratic organization, which depended on orderly communication. Many of the features that developed during the colonial period have remained even though the institutional basis has been altered. The twentieth century saw the addition of many more institutional arrangements for the control of the movement of people, commodities, and ideas.

In the late twentieth century, globalization revolved around the diminution of time and space to such an extent that the local and the global are powerfully connected. Instantaneous communication overcomes geography, accentuates interdependence, and enmeshes people in efforts to connect and resist interrelationships at one and the same time. An irony of globalization is its own working whereby the speed and impact of events in distant parts of the globe may create xenophobia, reactionary politics, and conflict (Held and McGrew, 2000). Communication was detached from physical distance with the advent of underwater cables in the nineteenth century, and during the twentieth century the trend continued toward what theorist Marshall McLuhan identified as a "global village" (McLuhan, 1962). Significant changes occurred in the late twentieth century in the extent of global networks. Quantum leaps in the intensity and magnitude of communication led to changes in the structures that organize the flow of information, goods, and services. These changes have affected human behavior and physical evolution in various ways. It is important to realize that although globalization integrates even greater numbers of the world's population, it also produces extensive divisions between and within nation-states. At the same time as there is global cooperation, there is also division. Furthermore, there is a pervasive feeling that all humans are living in a "runaway world," as Anthony Giddens (2000) put it, where decisions are taken by other people and citizens have lost their ability to effect change through civic action and the expression of democratic mechanisms.

The impact of globalization may be seen in international migration and travel, both of which are more extensive now than ever before in human history. Human migrations first took place in the *Homo erectus* period (1.8–.5m bp) and continued with the *Homo sapiens* migration, producing genetically distinct but not isolated populations over the past 100,000 years (Mascie-Taylor and Lasker, 1988). About ten thousand years ago, neolithic migrations resulted in population diffusion in Europe and Africa (Wijsman and Cavalli-Sforza, 1984) and initiated new populations in the Americas and Australia. Although such migrations were significant in a biological sense, they were gradual. On the other hand, the colonial slave trade from Africa to the Americas along with European migration to the same areas produced large-scale and long-distant relocation of distinct populations to new environments. It is important to note that the degree of mobility among migrant populations is closely connected with the stratification of the world's population into what Zygmunt Bauman called "tourists," who travel through life without restriction, and "vagabonds," who are tossed from place to place (1998). Migration is both cause and product of the inequalities accentuated by globalization. Diasporas of South Asians, Chinese, Mexicans, and others have resulted in communities with transnational identities, with migrants frequently maintaining ties with their places of origin. With such migrations increasing in size and geographical range, issues arise around identity politics, cultural hybridity, racism, human rights, and the nature of the relationship between the nation and the state. Refugee movements (e.g., the Palestinians) pose questions about the effects of statelessness and identity within and between nation-states. Arjun Appadurai (1996) coined the term *ethnoscape* to refer to those groups that appear to affect the politics between nations to an unprecedented degree and whose stable communities of kinship, friendship, work, and place are also being affected. The extent to which the early twenty-first century is undergoing a shift to a global culture and the role of distinct ethnic cultures are widely debated (Featherstone, 1990; Held and McGrew, 2000).

The biological effects of globalization are also significant. Ronald Barrett and colleagues (1998) identified three epidemiologic transitions. In the first transition with settled agriculture, pathogenic transmission occurred with the proximity of people in villages. There was also transmission of pathogens during the Neolithic period arising from trade and migration. More recently, the colonial migrations of Europeans to the Americas killed millions of people who had no prior exposure to European pathogens. The second transition featured a decline in the industrialized world in pathogenic diseases and an increase in chronic, noninfectious diseases such as cancer. In the third transition, there are new diseases, the reemergence of preexisting diseases, and new antimicrobial-resistant strains of pathogens. Although migration and trade as vectors in pathogenic transmission are not new, many diseases such as influenza, tuberculosis, and new diseases such as HIV can now spread rapidly from one continent to another. These diseases reflect global patterns of neocolonialism, seasonal migrant labor, international migration, tourism, and the gendered experience of poverty, which encourages sexual exploitation.

In the emerging field of evolutionary medicine, globalization presents another area of inquiry. According to Allison Jolly (1999), humanity is entering a new phase of evolution in which the world is a global organism akin to a superorganism. Despite some global efforts against poverty, disease, and environmental degradation, there is an ever-increasing gap between the rich and the poor. In accordance with Jolly's suggestion, it is clear that the privileged are reaping the benefits of medical practices such as immunization and disease control. Malaria, tuberculosis, and HIV are examples of such diseases reflecting medical inequalities on a global scale. New reproductive and genetic modification technologies pose yet other questions about how evolutionary practices are embedded in global inequalities. It may be that in a globalized world, those who control the wealth may survive because they are better able to direct their own evolution.

BIBLIOGRAPHY

Appadurai, A. *Modernity at Large: Cultural Dimensions of Globalization.* Minneapolis, 1996. Examines culture and ethnicity as transnational phenomena.

Barrett, R., C. W. Kuzawa, T. McDade, and G. Armelagos. "Emerging and Re-emerging Infectious Diseases: The Third Epidemiologic Transition." *Annual Review of Anthropology* 27 (1998): 247–271. A comprehensive introduction to diseases within the paradigm of evolutionary medicine.

Bauman, Z. *Globalization: The Human Consequences.* New York, 1998. An insightful and reflective examination of global inequalities.

Featherstone, M., ed. *Global Culture: Nationalism, Globalization and Modernity.* London, 1990. An early collection of significant writings on globalization and culture.

Frank, A. G., and B. K. Gills. *The World System: Five Hundred*

Years or Five Thousand? London, 1993. A collection of articles on the character and duration of world systems.

Friedman, J. *Cultural Identity and Global Process*. London, 1994. A significant contribution to the anthropology of identity.

Giddens, A. *Runaway World: How Globalization Is Reshaping Our Lives*. New York, 2000. An interesting review by a leading social commentator.

Held, D., and A. McGrew, eds. *The Global Transformation Reader: An Introduction to the Globalization Debate*. Cambridge, 2000. A comprehensive reader on the globalization debate between globalists and skeptics.

Jolly, A. *Lucy's Legacy: Sex and Intelligence in Human Evolution*. Cambridge, Mass., 1999. Introduces the notion of a global organism as a new stage of human evolution.

Kearney, M. "The Local and the Global: The Anthropology of Globalization and Transnationalism." *Annual Review of Anthropology* 24 (1995): 547–565. A succinct examination of the anthropology of globalization.

Mascie-Taylor, C. G. N., and G. W. Lasker, eds. *Biological Aspects of Human Migration*. Cambridge, 1988. A comprehensive collection on biological aspects of human migrations.

McLuhan, M. *The Gutenberg Galaxy*. Toronto, 1962. A classic study that introduced the idea that "[t]he new electronic interdependence recreates the world in the image of a global village" (p. 31).

Wallerstein, I. *The Modern World-System*. New York, 1974. The standard work outlining the development of the modern world system.

Wijsman, E. M., and L. L. Cavalli-Sforza. "Migration and Genetic Population Structure with Special Reference to Humans." *Annual Review of Ecology and Systematics* 15 (1984): 279–301. A technical review of migration and genetics.

— GAIL R. POOL

GROUP LIVING

Chances are that you see groups of animals every day. A flock of birds surrounding a rubbish bin, mosquitoes buzzing in a mating swarm, a pack of dogs running down the street, and worker ants marching along a pheromonal trail are all examples of animal groups. Groups are aggregations of individuals that range in size from a few to thousands, and include both relatively stable and ephemeral aggregations. This diversity in size and permanence makes the study of the evolution of grouping behavior a challenge, but several themes emerge from the contemporary literature. All involve quantifying the costs and benefits of aggregation as well as identifying constraints that may prevent individuals from living alone.

Aggregation often involves active decisions made by animals, but it also may occur passively. Consider the case of planktonic larvae, which are distributed by oceanic or tidal currents. Although such aggregations require no additional explanation, they offer the opportunity for more complex social relationships within them. The benefits for acquiring food and reducing the risk of predation are the most common explanations for group living, whereas increased competition, reproductive sup-

pression, and the increased risk of parasite or disease transmission are commonly cited costs.

Animals may aggregate around important resources. For example, patches of food may attract individuals of one or more species and create ephemeral foraging aggregations. Group members need not be related, nor even of the same species, to forage together on an important resource. Group size will be limited by the amount and defensibility of the food and by the presence of other known patches of food: larger groups will be seen around larger patches, and less defensible food (such as widely scattered seeds) will support larger groups than the same amount of food distributed in a defensible patch. In some cases, animals may distribute themselves according to the ideal free distribution, where food intake (and presumably fitness) is equivalent among all patches. [*See* Ideal Free Distribution.]

Competition and dominance relationships may make it less profitable for some individual to remain in a group. Dominant or "despotic" individuals—those that restrict free access to resources—increase the cost of remaining in a group for subordinates and, if dominants successfully defend resources, illustrate one reason why observed group sizes may not fit an ideal free distribution. In more permanent social groups, kinship may have an important effect on the relationship between despotism, which leads to biased fitness, and the benefits of sociality. Relatives, who can potentially gain indirect reproductive success by helping raise kin, should be more likely to reside in despotic groups than nonrelatives, who can gain direct fitness only by rearing their own offspring.

The payoffs of being in a group of a particular size vary depending on whether an individual is in the group or trying to join the group. If being in a group is beneficial, nongroup members will gain fitness by joining, even if the fitness of already-present members is reduced by the addition of another individual. Such "insider–outsider" conflicts remain to be better understood, but they may explain why group sizes are often larger than what is seemingly optimal.

There are other benefits from foraging in groups. Specifically, groups of insectivorous or carnivorous animals may flush more prey and have greater foraging success compared to solitary individuals. Such activity need not be highly coordinated, but group hunting, as reported in some social carnivores, involves the coordinated movement and behavior of several individuals (see Vignette). Additionally, group living may facilitate food finding for group members either through information centers (aggregations where information about the location of food is exchanged) or through more simple direct observations of the foraging success of others.

Conspecifics may form ephemeral aggregations around no identifiable resource. For example, males of lekking species (e.g., mosquitoes) aggregate and await visits by females, whereupon males will display and attempt to mate. [*See* Leks.]

Individuals may aggregate to reduce the risk of predation. Such antipredator benefits of aggregation may be obtained several ways.

Predation risk is "diluted" in larger groups because there are alternative prey. Assuming a predator will kill only one individual, the risk of predation declines with the addition of each individual and quickly asymptotes. Other functional relationships are possible, and the benefit may remain if predators take more than one individual. Marlin attacking tightly clumped schools of small fish kill many fish, but a solitary fish would make an easy target and the prey would continue to group despite ongoing attacks.

Animals in groups should have a greater ability to detect predators. Detecting predators with sufficient time to escape is a challenge many species face, and the presence of more eyes (and ears) to detect predators may be directly translated into survival.

Both detection and dilution models predict that individuals will allocate less time to antipredator vigilance and more time to foraging, as group size increases. These changes in time allocation are widely reported in birds and mammals. Scramble competition for a finite resource in which individuals forage more because competition increases with group size could produce a similar group size effect.

Assuming the case of antipredator vigilance, a question arises about whether animals modify their vigilance simply because other individuals are present or in response to the vigilance patterns of others. The original models assumed that there was perfect information transfer from the individual first detecting a predator and the rest of the group. Without perfect information, animals detecting predators may be safer than those relying on the information from others. Recent studies have questioned this assumption of perfect information transfer on both theoretical and empirical grounds and suggest that antipredator benefits are not likely to rely on "collective detection," and that the conceptually simple distinction between detection and dilution may be more complex and interrelated than originally envisioned.

Within a group, an individual's position with respect to others and the edge may affect predation risk. Centrally located animals are generally considered to be relatively safer, but this need not always be so. Predators can be viewed as being surrounded by a "zone of danger"—individuals within this zone of danger are likely to be killed. Fast-moving, agile predators (e.g., some raptors) may have a relatively large zone of danger, and centrally located individuals may be no safer than peripheral ones.

Groups of animals may be more effective at deterring predators or of warning conspecifics about predation

COOPERATIVE HUNTING IN SOCIAL CARNIVORES

There has been some controversy over the reasons large social carnivores form groups. Early analyses focused on the conspicuous and coordinated communal hunting seen in lions and wild dogs. Subsequent detailed analyses of the benefits of group hunting success in lions found that the per capita food intake was the lowest at the most common group size, seemingly refuting this obvious hypothesis. However, per capita intake ignores the costs of pursuing and capturing prey, which may be substantial in species that chase their prey over long distances. More recently, detailed studies of both the costs and the benefits of group hunting in wild dogs suggested that group members increased the amount of food acquired each day *per* kilometer they chased prey, and that viewed this way, wild dogs live in nearly optimally sized groups.

—DANIEL T. BLUMSTEIN

risk. Nesting colonial birds mob both aerial and terrestrial predators in or around the colony. Such communal defense, whether coordinated or not, may reduce the likelihood of predation or nest predation for all group members. Animals in groups may emit special signals designed to warn conspecifics about the risk of predation. [*See* Alarm Calls.]

Animals may aggregate for energetic benefits not directly related to food. Breeding male emperor penguins huddle together through the long Antarctic winter and move, in a more or less continuous trudge, from the colder periphery to the center of the group, and out again. Solitary penguins would freeze to death. Studies of alpine marmots have demonstrated that these hibernating rodents increase their chance of overwinter survival by hibernating socially in groups.

Perhaps the most interesting groups are those with more stable aggregations, for it is in these groups that animals often engage in apparently altruistic behavior, and some significant costs of sociality emerge. [*See* Social Evolution.] Stable groups include a remarkable variety of mating and spacing systems, which range from colonies of more or less solitary ground squirrels living in patchy habitat, to nesting birds that may engage in communal nest defense, to complex communal societies characterized by individuals' delaying dispersal from their natal group, having specific behavioral roles, and illustrating reproductive skew (i.e., not all individuals within the group breed), and to the truly eusocial Hymenoptera and thrips, with their fixed social castes. [*See* Eusociality, *article on* Eusociality in Mammals.]

Understanding the occurrence and dynamics of these more permanent and complex groups also generates the new problem of disentangling evolutionary origin from maintenance. [*See* Adaptation.] When the net benefit of remaining in more ephemeral groups changes, we expect individuals to behave in dynamic ways so as to maximize their fitness. However, some constraints may both foster the formation of more permanent societies and reduce or eliminate options for their residents. For example, sociality in ground-dwelling sciurid rodents (squirrels, prairie dogs, and marmots) is largely explained by their inability to reach adult body size in their first year of life. Females of relatively small species that mature quickly can disperse and reproduce independently. Some of these species live in colonies containing many breeding females. In contrast, the larger-bodied marmots are unable to mature in their first year of life and social groups (groups of animals that interact with each other and share a home range and burrows) contain litters from one or more years. Despotic breeders may suppress reproduction of their offspring, but related nonbreeders help thermoregulate their nondescendant kin and obtain an indirect fitness payoff for doing so.

Colonial life has other costs. Pathogens and parasites are likely to pass easily between individuals living in large colonies as seen in both black-tailed prairie dogs and bank swallows, where individuals in larger groups have more ectoparasites. Studies of the highly colonial bank swallow also illustrate the competitive costs associated with sociality: competition for nesting material and sites is greater in larger colonies, and animals interfere with each other more than is seen in less social species. And, even though animals may aggregate to reduce predation risk, studies of colonially nesting birds demonstrate that predator attack rates increase in highly conspicuous nesting colonies. Interestingly, this greater attack rate is offset by the collective defense and predator mobbing often seen in colonially nesting birds.

[*See also* Cooperation; Optimality Theory, *article on* Optimal Foraging; Reciprocal Altruism.]

BIBLIOGRAPHY

Bednekoff, P. A., and S. L. Lima. "Re-examining Safety in Numbers: Interactions between Risk Dilution and Collective Detection Depend upon Predator Targeting Behaviour." *Proceedings of the Royal Society of London B* 265 (1998): 2021–2026. Theoretically and empirically questions the distinction between detection and dilution models for explaining how antipredator vigilance changes as a function of group size.

Blumstein, D. T., and K. B. Armitage. "Cooperative Breeding in Marmots." *Oikos* 84 (1999): 369–382. Discusses evolution of complex sociality in sciurid rodents.

Creel, S., and N. M. Creel. "Communal Hunting and Pack Size in African Wild Dogs, *Lycaon pictus.*" *Animal Behaviour* 50 (1995): 1325–1339. Argues that after considering costs and benefits of communal hunting, wild dogs are found in nearly optimally sized groups.

Giraldeau, L.-A. "The Stable Group and the Determinants of For-
aging Group Size." In *The Ecology of Social Behavior*, edited by C. N. Slobodchikoff, pp. 33–53. San Diego, Calif., 1988. Discusses, and models, insider-outsider conflicts.

Krebs, J. R., and N. B. Davies. *An Introduction to Behavioural Ecology.* Oxford, 1993. Excellent introductory textbook that discusses the evolution of grouping.

Pulliam, H. R. "On the Advantages of Flocking." *Journal of Theoretical Biology* 38 (1973): 419–422. The model of antipredator vigilance, which discussed the benefits if members of a social group scanned for predators independently. Spawned considerable empirical research.

Vehrencamp, S. L. "A Model for the Evolution of Despotic versus Egalitarian Societies." *Animal Behaviour* 31 (1983): 667–682. Explains how kinship influences the benefits to being a subordinant in a group with dominant despots.

— DANIEL T. BLUMSTEIN

GROUP SELECTION

Throughout the history of biology, higher-level units such as social groups, species, and ecosystems have been described as well adapted to their environments in the same sense that individual organisms are well adapted. However, the process of natural selection does not automatically evolve adaptations at all levels of the biological hierarchy. Consider an individual-level adaptation such as cryptic coloration in birds. Individual birds vary in the degree to which they can be seen against their background, and in every generation the most conspicuous individuals are removed by predators. If phenotypic variation is heritable, individuals will evolve to be more cryptic over time. Now consider a social adaptation such as members of a bird flock warning each other about approaching predators. It is not obvious that callers will survive and reproduce better than noncallers. If uttering a cry attracts the attention of the predator, then callers place themselves at risk by warning others. In general, groups function best when members provide benefits to one another, but it is difficult to translate this kind of social organization into the currency of relative fitness upon which natural selection is based.

Darwin (1871) was aware of this problem and proposed a solution. Suppose there is not just one flock of birds but many flocks. Furthermore, suppose that the flocks vary in their proportion of callers. Even if a caller does not have a fitness advantage within its own flock, groups of callers will be more successful than groups of noncallers. In short, Darwin imagined a process of natural selection among groups in addition to a process of natural selection among individuals within groups. Groups can evolve into adaptive units, but only if there is a process of group-level selection that is stronger than the often opposing process of individual-level selection. This was the origin of an approach to evolutionary biology called multilevel selection theory (reviewed by Sober and Wilson, 1998). The general rule that emerges from multilevel selection theory is that "adaptation at level x of the biological hierarchy requires a corresponding process of natural selection at level x."

History of Multilevel Selection Theory. Despite Darwin's insight and cautious use of group selection to explain phenomena such as social insect colonies and human morality, many biologists continued to portray higher-level units as adaptive without attention to evolutionary mechanisms. Phrases such as *for the good of the group* and *for the good of the species* were used along with *for the good of the individual*, as if there were no need to distinguish among levels of selection. When authors demonstrated an awareness of multilevel selection, they often assumed without good reason that higher-level selection was strong enough to prevail over lower-level selection. These issues came to a head in the 1960s, resulting in a consensus that higher-level selection is almost invariably weak compared with lower-level selection and can be ignored for most purposes (Williams, 1966). It is important to emphasize that the theory of individual selection, as it was called, did not alter the basic logic of multilevel selection theory. Individual selectionists have always acknowledged the theoretical possibility of higher-level selection and have based their argument on the strength of individual-level selection compared with higher-level selection.

The rejection of higher-level selection was regarded as a major advance, similar to the rejection of Lamarckism, allowing evolutionary biologists to close the door on one set of possibilities and concentrate their attention elsewhere. Nevertheless, developments since the 1960s have shown that the rejection of higher-level selection was premature. Theoretical models, laboratory experiments, and field studies all point to adaptations that evolve by increasing the relative fitness of higher-level units, despite being selectively neutral or disadvantageous within those units. This does not mean that "for the good of the group" thinking can return in its earlier uncritical form, but it does mean that higher-level selection must be considered as a legitimate possibility in the evolution of many traits. If individual-level selection is not invariably stronger than higher-level selection, there is no recourse but to return to a fully multilevel view of the evolutionary process.

Defining units of selection. For multilevel selection theory to be a productive research program, it must clearly define units of selection. At first this might seem like a daunting task. Higher-level adaptations have been imagined everywhere, from bird flocks, ant colonies, and fish schools at one end of the spectrum to entire species and ecosystems at the other end. Nevertheless, this diversity exists only when we consider a diversity of traits. When we consider the evolution of a single trait, there is much less ambiguity about the appropriate unit(s) of selection.

Returning to our bird example, the choice of a single flock to study the evolution of warning calls appeared so natural as to be unquestioned. This is because it is assumed as part of the example that all members of a flock are at risk from the approaching predator and hear the cry of any member of the flock. Uttering a cry alters fitness within the flock and has no effect on other flocks. If it turned out that the predator targets several flocks at once and that a warning cry is heard by all the flocks, we would need to change our definition of groups. Similarly, if it turned out that a cry is heard only by the caller's nearest neighbor rather than the whole flock, we might need to change our definition of groups yet again. The reason that the definition of groups is so closely tied to the details of the trait is because we are trying to predict the evolution of the trait. When the trait is a nonsocial behavior that alters the fitness of the individual alone, we need not concern ourselves with groups. But when the trait is a social behavior, the fitness of an individual is often a function of its own trait and the traits of the individuals with whom it interacts. The set of individuals with whom it interacts constitutes its group, which must be identified accurately to calculate the fitnesses that determine the outcome of evolution. This is a requirement for all evolutionary theories of social behavior, not just multilevel selection theory.

It follows that groups must be defined separately for each and every trait. Suppose that we decided to study resource conservation in the same species of bird. Individuals that eat moderately have fewer babies than immoderate eaters, but moderate groups persist while immoderate groups overexploit their resources and become extinct (as proposed by Wynne-Edwards, 1962, 1986). This is the same problem of altruism and selfishness that exists for warning cries, but we need to find the appropriate grouping for resource conservation. Bird flocks may be appropriate if they live on exclusive territories, but not if many flocks share the same resources. If the birds live on an archipelago of islands, then perhaps a single island is the appropriate group for this particular trait. However, group selection may be less effective at the scale of islands than at the scale of bird flocks. If so, then the birds might behave altruistically with respect to warning cries but not with respect to feeding.

Sometimes the relevant groupings have discrete boundaries but sometimes they do not. Sharp boundaries exist for parasites in hosts but not for plants in fields. In both cases, however, interactions are localized and influence the fitness of a small subset of the total population. It is the localized nature of fitness-determining interactions, not sharp boundaries, that forms the basis of multilevel selection theory. The fundamental question is whether adaptations evolve, not by increasing the fitness of the individual relative to its immediate neighbors, but by increasing the fitness of larger collectives relative to other larger collectives.

Comparing fitnesses within and among units.
Once the relevant unit(s) of selection have been determined for a particular trait, the next task is to evaluate the direction and strength of natural selection at each level of the hierarchy. Natural selection at each level is based on the relative fitness of units within the next higher unit. Our bird example describes a two-tier hierarchy, with individuals in social groups and social groups in a metapopulation (a population of groups). First we compared the relative fitness of callers versus noncallers within single groups, concluding that natural selection at this level favored the trait of noncalling. Then we compared the relative fitness of calling groups versus noncalling groups in a metapopulation, concluding that natural selection at this level favored the trait of calling. The relative strength of these opposing evolutionary forces can be influenced by many factors. However, if calling is observed and if we correctly understand the facts of the matter, we can tentatively conclude that calling evolved by group-level selection, despite being selectively disadvantageous within groups.

It is possible for seemingly altruistic behaviors to be individually selfish upon closer inspection. Continuing with our bird example, suppose that uttering a call does not increase the risk of being attacked by an approaching predator. On the contrary, calling advertises to the predator that it has been spotted and that a less vigilant member of the group should be targeted. If these are the facts of the matter, then the evolution of so-called warning calls could be explained entirely by within-group selection. Calling individuals survive and reproduce better than noncalling individuals in the same group. The function of the adaptation is not to warn other members of the group but to communicate with the predator in a way that actually endangers other members of the group. We would be right to reject group selection in this case. Numerous examples of seemingly altruistic behaviors can probably be explained in this fashion. There is no substitute for knowing the facts of the matter, and the way to do this is to rigorously compare relative fitnesses within and among groups.

In addition, however, a subtle shift in perspective can make even group-selected traits appear individually selfish. The shift involves averaging the fitness of individuals across groups, rather than comparing their fitnesses within single groups. To pick an extreme example, imagine a flock of birds with one caller and nine noncallers. All members have a low fitness in this flock because only one bird is looking out for predators; however, the caller has the lowest fitness of all. Let us say that the chance of surviving predators is 50 percent for the noncallers and 25 percent for the caller. A second flock of birds has nine callers and one noncaller. All members in this flock

have a high fitness because so many members are looking out for predators; however, the shirking noncaller has the highest fitness of all. Let us say that the chance of surviving predators is 100 percent for the noncaller and 75 percent for the callers. When we compare the fitness of callers and noncallers within each group, we see that callers are the losers in both cases. However, the group with more callers fares better than the group with fewer callers. This is the classic group selection scenario that began with Darwin. Now let us calculate the average survival of callers and noncallers across the groups. One noncaller has a survival of 100 percent and nine have a survival of 50 percent, for an average survival of 55 percent. One caller has a survival of 25 percent and nine have a survival of 75 percent, for an average survival of 70 percent. The average caller is more fit than the average noncaller, so why not say that calling evolves by individual selection? Here, the need to invoke group selection appears to vanish. Of course, the disappearance is an illusion. The need for multiple groups and variation among groups is essential for the calling behavior to evolve.

To summarize, multilevel selection theory requires a certain kind of fitness comparison based on the relative fitness of lower-level units within higher-level units. Individual-level selection is defined by multilevel selection theory (and always has been) as the relative fitness of individuals within single groups. Group-level selection is the relative fitness of groups within a metapopulation. Extending the hierarchy downward, gene-level selection can be defined as the relative fitness of genes within a single individual organism. These definitions allow natural selection at each level of the hierarchy to be clearly identified and studied. However, averaging the fitness of individuals across groups makes group-level adaptations seem to disappear and to reappear as a form of individual selfishness. And averaging the fitness of genes across individuals and groups makes group- and individual-level adaptations seem to disappear and to reappear as a form of gene selfishness. Much of the disagreement about multilevel selection can be traced to this subtle shift of perspective.

Examples of higher-level adaptations.
Armed with a clear-cut definition of units and the appropriate fitness comparisons, multilevel selection becomes a practical method for studying evolution, with little reason for controversy. Once again, the fundamental question is whether adaptations evolve by increasing the relative fitness of higher-level units, despite being selectively neutral or disadvantageous at lower levels of the biological hierarchy. Many examples of higher-level adaptations have been documented, some of which will be listed here.

Sex ratios. It has been known since Fisher (1930) that natural selection within single groups favors an equal investment in males and females. However, groups with female-biased sex ratios grow faster than groups with male-biased sex ratios. Female-biased sex ratios exist in many species, and it has become uncontroversial to regard them as a product of among-group selection. For example, Herre (1999, p. 217) states that "the key to understanding how much greater or lesser degrees of female bias are favored lies in understanding how among- and within-group selection are balanced."

Disease evolution. Disease organisms obviously live in a multigroup population structure. The bacterial or viral population within a single host can persist for many hundreds of generations and can achieve densities greater than the human population on earth. The traits that maximize the relative fitness of a disease strain within a single host are very different from the traits that maximize the contribution of the whole population to the infection of future hosts. The use of multilevel selection to study disease evolution has become uncontroversial. Bull (1994, p. 1425) states that "both levels operate together in what may be regarded as a classic group selection hierarchy, the groups being the populations of parasites within hosts. . . . To understand the complexities of this two-level evolutionary process, it is easiest to dissect the process into separate between-host and within-host components, studying each component before assembling them."

Brave and cowardly lions. One of Darwin's key passages on group selection is about human bravery in battle as a group-selected trait. Recent research on lions has revealed individual differences in bravery, appearing to confirm Darwin's intuition. The following passage from Packer and Heinsohn (1996, p. 1216; see also Heinsohn and Packer, 1995) clearly describes the fitness consequences of bravery and cowardice for multilevel selection:

> Female lions share a common resource, the territory; but only a proportion of females pay the full costs of territorial defense. If too few females accept the responsibilities of leadership, the territory will be lost. If enough females cooperate to defend the range, their territory is maintained, but their collective effort is vulnerable to abuse by their companions. Leaders do not gain "additional benefits" from leading, but they do provide an opportunity for laggards to gain a free ride.

Building a mat and a tower. The bacteria *Pseudomonas fluorescens* lives on the surface of the water by producing a polymer that forms into a mat. Nonproducing "free-riders" evolve by mutation and replicate faster than the "altruistic" producers. When the proportion of cheaters becomes great enough, the mat disintegrates and the entire colony sinks to the bottom (Rainey and Travisano, 1998). This is a classic multilevel scenario involving fitness differences within and among groups.

Similarly, the amoeba *Dictyostelium discoideum* has a solitary foraging stage and a social stage in which groups form and differentiate into nonreproductive stalk cells and reproductive spores. The group-level advantage of this reproductive division of labor and the potential for cheating are obvious, and are currently being studied from a multilevel perspective (Strassman et al., 2000).

The major transitions of life. One of the most important developments in multilevel selection theory has been the interpretation of individual organisms as themselves higher-level units in the biological hierarchy (Maynard Smith and Szathmary, 1995). Evolution occurs not only by mutational change but by social groups becoming so functionally integrated that they become the new organism. It is likely that these coalescing events have occurred repeatedly throughout the history of life, including the origin of life itself as a social group of cooperating molecular reactions (Michod, 1999). Thinking of individual organisms as higher-level units vastly expands the domain of multilevel selection theory to include topics such as genetics and development, which involve the same potential conflicts between levels of selection originally envisioned for individuals in social groups.

Human social groups. It is likely that functionally organized human societies can be interpreted as a product of genetic and cultural group selection operating both in the distant past, shaping our innate psychology, and in the present, shaping our recently evolved cultural institutions (Boehm, 1999; Boyd and Richerson, 1985; Sober and Wilson, 1998; Wilson, 2002).

Ecosystem-level selection. It may seem extraordinary to think of whole ecosystems as units of selection. However, when replicate microbial ecosystems are established in the laboratory, they vary in their phenotypic properties despite the fact that they were initiated by hundreds of species numbering millions of individuals. Furthermore, when new ecosystems are formed from old ecosystems, in at least some cases the "offspring" share the phenotypic properties of the "parents" (Swenson, Wilson, and Elias, 2000; Swenson, Arendt, and Wilson, 2000; Goodnight, 2000). These and other multilevel selection experiments in the laboratory (reviewed by Goodnight and Stevens, 1997) reveal that even ecosystems can possess the fundamental ingredients of phenotypic variation and heritability that allows their properties to be molded by natural and artificial selection. With respect to natural selection at the level of microbial ecosystems, in an article on quorum sensing (communication among microbes), Decho (1999) concludes that "a new concept is emerging concerning the development and evolution of bacterial communities. The concept addresses bacteria, not as individuals, but rather as an 'organism.' The concept promotes 'self-organization' and cooperativity among cells as a driving force in community development, rather than the classical natural selection of individuals."

Future of Multilevel Selection Theory. Multilevel selection is a powerful theoretical framework that can be applied to an extraordinary range of phenomena, from the origin of life to human morality. Despite its turbulent past, it is increasingly being accepted as a part of mainstream evolutionary biology. Some authors refer to an "old" and "new" version of multilevel selection theory to distinguish between the ideas that were rejected in the 1960s and those that have become acceptable today. This is a simplistic view of scientific change, since the basic logic of multilevel selection theory has remained remarkably constant from Darwin to the present. The question has always been whether adaptations can evolve by increasing the fitness of higher-level collectives, despite being selectively neutral or disadvantageous within those collectives. The answer that has emerged is yes but only sometimes. Higher-level adaptations cannot be categorically accepted or rejected. The only recourse is to study evolution in all its multilevel complexity.

BIBLIOGRAPHY

Boehm, C. *Hierarchy in the Forest: Egalitarianism and the Evolution of Human Altruism.* Cambridge, Mass., 1999.
Boyd, R., and P. J. Richerson. *Culture and the Evolutionary Process.* Chicago, 1985.
Bull, J. J. "Virulence." *Evolution* 48 (1994): 1423–1437.
Darwin, C. *The Descent of Man and Selection in Relation to Sex.* New York, 1871.
Decho, A. W. "Chemical Communication within Microbial Biofilms: Chemotaxis and Quorum Sensing in Bacterial Cells." In *Microbial Extracellular Polymeric Substances*, edited by J. Wingender, T. Neu, and H.-K. Fleming. Berlin, 1999.
Fisher, R. A. *The Genetical Theory of Natural Selection.* Oxford, 1930. Repr. New York, 1958.
Goodnight, C. J. "Heritability at the Ecosystem Level." *Proceedings of the National Academy of Sciences* 97 (2000): 9365–9366.
Goodnight, C. J., and L. Stevens. "Experimental Studies of Group Selection: What Do They Tell Us about Group Selection in Nature?" *American Naturalist* 150 (1997): S59–S79.
Heinsohn, R., and C. Packer. "Complex Cooperative Strategies in Group-Territorial African Lions." *Science* 269 (1995): 1260–1262.
Herre, E. A. "Laws Governing Species Interactions? Encouragement and Caution from Figs and their Associates." In *Levels of Selection in Evolution*, edited by L. Keller, pp. 209–237. Princeton, 1999.
Maynard Smith, J., and E. Szathmáry. *The Major Transitions of Life.* New York, 1995.
Michod, R. E. *Darwinian Dynamics.* Princeton, 1999.
Packer, C., and R. Heinsohn "Lioness Leadership." *Science* 271 (1999): 1215–1216.
Rainey, P. B., and M. Travisano. "Adaptive Radiation in a Heterogeneous Environment." *Nature* 394 (1998): 69–72.
Sober, E., and D. S. Wilson. *Unto Others: The Evolution and Psychology of Unselfish Behavior.* Cambridge, Mass., 1998.
Strassman, J. E., Y. Zhu, and D. C. Queller. "Altruism and Social Cheating in the Social Amoeba Dictyostelium Discoideum." *Nature* 408 (2000): 965–967.
Swenson, W., J. Arendt, and D. S. Wilson. "Artificial Selection of Microbial Ecosystems for 3-Chloroaniline Biodegradation." *Environmental Microbiology* 2 (2000): 564–571.
Swenson, W., D. S. Wilson, and R. Elias. "Artificial Ecosystem Selection." *Proceedings of the National Academy of Sciences* 97 (2000): 9110–9114.
Williams, G. C. *Adaptation and Natural Selection: A Critique of Some Current Evolutionary Thought.* Princeton, 1966.
Wilson, D. S. *Darwin's Cathedral: Evolution, Religion, and the Nature of Society.* Chicago, 2002.
Waynne-Edwards, V. C. *Animal Disperson in Relation to Social Behavior.* Edinburgh, 1962.
Waynne-Edwards, V. C. *Evolution through Group Selection.* Oxford, 1986.

— DAVID SLOAN WILSON

GYMNOSPERMS. *See* Plants.

H

HALDANE, JOHN BURDON SANDERSON

J. B. S. Haldane (1892–1964), population geneticist, came from a Scottish family of considerable intellectual and political prominence. His father, the noted physiologist, J. S. Haldane, introduced him to biological research at an early age. Though formally trained at Oxford in mathematics and classics, Haldane became a physiologist and biochemist at Cambridge in the 1920s. During this period, he made important contributions to theoretical enzyme kinetics and pioneered the model that an enzyme acts on a substrate by exerting a strain at the active site. Independent of A. I. Oparin, in 1929 Haldane proposed that life originated in a "hot dilute soup" in a reducing atmosphere through the action of ultraviolet rays.

Haldane's most important scientific contributions were in mathematical population genetics, evolutionary theory, and human genetics. In 1919 he proposed a method to compute relative distances between genes along a chromosome (linkage map) using recombination frequencies. In 1922, he proposed what came to be known as Haldane's rule of interspecific hybrids: "When in the F_1 offspring of two different animal races [species] one sex is absent, rare, or sterile, that sex is the heterozygous [heterogametic] sex." In 1924, Haldane began publishing a series of ten papers, "A Mathematical Theory of Natural and Artificial Selection." The first one in this series presented the most comprehensive analysis of models of selection on populations to date and also laid down what became the standard framework for constructing such models. According to Haldane, the purpose of these models is to calculate the rate of change of a trait in a population based on the differences in reproduction between individuals with different values of the trait, that is, the rate of evolution based on the intensity of natural selection. From the observed rate of evolutionary change in the color of the peppered month (*Biston betularia*), Haldane used his mathematical models to predict that the dominant dark form reproduced one and a half times as much as the recessive light form. Such a huge reproductive difference was unexpected. However, this prediction was confirmed by H. B. D. Kettlewell in the 1950s, becoming the first significant confirmation of mathematical population genetics in the field.

The fifth paper in this series derived an equation for the balance between mutation and selection. In the early 1930s, Haldane used this equation to provide the first estimate of the spontaneous mutation rate of a human gene and prophetically argued that this rate differed in the two sexes. The eighth paper anticipated significant parts of Sewall Wright's shifting balance theory of evolution. Haldane collected his insights in a 1932 book, *The Causes of Evolution*. This was the first work to link population genetics with the work of T. H. Morgan and his students on the physical basis of heredity. The mathematical appendix collected together the results of Ronald Aylmer Fisher, Haldane, and Sewall Wright; it also provided the first analysis of kin selection. Fisher's *Genetical Theory of Natural Selection* had appeared in 1930; Wright's long paper "Evolution in Mendelian Populations" had appeared in 1931. With the publication of *Causes*, an important phase of the development of modern evolutionary theory was completed.

Moving to University College London, Haldane turned in the 1930s to human genetics. In 1936, he published the first linkage map of a human chromosome, placing six loci at relative positions on the X chromosome. In a 1937 paper he showed that it is the mutation rate and not the intensity of its effect that determines a mutation's effect on a population. This principle came to be called "mutation load" after it was rediscovered by H. J. Muller in 1950. Finally, in a 1957 paper, "The Cost of Selection," Haldane showed that the amount of selection required to substitute an allele in a population depended on its initial frequency, not on the intensity of selection. This principle was used by the geneticist Motoo Kimura in his initial argument for the neutral theory of molecular evolution.

Haldane emigrated to India in 1956, eventually becoming an Indian citizen. He died there of cancer in Bhubaneshwar in 1964.

Besides his scientific work, Haldane was a prolific science popularizer. In one such work, *Daedalus* (1922), he raised the possibility of ectopic birth (outside the body). For about ten years, he was a member of the Communist Party, eventually quitting over the Lysenko affair. Many of his short popular articles were written for the party's *Daily Worker* and were collected into ten volumes. These popular writings have had almost as much influence as Haldane's work in evolutionary theory.

[*See also* Genetic Load; Haldane's Rule; Senescence; Shifting Balance.]

BIBLIOGRAPHY

Clark, R. W. *J. B. S.: The Life and Work of J. B. S. Haldane.* New York, 1969.

Crow, J. F. "Centennial: J. B. S. Haldane, 1892–1964." *Genetics* 130 (1992): 1–6.

Dronamraju, K. R., ed. *Selected Genetic Papers of J. B. S. Haldane.* New York, 1990.

Haldane, J. B. S. *Daedalus.* London, 1922.

Haldane, J. B. S. *Possible Worlds and Other Papers.* New York, 1928.

Haldane, J. B. S. *Enzymes.* London, 1930.

Haldane, J. B. S. *The Causes of Evolution.* London, 1932.

Haldane, J. B. S. *The Inequality of Man and Other Essays.* London, 1932.

Haldane, J. B. S. *Heredity and Politics.* London, 1938.

Haldane, J. B. S. *The Marxist Philosophy and the Sciences.* New York, 1939.

Haldane, J. B. S. *New Paths in Genetics.* New York, 1942.

Haldane, J. B. S. *The Biochemistry of Genetics.* New York, 1954.

Provine, W. B. *The Origins of Theoretical Population Genetics.* Chicago, 1971.

Sarkar, S., ed. *The Founders of Evolutionary Genetics.* Dordrecht, The Netherlands, 1992.

Sarkar, S. "Science, Philosophy, and Politics in the Work of J. B. S. Haldane, 1922–1937." *Biology and Philosophy* 7 (1992): 385–409.

Wright, S. "Evolution in Mendelian Populations." *Genetics* 16 (1931): 97–159.

— SAHOTRA SARKAR

HALDANE'S RULE

"When in the F1 offspring of two different animal races one sex is absent, rare, or sterile, that sex is the heterozygous (= heterogametic) sex." This rule was formulated by J. B. S. Haldane, who made a seminal contribution to the study of speciation, as well as to many other fields of science. Nearly eighty years after Haldane's original proposal, his "rule" remains the best-known empirical generalization about the patterns of speciation, and it remains an active topic of research. According to this rule, unisexual sterility and inviability in first-generation (F1) hybrids will be more severe or more prevalent in the sex with two different sex chromosomes, called the heterogametic sex. In mammals, the heterogametic sex (symbolized XY) is male, and the homogametic sex (XX) is female. Most insects, including the fruit fly (*Drosophila* spp.), also have heterogametic males; in these species, sterility of hybrids is almost entirely confined to males. The pattern is reversed in birds and in the insect order Lepidoptera (butterflies and moths): the heterogametic sex is female (ZW), and the homogametic sex is male (ZZ). Haldane's rule applies in many different groups of animals, both those with heterogametic males and those with heterogametic females, and in vertebrates as well as invertebrates. Biologists studying speciation have been drawn to Haldane's rule precisely because of its generality; if the same patterns appear to apply in very different animals, then perhaps the process of speciation itself has general properties.

Since Haldane's original formulation, associated observations and phenomena have been found. In groups with heterogametic males, there are fewer exceptions to the sterility aspect of Haldane's rule than to the viability aspect; that is, a number of crosses produce viable but sterile males. Moreover, hybrid sterility generally evolves faster than hybrid inviability in these groups. There are also morphological and behavioral equivalents of Haldane's rule. For instance, in some crosses between species of flour beetles, male (heterogametic) hybrids more often have missing limbs and other morphological abnormalities than do their homogametic sisters. Hybrid males in some crosses between different *Drosophila* species sometimes display mating or other behavioral dysfunctions.

Haldane's Rule as a Composite Phenomenon. Biologists are interested not just in finding patterns, but also in formulating and testing explanations of those patterns. What explains Haldane's rule? Is there a single explanation, or several? Are the explanations for Haldane's rule the same or different for male versus female heterogametic taxa? Does the same mechanism account for the patterns of viability versus sterility? After a long lull, beginning around 1985 evolutionary geneticists pursued these questions with increasing intensity. During an exciting and occasionally rancorous decade-long period, various explanations were presented, tested, discarded, and sometimes resurrected.

By the mid-1990s, evolutionary geneticists studying Haldane's rule reached a general consensus: Haldane's rule is a composite phenomenon with several explanations. Although different researchers may emphasize different explanations, they generally agree that the patterns of Haldane's rule and associated phenomena result from three main processes:

1. The alleles contributing to hybrid dysfunction are at least somewhat recessive in their adverse effects in hybrids. This is called the "dominance theory." According to this model, the effects of recessive deleterious alleles in the homogametic F1 hybrids will be masked at least in part by alleles from the other parent. In the heterogametic hybrids, there is no masking—these deleterious alleles will manifest their full effects. Note that these alleles are deleterious and recessive only in the hybrid genetic background. In the pure species background, they are either neutral or advantageous, and their dominance relations are irrelevant. The dominance theory will promote Haldane's rule in both male and female heterogametic groups and will affect all aspects (sterility, viability, morphological, and behavioral) of the rule.

2. Genes contributing to hybrid male sterility evolve faster than do those contributing to hybrid female sterility. This is sometimes called the "faster male theory" and presumably arises from greater sexual selection operating on males than on females. According to this model, selection will fix alleles that affect male reproductive systems. These alleles will tend to have, in the hybrid genetic background, pleiotropic negative effects on the reproductive systems of male hybrids. The faster male theory will promote Haldane's rule in male heterogametic groups and work against Haldane's rule in female heterogametic groups. This process should affect only the sterility and not the viability aspect of Haldane's rule.

3. Negative interactions between the X chromosome and cytoplasm can contribute to dysfunction in female hybrids. X–cytoplasm interactions are not possible in male hybrids because they receive both their X and cytoplasm from their mother. In contrast, because female hybrids have one X chromosome from their mother and one from their father, interactions between their paternally derived X and the cytoplasm are possible. This process will attenuate Haldane's rule in groups with heterogametic males and promote the rule in groups with homogametic taxa. It generally affects the viability aspect of the rule more than the sterility aspect.

Both the dominance theory and the faster male theory have been subjected to sophisticated mathematical analysis. These analyses show that Haldane's rule can result even when the alleles causing hybrid incompatibility are dominant, provided that the evolution of male-specific incompatibility in hybrids is sufficiently faster than that of female-specific incompatibility. Haldane's rule is also likely to apply in female heterogametic groups, despite the faster evolution of male-specific incompatibility, if the incompatibility alleles are sufficiently recessive.

Support for the Explanations. Several lines of evidence, coming mostly from experimentalists working with *Drosophila*, support the dominance theory. One major line of evidence came from the use of special genetic strains in which females have their two X chromosomes attached to each other. These crosses allow for the production of "unbalanced" female hybrids that have both of their X chromosomes from a single parent. While normal female hybrids (with one X chromosome from each species) are viable, these "unbalanced" females are inviable, like their heterogametic brothers. Orr (1997) discusses other lines of support for the dominance theory.

A study by Presgraves and Orr (1998) of the patterns of speciation observed in two genera of mosquitoes has bolstered the dominance theory. The genus *Anopheles* has a sex-determination pattern like those of *Drosophila* and mammals, in which males possess degenerate Y chromosomes. In the other genus, *Aedes*, females are XX and males are XY, but the X and Y chromosomes share virtually all of their genetic loci, differing only in a small sex-determination region. Because of the genetic similarity of the X and Y chromosomes in *Aedes*, F1 males are not more unbalanced than F1 females with respect to chromosome composition, and the dominance theory should not contribute to the inviability aspect of Haldane's rule in this genus. In fact, whereas *Anopheles* follows Haldane's rule for inviability, *Aedes* does not—consistent with the expectations of the dominance theory.

Several lines of research also support the faster male theory. Both in mammals and in *Drosophila*, cases of hybrid male sterility greatly outnumber those of hybrid female sterility. In the best-studied crosses between different species of *Drosophila*, the number of regions of the genome contributing to hybrid male sterility is greater than those contributing to hybrid female sterility by a ratio of about 10:1. This is in stark contrast to the fact that mutagenesis within species results in far more mutations that cause lethality than in mutations that cause sterility. Moreover, genes that are expressed in the male reproductive tract sometimes evolve very quickly, suggesting the action of strong sexual selection. Unlike the dominance theory, the faster male theory is not affected by the mode of chromosomal sex determination. Whereas the *Aedes* mosquito genus does not follow Haldane's rule for inviability, both it and *Anopheles* follow Haldane's rule for sterility.

There are fewer cases of heterogametic inviability than of sterility in *Drosophila*, and also more cases that tend in the direction opposite that predicted by Haldane. Interactions between X chromosome and cytoplasm cause most of these exceptions, according to the consensus view. The best-known exception to Haldane's rule occurs in the cross between *D. melanogaster* males and *D. simulans* females, where F1 females are usually inviable but F1 males are usually viable. Sawamura and colleagues have shown that the inviability of these hybrid females is caused by a negative interaction between an X chromosome allele from *D. melanogaster* and *D. simulans* cytoplasm. The viability of F1 males from this cross is not affected because these males do not have a *D. melanogaster* X chromosome.

Unanswered Questions. What are the biochemical and physiological reasons for the dominance relationships seen among alleles in hybrids? At present, dominance relations can only be treated as a black box; virtually nothing is known about them at a functional level.

Why is there an acceleration of hybrid male sterility in groups with heterogametic males? Sexual selection, including cryptic sexual selection (e.g., sperm competition), is the generally accepted answer, but further study is needed.

[*See also* Haldane, John Burdon Sanderson.]

BIBLIOGRAPHY

Haldane, J. B. S. "Sex-ratio and Unisexual Sterility in Hybrid Animals." *Journal of Genetics* 12 (1922): 101–109.

Laurie, C. C. "The Weaker Sex Is Heterogametic: 75 Years of Haldane's Rule." *Genetics* 147 (1997): 937–951.

Noor, M. A. F. "Genetics of Sexual Isolation and Courtship Dysfunction in Male Hybrids of *Drosophila pseudoobscura* and *D. persimilis*." *Evolution* 51 (1997): 809–815.

Orr, H. A. "Haldane's Rule." *Annual Review of Ecology and Systematics* 28 (1997): 195–218.

Presgraves, D. C., and H. A. Orr. "Haldane's Rule in Taxa Lacking a Hemizygous X." Science 282 (1998): 952–954.

Sawamura, K., T. Taira, and T. K. Watanabe. "Hybrid Lethal Systems in the *Drosophila melanogaster* Species Complex. I. The *maternal hybrid rescue (mhr)* Gene of *Drosophila simulans*." *Genetics* 133 (1993): 299–305.

Turelli, M., and H. A. Orr. "Dominance, Epistasis, and the Genetics of Postzygotic Isolation." *Genetics* 154 (2000): 1663–1679.

Turelli, M., and H. A. Orr. "The Dominance Theory of Haldane's Rule." *Genetics* 140 (1995): 389–402.

Wade, M. J., N. A. Johnson, R. Jones, V. Siguel, and M. McNaughton. "Genetic Variation Segregating in Natural Populations of *Tribolium castaneum* Affecting Traits Observed in Hybrids with *T. freemani*." *Genetics* 147 (1997): 1235–1247.

Wu, C.-I., and A. W. Davis. "Evolution of Postmating Reproductive Isolation: The Composite Nature of Haldane's Rule and Its Genetic Bases." *American Naturalist* 142 (1993): 187–212.

Wu, C.-I., N. A. Johnson, and M. F. Palopoli. "Haldane's Rule and Its Legacy: Why Are There So Many Sterile Males?" *Trends in Ecological Evolution* 11 (1996): 281–284.

— NORMAN A. JOHNSON

HAMILTON, WILLIAM D.

W. D. Hamilton was one of the greatest evolutionary theorists since Darwin. Certainly, where social theory based on natural selection is concerned, he was easily our deepest and most original thinker.

His first work (1964)—his theory of inclusive fitness—was his most important, because it is the only true advance since Darwin in our understanding of natural selection. Hamilton's work is a natural and inevitable extension of Darwinian logic. In Darwin's system, natural selection refers to individual differences in reproductive success (RS) in nature, where RS is the number of surviving offspring produced. Hamilton enlarged the concept to include RS effects on other relatives; that is, not just fitness or reproductive success but inclusive fitness, defined (roughly) as an individual's RS plus effects on the RS of relatives, each devalued by the appropriate degree of relatedness (r).

This idea had been briefly advanced by R. A. Fisher and J. B. S. Haldane, but neither took it seriously and neither provided any kind of mathematical foundation. That foundation was not as obvious as it sounds. For a rare altruistic gene, it is clear that Br > C will give positive selection, where B is the benefit conferred and C the cost suffered; but the matter is not so obvious at intermediate gene frequencies. As the altruistic gene spreads, should not the criterion for positive selection be relaxed?

Hamilton showed that the answer is "no" and that his simple rule worked for all gene frequencies. He once told the story of sitting down as a doctoral student to write to Haldane, but to formulate each question precisely he had to do additional work and after a couple of years of such work he never sent the letter because by then he had worked out all the answers himself. A noteworthy implication of Hamilton's work was that in almost all species the individual was no longer expected to have a unitary self-interest, because genetic elements are inherited according to different rules (contrast paternal transmission of the Y chromosome with maternal transmission of mitochondrial DNA).

He soon followed this work with major advances in understanding selection acting on the sex ratio, the moulding of senescence by natural selection, the aggregation and dispersal of organisms, the evolution of the social insects, the evolution of dimorphic males, and the origin of higher taxonomic units in insects. For the latter he argued that the more-or-less closed spaces created by rotting wood imposed a system of small, inbred subpopulations in insects inhabiting it, leading to a great diversity of homozygous forms, often with arbitrary, novel characters (such as a second complete metamorphosis in many male scale insects). In 1981 with Robert Axelrod he laid the mathematical foundation for the study of reciprocal altruism, when they showed that the simple rule of tit-for-tat in playing iterated games of Prisoner's Dilemma was itself evolutionarily stable.

Twenty years ago Hamilton began to devote most of his time to the theory that parasites play a key role in generating sexual reproduction in their hosts, recombination being a defence against very rapidly and antagonistically co-evolving parasites. In his memorable phrase, sexual species are "guilds of genotypes committed to free, fair exchange of biochemical technology for parasite exclusion." He was not the first to advance this theory but he took it more seriously than others and he worked most successfully to define the form of the argument as well as its implications. Notable here was his work with Marlene Zuk on parasites as a key to mate choice. In 1982 they showed that species of birds with higher loads of blood parasites showed more colour and complex song, an unexpected finding unless parasite-rich environments favoured mate choice for these traits, thereby driving up their average value.

It is hard to capture on paper the beauty of the man

and the reason that so many evolutionists felt such a deep personal connection to Bill Hamilton. He had the most subtle, multi-layered mind I have ever encountered. What he said often had double and even triple meanings so that, while the rest of us speak and think in single notes, he thought in chords. He was modest in style, with a warm sense of humour. For example, he had no illusions about the clarity of his lecturing style, and once told a class we taught at Harvard that after hearing him lecture they would wonder whether he understood even his own ideas.

His letters were laced with humorous asides. He once sent me a news clipping of a human father-to-son testicle transplant, along with the comment, "New vistas for parent-offspring confict?" The last time I saw Bill, at Oxford in December 1998, he pointed with pride to the two, and possibly three, species of moss growing on his Volvo—indeed on its windows—and told me that this was a clear advantage of Oxford over Cambridge, the latter climate being too dry. (He had come to Oxford University in 1984, after seven years at the University of Michigan and 13 at Imperial College, London.)

Bill Hamilton was a naturalist of legendary knowledge, especially of insects, but he was also an acute observer of human behaviour, right down to the minutiae of your own actions in his presence. Had I noticed, he asked, that lopsided facial expressions in humans are almost exclusively male? (No, but I have seen it a hundred times since then!) He was an evolutionist to the core, and was always heartened by news of fellow evolutionists enjoying some reproductive success. In a similar spirit I take joy in the lives of his three daughters, Helen, Ruth and Rowena, not to mention his many surviving siblings. But the loss of this "gentle giant" is very great. Bill died at the age of 63 on 7 March 2000, from complications after contracting malaria during fieldwork in the Congo in January, work which was designed to locate more exactly the chimpanzee populations that donated HIV-1 to humans, as well as the mode of transmission. He had been strong in mind, body and spirit, with many new projects and thoughts under way. He will be sorely missed for many years to come.

— ROBERT TRIVERS

Reprinted by permission from *Nature* Vol. 404 (2000): 828. Copyright ©2000 Macmillan Magazines Ltd.

HAPLODIPLOIDY. *See* Eusociality.

HARDY–WEINBERG EQUATION

The Hardy–Weinberg law describes the expected proportions or frequencies of genotypes that arise with random mating in large diploid populations. The relationship was discovered independently in 1908 by G. H. Hardy, a British mathematician, and W. Weinberg, a German geneticist. Its value lies in describing a baseline expectation for gene frequencies in populations when it is assumed that selection is not operating. The assumptions underlying the relationship are that population size is infinite, mating is random, and there is no selection, migration, or mutation. The relationship is sensitive to deviations from random mating and to some but not all forms of selection, but robust to violations of the other assumptions. For example, the relationship provides accurate frequencies for populations in excess of 100 breeding individuals, and it is not biased by rates of mutation typical for most nuclear genes.

Consider a gene with two alleles, A and a, whose frequencies are p and q, so that $p + q = 1$. If these alleles are joined randomly (as would result from random mating in the population), then the probabilities of the diploid genotypes are given by the expansion of the binomial:

$$(p + q)^2 = p^2 + 2pq + q^2, \qquad (1)$$

indicating that the frequencies of the AA, Aa, and aa genotypes are expected to be p^2, $2pq$, and q^2, respectively. Thus, if allele A is found in 50 percent of individuals, it is expected that 25 percent ($.5^2$) of offspring will carry two copies (be homozygous) of allele A, and so on for the other genotypes. The Hardy–Weinberg law may be illustrated in a more intuitive manner with a contingency table. Imagine reproduction in the blue mussel, *Mytilus edulis*, which releases its gametes directly into the water. The gametes are mixed and joined randomly by the swirling ocean water. Once again, if allele A has frequency p, and allele a has frequency $1 - p = q$, then the probability of any genotype in the contingency table is given by the product of the frequencies in the appropriate row and column:

		Sperm	
		$F(A) = p$	$F(a) = q$
Eggs	$F(A) = p$	$P(AA) = p^2$	$P(Aa) = pq$
	$F(a) = q$	$P(Aa) = pq$	$P(aa) = q^2$

Thus, the probabilities of AA, Aa, and aa being formed by the random union of gametes are p^2, $2pq$, and q^2, respectively.

Random union of gametes with frequencies p and q produces genotypic proportions of p^2, $2pq$, and q^2, and production of haploid gametes from this population yields gametes with frequencies p and q. Thus, when the assumptions underlying the Hardy–Weinberg law are met, allelic frequencies are stable through time. This is often referred to as Hardy–Weinberg equilibrium.

Estimation of allelic frequencies from observed ge-

TABLE 1. A Chi-Square Test of the Fit of Observed Genotypes to the Number of Genotypes Expected under the Hardy–Weinberg Law

	Genotypes			Allelic frequencies		
	AA	As	aa	N	p	q
Observed	595	210	195	1000	.70	.30
Expected	490	420	90	1000	.70	.30
(Expected)	(p^2N)	$(2pqN)$	(q^2N)			
X^2	22.5	105.0	122.5	250.0		

NOTE: the value of X^2 exceeds the critical value of 3.68, indicating that the observed numbers of genotypes are significantly different than those expected by the Hardy–Weinberg law.

TABLE 2. The Hardy–Weinberg Law Is Not a Sensitive Test for Selection

In this example, selection on Hardy–Weinberg proportions moves the system to a different Hardy–Weinberg equilibrium.

	Genotypic frequencies			Allelic frequencies		
	AA	As	aa	Sum	p	q
Before selection	$p^2 = .250$	$2pq = .500$	$q^2 = .250$	1.0	.500	.500
Fitnesses	1	.5	.25			
After selection	.25	.25	.0625	.5625		
After selection	$p^2 = .445$	$2pq = .444$	$q^2 = .111$	1.0	.667	.333

notypic frequencies is simple when all three genotypes can be distinguished:

$$p = f(AA) + 0.5 * f(Aa) \quad (2)$$

and

$$q = f(aa) + 0.5 * f(aa) \quad (3)$$

where f is used to denote frequency.

This estimation is appropriate regardless of the fit of observed genotypic frequencies to Hardy–Weinberg proportions. However, this direct estimation of allelic frequencies is not an option when dominance between the alleles condenses three genotypic classes into two phenotypic classes.

Consider the variation in the shape of the human ear, which is influenced by two alleles. A dominant allele, D, causes individuals with DD and Dd genotypes to have a dangling earlobe, while the recessive genotype, dd, produces an attached earlobe. Suppose that a sample contains 91 individuals with dangling earlobes, and 9 with attached earlobes; the frequencies of the dangling and attached earlobe phenotypes are 0.91 and 0.09, respectively. However, because the numbers of DD and Dd genotypes among the 91 individuals with dangling earlobes is not known, the equations above cannot be used to estimate allelic frequencies. In this case, we must make the assumption that genotypic proportions conform to Hardy–Weinberg expectations, so that

$$f(dangling) = f(DD) + f(Dd) = p^2 + 2pq = 0.91 \quad (4)$$

and

$$f(attached) = f(dd) = q^2 = 0.09. \quad (5)$$

The frequency of the recessive allele is estimated as the square root of the observed frequency of the recessive phenotype, and thus, for this example,

$$q = \sqrt{f(dd)} = \sqrt{0.09} = 0.3. \quad (6)$$

The Hardy–Weinberg law is most easily illustrated with two alleles, but it applies to polymorphisms with two or more alleles. For example, imagine a polymorphism segregating three alleles, A, a, and α, with allelic frequencies p, q, and r, respectively, summing to 1.0. The appropriate binomial expansion, analogous to equation 1, is

$$(p + q + r)^2 = p^2 + 2pq + q^2 + 2pr + 2qr + r^2. \quad (7)$$

The fit of empirical, or observed, data to Hardy–Weinberg expectations is tested with a statistical test known as the Chi-squared test, which has the general form

$$\chi^2 = \frac{(O - E)^2}{E} \quad (8)$$

where O and E are observed and expected numbers of individuals. A test of the fit of observed to expected numbers of genotypes is presented in Table 1. From the observed numbers, p and q were estimated as in equations 2 and 3, above. The expected numbers are p^2N, $2pqN$, and q^2N for the AA, Aa, and aa genotypes respectively; N is the sample size in the observed sample. The value of X^2 is summed across the genotypes, and for a polymorphism with two alleles, the number of degrees of freedom is one, and the critical value of X^2 is 3.68. This is the largest value of the Chi-squared statistic that we would normally expect to see if the gene frequencies conform to Hardy–Weinberg expectations. That is, if the value of X^2 exceeds 3.68, the observed data deviate from Hardy–Weinberg expectations. In this particular example, $X^2 = 250.0$, indicating that the observed numbers of genotypes deviate markedly from the expected numbers. Deficiencies of heterozygotes are most commonly attributable to inbreeding, but they might be a consequence of mixing of populations with contrasting allelic frequencies (the Wahlund effect), or produced by selection. An excess of heterozygotes is most commonly attributable to selection, but it could also be produced if allelic frequencies differ between the sexes.

Although the Hardy–Weinberg law relies on the assumption of no selection, conformance of observed frequencies to Hardy–Weinberg expectations cannot be interpreted as evidence of no selection, or neutrality. For example, consider a case in which genotypic frequencies conform to Hardy–Weinberg expectations but are subjected to selection so that fitnesses are 1, $1 - s$, and $(1 - s)^2$ for the AA, Aa, and aa genotypes, where s is the proportion selected against. This type of selection will change both genotypic and allelic frequencies, but the genotypes remaining after the selective event will be in the proportions p^2, $2pq$, and q^2 for the new values of p and q. This point is illustrated with the selection coefficient, s, equal to 0.5 in Table 2.

[*See also* Inbreeding; Mutation, *article on* Evolution of Mutation Rates.]

— JEFFRY B. MITTON

HEART DISEASE. *See* Coronary Artery Disease.

HEMICHORDATES. *See* Animals.

HEREDITARY DISEASE. *See* Disease, *article on* Hereditary Disease.

HETEROCHRONY

Heterochrony (Greek for "different time") can be broadly defined as the change in the rate or timing of development of some trait. Because nearly any change to a trait's development involves some change in rate, this definition encompasses nearly all morphological evolution. A narrower definition requires a uniform change in rate or timing of some developmental process. Although some authors do treat all evolution as heterochrony, most see it as a particular kind of change that can be tested for empirically.

The term was coined by the German biologist Ernst Haeckel (1834–1919) to describe an exception to his biogenetic law ("Ontogeny is a brief and rapid recapitu-

lation of phylogeny"). Haeckel believed that the adult morphologies of an organism's ancestors are repeated in the stages through which that organism passes in developing from egg to adult. This means that the time and position at which a character first appears in a developing embryo should correspond to the historical time at which that character first appeared in evolution and its initial morphological position. Haeckel identified two processes that could confound the expected pattern: heterotopy, a shift in the location at which a character arises, and heterochrony, a shift in the timing of appearance of the character. Although the biogenic law has been discarded, the term *heterochrony* has persisted, though its meaning has changed.

Types of Heterochrony. In the 1930s, Gavin de Beer elevated heterochrony to a class of evolutionary mechanisms. De Beer (1940) described eight different kinds of heterochrony, corresponding to different ways in which the presence or absence of a character in juveniles or adults could change. For example, neoteny was defined as a case in which a character seen in the juvenile of the ancestor comes to be found in the adult of the descendant.

This approach of distinguishing different types of heterochrony was continued by S. J. Gould (1977), who modified de Beer's classification somewhat and rephrased the concept in terms of size and shape, rather than presence or absence. Gould's "clock model" defines different types of heterochrony by considering a particular growth stage (such as sexual maturity) and noting how far size and shape have developed in a descendant relative to its ancestor at the same stage. In this view, neoteny is a case in which the adult descendant is the same size as its ancestor but is shaped like the ancestor at a younger age.

Because the clock model considers size and shape at a particular time, it includes no direct representation of developmental change. The ontogenetic trajectory approach, devised by P. Alberch and colleagues (1979), remedies this by plotting curves that represent the value of some trait as a function of time (age) for ancestor and descendant (Figure 1). The different sorts of heterochrony correspond to different ways in which the ancestral trajectory would have to be transformed to produce the descendant trajectory.

In some instances, pairs of trajectories are not related by any uniform change in rate or timing (Figure 1E). Some authors argue that such cases should not be called heterochrony because they involve change in the nature of the developmental process, with changes in rate being secondary consequences. These cases are invisible under de Beer's classification or the clock model.

Because time is hard to measure in many cases, some authors have used allometric plots of shape versus body size. This is risky, because a change in an allometric

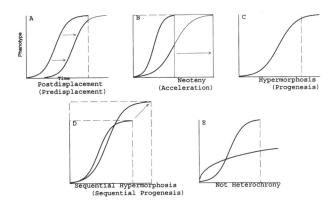

FIGURE 1. Ontogenetic Trajectories Showing Different Kinds of Heterochrony.
The first term is the case in which the solid line is the ancestor and the dashed line the descendant. The term in parentheses is the reverse transformation (dashed line is ancestral).   Drawing by Sean H. Rice.

curve could result from a change in the development of shape or size or both. Such plots can be suggestive of heterochrony, though, especially in cases of hypermorphosis or progenesis.

Most researchers now identify six kinds of heterochrony:

Hypermorphosis—Development of the character in the descendant continues beyond the point at which it stopped in the ancestor.

Progenesis—Development of the character in the descendant stops earlier than it did in the ancestor.

Acceleration—The character develops at a higher rate in the descendant than in the ancestor.

Neoteny—The character develops at a lower rate in the descendant than in the ancestor.

Postdisplacement—The character begins developing later in the descendant than it did in the ancestor.

Predisplacement—The character begins developing earlier in the descendant than it did in the ancestor.

The term *sequential hypermorphosis* (or *sequential progenesis*) refers to cases in which a trajectory is made up of multiple segments and each of these undergoes hypermorphosis (or progenesis).

In addition, two other terms are often used.

Paramorphosis—The adult of the ancestor resembles a juvenile of the descendant.

Paedomorphosis—The adult of the descendant resembles a juvenile of the ancestor.

These last two terms describe the outcome of development rather than the process itself. In contrast, the first six terms listed describe changes in the process of development.

Factor in Morphological and Human Evolution.
There is good evidence that heterochrony sometimes plays an important role in morphological evolution. In some cases, a complex life cycle is attenuated as reproduction shifts to what was previously a juvenile stage and the ancestral adult stage is lost. This pattern has been documented from a number of different organisms and may result from slowing of somatic development (neoteny, seen in some salamanders) or from predisplacement of gonadal maturation combined with progenesis of somatic development (seen in some ascidians), leading to a descendant that is smaller and shorter lived than its ancestor. Such a change in the entire life cycle constitutes global heterochrony. It is also possible for a change in timing to produce localized change in a particular trait.

Heterochrony has often been invoked as a factor in human evolution. In a number of ways, including facial features, humans resemble juvenile chimpanzees more closely than they resemble adult chimps (i.e., humans are paedomorphic). This observation led to the assertion that neoteny has played a role in human evolution. Analysis of ontogenetic trajectories for body size does show that, relative to chimps, human growth is both slowed down (neotenic) and extended (hypermorphic) up until the onset of puberty.

The ontogenetic trajectory for human brain size and that for chimpanzee brain size are related almost perfectly by sequential hypermorphosis. Interestingly, the human (or chimpanzee) brain growth trajectory is not related by any heterochronic transformations to the curve seen in monkeys. Thus, heterochrony (even under a restrictive definition) has clearly played a role in evolution, even though it does not account for all morphological change.

[*See also* Heterotopy.]

BIBLIOGRAPHY

Alberch, P., S. J. Gould, G. F. Oster, and D. B. Wake. "Size and Shape in Ontogeny and Phylogeny." *Paleobiology* 5 (1979): 296–317.

de Beer, G. *Embryos and Ancestors.* Oxford, 1940.

Gould, S. J. *Ontogeny and Phylogeny.* Cambridge, Mass., 1977. Extensively discusses the history of ideas of heterochrony. Also presents a new classification of heterochronic processes.

Haeckel, E. *The evolution of Man: A Popular Exposition of the Principal Points of Human Ontogeny and Phylogeny.* New York, 1879. English-language discussion of the author's ideas, including the biogenic law; first published in German.

McKinney, M. L., and K. J. McNamara. *Heterochrony: The Evolution of Ontogeny.* New York, 1991. Written from the viewpoint that nearly all morphological evolution is heterochrony.

Minugh-Purvis, N., and K. J. McNamara, eds. *Human Evolution through Developmental Change.* Baltimore, 2001. Multiauthored volume with many treatments of heterochrony in human evolution.

Raff, R. A. *The Shape of Life.* Chicago, 1996. Devotes a chapter to heterochrony from the standpoint of a developmental biologist.

Rice, S. H. "The Analysis of Ontogenetic Trajectories: When a

Change in Size or Shape Is Not Heterochrony." *Proceedings of the National Academy of Sciences* USA 94 (1997): 907–912. Argues for a restricted definition of heterochrony that can be tested for by comparing ontogenetic trajectories.

— SEAN H. RICE

HETEROTOPY

The term *heterotopy* refers to an evolutionary change to the location at which a feature arises or a process occurs in a descendant species. German biologist Ernst Haeckel (1875) coined the term along with the term *heterochrony*. Both were originally defined in the context of Haeckel's biogenetic law, which states that ontogeny recapitulates phylogeny. Both heterotopy and heterochrony are exceptions to that law, and both were seen as significant for the same reason—they complicate reconstructing evolutionary history. They provide what was considered "false history" because the evolutionary sequence of morphologies cannot be read directly from the developmental sequences of forms. Following the demise of the Biogenetic Law, it no longer provides a basis for reconstructing phylogenies, but the concepts of heterochrony and heterotopy have survived. Their significance, however, is dramatically altered. The primary significance of heterotopy in the modern literature is as one source of evolutionary novelties.

Two interpretations of heterotopy coexist in the literature. One emphasizes morphology, a rearrangement in the locations of anatomical parts. The other emphasizes developmental processes, a rearrangement in the locations of processes. The morphological concept of heterotopy concerns the products of development, whereas the process concept of heterotopy concerns the processes themselves. It is important to recognize that they are two different concepts, not two perspectives of the same phenomenon. There is no necessary causal relationship between them, although many morphological heterotopies might arise by heterotopies of process. Nonetheless, the defining criteria for the two concepts differ.

The morphological concept of heterochrony is especially prevalent in the medical literature, where it refers to abnormally located structures, for example, pancreatic tissue displaced to the stomach. Such cases, though not evolutionary, can still be informative about developmental causes of heterotopy and also about its effects on fitness. Several morphological heterotopies have been discovered in plants, such as the occurrence of flowers on the surface of leaf lamina (for other botanical examples, *see* Li and Johnston, 2000). Morphological heterotopies have also been found in animals, although they are not always called by that term. Brian Hall discusses several of these in his illuminating book *Evolutionary Developmental Biology* (1998). One example from vertebrates is the externalization of the ribs in the evolution of the turtle carapace. Primitively, tetrapods' ribs are internal to the pectoral and pelvic girdles, but ribs are externalized in turtles owing to the expansion and fusion of the ribs with each other and with dermal ossifications (dermal shields). Another example is the externalization of cheek pouches in rodents, which implicates a heterotopy of process (Brylski and Hall, 1988). Homeotic mutants might also be seen as cases of heterotopy; for example, *Aristopedia* mutation (in the fruit fly *Drosophila*) transforms an antenna into a leg. Some authors consider homeotic mutations to be partly distinct from heterotopy because they involve replacements rather than rearrangements of structures (Li and Johnston, 2000). However, they could be classed as heterotopies because they involve modifications in the spatial organization of morphology.

Heterotopies of process involve evolutionary changes in the location or spatial patterning of developmental processes. Among these are changes in location of inductive interactions, in positional information, or in the spatial distribution of morphogenetic signals or growth fields. A modification in the site of an inductive interaction is responsible for the fur-lined external cheek pouch of geomyoid rodents; the developing pouch grows into facial mesenchyme, bringing pouch epithelium into contact with hair-forming mesenchyme (rather than with mucous-forming epithelium, as in developing internal pouch). This displacement in the site of the inductive interaction is itself due to heterotopy, an anterior displacement of the buccal evagination (Brylski and Hall, 1988). Changes in positional information are exemplified by the evolution of mammalian limb morphology; this affects not only where limbs are located but also what kind of limbs develop (Lovejoy et al., 1999). Changes in the spatial distribution of proteins have been found in the evolution of echinoids (Wray and McClay, 1989), and modifications in the spatial distribution of growth fields are discerned in the evolution of hominid facial morphology (Lieberman, 2000), as well as in the generation of diverse body shapes in piranhas (Zelditch et al., 2000).

Heterotopy is significant because it generates novelties. This is also true of heterochrony, but in cases of heterochrony, the novelties resemble either ancestral features found at earlier ontogenetic stages or features that arise by extrapolation of the ancestral ontogeny (Gould, 1977, 2000). Heterochrony is an alteration in rate or timing of an otherwise conserved ontogeny, whereas heterotopy is an alteration of that ontogeny. The novelties need not be especially dramatic; for example, shifts in the locations of growth fields could simply alter proportions of bones. Nonetheless, heterochrony and heterotopy have strikingly different implications regarding the evolution of development. Heterochrony is widely viewed as a manifestation of developmental constraints;

morphological evolution is intrinsically channeled along a conservative ancestral ontogeny (Gould, 1977). In contrast, heterotopy implies that the ontogeny is redirected, so evolutionary change occurs along a novel direction, not that of the ancestral ontogeny. It may often be difficult to distinguish heterotopy from heterochrony because development may be modified in both timing and spatial patterning. Such combinations imply that neither spatial nor temporal patterning is constrained. Rather, they indicate the opposite—that both aspects of patterning are historically labile.

The frequency of heterotopy is still largely unknown. On empirical grounds, Klingenberg (1998) concludes that it is very rare in that late stages of development are typically modified by simply extending or truncating a conservative ontogeny (i.e., heterochrony). This conclusion, based on quantitative comparisons of ontogenetic allometries, requires more rigorous testing. On more conceptual grounds, some workers suspect that heterotopy is probably rare because it is a more profound change than heterochrony in that it generates novelties (see Li and Johnston, 2000). However, others argue that heterotopy is likely to be just as common as heterochrony because there is no good reason to suspect development is more conservative in spatial patterning than in rate or timing (Wray and McClay, 1989; Zelditch and Fink, 1996).

[See also Heterochrony.]

BIBLIOGRAPHY

Brylski, P., and B. K. Hall. "Ontogeny of a Macroevolutionary Phenotype: The External Cheek Pouches of Geomyoid Rodents." *Evolution* 42 (1988): 391–395.

Gould, S. J. *Ontogeny and Phylogeny*. Cambridge, Mass., 1977.

Gould, S. J. "Of Coiled Oysters and Big Brains: How to Rescue the Terminology of Heterochrony, Now Gone Astray." *Evolution and Development* 2(5, 2000): 241–248.

Haeckel, E. "Die Gastrula und die Eifurchung der Thiere." *Jenaische Zeitschrift für Naturwissenschaft* 9 (1875): 402–508.

Hall, B. K. *Evolutionary Developmental Biology*. London, 1998.

Klingenberg, Christian Peter. "Heterochrony and Allometry: The Analysis of Evolutionary Change in Ontogeny." *Biological Reviews* 72(1, 1998): 79–123.

Li, P., and M. O. Johnston. "Heterochrony in Plant Evolutionary Studies through the Twentieth Century." *Botanical Review* 66(1, 2000): 57–88.

Lieberman, D. E. "Ontogeny, Homology and Phylogeny in the Hominid Craniofacial Skeleton: The Problem of the Browridge." In *Development, Growth and Evolution: Implications for the Study of the Hominid Skeleton*, edited by Paul O'Higgins and Martin Cohen, pp. 85–122. San Diego, 2000.

Lovejoy, C. Owen, M. J. Cohn, and T. D. White. "Morphological Analysis of the Mammalian Postcranium: A Developmental Perspective." *Proceedings of the National Academy of Sciences* 96(23, 1999): 13247–13252.

Wray, G. A., and D. R. McClay. "Molecular Heterochronies and Heterotopies in the Evolution of Early Echinoid Development." *Evolution* 43(4, 1989): 803–813.

Zelditch, M. L., and W. L. Fink. "Heterochrony and Heterotopy: Stability and Innovation in the Evolution of Form." *Paleobiology* 22(2, 1996): 241–254.

Zelditch, M. L., H. D. Sheets, and W. L. Fink. "Spatiotemporal Reorganization of Growth Rates in the Evolution of Ontogeny." *Evolution* 54(4, 2000): 1363–1371.

— MIRIAM ZELDITCH

HETEROZYGOTE ADVANTAGE

[*This entry comprises two articles. The first article provides a summary overview of heterozygote advantage, starting with the definition of what constitutes a heterozygote/homozygote and discussing selection for a heterozygote as a cause for polymorphism; the second article provides a case study illuminating several important topics in evolutionary genetics.*]

An Overview

Heterozygous advantage arises when genetic heterozygotes have higher fitness than homozygotes. This mode of selection, like other forms of balancing selection, actively maintains genetic variation. Heterozygote advantage is most simply presented for a gene segregating two alleles in a diploid species with random mating and no inbreeding (Table 1). Consider a gene with two alleles, A and a, whose frequencies are p and q, so that $p + q = 1$. In a population with random mating, the Hardy–Weinberg Equilibrium (HWE) defines the relationship between allelic and genotypic frequencies. Assumptions underlying the HWE include infinite or very large population size, and negligible impact from natural selection, migration, and mutation. The genotypic proportions produced by random mating are p^2, $2pq$, and q^2, for AA, Aa, and aa, respectively (Table 1). These are the genotypic proportions prior to selective events.

The intensity of selection on a genotype is measured by selection coefficients (s and t in Table 1). Selection coefficients vary from zero to one, covering the range of neutrality to lethality, and they may represent either the total selection against the genotypes, or a component of selection such as mating success.

Relative fitnesses are the complements of the selection coefficients, scaled relative to the heterozygotes so that $w_{aa} = 1 - t$, $w_{Aa} = 1$ and $w_{AA} = 1 - s$. The total fitnesses of genotypes are measured with reproductive success, or the relative magnitude of their genetic contribution to the next generation. Relative fitnesses are scaled so that the highest fitness is 1.0 and complete failure is 0.0. A fitness of 1.0 does not indicate that selection does not act on this genotype, only that it is the most successful.

When two alleles are segregating at a locus, heterozygous advantage produces a stable polymorphism, so

TABLE 1. A Simple Model for Heterozygote Advantage at Equilibrium Allelic Frequencies, Demonstrating That Selection Modifies Genotypic Frequencies without Changing Allelic Frequencies

	Genotypes			*Alleles*	
	AA	AS	SS	A	S
Frequencies before selection	p^2	$2pq$	q^2	p	q
Selection coefficients	s	0	t		
Relative fitnesses	$1 - s$	1	$1 - t$		
Frequencies after selection	$\dfrac{p^2 (1 - s)}{1 - sp^2 - tq^2}$	$\dfrac{2pq}{1 - sp^2 - tq^2}$	$\dfrac{q^2}{1 - sp^2 - tq^2}$	p	q
Progeny frequencies	p^2	$2pq$	q^2	p	q

NOTE: Progeny frequencies are from the individuals in the next generation, before selection.

that the frequencies of the two alleles reach an equilibrium and return to equilibrium frequencies if they are displaced. A simple model illustrates the evolutionary dynamics of heterozygous advantage at equilibrium frequencies. To illustrate how selection modifies genotypic frequencies while stabilizing allelic frequencies, we can compare the allelic and genotypic frequencies before and after selection (Table 1). The genotypic proportions after selection are the product of the initial genotypic frequencies multiplied by the relative fitnesses, divided by the proportion of the population remaining after selection, $1 - sp^2 - tq^2$. To calculate the allelic frequencies, sum the proportion of a given homozygote with half of the proportion of the heterozygote:

$$p = \text{proportion (AA)} + 0.5*\text{proportion (Aa)};\quad (1)$$

similarly,

$$q = \text{proportion (aa)} + 0.5*\text{proportion (Aa)}.\quad (2)$$

To determine the rate and trajectory of evolution, this operation can be done repeatedly. For the case of heterozygous advantage, natural selection brings allelic frequencies to a stable equilibrium, where

$$p = t / (s + t)\quad (3)$$

and

$$q = s / (s + t).\quad (4)$$

The equilibrium allelic frequencies are dependent primarily on the relative fitnesses of the homozygotes, rather than on the intensity of selection. For example, if the selection coefficients, s and t, are .01 and .02 (weak selection) in one population, but .2 and .4 (strong selection) in another population, selection will bring both populations to the same equilibrium.

Sickle-Cell Hemoglobin. Sickle-cell anemia was the first fully described "molecular disease" (Neel, 1949) caused not by a bacterial, viral, or parasitic infection but by a genetic mutation with physiological consequences. In vertebrates, hemoglobin picks up oxygen in the lungs or gills and delivers it to tissues. A functional adult hemoglobin is a tetramer, composed of two subunits of α and two of β hemoglobin. The normal β subunit of adult hemoglobin is determined by the allele A, and sickle-cell subunit is produced by allele S, which has a point mutation that changes the sixth amino acid from glutamate to valine (Ingram, 1957). The AA homozygote has normal adult haemoglobin, AS is the sickle-cell heterozygote, and SS is the sickle-cell homozygote.

The SS genotype causes a severe form of anemia. The replacement of glutamate with valine makes that end of the hemoglobin form a ring of amino acids. Under conditions of low oxygen tension, such as strenuous exercise, the rings cause crystallinelike aggregates. The crystalline aggregates modify the shape of the red blood cell from a biconcave disk to a sickle, which is prone to lyse. The lysed cells can clog capillaries, restricting the blood flow and causing painful swelling. Without medical care, anemia and blood clots can be fatal. By comparison, sickle-cell heterozygotes have a mild anemia, but normally not a pathological condition.

The S form of β haemoglobin is rare or absent in most parts of the world, but it reaches moderate allelic frequencies, up to 0.16, in western and central Africa, the Arabian Peninsula, and southern India. The geographic areas in which S reaches moderate frequencies all have chronically high incidence of the form of malaria caused by *Plasmodium falciparum* (Allison, 1964). A series of field and laboratory studies have convincingly demonstrated that, in the presence of falciparum malaria, the sickle-cell heterozygote has the highest fitness. Direct experiments with volunteers in the field showed that, in comparison to normal homozygotes, heterozygotes have higher resistance to becoming infected with malaria. Once infected, heterozygotes sustain lower levels of parasitic infection. The red blood cells of heterozygotes turn over more quickly than those in normal homozygotes, so a heterozygore is a suboptimal host for malarial infection. Finally, hospital studies revealed that the probabilities of severe and fatal cases of malaria were lower in sickle-cell heterozygotes than in normal homozygotes.

Thus, in areas where falciparum malaria is a common

TABLE 2. Frequencies For the Sickle-Cell Polymorphism through One Generation of Balancing Selection

	Genotypes			Alleles	
	AA	AS	SS	A	S
Frequencies before selection	.709	.266	.025	.842	.158
Selection coefficients	.15	0	.8		
Relative fitnesses	.85	1.0	.20		
Frequencies after selection	.690	.304	.006	.842	.158
Progeny frequencies	.709	.266	.025	.842	.158

NOTE: Progeny frequencies are from the individuals in the next generation, before selection.

threat to health and life, heterozygotes have the highest fitness (Table 2); the disadvantage of their mild anemia is more than compensated for by their lower probabilities of contracting, suffering from, and dying from the disease. In comparison to heterozygotes, the normal homozygotes have a higher incidence of malaria, are more likely to have a severe infection, and are much more likely to die from malaria (Cavalli-Sforza and Bodmer, 1971). Selection on the sickle-cell polymorphism can be presented in a model that uses the approximate genotypic fitnesses and allelic frequencies for an environment with falciparum malaria (Table 2). The fitnesses are estimated to be .85, 1.0 and .20 for the AA, AS, and SS genotypes, respectively. These fitnesses, when used in equations 3 and 4 above, predict equilibrium frequencies of 0.842 for A and 0.158 for S. These frequencies are approximately those seen in areas plagued by falciparum malaria. The genotypic frequencies were produced from these allelic frequencies by using the relationship in the HWE (Table 1). This model shows that selection increases the frequency of heterozygotes, and decreases the frequencies of homozygotes. Random mating returns the genotypic frequencies to the initial, equilibrium frequencies. Neither selection nor random mating modifies the allelic frequencies.

The sickle cell polymorphism clearly reveals that fitness is an interaction between genotype and environment. The heterozygote has the highest fitness in areas with falciparum malaria, but the AA homozygote has the highest fitness elsewhere. Consequently, the S allele is extremely rare or absent in areas free of falciparum malaria.

The Number of Documented Examples Is Small.
The number of empirical examples of heterozygous advantage at a single gene is surprisingly small (see summaries by Endler, 1986; Mitton, 1997). This lack of empirical evidence for heterozygous advantage leads most evolutionary biologists to conclude that this form of balancing selection is rare, and other forms, such as frequency-dependent selection, are more common. A minority of evolutionary biologists contend that it is not that heterozygous advantage is so rare, but that it is

rarely documented to the exhaustive degree needed to be convincing.

One possible explanation for the paucity of cases of heterozygous advantage is in our expectation for a demonstration of heterozygous advantage. Because it is usually difficult and sometimes impossible to get lifetime estimates of fitness, evolutionary biologists often estimate fitness with a single component of fitness, such as growth rate or survival during a single selective event. It is possible that heterozygous advantage may be very difficult to detect during a single selective event, yet still relatively common.

Environmental conditions change over geographic space, and conditions also fluctuate in time. Because fitness is influenced by both genetic and environmental conditions, selection must also fluctuate in space and time. A simple model (Table 3) illustrates how fluctuation in environmental conditions, and therefore in selective regimes, can produce a cumulative heterozygous advantage even when there is no evidence of heterozygous advantage during any single selective event. Assume that in any single event, heterozygote fitness is perfectly intermediate between homozygote fitnesses, as they are in Table 3. These events might be alternating seasons in the year, or bright, warm days versus dark, chilly nights, or alternations of pure salt versus brackish water in a coastal marsh. An excellent empirical example of this reversal of fitnesses (Vrijenhoek, 1996) is seen

TABLE 3. Heterozygous Advantage Can Accumulate from a Series of Selective Events

Genotype	w_1	w_2	w_3	w_4	w_{cum}	w_{rel}
AA	1.0	.4	1.0	.8	.3200	.85
Aa	.7	.7	.8	.9	.3528	1.0
aa	.4	1.0	.6	1.0	.2400	.68

NOTE: w_1 through w_4 are fitnesses in selective events 1 through 4. Cumulative fitness, w_{cum}, is the product of w_1, w_2, w_3, and w_4. Relative fitness, w_{rel}, is obtained by dividing all of the w_{cum} by the highest w_{cum}. Note that the heterozygote has an intermediate fitness in each selective event, but emerges from the series of selective events with the highest relative fitness.

in the fitnesses of the fish *Peociliopsis monacha* between stresses imposed by cold and heat exacerbated by low oxygen (Figure 1). When two or more selective events are chained in this way and alternate homozygotes are favored during successive selective events, heterozygous advantage may accumulate (Mitton, 1997; Gillespie, 1991).

If heterozygous advantage is somewhat rare during single, selective events but commonly accumulates over a series of selective events, then heterozygous advantage may be more apparent if fitness is evaluated over an extended period of time. Many observations from population studies with allozymes are consistent with this hypothesis (Mitton, 1997). For example, when stands of conifers are sampled randomly, the genotypic proportions generally fit Hardy–Weinberg expectations; however, when only mature, only the largest, or only the oldest trees are sampled, analyses often reveal excesses of heterozygotes. Excess heterozygosity cannot be produced by selection against inbred individuals but must be a different mode of selection in which heterozygotes are favored. Artificial selection imposed by foresters has genetic consequences similar to those imposed by natural selection. When foresters established orchards of Engelmann spruce (*Picea engelmannii*), Norway spruce (*Picea abies*), Sitka spruce (*Picea sitchensis*), and Douglas fir (*Pseudotsuga menziesii*), they selected trees according to the normal criteria used by foresters, but they unknowingly chose trees with higher individual heterozygosities than would be found in a random sample of trees. The healthy, vigorous trees favored by foresters are highly heterozygous (Mitton, 1997).

Studies of enzyme kinetics, the biochemical performance of enzymes, have revealed some cases of heterozygous advantage reflected in a variety of components of fitness. For example, kinetic variation at an alcohol dehydrogenase polymorphism in tiger salamanders (*Ambystoma tigrinum*) influences oxygen consumption, growth rate, and success at metamorphosis to the adult form (Carter, 1997; Carter et al., 2000). One homozygote is favored in anoxic ponds, while the other homozygote is favored in ponds with high levels of dissolved oxygen. In areas with a mosaic of anoxic and hypoxic ponds, the heterozygote enjoys the highest fitness. Similarly, in one alpine butterfly, enzyme efficiency at the enzyme phosphoglucose isomerase varies among genotypes and is highest in the most common heterozygote. This genotype also flies over the greatest range of temperatures and has the highest survival, male mating success, and female fecundity (Watt, 1992).

It is a simple matter to detect natural selection with a genetic polymorphism, such as an enzyme polymorphism; however, because genes are embedded in chromosomes with hundreds or thousands of other genes, it is conceptually difficult and experimentally laborious to

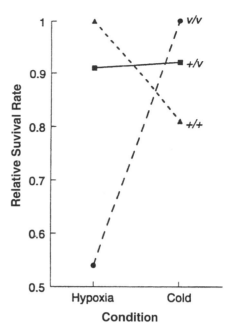

FIGURE 1. Relative Survival Rates of Ldh Genotypes in *Poeciliopsis monacha* During Hypoxic and Cold Stress. Relative survival is determined as the absolute survival rate for that genotype divided by the maximal survival rate under a particular type of stress. Because the heterozygote has intermediate survival in each environment, these data are similar to those modeled in Table 3.   From Vrijenhoek, 1966.

demonstrate that selection discriminates among the genotypes at a locus, and to characterize unambiguously and completely the form of that selection. Indirect evidence suggests that balancing selection, or selection that maintains genetic polymorphism, is not rare. Evolutionary biologists have theoretical considerations and some data that lead them to suspect that other forms of balancing selection, such as frequency-dependent selection, are more common than heterozygous advantage.

[*See also* Genetic Polymorphism; Hardy–Weinberg Equation; Malaria.]

BIBLIOGRAPHY

Allison, A. C. "Polymorphism and Natural Selection in Human Populations." *Symposia in Quantitative Biology* 29 (1964): 137–149.

Carter, P. A., J. B. Mitton, T. D. Kocher, and J. R. Coehlo. "Maintenance of the Adh Polymorphism in Tiger Salamanders II. Differences in Biochemical Function Among Allozymes." *Functional Ecology* 14 (2000): 70–76.

Carter, P. A. "Maintenance of the Adh Polymorphism in *Ambystoma tigrinum nebulosum* (Tiger Salamanders). I. Genotypic Differences in Time to Metamorphosis in Extreme Oxygen Environments." *Heredity* 78 (1997): 101–109.

Cavalli-Sforza, L. L., and W. F. Bodmer. *The Genetics of Human Populations*. San Francisco, 1971. An authoritative source on human genetics, and a thorough summary of studies of the sickle-cell polymorphism.

Endler, J. A. *Natural Selection in the Wild*. Princeton, 1986. A comprehensive compilation of studies of selection in natural populations.

Gillespie, J. H. *The Causes of Molecular Evolution*. Oxford and New York, 1991. Primarily a theoretical treatment of models of DNA sequence evolution.

Ingram, V. M. "Gene Mutations in Human Haemoglobin: The Chemical Differences Between Normal and Sickle Cell Haemoglobin." *Nature* 180 (1957): 326–328. This paper reports the single amino acid substitution in the sickle haemoglobin.

Mitton, J. B. *Selection in Natural Populations*. Oxford and New York, 1997. A compilation of empirical studies of natural selection on protein and enzyme polymorphisms, and discussion of modes and intensities of selection.

Neel, J. V. "The Inheritance of Sickle-Cell Anemia." *Science* 110 (1949): 64–66. The first demonstration of the Mendelian inheritance of sickle-cell anemia.

Vrijenhoek, R. C. "Conservation Genetics of North American Desert Fishes." In *Conservation Genetics: Case Histories from Nature*, edited by J. C. Avise, and J. L. Hamrick, pp. 367–397. New York, 1996. A clear case of genotypic fitnesses changing with environmental conditions.

Watt, W. B. "Eggs, Enzymes, and Evolution—Natural Genetic Variants Change Insect Fecundity." *Proceedings of the National Academy of Sciences USA* 89 (1992): 10,608–10,612. A summary of how the enzyme kinetics of an enzyme polymorphism exhibit heterozygous advantage, and result in heterozygous advantage for viability, male mating success, and female fecundity in an alpine butterfly.

— Jeffry B. Mitton

Sickle-Cell Anemia and Thalassemia

During normal human development, the structure of hemoglobin changes as an adaptive mechanism reflecting particular oxygen requirements at different ages. Normal adults have a major hemoglobin that consists of two different pairs of peptide chains called α and β, each of

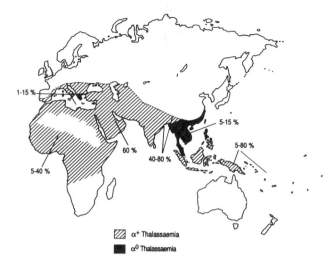

Figure 2. World Map of the Distribution of α Thalassemia. α° thalassemia from the deletion of both α genes; α+ thalassemia is due to the deletion of one of the pair of α globin genes.   From Weatherall and Clegg, 2001.

which is attached to a heme molecule to which oxygen binds. The inherited disorders of hemoglobin, the most common genetic diseases, fall into two main groups. First, there are the thalassemias (from the Greek word *thalassa*, meaning "sea"), originally believed to be restricted to those of Mediterranean origin. The thalassemias all result from defective synthesis of either the α or β globin chains and hence are classified into α and β thalassemias. The other group consists of inherited abnormalities of the structure of either the α or β globin chains. At first they were designated by letters of the alphabet; when there were no more letters available for designation, they were referred to by the name of the region in which they were first discovered. Although over 700 structural variants of this type have been described, only three, hemoglobins S, C, and E, all of which result from a single amino acid substitution in the β globin chain, are common. Nearly all the genetic disorders of hemoglobin follow an autosomal recessive inheritance.

The thalassemias occur at a high frequency in regions stretching through the Mediterranean, the Middle East, the Indian subcontinent, and Myanmar, and throughout Southeast Asia in a line starting in southern China and stretching down the Malay Peninsula into the island populations of Melanesia (Figure 1). The carrier frequency of β thalassemia in these populations ranges from 2 to 20 percent, while that for some forms of α thalassemia, which has a similar distribution (Figure 2), reaches as high as 80 percent in some regions. Hemoglobin S, which in the homozygous state causes sickle-cell anemia, is distributed throughout sub-Saharan Africa, the Mediter-

Figure 1. The Distribution of β Thalassemia.
The common mutations in different populations are shown.
From Weatherall and Clegg, 2001.

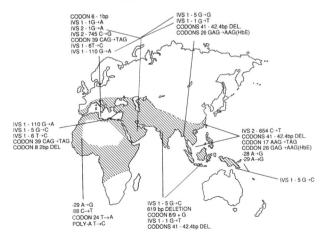

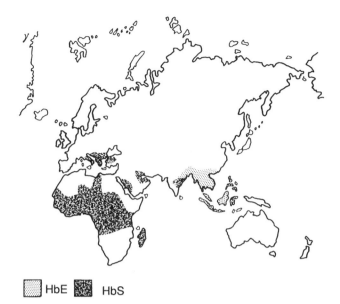

HbE HbS

FIGURE 3. The Distribution of Hemoglobins S and E.
From Weatherall and Clegg, 2001.

ranean, the Middle East, and parts of India (Figure 3); the frequency of the carrier state ranges from 5 to over 40 percent. Hemoglobin E, the most common structural hemoglobin in the world population, is confined to the eastern regions of the Indian subcontinent, Myanmar, and Southeast Asia (Figure 3). Its frequency varies; carrier rates of over 60 percent of the population occur in eastern Thailand and parts of Cambodia.

The extraordinarily high gene frequencies for these recessive disorders, many of which cause death in early life, puzzled population geneticists, particularly those who had become interested in human mutation rates as a result of their investigations of survivors of the atomic bombs that had been dropped on Japan during the Second World War. In their first population studies in the United States, the American geneticists James Neel and William Valentine calculated a mutation rate for thalassemia of 1 in 2,500. Their calculations assumed that the fitness of a β thalassemia homozygote is 0 and that the heterozygote is selectively neutral. An even higher rate was proposed by the Italian hematologist Ezio Silvestroni, who also suggested that some form of positive selection might have operated to maintain the frequency of thalassemia heterozygotes.

In 1948, the English geneticist and polymorph J. B. S. Haldane, in an address before the Eighth International Congress of Genetics in Stockholm, also agreed that heterozygote selection must be a major factor in maintaining the high frequency of thalassemia. Furthermore, Haldane suggested that the selective agent might be malaria (Figure 4). He said that "the corpuscles of the anaemic

heterozygotes are smaller than normal, and more resistant to hypotonic solutions. It is at least conceivable that they are also more resistant to attacks by the sporozoa which cause malaria" (Haldane, 1949).

It followed from Haldane's hypothesis that the high frequency of thalassemia might reflect a state of balanced polymorphism. The English geneticist E. B. Ford defined polymorphism as "the occurrence together in the same habitat at the same time of two or more distinctive forms of the species in such proportions that the rarest of them cannot be maintained merely by recurrent mutation." R. A. Fisher, one of the founders of population genetics, showed that the presence of the two forms, or alleles, must reflect a balance of selective forces maintaining each of them in a population. If this were not the case, one of the forms would increase in frequency to the exclusion of the other. Hence the term *balanced polymorphism* was invented to describe a situation in which the gene frequency for the advantageous heterozygous state of a condition like thalassemia will increase until it is balanced by the loss of homozygotes from the population.

The first evidence in support of what became known as the "malaria hypothesis" was obtained in the 1960s in an extensive series of studies by Anthony Allison, an English geneticist. He was able to obtain reasonably convincing evidence that the high frequency of the sickle-cell gene reflects heterozygote advantage against *P. falciparum* malaria, the most severe form of malarial infection. Many years later, this work was confirmed by another English geneticist, Adrian Hill, who, using the rigorous definition of "severe malaria" as defined by the World Health Organization was able to demonstrate that

FIGURE 4. The Distribution of Malaria in the Old World Before Eradication Programs.   From Weatherall and Clegg, 2001.

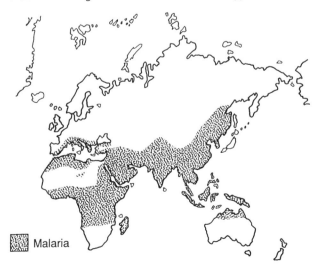

Malaria

the sickle-cell carriers have as high as 80 percent protection against a disease of this type.

Early attempts to test the malaria hypothesis in relationship to maintaining the high gene frequencies for different forms of thalassemia met with mixed fortunes. A reasonably clear correlation between the frequency of malaria and β thalassemia was observed in the Sardinian population, although when such associations were sought in other parts of the world they were not found. Furthermore, it was difficult to understand why the thalassemias were so widespread among so many different populations. There was much speculation about where the disease might have first arisen. One suggestion was in the ancient populations that inhabited Sicily, Greece, and parts of Italy, after which it spread eastward within the empire of Alexander the Great. Other writers who observed thalassemia in Persians proposed that the genes had arisen in the Orient and spread to the Mediterranean from the East.

The confirmation of Haldane's hypothesis regarding heterozygote advantage as the basis for the high frequency of thalassemia had to await the molecular era. Once it was possible to analyze the molecular basis for thalassemia, it was found that both the α and β thalassemias result from many different mutations of the α and β globin genes; over 200 different mutations have been found to cause β thalassemia and almost 100 to underlie α thalassemia. Remarkably, it turned out that every population in which thalassemia is common has its own particular set of mutations of the globin genes. Clearly, the thalassemias arose by mutation, after which their frequency was magnified by a locally acting selective agent. Proof that this was heterozygote advantage with respect to malarial infection has been obtained for α thalassemia in studies in the Vanuatuan islands and Papua New Guinea. Interestingly, it has been found recently that the selection for mild forms of α thalassemia may involve both heterozygotes and homozygotes. In this case, it seems likely that we are observing a transient rather than a balanced polymorphism, that is, if the selective factor (malaria) had persisted, the α thalassemias would have gone to fixation in the population. Although formal evidence of this type has not yet been obtained in the case of β thalassemia (and, incidentally, for hemoglobin E, which has the phenotype of a mild form of β thalassemia), it seems very likely that their high frequencies result from the same mechanism.

Although there is very good evidence that heterozygote advantage against the severe form of malaria has played a major role in maintaining the high gene frequency for these inherited disorders of hemoglobin one major problem remains to be resolved. Haldane's original hypothesis—that in the case of thalassemia, the small, underhemoglobinized red cells of heterozygotes might provide an inadequate environment for develop-

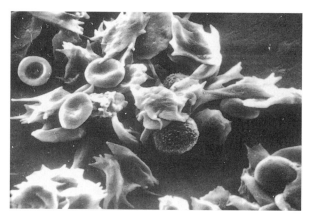

FIGURE 5. Normal and Sickled Red Blood Cells, Partial Sickling. Visuals Unlimited.

ment of the malarial parasite—has not stood up to experimental verification. In fact, parasites invade, grow, and multiply in these cells as well as in normal red blood cells. Current studies suggest that a much more subtle immune mechanism is involved in protection, although the details remain to be worked out.

Similarly, it is still not clear why the sickle-cell trait protects against malaria. Two mechanisms have been suggested, both of which are based on some experimental evidence. First, it has been found that if the red cells of individuals with the sickle-cell trait are maintained in the deoxygenated state, malarial parasites do not grow and develop in them as well as they do in normal red cells under the same conditions. It has also been suggested that parasitization of red cells containing hemoglobin S results in intravascular sickling (Figure 5) and premature destruction of the parasites in the spleen, hence preventing them from completing their life cycles. Currently, it is not clear which, if either, of these mechanisms are involved.

But whatever the cellular mechanisms involved, the thalassemias and sickle-cell anaemia are the best defined examples of heterozygote advantage as the basis for the persistence of high frequencies of serious genetic diseases in human populations.

[*See also* Adaptation; Disease, *articles on* Hereditary Disease, *and* Infectious Disease; Haldane, John Burdon Sanderson; Malaria; Natural Selection.]

BIBLIOGRAPHY

Allison, A. C. "Polymorphism and Natural Selection in Human Populations." *Cold Spring Harbor Symposium on Quantitative Biology* 29 (1964) 137–149.

Bodmer, W. F., and L. L. Cavalli-Sforza. *Genes, Evolution and Man.* San Francisco, 1976.

Haldane, J. B. S. "The Rate of Mutation of Human Genes." *Proceedings of the Eighth International Congress on Genetics Hereditas* 35 (1949) 267–273.

Hill, A. V. S., C. E. M. Allsop, and D. Kwiatkowski. "Common West African HLA Antigens Are Associated with Protection from Severe Malaria." *Nature* 352 (1991): 595–600.

Neel, J. V., and W. N. Valentine. 1947. "Further Studies on the Genetics of Thalassemia." *Genetics* 32 (1947): 38–63.

Silvestroni, E., and I. Bianco. "Sulla frequenza dei porta tori di malatia di morbo di Codey e primi observazioni sulla frequenza dei portatore di microcitemia nel Ferrarese e inakune regioni limitrofe." *Boll Atti Acad Med* 72 (1947): 32–.

Weatherall, D. J., and J. B. Clegg. *The Thalassaemia Syndromes.* 4th ed. Oxford, 2001.

Weatherall, D. J., J. B., Clegg, and D. Kwiatkowski. "The Role of Genomics in Studying Genetic Susceptibility to Infectious Disease." *Genome Research* 7 (1997): 967–973.

— DAVID WEATHERALL

HISTORY OF EVOLUTIONARY THOUGHT.

See Overview Essay *on* History of Evolutionary Thought *at the beginning of Volume 1.*

HITCHHIKING

The classical neutral theory of molecular variation and evolution assumes that events at a given locus can be considered in isolation from other loci in the genome. However, it has long been recognized that selection acting at one place in the genome can affect variability at nearby sites (Barton, 2000; Maynard Smith and Haigh, 1974). This theory of "genetic hitchhiking" has become important in interpreting evidence, initially from the fruit fly (*Drosophila melanogaster*), for reduced levels of DNA sequence variability in genomic regions where the rate of genetic recombination is low (Aguadé et al., 1989; Aquadro et al., 1994). In *Drosophila*, the possibility can be ruled out that recombination itself generates mutations, and thereby caused increased variation. If this were the explanation, divergence between the sequences of different species would also increase with the recombination rate experienced by the genes, but it does not (Aquadro et al., 1994). The relationship between variability and recombination rate thus probably arises from natural selection at sites in the genome that are linked to the loci concerned.

This effect has now been found in several other organisms, including mammals and plants (Charlesworth and Charlesworth, 1998). Interspecies comparisons also suggest that self-fertilizing plants, whose effective recombination rates are low, have reduced variability as a result of the same processes (Charlesworth and Charlesworth, 1998). The nature of the selection processes involved, however, is still controversial.

Hitchhiking by Adaptive Mutations. Variability at a neutral locus can be reduced by the spread of a favorable mutation at a linked locus (Barton, 2000; Maynard Smith and Haigh, 1974). In the most extreme case,

a unique favorable mutation arises at a site that is completely linked to a site segregating for neutral variants. Fixation of the favored allele in the population causes fixation of the variant at the neutral locus that was present in the chromosome in which the mutation arose (Figure 1). This is the effect that was originally termed *hitchhiking* (Maynard Smith and Haigh, 1974). It is now often referred to as a *selective sweep* (Berry et al., 1991).

The stronger the effect of a selective sweep on variability at a site in the genome, the stronger the selection at the site undergoing selection and the lower its frequency of recombination with the selected site. The most extreme effects occur when there is no recombination, as in the fourth chromosome of *D. melanogaster* (Berry et al., 1991). A selective sweep, however, can influence variability at any sufficiently closely linked site. An example has been detected at the maize *teosinte-branched* locus. The allele of this gene that gives maize its characteristic branching pattern is now fixed in maize and was probably under intense selection by humans during its domestication. In maize, part of the 5' noncoding region of this gene has very low sequence variation, compared to the rest of the gene, whereas this region has high diversity in the ancestral species, teosinte. Information on the amount of recombination per nucleotide in maize DNA was used to estimate the strength of this selection from the size of this low-variation region (Wang et al., 1999).

It is also possible to model an evolutionary steady state under a constant input of adaptively favorable mutations into a population, at sites scattered over the genome. The equilibrium level of nucleotide site diversity under mutation and drift at a neutral locus depends on the rate of favorable gene substitutions per nucleotide site, the product of the strength of selection and the effective population size, and the local rate of recombination per nucleotide site (Stephan, 1995). A neutral locus in a region of low recombination is thus expected to have reduced diversity, because it experiences greater effects of substitutions at closely linked sites than a gene in a more freely recombining region. The data of Aquadro and colleagues (1994) on loci on the *D. melanogaster* third chromosome fit the predictions of this model (Stephan, 1995); assuming a mean selective advantage of 0.01, the data suggest a substitution rate of favorable alleles in the genome of 1 per 4,000 generations.

Background Selection. The interpretation of the *Drosophila* data, in terms of successive selective sweeps, suggested that the low diversity of low recombination regions provided a molecular signature of pervasive adaptive evolution. An alternative explanation is hitchhiking arising from elimination of deleterious mutations from the population: background selection (Charlesworth et al., 1993). Deleterious mutations are usually removed from the population within a few generations of their origin. A neutral variant that is tightly linked to loci where mutations are occurring thus has a high chance of also being eliminated before it is uncoupled from the deleterious alleles by recombination, and so contributes little to the overall level of neutral variability in the population (see Figure 1). Genomes of higher organisms contain numerous loci subject to mutation to deleterious alleles, so that the relatively weak effects of many individual mutations may have large cumulative effects.

Assuming independence among loci, if the mean number of deleterious mutations per haploid genome per generation is $\bar{n}$, then the equilibrium frequency of haploid non-recombining genomes which are free of deleterious mutations is $f_0 = \exp(-\bar{n})$. Thus, even if $\bar{n}$ is only 3, about 5 percent of the population will be free of deleterious mutants. Only neutral variants that arise in this portion of the population contribute significantly to variability (Charlesworth et al., 1993). For a genomic region where recombination rarely occurs, the equilibrium level of neutral diversity is thus reduced by a factor of f_0, calculated for the region in question. If $\bar{n}$ is sufficiently large in such a region (i.e., if its mutation rate is large enough), genetic diversity may be greatly reduced by background selection. This effect may be important in the centromeric regions of chromosomes, or in asexual or highly self-fertilizing populations (Charlesworth et al., 1993).

The background selection theory can be extended to take into account the effects of recombination, generating predicted relative values of the DNA sequence variabilities for different positions on a chromosome that are quite similar to the patterns observed in *D. melanogaster* (Figure 2) (Charlesworth, 1996; Hudson and Kaplan, 1995). Frequent selective sweeps are thus not necessary to explain the observed relations between recombination rate and genetic diversity.

A way to discriminate between these two theories is to examine the departure of the frequency distribution of nucleotide site variants from neutral expectation in regions where variability is strongly reduced. During the slow recovery from a selective sweep that eliminates all or most variation at linked sites, segregating sites in a sample sequence should show an excess of low frequencies of variants, given the observed level of nucleotide site variation (Braverman et al., 1995). This effect is usually weak with background selection (Charlesworth et al., 1995). Genes in regions of reduced recombination in *D. melanogaster* only rarely depart from neutral expectation (Braverman et al., 1995; Charlesworth et al., 1995), suggesting that selective sweeps are not frequent in this species. This is consistent with background selection, but not conclusive evidence in favor of it, because a flux of mutations under weak directional or fluctuating selection can also cause reduced neutral variation at a closely linked site without greatly distorting its allele frequency distribution (Gillespie, 1997).

FIGURE 1. The Effects of a Selective Sweep and Background Selection with Complete Linkage.
A_1, A_2, and A_3 are neutral alleles at a polymorphic locus. B is a nearby locus with alleles that are selectively different. The left side of the figure shows an initial population fixed for allele b. In a selective sweep (the upper right hand part of the figure), the spread of the selectively favorable B allele causes the population to become fixed for allele A_1 at the neutral locus. Under background selection, mutation to a deleterious B allele leads to elimination of the associated allele A_1 at the neutral locus, so that diversity at the A locus is reduced to two alleles, A_2 and A_3.   Drawing by Deborah and Brian Charlesworth.

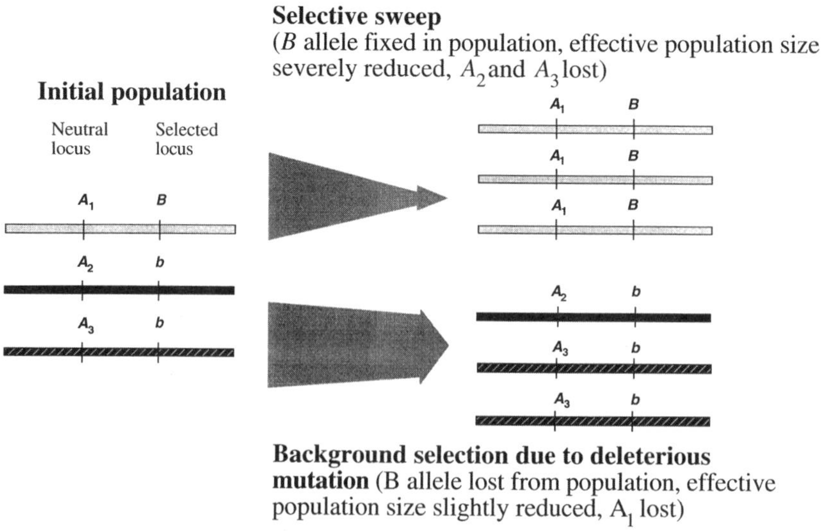

Selective sweep
(B allele fixed in population, effective population size severely reduced, A_2 and A_3 lost)

Initial population

Neutral   Selected
locus   locus

A_1 B

A_2 b

A_3 b

A_1 B
A_1 B
A_1 B

A_2 b
A_3 b
A_3 b

Background selection due to deleterious mutation (B allele lost from population, effective population size slightly reduced, A_1 lost)

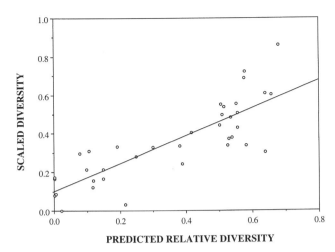

FIGURE 2. Comparison of the Observed Values of Nucleotide Site Diversities for *D. melanogaster* with the Values Predicted under the Background Selection Model.
The values predicted are expressed relative to the neutral values in the absence of background selection (Charlesworth, 1996). To enable comparisons throughout the genome, the observed value for each locus is scaled relative to the mean for the chromosome arm in which it is located, and adjusted so that this mean is the same for observed and predicted values.   Drawing by Deborah and Brian Charlesworth.

Effects of Selective Sweeps and Background Selection on Weakly Selected Sites. Silent changes to third coding positions in many species are not neutral, but are subject to very weak selection leading to biased codon usage. Because the selection is very weak, genetic drift and mutation are important in addition to selection (Akashi et al., 1998). In *D. melanogaster*, the level of codon bias is reduced in regions of restricted recombination (Comeron et al., 1999), probably also reflecting the effects of selective sweeps and background selection in reducing the effective size of the population in such regions. In addition, interference between the weakly selected sites themselves can considerably reduce the level of variability and codon usage bias in regions of low recombination (Comeron et al., 1999; McVean and Charlesworth, 2000). Sorting out the contributions of these various processes is a major research task for the future.

BIBLIOGRAPHY

Aguadé, M., N. Miyashita, and C. H. Langley. "Reduced Variation in the *Yellow-Achaete-Scute* Region in Natural Populations of *Drosophila melanogaster*." *Genetics* 122 (1989): 607–615.

Akashi, H., R. Kliman, and A. Eyre-Walker. "Mutation Pressure, Natural Selection, and the Evolution of Base Composition in *Drosophila*." *Genetica* 102/103 (1998): 49–60.

Aquadro, C. F., D. J. Begun, and E. C. Kindahl. "Selection, Recombination, and DNA Polymorphism in *Drosophila*." In *Non-neutral Evolution: Theories and Molecular Data*, edited by B. Golding, pp. 46–56. London: 1994.

Barton, N. H. "Genetic Hitchhiking." *Philosophical Transactions of the Royal Society of London B* 355 (2000): 1553–1562.

Berry, A. J., J. W. Ajioka, and M. Kreitman. "Lack of Polymorphism on the *Drosophila* Fourth Chromosome Resulting from Selection." *Genetics* 129 (1991): 1111–1117.

Braverman, J. M., R. R. Hudson, N. L. Kaplan, C. H. Langley, and W. Stephan. "The Hitchhiking Effect on the Site Frequency Spectrum of DNA Polymorphism." *Genetics* 140 (1995): 783–796.

Charlesworth, B. "Background Selection and Patterns of Genetic Diversity in *Drosophila melanogaster*." *Genetical Research* 68 (1996): 131–150.

Charlesworth, B., M. T. Morgan, and D. Charlesworth. "The Effect of Deleterious Mutations on Neutral Molecular Variation." *Genetics* 34 (1993): 1289–1303.

Charlesworth, D., and B. Charlesworth. "Sequence Variation: Looking for Effects of Genetic Linkage." *Current Biology* 8 (1998): R658–661.

Charlesworth, D., B. Charlesworth, and M. T. Morgan. "The Pattern of Neutral Molecular Variation under the Background Selection model." *Genetics* 141 (1995): 1619–1632.

Comeron, J. M., M. Kreitman, and M. Aguadé. "Natural Selection on Synonymous Sites Is Correlated with Gene Length and Recombination in *Drosophila*." *Genetics* 151 (1999): 239–249.

Gillespie, J. H. "Junk Ain't What Junk Does: Neutral Alleles in a Selected Context." *Gene* 205 (1997): 291–299.

Hudson, R. R., and N. L. Kaplan. "Deleterious Background Selection with Recombination." *Genetics* 141 (1995): 1605–1617.

Maynard Smith, J., and J. Haigh. "The Hitch-hiking Effect of a Favourable Gene." *Genetical Research* 23 (1974): 23–35.

McVean, G. A. T., and B. Charlesworth. "The Effects of Hill-Robertson Interference between Weakly Selected Mutations on Patterns of Molecular Evolution and Variation." *Genetics* 155 (2000): 929–944.

Stephan, W. "An Improved Method for Estimating the Rate of Fixation of Favorable Mutations Based on DNA Polymorphism Data." *Molecular Biology and Evolution* 12 (1995): 959–962.

Wang, R.-L., A. Stec, J. Hey, J. F. Doebley, and L. Lukens. "The Limits of Selection during Maize Domestication." *Nature* 398 (1999): 236–239.

— Brian Charlesworth and Deborah Charlesworth

HIV. *See* Acquired Immune Deficiency Syndrome.

HOMEOBOX

The homeobox was discovered in, and gained its name from, the homeotic genes of the fruit fly *Drosophila melanogaster* in 1984. It rapidly became a galvanizing force in the rejuvenated field of evolutionary developmental biology. The homeobox is a 180 base-pair motif encoding a 60 amino acid homeodomain. The homeodomain takes the form of three alpha-helices, with a turn between the second and third, so it is a type of "Helix-turn-Helix" protein. Homeodomains are sequence-specific DNA-binding domains in proteins that usually act as transcription factors, controlling the activity of downstream target genes. The levels of sequence similarity between homeodomains are most pronounced within the third helix, which is the section of the domain that makes the most contacts with the DNA. Indeed, the sequence-specific contacts are made via residues in this third helix.

Homeobox containing genes are found in plants and fungi but are most numerous and diverse in animals. The genes have prominent roles in the development of multicellular organisms, regulating cell position, fate, and differentiation. They often function at prominent early points in developmental programs, such that when the genes are mutated or knocked out, then whole organs or structures are affected. The genes are thus often cast as "master control genes."

The most famous examples of homeobox genes are the *Hox* genes. *Hox* genes, however, form only a subset of the large family of homeobox-containing genes, the term *Hox* being restricted to genes within the *Hox* gene clusters. *Hox*-like genes, such as the *Gsx*, *Xlox*, *Cdx*, *Nkx*, and *Mnx* classes, are as equally related to the *Hox* gene sequences as some of the different *Hox* classes are to each other, but these *Hox*-like genes do not reside in the *Hox* clusters. Together the *Hox* and *Hox*-like genes form the Antp-superclass, which is the largest group of homeobox genes, the other groups being the Prd/Prd-like, the POU, ZF, and LIM superclasses (for a classification, see Duboule, 1994). There are a handful of atypical homeodomains, with three amino acids more than the canonical 60. Where the protein structure has been determined, however, the atypical homeodomains still take the Helix-turn-Helix form.

Homologous homeobox genes are present across phyla. This allows comparisons of their roles in development across phyla, which both permits an assessment of the diversity and evolution of developmental mechanisms and enables the refinement of statements of homology of morphology. Care must be taken, however, with such comparisons owing to the frequent occurrence of the same homeobox gene in a diversity of developmental processes, even within the same organism. Expression of the same gene in two structures of two different species cannot alone be taken as evidence of homology of those structures. The genes do provide, however, a route into understanding how morphologies are built and whether they are built by homologous mechanisms, or are likely to be convergent, or are mechanistically easy to develop (and thus relatively easy to evolve, possibly repeatedly).

The homeobox gene *Pax6* and its essential function early in the development of eyes of animals as diverse and phylogenetically distant as flies and mice illustrates the potential of homeobox genes in understanding evolution of developmental mechanisms and animal morphology. *Pax6* is in the Prd superclass of homeobox

genes. Loss-of-function mutants in *Drosophila* fail to develop eyes. Loss-of-function mutants in mice and humans also fail to develop eyes. Early expression of *Pax6* during eye development seems to occur throughout bilaterian animals. Eye development is restored when the mouse gene is put into flies that lack their own copy of the gene. Ectopic expression of the gene can lead to development of ectopic eyes. *Pax6* is thus both essential and early-acting in eye development in animals. What does this mean for homology of eyes and for the evolution of developmental mechanisms? From animal phylogenies and comparison of the many different forms of eyes, it is thought that eyes evolved independently many times. The involvement of the homologous gene in the development of these supposedly nonhomologous (independently evolved) organs does not necessarily mean that the theory of multiple origins of eyes is wrong, and that all eyes are in fact homologous. Instead, the common ancestor of all animals with eyes clearly had a *Pax6* gene, and probably used it in the development of the evolutionary precursor(s) of eyes, probably a basic photoreceptor cell type (few animals have no photoreceptor cells). The evolution of the different forms of eyes then occurred from such photoreceptor cells, leaving *Pax6* operating early in the development of the photoreception systems while further developmental processes were added on to make the different nonhomologous forms of eyes. *Pax6* involvement in the development of photoreception systems and organs is thus homologous (monophyletic), but the advanced forms of photoreception apparatus that we call eyes arose independently in several lineages, all derived from the homologous eye-precursor photoreception system.

Pax6, then, with its essential function in the development of eyes across the animal phyla, provides the perfect entry point to dissecting how different forms of eyes are constructed, to understand the complexity and evolutionary history of these forms. Investigating *Pax6* can also show us how eyeless or blind forms of animals have evolved from sighted ancestors. Blind cave fish are blind owing to disruption of their eye development. Expression of their *Pax6* gene is altered accordingly. Rather than being strongly expressed in the optic primordia and across the forebrain (as in sighted surface fish), *Pax6* expression in blind cavefish is reduced and does not cross the forebrain. This modification of Pax6 expression and the evolution of blindness has occurred independently several times in cavefish.

Other examples exist of a homologous homeobox gene being conserved widely across animal phyla and being involved in the development of similar structures. The *Distalless* gene controls the development of appendages in flies and mice and is expressed in appendages or body wall outgrowths across many animal phyla. The same holds true for the *tinman* gene and "heart" (circulatory pump) development, *Otx* in the anterior central nervous system, and *Evx* and *Cdx* in patterning the posterior of animals. From such data, a hypothetical ancestor can be constructed, with homeobox genes controlling the development of the evolutionary precursors of diverse structures and organs.

Examples of homeobox genes playing prominent roles in development are not restricted to animals. The first homeobox gene to be cloned from a plant was *Knotted1* (now known to be in the KNOX class). This gene is involved in leaf development via exclusion from the leaf primordium. Ectopic expression in the primordium produces the "knotted" phenotype whereby a simple leaf becomes lobed. In tomatoes KNOX is expressed within the leaf primordia, and tomato leaves consist of leaflets, reminiscent of the knotted phenotype.

What these homeobox data show us is that homologous developmental genes can be deployed in, for example, the legs of flies and mice and tube feet of echinoderms, but these appendages are not necessarily homologous structures. Homology must be considered at different levels: morphology, developmental gene networks, and individual genes. Homology can be present in one level and not in others. Homologous genes, and even gene networks, can be coopted into the genesis of nonhomologous structures. Nevertheless, homeobox genes allow us to address these issues to get a better understanding of homology of structures and developmental mechanisms, which forms the core of understanding how different forms have evolved. Homeobox genes also indicate some of the molecular components of specific evolutionary events, such as blindness evolving in cavefish and multilobed leaf evolution in plants.

[*See also* Homeosis; *Hox* Genes.]

BIBLIOGRAPHY

Duboule, D., ed. *Guidebook to the Homeobox Genes.* New York, 1994. Useful general homeobox information and historical chapters. The coverage of the genes is now out of date but still provides a good overall impression of their diversity and functions.

Gehring, W. J. *Master Control Genes in Development and Evolution: The Homeobox Story.* New Haven and London, 1998. A personal view of the discovery of the homeobox and the subsequent development of the field, from one of the codiscoverers.

Gehring, W. J., and K. Ikeo. "Pax6: Mastering Eye Morphogenesis and Eye Evolution." *Trends in Genetics* 15 (1999): 371–377.

Panganiban, G., et al. "The Origin and Evolution of Animal Appendages." *Proceedings of the National Academy of Sciences USA* 94 (1997): 5162–5166. Comparative expression data on the expression of the Distalless protein in animal appendages, and a discussion of its significance.

Strickler, A. G., Y. Yamamoto, and W. R. Jeffery. "Early and Late Changes in *Pax6* Expression Accompany Eye Degeneration During Cavefish Development." *Development, Genes and Evolution* 211 (2001): 138–144.

Tsiantis, M. "Control of Shoot Cell Fate: Beyond Homeoboxes." *The Plant Cell* 13 (2001): 733–738. A review of the role of homeobox genes in plant leaf development, and their regulation.

Wray, G. A., and E. Abouheif. "When Is Homology Not Homology?" *Current Opinion in Genetics and Development* 8 (1998): 675–680. A review of the distinction of homology at different levels, of gene and morphology.

— DAVID E. K. FERRIER

HOMEOSIS

Homeosis is the production or occurrence of a structure, such as an organ, in a location on the individual that is atypical for that species. The term was coined in 1894 by William Bateson (1861–1926) in his book *Materials for the Study of Variation Treated with Especial Regard to Discontinuity in the Origin of Species*, which was a compendium of biological variations. Bateson distinguished between discontinuous variations, such as presence or absence of a structure, and continuous variations, such as depth of coloration. He believed, incorrectly, that these two kinds of variation had different causes and that new species arose primarily as a result of novel, discontinuous phenotypic variations. We now know that mutations cause both kinds of variation and that speciation probably usually occurs by the gradual accumulation of small changes.

Hundreds of homeotic mutations have been observed in plants and animals. These mutations are usually highly detrimental or fatal in early life. Some that are not fatal have been selected and cultivated by humans. The approximately 100 species of wild rose, for example, have five petals and a large number of stamens. The cultivated rose has a large number of petals, which develop in place of many of the stamens. Some other well-known homeotic mutations include such "double" flowers (extra petals) in many other horticultural varieties such as lilac; legs replacing antennae in insects; an extra pair of wings in *Drosophila*; extra vertebrae; and riblike structures extending from the anteriormost part of the vertebral column, which normally lacks ribs.

The study of homeotic mutations is currently an active research area that is particularly important because it connects evolutionary biology with molecular genetics and developmental biology. Homeosis and its genetic basis have been most thoroughly studied in three groups of organisms—invertebrates with segmented bodies, vertebrates, and flowering plants—in all of which the presence or absence of a particular organ at a particular location is controlled by genes encoding transcription factors. These transcription factors are homeotic genes: they turn on or off blocks of genes responsible for organ development. The best-studied homeotic genes are the MADS-box genes in plants and the *Hox* (homeobox) genes in animals. When these genes fail to function properly, structures can be produced in abnormal places, resulting in homeosis. The study of homeotic genes is important to evolutionary developmental biology not because organisms evolve by addition of well-formed structures in novel locations, but because the great diversity of forms results from evolutionary modifications of regulatory interactions among genes and not simply the acquisition of new structural genes. Examples of homeotic mutations can be found in the fruit fly (*Drosophila melanogaster*) and flowering plants.

The fruit fly, like all insects, has three main body regions: head, thorax, and abdomen. Each of these regions is further divided into segments that are modified in various ways. For example, the *Drosophila* thorax consists of three segments, each with a pair of legs. The middle segment also bears a pair of wings, and the posterior segment bears a pair of halteres, small balancing organs that resemble vestigial wings. In flies possessing the *ultrabithorax* (*Ubx*) mutation, the homeotic gene *Ubx* fails to function, and the third thoracic segment is converted to the second type (*Ubx* normally functions in the third but not the second segment). The formation of wings on the third thoracic segment is an example of homeosis.

A flower is a short stem with four concentric whorls of highly modified leaves. From outside to inside, these "leaves" are sepals, petals, stamens, and carpels (the pistil), the production of which requires the function of three classes of homeotic MADS-box genes. Each class of gene, termed A, B, and C, normally influences organ identity in two adjacent whorls. In addition, A and C are mutually antagonistic. As a result, a lack of function in one of these genes can permit expression of a transcription factor in an atypical whorl, causing inappropriate organs to form. Many "double" flowers, for example, consist of sepals, petals, petals, and sepals, and are examples of homeosis.

[*See also* Homeobox.]

BIBLIOGRAPHY

Carroll, S. B., J. K. Grenier, and S. D. Weatherbee. *From DNA to Diversity.* Oxford, 2001. An excellent, well-illustrated, and very readable exposition of how gene interactions have evolved to produce the variety of animal forms.

Howell, S. H. *Molecular Genetics of Plant Development.* Cambridge, 1998. The molecular genetics of nearly all aspects of plant development, including floral development.

Meyerowitz, E. M. "The Genetics of Flower Development." *Scientific American* 271 (November 1994): 56–65. Describes the "ABC" model of flower development and the homeotic phenotypes arising from mutations in these *MADS-box* genes.

Gerhart, J., and M. Kirschner. *Cells, Embryos, and Evolution.* Oxford, 1997. A richly informative postgraduate-level text on animal body plans, developmental regulation, and evolutionary diversity.

Gilbert, S. F. *Developmental Biology.* 6th ed. Sunderland, Mass.,

2000. A fine undergraduate textbook on animal development, with descriptions of *Hox* gene action and evolution.

— MARK O. JOHNSTON

HOMINID EVOLUTION

[*This entry comprises five articles:*

An Overview
Ardipithecus and Australopithecines
Early Homo
Archaic Homo sapiens
Neanderthals

The introductory article provides a critical discussion of the hominid radiations arising within hominoids; the companion articles focus on morphological variation, life history, geographical distribution, and phylogenetic relationships. For related discussions, see Archaeological Inference; Human Evolution; Modern *Homo sapiens; and* Neogene Climate Change.]

An Overview

Human evolution has played a major role in the field of evolutionary biology. At the inception of the Darwinian revolution, it was the implications of the general theory for one particular species—humans—that caught the public's imagination and had reverberations across the political and religious fabric of Victorian society. As the nineteenth and twentieth centuries progressed, each new fossil find that showed the link between humans and apes was hailed as a great discovery, and many have become parts of the English language—Neanderthals, Java man, Piltdown man, Lucy. In recent years, genetics has added to this vocabulary with terms such as *African Eve.* Furthermore, beyond the narrow confines of the fossil record, human evolution in a broader sense has been the central battleground for the debates between the social and biological sciences concerning human nature and behavior, as well as framework for developing extensions of classic Darwinian theory—memes and cultural evolution. The reasons for this state of affairs are fairly obvious. Humans are intrinsically interesting to themselves, and so their evolution is generally of greater concern than that of *Caenorhabditis elegans,* the nematode worm. Perhaps more important, their complexity, behavioral flexibility, diversity, and apparent free will represent the greatest challenges to evolutionary theory, and therefore the greatest prospect for refutation—and evolution has been a theory that many have been keen to demolish.

In light of this, the documentation of the pattern of human evolution is significant. Two reasons for this can

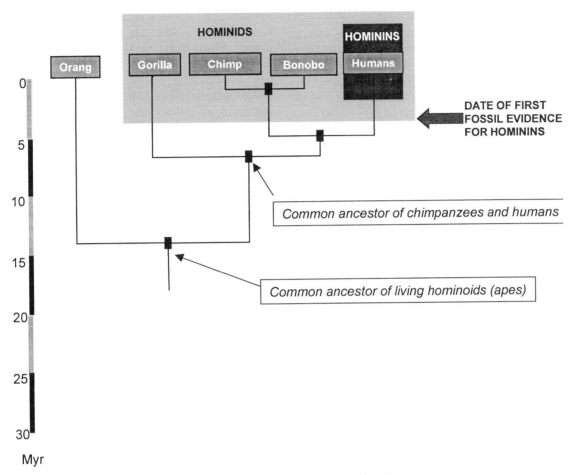

FIGURE 1. Relationships between Humans and Apes, Based on Molecular Genetic Evidence.
This approach has demonstrated the very close relationship between humans and chimpanzees, and the relatively recent age (approximately 5.0 Myr) of their last common ancestor. These results are consistent with the fossil record.   Drawing by Robert Foley.

be emphasized. The first is that such a documentation provides the answer to the question posed by T. H. Huxley in 1860: what is humanity's place in nature? The second is that this pattern can supply insights into the mechanisms by which human evolution has occurred, and thus how complex organisms might have emerged from more simple ones.

Human Lineage. The starting point for any examination of human evolution is to place the human lineage in context. This can be done without recourse to the fossil record. Humans are primates, that is, they belong to the order that includes monkeys and apes. More specifically, humans are hominoids, being more closely related to apes than to monkeys. Although this has been well established for a very long time, molecular evolution has revealed more: that chimpanzees are more closely related to humans than they are to gorillas, and that these three species—humans, chimps, and goril-

las—form, in effect, the African apes, a group separate from the other great ape, the orangutan. [*See* Molecular Evolution.] This, linked to the ability to use the molecular approach to provide chronological estimates, means that it is possible to say, with some certainty, that the first hominids occurred in Africa somewhere between six million and four million years ago (see Figure 1).

Hominin is a term used to define the lineage that evolved from the last common ancestor with the chimpanzees to humans. This term has replaced the term *hominid*, reflecting more accurately the close relationship between humans and African apes. There is now considerable fossil evidence that supports the notion of an African origin for the hominin lineage in the period around five million years ago. The earliest known fossils come from around four and a half million years ago and belong to a species called *Ardipithecus ramidus*. It was found in 1994 in the Middle Awash region of Ethiopia. It is fairly consistent with a very early hominin, having a mixture of both primitive, that is, apelike, and more derived characteristics. What makes it a hominin, and therefore more closely related to humans, is the reduced canine. Most apes have large, daggerlike canines, whereas humans and their ancestors have ones that are reduced and blunted. Although there are other details of the anatomy that are indicative of a hominin, there is little about this fossil that suggests that it is more than a very apelike creature. Other recently discovered material from Kenya, dated to over five million years ago and tentatively named *Orrorin tugenensis*, may throw further light on this early phase of hominin evolution.

From 4.2 million years ago there is far more extensive evidence for the pattern of human evolution. A range of species that are now known (see Table 1) document three major elements to the early phases. The first of these, shown most clearly in *A. anamensis*, *A. afarensis*, and a set of footprints from Laetoli in Tanzania, indicate that some form of bipedalism is present in the early hominins. Upright walking is one of the major, and most divergent, characteristics of humans and has resulted in a major reorganization of the skeleton. It is becoming increasingly clear that, although the early hominins may have retained certain aspects of an arboreal life, the shift to a terrestrial adaptation involving bipedalism was the major evolutionary trend. However, this occurred without any apparent change in other important elements of humans, such as toolmaking and enlargement of the brain.

The second element in the early phase of hominin evolution is that, rather than these first hominids evolving unerringly in a steplike progression toward the modern condition of humanity, they instead diversified or underwent an adaptive radiation. Between four million and two million years ago, there are as many as ten known species of hominin. They are all bipedal and share a number of basic characteristics, such as flatter faces than chimpanzees, smaller canines, and relatively large back teeth suitable for grinding coarse plant matter. This radiation is important, as it shows that the pattern of early hominin evolution is very similar to that of other animals, involving diversification of lineages in relation to local environmental conditions and isolation. There is very little about this radiation that can be seen as a precursor for later human evolution. Rather, it appears to be a radiation of bipedal apes, probably with behavioral systems not dissimilar to those of chimpanzees.

The third element is that this evolutionary phenomenon is, according to current evidence, entirely an African affair. At this time, hominins are known only from eastern and southern parts of Africa. Although this may represent the bias of preservation in the fossil record, it is also probable that it reflects the fact that being a hominin at this time was environmentally restricted to relatively dry and warm habitats.

Pattern of Evolution. This first phase of human evolution, which can be reasonably described as the australopithecine stage, lasts from over four million to close to one million years ago. In terms of evolutionary trends, it is characterized in particular by the evolution of more and more specialized and robust molar teeth, with changes in the associated musculature and anatomy of the skull and jaws. The hominins become increasingly robust, and it appears they were specializing in coarse, hard food, as would be appropriate in the context of the climatic changes that were occurring at this time.

At around two and a half million years ago, when this radiation was developing, and long prior to the extinction of the robust australopithecines, a new trend emerges within hominin evolution. This is characterized by a reversal of the pattern of molar enlargement, a reduction in jaw and face size, and an increase in the size of the brain (Figure 2). Below the head there is also a shift to the rather gracile, cylindrical anatomy that characterizes modern humans with a fully bipedal gait. Although the origins are obscure, this represents the evolution of the genus *Homo*, to which living humans belong.

In contrast to the australopithecines, this genus does show an evolutionary trend indicative of both modern human biology and behavior. Growth patterns become increasingly like those of modern humans, with a delayed pattern of maturation. Stone tool manufacture becomes more commonplace, first with simple chopping tools and flakes (mode 1 industries, by 2.5 million years ago), then with more complex bifacially and often symmetrical shaped tools (mode 2, hand axes by 1.4 million years ago), then with prepared flaking technologies (mode 3, by 300,000 years ago). Finally, brain size increases, reaching modern human levels by 250,000 years ago. With this, hominins move beyond the confines of the African continent and tropical habitats.

The species of *Homo* that spread out during the period after two million years ago, and especially after one million years ago, are known as *Homo ergaster* or *Homo erectus*. The descendants of these species gave rise to modern humans, sometimes about 200,000 years ago. There has been considerable controversy over whether this process occurred gradually and in a mosaic fashion over a wide geographical area, with the retention of regional characteristics of the archaic populations into their modern counterparts, or whether it took place in a restricted, bottlenecked population in Africa relatively quickly, with modern humans emerging and dispersing from this point about 150,000 years ago. Most of the evidence that has accumulated in the last decade has strongly supported the second of these models: the so-called Out of Africa model. Much of this evidence has come from evolutionary molecular genetics, which has

TABLE 1. Summary of the Known Fossil Hominins

Principal Fossil Hominin Groups	Approximate Time Range	Significance
Earliest hominins		
Orrorin tugenensis	>5.0 Myr	Recently discovered and highly controversial earliest hominin from Kenya
Ardipithecus ramidus	±4.4 Myr	Most primitive known hominin from Ethiopia, with strong evidence for link to chimpanzees
Kenyanthropus platyiops	±3.5 Myr	Recently discovered Kenyan hominin with complex mixture of derived and primitive traits
Australopithecines		
Australopithecus anamensis	4.2–3.9 Myr	From Kenya, earliest evidence for clear bipedalism and establishment of australopithecine traits
Australopithecus afarensis	3.7–2.9 Myr	Classic australopithecine features of bipedalism, molarization, and apelike brain size, known from Ethiopia and Tanzania
Australopithecus africanus	3.5–2.5 Myr	First (1924) evidence for the African origin for hominins, now significant for showing diversity in australopithecines and extension to southern Africa
Australopithecus bahrelghazali	±2.5 Myr	Poorly known species, similar to other early australopithecines, but showing extension of geographical range into the Sahel
Australopithecus garhi	±2.5 Myr	Recently discovered Ethiopian species showing mixed australopithecine features with some brain enlargement, possibly associated with stone tools
Robust australopithecines		
Australopithecus aethiopicus	2.6–2.3 Myr	Earliest known australopithecine showing full megadontic (enlarged molars) specializations
Australopithecus robustus	2.0–?1.0 Myr	Megadontic australopithecines from South Africa
Australopithecus boisei	2.3–1.4 Myr	Megadontic australopithecines from East Africa
Transitional forms		
Homo habilis	2.3–1.7 Myr	First hominin to show reduction in face and tooth size and some brain enlargement; controversially placed as the first *Homo*, probably associated with stone tools
Homo rudolfensis	2.3–1.7 Myr	An early hominin showing brain enlargement, but with large teeth; controversially placed in *Homo*
Archaic *Homo*		
Homo ergaster	1.9–?0.9 Myr	Earliest unquestioned *Homo* showing both brain enlargement and modern form of bipedal skeleton; first hominin known outside Africa (Georgia)
Homo erectus	1.9–?0.4 Myr	Form of archaic *Homo*, known principally in Asia.
Homo heidelbergensis	0.7–?0.2 Myr	Robust *Homo* from Africa and Europe, strongly associated with mode 2 (Acheulean hand axes) technologies, showing considerable brain enlargement
Homo antecessor	±0.8 Myr	Controversial hominin with some modern features known from Spain
Homo helmei	0.3–0.2 Myr	Controversial species with larger brains and mode 3 technologies proposed as more derived ancestor of Neanderthals and modern humans
Homo neanderthalensis	?0.2–0.027 Myr	Among the best known fossil hominins, this is a specialized European and western Asian hominin, with very large brain and cold climate specializations
Modern humans		
Homo sapiens	0.14 Myr–present	Fully modern humans, known from Africa from around 140,000 years ago, but elsewhere considerably later (less than 60,000 years ago)

Myr = million years ago

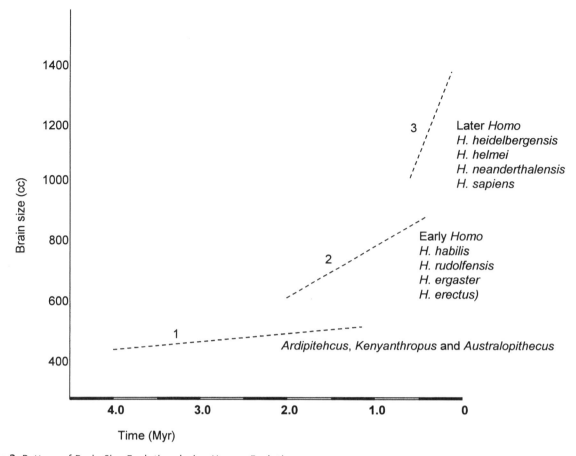

FIGURE 2. Pattern of Brain Size Evolution during Human Evolution.
The rate of encephalization is far greater in the later part of human evolution.   Drawing by Robert Foley.

shown that there is greater diversity in modern human population genetics in Africa than elsewhere, and this is compatible with an older African population. Furthermore, these studies have shown that, compared to other hominoids, humans lack genetic diversity. This is strongly consistent with a recent demographic bottleneck.

Modern humans therefore are thought to have had their evolutionary origins in the archaic populations of Africa about 250,000 years ago. From a small, restricted population, probably occurring in a period of glacial tropical aridity in Africa, there was an expansion of some of these populations, first across Africa, then beyond, such that eventually the world was colonized by modern humans. During this process, it appears that other populations of species of hominins, such as the Neanderthals, which lived in Europe and the Middle East during the last two glaciations or ice ages, became extinct. By 25,000 years ago, in contrast to virtually the whole hominid fossil record of the last five million years, there was only a single surviving species of hominin—*Homo sapiens*.

The evidence concerning human evolution, which is

drawn from comparative anatomy, paleoanthropology, archaeology, and genetics, shows that modern humans are on a continuum with other primates and are closely related to the African apes, thus confirming the implications of Charles Darwin's original evolutionary speculations and Huxley's question concerning the place of humanity in nature. The pattern shown in the fossil record is broadly one of gradual evolution, but this is not a completely constant process, in terms of either rate or direction. The earlier part of hominin evolution is characterized by the presence of a key human trait—bipedalism—but in other ways is markedly divergent. It shows both the retention of many conservative apelike characteristics of behavior and adaptation and the evolution of dental specializations that do not occur in modern humans. In that sense, the evolution of *Homo* represents only a part of the overall pattern of hominin evolution.

Within *Homo* there is a pattern of gradual emergence of modern human behavior, but again this is not a constant process. The anatomy and the archaeology related to *Homo ergaster* and *Homo erectus* show remarkable stability over a long period of time, and it is really only in the last 250,000 years that there is a rapid acceleration in evolutionary rate and the emergence of modern human behavior.

The issue of whether human evolution can throw light on evolutionary processes and their limits is harder to resolve. The pattern of the fossil record does show patterns that are consistent with adaptive principles in relation to changing climates and environments, as well as increased behavioral complexity. The fact that both australopithecines and *Homo* species undergo adaptive radiations also shows that they conform to the pattern of other lineages. What remains to be resolved is when key cognitive traits such as language, theory of mind, and complex social structures and institutions develop, and the impact they have on the evolution of modern human diversity. It is clear that these events occurred in the last 5 percent of hominin evolution, but whether this was a gradual process over 250,000 years or a major evolutionary leap occurring in a short-lived speciation event remains to be determined with any precision.

Explaining Human Evolution. The fossil and ge-

netic evidence provides the empirical basis on which it is possible to reconstruct the history of the human lineage—in other words, to provide a description of what happened in the past. This is important, because it provides one key element in any Darwinian explanation: the context for events, the environment in which particular evolutionary novelties may have proved advantageous. Another element, however, is to provide an explanation for why the hominin lineage evolved in the way it did, as well as human adaptations have been successful in those particular contexts.

There have been many such explanations for the evolution of humans. One such set of explanations often isolates one key hominin trait and sees that as the trigger for all other changes. Many such triggers have been identified and located in different hominin fossils. These would include bipedalism as the basis for carrying and food sharing, hunting as the basis for cooperation, and tool use and manufacture as the basis for complex cognition and planning. Perhaps one of the key empirical findings of the last half century is that none of these triggers seem to be sufficient, for the fossil record shows

FIGURE 3. Summary Diagram of Human Evolution, Showing the Two Main Radiations—Early Hominins and *Homo.* The principal pattern is one of diversification, with many species existing, often at the same time. The current situation with only one hominin species extant is very rare. Constructing a detailed phylogeny or set of evolutionary relationships between these species has proved extremely difficult, probably because of the high rate of convergent evolution among them.   Drawing by Robert Foley.

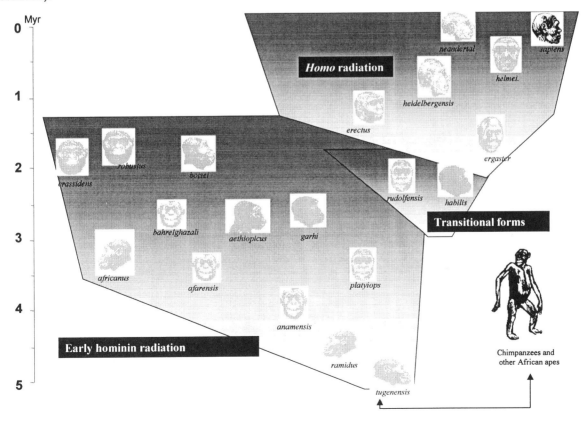

that humanity does not arrive as a package, but different elements can be dispersed over millions of years—bipedalism, for example, precedes both tool use and brain enlargement by several million years. Thus, there is a need for a more complex set of explanations.

One basis for current explanations relates to climatic change. The period in which the hominins evolved is one of marked climatic cooling—the onset of the ice ages or glaciation periods. The pattern of climatic change across the last five million years is one of progressive average cooling, with full glacial conditions being present by one million years ago. However, this is not a steady state; prior to the full glaciation periods, there were marked oscillations on a relatively long time scale, and after one million years ago there was a regular cyclicity to periods of full glaciation and interglacials. If climate is a driving force in human evolution, then it can operate both through actual changes in climate and environment and in increased variability.

Climatic change can be said to have had a number of marked effects on human evolution. The most profound of these is the evolution of bipedalism itself. The global cooling that occurred in the Pliocene epoch (5.0–1.6 million years ago) affected tropical Africa by making it more arid, and there was thus a reduction in forest distributions and an expansion of woodland and open habitats. It is apparently in these habitats that hominins first evolved specialized, terrestrial bipedalism, even if they still continued to feed, as some have argued, in the trees.

The second effect occurred in the Pleistocene epoch (1.6 million–10,000 years ago). During this epoch, the northern latitudes were subjected to periods of extreme glaciation, interspersed by shorter, warmer periods, such as the one we are living in at the moment. The warmer periods seem to have promoted population expansions and dispersals into higher latitudes, and thus the colonization of Eurasia. However, cold phases interrupted these expansions and caused widespread depopulation and extinction. It was these oscillations that created the demographic context in which small, isolated populations could adapt or drift, resulting in the complex evolutionary geography of the later hominins that we now recognize.

If the climate provides the context for any explanation of the pattern of hominin evolution, the other ingredient required is that of behavior. From the morphological and archaeological record, it is clear that the hominins changed across this time. Most paleoanthropologists agree that there are a number of key turning points in human behavioral evolution. The first of these is the switch to a more fully committed terrestriality among early hominids, resulting in megadontic specializations and range expansions in African savannas. The second probably occurs with the evolution of *Homo ergaster*, when a shift to a more human life history strategy resulted in larger brains, a more meat-based diet, more tool use, greater habitat tolerance, and more complex social organization. There has been considerable debate as to whether this occurs as a result of changes in male reproductive strategies in male–kin alliances, made possible by greater hunting abilities, or by changes in female strategies in response to cooperation between different generations of females (related to the evolution of menopause). Finally, there is a further major change within the last 250,000 years, associated with the acceleration of brain evolution. It has been argued that this change reflects the evolution of full modern cognition, associated with greater social complexity and the origins of language. Again, whether this occurs specifically among modern humans or is part of a broader evolutionary event, including Neanderthals, is debated.

What all these explanations have in common is that they envisage a close relationship between social and cognitive evolution—and hence cultural capacities—and ecological and biological circumstances. In other words, modern approaches to human evolution place much greater emphasis on the Darwinian package—selection favoring behaviors that enhance reproductive success, integrated fully into ecological and developmental systems—rather than seeing human cultural and cognitive development as a separate issue from the rest of evolutionary biology.

[*See also* Archaeological Inference; Geochronology; Modern *Homo sapiens, article on* Neanderthal–Modern Human Divergence; Neogene Climate Change.]

BIBLIOGRAPHY

Foley, R. A. *Humans Before Humanity: An Evolutionary Perspective.* Oxford, 1995.
Johanson, D., and B. Edgar. *From Lucy to Language.* London, 1996.
Klein, R. G. *The Human Career.* Chicago, 1999.
Lewin, R. *Principles of Human Evolution.* Oxford, 1999.

— ROBERT FOLEY

Ardipithecus and *Australopithecus*

The fossil record for the first half of human evolution is exclusively African (hominids are human ancestors and near-relatives that postdate our evolutionary divergence from chimpanzees). Since 1924, finds of African *Australopithecus* fossils and their contexts have pushed knowledge of human ancestry to 4.1 million years ago (mya). The fossil record for earlier exclusively human ancestors, however, was extremely poor until the 1990s. Recent discoveries have now extended knowledge of the earliest hominids almost to the last branching point between modern apes and humans at approximately 6 to 8 million years ago.

Ardipithecus is recognized as the earliest hominid because it possesses anatomical characters shared exclusively with later *Australopithecus* species. Several

Australopithecus species are recognized, in at least two lineages. The genus spans the period approximately 4.1 to 1.4 mya and is known from South Africa, Malawi, Tanzania, Kenya, Ethiopia, and Chad. All *Australopithecus* species exhibit postcanine megadontia and skeletons adapted to bipedality. Brain sizes were within ranges for the extant apes. *Australopithecus* is widely recognized as the genus ancestral to relatively more encephalized and mostly later species of *Homo*.

The first *Australopithecus* fossil was found in 1924. In 1936, the first recognized adult *Australopithecus* specimen was recovered from Sterkfontein, South Africa. Two years later, Robert Broom published some adult remains from Kromdraai as *Paranthropus robustus*, calling attention to the many specialized craniodental features of this species. By the 1950s, more complete fossils had been recovered from the South African sites, including postcranial fossils that demonstrated that *Australopithecus* was a hominid. It was already clear that the specialized *Australopithecus robustus* was not a human ancestor, but rather a separate hominid lineage.

In 1959, the focus of early hominid recovery shifted from southern to eastern Africa with the recovery of a young adult male skull by Mary Leakey at Tanzania's Olduvai Gorge. Louis Leakey named the new form *Zinjanthropus boisei* and first attributed to it the contemporary Oldowan lithic technology that lay in close spatial proximity. The application of the potassium-argon radioisotopic technique to nearby volcanic horizons dated this specimen at what was then a surprising 1.75 million years. The contemporary *Homo habilis* was found shortly thereafter, and a debate ensued about the taxonomic status of these fossils.

Work in Kenya's Turkana Basin during the 1970s dramatically increased sample sizes of both early hominid lineages and extended the record more deeply in time. In 1975, a very early cranium of *Homo erectus* (KNM-ER 3733) was recovered from Koobi Fora sediments that had also yielded *A. boisei*. This falsified the long-standing "single species" hypotheses about early human evolution. When hominid fossils older than 3.0 mya were found in the 1970s at Laetoli (Tanzania) and Hadar (Ethiopia), they were first assigned to *Homo* because they showed none of the specializations of *A. boisei*. However, subsequent analysis by Don Johanson and Tim White led to the 1978 description of the combined Laetoli and Hadar samples as belonging to a separate species within the genus *Australopithecus*, *A. afarensis*. The discovery of a fairly complete partial skeleton from Hadar in 1974 (A.L. 288-1; "Lucy") and three sets of fossilized footprints at Laetoli in 1978 represented important new evidence from a previously mostly unknown time.

Controversies about these fossils, dated to 3.0 to 3.5 mya and attributed to *Australopithecus afarensis*, ensued during the 1980s. These debates involved the distinctiveness of *A. africanus* from *A. afarensis*; whether Hadar *A. afarensis* was a single species; whether Laetoli and Hadar sampled the same taxon; and whether the locomotor habits of these earliest hominids involved a significant degree of arboreality. In 1986, A. Walker and colleagues described a cranium west of Lake Turkana as an early representative of *A. boisei* (the "black skull"). Most workers saw this fossil as validating the existence of an ancestral robust species, *A. aethiopicus*, distinct from *A. boisei*, but the link between *A. afarensis* and early *Homo* remained elusive.

In the early 1990s, work resumed at Hadar and in the Middle Awash of the Ethiopian Afar. This research generated greatly enlarged samples of *A. afarensis*. Work by Alun Hughes, Ron Clarke, and Phillip Tobias at Sterkfontein in South Africa also led to dramatically enlarged samples of *A. africanus*. The 1990s also saw the discovery of remains attributed to *A. boisei* in Malawi and at the site of Konso-Gardula, southern Ethiopia, where Gen Suwa and Yonas Beyene found the first associated mandible and cranium of this species. The important South African *A. robustus* site of Drimolen was opened by Andre Keyser in the 1990s. Overshadowing recovery of these additional samples of previously recognized *Australopithecus* species in the 1990s was a burst of discovery of previously unknown hominid species in Kenya and Ethiopia.

In 1994, the Middle Awash paleoanthropological project described a new, 4.39 mya hominid species, *A. ramidus*, based on its 1992 and 1993 discoveries at Aramis, Ethiopia. More fossils of this form were subsequently recovered, and the new genus name *Ardipithecus* was made available to receive this species. Later in 1995, Meave Leakey and colleagues announced the new species *Australopithecus anamensis*, based largely on material from Kanapoi, Kenya. In 1995, limited craniodental remains initially attributed to a new species of *Australopithecus* were recovered in Chad, central Africa, by a team led by Michel Brunet. In 1998, Ron Clarke announced the discovery of a partial skeleton of *Australopithecus* deep within the Sterkfontein cave complex. In 1999, a 2.5 mya partial cranium with dentition, from Bouri in the Middle Awash study area of Ethiopia, was named *Australopithecus garhi* by Asfaw and colleagues; simultaneously, evidence for contemporaneous butchery of large mammals with stone tools such as those found at Gona was announced. In early 2001, Meave Leakey and colleagues attributed a poorly preserved cranium to a new hominid genus and species, "Kenyanthropus platyops."

There are no fossil representatives of the lineages leading to the chimpanzee, bonobo, or gorilla; however, recently discovered fossils assigned to *Ardipithecus* from the Middle Awash, Ethiopia, date to over 5.6 mya, very close to the times of our common ancestor. These fossils pre-date Middle Awash remains of *Ardipithecus* dated to 4.39 mya, pushing back the earliest date for hominid fossils. This genus of hominid inhabited a wooded setting. Body size appears to have been within the *Australopithecus* range. The cranial remains are more primitive than those of *Australopithecus*; there is less postcanine megadontia than in *Australopithecus*; incisors are not expanded as in chimpanzees; and the adult canines are small and feminized, but large relative to the posterior dentition. The permanent third premolars and first deciduous premolars are unmolarized relative to *Australopithecus*, but the canine/premolar complex is non-honing. *Ardipithecus* tooth enamel is relatively and absolutely thin compared to *Australopithecus*.

The lack of a honing canine/premolar complex in *Ardipithecus* identifies it as a hominid. Aspects of its postcranial skeleton are consistent with regular bipedality. Phylogenetically, *Ardipithecus* could represent the direct ancestor of its slightly younger, more derived close relative, *Australopithecus anamensis*. This interpretation would imply a period of rapid anatomical change during the 220,000-year interval between 4.39 and 4.17 mya. Alternatively, *Ardipithecus ramidus* may represent a relict species (persistent mother species) little changed from the ancestor of *Australopithecus*. Resolution of this phylogenetic issue can be determined only by additional discoveries. There is currently nothing about the morphology of *Ardipithecus ramidus* that would exclude this taxon as the ancestor of later hominids.

The genus *Australopithecus* contains seven widely recognized species. The earliest, *A. anamensis*, is the most primitive but is not skeletally well known. Its direct lineal descendant is the well-known *A. afarensis*, which is documented from most of eastern Africa and dates between 3.6 and 2.9 mya. The anatomy of *A. afarensis* is better known than any other *Australopithecus* species, with most skeletal elements represented and sets of footprints available to show that it was habitually bipedal. The *A. afarensis* foot has longitudinal and transverse arches, an expanded heel, and an aligned big toe. The knee and pelvis are humanlike; the hand has a more gracile thumb than that of modern humans; and the foot and hand phalanges are slightly longer and more curved than in humans. *A. afarensis* limb proportions, known only from a single individual (A.L. 288-1; Lucy), feature a long forearm and a femur length intermediate between those of apes and humans. Body size in *A. afarensis* is moderately sexually dimorphic, and the body size range is large. This species is characterized by several apelike cranial and dental features. The palate is shallow; tooth rows range from parallel to convergent posteriorly (often with diastema); and cranial capacity ranges from about 380 to 430 cubic centimeters on a small sample.

There is strong sexual dimorphism in cranial size, cresting, and pneumatization. Compared with later *Australopithecus* species, upper and lower canines are large relative to postcanine teeth, and the lower third premolar is sometimes unicuspid. Various attempts have been made to divide the Hadar sample into separate taxa, and to link one or more of these subdivisions with later hominid species; however, such interpretations have proven contradictory among themselves, and most often they have been found to be based on fallacies in statistical manipulation, or on errors of basic observation or logic.

Australopithecus afarensis is known to have been distributed from Chad in the west to Ethiopia in the east and Tanzania to the south. The recent discovery of Sterkfontein's oldest fossil hominid, the StW 573 partial skeleton, may dramatically extend the known geographic range of *A. afarensis*. This may belong not to *A. africanus* but rather to *A. afarensis* or to a very close relative that may represent the immediate ancestor of *A. africanus*. Indeed, even before the discovery of this skeleton, it was widely held that *A. afarensis* was a suitable ancestor for all later hominid species. A Kenyan cranium dating to approximately 3.5 mya has recently been claimed to signal a new genus and species contemporary with *A. afarensis*; however, its morphology has been highly altered by distortion, and it most likely represents a small *A. afarensis* individual. Evidence for multiple hominid lineages before 3.0 mya is, therefore, still equivocal.

All later species of *Australopithecus* are variations on the basic theme revealed by the *A. afarensis* fossils. Where known, these species are represented by basically indistinguishable postcranial skeletons. The species are, therefore, recognized exclusively by craniodental features. The evolutionary relationships of these species are difficult to evaluate because of the fragmentary nature of the fossil record and the fact that there was considerable parallel evolution among them.

The only stratigraphic succession that currently bears directly on the limited early hominid diversification after 3.0 mya is Ethiopia's Omo Shungura sequence. Unfortunately, this succession does not reveal many details about the nature of the evolutionary processes through which at least two species lineages arose from populations of *A. afarensis*. One of these lineages was the time-successive series *A. aethiopicus* to *A. boisei*. The other was a lineage ancestral to the genus *Homo*. However, even the question of how many contemporary early *Homo* species existed is controversial. Based on teeth, the cladogenesis between robust and nonrobust lineages occurred before 2.7 mya. Specimens belonging to the robust lineage (*A. aethiopicus*) appear for the first time in Omo Shungura Member C, about 2.6 mya. The well-preserved cranium of this taxon, KNM-WT 17000 from West Turkana, is only slightly younger.

Many workers have adopted the position that the three robust *Australopithecus* species form a monophyletic group (sometimes called *Paranthropus*), with *A. aethiopicus* as the common ancestor. This position is based largely on extensive trait lists used in numerical cladistic analyses. However, these traits, or characters, are not necessarily independent: virtually all are related directly to basic masticatory adaptations involving megadontia, small anterior teeth, and large muscles of mastication. Similar dentognathic adaptations have independently arisen numerous times among primates, and even early *Homo* species evolved many of these characters in parallel. It remains a possibility that *A. robustus* evolved independently from *A. africanus* in southern Africa. It is also possible that *A. africanus* evolved from *A. afarensis* to become an endemic South African form that left no descendants.

The 1999 description of the megadont, nonrobust species *Australopithecus garhi* from the Middle Awash emphasized the still poor understanding of early hominid phylogenetic history after 3.0 mya. The species name, *garhi*, means "surprise" in the Afar language, and *A. garhi* was completely unpredicted by any phylogenetic analysis that preceded it. The discovery of *A. garhi* and the recognition of rampant homoplasy in early hominid masticatory systems have resulted in an unresolved polychotomy in early hominid phylogeny that involves *A. garhi*, *A. africanus*, *A. robustus*, *A. aethiopicus*, and early *Homo* (whether one or more species). It is likely that this uncertainty will persist until the fossil record is substantially augmented for the time period 2–3 mya.

[*See also* Geochronology; Neogene Climate Change.]

BIBLIOGRAPHY

Asfaw, B., T. D. White, C. O. Lovejoy, B. Latimer, and S. Simpson. "*Australopithecus garhi*: A New Species of Early Hominid from Ethiopia." *Science* 284 (1999): 629–634. The first publication on this species, and a discussion of early hominid evolutionary relationships.

Broom, R., and G. W. H. Schepers. "The South African Fossil Ape-men—the Australopithecinae." *Transvaal Museum Memoir* 2 (1946): 1–272. A review of the material available before *Australopithecus* was widely accepted as a hominid.

Ciochon, R. L., and J. G. Fleagle, eds. *The Human Evolution Source Book*. Englewood Cliffs, N.J., 1993. A collective volume that reprints classic papers across the entire span of human evolution.

Dart, R. A. "*Australopithecus africanus*: The Man-Ape of South Africa." *Nature* 115 (1925): 195–199. The classic first description.

Grine, F. E., ed. *Evolutionary History of the "Robust" Australopithecines*. New York, 1988. A collective volume with a wide variety of papers on the biology of these near-relatives of our ancestors.

Haile-Selassie, Y. "Late Miocene Hominids from the Middle Awash, Ethiopia." *Nature* 412 (2001): 178–181. Describes *Ardipithecus* fossils dating to 5.2–5.8 mya.

Johanson, D. C., et al. "Pliocene Hominids from Hadar, Ethiopia." *American Journal of Physical Anthropology* 57 (1982): 373–719. Detailed descriptions and illustrations of the original Hadar discoveries, including "Lucy."

Leakey, M. D., and J. M. Harris, eds. *Laetoli: A Pliocene Site in Northern Tanzania*. Oxford, 1987. The comprehensive monograph on the footprint site.

Leakey, M. G. "New Four-Million-Year-Old Hominid Species from Kanapoi and Allia Bay, Kenya." *Nature* 376 (1995): 565–571. Description of the species *Australopithecus anamensis*.

Leakey, M. G., F. Spoor, F. H. Brown, P. N. Gathogo, C. Kiarie, L. N. Leakey, and I. McDougall. "New Hominid Genus from Eastern Africa Shows Diverse Middle Pliocene Lineages." *Nature* 410 (2001): 433–440. The latest claim of diversity early in the hominid fossil record.

Meikle, W. E., and S. T. Parker, eds. *Naming Our Ancestors*. Prospect Heights, Ill., 1994. An excellent background to nomenclature of hominid fossils.

Reader, J. *Missing Links: The Hunt for Earliest Man*. Boston, 1981. Richly illustrated volume with an engaging narrative about the history of discovery and interpretation of hominid fossils.

Robinson, J. T. *The Dentition of the Australopithecinae*. (Transvaal Museum Memoir 9.) A classic description of the original collections; many new fossils have been found, many still not published.

Tobias, P. V. *Olduvai Gorge*. vol. 2, *The Cranium and Maxillary Dentition of* Australopithecus (Zinjanthropus) boisei. Cambridge, 1967. A classic monograph in the field.

White, T. D., G. Suwa, and B. Asfaw. "*Australopithecus ramidus*, a New Species of Early Hominid from Aramis, Ethiopia." *Nature* 371 (1994): 306–312. The original description of the species; more fossils have been found and are under analysis.

— TIM WHITE

Early *Homo*

Homo habilis, *Homo rudolfensis*, *Homo ergaster*, and *Homo erectus* form the early part of the genus *Homo* from the Late Pliocene epoch in Africa to the Late Pleistocene epoch in Indonesia. Some have suggested that the first two taxa should be placed in *Australopithecus* and that it is only the appearance of *H. ergaster* that marks the emergence of genus *Homo*. The *Homo* clade probably originated in Africa between two and three million years ago. The earliest specimen regularly attributed to the genus (broadly defined) is a maxilla from about 2.3 million years ago found in sediments in Ethiopia, but it is too fragmentary to be placed in a species.

Members of the genus *Homo* are distinguished from the ancestral genus *Australopithecus* by morphological characteristics such as larger brains and reduced dentition, as well as behavioral traits, such as the ability to manufacture and use stone tools. *H. ergaster* and *H. erectus* are also larger in body size than australopithecines. Like modern humans, their thorax shape and limb proportions indicate full commitment to walking/run-

THE TURKANA BOY

While surveying exposures on the western shore of Lake Turkana, Kenya, Kamoya Kimeu (a highly successful Kenyan fossil hunter) spotted a piece of hominid frontal bone the size of a matchbox. The rest of the associated fossils constituted the most complete early hominid skeleton ever collected (Figures 2 and 3). The bones of the juvenile *Homo erectus*, dubbed "Turkana boy," were situated on the south bank of the Nariokotome River in sediments dating to 1.53 million years.

This extraordinary discovery brought several novel insights to the study of early *Homo*. Based on tooth eruption and growth plate fusion in the long bones, the boy was estimated to have died at eleven years of age. Already at that age the boy stood 5'3" (160 cm) and was projected to grow to a height of 6'1" (185 cm), depending on the growth model chosen. Like stature, the limb proportions of the Turkana boy were within the range of modern humans. His distal limb segments (forearms and shanks) were long relative to his proximal limb segments (upper arms and thighs), similar to the pattern found in equatorial peoples today.

For the first time, researchers were able to examine the bones of *Homo erectus* for possible indicators of language. Evidence from the vertebrae of the Turkana boy suggest that language did not evolve until after the Middle Pleistocene epoch. The narrow vertebral canal of *H. erectus* suggests that the spinal cord was too small to contain the motor neurons needed to control breathing patterns in human speech. *Homo erectus* probably did not have language as we know it today.

—HOLLY M. DUNSWORTH AND ALAN WALKER

hominid) 7. The word *habilis* means "able" or "handy" and refers to somewhat controversial evidence of tool-making capabilities in the hand morphology of OH 7.

Many workers at the time contested the validity of the new species *H. habilis*. They argued that there simply was not enough morphological room in the hominid line for another species between *Australopithecus africanus* and *Homo erectus*. Since its establishment, the *H. habilis* hypodigm (the collection of known material belonging to a species) has grown to include many specimens from Olduvai such as a fragmentary, elderly skeleton (OH 62) and a cranium from Koobi Fora (KNM-ER 1813; Figure 1).

The discoverers of OH 62 argued that *H. habilis* was like *Australopithecus* in terms of size and limb proportions. However, the reliability of estimating the length of the lower limb from this fragmentary skeleton has been questioned. The postulated transition from a gracile *Australopithecus* to *H. habilis* is marked mainly by changes in craniofacial morphology. These include an increase in cranial capacity, less cranial buttressing, and smaller teeth.

Homo rudolfensis. Koobi Fora sediments also preserved the cranium KNM-ER 1470 (Figure 1), dated to about 1.9 million years. The mosaic of cranial features of 1470 is perplexing and has led to the designation of another *Homo* species, *H. rudolfensis*. The cranium has a larger volume (750 ml) than the contemporaneous KNM-ER 1813 *H. habilis* specimen (510 ml), to which it is most often compared. However, 1470 has some primitive *Australopithecus*-like features in the face and, as judged from the roots, had very large teeth.

Some experts lump all the fossils of early *Homo* from around 2.0 million years ago into *H. habilis*, thus deemphasizing the morphological disparity between the spec-

ning on two legs. Body size, tooth eruption patterns, and other aspects of skeletal development also indicate that, like modern humans, they probably matured later than did the australopithecines.

The genus *Homo* may have been the first hominid to exist in and be adapted to an open environment. It has an unprecedented geographic range for a primate. *Homo erectus* fossils are found across the Old World, marking the first extra-African dispersal of a species in the family Hominidae.

Homo habilis. *Homo habilis* is known best from Late Pliocene deposits at Olduvai Gorge on the western flank of the Rift Valley of Tanzania, and from the fossil-rich locality known as Koobi Fora on the eastern shores of Lake Turkana, Kenya. Louis Leakey and colleagues first described the species in 1964 based on fossils from Olduvai Gorge, including the type specimen OH (Olduvai

FIGURE 1. Frontal Views of Early *Homo* Crania. KNM-ER 1813 (left) is usually considered to be *H. habilis*, and KNM-ER 1470 (right) is *H. rudolfensis*. Photograph Alan Walker © National Museums of Kenya.

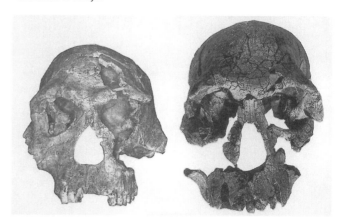

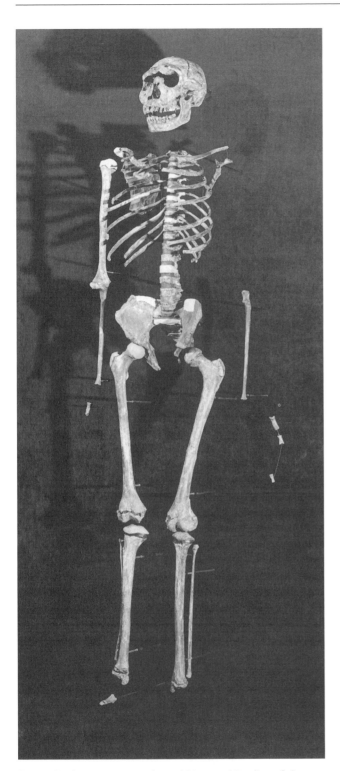

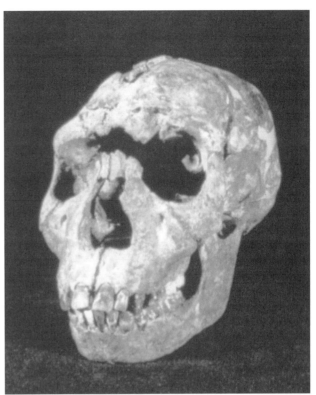

FIGURE 3. Three-Quarters View of Skull of the Turkana Boy.
Photograph Alan Walker © National Museums of Kenya.

FIGURE 2. Three-Quarters View of Mounted Replica of the Turkana Boy Skeleton KNM-WT 15000. Some researchers call this specimen an early *Homo erectus,* whereas others refer to it as *H. ergaster.*   Photograph Alan Walker © National Museums of Kenya.

imens. However, others distinguish specimens based on size and call the small individuals *H. habilis* and the large ones *H. rudolfensis*. In a similar fashion, isolated *Homo* limb bones from this time period are usually assigned to *H. rudolfensis* if they look like those of *H. erectus*. An earlier fossil specimen recently assigned to a new genus, *Kenyanthropus*, has some similarities with *rudolfensis* fossils, and it has been suggested that the *rudolfensis* material might be better accommodated by this new genus.

Homo erectus (Including Homo ergaster). The first known specimen of *Homo erectus* was discovered in Java, Indonesia. In 1891, Eugène Dubois, a Dutch anatomist, discovered a skullcap and femur at Trinil. He called the new species *Pithecanthropus erectus*, meaning "ape-man who walked erect," and considered it the "missing link" between apes and humans.

Today we include the Trinil fossils in the widespread species *Homo erectus*. Fossils of the species have been collected from South Africa, Tanzania, Kenya, Ethiopia, Algeria, Morocco, Italy, Germany, Georgia, India, China, and Indonesia. The oldest specimens come from Africa, the Caucasus, and Java and are dated at about 1.8 million years. These very early dates outside of Africa indicate that *H. erectus* dispersed across the Old World

FIGURE 4. A Collection of Acheulian Hand Axes Undisturbed Since Their Discovery in 1942 at Olorgesailie, Kenya.
Photograph Alan Walker © National Museums of Kenya.

almost instantaneously, as soon as the species arose in Africa. Africa is the most likely origin for *H. erectus* because it is the only continent with the remains of antecedent hominids. *Homo erectus* persisted very late in the Pleistocene epoch in Indonesia to possibly as late as 30,000 years ago, which suggests that the species survived in isolation while modern humans spread everywhere in the Old World.

Homo erectus is the first hominid species to reach modern human body size and limb proportions (see Vignette). The cranial capacity of *H. erectus* is on average about three quarters (900 cc) that of modern humans (1,300 cc), but the postcranial anatomy of *H. erectus* is barely distinguishable from that of modern humans.

The most common characteristics used to define *H. erectus* are located on the skull. These features include, for example, a long and low skull with the greatest breadth toward the base of the cranium, a flat, receding frontal bone, a pronounced supraorbital torus (browridge) with an associated supratoral sulcus (furrow) behind it, sagittal keeling, significant postorbital constriction, an occipital torus (a strong angle at the back of the skull), and thick cranial vault bones. Some experts argue that these features are unique to Asian *H. erectus*, thus advocating a separate species for the African specimens. When this taxonomic scheme is favored, *H. erectus* is used only for the Asian assemblage, and *H. ergaster* is given to the African and Georgian fossils.

Archaeology, Behavior, and Ecology. The oldest, most primitive tools belong to what is called the Oldowan tradition because much of the record consists of worked lavas and quartzites from Olduvai Gorge. The Oldowan industry, which arose as early as 2.6 million years ago, is also known from Koobi Fora and other sites in North and South Africa.

The Acheulean industry followed the Oldowan tra-

dition in Africa around 1.5 million years ago, and is credited as the handiwork of *Homo ergaster*. Large, bulky, bifacial tools such as hand axes, picks, and cleavers characterize the Acheulean tradition. With quite an abundant record of Early Stone Age material, archaeologists have investigated the functional use of early tools and why hominids took up toolmaking. Stone tools were most likely used to break open bones to get the high-quality marrow, to cut through skin as a set of second teeth, to disarticulate large animals for transport, or perhaps for making wooden tools.

The first bones that exhibit cut marks made by stone tools (distinguished microscopically from other damage) come from deposits in Ethiopia dated to 2.5 million years. There is a direct causal association between stone tools and cut-marked bovid fossils at Olduvai Gorge. Bones with both carnivore tooth marks and cut marks are found throughout Olduvai. Carnivore tooth marks superimposed over hominid cut marks show that carnivores scavenged from hominids, but cut marks have also been found over tooth marks, showing also that hominids scavenged from carnivores. There is general agreement among archaeologists that hominids would not have been effective primary predators of large ungulates in the Paleolithic period. But debate continues about whether they were successful at displacing large carnivores soon after a kill, or only later when most of the meat had been consumed.

Many authorities agree that this transitional scavenging period would have naturally led up to full-fledged hunting. With the emergence of *Homo erectus*, dietary dependence on meat increased and hunting may have become crucial. Mammalian predators have very large geographic ranges. The wide distribution of *H. erectus* fossils, combined with archaeological evidence, suggests that it was indeed a predator. Nevertheless, some suggest that other foraging strategies may have been key to this range expansion, and argue that clear evidence of full-fledged big game hunting emerges only within the last 500,000 years.

Evolution. We have not yet narrowed down a single contender for the apical ancestor of the genus *Homo*. *Australopithecus africanus* is often rooted just below *H. habilis*, but it has also been placed in the genus *Homo*. *Australopithecus garhi* is an equally appropriate candidate for the ancestor to *Homo* because of its *Homo*-like tooth and palate shape. The development of stone tools and the shift to a more carnivorous diet could have led to adaptations for small teeth.

[*See also* Archaeological Inference; Geochronology; Neogene Climate Change.]

BIBLIOGRAPHY

Capaldo, S. D. "Experimental Determinations of Carcass Processing by Plio-Pleistocene Hominids and Carnivores at FLK 22

(Zinjanthropus), Olduvai Gorge, Tanzania." *Journal of Human Evolution* 33 (1997): 555–597. Reviews the arguments and evidence for meat acquisition at an especially important early site.

Conroy, G. C. *Reconstructing Human Origins*. New York, 1997. Excellent for both introductory and advanced readers.

Johanson, D. C., et al. "New Partial Skeleton of *Homo habilis* from Olduvai Gorge, Tanzania." *Nature* 327 (1987): 205–209. Description and discussion of the OH 62 remains.

Leakey, L. S. B., P. V. Tobias, and J. R. Napier. "A New Species of the Genus *Homo* from Olduvai Gorge." *Nature* 202 (1964): 7–9. The authors name the new species *H. habilis*, as well as redefine the genus *Homo*.

Leakey, M. D. *Olduvai Gorge*, vol. 3. Cambridge, 1981. An authoritative account of the early tools from Olduvai Gorge.

O'Connell, J. F., K. Hawkes, and N. G. Blurton Jones. "Grandmothering and the Evolution of *Homo erectus*." *Journal of Human Evolution* 36 (1999): 461–485. Presents arguments and evidence for the hypothesis that changes in female foraging strategies and consequent life history changes explain the evolution of *H. erectus*.

Potts, R., and P. Shipman. "Cut Marks Made by Stone Tools on Bones from Olduvai Gorge, Tanzania." *Nature* 291 (1981): 577–580. A classic study investigating the evidence for scavenging versus hunting.

Rightmire, G. P. *The Evolution of* Homo erectus: *Comparative Anatomical Studies of an Extinct Human Species*. Cambridge, 1990.

Schick, K. D., and N. Toth. *Making Silent Stones Speak: Human Evolution and the Dawn of Technology*. New York, 1993. All about making, using, and interpreting stone tools.

Semaw, S., et al. "2.5-Million-Year-Old Stone Tools from Gona, Ethiopia." *Nature* 385 (1997): 333–336. The oldest stone tools on record.

Shipman, P. *The Man Who Found the Missing Link*. New York, 2001. A biography of Eugene Dubois, discoverer of the *Homo erectus* from Trinil, Java.

Shipman, P., and A. Walker. "The Costs of Becoming a Predator." *Journal of Human Evolution* 18 (1989): 373–392.

Walker, A., and R. Leakey. *The Nariokotome* Homo erectus *Skeleton*. Cambridge, Mass., 1993. A monograph dedicated to the Turkana boy remains.

Wood, B. A. "Origin and Evolution of the Genus *Homo*." *Nature* 355 (1992): 783–790.

Wood, B. A., and M. Collard. "The Human Genus." *Science* 284 (1999): 65–71.

— HOLLY M. DUNSWORTH AND ALAN WALKER

Archaic *Homo sapiens*

The term archaic *Homo sapiens* has been applied to a number of Middle Pleistocene fossils. This period spans from 780,000 years ago to 128,000 years ago. The lower limit corresponds to a change in magnetic polarity of the earth, that is, the transition from the Matuyama chron of reversed polarity to the Brunhes chron of normal polarity (the current polarity). The upper limit of the Middle Pleistocene coincides with the beginning of the last interglacial period and a marine transgression associated with the melting of the ice sheets.

In the 1980s, some authors started to sort the Middle Pleistocene human sample into two groups, with the fossils from East Asia assigned to the species *Homo erectus* (in a restricted sense). The hypodigm of this species included all the Javanese fossils (from the earliest specimens to the much later Ngandong sample), as well as the Chinese fossils from the sites of Lantian, Zhoukoudian, and Hexian. There are certainly fossils in Java older than one million years, and some of them (from Modjokerto and from the base of the Sangiran sequence) have been dated as 1.8 million to 1.6 million years old, although not without controversy because of the uncertainties in the stratigraphic position of the remains. In China, the skull from Gongwanling (Lantian) is perhaps one million years old, and artifact-bearing sediments at Xiaochangliang in the Nihewan Basin of northern China have recently been found to be 1.4 million years old; but there is no reliable evidence of a much earlier human presence, except for the controversial incisor and mandibular fragment (with two teeth) found at Longuppo with some problematic artifacts. Finally, the latest *Homo erectus (sensu stricto)* specimens would be the large sample of crania from Ngandong, considered to be from the Upper Pleistocene (less than 128,000 years old). This collection of Chinese and Javanese fossils was argued to exhibit a number of derived traits (or specializations) that would exclude them from modern human ancestry. In other words, they were too specialized in their own direction to evolve into *Homo sapiens*.

The second Middle Pleistocene human assemblage would include European specimens from the sites of Mauer, Bilzingsleben, Vértesszöllös, Arago, and Petralona, as well as some African fossils (Bodo, Ndutu, and Broken Hill being the most relevant), and perhaps a few Chinese specimens (notably, Jinniushan and Dali, and perhaps Yunxian). These fossils were said to be different from East Asian *Homo erectus* at the specific level. They lacked the specializations that would preclude them from being considered human ancestors and showed a range of cranial capacities intermediate between that of modern humans and that of *Homo erectus*. Furthermore, this group would represent the common stem from which Neanderthals and modern humans diverged, predating the Neanderthal/modern human split.

In summary, this interpretation of the Middle Pleistocene fossil record replaced a unilinear model of the evolution of *Homo* with a pattern with three branches, which would have coexisted in the late Middle Pleistocene and part of the Upper Pleistocene. It is also important to point out that in the Neanderthal and modern lineages, there was a marked (and independent) increase in encephalization.

The non-*erectus* Middle Pleistocene fossils that are supposed to constitute the common stock for Neanderthals and modern humans, were named archaic *Homo sapiens*. This name is only taxonomically valid if Neanderthals are considered part of the human species, with the status of subspecies: *Homo sapiens neanderthalensis*. However, in recent years, many paleoanthropologists have adopted the cladistic approach to systematics. This has led to an appreciation of the importance of derived characteristics in phylogenetics and has also produced an increase in the number of taxa recognized. The Neanderthals are now broadly considered a separate species: *Homo neanderthalensis*. Thus, the common stock for Neanderthals and modern humans could not be called archaic *Homo sapiens* anymore, and this name was replaced by the specific designation *Homo heidelbergensis* (Schoetensack, 1908).

Because the specimen for this species is a European fossil, the Mauer mandible (Germany), it is worth examining the European fossil record again. The site of Dmanisi (Georgia) has yielded three human skulls and two mandibles that are dated as 1.7 million years old. These fossils have been assigned to the species *Homo ergaster*, which, to some authors, is different from East Asian *Homo erectus* and to other experts is only an early African variant of this group. It is not yet clear, however, that the species found at Dmanisi represents the hominids that peopled Europe, because the fauna associated with the human remains has a strong African component. It is possible, then, that the Georgian population was simply an extension of the African stock, and they never really inhabited the seasonal ecosystems of Europe.

There have been many claims (especially of an archaeological nature) for a human presence in Europe during the Lower Pleistocene (more than 780,000 years ago), but so far the only reliable evidence is from the Gran Dolina cave site, in the Sierra de Atapuerca, in northern Spain. In 1994 and 1995, some eighty human fossils and around two hundred artifacts were recovered in a level dated biostratigraphically, as well as by paleomagnetism and ESR, as around 800,000 years old. Based on a unique combination of morphological characteristics, the fossils were assigned to a new human species: *Homo antecessor*. The tool complex lacks hand axes, cleavers, picks, blades, or well-elaborated retouched flakes, and can be assigned to the pre-Acheulean technology (or mode 1). Also in 1994, a large part of a human braincase was found in Ceprano (Italy), unfortunately, without any paleontological or archaeological context. However, the fossil was correlated geologically to other beds in the region that date to the late Lower Pleistocene and have yielded pre-Acheulean tools. The Ceprano specimen has primitive traits, such as a well-developed supraorbital torus and an angulated occipital bone, together with other more modern characteristics, such as a high temporal squama, and has recently been assigned to *Homo antecessor*.

It is difficult to accurately date most of the human fossils of the European Middle Pleistocene. Traditionally, they were assigned to the Alpine glacial and interglacial periods or to other climatic phases established in northern Europe or England. Although these climatic events do affect large continental areas, they are local occurrences, and it is presently preferable to use the ocean paleotemperature scale, which is based on the ratio between the stable oxygen isotopes ^{18}O and ^{16}O found in the shells of marine microfossils. This ratio is related to the total amount of water that is accumulated in ice, and thus expresses global marine temperatures.

The Middle Pleistocene human fossils of Europe can be sorted into three chronological assemblages. Two specimens of the earliest group are thought to belong to oxygen isotope stage (OIS) 13 (524,000–478,000 years ago). They are the Mauer mandible and the Boxgrove tibial shaft (England). In the old parlance, these fossils would be chronologically referred to as Cromerian. The Arago cave (France) has yielded three mandibles, a face with a parietal bone, a coxal bone, and some other remains, all of which are dated to around 450,000 years ago, that is, within OIS 12 (or Mindel, as it is still sometimes called).

There is a large sample of European Middle Pleistocene fossils of intermediate age, being broadly assigned to OIS 11 to 8 (423,000–245,000 years ago). The site of Swanscombe, in the Thames valley (England), which has yielded an occipital bone and the two associated parietal bones, lies above deposits that are assigned to the glaciation locally called Anglia, and correlated to OIS 12. Thus, the Swanscombe fossil would correspond to OIS 11 to and it is dated at around 400,000 years old. On the European continent, fossils belonging to the OIS 11 (423,000–362,000 years ago), 8 group include fragments of two braincases found in Bilzingsleben (Germany), the Steinheim cranium (also from Germany) and, above all, the extraordinary collection of human remains that is being recovered from the Sima de los Huesos, in the Sierra de Atapuerca (Spain) (see Vignette). The occipital bone from Vértesszöllös (Hungary) is morphologically consistent with the intermediate group, but the uranium-series datings place it within the late group. The same can be said of the very complete cranium from Petralona (Greece). The fragmentary braincase from Reilingen (Germany) has no context, but it resembles Swanscombe. It is important to note that, in sharp contrast with the European Lower Pleistocene sites, the earliest Middle Pleistocene sites contain an Acheulean industry (or mode 2).

Finally, there is a group of fossils that are commonly known as the late Middle Pleistocene group (or late Riss in the nomenclature of the Alpine glaciations) that is correlated today to oxygen stages 7 and 6 (245,000–128,000 years ago). Among this late Middle Pleistocene group, the most important fossils come from La Chaise-

Suard, Fontéchevade, Bau de l'Aubesier, and Biache-Saint-Vaast in France and Pontnewydd in Wales. There is also a good collection of skulls from Ehringsdorf (Germany) that certainly corresponds to a warm phase, which could be OIS 5 (the last interglacial, as was traditionally considered) or OIS 7 (as is today more widely believed). Finally, in the bottom of a shaft in the cave of Altamura (Italy) lies a complete skeleton (coated by calcite concretion) that could belong to the late Middle Pleistocene group, but also perhaps to the Upper Pleistocene.

The morphology of the European late Middle Pleistocene group is clearly Neanderthal, and these fossils could be included in the species *Homo neanderthalensis* as early forms. The last interglacial (OIS 5) fossils, such as the Krapina sample (Croatia) and the two Saccopastore crania (Italy), can also be considered early Neanderthals, because they are not as derived (specialized) as the last glaciation "classic" Neanderthals.

In the chronologically intermediate Middle Pleistocene group, Neanderthal traits (more or less developed) are common. For example, the Swanscombe fragmentary braincase is clearly Neanderthal reminiscent in the occipital bone, showing both a suprainiac fossa and a straight nucal torus with two lateral projections. This morphology is considered a Neanderthal specialization (derived trait or apomorphy in the cladistic jargon). However, the OIS 11–8 fossils cannot be characterized as fully Neanderthal, and would be better called "pre-Neandertals."

In the Mauer mandible and in the Arago sample, both from the early Middle Pleistocene group, some authors see incipient Neanderthal characteristics, whereas others do not. A new source of information has recently arrived to the field of paleoanthropology: the analysis of ancient DNA. Some Neanderthal mitochondrial DNA has already been obtained, and the genetic distance from modern populations is greater that expected, pointing to a Neanderthal/anatomically modern split about half a million years ago.

A gradual and linear (or anagenetic) model of human evolution in the Middle Pleistocene is a possibility to consider. However, there is no clear temporal sequence to the process of "Neanderthalization," and sorting the fossils from less to more Neanderthal derived does not coincide with earlier to later fossils. To give a good example, the Swanscombe morphology is more like Neanderthal than that of any of the Sima de los Huesos crania, which are supposed to be younger. However, the large uncertainties in the chronology of most of the European Middle Pleistocene fossils, as well as their fragmentary nature, make the accuracy of both the morphological and chronological sequences problematic.

On the other hand, the anagenetic model of evolution predicts only an increase in the frequencies of the de-

THE SIERRA DE ATAPUERCA SITES

The Sierra de Atapuerca consists of an isolated hill of cretaceous limestone, in the northern part of Spain. Numerous cave sites of great potential have been systematically excavated in the last twenty years.

One of them is the Dolina site, an 18-meter-deep cave infilling. During the field seasons of 1994 and 1995, a test pit reached level 6, which yielded a rich faunal assemblage, including nearly eighty human remains. Two hundred lithic artifacts were also found in this horizon.

The eighty human fragments belong to at least six individuals that were defleshed and eaten by other humans. Based on a unique combination of morphological traits present in these specimens, a new specific name (*Homo antecessor*) was given to the Gran Dolina fossils.

A second important site is the Sima de los Huesos, located at the foot of a 13-meter-deep shaft and placed more than 500 meters from the current opening of another cave.

Up to now, the Sima de los Huesos has yielded three thousand human fossils belonging to at least thirty individuals, including several well-preserved crania (one of them a complete skull) and an almost intact pelvis. The human and bear fossils have been directly dated by uranium series and ESR, yielding an age around 300,000 years old.

The presence in the Atapuerca collection of all the skeletal elements, including the smallest and most delicate ones (e.g., immature epiphyses or the hand and foot distal phalanges), leads us to think that there was an accumulation of human corpses in the cave. What was the origin of the accumulation of so many human bodies? The Sima de los Huesos could represent the earliest known funerary practice.

—JUAN LUIS ARSUAGA

rived traits with time, not that a given fossil must be necessarily more derived than any earlier one and less specialized than any later specimen. Unfortunately, the only good representation we have of a European Middle Pleistocene population is that of the Sima de los Huesos.

In summary, there is so much "noise" in the present European Middle Pleistocene fossil record that the "signal" is difficult to perceive. The discovery of new fossils in the future will undoubtedly help clarify the picture. Still, a hypothetical scenario that is consistent with the facts can currently be elaborated:

1. At or just prior to a half a million years ago, the European and African human stocks split.
2. In the period between half a million years ago and a

quarter million years ago, some Neanderthal specializations evolved in the European stock but not in the African stock. In Europe, these derived traits were present, with different frequencies in different regions. Some of these traits were independent of each other, whereas others should be considered together as morphological sets.

3. Around a quarter million years ago, the frequencies of these derived traits greatly increased, and the different sets tended to appear together, producing a more defined Neanderthal morphology. This frequency increase could be interpreted as the result of a population bottleneck, perhaps associated with a shrinkage of the European population and a retreat to the Mediterranean coast in southern Europe as a result of a cold period (possibly that of OIS 8). These bottlenecks produced a strong genetic drift.

4. In the succeeding cold phases, OIS 6 and OIS 4, new population shrinkages further exaggerated the specializations, producing the "classic" Neanderthals that eventually went extinct around 30,000 years ago.

Even with all the problems posed by a scarce fossil record, it appears that the African specimens suggest a different evolutionary pattern. There are a number of crania of late Middle Pleistocene (or early Upper Pleistocene) age that can be called "premodern." The most important of these are the specimens Djebel Irhoud 1 and 2 (Morocco), Ngaloba 18 (Tanzania), Florisbad (South Africa), Omo Kibish 2 (Ethiopia), and Eliye Springs and ER 3884 (Kenya). The fossils of Klasies River Mouth (in South Africa), dated to 120,000 years ago, are more clearly anatomically modern (or early modern), as well as the Omo Kibish 1 partial skeleton (of roughly the same age) and the Qafzeh and Skhul samples of skeletons from Israel (100,000 years old). The paleontological evidence is thus consistent with the hypothesis of the molecular biologists of an African origin of *Homo sapiens*.

The main earlier African Middle Pleistocene fossils are Bodo (Ethiopia), Broken Hill (or Kabwe, Zambia), Ndutu and Eyasi (Tanzania), Salé (Morocco), and Elandsfontein (South Africa). Whereas the fossils of the intermediate Middle Pleistocene European group show clear incipient Neanderthal traits, it is difficult to recognize "modern" traits in the contemporaneous African group. It is possible that the origin of anatomically modern humans was not the result of genetic drift but rather a "punctuational" event. In other words, it is possible that the morphology specific to the modern human species did not result from an increase in the frequencies of preexisting traits, but from the sudden appearance of new characteristics.

If the term archaic *Homo sapiens* is dropped, there are two different taxonomic alternatives. One is to lump

in *Homo heidelbergensis* fossils that predate the divergence of the Neanderthal branch (such as the Gran Dolina fossils from Atapuerca), together with the European pre-Neanderthals and their African contemporaries. This would essentially replace the invalid taxonomic category with a proper taxonomic classification, but would still include nearly all the same specimens as before. The other alternative is to use three different names for these three groups: *Homo antecessor* for the fossils prior to the split (currently Gran Dolina and Ceprano), *Homo heidelbergensis* for the European pre-Neanderthals, and *Homo rhodesiensis* for the earlier African Middle Pleistocene group.

[*See also* Archaeological Inference; Geochronology; Neogene Climate Change.]

BIBLIOGRAPHY

Andrews, P. "An Alternative Interpretation of the Characters Used to define *Homo erectus*." *Courier Forschungs-Institut Senckenberg* 69 (1984): 167–175. One of the first papers that excludes East Asian *Homo erectus* from modern human origins.

Arsuaga, J.-L., I. Martínez, A. Gracia, and C. Lorenzo. "The Sima de los Huesos Crania (Sierra de Atapuerca, Spain): A Comparative Study." *Journal of Human Evolution* 33 (1997): 219–281. On the Sima de los Huesos fossils and the European Middle Pleistocene origin of Neanderthals.

Bermúdez de Castro, J. M., J. L. Arsuaga, E. Carbonell, A. Rosas, I. Martínez, and M. Mosquera. "A Hominid from the Lower Pleistocene of Atapuerca: Possible Ancestor to Neanderthals and Modern Humans." *Science* 276 (1997): 1392–1395. On the European Pleistocene fossils from Gran Dolina in Atapuerca and the naming of *Homo antecessor*.

Cann, R. L., M. Stoneking, and A. C. Wilson. "Mitochondrial DNA and Human Evolution." *Nature* 325 (1987): 31–36. Pioneering work in ancient human DNA, which suggested a late African origin for modern humans.

Hublin, J.-J. "Climatic Changes, Paleogeography, and the Evolution of the Neandertals." In *Neandertals and Modern Humans in Western Asia*, edited by T. Akazawa, K. Aoki, and O. Bar-Yosef, pp. 295–310. New York, 1998. A recent perspective on the origin of the Neanderthals.

Krings, M., H. Geisert, R. W. Schmitz, H. Krainitzki, and S. Pääbo. "DNA Sequence of the Mitochondrial Hypervariable Region II from the Neanderthal Type Specimen." *Proceedings of the National Academy of Sciences USA* 96 (1999): 5581–5585. The first Neanderthal mitochondrial DNA recovered from a fossil specimen, suggesting an early split for the Neanderthal/modern human lineages.

Manzi, G., F. Mallegni, and A. Ascenzi. "A Cranium for the Earliest Europeans: Phylogenetic Position of the Hominid from Ceprano, Italy." *Proceedings of the National Academy of Sciences USA* 98 (2001): 10011–10016. A phylogenetic analysis of the Early Pleistocene human cranium from Italy.

Rak, Y. "Morphological Variation in *Homo neanderthalensis* and *Homo sapiens* in the Levant: A Biogeographic Model." In *Species, Species Concepts, and Primate Evolution*, edited by W. H. Kimbel and L. B. Martin, pp. 523–536. New York, 1993. A perspective on the different specializations within both Neanderthals and modern humans and their implications for phylogenetic analyses.

Rightmire, G. P. "Human Evolution in the Middle Pleistocene: The Role of *Homo heidelbergensis.*" *Evolutionary Anthropology* 6 (1998): 218–227. A current perspective on the significance and importance of *Homo heidelbergensis.*

Stringer, C. B. "The Evolution of *Homo sapiens.*" In *The Great Ideas Today,* pp. 42–94. Chicago, 1992. An exhaustive analysis of the fossil record of archaic *Homo sapiens.*

Stringer, C., and R. McKie. *African Exodus: The Origins of Modern Humanity.* New York, 1998. A popular account of the origins of modern humans.

Stringer, C. B., and J.-J. Hublin. "New Age Estimates for the Swanscombe Hominid, and Their Significance for Human Evolution." *Journal of Human Evolution* 37 (1999): 873–877. A recent article reevaluating the tempo and mode of human evolution during the European Middle Pleistocene.

Zhu, R. X., K. A. Hoffman, R. Potts, C. L. Deng, Y. X. Pan, B. Guo, C. D. Shi, Z. T. Guo, B. Y. Yuan, Y. M. Hou, and W. W. Huang. "Earliest Presence of Humans in Northeast Asia." *Nature* 413 (2001): 413–417.

— JUAN LUIS ARSUAGA

Neanderthals

Neanderthals (or Neandertals; see Vignette) are a group of premodern humans referred to taxonomically as *Homo neanderthalensis* or, in grade terms, as an archaic form of *Homo sapiens*, known from the Late Pleistocene epoch (c.130,000–28,000 years ago) of Europe and western Asia. The earliest Neanderthal remains found were that of a fragmentary child's cranium in 1830 (Engis, Belgium) and a well-preserved adult cranium in 1848 (Forbe's Quarry, Gibraltar). However, these were not recognized as distinct from modern humans at the time of their discovery. The finding of a partial skeleton in 1856 by miners working in the Feldhofer Cave in the Neander Valley near Düsseldorf, Germany, generated considerably more interest. The large browridges, low sloping forehead, and robustly built arms, legs, and ribs were the first premodern fossil human remains recognized and scientifically named (King, 1864).

Not all specialists initially concurred with King's naming of a new species, and instead argued that the peculiar anatomy of the Feldhofer remains was of pathological origin in a modern human. However, discoveries of additional Neanderthal remains in 1886 in caves at Spy, Belgium, and in 1899 at the Krapina Cave in Croatia, followed by a series of spectacular discoveries at several sites in southwestern France between 1908 and 1914, firmly established Neanderthals as a distinctive fossil human group. Moreover, these human fossils were found in direct association with stone tools and bones of extinct animals that indicated great antiquity.

Today more than 275 Neanderthal individuals have been recovered from more than seventy sites. These include several relatively complete skeletons found in burial contexts, including neonates, infants, juveniles, adolescents, and adults. The known geographical range

NEANDERTHAL OR NEANDERTAL?

The German word for *valley* is *thal* (pronounced with a silent *h*), which was later converted to *tal* with other changes in German orthography. This explains why both the spellings *Neanderthal* and *Neandertal* are commonly used, although there is only one correct pronunciation (*Neandertal*).

—ROBERT G. FRANCISCUS

inhabited by Neanderthals stretched from the Atlantic coastal lines of Europe in the west to the Uzbekistan Mountains of western Asia, and from Pontnewydd, Wales, in the north to the northern shores of the Mediterranean in the south. Many of the more complete and better known specimens come from colder time periods during the last glaciation period or ice age (c.70,000–40,000 years ago), however, Neanderthals existed successfully in a variety of climates including earlier, warmer interglacial phases as well. The anatomy and behavior of Neanderthals is today comparatively well understood relative to other premodern forms of humanity. This is partly due to the historical length of time that they have been studied, as well as the relative abundance of their sites and physical remains. The genetic relationship of Neanderthals to modern humans is the most enduring question in human evolutionary studies and is still vigorously debated today.

Morphology. The widely influential monograph by Marcellin Boule (1911–1913) of the famous *"old man"* of La Chapelle-aux-Saints, France, a virtually complete specimen from the last glaciation period, established a long-standing view of Neanderthals as rather primitive, brutish creatures with limited intelligence. Prior to the late 1970s, Neanderthals were also widely viewed as a global morphological and behavioral stage in human evolution with, for example, African and East Asian Neanderthal variants. It is now clear that the Neanderthal morphological pattern was confined exclusively to Europe, the Near East, and portions of western Asia. Neanderthals exhibit a mosaic mix of anatomical characteristics that fall into three basic categories: primitive traits shared with earlier and conspecific members of the genus *Homo* (symplesiomorphies), advanced or derived traits shared with modern humans (synapomorphies), and uniquely derived traits found exclusively in Neanderthals (autapomorphies).

Primitive traits include a cranial vault that is relatively long and low with well-developed central browridges and a wide cranial base; a large facial skeleton with a wide nasal breadth and large incisor teeth; and a mandible that lacks a true mental eminence or chin. Primitive features of the postcranial skeleton include robust, thickened shaft bones and well-developed muscle attachment sites.

Derived cranial traits that Neanderthals share with modern humans include a larger brain, thinner skull vault bones, and a reduction of the nuchal torus in the rear part of the cranium to a more rounded occipital profile. Shared derived facial features include reduced lateral facial projection and chewing muscle attachment areas, reduced lateral browridges, and smaller premolars and molars. A reduction in the robusticity of the iliac pillar of the pelvis directly above the hip joint is arguably one of the few shared derived postcranial features (Figure 1).

Uniquely derived cranial traits that most Neanderthals share, but are rarely found in other groups, include a spherical shape of the braincase when viewed from the rear and a distinct roughened depression (suprainiac fossa) in the midline of the occipital bone just above the attachment area of the neck muscles. The side of the cranium in the vicinity of the ear region also manifests a series of uniquely derived traits including small mastoid processes (the attachment area for muscles that balance and turn the head), a mastoid tubercle, and an enlarged area adjacent to the mastoid process known as the juxtamastoid eminence. The bony labyrinth of the inner ear has a relatively small and downwardly positioned posterior semicircular canal. The hole at the base of the cranium that provides passageway for the spinal cord (the foramen magnum) is elongated from back to front, especially in Neanderthal children, and the external base of the cranium (ectobasicranium) is unusually flattened. The flattened ectobasicranium has been linked to a higher position of the voicebox in Neanderthal throats, possibly precluding the full complement of vowel sound production present in modern humans, although this remains controversial.

Uniquely derived traits in the facial skeleton include a large nasal opening (relative to the overall size of the face) that is square-shaped rather than pear-shaped. The area of the cheekbones adjacent to the nasal opening is markedly everted, producing a rather projecting nasal prominence. The greater bony cheek region below the eye orbit is expanded and inflated rather than showing the more usual depression (a canine fossa) found in other groups. The cheekbones to the side of the face are swept back rather than angulated, and in conjunction with the nasal prominence produces a face that appears markedly projecting in midline. The large, projecting nasal region, in particular, has been interpreted as an adaptive response to living in extremely cold and arid glacial environments.

Neanderthal front teeth (incisors) are relatively large compared to their molars, and the lower jaw (mandible) of Neanderthals includes a large gap (retromolar space) between the back edge of their third molars and vertical front edge of the ascending ramus. Some of the other uniquely derived features in the lower jaw include an oval-horizontal configuration of the opening for the mandibular nerve (mandibular foramen), an enlarged tubercle for attachment of the medial pterygoid muscle on the internal side of the ascending ramus, and a lateral expansion of the mandibular condyle that forms part of the temporomandibular joint. Many of the derived dental, mandibular, and facial features noted here have been interpreted as an adaptive response among Neanderthals to high levels of bite forces and an unusual use of their front teeth as tools, as well as the result of unique patterns of growth and development (heterochrony).

From the neck down, Neanderthal skeletons exhibit several features that are thought to be reflections of cold-climate adaptation found in some recent modern populations. These include short stature coupled with a wide pelvis and barrel-chested thorax and relatively short lower arm and lower leg segment lengths. Uniquely derived features rarely found outside of Neanderthal samples include pelvic pubic bones that are thin in cross section and longest in males rather than females (a reversal of the usual pattern of sexual dimorphism), as well as shoulder blades that have an unusual positioning for the teres minor muscle attachment area (one of the rotator cuff muscles).

Behavior. The tool complex generally associated with Neanderthals is known as the Mousterian, often more generally referred to as the Middle Paleolithic. In Europe, the Mousterian begins as early as 200,000 years ago, although most sites are dated to between 130,000 and 36,000 years ago. Mousterian technology is dominated by retouched flake tools and included a more sophisticated method of controlling the size of flakes removed from cores (Levallois core technology) compared to earlier stone shaping techniques. Although Neanderthals were quite skilled in making a large variety of different stone tool types, their habitual use of raw material did not extend to other sources such as antler, bone, and ivory, as was the case for early modern people.

Although there is a general association between Neanderthals and Mousterian stone tool production, it is not an invariant relationship because early modern humans from the sites of Skhul and Qafzeh in present-day Israel are also directly associated with Mousterian tools. Moreover, Neanderthals have been found in post-Mousterian contexts such as the Châtelperronian industry, which does include non-stone-based artifacts and perforated pendants of shells, teeth, and bone (items found more frequently in modern human archaeological contexts). Neanderthals controlled and made frequent use of fire, although their hearths tended to be small, simple

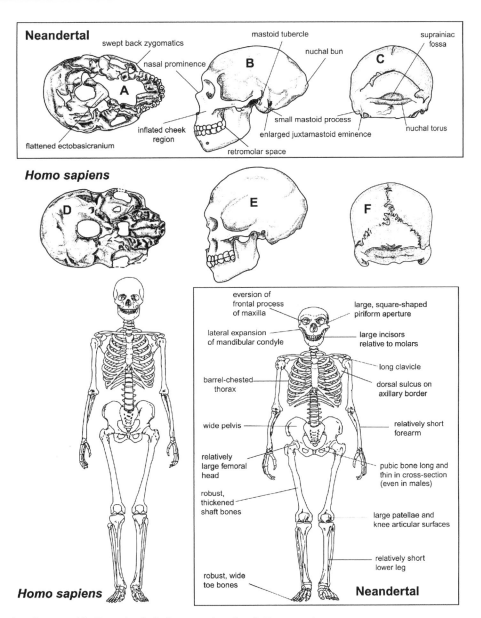

FIGURE 1. Selected Traits Found in Neanderthals Compared to Fossil *Homo sapiens*.
Crania A and C = La Ferrassie 1; B = La Chapelle-aux-Saints, (reconstructed); D and F = Cro-Magnon 1; E = Predmosti 3).
All figures modified from Klein, *The Human Career* 2d ed., University of Chicago Press, 1999; and Tattersall, *The Last Neanderthal*, Westview Press, 1999, by K. Mykris.

surface fires compared to the larger, deeper, and more centrally located permanent hearth structures found in many early modern living sites. Neanderthal sites, in general, seem to be somewhat less spatially organized, smaller, and more ephemeral than those of early modern humans.

Interpretations of Neanderthal subsistence behavior have varied widely, ranging from frequent, well-planned sophisticated hunting of large Ice Age megafauna, to more opportunistic foraging of smaller game and persistent scavenging of larger animals. There is now good evidence from animal remains, stone-tipped and wooden spears, and Neanderthal injury patterns indicating that they hunted midrange to large-sized mammals. Moreover, bone chemistry (stable isotope) analysis of Neanderthal skeletal remains from geographically dispersed sites indicates substantial protein intake, suggesting active predation as top-level carnivores (Richards et al., 2000).

Neanderthals provide the first unambiguous evidence for intentional burial of the dead, although the practice was not ubiquitous. Moreover, Neanderthal burials typically lack the associated grave goods and ceremonial adornments that are common in early modern burial practices. Art and personal adornments have been occasionally found in association with Neanderthal skeletal and cultural remains, especially in Châtelperronnian contexts; however, these are relatively rare compared to their much greater frequency in early modern human archaeological sites.

Cut marks and other damage on Neanderthal skeletons frequently have been attributed to cannibalism, although alternative explanations such as secondary burial and carnivore (hyena) damage also have been suggested. However, unambiguous evidence for dietary cannibalism among Neanderthals, though infrequent, does exist (Defleur et al., 1999). Neanderthal paleodemographic studies consistently indicate high prime age mortality and the lack of postmenopausal survivorship, suggesting high levels of subsistence mobility, adaptive stress, and frequent local demographic crises (Bocquet-Appel and Arsuaga, 1999; Trinkaus, 1995). [See Population Trends for errors in these estimates.]

Neanderthals were clearly behaviorally more sophisticated than earlier hominids. They were the first humans to successfully adapt to glacial and cold-steppe midlatitude environments in Europe. Nonetheless, many of the morphological and behavioral patterns described above suggest that Neanderthals relied more on bodily robusticity and strength, and less so on technological and cultural buffering to subsist in their environments compared to early modern humans.

Phylogenetic History. Neanderthal origins are clearly found in the late Middle Pleistocene epoch of Europe. As early as 450,000 years ago, uniquely derived features of the upper face and mandible appear in incipient form on several specimens from the site of Arago, France, and become more strongly established in the large sample from the Sima de los Huesos, Atapuerca, Spain, at about 300,000 years ago. Sometime later, Neanderthal features found at the rear (occipital) of the cranium become increasingly more frequent, followed by the uniquely derived features of the cranial vault and the temporal bone (Hublin, 1998). This mosaic pattern of Neanderthal trait evolution crystallizes by the Late Pleistocene epoch (c.130,000 years ago). The earliest evidence for Neanderthal traits occurs during a glacial period, and during most of the last 500,000 years European climatic conditions were cool to cold. Such conditions would have resulted in a degree of relative isolation of endemic populations. Random genetic drift and continued selection in colder environments likely played an important role in Neanderthal differentiation.

At some point, Neanderthals expanded their range from Europe into the Near East and western Asia, although precisely when this occurred remains uncertain. It is possible that this occurred during a warm intergla-

cial phase beginning about 127,000 years ago. This would have resulted in substantial temporal overlap with robust early modern humans in the Levant from the Skhul and Qafzeh sites. Alternatively, Neanderthal incursions from Europe into an area already abandoned by the Skhul/Qafzeh populations in the Near East could have occurred later as environments deteriorated during a cold glacial phase beginning about 71,000 years ago.

Approximately 40,000 years ago, modern humans arrived in Europe from western Asia and came into contact with the indigenous Neanderthal populations. The duration and nature of this contact are still debated. One view holds that contact lasted only a few millennia, with minimal genetic and cultural exchange. According to this view, the Neanderthals succumbed quickly to the technically and culturally more sophisticated adaptive pattern of modern humans, including perhaps the full capacity for language (Klein, 1999). Proponents of this notion also view genetic hybridization between Neanderthals and modern humans as minimal and evolutionarily insignificant, usually assigning Neanderthals to a separate species: *Homo neanderthalensis*. Others, citing late Neanderthals in Europe dated to 29,000 years ago, argue that Neanderthal–modern human coexistence lasted as long as 10,000 years, with Neanderthals slowly retreating into their last strongholds in southern Italy, Greece, and Iberia, and into the Balkans and Caucasus mountains. Proponents of this view cite evidence for a significant level of gene flow between Neanderthals and modern humans based on the persistence of some Neanderthal features in early modern European specimens (Duarte et al., 1999; Wolpoff, 1999). Châtelperronnian and other "transitional" tools manufactured by Neanderthals may be the outcome of Neanderthal–modern human interaction, through either imitation or actual Neanderthal inventions spurred on by ecological competition.

[See also Modern *Homo sapiens*, article on Neanderthal–Modern Human Divergence.]

BIBLIOGRAPHY

Adcock, G. J., et al. "Mitochondrial DNA Sequences in Ancient Australians: Implications for Modern Human Origins." *Proceedings of the National Academy of Sciences USA* 98 (2001): 537–542. Technical treatment of the first DNA recovered and published from an early modern human.

Bocquet-Appel, J.-P., and J.-L. Arsuaga. "Age Distributions of Hominid Samples at Atapuerca (SH) and Krapina indicate accumulation by catastrophe." *Journal of Archaeological Science* 26 (1999): 327–338.

Boule, M. "L'homme fossile de La Chapelle-aux-Saints." *Annales de Paleontologie* 6 (1911–1913): 111–172; 7: 21–56, 85–192; 8: 1–70.

Defleur, A., et al. "Neanderthal Cannibalism at Moula-Guercy, Ardèche, France." *Science* 286 (1999): 128–131.

Duarte, C., et al. "The Early Upper Paleolithic Human Skeleton from the Abrigo do Lagar Velho (Portugal) and Modern Human

Emergence in Iberia." *Proceedings of the National Academy of Sciences USA* 96 (1999): 7604–7609. Technical treatment of a 24,000-year-old modern human child argued to show Neanderthal–modern human interbreeding.

Hublin, J.-J. "Climatic Changes, Paleogeography, and the Evolution of the Neandertals." In *Neandertals and Modern Humans in Western Asia*, edited by T. Akazawa, K. Aoki, and O. Bar-Yosef. pp. 295–310. 1998.

King, W. "The Reputed Fossil Man of Neanderthal." *Quarterly Journal of Science* 1 (1864): 88–97.

Klein, R. G. *The Human Career: Human Biological and Cultural Origins.* 2d ed. Chicago, 1999. Contains comprehensive treatment of the Neanderthal debate from the perspective of genetic and behavioral replacement by modern humans.

Krings, M., et al. "Neandertal DNA Sequences and the Origin of Modern Humans." *Cell* 90 (1997): 19–30. Technical treatment of the first DNA extracted and published from a Neanderthal fossil.

Krings, M., et al. "DNA Sequence of the Mitochondrial Hypervariable Region II from the Neandertal Type Specimen." *Proceedings of the National Academy of Sciences USA* 96 (1999): 5581–5585.

Krings, M., et al. "A View of Neandertal Genetic Diversity." *Nature Genetics* 26 (2000): 144–146.

Ovchinnikov, I. V., et al. "Molecular Analysis of Neanderthal DNA from the Northern Caucasus." *Nature* 404 (2000): 490–493.

Richards, M. P., et al. "Neanderthal Diet at Vindija and Neanderthal Predation: The Evidence from Stable Isotopes." *Proceedings of the National Academy of Sciences USA* 97 (2000): 7663–7666.

Trinkaus, E. "Neanderthal Mortality Patterns." *Journal of Archaeological Science* 22 (1995): 121–142.

Wolpoff, M. H. *Paleoanthropology.* 2d ed. Boston, 1999. Contains comprehensive treatment of the Neanderthal debate from the perspective of genetic and behavioral continuity with modern humans.

— ROBERT G. FRANCISCUS

HOMO, GENUS. *See* Hominid Evolution, *article on* Early Homo.

HOMOLOGY

The term *homology* can be used to refer to any form of correspondence, whether between geometric objects, organic compounds, locations of tumors in the body, or positions on a chromosome during recombination (e.g., homologous recombination). The term is derived from the Greek word *homologos*, meaning "agreeing."

In biology, the primary use of the term *homology* is to describe the correspondence between body parts, characters, genes, or other biological features in different species. For example, the wing of a bird and the forelimb of a crocodile are considered homologous because they represent paired appendages of the same basic design despite having widely different shapes and functions. This form of correspondence is often contrasted with similarities among body parts that serve the same function but that are not equivalent in their structure. This form of similarity is often called analogy.

History. The distinction between homology and analogy was established by Richard Owen. These terms, however, were used before Owen (Panchen, 1994), and the basic ideas of homology and analogy can even be traced back to Aristotle (Rupke, 1994).

In 1827, Owen was appointed as curator to the Hunterian Museum in London and charged with preparing a catalog for the museum's extensive collections. To achieve this, he had to decide on a principle for arranging the specimens into a natural system. The intellectual environment in which Owen tried to achieve this was the highly publicized debate between Georges Cuvier (1769–1832) and Étienne Geoffroy de Saint-Hilaire (1772–1844) in Paris (1828–1829). The debate essentially was about whether the diversity of life is best understood as reflecting the functional needs of organisms (functionalism, Cuvier), or as reflecting an underlying plan of organization (unity of type, Geoffroy de Saint-Hilaire) independent of functional needs. The most striking argument for the reality of the unity of type came from the fact that the limbs of all tetrapods (four-legged vertebrates) are built according to the same basic plan with upper arm/leg, lower arm/leg, and hand/foot. This plan is recognizable regardless of the different functions limbs can serve, such as walking, climbing, swimming, digging, and flying. Based on this insight, Owen decided that the most natural way to describe and catalog the specimens in the Hunterian collection was to identify the corresponding parts in different species regardless of their function and to arrange the specimens according to the pattern of shared body parts (i.e., homologous parts). The result was a system in which homologous characters are shared among nested sets of species. This hierarchical pattern of shared characters between species was so striking that it needed to be explained. Owen proposed that homologous characters owe their identity to their derivation from a common body plan, the archetype. The archetype can be understood as an abstract principle or law of nature. According to Owen's interpretation, each species is a particular realization of this law. As such, the concept of the archetype was in line with contemporary scientific thinking, which sought to discover laws of nature from large collections of observations.

The concept of homology was thus developed prior to and independent of the theory of evolution. Using the distinction between homology and analogy in his anatomical work, Owen paved the way for the theory of evolution. The final step in this direction, however, was taken by Charles Darwin, who interpreted the archetype as a real biological ancestor. Darwin proposed that species share homologous characters because they are derived from an ancestor that had the same character. This leads to the most commonly used definition of homology in biology: characters in two species are homologous if they correspond to the same character in a common ancestor (see Definitions of Homology). With this conceptual leap, Darwin turned the results of preevolutionary comparative anatomy into the largest body of evidence supporting the theory of evolution. The theory of evolution, that is, the assumption that recent species are derived from common ancestors by a process of descent with modifications, is by far the most simple and convincing explanation for the vast number of facts accumulated by comparative anatomy.

With the Darwinian revolution, the homology concept assumed a new role. It became the main tool for the reconstruction of phylogenetic relationships (Hennig, 1966). Homologous characters found in two species are taken as evidence that the two species are related to each other. If, two species, say, A and B, share more homologous characters than either of them do with a third species C, one can infer that they may be more closely related to each other than each of them is to C. The study of all homologous characters found in a group of organisms is the basis for the reconstruction of phylogenetic relationships. Much of the theoretical development in morphology since Darwin has been dedicated to the development of criteria for recognizing homologous characters (see Criteria for Homology) (Remane, 1952). Homology is thus the conceptual basis for the extensive research program in comparative biology that followed the Darwinian revolution.

For much of the nineteenth and twentieth centuries, homology was accepted as a simple empirical fact with few theoretical implications. The recent progress of developmental biology in elucidating the molecular mechanisms of development have led to a renewed interest in the concept (Bock and Cardew, 1999). It is tempting to assume that homologous characters owe their identity to the genetic information inherited from a common ancestor. Although it is certainly the case that the same genetic information directs the development of homologous characters in closely related species, this is often not the case for distantly related species.

Many examples of homologous characters that are developmentally nonequivalent have been described (see Chapter 21 in Hall, 1998). The most striking examples can be found in insect development, in which the molecular biology of the developmental processes is best known. All insects have a segmented body plan, and there is no question that the segments of all insects are homologous in the sense that they are derived from the segments of the common ancestor of insects. In the fruit fly (*Drosophila melanogaster*), development of segments is controlled by three sets of genes: maternal effect, gap, and pair-rule genes. One of the pair-rule genes of *Drosophila* is *even-skipped*. The *even-skipped* gene in

grasshoppers, however, does not participate in segment development (Patel et al., 1992). Similarly, the development of the head region in *Drosophila* depends on the activity of a gene called *bicoid*, but this gene does not exist in any other insect order (Schmidt-Ott, 2001).

These recent experimental findings confirm an earlier suspicion that homology among morphological characters is not necessarily associated with a common genetic or developmental basis (DeBeer, 1971). The reason is that developmental processes can evolve after a character is established and while the identity of the adult character is maintained. What, then, is the mechanistic basis of homology if not shared developmental genetic information? This intriguing question is currently unresolved (for details, see Hall, 1998).

An attempt to solve this problem is the so-called biological homology concept (see Definitions of Homology). The basic idea behind the biological homology concept is that characters retain their recognizable identity during evolutionary change because of developmental constraints on their variability. For example, the mode of development of the limb skeleton always follows a stereotypic pattern (Shubin and Alberch, 1986). First, a clump of cells forms as the anlage (primordium) of the upper arm/leg bones (humerus/femur). Next, this clump of cells splits at its distal end to form two cell clumps that develop into the two lower arm or leg elements (radius and ulna, tibia and fibula, respectively). Finally, the complicated morphology of the hand/foot forms. A tetrapod limb is therefore always recognizable by the fact that it has one bone attached to the body, then two bones forming the lower limb and more bones at the distal end. It follows that the developmental processes and the genes that direct them can change without affecting character identity as long as the changes do not affect the developmental constraints of the character. For example, it is not necessary to know from where in the embryo the cells are recruited into the limb, as along as they follow the same morphogenetic rules during pattern formation.

Applications of Homology. Although the homology concept as developed by Owen and Darwin and their successors was originally applied only to morphological characters, it can be used for any biological character. These characters can be at any level of organization, including DNA, proteins, cells, tissues, organs, and even behaviors. The most important use of homology is in systematics, where it is used to reconstruct the phylogenetic relationships of species (Hennig, 1966).

An important nontraditional use of homology is in the comparative study of behavior. The Austrian ethologist Konrad Lorenz (1903–1989) was the first to recognize that some behavioral patterns fulfill the same criteria for homology as morphological characters (Lorenz, 1965) (see Criteria for Homology). Homology of behavioral patterns implies that behavior evolves by the same mechanisms as morphological characters. This insight led to the highly productive research program of ethology, the comparative study of animal behavior.

In molecular genetics, one speaks of homologous genes or proteins in the same sense as one speaks of homologous body parts if the genes are derived from the same gene in a common ancestor (Fitch, 1970). Because genes are frequently duplicated in evolution, it is important to distinguish between two forms of homology among genes. Genes that have diverged as a result of independent evolution following a speciation event are called orthologous. If genes that have diverged within a species as a result of a gene duplication event are called paralogous. Paralogy corresponds conceptually to serial homology of morphological characters because both reflect two structures in the same organism (see Definitions of Homology). Finally, genes that have diverged because of horizontal gene transfer between species are called xenologous. The comparison of the DNA sequence of orthologous genes is now the standard method for the reconstruction of phylogenetic relationships among living species (Hillis et al., 1996).

Criteria for Homology. Homologous characters are used to reconstruct phylogenetic relationships. To do this, it is necessary to establish standards for recognizing potentially homologous characters. The Darwinian definition of *homology* is not useful for this purpose because the Darwinian definition assumes that phylogenetic relationships are known. Criteria for recognizing homology therefore have to be independent of any knowledge of phylogenetic relationships. A systematic framework of "homology criteria" was established by Adolf Remane (Remane, 1952; Riedl, 1978). Remane distinguished between six criteria, three main criteria (1–3) and three ancillary criteria (4–6).

1. *Position.* Homology is likely if a structure is consistently found in the same relative position to other structures. Examples are bone plates of the skull that are not obviously individualized except that they tend to maintain their relative position to each other.
2. *Structure.* Similar structures may be homologous independent of their position if they share many features. The likelihood of homology increases with the complexity of the compared structures.
3. *Transition.* Even dissimilar structures in different positions can be homologous if one can find transitional forms. It is essential that all transitional forms are homologous according to the criteria of position and/or structure. The transitional forms can be ontogenetic stages or taxonomic intermediates. The most famous example is the homology between the ear bones of mammals and the primary jaw joint of reptiles. The homology of these very different struc-

tures was established when transitional forms were found in the fossil record.

4. *General congruence criterion.* Even simple structures may be homologous if they appear frequently in closely related species.
5. *Special congruence criterion.* The likelihood of homology among structures increases with the number of other structures that have the same distribution among species.
6. *Negative congruence criterion.* The likelihood of homology decreases with the frequency of occurrences of this character in unrelated species.

It is important to note that these criteria do not have the status of definitions, nor do they prove homology (Patterson, 1982). They are heuristic guidelines for the recognition of potential homologies that need to be tested by phylogenetic analysis. The final determination of homology is based on a comparison between the phylogenetic relationships and the distribution of the character. Only those hypothetical homologies that are congruent with the phylogenetic relationships can be true homologies. This criterion overlaps with Remane's ancillary criteria of congruence, but it is given more fundamental importance by Patterson (1982) than it had in Remane's system.

Definitions of Homology. We can distinguish between four definitions of *homology* that reflect different research traditions and goals. The original definition by Richard Owen can be called typological because it does not depend on evolutionary theory (see History). In contrast, Darwin's definition is historical because it is based on the historical or genealogical relationships between species. Finally, the concepts that emphasize the developmental basis of homology can be either informational or biological because they are concerned with the mechanistic basis for similarity.

The typological homology concept. Even though the theoretical implications of the typological homology concept are no longer part of modern biology, it still provides the intuitive basis for all practical applications of the concept. In 1843, Owen defined *homologue* as the same organ in different animals under every variety of form and function and *analogue* as a part or organ in one animal that has the same function as another part or organ in a different animal (cited after Owen, 1848). Owen also emphasized that homologues can have the same function, thus, the distinction is not exclusive. The definition needs to be extended to include all forms of life, not just animals.

Owen further distinguished between three forms of homology: special, general, and serial. Special homology is the relationship between two specific characters that are found in different organisms. This notion corresponds to the definition of *homology* that is still in use today. General homology is the relationship of a specific organ to the abstract concept of which it is a special case. In Owen's words, general homology is the relationship of an organ to its corresponding part in the "archetype" (for the archetype concept, see History). Finally, the relationship of sameness may apply to different parts of the same body, such as in the case of different vertebrae in the vertebral column of a fish. This form is called serial homology.

The weaknesses of the typological homology concept are that it does not explain why species have homologous characters and what exactly is meant by "sameness" of characters.

The historical homology concept. The evolutionary interpretation of homology relates patterns of anatomical similarity to the evolutionary process. According to this approach, a homologue is a character in two species that corresponds to the same character in the most recent common ancestor of these species (Mayr, 1982). This definition explains, by reference to a concrete natural process, why corresponding characters can be found in certain species but not in others. In particular, it explains why homologues tend to be shared among nested sets of species. This definition is also the basis for the application of the homology concept in comparative biology, in particular, phylogenetic research.

For the purpose of phylogenetic reconstruction, it is important to distinguish between two kinds of homologous characters in a given group of species (Henning, 1966). Homologous characters that are shared among a certain group of organisms but not with others are called apomorphic for this group. Apomorphic characters are defined as the shared derived characters of a group of species. In contrast, if the homologous characters have been inherited from a distant ancestor, the characters are called plesiomorphic. Only shared derived (apomorphic) characters allow us to infer the phylogenetic relationships.

The historical homology concept excludes serial homology, because the similarity of repeated parts is not caused by species diversification. Despite its strength and heuristic importance, however, this concept shares a weakness with the typological homology concept. Neither concept explains the notion of sameness among phenotypic characters. This is problematic, because homologous characters do not need to have the same genetic basis (see History). This weakness, though, does not apply to the homology of genes, which are transmitted between generations through a process of direct replication.

Informational homology concepts. In the early 1980s, two definitions were advanced that intended to explain the notion of sameness implicit in the historical homology concept. Both refer to the notion of information. Van Valen argued that "[h]omology is resemblance caused by a continuity of information. In biology it is a unified developmental process" (Van Valen, 1982, p. 305). Similarly, Osche wrote that "[h]omologies are non-random similarities of complex structures which are based on common genetic information (in the sense of instruction)" (Osche, 1982; cited in Haszprunar, 1992, p. 14). These definitions have the virtue to include both serial and historical homology. The weakness of these definitions is that it is not clear how the development of homologous characters can be different without affecting character identity. What exactly is the information that is shared between homologous characters that have differences in their development?

The biological homology concept. The biological homology concept complements the historical homology concept. The goal is to further explicate the notion of sameness underlying any notion of homology. "Structures from two individuals or from the same individual are homologous if they share a set of developmental constraints, caused by locally acting self-regulatory mechanisms of . . . differentiation" (Wagner, 1989, p. 62). The maintenance of character identity during evolution is explained by constraints that limit the evolution of the character to a bounded set of character states. This definition also allows for flexibility in the developmental pathways of homologous characters as long as these differences do not affect the developmental constraints of the structure. It is also applicable to serial homology.

BIBLIOGRAPHY

Bock, G. R., and G. Cardew, eds. *Homology*. London, 1999.
DeBeer, G. R. *Homology, an Unsolved Problem*. London, 1971.
Fitch, W. M. "Distinguishing Homologous from Analogous Proteins." Systematic Zoology 19 (1970): 99–113.
Hall, B. K. *Evolutionary Developmental Biology*. London, 1998.
Haszprunar, G. "The Types of Homology and Their Significance for Evolutionary Biology and Phylogenetics." *Evolutionary Biology* 5 (1992): 13–24.
Hennig, W. *Phylogenetic Systematics*. Urbana, Ill., 1966.
Hillis, D. M., C. Moritz, and B. K. Mable. *Molecular Systematics*. Sunderland, Mass., 1996.
Lorenz, K. *Evolution and Modification of Behavior*. Chicago, 1965.
Mayr, E. *The Growth of Biological Thought*. London, 1982.
Osche, G., "Rekapitulationentwicklung und ihre Bedeutung für die Phylogenetik—Wann gilt die 'Biogenetische Grundregel'?" *Verh. naturw. Ver. Hamburg (N.F.)* 25 (1982): 5–31.
Owen, R. *On the Archetype and Homologies of the Vertebrate Skeleton*. London, 1848.
Panchen, A. L. "Richard Owen and the Concept of Homology." In *Homology: The Hierarchical Basis of Comparative Biology*, edited by B. K. Hall. San Diego, Calif., 1994.
Patel, N. H., E. E. Ball, and C. S. Goodman. "Changing Role of Even-Skipped during the Evolution of Insect Pattern Formation." *Nature* 357 (1992): 339–342.
Patterson, C. "Morphological Characters and Homology." In *Problems of Phylogenetic Reconstruction*, edited by K. A. Joysey and A. E. Friday, pp. 21–74. London and New York, 1982.
Remane, A. *Die Grundlagen des natürlichen Systems, der ver-*

gleichenden Anatomie und der Phylogenetik. Leipzig, Germany, 1952.

Riedl, R. *Order in Living Organisms: A Systems Analysis of Evolution.* New York, 1978.

Rupke, N. A. *Richard Owen: Victorial Naturalist.* New Haven, 1994.

Schmidt-Ott, U. "Different Ways to Make a Head." *BioEssays* 23 (2001): 8–11.

Shubin, N., and P. Alberch. "A Morphogenetic Approach to the Origin and Basic Organization of the Tetrapod Limb." *Evolutionary Biology* 20 (1986): 319–387.

Van Valen, L. "Homology and Causes." *Journal of Morphology* 173 (1982): 305–312.

Wagner, G. P. "The Biological Homology Concept." *Annual Review of Ecology and Systematics* 20 (1989): 51–69.

— GÜNTER P. WAGNER

HOMOPLASY

The idea of evolution as "descent with modification" suggests a process of divergence between species over time. Some of the most evocative examples of this are rapid "adaptive radiations" on isolated oceanic islands, such as Darwin's finches on the Galapagos and Hawaiian silverswords. In each case, dramatic ecological specialization is accompanied by remarkable divergence in morphology among endemic species. However, although divergence among species is perhaps the overriding product of evolutionary processes, numerous cases of apparent "convergence" are also known. *Homoplasy*, a term coined by E. Ray Lankester in 1870, refers to any and all evolutionary processes that produce similarity in features between species by modification of ancestral species that were dissimilar in those features. It is thus distinguishable from *homology*, which refers to similarities that are merely retained as a consequence of common ancestry. Familiar examples of homoplasy are the repeated evolution of wings in tetrapod vertebrates (birds, bats, and pterosaurs) and succulent stems in plants living in arid environments (cacti, euphorbs). Charles Darwin, in his *Origin of Species* (1859), felt compelled to discuss these "analogical variations" as potential stumbling blocks to the development of his theory of natural selection. His examples included electric organs in divergent groups of fishes and elaborate pollen aggregations known as pollinia in orchids and milkweeds.

Some kinds of homoplasy are easy to spot; others are not. The dramatic elongation in the neck of giraffes is matched by several instances of greatly lengthened necks in staphylinid beetles of the subfamily Scaphidiinae, some of which are longer than the rest of their bodies. However, no biologist would mistake this obvious convergence between a mammal and an insect for a homology, because different tissues and structures have been modified to give rise to similarity of a very coarse type. The basic body plan of these two organisms is so different that any shared similarity in neck must have evolved in different ways. In contrast, the chemical compound caffeine has evolved independently several times in plants (as have biochemically related compounds that also have stimulating effects on the human nervous system), but the chemical structure of caffeine is identical in these different groups. Unlike the previous example, there is little ambiguity about the description of the trait, and the similarity is precise rather than coarse.

These examples illustrate the distinction between two types of homoplasy. Similarity of a coarse kind, involving different structures, usually evolving in very distantly related species, is convergence. Similarity of a much more precise nature, involving the same structures (to the limit of observational resolution) and often evolving in more closely related species, is parallelism. These definitions refer to both the degree of similarity and the degree of relationship, either or both of which may be hard to quantify. Of the two, closeness of relationship may be the more problematic. Even with perfect phylogenetic information, there is no nonarbitrary division with which to distinguish "closely" and "distantly" related. However, it is sometimes possible to distinguish whether traits are the same or not, using morphological, developmental, or genetic criteria, and in these cases it is straightforward to distinguish between convergence and parallelism.

A third kind of homoplasy is reversal, the reacquisition of a trait exhibited by an ancestor after that trait has been modified. Two different kinds of reversal can be distinguished along the lines described above, depending on the coarseness of the similarity to the ancestral form. Birds all have wings, but flightless birds have evolved perhaps as many as thirty times (often on islands). "Flightlessness" is a reversal to the ancestral condition in tetrapod vertebrates, but it is a coarse reversal, because the forelimbs of ancestral tetrapods are only vaguely similar to those of flightless birds, which usually retain some vestiges of their former muscle and skeletal adaptations. Among the many other examples are limb loss in vertebrates such as snakes, eyelessness (or at least blindness) in animals inhabiting caves, and the loss of photosynthesis in many parasitic plants.

Homoplasy of all three types is commonplace not just in morphological traits but throughout the organism. Homoplasy in molecular characters has been studied extensively. The degeneracy of the genetic code means that many of the same nucleotide substitutions can occur over and over in a DNA sequence without affecting the amino acid that is assembled into the mature protein. Such selectively neutral differences exhibit rampant homoplasy at the nucleotide level. Some amino acids can be replaced by others that have similar biochemical and biophysical properties without altering the three-dimensional structure of the protein or its interactions with other molecules, leading to homoplasy at the protein level. Large configurations in the genome can even exhibit homoplasy. Most plant chloroplast genomes contain a large stretch of genes that is duplicated and inverted. This complex inverted repeat has been lost independently within the flowering plant clade Leguminosae and within conifers, a clade outside of flowering plants.

The widespread occurrence of homoplasy across all taxa and all kinds of traits has prompted attempts to identify common regularities or "laws." Dollo's law is perhaps the most widely cited example. Dollo (1903; Gould, 1970, p. 196) argued that "[F]unctional or physiological reversal occurs; structural or morphological reversal does not occur." Flightless birds are a good example: the function of flight is lost, but vestiges of wings remain. Though much maligned and misunderstood, Dollo's "law" is really just an argument about complexity and the improbability that a lengthy series of evolutionary steps leading to a complex structure will be undone exactly by a series of reversals of each of these steps. Another apparent pattern is the existence of "parallel tendencies," in which parallelisms seem to be confined to very closely related species. For example, a bizarre parallelism of eyes on long stalks is seen sporadically in Diptera (true flies), but only in a group of closely related families. This pattern might be expected if the repeated origin of a trait is contingent on the existence of an underlying trait that evolved earlier in a more inclusive group of species.

Homoplasy in Phylogenetic Inference. In addition to its implications for evolutionary diversification, homoplasy assumes another role in modern comparative biology. It often provides misleading evidence about phylogenetic relationships. The telltale signature of evolutionary kinship is the presence of a shared novel trait between species—a homology. For example, a large fraction of land plants have a novel trait, seeds, that their relatives lack. Seeds are thought to have originated once in the common ancestor of these plants, providing evidence that all seed plants are more closely related to each other than to other plants. Had seeds originated more than once from distinct ancestors and subsequently converged, then it would be a mistake to regard seeds as evidence of close relationship. A supposed homology would actually be a homoplasy. Reconstructing phylogenetic relationships can thus be seen as an attempt to uncover homology and ignore homoplasy.

How can homoplasy's effects on the reconstruction of phylogenies be minimized? One strategy involves nothing more than patient, careful analysis of comparative data, to avoid making "obvious" mistakes that represent coarse convergences. Metaphorically speaking, increasing the "magnification on the microscope" often converts apparent similarities to evident dissimilarities. A second strategy is to deemphasize characteristics that are suspected homoplasies. Prior to the advent of phylogenetic systematics, most systematists explicitly made decisions about traits judged "unreliable" based on presumptions about levels of homoplasy. Despite a general movement away from this approach in recent decades, controversies persist about whether consistent differences in levels of homoplasy exist in different classes of characters. For example, are DNA sequences more or less prone to homoplasy than morphological traits? A third strategy is to remain as neutral as possible about levels of homoplasy prior to a phylogenetic analysis, and instead let the evidence contained in a large set of characters speak for itself. The presumption is that evidence in the homologies will generally outweigh evidence in the homoplasies, and the homoplasies will be discovered after the fact. Of course, all phylogenetic studies give less weight to some data implicitly by failing to include them at all, but at least among the data included, the weight of evidence rules. Phylogeneticists have championed this third strategy in the last several decades.

A fourth "strategy," which rears itself with historical regularity, is to assert that a newly discovered class of data is simply free from homoplasy. Chromosome inversions and genome rearrangements are two examples of data that were once argued to have special dispensations with respect to homoplasy. In these and similar cases, however, thorough phylogenetic analysis of multiple independent lines of evidence has always revealed homoplasy lurking in the background. Undoubtedly, this will continue to bedevil attempts to obtain an accurate picture of the tree of life.

[*See also* Dollo's Law; Homology.]

BIBLIOGRAPHY

Ashihara, H., and A. Crozier. "Biosynthesis and Metabolism of Caffeine and Related Purine Alkaloids in Plants." *Advances in Botanical Research* 30 (1999): 118–205.

Darwin, C. *The Origin of Species.* New York, 1859.

Fong, D. W., T. C. Kane, and D. C. Culver. "Vestigialization and Loss of Nonfunctional Characters." *Annual Review of Ecology and Systematics* 26 (1999): 249–268. Survey of evolutionary reversals, with special attention to flightlessness, limb loss, lung loss, and adaptations to subterranean environments.

Gould, S. J. "Dollo on Dollo's Law: Irreversibility and the Status of Evolutionary Laws." *Journal of the History of Biology* 3 (1970): 189–212. Convincing rehabilitation of Dollo in the context of modern views on macroevolution.

Land, M. F., and R. D. Fernald. "The Evolution of Eyes." *Annual Review of Neuroscience* 15 (1992): 1–29. Fascinating account of dramatic convergences in the evolution of animal eyes.

Leschen, R. A. B., and I. Lobl. "Phylogeny of Scaphidiinae with Redefinition of Tribal and Generic Limits (Coleoptera: Staphylinidae)." *Revue Suisse de Zoologie* 102 (1995): 425–474.

Sanderson, M. J., and L. Hufford. *Homoplasy: The Recurrence of Similarity in Evolution.* San Diego, 1996. Edited volume containing contributions on the theory of homoplasy, its role in phylogenetic analysis, and numerous examples from different groups of organisms.

— MICHAEL J. SANDERSON

HONEYBEES

Honeybees are a monophyletic group of some dozen species of social bees, all classified in the genus *Apis*. One species, *Apis mellifera*, is the familiar bee used for beekeeping.

Apis mellifera has been the subject of scientific observations since ancient times. Human fascination with honeybees reflects not only their astonishing industry as makers of honey and pollinators of crops but also their intricate societies. Whereas most animal groups (such as deer herds and wolf packs) are temporary assemblages in which each member can survive and reproduce independently, a honeybee colony is a thoroughly integrated unit composed of highly interdependent individuals.

There are two keys to understanding why honeybee colonies evolved a high degree of functional integration: their genetic structure and how a colony's members achieve genetic success. A colony's genetic structure is that of a family. Each colony derives from one mother (queen) and multiple fathers (drones). The queen mates with some dozen drones when she is young, stores their sperm, and uses it throughout her life. The vast majority of the queen's offspring is daughters that spend their lives as workers, performing all tasks within a colony (brood care, food collection, etc.) other than egg laying. The remaining offspring are a small number of seasonally produced sons (drones) and daughters (queens) that serve only to pass the colony's genes into the future.

Queens and drones, like most animals, achieve genetic success by giving rise to offspring, which themselves eventually reproduce. In honeybees, these reproductive offspring are the drones and queens produced by a colony. Workers achieve genetic success differently.

The worker bees of a honeybee colony and the somatic cells of a multicellular organism have a fundamental similarity: both lack direct reproduction. Nevertheless, both can foster the transmission of their genes into future generations by helping individuals carrying their genes to reproduce. Just as somatic cells toil selflessly to enable their body's germ cells to produce gametes, worker bees toil selflessly to enable their colony's queen to produce queen and drone offspring. Thus, the labor of a worker bee should be viewed as her striving to propagate her genes as they are represented in her mother queen's germ cells and stored sperm. Moreover, recent studies have shown that the genetic propagules of a honeybee colony—the queens and drones it produces—carry an unbiased sample of the genes in the colony's workers. This implies that the mother queen of a honeybee colony is not only a shared pathway but also a fair pathway for the propagation of the workers' genes. Workers therefore have equal genetic stakes in a colony's reproductive success and thus virtually congruent genetic interests.

Given the close overlap of the genetic interests of a colony's workers, it is not surprising that they cooperate for the common good. The most impressive form of their cooperation is the sharing of information about rich food sources by means of the waggle dance. This is a behavior performed deep inside the colony's nest, in which a successful forager reenacts a miniaturized version of her outward flight to a rewarding patch of flowers. Other workers learn the distance, direction, and scent of the flowers from the dance and translate this information into a flight to the specified flowers.

To examine how workers communicate using waggle dances, let us follow a bee upon her return from a rich food source. Her find is a cluster of flowers located a moderate distance from her nest, say, 1,500 meters (1,640 yards), and along a line 40° to the right of the line running between the sun and her nest (Figure 1). Excited by her discovery, she scrambles onto one of the vertical combs inside the nest, and here, amidst a massed throng of her sisters, she performs a waggle dance. This involves running through a small figure-eight pattern: a straight run followed by a right turn to circle back to the starting point, another straight run, followed by a left turn and circle to the left, and so on. The straight portion of the dance is the informative part and is given special emphasis by two actions. First, while walking straight ahead, the dancer vigorously vibrates (waggles) her body laterally at a low frequency (13 Hz). She further emphasizes the straight runs by emitting a buzzing sound. Produced by the flight muscles and characterized by a frequency (280 Hz) corresponding to wing beat vibrations, these buzzes evidently represent a ritualized flight-intention behavior. Usually several bees will crowd behind a dancer, their antennae extended toward her. The followers detect the low-frequency sound, transmitted through the comb, with vibration detectors in their legs, and they detect the high-frequency dance sound, transmitted through the air, with vibration detectors in their antennae. The distal portion of a worker's antennae has a resonant frequency of 280 Hz, matching the dance's high-frequency sound.

The direction and duration of straight runs correlate closely with the direction and distance of the flower patch. Flowers located in line with the sun are indicated by straight runs in an upward direction, and any angle to the right or left of the sun is coded by a corresponding angle to the right or left of straight up. The flowers' distance is indicated by the duration of the straight runs.

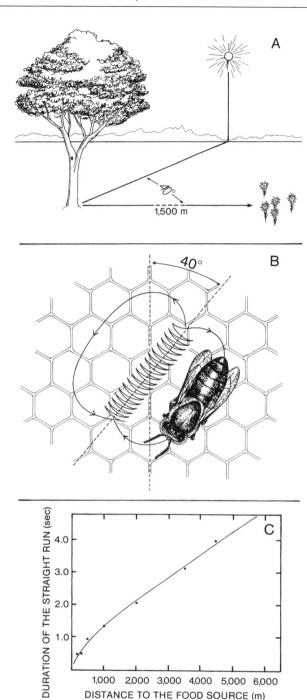

FIGURE 1. The Waggle Dance of the Honeybee. (A) In this example, the patch of flowers lies 1,500 meters (4,920 feet) out along a line 40° to the right of the sun as a bee leaves her colony's nest. (B) To advertise these flowers, the bee runs through a figure-eight pattern inside the nest, vibrating her body laterally as she passes through the central, straight portion of each dance circuit. The straight run is oriented on the vertical comb by transposing the angle between the flowers and the sun to the angle between the straight run and vertical. (C) Distance to the flowers is coded by the duration of the straight run.   Illustration prepared by Sandra Olenik. Seeley, Thomas D. *Honeybee Ecology*. Copyright © 1985 by PUP. Reprinted by permission of Princeton University Press.

Evidently, a dance follower determines the angle of the straight runs by aligning herself with the dancer during a straight run and noting her own body's orientation. A dance follower determines the duration of the straight runs by noting the duration of the substrate and airborne sounds produced during each dance circuit.

Besides information about direction and distance of a flower patch, a dancing bee shares information about the scent of her flowers. She carries home this scent both in her waxy cuticle and in her food—pollen on her hindlegs or nectar in her crop. Dance followers learn the flowers' scent and use this information to pinpoint the rewarding flowers after using the dance's vectorial information to arrive at their general location.

[*See also articles on* Eusociality; Kin Recognition; Kin Selection.]

BIBLIOGRAPHY

Frisch, K. von. *The Dance Language and Orientation of Bees*. Cambridge, Mass., 1967. A synthesis of Karl von Frisch's research on the honeybee's waggle dance, for which he received the 1973 Nobel Prize in physiology/medicine.

Lindauer, M. *Communication among Social Bees*. Cambridge, Mass., 1971. A nontechnical presentation on the evolutionary origins of the waggle dance and on its role in various activities such as water collection and nest-site selection.

Seeley, T. D. *The Wisdom of the Hive*. Cambridge, Mass., 1995. A summary of research on how the thousands of workers in a colony function as a coherent system in gathering their food.

Wilson, E. O. *The Insect Societies*. Cambridge, Mass., 1971. Still the best general review available on the biology of social insects, including honeybees.

— THOMAS D. SEELEY

HOST BEHAVIOR, MANIPULATION OF

A starling, a pillbug, and a worm can teach us much about the evolution of parasites, in particular, how parasites can manipulate host behavior. The thorny-headed worm (*Plagiorhynchus cylindraceus*) lives in the small intestine of the starling, releasing eggs with bird feces. When a terrestrial isopod (pillbug) eats these eggs, the parasite hatches, burrows through the pillbug intestine, and matures to a stage known as a cystacanth in the pillbug's body cavity. When the cystacanth is eaten (along with the pillbug) by a starling or other suitable avian host, it becomes an adult in the host intestine. It turns out that a pillbug containing a cystacanth is more likely to be eaten than an uninfected pillbug—a consequence of the fact that infected pillbugs do not behave like uninfected pillbugs. Infected pillbugs spend more time on light-colored surfaces and areas of lower humidity and ignore overhanging shelter. These behavioral aberrations increase their risk of predation and, in so doing, transmit the parasite more effectively.

Such manipulation is not restricted to worms and crustaceans, but runs the gamut of host–parasite associations and almost every type of behavior we can imagine. Thus, the virus that causes rabies invades areas of the brain that influence aggressive behavior, then uses the nervous system to access the salivary glands, all to the advantage of the virus: swallowing is impaired, virus-laden saliva collects, and the host bites other potential hosts. Mosquitoes transmitting the protist that causes malaria find themselves unable to feed efficiently. They probe, then probe again, injecting the microscopic pathogen into the bloodstream of one host, then another. The protist can be equally manipulative on its return trip from vertebrate to mosquito. In some associations, the vertebrate host is most ill—and least able to defend itself from biting flies—at the very time that the parasite is in the peripheral blood, ready to enter the insect vector.

The theme running through these and many other examples is that natural selection does not necessarily always favor a benign or a pathogenic parasite. Rather, it favors the parasite that gets transmitted, whether through benign or nefarious means. In some cases, transmission is enhanced by harming a host, making it available to vector or predator; in others, minimizing harm to a host protects a parasite's stake in the future, as the host disperses parasite propagules to other hosts or into the environment. Either outcome can be adaptive, depending on the transmission mode.

Parasites even meddle with reproductive behaviors—arguably the behaviors most closely linked to fitness. Bacteria in the genus *Wolbachia*, for example, are transmitted cytoplasmically, from insect mother to offspring. Given this mode of transmission, it is not surprising that the parasite manipulates dramatically the proportion of female offspring. Although this is not a behavioral manipulation, there is no shortage of such manipulation in the realm of host reproduction. Many insect-dwelling nematodes induce false oviposition, in which parasitized insects—male as well as female—fly to oviposition sites and "oviposit" the parasite in the vicinity of the next generation of young. This is one manifestation of a broader phenomenon, that of parasitic castration, in which a parasite either interferes with gonadal development or destroys existing gonads. Castrating parasites might benefit in ways ranging from the change in host energy allocation to the reduction of host risk associated with reproductive activities.

Such changes are actually more in the province of enhanced parasite survival than transmission, although neither would serve well without the other. As phenomena, however, they can be distinct. The most macabre example of survival enhancement can be found in the world of caterpillars that are hosts to parasitic wasps. The wasps feed on the innards of caterpillars, then emerge from their dying hosts to pupate and become adults (see Figure 1). It is not uncommon for parasitic

FIGURE 1. Parasitic Wasp Larvae Leaving White Caterpillar.
© Hans Pfletschinger/Peter Arnold, Inc.

wasps to manipulate the behavior and development of their caterpillars, but one species goes so far as to cause the caterpillar to spin a web over the emerging wasps and defend the pupae with its dying movements.

Like most behaviors, the right place and the right time are crucial to the success of these altered behaviors. Although it is often difficult to study behavior over the course of an infection, the timing of some behavioral changes has been shown to support the hypothesis that these changes benefit the parasite. For instance, another thorny-headed worm, this one infecting ducks, is a cystacanth in small crustaceans. Normally, these crustaceans dive and seek dark areas when the water is disturbed, but if infected, they will move toward the light and skim around on the surface, clinging to a variety of floating objects—including ducks. They are eaten in far greater numbers than uninfected counterparts, thus indicating that these behavioral shifts benefit the parasite, but even more remarkably, the behavioral change does not occur until the cystacanth is actually infective for a duck.

Behavioral changes should also take the parasite to the right place; we have already seen the utility of false oviposition in this regard. Even more striking is "tree-top" disease of insects. Produced by a variety of parasitic fungi, viruses, and bacteria, these pathogens cause elevation seeking on the part of dying hosts. Perched high above its conspecifics, the disintegrating host produces a rain of infective propagules that is dispersed much more effectively than it would be if the insect died in a normal, lower location.

There are innumerable additional examples of altered host behaviors that enhance parasite transmission, but the notion of elevation seeking can open a door into a new way of looking at altered behaviors—behaviors that benefit the host rather than the parasite. Houseflies parasitized by fungi, for example, seek warmth; if they can raise their body temperature high enough during the first days of the infection, they will kill the otherwise lethal fungus. In fact, many poikilotherms engage in behavioral fever. They fight pathogens by behavioral increases in body temperature. How do they do this? By elevation seeking, moving toward the light, basking, perhaps hyperactivity—in other words, the same behaviors that might also appear to enhance the transmission of some parasites, especially those transmitted in the food chain. Perhaps those trophically transmitted parasites (such as the cystacanths) have usurped the host's defensive behavior and turned it to their own advantage.

Indeed, a large suite of altered behaviors promotes the fitness of parasitized hosts. Behaviors associated with temperature shifts are important. Even our own metabolic fever, the fever of homeotherms, has behavioral correlates that enhance fitness. When we (or our pets) are feverish, we want to sleep, often to the exclusion of other activities. We say we "feel sick," dogs refuse to eat, normally fastidious cats look a mess. However, it is also possible that the almost mandatory rest imposed by fever is one way to conserve energy that is then devoted to the immune response and to the energetic demands of fever itself.

Defenses are not limited to temperature, but can also include chemicals. Thus, grooming not only mechanically removes ectoparasites, but saliva may be effective against microscopic ones as well. The saliva of some mammals is harmful to some bacteria, which may be why dogs were encouraged to lick human wounds in the Middle Ages. Moreover, the natural world contains medicinal compounds that we have long forgotton; ailing chimpanzees, however, make good use of vegetation that contains natural "worming" components, as well as soils that are analogous to the compounds we take for intestinal disorders. In all of these cases—moving to hot (or cold) locations, grooming, choosing special foods—infected animals behave differently than uninfected ones in ways that offer defenses against parasites.

Parasites can also affect the behavior of uninfected hosts, which may go to great lengths to avoid being parasitized. When we housebreak a puppy or provide a litter box for our cat, we are taking advantage of the fact that many animals are particular about eliminative areas; for instance, many herbivores avoid grazing in areas where they deposit feces—and perhaps parasites. As for ectoparasites, howler monkeys spend up to one-quarter of their metabolic budgets combating blood-feeding flies, and reindeer may migrate to get away from warble flies. If one doesn't move away, joining a group may be advantageous. In the so-called selfish herd, animals on the edges of the herd are at greater risk of attack from either flies or predators than animals in the center are. Thus, Holstein cows will "bunch" when in the presence of face flies, and stickleback fish will form larger shoals in the presence of fish lice. On the other hand, sociality exposes hosts to more directly transmitted pathogens than they would otherwise encounter; in fact, this may contribute to xenophobic behavior in many primates. Increasing human population density probably contributes to the appearance of seemingly new diseases that simply could not be maintained and transmitted in earlier, less crowded times. We now know that even mate selection is heavily influenced by parasites; many organisms will choose mates on the basis of traits that are consistent with good health—choices that not only may provide "good genes" for offspring but may also reduce the probability of contagion.

In summary, parasites can influence behavior in ways that benefit parasites or hosts, or in some cases, neither. Even more intriguing is the realization that the parasite-induced behavioral alterations we know best are those we see. However, parasites can also alter odor, sound, and other stimuli we perceive only dimly. Moreover, given the near ubiquity of parasites, one must wonder if "normal" animals really exist. In the presence of these hidden guests, things are not always as they seem.

[*See also* Sex Ratios; Virulence.]

BIBLIOGRAPHY

Anderson, R. M., and R. M. May. *Infectious Diseases of Humans: Dynamics and Control.* Oxford, 1992. A comprehensive, technical treatment of the evolution, population biology, and epidemiology of disease; it does not emphasize behavior, but provides valuable context.

Fialho, R. F., and L. Stevens. "Male-Killing *Wolbachia* in a Flour Beetle." *Proceedings of the Royal Society of London B* 247 (2000): 1469–1474. A recent exploration of one way in which *Wolbachia* skews insect sex ratio and a guide to other recent literature.

Hart, B. "Behavioral Adaptations to Parasites: An Ethological Approach." *Journal of Parasitology* 78 (1992): 256–265.

Moore, J. *Parasites and the Behavior of Animals.* New York, 2002.

— JANICE MOORE

HOX GENES

Hox genes are the architects of embryonic development. Their discovery in 1984 fundamentally altered the courses of evolutionary and developmental biology, and led to a molecular reunion of these two fields in the form of evolutionary developmental biology.

Naturalists had long been aware of the potential for development to occasionally generate bizarre mutants, in which one body part developed in place of another. These mutations ranged from development of an extra thumb to formation of entire limbs at ectopic positions in the body. These anatomical substitutions, or transformations, were originally labelled "homeotic" by the English naturalist William Bateson. Geneticists working with the fruit fly, *Drosophila*, also observed spontaneous mutants in which body parts developed in inappropriate positions, but unlike those oddities found in the gardens of Victorian naturalists, the mutant flies could be bred in the laboratory and isolated using genetics. A collection of homeotic mutants emerged from this approach, displaying such radical alterations to the body plan as formation of a second pair of wings in place of halteres, and development of legs in place of an antennae. [*See Body Plans.*] The mutants were named in accordance with their phenotypes, with the aforementioned four-winged fly being labelled "Bithorax" (because the wing-bearing second thoracic segment developed in place of the haltere-bearing third thoracic segment), and the fly with legs in place of antennae receiving the fitting title of "Antennapedia." A common feature of these homeotic mutants is that they exhibit changes in the *identity* of anatomical structures, as if a developmental fate switch has been thrown at an inappropriate position of the body. Once the switch is triggered, the cells seemingly follow the normal course of development, but in abnormal sites. For example, the antennapedia mutant flies form a morphologically sound leg, with the exquisite details of a fly leg, but this leg is situated in a position normally occupied by an antenna. At a cellular level, the cells that form the leg are the same cells that normally form the antenna, only in the mutant these cells follow the genetic program for leg development. Thus, homeotic transformations are caused by mutations in genes that determine the fate of particular cells during development. Accordingly, these genes were dubbed *homeotic genes.*

Physical mapping of the *Drosophila* homeotic genes revealed that they are physically linked in two complexes, termed the Bithorax and Antennapedia Complexes. The Bithorax Complex (BX-C) contains the *abdominal (abd)A, AbdB* and *Ultrabithorax (Ubx)* genes, and the Antennapedia Complex (ANT-C) contains the *labial (lab), proboscopedia (pb), Deformed (Dfd), Sex combs reduced (Scr)* and *Antennapedia (Antp)* genes. Collectively these genes act to draw a molecular map of the early embryo by specifying the positional identities of cells in different locations, such that cells in a particular segment of the fly know, based on the genes expressed within, whether they should form antennae, legs, wings, halteres, eyes, mouthparts, or genitalia (for a review of positional information, the reader is directed to the paper by Wolpert). Homeotic genes act cell autonomously, affecting differentiation only in the cells in which they are expressed (nonexpressing neighbors are not affected).

Cloning of the homeotic genes revealed the presence of a highly conserved sequence of 180 base pairs, known as the homeobox, which encodes a 60 amino acid DNA binding region known as the homeodomain. Homeobox genes encode transcription factors, which regulate expression of other genes in the same cell (hence their cell-autonomous activity). The homeobox is found in each of the fly homeotic genes described above. *Hox* genes belong to the larger family of homeobox genes, but the designation "*Hox*" is reserved for the eight linked genes found within the fly homeotic (HOM or *HOX*) cluster, and their homologues in other taxa.

Hox genes have been found in a remarkably wide range of animals. The evolutionary history of the *Hox* cluster is beginning to come to light, thanks to important new work in comparative genomics and comparative developmental biology. True *Hox* gene clusters are found throughout the *triploblasts* (higher animals with three germ layers) including vertebrates, arthropods, annelid worms, nematode worms, echinoderms and molluscs. In diploblasts (basal animals with two germ layers), the situation is somewhat more complex; genes that are quite closely related to the *Hox* genes have been found in the cnidarians (e.g., corals, sea anemones, hydroids, jellyfish) and ctenophores (comb jellyfish), but whether these genes are clustered remains largely unresolved (for a review of this topic, the reader is directed to the paper by Ferrier and Holland). At least some cnidarians appear to have linked *Hox*-like genes, suggesting that the origin of the *Hox* cluster could date back to the origin of eumetazoans, and in the absence of clear data from sponges, an origin at the base of metazoans cannot be excluded. Further work is needed to resolve these early nodes. The emerging picture of *Hox* gene evolution is now sufficiently detailed to allow for the proposal of a hypothetical scenario for *Hox* origins. According to the model, tandem duplications of a single ancestral homeobox gene gave rise to an ancestral cluster known as the Proto*Hox* cluster. This hypothetical ancestral cluster may have consisted of four *Hox*-like genes (a posterior gene, a middle gene, an *Xlox/Hox3* gene and an anterior gene). Subsequent duplication of the Proto*Hox* cluster gave rise to the *Hox* cluster and Para*Hox* cluster. The single *Hox* cluster was then expanded by tandem duplications, and descendents of this single *Hox* cluster can be found in all triploblastic animals. A single cluster has been retained by the invertebrate triploblasts (including *Drosophila* and the cephalochordate amphioxus), but in the vertebrates, partial or complete genome duplications have resulted in the evolution of multiple *Hox* clusters. Mammals, for example, have four *Hox* clusters (designated *HoxA, HoxB HoxC* and *HoxD*), and in teleost fishes, another round of genome duplication has resulted in seven clusters (the eighth cluster was not retained). Each of these clusters contains members of the thirteen *Hox* paralogy groups (though losses of individual genes during evolution means that no single cluster contains all 13 paralogues).

The fact that *Hox* genes are clustered in the genome is of paramount importance, because this genomic organization is causally linked to their expression patterns and functions during development. Genes situated at the 3′ end of the cluster are expressed in the anterior (head) region of the embryo, 5′ genes are expressed in the posterior (tail) end of the embryo, and genes in the middle of the cluster are expressed in the middle of the embryo. This correspondence between position of genes within the cluster and expression pattern in the embryo is termed spatial colinearity. The timing of expression is also related to the clustering of *Hox* genes. Temporal colinearity refers to the sequential cascade of *Hox* gene transcription, in which 3′ genes are expressed early and 5′ genes are expressed late.

Colinear expression of *Hox* genes during development results in molecular regionalization of the embryonic body, with cells at different positions along the anterior to posterior axis expressing different combinations of *Hox* genes, like a molecular coordinate system. These differences in *Hox* gene expression are translated by the cells to determine their relative positions within the embryo. Ultimately, qualitative and quantitative differences in *Hox* gene expression cause cells at different positions to follow different programs of development. In anatomical terms, this could mean cells at one position in the developing gut will give rise to the stomach and cells at a more posterior position will give rise to the rectum. *Hox* genes are involved in specifying regional identity throughout the body. In vertebrate embryos, for example, the central nervous system, axial skeleton (vertebrae), gut, skin, limb skeleton, muscle, and hair are patterned by *Hox* genes. If the same ancient set of genes is operating to control development of all organ systems in all animals, how has variation been introduced? This question lies at the heart of evolutionary developmental biology.

Introduction of heritable variation occurs during embryonic development. Embryogenesis therefore represents the window of opportunity for modification of developmental programs that can lead to the emergence of morphological novelties. The idea that evolution progresses by the generation of "hopeful monsters" is not new, but is it a viable mode of evolution? Probably not. Experiments in mouse and fly genetics have demonstrated quite clearly that mutations in *Hox* genes can result in radical alterations to the body, by altering the course of development. Although at first glance this may seem like a good way to generate rapid and radical diversification, mutations in *Hox* genes alter every organ system in which that gene operates. This means that a mutation in a single *Hox* gene could, for example, result in an animal with a longer neck, but this alteration is likely to be accompanied by changes to the cerebellum, gastrointestinal tract, urogenital system, limb skeleton and inner ear. Such a radical reorganization of the body is likely to be incompatible with life. However, changes

in *Hox* gene expression during development can be altered on a smaller scale, and in a modular pattern. Important work on the genetic regulation of *Hox* genes has revealed that their fine-scale expression patterns in different tissues are controlled by genetic modules, such as enhancers, situated adjacent to the gene. Mutations in these *cis*-regulatory elements can result in localized and often less catastrophic, changes to the body. Evolutionary modifications to the regulators of *Hox* genes, rather than to the *Hox* genes themselves, would allow more localized and subtle changes to the body. This process is known as *regulatory evolution*, and there is increasing evidence that modification of *cis*-regulatory elements has been a real mechanism of evolutionary change.

Boundaries are a key feature of *Hox* gene expression, and refer specifically to the spatial limits of a gene expression domain. *Hox* gene expression domains generally have sharply defined boundaries, and these molecular boundaries result in the specification of precise anatomical boundaries during development. A good example of anatomical boundaries can be seen in the vertebral column, where there are clear transitions between the different types of vertebrae. The transitions between cervical (neck), thoracic (chest), and lumbar (lower back) vertebrae are not graded, but are sharp transitions that occur over the space of one segment. A cursory glance at a vertebrate skeleton shows that on one side of an intervertebral disk lies a cervical vertebra, and on the other side lies a thoracic vertebra. These anatomical boundaries are established during development by boundaries of *Hox* gene expression. A one-segment shift in the position of a *Hox* expression boundary will result in a one-segment shift in vertebral identity. Importantly, these are changes in segment identity, but not segment number. For example, if the *Hox* gene expression boundary between thoracic and cervical vertebrae is shifted anteriorly by one segment, so that the most posterior cervical vertebra now lies within the thoracic *Hox* expression territory, the identity of that cervical vertebra will be posteriorized, and it will now develop as a thoracic vertebra. Although the thoracic region of the spine will be increased by one segment and the cervical region will be decreased by one segment, the total number of vertebral segments remains unchanged. This is an example of a homeotic transformation caused by a change in *Hox* gene expression. If the position of such a *Hox* expression boundary is controlled by a *cis*-regulatory element, then a precise, localized anatomical transformation could be achieved by changing its DNA sequence.

Although *Hox* genes are generally associated with regionalization of the primary body axis of the embryo, they are also involved in development of secondary axes, such as wings, legs, genitalia, and other appendages. Initially, the position at which appendages develop on an embryo is controlled by *Hox* gene expression in the trunk. For example, the fly wing develops only on the second thoracic segment and the haltere forms on the third. As described above, if the positional identity of a trunk segment is changed, then this will be accompanied by a concomitant change in the identity of the associated appendage (as in the bithorax mutant). *Hox* genes are also involved in later development of appendages. In the vertebrate limbs, 5′ genes of the *Hox*A and *Hox*D complexes play a crucial developmental role. As in the trunk, *Hox* genes are involved in specifying regional identities of limb structures. Detailed studies of *Hox* gene function in the limbs has also revealed that *Hox* genes control growth by regulating cell proliferation. Involvement of *Hox* genes in the regulation of growth is not exclusive to the limb, but seems to be a general role of *Hox* genes. Indeed, many of the homeotic transformations described above are the results of altered growth programs.

The molecular rebirth of evolutionary developmental biology has resulted in the identification of molecular genetic mechanisms that can account for evolutionary modifications to the body plan. *Hox* genes have been the most intensively studied candidates for these mechanisms, and there are already a number of examples in which evolutionary modifications of animal anatomy are associated with changes in the regulation of *Hox* gene expression. Snakes serve as an interesting example. Snakes lack paired limbs and have hundreds of morphologically similar, thoracic-like (i.e., ribbearing) vertebrae. Snakes evolved from lizardlike ancestors with complete forelimbs and hindlimbs, and a clearly regionalized axial skeleton. A study of *Hox* gene expression in snake embryos showed that the evolutionary respecification of cervical and lumbar vertebrae towards a thoracic identity was associated with an expansion of the *Hox* gene expression domain that controls development of the thorax. Similarly, loss of forelimbs is associated with an anterior shift in the expression boundary of a *Hox* gene that is normally expressed in the limbless body wall of tetrapods. Anterior expansions of *Hox* gene expression domains resulted in posteriorization of the anterior part of the snake trunk, transforming cervical vertebrae into thoracic vertebrae and the forelimb region into limbless body wall. Thus, the snake body plan has evolved by homeosis. Although the precise nature of the regulatory changes responsible for the modulation of gene expression is not yet know, the structure of this snake *Hox* protein is so similar to that of other tetrapods that the same antibody can recognize the protein in mouse, chicken and python embryos. This suggests that the change in *Hox* expression is caused by regulatory evolution and not by evolution of the structural *Hox* gene. Similar transformations in *Hox* expression are re-

sponsible for evolution of crustacean appendages. During crustacean evolution, the anterior limbs have been modified into feeding appendages that resemble mouthparts. Comparative studies of *Hox* gene expression in a diverse sample of crustacean embryos revealed that transformation of limbs to mouthparts was associated with retraction of a *Hox* gene expression boundary from the segment that bears the transformed limbs. Thus, by eliminating *Hox* expression from certain limb-bearing segments, the identities of the appendages on those segments were transformed. This is another example of evolution by homeosis. There are also examples of what appear to be homeotic mutants that have evolved by evolutionary changes to the downstream targets of the *Hox* genes. For example, butterflies resemble the *Drosophila* bithorax mutant, in that they have two pairs of wings (rather than one pair of wings and one pair of halteres). However, development of the second pair of wings is not associated with changes in expression of *Ultrabithorax* (*Ubx*) gene, which is responsible for the fly bithorax mutant. Instead, evolution of insect wing number was associated with changes in the Ubx-regulated target genes. Thus, homeotic mutations can arise from changes to the upsteam regulation of, or the downstream response to, *Hox* genes.

Our understanding of *Hox* gene regulation and targets is still at an early stage of development. As the mechanistic details of these genetic pathways are uncovered, our understanding of the role of *Hox* genes in evolution will be improved. Undoubtedly, morphological evolution does not occur solely by homeosis, and as the field develops, we can expect to discover more fine-scale anatomical changes that are the results of evolving *Hox* gene regulation.

[*See also* Homeobox; Homeosis.]

BIBLIOGRAPHY

Akam, M. "Hox Genes, Homeosis and the Evolution of Segment Identity: No Need for Hopeless Monsters." *International Journal of Developmental Biology*, 42 (1998): 455–451.

Averof, M., and N. H. Patel. "Crustacean Appendage Evolution Associated with Changes in Hox Gene Expression." *Nature* 388 (1997): 682–686.

Carroll, S. B., J. K. Grenier, and S. D. Weatherbee. *From DNA to Diversity: Molecular Genetics and the Evolution of Animal Design*. Malden, Mass., 2001.

Cohn, M. J., and C. Tickle. "Developmental Basis of Limblessness and Axial Patterning in Snakes." *Nature* 399 (1999): 474–479.

de Rosa, R., J. K. Grenier, T. Andreeva, C. E. Cook, A. Adoutte, M. Akam, S. B. Carroll, and G. Balavoine. "Hox Genes in Brachiopods and Priapulids and Protostome Evolution." *Nature* 399 (1999): 772–776.

Duboule, D., ed. Guidebook to the Homeobox genes. Oxford, 1994.

Ferrier, D. E. K., and P. W. H. Holland. "Ancient Origin of the Hox Gene Cluster." *Nature Reviews Genetics* 2 (2001): 33–38.

Kondo, T., and D. Duboule. "Breaking Colinearity in the Mouse HoxD Complex." *Cell* 97 (1999): 407–417.

Wolpert, L. "One Hundred Years of Positional Information." *Trends in Genetics* 12 (1996): 359–364.

— MARTIN J. COHN

HUMAN EVOLUTION

[*This entry comprises two articles. The first article provides an overview of the history of ideas in Western thought including sociobiological approaches to the past and the present; the second article offers a wide-ranging discussion of morphological and physiological adaptations in humans. For related discussions, see the* Overview Essay *on* History of Evolutionary Thought *at the beginning of Volume 1;* Hominid Evolution; Modern *Homo sapiens; and* Primates.]

History of Ideas

Western thought has always contained evolutionary ideas, but these were usually masked by the prevailing themes in religion, philosophy, and science. Perhaps the first explicitly evolutionary school of thought was developed by the Greek Epicurus (c.341–270 BCE) and his Roman follower Lucretius (c.95–52 BCE). Adopting an atomistic concept of matter and a materialist concept of living things, the Epicureans held that random atomic movements would create physical systems of all shapes and sizes, some of which were more stable than others. By this process, according to Lucretius, human society had evolved through ages of stone, bronze, and iron technology.

Such Greco-Roman evolutionism was all but lost during the Middle Ages, when neo-Platonic and Christian philosophers came to see each living kind as a divinely created essence or type. Because no type could be transformed into another type, the order of nature was fixed and immutable (unless God were to intervene, of course, as in the case of the Flood). All natural kinds, furthermore, were arranged along a scale from "lower" to "higher" forms. This *scala naturae*, which had been the organizing principle of Aristotle's biology, would be interpreted in Christian terms as a Great Chain of Being ascending from inorganic matter to plants, animals, humans, angels, and eventually to God. Each species had a unique position in the Chain, just above a slightly less perfect form and just below a form that was slightly more advanced. In this sense, the Chain would seem to suggest a progressive series of "links," but because the series was divinely created and unchanging, the higher links could hardly have "evolved" from the lower.

Although the human species occupied a distinct position on the Great Chain of Being, the various human "races" or "nations" were commonly classified according to their line of descent from Noah, whose offspring had repopulated the world after the Flood. This biblical ge-nealogy assumed not a linear, static hierarchy of human kinds but a branching tree of ethnological relationships—what might be called the Great Tree of Becoming. In contrast to the Chain, which could be said to exist outside both geography and history, the Tree was rooted in a specific place—Mount Ararat in Armenia, where the Ark had come to rest—and had developed in some definite (though unknown) historical sequence. In this sense, biblical ethnology possessed a historicity that was entirely lacking in contemporary zoological and botanical classification.

Yet biblical ethnology was hardly "evolutionary" in the Darwinian (or even the Epicurean) sense, in part because it assumed a chronology of thousands rather than millions of years. Any biblical approach was constrained by the age of the earth as calculated by Archbishop James Ussher (1581–1656) and other Christian chronographers. According to Ussher, the earth had been created in 4004 BCE, and the Flood had taken place in 2349 BCE. This meant that humans and other living things had dispersed from Mount Ararat to repopulate the world in just over four millennia. Racial variety, furthermore, was generally attributed not to progressive change but to degeneration from a superior starting point. Some races retained the blessings that God had bestowed upon Noah, whereas others carried the curse of Ham (or of Cain) and had drifted into idolatry or worse. If history had any general direction, then, it was from a "higher" to a "lower" condition, at least for the non-Christian nations.

In the eighteenth century, Carolus Linnaeus (1707–1778), Georges-Louis Leclerc Buffon (1707–1788), and most other naturalists remained committed to the doctrine of the fixity of species. The new advocates of progress, however, were ready to ignore biblical convention and argue instead for a dynamic universe in which both natural and human history followed a generally "upward" direction. Thus, by the 1750s, the Great Chain of Being was being "temporalized," allowing for the progress of organic forms to higher levels of perfection. At the same time, human societies were being ranked according to another scale—an evolutionary "ladder" indicating lower or higher stages of civilization. Adam Smith (1723–1790) and other leaders of the Scottish Enlightenment came to identify four stages of economic development: hunting, pastoralism, agriculture, and commerce. Similar schemes would multiply in the "conjectural histories" produced by French philosophers such as A. R. J. Turgot (1727–1781) and the Marquis de Condorcet (1743–1794). By the start of the French Revolution, this mode of progressive evolutionism had eclipsed the biblical account that had long provided the Western framework for world history.

In the wake of revolution, however, and with the spread of modern nationalism, the biblical style of history enjoyed a significant revival. The ethnologists of the early nineteenth century struggled to amass empirical evidence to reconstruct the human genealogy in all its ethnic detail. Linguistic evidence came to be seen as an especially valuable guide to historical relationships between contemporary "nations" or "races." The ethnological potential of such evidence had been dramatically demonstrated in 1786, when Sir William Jones (1746–1794) had posited an Indo-European family of languages, including Sanskrit, Greek, Latin, Gothic, Celtic, and Old Persian, based on their common descent from a lost ancestral tongue. During the same period, the English physician James Cowles Prichard (1786–1846) was marshaling linguistic, anatomical, and biogeographical data to reconstruct what he called "the physical history of man"—all within the traditional framework of biblical ethnology. In retrospect, the linguistics and the ethnology of the early nineteenth century are remarkable for their genealogical style of historical analysis. What Charles Darwin (1809–1882) would later call "propinquity of descent" (rather than a shared stage of development) was seen as the principal source of similarities between existing languages or peoples.

As a way of classifying species, however, a genealogical approach had yet to be developed. Even the evolutionists of the day lacked a truly genealogical framework and assumed instead that progressive development from simple to complex was the master pattern of natural history. Darwin's grandfather, Erasmus Darwin (1731–1802), was a freethinking physician and poet who speculated that "all warm-blooded animals have arisen from one living filament." Such progressive change was attributed to an organism's "faculty of continuing to improve," combined with the transmission of improvements to descendant generations. Later evolutionists such as Jean-Baptiste de Lamarck (1744–1829), Robert Chambers (1802–1871), and Herbert Spencer (1820–1903) continued to imagine that the history of life was governed by an inherent progressive trend in combination with the inheritance of acquired characteristics.

This kind of progressive evolutionism would be overturned by the work of Charles Darwin. In 1837, soon after his return from the voyage of the *Beagle*, Darwin became convinced that species are mutable and that similar species are somehow related by common descent. For months, however, he was at a loss to explain *how* species change. Before he had hit upon the mechanism of natural selection, Darwin was tempted to fall back on the idea of inherited improvements, and he would continue to invoke this "Lamarckian" principle even after publishing his own theory. What Darwin did *not* accept was the idea of an inevitable tendency toward progress in the history of life. Instead, Darwin set himself the task of explaining both adaptation and speciation in terms of external forces rather than internal

drives. This distinguishes Darwin's approach from the inherently teleological views propounded by the evolutionists before him.

When he first presented his theory in *On the Origin of Species* (1859), Darwin did not share his (already well-developed) thoughts on human evolution. In the twelve years that elapsed before he did publish on the topic, there were major shifts in scientific thinking about the age of the earth and the historical significance of fossils, apes, and "savages." The short biblical chronology had already been cast in doubt by the first discovery of a Neanderthal skull near Düsseldorf in 1857. The following year, in Brixham Cave on the Devon coast, scrapers and other artifacts were discovered in conjunction with the remains of extinct animals. Thus, by the time the *Origin* went to press, Charles Lyell (1795–1875) and other leading geologists were ready to accept that the earth and its inhabitants, including humans, were much older than the Christian chronographers had estimated. Now, in T. R. Trautmann's words (1987:221), "Time opened out indefinitely backward; we may truly say that the bottom dropped out of history."

The greatly expanded chronology allowed Darwin's followers to replace the biblical narrative with a naturalistic account of human origins. Thus, in *Man's Place in Nature* (1863), Thomas Henry Huxley (1825–1895) marshaled the anatomical evidence for a close evolutionary relationship between humans and other primates. Although he speculated that *Homo sapiens* had desended from "a man-like ape," Huxley remained cautious and was unwilling at this point to present a more detailed scenario of how the human species had evolved. The first such scenario was developed in 1868 by the leading German evolutionist, Ernst Haeckel (1834–1919), who argued that humans had evolved from apes by "becoming habituated to an upright walk."

At the same time, a chronology of millions of years presented anthropologists and historians with an unprecedented challenge: how could they ever recover what had happened so long ago? There seemed no longer any reasonable prospect of reconstructing the early migrations and lines of descent of the world's peoples, as Prichard and other early ethnologists had hoped to do. To fill the enormous gap between the "dawn of man" and the beginnings of written history, anthropologists increasingly turned to the "savages" in the ethnographic record as the best available models of prehistoric humanity. Though greatly simplifying the task at hand, this approach also tended to ignore both history and geography by arranging cases from various eras and from all over the world in the same category on the basis of their shared level of cultural complexity or progress.

In this way, the eighteenth-century style of social evolutionism was powerfully revived in the work of anthropologists such as Lewis Henry Morgan (1818–1881),

John Ferguson McLennan (1827–1881), and Edward Burnett Tylor (1832–1917). According to Morgan, for example, all human societies were progressing through three major stages—savagery, barbarism, and civilization—and existing "savages" could thus be taken to approximate the way of life followed by civilized peoples in the distant past. Though loosely associated with Darwin's circle, Morgan and his contemporaries made little use of the concept of natural selection and generally owed less to Darwin than they did to the philosophers of the Enlightenment and to pre-Darwinian evolutionists such as Lucretius and Lamarck. At the same time, Darwin himself was significantly influenced by the social evolutionists, as reflected in his major work on human beings, *The Descent of Man* (1871). Here the works of Morgan, McLennan, and Tylor were cited in support of the argument that "all civilized nations are the descendants of barbarians," and Darwin clearly accepted the idea that there were "higher" and "lower" nations or races. As a result, *The Descent of Man* is a curious mixture of Darwin's original genealogical thinking, which stressed the branching, open-ended pattern of evolutionary history, and the new social evolutionism that tended to rely instead on the notion of a universal scale or ladder of progress.

Despite all this, one of the major contributions of *The Descent of Man* was Darwin's attempt to account for human origins in nonteleological terms. He speculated that "a change in its manner of procuring subsistence" or "a change in the conditions of its native country" might have caused "some ancient member in the great series of Primates . . . to live somewhat less on trees and more on the ground." Increasing terrestriality would have led either to strictly quadrupedal locomotion, as in baboons, or to the human pattern of bipedalism. Once they were bipedal, people could use their hands to fashion weapons and other tools, which Darwin saw as crucial to the success of humans in the struggle for existence. Yet it is worth reiterating that this scenario proceeds from historically contingent and unpredictable changes in external conditions. For Darwin, then, the evolution of "higher" intellectual and moral traits has resulted not from an internal drive toward complexity or improvement, but from the vicissitudes of ecological history.

Throughout the nineteenth century, however, Darwin's emphasis on ecological and historical contingency was largely ignored as progressive evolutionism prevailed in both the biological and the social sciences. Most evolutionists in this period saw little need to identify specific ecological conditions that might have selected for the distinctive characteristics of humankind. The transition to erect posture in particular attracted little attention until the 1890s, after the discovery of "Java man" (*Pithecanthropus erectus*, now included in the genus *Homo*), a fully bipedal hominid with a brain size inter-

mediate between apes and modern humans. For more than forty years, evolutionists were divided on the question of whether terrestriality and bipedalism had preceded the enlargement of the brain, as Darwin had suggested, or large brains had enabled a more or less conscious decision to leave the trees and to begin walking upright. Although the accumulating fossil evidence—including the notorious Piltdown remains, discovered in 1912 and later determined to be fraudulent—would be marshaled in favor of one side or the other, the empirical debate seems to have masked a deeper philosophical split. Were human beings the fortuitous outcome of a truly contingent process, or was the development of humankind in some sense guided by intelligent foresight after all—if not that of God, then that of Man himself? In arguing for an intelligent "escape" from the arboreal world of our primate ancestors, evolutionists such as Grafton Elliot Smith (1871–1937) seemed eager to salvage the "mind first" notion of history that Darwin had sought to demolish.

When Elliot Smith, who was a leading authority on primate neuroanatomy, turned his attention to the history of human culture, he was much more inclined to recognize the role of historical contingency and chance. Because each society, he argued, had a unique history of migration, cultural contact, and diffusion, there could be no universal scale of social evolution, no sequence of stages through which all societies were likely to pass. Along with the psychologist and social anthropologist W. H. R. Rivers (1864–1922), Elliot Smith led a diffusionist movement in Britain, even as related forms of culture history were being developed by Fritz Graebner (1877–1934) in Germany and by Franz Boas (1858–1942) in the United States.

Yet the heyday of diffusionism was brief. In the 1940s, social evolutionism was revived in the work of an Australian archaeologist, V. Gordon Childe (1892–1957), and an American cultural anthropologist, Leslie A. White (1900–1975). Like Morgan and Tylor before them, these "neo-evolutionists" tended to think of human history as a unilinear rather than a branching process. If any one Victorian provided their inspiration, it was not Charles Darwin but Karl Marx (1818–1883), who had analyzed economic history as a sequence of stages set off from each other by revolutionary changes in modes of production. As a corrective to the unilinear schemes favored by Childe and White, the American anthropologist Julian Steward (1902–1972) proposed a theory of "multilinear evolution," according to which certain basic types of society, such as the hunting band or the agrarian civilization, tend to develop in similar ways under similar conditions, even though "few concrete aspects of culture will appear among all groups of mankind in a regular sequence."

Within biology, meanwhile, the teleological (and often Lamarckian) theories of evolution that had flourished in the early twentieth century were being swept aside. The "modern synthesis" of evolutionary theory and genetics reaffirmed Darwin's concept of natural selection and completed the scientific revolution that he had begun some ninety years earlier. As the new biology was incorporated into physical anthropology, the study of human evolution began to shift away from progressive typologies and racial classifications and toward the analysis of historical processes within variable populations. The movement's leader was Sherwood Washburn (1911–2000), who argued that the biological traits of any organism must be understood in the context of its ecological adaptation and way of life.

In the context of human evolution, Washburn elaborated ideas first proposed by Raymond Dart (1893–1988), an erstwhile student of Elliot Smith. In 1924, near the village of Taung in South Africa, Dart had discovered the "Taung child," a fossilized skull at least two million years old. As a specimen of *Australopithecus africanus*, the Taung skull had a small, ape-sized cranium that was balanced on the vertebral column, suggesting upright locomotion. Like Darwin, Dart proposed that bipedalism facilitated the use of tools, which in turn favored an expansion in cranial capacity. Yet Dart was particularly struck by the fact that australopithecine remains were found in association with numerous bones of antelopes, gazelles, and boars. Bipedal hominids, Dart reasoned, would have made especially effective hunters. In pursuing Dart's hypothesis, Washburn argued that a number of distinctively human characteristics, including bipedalism, large brains, and the prolonged period of infant and juvenile dependency, could all be traced to the hunting adaptation.

The growing influence of Washburn's views led in 1966 to a landmark conference on the theme of "Man the Hunter." Here biological anthropologists were brought together with social and cultural anthropologists to discuss and integrate their findings about hunter-gatherer populations, both in the ethnographic record and in the course of human evolution. At the conference, Washburn summarized the evidence for his "hunting hypothesis," which maintains that the provisioning of females and their young by hunting males transformed primate mating into human marriage. According to Washburn, then, the human family is a result of the addition of a provisioning father to the mother-plus-young social group found in other primates.

At the same time, the emphasis on paternal provisioning as the driving force of human social evolution provoked a backlash among feminist anthropologists, who answered the hunting hypothesis with their own "gathering hypothesis." In light of the growing evidence that foragers often rely on plant resources more than they do on game animals, Nancy Tanner, Adrienne Zihl-

man, and others argued that the productive activities of women had been systematically neglected or ignored in models of human evolution. Gathering rather than hunting was argued to be the key human adaptation leading to technological innovation and increasing brain size. In some versions of the gathering hypothesis, women were seen as largely self-sufficient producers, and the sexual division of labor that Washburn had cast as the hallmark of the family was all but eliminated from the story of human origins. Yet a renewed emphasis on conjugal cooperation appeared in many subsequent accounts. In 1978, for example, Glynn Isaac (1937–1985) applied Washburn's hypothesis to the archaeological record of early hominids. Isaac proposed that Plio-Pleistocene assemblages in East Africa were evidence of campsites, and that the different kinds of food produced by males and females were shared at a home base.

Variations on the hunting hypothesis dominated anthropological discussion about human social evolution for about twenty years, from the mid-1960s to the mid-1980s. By the end of this period, the fossil record of human evolution was being supplemented by new kinds of evidence. The development of molecular genetics in particular led to major breakthroughs in the reconstruction of the human genealogy. In 1984, Charles G. Sibley and Jon E. Ahlquist reconsidered the question of "man's place in nature" by analyzing the DNA of humans, gorillas, and chimpanzees. Their results suggest that the common ancestor of humans and the African apes lived only about five million years ago—much more recently than anyone had imagined. Chimpanzees, furthermore, were shown to be less closely related to gorillas than they are to humans—another startling result that suggests that people are, phylogenetically, nothing more and nothing less than apes.

The refinement of molecular techniques coincided with an explosion in scientific knowledge about the behavior of nonhuman primates in the wild. The coming of age of primatology is conveniently marked by the publication of Jane Goodall's *Chimpanzees of Gombe* (Cambridge, Mass., 1986), the product of a quarter century of field research at the same site in Tanzania. Such research has demonstrated that chimpanzees and humans share a number of unusual behavioral patterns, including hunting as a predominantly male activity, the fashioning and use of tools, elaborate political maneuvering within the social group, and lethal intergroup raiding conducted by male coalitions. Observations of chimpanzee "warfare" in particular led the primatologist Richard W. Wrangham to propose a phylogenetic hypothesis: the common ancestor of chimpanzees and humans probably lived in male-bonded groups and already exhibited a pattern of cooperative aggression against neighbors.

In contrast to the apes, however, modern humans are

extremely widely dispersed, having moved in successive waves out of Africa to populate most of the world. Here again, DNA sequencing and other molecular genetic techniques have opened new avenues of research. Analyses of genes in living human groups are now being used to track the movements of human populations over the last 100,000 years or more. In a sense, the ethnological project that was begun by Jones and Prichard, only to be abandoned in the 1860s when the long chronology made historical reconstruction seem impossible, has been revived in the work of L. Luca Cavalli-Sforza and his associates (*The History and Geography of Human Genes*, Princeton, 1994). By integrating the genetic data with the evidence of historical linguistics and archaeology, such work may yet complete the Great Tree of ethnological relationships.

[*See also* Hominid Evolution: An Overview; Primates, *article on* Primate Societies and Social Life.]

BIBLIOGRAPHY

Alter, S. J. *Darwinism and the Linguistic Image: Language, Race, and Natural Theology in the Nineteenth Century.* Baltimore, 1999. A revealing analysis of how comparative philology—the study of language families and their relationships—influenced Darwin, Haeckel, and other Victorian evolutionists.

Bowler, P. J. *Theories of Human Evolution: A Century of Debate, 1844–1944.* Baltimore, 1986. A leading historian of biology provides a masterful survey of ideas and discoveries in paleoanthropology.

Corbey, R., and W. Roebroeks, eds. *Studying Human Origins: Disciplinary History and Epistemology.* Amsterdam, 2001. State-of-the-art papers on the history and philosophy of paleoanthropology and related disciplines.

de Waal, F. B. M., ed. *Tree of Origin: What Primate Behavior Can Tell Us about Human Social Evolution.* Cambridge, Mass., 2001. In light of their findings about the behavior of monkeys and apes, nine leading primatologists reflect on human evolution, ecology, sexuality, and social organization.

Gruber, H. E. *Darwin on Man: A Psychological Study of Scientific Creativity.* 2d ed. Chicago, 1981. A fascinating analysis of Darwin's early notebooks, with special attention to his writings on human nature and mental evolution.

Haraway, D. J. *Primate Visions: Gender, Race, and Nature in the World of Modern Science.* New York, 1989. A feminist critique of primatology as a source of modern assumptions about human nature and sexuality. As the only comprehensive history of the field, it remains indispensable for an understanding of how the study of wild monkeys and apes came to be incorporated into the science of "man."

Kuper, A. *The Chosen Primate: Human Nature and Cultural Diversity.* Cambridge, Mass., 1994. Aiming to restore communication between cultural and biological approaches in the human sciences, a social anthropologist surveys the recent history of evolutionary biology, paleoanthropology, primatology, and human ecology.

Landau, M. *Narratives of Human Evolution.* New Haven, 1991. Uncovers the recurring motifs and narrative structures found in scientific accounts of human evolution and suggests that Darwin, Haeckel, Elliot Smith, and other evolutionists were accomplished storytellers as well as scientists.

Lovejoy, A. O. *The Great Chain of Being: A Study of the History of an Idea.* Cambridge, Mass., 1936. Remains the definitive study of how nature was conceptualized by most Western philosophers and scientists before the nineteenth century.

McCown, T. D., and K. A. R. Kennedy, eds. *Climbing Man's Family Tree: A Collection of Major Writings on Human Phylogeny, 1699 to 1971.* Englewood Cliffs, N.J., 1972. An excellent anthology of primary sources, including selections from Buffon, Lamarck, Lyell, Darwin, Huxley, Haeckel, and Elliot Smith.

Richards, R. J. *Darwin and the Emergence of Evolutionary Theories of Mind and Behavior.* Chicago, 1987. A trenchant analysis of how Darwinism has been used to understand the moral and instinctual dimensions of the psyche.

Spencer, F. *A History of American Physical Anthropology, 1930–1980.* New York, 1982. Surveys a crucial half century of research on human evolution and related topics.

Stocking, G. W., Jr. *Victorian Anthropology.* New York, 1987. The definitive study of nineteenth-century evolutionism by the leading authority on the history of anthropology.

Stocking, G. W., Jr., ed. *Bones, Bodies, Behavior: Essays on Biological Anthropology.* Madison, Wis., 1988. An excellent collection of studies in the history of biological anthropology, including paleoanthropology, race science, and behavioral primatology.

Trautmann, T. R. *Lewis Henry Morgan and the Invention of Kinship.* Berkeley, 1987. The single best biography of any of the social evolutionists. Explores the influence of comparative philology, the collapse of the biblical chronology, and the emergence of kinship as an object of anthropological study.

Trigger, B. G. *Sociocultural Evolution: Calculation and Contingency.* Oxford, 1998. An archaeologist traces the history of social and cultural evolutionism, from the late Middle Ages to the late twentieth century.

— LARS RODSETH

Morphological and Physiological Adaptations

Humans practice a unique mode of locomotion. In 1775, the German anatomist Johann Blumenbach became the first person to propose that bipedalism and its morphological correlates constitute the defining feature of the human species. About 225 years later, Blumenbach's pronouncement rings truer than ever. An increasingly deep record of fossil hominids clearly shows that bipedalism, along with the large brain, has been the core trademark of the human clade (Hominidae) since its inception in the Plio-Pleistocene. Moreover, the case can be made that a great part of what has followed within hominids has been shaped, constrained, or promoted by this peculiar form of locomotion. It is no wonder, then, that human locomotion, its origins, its structural and functional basis, and its historical transformation are dominant themes in the study of human evolution.

Adaptations for Bipedalism. Without the counterbalancing advantage of a tail, the primary challenge of hominid bipedalism is maintaining balance without excessive muscular effort. Several important structural modifications promote effective and efficient body stabilization during standing. Straight legs, strategic curvature in the vertebral column, and the orientation of the pelvis all position the center of gravity over the hip, knee, and foot. Stability is further enhanced by an unusually low center of gravity. Apes lack all these modifications. They stand bipedally with strongly flexed hip and knee joints, and a center of gravity that is aligned with neither joint. The massive arms, large pot-bellied torso, and relatively small legs give the ape a high, forward-placed center of gravity. An ape must expend considerable muscular effort to stabilize the hip and knee and to prevent the trunk from toppling forward. In humans, the metabolic cost of standing is not much more than lying down. This is ample testimony to exceptional balance mechanisms, including a significant degree of passive support.

Bipedal locomotion intensifies the problem of balance because of its alternating periods of single-limbed support. A key modification for hominid bipedalism is the valgus orientation of the femur, which brings the knees close together beneath the body. As a result, the body needs to shift only slightly with each step to reposition the center of gravity over the supporting foot. The trunk would otherwise have to be thrown more dramatically from side to side, making balance more difficult and increasing the mechanical work of locomotion. Chimps lack the valgus femur and exhibit precisely these sorts of gyrations when walking bipedally. Although a valgus femur reduces the problem of trunk balance on one leg, active stabilization of the hip joint is nonetheless required. The lesser gluteal muscles assume this strategic role. The large, distinctive gluteus maximus of modern humans is surprisingly inactive during erect standing and normal walking on flat surfaces; however, as a powerful extensor of the hip, it also keeps the trunk from pitching forward over the hips. The gluteus maximus is strongly utilized when a human is leaning over, walking up inclines, or running.

The striding gaits of modern humans are also dependent on a highly specialized foot that is stiff with relatively short, robust toes. The hallux (big toe) is especially strong and is fixed in a position parallel to the remaining toes. The entire structure forms an effective lever system for transferring the propulsive force generated by the calf muscles to the ground, much of it through the big toe. A well-formed connective tissue sheet, the plantar aponeurosis, is essential to the relay of force from the heel to the toes; it also helps to maintain the characteristic longitudinal arch of the human foot. The foot structure of apes reflects a wider range of mechanical functions. The hallux is divergent, mobile, and short relative to the adjacent toes, which are long, curved, and slender. The ape's foot is operationally prehensile and is used extensively to grasp branches and provide other footholds when the animal is moving in trees. Without a well-formed plantar aponeurosis, the ape foot tends to deform in the tarsal region during the push-off phase of walking; it is therefore incapable of delivering the strong propulsive thrusts at the toes that characterize the human gait.

Australopithecines: Out of the Trees—Halfway. The extinct australopithecines, or simply australopiths, include both gracile (*Aridipithecus*, *Australopithecus*) and robust (*Paranthropus*) species. The gracile australopiths are the oldest and most primitive hominids and include within their ranks the ancestors of *Homo*. Australopith anatomy offers the most direct evidence about the earliest stages of bipedalism.

Raymond Dart's 1925 description of the famous Taung child, a juvenile *Australopithecus africanus* from South Africa, offered the first tangible clue that a bipedal gait was not confined to *Homo*. Dart used indirect evidence to posit a fully upright gait for *A. africanus*, but later discoveries of remains of pelvis and leg bones (including a distinctly valgus femur) confirmed his suspicions. More recently, an exceptionally complete skeleton of *A. afarenesis* (AL 288; "Lucy") from the Hardar region of Ethiopia has provided clear documentation of bipedal walking in very early hominids. The spectacular discovery of fossilized trackways at Laetoli, Tanzania, in the early 1990s provided direct evidence for the existence of a fully bipedal hominid dating to approximately 3.6 million years ago. These footprints are widely believed to have been made by *A. afarensis*. There is general agreement that some form of habitual bipedal locomotion was practiced by all the australopiths. The evidence suggests that the origin of hominids may, in fact, be associated with this key shift in locomotor behavior.

Still, there is dispute over the exact nature of australopithecine bipedalism and the extent to which the overall locomotor repertoire of these ancestral hominids may have differed from that of modern humans and apes. Owen Lovejoy, Bruce Latimer, and others undertook the initial functional analyses of the AL 288 materials. They concluded, based primarily on the anatomy of the pelvis and lower limbs, that *A. afarensis* had a gait and overall locomotor abilities not meaningfully different from that of modern humans. This interpretation was unexpected and widely challenged. Subsequent studies of "Lucy" and related materials have convinced most workers that *A. afarensis*, and probably all australopiths, exhibits a unique combination of locomotor adaptations distinctly different from *Homo*. For example, *A. afarensis* has relatively long, curved finger and toe bones, coupled with a shoulder socket that points upward, as in apes, rather than outward, as in humans. The arms are also relatively long and the legs short in australopiths, giving them body proportions distinctly more like apes than like humans.

The supposed humanness of the pelvis, thigh, and

foot morphology of australopiths is also disputed. The pelvis is unusually wide across the ilia and is combined with a femur that has a very small head and long neck compared to *Homo* and to apes. The australopith arrangement gives the gluteal muscles more leverage to resist the lateral tipping of the trunk. In turn, this may reflect an overall mechanism of hip stabilization less effective than that of *Homo*, and a resulting need to compensate. The small size of the joint surfaces within Lucy's leg implies that they were not normally subjected to the level of force commonly encountered in humans. Indeed, the relatively small size of the lumbar and sacral vertebrae has prompted the suggestion that *A. afarensis* may have walked with somewhat flexed hip and knee joints (so-called Groucho walking) to avoid the large shocks associated with the straight-legged gait of modern humans. Computerized modeling of such a hypothetical gait has shown it to be inconsistent with the anatomy of *A. afarensis* and energetically quite expensive. Still, it is unlikely these hominids engaged in the behaviors that routinely produce large, repetitive skeletal loads in modern humans, such as running or jumping.

New materials of *A. africanus* likewise question the humanlike function of the australopith foot. They indicate that the hallux may have been capable of significant lateral movement and divergence—less than in chimps, but very unlike *Homo*. Moreover, chimps trained to walk bipedally across damp sand are now known sometimes to produce footprints remarkably similar to the tracks at Laetoli. There are also suspicions that the foot of *A. afarensis* lacked a fully formed longitudinal arch and, by extension, a plantar aponeurosis. These interpretations point to a foot incapable of generating the powerful thrusts seen in human locomotion, but still quite serviceable in arboreal climbing and grasping.

The australopiths offer a superb example of mosaic evolution in which different features became humanlike at different times and at different rates. The unique combination of human and apelike morphologies makes the australopiths without close analogue in the modern world, and therefore extremely difficult to reconstruct with respect to function and behavior. Collectively, however, the available evidence strongly implies that these primitive hominids continued to make frequent use of arboreal habitats, even if they were fully committed bipeds on the ground. Some researchers have argued that the climbing abilities of *A. afarensis* may well have exceeded those of modern chimps, which routinely occupy trees in order to secure food and protective shelter.

***Homo*: Into the Savanna—All the Way.** *Homo habilis* is still too poorly understood to conclude anything concrete about its locomotor abilities. Thus, an adolescent *H. ergaster* (WT 15000) from Kenya, dated approximately 1.5 million years ago, is the first known hominid with a truly human postcranial skeleton, including the characteristic long legs and relatively short arms. The pelvis is modern, with an expanded, laterally oriented iliac blade above the hip socket, signaling large, powerful gluteal muscles effectively positioned for hip stabilization. The pelvis is relatively narrower than in australopiths, and the neck of the femur is short by comparison. Importantly, the load-bearing surfaces of the legs and lower back are both absolutely and relatively much more robust than in the australopiths. In short, there is nothing in *H. ergaster* to suggest that its basic locomotor abilities departed significantly from those of modern *Homo*. It almost surely used a striding, stiff-legged walk and was capable of sustained running. WT 15000 also documents the large body size attained by early *Homo*, with an estimated adult stature of 5 feet 10 inches to 6 feet 2 inches.

Most paleoanthropologists believe that the basic human body type emerged in response to the rigors of life in an open tropical savanna. Central to this model is the notion that humans are adapted to high, sustained levels of aerobic exercise, long-distance travel, and effective temperature control during such activities. A serious problem facing any active hominid moving from a mesic, forested environment to hotter, drier, open habitats is that of maintaining acceptable core body temperatures. The added heat loads arise from two sources: incident solar radiation, and the endogenous heat generated by muscular activity. Humans, with a relatively naked skin and many widely distributed sweat glands, possess the most sophisticated of all mammalian evaporative cooling mechanisms. Nothing remotely comparable exists in other primates, including the apes. In all likelihood, australopiths lacked a comparable cooling system. Because the secretions are from eccrine sweat glands and are thus quite watery, human sweating is extremely effective in evaporative heat loss. The primary disadvantage is obvious: intense, prolonged sweating may lead to large loss of water and dangerous dehydration. Sweating is also of limited value under humid conditions.

The dissipation of heat, both evaporative and convective, is favored by a high surface-to-volume ratio. Ruff suggests that heat dissipation may help to explain the emergence of long, slender limbs as well as narrow, elongate torsos in hominids like *H. ergaster*. The sub-Saharan peoples of eastern and northern Africa exhibit similar body proportions, as do those of several other regions with hot, dry climates. In regard to limb proportions, humans thus conform to a well-known ecogeographic pattern among homeothermic vertebrates, Allen's Rule, which postulates a positive correlation between length of extremities (limbs, tails, ears) and mean environmental temperature. This same relationship provides a plausible explanation for the characteristically short distal limb segments of Neanderthals (*H. neanderthal-*

ensis), a species with a decidedly temperate distribution.

In modern humans, the width across the top of the pelvis (bi-iliac width) is well correlated with total body surface area. Bi-iliac width is also highly constrained, varying little with gender and adult stature within local populations. There is, however, predictable variation in the relationship of pelvic width to adult stature according to regional mean annual temperature. It seems that climatically mediated adjustments in body surface area relative to body mass are effected chiefly by changes in the linear dimension of the torso, rather than in the width. Accordingly, the bi-iliac/stature ratio is low in people inhabiting hot, dry climates, and high in those from cool, temperate regions (or high altitudes). This climatic influence on hominid body proportions might offer a partial explanation for the relatively wide bi-iliac span of the smaller australopithecines. Although this interpretation does not negate the biomechanical influences on the australopith pelvis, it underscores the important fact that morphologies are often shaped by multiple selective influences.

Humans utilize two distinct, speed-dependent gaits: the walk and run. Transition between the two gaits occurs at speeds of 2.0—2.4 meters per second and serves to minimize the energetic cost of transport. Both fast and slow walking are much more costly than jogging; intermediate walking speeds are the most economical. The large contrasts in the cost of walking and running reflect fundamental differences in gait mechanics. Walking utilizes a relatively straight, stiff leg to promote the exchange of the body's kinetic and potential energy on every step. The mechanism has been likened to an egg rolling end over end: it works well only at modest speeds. Running is more like a bouncing ball. The supporting leg is more compliant than in walking, and it makes significant use of stored elastic strain energy through the stretch and recoil of springy structures within the leg. This gait operates effectively over a much greater range of speed.

As a new addition to the hominid behavioral repertoire, sustained running could account for several additional structural and physiological modifications unique to *Homo*. Certain aspects of leg design (e.g., Achilles tendon) are consistent with specialization for running, but they make little sense in terms of walking. Running also enhances the efficiency of the sweat-based evaporative cooling system of humans, especially in still air. However, because it produces larger endogenous heat loads than walking, running also demands a more effective cooling mechanism. The significantly larger body size of Pleistocene *Homo*, compared to australopiths (especially graciles), could likewise reflect specialization for endurance running. The reason is that the cost of transport in running mammals scales with strong negative allometry ($Mbody^{-0.33}$), thus placing a decided premium on larger body size. This scaling advantage exceeds that of the expected decline in basal metabolic rate ($Mbody^{-0.25}$) according to Klieber's Rule.

As a driver of morphologic change, human running has taken a distant back seat to walking in discussions of hominid evolution, mostly because it is both slower and relatively more expensive than in quadrupedal mammals. The inferiority of human running performance is evolutionarily relevant, however, only if humans have competed primarily with quadrupeds rather than with other hominids. Furthermore, among mammals in general, humans excel at endurance running, especially in hot, dry environments. Recent hunter-gatherer societies have sometimes exploited this advantage to run faster game animals to exhaustion.

[*See also* Hominid Evolution, *articles on Ardipithecus* and *Australopithecines*, Early *Homo, and* Archaic *Homo sapiens; articles on* Race.]

BIBLIOGRAPHY

Aiello, L., and C. Dean. *An Introduction to Human Evolutionary Anatomy*. New York, 1990. The best survey of hominoid functional and evolutionary morphology; extensively referenced; requires only a minimal background in human anatomy.

Alexander, R., McN. *The Human Machine*. New York, 1992. A very good introduction to the physics of human musculo-skeletal design and movement, by the preeminent biomechanist of our time.

Carrier, D. R.. "The Energetic Paradox of Human Running and Hominid Evolution." *Current Anthropology* 25 1984: 483–495. One of the few papers to emphasize the possible importance of endurance running in human evolution, chiefly from the viewpoint of temperature regulation during exercise.

Inman, V. T., H. J. Ralston, and F. Todd. *Human Walking*. Baltimore, 1981. The classic work on the mechanics of human walking, with many exceptional (and much reproduced) illustrations.

Lovejoy, C. O.. "Evolution of Human Walking." *Scientific American* 259 (1988): 118–125. A readable account of a leading anthropologist's interpretation of the early stages of hominid bipedalism from a functional perspective.

Ruff, C. B.. "Climate and Body Shape in Hominid Evolution." *Journal of Human Evolution* 21 (1991): 81–105.

— Dennis M. Bramble

HUMAN FAMILIES AND KIN GROUPS

The nuclear family is often presumed to be the basic social unit of human society, consisting of a husband, a wife, and their unmarried children, living together in one household. However, it is also common for parents and siblings of either spouse, and their own children and spouses, to unite with the members of the nuclear family to form an extended family.

Most other primates, longer lived and slower maturing than other mammals of similar body size, also live in groups composed largely of relatives—groups that

FIGURE 1. Large, Extended Family, Queens, New York.
© Catherine Ursillo/Photo Researchers, Inc.

resemble collections of extended human families in long-term cohesion and coresidence, in collaborative care for juveniles, defense against enemies, and in the complicated mix of affection, competition, and conflicts of interest common in social animals. Communities founded on such social relationships are fundamental to our evolutionary heritage.

The defining characteristic of the human nuclear family, however, is the species-typical behavior of marriage, found in every known society (the Na of China are the single possible exception). Ethnologists continue to argue about definitions that can accommodate every case, but the core of the institution consists of a socially approved, nonephemeral mating and child rearing relationship between a man and a woman (monogamy), a man and several women (polygyny), a woman and several men (polyandry), or several men and several women (polygynandry or group marriage) in which the man gains special sexual rights over a woman, and the woman gains access to resources under male control, protection from harassment by other men, and a designated social father for her children.

Marriage also links men to their wives' children. Some cultures hold that the child is wholly the product of the father or fathers, with the mother serving only as a receptacle. Others assert that both father(s) and mother donate substance to the formation of the fetus. Still others have ideologies that minimize the father's contribution to conception and gestation, while still holding that the child is his in a genealogical sense. Whatever the beliefs about conception and fetal growth, the crucial fact is that marriage designates the mother's husband as the primary social father of the child. A consistent finding of cross-cultural research is that children without social fathers, whether because of death, desertion, divorce, or bastardy, are handicapped in survival, reproduction, and many culturally specific measures of achievement.

Social Approval. It is social approval that distinguishes human marriage from analogous animal bonds having to do with mating, juvenile care, and related interdependencies. In other primate species, third parties, especially other males, usually take an active interest in mating relationships, because they are in competition for the paternity of offspring. In human societies, that mating competition is largely decided (although never absolutely) by public agreement that a husband has special rights. Marriage as well as patterns of extramarital sex that establish special relationships between particular men and women may reduce the dangers that other men can pose to women and their offspring.

The social approval of a marriage usually includes rituals wherein third parties such as parents, kin group elders, clergy, or government officials give public consent to the relationship. Often an exchange of gifts between the bride and groom or their families precedes the marriage. In the custom of bride service, a husband (or prospective husband, as in the biblical story of Jacob) works for his father-in-law in return for receiving his daughter as a wife. Where males control durable wealth, a bride-price payment, often in livestock, may be given by the prospective husband (usually with contributions from his kin) to the bride's family. Dowry, a marriage payment made by the bride's family, customarily goes not to the husband's family of origin, but to the new couple.

Some kinds of unions are inherently ineligible for social approval. Almost everywhere, it is forbidden to have sex with, and hence to marry, a primary relative—a mother, father, sister, brother, daughter, or son. Incest taboos are usually extended to cover more than the primary relatives, but which secondary and tertiary (and often more distant) relatives the taboo covers vary from culture to culture. In many societies, individuals who are not biologically related are linked through fictive kinship ties and are also subject to incest taboos. Where societies are stratified, sometimes the wealthy and powerful marry close kin, concentrating assets at the top. In a few archaic states—Egypt, Peru under the Incas—royal siblings married each other.

Although incest-averse behavior is common among primates, socially prescribed kin categories of marriage partners are confined to humans. In many societies, marriage with a certain category of relative (e.g., a cross-cousin) is preferred or required. There may be ongoing interchange between families or descent groups (see below), such that when a daughter of family or clan A marries a son of family or clan B, then a daughter of B will marry a son of A.

Residence Classification. People are like other primates in that either males or females tend to leave home when they grow up. In the vast majority of human societies, these moves are associated with marriage, either husband or wife (sometimes both) moving from his or her natal family (the nuclear family of orientation) to join with the other in forming a new unit (the nuclear family of procreation). Cultural preferences in these movements are called postmarital residence rules. Virilocality or patrilocality specifies residence with or near the husband's family, uxorilocality or matrilocality with or near the wife's family. When the newlyweds move back and forth, the pattern is called bilocality or ambilocality; when they establish residence in a new place, neolocality. Avunculocality occurs in a few societies, where residence is with the wife's mother's brother. Residence choices may vary through the life cycle.

The residence classification implies that marriages are long-lasting unions in which spouses always sleep under the same roof, but divorce and remarriage are common in many societies. Even where marriages are long lasting, husbands and wives may not reside together, as in much of Highland New Guinea, parts of aboriginal Australia, and in most polygynous societies in Africa. The Guajiro Indians of Colombia and Venezuela traditionally followed the custom of the "visiting husband," in which husband and wife did not even reside in the same community.

Unusual forms of marriage are found in some societies. Among some cattle-raising peoples of East and South Africa (e.g., the Nuer), a woman without brothers could take a wife for the purpose of providing male heirs for her patrilineal descent group (see below). The woman became a "social man" insofar as her role in this marriage was concerned, her wife impregnated by a male from the patrilineage. Among some of the Plains Indians of the United States, a man could become a berdache, a transvestite social woman, and might in this status be taken as a second wife by a warrior. In European history, royal marriages were sometimes contracted for political alliances only, without intent of sexual relations. In Catholic doctrine, celibate nuns are considered to be married to the deity. Among the matrilineal Nayar of India, a girl went through a marriage ceremony with a man she might never see again, before beginning her sexual and reproductive life by taking lovers. In the contemporary United States, lobbying for "gay marriage" of same-sex couples, without one partner necessarily taking the social role of the opposite sex, has resulted in some jurisdictions recognizing a legal union, although distinguishing it from traditional marriage.

Divorce tends to be more common in uxorilocal societies, where a wife lives among her close kin and her father directs the activities of the junior men, distributing the products of their labor. Aside from this contrast between uxorilocal and virilocal societies, there exists a great deal of intercultural variation in the frequency of divorce and the average number of spouses per individual. In a handful of well-studied hunter-gatherer communities, the frequency of divorce is lower where men have fewer mating opportunities, but not where fathers have larger effects on their children's welfare. It is a finding that throws into relief the endless conflict between male and female reproductive interests, men often serving their own interests best by maximizing the numbers of children they sire, whereas women serve theirs by maximizing the survivorship of their children. The human family can be seen as both a treaty and a battleground in this conflict.

Descent Groups. In a large number of human societies, people belong to groups of kin larger than the family on the basis of their common descent from a founding ancestor. Descent may be unilineal, traced exclusively through either the male or the female line. When all the genealogical links from each group member back to the founding ancestor can be specified, these descent groups are known as patrilineages or matrilineages, respectively. When only the ideology of common descent remains, but the genealogical links are forgotten, they are called patriclans or matriclans. In about a third of societies with ancestor-focused groups, descent may in principle be traced through any combination of male and female ancestors.

Descent groups often hold land and other assets, as well as engaging in collective efforts in warfare and religious practice. When descent groups are prominent, responsibility for the actions of any member is usually shared by all. Most primary loyalties attach to fellow descent group members. Relations between and within descent groups are usually modeled on family relations; for example, two patrilineages may be considered "like brothers." Unilineal descent groups are often exogamous, for members of the same descent group to marry each other would be incestuous. Thus, people have close kin who are not in their descent group. With exogamous patrilineages, for example, one's mother and all relatives on her side belong to different lineages, as do a woman's own children and those of her sisters.

Basis of Marriage. Human communities and families differ from other primate social groups in that males and females engage in different economic activities. Among hunting and gathering peoples, where economics has mainly to do with getting and distributing food, the general pattern is that the men do most of the hunting and the women most of the gathering. This evidence has led to speculation that the origin of marriage and the nuclear family lies in the provisioning of wives by husbands. The idea is that marriage is a bargain in which husbands provide meat that their wives cannot hunt for themselves because they are encumbered by the highly dependent juveniles characteristic of *Homo sapiens*. In return, wives restrict their sexual relations to their husbands, increasing the husband's confidence that he is feeding his own offspring.

PARTIBLE PATERNITY

The Barí Indians of Venezuela are tropical forest horticulturalists who supplement their root crops with fish and game. They traditionally lived in communal longhouses holding about fifty people. Within these houses, people were clustered into up to a dozen nuclear or small extended families that cooked at the same hearth, hung their hammocks in the same section of the house, and maintained their own separate household economy.

The Barí believed that all men who had sex with a woman around the beginning of her pregnancy, or at any time during the pregnancy, contributed to the development of the fetus. Two thirds of Barí woman had a lover for at least one of their pregnancies (median number of pregnancies = 8), a practice to which their husbands did not object. As a result, about 23 percent of all Barí children had two or more social fathers, the mother's husband, and her lover(s), who, as secondary fathers, had also recognized obligations to the child (although there was no mechanism to enforce them).

Analysis of demographic data gathered on 916 remembered pregnancies of 114 Barí women showed that pregnancies involving secondary fathers had a lower rate of miscarriage than pregnancies attributed only to the woman's husband. Further, among live-born children, those with secondary fathers had a higher rate of survival to age 15. These differences are probably attributable to gifts of fish and game provided by the lovers, first to the pregnant mother and later to her child.

—STEPHEN BECKERMAN

However, in a number of hunting and gathering societies (e.g., the Ache of Paraguay, the Hadza of Tanzania), women get about the same amount of meat from a hunter whether or not they are married to him. In some societies (e.g., the Warao of Venezuela), a woman's husband may provide her with less meat than do her father and brothers. In much of traditional horticultural Africa, women supply virtually all the subsistence for their children, with their husbands' contribution limited to clearing fields for the wives.

A further challenge to the idea that marriage is in essence a meat-for-fidelity contract between husbands and wives arises from ethnographic evidence about extramarital sex. Although marriage regularly involves sexual access between husband and wife, it is less common for it to require sexual exclusivity. In various parts of the world, a husband's brothers have legitimate sexual access to his wife, and the husband has legitimate

access to his wife's sisters. Among many peoples of lowland South America (e.g., the Barí of Venezuela), it is common for a married woman to take lovers who are considered to share the paternity of her children with her husband. In large areas of aboriginal central Australia, there were multiple ceremonial occasions for extramarital sexual relations. These cases underline the complexity and subtlety of human reproductive strategies and the multiple functions of the family in playing out those strategies.

[*See also* Hominid Evolution, *articles on* Early *Homo*, Archaic *Homo sapiens, and* Neanderthals; Human Foraging Strategies, *articles on* Subsistence Strategies and Subsistence Transitions, *and* Human Diet and Food Practices; Human Sociobiology and Behavior, *articles on* Human Sociobiology, Evolutionary Psychology, *and* Behavioral Ecology; Mate Choice, *article on* Human Mate Choice; Primates, *article on* Primate Societies and Social Life.]

BIBLIOGRAPHY

Baker, R., and M. A. Bellis. *Human Sperm Competition: Copulation, Masturbation and Infidelity.* New York, 1996. Synthesis of recent findings from several domains regarding human sexual behavior; some of this research has been challenged by other scholars.

Beckerman, S., and P. Valentine. "Introduction." In *Cultures of Multiple Fathers: The Theory and Practice of Partible Paternity in South America,* edited by S. Beckerman and P. Valentine. Gainesville, Fla., 2001. A review of cultural variation in the construal of fatherhood, especially in South America.

Blurton Jones, N., F. W. Marlowe, K. Hawkes, and J. F. O'Connell. "Paternal Investment and Hunter Gatherer Divorce Rates." In *Adaptation and Human Behavior: An Anthropological Perspective,* edited by L. Cronk, N. Chagnon, and W. Irons, pp. 69–90, New York, 2000.

Borgerhoff-Mulder, M. "Reproductive Decisions." In *Evolutionary Ecology and Human Behavior,* edited by B. Winterhalder, and E. A. Smith, pp. 339–374, 1992. An overview of evolutionary ecological theory with respect to human reproductive behavior.

Hrdy, S. *Mother Nature: A History of Mothers, Infants, and Natural Selection.* New York, 2000. A feminist sociobiological treatment of behavioral ecology and evolution, emphasizing female behavior.

Keesing, R. *Kin Groups and Social Structure.* New York, 1975. A highly readable and level headed introduction to main line anthropological thinking on the subject.

Murdock, G. P. *Social Structure.* New York, 1949. Classic comparative and statistical treatment of cross culture variation in the composition and structure of human families and larger kin groups.

Radcliffe-Brown, A. R. "Introduction." In *African Systems of Kinship and Marriage,* edited by A. R. Radcliffe-Brown and D. Forde, pp. 1–85, London, 1950. An elegant statement of the structuralist functionalist view of kinship and marriage; traditional social anthropology at a high point.

Washburn, S., and C. S. Lancaster. "The Evolution of Hunting." In *Man the Hunter,* edited by R. B. Lee and I. DeVore, pp. 293–

303, Chicago, 1968. The beginnings of modern thinking about the origin of the human family.

— STEPHEN BECKERMAN

HUMAN FORAGING STRATEGIES

[*This entry comprises three articles:*

An Overview
Subsistence Strategies and Subsistence Transitions
Human Diet and Food Practices

The first article discusses the anthropological applications of foraging and the transition from foraging to cultivation; the second article provides an account of the economics of subsistence systems; the third article offers a wide-ranging account of foraging strategies including the hunting hypothesis of human evolution, cooking and its consequences, and the recent addition of grain and dairy components to human diets. For related discussions, see Agriculture; *and* Optimality Theory, *article on* Optimal Foraging.]

An Overview

People who hunt and gather, or forage, for a living are confronted with many decisions in acquiring food. Where should they go to look for food? Which foods should they choose once they get there? Optimal foraging theory (OFT) consists of models that provide answers to these and other questions that foragers face. The models and the answers they provide assume that natural selection has designed all animals—from spiders to people— to choose foraging options that produce the highest possible benefit, usually the rate of energy gain, for the cost of time expended. Applications of OFT to human behavior have provided insight into a number of important issues, from models of human evolution and key subsistence transitions to wilderness conservation.

Broad Spectrum Transitions. A dramatic increase in the diversity of foods selected by foragers occurred in many parts of the world after about eighteen thousand years ago. The resources added or selected with increasing frequency at this time are ones that yield low-calorie benefits for the amount of time spent catching and processing them. These include smaller, more difficult-to-catch mammals, fish, and birds and plants that require considerable processing to remove toxins or improve digestibility. The increasing use of low-return resources that characterized this so-called broad spectrum revolution implies a decline in overall return rates or foraging efficiency. This trend is predicted from mathematical models of optimal foraging that focus on how a forager should choose among a range of resources that vary in

the rate of energy earned for the time spent in pursuing and processing (i.e., "handling") them. These "prey choice" models show that when the rate at which a forager finds high-return resources declines, foragers should be less selective, and begin to pursue lower return resources as well. Increasing human population densities and environmental changes involving extinctions of large game species are likely catalysts for this change.

The broad spectrum transition also entails a shift in how foraging time is spent: from a case where most foraging effort involves the search for high-return resources to one where most costs are devoted to handling more abundant but lower return ones. When handling costs are high, much can be gained by reducing them with technology. Accordingly, broad spectrum foragers developed increasingly sophisticated and diversified tool kits (e.g., nets and millingstones) to reduce those costs.

Similar processes may account for other broad spectrum shifts. For example, the prey remains found with Neanderthals and other archaic *Homo sapiens* fossils (150,000–40,000 years ago) are often dominated by a single species of large mammal. [*See* Human Evolution, *article on* Morphological and Physiological Adaptations.] In contrast, fossils of later anatomically modern peoples are associated with higher diversities of larger mammals as well as variable, but often appreciable, quantities of more difficult-to-catch fish, rabbits, and birds. Although this pattern may reflect a limited hunting capacity of Neanderthals, it is more simply read as an expansion of diet breadth, driven by population growth and declining encounter rates with the larger, high-ranked prey.

Declining foraging efficiencies have been well documented in more recent prehistoric settings. For example, in the face of ever-increasing human population densities in California during the Late Holocene epoch (2,500 years ago up to the time of contact with Europeans), the abundance of larger animals such as elk and deer declined significantly compared to smaller vertebrates. Increasing use of labor-intensive plant resources parallels this trend, as do human skeletal indicators of increasing dietary/disease stress. Later increases in deer hunting in some settings reflect the exhaustion of local resources and an increasing use of less-depleted patches located far from residences. Evidence for this includes differential transport and discard of deer body parts, behavior that has received recent modeling and ethnographic analysis.

Domestication and Agriculture. The degree to which foragers rely on expensive resources also affects the process of domestication. Only when foragers spend far more time pursuing and processing resources than searching for them do they have much to gain by reducing handling costs. The cultivation, weeding, and harvesting of wild cereal grass seeds might, for example,

WHY DIDN'T NATIVE CALIFORNIANS FARM?

California has the largest agricultural economy in the United States and produces nearly 40 million tons of fruits and vegetables a year. So why didn't aboriginal Californians farm when domesticated corn, beans, and squash were the staff of life for peoples living in nearby Mexico and the American Southwest? The perspective of optimal foraging theory may shed light on this question. The diet breadth model predicts that whether or not a given resource is utilized depends on the availability of resources that produce *higher* net energy gains for the time invested in procuring them. The fact that California Indians never adopted domesticated crops suggests, that other resources yielding higher return rates were available. Acorns may have been one such resource. The species diversity and abundance of California oaks is legendary and acorns were a staple to native peoples living there at contact. Ethnographically derived return rate data for acorns and corn are very similar, ranging from about 300 to 1,800 kcal/hr. The rates vary positively with nut size in acorns and negatively with time spent in cultivation for corn. Return rates for corn would likely have been at the low end of their range in prehistoric California, however, because insufficient rainfall occurs during the growing season and, in most cases, intensive investment in irrigation would have been required to grow it there. If so, all but the smallest species of acorns would have been a better economic option than corn.

—Jack M. Broughton

lead to their proliferation. Increased density of the grasses might then yield reduced handling costs. Broad spectrum diets may be a necessary precursor to the domestication and farming that emerged in some regions of the Old World as early as ten thousand years ago. Not all broad spectrum foragers shifted to farming, simply because few plants and animals possess the life history and genetic characteristics that permit rapid evolution and domestication. Domesticated animals such as cattle and goats probably evolved as processing machines that reduced the handling costs of low-ranked plants by converting them to milk.

The optimal foraging perspective raises questions about agriculture as an ideal subsistence strategy that should have been adopted whenever and wherever people were smart enough to recognize its virtues. Indeed, recent ethnographic research shows that agriculture is a very low-return enterprise. Although some increases in return rates likely resulted from the domestication process (e.g., larger seeds), farming appears to represent the efforts of people to make the best of situations where higher return wild resources became scarce.

This may be supported by analyses of human skeletons representing hunter-gatherers and agriculturalists; agriculturalists display increased frequencies of disease and/or dietary stress markers, lower average age at death, reduced stature, and increased evidence of violent trauma. [*See* Population Trends *for problems in interpreting these data.*] The high population densities and low return rates that characterize farming thus seem to go hand in hand with famine, disease, and war. Declining health and increasing violence accompany an increase in human population densities and declining return rates through time among some hunter-gatherer populations as well (e.g., California).

Foraging Differences by Sex and Age. One of the most extensive applications of optimal foraging models to living people was conducted among the Ache of Paraguay. The behavior of the Aché matched several predictions of the prey-choice model: (1) all of the resources selected produced higher caloric returns than the rate of return yielded for the habitat as a whole, and (2) all of the resources ignored had return rates lower than the overall rate. But these results apply only to the average Aché forager; men and women actually follow very different foraging strategies, neither obeying expectations of the model. Men ignore many resources, especially plants that would cause them to gain higher overall caloric return rates, and women ignore high-ranked large game and focus on plant foods.

These failures of the model's predictions have stimulated research into sex differences in foraging. Although some researchers suggest that differences in male and female foraging may serve to maximize the production of balanced household diets, others argue that the sexes forage with very different goals in mind. Women may forage to provision their offspring and are constrained to pursue resources that do not conflict with child care. Men may forage to elicit social attention and increased extramarital mating access by focusing on large and impressive resources (big game) that are more irregularly found and more widely shared beyond the family. Men's hunting may more often be about status competition than parenting. This suggestion has implications for models of human evolution that place emphasis on male provisioning of mates and offspring.

Recent research has also focused on the foraging behavior of children. Children can be surprisingly productive foragers; the extent to which they contribute varies with the nature of the resources available and factors that affect their safety. Parents will even adjust their own foraging strategies to take advantage of children. Mothers, for example, will accept lower personal return rates to maximize the returns that they achieve in working with their children as a team.

Juveniles are also less selective than adults about the foods they take. Although some experts have argued that this may reflect a lack of experience or lower foraging capabilities of children, recent ethnographic tests of the prey-choice model show that children are actually maximizing returns. Their wider diet breadths follow predictably from their slower walking speeds, reduced encounter rates, and lower overall return rates. These studies have implications for models of human evolution that assume our long juvenile period is dictated by a major investment required to learn how to forage efficiently.

Conservation Biology. Many writers portray indigenous peoples as "ecologically noble savages" living in harmony with the natural world. In North America, this perception underlies wilderness conservation policies; restoring or maintaining ecosystems to their "pristine" condition simply requires the elimination of European influences. Applications of optimal foraging models to native foragers have challenged this perception. Patterns in prehistoric animal exploitation in California (see above), for example, follow predictions of foraging models and suggest that the prehistoric faunal landscape was fundamentally anthropogenic. Similarly, studies of living Amazonian foragers show that hunting decisions follow immediate concerns of caloric gains versus costs, not strategies that would preserve game populations for future use; hunters do not shy from killing highly profitable but nearly depleted prey. In general, archaeological and ethnographic evidence shows that the scale of impact foragers have on their environments is quite variable but linked positively to population densities, not cultural affiliation.

[*See also articles on* Agriculture; Human Sociobiology and Behavior, *article on* Human Sociobiology; Optimality Theory, *article on* Optimal Foraging.]

BIBLIOGRAPHY

Alvard, M. "Intraspecific Prey Choice by Amazonian Hunters." *Current Anthropology* 36 (1995): 789–818. A study of modern Amazonian hunter-gatherers that suggests they do not follow conservationist hunting strategies. Includes extended debate from other scholars.

Bird, D. W., and R. Bliege Bird. "The Ethnoarchaeology of Juvenile Foragers: Shellfishing Strategies among Meriam Children." *Journal of Anthropological Archaeology* 19 (2000): 461–476. Shows that children forage optimally for shellfish but differently than adults. Also shows the importance of differential transport of shells back home, a factor that would confuse archaeological reconstructions of diet.

Broughton, J. M., and J. F. O'Connell. "On Evolutionary Ecology, Selectionist Archaeology, and Behavioral Archaeology." *American Antiquity* 64 (1999): 153–165. Summarizes archaeological implications of foraging models and the evidence and bibliographical references for declining foraging efficiencies and related phenomena in the Late Holocene epoch of California.

Cohen, M. N. "The Significance of Long-term Changes in Human Diet and Food Economy." In *Food and Evolution: Toward a Theory of Human Food Habits*, edited by M. Harris and E. B. Ross, pp. 261–283. Philadelphia, 1987. Discusses the broad spectrum revolution, including possible preservation biases, and the adoption agriculture from a very general optimal foraging perspective. Provides a summary of the human skeletal evidence for differences in dietary stress between hunter-gatherers and early agriculturalists. Written for a general audience.

Hawkes, K., and J. F. O'Connell. "On Optimal Foraging Models and Subsistence Transitions." *Current Anthropology* 33 (1992): 63–66. An important theoretical paper that discusses how optimal foraging models have implications for transitions from hunting and gathering to agriculture and associated technological changes.

Hawkes, K., J. F. O'Connell, and L. Rogers. "The Behavioral Ecology of Modern Hunter-Gatherers, and Human Evolution." *Trends in Ecology and Evolution* 12 (1997): 29–32. A brief summary of human behavioral ecology, including optimal foraging theory, that highlights the implications for models of human evolution. Written for a scientifically educated but nonspecialized audience.

Kaplan, H., and K. Hill. "The Evolutionary Ecology of Food Acquisition." In *Evolutionary Ecology and Human Behavior*, edited by E. A. Smith and B. Winterhalder, pp. 167–202. New York, 1992. A summary of applications of optimal foraging theory to living foragers.

Kelly, R. L. *The Foraging Spectrum: Diversity in Hunter-Gatherer Lifeways.* Washington, D.C., 1995. A detailed, comparative examination of variation in foraging related behavior of ethnographically documented hunter-gatherers from the perspective of optimal foraging theory. Treatment includes mobility, sharing, territoriality, group size, and social hierarchies.

Marean, C. W., and Z. Assefa. "Zooarchaeological Evidence for the Faunal Exploitation Behavior of Neandertals and Early Modern Humans." *Evolutionary Anthropology* 8 (1999): 22–37. Discusses the animal bone evidence for broadening diets between Neanderthals and early modern humans.

Winterhalder, B., and E. A. Smith. *Hunter-Gatherer Foraging Strategies: Ethnographic and Archaeological Analyses.* Chicago, 1981. This is an edited volume that includes a summary of the most important optimal foraging models and eight early case studies.

Winterhalder, B., and E. A. Smith. "Analyzing Adaptive Strategies: Human Behavioral Ecology at Twenty-five." *Evolutionary Anthropology* 9 (2000): 51–72. A comprehensive review of applications of behavioral ecology, including optimal foraging theory, to people. Includes extensive bibliographic references.

— JACK M. BROUGHTON

Subsistence Strategies and Subsistence Transitions

Until about twelve thousand years ago, all humans lived as hunters and gatherers; subsistence was based on hunting wild animals, gathering wild plant foods, and fishing, with no domestication of plants or animals, except the dog. Very few people follow this form of subsistence today, and they are mostly restricted to ranges in harsh and marginal habitats, so it is not clear how representative they are of foragers from our prehistory.

From what we observe, forager subsistence strategies are diverse but typically involve living in small groups in which food is shared and the groups are nomadic, without centralized authority, and with a clear division of labor between the sexes: men focus on hunting and women on gathering.

Nineteenth-century anthropologists viewed hunter-gatherers as relics of a former, primitive lifestyle, at the bottom of a hierarchy in which other forms of subsistence, such as horticulture, then agriculture and stock rearing (pastoralism), then urbanization represented higher forms. This view of social progression over time is known in social anthropology as evolutionist. Evolutionist views are a legacy of Charles Darwin's contemporary, Herbert Spencer, and perhaps best characterized by the U.S. anthropologist Lewis Henry Morgan's scheme of "savagery, barbarism, and civilization," in which foragers were classified as savages. Such classifications rest on the proposition that social evolution is progressive change from simple to complex societies. This contrasts with the Darwinian view held by contemporary human behavioral ecologists that people generally adjust their behavior in response to their environment, and do so in ways likely to promote their own individual fitness. It is now clear that individuals and whole groups may make transitions between different modes of subsistence in any direction, depending on current costs, benefits, and constraints, which vary with changing climate, human effects on local environments, and the practices of neighboring populations.

Many foragers (hunter-gatherers) choose to keep small farms or gardens for part of the year, or work on other people's farms for cash or food, returning to hunting and gathering at times of the year when hunting returns are good. Many had (and some still have) strong links with farmers and herders, either through trading relationships (in which foragers bring goods such as honey) or through stronger patron–client relationships, in which foragers are more exploited (such as the Pygmies of central Africa). There is evidence that some foragers in tropical South America returned to foraging as a secondary adaptation, after millennia of farming, because the European conquests in the sixteenth to eighteenth centuries made them vulnerable to exploitation and because mobile hunting and gathering offered greater security. Some researchers believe that San foragers in southern Africa were formerly pastoralists who lost their stock, but this view is controversial.

It has been suggested that the framework of optimal foraging theory can help explain such transitions. To date, mostly verbal rather than formal models have been applied to the transition between foraging and agriculture. It is possible that the extinction of many large prey species, as a result of hunting, increased the profitability of an agricultural lifestyle. This could have hastened the transition to agriculture, or agriculture may have been more profitable anyway in many areas under the equilibrating climatic conditions of the Holocene epoch, and people who were predominantly farmers continued to hunt and drive prey extinct, as they do in many parts of the world today. The adoption of farming undoubtedly increased fertility and hence the human population density increased dramatically, although some have argued that this was associated with the onset of many new diseases and possibly generally lower levels of health. [See Disease: Demography and Human Disease.]

Transitions between farming and pastoralism are also very common. Groups speaking very closely related languages can include both specialized pastoralists and farmers. Especially in Africa, there are many examples of families combining cattle herding with farming for part of the year, a mode of subsistence known as agro-pastoralism. Dynamic optimality models, in which decisions can depend on the state of the decision maker, have been used to explore transitions between farming, herding and agropastoralism in sub-Saharan Africa. The subsistence strategy that best promotes long-term survival can depend both on the productivity of the land and on household wealth. Herding requires wealth in the form of animals. This is not necessarily true of farming, as land is not limiting in many areas—it is the labor needed to cultivate it that frequently limits production. A decision to combine herding and farming constrains the productivity of herding also largely because of labor constraints. Dynamic modeling shows that specialized pastoralism is optimal even in areas with good agricultural productivity, but only for the wealthy, that is, for those with large herds. If herd sizes are too small, the risk of having to eat into the herd at times of drought makes pure pastoralism unviable. At intermediate levels of wealth and land productivity, agropastoralism is optimal, especially as it can be combined with grain storage for use in lean periods. Only in areas of very high agricultural productivity is a strategy of pure agriculture optimal. Such modeling can help explain why, when wealth or land productivity fluctuates over time, either within a household or over the whole population, people will shift between farming and pastoralism, or indeed into foraging if wealth is low. Increasing wealth will frequently favor a transition into pastoralism—development projects that have encouraged nomadic peoples to settle and take up farming have sometimes been surprised to see people moving back into nomadic life when times are good. As the wealth of pastoralists declines owing to competition with farmers for their grazing lands (which formerly were owned communally), a cycle develops in which the more pastoralists settle as farmers, the more prevalent are the conditions in which farming is optimal. Soon pastoralism, like foraging, becomes restricted almost entirely to land so marginal that it is impossible to grow crops.

Many other aspects of culture may change dramatically when subsistence strategies change. For example, some Somali groups in the Horn of Africa are now farming but were recently pastoralists, and they can no longer pay the traditional bride price in camels; bride price in grain is now considered appropriate. This transition is also associated with a less nomadic lifestyle, so house construction practices change to more permanent structures, and land tenure rules change from communal to individual ownership. Pastoralism is frequently associated with high degrees of polygyny, and also patrilineal kinship systems, so it is possible that these features too could change over time.

What is clear is that human subsistence strategies are dynamic, and transitions between different modes of subsistence will depend on current benefits, costs, and constraints. Subsistence transitions trigger changes in diverse aspects of the culture, although other aspects, like language, may remain relatively unchanged, which can then be used to help us to untangle our dynamic history.

Both the numbers of individuals practicing cultivation and their geographic range have grown rapidly over the last ten thousand years as agriculturalists and pastoralists came to dominate Asia and Europe, then Africa. There is debate as to the extent to which farming and herding spread through the Holocene as local populations adjusted to changing opportunities and constraints, or whether farming and herding peoples expanded into new areas where they simply outreproduced and outcompeted resident foragers, driving them to extinction.

The study of both languages and genes is helping to throw light on the relative importance of these processes. There are places where linguistic diversity is high, language families cover small areas, and unusual "isolate" languages are common (such zones are common in the Americas and Australia). In such cases, languages are thought to be of great antiquity and the patterns of linguistic diversity may represent the settlement patterns of Pleistocene populations, prior to the invention of agriculture. If farming has been adopted in these areas, it is probable that local populations adjusted to changing opportunities rather than the arrival of new populations.

There are other zones where a large language family may dominate a wide geographical range, and related languages branch from a common root, like species on a phylogenetic tree. This pattern is thought to reflect the spread of an expanding population from an area, normally associated with a technological innovation such as the cultivation of a new crop. The Indo-European lan-

FIGURE 1. Preparing a Meal of Pandanus Nuts Gathered from the Eastern Highlands of Papua New Guinea. © David Gillison/ Peter Arnold, Inc.

guages follow this pattern and appear to have spread from the Middle East northwest across Europe. The Italian geneticist Luca Cavalli-Sforza and colleague (1994) find clines of genetic alleles following roughly the same trajectory. They propose a "wave of advance" model in which an expanding population of dispersing farmers replaced existing forager populations—a process sometimes described as demic diffusion. The Bantu expansion from a small area of western Africa, which led Bantu-speaking people to dominate western, eastern, and some of southern Africa, is thought to reflect another farming dispersal, based on yam cultivation in the west and cattle herding in the east. Horses and ploughs may be behind other population and linguistic expansions in central Asia.

Genes from forager populations could survive such farming dispersals if foragers interbred with newcomers. Anthropological studies of present-day populations include examples of forager populations losing women to wealthier farmers and pastoralists. In one case, foragers were seen to use cattle obtained as the bride price for marrying their daughters to herders, to make the transition from foraging into pastoralism. Where farming was simply more productive than foraging, it is possible that forager women were taken into farming villages as wives, whereas forager males would not be successful in attracting mates. If this were a common process during farming dispersals, we might expect DNA in the mitochondria (which is inherited only down the female line) to represent older populations than the DNA on the Y chromosome (which is inherited only down the male line). However, there is little evidence for this—some genetic evidence suggests that females are actually more mobile than males, but this could reflect more recent migration patterns. The long-distance maritime colonizations carried out by Europeans in the last few centuries were especially devastating for indigenous populations in the Americas and Australia. In these cases, both the introduction of new diseases and competition played a role in decimating, native populations. New farming practices spread mainly by demic diffusion, generally without widespread interbreeding in Australia and much of North America. Colonization patterns in Latin America preserved more indigenous genes, which, under current economic circumstances, are now moving north.

[*See also* Agriculture, *article on* Spread of Agriculture; Human Sociobiology and Behavior, *articles on* Human Sociobiology *and* Behavioral Ecology; Optimality Theory, *article on* Optimal Foraging.]

BIBLIOGRAPHY

Cavalli-Sforza, L. L., P. Menozzi, and A. Piazza. *The History and Geography of Human Genes.* Princeton, 1994. Compilation of work on how classical genetics informs our understanding of human migrations worldwide.

Cronk, L. "From Hunters to Herders: Subsistence Change as a Reproductive Strategy." *Current Anthropology* 30 (1989): 224–234. A case study of a transition from foraging to herding involving bride price.

Diamond, J. M. *Guns, Germs and Steel: A Short History of Everybody for the Last 13,000 Years.* London, 1998. Popular account of how technological advances underly population expansions and colonization.

Gray, R. D., and F. M. Jordan. "Language Trees Support the Express-Train Sequence of Austronesian Expansion." *Nature* 405 (2000): 1052–1055.

Hawkes, K., and J. O'Connell. "On Optimal Foraging Models and Subsistence Transitions." *Current Anthropology* 33 (1992): 63–66. Note on the use and misuse of optimal foraging models to understand transitions between foraging and farming.

Mace, R. "Transitions between Cultivation and Pastorlism." *Current Anthropology* 34 (1993): 363–382. Dynamic optimality model comparing herding, farming, and agropastoralism as alternative subsistence strategies.

Lee, R. B., and R. Daly, eds. *The Cambridge Encyclopedia of Hunters and Gatherers.* Cambridge, 1999. Comprehensive account of foraging lifestyles and current debates, including detailed case studies of over forty forager groups.

Renfrew, C. "World Linguistic Diversity and Farming Dispersals." In *Archaeology and Language*, edited by R. Blench and M. Spriggs, vol. 1, pp. 83–90. New York, 1996.

— RUTH MACE

Human Diet and Food Practices

The decisions that we make to provision ourselves and others with food determine whether we will survive, reproduce, or assist others to do so. Each one of us is the product of an ancestry of only those who successfully solved the problems involved in acquiring and handling food. We have good reason, then, to suspect that we have strong inclinations and finely tuned capacities to assess the merits of alternative diets and foraging strategies.

Being Choosy. What kinds of resources are humans designed to acquire and eat? The answer, of course, depends on who and where you are: "food" is a socially specific category that varies between and within societies. For example, for most people of European decent, insect larvae are not food. But for many Australian Aborigines, cossid larvae are relished and often an important part of daily subsistence. The texture, flavor, and means of acquiring cossid larvae are foreign only to foreigners. It is clear that people learn many food preferences as they grow up. We may even develop preferences for certain tastes that flavor our mothers' milk.

Despite great diversity across cultures, there are also some striking regularities. Within societies, what people consider to be food is always only a small subset of the edible resources locally available, and often only a few of these are regularly exploited. Within societies as well, and even within the same family, individual members usually focus on acquiring different types of food depending on their sex and age. What structures such variability in our selectivity? Cultural and individual preferences are the immediate causes, but what influences the patterning in these within and across human societies?

Much of the work toward answering the question of human food selectivity has focused on evaluating the costs and benefits of foraging among people who hunt and gather for a living. In societies as different as the Alyawara in central Australia, Inujjuamiut of Arctic Canada, the Ache of neotropical Paraguay, and the Cree in North American boreal forests, anthropologists throughout the 1980s demonstrated that people chose the array of wild resources to exploit that, on average, increased the rate at which they acquired energy (energy captured per unit of time spent searching for and handling resources). Moreover, in many of these cases, as overall rate increased, foragers focused more narrowly on the more profitable prey types; conversely, as income decreased, their "diet breadth" broadened to include more resources. [*See* Human Foraging Strategies, *article on* Subsistence Strategies and Subsistence Transitions.]

The results of these studies on traditional societies also highlight tremendous variability in human foraging, especially with regard to sex and age. Men, women, and children often make very different decisions about how to allocate foraging time and choose resources. For example, among the Ache, with whom anthropologists have done extensive studies of foraging, only men acquire meat and honey, women specialize in plants (especially palm starch), and juveniles rely heavily on the foraging efforts of others. Although Ache men maximize the amount of meat acquired per hour of hunting, they decrease their total rate of energy production by passing over the opportunity to procure palm starch. Women, in contrast, do not pursue highly profitable game even though doing so would boost nutrient acquisition rate.

These differences illustrate a common tendency for sexual division of labor in human societies. Although this tendency varies across societies in both kind and degree (Australian women hunt large lizards, and Australian men gather plant foods for their own consumption when they are moving over the country), men and women almost always forage differently. There are some sex differences in foraging strategies in other primates, but never as extreme as the human differences. Even in comparison to chimpanzees, where only males hunt and females do more termite fishing, the degree to which human food practices are consistently linked to the sex of the forager is striking.

Some of the most influential scenarios to explain human evolution have focused on sex and age-linked foraging practices and their possible role in structuring our unique constellation of life history traits and social behavior. Compared to chimpanzees, humans mature late, have shorter interbirth intervals, wean infants early, and generally live much longer even where mortality rates are high. [See Life History Theory, *article on* Human Life Histories.] Our comparatively high birth rate suggests that a human mother usually gets help supporting offspring from someone else. The father's hunting and sharing of meat has often been nominated to be the primary component of this help. Anthropologists have suggested that along with a change in climatic conditions at the beginning of the Pleistocene epoch (about 1.8 million years ago) came new opportunities for some hominids to focus on hunting. Many researchers have argued that this unique foraging niche is the catalyst that propelled human evolution: it increased the benefits of larger brains, long juvenile periods, pair bonding, food sharing, and a complementary sexual division of labor. Attractive as this hypothesis seems, it has proved difficult to confirm. The resources taken by men, especially meat, are often so widely shared that others get as much as wives and offspring do. This wide sharing means that hunters' rate of nutrient income for their own households is much lower than their total acquisition rate.

Anthropologists have often assumed that shares of meat distributed to others from a hunter's kill are repaid to him later. If so, men might serve the nutritional welfare of their households by participating in this exchange. However, indications of reciprocal meat exchange like this have been difficult to find: households of those who capture widely shared meat more frequently do not receive more frequent shares. There is some evidence of reciprocity for food in general; however, the common pattern for big game is not one of exchange, but of unconditional distribution.

Among the Meriam of western Melanesia, women fish

"THRIFTY" GENES AND THRIFTY METABOLISM

The stereotypical Westernized diet that we luxuriate in is, obviously, a very recent evolutionary development, and human diet has traditionally been far lower in sugar, fat, salt, and cholesterol, and has been punctuated by frequent periods of famine. Because of this, humans have traditionally been under more selective pressure than those in the West to have an efficient or "thrifty" metabolism to utilize every available calorie and smidgen of salt. This has involved a number of physiological adaptations—for example, rapid insulin release to enhance energy storage, or better renal retention of salt.

And what happens as traditional societies transition into Western diets, when "thrifty" physiology meets supermarkets? High rates of diseases of plenty—obesity, adult-onset diabetes, and renal hypertension. For example, some Pacific Islander populations in their first generation of Western diets have fifteen times the rate of diabetes as in America. With a Western diet, a thrifty metabolism increases the risk of early death, and most in the West descend from those who, with the emergence of Western diets, had the good luck of having the *least* thrifty metabolism. Thus, what is adaptive for a hunter-gatherer can be pathogenic for a business executive, and vice-versa.

The degree of "thriftiness" of metabolism has strong genetic components (for example, different genetic variants of insulin receptors), and is thus subject to evolutionary selection. In addition, food availability during fetal life is also a factor. Thus, fetal malnutrition causes a lifelong thriftiness of metabolism, increasing the risk of obesity, diabetes, and cardiovascular disease (a new area of science called "Fetal Origins of Adult Disease").

—ROBERT SAPOLSKY

from shore, collect shellfish, and acquire nesting sea turtles. Men also do this, but they specialize in hunting offshore turtle and fishing for big game. Although men supply a lot of food for the community, the large game they acquire is unreliable and shared so widely that a hunter's own energetic productivity and his household contribution are very low. Women, in contrast, focus on hunting and fishing activities that provide a steady stream of household provisions. This is a pan-human theme: when it comes to big game, men's efforts have a high risk of failure, and little of the meat goes to a hunter's own household.

An alternative to the hunting as paternal provisioning and reciprocal food sharing hypotheses has been suggested: large animal acquisition advances men's sociopolitical goals; men's hunting is a competitive strategy for achieving prestige, social attention, and political advantage. This seems clearly to be the case with the Meriam.

According to this hypothesis, foraging differences between men and women reflect not only the constraints of child care on females but also sexual selection on reproductive strategies—men competing with each other to display qualities reflected in hunting large prey. Work over the last decade has underlined variation in the payoffs that human foragers earn for their effort, and the importance of social as well as ecological factors in shaping human resource choices.

Children's Foraging. Unlike other juvenile primates, humans are provisioned for many years after weaning. Even Hadza children in Tanzania, who are renowned for their productive foraging, rely on food from others, often grandmothers. Resources that adult men and women gather are difficult for children to acquire and prepare. Adults commonly gather foods, such as tubers, that require extraction and extensive cooking. In many areas of the world, tubers are an important component of human diets, and have probably been so throughout much of the past two million years.

Where children help provision themselves, their strategies and choices are very different from those of adults. Conventional wisdom among parents and some anthropologists has it that these differences result from the long time it takes to learn adult foraging. Accordingly, our long childhoods are thought to be designed to accommodate such learning. But recent work demonstrates that children's food practices are not simply about learning to become an adult. Children's tastes, food acquisition, and time allocation are finely tuned to suit the needs of small bodies that face different health insults and energetic constraints. For example, Hadza children are more constrained in their foraging by strength than by learning and experience, whereas Meriam children are limited by energetic constraints on walking speeds. In other cases, age differences in diet breadth and receptiveness to new foods fluctuate with overall foraging return rates and children's special sensitivity to bitter or toxic secondary compounds in certain foods. These results turn the usual notion of childhood on its head: maybe our long childhood did not evolve because of the benefits of more learning, but more learning occurs simply because we have such a long juvenile period.

Much of what it means to be human revolves around differences between us and other primates in the way we forage for food. People differing in age and sex differ in their foraging practices, and this variation differs in turn with ecological circumstances. Important clues about human evolution can be found by paying close attention to details of food acquisition and distribution, as well as to the different goals associated with different food practices.

[*See also* Archaeological Inference; Hominid Evolution, *articles on* Early *Homo*, Archaic *Homo sapiens*, *and* Neanderthals; Primates, *article on* Primate Societies and Social Life.]

BIBLIOGRAPHY

Altman, J. C. *Hunter-Gatherers Today: An Aboriginal Economy in North Australia.* Canberra, Australia, 1987. A detailed account of production, time allocation, and sharing among Gunwinggu speakers. An entire chapter is devoted to meat distributions.

Bird, D. W., and R. Bliege Bird. "The Ethnoarchaeology of Juvenile Foragers: Shellfishing Strategies among Meriam Children." *Journal of Anthropological Archaeology* 19 (2000): 461–476. Provides an introduction to children's subsistence and prey choice, with a focus on intertidal activities.

Bliege Bird, R. "Cooperation and Conflict: The Behavioral Ecology of the Sexual Division of Labor." *Evolutionary Anthropology* 8 (1999): 65–75. A review of variability in men's and women's subsistence strategies.

Bliege Bird, R., E. A. Smith, and D. W. Bird. "The Hunting Handicap: Costly Signaling in Human Foraging Strategies." *Behavioral Ecology and Sociobiology* (2001). A technical article on the sociopolitical value of Meriam hunting and fishing.

Blurton Jones, N. G., K. Hawkes, and J. F. O'Connell. "Why do Hadza Children Forage?" *Uniting Psychology and Biology: Integrative Perspectives on Human Development*, edited N. L. Segal, G. E. Weisfeld, and C. C. Weisfeld. New York, 1997. Argues that children select different resources because of less strength, not learning constraints.

Cashdan, E. "Adaptiveness of Food Learning and Food Aversions in Children." *Anthropology of Food* 37 (1998): 613–632. An evolutionary perspective of children's learned food tastes and dietary norms.

Gurven, M., K. Hill, A. Hurtado, and R. Lyles. "Food Transfers among Hiwi Foragers of Venezuela: Tests of Reciprocity." *Human Ecology* 28 (2000): 171–214. A good example of research on contingency in aspects of food sharing.

Hawkes, K. "Showing-off: Tests of an Hypothesis about Men's Foraging Goals." *Ethology and Sociobiology* 12 (1991): 29–54. A provocative and controversial article that develops the hypothesis that men's subsistence often maximizes social attention but not paternal investment.

Hrdy, S. B. *Mother Nature: A History of Mothers, Infants, and Natural Selection.* New York, 1999. An engaging book on the real trade-offs involved in human parenting and childhood.

Kaplan, H., and K. Hill. "The Evolutionary Ecology of Food Acquisition." In *Evolutionary Ecology and Human Behavior*, edited by E. A. Smith and B. Winterhalder. New York, 1992. A very good, somewhat technical, summary of theory and applications of foraging models, along with a useful overview of Ache subsistence.

Kaplan, H., K. Hill, J. Lancaster, and A. M. Hurtado. "A Theory of Human Life History Evolution: Diet, Intelligence, and Longevity." *Evolutionary Anthropology* 9 (2000): 156–185. Comprehensive treatment of learning and hunting, along with their possible role in human evolution.

O'Connell, J. F., K. Hawkes, and N. G. Blurton Jones. "Grandmothering and the Evolution of *Homo erectus*." *Journal of Human Evolution* 36 (1999): 461–485. An alternative to the hypothesis that the evolution of the human genus is driven by male provisioning.

Stanford, C. B. *The Hunting Apes: Meat Eating and the Origins of Human Behavior*. Princeton, 1999. A very readable overview of arguments about the role of meat in human evolution; compares human and chimpanzee hunting.

Winterhalder, B., and E. A. Smith, eds. *Hunter-Gatherer Foraging Strategies*. Chicago, 1981. A classic volume on variability in resource exploitation and foraging groups among people around the world.

Winterhalder, B., and E. A. Smith. "Analyzing Adaptive Strategies: Human Behavioral Ecology at Twenty-five." *Evolutionary Anthropology* 9 (2000): 51–72. One of the best reviews of evolutionary approaches to human resource choice and sharing.

Wrangham, R. W., J. H. Jones, G. Laden, D. Pilbeam, and N. L. Conklin-Brittain. "The Raw and the Stolen: Cooking and the Ecology of Human Origins." *Current Anthropology* 40 (1999): 567–594. Stresses the importance of resources that require extraction and cooking in hominid evolution.

— DOUGLAS BIRD

HUMAN GENOME PROJECT. *See* Overview Essay *on* Genomics and the Dawn of Proteomics *at the beginning of Volume 1*; Genome Sequencing Projects.

HUMAN RACES. *See* Race.

HUMAN SOCIETIES, EVOLUTION OF

Human societies have evolved from the small-scale and intimate groups of the Stone Age to the large and complex urban states of today. For more than a hundred thousand years, humans were organized as flexible bands of perhaps twenty to fifty members, subsisting by foraging. These foragers colonized the world's continents and adapted to extreme differences in climate, ecology, and food availability, but the scale of their social units remained little changed. Then, about ten thousand years ago, societies began to change substantially. World population grew in size a thousand times, from about four million people to four billion. People established first villages, then larger settlements and eventually cities, and the sizes of polities grew to the thousands, tens of thousands, and then millions and more. Agriculture was developed independently in at least six regions and was intensified using irrigation, terracing, and drained fields. New technologies were devised and elaborated, including the independent development of the synthetic materials, ceramics and metals. In region after region, archaeological research documents evolutionary trajectories, at first largely independent and then increasingly dependent. These trajectories demonstrate linear relationships between cultural complexity and time.

Since the nineteenth century, evolution has been recognized as systematic change in molecular structures, biological organisms, ecosystems, social systems, cultures, and other organic forms. The processes and outcomes of these evolutionary developments are related but not necessarily reducible to a few common mechanisms of change. Social evolution, unlike biological evolution, appears to be both rapid and directional. Although changes in social forms are neither irreversible nor inevitable, they are to some degree systematic and parallel in one location after another. To understand social evolution, we must clarify the specific nature of change.

Units of Analysis for Social Evolution. In biology, selection acts on organisms as the vessels for genes that code biological forms. Because characteristics that increase reproductive success are "selected for," it is possible to speak of an organism as having an inherent goal to maximize its contribution of genes to the next generation. Does a similar process apply to social evolution?

Much of social evolutionary theory has assumed that societies are adapted as wholes to specific biocultural environments. Societies have been described with analogies to organisms, such that cultural traits are said to function to maintain or reproduce that society. Presumably, societies would need to be selected for as whole systems, but the processes for change in such theories have rarely been clarified. Critiques of social evolutionary theory and of related ideas in cultural ecology often focus on inadequate conceptualization of their functionalist logic and on their inadequate selection process. Attempts to investigate the conditions that might cause group selection have mainly given negative results (see, however, the Vignette concerning the Inca empire for an example of group selection). An effective theory of social evolution may require mechanisms for change that are different from those of the Darwinian synthesis for organic evolution.

The units of social evolution are societies, which are open systems of humans and the social groups that they form. Social systems are structured by cultural relationships, both formal and informal, that include kinship, friendship, politics, and religion. Unlike an organism, societies are not single entities and do not have unified goals. They are multifaceted and highly dynamic. The interests of individuals and groups within a society combine and conflict, as groups form in opposition to other groups and dissolve with internal conflict.

Societies are like ecosystems, structured through flows of energy and matter. Human organizations distribute these flows of food, technology, and wealth used by their populations. The evolution of complex societies required technological developments that generate surpluses, and these surpluses were channeled to finance emerging institutions.

Social Evolution: Specific and General. Social evolution is both specific and general. Specific evolution is about local changes to a particular society. It de-

THE INCA EMPIRE

The Inca was the largest and most complex polity of the New World. Its empire extended from Chile through Peru to Ecuador, incorporating more than 500,000 square kilometers with perhaps 8 to 14 million subjects. The empire exercised power over more than 100 ethnic groups, themselves formerly fragmented into many autonomous polities. The Inca rose to power rapidly. About 600 years ago, the Andean region was divided among warring chiefdoms in the highlands and larger states on the coast. In the highlands, the Inca successfully consolidated power around Cuzco and expanded rapidly by conquest to the north and south. Within a few generations, its army had conquered a vast empire that was held together by elaborate institutions of military, religion, administration, communication, and finance. These institutions consolidated and expanded the primary sources of power on which the state depended. Then came the Spanish in 1532, and the powerful Inca empire was itself defeated and transformed into a colony. In both cases of conquest, differential military power effected major group selection as envisioned by Spencer.

Illustrating the historical nature of social evolution, the Inca state was structurally like the chiefdoms from which it evolved. Most chiefdoms depend on staple finance through which surpluses are mobilized to support ruling institutions. The Inca owned all lands, and they mobilized community corvée labor to farm and to fill the state's storage. Stored staples support laborers, specialized administrators, warriors, and crafters. The Inca state eventually expanded its system of redistribution by establishing state farms, wealth finance, and fulltime craft specialists. Inca facilities built in its short history included 40,000 kilometers of roads, many administrative centers, large storage facilities, fortifications, estates, and religious shrines. The tourist in the Andes is still impressed by the empire's great accomplishments.

—TIMOTHY EARLE

scribes the history of a society's expansion and its adaptation to changing conditions. Steward (1955) argued that cultures adapt to specific environments, creating multiple parallel evolutionary pathways to complexity. General evolution, in contrast, involves prevailing developments of societies from simple to complex. Derived from the writings of White (1959) and his nineteenth-century forebears, human macro-history has a directionality toward complexity.

Specific social evolution is studied now primarily by archaeologists, who record changes in prehistoric and historical sequences of particular regions. By their surveys and excavations of prehistoric settlements, archaeologists are able to study changes in population, subsistence, settlement, property relationships, social and political organization, exchange, and religious practices. For example, the number and size of settlements help measure population sizes, the mass of monuments helps describe the organization of labor, and wealth in burials and houses suggests degrees of stratification. Researchers attempt to correlate changes in key variables to identify processes for social evolution.

Specific evolution corresponds to Darwin's concept of biological evolution as a historical science. Thus, researchers can understand evolutionary changes in bluebirds or in Danish Neolithic societies, but not the specifics of those birds or societies that are outcomes of long chains of historical events. Within human societies, unique cultural factors, which involve human intentions and values, increase the uniqueness of each historical sequence. Attempts have been made to compare archaeologically the long-term evolutionary changes, as in the development of states in Mesopotamia and Mesoamerica and of chiefdoms in the Hawaiian Islands, the Andes, and Denmark. These studies help identify common processes at work, but with considerable historical variation from one sequence to the next.

A successful approach to social evolution has been to study cultural divergence among groups with the same basic phylogenetic history. Such work was pioneered by cultural anthropologists. Particularly well known is Polynesia, where a common ancestral population colonized the islands of the deep Pacific and societies diverged as different historical lines arose and developed with contrasting island ecologies and configurations. Other examples include the variation among Amerindian societies in the Southwest corresponding to difference in subsistence agriculture, and political evolution among highland New Guinea societies reflecting the timing of sweet potato introduction. As archaeology comes of age in Polynesia, Mesoamerica, North America, Europe, and elsewhere, the potential for systematic research on social radiation seems very promising.

General social evolution has been used by cultural anthropologists as a simple framework to organize world societies. Since Linnaeus, scientists have organized and classified the natural world. In the nineteenth century, an interest in general social evolution was central to social theorists such as Herbert Spencer and Karl Marx. Anthropologists were then beginning to collect information on and to make sense of the incredible variation in human societies that were being brought within European colonial orbits. Western museums were amassing large collections of material culture from around the world and organizing them for their urban patrons to see foreign cultures. The schemes used to classify collections were regional, as in North America or Africa, and evolutionary, as from stone to bronze to iron, or from foraging to horticultural to agricultural societies.

Pioneering the comparative study of cultures, Lewis Henry Morgan collected descriptions from around the world and classified them according to his typology from savagery to barbarism to civilization. In the twentieth century, following an anti-evolutionary critique by Franz Boas and his students, neo-evolutionists, again largely cultural anthropologists, refined the classification schemes based on levels (scale) and centrality of social integration. Most famous was Elman Service's four-part classification into band, tribe, chiefdom, and state, to which were added Marshall Sahlins's distinction between head men and Big Men in tribes and Timothy Earle's division between simple and complex chiefdoms. An alternative, but comparable scheme was proposed by Fried (1967).

Johnson and Earle present a third-generation revision of Service's classification as four levels of social integration: the family level, the local group, the chiefdom, and the state. Each new level of integration is created by embedding lower-level units. The family level (bands or smaller egalitarian societies) comprises small and changing groups of five or so families (perhaps 25 people) who are coresidential in camps or hamlets. The local group (tribes with both head men and Big Men) organizes a number of family-level-sized units to create villages or hamlet clusters with several hundred members. Then chiefdoms (ranked or stratified societies) are regional polities with institutionalized governance and some social stratification that embed several local groups to create regional polities with populations in the low thousands (simple chiefdoms) to tens of thousands (complex chiefdoms). States are regional and supraregional polities created through conquest and incorporation of a number of chiefdom-sized groups. Each new level of integration must have new institutional characteristics in order effectively to organize the enlarged scale of societies. Explanations of social evolution focus on why larger scale and more centrally organized societies should be developed.

Many researchers have criticized evolutionary typologies and their supposed simplicity, unilinearity, and progressiveness. Those who use such classification now do so with caution, avoiding types as natural (essential) categories. No typical chiefdom or state can be said to exist, and researchers no longer search for the origins of a type (like a chiefdom). Rather, the types are used to establish clusters of societies of similar scale within which we can investigate the causes of variation. Analyses across stages of complexity are usually limited to specific historical sequences. We attempt to make clear that parallel evolutionary trajectories abound. The typologies have at least heuristic value; they emphasize that the nature of human societies has changed fundamentally, and that we as anthropologists need to investigate those changes.

Mechanism for Evolutionary Change. While the Darwinian synthesis has isolated the mechanisms for biological evolution, most work on social evolution has been descriptive and classificatory. Cultural anthropologists and archaeologists have shown that systematic changes have taken place, but the mechanisms of change are inadequately understood. Some attention has focused on trying to bring a Darwinian selectionist approach to bear on cultural evolution. Although these approaches have promise, especially for foraging societies, so far they have been inadequate to understand the tempo and processes that result in more complex societies. Other evolutionary mechanisms that have been discussed involve adaptational and political processes.

Adaptational processes. Steward and Service envision the evolution of human societies as involving adaptational processes. Human institutions are thought to be organizational solutions to critical problems of adaptation. Analogous to technological innovation, the evolution of central leadership is conceived as the development of management solutions that benefit the group. This viewpoint sees the flexible organization of family-level society as an adaptation to the fluctuating resource base of foraging economies in marginal environments. The subsequent development of decentralized regional ceremonial complexes and political collectivities between local groups could then be seen as managing risk, warfare, exchange, and exogamy. Chiefs then are seen as managers of the redistribution and decision-making required to solve emerging regional problems. For example, the possibility for more productive and secure agriculture could encourage the construction of irrigation, requiring people to accept leaders to coordinate construction and maintenance. In all cases, the logic is similar: a need to adopt a particular solution for group survival, or the advantage that it confers, encourages people to create more complex social arrangements, ultimately with strong leaders and complex political organizations.

The selective mechanism at work has usually been left vague. Perhaps groups with more effective means of adaptation could exclude others with simpler and less effective organizations. Spencer's model of social evolution as "the survival of the fittest" rested on such social Darwinism (although Darwin himself never embraced such ideas). Group selection continues to be discussed, although the situations where it can be shown to be important are quite limited. An example is the role of warfare in group selection, as groups with more central organization and correspondingly more effective military strategies expand against others. As in the Inca case, the formation of complex societies often requires conquest and forceful incorporation. Although group selection may cause the evolution of more powerful military strategies, it seems unlikely that group selection could be frequent enough to fine-tune the organizational structure of societies to the adaptational challenges that subsistence economies face.

Political processes. An alternative approach emphasizes the political processes for the evolution of leadership and stratification in societies. Deriving from the historical materialism of Marx, social stratification is seen as emerging from economic conditions of control. Intensification creates opportunities to produce a surplus that can be used to support a ruling elite and its associated institutions. For example, intensive agriculture circumscribes populations by tethering them to lands with improved facilities. The construction of irrigation systems can allow their ownership by elites, who can then distribute surplus received as rent. The redistributive and market systems of chiefdoms and states are seen as mechanisms of institutional finance, rather than as means of adaptation. The commoner is caught in asymmetrical relationships as elites fashion ruling institutions.

The political economy comprises the channels that support the ruling elite and their institutions of control. Unlike the subsistence economy, the political economy is inherently growth-oriented. With intense competition for leadership, individuals must maximize their access to labor and resources used to fund competitive strategies. Growth in a political economy can be rapid, constrained only by the ability to mobilize resources from commoners. But what confers and limits this ability to mobilize surplus?

Long-term intensification of the subsistence economy creates the conditions of potential control. As Lenski (1966) argues, institutional privilege derives from a fragile balance of power and need. Although state coercion never lies far below the surface, stable systems of domination depend on an elite's ability to provide (or deny) essential products or services. The evolution of complex societies can be seen as dependent on creating the conditions that allow power to be centrally controlled.

Three fundamental sources of power are important. Economic power derives from control over the production and distribution of goods needed and desired. Military power is based on might and intimidation—the ability to exclude others from those goods. Ideological power is based on belief that a natural order of inequality and leadership give inalienable rights to some and not others. Control over the flow of energy and materials in the economy fuels economic, military, and ideological power and allows for a progressive centralization of social institutions. Where it is possible to control the production of surplus, as in the construction of irrigation systems,

the surplus is reinvested in further intensification of the subsistence economy where control is maximized, in warriors who defend unequal ownership rights, and in an ideology that legitimizes inequality.

Alternative means exist to exert power over people, and power may well become distributed quite broadly and central control may be highly problematic. While encompassing complicated and conflicting power relationships, social evolution is based on the strategic channeling of the energy and material flows that permit control over multiple power sources. We can conclude that the evolution of complex societies involves the building of a productive economy from which a surplus is mobilized and channeled to build the institutions of complexity involving new scales of integration and central control.

[*See also* Agriculture, *articles on* Origins of Agriculture *and* Spread of Agriculture.]

BIBLIOGRAPHY

Childe, V. G. *Social Evolution*. London, 1951. Seminal study of social evolution using archaeological evidence.

Earle, T. *How Chiefs Come to Power*. Stanford, 1997. Comparison of evolutionary change in chiefdom in three independent world regions.

Feil, D. K. *The Evolution of Highland Papuan New Guinea Societies*. Cambridge, 1987. Ethnographic descriptions of divergence in related societies.

Fried, M. *The Evolution of Political Society*. New York, 1967. Classic synthesis of social evolution using ethnographic cases.

Johnson, A., and T. Earle. *The Evolution of Human Societies*. Stanford, 2000. Most recent synthesis of social evolution using ethnographic cases.

Kelly, R. *The Forager Spectrum*. Washington, D.C., 1995. Synthesizes worldwide variation in ethnographic forager-based societies.

Kirch, P. *The Wet and the Dry*. Chicago, 1994. A comparative study of the evolution of Polynesian societies, emphasizing contrasting subsistence economies and political development.

Lenski, G. *Power and Privilege: A Theory of Social Stratification*. New York, 1966. Influential sociological study of the importance of power with immediate relevance to social evolution.

Mayr, E. "Darwin's Influence on Modern Thought." *Scientific American* July (2000): 79–83. Thoughtful discussion of the broadly philosophical implications of evolutionary theory as historical science.

Morgan, L. H. *Ancient Society*. Chicago, 1877. The first major comparative study of human societies using an evolutionary approach in anthropology. Influential on Marx and Engels.

Peregrine, P. "Cross-Cultural Comparative Approaches in Archaeology." *Annual Review in Anthropology*, 2001. Review of the use of comparison in archaeology, emphasizing evolutionary approaches.

Sahlins, M. *Social Stratification in Polynesia*. Seattle, 1958. Influential study of adaptive radiation in Polynesia.

Sahlins, M. "Evolution: Specific and General." In *Evolution and Culture*, edited by M. Sahlins and E. Service, pp. 12–44. Ann Arbor, 1960. A key article partly resolving the debate between Steward and White.

Service, E. *Primitive Social Organization*. New York, 1962. The frequently cited typology in general evolution, using an adaptational perspective typical of cultural ecology in the 1960s.

Steward, J. *Theory of Culture Change.* Urbana, 1955. Development of cultural ecology and the concept of parallel social evolution.

Trigger, B. *Sociocultural Evolution.* Oxford, 1998. An elegant history and defense of evolutionary approaches to long-term history.

White, L. *The Evolution of Culture.* New York, 1959. Classic study of general cultural evolution, emphasizing the roles of technology and energy in social evolution.

Wittfogel, K. *Oriental Despotism.* 1957. A comparative study of the development of complex societies resulting from the use of complex irrigation systems.

— Timothy Earle

HUMAN SOCIOBIOLOGY AND BEHAVIOR

[*This entry comprises four articles:*

> An Overview
> Human Sociobiology
> Evolutionary Psychology
> Behavioral Ecology

The introductory article provides a general overview of human sociobiology and behavior; the second article discusses the theoretical arguments for a genetic theory of social behavior in humans; the third article focuses on the evolution of the human mind and current developments in cognitive neuroscience; the fourth article introduces the idea that human behavior can be studied from an adaptationist perspective and provides a review of the key topics that human behavioral ecologists investigate. For related discussions, see Human Foraging Strategies; *and* Parental Care.]

An Overview

Until recently, evolutionary thinking has played little role in the study of human behavior. The causes of this neglect are complex, but one important factor is that biology did not have an adequate theory for the evolution of behavior until the 1960s, and therefore had little of interest to say to social scientists about human behavior. Biologists from Charles Darwin to Theodosius Dobzhansky had applied evolutionary ideas to human behavior, but their efforts were largely ignored by social scientists. In the early 1960s, William D. Hamilton (1964) published the first of a series of papers that radically transformed our understanding of the evolution of behavior. Hamilton's work, and subsequent work by Eric Charnov, John Maynard Smith, Robert Trivers, and George Williams, showed how the systematic application of adaptationist thinking could explain observed patterns of competition, cooperation, conflict, parental investment, sex ratios, signaling, food choice, and dispersal in nature. This new theory was based on the idea

that natural selection favors behaviors that enhance the relative reproductive success of individuals, not behaviors that enhance the welfare of the species. These theoretical contributions transformed our understanding of behavior and generated a tidal wave of new empirical and theoretical work. (See Krebs and Davies, 1993, and Alcock, 1997, for textbook treatments.)

It was not long before these new ideas in biology were applied to human behavior. The first salvo came in the mid-1970s, when Richard Alexander (1974) published a long article in the *Annual Review of Ecology and Systematics* and Edmund O. Wilson's book, *Sociobiology: The New Synthesis* (1975), appeared. Although both of these works focused mainly on the evolution of behavior in nonhuman animals, they included discussions of how evolution shaped human behavior. Over the next few years a number of researchers in anthropology and psychology followed their lead and produced influential work. In 1978, Martin Daly and Margo Wilson published the first edition of their authoritative text *Sex, Evolution, and Behavior.* In 1979, Donald Symons published *The Evolution of Human Sexuality.* Napoleon Chagnon and William Irons collected much of the best new empirical work on the evolutionary basis of human behavior in an edited volume published in 1979. Around the same time, programs of anthropological field work rooted in evolutionary theory were launched by Chagnon and Irons, then at Northwestern University, and by Irven DeVore at Harvard University. The programs yielded a number of empirical projects that focused on social patterns in small-scale societies.

These empirical projects borrowed both theory and empirical methods from evolutionary biologists. Adaptationist reasoning was used to generate hypotheses about human behavior that could be tested empirically. Monique Borgerhoff Mulder's (1988) study of bride-wealth payments among the Kipsigis, a Kalenjin-speaking group living in Kenya, provides a good example of this approach. Among the Kipsigis, a groom's father makes a payment of cash and livestock to the father of the bride at the time of marriage. Borgerhoff Mulder reasoned that Kipsigis bride-wealth payments provide an observable index of the qualities that are valued in prospective spouses, and predicted that bride-wealth offers should reflect the potential reproductive value of the prospective bride. In fact, the highest bride-wealths were paid for the women who were youngest when they reached menarche. Among the Kipsigis, age at menarche is a reliable index of women's reproductive potential. Kipsigis women who reach menarche early have longer reproductive life spans, higher annual fertility, and higher survivorship among their offspring than do women who mature at later ages.

Research applying evolutionary theory to human behavior was strongly criticized from a wide variety of perspectives, ranging from the frankly political to the rigorously scientific. The critiques, which continue to the present, are too diverse to descibe here. (See Segresträle, 2000, for a history of these controversies.) However, many critiques are rooted in the common belief that evolutionary explanations entail "genetic determinism." This argument is still quite common, and it is important to see why it is wrong.

Many social scientists mistakenly believe that evolutionary explanations imply that behavioral differences between individuals are caused by genetic differences because natural selection cannot create adaptations unless genetic differences between individuals exist. Because there is considerable evidence that human differences are the product of learning and culture, adaptive explanations must not be relevant to human behavior. The flaw in this argument is that natural selection can shape the mechanisms that control the way animals assess and respond to opportunities and dangers. Individuals inheriting the same control mechanisms then vary their behavior adaptively with circumstances. In humans, natural selection has shaped the mechanisms that control-learning and the acquisition of culture so that individuals respond adaptively to environmental contingencies. (See Smith and Winterhalder, 1992, for an extended version of this argument.)

Contemporary Research. During the 1980s, research on evolution and human behavior split into two distinct, but related, threads. Evolutionary psychology is concerned with how the past environments in which humans evolved shaped the human mind. Human behavioral ecology focuses on how evolution shapes behavioral strategies and reproductive outcomes that can be observed today and in the historical record.

Evolutionary psychology. Evolutionary psychologists use evolutionary theory to understand human psychology. This research program is based on three precepts. First, most learning is built on information-rich, modular psychology. Because brains are expensive to build and to maintain, organisms cannot excel at all cognitive tasks. Instead, minds are built up out of a large number of special-purpose mechanisms that solve particular kinds of problems. Second, natural selection determines the kinds of problems that the brains of particular species are good at solving. To understand the psychology of any animal species, we must know what kinds of problems these animals need to solve in nature. Third, complex adaptations evolve relatively slowly. We know that people lived in in small-scale foraging societies until very recently. Thus, to understand human psychology, we must identify the kinds of problems humans needed to solve when they lived in small groups as Pleistocene foragers.

The work of Leda Cosmides and John Tooby on "social contract" reasoning provides a good example of this approach (Cosmides and Tooby, 1992). Beginning in the 1960s, psychologists had demonstrated that people's ability to reason depends on content. Laboratory subjects were much more likely to make errors with some problems than others even when the problems had the same logical structure. Moreover, psychologists were unable to discern why some content made problems easy and others hard. Cosmides and Tooby reasoned that solving problems involving reciprocal altruism was crucial for success in the small groups that have characterized human societies for most of our history, and because the big problem with reciprocal altruism is being exploited, they predicted that human cognition should be tuned to look for cheaters in social exchanges. They then executed a series of experiments that show that this prediction was correct—problems in which the logically correct answer is equivalent to detecting cheaters are much easier to solve than many other kinds of problems. Even more compelling, problems in which social contract reasoning opposed logical reasoning were harder than abstract problems. This work has been tested cross-culturally (Sugiyama et al., 1995).

The reasoning of evolutionary psychology can also be applied to data collected outside the psychology laboratory. For example, Martin Daly and Margo Wilson (1993) have studied the psychology of parental investment in contemporary Western populations. In most ancestral environments, parents' ability to invest in their offspring would have been limited, and natural selection would have favored psychological mechanisms that cause parents to terminate investment in offspring that are not likely to survive and to channel investment toward the children that are most likely to live to maturity and to reproduce successfully. We expect these psychological mechanisms to be sensitive to the child's condition, the parents' economic circumstances, and the parents' alternative opportunities to pass on their genetic material. Cross-cultural ethnographic data on infanticide in traditional societies largely fit these predictions. Although it is unlikely that child abuse in wealthy industrialized nations is actually adaptive, it is plausible that the same psychological mechanisms that lead to infanticide in simpler, more resource-constrained societies may also lead to child abuse in our own society. Thus, Daly and Wilson reasoned that child abuse was most likely to occur when families are stressed, when the child is sickly or handicapped, or when the child's paternity is uncertain. Crime statistics and child abuse reports from the United States and Canada are largely consistent with these predictions. For example, young children in families with one stepparent are almost forty times more likely to experience child abuse than children in families with two biological parents, even when other social risk factors are controlled.

The logic of evolutionary psychology can also be

used to study the interaction of cultural and genetic evolution. Although social learning is not unique to humans, humans aquire much more of their behavior through social learning than any other animals do. Peter Richerson, Robert Boyd, and their coworkers have studied how natural selection should shape the psychology of social learning to reduce the costs of individual learning and experimentation. For example, Joseph Henrich and Robert Boyd (1998) analyzed mathematical models that suggest that in a wide range of temporally and spatially varying environments natural selection will favor a tendency to copy the more common behavior. Such "content independent" social learning mechanisms provide a means of rapidly sifting through the plethora of information available in the social world and efficiently extracting adaptive behaviors. These social learning shortcuts do not always result in the best behaviors, nor do they prevent the acquisition of maladaptive behaviors. Nevertheless, averaged over many environments and behavioral domains (e.g., foraging, hunting, and social interaction), these cultural transmission mechanisms provide fast and frugal means to acquire complex, highly adaptive behavioral repertoires. Empirical research by psychologists, economists, and sociologists confirms that people are likely to adopt common behaviors across a wide range of decision domains. Although much of this work confounds normative conformity (going with the popular choice to avoid appearing deviant) with conformist transmission (using the popularity of a choice as an indirect measure of its worth), recent work shows that conformist transmission is important in a variety of social learning problems.

Although the logic underlying evolutionary psychology is compelling and the paradigm has generated a substantial body of valuable empirical work, it is vulnerable to at least one objection: it is hard to determine the kind of environment that has shaped the evolution of human reasoning. Although evolutionary psychologists assume that complex adaptations evolve slowly, we actually know very little about the rate at which complex adaptations like these special-purpose mental mechanisms evolve. It is possible that new mental mechanisms might have evolved since the origin of agriculture just 10,000 years ago, but it is also possible that some mental mechanisms might have been shaped by selection long before the origin of modern humans 200,000 years ago. There is also great uncertainty about the ecology and behavior of extinct hominids. Some authorities believe that hominids have lived for two million years in societies that resemble those of contemporary human foragers, whereas others think that such societies did not arise until 50,000 years ago. If early hominids lived lives that resembled the lives of contemporary foragers, then it is reasonable to think that the human brain has evolved to solve the kinds of problems that confront modern foragers.

Human behavioral ecology. Human behavioral ecologists generally adopt the same kinds of tactics and methods that evolutionary biologists use when they want to understand particular aspects of the morphology and behavior of other organisms. Evolutionary biologists assume that phenotypes have been shaped by natural selection, and rely on this assumption to explain features of their morphology, physiology, and behavior. For example, primates that eat leaves have different kinds of teeth and guts than primates that eat mainly ripe fruit, primate females that cache their young while they are foraging and feed them at long intervals have richer milk than females that carry their young with them, and male monkeys that compete actively for access to mates have larger bodies and longer canines than male monkeys that form long-term pair-bonds and face little competition from other males for access to females. Of course, there are many reasons why evolution may not yield adaptations in every case: selection acting on one characteristic may affect others, genetic drift may lead to maladaptation, or constraints may prevent adaptation from occurring. To be certain that adaptive reasoning is justified in any particular case, a biologist needs to know enough about the genetics, development, and history of the organism of interest to discount maladaptive processes. In practice, biologists almost never know much about these things. Many biologists simply ignore these complications and assume that observed phenotypes are adaptive, a research tactic that Oxford biologist Alan Grafen calls the phenotypic gambit. Biologists use the phenotypic gambit, knowing that it may not be correct in every case, because experience in many areas of biology suggests that adaptive reasoning that ignores the complexities of particular control mechanisms is often a useful means to understand why animals behave the way they do. Conversely, explanations of how those mechanisms work may say little about why they evolved to work as they do. The distinction between these two kinds of inquiry is captured by Ernst Mayr's distinction between "ultimate" and "proximate" explanations. When human behavioral ecologists apply the phenotypic gambit to the human species—they usually ignore proximate questions about how human decision mechanisms work. They assume that, like other animals, humans act as if they are attempting to maximize their genetic fitness. The assumption provides human behavioral ecologists with a rich source of hypotheses about human behavior.

What distinguishes behavioral ecology from evolutionary psychology is that behavioral ecologists look first to the ways that ecological forces influence human behavior. For example, A number of human behavioral ecologists have used optimal foraging theory to investigate subsistence strategies of human foragers (reviewed by Kaplan and Hill, 1992). Optimal foraging theory assumes that foragers generally maximize their fitness by maximizing their rate of nutrient acquisition. Models then predicts that foragers will selectively exploit resources that provide favorable return rates. Among the Ache, a group of people who forage in the forests of the Amazon basin, subsistence behavior largely conforms to predictions derived from these models. In general, the Ache concentrated their foraging efforts on resources that generated a higher rate of return than the mean, as they tended to take the most productive resources most often. However, this fails to account for two aspects of Ache foraging behavior. First, the men routinely ignore plant foods, even the ones that generate relatively high rates of return. Men could greatly increase the number of calories they obtain per hour if they devoted themselves to extracting and processing starch from the trunks of palm trees. Instead, men mainly hunt. Second, women routinely ignore resources, such as honey and meat, that bring high rates of return and concentrate their efforts on plant foods with lower return rates. These discrepancies suggest that human foraging strategies are not simply designed to maximize caloric returns; foraging strategies may be adjusted to accommodate needs for specific nutrients or to balance competing constraints such as child care. An alternative hypothesis is that the sex differences result from different foraging goals. Women pursue resources they can count on getting and keeping for their own families. Men specialize in risky resources that are widely shared. Women may be foraging to provision their children, whereas men hunt to signal qualities that affect their relative standing among men. Hunting success supplies valuable common goods, so this hypothesis also implies a pathway to cooperation through costly signaling.

Human behavioral ecologists have used an evolutionary perspective to both uncover and explain such behavioral differences between the sexes. Some of this work focuses on factors that are associated with variation in reproductive success and that shape reproductive strategies within societies. For example, in traditional and historical populations, wealth is consistently associated with men's reproductive success (Cronk, 1991, Irons, 1979;) because wealthy men tend to have younger and more fertile wives, are more likely to have multiple wives, have greater survivorship among their offspring, and are able to secure better marriage prospects for their offspring (Voland, 1998).

Behavioral ecologists also ask how natural selection shapes individuals' decisions about who to marry, how many children to have, and how much to invest in each of their offspring. Eckart Voland and Robin Dunbar (1995) used parish records from the Krummhörn region of northwestern Germany to investigate parental investment strategies. The fertile soil of the Krummhörn supported productive agriculture and dairy farming and

considerable concentration of wealth in certain lineages. However, population growth and economic opportunities were constrained because the region is bounded on three sides by the sea and on the fourth by unproductive moorland. Wealthy, landed men who married achieved higher reproductive success than other men (Klindworth and Voland, 1995), but not all sons from landed families were equally advantaged. Primogeniture, practiced by landed families in the Krummhörn, prevented the fragmentation of land holdings and maintained lineage wealth, but seriously limited sons' prospects. Thus, in landed families, the number of surviving sons was negatively related to their chance of marrying and positively related to their chance of emigrating from the region (Voland and Dunbar, 1995). These conditions were also reflected in the mortality statistics for young men—sons were uniformly less likely to survive than daughters, and the presence of surviving brothers substantially reduced a young man's chance of surviving to the age of fifteen. Several lines of evidence suggest that the differential survivorship of sons and daughters in landed families was the result of parental manipulation of the sex ratio among their progeny. Parents limited investment in offspring who were unlikely to become reproductively successful.

Human behavioral ecology provides insight about how and why people in traditional societies make certain kinds of decisions in certain contexts, but it has been less successful in explaining the broader patterns of human social organization and behavior. Looking back on her own work on Kipsigis women's preferences for wealthy men, Borgerhoff Mulder (1997) wrote, "We still have a very poor understanding of the social and ecological constraints that shape behavioral options," and her own analysis "fails to tackle seriously the question of why different marriage systems emerge. As always, we trade comparative breadth for intrasocietal depth."

BIBLIOGRAPHY

Alcock, John. *Animal Behavior: An Evolutionary Approach.* 6th ed. Sunderland, Mass., 1998.

Alexander, Richard D. "The Evolution of Social Behavior." *Annual Review of Ecology and Systematics* 5 (1974): 325–383.

Borgerhoff Mulder, Monique. "Early Maturing Kipsigis Women Have Higher Reproductive Success than Later Maturing Women and Cost More to Marry." *Behavioral Ecology and Sociobiology* 24 (1988): 145–153.

Borgerhoff Mulder, Monique. "Marrying a Married Man: A Postscript." In *Human Nature*, edited by L. Betzig, pp. 115–117. New York, 1997.

Chagnon, Napoleon A., and William Irons. *Evolutionary Biology and Human Social Behavior: An Anthropological Perspective.* North Scituate, Mass., 1979.

Cosmides, Leda, and John Tooby. "Cognitive Adaptations for Social Exchange." In *The Adapted Mind*, edited by Jerome Barkow, Leda Cosmides, and John Tooby. New York, 1992.

Cronk, Lee. "Wealth, Status, and Reproductive Success among the

Mukogodo of Kenya." *American Anthropologist* 93 (1991): 345–360.

Daly, Martin, and Margo Wilson. *Sex, Evolution, and Behavior.* Boston, 1983.

Daly, Martin, and Margo Wilson. *Homicide.* New York, 1988.

Hamilton, W. D. "The Genetical Evolution of Social Behavior." *Journal of Theoretical Biology* (1964): 1–52.

Henrich, Joseph, and Robert Boyd. "The Evolution of Conformist Transmission and the Emergence of Between-Group Differences." *Evolution and Human Behavior* 19 (1998): 215–242.

Irons, William. "Cultural and Biological Success." In *Evolutionary Behavior and Human Social Behavior*, edited by N. A. Chagnon and W. Irons. pp. 257–272. North Scituate, Mass., 1979.

Kaplan, Hillard, and Kim Hill. "Human Subsistence Behavior." In *Evolution, Ecology, and Human Behavior*, edited by E. A. Smith and B. Winterhalder, pp. 167–202. Chicago, 1992.

Kaplan, Hillard, Kim Hill, J. Lancaster, and A. Madgalena Hurtado. "A Theory of Human Life History Evolution: Diet, Intelligence, and Longevity." *Evolutionary Anthropology* 9 (2000): 156–185.

Klindworth, Heike, and Eckart Voland. "How Did the Krummhörn Elite Males Achieve Above-Average Reproductive Success?" *Human Nature* 6 (1995): 221–240.

Krebs, John R., and Nicholas B. Davies. *An Introduction to Behavioral Ecology.* Oxford, 1993.

Segrestråle, Ullica Christina Olofsdotter. *Defenders of the Truth: The Battle for Science in the Sociobiology Debate and Beyond.* New York, 2000.

Smith, Eric A., and Bruce Winterhalder. *Evolutionary Ecology and Human Behavior.* New York, 1992.

Sugiyama, Lawrence, John Tooby, and Leda Cosmides. "Testing for Universality: Reasoning Adaptations among the Achuar of Amazonia." In *Meetings of the Human Behavior and Evolution Society.* Santa Barbara, Calif., 1995.

Symons, Donald. *The Evolution of Human Sexuality.* New York, 1979.

Voland, Eckart. "Evolutionary Ecology of Human Reproduction." *Annual Review of Anthropology* 27 (1998): 347–374.

Voland, Eckart, and Robin I. M. Dunbar. "Resource Competition and Reproduction: The Relationship between Economic and Parental Strategies in the Krummhörn Population (1720–1874)." *Human Nature* 6 (1995): 33–49.

Wilson, Edward O. *Sociobiology: The New Synthesis.* Cambridge, Mass., 1975.

— ROBERT BOYD AND JOAN B. SILK

Human Sociobiology

Human sociobiology can be defined as the study of human social behavior from the perspective of evolutionary theory. As John Alcock (2001) points out, the term "sociobiology" is usually restricted, in practice, to studies in which the main goal is to explain the role of natural selection in shaping social behaviors and societies. This is true whether the subject is human behavior or nonhuman animal behavior. The central theoretical issue is how natural selection has favored particular forms of social behavior. Sociobiology is closely related to behavioral ecology, which can be defined as the study of behavior as an adaptation to an animal's environment; sociobiology can be defined as the part of behavioral ecology that is concerned with adaptation to the social environment.

The topics studied in human sociobiology are, for the most part, the same topics that sociobiologists address when studying nonhuman animals. These include kin altruism, mating strategies, parenting strategies, life-history strategies, status or dominance hierarchies, reciprocal altruism, and the formation of cooperative groups, and communication. Much of the work in human sociobiology has been done in more traditional societies or has incorporated a cross-cultural element, on the assumption that human social behavior reflects an interaction of genetic and cultural influences.

Kin Altruism. W. D. Hamilton's theory that organisms tend to behave as if they valued their own fitness over that of their neighbors according to the portion of genes they share by recent common descent has been tested extensively with human data and has generally been found to be strongly supported (see Cronk, 1991). For the most part, the predictions tested have been that a human being will be more or less helpful to another human being in direct relation to the portion of genes the two share by recent common descent.

Martin Daly and Margo Wilson (1988), in a classic study showed that the risk of homicide is eleven times greater for individuals living in the same household who are not genetically related than for genetic kin. Members of the same household were compared because people who reside in the same household have, on average, equal opportunities to annoy and provoke one another to hostile action. Individuals not living in the same household are less likely to provoke aggressive behavior toward one another. They also were able to show, using nation-wide Canadian statistics, that a child's risk of being killed by its caretakers, when the child is two years of age or younger is approximately 70 times higher when the child lives with one natural parent and one stepparent rather than two natural parents. For the most part the caretakers responsible for the killing are step fathers.

N. A. Chagnon has been able to show that close genetic ties play an important role in many aspect of the lives of the Yanomamo, an indigenous group living in the Amazon rainforest of southern Venezuela, in villages of 40 to 300 individuals. These villages are separated by several hours' to several days' walk and are frequently hostile toward one another. Hostile villages raid and kill members of enemy villages. Members of the same village support one another in this inter-village warfare, and residents of different villages are likely to be involved in lethal encounters. In this situation, the Yanomamo strongly prefer to reside with close relatives. Chagnon has also shown that Yanomamo villages tend to grow in population over time and, after reaching a certain size, to split into two smaller villages. Close kin tend to stay together in the same new village, and distant kin join separate villages. Staying with someone in the same village is an act of altruism toward that person since it carries with it a high probability that co-residents will support one another in the future when their village is attacked. Taking up residence in separate villages from someone else is an act of selfishness since the separate villages are likely to go to war with one another in the future. Chagnon used statistics drawn from a large number of villages to show that the average genetic relatedness among village residents was a strong predictor of whether the village was about to fission. Low relatedness tended to lead to fissioning into smaller villages, each with higher average relatedness. Chagnon's analysis showed that a phenomenon that many anthropologists had assumed was shaped completely by culture strongly reflects an evolved propensity toward kin altruism of the sort predicted by Hamilton's theory of kinship.

Mating Strategies. Human sociobiologists have studied mating strategies extensively and have learned that, like kin altruism, this aspect of human behavior reflects the evolutionary heritage of our species. Borgerhoff Mulder's analysis (1988) of variation in bridewealth among the Kipsigis of Kenya shows that higher bridewealth is paid for brides with a higher reproductive value—that is, a greater probability of future high fertility. Early-maturing brides fetch higher bridewealth, on average, and have higher fertility after marriage. Plumper and healthier brides also bring higher bridewealth; brides who have already had children and therefore are likely to have fewer in the future bring lower bridewealth.

The majority of the more traditional societies that have been the focus of anthropological research allow polygyny. Having multiple wives offers clear reproductive benefits to men, but it is less obvious why a female would choose to be a second or later mate, rather than the first mate of a previously unmarried man. Models worked out among nonhuman animals suggest that polygyny can result from female choice when males possess resources useful to females for rearing offspring. The position of second or later mate of a male with extensive resources may give a female access to more resources for offspring care than would the position of the sole mate of a male with few resources. However, recent research has shown that, in some human societies, polygyny is beneficial to men but harmful to the reproductive interests of women. Strassmann (2000) reports that, among the Dogon of Mali, polygyny adversely affects the survivorship of children and hence the lifetime reproductive success of women. Strassmann's research involved the use of anthropometric measures (weight and height of children of specific ages) to demonstrate that the children of polygynously married women suffer a disadvantage in diet. The fact that resources per child

are clearly reduced for these women in the Dogon environment makes clear the proximate cause of higher mortality for their children.

Available studies on polygyny suggest that some forms of human polygyny are beneficial to women and fit the female choice model, while others are harmful to women and fit the male coercion model. Why it is that the marriage systems of some societies serve male interests at the expense of female interests, while others seem to serve both, is an important question for future research. As research in human sociobiology answers some questions, it raises new ones.

Parenting Strategies. The concept of parental investment formulated by R. L. Trivers has provided inspiration for productive lines of research on human parenting behavior. Trivers's central insight was that human parents, like all other parents, have limited resources for nurturing offspring and must make choices concerning how much to invest in each offspring. One issue is how much to invest in sons versus daughters. Trivers, working with D. E. Willard, suggested that when it is predictable from the parent's condition that offspring of one sex will have higher average reproductive success than offspring of the other sex, natural selection should favor the ability to shift parental investment to emphasize nurturing the sex with the highest probable reproductive success.

Early research by Mildred Dickemann suggested that, in the human species, rank in a social hierarchy could predict the probable future reproductive success of sons versus daughters. Sons of families who rank high tend to have higher reproductive success than their sisters, and daughters of families who rank low tend to have higher reproductive success than their brothers. She documented this with data from British India. This difference in reproductive success between sons and daughters is greatest when women tend to marry men of higher social status than their natal families, and men at the top of the hierarchy are polygynous. Dickemann used the Trivers–Willard hypothesis to predict higher rates of female infanticide among higher castes in the early colonial period of British India, and she got confirming results from census data collected by the colonial government.

It is commonly assumed that in traditional societies, parents prefer sons over daughters. The Trivers–Willard hypothesis, however, suggests the counterintuitive prediction that at the bottom of social hierarchies in which women marry up and men can be polygynous, parents would favor daughters over sons. Cronk tested this using ethnographic and demographic data along with anthropometric measures of health and nutritional status (weight, height, and skinfold measure of fat reserves). He did focal follows—that is, he followed particular parents for a day or so and recorded all of their parenting activities. These data confirmed that among the low-status group he was studying, the Mukogodo of Kenya, parents did more to nurture daughters than sons, and, as a consequence, daughters were more likely to survive to adulthood. Cronk was also able to demonstrate with demographic data that Mukogodo daughters have more children, on average, than do Mukogodo sons. Mukogodo women have an expanded mating pool because they can marry both Mukogodo men and men of higher-status groups; Mukogodo men, in contrast, usually cannot marry women outside their group, and the fact that many Mukogodo women marry into higher-ranking groups limits the number remaining for Mukogodo men.

This result was all the more surprising because Mukogodo, when asked, usually say they prefer sons over daughters; Cronk uncovered strong evidence of a gap between Mukogodo culture (what they consciously claim) and their behavior. This led to interesting suggestions about the nature of culture (defined as socially transmitted information) and behavior. The gap between culture and behavior, in this case, appears to result from the fact that the Mukogodo are former hunters and gatherers who have been assimilated to the pastoralist Maasai culture, and their poverty and the low status of their former lifestyle places them at the bottom of the Maasai social hierarchy. Maasai in general prefer sons, and the Mukogodo prefer to claim an attitude that makes them seem more Maasai, even though it does not reflect their actual behavior.

One issue in parenting strategy is the trade-off between number of offspring and the extent of nurturance devoted to each individual offspring. In the human species, few offspring are produced and great effort is usually put into the rearing of each individual offspring. This continues a long-term trend in primate evolution. In the human case, nurturing children includes ensuring that they acquire the social and technological skills required by their society, so the optimal trade-off between number and nurturance can vary from one social setting to another—one of many examples showing that human beings are flexible in response to the demands of different cultural and social environments.

Some sociobiologists studying humans have used this observation as a starting point for building a theory of the transition to very low fertility that has accompanied modernization. J. W. Wood (1994) has presented evidence that human populations tend to fall into two distinct groups, natural fertility populations and controlled fertility populations. The first group makes little or no use of any form of artificial birth control, whereas the second group makes extensive use of it to limit fertility. A woman who lives through her reproductive years in a natural-fertility population has 6.1 children, on average; a woman who spends her reproductive years in a controlled-fertility population has 2.6 children. The averages vary among societies, but there is little overlap between natural and controlled-fertility populations in terms of lifetime fertility.

The shift from natural fertility to controlled fertility is a recent historical phenomenon associated with industrialization, urbanization, and extensive education. On the surface, this shift seems to contradict the expectations of evolutionary theory. Why should people raise fewer children when they have much greater resources for nurturing children than was typical in ancestral populations? Several researchers have suggested that a perceived need for an increase in nurturance per child underlies this shift. Kaplan and Lancaster (2000) analyzed this issue by combining life-history theory and economic theory. They argue that the psychological and physiological mechanisms that evolved in the human descent line tend to maximize not only the acquisition of skills and knowledge that can increase the effectiveness of resource acquisition, but also the biological growth of an adult phenotype. Their models show that when additional investment in children's skills and knowledge has a large effect on the children's ability to acquire resources, parents invest more in fewer children. Ruth Mace (2000) using data from a nomadic pastoral population, has constructed models, that look at a combination of reproduction and the acquisition of wealth rather than skills embodied capital; her research produced similar results, suggesting that the psychological mechanisms that optimized reproduction in traditional societies lead people in modern societies to increase nurturance per child and to lower fertility.

Expanding Scope of Human Sociobiology. The first topics studied in this field were those that had occupied the attention of sociobiologists studying animals—kin altruism, mating strategies, parenting strategies, and closely related topics such as infanticide, parent–offspring conflict, and competition for resources and status. There is now strong evidence that, despite cultural variation, humans beings everywhere are nepotistic, that male and female mating strategies differ in predictable ways, that parents tend to bias their parental investment toward offspring with better prospects for eventual reproductive success, and that the goals of human beings are generally those that in ancestral environments led to reproductive success and aided kin.

At the same time, sociobiologists studying humans have expanded their range of topics. Many of the new areas of inquiry are connected with the question of why human beings became ultrasocial—that is, why they form such large cooperative groups characterized by complex divisions of labor. R. D. Alexander (1987) attacked the question of morality. Why do human beings everywhere have the notion that certain forms of behavior are commendable and worthy of reward, and other forms are reprehensible and deserving of punishment?

Why do they attach great importance to these moral judgments even when they themselves are not directly affected by the behaviors in question? Alexander's answer is that human evolution has been driven primarily by competition between groups, and this has favored traits that allowed human beings to form larger, better-united groups. Kin altruism could not serve as a foundation for such large groups. Alexander suggested indirect reciprocity as one mechanism for forming larger groups. Indirect reciprocity consists of observing others and rewarding those who act altruistically, while punishing those who act selfishly—in very simple terms, being nice to nice people and nasty to nasty ones. Moral systems, which are systems of indirect reciprocity, made the formation of larger cooperative groups possible, and so the psychological mechanisms underlying these systems were favored in human evolution.

Another promising line of exploration of ways in which human beings evolved traits that aid in expanding the sphere of cooperation is to explore the role of commitments as social strategies. This idea was developed initially by game theorists who were not concerned with evolution. Thomas Schelling, analyzing the Cold War, noted that a firm commitment to retaliate against a preemptive first strike could deter such a first strike even though it could not save the victim and might even, in a nuclear exchange, make the victim's situation worse. A commitment to retaliate, even if it were destructive to the retaliator, could deter an attack. Paradoxically, this meant that a contingent commitment to do something contrary to one's self-interest could serve that self-interest by making the triggering of the commitment unlikely and, at the same time, making it more likely that other parties would act in one's interest. Robert Frank, an economist, explored these issues (1988) and pointed out that the use of commitments as social strategies makes sense in terms of human evolution. These new lines of research have expanded our understanding of how human beings have evolved to form very large cooperating groups and networks. They have also added to the list of topics that human sociobiologists study, using evolutionary theory as a guide. The new areas explored include morality, religion, law, economics, and even the basic nature of culture.

BIBLIOGRAPHY

Many empirical studies in human sociobiology are published in two journals devoted primarily to research on human behavior guided by evolutionary theory: *Evolution and Human Behavior* and *Human Nature*.

Alcock, J. *The Triumph of Sociobiology.* Oxford, 2001.

Alexander, R. D. *The Biology of Moral Systems.* New York, 1987. Presents the theory that group–group competition caused runaway selection for traits that aided in the formation of larger, better-united groups in the human descent line and introduces the concept of indirect reciprocity.

Alexander, R. D. *Darwinism and Human Affairs.* Seattle, 1979. An early overview emphasizing the use of anthropological and cross-cultural data to evaluate evolutionary hypotheses about human behavior.

Betzig, L. L., ed. *Human Nature: A Critical Reader.* New York, 1996. Reprints classical empirical studies in human sociobiology and evolutionary psychology.

Betzig, L. L. *Despotism and Differential Reproduction: A Darwinian View of History.* Hawthorne, N.Y., 1986. Presents cross-cultural data supporting the hypothesis that human beings use political power for reproductive ends in traditional societies, especially for the pursuit of multiple mates by men.

Betzig, L. L., M. Borgerhoff Mulder, and P. Turke, eds. *Human Reproductive Behaviour: A Darwinian Perspective.* Cambridge: Cambridge Univ. Press, 1988. A collection of state-of-the-art empirical studies.

Borgerhoff Mulder, M. "Kipsigis Bridewealth Payments." In L. Betzig et al. (1988), pp. 65–82.

Chagnon, N. A., and W. Irons, eds. *Evolutionary Biology and Human Social Behavior: An Anthropological Perspective.* North Scituate, Mass., 1979. The first attempt to present the use of sociobiological theory in anthropology; set the agenda for early sociobiological research in anthropology.

Cronk, L. *That Complex Whole: Culture and the Evolution of Human Behavior.* Boulder, Colo., 1999. Addresses the issue of why human beings have elaborate cultural traditions and finds the answer in the use of culture to manipulate the behavior of other human beings.

Cronk, L. "Human Behavioral Ecology." *Annual Reviews of Anthropology* 20 (1991): 41–75.

Cronk, L., N. A. Chagnon, and W. Irons, eds. *Adaptation and Human Behavior: An Anthropological Perspective.* Hawthorne, N.Y., 2000. Presents state-of-the-art empirical papers commemorating the 1979 Chagnon and Irons edited volume, showcasing progress in human behavioral ecology (human sociobiology) since 1979.

Daly, M., and M. Wilson. *Homicide.* New York, 1988. Uses data from the United States, Canada and other countries to test evolutionary hypotheses about violence and homicide; makes a persuasive case that human violence cannot be understood without reference to human evolution.

Dickemann, M. "Female Infanticide, Reproductive Strategies, and Social Stratification: A Preliminary Model." In Chagnon and Irons (1979), pp. 321–367.

Dunbar, R. *Grooming, Gossip, and the Evolution of Language.* Cambridge, Mass., 1996. Presents the hypothesis that speech, gossip, and small talk have replaced primate grooming in the human species and facilitate the formation of larger social groups.

Frank, R. *Passions within Reason: The Strategic Role of Emotions.* New York, 1988.

Hill, K., and M. Hurtado. *Ache Life History; The Ecology and Demography of a Foraging People.* Hawthorne, N.Y., 1996. A classic evolutionary analysis of the lives of a group of foragers using ethnography and demographic data, organized around life-history theory.

Hrdy, S. B. *Mother Nature.* New York, 1999. Thorough and insightful analysis of the many dimensions of mothering which shows that evolution explains complexities of mothering behavior in both human beings and other animals.

Kaplan, H., and J. Lancaster. "The Evolutionary Economics and Psychology of the Demographic Transition to Low Fertility." In Cronk et al. (2000), pp. 283–322.

Low, B. S. *Why Sex Matters: A Darwinian Look at Human Behavior*. Princeton, 2000. Excellent overview of human behavioral ecology (human sociobiology) over the past two decades; organized around the fact that men and women reproduce in different ways and therefore, use different strategies to acquire resources, mates, and to parent.

Mace, Ruth. "An Adaptive Model of Human Reproductive Rate Where Wealth Is Inherited: Why People Have Small Families." In Cronk et al. 2000, pp. 261–281.

Nesse, R., ed. *Natural Selection and the Capacity for Commitment*. New York, 2002. Explores the game-theory analysis of commitment as a social strategy and its implications for law, economics, religion, and emotional adjustment, among other subjects.

Nesse, R. M., and G. C. Williams. *Why We Get Sick: The New Science of Darwinian Medicine*. New York, 1994. Excellent introduction to the new field of evolutionary medicine; clarifies that disease cannot be understood and controlled without evolutionary theory.

Smith, E. A., and B. Winterhalder, eds. *Evolutionary Ecology and Human Behavior*. Hawthorne, N. Y., 1992. Rigorous presentation of human behavioral ecology, including adaptation to both natural and human social environments.

Strassman, Beverly. "Polygyny, Family Structure, and Child Mortality: A Prospective Study among the Dogon of Mali." In Cronk et al. (2000), pp. 49–67.

Wilson, E. O. *On Human Nature*. Cambridge, Mass., 1978. Early exploration of human sociobiology by the author of the watershed volume *Sociobiology: The New Synthesis* Cambridge, Mass., 1975.

Winterhalder, B., and E. A. Smith. "Analyzing Adaptive Strategies: Human Behavioral Ecology at Twenty-Five." *Evolutionary Anthropology* 9.2 (2000): 51–72. Review article covering all of behavioral ecology, including human sociobiology.

Wood, J. W. *Dynamics of Human Reproduction: Biology, Biometry, Demography*. New York, 1994.

— WILLIAM IRONS

Evolutionary Psychology

Humans exhibit many of the same anatomical and physiological adaptations as other primates. When it comes to psychology and behavior, however, humans are much more complicated. Although we share with other primates basic emotions—rage, fear, sexual desire, and mother love—humans alone can communicate symbolically through written and spoken language, create and disseminate works of mathematics, science, and art, and manipulate the behavior of others with appeals to duty, religion, laws, and promises. And although other primates may be able to deceive one another, language allows us to devise elaborate stories, excuses, and lies. Evolutionary psychologists study the human mental functions that underlie these capacities by drawing on cross-species comparisons, cross-cultural comparisons, case studies (e.g., of brain damage), and laboratory experiments.

Like other evolutionary scientists, evolutionary psychologists generally assume that the attributes of the species they study (in this case, humans) have evolved through natural selection. Specifically, evolutionary psychologists assume that the brain circuitry and processes underlying human behavior include features maintained from our primate ancestry and modifications that were selected because, at the time they evolved, they led to adaptive behavior. It is thus sometimes said that the human brain houses an "adapted mind." However, evolutionary psychologists do not assume that all human behavior is adaptive. Because the environments we live in today are in many ways dissimilar to the environments that our ancestors were designed to cope with, many of our behaviors may be anachronistic—once adaptive, but now neutral or even maladaptive. It is thus also sometimes said that the human brain houses a "maladapted mind."

The Brain. Physically, the human brain is much like that of all other mammals, consisting of a brain stem, which regulates mostly unconscious physical activity; a subcortical system, which modulates drives, emotions, physical sensations, and memory; and a convoluted cortex, which allows for the handling of greater amounts and complexity of information than is possible by members of other taxa. Primates as an order have brains relatively larger than other mammals of similar body sizes, and the human brain is significantly larger—even in proportion to body mass—than those of other primates. [*See* Brain Size Evolution.] As mammalian brains increased in size, the cerebral cortex increased relatively more than other parts. It is this expanded cortex that confers on humans great intelligence, a capacity for symbolic representation, and a reflective self-awareness.

Some parts of the human brain are physically distinguishable from adjacent areas, allowing for three-dimensional mapping of subcortical structures as well as two-dimensional mapping of the cortical surface. Boundaries of these structurally distinct areas often coincide with boundaries of functionally distinct areas, leading some theorists to conceive of the brain as not just a single organ, but as many organs or functional "modules" bundled together. These different organs or modules interconnect with one another to varying degrees, allowing both serial and parallel information processing at multiple levels of integration. Consequently, a person can engage in several activities at the same time, and can even experience two or more seemingly contradictory thoughts or feelings simultaneously—such as love and hate or curiosity and fear. Another consequence of this design is that, although some mental activity is available to consciousness, most occurs at a level of physical automaticity. We do not "know," for example, how we change the focus of our eyes from near to far, or how we retrieve an old phone number out of memory, or how we come to distinguish and prefer chocolate as compared to vanilla—we just do.

Natural selection has responded to the existence of static and predictable stimuli with the evolution of simple low-level processing mechanisms. Collectively, these confer a set of panhuman "instinctive" behavioral responses and psychological preferences. The existence of spatially diverse and temporally dynamic stimuli, in contrast, has led to the evolution of more general and more complex learning mechanisms designed to generate what is statistically likely to be the most adaptive response in each unique or new circumstance. The expanded human cortex, with its increased capacity for associative learning, abstract generalization, and transitive reasoning, is what allows for this adaptive, functional plasticity, and is what underlies the great cultural and individual diversity of the human species.

The Adapted Mind. In stark contrast to Skinnerian behaviorists, who view the infant mind as a tabula rasa, or blank slate, evolutionary psychologists are quick to point out that baby brains are not, in fact, blank or empty; infants come into the world with rudimentary knowledge and specialized information processing systems already "wired in." Furthermore, although knowledge does increase with experience, certain types of information are absorbed quite readily by the human brain, whereas other types are acquired only with difficulty. From an evolutionary perspective, the differential ease with which we learn about different aspects of our experience reflects their relative importance for our ancestors' survival and reproductive success.

Emotion. Perhaps the most important psychological adaptation for survival and reproduction is the sense of emotional attachment that develops between an infant and its mother. This is true for other primates as well as for humans, but humans develop motor skills more slowly as infants. Although we wean infants a bit earlier than chimpanzees do, human offspring require intensive investment and supervision long after weaning. Like other primates, young human children left on their own would have been vulnerable to starvation, exposure, predators, and the abuse of intolerant adults throughout evolutionary history. Close bonds between mother and child have thus been in the genetic interest of both mother and child—a bond to keep the child from wandering away from its primary protector and to keep the mother motivated to address the constant demands of what must often have been a distracting and irritating burden.

To this end, infants can recognize their mother's voice and smell soon after birth, and as soon as their eyes are able to focus enough to discriminate individual faces, they can recognize—and show preference for—that of their own mother. Once human children are old enough to crawl (potentially into danger), babies develop an intense desire to be within sight of their mother, and when temporarily separated, they experience and communicate great distress. Mothers, reciprocally, develop an intense attachment to their child above all others, and they too experience distress upon separation. Attachment, separation anxiety, and the many forms of preverbal communication that mutually reward mother–infant interactions are features of the human psyche that have deep roots in our primate heritage and are both universal and compelling.

Other emotions appear early in life, across cultures, and without having to be learned. These so-called primary emotions include fear, anger, happiness, sadness, surprise, disgust and, according to some, curiosity. The primary emotions are universal not only in terms of our phenomenological experience but also in terms of their outward expression; facial expressions of the primary emotions are consistently performed across cultures and ages even in children blind from birth, and (among the sighted) are recognized quickly and reliably. Babies communicate with facial expression long before they can see the expressions of others, and at about seven months of age when their visual system has developed enough to discriminate facial detail, they are differentially and appropriately responsive to the facial expressions of those around them.

Later in development, children acquire the so-called social emotions, such as guilt, shame, allegiance, vengeance, sympathy, remorse, and gratitude. Unlike the primary emotions, the social emotions require an awareness and appreciation of social context, and so routinely appear only after a child is physically and mentally capable of navigating the social world—around ages four, five, and six. The social emotions are critical for cooperating with nonrelatives and for getting along in large social groups in which people must often depend upon one another. The social emotions motivate us to repeat and to reward mutually beneficial interactions, and to avoid and to punish those who take advantage.

Knowing whom to trust and whom to avoid requires the ability to discriminate between and remember unique individuals, as well as to remember how they behaved in previous encounters. To this end, natural selection has honed in us a rather amazing capacity to recognize individual faces and voices—even after as little as a single exposure and sometimes after as long as decades between encounters. Furthermore, recognition of specific individuals is often associated with an automatically generated emotion—fear, anger, pleasure, anxiety—depending on our past experience with that person. In fact, one of the first human universals to be documented was the facial expression of anxiety upon first seeing a stranger approach, followed by an "eyebrow flash" and a smile of relief and pleasure when he or she appeared close enough to be recognized as a friend.

We even have emotional responses to strangers who look or act like people we have previously met. These "gut feelings" about people (and situations) are, essentially, classically conditioned visceral responses that send us a message that if experience is anything to go by, we should pay special attention because these same conditions were, in our own past, associated with salient outcomes. Emotion, even in the form of such vague "intuitions," is designed to guide us away from situations that are statistically likely to be harmful, and toward situations that are statistically likely to be beneficial. For evolutionary psychologists, the fact that our emotions and visceral responses are so automatic and so easily conditioned is a legacy of our past that has been built into our adapted brain and our adapted mind.

Cognition. Important aspects of the behavior of the physical world also seem to be innately wired into, or easily acquired by, our adapted brain. Universally, babies appear to experience anxiety about steep drop-offs as soon as they can see them—without having to learn by experiencing a fall. They also flinch or move away from objects that are getting larger on a projection screen and therefore appear to be coming toward them. Although babies cannot count or do mathematics, they very quickly internalize the fundamental concepts of Newtonian mechanics (length, mass, speed, and gravity), as well as the concepts of more and less, larger and smaller. Without having to be taught, all but the most profoundly retarded of individuals automatically acquires, and communicates about, one of the most abstract concepts of all: the relation between past, present, and future.

Language is another kind of complex and abstract knowledge that is second nature to very young children. Across all cultures and language groups, children progress through a regularized series of stages of language development, acquiring the ability to both understand and produce grammatical speech (or, in the case of deaf children, visual signs). To acquire the 60,000-word vocabulary typical of an eighteen-year-old, a person must learn, on average, ten words a day; at their peak, children acquire several new words an hour. That this is a highly specialized kind of learning that is prewired into our adapted mind is evident by the fact that most adults have to work extremely hard to acquire a second or third language, and if they succeed in doing so, they are rarely as fluent as in their first language. Furthermore, it is exceedingly difficult to "teach" the smartest computers and robots how to understand even very elementary forms of speech, and although other animals can be taught the meanings of certain word or symbol sequences, they are unable to acquire the rules of grammar and syntax that come so naturally to human children. Linguistic skill is a human-specific attribute that develops easily and quickly during a critical period of brain development, and once achieved, it is never forgotten. For evolutionary psychologists, language facility is a paradigmatic feature of our adapted mind.

Another feature of our adapted mind is an innate preference for things that were "good" for our ancestors and a distaste of things that were "bad." We naturally like sweet foods that provide us with the necessary glucose for our calorie-hungry brain and salty foods that provide us with the minerals to run our neuronal sodium pump; yet we have to acquire (and in some cases may never acquire) a taste for bitter foods and for foul-tasting or foul-smelling foods, which signal our brain that they may contain toxins. Even our taste for spices is adaptive: the most commonly used seasonings in cuisines around the world (lemon, garlic, various peppers, and oregano) are those that protect us with their potent antibacterial properties. Our brain also automatically causes us to develop highly ingrained aversions to foods that we ingested several hours before becoming ill. Even in cases when we consciously know that it was not the salad dressing or the broccoli au gratin that actually made us sick, we may still feel nauseous just thinking about the item ten years later. Our adapted mind is watching out for us—motivating us to eat things that we can use and to avoid things that might poison us.

Our adapted mind also demonstrates other kinds of "taste." Children around the world show a decided preference for parklike landscapes that provide plenty of water, flowers, and foliage, and for forms of play that provide exercise, strengthen muscles, and increase physical coordination. Adults particularly admire the beautiful faces and shapely bodies of the young, healthy, and disease-free—those who are the safest to befriend and most profitable to mate with. And people of all ages have a predisposition to pay attention to, admire, and emulate the practices and preferences of successful peers. These aspects of our psychology are obvious to advertisers who today use outdoor scenes, sports, and sex to get our attention, and who pay movie stars and sporting idols to promote their products, but the reason they exist in the first place is because they helped our ancestors to survive, acquire physical and social skills, find a mate, reproduce, and successfully raise children.

Even our facility at logical reasoning—an aptitude that many philosophers consider to be the hallmark of our species and a measure of our mental flexibility—seems to be in some ways programmed. For example, just as we so readily develop emotional "intuitions" about certain kinds of people and places, we also readily develop mental stereotypes. Stereotypes are basically abstract concepts that arise from the mental integration of experience and that are designed to represent a statistically accurate synopsis of reality. Stereotypes are formed by a process that parallels what we call "inductive reasoning," but which occurs at an automated rather than accessible level. The automaticity of stereotypes allows us to respond quickly to new stimuli so that we do not waste precious time assessing each nuance of each new minute of our day. In this way, having a mental stereotype is much like having the concept of "more versus less" or "sweet versus bitter": it is a very gross assessment, but it is often all that is needed.

Deductive logic, too, exhibits features that signal the hand of natural selection. Solution of logic problems involving the classic "if p, then q" form may seem either simplistic or deviously difficult, depending on the specific content of the statements p and q. When problems are worded such that the task is to identify people who may be breaking a social rule (e.g., "If you take cake, then you must pay for it"), the solutions are readily apparent: anyone who takes cake (p) must be checked, and anyone who does not pay ($not\ q$) must be checked. When, however, problems are worded such that the task is to identify inanimate objects that may be breaking an arbitrary physical or numerical rule (e.g., "If a card is even-numbered, then it must go in a green folder"), the solutions prove to be quite elusive: everyone realizes that they must check for even-numbered cards (p), but a large majority fail to realize that they must check cards in the nongreen ($not\ q$) rather than the green (q) folders. That is, despite the identical format of the problems, success is achieved easily only on those that have clear relevance for human social interaction. Our facility for deductive logic is, like our memory and our emotional responsivity, selectively tuned to protect us from situations in which we might be harmed or cheated by other people.

In fact, it might have been the complexity of human social groups and interactions that created the selection pressures that led to our great intelligence. Populations of social species typically have some form of hierarchical organization involving complex, sometimes dynamic, relationships, and youngsters in such groups must learn not simply who is their mother, but who is their mother's friend and who is their mother's enemy, who leads the group and who follows, who dominates and who can be dominated. Furthermore, each individual's status in the group must be continually monitored because relationships may change with age or reproductive status and as enmities and alliances form, dissolve, and reconfigure. Keeping track of all these identities, relationships, and their history requires keen faculties of discrimination, memory, and inference.

Although our highly social primate relatives can do these things to some extent, humans have taken this kind of mental social register to another level altogether. To be able to infer how other people might respond to a contemplated action, and thus to plan or strategize about our own behavior, we rely not only on our knowledge about particular others but also on a "theory of mind." At about age four, children begin to understand that other people have independent thoughts, feelings, and beliefs—that each person has his or her own mind—

and that the contents of each mind are, in fact, different. This critical insight completely changes how the child behaves, for now he or she can attempt to "manage" others by manipulating their feelings and beliefs. It is at this age that human children first start to show sympathy and empathy for others who are in distress, first start to use gifts, promises, and friendship to create alliances, and first start to lie to protect their own interests.

In many ways, our intelligence is really a specialized social intelligence. We devote much of our everyday conscious thought not to how to walk or focus our eyes or do science, but to figuring out how best to deal with others and how to improve our own image and status in their thoughts. We use our amazing gift for language not primarily to discuss philosophy or the structure of matter, but to gossip, seduce, insult, boast, and assign credit and blame. Our television, magazines, books, and newspapers cater to our hunger for knowledge about other people—even make-believe people—over that about other creatures or the physical world. In sum, just as coral polyps have evolved to be natural architects, sheep to be natural lawnmowers, and cats to be natural mousetraps, humans have evolved to be natural psychologists.

Personality. Because predicting other people's behavior is so important to us, any aspect of a person that is consistent and can help us to predict accurately becomes salient. Such relatively stable behavioral predispositions are what we refer to in everyday parlance as "personality." We care about others' personality not only when it comes to important decisions about marriage, politics, and business partnerships, but also when making decisions about everyday social and economic transactions such as with whom to carpool or eat lunch. We have created an expansive lexicon describing intricate nuances of personality, both normal and abnormal, and we use these to label ourselves and others—sometimes for the sake of accurate prediction, and sometimes for more nefarious purposes.

There are many systems that modern psychologists use to discuss variation in personality, but from an evolutionary perspective there are two major dimensions that repeatedly emerge as most important: dominance/submissiveness and friendliness/hostility. Our special attention to these two dimensions of personality seems to highlight once again the importance of social hierarchies, factions, and alliances to human life.

Also salient to us are some consistent differences between the sexes. Boys and men around the world are, on average, more physically aggressive, more competitive, more impulsive, and more risk-prone than girls and women, who are, on average, more nurturant, more empathetic, more cooperative, and more harm-avoidant than boys and men. These differences manifest in multitudinous ways, including differential representation of the sexes in certain activities and careers, greater participation by males in war and both violent and nonviolent crime, and a longer life expectancy for females with higher death rates for males at every age.

Although human politics and sex role relations are vastly more complicated than those of our primate (and other mammalian) kin, the physical and temperamental differences that underlie human sex differences are also found in other species. Male mammals generally have a higher metabolic rate than females, reach sexual maturity at a later age but at a larger size, are more aggressive, and have a greater ratio of muscle to fat than do females; like human males, they have a shorter life expectancy and higher death rates at all ages. Female mammals generally are more sexually discriminating than males, and in addition to being the sole provisioner of nursing young, are typically the sole social parent as well. These taxon-wide sex differences have evolved as a consequence of intrasexual competition among males and gestation-driven selection in females. Although we humans may exaggerate them or mask them with our cultural inventions, they are a fundamental aspect of our evolved nature.

The Maladapted Mind. Psychologists and psychiatrists are now beginning to apply evolutionary theory to human emotion, cognition, and personality to try to understand how our mental adaptations sometimes go awry in modern society, causing psychosocial dysfunction and pathology.

Of the primary emotions, the one that is most easily recognizable in pathological form is sadness. Clinical depression can become so severe as to culminate in suicide—clearly not an adaptive outcome. The symptoms of depression, however, can be seen to be extensions of normal sadness—an emotion that is designed to orient us away from misadventure after having experienced a loss. It is after having experienced a loss, of course, that further losses would be most costly. Thus, depression may serve as a protective buffer when we are at our weakest. The pathology of modern depression is not that it exists, but that we do not regroup and come out of it in a timely and functional manner. The selective process that fine-tuned our emotional responsivity could not have foreseen features of today's existence such as the absence of extended kin support networks and the stress of constantly having to be alert to the behavior and intentions of strangers and modern "institutional entities" (e.g., law, big business, and the stock market). Our fears and anxieties are attuned to an era that is long gone: our (reasonable) fear of steep drop-offs can prevent us from flying, and our (reasonable) anxiety about strangers can give us disabling stage fright and contribute to xenophobia.

Reason itself can also be led astray. As heuristics, stereotypes are designed to work to our advantage, but the "experiences" that now go into our mental computations include not just real experiences that reflect true probabilities but also thousands upon thousands of images of murder and mayhem from movies, newspapers, and television. In fact, the more media a person consumes, the more likely he or she is to have exaggerated fears and inflated estimates of the likelihood of personally experiencing a frequently portrayed tragedy. Similarly, our judgments about our own physical attractiveness and the attractiveness of our partners is radically (negatively) altered by the fact that we are constantly surrounded by (artificially enhanced) images of the most beautiful of the beautiful.

Even personality attributes that may have been adaptive in the past may no longer be so. Aggressive and narcissistic men who are today diagnosed with antisocial personality disorder may have been the successful warrior-kings of old. Harm-avoidant and compulsive women who today succumb to anorexia may have been the best midwives and mothers.

Over a century ago, Charles Darwin predicted that "In the distant future . . . psychology will be based on a new foundation" (*The Origin of Species*, 1859). That time has come.

[*See also* Emotions and Self-Knowledge; Human Families and Kin Groups; Mate Choice, *article on* Human Mate Choice; Parental Care; Primates, *article on* Primate Societies and Social Life.]

BIBLIOGRAPHY

Barkow, J., L. Cosmides, and J. Tooby, eds. *The Adapted Mind*. New York, 1992. A rather technical book that introduced the term *evolutionary psychology* and first defined this research program.

Baron-Cohen, S., ed. *The Maladapted Mind*. East Sussex, England, 1997. Each chapter discusses some form of modern psychopathology from an evolutionary perspective.

Buss, D. M. "Evolutionary Personality Psychology." *Annual Review of Psychology* 42 (1991): 459–491.

Buss, D. M. "Evolutionary Psychology: A New Paradigm for Psychological Science." *Psychological Inquiry* 6 (1995): 1–30. A paper on philosophy and method of evolutionary psychology, followed by peer commentaries and a rejoinder by the author.

Byrne, R. W., and A. Whiten. *Machiavellian Intelligence: Social Expertise and the Evolution of Intellect in Monkeys, Apes, and Humans*. Oxford, 1988.

Crawford, C., and D. L. Krebs, eds. *Handbook of Evolutionary Psychology: Ideas, Issues, and Applications*. Mahwah, N.J., 1998.

Cummins, D. D., and C. Allen, eds. *The Evolution of Mind*. Oxford, 1998.

Damasio, A. R. *Descarte's Error: Emotion, Reason and the Human Brain*. New York, 1994. A premier neuroscientist discusses his compelling work on emotion and cognition.

Eibl-Eibesfeldt, I. *Human Ethology*. New York, 1989. A comprehensive and detailed documentary of human behavior in comparative perspective. Includes many classic photos and diagrams.

Frank, R. H. *Passions within Reason: The Strategic Role of the Emotions*. New York, 1988. An evolutionary economist shows how and why humans are not simply "rational" beings.

Gaulin, S. J. C., and D. H. McBurney. *Psychology: An Evolutionary*

Approach. Upper Saddle River, N.J., 2001. An excellent textbook; the most reader-friendly item on this list.

Gazzaniga, M. S., ed. *The Cognitive Neurosciences.* Cambridge, Mass., 1995. This advanced specialty encyclopedia includes several entries on evolutionary psychology, including those by Premack and Premack (social awareness and theory of mind), Brothers (social awareness and theory of mind), Tooby and Cosmides (functional organization of the brain/mind), Cosmides and Tooby (cognition as mental computation), Gaulin (sex differences in cognition), Preuss (comparative cognitive neuroscience), Gallistel (cognitive modularity), and Daly and Wilson (parental motivations).

Griffiths, P. E. "Modularity, and the Psychoevolutionary Theory of Emotion." *Biology and Philosophy* 5 (1990): 175–196.

Hrdy, S. *Mother Nature: A History of Mothers, Infants, and Natural Selection.* New York, 1999. A best-seller by a premier anthropologist focusing on human and nonhuman primate mother–infant interactions from an evolutionary perspective.

Irons, W. "How Did Morality Evolve?" *Zygon: The Journal of Science and Religion,* 26 (1991): 49–89.

Katz, L., ed. *Evolutionary Origins of Morality: Cross-Disciplinary Perspectives.* Essex, England, 2000. Four papers on the evolution of morality, each followed by a series of expert commentaries and a rejoinder by the author. Reprinted in its entirety from a special issue of the *Journal of Consciousness Studies* (7.1–2 2000).

Lamb, M. E., R. A. Thompson, W. P. Gardner, E. L. Charnov, and D. Estes. "Security of Infantile Attachment as Assessed in the 'Strange Situation': Its Study and Biological Interpretation." *Behavioral and Brain Sciences* 7 (1984): 127–171. A technical review on attachment followed by a series of expert commentaries and a rejoinder by the authors.

MacDonald, K. B., ed. *Sociobiological Perspectives on Human Development.* New York, 1988. Includes chapters on adolescence as a life stage, development of cognition, and family interactions.

Mealey, L. *Sex Differences: Developmental and Evolutionary Strategies.* San Diego, Calif., 2000. Compares human and nonhuman behavioral and psychological sex differences from an evolutionary perspective.

Nesse, R. M. "Evolutionary Explanations of Emotions." *Human Nature* 1 (1990): 261–289. A lucid presentation by an evolutionary psychiatrist; hard to find, but worth reading.

Panksepp, J. *Affective Neuroscience.* New York, 1998. Advanced text intergrating comparative neuroscience with theories of emotion.

Pinker, S. *The Language Instinct.* Cambridge, Mass., 1994. The best of the books on the evolution of language.

Scheibel, A. B., and J. W. Schopf, eds. *The Origin and Evolution of Intelligence.* Boston, 1997. Written at a somewhat less technical level than *The Cognitive Neurosciences* (above), this collection is also shorter and restricted to evolutionary angles, but authors and topics overlap. Includes Seyfarth and Cheney on monkey minds, Savage-Rumbaugh on ape language, Cosmides and Tooby on modularity, and Pinker on language.

Simpson, J. A., and D. T. Kenrick, eds. *Evolutionary Social Psychology.* Hillsdale, N.J., 1996. Many chapters discuss cognition in the context of social relationships.

Wilson, M., and M. Daly, "Competitiveness, Risk-taking, and Violence: The Young Male Syndrome." *Ethology and Sociobiology* 6 (1985): 59–73. A classic discussion of male risk taking from an evolutionary perspective.

— LINDA MEALEY

Behavioral Ecology

Human behavioral ecology is the study of how human behavior can be understood as an adaptive response to an individual's environment. Central to this approach is the hypothesis-driven framework that uses models derived from neo-Darwinian theory. Behavioral strategies are assumed to be adapted to maximize individual fitness. Formal hypotheses are erected to explain behavior based on the costs, benefits, and constraints associated with each behavioral option in a given environment. Costs and benefits are measured in currencies—such as energy intake—that are presumed to be related to fitness. Hypotheses can be tested by examining individual variation in behavior; sometimes the variation across different human cultures can also be used.

This emphasis on explaining the variability in human behavior and using formal optimality models to pose hypotheses to be tested contrasts with the approach of either classical sociobiology or evolutionary psychology, in which more emphasis is placed on human universals. Understanding the details of causation (e.g., psychological mechanisms) tends to be secondary to the task of ascertaining whether behavior is following the predicted pattern. Possible constraints of a genetic, phylogenetic, or cognitive nature are generally not explicitly considered—the power of evolution to reach an adaptive solution is considered strong. This approach, which is sometimes described as reductionist, helps to reduce the daunting complexity of some problems in the evolution of human behavior and thus generate testable predictions. The powerful combination of an empirical tradition and a theoretical framework is new to most social sciences, and the framework of evolutionary ecology has now been fruitfully applied to several areas of human behavior.

Foraging and Resource Exploitation. Formal optimal foraging models have been used to understand the dietary choices of traditional populations, particularly of hunter-gatherers. The most widely used model is known as the diet breadth model, which calculates the varying rates at which foragers acquire nutrients depending on which prey items they pursue, and hence predicts which prey should and should not be pursued. Some use has been made of patch choice models, which show that traditional hunters may leave areas when prey densities are low, not because they are conservationists, but because it pays to move on to more profitable areas. All optimal foraging theories specify particular currencies in which to measure fitness-related benefits, usually the rate of energy intake. They specify a particular goal— often maximizing the long-term mean, and particular trade-offs and constraints, for example, time spent searching for prey versus time spent pursuing and processing them. [*See* Optimality Theory: Optimal Foraging.] Most optimal foraging studies on animals show that they make prey choices that serve the goal of maximizing mean energy intake rate, not other plausible currencies such as energetic efficiency (i.e., the ratio of energy intake to energy expended).

The diet breadth model has proven to be a good predictor of human forager diet, but studies reveal that males, females, and children generally make different choices. These differences may largely arise from different constraints rather than different currencies, for example, children's walking speeds and acquisition rates may differ from those of adults, so they maximize their energetic returns within those constraints. In most foraging societies, men tend to hunt and gather, whereas women concentrate exclusively on gathering. This likely arises from women being constrained by the demands of caring for children. However, calculations of the energy intake rate of Ache foragers in Paraguay suggest that although women are maximizing the energy intake rate by gathering, men's returns from hunting are lower. This may be because the protein and iron content of meat give it a nutritional value that exceeds that calculated from energy content alone. But this finding has stimulated alternative hypotheses, such as that men's foraging may often aim at a different goal. They may be hunting to show off and impress women, as well as other men, at the expense of some calorific returns. Aspects of both explanations may be correct, in that by obtaining a valuable nutritional resource that women cannot themselves easily acquire (i.e., meat), men gain the potential to trade meat for sex, and hence higher reproductive success.

Most foragers share the catch among the hunters or among the whole camp. A range of explanations from kin selection, tolerated theft (large prey items are hard to defend from group members and cannot be easily stored), to simply reciprocal altruism may apply. However, certain successful hunters do appear to contribute a disproportionate amount to the group, which remains a puzzle. The high status that results might underly their apparent generosity. It is clear the humans evolved in small groups that probably had frequent aggressive, competitive interactions with other groups—conditions that may have favored the evolution of collaborative coalitions between men. Game theory has been useful in attempting to understand how such apparently altruistic behavior might emerge.

Foraging strategies will often depend on the state of the decision maker, such as their level of resources or risk of starvation. Dynamic optimality models that allow decisions to be contingent on the state of the decision maker have been useful in understanding the resource choices of farmers and herders. Gabbra camel herders in northern Kenya keep camels, which are drought resistant but slow breeders, and also goats. The species

that would maximize energy intake rate is goats, but they are risky, because a drought could reduce the herd to a level below that needed to support the family. State-dependent, dynamic optimality models can predict the proportion of the herd that could be kept to maximize long-term household survival, and how that optimum varies with total herd size. This approach can be extended to examine the circumstances that predict major transitions from one mode of subsistence to another. [*See articles on* Human Foraging Strategies.] In theory, trade-offs between two or more very different behaviors (e.g., seeking mates, seeking food, and caring for children) could also be examined using these models, as long as all behavioral options can be measured in a common currency.

Mating and Marriage. Most of the behavioral ecological studies of human mating and marriage patterns have examined how control of resources influences mate choice. In hunter-gatherer societies, there is little "ownership" as such, so prowess at hunting appears to be a good determinant of male reproductive success. In farming and herding societies, land or animals are usually owned, and males, in particular, can use that ownership to attract and control mates. The resource holding potential model of polygyny predicts that women will marry polygynously only when a share of a wealthy man's resources exceeds exclusive access to the resources of a poor, but monogamous, man. This model has been tested in pastoralist and agropastoralist societies. As predicted, males with large holdings of land or livestock attract more wives and thus have higher reproductive success, whereas the variation in female reproductive success is lower. However, there is some evidence that the children of polygynously married women may not always fare as well as those from monogamous unions, a finding inconsistent with this female choice model of polygyny.

At the opposite extreme, where good farming land is extremely limited, as in Tibet, the rare, polyandrous marriage system may arise, in which several brothers share one wife, because the family farm could not support more than one woman's children.

The strength of the marriage bond shows great cultural variation. In farming and herding societies, patrilineal control of resources (wealth is passed from father to son) can limit women's freedom to leave marriages, as they would have to abandon both their means to live and their children in the case of divorce. A significant minority of cultures show matrilineal control of resources (wealth tends to pass from mother to daughter, or from a man to his sister's sons). How such societies arose is not clear, although they tend to be associated with farming systems where much labor is required and tend not to be associated with pastoralism. It may be that in such circumstances land is not worth defending, whereas livestock, in contrast, is a valuable resource worth fighting for and passing on to your sons. However matriliny arose, marriage bonds in matrilineal societies are relatively weak—men and women tend to show serial monogamy or polygamy. Women live and work on their farms with their mothers, sisters, and daughters, so the woman's family economy is less disrupted by marital changes than it would be in a patrilineal system.

Parental Investment. Parental investment is anything that a parent invests in offspring that benefit that offspring at some cost to the parent's future reproduction. As such, it can range from a long interval to the next birth, to paying for a child's schooling. It is closely linked with marriage and mating opportunities, especially for men. The likelihood that a man will invest in his children is closely related to the stability of marriage. When the benefits of new breeding opportunities exceed the benefits of paternal care of offspring, males are more likely to leave their existing children and enter into new marriages. In addition to the gains from pursuing other mating opportunities, men can rarely be certain of their paternity. Paternity uncertainty means that, on balance, males may be less inclined to invest in offspring as heavily as females, as they are on average slightly less related to them. There is evidence from modern Western societies that child abuse is closely linked with paternity uncertainty and marital instability. Stepfathers have been associated with increased risk of harm to children across both traditional and modern societies, although this does depend on costs and benefits. In many cases, care directed toward a new mate's existing children may be an essential part of maintaining a new partnership. Mothers also occasionally harm children from marriages that have ended. In such cases, infanticide could enhance her prospects of remarriage or terminate investment in a child that is unlikely to be reared successfully without paternal help.

Although overall parental investment in sons and daughters is often about equal, sex-biases in parental investment may arise when the costs and opportunities for children vary with sex. Nutrition in the womb can have long-term influences on body size, which can effect reproductive success in later life; there is evidence that tall stature increases men's attractiveness and reproductive success, and that low birth weight males are especially likely to die. If low birth weight is more harmful to a male's than a female's future reproductive success, then mothers in relatively good condition should bear more sons. However, as with other mammals, human sex ratios at birth do not deviate much from 50:50. Most sex biases in parental investment occur after birth.

Parents can increase the success of their children in the marriage market by endowing sons with bride price (paid by the groom's family to the bride's) or daughters with dowry (paid by the bride's family to the groom's).

From evolutionary principles, we would expect bride price to be seen when there is competition between males for females and dowry when there is competition between females for males. This is broadly borne out. Bride price is more common where there is resource-holding polygyny (societies in which there is no real advantage gained by a woman by marrying a wealthy man because she is likely to become one of many wives). Where monogamy is socially imposed, as in Western and some Asian cultures, there are clear advantages to females marrying wealthy men, because they and their children are the sole beneficiaries of that wealth. In these societies, female competition is intense and dowry is common.

These marriage costs can be a major part of raising a child. Strong sex preferences can emerge when these costs are asymmetric between the sexes. This then influences other areas of parental investment. For example, female infanticide has been associated with high dowry costs in some Asian societies (i.e., parents with daughters will suffer high costs when those daughters marry), and higher male childhood mortality is found in matrilineal areas of southern Africa. In poor families, the opportunity for girls to "marry up" the social scale often exceeds that of boys, so girls may be given more food, medical care or attention than boys. Sex preferences can influence interbirth intervals; parents without a child of the desired sex have been shown to move on to the next pregnancy faster. Female-biased infanticide historically was not uncommon in several hunter-gatherer populations; it can arise when female's marriage prospects are limited by a heavy reliance on male hunting for food. This can lead to ecologically imposed monogamy and high adult male mortality, such as among the Inuit of North America.

Rapid reproduction is associated with higher child mortality. However, there is not much evidence that this is maladaptive to the mother. Faster breeding mothers are generally those found to have the highest reproductive success, despite possible higher rates of infant mortality; mothers of twins are typically the most fertile of all. A study of birth spacing in the nomadic !Kung foragers of southern Africa is one of the few that found evidence that a four-year birth interval (which is long by human standards) is adaptive, and explained it as enabling the mother to carry large weights of food and the child itself for long distances. However, other researchers have attributed this long birth interval to sexually transmitted disease, and still others have argued that differences in the trade-offs faced by different women were not suitably accounted for in this study. The recent decline in fertility around the world, known as the demographic transition, could be understood as an example of hyperinvestment in individual children leading to very low fertility; however, the balance of evidence suggests that this is actually maladaptive. [*See* Demographic Transition.]

Life History. Life history concerns the timing of life events and how this can be understood in the context of trading off energy put into growth and avoiding mortality, on the one hand, against reproduction, on the other. Thus, it concerns patterns of development, age at first reproduction, rate of giving birth, and aging. An example of a trade-off might be that later age at first reproduction is associated with taller stature in females. Although this may temporarily reduce a female's reproductive success relative to her peers when in her teens there is evidence that it may be beneficial to her reproductive success in adulthood.

There is considerable interest in how the human life history differs from that of the other great apes. Compared to our closest living primate relatives, childhood is long and puberty is accompanied by a growth spurt. Humans reproduce with birth intervals unusually short for average body size (three years, compared to eight years for orangutan females, which are of similar weight). Humans can also live much longer than any other primate. Chimpanzees that live to the age of fifty years display indications of menopause, but chimpanzees very rarely live that long. General health and vigor and extended survival after menopause appear to be uniquely human—twenty years of postreproductive life is common even in traditional societies, and nothing comparable has been observed in other primates. Many of these unusual characteristics have been attributed to our large brains and social lifestyle. Childhood may be necessary to learn how to use our mental abilities to the full, or, alternatively, our long childhood may be an aspect of other life history adjustments.

The high human birth rate is almost universally attributed to the division of labor in human society. Nonhuman primate females have to rear offspring alone, but human females are fed by males and may also be fed and assisted with child rearing by matrilineal kin. The relative importance of male versus matrilineal help probably varies from culture to culture. Among Ache foragers in Paraguay, male help is most important, whereas among Hadza foragers in Tanzania, female kin are important. Female menopause substantially precedes age-related termination of other aspects of physiology. It has been argued that menopause evolved because in human ancestors the benefits to a female of caring for grandchildren exceeded the benefits of reproducing in older age. Fitness benefits to children with grandmothers have been shown among Mandinka farmers in Gambia, where maternal grandmothers did improve the survival of toddlers; but models suggest this effect is not large enough to favor menopause. Other models propose that the rate of ovarian senescence has not changed much in our lineage (i.e., humans do not differ much from other apes on

this measure), whereas our rate of somatic senescence has slowed. These models require that the effect of grandmothers' help be great enough to maintain our longevity. However, the details of life history modeling are complex, with new models appearing all the time, so this question is far from resolved.

Problems and Prospects. The discipline draws on the same theoretical framework and many of the same methodological tools that are used by behavioral and evolutionary ecologists studying animal behavior. However, there are major distinctions between human and animal behavioral ecology that are both theoretical and methodological. Human behavioral ecologists have to incorporate culture far more than do animal behaviorists. The cultural rules of the society in which one lives are likely to be a very important feature of the environment. The approach of human behavioral ecologists, in contrast to some other disciplines examining the evolution of human behavior, is generally to consider culture as it would any other aspects of the environment. All behavioral ecologists consider aspects of behavior that are at least partially learned, and the extent to which behavioral responses are learned is not directly relevant to testing whether or not they maximize fitness. Genetically and culturally determined traits can both show great phenotypic plasticity—variation that is a conditional response to environmental circumstances. Behaviors that are entirely cultural in origin, such as the amount of money parents give in bride price to enable offspring to marry, have been shown to fit patterns that maximize reproductive success. Models exploring how cultural traits are transmitted through populations, and how they coevolve with genes, have produced theoretical scenarios in which genetically maladaptive cultural behavior can arise. [*See articles on* Cultural Evolution.] However, clear empirical examples of behaviors that fit such models are lacking, weakening the case that such circumstances are common. Even if extraordinary cultural systems do emerge, the behavioral strategies of individuals within those systems can often be understood as maximizing fitness within the constraints imposed by the system. Hence, aspects of culture are considered both as part of the environment to which behavior is adapted and as adaptive strategies in themselves.

Human behavioral ecologists face the methodological problem that it is often not possible, for ethical and practical reasons, to conduct carefully controlled experiments of the kind that have been performed in studies of animals. Particularly in traditional societies, behavioral decisions really do influence the health and well-being of people, so natural variation in human behavior is frequently all that can be used to infer the nature of people's responses to different conditions. This means that some arguments about the causation or currencies used in human decision making cannot be easily resolved. Particularly in the case of life history variation, it can be hard to untangle whether observed trends are phenotypic correlation (trends attributable to the different states of individuals', such as their body condition) or responses of individuals of equal phenotypic condition facing different costs and benefits. For example, life history theory predicts that females who raise many children will experience greater costs than those who raise only a few children, probably paid for by a shorter life span. In birds, the number of eggs in a nest can be increased or decreased experimentally, and where this manipulation has been done, future fitness costs associated with raising larger broods (including earlier maternal death) have been found. In humans, at best only weak correlation between female longevity and the number of offspring have been found, but this is probably because only those women who are phenotypically the fittest are producing the largest families. We cannot manipulate family size, so it is hard to estimate exactly what the costs of raising a large family really are. Although the inability to conduct such experiments can raise formidable problems, the field is helped by other characteristics of humans not found in animals relating to the detail of information that can often be obtained. In particular, people's own descriptions of their strategies and life histories, or the existence of historical demographic, legal, or medical records, can provide valuable data of a quality that could only ever be obtained in animal studies by years or decades of continuous observation.

Behavioral ecology not only has introduced fundamental differences in approach from the other human sciences but also has highlighted new problems in the field of understanding human behavior. Some of these problems have been solved, others have simply alerted us to novel and often important alternative explanations. Marital stability, child abuse, and rates of fertility are some of the areas relevant to human health and well-being, to which human behavioral ecology has made significant contributions to our understanding.

[See also Human Families and Kin Groups; Human Foraging Strategies, articles on Subsistence Strategies and Subsistence Transitions and Human Diet and Food Practices; Mate Choice, article on Human Mate Choice.]

BIBLIOGRAPHY

Betzig, L., ed. Human Nature. New York, 1997. A reader containing several classic articles in human behavioral ecology, including key studies on child abuse by stepparents and hunter-gatherer interbirth intervals.

Borgerhoff Mulder, M. "Reproductive Success in Three Kipsigis Cohorts." In Reproductive Success: Studies of Individual Variation in Contrasting Breeding Systems, edited by T. Clutton-Brock, pp. 419–435. Chicago, 1988. Study of polygyny in agropastoralists.

Cronk, L., N. Chagnon, and W. Irons, eds. Adaptation and Human

Behavior: An Anthropological Perspective. New York, 2000. Collection of current work across a range of fields in human evolutionary ecology.

Daly, M., and M. Wilson. Homicide. New York, 1988. Classic book on sociobiology of homicide, including infanticide.

Hartung, J. "Matrilineal Inheritance: New Theory and Analysis." Behavioral and Brain Sciences 8 (1985): 661–688.

Hawkes, K. "Why Do Men Hunt? Some Benefits for Risky Strategies." In Risk and Uncertainty in Tirbal and Peasant Economies, edited by E. Cashdan, pp. 145–166. Boulder, 1990. Proposes that male foragers hunt to show off.

Hawkes, K., J. F. O'Connell, N. G., Blurton Jones, H. Alvares, and E. L. Charnov. "Grandmothering, Menopause and the Evolution of Human Life Histories." Proceedings of the National Academy of Sciences USA 95 (1998): 1336–1339.

Hill, K. "Life History Theory and Evolutionary Theory Anthropology." Evolution and Anthropology: 2 (1993): 78–88. Review.

Hill, K., and M. Hurtado. Ache Life History: The Ecology and Demography of a Foraging People. New York, 1997. A unique account of the life history of a foraging group from an evolutionary ecological perspective.

Mace, R., and A. I. Houston. "Pastoralist Strategies for Survival in Unpredictable Environments: A Model of Herd Composition That Maximises Household Viability." Agricultural Systems 31 (1989): 185–204. Use of dynamic optimality modeling to predict herder decision making.

Sear, R., R. Mace, and I. A. McGregor. "Maternal Grandmothers Improve Nutritional Status and Survival of Children in Rural Gambia." Proceedings of the Royal Society of London B 267 (2000): 1641–1647.

Smith, E. A., and B. Winterhalder. Evolutionary Ecology and Human Behavior. New York, 1992. Textbook.

Strassman, B. I. "Polygyny as a Risk Factor for Child Mortality among the Dogon." Current Anthropology 38 (1997): 688–695.

Voland, E. "Evolutionary Ecology of Human Reproduction." Annual Review of Anthropology 27 (1998): 347–374. Review, reporting many studies on human parental investment.

Winterhalder, B., and E. A. Smith. "Analyzing Adaptive Strategies: Human Behavioural Ecology at Twenty-five." Evolutionary Anthropology 9 (2000): 51–72. Review.

— RUTH MACE

HUXLEY, THOMAS HENRY

The English biologist Thomas Henry Huxley (1825–1895), dubbed "Darwin's bulldog," was a prominent champion of evolution. Known for his witticisms, Huxley is celebrated in partisan accounts for rebuffing Anglican Bishop Samuel Wilberforce at Oxford University in 1860 (where Huxley, legend has it, said he would rather have an ape for an ancestor than a bishop who prostituted his gifts). But this one-dimensional reminiscence belies the importance of Huxley's life. Even his "evolution versus Creation" polarization, which shaped a century of debate, was symptomatic of deeper social events. In Victorian times, the rising nonconformist industrialists were usurping church authority. Huxley capitalized on this movement; he championed the new laboratory researcher as a cultural leader who could undermine the miraculous stays of the old-style Tory-Anglicanism opposed to reform. He coined the word *agnosis* in 1869, and made agnosticism, the idea that the existence in God is unknown and probably unknowable, integral to his Darwinian proselytizing. By contrasting naturalistic science to Genesis, as taught by the pulpit power holders in English society, he helped define the rival professionalizing drive in science.

Huxley had not always supported evolution. As assistant surgeon on the HMS *Rattlesnake*'s voyage to Australia and New Guinea (1846–1850), he had made his name studying coelenterates, showing that they were composed of two membranes, later called ectoderm and endoderm. For this he received the Royal Society's Royal Medal in 1852. Two years later, he panned the tenth edition of the anonymous evolutionary *Vestiges of the Natural History of Creation* so savagely that one critic later wondered how Huxley could jump so effortlessly from rejecting evolution to accepting a bestial ape ancestry. In truth, Huxley caviled less at the evolution in *Vestiges* than at the book's use of divinely driven laws to accomplish it. By contrast, the naturalism (or rejection of miraculous mechanism) of Charles Darwin's *Origin of Species* (1859) suited Huxley's secular ends.

Huxley, teaching at the School of Mines in London, first visited Darwin at Downe in 1856. Darwin discussed his project, and immediately Huxley returned to his classroom to lampoon supernatural creation. He quickly pushed into the human arena (although Darwin had decided to leave humans out of the *Origin*). By 1858, Huxley was arguing at the Royal Institution that humans scarcely departed more from gorillas than gorillas did from baboons, anatomically speaking. Further inflaming Tory sensibilities, he informed the working classes in 1859 that humans had emerged from animals, playing up the lowly-origins-to-lofty-future scenario that appealed to the underclass (with no natural Anglican Oxbridge constituency, Huxley had to canvass widely for his support).

His rousing defenses of Darwin's *Origin of Species* actually did scant justice to natural selection. Huxley preferred some inner fount of variation that could generate new species at once. But for allowing the secular intellectual to rival the pulpit's power, he hailed the book a "Whitworth gun in the armoury of liberalism"; Huxley's attack on theology always had a social dimension. His retort to Wilberforce's ribaldry at Oxford in 1860—was the ape on his grandfather's or grandmother's side?—contrasted the earnest Puritan of science with purple-vested church frivolity. Science was made to look more serious in an increasingly straitlaced age. Whatever was said, and no one could quite remember, the event became a propagandist coup in the Darwinians' retelling.

Never were two naturalists more complementary than the "plebeian" Huxley and gentlemanly Darwin: one needed a champion, the other a cause. Huxley's cause

was a "new Reformation," and Darwinism allowed him to ask taboo questions about human origins and the meaning of morality. With his wife, Henrietta, recuperating at Darwin's home in 1861, after the death of their son Noel, Huxley challenged Richard Owen's claim that human brains were unique. Their cerebral hemispheres extend back into a new lobe with a singular structure on its floor, the hippocampus minor; or so Owen said, and he erected a subclass solely for humans on its basis. Huxley insisted that apes also possessed a hippocampus minor. In newspapers, Owen tried to discredit Huxley as an "advocate of man's origin from a transmuted ape," thereby setting the evolutionary terms of the debate. Nothing brought human evolution so sharply into public focus. Huxley, having taken over the *Natural History Review*, ran articles proving Owen wrong. Simultaneously, Huxley examined the recently discovered Neanderthal man. With its sloped forehead and prominent browridges, this was the only known human fossil to depart from modern anatomy, although he proved that it was not the "ape man" that some suggested. The result was *Evidence as to Man's Place in Nature* (1863), the first post-Darwinian book to concentrate on human–ape relations.

Huxley turned to paleontology. From Britain's coal mines came Carboniferous amphibians—crocodile-shaped labyrinthodonts—and he had named eleven new genera by 1865. He called the Devonian lobe-fin fishes "Crossopterygians" and suspected that they too had been air breathers. Although these had often been confused with labyrinthodont fossils, Huxley made no evolutionary connection. Much of his academic work in fact remained in the morphological mainstream—structural description was his forte. However, by 1867, he was influenced by Ernst Haeckel's speculative "phylogenies" (Haeckel's neologism), or ancestral pathways. Huxley began reclassifying birds according to their presumed derivation. More spectacularly, in popular talks from 1868, he suggested that ostrichlike ratites had evolved from small dinosaurs such as *Compsognathus*. He accepted that many dinosaurs were themselves bipedal: whereas Owen had reconstructed them like rhinoceroses, Huxley reinvented dinosaurs as the "Ornithoscelida" ("bird leg"). He even floated the idea that *Compsognathus* might have been feathered like its descendants. But the "demonstrative evidence" of evolution, he said during a tour of the United States in 1876, came in the form of the Yale paleontologist Othniel C. Marsh's fossil horses from Nebraska. Huxley's chart of horse ancestry, from today's single-toed steeds back to their dog-sized three-toed ancestors—published in *American Addresses* (1877)—became the paradigm fossil proof for evolution.

In the first article on "Evolution" in the *Encyclopaedia Britannica*, which Huxley wrote in 1878, he mentioned natural selection only once. He was less interested in evolution's "physiological causes" than its philosophical import. He did not even discuss evolution in the classroom. That was because he had started training educators in his laboratory, relocated to South Kensington in 1871; he was kick-starting science in what had been Classics-based schools, and he restricted his course to a simplified biology, based on discrete animal and plant types, that the teachers could take back to their classrooms. All the same, the words made fashionable by Huxley from 1870—*evolution, biology,* and *agnosticism*—show his success in investing nature with new meaning. Evolution was to be naturalistic, in contrast to the images propagated by rival Darwinians such as the Catholic St. George Mivart and the spiritualist Alfred Russel Wallace. It was to be the prerogative of trained biologists, who were agnostic on moral issues—they were the neutral professionals. Huxley was fashioning the late Victorian image of the scientist.

Not that that stopped him spending his retirement in theological controversy, decrying the "sin of faith." His pronouncement on *Evolution and Ethics* in 1893 was a final denial of a weak-to-the-wall Darwinism in the ethical realm. Rather, humans had created a caring society, where a wall was built around the weak. Our ethics were undeniably a product of natural processes, but he left the problem of explaining them to the twentieth century. *Evolution and Ethics,* Huxley admitted, was as much a political statement. It was his attempt to strike a middle way between the socialists who denounced Darwin's competitive Malthusianism and Herbert Spencer's laissez-faire extremists, who refused to sanction state welfare. By dissociating ethics from a heartless Darwinism, Huxley pushed evolution further into the liberal political mainstream.

Huxley was the foremost publicist for a naturalistic worldview, based on the supremacy of science, evolution, and agnosticism, in the English-speaking realm. He died in 1895, as he would have wanted, in mid polemic, defending his naturalistic principles.

[*See also* Darwin, Charles; *articles on* Hominid Evolution.]

BIBLIOGRAPHY

Barr, A. P. ed. *Thomas Henry Huxley's Place in Science and Letters: Centenary Essays.* Athens, Ga., and London, 1997. Essays commemorating the centenary of Huxley's death.

Bibby, C. *T. H. Huxley: Scientist, Humanist and Educator.* New York, 1960. Older but valuable study of Huxley the teacher.

Desmond, A. *Huxley: From Devil's Disciple to Evolution's High Priest.* Reading, Mass., 1997. Full-blooded biography explaining Huxley's evolutionary and agnostic propagandism in its Victorian setting.

Desmond, A. *Archetypes and Ancestors: Palaeontology in Victorian London, 1850–1875.* Chicago and London, 1984. Huxley's use of fossils in the evolutionary debate.

Di Gregorio, M. *T. H. Huxley's Place in Natural Science.* New Haven and London, 1984. The internal workings of Huxley's biology, with a detailed bibliography of Huxley's published writings.

Huxley, L., ed. *Life and Letters of Thomas Henry Huxley.* 2 vols. London, 1900. Original filial "life," essential but dated.

Jensen, J. V. *Thomas Henry Huxley: Communicating for Science.* Newark, Del., 1991. Manichean approach to Huxley's often anti-Anglican rhetoric.

Lyons, S. L. *Thomas Henry Huxley: The Evolution of a Scientist.* Amherst, N.Y., 1999. Huxley as the Victorian morphologist.

Paradis, J. G., and G. C. Williams, eds. *Evolution and Ethics.* Princeton, 1989. On Huxley's denial of a Darwinian basis for ethics.

ONLINE RESOURCES

http://aleph0.clarku.edu/huxley/. The Huxley File. Created by Charles Blinderman and David Joyce at Clark University. A vast library, comprising 680 articles, letters, and books published by Huxley (and some unpublished), 148 illustrations, and 120 commentaries. Well laid out, with a full bibliography by James Paradis (based on Mario di Gregorio's published bibliography) and four indexes.

— ADRIAN DESMOND

HYBRID ZONES

A hybrid zone occurs where distinct genetic forms meet, mate, and produce genetically mixed offspring. Such zones may be found between species, subspecies, races, and forms, and they may vary in width, length, and fragmentation. Typically, they are relatively narrow and divide a species distribution into a patchwork of parapatric races and subspecies.

Many Examples. There are now several hundred examples in the literature. They are reported particularly in birds, mammals, salamanders, frogs, fishes, butterflies, grasshoppers, and plants. However, the scientific cover is uneven both taxonomically and geographically. The advent of genetic methods for chromosome, protein, and DNA analysis, in addition to traditional morphological and behavioral characters, has shown that hybrid zones can cryptically subdivide the genome of a species across its range into a patchwork of genetic forms.

Birds with distinct plumage provide classic examples of hybrid zones. In Europe, the eastern hooded and western carrion crows hybridize down the center, while in North America the eastern yellow and western red-shafted flickers hybridize in the Great Plains. A number of recent multifaceted studies in small mammal, bird, reptile, amphibian, insect, and plant species provide instructive case studies. Such investigations provide evidence on the mechanisms maintaining hybrid zones and on the forces concerned in their origins. The following will serve as an exemplar.

Chorthippus parallelus, the meadow grasshopper. This species is widespread across Europe and in

Asia, with several subspecies distinguished morphologically (Figure 1). A hybrid zone exists along the Pyrenees between *C. p. parallelus* from north and central Europe and *C. p. erythropus* from Iberia. These two subspecies differ in many morphological and behavioral characters, including color, stridulatory pegs, and song, as well as in chromosomal and molecular ones. They meet and hybridize along the mountain divide in passes and hills below 2100 meters. The zone is fairly narrow but varies in width from place to place. The cline width, in which a character changes gradually from pure *parallelus* to pure *erythropus*, varies among characters (e.g., that for an X-chromosome C-band is <2km, while that for song syllable length is >20km). Most characters have broadly coincident clines, but a few are displaced from the center (Figure 2). These details vary among transects in different places across the Pyrenees, which reflects the complex interaction of many genes hybridizing over many generations in different habitat structures.

When the two subspecies are crossed in the laboratory, the F1 hybrid males are sterile owing to dysfunctional testes, while the F1 females are fit and fertile. This allows gene flow in the zone through female back-

crosses and derived crosses. Most hybrid zones that have been analyzed in detail show some form of hybrid unfitness or sterility, and the balance between gene flow by dispersal of parental forms into the zone and selection against hybrid genomes determine the width of the zone. These are called *tension zones*.

Although *parallelus* and *erythropus* meet along a major mountain ridge with somewhat different environments and biota on either side, they are not specially adapted to or limited by these differences. As its name suggests, the species's habitat is temperate meadows, which are patchily distributed along valleys, over hills, and across plains. For this species, the Pyrenees do not pose an ecotonal divide, and its habitat is not distributed as a mosaic of alternative types. However, for some hybrid zones the close adaptation of their parapatric taxa to particular habitats is important for their structure and location.

The two *Chorthippus* subspecies have somewhat divergent mate recognition systems, with differences in the male calling and courtship songs, and in cuticular hydrocarbon composition, which provides olfactory signals. Individuals from pure populations show significant

FIGURE 1. The Postglacial Colonization Routes of the Meadow Grasshopper, *C. parallelus*, as Deduced from DNA Similarities (Path of Arrows).
At least five major southern refugia existed and hybrid zones (heavy dashed lines) have formed between the distinct genomic races from these.   Drawing by Godfrey M. Hewitt.

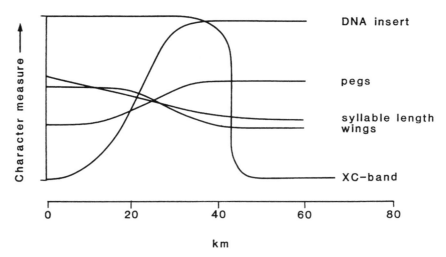

Figure 2. Clines in Some Characters in *C. parallelus* That Change from North to South Through the Col de Portalet in the West Pyrenees, Including a DNA Insertion, an X-Chromosome C-Band, Stridulatory Pegs, Wing Ratio and Song Syllable Length. Most are coincident with varying widths.   Drawing by Godfrey M. Hewitt.

levels of assortataive mating when mixed in laboratory experiments. The development of such pre-mating isolation is part of the process of speciation.

Recent studies of nuclear and mitochondrial DNA sequences in *C. parallelus* across Europe have shown that the *parallelus* and *erythropus* from north and south of the Pyrenees form distinct clades of some age. Furthermore, they reveal distinct genomes in Iberia, Italy, Greece, Turkey, and the Balkans (Figure 1). The rest of the northern European population is similar to and apparently derived from the Balkan genome. The diverged Italian and northern genomic races have subsequently been shown to form a strong hybrid zone along the Alps where they meet, and it seems likely that other such morphologically cryptic zones exist in the southern parts of the range.

Types of Hybrid Zones. The preceding study demonstrates that many character differences are involved in the hybrid zone, and their controlling genes introgress various distances into the other genome. Hybrids are less fit than parental types, which also have divergent mate recognition systems. Along with other case studies, three main models for hybrid zones are apparent.

Tension zone model. Where two diverged genomes abut, individual dispersal will carry alleles into the other form, and they will mix progressively over generations, producing wider clines. If, however, the heterozygotes or hybrids are less fit, then this will reduce their frequency and produce a balance between dispersal (δ) and selection (s). The sigmoidal cline produced has a width $w \propto \delta/\sqrt{s}$. This is a pure tension zone where the genotype selection does not depend on the environment.

Ecotonal zone model. An ecotone exists where two different environments meet. Where each parental taxon

is better adapted to one habitat, allele frequencies will change across the hybrid zone between them. The balance of this environmental selection and dispersal will again produce clines. The mathematics of both types of clines produced by dispersal and selection has been thoroughly investigated and can help to distinguish them, in conjunction with pertinent experimentation.

Mosaic zone model. If the environments of the two forms are fragmented in a mosaic pattern where they meet, rather than a simple parapatry, then strong selection for each genotype will cause the genotypes to be likewise distributed. Hybrids will be produced at the edges of neighboring individual patches of the two habitats, and dispersal will generate clines, but overall a more complex pattern will ensue. A number of variant models have been proposed in particular situations, the most recurrent being the Bounded Hybrid Superiority Zone. This argues that if hybrids were fitter in the ecotonal environment, then this would produce a hybrid zone. Although individual hybrid genotypes may occasionally be fitter, recombination in the zone would tend to dissemble favored genotypes. There is little evidence for this model.

There is evidence for selection against hybrids in many hybrid zones. In addition to the sterility seen in *C. parallelus*, the mountain grasshopper *Podisma pedestris* shows high embryo mortality, the toad *Bombina* increased tadpole abnormalities, and the Australian grasshopper *Caledia captiva* F2 hybrid breakdown. In a number of cases, field mark-recapture estimates of dispersal, combined with such hybrid unfitness, provide satisfactory explanations of zone width under a tension zone model.

Perhaps a quarter of hybrid zones are reported near ecotones and environmental transitions, and this indicates that the parapatric taxa may have genotypes that are better adapted to the conditions on their side. However, as in the case of *C. parallelus* along the Pyrenees, hybrid unfitness may also be involved, and the determining balance may be a tension zone. Furthermore, most environmental gradients are broader than their hybrid zones, which indicates that they are not strongly determining allele frequency and clines. Also, environmental gradients vary across an ecotone, and yet hybrid zones do not seem to respond to this. For example, the toad *Bombina* shows similar clines for allozymes and morphology in zone transects hundreds of kilometers apart. Another phenomenon difficult to explain purely in terms of environmental gradients is the coincidence of clines for many characters in the zone. Genes for different functions would not be expected to have selection change at the same place.

There are a few well-researched cases where particular genotypes are found in distinct environments. In the northeastern United States, there is a hybrid zone between the two crickets *Gryllus firmus* and *G. pennsylvanicus*, where each is associated with either sandy or loamy soil. These types of soil meet and intermingle in patches. This has produced a classic mosaic hybrid zone, and selection by the two habitats on the species must be relatively strong. Species of Louisiana *Iris*, such as *I. fulva* and *I. brevicaulis*, also live in different habitats among the forests and bayous, and hybrid zones occur in the intermediate mixed habitat. Such habitat structuring may be especially prevalent in plants and plant-specific insects.

The forces maintaining a narrow hybrid zone in the face of dispersal, gene flow, and introgression vary from case to case. Hybrid unfitness in some form is usually involved, and there may often be an environmental component, which sometimes is strong enough to structure the zone.

Origins of Hybrid Zones. There has been considerable debate as to whether hybrid zones have primary or secondary origins. This question arises because the patterns of genetic variation produced in the past are similar under both hypotheses. However, advances in paleobiology and phylogenetics now provide discriminating evidence.

A primary zone is one in which differences evolved in a continuous distribution, with different alleles being selected toward opposite ends of an environmental gradient. The clines so generated are steepened by selection for modifying genes to produce a narrow hybrid zone between two internally coadapted genotypes. A secondary zone is the result of contact between two already diverged genomes that were previously separate and have changed their ranges. Dispersal and hybridization mixes the genomes, producing clines across the zone.

Although there is good evidence for primary divergence, in that steep clines can be generated across sharp environmental changes, several strong arguments favor secondary contact as the proximate origin of most hybrid zones. First, major climatic changes have repeatedly modified species ranges globally, particularly in the last 1 million years. This is particularly well established in Europe and North America, where current temperate species hybrid zones were formed as the ice and tundra retreated after the last ice age, some 16,000–6,000 years ago. There is growing evidence for changes in distribution of tropical species also through these climatic oscillations. Second, different patterns of DNA sequence phylogeny are expected under primary and secondary models. Two taxa that diverged allopatrically should form distinct diverged DNA clades that are preserved on secondary contact; under primary divergence, however, neighboring populations on either side of the zone should have similar DNA sequences. Third, a composite argument is that, given the recent postglacial origin of most zones, it seems not possible that such large differences and many coincident clines could have arisen so quickly between the two taxa.

Molecular Phylogeographic Evidence. The phylogenetic test of distinct DNA clades argues for allopatric divergence and secondary contact in parapatric species of the Andean mouse, *Akodon*, and also for the subspecies of *C. parallelus* in the Pyrenees. The depth of DNA divergence indicates that their hybridizing lineages were effectively separated in the Pleistocene. Further examples come in the toad *Bombina* at some 5 million years, the hedgehog *Erinaceus* at 3–5 million years, and the mouse *Mus* at 1 million years. For the *C. parallelus* subspecies, a 1 percent mtDNA divergence requires isolation over about 0.5 million years, or 3–5 ice ages, and the glacial refugia were in southern peninsulas of Spain, Italy, Greece, Turkey, and the Balkans. One may deduce from DNA similarities that after the last ice age about 16,000 years ago, *C. parallelus* from the Balkans colonized most of Europe, while the Iberian genomes reached the Pyrenees and the Italian ones reached the Alps. As these mountain icecaps melted some 8,000–9,000 years ago, the hybrid zones formed as the Balkan/Iberian and Balkan/Italian genomes met along the ridges (Figure 1).

Using recent molecular data, we can deduce the colonization routes of different genomes from southern refugia for several species. Most patterns are similar to that of *C. parallelus*, but in a number of cases (like the brown bear, *Ursus arctos*), the Iberian and Balkan genomes expand and form hybrid zones in the middle of northern Europe. In only a few species, like the hedgehog *Erinaceus*, does the Italian genome also expand beyond the Alps and colonize the north. Hybrid zones result from these secondary genomic contacts, and clusters for several species are found particularly in the Alps, in central

Scandinavia and Europe, and in the Pyrenees. These have been called *suture zones.*

This phenomenon was first noted for morphological characters in North America. The use of modern DNA methods provides similar phylogeographic deductions of subspecific divergence, postglacial colonization, and hybrid zone formation for this continent's biota as well. Less is known about climatically induced range changes in tropical parts, but research in the rain forest of northeastern Australia and the northern Andes suggests that similar processes operate, although perhaps over shorter distances.

Age and Stability of Zones. The preceding evidence supports a postglacial secondary origin for most hybrid zones, and their locations suggest that they have not moved far from their original contact positions. Once formed, a tension zone will tend to be trapped in regions of low density for the species, where dispersal and gene flow from both sides is equally low. Such regions are many, and they may coincide with an environmental transition and combine with ecological adaptation at an ecotone or topographical barrier.

However, the genetic divergence between the component taxa making secondary contact has built up over several to many major climatic oscillations and range changes in the Pleistocene. Hybrid zones must have formed and been destroyed many times, and not necessarily in the same places. Changes in climate are the main controllers of hybrid zones. Species ranges are still changing naturally, and man is also having effects. For example, in North America several bird species (like the warblers *Vermivora*) now hybridize as a result of colonial agriculture, and in the southern United States, introduced fire ants (*Solenopsis*) form a moving hybrid zone.

Hybrid Zones as Evolutionary Laboratories. DNA sequences strongly confirm that the component genomes forming different hybrid zones have diverged over various lengths of time, sometimes for millions of years. Over such a time scale, populations in a species may diverge by chance and selection into races, subspecies, and species. Hybrid zones thus provide comparisons of genomes at different states and stages in this process of evolution and speciation.

Hybrid zones are particularly involved in certain proposed modes of speciation, such as reinforcement and parapatric and accumulated allopatric speciation. They have been used to provide experimental evidence to test these, particularly in challenging the feasibility of reinforcement and in clarifying the ecology of hybrid speciation.

Natural hybridization provides recombinant genotypes with a range of contributions from the parents. With modern genetic technology, the genes concerned in the differing adaptations can be located, sequenced, and studied. One may ask how fast characters change, when and where pre- and post-mating isolation evolve, and what genetic changes are involved.

Hybrid zones identify adjacent genomes and races, and with molecular phylogeography their Pleistocene history may be deduced. This indicates that present biodiversity is geographically structured at several genetic levels. Furthermore, species communities are formed as mixtures of genotypes from several refugia, and these have not been constant over the Pleistocene ice ages.

BIBLIOGRAPHY

Arnold, M. L. *Natural Hybridization and Evolution.* Oxford and New York, 1997. Most hybrid zone studies concern animals and this view from a plant biologist is useful in highlighting the role of hybrids in speciation.

Barton, N. H., and G. M. Hewitt. "Analysis of Hybrid Zones." *Annual Review of Ecology and Systematics* 16 (1985): 113–148. A source review listing and analyzing data from some 150 hybrid zones then available.

Butlin, R. K. "Reinforcement—An Idea Evolving." *Trends in Ecology and Evolution* 10 (1995): 432–434. A balanced report on this much debated hypothesis of speciation in which hybrid zones are central.

Harrison, R. G., ed. *Hybrid Zones and the Evolutionary Process.* Oxford and New York, 1993. An integrated collection of major articles by leading workers on the concepts and case studies of hybrid zones.

Hewitt, G. M. "Hybrid Zones—Natural Laboratories for Evolutionary Studies." *Trends in Ecology and Evolution* 3 (1988): 158–167. A broad introduction to the evidence and concepts; a useful place to start in understanding the more recent literature.

Hewitt, G. M. "The Genetic Legacy of the Quaternary Ice Ages." *Nature* 405 (2000): 907–913. This joins recent advances in paleoclimatic and phylogeographic studies to understand present genetic structure of species, including hybrid zone formation and significance.

Howard, D. J., and S. H. Berlocher, eds. *Endless Forms: Species and Speciation.* Oxford and New York, 1998. The major modern book on speciation, containing articles on recent advances and a section on hybrid zones.

Kim, S-C., and L. H. Rieseberg. "The Contribution of Epistasis to Species Differences in Annual Sunflowers." *Molecular Ecology* 10 (2001): 683–690. An example of research using modern genetic techniques to ask fundamental questions about adaptive evolution and introgression across a hybrid zone.

Patton, J. L., and M. F. Smith. "MtDNA Phylogeny of Andean Mice: A Test of Diversification Across Ecological Gradients." *Evolution* 46 (1992): 174–183. A classic paper using modern DNA approaches to distinguish hybrid zone hypotheses of divergence and speciation.

Shoemaker, D. D., et al. "Genetic Structure and Evolution of a Fire Ant Hybrid Zone." *Evolution* 48 (1996): 392–407. A manmade hybrid zone between introduced species that provides a rare chance to study movement and evolution in progress.

Virdee, S. R., and G. M. Hewitt. "Clines for Hybrid Dysfunction in a Grasshopper Hybrid Zone." *Evolution* 48 (1994): 392–407. Hybrid zones may evolve in several ways, and this examines some consequences of interaction among dispersal, hybrid sterility and selection over some 9,000 generations.

— GODFREY M. HEWITT

R
576.803
P1339
v.1

LINCOLN CHRISTIAN COLLEGE AND SEMINARY   104695

3 4711 00169 6758